VOLUME

A millilitre is about
one-fifth of a teaspoon

A litre is becoming the standard
size for soft drink, wine, and
liquor bottles in the U.S. If
its price is the same as for the
old quart bottle, you're in luck
because a litre is equal to
slightly more than a quart

The gas tank of a large car
holds 64 to 75 litres of gasoline

Volume Conversion

1 tsp = 5 ml	1 ml = 0.03 fl oz
1 tbsp = 15 ml	1 l = 2.1 pt
1 fl oz = 30 ml	1 l = 1.06 qt
1 cup = 0.24 l	1 l = 0.26 gal
1 pt = 0.47 l	
1 qt = 0.95 l	
1 gal = 3.8 l	

TEMPERATURE

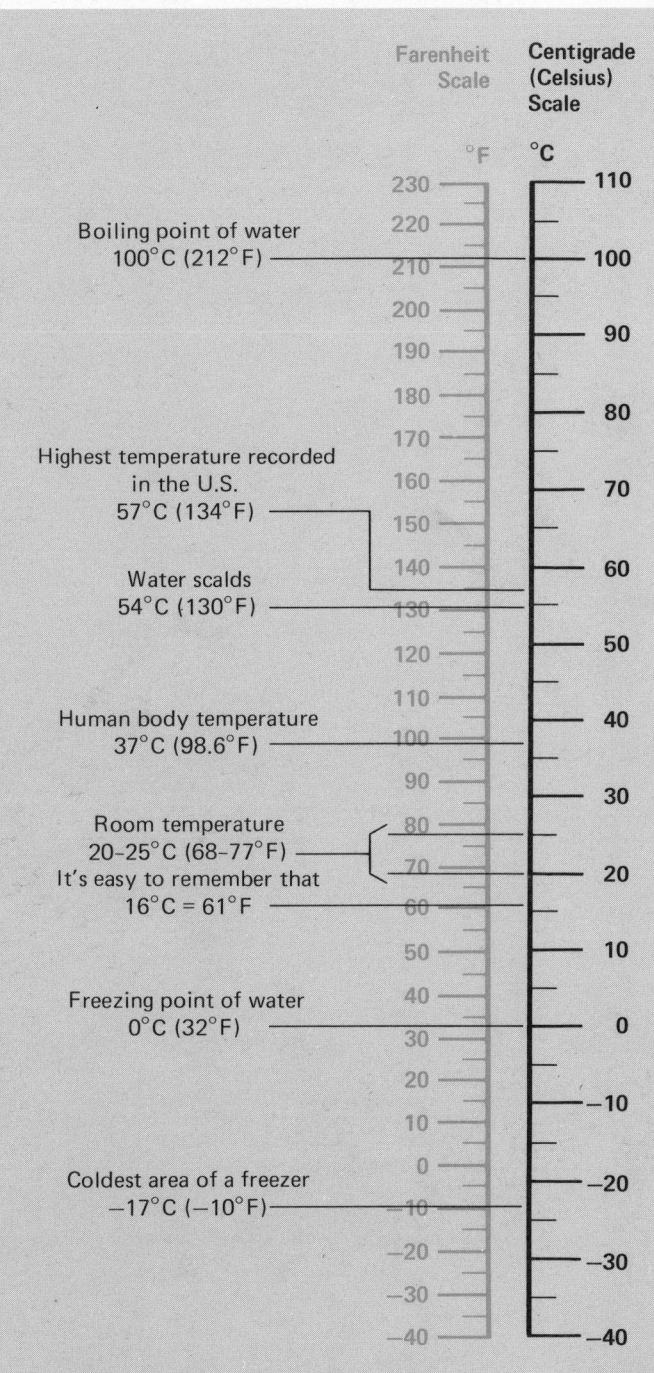

Boiling point of water
100°C (212°F)

Highest temperature recorded
in the U.S.
57°C (134°F)

Water scalds
54°C (130°F)

Human body temperature
37°C (98.6°F)

Room temperature
20–25°C (68–77°F)
It's easy to remember that
16°C = 61°F

Freezing point of water
0°C (32°F)

Coldest area of a freezer
−17°C (−10°F)

Temperature Conversion

$$°C = \frac{(°F - 32) \times 5}{9}$$

$$°F = \frac{°C \times 9}{5} + 32$$

Interval Equivalents

°C	°F
1° =	1.8°
5° =	9°
10° =	18°

Biology

Second Edition

SAUNDERS COLLEGE PUBLISHING

Philadelphia   New York   Chicago
San Francisco   Montreal   Toronto
London   Sydney   Tokyo   Mexico City
Rio de Janeiro   Madrid

Biology

Second Edition

Karen Arms
Pamela S. Camp

Address orders to:
383 Madison Avenue
New York, NY 10017

Address editorial correspondence to:
West Washington Square
Philadelphia, PA 19105

This book was set in Primer by York Graphic Services.
The editors were Michael Brown, Lee Walters, Carol Field, and Elizabeth Galbraith.
The art & design director was Richard L. Moore.
The text design was done by Nancy E. J. Grossman.
The cover design was done by Richard L. Moore.
The artwork was drawn by Vantage Art, Inc.
The production manager was Tom O'Connor.
This book was printed by Kingsport Press.

Cover credit: Scarlet Ibis in Trinidad Swamp, © Towsend Dickinson.

BIOLOGY   ISBN 0-03-059961-X

1234   071   987654321

CBS COLLEGE PUBLISHING
Saunders College Publishing
Holt, Rinehart and Winston
The Dryden Press

To Paul and Bill,
the critics on the hearth,
with love

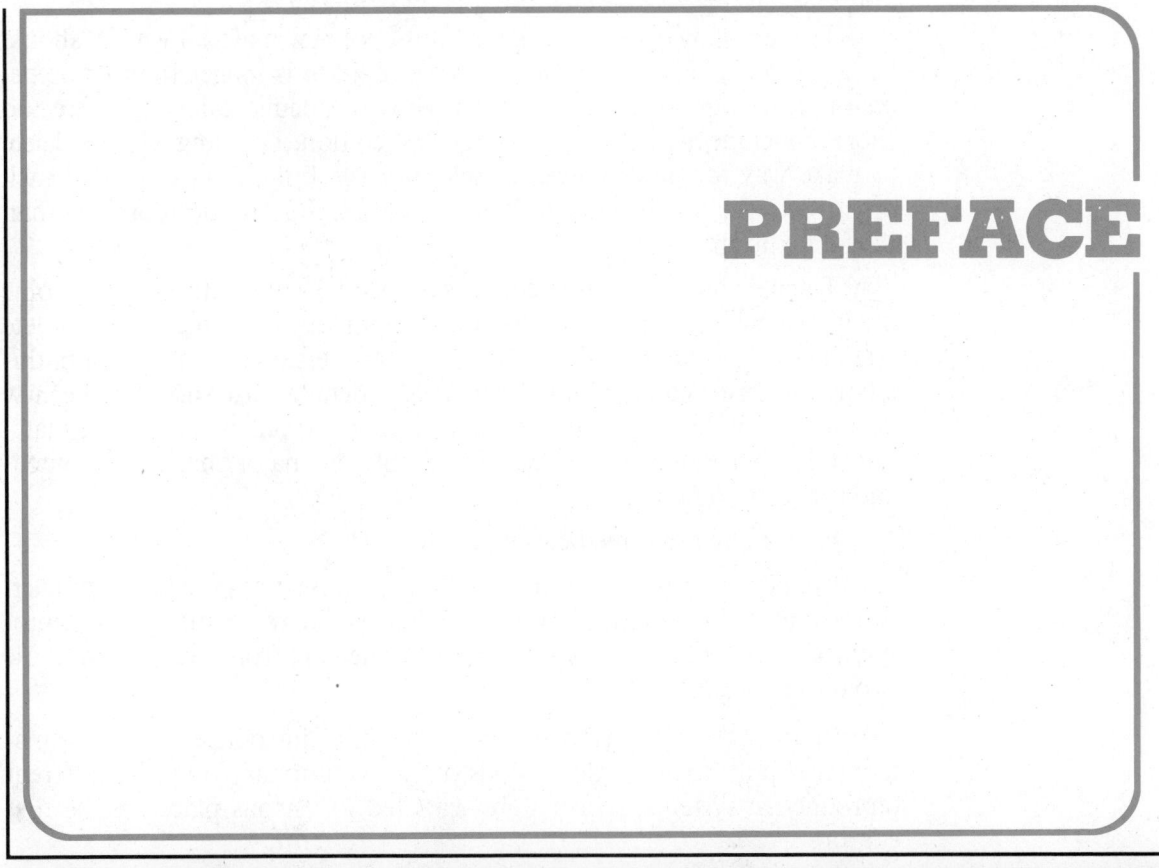

PREFACE

Preparing this second edition of *Biology* has been an exciting task — a chance to bring in new ideas, incorporate biological discoveries of the last few years, and make material easier to understand by rewriting, redrawing, and finding new photographs. Despite these opportunities, many pitfalls gape for the unwary authors of a new edition. Our first impulse was to start from scratch. But our editors and users kept us in check with frequent reminders that the idea was to revise, and not to destroy what people liked in the first edition.

Although it is a relatively small part of preparing a new edition, recording the advances of the last few years' research is the greatest thrill. How exciting it is to report that we now know the nucleotide sequences of dozens of genes, when we remember hours spent listening to chemists explain why it was unlikely that anyone would analyze the nucleotide sequence of a DNA or RNA molecule in our lifetimes!

The temptation to add large quantities of new material which "should be in the book" can mean that each new edition is longer than its predecessor, to the deep distress of users whose academic calendars were too short to include all the material in the first edition. Deciding what to delete to make way for new material is seldom easy, but we have goaded each other on to the disagreeable task and this book contains no more text than the first edition.

Teachers often greet new editions of their favorite textbooks with some dread: a new edition means the drudgery of updating the now-obsolete reading lists and handouts used with the older version. To save most of this labor, the Instructor's Manual provides a detailed list showing the new location of all the material in the first edition, as well as identifying material added or deleted. Here we mention only the major changes in content and organization.

First, there are three new chapters:

Chapter 6, "Energy and Living Cells," summarizes the basics of energetics that a student needs for the chapters on respiration and photosynthesis, and, we hope, removes some of the pain from the study of these two difficult chapters.

Chapter 18, "Evolution and Reproduction," introduces some new material and pulls together discussions of the evolutionary origin of different reproductive systems, which were scattered in various places in the first edition.

Chapter 34, "Defenses Against Disease," is largely an introduction to immunology, much expanded (and, of course, updated) from the coverage in the first edition.

In addition, the kingdom Protista now has a chapter all to itself.

Second, several chapters have disappeared. Most of these are from the part of the book on ecology and evolution. Most users of the first edition did not have time for all of these chapters. Rather than see many chapters unused, we have tried to cut out the less-essential or repetitive material, and have merged the more interesting and important topics from the deleted chapters into those that remain. Parts of the chapter on communities have been merged into Chapter 43, "Biomes," and Chapter 44, "Ecosystems," and sections from other chapters have been moved forward into the Parts on plant or animal biology. We have also combined the two chapters on animal transport systems with the chapter on thermoregulation.

Third, the chapters on evolution are now placed nearer the front of the book, immediately after the related topics of molecular biology and genetics, because many users taught the topics in this order in their courses. Our choice of chapter order may not matter much; biology is thoroughly web-like, and anything is easier to understand when you already know everything else. As far as we know, all users of the first edition reordered the chapters to some extent in their teaching. The feedback we received confirms that the chapters are independent enough to be taught in various orders, according to teachers' preferences.

The running glossary and cross references to sections in other chapters are especially useful when the chapters are taught in a different order from that in the book. In this edition, we have incorporated more of the running glossary into the text to avoid what many users felt were too many footnotes; as before, the terms in the running glossary are indicated by asterisks. The text is again divided into short, numbered sections for easy identification, and we have now added subheadings to make an easier-to-follow outline. Many of the first edition's lettered sections are now subsections, so that the number of lettered sections is smaller. The extensive index/glossary of course remains, for we anticipate that this will be used as a reference book as well as a textbook.

All the main features of the first edition are still present — each chapter contains a summary, objectives, a self-quiz, questions for discussion, and a list of references and further reading. We have moved the objectives to the end of the chapters; we find that many students are intimidated if they read objectives with a lot of long, unfamiliar words before they start a chapter; they tend to skip the objectives and they may or may not remember to study the objectives after they have read the chapter. If students come to the objectives *after* reading the text for the first time, they find that they can already do most of the objectives but must go back and read a little more carefully to master the others before they are ready to tackle the self-quiz. Students with stronger backgrounds and vocabularies will find it worthwhile to skim the objectives before reading the chapter so that they know what to look for in the chapter. Having chosen to place the objectives at the back, we have rewritten the introductions to the chapters with special attention to preparing the students for the material in the chapter.

Objectives are a list of things so important that each student is required to know or do them in order to pass the course. The lists we provide are our attempt to point out important and well-established concepts and vocabulary and necessary skills. We do not include in the objectives conjectural material or much detail, even though we feel that it is worth presenting this material in the text of the chapter for the sake of background, completeness, or current interest. New teachers are reminded that they will often disagree with our selections, and should alter the list of objectives to suit their own courses. A more detailed discussion of the preparation and use of objectives is included in the Instructor's Manual.

The self-quiz permits students to test their mastery of some of the more important objectives; answers are in the back of the book. There is not enough space to test all objectives exhaustively, and students should treat the list of objectives itself as an essay-type examination.

The questions for discussion should not be treated as part of the self-quiz. They are just what their name implies: questions to bat around in discussion groups or over lunch. Students should be able to work out the answers to some of them from the information in the chapter. However, many are unanswerable — questions that researchers are addressing today, questions that have puzzled people for centuries, or questions of ethics or opinion that we must all answer for ourselves.

The suggested readings offer great variety. They include original work referred to in the text or figures, reference books that we find reliable, bedtime reading with a biological slant, and works that approach difficult material from a different viewpoint.

At the behest of the United States Metric Association, we use the spellings "litre" and "metre" (instead of "meter" and "liter"), as used by scientists of all nations and by international businesses. We have also adopted the term "bisphosphate" now used by biochemists in place of "diphosphate" to indicate sugars with two phosphate groups, and have replaced the terms "blue-green algae" and "cyanophytes" with the increasingly favored "blue-green bacteria" and "cyanobacteria."

Possibly the greatest strength of this book is that it has been read and criticized by literally hundreds of biology students, teaching assistants, and experienced teachers of introductory biology. With the benefit of their experience to add to our own, we have identified difficult areas, devoted more space to them than is usual and, we hope, helped students over the conceptual hurdles involved. We hope this book presents biology as the fascinating and human subject that we ourselves enjoy so much.

KAREN ARMS

PAMELA S. CAMP

ACKNOWLEDGMENTS

The number of people who have helped to make and improve this book grows with every year that passes. The generous contributions of our teachers and students and friends and relatives have made our job much easier and cheered us on our way.

To ensure that the book is accurate and up to date, researchers in almost every area of biology have scanned chapters in their own specialties for mistakes and omissions. Dozens of experienced biology teachers also agreed to tell us what they thought was good and what was bad about the first edition, and to read the manuscript for this edition. They have helped to eliminate mistakes and, more important, have contributed their experience of areas students find difficult, additional material they have found useful and the hundred and one details that are important to students, and so to teachers. For their contributions to the second edition, our thanks to:

Robert L. Amy
Southwestern University

Carol M. Bailey
Dean Junior College

William E. Barstow
University of Georgia

Sally Bauer
Hudson Valley Community College

William M. Bethal
Saint Louis University

Charles K. Biernbaum
The College of Charleston

Lois M. Borgman
Mount Ida Junior College

Edmund D. Brodie, Jr.
Adelphi University

Mac A. Callaham
North Georgia College

C. Blaine Carpenter
Clayton Junior College

Simon L. Chung
Trinity College

Mildred A. Collins
Stillman College

Richard Collins
Louisiana State University

Deborah F. Cooperstein
Adelphi University

Roland Corey
U.S. Naval Academy

James R. Darwood
Westchester Community College

Peter Davies
Cornell University

D. G. Davis
University of Alabama

John D. Davis
Mississippi University for Women

Margaret L. S. DeSaiz
University of North Carolina at Chapel Hill

Al Dibold
Macon Junior College

James K. Dooley
Adelphi University

Robert Egan
University of Nebraska at Omaha

John Farmer
University of Oklahoma

Gerald Fassell
Cayuga Community College

Albert E. Feldman
Dutchess Community College

Alfred F. Finocchio
St. Bonaventure University

Carl Frankel
Pennsylvania State University at Hazleton

E. C. Franks
West Illinois University

C. W. Gaddis
University of Arizona

William H. Gilbert
Simpson College

Julius Hand Gooden
Bowie State College

John R. Gregg
Duke University

David A. Haskell
Smith College

Stephen C. Hedman
University of Minnesota at Duluth

Frank Heppner
University of Rhode Island

Terry Hill
Southwestern University

Peter Hinkle
Cornell University

Kenneth M. Hoff
Cleveland State University

Linda-Margaret Hunt
University of Notre Dame

Daniel R. Hystrom
Bakersfield College

Andre Jagendorf
Cornell University

Henry M. Knizeski, Jr.
Mercy College

Shirley J. Kurtzberg
Westchester Community College

Richard J. Lacey
Adelphi University

Arthur Lavigne
Middlesex Community College

Thomas Lonergan
University of New Orleans

William F. Loomis, Jr.
University of California at San Diego

John D. Lyon
Adirondack Community College

James Maranaccio
Somerset County College

W. H. Mason
Auburn University

Leathem Mehaffey
Vassar College

Michael C. Mix
Oregon State University

Louise M. Morgan
Mercer University

J. Thomas Mullins
University of Florida

James T. Murrell
Mississippi University for Women

Henry M. Muschio
Dutchess Community College

Lynne J. Osborn
Middlesex Community College

Marlene K. Palmer
Vassar College

Lloyd M. Pederson
San Joaquin Delta College

Ruthanne B. Pitkin
Allegheny College

Michael V. Plummer
Harding University

C. J. Probst, Jr.
University of New Orleans

Dennis C. Radabaugh
Ohio Wesleyan University

Francis V. Ranzoni
Vassar College

Basil Robinson
Holyoke Community College

Charles F. Rodell
St. John's University

Mitchel Sayare
University of Connecticut

John W. Sechrist
Wheaton College

John L. Shane
Middlesex Community College

Gertrude D. Shay
University of Dayton

Harry L. Sherman
Mississippi University for Women

Charles G. Smith
Lake Erie College

Marga H. Smith
Wittenberg University

Norman Smith
San Joaquin Delta College

Arlien Steiner
Tidewater Community College

David Stetler
Virginia Polytechnic Institute and State University

Darrell R. Stokes
Emory University

Carl A. Strang
Dickinson College

Stanley Szarek
Arizona State University

Jane B. Taylor
Northern Virginia Community College

Dale Therrieu
Pennsylvania State University

John Thornton
Oklahoma State University

Elizabeth K. Tomaszewski
Texas A & M University

Bik-Kwoon Tye
Cornell University

John M. Walker
Vincennes University

Eileen Walsh
Westchester Community College

Joseph Wood
University of Missouri at Columbia

Theo Zemek
College of Du Page

We would also like to acknowledge the considerable contributions of those who reviewed and otherwise helped to prepare the first edition:

Kraig Adler
Cornell University

John Alcock
Arizona State University

Betty Allamong
Ball State University

John C. Belton
California State University, Hayward

John P. Bihn
LaGuardia Community College

Antonie W. Blackler
Cornell University

Patricia Bonamo
State University of New York at Binghamton

Roger R. Bowers
California State University, L.A.

Peter F. Brussard
Cornell University

Neal D. Buffaloe
University of Central Arkansas

Brian Capon
California State University, L.A

Arthur G. Carroll
Oklahoma State University

Robert H. Catlett
University of Colorado

Shepley S. C. Chen
University of Illinois

Frances S. Chew
Tufts University

Norman T. Davis
University of Connecticut

F. Paul Doerder
University of Pittsburgh

Walter Rudd Douglass
Fordham University

Marvin Druger
Syracuse University

George F. Estabrook
University of Michigan

William L. Evans
University of Arkansas

Paul Feeny
Cornell University

Allen D. Forsythe
Holyoke Community College

Charles E. Foster
State University College at Potsdam

Tim Gaskin
Cuyahoga Community College

Lawrence Gilbert
University of Texas at Austin

Jack L. Gottschang
University of Cincinnati

James L. Greco
State University of New York at Albany

Arnold J. Greer
St. Louis Community College at Meramec

Steven N. Handel
Yale University

Stephen C. Hedman
University of Minnesota

Richard F. Heller
Bronx Community College

Kenneth M. Hoff
The Cleveland State University

Richard J. Hoffmann
University of Pittsburgh

Russell C. Hollingsworth
Tarrant County Junior College

Andre Jagendorf
Cornell University

Alan J. Jaworski
University of Georgia

Edmund D. Keiser
University of Mississippi

George H. Kieffer
University of Illinois

John A. W. Kirsch
Yale University

David Klingener
University of Massachusetts

Clifford E. LaMotte
Iowa State University

Laurence A. Larson
Ohio University

Georgia E. Lesh-Laurie
The Cleveland State University

Carmita E. Love
Community College of Philadelphia

Henry Merchant
George Washington University

Keith Moffat
Cornell University

Joseph J. Napolitano
Adelphi University

Milton Nathanson
Queens College

Alma Moon Novotny
University of Houston

John D. O'Connor
University of California, L.A.

Lowell P. Orr
Kent State University

Martha Constantine Paton
Princeton University

Donald I. Patt
Boston University

L. Jack Pierce
Mountain View College

Caroline M. Pond
The Open University

Gene A. Pratt
University of Wyoming

Ralph S. Quatrano
Oregon State University

Rudolf A. Raff
Indiana University

Martin Sacks
City University of New York, City College

Roger H. Sawyer
University of South Carolina

Howard A. Schneiderman
University of California, Irvine

Allan A. Schoenherr
Fullerton College

Gary A. Smith
Tarrant County Junior College

Boyd R. Strain
Duke University

Daryl Sweeney
University of Illinois

Joseph W. Vanable, Jr.
Purdue University

Marvalee H. Wake
University of California, Berkeley

Norman K. Wessells
Stanford University

James R. Willman
St. Louis Community College at Meramec

Norman Williams
University of Iowa

John T. Windell
University of Colorado

Clarence C. Wolfe
Northern Virginia Community College

Joseph M. Wood
University of Missouri, Columbia

Newell A. Younggren
University of Arizona

John L. Zimmerman
Kansas State University

In addition to these reviewers are people who have helped us in a variety of ways. They transported, fed and housed us, argued with us about biology and grammar, read chapters, and contributed ideas, reprints, and moral support. Without all these people this book would be but a pale shadow of itself: Mary Ahl, Nan Arms, Frank Avedon, Amy Avnet, Robert Blakemore, William Blau, Clive Bransom, David Campbell, James Case, Brian Chabot, Anna Chao, John and Roberta Christy, Edward Cox, Catherine Craig, Karen Crassweller, Lucian Del Priore, Debbie DenHerder, Bernard Dethier, Carolyn Eberhard, Sheldon Freedman, Sidney Fox, Jane Gibson, Susan Giffen, David Gosling, Bruce Gray, Dan Gray, George Gutman, JoAnn Haick, William Hallahan, John Heiser, Ronald Hoy, Robert Ingoglia, Andrew Jones, Janis Kelly, Donald Kennedy, Harold Kipperman, Randi Lebar, Ellis Loew, Jane and John Lytle, Ross MacIntyre, Allen MacNeill, Peter Marks, Robert May, Muriel Milks, John Muller, Thomas Mullins, Nancy Ostman, Dominick Paolillo, Marie Paolillo, Richard Pflanzer, Harold Pusey, Mark Rauscher, Thomas Roberts, Monika Robke, Richard Root, Kenneth Sandlan, Heidi Schallenberg, David Schindler, Stephanie Seramitis, Jo Shapiro, Philippa Shepherd, Daniel Simberloff, Janice and Ian Skidmore, Stephanie and Stephen Sutton, Barbara and David Usher, Nicholas Wade, Watt Webb, Steven Webster, Robert Whittaker, and Kathryn Wunderlich.

A special word of thanks to Biophoto Associates, a cooperative of camera-clicking biologists run by Gordon Leedale at Leeds University in England. Gordon and his assistant, Helena Cmiech, patiently catered to our changing whims and presented us with so many gorgeous photographs that our only difficulty was deciding which we could bear not to use in this book. Without such a wonderful source of photographs, this book would be much less beautiful and could not have been published for at least another year.

Most of the cartoons scattered through the book are from the lively pen of Rosemary Smith who produced her comment on the genetic code (Chapter 10) in 1965 and has continued her frivolous commentary on biology and biologists ever since. May Berenbaum produced the scatological essay in Chapter 49 for a newsletter that accompanied the first edition of the book; we liked it so much that we borrowed it for this edition.

We are lucky enough to have found a publisher with whom it is stimulating, infuriating, amusing, and a continuing pleasure to work. Kendall Getman, who roped us into this business in the first place, has now retired and we miss him very much. This is our second book with publisher Don Jackson and editor Lee Walters. With them we have suffered much and celebrated mightily; long may it last! Our gratitude for their patience and hard work goes also to Lyn Peters of Holt, Rinehart and Winston, who edited the first edition, and to Michael Brown, Tom O'Connor, Carol Field, Nancy Grossman, Leesa Massey, and Rick Moore at Saunders College Publishing who designed, edited, produced, and kept track of the million and one bits of paper that a book of this complexity generates. We should make special mention, too, of John Tugman, our marketing manager, and his magnificent sales force. Without them, this book would be born to blush unseen. Over the years all these people have become our faithful friends and allies; we even take their advice occasionally!

Our husbands, Paul Feeny and Bill Camp, have earned our special gratitude. Those who know his work will recognize Paul's hand in the chapters on ecology and evolution. Bill kept us on the straight and narrow where the text strays into the physical sciences. In addition, they typed, ran errands, suffered neglect and malnutrition, and generally immolated themselves on the twin altars of science and marital devotion.

Finally, to each other — gratitude and affection for a partnership that has lasted a remarkable ten years and brought both of us a job we love and lots of fun and satisfaction.

Ithaca, N.Y.   KAREN ARMS

September 1981   PAMELA S. CAMP

CONTENTS OVERVIEW

CONTENTS

PART I: CELLS

PART II: INFORMATION CODING AND TRANSFER

PART V: ANIMAL BIOLOGY

Biology

CHAPTER 1

INTRODUCTION

Living things first appeared on earth some 3.5 billion years ago. Their descendants have diversified into the several million kinds of living things (plants, animals, bacteria, and fungi) alive today. Throughout the ages people have been interested in the plants and animals that surround them. At first, this interest took the form of observation: people examined and named living things, classified and made lists of what lived where, and collected organisms as some people collect stamps. During this time, the study of living things was generally known as natural history, which is not a science; this was the forerunner of biology, which is a true science. The main thing that distinguishes a science, such as biology, chemistry, or physics, from the "humanities," such as art and literature, is the use of experiments to answer questions.

In this chapter we shall consider how scientists ask and answer questions. We then point out that science is only a limited way of looking at the world, and con-

Opposite: Fossil ammonites, relatives of modern
nautiluses and squids. (Biophoto Associates)

sider why and how it is limited. Finally, we turn to the characteristics of living organisms—which is really a brief synopsis of the concepts covered in this book.

1-A Science and Society

Science has assumed a position of enormous importance in modern society. Many decisions affecting our future depend upon appropriate interpretations of scientific discoveries. Democratic government demands that everyone participate in decisions on such subjects as population control, protection of wildlife, and compulsory immunization. The body of scientific knowledge and its practical applications, or technology, are so vast today that no one person, and even no one university of hundreds of scholars, can understand all of it. As responsible citizens, however, we can try to follow some of the important scientific studies that bear on public issues, and we can apply scientific reasoning in arriving at our own positions on these issues.

There is nothing mysterious about scientific reasoning or experiments. They are merely logical ways of trying to solve problems such as are used by business people, historians, and each of us in our everyday lives. We do not need specialized scientific training or knowledge to decide whether conclusions are justified from the data presented. We can request further tests of a theory that does not appear to be well supported by the evidence, and we can agree or disagree with predictions from a theory. We can improve the way we do these things ourselves if we first understand how a scientist arrives at conclusions about natural phenomena through the same kinds of processes.

1-B Scientific Method

You may never have thought about how you solve problems, test theories, or decide upon a plan of action. Let us consider how a biologist attacks a problem so that we can examine the main types of thinking involved.

Science usually starts with observations of the natural world and makes a generalization from those observations. For instance, if you are collecting insects, you may notice that many have black and yellow stripes. As you catch them, you probably think they are all bees or wasps and treat them with due caution. However, as you examine them more carefully, you may find that some have features clearly showing that they are flies rather than bees.

Is it merely coincidence that these flies look unlike their drab housefly cousins and resemble unrelated bees? Or do the striped flies gain some advantage from looking like bees? To answer this question, you must think of some ideas, or **hypotheses**, that will account for your observations. You may think of hypotheses such as "a fly's resemblance to bees protects it from predation," or "the flies fool bees into accepting them as members of the hive, and sneak in and steal honey."

The next step is to design and perform **experiments** to test the hypothesis. Because hypotheses usually cannot be tested directly, you must first develop a testable prediction from the hypothesis. Some hypotheses are of no use to science because they cannot be tested directly or even indirectly. For instance, the hypothesis "predators think flies are bees" is untestable because you can never know what an animal thinks.

But suppose you use your hypothesis to predict that a predator will not eat a fly that looks like a bee if it has first learned not to eat stinging bees. This prediction *can* be tested.

For this experiment, you need predators that eat insects—toads will do. Toads eat by catching insects flying or crawling nearby. If you put bees into the cage of a naïve toad (one not acquainted with bees), it will catch a few, learn that they sting, and refuse to catch any more. You next put a black-and-yellow striped fly into the cage and see if the toad also refuses the fly. If so, the hypothesis that the fly's resemblance to bees protects it from predation is supported. You should check, though, to be sure the toad is still hungry by offering it a harmless housefly for

Figure 1-2 This insect looks like a bee because it has a hairy body and black and orange stripes; it is actually a harmless fly. (Biophoto Associates)

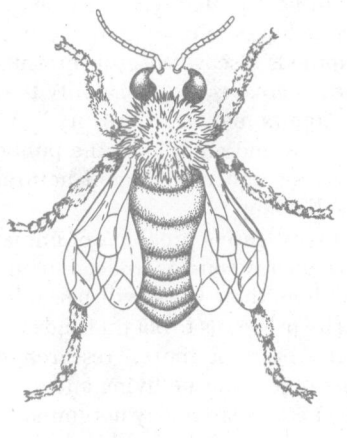

(a) Bee (b) Fly

Figure 1-3 Resemblance of a mimic fly to a bee is good enough to fool some of the people some of the time. However, there are differences: the fly has large eyes and stubby, club-shaped antennae. The bee's eyes are smaller and her antennae more slender. A bee also has two pairs of wings to a fly's one pair, but this may be hard to see.

dessert. But perhaps the bee has nothing to do with it; maybe toads just do not eat striped flies. To test this, you must use a second naïve toad. If this toad cheerfully devours black-and-yellow striped flies, you have gained additional support for the hypothesis that the fly's striped suit is advantageous because it resembles the bee's.

A valid experiment always includes a **control treatment**, such as the second toad above, as well as an **experimental treatment**, in which one (*and only one*) factor is varied—in this case, the opportunity to eat bees. If you had not used a control treatment, the toads might have refused the flies because of some factor that you did not notice, such as not being hungry, and you might have concluded, wrongly, that the resemblance to the bees was responsible.

It is disappointing to realize that a hypothesis can never formally be proved, but can only be disproved. You can never "prove" that the flies' stripes protect them from predation; you can only disprove the hypotheses that you think of as alternative explanations for the resemblance and show that the coloration always seems to be effective in discouraging predators of a wide range of species.

A hypothesis supported by many different lines of evidence from repeated experiments is promoted to the status of a **theory** and eventually comes to be regarded as a scientific "fact" or "law."

1-C Fact and Fiction

"It's a scientific fact" is often presented as the clincher to an argument. Most scientists, however, would argue that there is no such thing as a fact. The doubts and uncertainties inherent in scientific method make a mockery of the belief that, if it is scientific, it must be right.

"Facts" are usually thought of as things or events which are repeated in identical fashion or about which we have unambiguous records. "The sun rises every morning" looks like a fact on first glance, but it is really a prediction about something that will happen in the future. Another reason that "facts" are less sure than they seem is that they depend upon our faith in our senses or in some extension of our senses. A group of people may agree that a particular object is a table. Is that a fact? No, it is a statement resulting from a convention or definition; all have agreed to call that sort of object a table. Suppose several people look at two photographs, one of a table and one of an object floating in a lake. Everyone may agree, if the first photograph is a clear one, that the object portrayed is a table. When they look at the second photograph, one of them may say, "That is a Loch Ness monster," but the others may legitimately disagree. When technology, in the form of a camera, microscope, or oscilloscope intervenes between our senses and an object, as it often must in scientific research, the problem of interpreting what we see or hear or smell becomes even more subject to doubt. Thus a "fact" is really a piece of information that we, for present purposes, choose to call a fact, probably because we believe in it strongly, or because it seems highly likely that it will be repeated without change.

Although scientific facts, laws, and truths are much less reliable than is generally believed, most scientists do think that their methods discover useful information about objects and events, and that careful study increases the probability that science's generalizations about nature are a close approximation to "reality." Although scientists are motivated by many different goals and ambitions, the public support given to science rests on the belief that a better understanding of natural phenomena increases our ability to promote our well-being.

Much public support goes to scientific projects investigating problems of immediate concern, such as cancer research or alternative energy sources. Indeed, many scientists go through elaborate contortions to show how their work will be "relevant" to such areas of human concern when they write proposals to obtain funds for a research project. However, there is still a great deal of basic, or "pure," research to be done to discover the underlying principles of why objects in the living and non-living world around us behave as they do. Although such research may not immediately benefit the human race, it adds to our understanding of the world and will almost inevitably be put to use sooner or later. And even work that does not find an

Figure 1-4 When is a fact not a fact? Nineteenth century doctors were taught that men and women breathed differently: men used their diaphragms (the sheet of muscle below the rib cage) to expand their chests, whereas women raised the ribs near the top of the chest. Finally, a woman doctor found that women breathed in this way because their clothes were so fashionably tight that the diaphragm could not move far enough to admit air into the lungs. Some women, like the one in this drawing of 1870 styles, even had their lower ribs removed surgically so that they could lace their waists more tightly.

Figure 1-5 Science often becomes embroiled in political controversy. We have awakened to the fascinating intelligence of whales at a time when many whale species are in danger of extinction because so many of them have been caught and turned into dog food by whaling ships like this one. (Biophoto Associates)

immediate application can be intellectually satisfying, just as is the painting of a picture or the writing of a play. It is interesting that people will accept "art for art's sake," but often will not grant the same privilege to science, which must work for its keep.

1-D The Limitations of Science

Science is only one way of exploring the world around us. Historians try to understand the past (and sometimes to predict the future) by studying what people have done in the past. Religion attempts to explain truths about the human spirit, and philosophy collects information from many sources to draw conclusions about reality and human life.

Because it deals only with things that can be experienced directly or indirectly through the senses (sight and hearing, for example), science is completely excluded from phenomena that cannot be experienced in this way. Thus, by definition, science has nothing to say about the supernatural. As the biologist George Gaylord Simpson put it: "This is not to say that science necessarily denies the existence of immaterial or supernatural relationships, but only that, whether or not they exist, they are not the business of science." A recent Vatican Council report argued that religion should not censor science: "research performed in a truly scientific manner can never be in contrast with faith because both profane and religious realities have their origin in the same God."

That science does not deal with the supernatural or with the illogical does not mean that scientists are any less religious, political, or illogical than anyone else. Furthermore, the social environment strongly influences how scientists think and what projects they work on. Not many scientists today study the physics of how to build pyramids to last for a thousand years, because we do not want to build pyramids for the burial of our rulers. Similarly, politics rules science in that funds from government agencies provide most of the financial support for scientific research.

The history of science is replete with scientific dogmas that turned out to be wrong, although for a time they were widely accepted by other scientists. This is one reason why the cautious person—or society—will not place too much faith or invest heavily in a new scientific discovery until it is fairly clear that the theory will stand the test of time.

Figure 1-6 Living things use energy. These fireflies use some of the energy in their food to produce flashes of light used to attract mates. (Biophoto Associates, N.H.P.A.)

Figure 1-7 Living things respond actively to their environments. This swan is chasing away a duck that has come too close for the swan's comfort.

Scientists are fond of saying that science is never good or bad; only society's use of science has moral consequences. From a purist's point of view, this is true. The discovery that the atom could be split was merely a scientific discovery with no moral implications. It was society's decision to use this knowledge to build an atom bomb that produced the moral dilemma of whether or not it was ever right to use such a devastating weapon.

Despite the traditionally ostrich-like approach of scientists to the moral implications of their work, more and more scientists now feel that they must become involved in society's moral decisions about science, if only to make sure that the people responsible for the decisions are basing them on valid information. Some scientists even go so far as to say that certain sorts of research should not be done until society has worked out its moral position on the consequences. For instance, research on producing human babies outside the womb was banned in the United States during the 1970s because it was unclear whether or not a researcher might be held legally responsible for human life produced in this way. Some scientists now feel that they must take, and that society may force them to take, more moral responsibility for the consequences of their research. This attitude, carried to extremes, can destroy science. The Western world experienced about 500 years in which very few scientific discoveries were made because particular sorts of research or findings violated religious teachings. Few of us want to return to the Dark Ages.

There is no simple solution to this dilemma. The peaceful coexistence of science with society depends on citizens who understand what science is and what it can and cannot do, and who do not confuse scientific with moral, economic, or political values.

1-E   What Is Life?

This book is an introduction to biology, one of the branches of science. Although biology is the study of living things, it is notoriously difficult to define life. Living organisms have a number of characteristics that most nonliving systems do not have, yet none of these characteristics is unique to life. For each characteristic, we can think of some nonliving system that also shows the property. With this in mind, however, we may list the main features of living organisms:

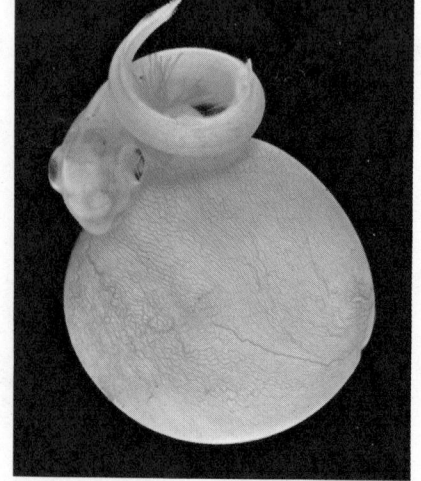

Figure 1-8 Living things develop. This dogfish embryo, lying on top of the yolk sac from which it draws nourishment, will develop into one of the smaller sharks. (Biophoto Associates)

1. *Living things are highly ordered.* The chemicals that make up living organisms are much more highly organized on a molecular level than are the chemicals that make up most nonliving systems.

Figure 1-9 Living things reproduce themselves. These flowers are the reproductive structures of pansy plants and will form the seeds that develop into new plants. (William Camp)

Figure 1-10 The information an organism needs is passed from each organism to its offspring. These mushrooms look alike because they have all inherited similar information from their parents. (Biophoto Associates, N.H.P.A.)

2. *Living things take energy from their environment.* Most organisms depend, directly or indirectly, on energy from the sun; green plants use this energy to make the food that supports virtually all other organisms on earth. All organisms use the energy from their food to maintain the orderliness of their bodies, and to grow and reproduce.

3. *Living things respond actively to their environment.* When you push a stone it may move passively. When you push an animal it may respond actively by running away, bending towards you, or rolling up into a ball. The stem of a plant bends toward light; a root grows downward. The capacity to respond to outside stimuli is universal to living things.

4. *Living things are adapted to their environment.* Living organisms and their component parts are well suited to their ways of life. Fish, earthworms, and frogs are all constructed in such a way that we can predict roughly how they live merely by looking at them. The fact that organisms are adapted to their environments in this way is one result of evolution (Section 1-F).

5. *Living things develop.* Everything changes with time, but living organisms change in complex ways that we call development. A crystal may grow by the addition of identical or similar molecules, but a plant or animal will develop new branches and organs, which may be chemically and structurally different from the chemicals and structures that produced it.

6. *Living things reproduce themselves.* New animals and plants arise only from the reproduction of other animals and plants. Living things produce offspring that are identical or nearly identical to themselves.

7. *The information that each organism needs to survive, develop, and reproduce is segregated within the organism and passed from each organism to its offspring.* The only information in a rock is the whole rock. Living things, on the other hand, contain a separate information store, their genetic material, which specifies the range of activities and structures that the organism can produce.

Figure 1-11 This sheep belongs to a breed with particularly heavy wool; it is adapted to cold, windy winters on high moorland pastures in Europe.

1-F   Evolution and Natural Selection

The characteristics of a living organism result from the interactions between its inherited genetic information and its environment. This interaction is the basis for the most important generalization in biology, that organisms evolve by means of natural selection. This theory can account for all of the properties of life listed above, and it predicts that these properties will be further enhanced in the future.

Evolution is the process by which organisms of one generation come to differ from those of preceding generations. For instance, most biologists believe that human beings have evolved from now-extinct animals that looked much like chimpanzees, and that this happened through a gradual accumulation of changes from generation to generation. Evolution occurs because individuals differ from each other in their genetic material, the material that determines how an organism obtains and uses energy, grows, develops, responds to its environment, and reproduces. An individual's chances of successfully coping with its environment and producing offspring will depend to some extent on the information carried by its genetic material. Individuals whose genetic material makes them well-equipped to deal with their particular environment will be more likely to survive and reproduce than will individuals with less favorable genetic material.

Figure 1-12 Living things are adapted to their environments. This delicate shrimp lives on a sea anemone's tentacles surrounded by stinging organs that kill or damage any other organism that touches them. The shrimp is immune to the stings, adapted to this way of life. (Steven Webster)

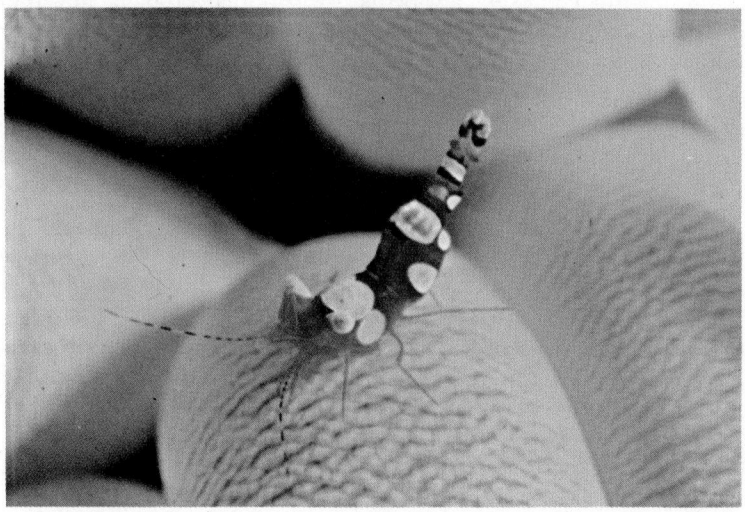

Figure 1-13 A tadpole, which will develop into an adult frog. (Biophoto Associates)

A fundamental reason for evolution is that in each generation some individuals of a species reproduce and some do not, and of those that reproduce, some produce more offspring than others. For instance, the length of an animal's hair is largely determined by its genes. A very cold winter may kill many individuals with short, sparse hair. Individuals with longer, thicker fur are more likely to survive the winter and reproduce the following spring. Because more animals with thicker fur breed and pass on the genetic material that dictates the growth of thick hair, a larger proportion of individuals in the next generation of the population have thick fur. **Natural selection** is the mechanism that allows a higher proportion of individuals with certain characteristics to reproduce, compared to individuals with other characteristics. In this example, low temperature is the agent of natural selection; in other words, cold acts as the **selective pressure** against individuals with scanty coats. As a result of natural selection, the genetic makeup of a population will change with time, and this change is evolution. The **evolutionary success** of an individual is measured by its contribution to the genetic makeup of future generations; the greater the proportion of individuals in future generations that are its descendants, the more evolutionarily successful the individual. Thus it does not really matter how long an organism lives; only its reproduction counts.

The result of natural selection is that populations undergo **adaptation**, or changes appropriate to their environments, over a period of time. The sum of all the selective pressures that a population encounters in its environment operates to select those genetic characteristics that are **adapted**, or well-suited, to the environment. For instance, through selection, populations living in cold areas will, during the course of their evolution, become better adapted to withstand the cold.

When we say that selection causes organisms to become adapted to their environments, we should note that "environment" in this evolutionary context is a catchall word that means much more than merely whether an organism lives in a forest rather than a desert and whether or not it can obtain enough food. It includes all the external factors that can affect whether the organism lives to reproduce.

An organism's environment includes its external environment as an embryo, juvenile, and adult. Let us, for example, consider a frog. Whether it successfully meets the "pressure" of its environment depends on the speed and normality of its embryonic development, whether bacteria penetrate the jelly coat of the egg and destroy it during development, whether as a tadpole it can find sufficient food for rapid growth and avoid being eaten by a predator, whether the pond in which it lives as a tadpole dries up before it becomes a frog, and whether as a small frog it avoids death by disease and predators. To make things more complicated, environmental pressures are frequently contradictory. For instance, a hot summer is in one way advantageous to our hypothetical frog because frog embryos develop faster at higher temperatures, but a hot summer also increases the chance that the tadpole's pond will dry up before it is ready to live on land. Even worse, environmental pressures are subject to frequent change. The frog must have characteristics that allow it to withstand both the heat of summer and the cold of winter; it should remain still to be safe from some predators and move quickly to escape from others, and so forth. Thus the genetic makeup of a frog is a finely balanced compromise brought about by selection for a number of opposing characteristics.

Adaptations

We have just defined adaptation as the process by which populations evolve to become better and better suited to their environments as a result of natural selection. Biologists use the word in a second way. An **adaptation** is any genetically determined trait that has been selected for and that occurs in a large part of the population because it increases an individual's chance of reproducing successfully. Much of biology involves the study of the adaptations of various organisms and how these adaptations work.

Adaptations may be broadly classed as anatomical, physiological, or behavioral. **Anatomical** adaptations are those involving the physical structure of the organism. For instance, a penguin's flippers are an anatomical adaptation that permits it to swim. An organism's **physiology** is all of the internal workings of its body: the biochemistry of its cells and the processes that allow it to digest food, exchange gases, excrete wastes, reproduce, move, and sense and respond to the outside world. An example of an extreme physiological adaptation to temperature is seen in the ability of some bacteria to live in hot springs at temperatures up to 80°C (175°F), which would destroy all biochemical activity in most other organisms. An example of an impressive behavioral adaptation is the ability of a kangaroo rat to eat the leaves of the desert saltbush. No other animal can eat this plant because its leaves are full of salt crystals. The kangaroo rat flakes off the salt-filled outer layer of the leaf with its front teeth, and then eats the salt-free inner part. This ability to prepare its food is a behavioral adaptation allowing the kangaroo rat to eat a food that is completely unavailable to other animals.

Energy and Natural Selection

We have seen that all living things must take in and use energy to maintain their bodies, to grow, to obtain more energy, and to reproduce. Since the evolutionarily successful individual is one that leaves descendants in future generations, natural selection favors those individuals that can channel the most energy into

Figure 1-14 Always in hot water: the pastel bands of color in the runoff from this geyser are colonies of various species of bacteria.

producing offspring. The use of energy in other activities such as feeding, fighting, or growing is selectively advantageous only so far as these activities result in the organism's accumulating more energy to produce offspring.

Each individual has an "energy income," all of the energy that it acquires during its lifetime. It also has an "energy budget," its allotment of different amounts of energy to various activities. The most evolutionarily successful organisms are those most effective in conversion of energy to offspring. This does not mean that organisms use all their energy directly to produce offspring. For example, suppose that a tree converts some of its energy into growing a large root system. The energy thus spent cannot be used to produce offspring. Its large root system may enable the tree to obtain a great deal of water and minerals from the soil and so to produce more leaves, another diversion of energy away from the production of offspring. However, its many leaves may enable the tree to synthesize more food than it would have otherwise, and so allow it to recoup some of its previous energy expenditure by producing more offspring in the end. Thus organisms make energy investments that may ultimately yield energy gains that can be reinvested in the production of offspring. Sometimes these investments will turn out to be selectively disadvantageous because they postpone production of offspring. If the organism meets an early death, it will never get a chance to reproduce. So any item in an organism's energy budget must have the potential to confer an ultimate reproductive gain that is commensurate with the risks involved in diverting energy away from the immediate production of offspring.

This discussion suggests that organisms should evolve in such a way that they use energy efficiently, saving as much energy as possible to put into offspring. This is sometimes true. However, natural selection also favors those individuals that grow and reproduce most rapidly. While we shall encounter many adaptations that seem to have evolved because they save their owners energy, we shall also encounter situations in which organisms expend energy very inefficiently on rapid growth or reproduction.

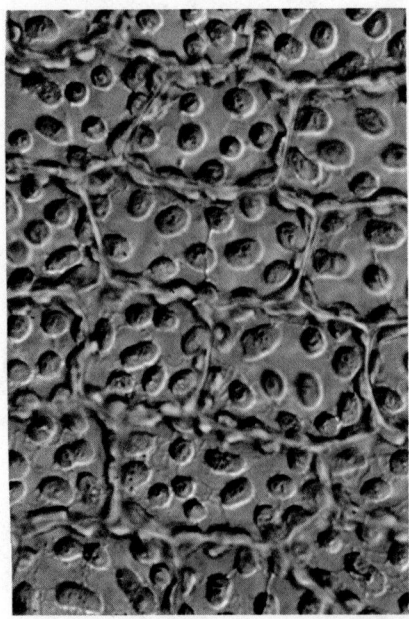

Figure 1-15 Energy conversion. The green objects in these cells of a moss are chloroplasts, structures used by mosses, trees and other plants to take in solar energy and convert it into food. (Biophoto Associates)

SUMMARY

Scientific knowledge develops by subjecting problems to the scientific method. First, observations are made on the problem. Alternate hypotheses that might explain the observations are then formulated, and the hypotheses are tested by experiments that will eliminate one or more hypotheses and strengthen those that remain. Scientific facts and theories are useful generalizations, but they are always open to question.

Biology is the science that studies living things. Life is difficult to define, but in general it can be said that living things take energy from their environments and use it to maintain a high degree of molecular order; they respond actively to stimuli, are adapted to their environments, and contain all the genetic information they need to develop, survive, and reproduce.

The theory of evolution states that populations of organisms change from one generation to another as a result of natural selection, the process by which only individuals with certain traits reproduce in each generation. Evolution ensures that a population of organisms will become increasingly well adapted to its environment. An adaptation that increases the ability to survive and reproduce becomes more common in the population as members who lack the adaptation are eliminated by natural selection.

Natural selection ensures that those organisms that are most effective in converting energy into offspring will be evolutionarily successful.

In the remainder of this book we will use the theory of evolution by natural selection and its resulting adaptations as a framework for the study of biology. The theory has been presented here in simplified terms; it will be developed with more sophistication in Chapter 16.

OBJECTIVES

From your study of this chapter you should be able to:

1. List three steps in the scientific method, and apply them to investigating a sample scientific problem.
2. List seven characteristics of living things, and state why it is difficult to define life.
3. Define evolution, natural selection, selective pressure, evolutionary success, adaptation, and energy budget.

QUESTIONS FOR DISCUSSION

1. How would you test the hypothesis that a fly's black and yellow stripes allow it to enter a beehive and steal honey?
2. After every hard rain you find dead earthworms lying on the sidewalk. What experiments would you perform to show the cause of death?
3. Each characteristic of life can be found in some nonliving thing. Can you think of examples of these?
4. To what extent do you think scientists should be held responsible for the social and moral consequences of their discoveries?
5. What might you expect was the selective pressure that resulted in each of the following adaptations?
   an elephant's trunk
   a leopard's spots
   human language
   the scent of honeysuckle
   the bark of a tree

REFERENCES AND FURTHER READING

Ayala, F. J., and T. Dobzhansky (eds.). *Studies in the Philosophy of Biology*. Berkeley: University of California Press, 1974. An excellent collection of essays on biology, its methods and impact on society.

Roszak, T. *Where the Wasteland Ends*. Garden City, N.Y.: Doubleday, 1973. An ambitious critique of modern science by a man who believes that a scientific view of the world dominates western society and is responsible for much of its malaise.

Stent, G. S. "Prematurity and uniqueness in scientific discovery." *Scientific American*, September 1972. Compares scientific discoveries with artistic creations and discusses what it means to say that a discovery is "ahead of its time."

Wallace, B., and A. M. Srb. *Adaptation*, 2nd ed. Englewood Cliffs, N.J.: Prentice-Hall, 1964. A rigorous discussion of adaptation, what it is and how it is brought about, distinguishing an adaptation from the process of adaptation.

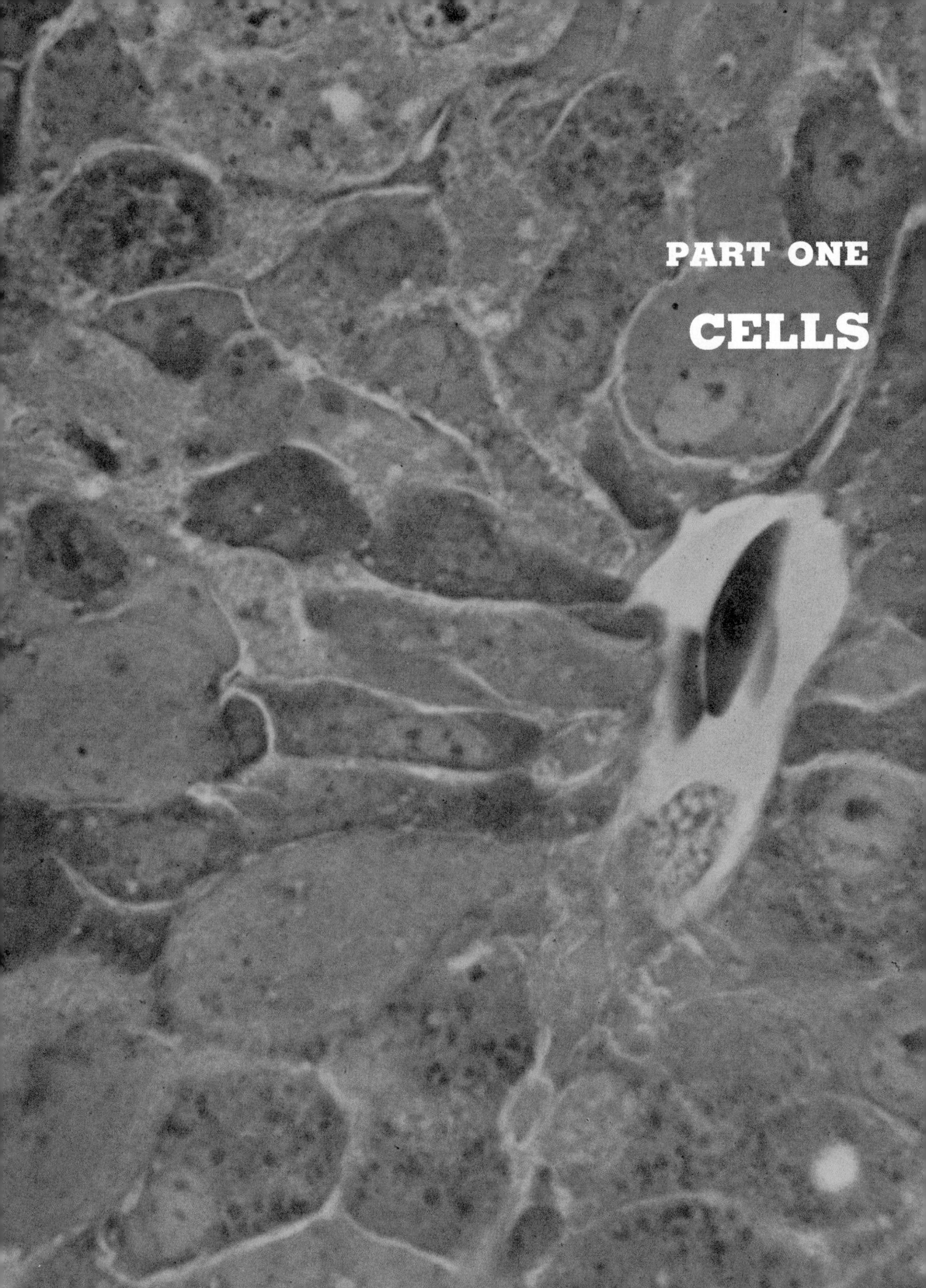

PART ONE

CELLS

CHAPTER 2

SOME BASIC CHEMISTRY

A human being has been defined as "twenty gallons of water and $5.00 worth of assorted chemicals." All organisms are composed of chemicals and their environment surrounds them with more chemicals. In this chapter we shall consider some aspects of chemistry that are needed for the study of biology.

2-A   Chemical Elements of Life

A chemical element is a substance that does not come apart into substances of other kinds during ordinary chemical reactions. Some familiar examples are copper, oxygen, and carbon; carbon occurs in two common forms, graphite (the "lead" in a pencil) and diamond.

Chemists have discovered more than 100 chemical elements, but living organisms contain only about 20. However, these are not the 20 most common elements. For example, silicon, the most common element in the earth's crust, is over 300 times more abundant (by weight) than carbon, yet carbon is an indispensable component of every living thing, whereas silicon is a major constituent of very few. This is because carbon has unique properties that make it peculiarly suitable to form biological structures, and evolution has selected those organisms whose chemical compositions are best at carrying on the processes of growth, activity, and reproduction.

Organisms select and take in the chemicals they need, rejecting others that are present in the environment. The roots of a plant, for example, take up certain minerals and exclude others. Animals have complex behavior patterns that enable them to obtain necessary chemicals. For instance, animals will travel many miles to visit salt deposits where they can replenish their supplies of sodium.

2-B Structure of Atoms

An **atom** is the smallest unit of an element that retains all the properties of the element. In the center of an atom, occupying a space equivalent to that of a marble in a football field, is a **nucleus**, which contains two kinds of particles, **protons** and **neutrons**. Protons and neutrons both have a mass* of about 1 atomic mass unit. Each proton, however, bears one positive electric charge, whereas a neutron has no charge. A third type of atomic particle is the electron. **Electrons** are so light that their mass is usually considered to be zero; each electron bears one negative electric charge. Electrons move so rapidly through the area around the nucleus that they occupy most of the space in an atom; this space is referred to as the atom's **electron orbitals** or **electron shells**.

The number of protons equals the number of electrons in an atom, so that an atom has a net electric charge of zero. The smallest atom, an atom of hydrogen, consists of one proton and one electron (Figure 2–1). All the atoms of a particular

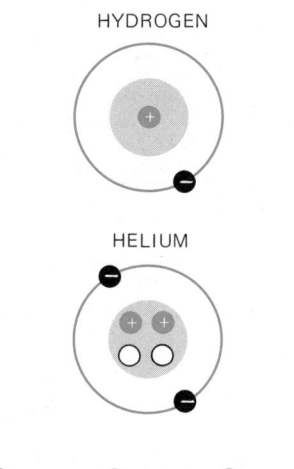

HYDROGEN

HELIUM

🔴 Electron ⚪ Neutron ⊕ Proton

Figure 2-1 Atomic models of hydrogen and helium atoms as Niels Bohr imagined them.

---

* Mass A measure of the quantity of matter; differs from weight, which is a measure of how strongly a body is attracted to earth and involves the interaction of mass with gravity.

TABLE 2-1 **CHEMICAL ELEMENTS FOUND IN ANIMALS, THEIR APPROXIMATE ABUNDANCE BY WEIGHT AND THEIR ATOMIC MASSES**

ELEMENT	SYMBOL[1]	WEIGHT, PERCENT	ATOMIC MASS[2]
Oxygen	O	62	16.0
Carbon	C	20	12.0
Hydrogen	H	10	1.0
Nitrogen	N	3.3	14.0
Calcium	Ca	2.5	40.0
Phosphorus	P	1.0	31.0
Sulfur	S	0.25	32.0
Potassium	K	0.25	39.0
Chlorine	Cl	0.2	35.5
Sodium	Na	0.10	23.0
Magnesium	Mg	0.07	24.5
Iodine	I	0.01	127.0
Iron	Fe	0.01	56.0
		99.59	

TRACE ELEMENTS (needed in very small amounts)

Copper	Cu		63.5
Manganese	Mn		55.0
Molybdenum	Mo		96.0
Cobalt	Co		59.0
Boron	B		11.0
Zinc	Zn		65.5
Fluorine	F		19.0
Selenium	Se		79.0
Chromium	Cr		52.0

[1]Each element is assigned a one- or two-letter symbol that is used as a chemical "shorthand" in writing chemical formulas and equations.

[2]Where atomic mass is not an integer, the mass given represents the average between commonly occurring isotopes of the element.

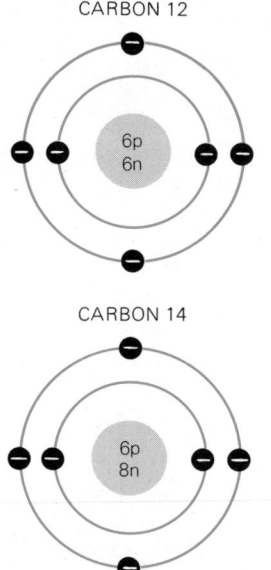

CARBON 12

CARBON 14

Figure 2-2 Isotopes of carbon. Carbon 12 is the most common isotope of carbon. Carbon 14 is radioactive carbon, often used in biological experiments to trace what happens to a carbon atom in the biochemistry or physiology of an organism.

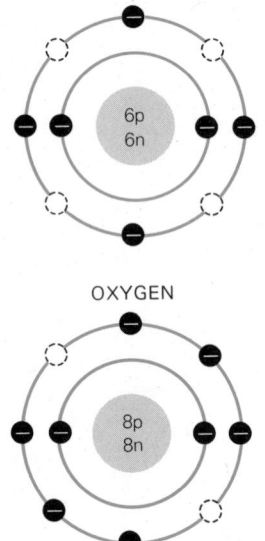

CARBON

OXYGEN

Figure 2-3 The Bohr models for carbon and oxygen. Carbon has room for four more electrons in its outer shell; oxygen can accommodate two more electrons before its outer shell has a full complement of eight.

element have the same number of protons in their nuclei; this number is the **atomic number** of the element. Because each element has a different characteristic number of protons, it also has a different atomic number from all other elements. For example, the atomic number of hydrogen is one, and of helium two.

The mass of an atom, the **atomic mass**, is determined by adding the number of its protons and the number of its neutrons, since protons and neutrons have the same mass. Some elements occur in two or more different forms, or **isotopes**; two isotopes of the same element have different numbers of neutrons, but the same number of protons (Figure 2-2). Thus they have the same atomic number but a different atomic mass.

Although atoms are too small to be examined directly, chemists and physicists have deduced many of their properties and developed models of their structure.

A model of atomic structure was advanced in 1913 by the physicist Niels Bohr (Figures 2-1, 2-2, 2-3). Since electrons are negatively charged, they are attracted to the positively charged nucleus. Bohr visualized electrons as moving in spherical spaces, called electron shells or **energy levels**, around the nucleus. The innermost shell can contain only two electrons; in the common elements, the next two outer shells can hold up to eight electrons each. Atoms tend to fill their outer shells with electrons through chemical reactions in which they gain, lose, or share electrons. For example, carbon has six protons, six neutrons, and six electrons. Hence it has two electrons in its first energy level, but only four in its second energy level, or outer shell. It can form a stable outer shell of eight electrons by gaining, losing, or sharing four electrons (Figure 2-3). The number of electrons that an element must share, lose, or gain to complete a stable outer electron shell is known as its **valence**. Thus carbon has a valence of four and oxygen has a valence of two. The valence of an atom determines how many connections an atom can form when it combines with other atoms.

A more realistic, but still simplified, way to look at an atom is to consider electrons as particles in constant, random motion. In this model of the atom, an electron's orbital is an imaginary sheath surrounding the space in which the electron passes most of its time (Figure 2-4). Different electrons have different amounts of energy which determine the orbitals in which they occur. An electron in the first energy level, closest to the nucleus, has a spherical orbital known as a *1s* orbital. This *1s* orbital can hold two electrons and no more.

The first two electrons in the second energy level (and in all higher energy levels) also occupy a spherical orbital, in this case the *2s* orbital. However, the second energy level can also have up to six more electrons. These electrons occupy three dumbbell-shaped orbitals, called *2p* orbitals, each of which contains two electrons (Figure 2-4). The orbitals of an atom define **electron charge clouds**, volumes of space in which the electrons can move. The shape of the charge clouds can be distorted by the presence of other atoms.

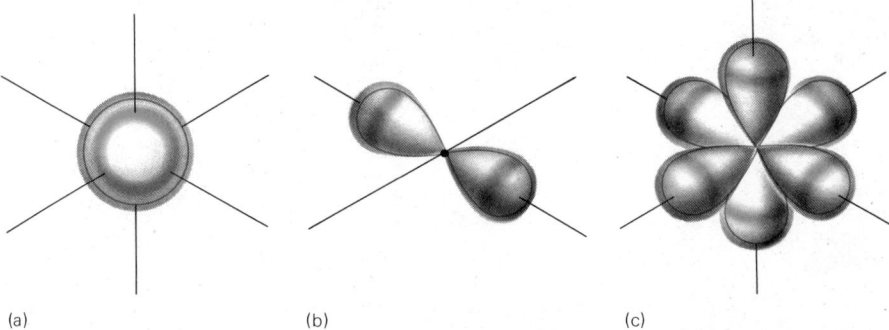

(a)   (b)   (c)

Figure 2-4 Electron clouds. (a) A *1s* orbital. Density of colored shading indicates the proportion of time spent by *1s* electrons in each part of the space around the nucleus. Although the positively charged nucleus attracts the negatively charged electrons, the electrons spend more time some distance from the nucleus, where a greater volume of space is available. The colored line surrounds the electron charge cloud, the space in which an electron spends 90 percent of its time. (b) A *2p* orbital; it can contain 2 electrons. (c) There are three *2p* orbitals, arranged at right angles to one another.

2-C Bonds Between Atoms

There are very few free atoms in nature. Since most atoms do not have stable electron configurations, they undergo chemical reactions in which they join, or **bond**, with one or more other atoms to achieve stable outer electron shells.

Three types of bonds which can be formed between atoms are important in living things. An **ionic bond** is formed when one atom gives up one or more of its outermost electrons, which is taken by another atom. As a result of this atomic give and take, both atoms end up with stable outermost energy shells. Atoms with fewer than four outer electrons usually lose electrons when they form ionic bonds; those with more than four gain electrons. For instance, sodium has one electron in its outermost shell. If that electron leaves, the atom will be left with a stable outer shell containing eight electrons. If sodium's outermost electron is given to a chlorine atom, which contains seven electrons in its outer shell, the chlorine atom will become stable, with eight electrons in its outermost shell (Figure 2-5). As a result of gaining or losing an electron, each newly formed particle now carries an electric charge and is known as an **ion**. The sodium ion has 11 protons but only 10 electrons, for a net charge of $+1$. The ionic form of chlorine is called a chlor*ide* ion; it has 18 electrons and only 17 protons, for a net charge of -1.

In a **covalent bond**, two atoms share a pair of electrons so that each has a stable, complete outer energy shell (Figure 2-6). In a double covalent bond, the bonded atoms share two pairs of electrons (Figure 2-7).

When two atoms of the same element bond covalently, the bond is electrically symmetrical (Figures 2-6, 2-7). Because the two atoms attract electrons equally, the average position of the shared electrons will be midway between the two atoms. If the two atoms are of different elements, one atom will usually be more **electronegative**; that is, it will attract electrons more strongly than the other, and the

HYDROGEN GAS (H₂)

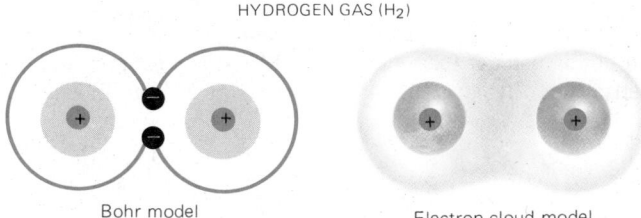

Bohr model   Electron cloud model

Figure 2-6 Two ways of drawing covalent bonding between two hydrogen atoms. Each hydrogen atom has one electron and so tends to gain another electron to complete its outer shell. By sharing their electrons, two hydrogen atoms both complete their outer shells.

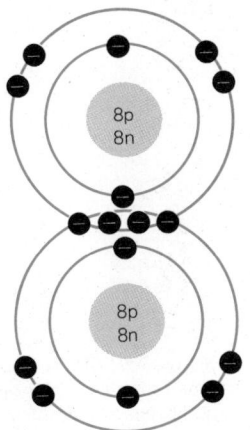

Figure 2-7 A double covalent bond between two oxygen atoms. The atoms share two pairs of electrons, so that each completes its outer electron shell.

SODIUM ATOM

CHLORINE ATOM

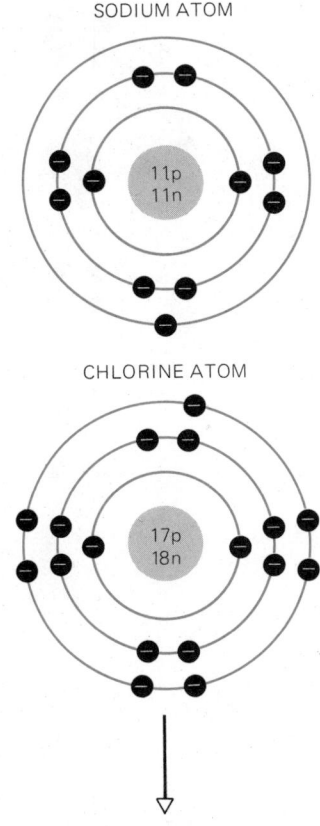

SODIUM ION (Na⁺)

(+)

CHLORIDE ION (Cl⁻)

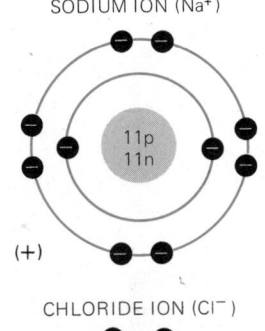

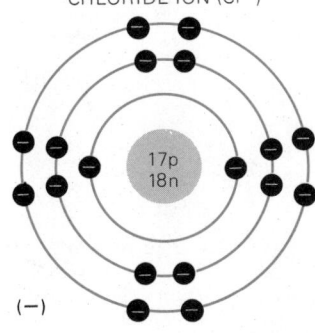

(−)

Figure 2-5 Formation of an ionic bond.

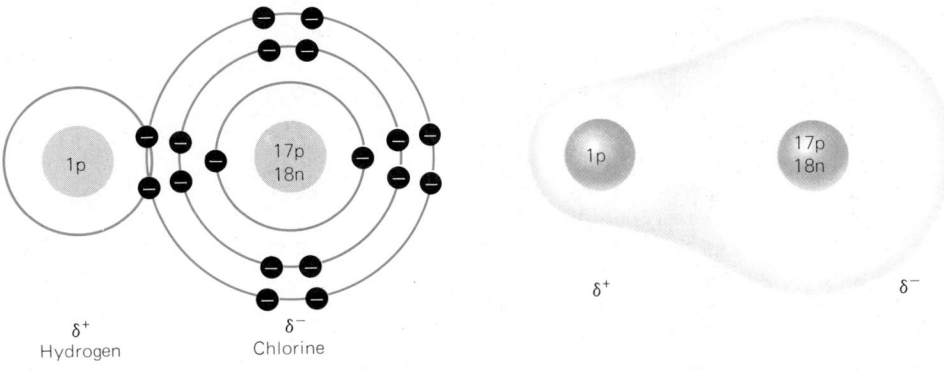

δ^+
Hydrogen

δ^-
Chlorine

δ^+

δ^-

Figure 2-8 Two representations of the polar molecule of hydrogen chloride. The shared pair of electrons in the covalent bond is more strongly attracted to the chlorine nucleus than to the hydrogen nucleus. As a result, the chlorine has a partial negative charge (δ^-) and the hydrogen atom has partly lost its electron, leaving it with a partial positive charge (δ^+).

shared electrons will spend more time near this atom. The resulting covalent bond is **polar**, that is, electrically asymmetrical (Figure 2–8). Such a bond, with some areas that tend to be electrically more negative and other areas that tend to be more positive, is known as a **polar bond**. The degree of polarity of bonds can be arranged in a continuum. In **nonpolar** covalent bonds, the position of the shared electrons, averaged over time, is symmetrical. In polar covalent bonds, the electrons tend to spend more time near one end of the bond; the greater the difference in electronegativity between two atoms, the more pronounced is this tendency. Ionic bonding is the extreme case of electrical polarity. The electrons spend almost all their time near one of the nuclei, and one atom essentially gives up one or more electrons to other atoms.

Besides covalent and ionic bonds, the third type of bond between atoms that is important in the chemistry of living organisms is the **hydrogen bond** (Figure 2–9). In this type of bond, one member of the interaction is a hydrogen atom that has a polar covalent bond to a strongly electronegative atom, usually oxygen or nitrogen. The electronegative atom pulls the shared electrons away from the hydrogen nucleus, leaving it somewhat exposed, and with a partial positive charge. Because of this partial positive charge, the hydrogen nucleus is attracted to another atom, with a partial negative charge (again usually oxygen or nitrogen), and this forms the hydrogen bond. Hydrogen bonds usually form between atoms that belong to different molecules, or between atoms that are on different parts of a large molecule.

2-D Compounds and Molecules

Because each type of atom gains, loses, or shares a particular number of electrons when it forms stable energy levels, any atom can form only limited and predictable types of bonds. A **compound** consists of two or more atoms of different elements, combined in specific proportions, with a specific pattern of bonds. Each compound, like each chemical element, has a definite formula and a definite set of properties, which differ from those of its component elements. A **molecule** is the

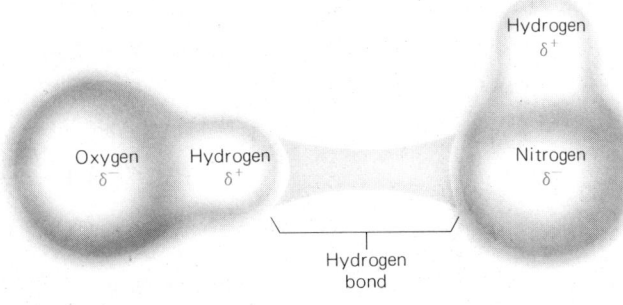

Hydrogen
δ^+

Oxygen
δ^-

Hydrogen
δ^+

Nitrogen
δ^-

Hydrogen
bond

Figure 2-9 A hydrogen bond (color) is a weak attraction between a polar-bonded hydrogen with a partial positive charge (δ^+) and a polar-bonded atom of nitrogen or oxygen with a partial negative charge (δ^-).

smallest part of a compound that retains all the properties of that compound, just as an atom is the smallest part of an element that retains all the properties of the element. Ionically bonded substances are usually referred to as being composed of ions rather than of molecules.

A **molecular formula** is a shorthand way of representing the kinds and amounts of different atoms present in a molecule, using the symbols for the elements. The symbol for a helium atom, He, for example, is also the symbol for a helium molecule, since helium has a filled outer shell, making it so stable that it will not bond to other atoms. NaCl represents sodium chloride or table salt. The formula NaCl indicates that table salt contains sodium and chloride ions in a 1:1 ratio. Water is H_2O. The subscript 2 indicates that there are two hydrogen atoms (H) for every oxygen atom (O) in a water molecule. Likewise, in CO_2—carbon dioxide—there are two oxygen atoms for each atom of carbon (C); and a molecule of oxygen gas, O_2, is made up of two oxygen atoms. Thus it is called molecular oxygen.

A **structural formula** takes up more space than a molecular formula, but it gives information about the arrangement of atoms in the molecule, as well as the numbers of each type of atom. Structurally, water is written H—O—H, indicating that each hydrogen is attached to the oxygen atom, but that the hydrogens are not bonded to each other. In carbon dioxide, each oxygen is double-bonded to the carbon atom: O=C=O.

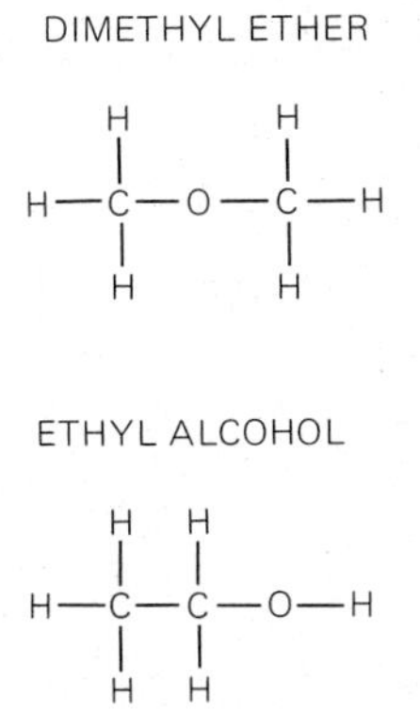

Structural formulas are indispensable in cases where the molecular formulas for two different molecules are the same. For example, both dimethyl ether and ethyl alcohol* have the molecular formulas C_2H_6O. Their structural formulas are different, though, since the same atoms can be arranged in different ways (Figure 2–10).

The **molecular weight** of a molecule is the sum of the atomic weights of all its atoms. (For the sake of convenience, we will consider mass and weight as equal, although this is not always the case. A man on the moon, for instance, has the same mass as he had on earth, but he weighs a lot less. Weight depends on the force of gravity; mass does not.) Using the atomic masses from Table 2–1, the molecular weight of water, H_2O, can be found as 2×1 for the two hydrogens, $+16$ for the oxygen, or 18 atomic mass units. The **gram molecular weight** (1 **mole**) of a substance is the molecular weight of the substance in grams. For example, a mole of water weighs 18 grams. A mole of any substance contains 6.023×10^{23} molecules. The mole is a useful quantity since it is based on the number of molecules; a mole of table sugar (358 grams) and a mole of ethanol (46 grams) contain the same number of molecules, although one gram of ethanol contains more than seven times as many molecules as a gram of sugar.

A **solution** consists of a **solvent** plus the substances dissolved in it, called **solutes**. The concentration of a solution is almost invariably expressed in molarity in biological experiments; for example, a **molar** (**1M**) solution contains a mole of solute in a total of 1 litre of solution.

DIMETHYL ETHER

ETHYL ALCOHOL

Figure 2-10 Structures of two molecules which both have the molecular formula C_2H_6O.

2-E   Chemical Reactions

During a **chemical reaction**, molecules or ions interact to form new substances. Reactions are often written in the form of chemical equations like this:

$$\underbrace{CH_4 + 2O_2}_{\text{reactants}} \longrightarrow \underbrace{CO_2 + 2H_2O}_{\text{products}}$$

methane*   oxygen   carbon dioxide   water

The starting materials (**reactants**) are written on the left and the **products** on the right, after the arrow. This particular equation indicates that two molecules of oxygen will combine with a single molecule of methane, and that for each carbon

---

* Ethyl alcohol   The type of alcohol in alcoholic beverages (beer, wine, etc.). Also known as ethanol.
* Methane   The main constituent of marsh gas.

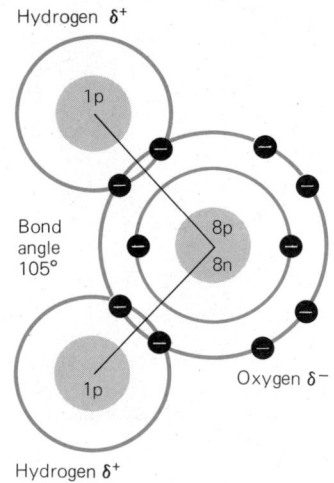

Hydrogen δ^+

1p

Bond angle 105°

8p
8n

Oxygen δ^-

1p

Hydrogen δ^+

(a) Bohr model

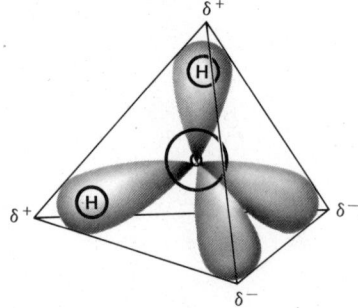

δ^+

(H)

δ^+ (H)

δ^-

δ^-

(b) Electron cloud model

Figure 2-11 Two ways of depicting a molecule of water.

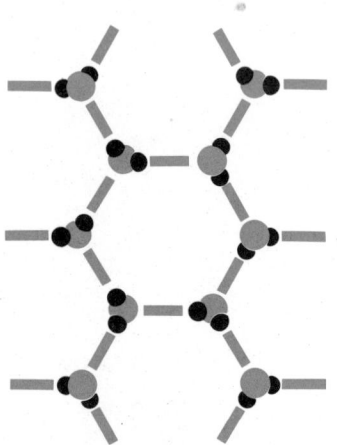

Figure 2-12 The arrangement of water molecules in an ice crystal. Each molecule becomes hydrogen-bonded to four others in the crystal structure. Only three connections are shown here; the fourth would go to a molecule above or below the plane of the page. Colored spheres = oxygens; black spheres = hydrogens; colored lines = hydrogen bonds.

dioxide molecule produced, two water molecules are produced as well. The number of molecules is given to balance the equation, that is, to indicate the proportions of reactants and products: two moles of oxygen are required for the complete combustion of one mole of methane, and two moles of water are produced for each mole of carbon dioxide. Sometimes the arrows in a chemical equation point in both directions, like this:

$$CO_2 + H_2O \rightleftharpoons H_2CO_3$$

This means that the reaction can go either from left to right (forward) or from right to left (backward), depending on the conditions. Such a reaction is said to be reversible.

The **equilibrium point** of a reaction is the point at which the relative concentrations of reactants and products are such that the rates of the forward and backward reactions are equal, and there is no net change in the concentrations of reactants and products. If the relative concentrations of reactants and products change, such as happens when another reaction is steadily creating reactants or removing products, the reaction tends to go more in the direction that will restore equilibrium. In an irreversible reaction, the equilibrium point is reached when all the reactants have been converted to products, whereas the equilibrium points of reversible reactions vary.

2-F Water

Water makes up the bulk of most living organisms, and it is impossible to imagine that life as we know it will be found on any planet that lacks an abundant supply of water. Water is vital to life because it has several extraordinary properties, based on the structure of its molecules.

In a molecule of water, an atom of oxygen is covalently bonded to two atoms of hydrogen (Figure 2–11). The molecule is bent, with an angle of 105° between the two bonds. Because oxygen is more electronegative than hydrogen, the molecule is polar, with a partial negative charge (δ^-) on the oxygen atom, and a partial positive charge on each hydrogen atom (δ^+). These partial charges are distributed in a roughly tetrahedral* configuration, with the oxygen nucleus at the center, the two partially positive hydrogens at two corners, and oxygen's two partially negative unshared electron pairs at the other two corners (Figure 2–11b).

As a result of this electrical polarity, water molecules form hydrogen bonds with one another. In ice crystals, a regular latticework forms as each molecule bonds to four others at the corners of its tetrahedron (Figure 2–12). Liquid water has some paracrystalline areas, with molecules in a regular tetrahedral array but closer together than in ice; at increasing temperatures, molecules in more and more areas of the water become randomly oriented, with rapid formation and breakage of hydrogen bonds. Although each hydrogen bond has a very short lifetime (about 10^{-11} second), large numbers of them acting together can contribute considerable stability to a group of molecules.

The structure of its molecules endows water with a number of properties that are essential to life:

1. Water is a solvent. More substances are soluble in water than in any other known liquid. Because it is polar, water can dissolve ions and polar molecules (Figure 2–13).

Nonpolar substances, such as those composed mainly of carbon and hydrogen atoms, will not dissolve in water, but instead form interfaces with it, such as the interfaces formed when oil and water are placed together. Interfaces between polar and nonpolar substances are the sites for many chemical reactions in living organisms; thus the inability of water to dissolve nonpolar substances is also necessary to life.

---

*Tetrahedral   In the shape of a tetrahedron, a geometric solid having four sides, each a triangle.

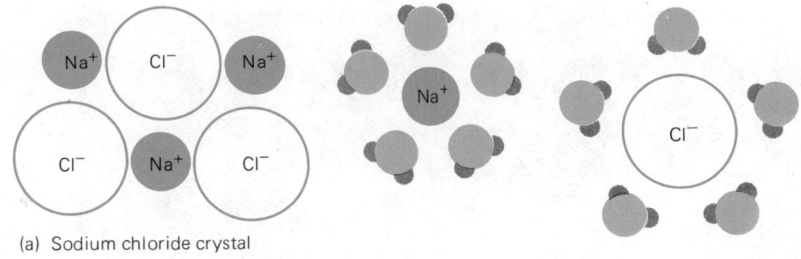

Figure 2-13 Sodium chloride (NaCl) dissolving in water. (a) Their opposite electrical charges attract Na⁺ and Cl⁻ to one another. (b) When NaCl and water meet, Na⁺ attracts the partial negative charges (δ^-) on the oxygens of water molecules, and Cl⁻ attracts the δ^+ charges of hydrogens. Altogether, the tiny electric tugs exerted by water molecules pull Na⁺ and Cl⁻ away from each other.

(a) Sodium chloride crystal

(b) Sodium (Na⁺) and chloride (Cl⁻) ions dissolved in water

2. Water is cohesive and adhesive. Cohesion is the holding together of like substances; **adhesion** is the holding of one substance to a different one. You can fill a glass with water to above the level of the rim of the glass; a water strider can run across the surface of a pond. Both these facts are a result of the **surface tension** of water, which means that a water surface appears to be covered by a "skin." Surface tension results from the cohesion of water molecules to one another by their hydrogen bonds. Water is more cohesive than any other liquid except mercury. That it is both adhesive and cohesive accounts for its **capillarity**, the ability of water to move up a piece of porous paper or to creep through the fine pores in the soil or in a leaf (Figure 2–14). Because it is highly polar, water also adheres strongly to any surface that bears an electric charge.

3. Water has a high specific heat. This means that it takes a lot of heat to raise the temperature of water, and much heat must be lost to lower its temperature. Compared with the air above it, a body of water warms up more slowly in the spring and cools down more slowly in the fall. Since living organisms are made up largely of water, they also gain and lose heat relatively slowly. Organisms produce enormous amounts of heat from chemical reactions in their bodies. Were it not for the high capacity of water to absorb heat, the temperature of an organism would rise to the point where life would be impossible.

Water also has a high **thermal conductivity**, which means that if one part of a body of water is heated, the heat is rapidly conducted throughout the rest of the water. Since all parts of an organism are usually linked to all other parts by water, heat produced in one part of an organism is transferred rapidly all over the body.

4. Water has a high boiling point. Because there is so much attraction between water molecules, it takes a great deal of heat to overcome this attraction, and to change water from a liquid to a gas. Temperatures on earth seldom reach the boiling point of water, and so living organisms need not contend with the problem of boiling away.

5. Water is a good evaporative coolant. Water molecules are attracted to each other so strongly that it takes a lot of energy to vaporize a water molecule. Many land-dwelling organisms make use of this property of water to cool their bodies when they overheat. Water evaporates from their surfaces using heat from the body, so cooling the organism.

6. Water has a high freezing point, and a lower density as a solid than as a liquid. These properties have both advantages and disadvantages for living things. The freezing point of water (or the melting point of ice), 0°C, is frequently

Figure 2-14 Capillarity of water. (a) Fibers in absorbent paper are touched to the surface of water. (b) Water adheres to the fibers and creeps up them. (c) Water that has crept up the fibers coheres to other water molecules still at the surface, pulling them up and filling the space between the fibers with water.

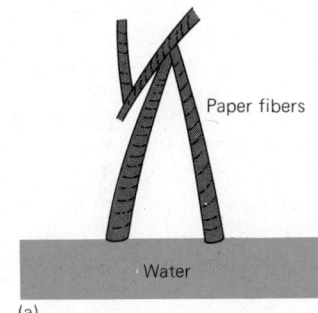
Paper fibers
Water
(a)

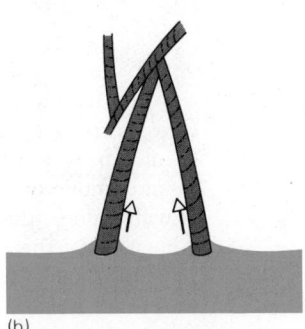
(b)

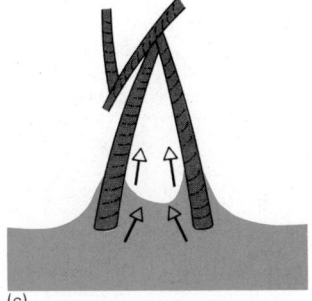
(c)

Figure 2-15 An iceberg in the Antarctic. An iceberg floats because it is composed of ice, which is less dense than water. (U.S. Navy)

reached in many climates. Water also contracts, becoming more dense as it cools. However, it reaches its greatest density at 4°C. As it cools below that point, water begins to expand again, becoming less dense. This is because the molecules begin to form a crystal lattice in which they are not packed so closely as they are in liquid water (Figure 2–12). As a result, ice is less dense than water and it floats. In natural bodies of water, the formation of ice on the top of the water creates a blanket of insulation that shields the underlying water and its inhabitants from further cooling.

When organisms freeze, ice crystals form within their bodies. Because water must expand to form these crystals, the internal structure of the body is destroyed, and subsequent thawing cannot restore its delicate architecture. Some organisms produce "antifreeze" substances that prevent the formation of ice crystals. Winter wheat and many insects and insect eggs have such an adaptation. Other organisms, such as tomato plants, are killed when they freeze and must complete a generation of growth and reproduction in the summer months between frosts.

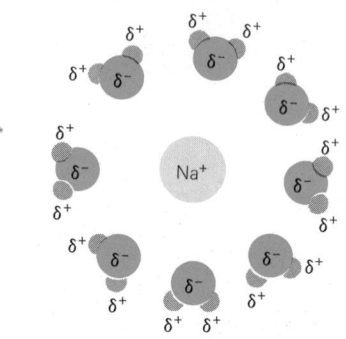

Figure 2-16 A hydrated sodium ion. A shell of water molecules forms because the positively charged sodium ion (Na^+) attracts the partial negative charges of oxygen atoms in water molecules. In shells around negative ions, water molecules orient with their hydrogens (δ^+) nearer the ion.

2-G   Dissociation and the pH Scale

In order to understand many biochemical processes, you must be aware that many substances found in living organisms dissociate into ions when they dissolve in water. Some compounds dissociate completely, whereas others dissociate only to a limited extent, so that, in solution, some of their molecules are intact and some are ionized. Water itself dissociates partially; the most common and important particles so formed are hydrogen ions (H^+) and hydroxyl ions (OH^-):

$$H_2O \longrightarrow H^+ + OH^-$$

Because water molecules carry both partial negative and partial positive charges, they can assist dissociation by forming "shells" around ions, shielding them from the attraction of oppositely charged ions in the solution and allowing them to move independently of one another. These hydrated ions, with their shells of water molecules, behave as if they were larger and move more slowly than would otherwise be the case (Figures 2–13 and 2–16).

Substances may be classified by the particles they yield when they dissociate in water. An **acid** is a substance that releases H^+ when it dissociates in water. For

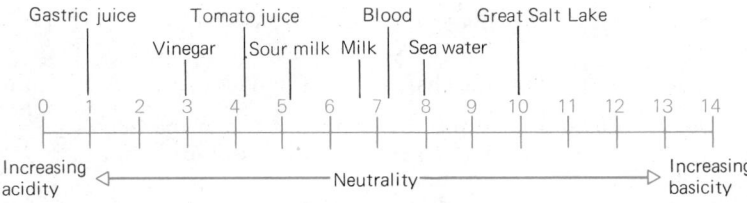

Figure 2-17   The pH scale, with some familiar markers.

instance, hydrogen chloride (HCl) yields hydrogen ions (H⁺) and chloride ions (Cl⁻) in solution, forming hydrochloric acid. A **base** is a substance that releases OH⁻ in water, or that accepts H⁺. Thus the base sodium hydroxide (NaOH)* dissociates into sodium ions (Na⁺) and hydroxyl ions (OH⁻). (Note that by these definitions water is both an acid and a base!) A **salt** is a substance in which the H⁺ of an acid is replaced by other positively charged ions. A salt dissociates into oppositely charged ions; for example, the salt sodium chloride (NaCl) dissociates into sodium and chloride ions.

The acidity or alkalinity (basicity) of a solution is indicated by a figure known as pH. **pH** is the negative logarithm of the molar concentration of hydrogen ions in the solution: $pH = -\log[H^+]$ ([ ] means concentration of the substance enclosed in the brackets). The pH scale runs from 0 to 14. The midpoint of the scale is 7, and a solution of pH 7 is neutral, neither acidic nor basic. This means that it contains equal numbers of hydroxyl and hydrogen ions. Pure water has a pH of about 7. A pH between 0 and 7 is acidic; a pH between 7 and 14 is basic. Since the pH scale is logarithmic, a solution of pH 5 is ten times as acidic as a solution of pH 6, and so forth.

Most of the chemical reactions of life occur most rapidly at a pH near the neutral point. In our own bodies, the blood and most body fluids have pH values of about 7.4. An empty stomach contains hydrochloric acid, which dissociates readily, and the fluid in the stomach is about pH 2.

SUMMARY

More than 100 different chemical elements are known, each with a unique set of chemical properties. About twenty elements are essential to living things. The special properties of those elements found in biological systems are necessary to the processes involved in growing, living, and reproducing.

Living organisms are subject to the same physical and chemical laws that govern nonliving systems. Like nonliving matter, they are made up of atoms, which bond in various ways to form compounds. Ionic bonds form when one atom takes one or more electrons given up by another atom, whereas covalent bonds form when atoms share electron pairs. Hydrogen bonds are much weaker attractions between partial positive and negative charges on polarly bonded atoms of different molecules. Chemical reactions rearrange the bonding of atoms and molecules and form different compounds.

Water is the most abundant substance in living things, and is absolutely necessary for life. The water molecule's structure and hydrogen-bonding ability make water unique among chemical compounds: it is a solvent of ionic and polar substances, it forms interfaces with nonpolar substances, it has high melting and boiling points, it has high heat capacity and thermal conductivity, and it is less dense as a solid than as a liquid. These properties make water vital to the economy of living organisms.

The acidity or alkalinity of a solution is expressed as its pH, a measure of its hydrogen ion concentration.

---

*Sodium hydroxide   The active ingredient in Drano drain cleaner.

OBJECTIVES

From your study of this chapter, you should be able to:

1. Define, use, and recognize the characteristics of the following: atom, proton, neutron, electron, isotope, ion, molecule, single bond, double bond, polar, nonpolar, mole, molecular weight, dissociation, acid, base, salt, pH scale.

2. Recognize examples of ionic, covalent, and hydrogen bonds and explain the differences between them.

3. Write the correct molecular formulas for water, carbon dioxide, oxygen gas, and table salt.

4. List and discuss six reasons why water plays an important role in living systems.

SELF-QUIZ

As appropriate, choose the correct alternative from words in parentheses, or select the lettered answer that correctly completes the sentence.

1. If the pH of a solution changes from 2 to 5, it has become more (acidic, basic); its hydrogen ion concentration has (increased, decreased, remained constant).

2. A positive ion has:
   a. more protons than electrons
   b. more electrons than protons
   c. equal numbers of neutrons and electrons
   d. equal numbers of protons and electrons
   e. more neutrons than electrons

3. Atoms of the same element that contain different numbers of neutrons are known as:
   a. nuclei
   b. citrons
   c. isotopes
   d. ions
   e. protons

4. What kind of bond involves the sharing of electrons between atoms such that each atom completes its outer electron shell? (covalent, ionic)

5. A water molecule is held together by (ionic, covalent, hydrogen) bonds.

6. In the dissociation of NaCl in water:
   a. water exerts forces that induce dissociation
   b. water is a passive solvent, accepting particles that dissociate because of their own internal forces
   c. water molecules lose hydrogen ions
   d. twice as many H^+ ions are formed as Na^+ ions
   e. equal numbers of H^+ and Na^+ ions are formed

7. The high boiling point of water is advantageous to living organisms because:
   a. the environment seldom reaches the boiling point of water
   b. organisms can easily boil off enough water to keep themselves cool
   c. it allows organisms to spread heat evenly throughout their bodies
   d. organisms can absorb a great deal of heat before they reach the boiling point
   e. ice can form as an insulating barrier at environmental temperatures that are frequently encountered

QUESTIONS FOR DISCUSSION

1. Mercury is even more cohesive than water. Why could living organisms not have mercury instead of water as the main fluid in their bodies?

2. Why does oxygen make up 62% of the weight of an organism? Why does hydrogen account for only 10% of an organism's weight?

REFERENCES AND FURTHER READING

Henderson, L. S. *The Fitness of the Environment.* Boston: Beacon Press, 1958. An interesting discussion of the ability of the physical and chemical conditions on earth to support life. Chapter 3, on water and its relationship to life, is especially good.

Hill, J. W. *Chemistry for Changing Times,* 3d ed. Minneapolis: Burgess, 1979. An excellent and entertaining introduction to chemistry written for people who know nothing about the subject.

Rodella, T. D., et al. *Through the Molecular Maze: A Helpful Guide to Chemistry for Beginning Life Science Students.* W. Kaufman, 1975.

White, E. H. *Chemical Background for the Biological Sciences.* 2d ed. Englewood Cliffs, N.J.: Prentice-Hall, 1970. An interesting book on the structures of atoms and molecules and the nature of chemical reactions.

CHAPTER 3

BIOLOGICAL CHEMISTRY

Biological chemistry is based on molecules that contain carbon; carbon compounds occur mainly in living organisms (and their remains) and are, therefore, called **organic** molecules. Only the simplest carbon compounds, such as carbon dioxide (CO_2) and carbonate compounds (containing CO_3^{2-}), are considered inorganic. All substances that do not contain carbon are also classed as inorganic.

Organisms form various types of small organic molecules called **monomers**. These monomers may then be linked together to form **polymers**, also called **macromolecules** (macro = big), which may be long, straight or branched chains. The unique properties of life depend upon the enormous array of different polymers that can be made up from a relatively small number of common biological monomers.

There are four main classes of organic compounds: carbohydrates, lipids, proteins, and nucleic acids.

In this chapter we shall look at the structures of members of these four groups of compounds. We shall also examine the functions of one group of proteins, the enzymes, in some detail; in later chapters we shall delve more deeply into the functions of carbohydrates, lipids, nucleic acids, and various kinds of proteins.

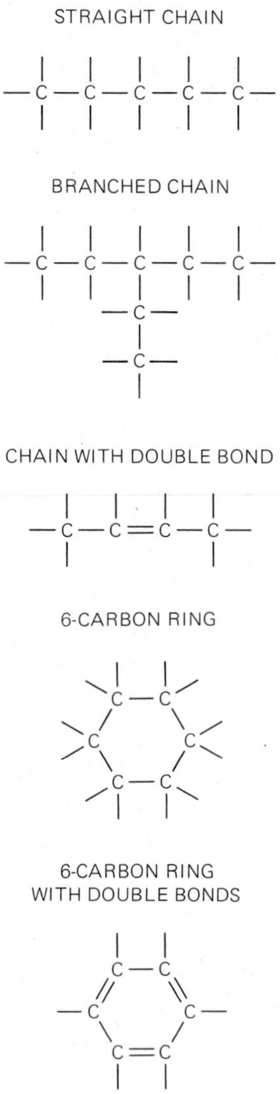

STRAIGHT CHAIN

$$-C-C-C-C-C-$$

BRANCHED CHAIN

$$-C-C-C-C-C-$$
$$-C-$$
$$-C-$$

CHAIN WITH DOUBLE BOND

$$-C-C=C-C-$$

6-CARBON RING

6-CARBON RING
WITH DOUBLE BONDS

Figure 3-2 Carbon atoms can bond together in various ways. The unconnected lines protruding from the carbon atoms indicate that any one of a number of elements can bond covalently in these positions. (Because a carbon atom forms tetrahedral bonds, the "straight" chain actually forms a zigzag in space.)

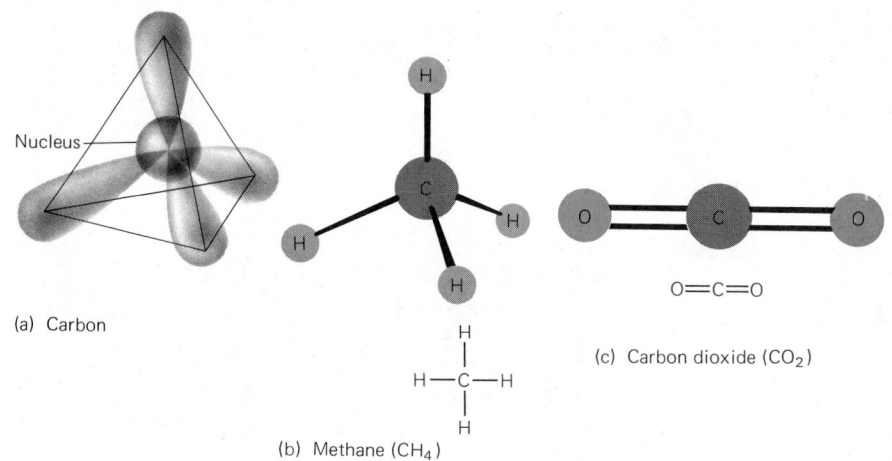

(a) Carbon

(b) Methane (CH_4)

(c) Carbon dioxide (CO_2)

Figure 3-1 Carbon and the bonds it forms. (a) A carbon atom can form four covalent bonds, which point to the four corners of a tetrahedron. (b) Methane (CH_4). The four hydrogens are bonded symmetrically around the carbon at the points of a tetrahedron. (c) Carbon dioxide (CO_2). Each oxygen atom is joined to the carbon atom by a double bond. The bonds move so that they are parallel, and the molecule is linear.

3-A Properties of Carbon

The properties of carbon atoms make them uniquely suited to be the basis of the chemistry of life. A carbon atom can form four covalent bonds (see Section 2-C); when it does, the four atoms bonded to carbon are as far from each other as possible, at the corners of a tetrahedron,* with the carbon nucleus at the center of the tetrahedron (Figure 3-1). Carbon atoms can also join with each other to form long chains, and these chains may also have carbon chains branching off from them. The ends of carbon chains may join to form ring structures (Figure 3-2).

Carbon is the only element that forms enough different, complex, stable compounds to make up the variety of molecules found in living things.

---

*Tetrahedron A four-sided solid whose sides are all triangles.

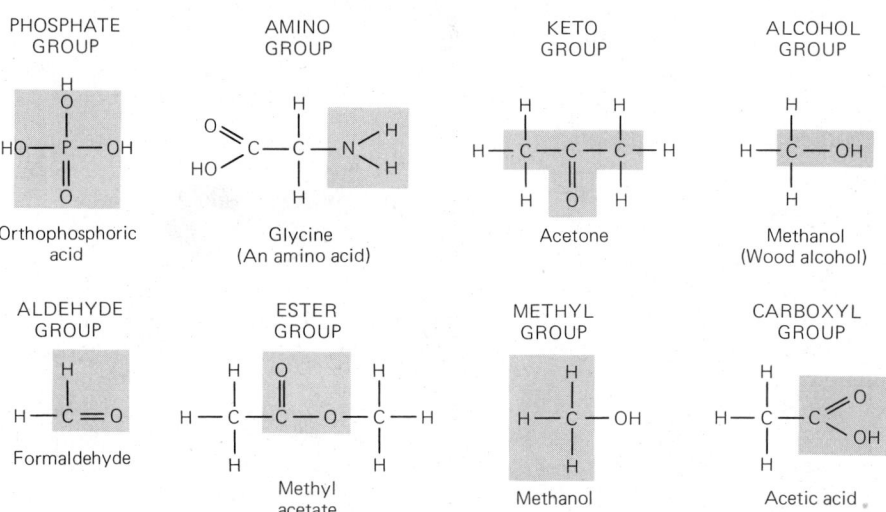

Figure 3-3 Some biologically important functional groups and examples of molecules in which they occur. Group names appear above each part; the group itself lies in the colored box, and the name of the entire molecule in which the group appears is given below the molecule.

3-B Functional Groups

Organic compounds occur in great variety, but certain **functional groups** are found repeatedly and offer a way of sorting organic molecules into a small number of molecular "families." For instance, if there is an alcohol (C—OH) group on a molecule, the molecule will tend to take part in certain chemical reactions and will behave in certain definite ways, no matter what the rest of the molecule is like. Compounds that contain carboxyl groups (COOH), such as acetic acid and amino acids, are called **organic acids** because they release hydrogen ions in solution (see Section 2-G). Some common functional groups are shown in Figure 3–3.

3-C Carbohydrates

Carbohydrates are the sugars, starches, and related molecules. The most familiar carbohydrates are energy sources, such as the sugars and starches, but cellulose, a structural molecule in plants, is also a carbohydrate.

Monosaccharides or "simple sugars," such as glucose, fructose, and galactose, are the simplest carbohydrates; they are the monomers, or subunits, from which more complex carbohydrate polymers are synthesized. A monosaccharide contains carbon, hydrogen, and oxygen in about a 1:2:1 ratio (represented as $[CH_2O]_n$), from three to nine carbon atoms in an unbranched chain (or ring), and a carbonyl (C=O) group (Figure 3–4). Monosaccharides may be classified by the number of carbon atoms they contain. A **triose** has three carbons (tri = three), a **tetrose** has four (tetr = four), a **pentose** five (pent = five), and a **hexose**, such as glucose, has six (hex = six) (Figure 3–5). When a monosaccharide containing five or more carbon atoms is dissolved in water (as it always is in a living system), most of the molecules take on a ring structure (Figure 3–6).

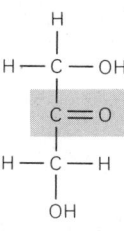

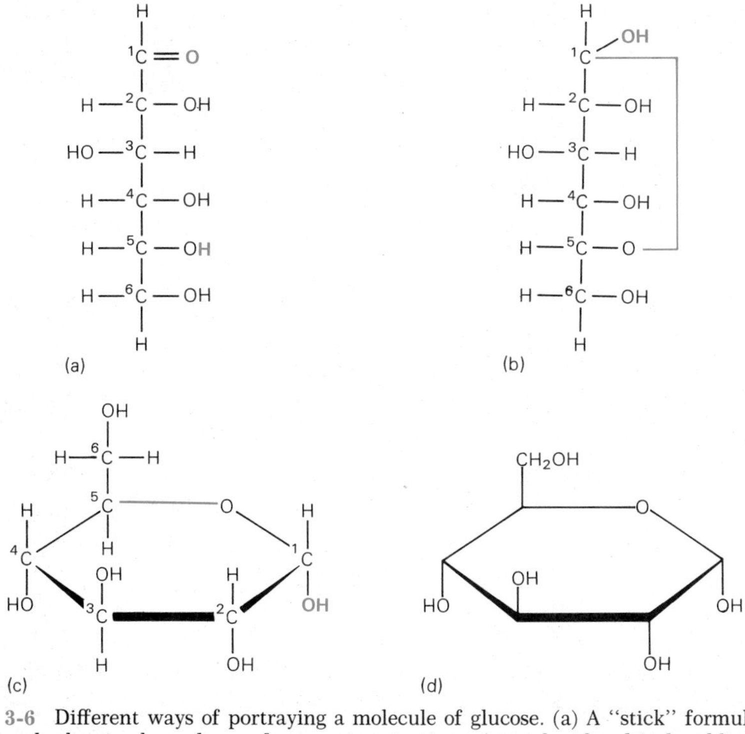

GLYCERALDEHYDE ($C_3H_6O_3$)

DIHYDROXYACETONE ($C_3H_6O_3$)

Figure 3-4 Triose sugars. These are the only two trioses. The C=O group (colored), characteristic of a monosaccharide, is on an end carbon in one sugar and on the middle carbon in the other.

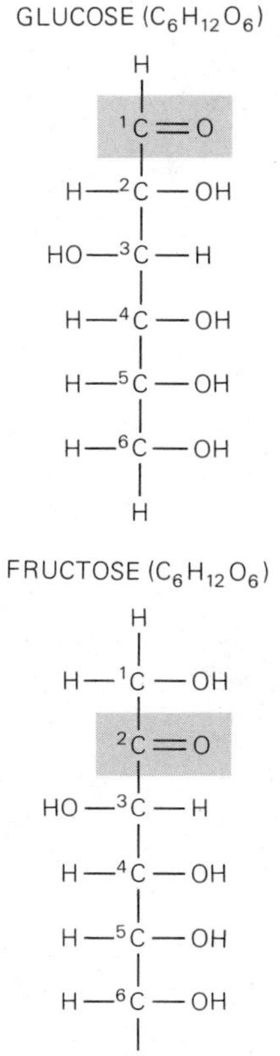

GLUCOSE ($C_6H_{12}O_6$)

FRUCTOSE ($C_6H_{12}O_6$)

Figure 3-5 Glucose and fructose, two hexose monosaccharides. Their carbonyl groups are colored. By convention, the carbon atoms are numbered so that the carbon at that end of the molecule which is more oxidized (that is, contains the higher ratio of oxygen to [carbon + hydrogen]) has the lowest number.

(a)

(b)

(c)

(d)

Figure 3-6 Different ways of portraying a molecule of glucose. (a) A "stick" formula. (b) A stick formula showing how glucose forms a ring structure. A covalent bond (colored line) forms between carbon #1 and the oxygen on carbon #5. The hydrogen previously attached to the oxygen on carbon #5 joins the oxygen on carbon #1 to form a hydroxyl (OH) group. (c) The "ring" formula gives a better idea of how the atoms in (b) are arranged in space. The bonds drawn with heavier lines are visualized as projecting out of the paper toward the observer. (d) This skeleton formula permits more information to be packed into less space. Carbon atoms are assumed at all angles unless another element is shown. Any carbon atom shown with fewer than four bonds to other atoms is assumed to be bonded to enough hydrogen atoms to fill its valence of four.

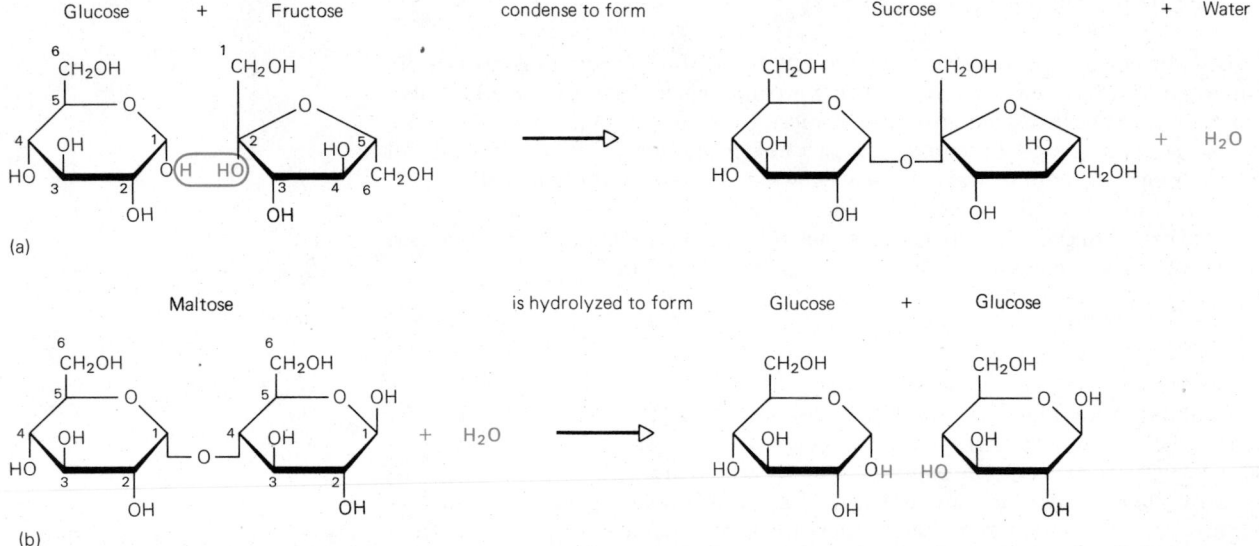

Figure 3-7 Condensation and hydrolysis. (a) A condensation between the monosaccharides glucose and fructose produces the disaccharide sucrose, and water. (b) The disaccharide maltose hydrolyzes upon addition of water to yield two molecules of glucose.

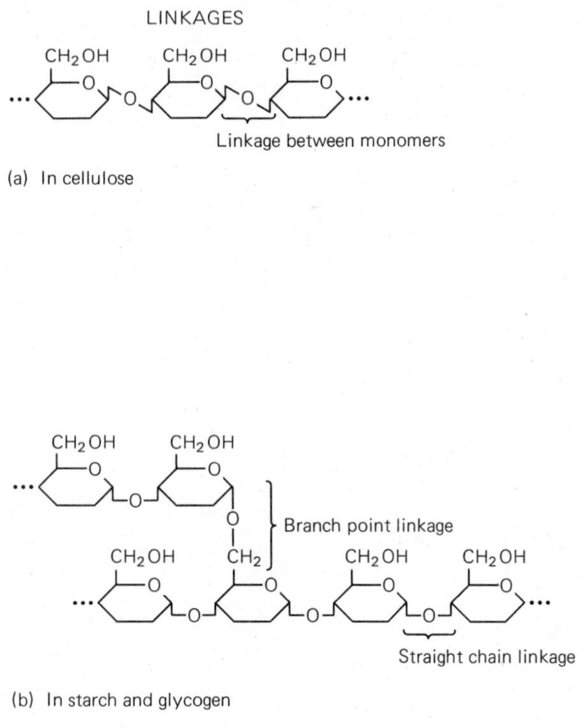

LINKAGES

Linkage between monomers

(a) In cellulose

MOLECULAR FORM

(c) Cellulose

Branch point linkage

Straight chain linkage

(b) In starch and glycogen

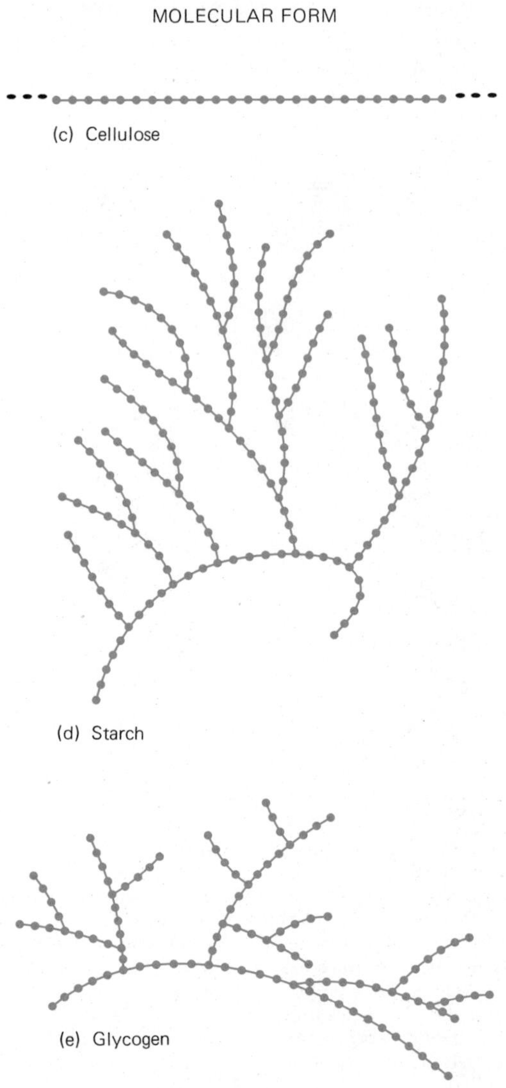

(d) Starch

(e) Glycogen

Figure 3-8 Linkages and molecular form of three important polysaccharides. Cellulose, starch, and glycogen are all made up of glucose monomers, but note the differences in the way these are joined. Cellulose molecules are unbranched. Glycogen branches more frequently than starch.

Two monosaccharides (mono = one) may bond together to form a **disaccharide** (di = two). Sucrose, maltose, and lactose are familiar disaccharides. A disaccharide is formed by a **condensation** or **dehydration synthesis reaction**, in which two smaller molecules become joined by losing hydrogen and oxygen atoms that go into formation of a water molecule (Figure 3–7). After each monomer loses an atom or two, the remaining part of the monomer is called a **residue**. For example, the components of a sucrose molecule are a glucose residue and a fructose residue.

A condensation is readily reversible. A molecule of water added to a disaccharide between the monosaccharide residues will split it by a **hydrolysis** reaction (hydro = water; lysis = breaking) into its component monosaccharides.

Condensation and hydrolysis reactions are very important in biological systems. Condensation reactions join organic monomers to make larger molecules; hydrolysis reactions reverse the process and release the original monomers. Food is digested by hydrolysis reactions.

By a series of condensation reactions, many monosaccharides can be joined to form a **polysaccharide**. **Starch** is a very common polysaccharide; it is the form in which many plants store excess sugar molecules; **cellulose** is a structural polysaccharide that stiffens and supports the bodies of plants, and **glycogen** is a polysaccharide made and stored in animal liver or muscle tissue, where it is available as a ready reserve of energy. These three important polysaccharides are all made up of glucose residues; they differ in the arrangements of the bonds between their glucose subunits, in the branching patterns of the polymer, and in the total number of glucose residues per chain (Figure 3–8).

3-D Lipids

Whereas carbohydrates are defined by their composition and are chemically related to one another, lipids are a heterogeneous group of compounds defined by their solubility. The word **lipid** is a general term for organic compounds that are soluble in nonpolar organic solvents such as ether, chloroform and benzene, and are essentially insoluble in water. Lipids contain carbon, hydrogen, and oxygen, but the proportion of oxygen in lipids is much lower than in carbohydrates; lipids may also contain other elements such as phosphorus and nitrogen. Because they are insoluble in water, lipids are vital components of the membranes that divide one aqueous (aqua = water) compartment in the body from another. They are also compounds in which energy can be stored in a concentrated form that does not interfere with the body's water balance.

We shall consider four groups of lipids: fatty acids, triglycerides, phospholipids, and steroids.

Fatty Acids

Fatty acids are the simplest lipids. A **fatty acid** is a long chain of carbon and hydrogen atoms with a carboxyl group on one end (Figure 3–9). If the hydrocarbon chain contains no double bonds between any of its carbon atoms (see Section 2-C), it is said to be **saturated**, as it is in palmitic acid; if it contains one or more double bonds, it is **unsaturated**, and has room for more hydrogen atoms.

The carboxyl end of a fatty acid molecule is **hydrophilic** (hydro = water; philic = loving), while the other end of the molecule is **hydrophobic** (phobic = hating). The hydrophilic end will dissolve in aqueous solutions in the cell, and the hydrophobic end, which is soluble in nonpolar organic solvents, will attach to or dissolve in nonpolar organic compounds. The tendency of fatty acids to lie along the interfaces between nonpolar and aqueous environments makes them important parts of the membranes that divide living systems into compartments. This property is also the source of the cleaning power of soaps and detergents (Figure 3–10).

Fatty acids seldom occur free, but are usually combined with other molecules to form substances such as **glycolipids** (carbohydrate + lipid) and **lipoproteins** (lipid + protein). They are also monomers of many more complex lipids.

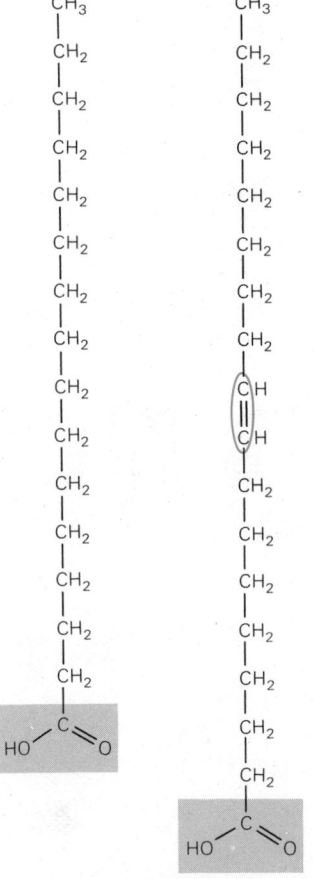

Figure 3-9 Fatty acids. The carboxyl group, which makes these compounds acids, is boxed in. The double bond that makes oleic acid unsaturated is circled. The most common fatty acids have even numbers of carbon atoms in chains 14 to 22 carbon atoms long. Acetic acid is the acid component of vinegar. Butyric acid is the active ingredient in rancid butter and in "body odor." Bloodhounds use it to track human scent.

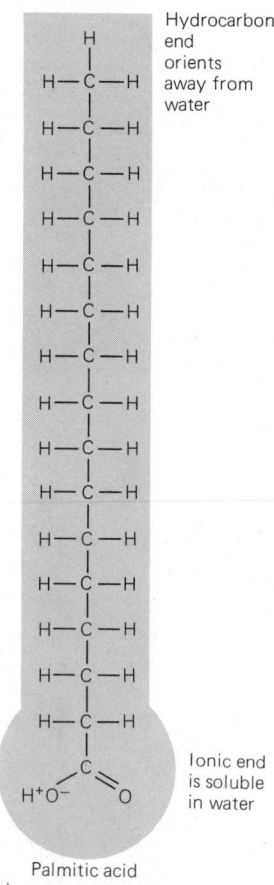

Hydrocarbon end orients away from water

Ionic end is soluble in water

Palmitic acid

(a)

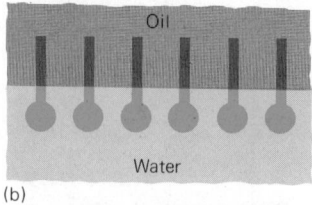

Oil

Water

(b)

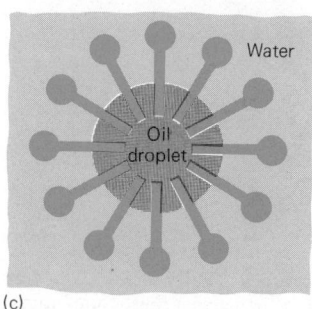

Water

Oil droplet

(c)

Figure 3-10 (a) Fatty acids have hydrophobic and hydrophilic ends. The ionic carboxyl group is soluble in water, while the hydrocarbon chain is not. (b) The molecules orient at interfaces between water and organic substances, or water and air, with their ionic ends in the water. (c) Soaps and detergents clean by removing droplets of oil or other organic substances from dirty surfaces. Sodium palmitate $(CH_3(CH_2)_{14}COO^-Na^+)$ is a soap. It surrounds an oil droplet, in which its hydrophobic chains dissolve. The hydrophilic carboxyl ends dissolve in water, so that the whole tiny droplet floats and can be washed away.

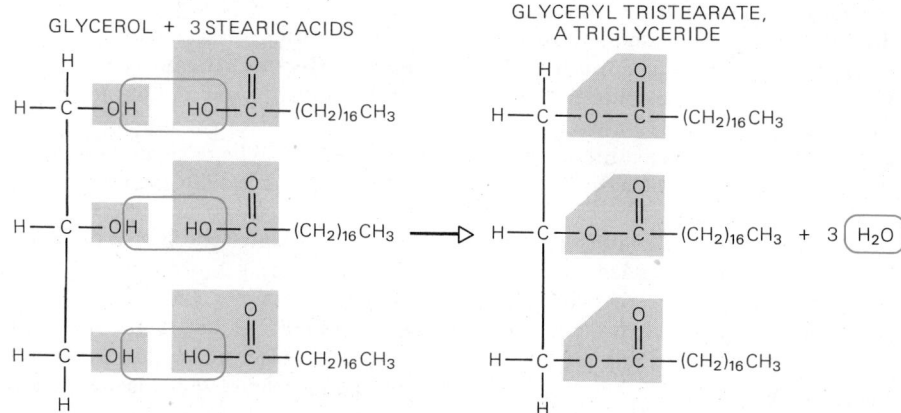

Figure 3-11 Formation of a triglyceride. Three molecules of fatty acids (of the same or different kinds) condense with glycerol, with the removal of water (ringed). An ester group (colored shapes at right) is formed between each alcohol group (—OH) of glycerol and the carboxyl group (—COOH) of each fatty acid.

Triglycerides

A **triglyceride** consists of a molecule of the alcohol glycerol to which three fatty acid molecules have been attached. Ester linkages (see Figure 3-3) join the carboxyl groups of the fatty acids with the alcohol groups of the glycerol (Figure 3-11).

Triglycerides that are solid at room temperatures are often referred to as "**fats**," whereas those that are liquid are called "**oils**." Oils generally contain more unsaturated fatty acids than do fats. Animals that live in cold habitats, such as fish in the Arctic and Antarctic, usually contain triglycerides that are less saturated than those of animals from warmer areas. If they did not, their bodies would become rigid at the low temperatures they encounter.

Waxes are esters formed from fatty acids and alcohols other than glycerol. Most waxes melt at higher temperatures than triglycerides.

Phospholipids

Phospholipids are similar to triglycerides except that in phospholipids one or two of the fatty acids are replaced by functional groups containing phosphorus, nitrogen, or both (Figure 3-12). They serve many important functions in cells. Phospholipids are generally found in combination with other substances, particularly proteins, and are important components of many biological membranes.

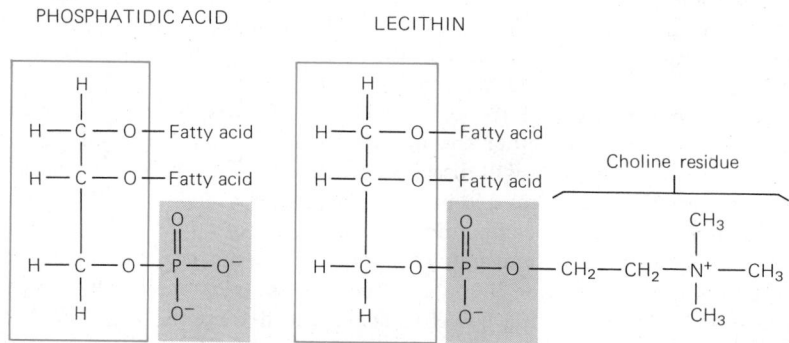

Figure 3-12 Many phospholipids are derivatives of phosphatidic acid, in which a glycerol (colored outline) has condensed with two fatty acids and a phosphate group (shaded box). Lecithin, a phospholipid found in the membranes of all cells, forms by condensation of phosphatidic acid with choline.

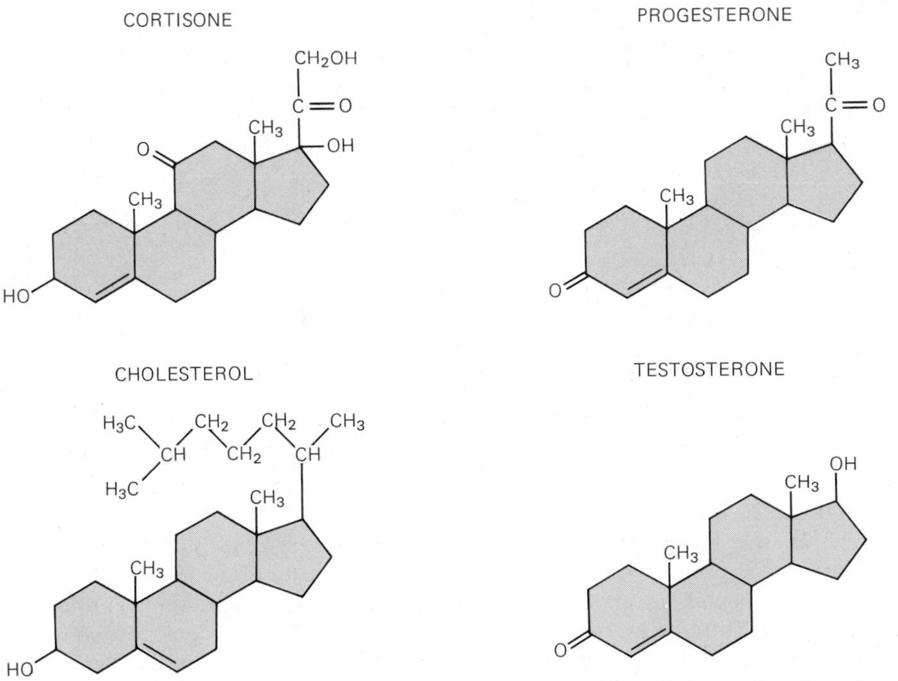

CORTISONE

PROGESTERONE

CHOLESTEROL

TESTOSTERONE

Figure 3-13 Some steroids. Notice that all have the same basic skeleton of carbon rings (colored), with various other groups added on.

Steroids

Steroids are fairly large lipids with a basic structure consisting of four carbon rings joined to one another. Various other groups are joined to these rings in different steroids. The sex hormones, as well as cortisone and cholesterol, are examples of biologically important steroids (Figure 3–13).

3-E   Proteins

Proteins are by far the most abundant organic components of a cell. They may perform a variety of functions; the most common proteins are catalysts of biological reactions or parts of structural molecules.

In addition to the carbon, hydrogen, and oxygen found in most organic molecules, proteins always contain nitrogen, and usually some sulfur.

Proteins are composed of amino acid monomers. An **amino acid** has a carboxyl group and an amino group attached to the same carbon atom. A variety of different functional groups or side chains (R in Figure 3–14) are also attached to this carbon. Their different side chains determine the differences between the twenty amino acids that commonly occur in proteins (Figure 3–15). From these twenty common

(a) Generalized amino acid   (b) Amino acid ionized in water

Figure 3-14 An amino acid. (a) Amino and carboxyl groups (shaded color) are always attached to the same carbon atom (colored ring), called the α carbon. The α carbon bears another group (R), which differs from one amino acid to another. (b) When dissolved in water, an amino acid may dissociate; hydrogen ions released by the carboxyl group are picked up by the amino group, so that the molecule has two oppositely charged areas. The molecule may also have other charges formed by dissociation of ions from the R group.

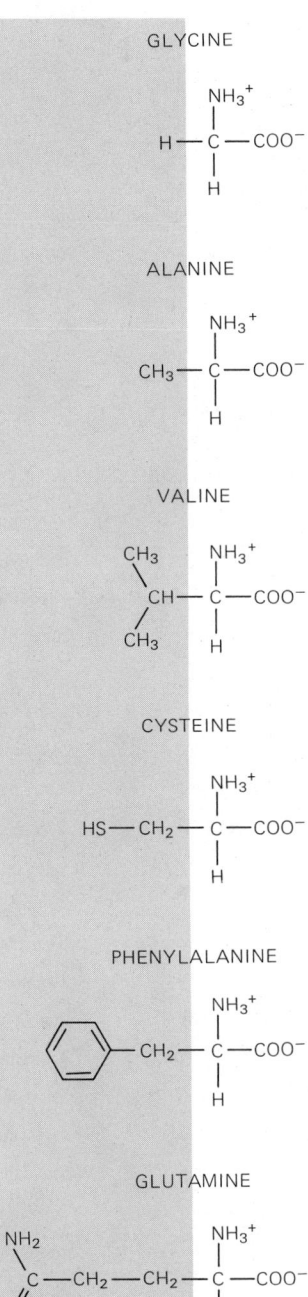

GLYCINE

ALANINE

VALINE

CYSTEINE

PHENYLALANINE

GLUTAMINE

Figure 3-15 Some amino acids. The R groups are shaded. The α carbon, carboxyl, and amino groups that define the amino acid are on the white background. Note the sulfur, amino groups, and carbon rings in some of the R groups.

**CHAPTER 3
BIOLOGICAL
CHEMISTRY** 33

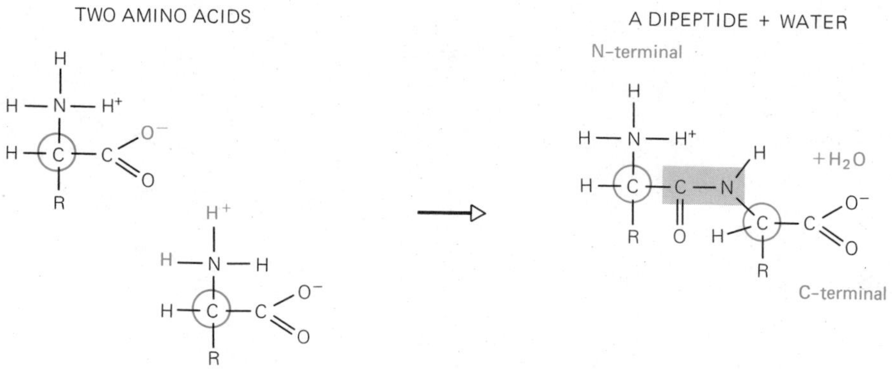

TWO AMINO ACIDS

A DIPEPTIDE + WATER

N-terminal

C-terminal

N-terminal ends

Glycine — Phenylalanine
Isoleucine — Valine
Valine — Asparagine
Glutamic acid — Glutamine
Glutamine — Histidine
Cysteine — Leucine
Cysteine—S—S—Cysteine
Alanine — Glycine
Serine — Serine
Valine — Histidine
Cysteine — Leucine
Serine — Valine
Leucine — Glutamic acid
Tyrosine — Alanine
Glutamine — Leucine
Leucine — Tyrosine
Glutamic acid — Leucine
Asparagine — Valine
Tyrosine — Cysteine
Cysteine—S—S—Glycine
Asparagine — Glutamic acid
Arginine
Glycine
Phenylalanine
Phenylalanine
Tyrosine
Threonine
Proline
Lysine
Alanine

A chain

B chain

C-terminal ends

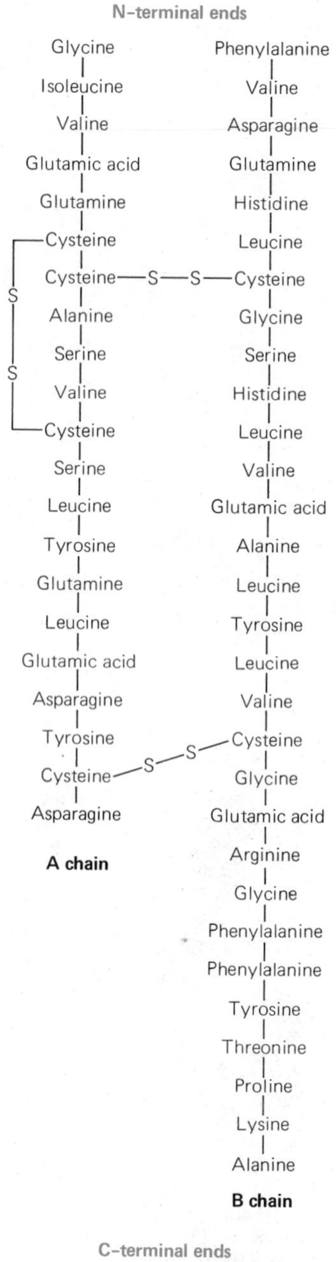

Figure 3-17 The amino acid sequence of insulin from cattle. The protein is made up of two polypeptide chains (A and B).

amino acid monomers, living things build thousands of kinds of different proteins. The vast diversity of living things is largely a result of the different sequences in which the 20 common amino acids can be arranged to form proteins.

Like other biological macromolecules, proteins are synthesized by condensation reactions that join the amino acid monomers. A **peptide bond** forms between the carboxyl carbon of one amino acid and the amino nitrogen of another. The condensation of two amino acids forms a **dipeptide** (Figure 3–16). Condensing more amino acids onto this dipeptide by the formation of peptide bonds between them would form a **polypeptide**. When two or more amino acids are linked, the amino acid residue at one end of the chain still has an intact amino group which has not taken part in formation of a peptide bond, and the residue at the other end has an intact carboxyl group. The end of the chain with the free amino group is called the N-terminal end, and the end with the free carboxyl group is called the C-terminal end (Figure 3–16).

A **protein** is a functional unit composed of one or more polypeptides. The protein insulin, made up of two linked polypeptide chains, is shown in Figure 3–17. Insulin is a hormone that stimulates the removal of glucose from the bloodstream. The insulin molecule is made up of 51 amino acid residues, which is small for a protein.

Protein Structure

Protein molecules are long and they may assume complex shapes. In order to analyze the structure of proteins in a systematic way, biochemists divide protein structure into four main components.

The **primary structure** of a protein is its unique sequence of amino acids. Its **secondary structure** is a regular pattern that is repeated along the polypeptide chain. The most common type of secondary structure is the alpha helix (also written α-helix) assumed by some polypeptide chains or parts of chains (Figure 3–18). An α-helix forms when hydrogen bonds arise between the amide ($>$NH) group in one peptide bond and the carbonyl (C$=$O) group of another (Figure 3–19).

Each protein has a characteristic **tertiary structure**, an overall shape assumed by the polypeptide chain. This tertiary structure is strongly influenced by the interactions of R groups in different parts of the chain; these interactions may be in the form of ionic bonds,* hydrogen bonds,* disulfide bonds (see Figure 3–17), or hydrophobic interactions between different R groups. A **hydrophobic interaction** is the tendency of nonpolar* R groups to stay close together because they are repelled by the polar* substance, water, that normally surrounds proteins. Two amino

---

*Ionic bond   The bond formed between two atoms when one gives up one or more electrons to the other.
*Hydrogen bond   Weak bond between two molecules or two parts of the same molecule due to the attraction of a hydrogen with a partial positive charge to an oxygen or nitrogen with a partial negative charge.
*Nonpolar   Electrically symmetrical (see Section 2-C).
*Polar   Electrically asymmetrical.

acids that affect tertiary structure profoundly are **cysteine** and **proline**. Two cysteine residues (Figure 3–15) can interact to form a covalent **disulfide bond** (S—S) between them. This may join two parts of a polypeptide chain, or may join one chain to another (see Figures 3–17 and 3–20). Proline has a different effect. It contains an inflexible ring structure that causes a kink in the molecule wherever a proline residue occurs (Figure 3–20).

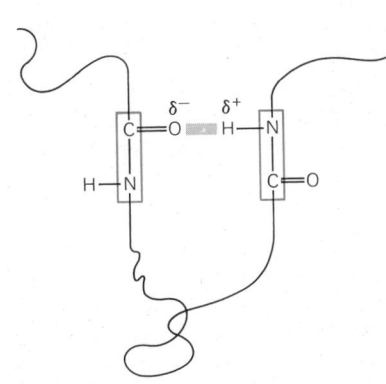

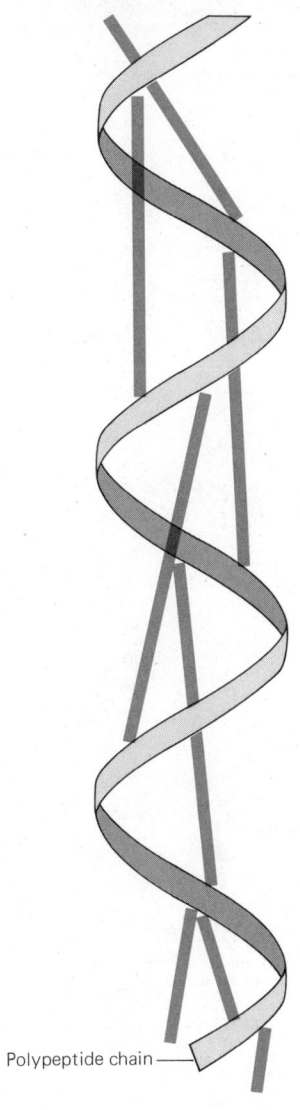

Figure 3-19 A hydrogen bond between two regions of a polypeptide chain. The nitrogen in a peptide bond (boxed) attracts the electron of its hydrogen, leaving a partial positive charge (δ^+) on the hydrogen. The double-bonded oxygen of another peptide bond has a partial negative charge (δ^-), which will attract this hydrogen weakly to form a hydrogen bond (colored line).

The term **quaternary structure** refers to the fact that some proteins are made up of two or more polypeptide chains that must fit together in a specific arrangement to form a complete, functional protein. For example, a molecule of the hemoglobin that makes our blood red is made up of four polypeptide strands.

Secondary, tertiary, and quaternary structures of proteins are not random. They are highly specific, and they are precisely dictated by the primary structure—that is, by the sequence of amino acids in the chain. Once the amino acids are linked in the proper order the chain automatically coils, loops, and folds into its correct secondary and tertiary structures. Likewise, the different polypeptide molecules that must come together to give a complex protein its quaternary structure are made so that their shapes and forces of attraction automatically fit them together. Polypeptide molecules can be made to lose their shape by gentle heating or by certain chemical treatments. When they are returned to more normal surroundings, they coil back up again and join together to re-form their original structure. Protein structures thus represent masterpieces of molecular engineering, fashioned by hundreds of millions of years of natural selection.

Polypeptide chain

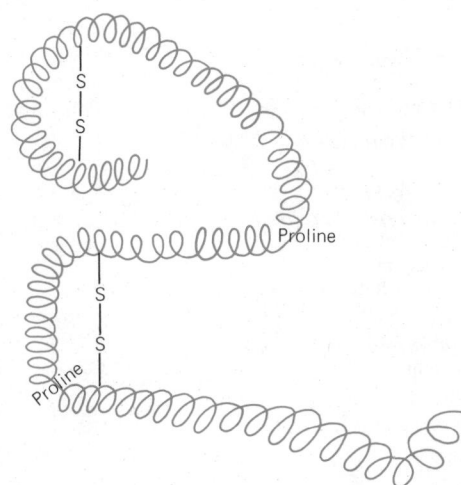

Figure 3-20 Three-dimensional structure of a hypothetical protein. Secondary structure is the helical coiling that results from hydrogen bonding between peptide bonds. Tertiary structure is a result of interactions between specific amino acids. For instance, wherever the amino acid proline occurs, there is a crick in the helix; disulfide bonds (—S—S—) between two cysteine residues join parts of the structure firmly together.

**CHAPTER 3
BIOLOGICAL
CHEMISTRY** **35**

Functions of Proteins

Proteins perform many important functions in living organisms (Table 3–1). Hair and fingernails are composed of **structural proteins**. Many types of proteins are also found in body fluids. Hemoglobin is a protein that carries oxygen in the blood; other **blood proteins** detoxify* poisonous substances or help combat disease. Proteins are also components of nearly all biological membranes, and they play important roles in regulating the passage of substances through membranes; **glyco-proteins**, composed of proteins linked to short carbohydrate chains, are especially important as membrane "markers" identifying particular types of cells. Other proteins are responsible for the ability of muscles to contract and for the detection of chemical substances that we can smell or taste. Insulin is one of a number of small protein **hormones**, chemical messengers between different parts of the body. But the most important proteins are the enzymes.

Enzymes

Most reactions between molecules in living organisms are controlled by **enzymes**, protein molecules that act as catalysts. A **catalyst** is a substance that increases the rate of a chemical reaction which would normally occur at a slower rate. The catalyst is not permanently changed by the reaction, nor does it alter the **equilibrium position** of the reaction—the relative amounts of reactants that are used up and of products that are formed.

As an example of the action of an enzyme, let us consider a reaction familiar to anyone who owns a cat, the hydrolysis of urea from the cat's urine into carbon

---

*Detoxify   Make less poisonous to a living organism.

TABLE 3–1   **SOME FUNCTIONS OF PROTEINS**

EXAMPLE	FUNCTION
Enzymes	
Amylase	Converts starch to glucose
DNA polymerase I	Repairs DNA molecule
Transaminase	Transfers amino group from one amino acid to another
Structural proteins	
Viral coat proteins	Outer covering of virus
Keratin	Hair, nails, horns, hoofs
Collagen	Tendons, cartilage
Hormones	
Insulin ⎫	
Glucagon ⎭	Regulate glucose metabolism
Oxytocin	Regulates milk production in female mammals
Vasopressin	Increases retention of water by kidney
Contractile proteins	
Actin	Thin contractile filaments in muscle
Myosin	Thick filaments in muscle
Storage proteins	
Casein	A nutrient protein in milk
Ferritin	Stores iron in spleen and egg yolk
Transport proteins	
Hemoglobin	Carries O_2 in blood
Myoglobin	Carries O_2 in muscle
Serum albumin	Carries fatty acids in blood
Cytochrome c	Transfers electrons
Immunological proteins	
γ-globulins	Form complexes with foreign proteins
Toxins	
Neurotoxin	Cobra venom blocker of nerve function

dioxide and ammonia, which gives its characteristic odor to a litterbox in need of cleaning:

$$H_2N—CO—NH_2 + H_2O \longrightarrow CO_2 + 2NH_3$$

urea   water   carbon   ammonia
                           dioxide

reactants   products

This reaction is mediated by an enzyme called **urease**, produced by bacteria that fall out of the air and grow in the cat's box. At room temperature and at pH 8,* a molecule of urease can catalyze the hydrolysis of about 30,000 molecules of urea in one second. In the absence of the enzyme, this reaction would take about 3 million years to complete. So in the presence of urease the reaction happens at more than a trillion times its natural rate. Some enzymes work more quickly than urease; others work more slowly.

An enzyme speeds up a reaction by lowering its **activation energy**. What does this mean? The hydrolysis of urea will occur without a catalyst, but very slowly. Urea and water molecules must have a great deal of energy, and must collide with great force, in order to react to form carbon dioxide and water. At room temperature, very few molecules of urea or water have sufficient energy to react, and collisions between molecules that have enough energy will be extremely infrequent. If the enzyme urease is present, it holds the urea and water molecules in such a way that they will react even when they have much lower energy; that is, the enzyme has lowered the energy threshold or activation energy of the reaction (Figure 3–21). Thus, enzymes permit organisms to carry out reactions rapidly, at the relatively low temperatures found in the body.

Any reaction can proceed in both the forward and backward directions. If neither the reactants nor the products are removed from the reaction system, the reaction will continue until it reaches equilibrium, at which point the original reactants are being formed by the backward reaction as fast as they are being taken up to form products by the forward reaction. The equilibrium position for any reaction is determined by energy changes during the reaction. An enzyme lowers the energy barriers to both the forward and reverse reactions to the same extent; thus an enzyme does not alter the equilibrium point of a reaction, but merely increases the speed with which the reaction reaches equilibrium. Furthermore, since a catalyst—such as an enzyme—is not permanently changed by participation in a reaction, it emerges exactly as it started, ready to catalyze another reaction.

ENZYME-SUBSTRATE COMPLEXES

Enzymes are very specific in that each catalyzes only one or a very few reactions. This specificity occurs because an enzyme actually binds to its **substrates**, the reactants which it affects, to form an **enzyme-substrate complex**.

The specific binding of a substrate to an enzyme is a result of a cluster of chemical groupings that make up the enzyme's **active site**. If an enzyme is steadily broken down by chemical digestion, it may retain the ability to bind its substrates even after quite a lot of the enzyme molecule has disappeared. This indicates that the active site is only a small fraction of the whole enzyme molecule.

The active site binds specifically with certain parts of the substrate molecule. For instance, if the substrate molecule has a hydrophobic and a positively charged group on one surface, the active site of the enzyme would be expected to have a hydrophobic and a negatively charged group at its active site, oriented in space so that they can interact with the hydrophobic and positively charged groups on the substrate. As a result of this "matching," enzymes and substrates bind by specific point-to-point interactions, so that the enzyme-substrate complex is always exactly the same every time it is formed. This means that each enzyme can bind only one or a very few kinds of substrates, and it accounts for the specificity of enzyme-medi-

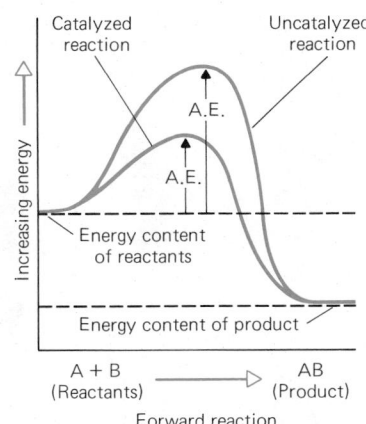

Figure 3-21 Lowering of activation energy by a catalyst. The activation energy (A.E.) is an energy barrier which the reactants A and B must overcome as they react to form the product AB. The reaction proceeds because the products have lower energy than the reactants.

---

*pH A measure of acidity or alkalinity; a pH of 8 is slightly basic (see Section 2-G).

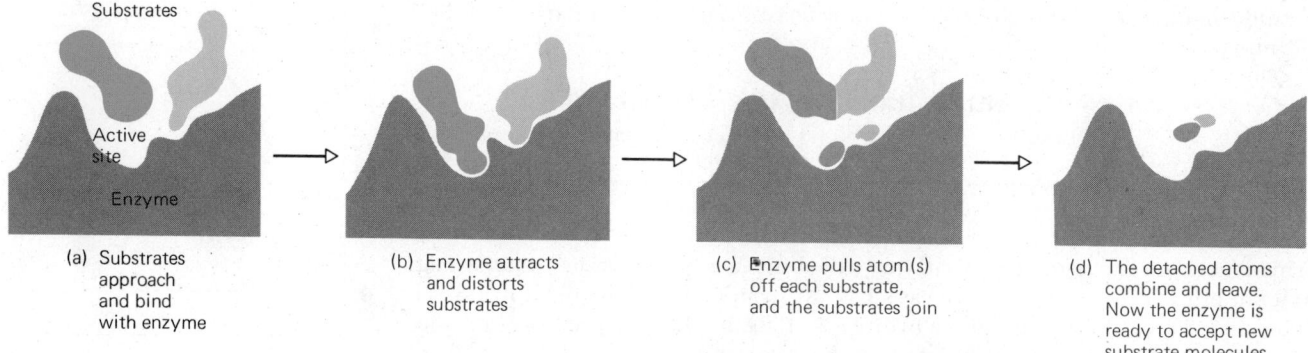

(a) Substrates approach and bind with enzyme

(b) Enzyme attracts and distorts substrates

(c) Enzyme pulls atom(s) off each substrate, and the substrates join

(d) The detached atoms combine and leave. Now the enzyme is ready to accept new substrate molecules

Figure 3-22 How an enzyme-catalyzed reaction between two substrates might look. Note that both the enzyme and substrates change shape somewhat when the substrates bind to the enzyme's active site. The enzyme probably loosens bonds in the substrate molecules, allowing them to release some of their atoms more easily and re-bond to each other.

ated reactions. Both enzyme and substrate(s) change shape slightly when they combine; this is known as "induced fit."

When an enzyme catalyzes the reaction of one substrate with another, it binds the two closely together, so that the parts of the substrates that will react are next to one another (Figure 3–22). The enzyme may alter its substrate by attracting electrons or atomic nuclei in such a way as to put tension on the bonds holding the different atoms of the substrate together. This makes the substrate more reactive, and is thought to be how an enzyme lowers the activation energy of a reaction.

Before some enzymes can bind their substrates, their active sites must contain additional ions or molecules which are not proteins. If these additional nonprotein substances are covalently bound to the enzyme, they are called **prosthetic groups**; if they are held by other kinds of bonds, they are called enzyme **cofactors**. Some cofactors are ions, such as calcium (Ca^{2+}) or magnesium (Mg^{2+}); others are nonprotein organic molecules called **coenzymes**.

FACTORS THAT AFFECT ENZYME ACTIVITY

Because substrate molecules must bind to specific sites on the enzyme, each enzyme molecule can catalyze only a certain number of reactions in a given time. Thus enzyme-mediated reactions proceed faster as more enzyme molecules are added. This is true only as long as all the enzyme molecules are occupied all the time. If the reaction reaches the point at which there are more enzyme than substrate molecules, more substrate will have to be added to speed it up again (Figure 3–23).

Because many of their amino acid residues bear R groups which ionize when they are dissolved in water, enzymes are electrically charged. The pH of the solution will determine the charges that an enzyme bears. For instance, in an acid solution, negatively charged R groups on the enzyme will tend to combine with H^+ ions in the solution; thus neutralized, the negative R groups can no longer bind to positively

Figure 3-23 Effect of enzyme and substrate concentration on the rate of an enzyme-mediated reaction. Raising the concentration of either increases the rate at which substrate is converted to product, but at high substrate concentrations the enzyme molecules become saturated. V_{max} (part b) is the maximum velocity of the reaction at a given enzyme concentration.

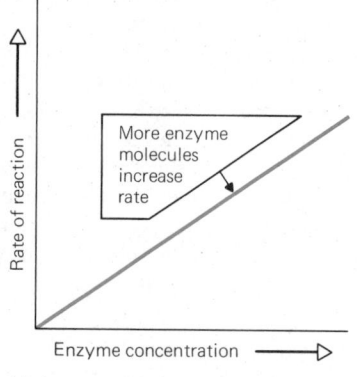

(a) Large excess of substrate present

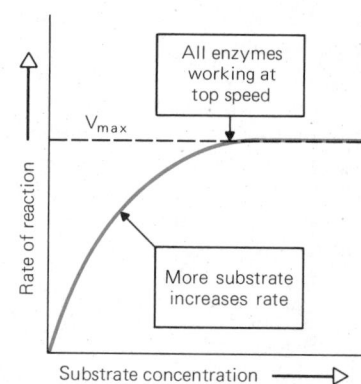

(b) Constant enzyme concentration

charged groups. Only if an enzyme bears the appropriate charged groups at its active site will it react with its substrate. This is why pH affects the rate of an enzyme-mediated reaction (Figure 3–24).

Most proteins work best when they are in a medium of approximately neutral pH; however, some organisms live in acidic or basic environments. For instance, some organisms live in the mineral springs of Yellowstone National Park (where both acidic and basic waters are found) and these organisms have enzymes that work at the pHs of their particular environment. Digestive enzymes of the human stomach work best at an extremely acidic pH, around 1.5–2.0. These enzymes become relatively inactive when they pass with the food into the small intestine, where stomach acid is neutralized and the pH is raised to about 8.

Temperature also affects the rate of enzyme reactions. Temperatures above about 60°C **denature** most proteins, or disrupt their structure permanently so that they lose all ability to function. (Heating food preserves it by destroying the enzyme activity of organisms that cause decay.) At the opposite extreme, chemical reactions proceed slowly at low temperatures because molecules move so slowly that few collisions occur between enzyme and substrate molecules. (Refrigeration preserves food by slowing the action of enzymes in organisms which cause decay, or enzymes in the food itself.) Through the mechanism of natural selection, enzymes adapted to function at a particular range of temperatures have been selected (Figure 3–25).

Organisms that live at extreme temperatures have enzymes that will function under these conditions. Some of the organisms known as blue-green bacteria live on the surfaces of glacial ice, and are adapted to temperatures close to the freezing point of water (0°C). Other members of this group inhabit the hot springs of Yellowstone, which, in addition to having unusually high or low pH values, may be at temperatures of 80 to 85°C.

3-F   Nucleic Acids

The **nucleic acids** include the largest molecules formed by organisms. There are two kinds of nucleic acids: **deoxyribonucleic acid (DNA)**, which is the genetic material, containing instructions for the order of amino acids in polypeptides, and **ribonucleic acid (RNA)**, which participates in polypeptide synthesis.

Nucleic acid monomers are called **nucleotides**. Each nucleotide, in turn, is composed of three distinct units: a phosphate group, a pentose sugar, and a ring-shaped nitrogenous base.* The bases are derivatives of either the single-ring base **pyrimidine** or the double-ring base **purine** (Figure 3–26). The sugar may be either

---

*Base   In this usage, a substance that can accept H+ in solution (see Section 2-G).

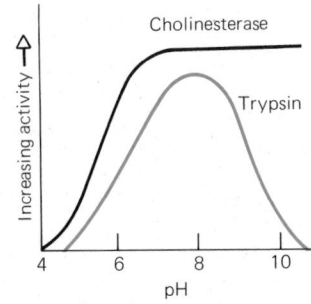

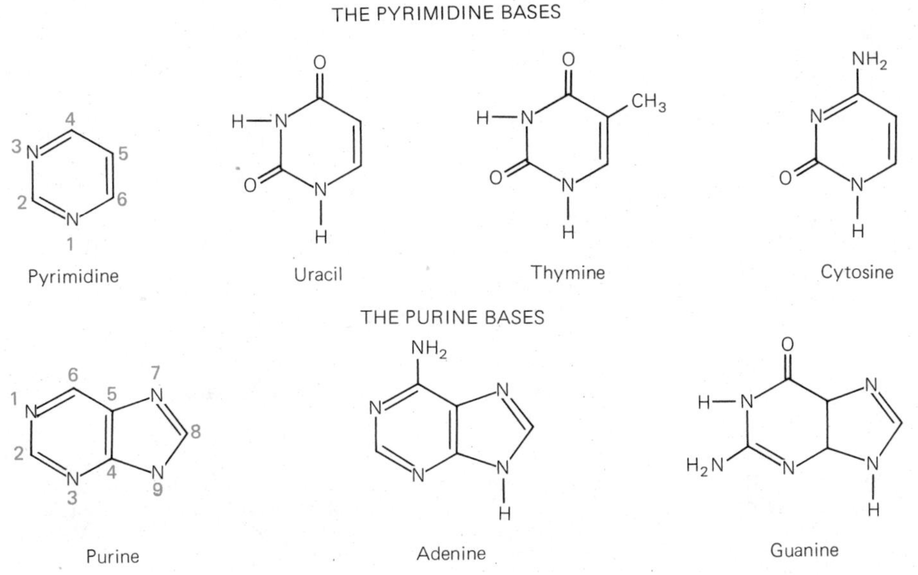

Figure 3-24 Effect of pH on the activity of two enzymes of mice. The pH-sensitivity of an enzyme depends on how many ionizable R groups are necessary to its activity, and their charge. Trypsin is an enzyme that hydrolyzes proteins; cholinesterase hydrolyzes substances that are important in the nervous system.

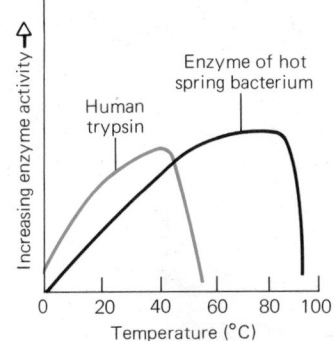

Figure 3-25 Effect of temperature on two enzymes that normally function at two different temperatures. The human body is normally at 37.6°C; the hot spring bacterium lives at about 80°C. The curve falls off steeply at high temperatures as the enzyme is denatured and its function completely destroyed.

THE PYRIMIDINE BASES

Pyrimidine   Uracil   Thymine   Cytosine

THE PURINE BASES

Figure 3-26 The common nitrogenous bases found in nucleic acids.

Purine   Adenine   Guanine

RIBOSE

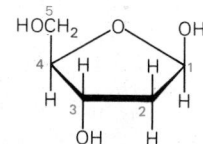

DEOXYRIBOSE

Figure 3-27 Ribose and deoxyribose, the pentose sugars found in nucleic acids. "Deoxy" means "deprived of oxygen"; note that deoxyribose has one less oxygen atom than ribose (colored atom).

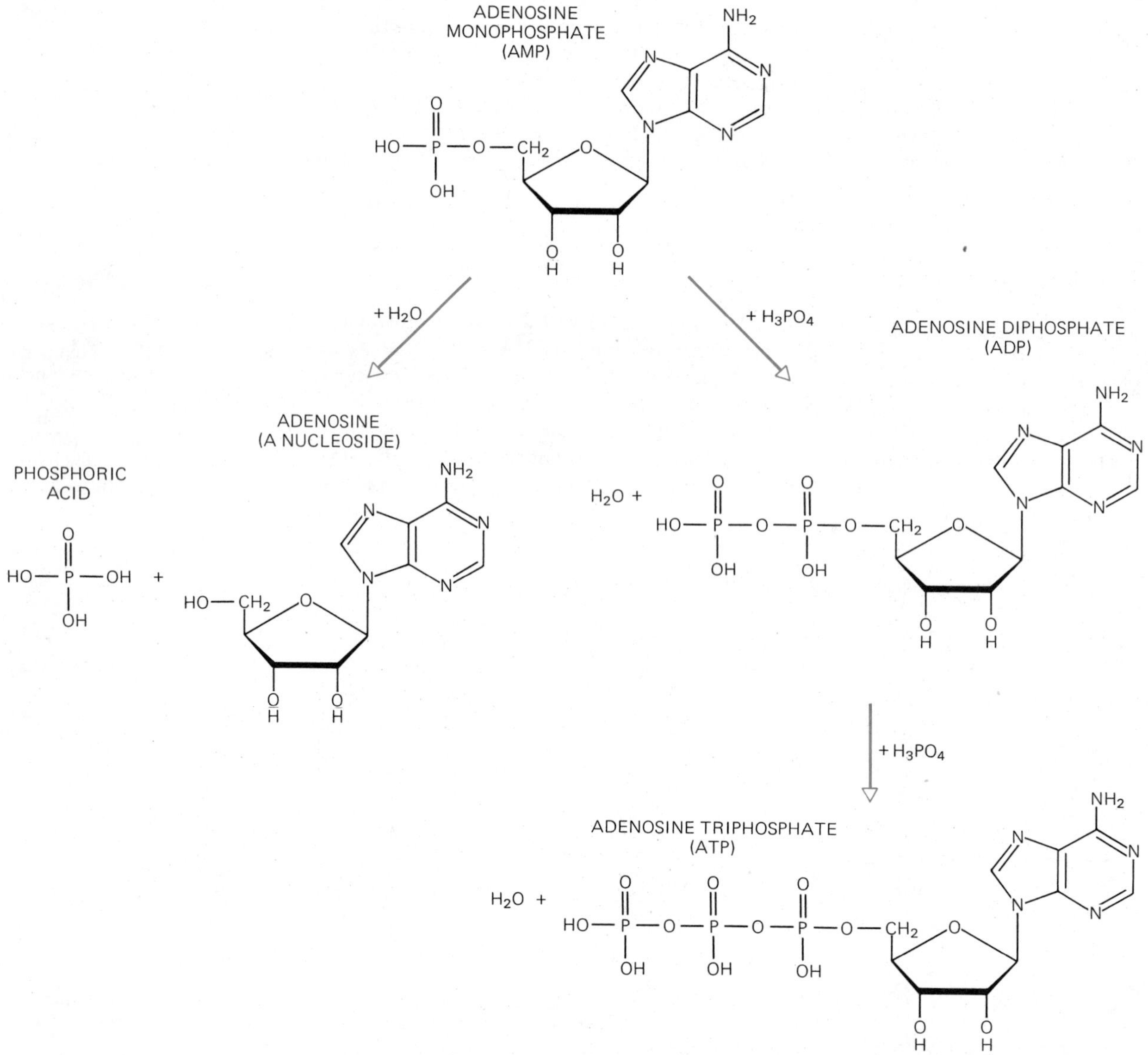

ADENOSINE
MONOPHOSPHATE
(AMP)

+ H₂O

+ H₃PO₄

ADENOSINE DIPHOSPHATE
(ADP)

PHOSPHORIC
ACID

ADENOSINE
(A NUCLEOSIDE)

H₂O +

+ H₃PO₄

ADENOSINE TRIPHOSPHATE
(ATP)

H₂O +

Figure 3-28 The nucleotide adenosine monophosphate (top) may lose its phosphate group to become a nucleoside (left) or may gain a phosphate to become adenosine diphosphate (right). Addition of a third phosphate yields adenosine triphosphate, used in synthesis of RNA and as an energy-donating molecule for a number of enzymatic reactions.

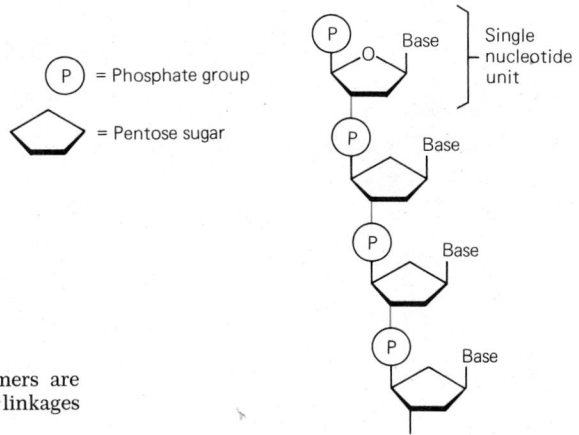

P = Phosphate group

= Pentose sugar

Single nucleotide unit

Base

Base

Base

Base

Figure 3-29 Nucleotide monomers are attached to each other by ester linkages (color).

ribose, which is found in the nucleotides that make up RNA (ribonucleic acid), or **deoxyribose**, found in the nucleotides that make up DNA (deoxyribonucleic acid) (Figure 3-27). The phosphate group is linked to the fifth carbon atom of the sugar by an ester linkage (Figure 3-28, top).

Hydrolysis of the phosphate group of a nucleotide yields a **nucleoside**, a substance composed of only a sugar and a nitrogenous base (Figure 3-28). Thus a nucleotide is the phosphate ester of a nucleoside. Cells contain nucleotides in the form of nucleoside monophosphates, diphosphates, and triphosphates—nucleosides linked to one, two, or three phosphate groups. These nucleotide molecules are named according to which sugar and which base they contain, and their number of phosphate groups (Table 3-2).

Besides their importance as the monomers from which nucleic acids are synthesized, a number of nucleotides have other important functions in living things. Some serve as coenzymes, or as the precursors from which coenzymes are synthesized. In addition, adenosine triphosphate (ATP) is an immediate source of energy for most living organisms, and guanosine triphosphate (GTP) fulfills a similar role in other organisms (see Chapters 6, 7, and 8). AMP is an important intermediary in the function of some hormones (Section 41-A).

When nucleotides are polymerized into nucleic acids, the phosphate group of one nucleotide is linked to the sugar of the next nucleotide to form a long backbone of alternating sugar and phosphate residues from which the bases extend along one side (Figure 3-29). We shall consider the structure of nucleic acids in more detail in Chapter 9.

TABLE 3-2 **THE COMMON NUCLEOSIDE TRIPHOSPHATES***

Needed in the Synthesis of RNA:	Needed in the Synthesis of DNA:
ATP (adenosine triphosphate)	dATP (deoxyadenosine triphosphate)
UTP (uridine triphosphate)	dTTP (deoxythymidine triphosphate)
CTP (cytidine triphosphate)	dCTP (deoxycytosine triphosphate)
GTP (guanosine triphosphate)	dGTP (deoxyguanosine triphosphate)

*Each triphosphate can give up one or two of its phosphate groups to become a diphosphate or monophosphate. E.g., uridine triphosphate (UTP) can become uridine monophosphate (UMP).

3-G Quantitative Comparisons of Classes of Biological Molecules

The sizes of the various kinds of molecules discussed in this chapter vary over several orders of magnitude. The monomers—monosaccharides, fatty acids, amino acids, and nucleotides—all fall below a molecular weight of 400 (found by adding the atomic masses of all their atoms; see Table 2-1). The common lipids are quite

TABLE 3-3 RELATIVE SIZES OF SOME BIOLOGICAL MOLECULES

SUBSTANCE	MOLECULAR WEIGHT	REFERENCES OR COMMENTS
Carbohydrates		
Glucose	180	Common monosaccharide (Fig. 3-6)
Sucrose	342	Table sugar (a disaccharide; Fig. 3-7)
Starch	up to 100 million	A storage polysaccharide (Fig. 3-8)
Cellulose	50,000-2,500,000	A structural polysaccharide, 300-15,000 glucose residues (Fig. 3-8)
Lipids		
Glyceryl tristearate	890	A triglyceride (Fig. 3-11)
Cholesterol	373	A steroid (Fig. 3-13)
Proteins		
Amino acids	75-204	
Insulin (cattle)	5700	Two polypeptide chains (Fig. 3-17)
Hemoglobin (human)	64,500	Four polypeptide chains + four prosthetic groups; carries O_2 in blood
DNA polymerase I	109,000	Single polypeptide chain; enzyme that repairs DNA molecules
Ribulose bisphosphate carboxylase	550,000	Sixteen subunits; enzyme of photosynthesis (see Section 8-E)
Nucleic Acids		
Adenosine monophosphate	347	See Figure 3-28, top
DNA	1.7 million-hundreds of millions	Only small viruses' DNA can be isolated unbroken for weight determination

small, not even deserving to be called polymers or macromolecules because they are made up of very few subunits. Polysaccharides are much larger and are very variable in size. A particular kind of DNA or protein, however, is always the same size, although the DNAs of different species can be of different sizes, and the proteins even within one species are of many sizes (Table 3-3).

The four classes of biological molecules are also found in different amounts in living organisms. In keeping with their many and crucial functions, it should be no surprise that the proteins are the most abundant class in a "typical" cell. As we become more familiar with the roles of the many substances in living organisms in succeeding chapters, we shall gain a better understanding of the relative abundances presented in Table 3-4.

TABLE 3-4 CHEMICAL COMPOSITION (EXCLUDING WATER) OF SOME COMMON BACTERIA

TYPE OF MOLECULE	PERCENT OF TOTAL DRY WEIGHT	COMMENTS
Small molecules	10	Inorganic ions, monomers, coenzymes
Polysaccharides and lipids	16	Protective outer wall and membrane; some glycogen stored inside bacterium
DNA	4	One or two molecules per bacterium; each molecule is about 1 mm[1] long and highly folded; the bacterium itself is only about 0.002 mm long
RNA	20	About 3000 different kinds
Proteins	50	About 2500 different kinds: about $\frac{1}{3}$ structural protein, $\frac{2}{3}$ enzymes

[1]mm = millimetre; there are about 25 mm in 1 inch (see "Metric System" end paper of this book).

SUMMARY

Living organisms are made up of water, mineral salts, and organic molecules. Organic molecules are made up of chains of carbon atoms with atoms of other elements attached. It is possible to construct an endless variety of "carbon skeletons" of various sizes and shapes for organic molecules.

Although we know of a myriad of organic substances, we can simplify our study of those that occur in living things by grouping them into four main categories: carbohydrates, lipids, proteins, and nucleic acids.

Carbohydrates are composed entirely, and lipids mainly, of carbon, hydrogen, and oxygen. Lipids are not soluble in water and are vital components of all biological membranes. Some carbohydrates and lipids are important energy-storage compounds that may be broken down to release energy. Plants make the polysaccharide cellulose, which gives them support and protection.

Proteins and nucleic acids are macromolecules that play a vital part in directing the growth, activity, and reproduction of cells. Proteins contain carbon, hydrogen, oxygen, nitrogen, and some sulfur; nucleic acids contain carbon, hydrogen, oxygen, nitrogen, and phosphorus.

Each organic macromolecule is a polymer made up of a number of monomeric subunits, and the types of subunits in each macromolecule are limited in number.

Monomers are joined together to form polymers by condensation reactions in which a bond is formed by removing a water molecule between the subunits. Macromolecules are broken down by hydrolysis, the addition of a water molecule between the subunits, which thus become separated.

Monomers		*Polymers*
Monosaccharides	$\xrightleftharpoons[+H_2O]{-H_2O}$	Polysaccharides
Glycerol and fatty acids	$\xrightleftharpoons[+H_2O]{-H_2O}$	Triglycerides (not true polymers)
Amino acids	$\xrightleftharpoons[+H_2O]{-H_2O}$	Proteins
Nucleotides	$\xrightleftharpoons[+H_2O]{-H_2O}$	Nucleic acids

Thousands of kinds of enzymes speed up the chemical reactions necessary to life. Enzymes are proteins adapted to fit specific substrate molecules. By lowering the activation energy for the reactions they mediate, enzymes enable organisms to carry out chemical reactions quickly at relatively low temperatures.

OBJECTIVES

From your study of this chapter, you should be able to:

1. Define or recognize examples of the following: organic compound, carboxyl group, amino group, polymer, monomer, carbohydrate, monosaccharide, disaccharide, hexose sugar, polysaccharide, polypeptide, lipid, triglyceride, fatty acid, amino acid, dipeptide, peptide bond, protein, enzyme, substrate, condensation reaction, dehydration synthesis, hydrolysis, catalyst, equilibrium position, pentose sugar, nucleic acid, nucleotide.

2. Classify the following in the appropriate classes of molecules listed in Objective 1, and briefly state their function in living organisms: glucose, cellulose, starch, glycogen, glycerol, enzyme, ATP.

3. List four main classes of biological macromolecules, state the role of each in living organisms, and name the type(s) of subunits from which each is synthesized and the chemical elements typical of each class of molecules.

4. Given a diagram of the structures of two sugars, or two amino acids, or glycerol and a fatty acid, draw and explain a condensation reaction between them, and name the classes of compounds to which the reactant(s) and products belong, or, given the products of such a condensation reaction, draw a diagram of the hydrolysis reaction that each would undergo.

5. Explain the effect of enzyme or substrate concentration, temperature, and pH on the rate of an enzyme-mediated reaction.

SELF-QUIZ

1. Which of the following is an amino acid?

   a.

   b. $H_2N—C—NH_2$
   (with O double-bonded to C)

   c. $H_3C—C—C$
   (with NH_2 and H on middle carbon, O and OH on end carbon)

   d. $H—C—C—OH$
   (with NH_2 and H on first carbon, H and H on second, O on the carbonyl)

   e. $H—C—C—C—C$
   (with H atoms and O and OH)

2. Pick out the R group on the amino acid from question #1.

3. Which of the following is *not* made up of hexose sugar subunits?
   a. sucrose   d. insulin
   b. starch   e. cellulose
   c. glycogen

4. Draw a hydrolysis reaction using the molecule below:

5. The molecule above is a:
   a. fatty acid   d. hexose sugar
   b. dipeptide   e. pentose sugar
   c. disaccharide

6. You have a solution of an enzyme. You put half the solution into each of two test tubes containing identical substrate at equal concentration. After waiting awhile, you test both solutions and find that the substrate in test tube A has been changed but the substrate in tube B has not been acted on by the enzyme. Suddenly you notice that tube B has been sitting on a hot plate with the switch turned to "high." The enzyme in tube B probably did not work because it had been:
   a. hydrolyzed
   b. denatured
   c. condensed
   d. catalyzed
   e. dehydrated

7. Which of the following statements about enzymes is *true*? Enzymes
   a. are altered permanently in the reaction they catalyze
   b. make the equilibrium of the reaction more favorable for the organism
   c. increase the energy of the reactant molecules
   d. lower the energy of activation of a reaction
   e. lower the energy of the products

8. Which of the following sets of elements is found in nucleic acids?
   a. carbon, hydrogen, oxygen, nitrogen, and phosphorus
   b. carbon, hydrogen, oxygen, nitrogen, and sulfur
   c. carbon, hydrogen, oxygen, nitrogen, sulfur, and phosphorus
   d. carbon, hydrogen, oxygen, and phosphorus
   e. carbon, hydrogen, oxygen, and nitrogen

QUESTIONS FOR DISCUSSION

1. Do you think life on another planet could be based on the chemistry of an element other than carbon?

2. An organism contains different amounts of the various enzymes it makes. Based on your own understanding of natural selection from Chapter 1, explain how this variation may have arisen.

REFERENCES AND FURTHER READING

Barker, R. *Organic Chemistry of Biological Compounds*. Englewood Cliffs, N.J.: Prentice-Hall, 1971. A good reference text on biologically important compounds.

Calvin, M., and W. A. Pryor. *Organic Chemistry of Life: Readings from Scientific American*. San Francisco: W. H. Freeman, 1973. A collection of articles on biologically important organic compounds; readable and well illustrated.

Dickerson, R. E., and I. Geis. *The Structure and Action of Proteins*. New York: Harper and Row, 1969. Introduces the chemistry and structure of various kinds of proteins; excellent illustrations, highly readable.

Sharon, N. "Carbohydrates." *Scientific American*, November 1980. Covers historic and modern discoveries about carbohydrate roles in organisms.

CHAPTER 4

STRUCTURE AND FUNCTION OF CELLS

Cells were first described by seventeenth-century microscopists. In 1665, Robert Hooke applied the word "cell" to the chambers in a thin section of cork, which we now know are the empty walls of dead cells in the bark of the cork oak. Matthias Schleiden, a botanist, and Theodor Schwann, a zoologist, formulated the **cell theory** during 1838 and 1839. The theory states (1) that cells are the fundamental units of life, and (2) that all organisms are made up of cells. In 1855, Rudolf Virchow added a third statement, (3) that cells arise only by division of other cells.

In this chapter we shall examine the structure of cells, the functions of their parts, and the differences between plant, animal, and bacterial cells.

Although cells vary considerably in size, most plant and animal cells are very small, between 5 and 40 μm in diameter (Tables 4–1 and 4–2). Most cells can be seen only if microscopes are used to magnify them, and most of our knowledge about cells has depended upon the gradual improvement of microscopes over the centuries.

TABLE 4-1 **UNITS OF SIZE USED IN MICROSCOPY**

1 metre (m) = 100 centimetres (cm) = 39.4 inches
1 cm = 10 millimetres (mm)
1 mm = 10^{-3} m† = 10^3 micrometres (μm)
(Micrometres were formerly known as microns, denoted by the Greek letter μ, pronounced "mew.")
1 μm = 10^{-6} m = 10^{-4} cm = 10^3 nanometres (nm)
1 nm = 10^{-9} m = 10^{-7} cm = 10 Angstroms (Å)
1 Å = 10^{-10} m = 10^{-8} cm = 10^{-1} nm
(Most biologists have now abandoned the use of Angstrom units, which will still, however, be encountered in older literature.)

The prefix "centi" means one-hundredth
"milli" means one-thousandth
"micro" means one-millionth
"nano" means one-billionth

† 1/10 = 0.1 = 10^{-1}; 10^{-3} = 1/1,000; 10^6 = 1,000,000

4-A Microscopy

Light Microscopes

Though we usually think of the lenses used in microscopes as relatively modern inventions, Italian monks developed the art of grinding lenses in the fourteenth century. These lenses were made into spectacles to improve the failing eyesight of monks who printed and illustrated exquisite manuscripts. By 1590, the Dutch lens grinders Hans and Zacharias Janssen had mounted two lenses in a tube to produce the first **compound microscope** (one with two main lenses). The subsequent improvement of light microscopes, and the invention of electron microscopes in the twentieth century, have permitted the viewing of smaller and smaller objects.

The compound light microscopes used today contain an **objective lens**, close to the object or specimen being viewed; this lens forms a magnified image of the specimen. The image is further magnified by the **ocular lens**, which lies close to the viewer's eye (Figure 4-1).

Figure 4-1 Comparison of a light microscope with an electron microscope. The electron microscope is shown upside down so that the parts correspond with those of the light microscope.

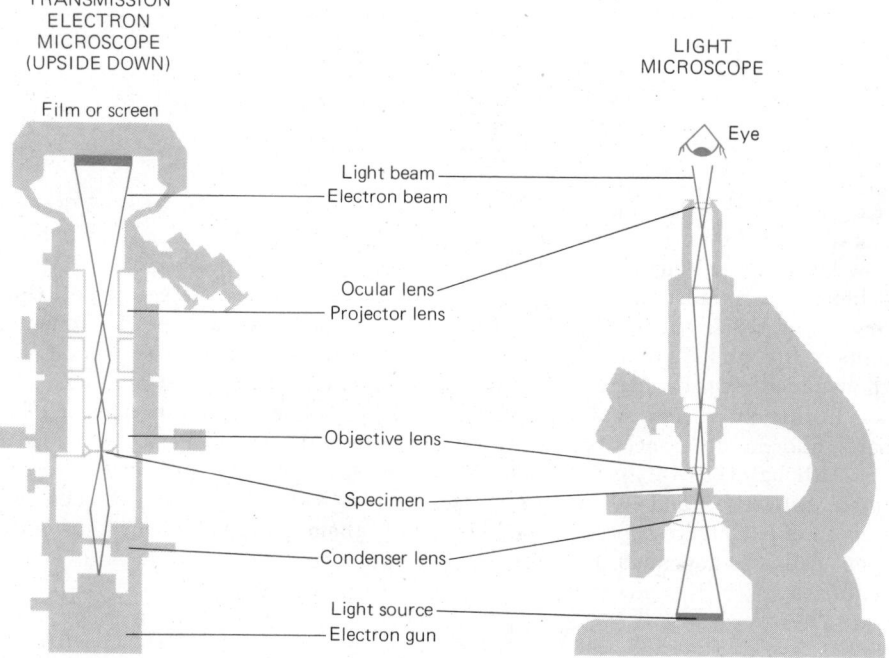

TRANSMISSION
ELECTRON
MICROSCOPE
(UPSIDE DOWN)

Film or screen

LIGHT
MICROSCOPE

Eye

Light beam
Electron beam

Ocular lens
Projector lens

Objective lens

Specimen

Condenser lens

Light source
Electron gun

TABLE 4-2 SOME BIOLOGICAL SIZES

Nerve cell	Up to 2 m long (but *very* thin)
Average body cell of an animal	10–20 μm in diameter
Average body cell of a plant	30–50 μm in diameter
Chloroplast of a higher plant	5–10 μm long
Mitochondrion	Up to 7 μm long
Escherichia coli (bacterium)	2 μm long
Ribosome	25 nm (= 0.025 μm) in diameter
DNA molecule	2 nm thick
Hydrogen atom	0.1 nm in diameter

Surprisingly, the magnifying power of a microscope is not the most important factor determining the size of the object that can be viewed. The most essential feature of a microscope is its **resolving power**, its ability to distinguish the separateness of two objects that are close together (Table 4–3). The resolving power of the human eye, for instance, is about 0.1 mm; we can distinguish two dots as being separate from each other when they are as little as 0.1 mm apart. Magnification is of no use without resolving power. If the lenses cannot distinguish detail (separateness of parts) in the object, they will produce a fuzzy image, and magnification will produce only a larger fuzzy image. The better the resolving power of a microscope, however, the greater the magnification which can usefully enlarge the detail distinguished by the lenses. A good light microscope can resolve two objects separated by 0.4 μm, and a magnification of about ×2000 will be needed so that the eye can also resolve the projected image. For any greater magnification to be of use, the resolving power must be increased.

The resolving power of a lens system is limited by the **diffraction**, or scattering, of light as it passes through the lens opening. Since diffraction at the objective lens produces an image that is larger than the specimen, two small specimens close together will have overlapping images which cannot be resolved as separate. Resolving power can be increased, therefore, by reducing diffraction as the light passes through the objective lens. This can be done by increasing the opening of the lens, since light is diffracted less as it passes through a wider opening, or by decreasing the wavelength* of light used, since shorter wavelengths of light are diffracted less than long wavelengths. For example, the ultraviolet microscope, using a wavelength about one quarter the average wavelength of white light, has four times better resolving power than a white light microscope.

So far we have considered only the generalized situation of distinguishing two different objects or points. If the "objects" are the two sides of one particle, then the particle itself will be resolved, or visible, only if it is at least as big as the resolving power of the microscope. Thus, an object smaller than 0.4 μm will not be visible through a light microscope. Although entire cells and their larger components can

---

* Wavelength Light may be considered as traveling in wavy lines. The distance between adjacent peaks of the line is the wavelength.

TABLE 4-3 TYPICAL RESOLVING POWERS OF VARIOUS OPTICAL SYSTEMS

SYSTEM	RESOLVING POWER
Human eye	0.1 mm = 100 μm at a distance of 10 cm
Light microscope	0.4 μm
Ultraviolet microscope	0.1 μm
Scanning electron microscope	5 nm (= 0.005 μm)
Transmission electron microscope	0.2 nm (= 0.0002 μm)

be seen with a light microscope, many smaller structures within a cell demand higher resolving power. This requires microscopes that use shorter wavelengths, such as electron microscopes.

Electron Microscopes

TRANSMISSION ELECTRON MICROSCOPES

In **transmission electron microscopes**, invented in the 1930s, electrons pass through the specimen, just as light passes through the specimen in a light microscope. When we visualize chemical reactions it is convenient to think of electrons as particles, but electrons can also be considered as electromagnetic waves. The wavelength of a typical beam of electrons is about 0.005 nm. This gives an electron microscope a theoretical resolving power of about 0.0025 nm. In fact, because of technical difficulties, the best transmission electron microscopes today have a resolution of about 0.2 nm.

The features of a transmission electron microscope are similar to those of a light microscope except that it is assembled upside down (see Figure 4–1). The lenses of an electron microscope are not glass but electromagnets, which can deflect the negatively charged electrons. A beam of electrons is produced by heating a tungsten filament in the electron gun. The beam is accelerated through the **condenser lens**, which focuses it, and then passes the focused beam through the specimen and the objective lens. In an electron microscope, the equivalent of the ocular lens in a light microscope is the **projector lens**. Since the human eye cannot detect electrons, the projector lens focuses the final electron beam on a photographic plate or fluorescent screen, where it will produce a visible image.

One disadvantage of transmission electron microscopes is that electrons are very easily deflected or absorbed by molecules in the air or by the specimen itself. For this reason specimens must be observed under a high vacuum, and must also be sliced very thin (50 to 100 nm). In contrast, specimens for light microscopy are seldom sliced thinner than 4000 nm.

SCANNING ELECTRON MICROSCOPE

A later invention than the transmission electron microscope is the **scanning electron microscope**, in which a beam of electrons bombards the surface of the object and causes lower-energy **secondary electrons** to be emitted from atoms in the surface of the specimen. These secondary electrons are collected and used to vary the intensity of a spot on a television screen that scans in synchrony with the electron beam. The resolving power, about 5 nm, is less than that of a transmission electron microscope, but the scanning electron microscope has other advantages. First, less preparation of the specimen is needed. Second, since the microscope has an extraordinary depth of focus, the surface of an intact specimen can be observed in great detail (Figure 4–2). In addition, some living organisms, such as hardy insects, can withstand the high vacuum of a scanning electron microscope and be viewed alive. However, as is the case with transmission electron microscopes, higher resolving powers require higher vacuums, so that only completely dehydrated, and therefore dead, specimens can be viewed if maximum resolution is to be obtained.

Specimen Preparation and Staining

In any microscope, a visible image is produced because certain parts of the specimen absorb or deflect light or electrons and other parts permit light or electrons to pass through. The final image thus shows light and dark areas in the specimen. To increase the contrast in the final image, most specimens are specially prepared and stained for microscopy. Tissues are first fixed (killed and preserved) in fixatives such as formaldehyde for light microscopy or glutaraldehyde and osmium tetroxide for electron microscopy. Then they are embedded in wax or resin and **sectioned** into thin slices with a glass or metal knife.

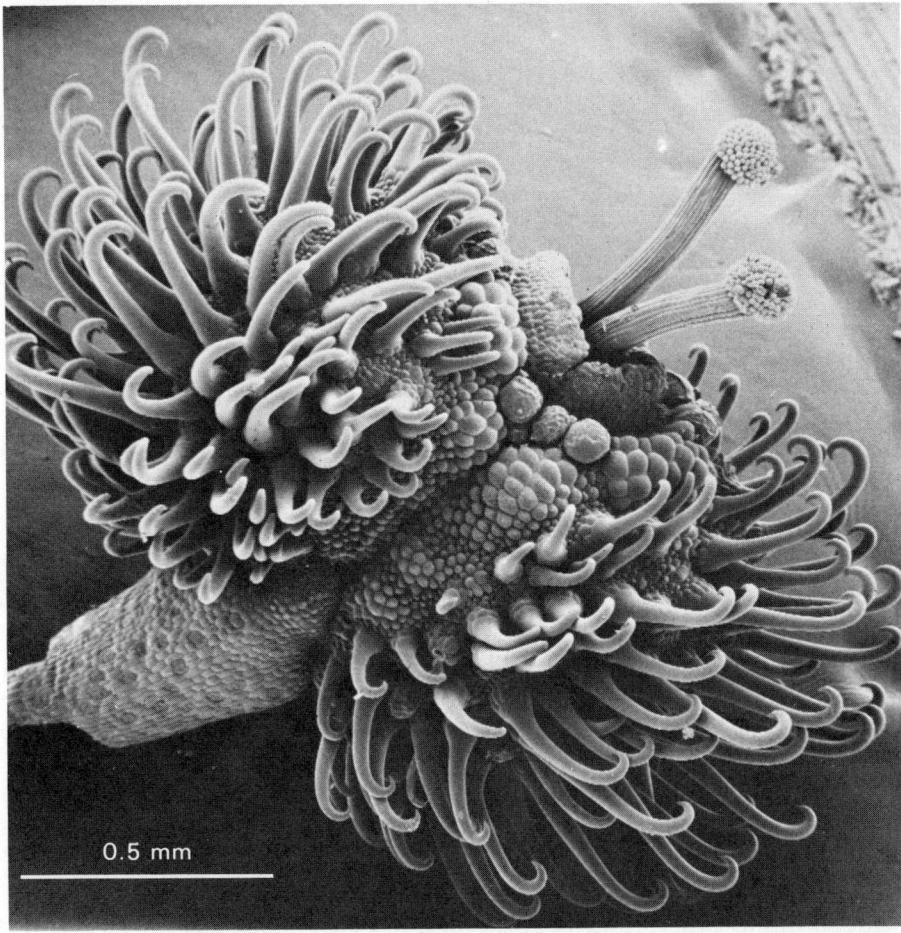

Figure 4-2 A scanning electron micrograph of a goosegrass flower with a pair of developing fruits. The fruits disperse by hooking onto an animal or onto our clothes when we walk through a meadow in late summer. (Biophoto Associates)

Stains for microscopy give contrast to the image by differentially absorbing light or electrons. For instance, alkaline structures, such as the chromosomes, which contain the genetic material, react with a blue stain called hematoxylin. The transmission electron microscope cannot produce colored pictures, but structures can be stained with heavy metal ions, which absorb electrons and so produce dark areas in the final image. A specimen might be stained with a lead solution, which reacts with acid structures to leave a deposit of lead that will absorb electrons. For scanning electron microscopy, the whole surface of the dried specimen is usually coated with a very thin layer of gold or some other good emitter of secondary electrons. In effect, the viewer "sees" the gold coating, not the specimen itself.

4-B Cell Function and Organelles

The cell theory states that a cell is the structural and functional unit of life. Many organisms are **unicellular**, composed of only one cell. Most plants, fungi, and animals, however, are **multicellular**, composed of many cells. All cells must carry out certain activities basic to life; in addition, each cell of a multicellular organism makes a specialized contribution to the functioning of the body as a whole. Some cells transport food, gases, and waste products; others sense changes in the external and internal environments, control movement, or perform other special tasks. Thus there is a division of labor among the cells of a multicellular organism. A specialized cell does not usually have structures or chemicals that most other cells lack; rather,

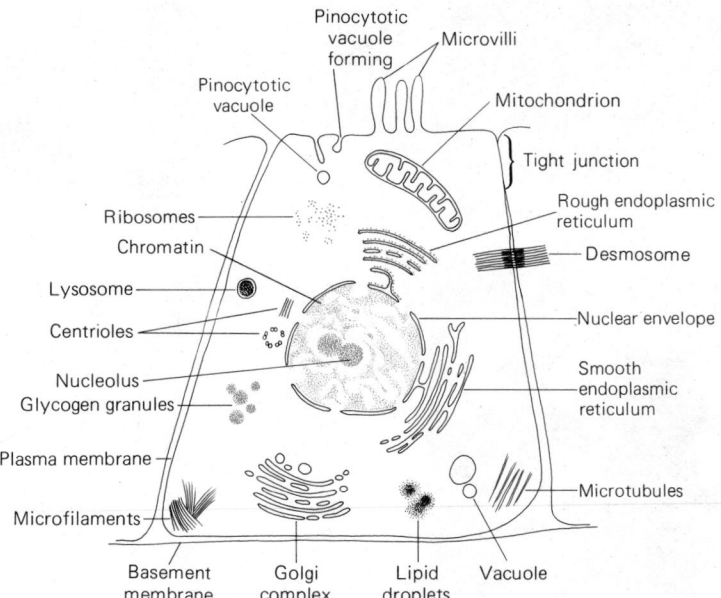

Figure 4-3 A generalized animal cell. This cell has organelles that would not usually occur in the same cell (e.g., microvilli are characteristic of cells lining the intestine; glycogen granules are found in liver cells).

it is specialized by the exaggeration of a particular characteristic or set of characteristics common to most types of cells. Specialized cells can be regarded as variations on a basic theme of cell structure and function (Figure 4–3).

Every cell is made up of (1) **cytoplasm**, the "living material" in which various activities take place and in which many discrete structures with particular functions are found; (2) a **plasma membrane**, which surrounds the cytoplasm; and (3) a **nucleus** (or a **nuclear area**), which contains the genetic material. In addition to these three universal components, the cells of plants and unicellular photosynthetic organisms have **cell walls** outside the plasma membrane and usually contain a fluid-filled **vacuole** surrounded by a membrane (Figure 4–4). The nucleus, vacuoles, and various other structures inside cells are called **organelles**.

Some organelles are found in essentially all cells since they perform functions that no cell can do without. What are these functions? Each cell must obtain or manufacture food, break down its food to release energy, and eliminate wastes into the environment. In addition, most cells reproduce themselves by dividing. A cell must also synthesize the various macromolecules that are necessary for all these processes.

Figure 4-4 Diagram of a generalized plant cell.

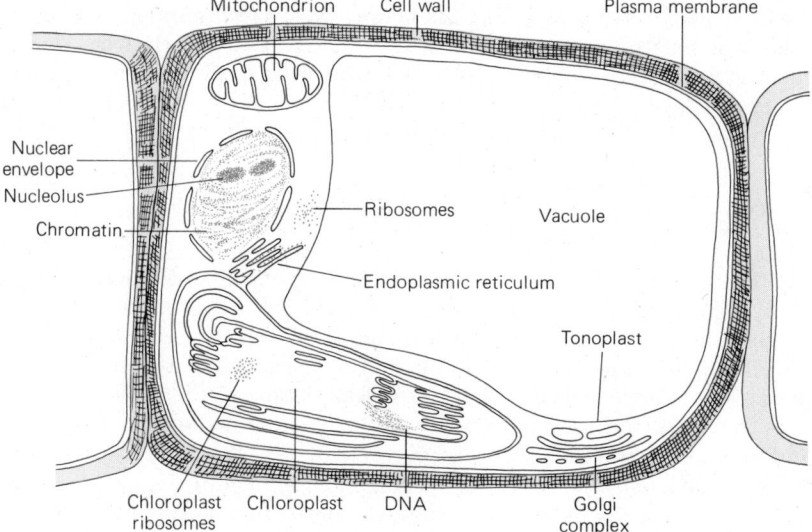

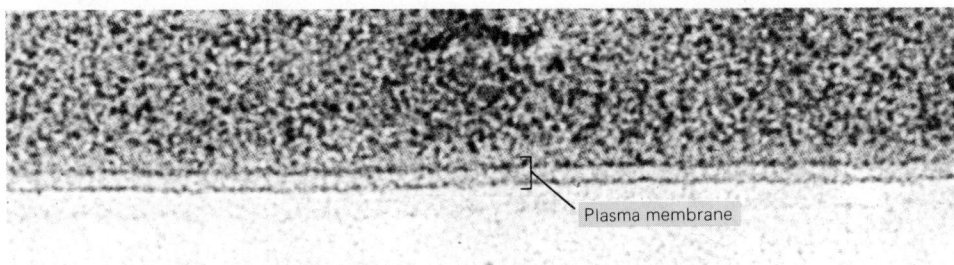

10 nm

Figure 4-5 This electron micrograph shows the two-layered structure of the plasma membrane. (Biophoto Associates)

There are two basic types of cells, prokaryotic and eukaryotic. **Prokaryotic** cells are simple in structure, and for this reason are believed to be evolutionarily more primitive than eukaryotic cells. Prokaryotic organisms are the bacteria, including the blue-green bacteria, which were formerly called blue-green algae. **Eukaryotic** cells contain many membrane-bound organelles, which are not found in prokaryotic cells. Eukaryotes include the **protists**, which are unicellular, and the multicellular plants, fungi, and animals.

Most of the rest of this chapter will examine the structure of eukaryotic cells. We shall examine organelles common to all eukaryotic cells and then turn to organelles that are found less universally. Finally, we shall consider the structure of prokaryotic cells.

The Plasma Membrane

Every cell is surrounded by a **plasma membrane**, whose primary function is to control what enters and leaves the cell. The membrane (sometimes called a **cell membrane** or **plasmalemma**) is only about 7.5 to 10 nm thick and is therefore too thin to be seen with a light microscope. However, many biologists of the early 1900s believed in its existence because of the way cells respond to certain chemical treatments. Electron microscopy finally showed that the plasma membrane is a real structure. An electron micrograph shows the membrane as a continuous double line surrounding the cell (Figure 4–5).

Biological membranes are so thin that they are very difficult to study, and their structure has been debated for many years. In 1972, S. J. Singer and Garth Nicolson proposed a model of the plasma membrane that is widely accepted today. According to this "fluid mosaic" model, the membrane consists largely of two layers of phospholipid molecules in which protein and glycoprotein (molecules made of carbohydrate and protein) molecules are embedded (Figure 4–6). In animal cells, cholesterol

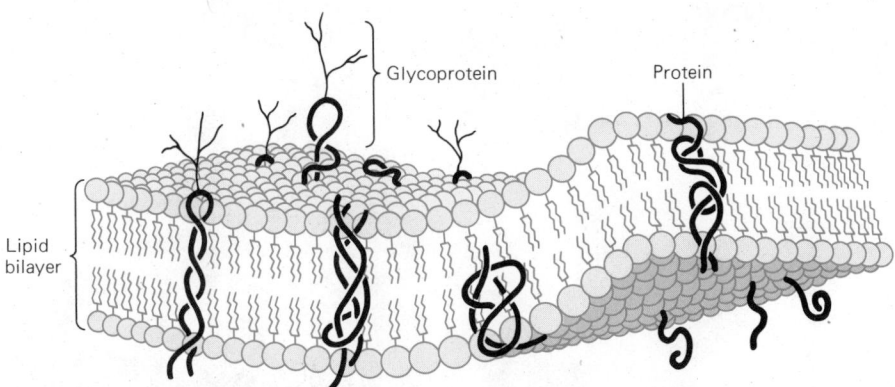

Figure 4-6 Structure of a plasma membrane according to the fluid mosaic model. Proteins and glycoproteins are embedded in a double layer of lipid molecules, oriented with their polar ends (colored spheres) at the outside of the membrane and their nonpolar ends (wavy colored lines) in the membrane's interior.

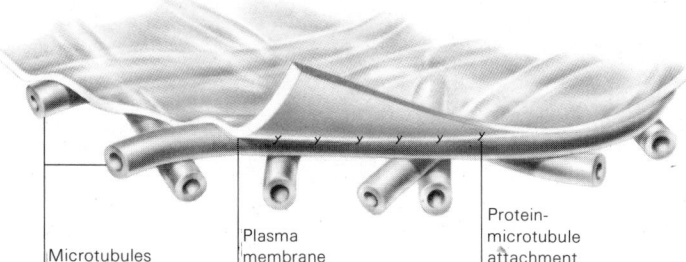

Microtubules Plasma membrane Protein-microtubule attachment

Figure 4-7 The tubular cytoskeleton found beneath the plasma membrane in some cells.

(Figure 3–13) is an important part of the membrane. It reduces the permeability of the lipid layer and strengthens the membrane. In some cells, the proteins in the membrane form an orderly pattern; in others their distribution looks more or less random. The actual proteins, glycoproteins, and lipids which make up the membrane vary from one type of cell to another. As a result, each type of cell has a unique "signature," written mainly in the glycoproteins that protrude from its surface. For example, human red blood cells of different types (type O, type A, etc.) differ from each other by the sugars that are attached to otherwise identical proteins in their plasma membranes. As we shall see in later chapters, cells recognize and interact with each other, and with specific substances such as hormones, by way of the signature glycoproteins on their plasma membranes.

The plasma membrane is considered fluid because the lipids can move about relatively freely in the membrane; many of the proteins can also move about. The fluidity is advantageous because it permits the membrane to seal itself if it is broken and to change in various ways, depending on the cell's specific requirements. About half the proteins in the membrane are relatively immobile; for some of them this is probably because they are attached to a **cytoskeleton** (cyto = cell). The cytoskeleton is made up of long, thin tubules which lie in a somewhat irregular network under the plasma membrane (Figure 4–7). In some cells, there is also an **exoskeleton** (= outside skeleton) outside the plasma membrane, and this may also bind some of the membrane proteins in position.

The cytoskeleton and exoskeleton are found only near plasma membranes, and they are not present in all cells. The basic fluid mosaic structure, however, appears to be common to all the membranes in a cell.

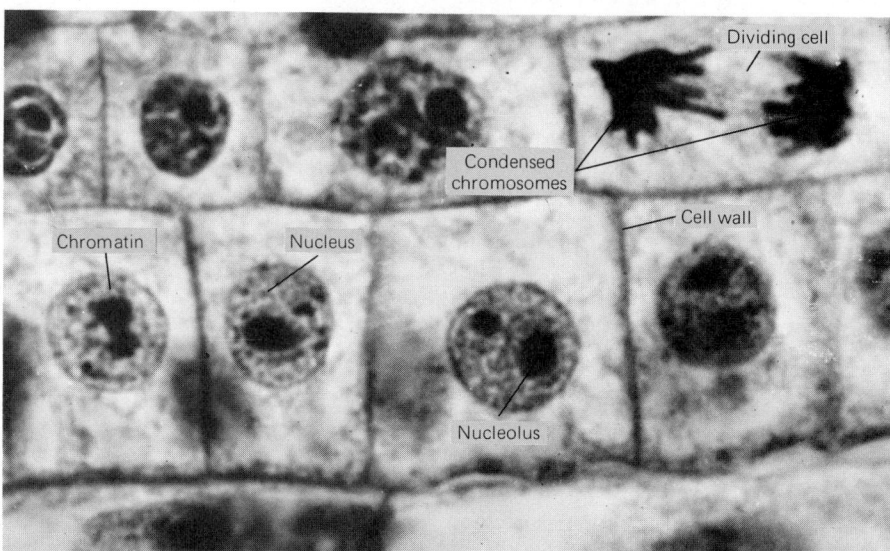

Dividing cell

Condensed chromosomes

Cell wall

Chromatin Nucleus

Nucleolus

Figure 4-8 Nuclei and chromosomes in onion root cells. (Biophoto Associates)

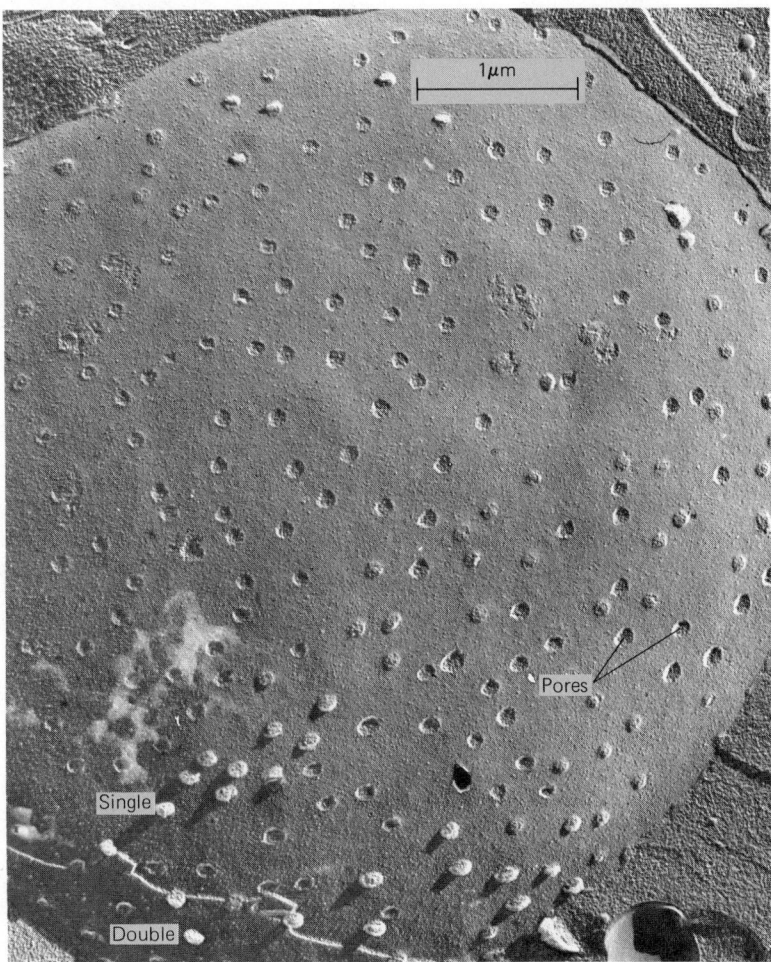

Figure 4-9 Electron micrograph of the surface of a nucleus. The specimen has fractured in such a way that the double layer of the nuclear envelope shows at the bottom but only the single inner layer remains in most of the photograph. Nuclear pores are visible, some of them apparently clogged with material. (Biophoto Associates)

The Nucleus

The nucleus is sometimes called the "control center" of the cell. It contains the **chromosomes**. The chromosomes consist of strands of DNA, the genetic material of the cell, with associated RNA and protein. The chromosomal DNA dictates the structure of the cell's proteins, and by this means controls most of the activities of the cell. When a cell is not dividing, its chromosomes are visible only as a tangle of threads called **chromatin**. When a cell divides, however, the chromosomes contract, and become distinct from one another (Figure 4–8).

The only structures normally visible in the nucleus of a nondividing cell are the **nucleoli**. These appear as one or more dark areas (Figures 4–3, 4–4), which disappear when the cell divides and reappear afterwards. The nucleoli are rich in protein and RNA and are the site at which ribosomes (discussed in the next section) are synthesized.

The nucleus is surrounded by a double membrane, the **nuclear membrane** or **nuclear envelope**. The nuclear membrane is perforated by fairly large pores, which serve as channels for the passage of materials between the nucleus and the cytoplasm of the cell (Figure 4–9).

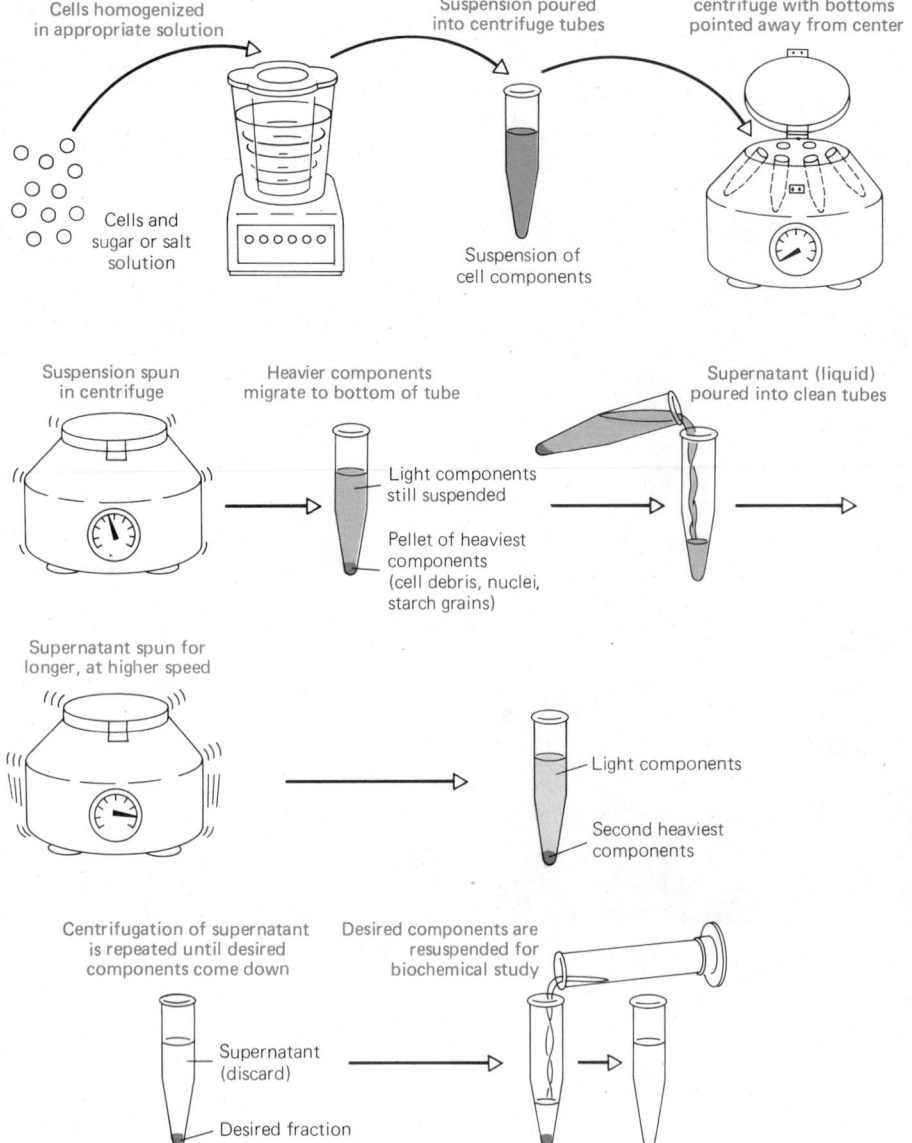

Cells homogenized in appropriate solution

Cells and sugar or salt solution

Suspension poured into centrifuge tubes

Suspension of cell components

Tubes placed in centrifuge with bottoms pointed away from center

Suspension spun in centrifuge

Heavier components migrate to bottom of tube

Light components still suspended

Pellet of heaviest components (cell debris, nuclei, starch grains)

Supernatant (liquid) poured into clean tubes

Supernatant spun for longer, at higher speed

Light components

Second heaviest components

Centrifugation of supernatant is repeated until desired components come down

Supernatant (discard)

Desired fraction

Desired components are resuspended for biochemical study

Figure 4-10 Differential centrifugation, a technique used to separate cells into their component parts.

Ribosomes

Ribosomes are the site of protein synthesis (see Chapter 10) and are therefore found in almost all cells. They are less than 30 nm in diameter, so that even electron micrographs show little detail of their structure. Most of our knowledge of ribosomes comes from work performed on ribosomes isolated from the rest of the cell by means of differential centrifugation (Figure 4–10) and studied in a test tube.

Some ribosomes are attached to membranes in the cell, some are free in the cytoplasm, and some are inside other organelles. A cell may contain half a million ribosomes, the number varying with the amount of protein the cell manufactures.

Mitochondria

Mitochondria (singular: **mitochondrion**) are organelles that supply the cell with its chemical energy store, adenosine triphosphate (ATP) (see Chapter 7). Mitochondria are most common in those areas with the greatest energy needs, such as a muscle in an animal or a growing root tip in a plant. A liver cell may contain as many as 2500 mitochondria.

Mitochondria have two membranes, an outer membrane separating the mitochondrion from the cytoplasm, and a highly folded inner membrane (Figures 4–3, 4–4). Mitochondria contain their own DNA, RNA, and ribosomes, and are self-replicating; that is, they divide and reproduce themselves. New mitochondria arise only by the division of existing ones; a cell cannot make them from raw materials.

The plasma membrane, nucleus, ribosomes, and mitochondria are found in virtually all eukaryotic cells. The rest of the cell components discussed in this chapter are found in many, but not all, cells.

The Cell Wall

A plant cell has a relatively rigid **cell wall** outside its flexible plasma membrane. The cell wall is quite porous, allowing water and dissolved substances to pass freely through it.

When a plant cell divides, the two newly formed cells build a common partition, called a **middle lamella**. This middle lamella is composed largely of polysaccharides called **pectins**, the substances that make fruit set into a jam or jelly. Pectins can be broken down by pectinase, an enzyme which causes the softening of fruit as it ripens. (Extra pectin must be added to jam made with overripe fruit.)

Each plant cell builds an elastic **primary cell wall** on its own side of the middle lamella. The primary cell wall is composed of cellulose molecules organized into thin fibers which can slide past each other, allowing the wall to stretch as the cell grows (Figure 4–11).

Many cells lay down a more rigid **secondary cell wall** when they are fully grown. The substance commonly called ''wood'' is composed mainly of secondary cell walls in which the cellulose has been reinforced with a strengthening material called **lignin**. Cells whose main function is support, such as those of a tree trunk, contain more reinforcing lignin in their secondary cell walls than do other types of cells.

The relative rigidity of cell walls accounts for many of the special characteristics of plants and plant cells. Cell walls contribute largely to the structural support of a plant's cells and of the entire plant body (stems, leaves, and roots). Because every plant cell is irrevocably cemented to its neighbors, the cells and organs of plants cannot move appreciably with respect to one another.

Plastids

Plastids are a prominent group of organelles found in plants and in photosynthetic protists. Like mitochondria, plastids contain DNA, RNA, and ribosomes, and replicate themselves. They have two outer membranes that separate them from the cytoplasm, and a separate system of internal membranes. Depending on the type of cell in which it occurs, a plastid may develop into a chloroplast, a chromoplast, or a leucoplast.

Chloroplasts are the site of photosynthesis, the process that uses the energy of sunlight to synthesize food molecules. Chloroplasts are green because they contain the green pigment chlorophyll (see Figures 4–4 and 1–15).

Chromoplasts are plastids that synthesize and store the red, yellow, and orange pigments responsible for the colors of many flowers and fruits. The most common leucoplasts are **amyloplasts**, colorless plastids that take in glucose and store it in the form of starch until it is needed by the plant. Amyloplasts in the root are also used in detecting the direction of gravity. This information is used to ensure that the root grows down into the soil.

Figure 4-11 Cellulose fibers in a plant cell wall. (Biophoto Associates)

1 µm

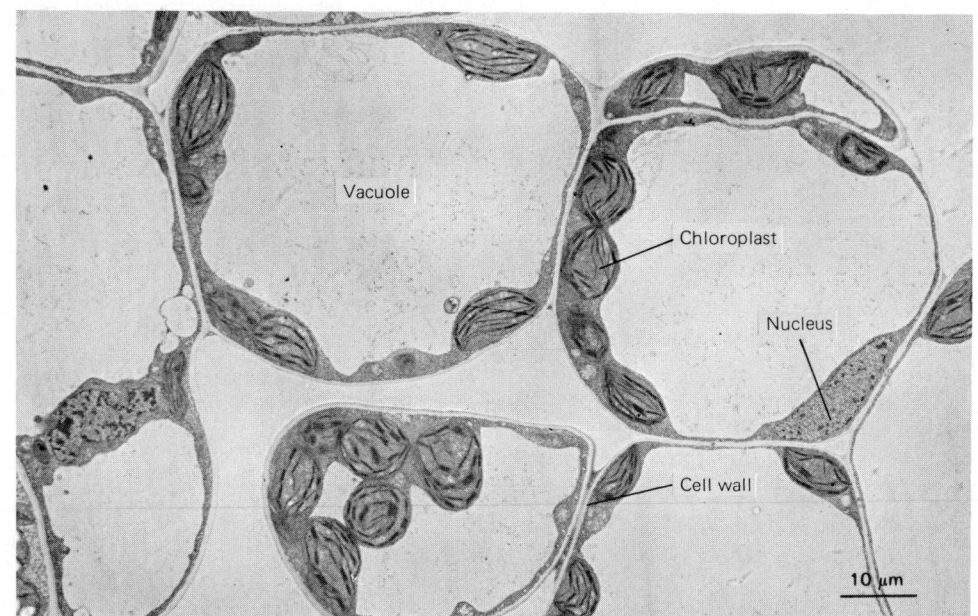

Figure 4-12 Cells from a spinach leaf as they appear at low power under the electron microscope. (Biophoto Associates)

Vacuoles

A **vacuole** is a fluid-filled space surrounded by a membrane. Vacuoles are present in many cells but are particularly prominent in plant cells (Figure 4–12). Typically, most of the volume of a plant cell is made up of a single vacuole surrounded by its membrane, the **tonoplast**. The tonoplast holds stored food, anthocyanin pigments (deep red, blue, or purple), salts, and other substances inside the vacuole. The vacuole is also convenient as a compartment in which toxic substances can be segregated from the rest of the cell. For instance, some acacia trees store cyanides—which make them poisonous to plant-eating animals—in their vacuoles. If the cyanides were in the cytoplasm, they would poison the rest of the cell. Unfortunately, the vacuole has proved very difficult to study because the tonoplast usually breaks when a plant cell is disrupted. When this happens, the vacuolar fluid mixes with the cytoplasm and the properties of the tonoplast cannot be determined.

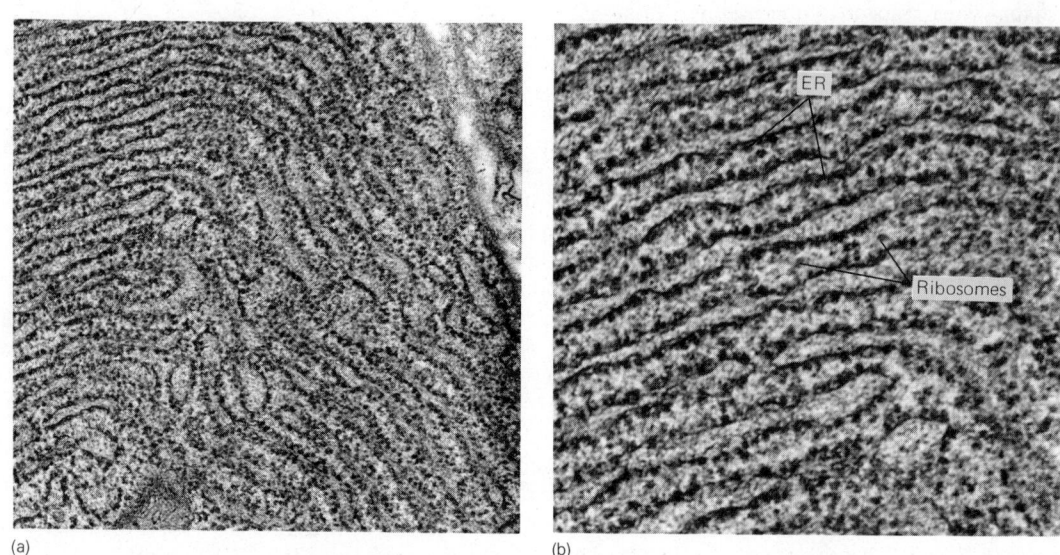

(a)   (b)

Figure 4-13 Endoplasmic reticulum with many ribosomes attached. This cell is from the pancreas of a mouse. (b) is a higher power view of (a).

Some animal cells take in substances from their surroundings by engulfing them in part of the plasma membrane, which then pinches off as a small vacuole called a **pinocytotic vacuole** (Figure 4–3).

Endoplasmic Reticulum

The **endoplasmic reticulum** (**ER**) is a collection of cytoplasmic membranes which is usually prominent in cells that synthesize a lot of protein and lipid. ER was first described by Keith Porter with electron micrographs published in 1945. In most cells it is continuous with the outer membrane of the nuclear envelope. The endoplasmic reticulum is highly variable in structure; it consists of a system of tubules and channels connected to one another (Figure 4–13). Since many enzymes work best when they are attached to membranes, the many folds of the endoplasmic reticulum serve as important sites for chemical reactions in the cell.

Ribosomes are frequently found attached to the endoplasmic reticulum, to form **rough endoplasmic reticulum** (Figure 4–13). Ribosomes need not be attached in this way in order to synthesize protein; they function in the same way if they are free in the cytoplasm. Cells that synthesize large amounts of protein, however, invariably contain considerable areas of rough ER. Similarly, cells that synthesize large amounts of lipids (e.g., cells that synthesize steroid hormones) contain a lot of smooth ER since most of a cell's lipid-synthesizing enzymes are also part of the ER.

Besides protein and lipid synthesis, the most general function of the endoplasmic reticulum is to divide the cell into two main compartments, one lying inside the reticulum and the other outside these membranes. This division into compartments performs a vital function, since many chemical processes take place within a cell at the same time, and it is often essential that they be separated so that they do not interfere with one another nor undo one another's accomplishments.

Since the nuclear membrane, ER, and plasma membrane are continuous in some electron micrographs, it has long been suggested that all these membranes have a common origin. Perhaps they are all synthesized by the ER? Some people think that this is the case because heavy metal stain introduced into the ER membranes as a marker sometimes turns up later in the nuclear membrane or the plasma membrane. Such an experiment is not conclusive, however, and at the moment we can only say that little is known about where and how a cell's membranes are synthesized.

Golgi Complex

The **Golgi complex** is an organelle made up of small membrane-bound sacs, or **vesicles**, which is often associated with the ER. (In plants it is sometimes known as a **dictyosome**.) It functions essentially as a packing station. In the 1960s it was found that when protein is synthesized in pancreas* cells, the newly formed protein moves from the ribosomes of the rough endoplasmic reticulum where it is synthesized into the cavity of the reticulum. From here it travels to the Golgi complex, where it is incorporated into vesicles. These vesicles leave the Golgi complex and travel to the surface of the cell, where they release the protein (Figure 4–14).

Lysosomes

A **lysosome** is a membrane-bound sac full of hydrolytic enzymes. Lysosomes are very variable in size and shape. They arise from the Golgi bodies. If the lysosome membrane is disrupted, its enzymes are released and they digest the rest of the cell. Lysosomes were first discovered in rat liver cells in 1952 and have since been found to be very common in animal cells. Little is known about their prevalence or function in plant cells. Lysosomes break down old and broken parts of the cell into small organic molecules that can be reused. They are especially active in breaking down the tail when a tadpole changes into a frog.

---

*Pancreas   An organ near the stomach that produces digestive enzymes and proteinaceous hormones.

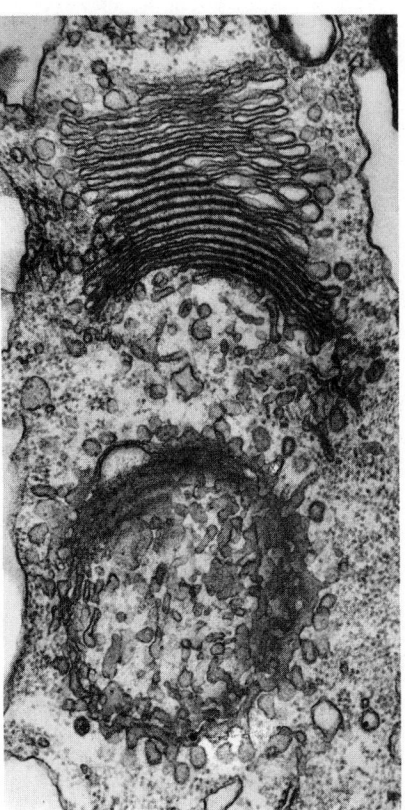

Figure 4-14   Two Golgi complexes at right angles to each other. The membranes looking like a pile of pancakes (top) are a stack of Golgi vesicles sectioned longitudinally (that is, cut from top to bottom of the stack). The circular structure at the bottom is one pancake seen from above. Note the characteristic smaller vesicles round the edges of the large "pancakes." (Biophoto Associates)

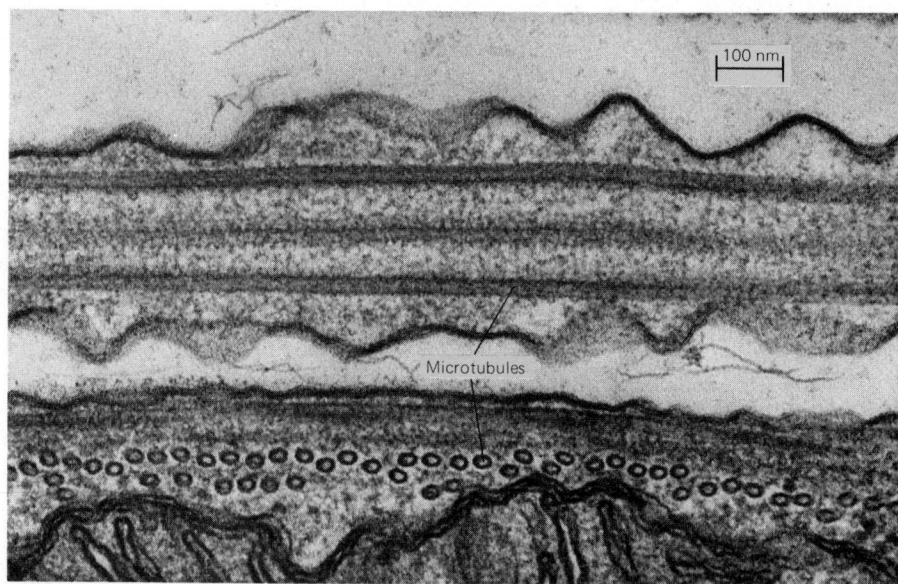

Figure 4-15 Microtubules in longitudinal section (top) and cross section (bottom). (Biophoto Associates)

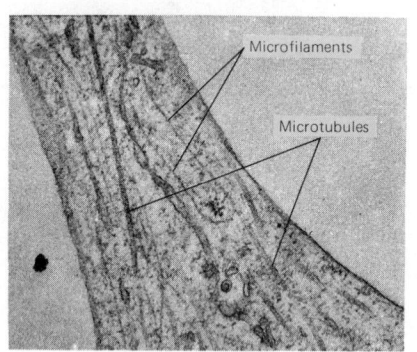

Figure 4-16 Microtubules and microfilaments in a chicken nerve cell. (Norman Wessells)

Microtubules and Microfilaments

Microtubules are long, slender tubes, 20 to 25 nm in diameter (Figure 4–15). They are made of protein subunits and appear to form an internal skeleton in both plant and animal cells. On occasion they act as tracks upon which other organelles move. The protein subunits from which microtubules form are apparently always present in cells. The microtubules are often assembled from these subunits to perform particular functions and then disassembled into subunits that can be used again. Cilia, flagella, and sperm tails (see below) are composed largely of microtubules.

Microfilaments are thin protein fibers that are more slender than microtubules—about 6 nm in diameter (Figure 4–16). They are believed to be a major part of the systems of contractile filaments which occur in some cells in both plants and animals. In simple form, they may help to cause cytoplasmic streaming and cell movement; in more complex form, microfilaments may account for the contraction of various types of muscle cells. Like microtubules, they are highly changeable and may exist as subunits or filaments, depending on localized conditions within the cell.

Cilia and Flagella

Cilia (singular: **cilium**) are short, thread-like organelles, present in large numbers on the surface of some protist, animal, and lower plant cells. **Flagella** (singular: **flagellum**) have the same basic structure as cilia, but they are longer and less numerous.

Cilia and flagella are organelles of locomotion; they function either to propel the cell to which they are attached or to move something past the cell. For example, the locomotory ability of a human sperm is provided by the single flagellum that forms its tail. Cilia move mucus and debris out of the air passages leading to the lungs; many unicellular organisms move by lashing their cilia or flagella. In all these cases it is believed that microtubules in the cilium or flagellum interact with other proteins to generate the forces that move the organelle.

Electron micrographs of cilia and flagella show that they contain a circle of nine pairs of microtubules surrounding another pair in the center of the structure (Figure 4–17). This so-called 9 + 2 arrangement is characteristic of eukaryotic flagella but different from the flagella of prokaryotes. A **basal body** is found near the site

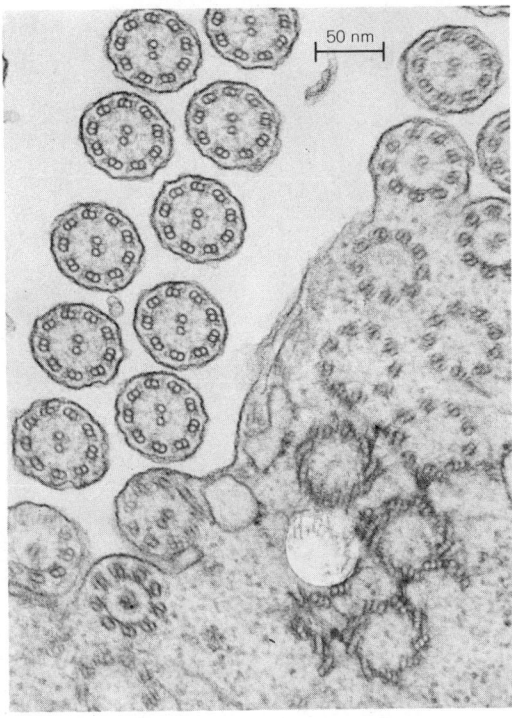

50 nm

where a cilium or flagellum attaches to the cell body (Figure 4–18). The basal body consists of nine triplets of microtubules in a circle. The function of basal bodies is not well understood. They appear to be sites where the protein subunits assemble to form the microtubules of cilia and flagella.

Centrioles

Two **centrioles** lie just outside the nucleus in many eukaryotic cells (see Figure 4–3). The electron microscope has revealed that each centriole is basically cylindrical in shape and is made up of nine groups of microtubules, with three microtubules in each group.

One function of centrioles is that they give rise to basal bodies. The circle of nine triple microtubules characteristic of centrioles is also found in the basal bodies of cilia (and flagella); two strands of each triplet extend from the basal body into the cilium, forming the circle of nine double tubules (Figure 4–18).

Before cell division, centrioles duplicate and move so that there is a pair at each end of the cell. For a long time it was also believed that they determined the plane in which a cell divides. On the other hand, cells of higher plants can divide but do not contain centrioles. If centrioles are vital to cell division, higher plant cells would be expected to contain equivalent structures, but if these exist, they have not so far been found.

Extracellular Space

Every cell in a multicellular animal is surrounded by a space which contains **extracellular fluid** (extra = outside) and the substances that hold cells together. The extracellular fluid is the immediate environment of the cell; it is the source of nearly all the molecules the cell takes in, as well as being the immediate sink for any waste products the cell excretes.

Adjacent cells may be attached to one another by several different types of junctions whose functions are poorly understood. In a **desmosome**, filamentous material is found in the space between the cells and in the cytoplasm of the cells on either side of the junction (see Figure 4–3); these contacts strengthen the cells they

Microtubules

Cilium

Plasma membrane

Basal body

Figure 4-18 A cilium with its basal body. The structure in the middle of the basal body is not a microtubule; its makeup is unknown.

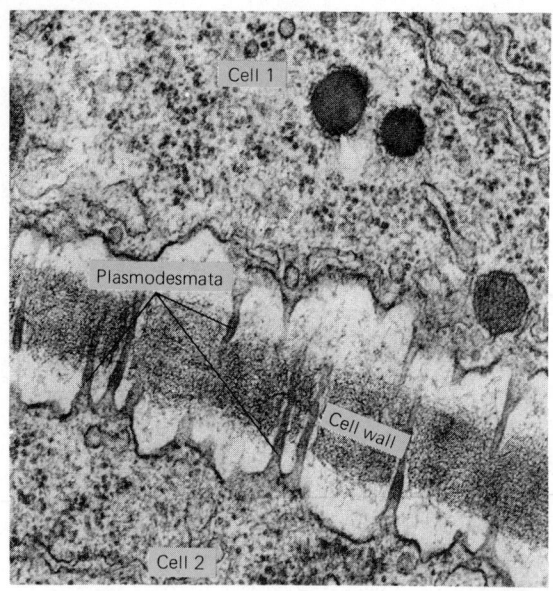

Figure 4-19 Plasmodesmata between two plant cells. (Biophoto Associates)

hold together. In a tight junction, the cell membranes of adjacent cells appear to be fused to one another, so that the space which usually lies between two cells is completely closed (see Figure 4–3). Tight junctions are found particularly on the outsides of organs and they prevent substances from leaking out between cells.

4-C Cell Communication

For an organism to function, the activities of its cells must be coordinated in many ways. For instance, chemical messengers carry information from one cell to another. (These messengers are considered in the chapters on hormones—Chapters 41 and 47.) Cells also communicate by the direct transfer of substances from one cell to another. In recent years, considerable interest has focused on channels by which substances can pass directly from one cell to another. These channels are very tiny and have been studied by electron microscopy and electrophysiological* techniques. If two microelectrodes* are placed, one inside and the other outside a cell, the electrical resistance between the two electrodes is found to be high, indicating that current cannot flow from inside the cell to the outside across the plasma membrane; the membrane is relatively impermeable to electrically charged particles. In contrast, if electrodes are placed inside two adjacent cells, the electrical resistance between them may be very low, indicating that electrically charged substances can move freely between the insides of the cells.

The main electrically charged particles in biological systems are protein molecules that are too big to leave the cell, and small ions such as Na^+, K^+, and Cl^-. The low electrical resistance between adjacent cells suggests that ions can move from one cell to another, probably by way of intercellular channels. It is sometimes possible to confirm such a finding by tracing the movement of fluorescent or radioactive substances from one cell to another.

Plasmodesmata (singular: **plasmodesma**) are another type of channel, connecting the cytoplasm of one plant cell with that of its neighbor (Figure 4–19). **Gap junctions** (sometimes called **nexus junctions**) are areas of low electrical resistance between animal cells, which appear to function like plasmodesmata in plants. Small ions and molecules can pass freely across gap junctions. These junctions are particularly common between cells in early embryos, and in the heart, where they speed the transmission of electrical impulses from one cell to the next.

---

* Electrophysiology The study of electrical properties of living systems.
* Microelectrode A very small probe used to conduct electric current into or out of a small area being studied.

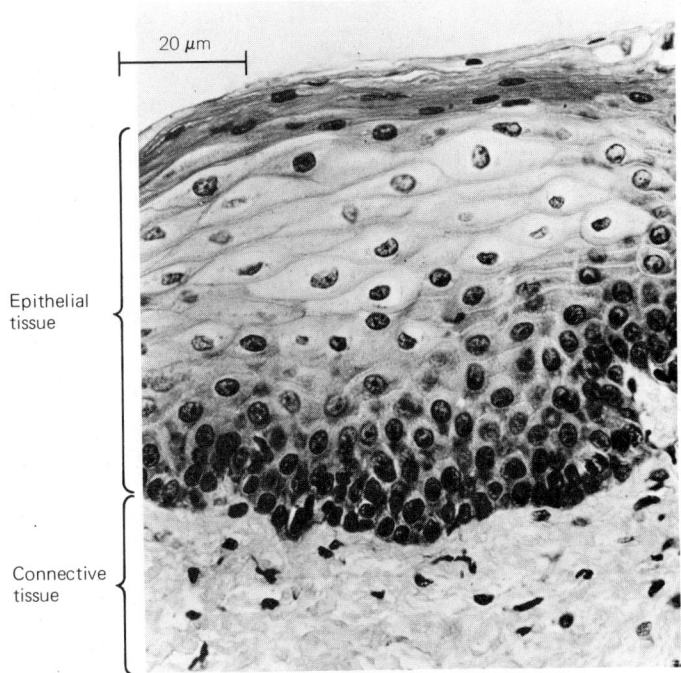

20 μm

Epithelial tissue

Connective tissue

Figure 4-20 A light microscope section through the human palate. (Biophoto Associates)

4-D Tissues and Organs

All but the simplest multicellular organisms contain a range of different types of cells, many of them specialized for different functions, and hence with different shapes, sizes, and cell chemistry. Cells of one or a few types form groups called **tissues**, held together in characteristic patterns by substances secreted by the cells themselves. **Organs** are functional units of the body made up of more than one type of tissue; examples are eyes, kidneys, muscles, leaves, or roots.

Animal Tissues

Animal tissues are generally divided into four main types:

Epithelial tissue forms coverings and linings, over the outside of the body or of an organ, or inside a cavity or tube (Figure 4–20). In keeping with its function, this type of tissue forms sheets one to several cells thick, with the cells tightly packed together. Substances may enter or leave the body through epithelia—gases through the lung epithelium and food through the digestive tract epithelium, for example. Epithelial tissues may also be specialized for secretion; the lining of the human digestive tract secretes mucus, and the epithelium of a snail secretes its shell.

Nervous tissue contains nerve cells with the special property of irritability, the ability to conduct electrical impulses in response to various stimuli (see Chapters 37 and 38). Various other types of cells closely associated with the nerve cells are also found in nervous tissue.

Muscle tissue is made up of cells with both irritability and the ability to contract (see Chapter 40).

Connective tissue is generally the most abundant type of animal tissue. It characteristically has a great deal of intercellular (= between cells) material secreted by the cells, usually in the form of fibers, gelatinous substance, or both. Cartilage, bone, and adipose (fat) tissue are the most familiar examples (see Chapter 40 for bone and cartilage).

Plant Tissues

Various systems are in use for classifying plant tissues, and the types recognized are too numerous to list them all here. Below are a few we shall meet in this book (Chapters 43 and 44).

Epidermis, the tissue equivalent to epithelium in animal cells, covers the outside of leaves and of non-woody stems and roots.

Vascular tissue transports water, food, hormones, and so forth between different parts of the plant; it is perhaps most familiar as the veins in leaves and the wood of trees. A plant is "woody" because cells in the vascular tissue have deposited a great deal of strengthening material in their cell walls.

Parenchyma cells form various tissues that make up most of the inside of leaves and of non-woody stems and roots. These cells have thin cell walls and they are often loosely packed, with many intercellular spaces.

The presence of many chloroplasts or storage plastids in some kinds of parenchyma shows that these cells are specialized for photosynthesis or food storage.

4-E Prokaryotic Cells

There are two basic types of cell organization. The cells considered so far in this chapter are the eukaryotic cells found in animals, plants, fungi, and unicellular organisms. Eukaryotic cells contain organelles that are bounded by membranes: nuclei, mitochondria, plastids, lysosomes, and so forth.

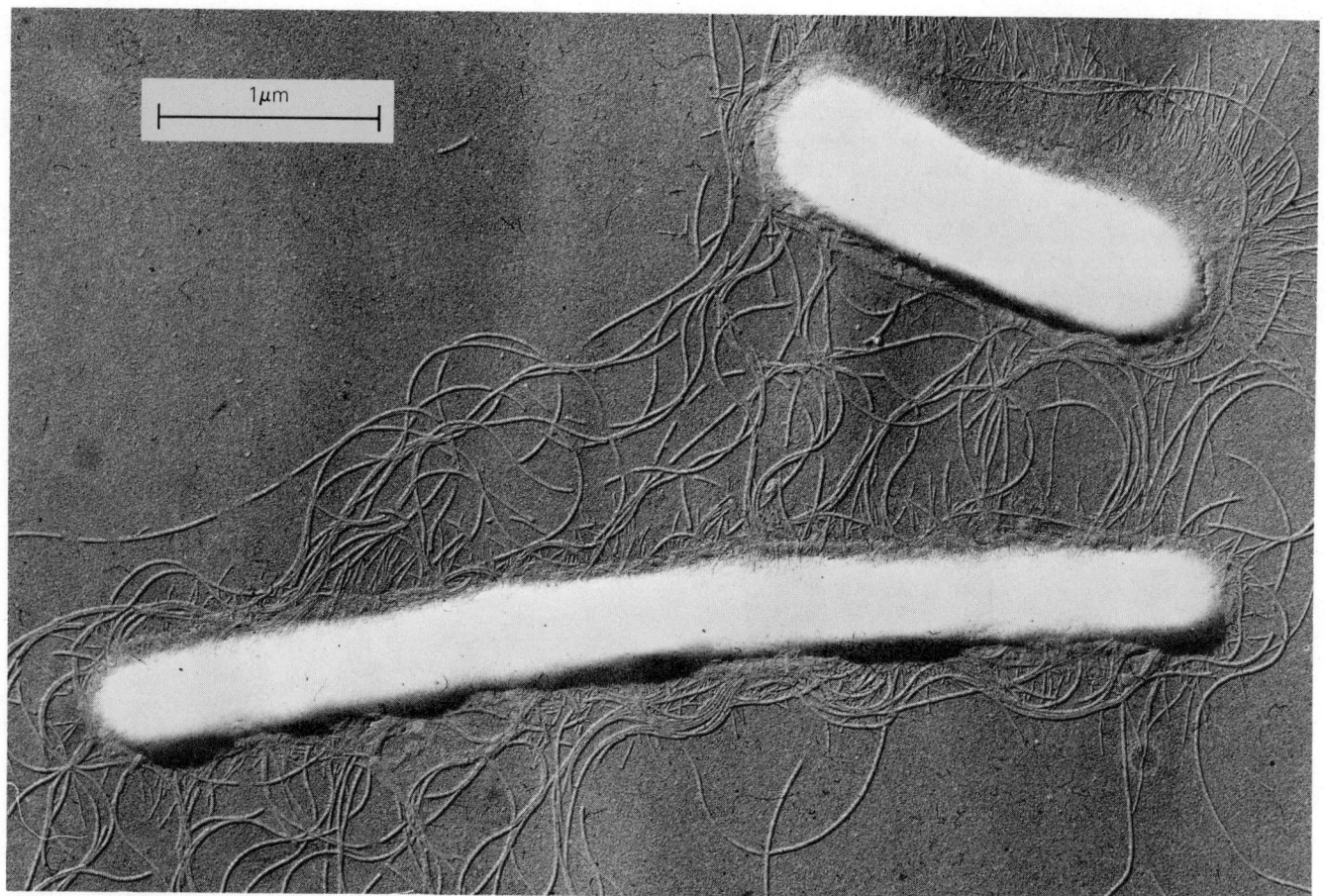

Figure 4-21 Prokaryotic cells. Two rod-shaped bacteria surrounded by flagella-like "pili." (Biophoto Associates)

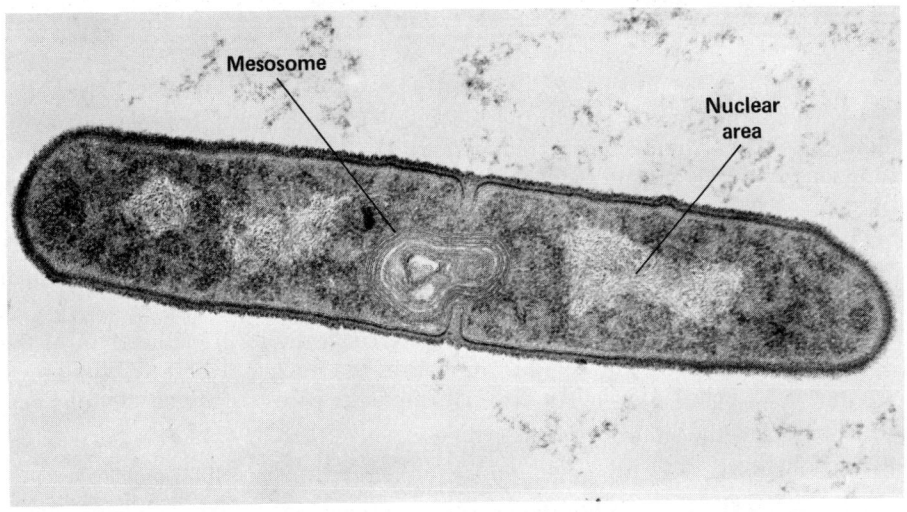

Figure 4-22 A bacterium dividing. Note the mesosome and nuclear area. (Biophoto Associates)

Prokaryotic cells, found only in bacteria, do not contain membrane-bound organelles (Figure 4–22), and must carry on all the essential life processes without them. Since it has no nuclear membrane, a prokaryotic cell has no distinct nucleus; its genetic material is contained in a single, circular molecule of double-stranded DNA in a "nuclear area." Prokaryotic DNA lacks the proteins associated with eukaryotic DNA. Prokaryotes do have ribosomes and a cell wall, but the ribosomes are somewhat smaller, and the cell wall is usually made up of carbohydrates and polypeptides rather than the cellulose that comprises the cell walls of eukaryotic plants.

Aside from the plasma membrane, which is present in prokaryotic as well as in eukaryotic cells, the only membranous structure found in many prokaryotes is the **mesosome**, attached to the plasma membrane. Its function is little understood. Various theories hold that it plays a role in producing new cell wall material following cell division; that it is involved in replication of the DNA prior to cell division; or that it is involved in the breakdown of food molecules to provide energy. In addition to the plasma membrane and mesosome, most photosynthetic prokaryotes also contain photosynthetic membranes, but these are not enclosed in chloroplasts as they are in the eukaryotic cells of plants.

TABLE 4-4 **COMPARISON OF EUKARYOTIC AND PROKARYOTIC CELLS**

STRUCTURE	EUKARYOTIC CELL	PROKARYOTIC CELL
Cell wall	Absent in animals	Present (different chemical composition)
Plasma membrane	Present	Present
Nucleus	Surrounded by a membrane	Nuclear area, not surrounded by a membrane
Chromosomes	Linear, with proteins	Circular, no protein component
Endoplasmic reticulum	Usually present	Absent
Ribosomes	Present	Present (different type)
Golgi complex	Present	Absent
Lysosomes	Present in many cells	Absent
Mitochondria	Present	Absent
Vacuoles	Present in most plant and some animal cells	Absent
Centrioles, cilia and flagella	Present in all except higher plants	Flagella of different type present in some bacteria

SUMMARY

Each cell must carry out all the processes of life: obtain food, release energy, eliminate wastes, divide and reproduce itself. In addition to carrying out its own functions, each cell of a multicellular organism must also carry out some specialized function as its particular contribution to the body's overall economy.

Important structures found in eukaryotic cells are:

1. The plasma membrane, a double layer of lipid, protein, and glycoprotein, which regulates the movement of substances into and out of the cell.

2. The nucleus, containing the genetic material in the form of the DNA of the chromosomes. The nucleolus is an area in the nucleus where ribosomes are made. A nuclear envelope with relatively large pores surrounds the nucleus.

3. Ribosomes, the sites of protein synthesis.

4. Mitochondria, which produce the cell's energy supply of ATP molecules.

5. The cell wall, a porous but rigid supporting structure outside the plasma membranes of plant cells.

6. Plastids, found only in plant cells. The most important plastids are the chloroplasts, which carry out photosynthesis.

7. Vacuoles, especially prominent in plant cells.

8. Endoplasmic reticulum, which divides the cell into compartments, forms the surface upon which many chemical reactions take place, and synthesizes the cell's membranes.

9. Golgi complexes, the areas where proteins are packaged for secretion to the exterior of the cell.

10. Lysosomes, sacs of hydrolytic enzymes involved in degradation of intracellular structures and substances.

11. Microtubules and microfilaments, structures concerned with movement of the cell and its organelles.

12. Cilia and flagella, organelles projecting from the cell and concerned with cell locomotion or with movement of external substances past the cell surface.

13. Centrioles, structures similar to basal bodies of cilia and flagella; their function is unclear.

In multicellular organisms, cells are organized into tissues, such as epithelium and connective tissue, and organs, such as kidneys and stems, each with its own particular function.

The prokaryotic cells of bacteria also have cell walls, plasma membranes, cytoplasm, and DNA; they differ from the eukaryotic cells of plants, animals, fungi, and protists in containing none of the major cytoplasmic organelles except ribosomes. Their main internal membrane system is the mesosome.

OBJECTIVES

From your study of this chapter, you should be able to:

1. Outline how a light microscope, a transmission electron microscope, and a scanning electron microscope work.

2. List at least four components of cells that can be seen with a light microscope and four that can be seen with an electron microscope but not with a light microscope.

3. Give the function of each of the following structures and state whether each would be found in plant, animal or prokaryotic cells: *plasma membrane, *nucleus, *nuclear membrane, nuclear area, *chromosome, *nucleolus, *ribosome, *mitochondrion, *cell wall, plastid, *chloroplast, *vacuole, *endoplasmic reticulum, *Golgi complex, lysosome, microtubule, microfilament, cilium, flagellum, centriole, gap junction, *plasmodesma, mesosome.

4. Identify any structure marked * in Objective #3 that appears in a photograph taken using a light microscope or transmission electron microscope.

5. Describe the main differences between a eukaryotic and a prokaryotic cell.

Match the structures listed on the right with the property given on the left.

_____ 1. Sites of protein synthesis.

_____ 2. Used to propel a cell through a fluid, or to move a fluid past the surface of a cell.

_____ 3. Rigid protective covering of some cells.

_____ 4. Carry hereditary information of cell.

_____ 5. Secretory apparatus of cells.

_____ 6. Contains digestive enzymes of cell.

_____ 7. Imparts color to leaves, flowers, and fruits of plants.

_____ 8. Storage compartment in plant cells.

_____ 9. Involved in synthesis of ribosomes.

_____ 10. Internal membranous structure in prokaryotic cells.

a. plasma membrane
b. cell wall
c. centrioles
d. chromosomes
e. cilia
f. endoplasmic reticulum
g. flagella
h. Golgi complex
i. lysosome
j. mesosome
k. mitochondrion
l. nucleolus
m. plastids
n. ribosomes
o. vacuole

For each item listed below, check whether it is found in cells of animals, cells of plants, prokaryotic cells, or cells of all types of organisms.

	Animal	Plant	Prokaryote	All
11. ribosome	_____	_____	_____	_____
12. flagella	_____	_____	_____	_____
13. cell wall	_____	_____	_____	_____
14. chromosome	_____	_____	_____	_____
15. mitochondrion	_____	_____	_____	_____

16. Which of the following is *not* found in the cells of higher plants?
   a. plasma membrane
   b. cell wall
   c. chloroplast
   d. ribosome
   e. centriole

17. Louis Pasteur placed sterile broth in two sterile containers, A and B. Container A had a straight mouth open to the atmosphere, while container B, which was also open, had a bent neck to prevent any particles entering the container from the air (see Figure 19-2). Which of these results would support the cell theory?
   a. After several weeks both A and B were teeming with organisms
   b. After several weeks A was teeming with organisms, but B contained no life.
   c. After several weeks B was teeming with organisms, but A contained no life.

QUESTIONS FOR DISCUSSION

1. What useful applications of the scanning electron microscope can you think of?

2. Would you expect the cells of your hair follicles to contain more ribosomes than a cell that stores fat? Why?

3. It has been said that animals, as we know them, could not exist if they had cell walls. Why not?

4. It has been suggested that plastids and mitochondria were originally free-living prokaryotes which, in the course of evolution, took up residence inside eukaryotic cells. What evidence (similarities, differences) can you muster for or against this theory?

REFERENCES AND FURTHER READING

DeRobertis, E. D. P., and E. M. F. DeRobertis. *Cell Biology,* 7th ed. Philadelphia: W. B. Saunders, 1979. An excellent text describing recent work on the features of cells; frequent summaries of information and conclusions.

Dyson, R. D. *Cell Biology,* 2d ed. Boston: Allyn and Bacon, 1978. A clearly written modern text.

Dustin, P. "Microtubules." *Scientific American,* August 1980.

Fawcett, D. W. *An Atlas of Fine Structure.* Philadelphia: W. B. Saunders, 1966. A collection of well-chosen electron micrographs of different kinds of cells and organelles with a brief description of each.

Giese, A. C. *Cell Physiology,* 5th ed. Philadelphia: W. B. Saunders, 1979. A well-illustrated, fairly advanced text on how cells do what they do; with excellent sections on the evolution of cells.

Singer, S. J., and G. Nicolson. "The fluid mosaic model of the structure of cell membranes." *Science* 175:720, 1972. The original description of the fluid mosaic model.

Staehelin, L. A., and B. E. Hull. "Junctions between living cells." *Scientific American,* May 1978.

CHAPTER 5

HOW THINGS ENTER AND LEAVE CELLS

An organism's chemical processes are carried out by its cells and organelles. Biochemical reactions require raw materials from outside the cell and generate waste products that the cell must discharge back into its environment. In order to remain alive, a cell must maintain **chemical homeostasis**; that is, it must keep its chemical composition constant within narrow limits. This is difficult because the interior of the cell and the environment outside the organism are very different in chemical composition (Figure 5–1). The plasma membrane performs the vital role of determining what substances enter and leave the cell; thus it plays a major part in determining the cell's internal composition. In this chapter we consider the various ways in which substances cross the plasma membrane.

Figure 5-1 Living things selectively accumulate some chemicals. The intensity of color represents the concentration of carbon, which the plant accumulates at much higher concentrations than those found in the environment. Black dots represent silicon, a plentiful element in the soil which is found only in minute amounts in living organisms.

5-A Movement in Liquids

Diffusion

Diffusion is one of the processes by which substances become randomly distributed. All substances are made up of particles, which are in constant, random motion. In any substance, some molecules move faster and some move more slowly. The faster a particle moves, the more **kinetic energy**, or energy of motion, it has. Heating a substance imparts more energy to its molecules, so that, on average, they move faster.

This spontaneous movement of molecules is responsible for **diffusion**, the process by which molecules of two or more substances move about until they become evenly mixed. Consider a drop of food coloring in a glass of water. A molecule of food coloring will move in one direction until it collides with some other molecule, either another food coloring molecule or a water molecule. When the two molecules collide, they both bounce off in new directions. When a molecule of food coloring is hit, there is a better chance that it will bounce off in a new direction which will carry it away from the drop than in the one direction that will take it back to the drop.

The molecules moving away from a drop of food coloring will set up a **concentration gradient** in which the concentration of food coloring molecules falls off gradually with distance from the center of the drop (Figure 5-2). On the whole, food coloring molecules will tend to move down this concentration gradient, or toward the area where they are less concentrated, until they are evenly dispersed throughout the glass (Figure 5-3). At the same time that the food coloring molecules are moving down their concentration gradient, the water molecules in the glass are also moving until they spread evenly through the glass, including the part where the food coloring was originally introduced.

Currents

The even mixing of food coloring and water molecules by diffusion takes place extremely slowly at room temperature. Other processes mix molecules much more quickly. Stirring the glass of water would speed the mixing of water and coloring molecules. Stirring introduces an outside force that sets up **currents**, which are large numbers of molecules moving together in the same direction. Examples of currents in nature include wind, running water, and smoke carried up a chimney by the heat of a fire. Organisms usually have internal currents which speed the movement of molecules; currents are created by the pumping of the heart, by breathing movements, and by the streaming of cytoplasm within a cell.

If currents are so much more effective at moving and mixing molecules, why do we pay any attention to diffusion? The answer is that substances moving into and out of cells are traveling very small distances, and in this situation diffusion is important. When a current moves past an object, it encounters resistance from the surface of the object. There is a **boundary layer** between the surface and the main body of the current, and the molecules in the boundary layer move much more slowly than the main current. A current may move molecules to the vicinity of a cell, but the molecules must traverse the remaining distance, across the boundary layer and to the cell, by diffusion (Figure 5-5).

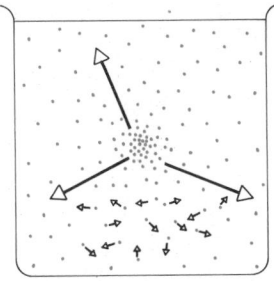
Figure 5-2 Although individual molecules may move towards their source (some short arrows), the average movement for all molecules is away from the source (long arrows), or down a concentration gradient.

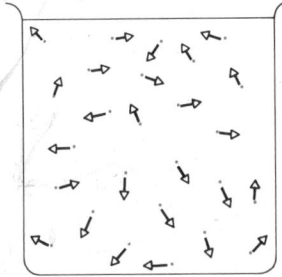
Figure 5-3 Eventually, the molecules of food coloring become evenly dispersed among the water molecules (not shown). The molecules remain in random motion and tend to retain a more-or-less even spacing.

Figure 5-4 Currents force large numbers of molecules to move and mix with other molecules they encounter.

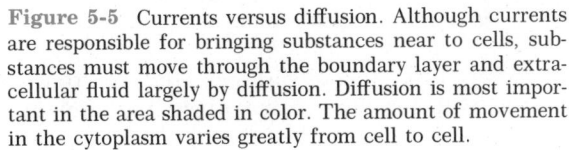

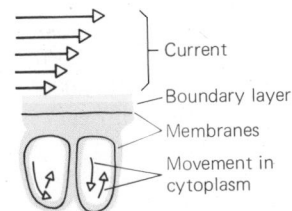

Figure 5-5 Currents versus diffusion. Although currents are responsible for bringing substances near to cells, substances must move through the boundary layer and extracellular fluid largely by diffusion. Diffusion is most important in the area shaded in color. The amount of movement in the cytoplasm varies greatly from cell to cell.

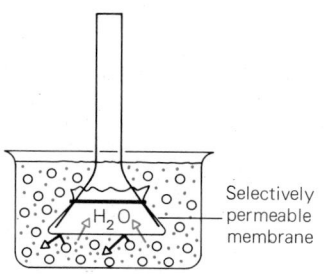

o Sugar molecule ·
· Salt ion

Selectively permeable membrane

(a) Salt ions can pass through the membrane into the water, but sugar molecules cannot.

Later

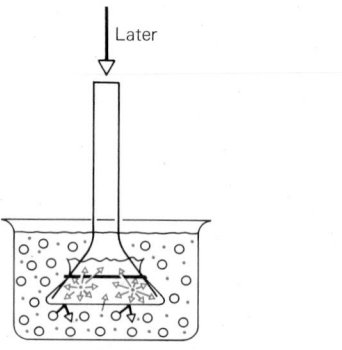

(b) There are very few salt ions inside the membrane, so there is a greater chance that an ion from outside will move in than that an ion from inside will move out.

Later

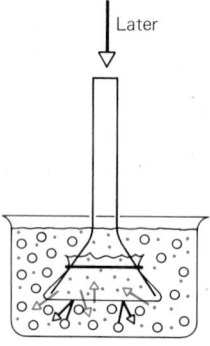

(c) At equilibrium, the concentration of salt is about equal on both sides of the membrane, and the movement of salt is equally likely in either direction. All the sugar remains outside; salt diffuses independently of sugar.

Figure 5-6 Diffusion across a selectively permeable membrane. The membrane is permeable to salt ions but not to sugar molecules. (Water molecules are disregarded for the sake of simplicity.)

5-B   Membrane Permeability

The plasma membrane is the main reason that living organisms differ from their environments in the proportions of a substance they contain.

The plasma membrane of a cell is **selectively permeable**; that is, it will allow some substances, to which it is **permeable**, to pass through, but will bar the passage of other substances, to which it is **impermeable** (Figure 5–6). Molecules of different substances diffuse independently of each other; in Figure 5–6, for example, the salt will diffuse across the membrane even though the sugar cannot.

The rate at which a substance can diffuse into the cell through the lipid of the plasma membrane depends on its solubility in lipids and on its molecular size. Small, highly lipid-soluble (nonpolar) substances such as ethylene and ether enter cells readily; we can visualize the process by imagining that molecules dissolve in the lipid of the plasma membrane and eventually find themselves inside the cell. Nonpolar substances usually enter cells quickly even when they are large molecules with high molecular weights and sizable diameters. As polar molecules or ions increase in size, their ability to penetrate the plasma membrane decreases more rapidly.

Water enters and leaves cells readily, probably through minute pores in the membrane. Many small ions such as sodium, chloride, and potassium also enter cells much more rapidly than would be predicted by their solubility in lipids. Small organic monomers such as glucose, glycerol, and amino acids can also enter cells. As we shall see shortly, special mechanisms permit cells to take in glucose, amino acids, and other substances more rapidly than would occur by diffusion alone.

Many large organic molecules are effectively barred from entrance into the cell. This may seem like an evolutionary mistake until you realize that it also means they cannot leave the cell. For instance, only the smallest carbohydrate molecules can cross the plasma membrane. This means that a cell can form polysaccharides, such as starch or glycogen, and these molecules will stay inside the cell, ready to be broken down as they are needed. If another cell needs stored carbohydrates, the storage cell must break the polysaccharide down to its constituent monosaccharide units, which can leave the cell and be transported to where they are needed.

The concentration of small molecules and ions in a cell results from the presence in the plasma membrane of protein **carrier molecules** whose job is to usher particular substances into or out of the cell. There are two main ways in which this can happen: facilitated diffusion and active transport.

5-C   Facilitated Diffusion

In **facilitated diffusion** a carrier in the plasma membrane combines with the molecule to be transported on one side of the membrane and releases it on the other side. The carrier increases the membrane's permeability to the substance but moves it only in the direction in which the substance would tend to travel on its own. In all cases that have been studied there is a specific carrier for each type of molecule.

An example of facilitated diffusion is the system that permits almost all cells to take up glucose more rapidly than they otherwise would. In many vertebrate tissues that have been studied, such as liver, red blood cells, and the lens of the eye, facilitated diffusion moves glucose across the plasma membrane in both directions by means of a carrier molecule. The carrier molecule is more likely to encounter and pick up a glucose molecule on that side of the membrane where glucose is more plentiful. When the cell is using up glucose quickly, the concentration inside the cell will fall; glucose will then be more plentiful outside the cell, and it will be moved into the cell more rapidly than it will move out. Facilitated diffusion is just as important in increasing the rate at which glucose leaves a cell as the rate at which it enters. Cells in the liver, for instance, not only remove glucose from the bloodstream when the blood glucose level is high, but also replenish the blood glucose when its level drops.

Muscle is a tissue in which the need for glucose varies greatly from time to time. Lack of sufficient glucose may actually limit the amount of work a muscle can do.

Activity speeds up facilitated diffusion in muscle, and some hormones have the same effect in other tissues. For instance, the hormone insulin increases the rate of glucose uptake by some human tissues.

In most tissues, it is clear that glucose moves along a concentration gradient. Thus facilitated diffusion, which merely increases the rate at which glucose crosses the plasma membrane, is adequate for the tissues' needs. However, there are cases where glucose (and other substances) must move against a concentration gradient. In such a case, active transport is necessary.

5-D Active Transport

Two things distinguish **active transport** from facilitated diffusion: first, active transport can move substances either with or against their concentration gradients; second, active transport requires the expenditure of energy.

Among the many active transport systems are the uptake of amino acids, peptides, nucleosides, and potassium by the bacterium *Escherichia coli*. These substances move into the bacterium only in the presence of the appropriate carriers and of a source of energy.

The best-studied active transport systems in higher organisms are those that move glucose from the gut into the cells of the intestine, and the sodium-potassium "pump." Both these systems use the high-energy molecule ATP as their source of energy.

The sodium-potassium pump is so called because it ejects sodium from a cell and brings in potassium. Nobody knows precisely how the pump works. It is found in most eukaryotic cells. An enzyme called sodium-potassium-ATPase breaks down ATP to release the energy required by the sodium-potassium pump.

There is more sodium outside than inside a cell, but this does not prove that all cells have sodium pumps. It might mean only that the plasma membrane is very impermeable to sodium and so excludes most sodium. How do we know that there is such a thing as the sodium pump? First, cells can be poisoned so that their ATP-manufacturing machinery cannot work. Such poisoned cells lose their ability to keep sodium out and potassium in; they leak potassium to the outside, and their internal sodium content rises. Second, cells can be incubated in a solution with radioactive sodium, so that their internal sodium gradually becomes radioactive. If they are then transferred to a medium containing nonradioactive sodium, they gradually lose the radioactive sodium to the outside and at the same time take in nonradioactive sodium. This indicates that sodium on the outside is constantly leaking in, and that the sodium that was originally inside the cell is leaving (presumably ejected by the carrier molecule). Thus the plasma membrane is not impermeable to sodium.

How do carrier molecules work? It seems that carriers are proteins and related substances in the plasma membrane. When a molecule to be transported across the membrane combines with the carrier, the carrier changes shape; this shape change may open a temporary pore in the membrane (the pore can be visualized not as a real hole but as a patch of membrane with temporarily altered shape such that the transported molecule can pass through easily). In the case of active transport, the energy from ATP may help to detach the transported molecule from its carrier on one side of the membrane, or may cause a change in the conformation of the carrier so that it can pick up another molecule for the return trip.

Recently a number of carrier molecules have been isolated from microorganisms. One of these is **valinomycin**, a potassium carrier. The valinomycin molecule resembles a hollow ball that stores the potassium ion inside. The outside of the ball is nonpolar and, therefore, soluble in the lipid of the plasma membrane. There is still much to learn about this system, such as precisely how valinomycin picks up and releases potassium, and why it is so specific for potassium.

5-E Exchange of Large Particles Across the Plasma Membrane

The plasma membrane is not a static structure; it is easily pushed or pulled, it fuses readily with the membranes of various internal organelles, and it is continu-

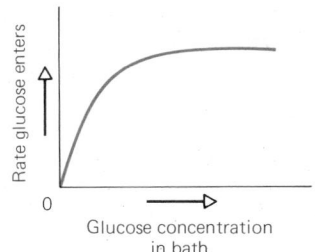

Figure 5-7 Influx of glucose into human red blood cells. The higher the concentration of glucose in the fluid bathing the cells, the faster glucose enters the cells by facilitated diffusion until all the glucose carrier molecules are occupied at any one time and glucose can move no faster.

Figure 5-8 Claude Bernard, the nineteenth-century physiologist who coined the aphorism "la fixité du milieu intérieur c'est la condition de la vie libre" (free life depends on the constancy of the internal environment). The plasma membrane plays an important role in homeostasis, the constancy of the internal environment, at the cellular level. (Smithsonian Institution)

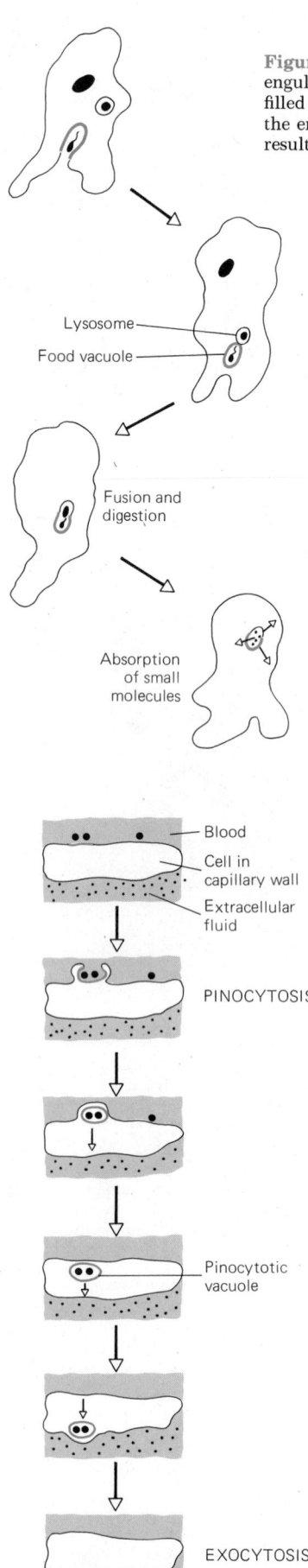

Figure 5-9 Phagocytosis, a form of endocytosis. An amoeba engulfs its prey and encloses it in a food vacuole. A lysosome filled with digestive enzymes fuses with the food vacuole and the engulfed organism is digested. The small molecules that result are absorbed into the cytoplasm.

Lysosome

Food vacuole

Fusion and digestion

Absorption of small molecules

Blood

Cell in capillary wall

Extracellular fluid

PINOCYTOSIS

Pinocytotic vacuole

EXOCYTOSIS

ous with various other membrane systems, such as the endoplasmic reticulum. The plasma membrane is thus a dynamic, fluid structure; its ability to change shape, and to fuse with or pinch off small pieces of membrane, allows it to engulf or export large particles.

In the process of **endocytosis**, cells incorporate material from their surroundings. The single-celled organism *Amoeba* engulfs its food by extending projections from the main body of its cell, surrounding the food, and then fusing the projections so that the food becomes enclosed in a membrane-bound sac in the amoeba's interior. This kind of endocytosis is known as **phagocytosis**—the engulfing of large particles, such as an entire bacterium or a fragment of a disintegrating cell, into a sac or vacuole in the cell (Figure 5-9). Phagocytosis is a major feeding method of many unicellular and simple multicellular animals.

In vertebrates (animals with backbones) and other higher animals, phagocytosis plays a part in defense against bacteria. Some types of white blood cells (phagocytes) engulf invading bacteria and digest them, thus serving as an important barrier to disease. Some diseases, such as leprosy and tuberculosis, are caused by bacteria that have adaptations making them resistant to digestion by phagocytes.

Phagocytosis also has a role in the normal physiology of vertebrates. For example, red blood cells and other cells undergo "programmed death" and their remains are cleaned up by phagocytes.

Endocytosis does not occur in most plant cells, because they are surrounded by thick cell walls which prevent large-sized particles from coming into contact with the plasma membrane.

Pinocytosis is a term sometimes used for endocytosis in which a cell takes in liquid or macromolecules rather than large particles (Figure 5-10).

The changes in the shape of the plasma membrane during endocytosis are related to local changes in the microfilaments that underlie the membrane. The mechanism is not yet well understood.

Materials can be discharged from a cell as well as engulfed In **exocytosis** cells discharge substances by fusing the membrane of an internal vesicle or vacuole with the plasma membrane, which then opens up and allows the substances in the vesicle or vacuole to escape from the cell (Figure 5–10). Substances released in this way may be indigestible food particles, or secretions such as hormones.

5-F   Osmosis

So far, we have neglected transport of the one substance most crucial to life: water. In order for a cell to maintain homeostasis, it must have strict control over its chemical content, which includes not only the absolute amounts of solutes but also their concentration. Thus the content of the solvent, water, in a cell must also be precisely regulated. Vital as water is to living cells, cells have no known carriers or other direct means of transporting water in or out. Water seems to travel through the cell membrane quite freely—faster, in fact, than any other substance.

Osmosis, the process by which water moves through a selectively permeable membrane, is a special case of diffusion. It involves the diffusion of a solvent, such as water, rather than the diffusion of substances dissolved in the solvent. In osmosis

Figure 5-10 Movement of fluid and macromolecules from the blood into the extracellular (outside the cells) fluid. The cells that form the capillary walls take in materials from the blood by pinocytosis. The pinocytotic vacuoles then move across the short distance to the opposite side of the cell, which faces the extracellular fluid, and there fuse with the cell membrane and discharge their contents.

in living cells, water moves across a membrane from a weak, or dilute, solution into a strong, or concentrated solution.

A simple way to demonstrate osmosis is to separate pure water from an aqueous solution* by a membrane that is permeable to water but not to the solute (Figure 5–11). After a time, the solution will increase in volume and the pure water will decrease in volume, as water passes by osmosis from the pure water, across the membrane, and into the solution.

Why does this happen? Water molecules can pass in either direction across the membrane. However, the water molecules in the solution bump into the solutes and also experience forces attracting them to solute particles (Figure 5–12); this retards the movement of the water molecules in the solution, and so water moves into the solution faster than it moves out.

In an osmotic system such as that shown in Figure 5–11, the net movement of water into the solution increases the height of the solution in the tube, and the weight of the column of solution exerts **hydrostatic pressure**. This pressure builds up as water enters the solution until it is pushing water molecules out as fast as they enter. The solution will remain at this level.

The extent of movement of water across a membrane can be predicted by knowing the osmotic potentials of the two solutions separated by the membrane. The **osmotic potential** of a solution is the tendency of the solution to gain water when it is separated from pure water by an ideal selectively permeable membrane. A stricter definition of osmotic potential is that it is the negative of the **osmotic pressure**, where the latter is the minimum pressure that must be applied to a solution to prevent it from gaining water when it is separated from pure water by an ideal selectively permeable membrane. The osmotic potential of pure water is zero; relative to water, any solution has an osmotic potential expressed in negative terms (Figure 5–13). The more concentrated the solution, the lower (*more negative*) its osmotic potential, and the greater its tendency to gain water from a solution with a higher (*less negative*) osmotic potential. The osmotic potential is the driving force of osmosis, since water tends to move in an energetically downhill direction, that is, in the direction of the lower osmotic potential.

The osmotic potential in a system depends on the concentration of particles in the solution, and on the attraction of water molecules to the particles, which slows the movement of the water molecules. There may be only one type of molecule dissolved in a solution, or there may be many, as in a living cell. Each molecule of an ionic substance dissociates into more than one particle in aqueous solution; NaCl dissociates into two particles, $MgCl_2$ into three, and so forth. The more particles there are in a solution, the lower the osmotic potential. If the solute particles are able to pass through the membrane, then the osmotic potential of the solution will gradually change as solute particles enter or leave it.

Now we can see how a cell can control its water content. The cell can create a difference in osmotic potential across its membrane by the active transport of solutes; water will then move by osmosis toward the side of the membrane where the osmotic potential is lower.

---

* Aqueous solution   Solution in which water is the solvent.

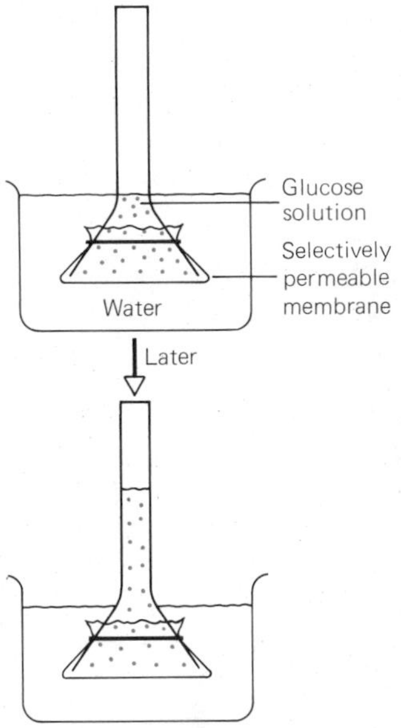

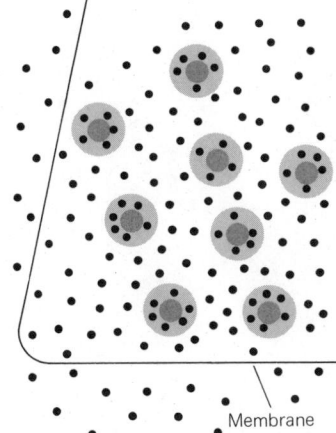

Figure 5-11 An osmotic system. The tube is open at both ends; the bottom of the tube is covered with a membrane permeable to water but not to glucose. The wide end of the tube is filled with a solution of glucose and immersed in pure water to the level of the solution. Later, the solution has risen in the tube, while the level of water in the outer container has fallen. Water has moved from the container into the tube by osmosis.

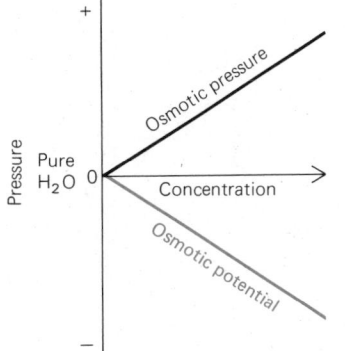

Figure 5-12 Solute molecules (full color) attract water molecules (black dots). Light color represents the drag exerted by the attraction to solute molecules, which slows the movement of these water molecules.

Figure 5-13 Osmotic potential and osmotic pressure. The osmotic potential is equal to the negative of the osmotic pressure (osmotic potential = −osmotic pressure). As the concentration of a particular substance in a solution increases, the osmotic potential of the solution decreases.

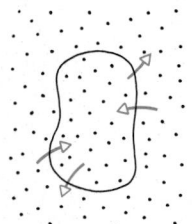

(a) Isotonic solution: The osmotic pressures of the cell and its surroundings are equal and the cell gains and loses equal amounts of water

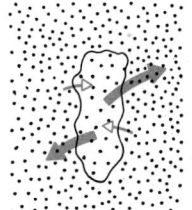

(b) Hypertonic solution: More water moves out of the cell than in; the cell shrinks as it loses water

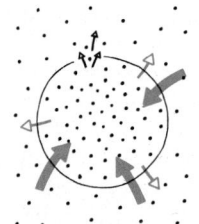

(c) Hypotonic solution: The cell gains more water than it loses; it swells and may eventually burst

Figure 5-14 Responses of an animal cell immersed in solutions of different osmotic characteristics.

Cells as Osmotic Systems

Cells behave as osmotic systems. A living cell contains a solution of various particles dissolved in water. It has a selectively permeable membrane and it must, in order to be living, be covered by some quantity of water, which also has solutes dissolved in it. If this extracellular solution has the same osmotic pressure as the intracellular (intra = within) solution, the cell is said to be living in an **isotonic** solution. If the external solution has a higher osmotic pressure than the internal solution, it is said to be **hypertonic** to the interior of the cell, and there will be a net flow of water out of the cell. And if the exterior solution is less concentrated than the interior of the cell, the exterior solution is said to be **hypotonic** to the interior of the cell, and water will tend to move into the cell (Figure 5–14). Some animal cells in dilute solutions may take in so much water that their internal pressure ruptures the plasma membrane, allowing the cell contents to escape. This process is known as **lysis** (bursting) of the cell. In the same situation, the rigid wall of a plant cell produces a **wall pressure** which opposes the outward pressure of swelling and makes most plant cells more resistant to swelling in a hypotonic solution (Figure 5–15).

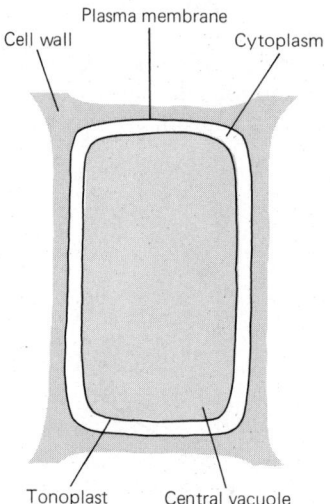

(a) The central vacuole, surrounded by its membrane, takes up most of the volume of the cell.

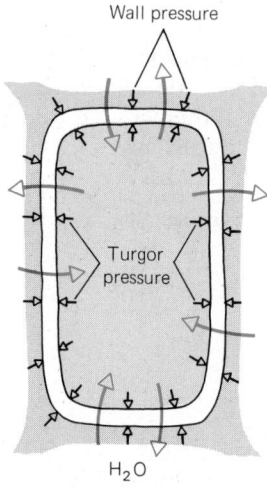

(b) As the cell swells, its internal turgor pressure is opposed by the pressure of the cell wall, which resists stretching (black arrows). When turgor pressure equals wall pressure, water is squeezed out by wall pressure as quickly as it enters the cell (colored arrows).

Figure 5-15 Contributions of the central vacuole and cell wall to plant cell shape.

Water Relations in Plant Cells

Most plant cells swell and shrink under osmotic stress by gaining or losing water from their central vacuoles.

The central vacuole is surrounded by a membrane, the **tonoplast**, which, like the plasma membrane, is selectively permeable and which also engages in active transport. The **cell sap** that fills the vacuole is a concentrated solution of salts, sugar, and various proteins. The low osmotic potential of the sap allows the vacuole to take up water. The vacuole thus serves as a biochemical reservoir for the cell and also plays an extremely important part in supporting the plant. A plant that has lost a lot of water will have shrunken central vacuoles in its cells. There will be no internal pressure in the cells, and they will hang limply together (Figure 5–16). The cells are said to be **plasmolyzed**, and a plant in this condition appears wilted. Even though the cell walls are strong, they cannot hold the plant up by themselves. However, if the plant has enough water to fill its vacuoles, the internal fluid pressure of the cells, pushing on the insides of all the cell walls, supports them in the proper configuration, and the plant appears fresh and healthy. The internal pressure of the cells is called **turgor pressure**, and a plant cell exerting pressure against its wall is said to be **turgid**.

5-G How Good Is the Plasma Membrane's Control of Permeability?

A cell's control over what enters and leaves seems, at first glance, to be imperfect since poisons like cyanide, carbon monoxide, and some drugs penetrate most plasma membranes readily and may kill the cell. A cell cannot, however, defend itself against unwanted substances by being impermeable. It must be somewhat permeable if it is to exchange necessary substances with its environment. Furthermore, cells cannot evolve defenses against toxins they have never encountered. It is clear that cells generally do evolve defenses against toxins which are common enough to act as effective selective pressures. It is extremely inconvenient to us, for instance, that many bacteria have evolved an impermeability to various antibiotics in the course of only a few years.

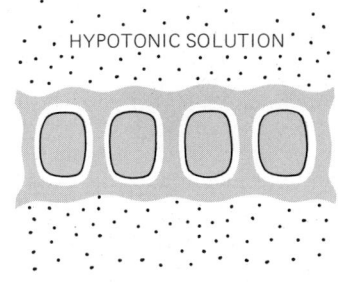

(a) Cells placed in a hypotonic solution are turgid, pressing against their cell walls.

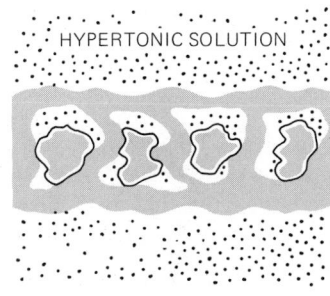

(b) Cells placed in a hypertonic solution become plasmolyzed (shrink), losing water from their vacuoles; the cell walls sag when they are not held in place by turgor pressure.

Figure 5-16 Osmotic relations in plant cells.

SUMMARY

Cells must maintain chemical homeostasis, the condition in which the concentrations of all substances are kept within the very narrow limits necessary for the cell's survival. Since cellular activities constantly demand new raw materials and produce waste products which must not be allowed to build up in the cell, a cell must maintain a lively commerce with its environment: it cannot simply shut itself up once it has achieved the desirable level of all its chemical constituents.

The primary responsibility for regulating what gets into or out of the cell rests with the plasma membrane. The membrane's lipid layer admits lipid-soluble molecules; its carrier molecules provide for the entry and exit of many polar molecules and ions. It is permeable to many types of small organic monomers, yet sufficiently impermeable to such vital cellular materials as nucleic acids, proteins, and polysaccharides.

When large particles must move into the cell, the membrane surrounds them and pinches off to become a vesicle or vacuole inside the cell, by the process of endocytosis. Substances can be discharged from many cells by the opposite process of exocytosis.

Cells gain or lose water by osmosis. Cells cannot control osmosis by influencing the movement of water molecules directly; rather they rely on their ability to perform active transport and to create an osmotic potential difference that will induce osmosis in the appropriate direction. The cell wall of a plant cell exerts a pressure that limits the cell's water content.

OBJECTIVES

From your study of this chapter, you should be able to:

1. Define and explain the following processes, and state or identify the characteristics that distinguish them from one another: diffusion, osmosis, active transport, facilitated diffusion, endocytosis, exocytosis.

2. Describe the structure of the plasma membrane, and relate its structure to the ability to carry on the processes mentioned in Objective #1.

3. Name the types of substances that enter the cell by each of the processes listed in Objective #1.

4. List four important features of the sodium pump and give reasons why the pump is important to a living cell.

5. Define and use the following terms correctly: gradient, permeable, osmotic potential, osmotic pressure, isotonic, hypertonic, hypotonic, lysis, plasmolysis, turgor.

6. Describe, using the appropriate vocabulary from Objective #5, the effect of placing a plant or animal cell in distilled water or in a concentrated salt solution.

7. Given (1) a description or picture of two solutions separated by a membrane, and (2) the permeability properties of the membrane, predict (a) which solution will have the higher osmotic potential, and (b) the total and net movement of water, solutes, or both in the system.

8. State the functions of the central vacuole and tonoplast in plant cells.

SELF-QUIZ

As appropriate, choose the correct word from each set in parentheses (in Questions 1, 4, and 6).

1. The higher the concentration of a solution, the (higher, lower) its osmotic potential, and the (more, less) water will enter it across a membrane permeable to water but not to solutes.

2. In the tube in the figure below, the membrane is permeable to both water and glucose. The tube is filled with a solution of glucose and placed in a beaker of water. Describe the sequence of events that will take place in this system, and draw the *final* water level in the tube and in the beaker.

   (Hint: the water molecules are smaller than glucose molecules and diffuse faster.)

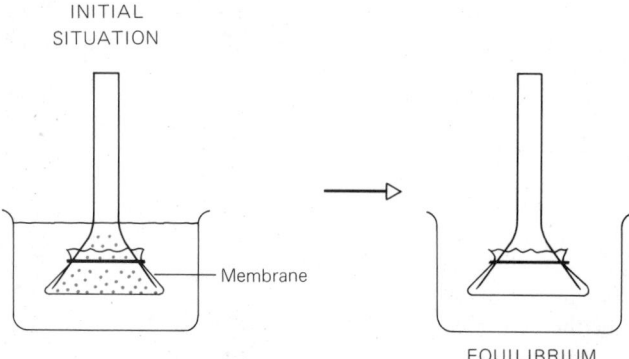

3. The U-tube in the figure below is divided in the middle by a membrane that is impermeable to starch but permeable to water. A 10% starch solution is put into the right-hand half of the tube and an equal amount of 6% starch solution is put into the left-hand half of the tube.

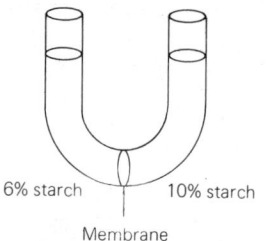

In this situation:
   a. water will move from the right to the left
   b. water will move from the left to the right
   c. starch will move from the right to the left
   d. water will move in both directions, but more from left to right than right to left
   e. water will move in both directions, but more from right to left than left to right

4. *Amoeba*, a single-celled organism, has a specialized structure called a contractile vacuole whose function is to collect excess water from the cell and discharge this water into the environment. From this information you can deduce that *Amoeba* lives in an environment that is (hypertonic, isotonic, hypotonic) to the interior solution of *Amoeba*. Thus, you would expect to find *Amoeba* living in:
   a. a freshwater pond
   b. the ocean
   c. Great Salt Lake
   d. a bottle of unpasteurized milk

5. In Chapter 4, we saw that the Golgi body discharges substances from the cell by fusing of vesicles with the plasma membrane and subsequent opening of what was the interior of the vesicle to the outside environment. This is an example of:
   a. exocytosis
   b. endocytosis
   c. active transport
   d. facilitated diffusion

6. At the produce counter in the supermarket, you pick up a head of lettuce whose outer leaves are wilted. You take it home and place it in ice water; the outer leaves become crisper. The vacuoles in the cells of these leaves are now (larger, smaller) than they were when you put the lettuce into your shopping cart.

QUESTIONS FOR DISCUSSION

1. Why don't cells just shut themselves off from the outside world and make their membranes impermeable so that nothing goes in or out?
2. Why is it important for cells to maintain chemical homeostasis?
3. Hydrogen cyanide (HCN) and carbon monoxide (CO) are poisons that penetrate cell membranes readily. Can you think of an explanation for the fact that no cells have evolved adaptations to keep these molecules out?
4. If cells act as osmotic systems, why don't we swell up and burst when we go swimming in fresh water, which is hypotonic to our blood, extracellular fluids, and intracellular solutions?

REFERENCES AND FURTHER READING

(Most of the references for Chapter 4 will also be useful here.)

Christensen, H. N. *Biological Transport,* 2nd ed. Reading, Massachusetts: W. A. Benjamin Inc., 1975. An advanced book covering many aspects of transport on the cellular and molecular levels.

Giese, A. C. *Cell Physiology*, 5th ed. Philadelphia: W. B. Saunders, 1979. A useful reference.

Kennedy, D., ed. *Cellular and Organismal Biology*. San Francisco: W. H. Freeman, 1974. Collection of articles on cell structure and function that originally appeared in *Scientific American;* with introductory essays.

CHAPTER

6

ENERGY AND LIVING CELLS

Pogo the 'Possum, hero of a comic strip popular in your authors' youth, once declared that he did not like bombs because "they put everything too everywhere." What he perhaps did not realize is that bombs are only a faster way to arrive at an inevitable end result. Left to its own devices, everything will eventually go everywhere anyway. This principle is called the **second law of thermodynamics**; it states that the amount of disorder, or **entropy**, in the universe must increase, and as entropy increases, the **free energy**, energy that can do useful work, decreases. In other words, the universe is going "downhill," in terms of free energy, toward the lowest possible level of free energy.

Living organisms appear to run directly counter to this law of increasing disorder in that they increase the molecular orderliness inside their bodies. As we saw in Chapter 5, an organism maintains a chemical composition very different from that of its environment, and it permits its internal chemical composition to vary only within narrow limits. An organism also increases orderliness by growing and so increasing the amount of matter which is organized into living cytoplasm.

Living things are not exempt from the second law of thermodynamics. Instead, organisms use energy from various sources to combat the universal tendency to increasing disorganization and to synthesize the molecules needed for growth, repair, and reproduction. In addition, organisms must use energy to move substances within their own bodies, and often to move the body itself.

Organisms obtain the energy they need for these activities mainly from food. During cellular respiration, a cell breaks down food molecules via a series of steps and releases the food energy in small packets that can be used to drive energy-requiring processes. Almost all the food in the world is made by photosynthesis, another step-by-step process, in which the energy of sunlight is stored in chemical bonds as plants build food molecules from smaller components—carbon dioxide, water, and various minerals.

In this chapter we introduce the rules of energy transactions and show how they apply to the chemical reactions—including the steps of respiration and photosynthesis—carried out by cells. We shall see that all cells have a few versatile tricks that they use to provide themselves with energy in usable doses. Since photosynthesis is the production of food, and respiration the breakdown of food, we often think of respiration and photosynthesis as opposites. But, as we shall see in this chapter, the two have many features in common.

In the next two chapters, we shall cover respiration and photosynthesis in greater detail.

6-A   Energy Transformations

Energy occurs in many familiar forms, such as heat, light, chemical, and electrical energy. Energy can be **transformed**, or changed from one form into another. For example, the filament of a light bulb converts the energy of an electric current into light energy and heat energy.

We often find it useful to distinguish between potential and kinetic energy. **Kinetic energy** involves motion—a rolling stone, a moving current of electrically charged particles, or a vibrating molecule, for example. **Potential energy** is associated with position; it is the possible energy that can be released if the position of something is permitted to change. A rock perched on a cliff has potential energy, which can be changed to kinetic energy if the rock is pushed over the edge. There is also potential energy in a bond between atoms in a molecule or in an electrical potential, in which oppositely charged particles, which are attracted to each other, are held apart by some barrier (Figure 6–1).

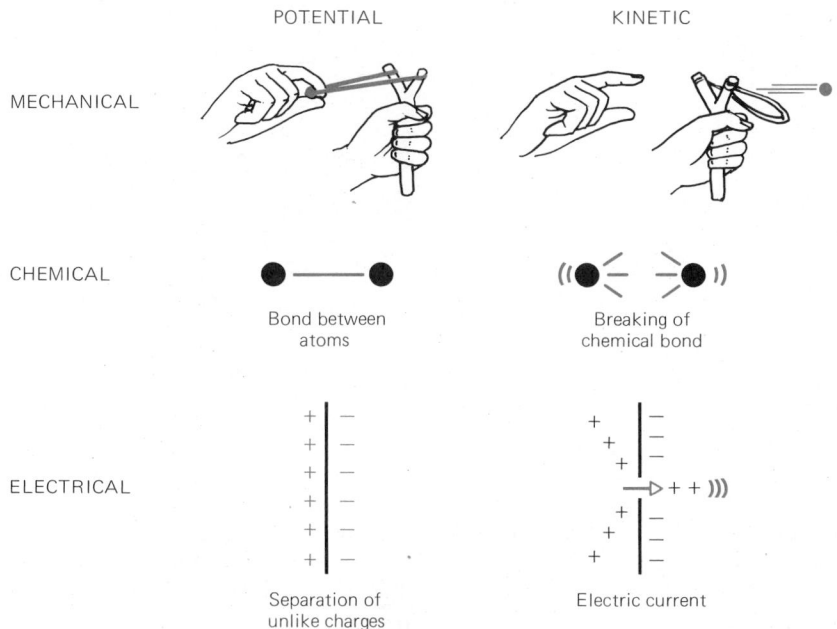

Figure 6-1 Some important forms of energy found in biological systems. Both potential and kinetic examples of mechanical, chemical, and electrical energy are shown.

Thermodynamics is the study of energy transformations. The first law of thermodynamics states that you can't win: energy can be neither created nor destroyed, it can only be transformed from one form to another. The second law of thermodynamics states that you can't even break even: in any change of energy from one form to another, the free energy decreases. Therefore, in a series of energy conversions, a smaller amount of free energy is transferred at each successive step. As the free energy decreases, there is an increase in the entropy in the system.

It is the second law of thermodynamics that poses problems, both for our energy-hungry society and for the world of living organisms. Each time an energy transformation releases usable free energy, some is lost as useless heat; for energy-using processes to continue, new sources of energy must be supplied constantly.

6-B Photosynthesis and Respiration

The immediate energy source for most cellular processes is **food molecules**— organic molecules that contain potential chemical energy in the form of the bonds between their atoms. This energy can be released by breaking the bonds, and the free energy so obtained (which is less than the free energy it took to make the bonds in the first place) can be used to meet the cell's energy requirements.

Life on earth today is utterly dependent on the sun—an energy source that will not be exhausted for perhaps 40 billion years. During **photosynthesis**, solar energy is stored by forming new bonds in food molecules. The raw materials of photosynthesis are the simple, low-energy inorganic molecules of carbon dioxide and water, which are assembled to form more complex food molecules; oxygen is given off as a by-product. Photosynthetic organisms are called **autotrophs** (auto = self; troph = food) because they do not require food molecules from other organisms to meet their energy needs. (There are other kinds of autotrophs, which use sources other than the sun for energy [Section 22-E], but their contribution to the world's food supply is negligible.)

In **cellular respiration**, food molecules are broken down, releasing energy that can be used by cells. Respiration uses oxygen in the complete breakdown of

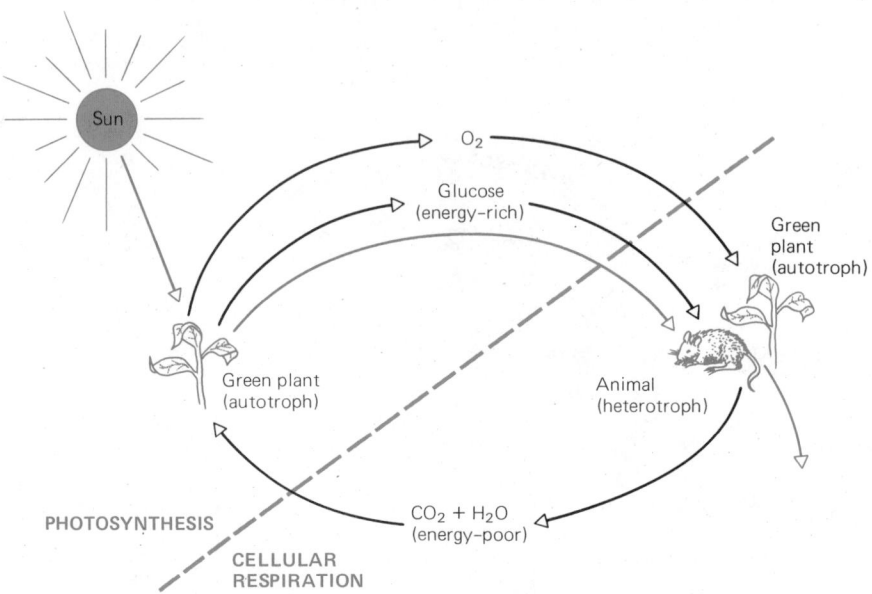

Figure 6-2 Flow of matter (black arrows) and energy (colored arrows) in living things. Green plants capture the sun's radiant energy and use it to synthesize organic molecules, such as glucose, from water and carbon dioxide. Oxygen is released as a by-product of photosynthesis. All organisms break down organic molecules to release energy during the process of cellular respiration. Respiration uses oxygen and produces carbon dioxide and water, which can be reused in photosynthesis. Although matter cycles indefinitely between photosynthesis and respiration, energy does not; the sun must constantly supply fresh energy.

food molecules; carbon dioxide and water (plus the released energy) are the end products. Thus, respiration has the overall effect of undoing photosynthesis; the carbon dioxide and water that went into photosynthesis are returned to the environment, and the energy released (unhappily for living organisms, much less than the energy originally trapped during photosynthesis) is used to drive the energy-requiring processes of the cell. Autotrophic organisms use respiration to break down the food they have made as its stored energy is needed. Organisms that cannot make their own food molecules, but must obtain the food they respire from other organisms, are called **heterotrophs** (hetero = other) (Figure 6–2).

The overall equations for photosynthesis and respiration are:

photosynthesis: $CO_2 + H_2O$ + energy $\longrightarrow$ organic molecules + O_2

respiration: organic molecules + $O_2 \longrightarrow CO_2 + H_2O$ + energy

These equations appear to be direct opposites: the raw materials of each are the end products of the other; and photosynthesis uses energy, whereas respiration releases it. However, both processes share some of the same kinds of chemical reactions and energy-transfer strategies. In the rest of this chapter we shall examine some of the features that these central energy-processing pathways of living cells have in common.

6-C   Chemical Reactions and Energy

The free energy of a chemical compound is a function of the bonds between its atoms. Each bond has a characteristic energy, usually defined as the amount of heat energy required to break the bond. A chemical reaction will tend to proceed in the direction in which the resulting molecules have the lowest possible free energy; this may be either toward the right or toward the left side in a reaction equation. In a chemical reaction, if the total bond energy in the reactant* molecules is greater than that in the product* molecules, energy is released as the reaction proceeds and the reaction is said to be **exergonic**. The oxidation of glucose is an exergonic reaction:

$$C_6H_{12}O_6 + 6O_2 \longrightarrow 6CO_2 + 6H_2O + \text{energy}$$

In an **endergonic** reaction, the energy of the reactants is less than that of the products, and energy must be added to the system to make the reaction proceed. For example, energy is needed to join a phosphate group to a molecule of glucose:

glucose + phosphate + energy $\longrightarrow$ glucose-phosphate + water

Reactions that are strongly exergonic proceed until almost all the reactants are used up. Where the free energy change in a reaction is not great, however, the reaction will have a tendency to go in both directions, as in the following conversion of glucose-phosphate to another sugar phosphate, fructose-phosphate:

glucose-phosphate $\rightleftharpoons$ fructose-phosphate

The longer arrow shows the exergonic reaction.

Every reaction has a characteristic equilibrium point,* which depends on the reactants and on conditions such as the temperature and pH of the environment. Reactions seldom reach equilibrium in living systems because other reactions constantly add new reactants or remove the products; this forces the reaction to keep going. In the reaction above, if fructose-phosphate is constantly used up in another reaction, glucose-phosphate will continue to be converted to fructose-phosphate.

---

* Reactant A substance that enters a chemical reaction and is changed by it.
* Product A substance that results from a chemical reaction.
* Equilibrium point The balance point in a chemical reaction at which the rate of the forward reaction, which changes the reactants to products, is equal to the rate of the back reaction, which changes products into reactants.

Oxidation-Reduction Reactions

Oxidation is a very common type of exergonic reaction in living organisms, involving transfer of an electron (e^-) from an electron donor molecule to an electron acceptor molecule. The molecule that loses the electron is oxidized, and the molecule that gains the electron is **reduced**. Oxidation and reduction are complementary reactions; for every oxidation, there is a corresponding reduction. Such redox (nickname for *red*uction-*ox*idation) reactions are often written:

$$Ae^- + B \longrightarrow A + Be^-$$

| electron donor | electron acceptor | has been oxidized (lower free energy than Ae^-) | has been reduced (higher free energy than B) |

or, written in the coupled form:

$$Ae^- \quad \overset{}{\diagdown} \quad B \quad \text{(reactants)}$$
$$A \quad \diagup \quad Be^- \quad \text{(products)}$$

oxidation (exergonic)   reduction (endergonic)

Oxidation and reduction may be accomplished in several ways. Substances are oxidized if they lose an entire hydrogen atom, which contains an electron. Adding oxygen also oxidizes a substance. Electrons are not actually removed from the molecule in this type of oxidation; however, since oxygen attracts electrons very strongly, electrons will spend more of their time near the oxygen nucleus and less time near the rest of the molecule. This constitutes partial removal of electrons from the original molecule.

Energy Intermediates

Chemical reactions in cells are catalyzed by enzymes, but enzymes cannot alter the energy changes that occur during a reaction nor make an endergonic reaction proceed without the necessary added energy. How then do organisms power endergonic reactions? The answer is that they use the energy released from an exergonic reaction to drive an endergonic reaction. For this to occur, the two reactions must have a common energy intermediate.

The high-energy compounds that act as energy intermediates in living cells have phosphate groups attached by so-called **high-energy phosphate bonds**; these release considerably more energy than that released by most other chemical bonds when they are hydrolyzed. In the synthesis of the polysaccharide glycogen from glucose subunits, the energy donor is the high-energy compound uridine triphosphate (UTP). Guanosine triphosphate (GTP) is another energy intermediate in animals. By far the most common energy intermediate in all living things, however, is adenosine triphosphate, or ATP; indeed, ATP is believed to be an energy intermediate in every living organism.

Not all energy intermediates are high-energy molecules. Energy can also be stored in living organisms in the form of a difference in the electrical potential energy between the two sides of a membrane (see Section 6-D).

How does an energy intermediate such as ATP permit a cell to drive an endergonic reaction? As a specific example, the removal of a phosphate group from phosphoenolpyruvate is an exergonic reaction which releases energy that can be used to power the endergonic addition of more glucose subunits to starch:

(1) phosphoenolpyruvate $\rightleftharpoons$ pyruvate + phosphate   (exergonic)

(2) glucose + (glucose)$_n$ $\rightleftharpoons$ (glucose)$_{n+1}$ + H_2O (endergonic)
         small   larger
    starch polymer   starch polymer

Reaction (2) is not energetically feasible as it is written. As can be seen from the arrows, it has a greater spontaneous tendency to proceed to the left than to the right. The energy-carrying intermediate ATP is used to permit reaction (2) to occur:

$$\text{phosphoenolpyruvate} + \text{ADP} \rightleftharpoons \text{pyruvate} + \text{ATP} \qquad \text{(exergonic)}$$
$$\text{ATP} + \text{glucose} \rightleftharpoons \text{(ADP-glucose)} + \text{phosphate} \qquad \text{(exergonic)}$$
$$\text{(ADP-glucose)} + \underset{\substack{\text{small} \\ \text{starch polymer}}}{\text{(glucose)}_n} \rightleftharpoons \underset{\substack{\text{larger} \\ \text{starch polymer}}}{\text{(glucose)}_{n+1}} + \text{ADP} \qquad \text{(exergonic)}$$

All of these reactions are energetically feasible because their products have lower free energies than do their reactants. The entire series is energetically "downhill" and can occur, catalyzed by enzymes, in a cell. An overall endergonic reaction is thus carried out by executing a series of exergonic reactions, making use of energy intermediates.

6-D Making ATP

The role of ATP in the cell's energy economy has been compared to that of cash in the human economy. Many enzymes hydrolyze ATP to provide the energy for a variety of activities—muscle contraction, active transport, waste excretion, and synthesis of new macromolecules, to name a few. ATP is also used during photosynthesis as an energy intermediate in the building of food molecules, which are the cell's savings account of stored energy, broken down at need to make the ready cash of ATP. The central position of ATP in the economy of life can be illustrated like this:

$$\text{Solar energy} \longrightarrow \text{ATP} \longleftrightarrow \text{Food molecules}$$
$$\downarrow$$
$$\text{Growth, Reproduction, Movement, etc.}$$

Both photosynthesis and respiration, then, can be regarded as means to the same end, the synthesis of ATP.

An ATP molecule consists of an adenine group, a ribose sugar group, and three phosphate groups (Figure 6–3). The bonds attaching the last two phosphate groups are high-energy phosphate bonds. Energy is usually released from ATP by hydrolyzing the terminal phosphate group, yielding ADP (adenosine diphosphate), an inorganic phosphate group (abbreviated as P_i), and about 7.3 kilocalories* of energy per mole of ATP hydrolyzed. ADP may be further hydrolyzed to AMP (adenosine monophosphate) plus P_i, releasing another 7.3 kcal per mole. At times, the two terminal phosphates of ATP are broken off together (see Figure 10–10).

The number of ATP molecules in a cell is relatively small, and ATP must be

---

* Kilocalorie 1 kilocalorie or kcal = 1 Calorie = 1000 calories. A calorie is the amount of heat needed to raise one gram of water from 14.5°C to 15.5°C. The energy values of foods are expressed in Calories.

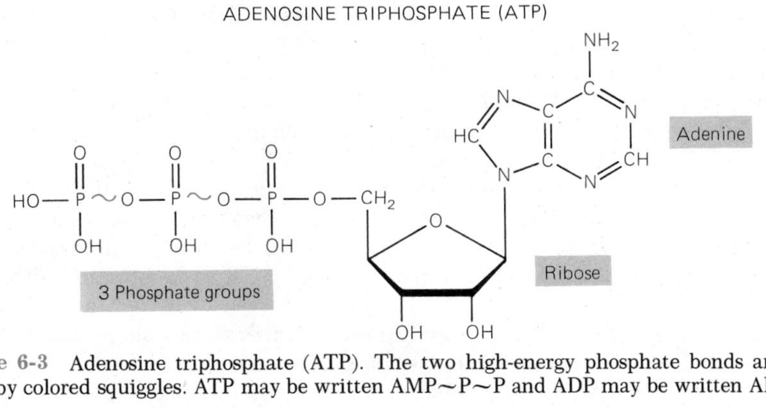

ADENOSINE TRIPHOSPHATE (ATP)

Figure 6-3 Adenosine triphosphate (ATP). The two high-energy phosphate bonds are indicated by colored squiggles. ATP may be written AMP~P~P and ADP may be written AMP~P.

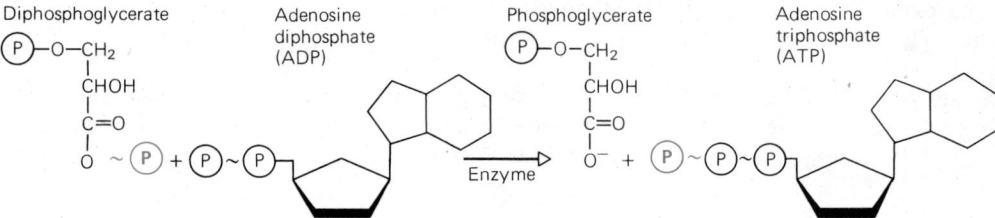

Diphosphoglycerate

Adenosine diphosphate (ADP)

Phosphoglycerate

Adenosine triphosphate (ATP)

Figure 6-4 Substrate-level phosphorylation of ADP to ATP, using diphosphoglycerate as a phosphate-group donor.

resynthesized as fast as it is broken down. Formation of ATP involves the **phosphorylation** of (that is, the addition of a phosphate group to) ADP. This may occur by two mechanisms.

Substrate-Level Phosphorylation

In **substrate-level phosphorylation**, a molecule with a high-energy phosphate bond which has a higher free energy of hydrolysis than that of ATP donates its phosphate group to ADP, forming ATP (Figure 6–4). The phosphorylation of ADP by phosphoenolpyruvate (see Section 6-C) is an example.

Chemiosmotic ATP Synthesis

In most cells, substrate-level phosphorylation produces only a minor fraction of the total ATP. Far more ATP is made by a second mechanism, involving membranes, which has been little understood until recently. It has long been known that the plasma membranes of bacteria and the inner membranes of mitochondria are involved in ATP production during respiration, and that the inner membranes of chloroplasts play some similar role in photosynthesis. This membrane-associated phosphorylation occurs by a process now known as **chemiosmosis**.

The chemiosmotic theory of ATP synthesis was proposed by Peter Mitchell in 1961. Although the theory is not yet complete in all its details, the available evidence for it is so convincing that Mitchell received a Nobel Prize for this work in 1978. Chemiosmotic ATP synthesis requires two steps: first, an energy supply is stockpiled, and then this energy is used to power the endergonic reaction of ATP synthesis. In this section we shall examine how the energy source is built up and how it is then used to drive ATP synthesis, the evidence that the theory is correct, and how the theory applies to some other situations.

ENERGY FOR CHEMIOSMOSIS. The energy for chemiosmosis exists in the form of an **electrochemical potential gradient**, built up by separating particles with positive and negative electric charges—in this case, the protons (H^+) and electrons (e^-) of hydrogen atoms—and accumulating the H^+ in a reservoir on one side of a membrane. The term "electrochemical" signifies that this potential gradient has both an electrical component (the difference in charges across the membrane) and a chemical component (the difference in H^+ concentration [pH]); each of these has potential energy in its own right. For this potential to exist, the membrane must be intact and relatively impermeable to H^+; the potential would dissipate rapidly if H^+ could move freely across the membrane.

The membrane is not simply an inert barrier to the movement of H^+; it also houses the **electron transport system**, a set of molecules responsible for setting up the electrochemical gradient in the first place. Electron transport systems are found in the plasma membranes of most bacteria. They are also found in the inner membranes of chloroplasts and mitochondria, the sites of photosynthesis and respiration, respectively, in eukaryotes (Figure 6–5). All of these membranes housing electron transport systems will be referred to simply as "the membranes" in the rest of this chapter.

The molecules of the electron transport system are arranged in precise positions in the membrane, in such a way that they pass electrons, sometimes accompanied by protons, very rapidly from one transport molecule to the next. As each molecule

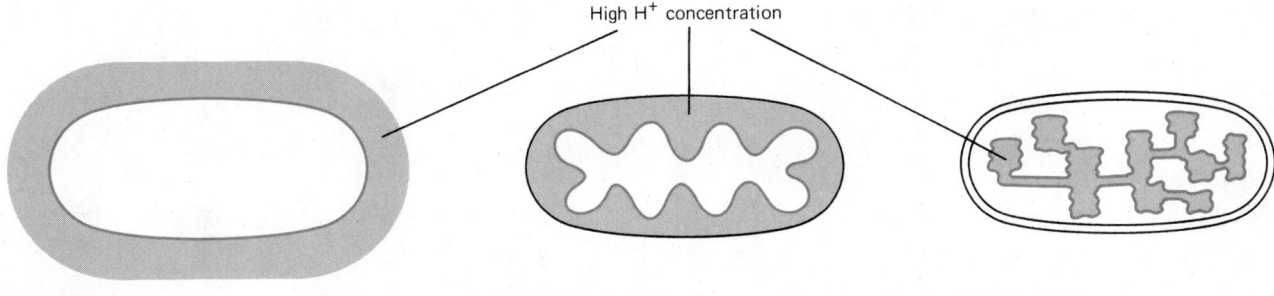

High H⁺ concentration

(a) Bacterial cell　　　　　　　(b) Mitochondrion　　　　　　　(c) Chloroplast

Figure 6-5 Positions of membranes and H⁺ reservoirs involved in chemiosmosis. Membranes in solid color contain electron transport systems: the plasma membranes of bacterial cells, and the inner membranes in mitochondria and chloroplasts. H⁺ accumulates in areas shown in pale color; note that this is outside the electron-transporting membrane in bacteria and mitochondria, but inside in chloroplasts.

accepts an electron or a hydrogen atom, it is temporarily reduced, and then it becomes oxidized once again as it passes its burden on to the next molecule. Each redox reaction in the series releases some free energy; electrons or hydrogen atoms start at the beginning of the system with relatively great energy, which is reduced somewhat at each step.

How is the buildup of H⁺ on one side of the membrane accomplished? The basic trick in both photosynthesis and respiration lies in the arrangement of the electron transport molecules: those that can carry hydrogen atoms alternate with those that can carry only electrons. A hydrogen atom carrier picks up two hydrogen atoms at one side of the membrane, carries them to the other side, and is oxidized by molecules that can carry only electrons. The H⁺ is dumped into the reservoir on the side of the membrane opposite that from which it started. Meanwhile, the electrons are passed by the electron transport system back to the first side of the membrane, thus preventing them from recombining with the H⁺, which would destroy the electrical potential created by separating the two oppositely charged components (Figure 6–6). The electrons may make several such round trips before all their energy is spent. (The electrons eventually join with other H⁺ ions; we shall see where these H⁺ come from in Chapters 7 and 8.)

The electrochemical potential gradient is built up as electrons weave back and forth across the membrane, in their passage from molecule to molecule in the electron transport chain. In turn, the potential energy is used to make ATP.

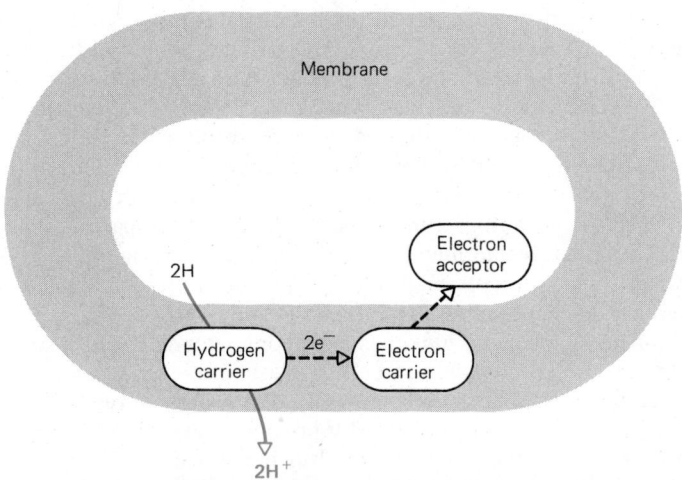

Membrane

Electron
acceptor

2H

Hydrogen
carrier   2e⁻   Electron
carrier

2H⁺

Figure 6-6 A simplified scheme to show how mitochondria build up an H⁺ reservoir outside a membrane. A hydrogen carrier molecule in the membrane transports two hydrogen atoms across to the other side, where it hands its burden on. But because the next molecules can carry only electrons, the H⁺ is released, having moved across the membrane. The electrons are carried back to the first side of the membrane, and eventually combine with an electron acceptor.

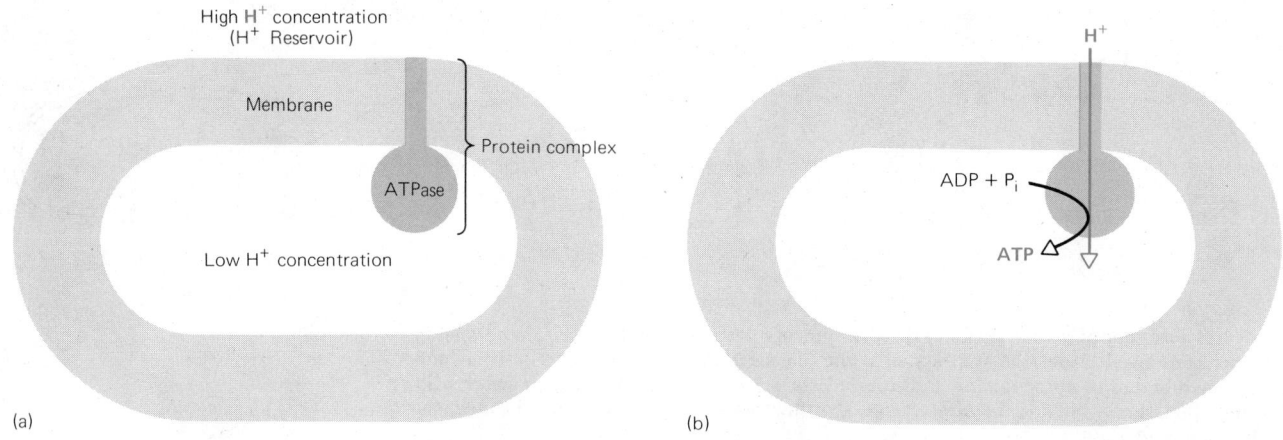

Figure 6-7 ATP synthesis powered by an electrochemical gradient in mitochondria or bacteria. (a) An ATPase enzyme, part of a protein complex, lies on the face of the membrane opposite the face where the H^+ has accumulated. (b) ATP synthesis occurs when H^+ crosses the membrane, going down its electrochemical gradient. The energy so released is used to form ATP from $ADP + P_i$.

HOW THE ELECTROCHEMICAL GRADIENT POWERS ATP SYNTHESIS. In addition to the molecules of the electron transport system, the membrane also contains protein complexes with ATPase enzymes (enzymes that can synthesize or hydrolyze ATP) sticking out on one side of the membrane (Figure 6–7). When the electrochemical gradient across the membrane has been built up, ATPase enzymes can use the energy to synthesize ATP. In the process, H^+ is thought to re-cross the membrane (2 H^+ per ATP synthesized in mitochondria, 3 H^+ per ATP in chloroplasts), and all the stored electrochemical potential energy of that H^+ is spent.

The electrochemical gradient is reduced as some of its H^+ returns to the original side of the membrane during ATP synthesis. Thus the electron transport system must keep working to maintain the electrochemical gradient.

EVIDENCE FOR THE CHEMIOSMOTIC THEORY. The available evidence makes the chemiosmotic theory more satisfying than alternative theories for several reasons. First, the only reasonable alternative to the chemiosmotic theory is that ATP is formed in electron transport systems, as it is in the cytoplasm, by the transfer of a phosphate group from some high-energy phosphate compound to ADP. Such a compound has never been discovered despite 30 years' work on the biochemistry of electron transport systems. On a more positive note, researchers have found that a respiring mitochondrion or bacterium (or a photosynthesizing chloroplast or bacterium) does in fact produce a change in the pH of the solution around it. In chloroplasts there is a difference of 3.5 pH units (more than a thousand-fold difference in H^+ concentration) across the membrane. Furthermore, chloroplasts in the dark can make ATP if experimentally provided with an artificial pH gradient. In such an experiment, the chloroplasts are allowed to reach equilibrium with a solution of a particular pH in the dark. When they are suddenly transferred, still in the dark, to a solution with a pH 2 or 3 units higher, the outer compartment soon reaches equilibrium with the new environment, while the inner compartment, surrounded by its relatively impermeable membrane, remains at the original pH. For a brief time thereafter, the chloroplasts produce ATP, even though ATP synthesis in chloroplasts usually occurs only in the light.

In addition, no alternative hypothesis explains why ATP synthesis stops if the inner membranes of mitochondria and chloroplasts are broken. Under such conditions, the electron transport reactions proceed, but they do not result in ATP synthesis. This phenomenon is readily explained if breaking the membrane permits hydrogen ions to move freely and destroys the electrochemical gradient. Similarly, the chemiosmotic hypothesis explains why detergents, which make membranes more permeable, prevent ATP synthesis.

Several substances that make mitochondrial membranes more permeable to H^+ have medical importance. If the membrane allows H^+ to leak through without being used to make ATP (that is, if H^+ movement is "uncoupled" from ATP synthesis), a well-nourished cell can literally starve to death. Dinitrophenols, yellow substances once used as food additives to make baked goods look as if they contained more eggs than they really did, make mitochondria leaky to H^+. These substances were prescribed for a time as a cure for obesity, but were abandoned after several patients ran out of ATP and died during treatment. Such uncoupling also occurs in normal animals at times. For example, the mitochondrial membranes in fat cells of hibernating animals contain an uncoupling protein, and most of the energy released during electron transport is used to make heat rather than ATP.

THE ELECTROCHEMICAL GRADIENT AS AN ENERGY INTERMEDIATE.   The energy of the electrochemical gradient can be used for tasks other than ATP synthesis. The flow of H^+ from the external H^+ reservoir into the bacterium *Escherichia coli* powers the flagellum during locomotion. In addition, movement of H^+ across the membrane in these bacteria and in mitochondria provides the energy needed for the active transport of sodium and calcium ions and of certain organic molecules. It seems probable that electrochemical energy is a much more common energy intermediate in living organisms than we now appreciate.

SUMMARY

Living organisms require energy in order to maintain their chemical composition, move, repair damage, grow, and reproduce. Photosynthesis and respiration are the central energy-processing pathways of life. Photosynthetic organisms capture the sun's energy and store it in the chemical bonds of food molecules, which can later be broken down during respiration to release the trapped energy.

The ultimate task of both photosynthesis and respiration is to produce energy intermediates to drive endergonic reactions. The most common energy intermediate is ATP; although some ATP is formed by substrate-level phosphorylation, most is believed to be made by chemiosmosis, using an electrochemical gradient as an energy source. This gradient consists of a reservoir of H^+ trapped on one side of an H^+-impermeable membrane. The electron transport system in the membrane builds up the gradient. Hydrogen carriers in the system carry pairs of hydrogen atoms across the membrane, where H^+ is dumped when the next transport molecules accept only the electrons. Transport of the electrons back to the first side of the membrane prevents them from recombining with their protons.

The electrochemical potential gradient is used to make ATP, possibly by allowing H^+ to leak back through the membrane by way of protein complexes attached to ATPase enzymes. The chemiosmotic theory is supported by experimental findings that ATP synthesis occurs in the presence of pH gradients across intact membranes, and that destroying the gradient or the membrane virtually halts ATP formation.

OBJECTIVES

From your study of this chapter, you should be able to:

1. In addition to the terms mentioned in the other objectives, use or interpret the following terms correctly:

potential energy	autotroph
free energy	heterotroph
entropy	ADP
	P_i

2. State the first and second laws of thermodynamics and explain how living organisms carry out endergonic chemical reactions.

3. List the basic starting materials and end products of photosynthesis and respiration, and describe the roles of photosynthesis and respiration in the energy economy of the living world.

4. Recognize examples of oxidation and reduction reactions.

5. Discuss the role of ATP in the energy economy of living organisms.

6. Distinguish between substrate-level phosphorylation and chemiosmotic phosphorylation.

7. Explain the importance of electron transport systems to cells.

8. Describe the chemiosmotic theory of ATP synthesis, including the role of the electron transport system, the membrane housing this system, the electrochemical gradient, and ATPase.

SELF-QUIZ

1. Does the following equation describe photosynthesis or respiration?

$$CO_2 + H_2O \longrightarrow C_6H_{12}O_6 + O_2$$

2. In the above equation, is carbon oxidized or reduced as a result of the reaction?

3. Cells are lysed, or broken up, and centrifuged, and the fraction containing their membranes is discarded. When both P_i containing radioactive phosphorus and ADP are added to the suspension of nonmembranous cell contents, some radioactive ATP is formed. This probably occurs by (substrate-level, chemiosmotic) phosphorylation.

4. A cell's electron transport system:
   a. makes ATP
   b. contains ATPase
   c. is responsible for separating hydrogen atoms into their components
   d. can work against the second law of thermodynamics
   e. can be destroyed by breaking the membrane in which it is embedded

5. The inner membranes of mitochondria and chloroplasts:
   a. are relatively impermeable to H^+
   b. have ATPase enzymes attached to only one face
   c. contain molecules of the electron transport system
   d. completely enclose the internal compartments
   e. all of the above

QUESTIONS FOR DISCUSSION

1. List as many energy-requiring activities carried out by living organisms as you can.

2. Why does a photosynthetic plant need to carry on cellular respiration?

3. The energy trapped in a glucose molecule ($C_6H_{12}O_6$) is only about 1% of the energy in the light quanta (units of light energy) needed to make that molecule from carbon dioxide. State the first and second laws of thermodynamics and show how they can be used to explain this. What happens to the remainder of the energy?

REFERENCES AND FURTHER READING

Hinkle, P. C., and R. E. McCarty. "How cells make ATP." *Scientific American*, March 1978. Presents the hypothesis of the chemiosmotic mechanism for ATP synthesis and gives a comparison for how it is thought to work in bacteria, mitochondria, and chloroplasts.

Racker, E. *A New Look at Mechanisms in Bioenergetics.* New York: Academic Press, 1976.

CHAPTER 7

FOOD AS FUEL

In Chapter 6, we saw that cells must constantly supply themselves with energy intermediates, principally ATP, to drive a multitude of energy-requiring activities. All organisms can make energy intermediates by breaking down organic food molecules. Most break down their food by **aerobic respiration**, a process that uses molecular oxygen in the oxidation* of high-energy food molecules; respiration produces the low-energy molecules carbon dioxide and water as by-products. Oxidation is familiar as the combustion, or burning, of wood or paper. But whereas combustion oxidizes organic molecules all at once, respiration oxidizes food in a series of controlled steps, each releasing a little of the food molecule's energy. This permits cells to recapture more energy in the form of energy intermediates than would be possible if the energy were released in one big burst.

Some cells make energy intermediates by **fermentation**, the breakdown of food molecules in the absence of molecular oxygen.

---

*Oxidation A chemical reaction that decreases the energy of a compound by adding oxygen to it or by removing electrons or hydrogen atoms from it.

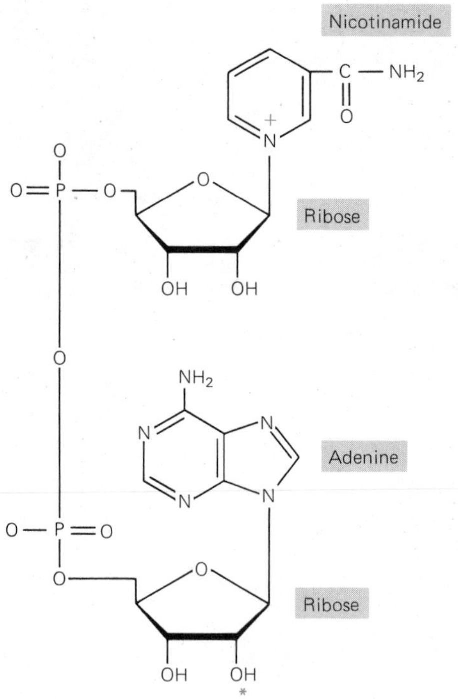

Figure 7-1 Nicotinamide adenine dinucleotide, NAD⁺, a coenzyme. The nicotinamide portion of the molecule is derived from the B vitamin niacin. Adding
another phosphate group at the colored *
gives NADP⁺, a coenzyme important in
photosynthesis (Chapter 8).

In this chapter we start by meeting several small, but vital, accessory molecules
that play important roles in helping to capture the energy of food molecules during
respiration. Then we shall see how cells break down the most commonly used food
molecule, the monosaccharide glucose. After this we shall see how some cells carry
out fermentation, and finally we shall see how organisms can obtain energy from
molecules other than glucose, and how the pathways of respiration connect with
other biochemical pathways in the cell.

7-A   Coenzymes

Many enzymes require additional molecules besides their substrates* before
they can act. These additional molecules are called **cofactors** or **coenzymes**. Many
of them are synthesized from the vitamins in our diets.

In this chapter, we shall meet three important coenzymes: nicotinamide adenine dinucleotide, commonly known as NAD⁺ (Figure 7–1); flavin adenine dinucleotide, or FAD; and coenzyme A, or CoA. (Niacin, riboflavin, and pantothenic acid,
respectively, are the vitamins from which these coenzymes are synthesized.) These
coenzymes may be considered as carrier molecules because they carry substances to
or from enzyme-catalyzed reactions; NAD⁺ and FAD, as we shall see, carry hydrogen, and CoA carries a 2-carbon acetyl group. Coenzymes last for a long time and can
be used over and over again, so they are required only in very small quantities by a
cell.

7-B   Breakdown of Glucose Under Aerobic
Conditions

In this section we shall follow the stepwise breakdown of glucose by a series of
exergonic* reactions; energy released at some steps in this breakdown process can be
used to synthesize ATP. We use the food molecule glucose to illustrate this breakdown because most cells use glucose rather than other available food molecules.

---

*Substrates The reactants in an enzyme-mediated chemical reaction.
*Exergonic In which the energy of the reactants is greater than the energy of the products.

Glucose may be broken down either **aerobically** (in the presence of oxygen) or **anaerobically** (when no oxygen is available). Both processes start out the same way, using the anaerobic reaction pathway called glycolysis. Under aerobic conditions, the products of glycolysis proceed to cellular respiration, made up of two pathways, the tricarboxylic acid cycle and oxidative phosphorylation.

1. **Glycolysis**. Glycolysis is a reaction series in which the 6-carbon glucose molecule is converted into two 3-carbon pyruvate molecules. During glycolysis two important products are synthesized: ATP and the reduced (hydrogen-laden) form of the coenzyme nicotinamide adenine dinucleotide, which may be written $NADH + H^+$ (Figure 7–2).

2. The **tricarboxylic acid cycle** or **citric acid cycle**. During the citric acid cycle, 2-carbon groups from pyruvate are broken down into carbon dioxide. ATP and the reduced coenzymes $NADH + H^+$ and $FADH_2$ are the important products of the cycle.

3. **Oxidative phosphorylation**. In this series of steps, reduced coenzymes from glycolysis and the citric acid cycle are oxidized as they pass the hydrogen atoms they carry to an electron transport chain embedded in a membrane. Here, the protons (H^+) of the hydrogen atoms are pushed to the outside of the membrane, setting up an electrochemical gradient across the membrane. This gradient eventually furnishes energy for ATP synthesis (Section 6-D). Electrons leaving the electron transport chain combine with oxygen and with hydrogen ions to form water. The term "oxidative phosphorylation" refers to the fact that the phosphorylation of ADP to form ATP is coupled with the oxidation of hydrogen-carrying coenzymes.

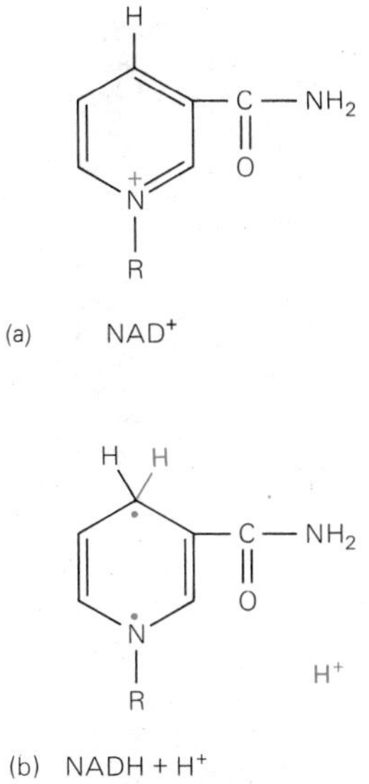

(a)   NAD⁺

(b) $NADH + H^+$

Figure 7-2 Oxidized and reduced forms of the coenzyme nicotinamide adenine dinucleotide. (a) The oxidized form, NAD^+, and (b) the reduced form, $NADH + H^+$. Only the nicotinamide portion of the molecule is shown; the rest of the molecule is indicated by R. When a substrate is oxidized by NAD^+, it loses two hydrogen ions plus two electrons (colored). One hydrogen ion and both electrons combine with NAD^+ to form NADH, and the other hydrogen ion is released into solution.

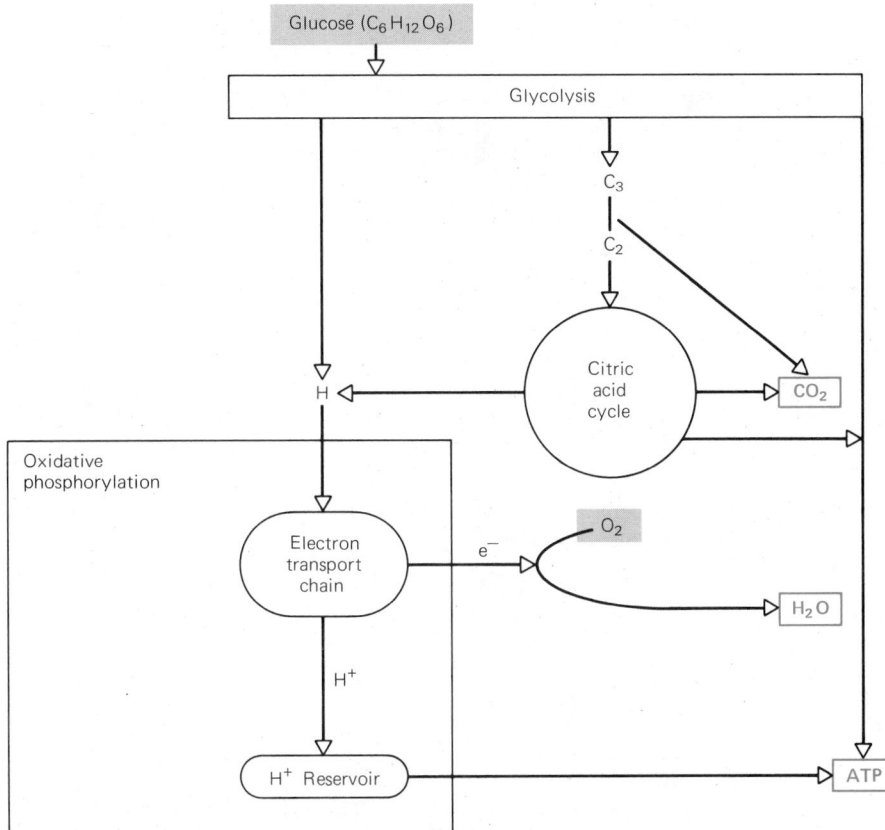

Figure 7-3 Summary of the aerobic respiration of glucose. The raw materials are in shaded boxes; the end products are in colored outlines. Glycolysis and the citric acid cycle make some ATP and also send hydrogen to the electron transport chain, which sets up a hydrogen ion reservoir to drive most of the cell's ATP synthesis.

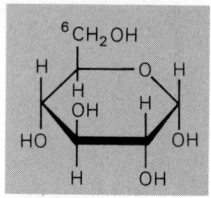

Glucose

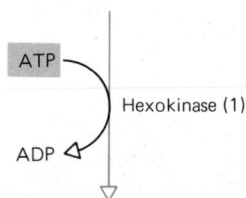

Glucose-6-phosphate

Phosphogluco-
isomerase
(2)

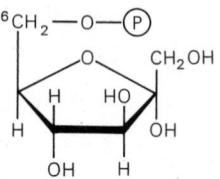

Fructose-6-phosphate

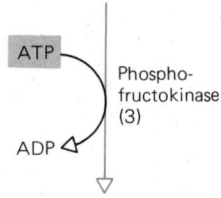

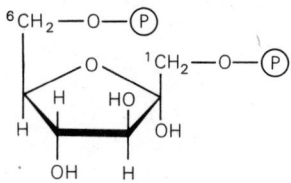

Fructose 1, 6-bisphosphate

Figure 7-4 The stages of glycolysis. Important molecules entering the pathway are enclosed in shaded boxes. Important products are in colored outlines. The enzymes that mediate the reactions are indicated beside the colored arrows.

Glycolysis

Glycolysis is an anaerobic reaction series that acts as the sole source of ATP synthesis in anaerobic fermentation (discussed in Section 7-C). Under aerobic conditions, glycolysis functions to break carbohydrates down into a form that, with slight modification, can enter the citric acid cycle.

The enzymes of glycolysis are dissolved in the cytoplasm. Follow the steps of glycolysis (Figure 7–4) as they are discussed:

1. The first step of glycolysis is the donation of a phosphate group from ATP to the sixth carbon atom of glucose. This gives the glucose molecule a negatively-charged phosphate group, which both prevents it from diffusing out of the cell through the plasma membrane and provides a recognition site by which the molecule binds to the enzyme that catalyzes the next step of glycolysis. The phosphorylation of glucose by ATP is a "pump priming" process, whereby some ATP must be used to initiate the series of energy-releasing processes that will yield an ATP profit. Other carbohydrates, such as glycogen, sucrose, fructose, galactose, etc., can be converted into glucose-6-phosphate and enter the glycolytic pathway at this point.

2. The next step involves an internal rearrangement of the glucose-6-phosphate molecule to form another 6-carbon compound, fructose-6-phosphate.

3. Addition of still another phosphate, again furnished by ATP, yields a molecule of fructose bisphosphate.

4. Fructose bisphosphate is split to form two 3-carbon molecules, each with a phosphate group attached to one end. One of these 3-carbon molecules, dihydroxyacetone phosphate, is eventually converted into the other, phosphoglyceraldehyde (PGAL). Up to this point, two ATPs have been used up during glycolysis.

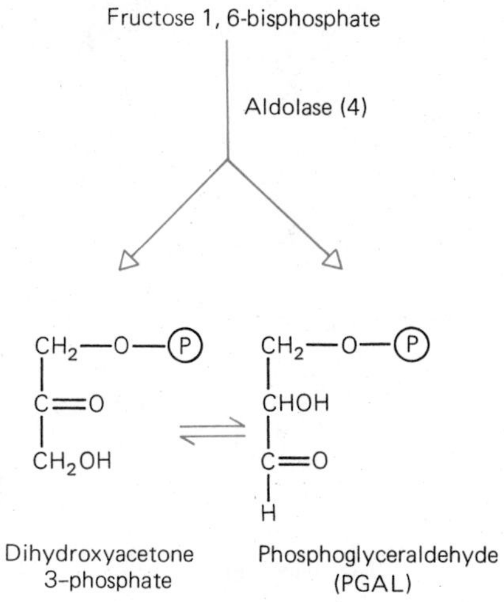

Fructose 1, 6-bisphosphate

Aldolase (4)

CH_2—O—(P)
|
C=O
|
CH_2OH

CH_2—O—(P)
|
CHOH
|
C=O
|
H

Dihydroxyacetone 3–phosphate

Phosphoglyceraldehyde (PGAL)

5. In the next step two phosphoglyceraldehyde molecules are oxidized and phosphorylated to form two diphosphoglycerate molecules. This requires two inorganic phosphate ions and two molecules of the coenzyme NAD^+. A phosphate group is added to each molecule of phosphoglyceraldehyde, and two hydrogen atoms are removed from it and used to reduce NAD^+ to $NADH + H^+$. Phosphoglyceraldehyde has become diphosphoglycerate, and its newly acquired phosphate group is attached with a high-energy bond.

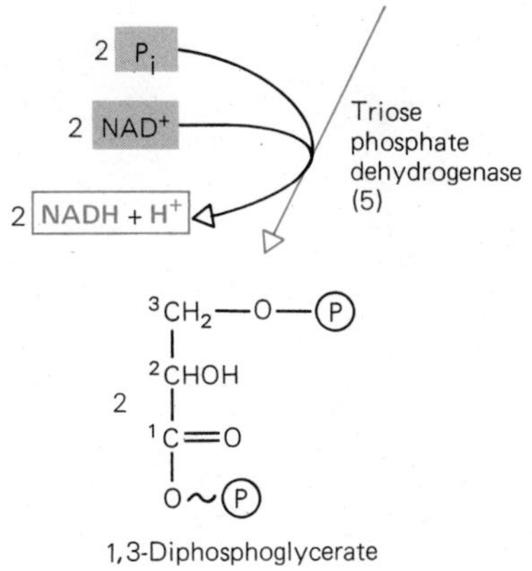

2 P_i

2 NAD^+

2 $NADH + H^+$

Triose phosphate dehydrogenase (5)

$^3CH_2$—O—(P)
|
^2CHOH
|
^1C=O
|
O~(P)

2

1,3-Diphosphoglycerate

6. The high-energy phosphates from two diphosphoglycerate molecules are transferred to two ADP molecules to make two ATP molecules. This transfer of a phosphate group from a high-energy molecule to ADP is a substrate-level phosphorylation (Section 6-D).

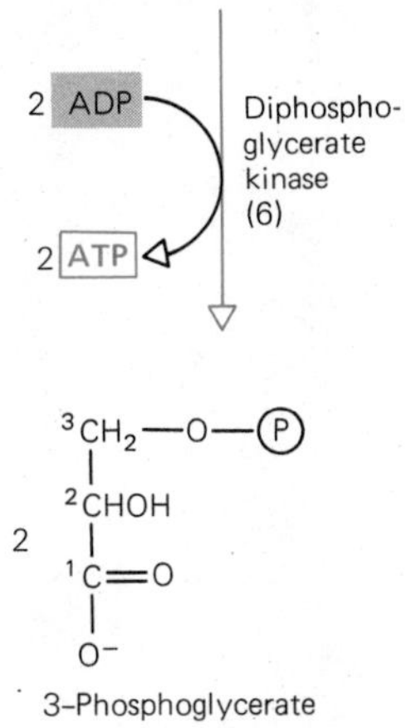

2 ADP

Diphosphoglycerate kinase (6)

2 ATP

$^3CH_2$—O—(P)
|
^2CHOH
|
^1C=O
|
O^-

2

3–Phosphoglycerate

Phosphoglyceromutase (7)

7. The remaining phosphate group of each phosphoglycerate molecule is transferred from the 3- to the 2-carbon position.

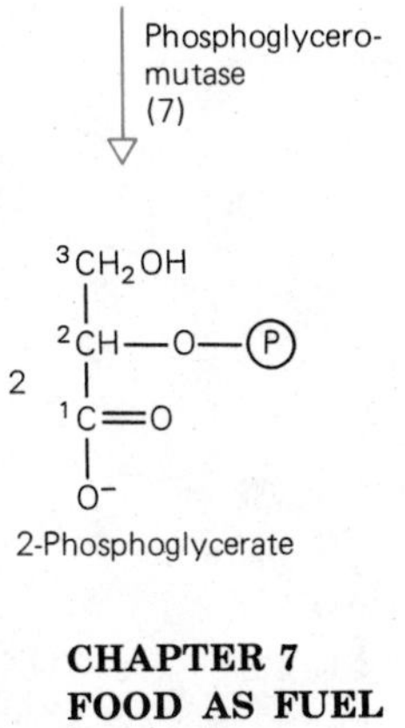

$^3CH_2OH$
|
^2CH—O—(P)
|
^1C=O
|
O^-

2

2-Phosphoglycerate

2-Phosphoglycerate

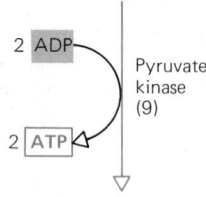

Enolase (8)

$2\,H_2O$

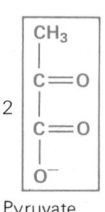

Phosphoenolpyruvate

$2\,\boxed{ADP}$

Pyruvate kinase (9)

$2\,\boxed{ATP}$

Pyruvate

8. A molecule of water is removed from each molecule of 2-phosphoglycerate, leading to formation of phosphoenolpyruvate; this conversion changes the phosphate bond into a high-energy phosphate bond.

9. The remaining phosphate group is used to phosphorylate ADP to ATP (another substrate-level phosphorylation), converting phosphoenolpyruvate into the 3-carbon compound pyruvate.

To summarize, glycolysis involves the conversion of a 6-carbon molecule (glucose) into two 3-carbon molecules (pyruvate):

$$C_6H_{12}O_6 + 2ATP + 4ADP + 2P_i + 2NAD^+ \longrightarrow$$
$$2C_3H_4O_3 + 2ADP + 4ATP + 2NADH + 2H^+$$

For each molecule of glucose broken down, glycolysis results in a net energy production of two molecules of ATP (four ATPs have been formed but two have been used up) and of two NADH + H^+. These NADH + H^+ will later donate their electrons to the electron transport chain to produce ATP during oxidative phosphorylation. The functions of glycolysis in the economy of the cell include the conversion of glucose into pyruvate, which goes into the citric acid cycle, and the production of a small amount of usable energy in the form of ATP.

Into the Mitochondrion

Glycolysis occurs in the cytoplasm. In eukaryotes, pyruvate formed during glycolysis enters a mitochondrion, where it is converted into a form that can enter the citric acid cycle. Because all of respiration except glycolysis occurs inside mitochondria, we shall take time out to examine the structure of these organelles before we proceed.

A mitochondrion is separated from the cytoplasm by its outer membrane; this membrane is permeable to a variety of molecules because it contains proteins that form pores. (In fact this membrane's existence is often disregarded; when biochemists speak of something crossing "the mitochondrial membrane" they mean the inner membrane.) The inner membrane is less permeable. It encloses a compartment containing a protein-rich solution called the **mitochondrial matrix**. Many of the enzymes of the citric acid cycle are dissolved in the matrix; the rest are attached to the inner face of the inner mitochondrial membrane. This membrane also contains the electron transport molecules, which create a reservoir of H^+ by dumping H^+ ions at the outside face of the inner membrane. In addition, the inner membrane contains a protein complex that attaches to ATPase enzymes lying on the inner face of the membrane (Figure 7–5). In prokaryotes, which lack mitochondria, the plasma membrane carries out the functions performed by the inner mitochondrial membrane in eukaryotes.

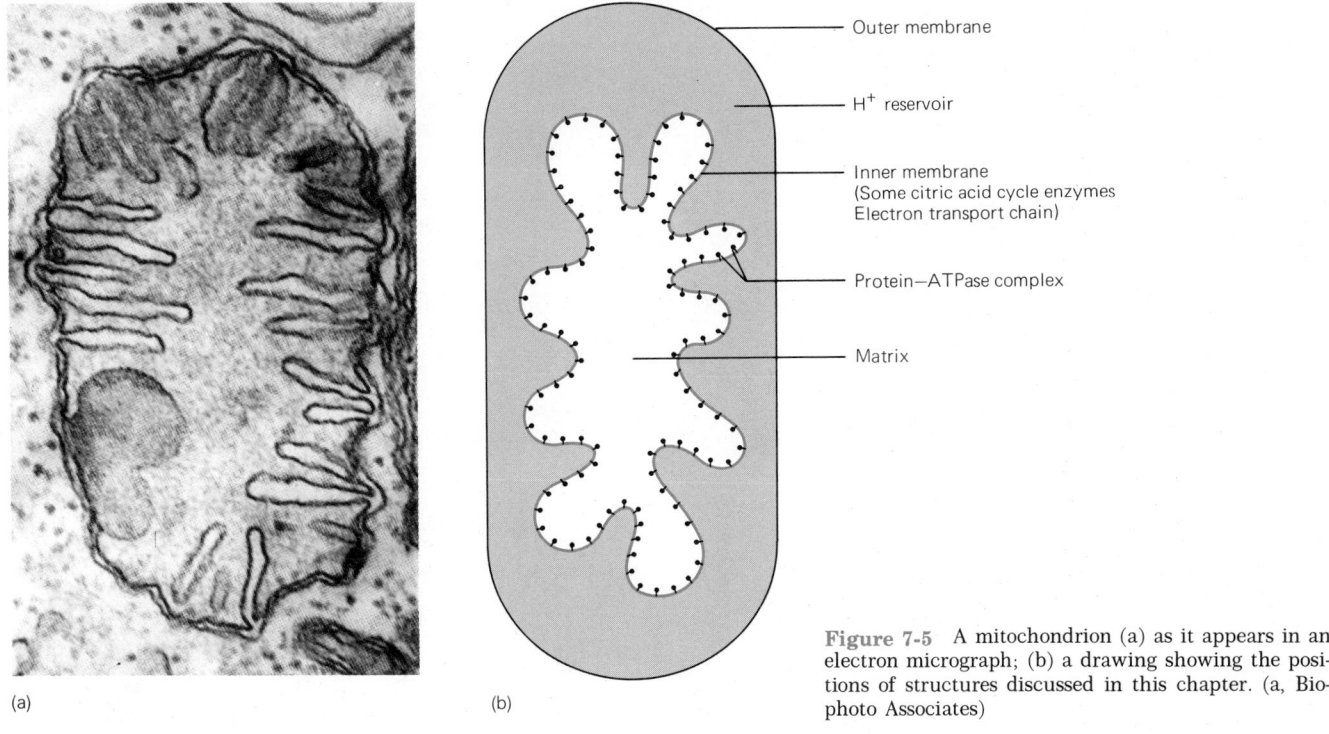

Figure 7-5 A mitochondrion (a) as it appears in an electron micrograph; (b) a drawing showing the positions of structures discussed in this chapter. (a, Biophoto Associates)

Outer membrane

H⁺ reservoir

Inner membrane
(Some citric acid cycle enzymes
Electron transport chain)

Protein—ATPase complex

Matrix

(a)

(b)

Now let us return to pyruvate, which we left in the mitochondrial matrix. Here a large complex of enzymes carries out a series of reactions summarized in Figure 7–6: the pyruvate loses one carbon in the form of carbon dioxide and ends up as a 2-carbon acetyl group attached to a coenzyme A molecule, forming acetyl CoA. Meanwhile, two hydrogens, one from CoA and one associated with pyruvate, reduce a molecule of NAD^+ to $NADH + H^+$. The acetyl group, once attached to CoA, is now ready to enter the citric acid cycle. In Figure 7–6 pyruvate, which is an ion, is shown attracting a hydrogen ion. If the hydrogen were bound by the oxygen atom shown, the molecule would be pyruvic acid. Many of the intermediates of glycolysis and the citric acid cycle exist, like pyruvate, as anions (negatively charged ions) at the pH found in cells, but they sometimes associate with H^+ to form acids (such as pyruvic acid) and may be represented as either anions or acids. In this chapter, the intermediates in glycolysis are shown as anions and those of the citric acid cycle as acids so that you can see examples of each form.

Citric Acid Cycle

The **citric acid** or **tricarboxylic acid cycle** is also known as the **Krebs cycle** after Sir Hans Krebs, who first described the cycle in 1937 and received a Nobel Prize for this work in 1953.

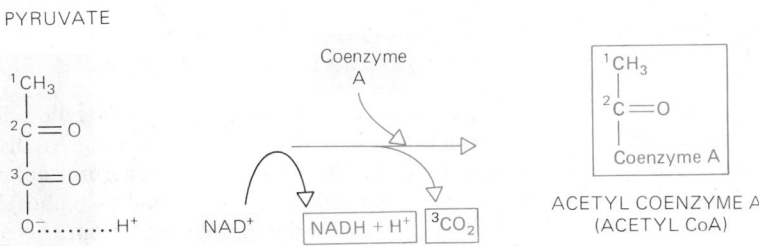

PYRUVATE

ACETYL COENZYME A
(ACETYL CoA)

Figure 7-6 Pyruvate is oxidized to an acetyl group which combines with coenzyme A to form acetyl CoA. The oxidizing agent is NAD^+, which is reduced to $NADH + H^+$ in the reaction. One of pyruvate's carbons is lost as carbon dioxide. Acetyl CoA enters the citric acid cycle.

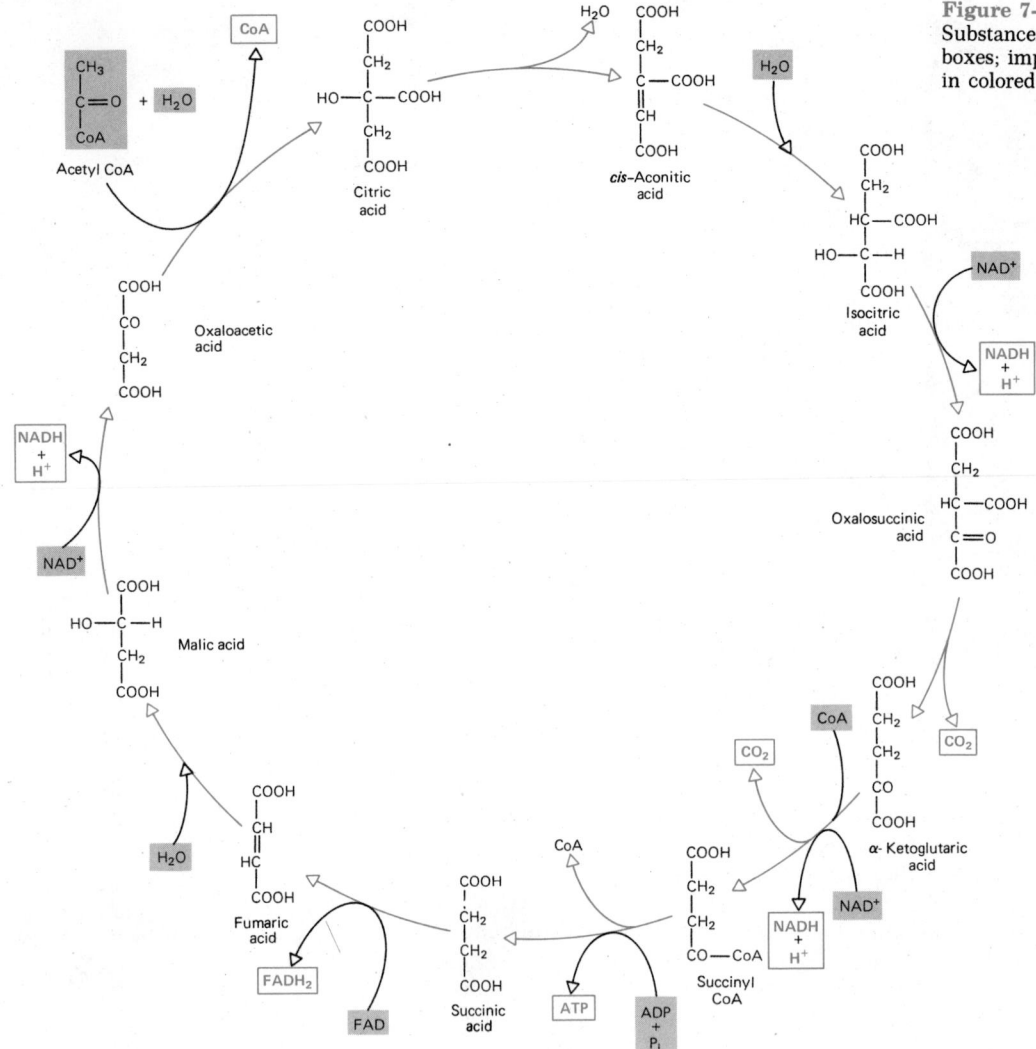

Figure 7-7 The tricarboxylic acid cycle. Substances entering the cycle are in shaded boxes; important products of the cycle are in colored outlines.

The reactions of the citric acid cycle are shown in Figure 7–7. Coenzyme A transfers its 2-carbon acetyl group to the 4-carbon compound oxaloacetic acid, forming a 6-carbon compound, citric acid. In the transfer, CoA is released and goes back to pick up another acetyl group. A series of enzymes next removes two of the six carbons, in the form of carbon dioxide, from the citric acid molecule; the remaining 4-carbon compound is converted into a new molecule of oxaloacetic acid, ready to accept another 2-carbon acetyl group from acetyl CoA.

The most important features of the citric acid cycle are these:

1. The carbon dioxide that we breathe out is a waste product. For each molecule of pyruvate oxidized, all three carbon atoms are given off as molecules of carbon dioxide; one carbon dioxide is produced in the conversion of pyruvate to acetyl CoA, and the other two in the citric acid cycle.

2. Hydrogen atoms are removed at various stages of the tricarboxylic acid cycle and passed to the electron acceptors NAD⁺ and FAD, forming NADH + H⁺ and FADH₂. Energetically speaking, this is the most important outcome of the cycle, because the reduced coenzymes are used in oxidative phosphorylation to produce the greatest part of the energy derived from the original glucose molecule.

3. One molecule of ATP is formed by substrate-level phosphorylation during each turn of the tricarboxylic acid cycle.

4. The citric acid cycle may be summarized:

$$\text{oxaloacetic acid} + \text{acetyl CoA} + \text{ADP} + P_i + 3NAD^+ + FAD \longrightarrow$$
$$\text{oxaloacetic acid} + \text{CoA} + 2CO_2 + ATP + 3NADH + 3H^+ + FADH_2$$

5. Because each glucose molecule gives rise to two pyruvates, it takes two trips around the citric acid cycle to break down one molecule of glucose.

Oxidative Phosphorylation

One molecule of glucose has now been completely oxidized. Some of the energy released has been used to make ATP by substrate-level phosphorylation, but most of it remains in the electrons carried by NADH + H$^+$ and FADH$_2$ and is ultimately used to synthesize ATP by oxidative phosphorylation.

Oxidative phosphorylation begins as NADH + H$^+$ and FADH$_2$ pass their electrons to the electron transport chain, or cytochrome chain, the main energy-releasing system of a cell. The system consists of a series of electron carrier molecules in precise arrangements in the inner membrane of the mitochondrion. These molecules pass electrons in a series of redox reactions,* ending with oxygen, the final electron acceptor. Some of the electron carriers are various **cytochromes**, protein molecules containing complex ring structures with iron in the center (see Figure 32–15). Some electron carriers, such as NADH + H$^+$, carry electrons in the form of hydrogen atoms; the cytochromes, however, carry electrons alone.

The details of electron transport have not been fully worked out, but we shall outline a proposed sequence of events. Each pair of hydrogen atoms passed to the electron transport chain by NADH + H$^+$ from the mitochondrial matrix is carried across the inner membrane. Here the next molecule in the chain is an iron–sulfur protein which, like a cytochrome, will accept only electrons and so releases the H$^+$ it has received to the outside of the inner mitochondrial membrane. The electron pair crosses back to the inner edge of the membrane to another carrier. This is a hydrogen carrier; it picks up two hydrogen ions (H$^+$), formed by the dissociation of water molecules in the mitochondrial matrix, and the whole process is repeated. After a third, similar round trip across the membrane, the electrons, now on the inner surface, react with hydrogen ions and oxygen to form water:

$$\tfrac{1}{2}O_2 + 2e^- + 2H^+ \longrightarrow H_2O$$

The electron pair has made three round trips across the membrane, accompanying three pairs of hydrogen ions that have made only one-way trips from the inner to the outer sides of the inner mitochondrial membrane. The hydrogen ion reservoir outside the membrane has gained six hydrogen ions from these maneuvers.

(A slight wrinkle in this story comes from the fact that when NADH + H$^+$ comes to the electron transport chain from glycolysis instead of from the citric acid cycle, and when FADH$_2$ comes from the citric acid cycle, they pass their electrons to the carrier at the beginning of the *second* round trip across the membrane, not to the molecule at the beginning of the first trip. Hence these electron pairs make fewer trips across the membrane, with fewer H$^+$ transported.)

The combination of electrons from the transport chain with oxygen and hydrogen ions completes the oxidation part of oxidative phosphorylation. The phosphorylation part produces ATP from ADP and depends on the use of energy from the electrochemical gradient that has been established by transporting H$^+$ out of the mitochondrial matrix.

The inner mitochondrial membrane contains protein complexes that attach to

---

*Redox reactions Reactions in which one reactant is reduced and another oxidized (see Section 6-C).

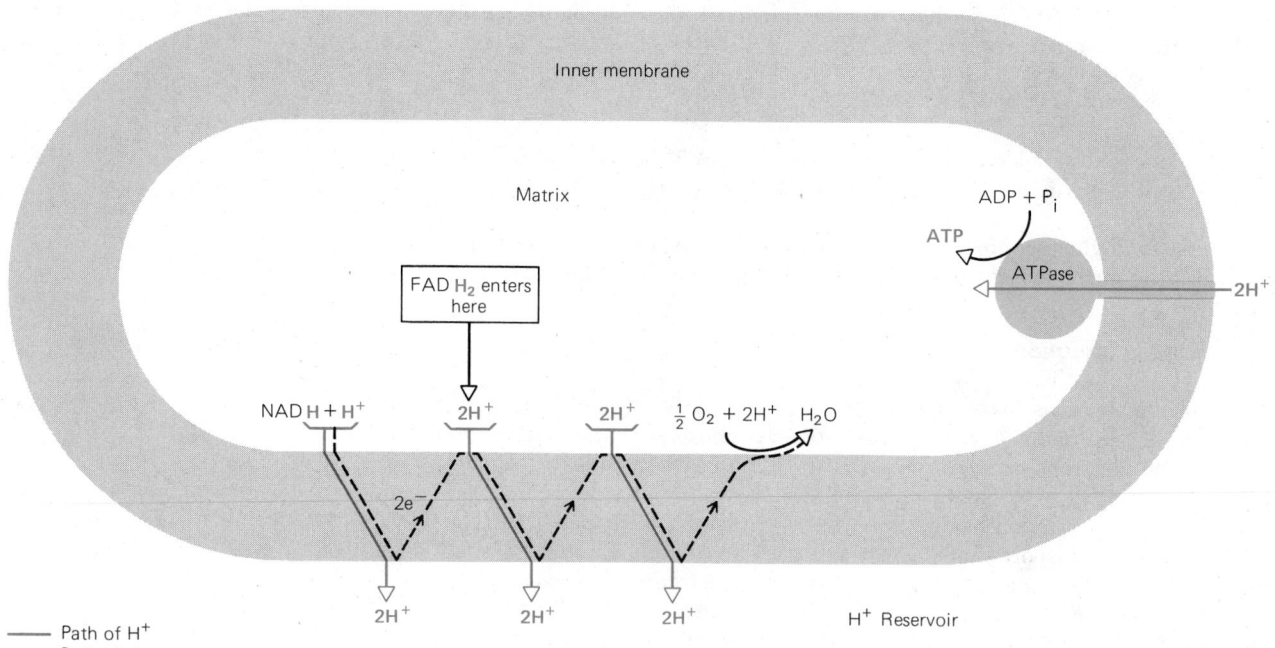

Figure 7-8 Simplified scheme of oxidative phosphorylation in mitochondria. At left, NADH + H$^+$ passes two hydrogen atoms (2H$^+$ + 2e$^-$) to the electron transport chain in the inner membrane. The electrons make three round trips, with 2H$^+$ being dropped off into the H$^+$ reservoir each time the hydrogen atoms reach the outer face of the membrane. On trips two and three, H$^+$ derived from dissociation of water replaces the original H$^+$ from NADH + H$^+$. 2H$^+$ from the reservoir are eventually used to drive phosphorylation of each ADP to ATP by ATPase enzymes associated with the membrane.

ATPase enzymes lying at the inner surface of the membrane (Figure 7–8). These enzymes use the power of the electrochemical gradient across the membrane to make ATP, eliminating 2 H$^+$ from the gradient for each ATP synthesized. (The exact mechanism of this reaction is not yet known.)

A total of about 64 H$^+$ is moved to the outside of the inner membrane for each original glucose molecule. From a ratio of 2 H$^+$ per ATP synthesized, we can calculate that a molecule of glucose will yield 32 ATP by oxidative phosphorylation, plus 4 ATP by substrate-level phosphorylation, for a total of 36 ATP per glucose molecule. This is a maximum estimate; the actual number of ATP molecules formed by oxidative phosphorylation may (like your estimated mpg) be lower for several reasons:

1. The 2 H$^+$ per ATP figure supposes a membrane perfectly impermeable to H$^+$, which is impossible. A slow rate of H$^+$ leakage back across the membrane without ATP synthesis is inevitable.

2. Some of the H$^+$ reservoir's energy is used, not to make ATP, but for other functions such as the active transport of sodium and calcium across the inner mitochondrial membrane.

3. It takes one H$^+$ to transport each ATP out of the mitochondrion, for a total of 3 H$^+$ per ATP formed and sent to the cytoplasm. Hence, if all 64 H$^+$ were used to make ATP that was sent to the cytoplasm, each glucose molecule would yield only about 21 ATP by oxidative phosphorylation, plus the 4 ATP from substrate-level phosphorylation, for a total of only 25 ATP.

Table 7–1 shows a balance sheet for ATP synthesized using energy from one molecule of glucose.

TABLE 7-1 ENERGY YIELD FROM THE BREAKDOWN OF ONE MOLECULE OF GLUCOSE

PROCESS	NUMBER (NET) OF ATP FROM SUBSTRATE-LEVEL PHOSPHORYLATION	NUMBER OF REDUCED HYDROGEN CARRIERS	NUMBER OF H⁺ DELIVERED TO H⁺ GRADIENT
Under Anaerobic Conditions			
Alcoholic fermentation	2 ATP		
Lactic acid fermentation	2 ATP		
Under Aerobic Conditions			
Glycolysis	2 ATP	2 NADH + H⁺	8 H⁺
Pyruvate to acetyl CoA		2 NADH + H⁺	12 H⁺
Citric acid cycle	2 ATP	6 NADH + H⁺	36 H⁺
		2 FADH₂	8 H⁺
			Total: 64 H⁺†

† These 64 H⁺ can be used to make 64 ÷ 2 = 32 ATP in the mitochondrion or 64 ÷ 3 = 21 ATP made and transported to the cytoplasm. Add 4 ATP from substrate-level phosphorylation under aerobic conditions to obtain the total number of ATP per molecule of glucose.

7-C Anaerobic Fermentation

Many cells can obtain energy in the absence of oxygen by the anaerobic process of **fermentation**. The most common type of fermentation uses the pathway of glycolysis, making pyruvate from glucose.

We have seen that oxygen serves as the final electron acceptor in the electron transport chain. If there is not enough oxygen in a cell to pick up electrons as fast as they are passed down the electron transport chain, NADH will be unable to pass its electrons into the chain. Since NAD⁺ is present in very small quantities, all the molecules of NAD⁺ will soon be converted to NADH + H⁺. Without NAD⁺ to accept electrons, both glycolysis and the citric acid cycle will slow down and stop. During anaerobic fermentation, a cell uses pyruvate as an alternative electron acceptor to which NADH can donate its electrons. The NAD⁺ freed in this way continues to act as an electron acceptor that will permit glycolysis, at least, to proceed, with the net production of two molecules of ATP for each molecule of glucose broken down. Much less ATP is produced by anaerobic fermentation than by aerobic respiration of a glucose molecule (Table 7–1), but this system provides a source of ATP when insufficient oxygen is available.

The citric acid cycle does not require oxygen directly. However, it is usually considered an aerobic pathway because it requires NAD⁺ as a coenzyme and, unlike glycolysis, it has no alternative mechanism to free NAD⁺ that has been reduced to NADH + H⁺. As a result, when the electron transport chain stops from lack of oxygen, the citric acid cycle stops as well.

Some types of cell, notably those of the brain, die rapidly if deprived of oxygen because they cannot survive for long if oxidative phosphorylation slows down significantly. Other cells are more tolerant of oxygen deprivation and can make do with the ATP produced by anerobic fermentation. Many invertebrates* and microorganisms (e.g., bacteria, fungi) can survive indefinitely in the absence of oxygen.

The first example of anaerobic fermentation to be worked out was the alcoholic fermentation of grape sugars by yeasts, unicellular fungi used to convert grape juice to wine. In alcoholic fermentation, pyruvate formed during glycolysis is converted to acetaldehyde and then to ethanol by NADH + H⁺ formed during glycolysis (Figure 7–9). The regenerated NAD⁺ can then be used again as a hydrogen acceptor in glycolysis. Thus when oxygen is not available, a yeast obtains ATP from the sub-

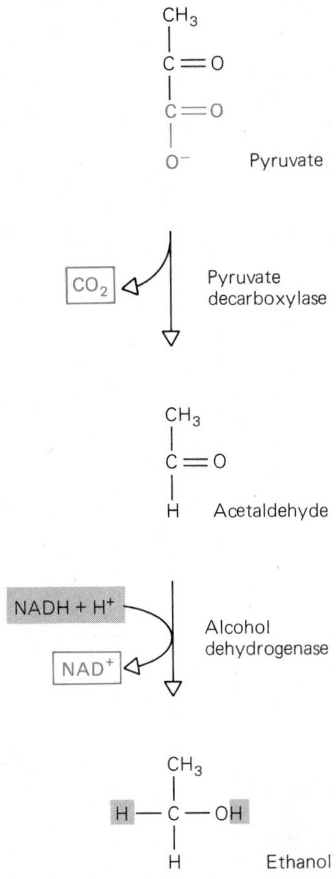

CH₃
|
C=O
|
C=O
|
O⁻   Pyruvate

CO₂   Pyruvate decarboxylase

CH₃
|
C=O
|
H   Acetaldehyde

NADH + H⁺   Alcohol dehydrogenase
NAD⁺

CH₃
|
H—C—OH
|
H   Ethanol

Figure 7-9 Alcoholic fermentation, as found in yeast. In this reaction, pyruvate produced in glycolysis acts as a hydrogen acceptor, releasing NAD⁺, which is needed for glycolysis to continue and produce ATP at a low rate while the cell is deprived of oxygen.

*Invertebrate   An animal that lacks a backbone (e.g., earthworm, snail).

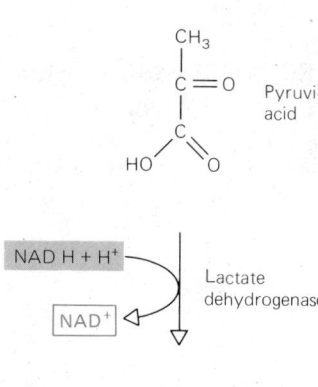

Figure 7-10 The conversion of pyruvic acid to lactic acid in muscle. When the cell lacks oxygen, NADH accumulates; pyruvic acid produced by glycolysis acts as a hydrogen acceptor and becomes lactic acid, thus releasing NAD$^+$ to continue glycolysis.

strate-level phosphorylation reactions of glycolysis. Variations on this sort of fermentation pathway occur in some yeasts, in other microorganisms, and in many invertebrates.

Anaerobic fermentation in higher animals is most common in muscle tissue during strenuous exercise. The muscle carries on aerobic respiration as fast as possible, but the oxygen supply is inadequate and NADH builds up. Such a buildup triggers anaerobic fermentation to supply a small additional amount of ATP. Pyruvate acts as an electron acceptor and is converted into lactic acid (Figure 7–10).

When a muscle operates anaerobically, it incurs an **oxygen debt**. This is because the lactic acid that accumulates during anaerobic fermentation must ultimately be oxidized back to pyruvate, and the electrons so recovered must be passed to the electron transport chain (Figure 7–11). We still breathe hard after we have stopped using our muscles strenuously because we are repaying the oxygen debt by taking in enough extra oxygen to combine with these electrons.

Most (although possibly not all) cells can carry on glycolysis, but not all organisms are capable of aerobic respiration. For this reason, and because the atmosphere of the earth probably contained little oxygen when life first evolved, it is generally believed that glycolysis is the more primitive pathway of food breakdown and that aerobic pathways evolved at a later date.

7-D   Alternative Food Molecules

The pathways we have been looking at are those by which carbohydrate molecules are broken down and their stored energy transferred into the high-energy phosphate bonds of ATP molecules. Polysaccharides eaten or stored in the body can be broken down to form glucose; monosaccharides other than glucose can be converted to glucose or fructose and so eventually end up as pyruvate. Thus the claim that other "natural" sugars are less fattening than sucrose or glucose is untrue; a cell treats them all alike.

Two other important classes of nutrients, fats and proteins, are metabolized in various ways, but they too ultimately reach either glycolysis or the citric acid cycle (Figure 7–12). Fats are broken down to glycerol and fatty acids. Glycerol is converted to phosphoglyceraldehyde and fed into the glycolytic pathway at that point. Fatty acids are broken down into 2-carbon fractions, which are acetyl groups; these combine with coenzyme A to enter the citric acid cycle as acetyl CoA.

Proteins in food are hydrolyzed into amino acids. An amino acid that is not needed to synthesize more proteins is **deaminated**; that is, its amino group is removed, and the rest of the molecule is converted into pyruvate, acetyl CoA, or one of the intermediates of the citric acid cycle, depending on the structure of the original amino acid. It then enters the pathway at the appropriate place.

If a cell receives more food than it needs for building macromolecules and releasing energy, the excess is converted to fat. As we have just seen, both carbohydrates and proteins can be channeled into pathways where they are converted to acetyl CoA; acetyl CoA, in turn, is the basic building block for synthesis of fatty acids (follow these pathways in Figure 7–12). The fatty acids are then incorporated into triglycerides (Figure 3–11) and stored as body fat. This is why eating more food than we need—whether carbohydrate, protein, or fat—results in gaining weight.

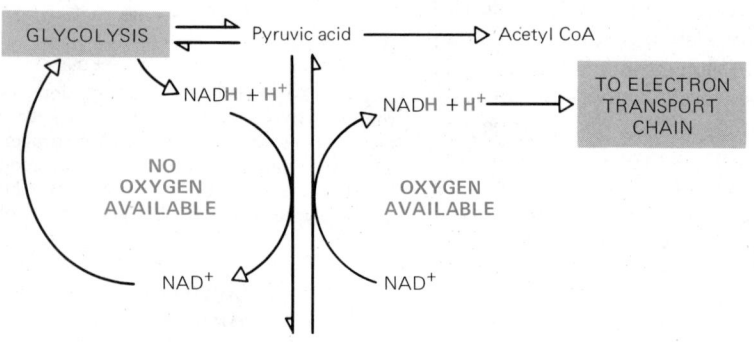

Figure 7-11 Relationship between aerobic respiration (on the right) and anaerobic fermentation (on the left) in muscle. When no oxygen is available, cells convert pyruvic acid to lactic acid. When oxygen again becomes available, the reaction is reversed.

Essay: Pasteur and Yeasts

The word enzyme means "in yeast." Most of the early studies on enzymes and their actions were attempts to understand the alcoholic fermentation by which wine is made. As early as 1785 the Academy of Florence offered a prize for a theory of fermentation that could be applied to keeping wine in better condition while it was transported. However, no real light was shed on the subject until the French wine industry asked Louis Pasteur to investigate the condition called "l'amer" that destroyed large quantities of the best Burgundy every year.

From his experiments, Pasteur concluded that fermentation occurred only when living yeast was present. Liebig, an influential chemist, thought otherwise and performed many experiments in which he killed yeast cells by boiling them and then tested them to see if they would ferment sugar. They would not, and enzymes, also called "ferments," came to be considered as catalysts that would not function outside a living cell. We now know that enzymes can function outside cells and that Liebig, in boiling the yeast cells, had also destroyed the tertiary structure*

of their enzymes so that they could no longer function. Pasteur discovered that *l'amer*, which turned wine sour, was caused by bacteria. Microscopic examination showed that the wine turned sour when it contained more bacteria than yeast cells. *L'amer* could be prevented by greater cleanliness, including sulfur sterilization, which is now standard practice in many winemaking steps. Pasteur also showed why it is important to exclude air during fermentation: wine yeasts produce alcohol under anaerobic conditions, but if oxygen is present, other yeasts and bacteria that convert alcohol into acetic acid ($CH_3CH_2OH \rightarrow CH_3COOH$) will outcompete the wine yeasts, and will turn the wine to vinegar.

Pasteur loved good wines and devoted many years to studies of their fermentation and aging. His book *Etudes sur le Vin*, published in 1866, revolutionized winemaking, giving it a scientific basis for the first time.

---

* Tertiary structure The folding pattern of a polypeptide chain.

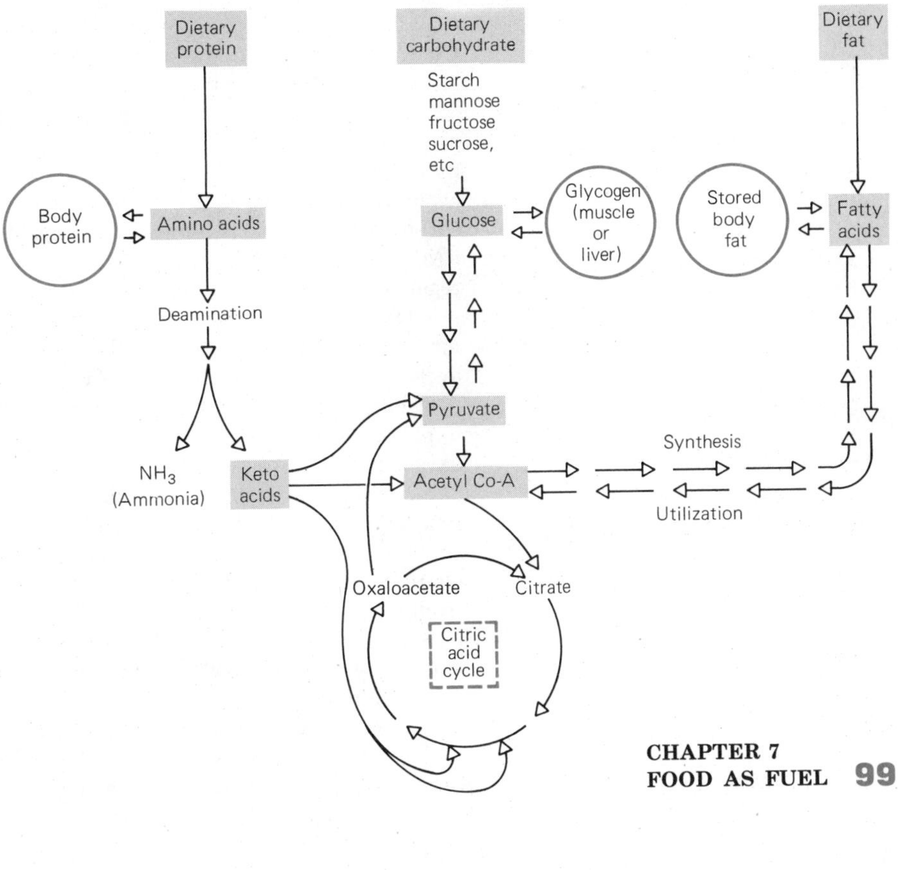

Figure 7-A Louis Pasteur (1822–1895) studied many aspects of chemistry and microbiology during an extremely productive scientific career. (Smithsonian Institution)

Figure 7-12 Proteins, fats, and carbohydrates in the diet may become part of the body (colored circles) or may be processed to release energy. The most usual way for a cell to obtain energy is through the process of glycolysis (from glucose to pyruvate, center), which breaks down carbohydrates to release energy. Fatty acids and some keto acids from the deamination of amino acids enter cellular respiration at the level of acetyl CoA; other keto acids enter as pyruvate or as intermediates in the citric acid cycle. Both proteins and carbohydrates can also follow pathways that result in accumulation of body fat.

Dietary protein

Body protein

Amino acids

Deamination

NH_3 (Ammonia)

Keto acids

Dietary carbohydrate

Starch mannose fructose sucrose, etc

Glucose

Glycogen (muscle or liver)

Pyruvate

Acetyl Co-A

Oxaloacetate

Citrate

Citric acid cycle

Synthesis

Utilization

Stored body fat

Dietary fat

Fatty acids

TABLE 7-2 ENERGY YIELD OF MAJOR FOOD COMPONENTS

CLASS	COMPOSITION	ENERGY YIELD	STORAGE
Carbohydrates	CH_2O	4 kcal/gram	Hydrated: attracts much water
Fats	CHO	9 kcal/gram	Hydrophobic: concentrated fat droplets
Proteins	CHON(S)	~4 kcal/gram	None

Table 7-2 shows the energy yield per unit of weight for carbohydrates, fats, and proteins. Fats give the highest energy yield per unit weight because fats contain a higher hydrogen : oxygen ratio than do carbohydrates. Since most of the energy from food molecules is obtained by passing hydrogens along the electron transport chain, the higher the proportion of hydrogen, by weight, in the molecule, the more energy is stored in a given weight of that substance.

The hydroxyl groups in carbohydrates make carbohydrate molecules polar; thus they attract and hold water around them. Since this water adds bulk and weight but not energy, stored carbohydrates yield little energy per unit of weight. Fats, on the other hand, are nonpolar; they repel water and arrange themselves into concentrated fat droplets. Thus fats can be stored much more compactly than carbohydrates. Animals usually store most of their excess food in the form of fats, while plants tend to store the polysaccharide starch.

SUMMARY

Cellular respiration is the process by which cells extract free energy from the energy stored in the chemical bonds of food molecules (usually glucose) and use this energy to regenerate their supply of ATP.

During glycolysis, glucose is broken down anaerobically to two molecules of pyruvate, and NAD^+ is reduced to $NADH + H^+$. Glycolysis also yields ATP by substrate-level phosphorylation.

Each pyruvate formed during glycolysis loses a carbon dioxide to become an acetyl group that combines with coenzyme A. Acetyl CoA enters the citric acid cycle by combining with the 4-carbon compound oxaloacetate. During one turn of the cycle, the equivalents of the two carbon atoms of the acetyl group are oxidized to carbon dioxide. ATP is produced by substrate-level phosphorylation, and NAD^+ and FAD are reduced to $NADH + H^+$ and $FADH_2$. Oxaloacetate and coenzyme A are regenerated and can go through the cycle again.

Oxidative phosphorylation produces most of the ATP derived from aerobic respiration of glucose. $NADH + H^+$ and $FADH_2$ from glycolysis and the citric acid cycle pass pairs of hydrogen atoms to the electron transport chain. The electron transport chain forms an electrochemical gradient of hydrogen ions that can be used to phosphorylate ADP to ATP. At the end of the chain, the electrons combine with oxygen to form water.

Neither glycolysis nor the citric acid cycle requires oxygen directly. However, in a cell that is short of the final electron acceptor, oxygen, most of the NAD^+ in the cell will be tied up as NADH, unable to release its electrons to the electron transport chain. Some cells have an adaptation that permits them to continue to produce ATP from glycolysis under such anaerobic conditions; pyruvate accepts electrons from $NADH + H^+$, releasing NAD^+ so that glycolysis can continue. No such mechanism exists for the tricarboxylic acid cycle, which therefore cannot function under anaerobic conditions.

Many other metabolic pathways feed into glycolysis, the citric acid cycle, and the electron transport chain, enabling cells to use many organic compounds other than glucose as food sources to generate usable energy in the form of ATP.

OBJECTIVES

From your study of this chapter, you should be able to:

1. Recognize examples of coenzymes; explain their functions in living organisms.

2. Explain why most organisms need oxygen and state how carbon dioxide and water are produced as waste products of cellular respiration.

3. Name the starting materials and the important end products of:
   a. glycolysis
   b. the citric acid cycle
   c. oxidative phosphorylation

4. Explain the functions of the three processes listed in Objective 3 in the scheme of cellular respiration.

5. State where in the cell glycolysis; the citric acid cycle, and oxidative phosphorylation occur in eukaryotes and in prokaryotes.

6. Explain the importance of the electron transport system to the cell.

7. Describe how mitochondria carry out ATP synthesis.

8. Compare and contrast the process of alcoholic fermentation by a wine yeast with the process of lactic acid fermentation by a muscle.

9. Explain how the ability to undergo anaerobic fermentation, and to acquire an oxygen debt, is a useful adaptation for a muscle.

10. Compare aerobic respiration and anaerobic fermentation with respect to the amount of ATP regenerated by each process.

11. Explain why we get fat when we eat more food than we need.

SELF-QUIZ

1. NAD^+ functions in cell respiration as a(n):
   a. energy intermediate
   b. enzyme
   c. coenzyme
   d. oxidizable substrate
   e. hydrogen donor

2. Which of the following statements about oxidative phosphorylation is *not* true?
   a. More of the ATP in a normal cell is formed by oxidative phosphorylation via the electron transport chain than by substrate-level phosphorylation.
   b. In eukaryotes, the formation of ATP by oxidative phosphorylation requires that the inner mitochondrial membrane remain intact.
   c. NAD^+ is a carrier molecule that travels down the electron transport chain to release ATP during oxidative phosphorylation.
   d. In eukaryotes, the electron transport chain and the enzymes of the citric acid cycle are located in mitochondria whereas the enzymes of glycolysis are located in the cytoplasm.

   e. The role of oxygen is to act as an acceptor for electrons.

3. Give the end products of the following reaction sequences:
   a. the citric acid cycle
   b. yeast fermentation
   c. electron transport chain

4. While a muscle is in the process of reducing an oxygen debt:
   a. lactate is converted into pyruvate
   b. all the NAD^+ is in the reduced form
   c. pyruvate is converted into lactate
   d. NADH acts as an oxygen acceptor

5. True or False? Both yeast cells and muscle make 2 ATP per glucose molecule fermented anaerobically.

6. True or False? Carbohydrate is unnecessary in the human diet since the products of fat and protein breakdown can enter the citric acid cycle to generate energy.

QUESTIONS FOR DISCUSSION

1. Vitamins are substances that organisms need in small amounts in the diet because they cannot synthesize them. You have encountered several vitamins in this chapter (e.g., niacin and pantothenic acid). What do they have in common? Why are they vital, and why are they required only in very small amounts?

2. Cyanide inactivates cytochromes. Why is it poisonous?

3. How many hydrogen atoms are there in a glucose molecule? How many hydrogen atoms are transported to the electron transport chain by coenzymes during respiration of one molecule of glucose? Where do the additional hydrogen atoms come from?

4. In the bacterium *Escherichia coli*, only $4 H^+$ leave the cell for each pair of hydrogen atoms passed to the electron transport chain. Two H^+ are required to phosphorylate each ADP to ATP. Calculate the total number of ATP's per glucose molecule made by these cells. (Remember, there are no mitochondria!)

5. Can you think of any possible reasons for the fact that the brain is so susceptible to oxygen shortage and cannot acquire an oxygen debt?

6. Why do foods that are rich in fat tend to be expensive (in dollars and cents) compared with carbohydrates?

REFERENCES AND FURTHER READING

Dyson, R. D. *Cell Biology: A Molecular Approach*, 2d ed. Boston: Allyn and Bacon, 1978. A cell biology book with an excellent molecular treatment of the function of mitochondria.

Krebs, H. A. "The history of the tricarboxylic acid cycle." *Perspectives in Biology and Medicine* 14:154, 1970. An engaging account of how the citric acid cycle was worked out by Sir Hans Krebs, who won a Nobel Prize for this work.

Lehninger, A. L. *Biochemistry*, 2d ed. New York: Worth, 1975. The serious student will find several chapters devoted to the biochemical events of respiration, including variation in different tissues and organisms, and the structure and function of mitochondria.

Stryer, L. *Biochemistry*, 2d ed. San Francisco: W. H. Freeman, 1980. A standard biochemistry textbook; excellent description of cellular respiration.

CHAPTER

PHOTOSYNTHESIS

Photosynthesis is the process whereby plants capture and store the energy of sunlight in the form of chemical bonds in carbohydrate molecules. **Green plants** perform photosynthesis using the green pigment **chlorophyll** to trap light energy, and using water as a source of hydrogen to reduce carbon dioxide as it is incorporated into organic molecules:

$$CO_2 + H_2O \xrightarrow{\substack{\text{light,} \\ \text{chlorophyll}}} CH_2O + O_2$$

The green plants are the prokaryotic blue-green bacteria, the eukaryotic unicellular and multicellular algae,* and the more familiar land plants. Photosynthetic bacteria also exist, but they use different light-trapping pigments and hydrogen sources (see Chapter 22).

---

*Algae   Eukaryotic photosynthetic organisms that do not form an embryo protected in the (female) parent's body; most algae are aquatic.

Hydrogen is an important medium of energy exchange in living cells. Adding hydrogen (reduction) increases most molecules' free energy (O_2 is a notable exception); the subsequent removal of the hydrogen (oxidation) releases energy as the molecule's free energy declines. In our study of respiration, we saw a series of such redox reactions occurring as pairs of hydrogen atoms passed down the electron transport chain. Their energy was bled off gradually before the hydrogens finally combined with oxygen to form water, a low-energy molecule. The diverted energy powered the synthesis of three molecules of ATP per hydrogen pair passed down the chain. The opposite reaction, splitting water into hydrogen and oxygen, thus requires an appreciable energy input. This, ultimately, is the reaction that light energy drives in photosynthesis. The hydrogen so released then supplies the energy to make ATP and to reduce carbon dioxide in the formation of carbohydrates.

During photosynthesis, green plants trap a fleeting source of energy, light, and transform it in several stages until, finally, it is stored as potential chemical energy in the bonds of organic molecules, which can be stored indefinitely. (The organic molecules of the coal and oil we burn for energy were made during photosynthesis hundreds of millions of years ago.)

All the oxygen in the air today—about 20% of our atmosphere—comes from photosynthesis. In addition, all our food comes, directly or indirectly, from photosynthesis. Our dependence on plants for food has stimulated research on how photosynthesis works and how we can manipulate plants and their environments to produce more food. Although we know much more about photosynthesis than most beginning students feel they want to learn, plants still hold many secrets that our most sophisticated equipment has yet to unravel, much less duplicate.

In this chapter, we shall first take a look at the structure of chloroplasts, the organelles that carry out photosynthesis in eukaryotic green plants. Then we shall examine the structure and properties of chlorophyll and other photosynthetic pigments, whose role is to absorb light energy and transform it into a form that the chloroplast's chemical machinery can use to drive the reactions of photosynthesis. We shall then study the mechanism of these photosynthetic reactions in some detail. Finally, we shall study factors that affect the rate of photosynthesis and see adaptations that enable plants in different habitats to flourish even though they must carry out photosynthesis under less than ideal conditions.

8-A   Structure of Chloroplasts

In our study of the chemical reactions of photosynthesis, it will be important to understand the locations of various molecules in the chloroplast, especially with respect to the complex system of internal membranes (thylakoids). In this section we shall introduce the basic chloroplast structure, and in later sections we shall see how the photosynthetic reactions relate to this structure.

A chloroplast is separated from the cytoplasm by its outer envelope, made up of two membranes. Inside the chloroplast is a series of **photosynthetic membranes** arranged in flattened sacs called **thylakoids** or **lamellae** (Figure 8–1[b]). All of the photosynthetic membranes are thought to be continuous, and to surround and enclose a continuous interior space; the hydrogen ion reservoir used in chemiosmotic synthesis of ATP accumulates inside this interior space (Figure 8–1[c]). Embedded in the thylakoid membranes are three important kinds of molecules: chlorophyll and other light-trapping photosynthetic pigments; the electron transport chain; and the ATP-synthesizing complexes. The arrangement of the photosynthetic thylakoid membranes provides a great deal of membrane surface area relative to the enclosed volume. This is important because the small volume of the H^+ reservoir can be built up to a high H^+ concentration rapidly by the activities of the many membrane-bound transport molecules surrounding it. In addition, the greater the surface area spread out to the light, the more light the membrane's pigment molecules can intercept.

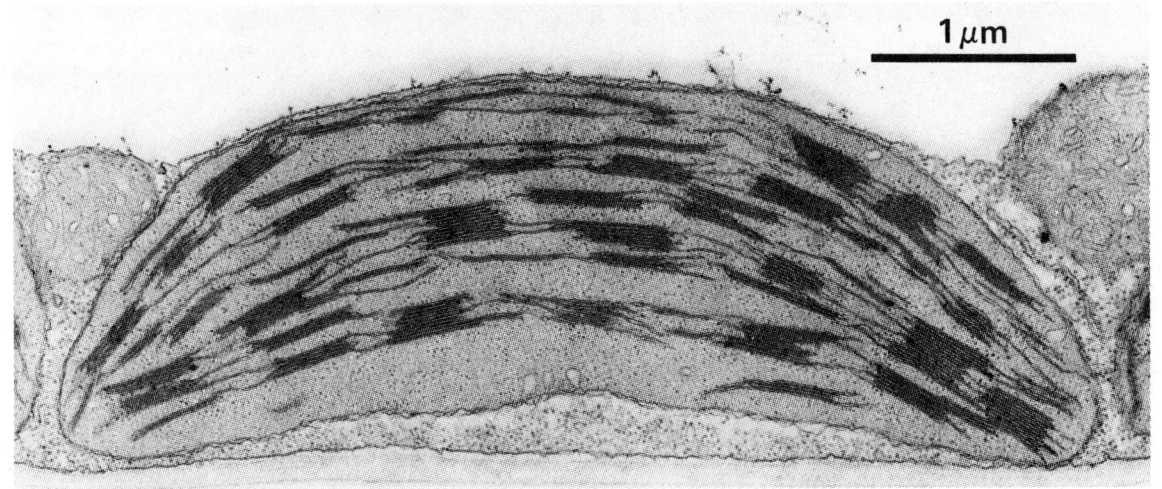

(a)

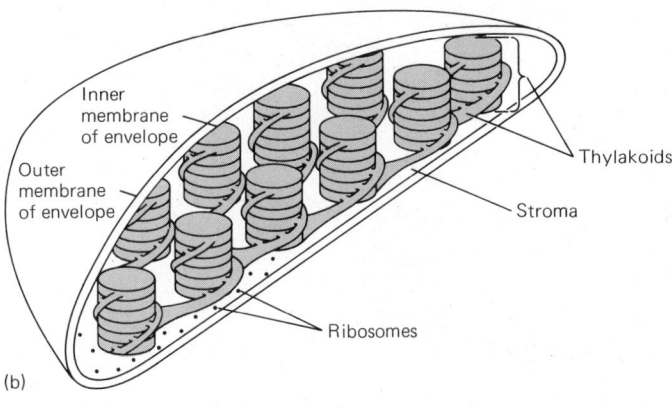

Inner membrane of envelope

Outer membrane of envelope

Thylakoids

Stroma

Ribosomes

(b)

Figure 8-1 (a) Electron micrograph of a chloroplast of tobacco (*Nicotiana tabacum*). (b) Interpretive drawing of the three-dimensional structure of a chloroplast. The photosynthetic membranes are colored. (c) Cross section through a single thylakoid. (a, Herbert W. Israel, Cornell University)

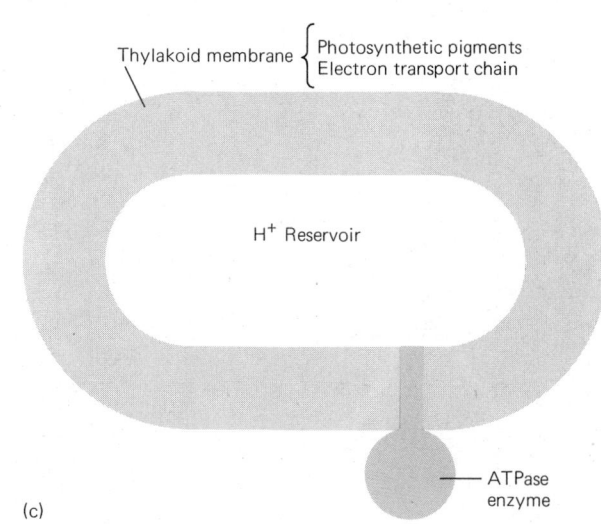

Thylakoid membrane

Photosynthetic pigments
Electron transport chain

H⁺ Reservoir

ATPase enzyme

(c)

The protein-rich solution surrounding the thylakoids in the chloroplast is called the **stroma**; it contains the enzymes that incorporate carbon dioxide into carbohydrate molecules. The stroma also contains the chloroplast's ribosomes and its DNA.

A eukaryotic green cell may have from one to 40 or more chloroplasts, depending on the species, age, and health of the cell. *Chlamydomonas*, a single-celled, flagellated green alga widely used in studying photosynthesis, has only one chloroplast. Spinach, another widely used plant, has many chloroplasts per cell.

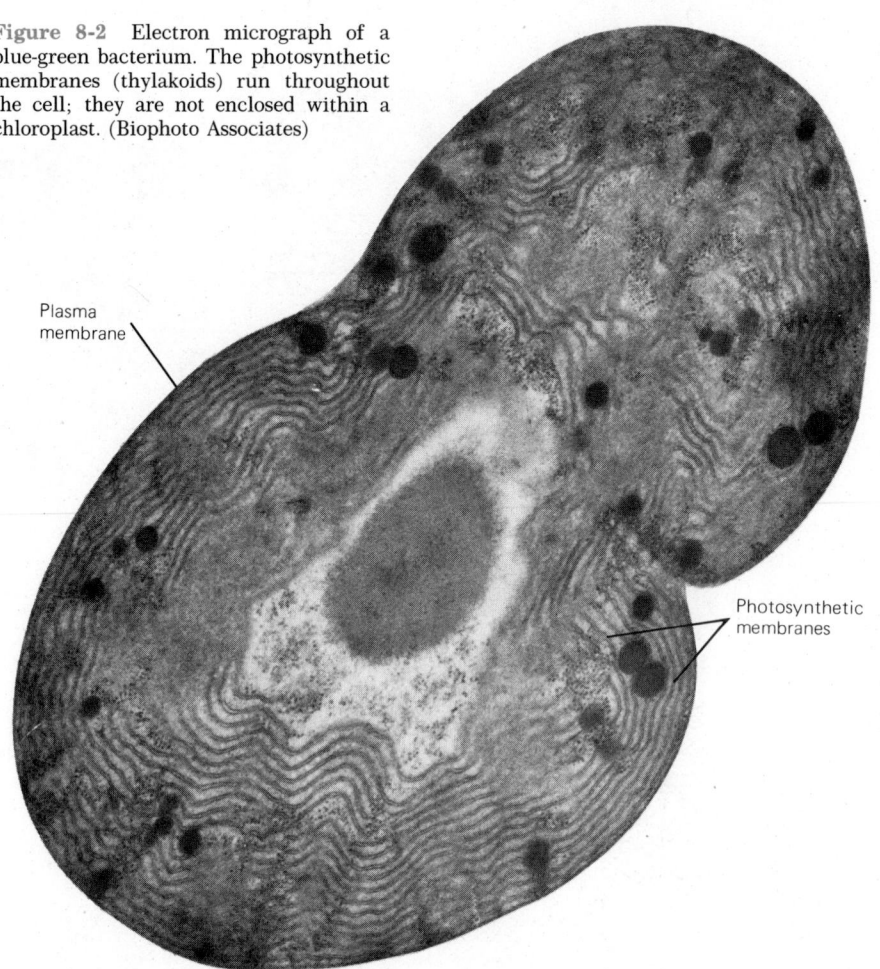

Figure 8-2 Electron micrograph of a blue-green bacterium. The photosynthetic membranes (thylakoids) run throughout the cell; they are not enclosed within a chloroplast. (Biophoto Associates)

Plasma membrane

Photosynthetic membranes

In blue-green bacteria, which are prokaryotes, there are no chloroplasts. Instead, photosynthetic membranes run through the interior of the cell (Figure 8–2).

8-B   Trapping Light Energy

Light is electromagnetic radiation in a particular band of wavelengths; part of this range is visible to our eyes (Figure 8–3). The shorter the wavelength of light, the more energy it has. Thus a quantum* of blue light has more energy than does a quantum of red light.

Objects appear colored when they reflect or transmit some wavelengths of light more than others. A red object reflects red light but absorbs light of other colors. A white object reflects all visible light, and a black object absorbs all colors.

Photosynthetic Pigments

When you look at leaf cells through a light microscope, you can see that the leaf's green color comes from the green chloroplasts in its cells; the rest of a leaf is almost colorless. A chloroplast is green because it contains the **pigment** (light-absorbing molecule) chlorophyll (chloro = green; phyll = leaf), which reflects and

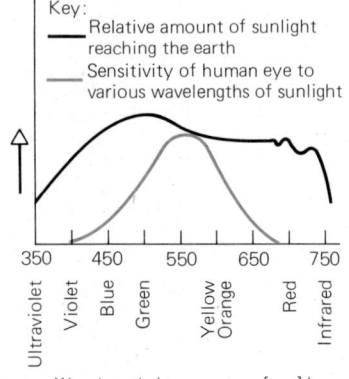

Key:
Relative amount of sunlight reaching the earth
Sensitivity of human eye to various wavelengths of sunlight

350   450   550   650   750

Ultraviolet   Violet   Blue   Green   Yellow   Orange   Red   Infrared

Wavelength (nanometres [nm])

Figure 8-3 Wavelengths of sunlight reaching the earth (black). Shorter wavelengths are known as ultraviolet; longer wavelengths are known as infrared. Our eyes can detect light with wavelengths of about 400–700 nm (color).

106   **CELLS**

*Quantum   Smallest possible unit of light energy.

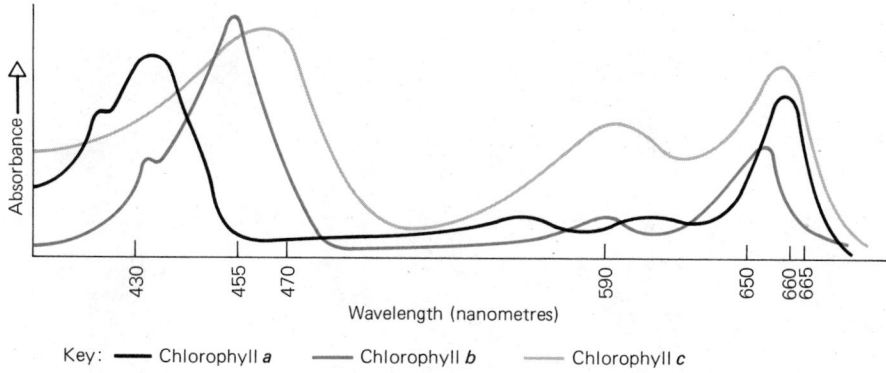

Key: —— Chlorophyll *a*   —— Chlorophyll *b*   —— Chlorophyll *c*

Figure 8-4 Absorption spectra of three chlorophylls.

transmits green light and absorbs light of other colors. Other pigments present in the chloroplast reflect different wavelengths and tinge the green with various hues.

Light energy must be absorbed before it can be used. The **absorption spectrum** of a pigment (Figure 8–4) is a graph in which the absorbance* of light by the purified pigment, in solution, is plotted against different wavelengths of light. The peaks in the absorption spectrum indicate which wavelengths are most strongly absorbed by the pigment. Assuming that the pigment in solution absorbs the same wavelengths that it absorbs in the intact plant (which is approximately true), a plant might be expected to carry out photosynthesis only when it is exposed to the wavelengths of light that its pigments can capture.

In 1883, a German, T. W. Engelmann, provided circumstantial evidence that chlorophyll plays an important role in photosynthesis. He used *Spirogyra*, an alga that has long, spiral chloroplasts. He put the alga on a microscope slide with some aerobic* bacteria. By using a prism to disperse the beam of light into a spectrum, he was able to expose different parts of the chloroplast to different wavelengths of light (Figure 8–5). Engelmann expected that if some wavelengths supported photosynthesis better than others, more oxygen would be given off by the parts of the alga exposed to those wavelengths, and the bacteria would cluster near those parts. The bacteria clustered in the bands of red and blue light, indicating that these wavelengths support photosynthesis most effectively. Since red and blue light are absorbed by chlorophyll, this result strongly suggested that the green chlorophylls, present in the algal chloroplast, were very important in photosynthesis.

---

* Absorbance The degree to which light is absorbed; measured as absorbance = 2 − log (% transmittance), where % transmittance is the percent of light shined into the solution that passes through it and can be detected on the other side.
* Aerobic Requiring molecular oxygen (O_2).

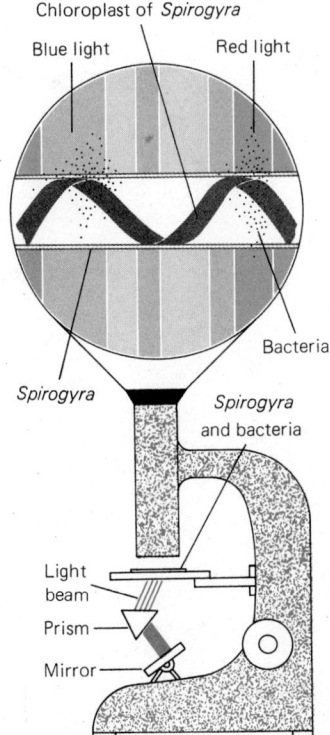

Chloroplast of *Spirogyra*
Blue light   Red light

Bacteria

Spirogyra   *Spirogyra* and bacteria

Light beam

Prism

Mirror

Figure 8-5 Engelmann's experiment to determine which wavelengths of light best supported photosynthesis. Engelmann inserted a prism in the beam of light reflected onto the stage by the mirror of a microscope. The prism dispersed the light into a spectrum as it reached the chloroplast of *Spirogyra* on the microscope stage. Oxygen released during photosynthesis attracted aerobic bacteria, which congregated where red and blue light fell on the alga. Engelmann concluded that red and blue light support higher rates of photosynthesis than do other wavelengths.

Figure 8-6 Molecular structure of one form of chlorophyll *a*. In chlorophyll *b*, the methyl group in the shaded circle is replaced by the aldehyde group in the colored ring.

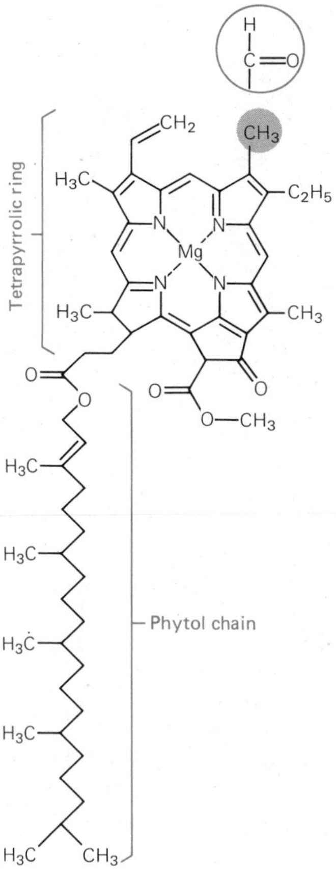

β–CAROTENE

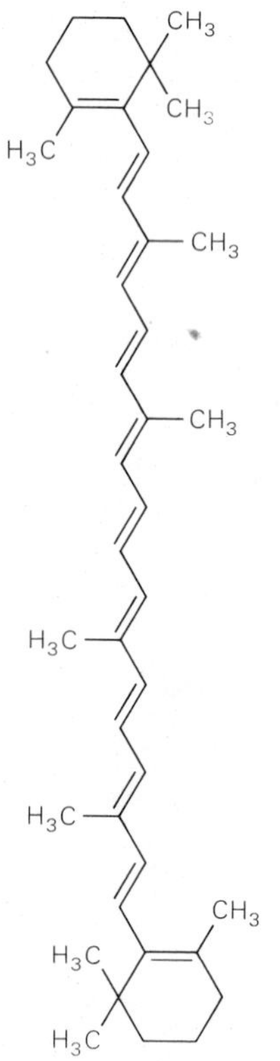

Figure 8-7 Structure of β-carotene, a carotenoid found in many higher plants.

The chlorophyll molecule has two distinct parts. At one end is a complex ring structure with a magnesium ion (Mg^{2+}) bound in the center. This is the active site, where light energy is trapped. The molecule's long nonpolar "tail" anchors it in the lipid of the photosynthetic membrane (Figure 8–6).

Chlorophyll *a*, the primary pigment involved in photosynthesis, is found in blue-green bacteria and in all photosynthetic eukaryotes. Recent research has shown that there are actually three or four variations on the structure shown in Figure 8–6, with slightly different absorption spectra, all of which are forms of chlorophyll *a*. In addition, there are other, still more different, chlorophylls—*b*, *c*, and *d*—which are characteristic of various evolutionary branches of algae and higher plants.

The light that plants receive contains more wavelengths than the blue and red ones absorbed by chlorophyll. These wavelengths are not wasted; plants have various **accessory photosynthetic pigments**, which absorb the energy and transfer it to chlorophyll *a*. Chlorophylls *b*, *c*, and *d* serve this function in many plants.

The **carotenoids** are an important group of accessory pigments found in all green plants. Carotenoids usually consist of long hydrocarbon chains with a ring structure at each end (Figure 8–7). Under high light intensities, the energy absorbed by chlorophyll cannot all be used productively; then oxygen molecules may be partly reduced, forming "radicals" that react with, and destroy, other molecules. These radicals bind readily to the double bonds in carotenoid molecules; the carotenoids thus prevent the radicals from destroying chlorophyll. There is also good evidence that energy absorbed by carotenoids is passed on to chlorophyll and used in photosynthesis.

Some plants break down their chlorophyll in the autumn and store its magnesium and nitrogen in the plant body before the leaves drop; this conserves scarce elements for reuse the following spring. The breakdown of chlorophyll makes the yellow, orange, or brown colors of the still-present carotenoids visible in autumn leaves.

Red algae and blue-green bacteria contain accessory pigments of another group, the **phycobilins**, in addition to chlorophyll *a* and carotenoids. Phycoerythrin is a

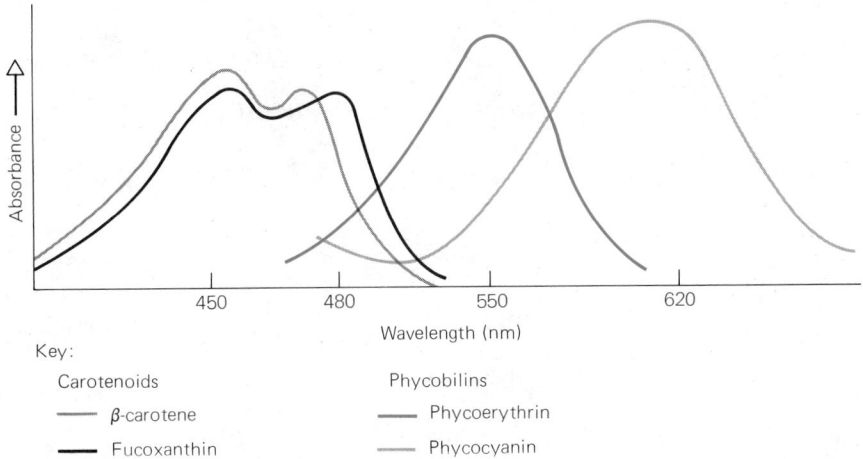

Key:

Carotenoids   Phycobilins

— β-carotene   — Phycoerythrin

— Fucoxanthin   — Phycocyanin

Figure 8-8 Absorption spectra for carotenoids and phycobilins, accessory photosynthetic pigments. β-carotene is common in higher plants; fucoxanthin is found in brown algae. Phycocyanin and phycoerythrin are common in red algae and blue-green bacteria. (See Chapter 25 for red and brown algae.)

red phycobilin, and phycocyanin looks blue. These pigments absorb light of wavelengths that chlorophyll does not absorb strongly, and transfer the energy to chlorophyll *a* for use in photosynthesis (Figure 8–8).

Photosystems

The thylakoid membranes contain photosynthetic pigment molecules arranged in clusters called photosystems. In each photosystem, there is a light-gathering "antenna" of chlorophyll *a*, carotenoids, and other pigments, which capture light energy from a broad range of the spectrum. This energy passes to a special chlorophyll *a* molecule in the photosystem's **light trap**, or **reaction center**, which also contains molecules that use the captured energy to start the next part of photosynthesis.

There are two kinds of photosystems, with different kinds of light traps. **Photosystem I** traps contain a form of chlorophyll *a* called P700 (short for pigment 700; this pigment's absorption peak is in the long red wavelengths of about 700 nm). **Photosystem II** traps use a form of chlorophyll *a* called P680, with a maximum absorption of shorter (680 nm) red wavelengths.

Because these absorption peaks are at the long-wavelength, lower-energy end of the visible spectrum (Figure 8–3), light of shorter wavelengths and higher energy can be absorbed by other photosystem pigments and passed downhill, energetically speaking, to a light trap. Furthermore, if a plant is given only light of wavelengths longer than 680 nm, it cannot carry on complete photosynthesis because these long wavelengths do not contain enough energy to activate the photosystem II P680.

8-C   What Happens in Photosynthesis?

Now that we have studied the architecture of chloroplasts, the organelles that house the photosynthetic machinery, and the structure and arrangement of the light-trapping pigment molecules, we are ready to outline the events of photosynthesis. In our discussion, we shall meet new examples of processes we saw in Chapters 6 and 7: reduction of hydrogen-carrying coenzymes, electron transport to set up a hydrogen ion reservoir, and chemiosmotic ATP synthesis. The one aspect of photosynthesis that we have not yet encountered is its photochemical reactions. A **photochemical reaction** is one activated by light energy; it occurs at a greater rate when light is brighter. A familiar example occurs when we take photographs: the more light there is, the more quickly the chemicals in the film react to produce the image. Once the initial photochemical reactions have taken place, the other reactions of photosynthesis can occur; they are the kind of reactions we have encountered before, sometimes called **thermochemical reactions** because their rates are increased by heat.

TABLE 8–1 BASIC CHEMICAL REACTIONS OF PHOTOSYNTHESIS, SHOWING ONLY RAW MATERIALS AND END PRODUCTS

REACTION OR REACTION SERIES	RAW MATERIALS	END PRODUCTS
Photochemical reactions	Light energy Pigments	Electrons
Electron transport	Electrons $NADP^+$	$NADPH + H^+$
	H_2O	$\begin{cases} O_2 \\ H^+ \text{ reservoir} \end{cases}$
Chemiosmosis	H^+ reservoir $ADP + P_i$	ATP
Carbon fixation	Ribulose bisphosphate $\Big\}$ CO_2	Sugars
	ATP	$ADP + P_i$
	$NADPH + H^+$	$NADP^+$

Photosynthesis may be broken down into four sets of events. We shall outline them here, and then elaborate them in more detail in the following sections. Table 8–1 and Figure 8–9 should serve as convenient summaries for future reference.

1. **Photochemical reactions.** Light energy knocks electrons off chlorophyll *a* molecules in photosynthetic reaction centers in the thylakoid membranes.

2. **Electron transport.** The electrons are picked up by the electron transport chain molecules in the thylakoid membranes. Eventually the electrons reach the outer (stroma) surface of the membrane, where they reduce a molecule of a hydrogen-carrying coenzyme, $NADP^+$, to $NADPH + H^+$ (we shall see the source of the H^+ needed to do this in the detailed discussion). In addition, the electron transport chain accumulates hydrogen ions in the H^+ reservoir, in the interior space of the thylakoid membranes. The splitting of water provides some of this H^+, and also produces the photosynthetic waste product O_2.

3. **Chemiosmosis.** Next, the H^+ reservoir powers the joining of ADP and P_i (inorganic phosphate ion) to form ATP. H^+ is allowed to leak through special channels in the thylakoid membrane associated with ATP-synthesizing enzymes. These enzymes jut into the stroma, from which they take ADP and P_i and into which they release ATP.[1]

4. **Carbon fixation.** The attachment of carbon dioxide to larger organic molecules (in photosynthesis, to molecules of ribulose bisphosphate) is called **carbon fixation.** This process occurs in the stroma, mediated by enzymes and using hydrogens from $NADPH + H^+$, and phosphate groups from ATP, produced by the reactions described above. Sugars are the products of this set of reactions.[2]

---

[1] Early studies of photosynthesis showed that an increase in light intensity increased the rate of the reactions producing ATP and $NADPH + H^+$. Accordingly, these were known as the *light reactions.* We now know that these are driven by the photochemical reactions, whose rate does increase with increasing light intensity, but that they are not themselves photochemical reactions. Accordingly, many researchers now prefer not to use the misleading term "light reactions." In this book, the former "light reactions" are divided into light absorption, electron transport, and chemiosmotic ATP synthesis.

[2] Carbon fixation and related reactions were formerly called the *dark reactions* because they do not speed up when light intensity increases. This term is also misleading; carbon fixation occurs in either light or darkness, and in fact, the enzymes involved seem to be activated by light. In this book we simply substitute the term "carbon fixation."

Figure 8-9 A summary of what happens in photosynthesis.

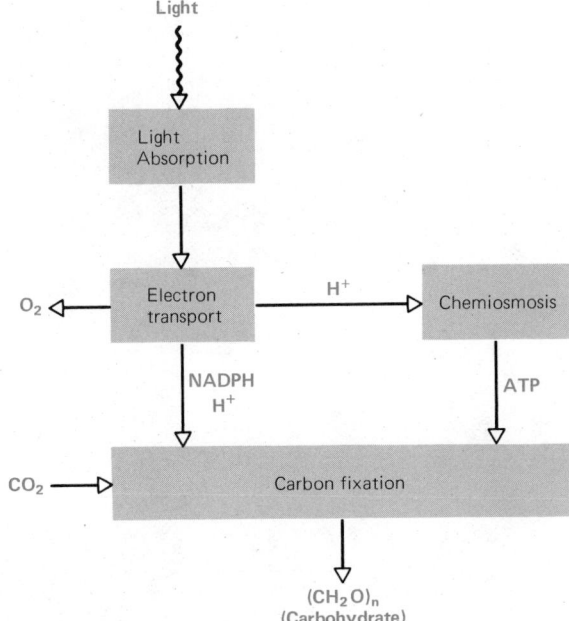

Keep this brief outline in mind as we describe these reactions in more detail. We shall treat parts 1 to 3 together because they interconnect intimately and involve the activities of molecules in the photosynthetic membranes; then we shall look at carbon fixation.

8-D Reactions Involving the Thylakoid Membranes

Light Absorption

The photosynthetic "kickoff" is the energetically uphill boost imparted to a chlorophyll electron by light energy; subsequent reactions proceed downhill, energetically speaking, transforming this electron's energy from one form to another (with energy loss at each step) until it is finally stored in the form of the chemical bonds of food molecules.

When light strikes a photosystem, any excited pigment passes the energy on to chlorophyll a (P700 or P680, depending on whether light has struck a photosystem I or II area) in the reaction center. Here the energy boosts one of the chlorophyll's electrons to such a high energy level that it reaches "escape velocity" and passes to an electron-accepting molecule in the reaction center. Such events happen twice, once in photosystem I and once in photosystem II, during each complete photosynthetic sequence.

Noncyclic Electron Flow

The passing of an electron from a chlorophyll molecule to its neighbor starts a series of redox reactions in the electron transport system, as each molecule is temporarily reduced by accepting electrons and then re-oxidized by passing them on to the next molecule. Electron transport eventually stores the energy derived from light in two forms:

1. The **reducing potential** of NADPH. When NADPH reduces another molecule by donating its hydrogen, it also gives off about 50 kcal of energy per mole (compare this with the 7.3 kcal per mole released by hydrolysis of ATP to ADP and P_i).

2. The electrochemical potential gradient of a membrane-bound H^+ reservoir, which is later used to power chemiosmotic ATP synthesis. The generation of the reservoir and subsequent ATP synthesis together are called **photosynthetic phosphorylation.**

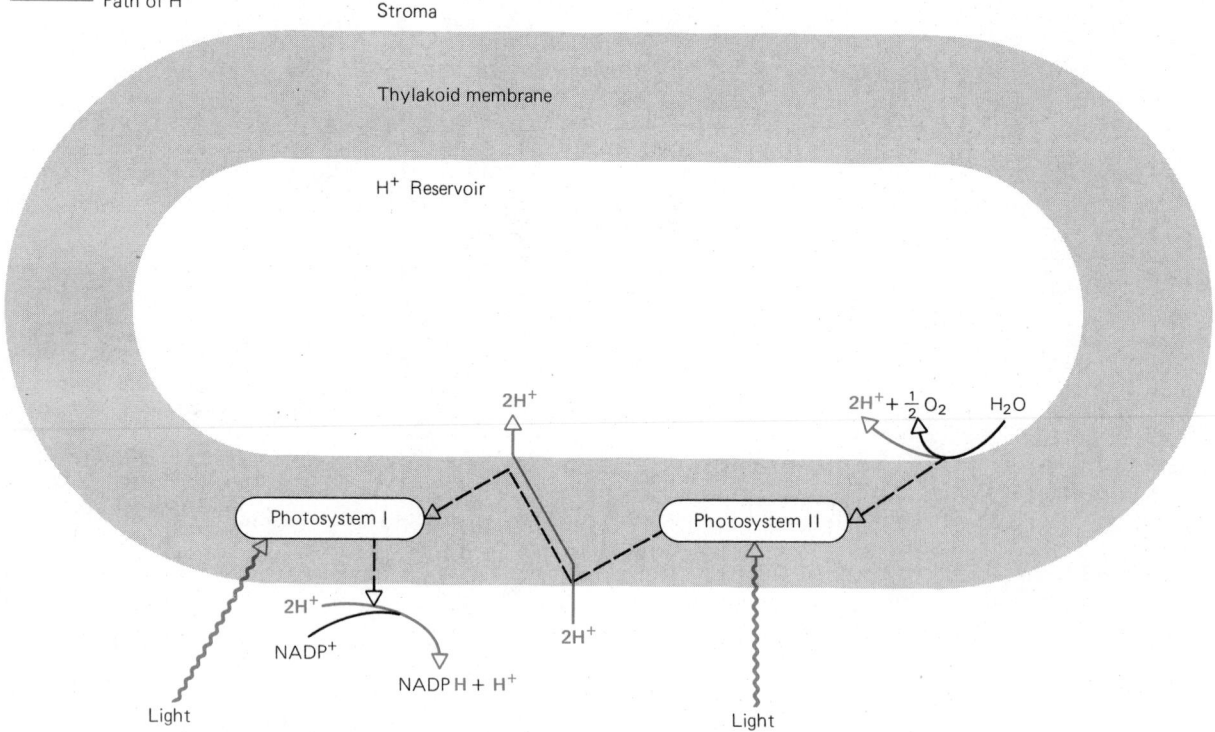

Figure 8-10 Summary of the events of electron flow in thylakoid membranes. Light initiates electron flow in photosystems I and II; as a result of electron flow, NADPH + H⁺, a hydrogen ion reservoir, and oxygen are formed.

Electrons journey through the photosynthetic membranes in three stages (Figure 8–10):

First Stage. When light energy enters a photosystem I trap, a P700 electron becomes excited, leaves the molecule, and is eventually passed to NADP⁺. NADP⁺ becomes NADPH by accepting two electrons from photosystem I and H⁺ from dissociated water in the stroma. A second such H⁺ associates with NADPH. Two P700 molecules are now short of an electron each, and these are replaced by electrons from photosystem II, by a circuitous pathway:

Second Stage. When light strikes photosystem II, P680 passes a high-energy electron to a shuttle molecule in the thylakoid membrane. This molecule is a hydrogen carrier; it will not carry electrons alone. It picks up two electrons from photosystem II and two hydrogen ions from dissociated water in the stroma and then moves across the membrane to the inner surface. Here the shuttle molecule reduces the next molecule in the electron transport chain, which is a cytochrome. From our study of respiration, we know that cytochromes carry only electrons; the two hydrogen ions are released into the H⁺ reservoir in the thylakoid interior space. The electrons, meanwhile, move along the chain to P700, returning it to its original state; it can now absorb more light and initiate the steps outlined in the first stage again.

Third Stage. Now two molecules of P680 lack an electron apiece. Each receives one from a molecule that can remove electrons from water, by the following reaction:

$$H_2O \longrightarrow \underset{\text{to P680}}{2e^-} + \underset{\substack{\text{to } H^+ \\ \text{reservoir}}}{2H^+} + \underset{\text{waste}}{\tfrac{1}{2}O_2}$$

This reaction also adds more H⁺ to the reservoir, and generates oxygen as a waste product; the oxygen diffuses away and eventually leaves the plant.

The mystery molecule(s) performing this reaction is not yet identified. It must be a very powerful oxidizing agent to split water into hydrogen and oxygen. Unlike the spontaneous dissociation of water into H^+ and OH^-, this reaction takes a lot of energy. So far, all we know is that this molecule is a protein associated with manganese (Mn^{2+}).

Chemiosmotic ATP Synthesis

The steps outlined in the second and third stages above build up a hydrogen ion reservoir inside the thylakoids, with a difference of about 3.5 pH units between the two sides of the membrane. Since the pH scale is logarithmic, this means that the hydrogen ion concentration inside is $10^{3.5}$ times as great as that outside. The electrochemical potential provided by this pH difference across the membrane supplies energy to synthesize ATP: ATPase enzymes phosphorylate ADP to ATP, using the energy of hydrogen ions moving out of the reservoir (Figure 8–11). Because ATP synthesis depletes the H^+ reservoir and reduces the pH difference across the membrane, light must continue to strike the chloroplast for ATP synthesis (and hence photosynthesis) to continue.

Much of this should sound familiar after our study of respiration. Note the differences, however: whereas in respiration electrons are passed *away from* hydrogen carriers and *toward* oxygen, in photosynthesis electrons pass *to* $NADP^+$ and *away from* oxygen. Furthermore, in respiration the H^+ reservoir accumulates outside the inner mitochondrial membrane and ATPase lies on the *inner* surface of the inner membrane, whereas in photosynthesis the H^+ reservoir accumulates in the interior of the thylakoid membranes and ATPase lies on the *outer* surface of these membranes, and it releases ATP directly into the stroma (Figure 8–11).

Cyclic Electron Flow: A Brief Digression

Electron flow can transport additional hydrogen ions into the reservoir by a pathway different from that already described. This process uses some of the same electron carriers, but in a different sequence. It is called **cyclic electron flow** because chlorophyll P700, in photosystem I, serves as the original electron donor as well as the final electron acceptor (Figure 8–12). By building up the hydrogen reservoir, cyclic electron flow stimulates the synthesis of ATP. However, the cyclic pathway does not create additional NADPH nor generate oxygen. It is thought that cyclic flow occurs when most of the $NADP^+$ has already been reduced, leaving a shortage of electron acceptors.

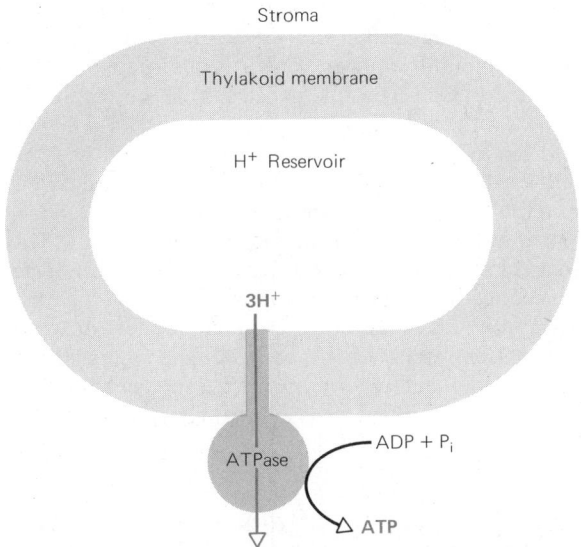

Figure 8-11 Chemiosmotic ATP synthesis in chloroplasts. The H^+ reservoir inside the thylakoid powers the phosphorylation of ADP to ATP by the ATPase enzyme lying on the stroma (outer) face of the membrane.

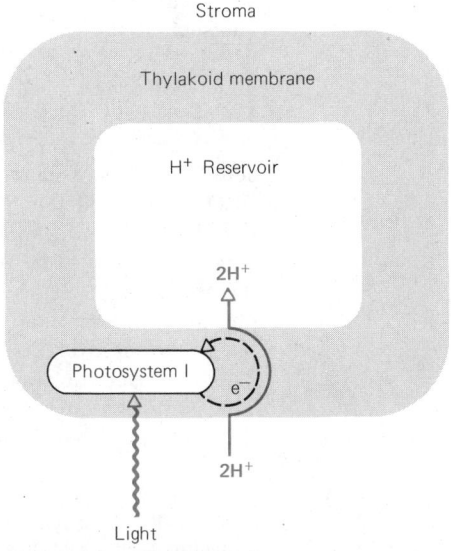

Figure 8-12 In cyclic electron flow (dashed black line), photosystem I acts as both donor and recipient of electrons. Cyclic flow adds hydrogen ions to the reservoir.

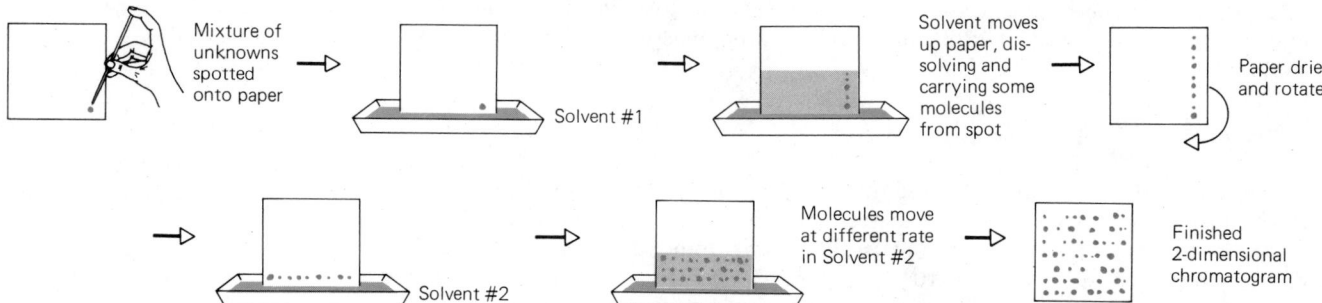

Mixture of unknowns spotted onto paper

Solvent #1

Solvent moves up paper, dissolving and carrying some molecules from spot

Paper dried and rotated

Solvent #2

Molecules move at different rate in Solvent #2

Finished 2-dimensional chromatogram

Figure 8-13 Calvin worked out the pathway of carbon dioxide fixation using two-dimensional paper chromatography to determine the kinds and amounts of substances labeled with ^{14}C in a spot of solution made from photosynthesizing cells. Substances that are most soluble in the solvents used move the farthest up the paper. In the finished chromatogram, each spot contains a different compound from the original mixture. After the paper has dried, it is placed on a sheet of unexposed photographic film, and left in a dark place. As the radioactive carbon in each test spot decays, the particles emitted expose the film, so that the spots containing carbon 14 take their own pictures, a process known as autoradiography. The locations of these spots are then compared to the locations of spots of known compounds subjected to the same procedure of paper chromatography that was used on the photosynthetic mixture; spots in the same location are presumed to be the same compound; this can be checked by cutting out the spots, redissolving them, and subjecting the minute quantities so obtained to sensitive chemical analysis.

8-E Carbon Fixation

Carbon is fixed in the stroma of chloroplasts (or in the cytoplasm of photosynthetic prokaryotes), using NADPH + H$^+$ and ATP produced as we have just seen. The stroma contains a number of different enzymes that can attach carbon dioxide to a preexisting organic molecule, and then pass the resulting molecule through a variety of different metabolic pathways.

Melvin Calvin and his associates elucidated the **C$_3$** or **Calvin cycle**, by which plants incorporate carbon dioxide into various carbohydrate molecules. These workers used carbon-14, a radioactive isotope of carbon, to trace the fate of the carbon atom from carbon dioxide. In such studies, a preparation of plant material is deprived of carbon dioxide and subjected to conditions that induce photosynthesis. $^{14}CO_2$ is then injected into the reaction vessel. At very short time intervals after the introduction of $^{14}CO_2$, some of the photosynthesizing cells are removed directly into boiling alcohol, which immediately denatures all of the cell enzymes and prevents any further reaction. The small organic molecules are then extracted from the cells and separated by paper chromatography (Figure 8–13). By determining the order in which the radioactively labeled compounds appear, Calvin constructed the sequence of reactions during and after carbon fixation; for this work he received a Nobel Prize in 1961.

When carbon dioxide is eliminated from a suspension of photosynthesizing cells, a 5-carbon compound called ribulose bisphosphate (RuBP) accumulates. If $^{14}CO_2$ is then added to the system, the ribulose bisphosphate disappears and radioactive, ^{14}C-containing molecules of the 3-carbon compound phosphoglycerate (PGA) appear (Figure 8–14). What seems to happen is that $^{14}CO_2$ is joined to RuBP, forming an unstable 6-carbon intermediate that immediately breaks into two 3-carbon phosphoglycerate molecules. These two PGA molecules are next reduced to phosphoglyceraldehyde (PGAL). This takes two steps. First each PGA molecule receives a second phosphate, this time with a "high-energy" bond, from a molecule of ATP. This new, high-energy phosphate bond is then broken as PGA is reduced to PGAL by the addition of hydrogen from NADPH.

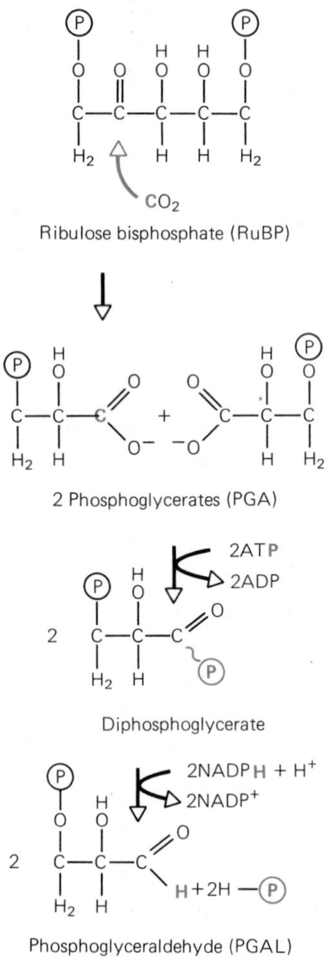

Ribulose bisphosphate (RuBP)

CO_2

2 Phosphoglycerates (PGA)

2ATP
2ADP

Diphosphoglycerate

2NADPH + H$^+$
2NADP$^+$

Phosphoglyceraldehyde (PGAL)

Figure 8-14 First reactions of carbon fixation (the Calvin cycle). The six-carbon compound formed by joining carbon dioxide to ribulose bisphosphate breaks into two molecules of phosphoglycerate (shown oriented in opposite directions). Using energy from ATP, these phosphoglycerate molecules are phosphorylated to form diphosphoglycerates, and are then reduced to phosphoglyceraldehyde, using NADPH + H$^+$. (Compare this sequence of reactions with those in Figure 7-4.)

The 3-carbon PGAL can be channeled into the formation of 6-carbon sugars by the joining together of two 3-carbon molecules, and once such hexose sugars are formed, they can be polymerized into starch, an energy-storage compound, or cellulose, which makes up the cell wall. Some PGAL must also be channeled into the formation of more of the 5-carbon sugar, ribulose, which is then put back into the Calvin cycle to receive more carbon dioxide, enabling the cycle to continue (Figure 8–15). PGAL can also enter biochemical pathways that produce amino acids.

It is interesting to note that the enzyme responsible for attaching carbon dioxide to ribulose bisphosphate, RuBP carboxylase, makes up about 25% of the protein in chloroplasts. This makes it far and away the most abundant protein in green tissue, and possibly in the world!

In summary, the Calvin cycle begins with carbon dioxide and a 5-carbon molecule, which react to form two 3-carbon molecules (PGA). Each of these PGA molecules uses an ATP and an NADPH + H$^+$ as it is converted to PGAL. A third ATP is required to phosphorylate ribulose phosphate to regenerate the starting molecule, ribulose bisphosphate. The overall equation is thus:

$$RuBP + CO_2 + 2NADPH + 2H^+ + 3ATP \longrightarrow$$
$$RuBP + CH_2O + 2NADP^+ + 3ADP + 3P_i$$

Six turns of the Calvin cycle are required to produce the equivalent of one (6-carbon) glucose molecule, that is, to fix six carbon atoms into organic form.

The ADP, P$_i$, and NADP$^+$ released by the C$_3$ cycle are recycled to form more ATP and NADPH. These substances are present in very small amounts, and so if either electron transport or the C$_3$ cycle is stopped, the other soon stops as well. The stockpile of ATP and NADPH, for instance, will last only a matter of seconds once the light is turned off; after the supply is exhausted, carbon fixation can no longer proceed.

8-F   What Controls the Rate of Photosynthesis?

Like all chemical reactions, photosynthesis proceeds faster under some conditions than under others. Because photosynthesis is made up of both photochemical and thermochemical reactions, both light and temperature influence its rate. In fact, observations of the rate of photosynthesis at different temperatures and light intensities led F. F. Blackman to deduce the existence of both photochemical and thermochemical reactions in 1905, long before any of the actual reactions were identified.

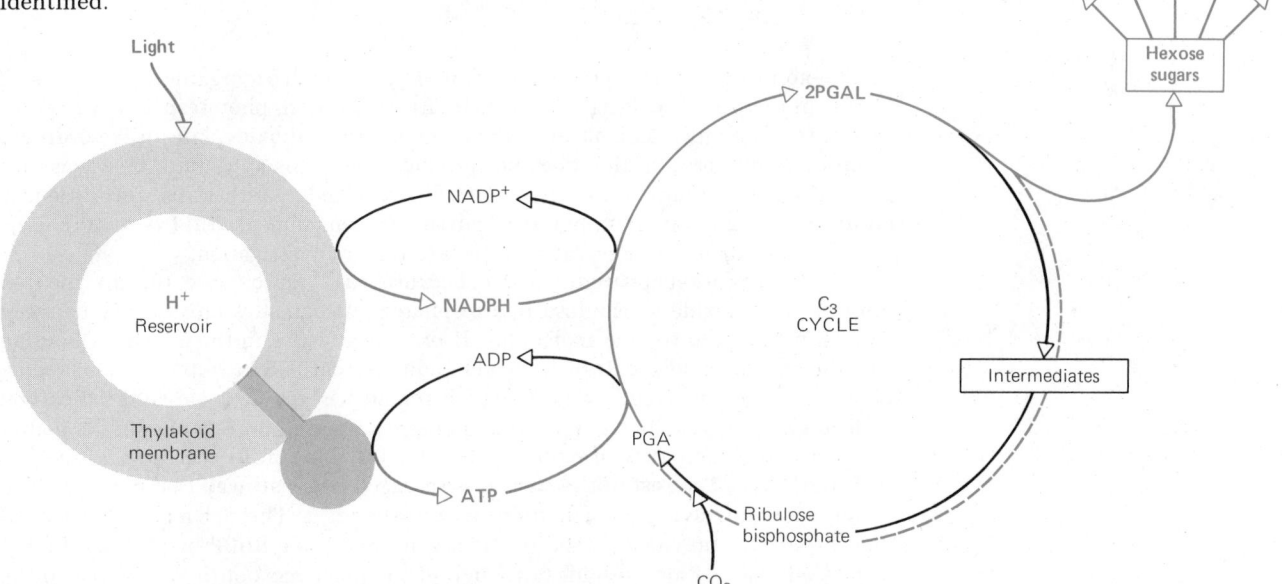

Figure 8-15 The C$_3$ cycle and its connections with the electron transport chain in the thylakoid membranes. Colored arrows trace the course of energy from sunlight through temporary energy intermediates to the more permanent energy-storage form of hexose sugars and their polymers.

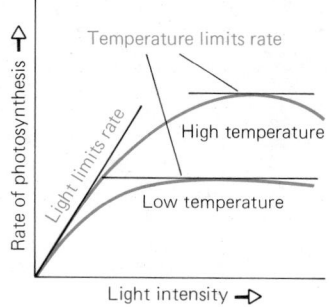

Figure 8-16 Rate of photosynthesis under different conditions of light intensity and temperature. At low light intensities the rate of photochemical reactions limits the overall rate of photosynthesis. At higher light intensities the rate of thermochemical reactions fails to keep up with the increased rate of the light-driven reactions.

In dim light, light limits the rate of photosynthesis; photochemical reactions that initiate electron transport, and hence chemiosmosis, go too slowly to produce NADPH and ATP as fast as carbon fixation uses them up. As light intensity increases, the rate of photosynthesis increases until ATP and NADPH become so plentiful that carbon fixation cannot keep up, and the curve levels off (Figure 8–16).

Photosynthesis also proceeds more slowly at low temperatures than at moderate temperatures because thermochemical reactions occur more slowly when there are fewer collisions between molecules per unit of time. At still higher temperatures, however, enzymes are denatured; this eventually stops all biochemical reactions. At low temperatures but high light intensities, therefore, carbon fixation does not use the reactants produced by the light-initiated reactions as quickly as they are produced (Figure 8–16).

The rate of a reaction can also be decreased by scarcity of raw materials or excess of products. On warm, bright days, light and temperature are optimum for photosynthesis, but a low concentration of the raw material carbon dioxide often limits its rate. Adding more carbon dioxide to the air in a greenhouse increases the rate of photosynthesis (and so of plant growth) of some greenhouse crops. Above a certain concentration, however, carbon dioxide inhibits photosynthesis.

The other raw material of photosynthesis, water, is so abundant in living tissue that the amount used in photosynthesis is negligible; variations in the plant's water content affect photosynthesis only indirectly. A plant's leaves lose water by evaporation through pores called **stomata**, which must be open for the plant to obtain carbon dioxide from the air. Under conditions of water stress, the stomata may be partly or completely closed; this conserves water but reduces access to carbon dioxide. During periods of water shortage, therefore, the decrease in photosynthetic rate is actually due to low carbon dioxide concentration.

The carbohydrate end products of photosynthesis are quickly changed into other chemical forms or removed from the chloroplast, and so they do not stay around to inhibit the reaction.

A high concentration of the other end product, oxygen, limits photosynthesis in several ways. Oxygen combines with electron transport molecules and so interferes with electron transport. It also oxidizes and destroys photosynthetic pigments. Oxygen's most important effect, however, is its role in photorespiration, discussed in the next section.

8-G   Photorespiration

Photorespiration is a process whereby oxygen oxidizes organic compounds in plants, in the presence of light. Like ordinary respiration, photorespiration releases carbon from organic compounds in the form of carbon dioxide, but unlike ordinary respiration, photorespiration does not produce ATP. This appears to be a wasteful process, but we assume that the state of our knowledge, rather than evolutionary failure, is at fault, and that photorespiration performs some useful function that we have not yet identified. Several theories are under investigation.

Photorespiration apparently occurs because RuBP carboxylase, the enzyme that joins carbon dioxide to ribulose bisphosphate, distinguishes only poorly between carbon dioxide and oxygen molecules. If oxygen attaches to the enzyme's binding site, the enzyme oxidizes RuBP instead of adding a carboxyl group to it. This oxidation releases one molecule of PGA, which remains in the C_3 cycle, and a 2-carbon molecule that leaves the chloroplast and enters a **peroxisome**, a membrane-bound sac of enzymes. Here further oxidation occurs, and some of the carbon is released as carbon dioxide. The rest of the carbon is salvaged by a pathway that converts it to amino acids, involving participation by mitochondria. The net result is a loss of carbon dioxide (previously fixed in the organic molecule RuBP by photosynthesis) and a slight gain of amino acids (although plants make most of their amino acids by other routes).

Photorespiration is enhanced by bright light, high temperatures, high oxygen concentrations, and low carbon dioxide concentrations. On a warm, bright day, when a plant is undergoing rapid photosynthesis, up to 50% of the carbon dioxide fixed in photosynthesis may be lost again in photorespiration. Because of its obvious importance to the productivity of agricultural crops, there is a great deal of research into this odd process.

8-H Ecological Aspects of Photosynthesis

Green plants live in many different kinds of habitats, and many of them have evolved variations on the photosynthetic pathways that adapt them to particular environmental conditions. Here we shall outline a few important photosynthetic adaptations of land plants.

The C_4 or Hatch–Slack Pathway

The C_3 cycle is not the only way plants can fix carbon dioxide. Most, if not all, plants can also add carbon dioxide to the 3-carbon compound phosphoenolpyruvate (PEP)* to form the 4-carbon compound oxaloacetate.* In the 1960s botanists H. P. Kortschak in Hawaii and M. D. Hatch and C. R. Slack in Australia found that this ability is peculiarly well developed in some plants, now called C_4 plants, with especially high rates of photosynthesis. These C_4 plants include crabgrass and other important weeds, as well as some of our most important agricultural plants, including sugarcane, corn, millet, and sorghum.

The leaves of C_4 plants contain two types of photosynthetic cells; **bundle-sheath cells** are packed tightly in a circle and surrounded by **mesophyll cells**. As in all leaves, the mesophyll cells lie near air spaces, which exchange gases with the atmosphere through stomata in the leaf surface (Figure 8–17). Carbon dioxide is fixed in the mesophyll cells; the resulting oxaloacetate is converted into other compounds, which travel to the bundle-sheath cells. Here the carbon dioxide is removed and re-fixed by the C_3 cycle.

The net effect of the Hatch-Slack pathway is to transfer carbon dioxide from one cell to another. It has been proposed that this "supercharges" the bundle-sheath cells with carbon dioxide, increasing the chances that CO_2, rather than O_2, will combine with RuBP carboxylase molecules. This results in increased photosynthesis and decreased photorespiration.

The C_4 pathway uses 5 ATP per molecule of carbon dioxide fixed instead of the 3 ATP used in C_3 photosynthesis alone. Thus, C_4 plants use energy less efficiently. However, their photosynthetic rate is faster than that of C_3 plants under certain conditions. The sacrifice of efficiency for speed is advantageous because it allows organisms to grow and reproduce faster, as we shall see in several cases in this book. However, this tradeoff works only when energy is abundant, because it requires profligate use of energy. C_4 photosynthesis has evolved in plants of somewhat dry tropical and subtropical* climates, such as grasslands, where there is plenty of light energy to drive ATP synthesis, but where the plant may experience water stress. In such a situation, the C_4 plant's stomata may close partially, reducing loss of water vapor from the leaves. This also reduces uptake of CO_2 from the air, but possessing

---

* Phosphoenolpyruvate See Figure 7–4.
* Oxaloacetate See Figure 7–7.
* Subtropical Just north or south of tropical areas; warm; e.g., Florida, northern Mexico.

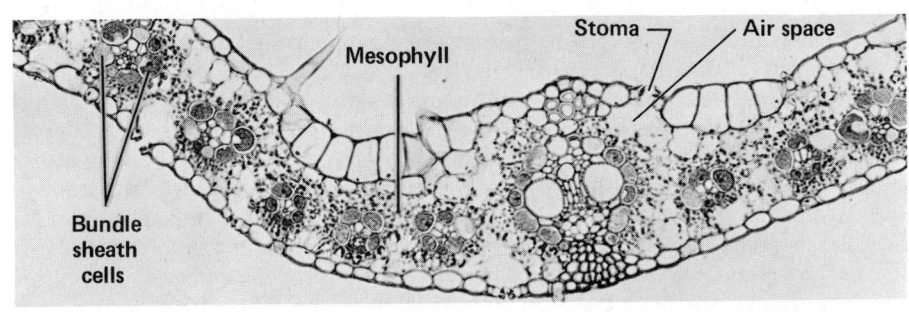

Figure 8-17 Cross section of a leaf of corn, a C_4 plant, showing stomata, air spaces, and bundle-sheath and mesophyll cells. (Carolina Biological Supply Company)

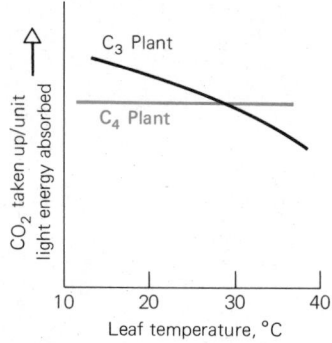

Figure 8-18 Comparison of net photo-synthesis in a C_3 and a C_4 plant at increasing temperatures. Photosynthesis measured as CO_2 taken up per unit of light energy absorbed. (Redrawn after J. R. Ehleringer, *Oecologia* 31(3):255, Springer-Verlag, Berlin, Heidelberg, New York.)

the C_4 pathway allows the plant to capture enough carbon dioxide for a high rate of photosynthesis anyway.

The rapid growth of C_4 plants has attracted the attention of those interested in increasing the productivity of crops. Because this rapid growth depends on high rates of photosynthesis in bright light, however, C_4 plants are at a relative disadvantage in cool or shady situations, and C_3 and C_4 plants coexist, with neither having a noticeable edge, in many habitats (Figure 8–18).

Crassulacean Acid Metabolism

Crassulacean acid metabolism (CAM) is a variation of C_4 photosynthesis that has evolved in several different plant families primarily adapted to desert living (the families Crassulaceae, e.g., hens-and-chicks; Cactaceae, the cacti; and Euphorbiaceae, the euphorbs). Most species with CAM are succulent; that is, they store water in fleshy stems or leaves. The stomata of these plants are closed during the day, reducing water loss. CAM plants obtain carbon dioxide by opening the stomata at night and fixing it in organic acids. During the day, when the stomata are closed, carbon dioxide is removed from the acids, and light energy is used to re-fix the carbon dioxide by way of the C_3 pathway. Both C_4 photosynthesis and CAM store carbon dioxide for later use, but in CAM, activity is divided between day and night instead of between different types of cells. Although CAM plants receive plenty of sun, they must conserve water very strictly, and so their photosynthetic rates are low. However, their ability to conserve water enables them to survive in habitats too dry for the "fast food" plants that would otherwise crowd them out.

Sun Plants and Shade Plants

The biochemical aspects of photosynthesis are only part of the way plants are adapted to different environments. Very often, changes in leaf anatomy and shape, and the arrangement of leaves with respect to each other, affect rates of photosynthesis. One example of this is in adaptation to shade as opposed to intense light. Plants can be separated into two groups, sun plants and shade plants. While **sun plants**, such as soybean, cotton, and tomato, have increasing rates of photosynthesis as light intensity increases, **shade plants** do not. Shade plants, including many ferns, African violets, and philodendron, simply do not photosynthesize at a very high rate, even when much light is available. On the other hand, shade plants are much more efficient at utilizing very low light levels (Figure 8–19). Some species have both shade leaves and sun leaves on the same plant. For example, many forest trees, such as the oaks, will have sun leaves near the top and shade leaves near the bottom. Sun leaves tend to be smaller and thicker, with more pronounced lobing than shade leaves. Beyond these differences in overall form, sun leaves have more photosynthetic cells, RuBP carboxylase, and stomata per unit of surface area; these permit their high rates of photosynthesis. Shade leaves are thinner, with fewer photosynthetic cells in a unit of leaf area, but their ratio of chlorophyll to carboxylase is higher than in sun leaves.

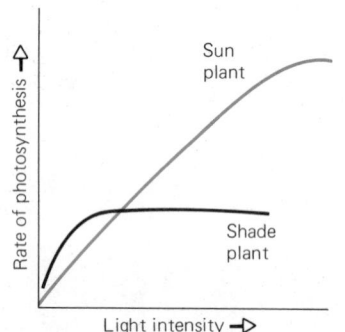

Figure 8-19 Photosynthetic rates of sun plants and shade plants at different light intensities. Sun plants can utilize light at higher intensities, but are not so efficient at low light intensities.

SUMMARY

Photosynthesis is the process by which green plants store the energy of sunlight by converting carbon dioxide and water into organic compounds. These organic compounds are used in turn, by plants and by the animals that eat plants, to build cells and to power other energy-requiring processes. Respiration is, in many ways, the reverse of photosynthesis, since it breaks down the end products of photosynthesis and releases their stored energy. By the same token, respiratory end products, carbon dioxide and water, which are compounds with very little energy, can be recycled by green plants in photosynthesis. Since energy cannot be recycled, however, the sunlight that drives photosynthesis is the ultimate source of energy for nearly all life on earth.

Photosynthesis may be considered in three parts:

1. The energy of sunlight is trapped by molecules of chlorophyll and other photosynthetic pigments in the thylakoid membranes of chloroplasts, initiating a flow of electrons through the membrane's electron transport chain.

2. The flow of electrons reduces $NADP^+$ to $NADPH + H^+$ and creates the H^+ reservoir used by ATPase enzymes to phosphorylate ADP to ATP. Oxygen from water is released as a by-product. The $NADPH + H^+$ and ATP are released into the stroma of chloroplasts, where their energy is used to fix carbon dioxide.

3. During the C_3 cycle carbon dioxide becomes attached to a 5-carbon sugar, ribulose bisphosphate, which then breaks to yield two 3-carbon phosphoglycerate molecules; these are then phosphorylated by ATP and reduced by hydrogens from $NADPH + H^+$ to make phosphoglyceraldehyde. The resulting ADP and $NADP^+$ are recycled. The 3-carbon PGAL molecules may be made into structural or energy-storing molecules, or may be processed, with the use of more ATP, to make more ribulose for carbon fixation. In C_4 and CAM plants, carbon dioxide is temporarily added to a 3-carbon molecule, from which it is later removed and re-fixed by the C_3 pathway.

A photosynthesizing plant captures light energy in two parts of the reaction sequence. In each case, the light energy boosts an electron to a high energy level, and the energy so transferred is then channeled into increasingly more permanent forms of energy storage:

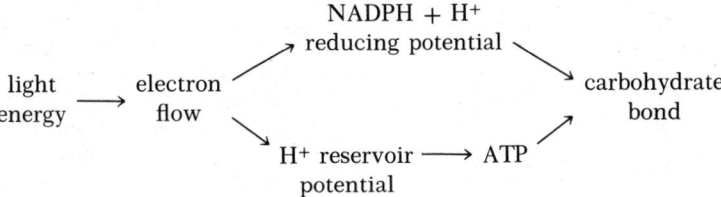

OBJECTIVES

From your study of this chapter, you should be able to:

1. Name or recognize the necessary raw materials of photosynthesis and the important end products.

2. Describe or sketch the structure of a chloroplast; explain where the following are to be found and the importance of their location to their roles in photosynthesis: chlorophyll, electron transport system, ATPase enzymes, hydrogen ion reservoir, C_3 cycle enzymes.

3. Name the three main groups of photosynthetic pigments and state the functions they perform.

4. State which colors of light are most effective in promoting photosynthesis, and explain why.

5. Name the raw materials and end products of the electron transport reactions, chemiosmotic ATP synthesis, and the C_3 cycle. Predict how altering light intensity or temperature will affect each.

6. State what drives carbon fixation and explain what happens in the C_3 cycle.

7. Summarize the important steps in energy transfer during photosynthesis.

SELF-QUIZ

Match:

___ 1. Location of chlorophyll

___ 2. Location of enzymes for carbon fixation

___ 3. Location of ATPase in chloroplasts

___ 4. Location of H^+ reservoir for ATP synthesis

a. stroma
b. chloroplast envelope
c. photosynthetic membranes (thylakoid membranes)
d. thylakoid interior

Match (give all correct answers):

___ 5. ribulose bisphosphate

___ 6. $NADP^+$

___ 7. PGAL (phosphoglyceraldehyde)

___ 8. O_2

___ 9. CO_2

___ 10. $ADP + P_i$

a. raw material of electron transport reactions
b. end product of electron transport reactions
c. raw material for chemiosmotic ATP synthesis
d. end product of chemiosmotic ATP synthesis
e. raw material of C_3 cycle
f. end product of C_3 cycle

119

11. The oxygen from H_2O is incorporated into:
   a. oxygen gas
   b. water
   c. carbohydrates
   d. NADPH + H^+
   e. ATP

12. Engelmann's experiment with *Spirogyra* demonstrated that:
   a. the full spectrum of sunlight is needed for photosynthesis
   b. only red wavelengths are effective in causing photosynthesis
   c. only blue wavelengths are effective
   d. both red and blue wavelengths are effective
   e. only green wavelengths are effective
   f. both green and red wavelengths are effective

13. Red and blue light support the highest rates of photosynthesis because:
   a. these are the only wavelengths reaching the earth from the sun
   b. these are the only wavelengths that carotenoids cannot absorb
   c. chlorophyll absorbs these wavelengths more than other wavelengths
   d. these wavelengths have the highest energy in the visible spectrum
   e. these wavelengths activate the ATPase enzyme

14. The role of phycobilins in photosynthesis is to:
   a. absorb and pass energy to chlorophyll *a*
   b. donate electrons to the electron transport chain
   c. fix carbon dioxide
   d. carry hydrogen or electrons
   e. all of the above

QUESTIONS FOR DISCUSSION

1. Why is photosynthesis only about 1% efficient in converting the energy in sunlight that strikes a leaf into energy in organic molecules? What happens to the rest of the energy?

2. In the early 1930s, C. B. van Niel found that purple sulfur bacteria use light to make carbohydrates. They require hydrogen sulfide as a raw material and give off sulfur. Van Niel speculated that the reactions in sulfur bacteria were analogous to photosynthesis in green plants:

General reaction for green plants:

$$CO_2 + H_2O \xrightarrow{\text{light}} (CH_2O) + H_2O + O_2$$

General reaction for sulfur bacteria:

$$CO_2 + H_2S \xrightarrow{\text{light}} (CH_2O) + H_2O + 2S$$

(This was the first indication that the oxygen produced in photosynthesis probably came from water rather than from carbon dioxide.) Other kinds of bacteria produce food by processes called chemosynthesis, using energy from inorganic chemical reactions rather than light energy. Why is the kind of photosynthesis outlined in this chapter so much more prevalent among living organisms today than the kind of photosynthesis used by sulfur bacteria or the chemosynthesis used by bacteria?

3. Oxygen exists in two forms: ^{18}O has two extra neutrons in its nucleus, making it heavier than the more common ^{16}O. In 1941, Samuel Ruben and Martin Kamen used ^{18}O to confirm van Niel's proposal that the oxygen produced in photosynthesis comes from water. Complete the following equations to show the results you would expect to find from this experiment.

$$C\ ^{16}O_2 + H_2\ ^{18}O \longrightarrow$$

$$C\ ^{18}O_2 + H_2\ ^{16}O \longrightarrow$$

4. Why do trees reabsorb their chlorophylls, but not their carotenoids, in the autumn?

5. If oxygen is given off as a result of electron transport in photosynthetic membranes, why does the rate of oxygen emission level off when the rate of the C_3 cycle limits the rate of photosynthesis?

REFERENCES AND FURTHER READING

Bidwell, R. G. S. *Plant Physiology*, 2d ed. New York: Macmillan Publishing Co., Inc., 1979. Presents a modern view of photosynthesis in more detail.

Ehleringer, J. R. "Implications of quantum yield differences on the distributions of C_3 and C_4 grasses." *Oecologia* 31(3):255, 1978.

Galston, A. W., P. J. Davies, and R. L. Satter. *The Life of the Green Plant*, 3d ed. Englewood Cliffs, NJ: Prentice-Hall, Inc., 1980. A modern text with many fascinating tidbits.

Govindjee, and R. Govindjee. "Primary events of photosynthesis." *Scientific American*, December 1974. The absorption of light energy and excitation of pigment molecules are covered at an advanced level.

Lehninger, A. L. *Biochemistry*, 2d ed. New York: Worth, 1975. Chapters 22 and 23 give a comprehensive summary for the serious student.

Levine, R. P. "The mechanism of photosynthesis." *Scientific American*, December 1969. Covers the two photosystems, the absorption of light energy, and its use to initiate the oxidation-reduction steps of electron transfer.

Rabinowitch, E. I., and Govindjee. *Photosynthesis*. New York: John Wiley and Sons, Inc., 1969. Presents the historical and experimental basis of our knowledge of photosynthesis.

Trebst, A. "Energy conversion in photosynthetic electron transport of chloroplasts." *Annual Review of Plant Physiology* 25:423, 1974. Reviews work on electron transport in chloroplast membranes.

PART TWO

INFORMATION CODING
AND TRANSFER

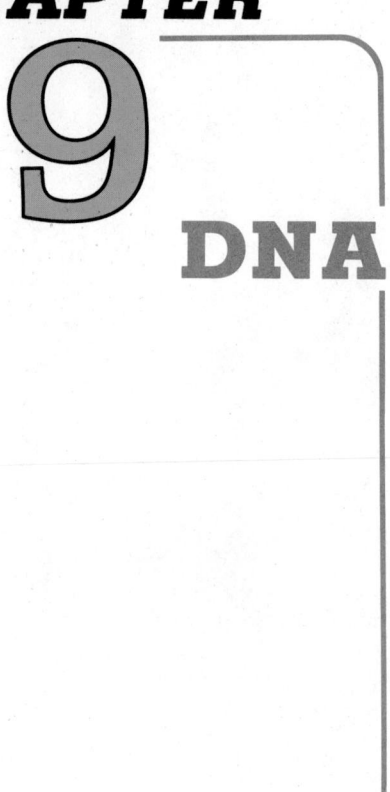

CHAPTER

9

DNA

In Chapters 9 to 14 we address the central mystery of life. What organizes a tiny mass of material into a functioning cell capable of regulating its internal chemical composition, growing, and reproducing? What directed the single fertilized egg from which each of us originated to divide, and the resulting mass of cells to move around, grow, absorb nourishment, and take shape as a unique individual? What makes each of us distinct from other individuals but gives all of us a basic similarity as members of the human species? And what allows discerning relatives hovering over a new arrival to proclaim that it has its father's nose and its mother's crooked smile? The answer to all these questions is **genetic information**. The various different units that make up the total of an organism's genetic information (units governing characteristics such as hair color, nose size, and blood type) are called **genes**.

Each of us is a unique individual because each of us has a unique combination of genes. Genes contain information dictating what proteins our cells make, and in what proportions, and how the production and interaction of our proteins can be influenced by the environment in which we develop and live. Identical twins, who

"Dr. Wilshaw here has a special talent for recombinant DNA . . ."
(Rosemary Smith)

share identical genetic information, differ from each other only to the extent that they are exposed to different environments.

We received our genetic information from our parents: half from our mother's egg and half from our father's sperm. During the last century, biologists studied cells as they divided and formed eggs and sperm. Before cell division, the chromosomes separated from each other in such a way that half of the chromosomes from the original cell ended up in each egg or sperm. This pattern convinced biologists that chromosomes were the bearers of genetic information. One problem remained, however; chromosomes were composed of two substances, DNA (Section 3-F) and proteins. Which carried the genetic information?

Obviously genes must contain a variety of information. Scientists had already discovered that proteins were a diverse and complex group of molecules, but no one knew much about DNA beyond the fact that it consists of four types of nucleotide monomers, or subunits, in contrast to the 20 different amino acid monomers commonly found in proteins. DNA was assumed to be built of monotonously repeating units of the four nucleotides. Because the structure of chromosomal proteins is more complicated than that of DNA, most scientists thought the proteins carried the genetic information, and it was not until the 1940s that this was shown to be wrong.

In this chapter we shall examine the evidence showing that the genetic material is in fact DNA, and we shall see how scientists worked out the structure of DNA. This structure is one of the simplest, and at the same time the most elegant, of evolutionary achievements; it dictates the duplication of DNA in exact copies, which are then passed on to future generations of cells.

The elucidation of the structure of DNA touched off a flurry of experiments on the molecular basis of inheritance, often called **molecular biology**—one of the most exciting fields of modern science. In less than 40 years, molecular biologists have progressed from not even knowing that genes were made of DNA to the point where they can place a catalogue order for the ingredients needed to synthesize particular genes in the laboratory. New discoveries are announced with bewildering frequency, discoveries still so new that precisely what they mean is anyone's guess.

We cover the story of DNA roughly in the order in which it unfolded. In this way, we can follow the logical progression from one discovery to the next, which is more obvious in this area than in most other areas of biology.

9-A   Evidence That DNA Is the Genetic Material

Bacterial Transformation

The first evidence to cast doubt on the belief that genes are made of proteins came from studies of **bacterial transformation**, the transfer of genetic characteristics from one bacterial cell to another.

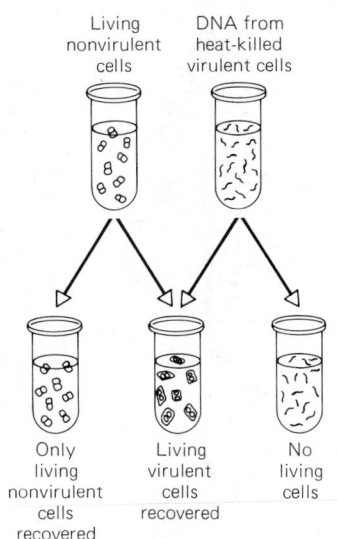

Living nonvirulent cells

DNA from heat-killed virulent cells

Only living nonvirulent cells recovered

Living virulent cells recovered

No living cells

Figure 9-1 Bacterial transformation by DNA from heat-killed virulent cells.

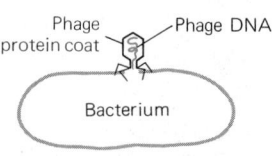

PHAGE INJECTS DNA INTO BACTERIUM; PROTEIN COAT REMAINS OUTSIDE

Phage protein coat

Phage DNA

Bacterium

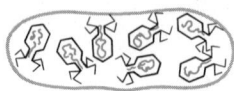

BACTERIUM PRODUCES MORE PHAGE

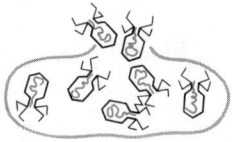

BACTERIUM RUPTURES, RELEASING PHAGE

Figure 9-2 Infection of a bacterium by a bacteriophage. The virus injects its DNA, but its protein coat remains outside the bacterium. Many new phages are produced inside the bacterium, indicating that DNA, not protein, is the phage hereditary material.

In 1944, O. T. Avery, C. M. MacLeod, and M. McCarty investigated the chemical nature of the substance that could alter the genetic characteristics of the recipient when transferred from one bacterium to another. They separated carbohydrate, lipid, protein, DNA, and RNA from heat-killed virulent (disease-causing) bacteria, and added each of these fractions to a separate suspension of living, nonvirulent bacteria. They then cultured the bacteria in each suspension (Figure 9–1) and found that some of the living cells exposed to DNA from the dead bacteria had become virulent, whereas those treated with the other fractions from the dead bacteria were unchanged. Treatment with DNA had endowed the living bacterial cells with genetic characteristics that they had not possessed before.

The obvious conclusion—that DNA is the genetic material—was not widely accepted at the time. This was mainly because few biologists at the time considered that bacteria were "real" organisms. What did it matter if DNA caused genetic transformation in bacteria? It was nearly ten years before biologists realized that genetic information takes essentially the same form in viruses and in all organisms, from bacteria to plants and animals. We now know that, during bacterial transformation, transforming DNA becomes incorporated into the genetic material (DNA) of the recipient bacterium, and it endows the recipient cell with any genetic traits carried by the transforming DNA. Ironically, most of the major discoveries in molecular biology have been made using bacteria. Good techniques for working with eukaryote* DNA were not developed until the 1970s.

Bacteriophages

In 1952, Alfred Hershey and Martha Chase reported experiments with bacteriophages that lent strong support to the theory that DNA is the genetic material. **Bacteriophages** (known to their intimates as **phages**) are viruses that attack bacteria. A virus particle consists of a molecule of DNA (or RNA) inside a protein coat. A phage attacks a bacterium by injecting its genetic material into the bacterial cell; the phage genetic material then somehow "takes over" the genetic machinery of the bacterium, with the result that the bacterium makes anywhere from a few to hundreds of new phages. Eventually the bacterium bursts, releasing these newly formed viruses.

Hershey and Chase used radioactive isotopes* that permitted them to distinguish between the protein coat and the DNA of phages. Most protein contains sulfur but not phosphorus, whereas DNA contains phosphorus but not sulfur. Hershey and Chase labeled phage protein with radioactive sulfur and phage DNA with radioactive phosphorus and found that when bacteria were infected with phage, the radioactive phosphorus in the phage always entered the bacterium, while all of the radioactive sulfur remained outside (Figure 9–2). This was strong evidence that the phage genetic material contained only DNA and no protein. This experiment shows how microorganisms can often provide information that cannot be obtained easily from higher organisms. Plant or animal cells could not have been used because considerable quantities of protein are tightly bound to the DNA of the chromosomes of these higher organisms.

Quantity of DNA in Cells

Circumstantial evidence that DNA is also the genetic material in higher organisms came from measurements of the amount of DNA in different cells. For example, the body cells from the liver and kidneys of a chicken contain the same amount of DNA as each other and twice as much as sperm, which are reproductive cells. (Since two reproductive cells—sperm and egg—combine to form the new individ-

---

*Eukaryotes All organisms other than bacteria; fungi, plants, protists, and animals.

*Isotopes Atoms of an element that have different numbers of neutrons in the nucleus and hence different atomic weights. A radioactive isotope emits radioactive particles because it is unstable.

ual, each reproductive cell must have half as much genetic information as a body cell, or else the amount of genetic information would double in each generation.) This distribution of DNA is what one would expect of the genetic material. The distribution of protein in cells, on the other hand, varies considerably from one tissue to another and is not necessarily lower in reproductive cells. This makes it less reasonable to suppose that protein is the genetic material.

Base Content of DNA

Possibly the most convincing biochemical evidence that DNA is the genetic material came from the discovery that all members of any species of organism contain DNA with almost exactly the same chemical composition, which is not true of the proteins.

DNA is made up of nucleotides. Each nucleotide contains (a) a nitrogenous base derived from either pyrimidine or purine; (b) a pentose sugar, deoxyribose; and (c) a phosphate group (see Figure 9–3). DNA contains four types of nucleotides, which are identical in their sugar and phosphate groups. They differ only in their nitrogenous bases, which may be either the double-ring purine derivatives adenine (A) or guanine (G), or the single-ring pyrimidine derivatives cytosine (C) or thymine (T) (see Figure 3–26).

Until 1949, it appeared that DNA contained the four nitrogenous bases in roughly equal amounts. Then, however, Erwin Chargaff and his coworkers reported that the DNA of different species has a different content of bases (Table 9–1). On the other hand, DNA from different members of a single species, or from different tissues of a single individual, has the same base composition. Table 9–1 also shows that, within the limits of experimental error, the DNA of a given species contains equal numbers of adenine and thymine nucleotides and equal numbers of guanine and cytosine nucleotides. This finding eventually became a major clue in the elucidation of the molecular structure of DNA.

TABLE 9–1 **APPROXIMATE AMOUNTS AND RATIOS OF NITROGENOUS BASES FROM DIFFERENT ORGANISMS**

| | BASE COMPOSITION IN MOLE* PERCENT | | | | BASE RATIOS | |
| | Purines | | Pyrimidines | | | |
	A	G	C	T	A/T	G/C
Animals						
Human	30.9	19.9	19.8	29.4	1.05	1.00
Chicken	28.8	20.5	21.5	29.2	1.02	0.95
Locust	29.3	20.5	20.7	29.3	1.00	1.00
Plant						
Wheat	27.3	22.7	22.8	27.1	1.01	1.00
Fungus						
Yeast	31.3	18.7	17.1	32.9	0.95	1.09
Bacterium						
Escherichia coli	24.7	26.0	25.7	23.6	1.04	1.01
Bacteriophage						
Phage T$_7$	26.0	24.0	24.0	26.0	1.00	1.00
Phage λ	21.3	28.6	27.2	22.9	0.92	1.05

*Mole Gram molecular weight; a way of measuring a substance in terms of the number of molecules it contains (Section 2-D). Equal numbers of moles contain equal numbers of molecules regardless of the molecular weight of a substance, and so molar measurements are used when comparing the ratios of numbers of molecules of various substances present.

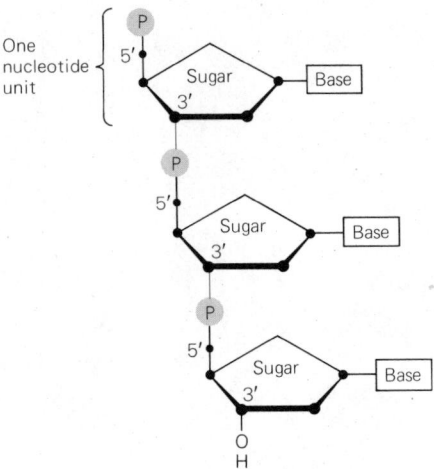

One nucleotide unit

Figure 9-3 Linkage of nucleotides into a strand. Phosphate groups are indicated by P in a colored circle.

9-B The Structure of DNA

By the early 1950s there was compelling evidence that DNA carries a cell's genetic information, and many people were trying to work out how this large molecule was constructed. Any model of the structure of DNA had to take into account the following facts:

1. DNA is made up of nucleotides.

2. When the nucleotides are linked together in a strand of DNA, the phosphate group attached to the 5′ (pronounced "five prime") carbon of the deoxyribose of one nucleotide becomes joined to the 3′ carbon on the deoxyribose ring of an adjacent nucleotide (Figure 9–3). The string of alternating phosphate and sugar groups, called the sugar-phosphate backbone, is held together by covalent bonds. The backbone is not symmetrical; it has a definite orientation, with a free 3′ hydroxyl group at one end and a free 5′ phosphate group at the other. The purine and pyrimidine bases stick out to one side of the sugar-phosphate backbone.

3. In each DNA molecule the number of nucleotides containing adenine equals the number of nucleotides containing thymine (A = T), and the number of nucleotides containing cytosine equals the number of nucleotides containing guanine (C = G) (Table 9–1).

4. There is so much DNA in a eukaryotic cell that if the nucleotides composing a single chromosome were strung out in a straight line, the DNA molecule would be more than a metre long. DNA molecules must be compressed in some way to permit them to fit into a cell's nucleus.

5. Most molecules with regular structure produce specific x-ray diffraction pictures. The most direct evidence for the structure of DNA came from studies of the way in which highly purified fibers of DNA diffracted x-rays. In 1952, Rosalind Franklin produced x-ray diffraction photographs of DNA which indicated that DNA was arranged in a spiral, or helix, with the bases oriented perpendicular to the fiber. The photographs also provided evidence that the sugar-phosphate strands which formed the backbone of the DNA fiber were on the outside of the molecule, and the bases were on the inside. Reflections along the fiber axis indicated that one turn of the helix contained 10 nucleotides.

At the time Franklin made these pictures, there was a frantic race going on to solve the mystery of the structure of DNA. Linus Pauling, an American chemist, had already published a model of the structure that turned out to have an embarrassingly elementary flaw, and he was bent on retrieving his reputation. Maurice Wilkins, head of the laboratory where Franklin worked in London, was also keenly on the scent. The first people to fit all the evidence together in a workable form were James Watson, a postdoctoral researcher, and Francis Crick, an erstwhile physicist; they used a set of scale models of nucleotides to build and rebuild possible structures until they found one that fitted all the data.

The structure worked out by Watson and Crick consists of two strands of DNA; the strands are arranged like a ladder, with the sides being the sugar-phosphate backbones of the two strands and the rungs being the bases. Each rung consists of a single-ring base attached to one DNA strand and a double-ring base on the other strand. A rung may consist of either an adenine opposite a thymine, or a guanine opposite a cytosine; in each rung, either base may be on either strand. The pair of bases in each rung is held together by hydrogen bonds. Two hydrogen bonds hold an adenine-thymine pair together, and three bonds hold a guanine-cytosine pair together (Figure 9–4). A and T, and G and C, form the most stable combinations of hydrogen bonds, and this explains Chargaff's finding that A = T and G = C in the DNA of any species. Since each pair consists of one single-ring and one double-ring

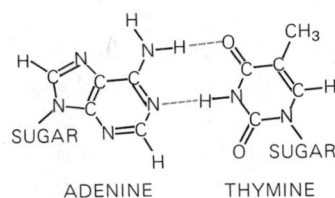

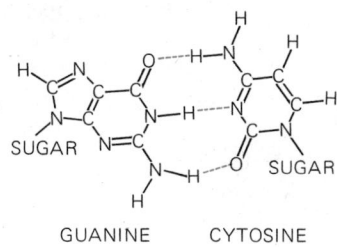

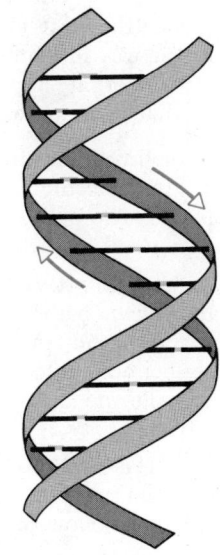

Figure 9-4 Hydrogen bonding (color) between base pairs of DNA.

Figure 9-5 Structure of DNA. Colored arrows indicate that the two strands are antiparallel. Light color indicates the hydrogen bonds between the bases that hold the two strands together.

base, all the rungs of the ladder are the same width, and the backbones of the two DNA strands are always the same distance from one another, instead of bending in and out. Watson and Crick also saw that for hydrogen bonds to form properly between the base pairs in DNA, the two nucleotide strands of the DNA molecule had to run in opposite directions; that is, they are antiparallel, with the free 5'-phosphate groups of the two strands at opposite ends of the molecule. Finally, the whole ladder is twisted, with ten bases per turn, to form the spiral detected by the x-ray photographs; because the spiral is composed of two strands wound around each other, the DNA molecule is referred to as a double helix (Figure 9–5).

Tremendously excited by the simple yet elegant structure they had worked out, Watson and Crick hurriedly wrote up a two-page paper for quick publication. As Watson remarked, "It was too pretty not to be true." Watson, Crick, and Wilkins received a Nobel Prize for their work on DNA in 1962.

9-C DNA Replication

Before a cell divides, DNA replicates, or duplicates itself, and both the offspring of the cell receive the same genetic information as that in the parent cell. Watson and Crick pointed out that the double-stranded, base-paired structure of the DNA molecule suggested a means whereby the genetic information could be replicated accurately. The two strands of the DNA helix contain complementary information in that the nucleotide sequence of one strand supplies the information needed to produce its partner strand (e.g., a strand that runs A—T—G—C—A—A must have a partner that runs T—A—C—G—T—T). If the two strands of a DNA molecule are separated, each strand can be used as a mold, or template, to produce a complementary strand. The template and its complement together then form a new DNA molecule, identical with the original molecule. Watson and Crick suggested that this was in fact how DNA replicated. Their paper stimulated a flurry of experiments designed to test this hypothesis.

In 1958, Matthew Meselson and Franklin Stahl published the first convincing evidence for the way in which DNA replicates. Meselson and Stahl formulated the following three alternate hypotheses (Figure 9–6), any one of which would account for the facts that were known about DNA replication:

1. **Conservative replication.** In this hypothesis, the two strands of the parent DNA molecule act as templates for a completely new double-stranded molecule. The parent DNA molecule is preserved intact and goes into one daughter cell, and the new molecule goes into the other daughter cell.

2. **Semi-conservative replication.** According to this hypothesis, the two strands of the parental molecule separate like the two sides of a zipper, and each acts as the template for formation of one new strand, which then becomes bound to it to form a complete molecule. Each daughter cell inherits a DNA molecule that is a hybrid, consisting of one new and one parental strand.

3. **Dispersive replication.** The parental DNA is broken up into short segments used as templates for the formation of two new double helices, which are then somehow joined together.

In a beautifully simple series of experiments, Meselson and Stahl showed that the first and third of these hypotheses could be disproved, and provided strong support for the theory of semi-conservative replication. (Note that a hypothesis can never be proved, because we can never be certain we have taken all possible factors into account; however, it is possible to disprove a hypothesis if the data do not agree with predictions of what should happen if the hypothesis is true.)

Meselson and Stahl needed a way to distinguish between "old" and "new" DNA. They did this using two different isotopes of nitrogen: ^{14}N, the more common isotope, and ^{15}N, a heavier isotope. First Meselson and Stahl grew bacteria in a nutrient medium containing ^{15}N for several generations; now virtually all the bacteria had DNA containing ^{15}N. They next transferred the bacteria to a nutrient me-

Figure 9-6 Alternative hypotheses of DNA replication tested by Meselson and Stahl.

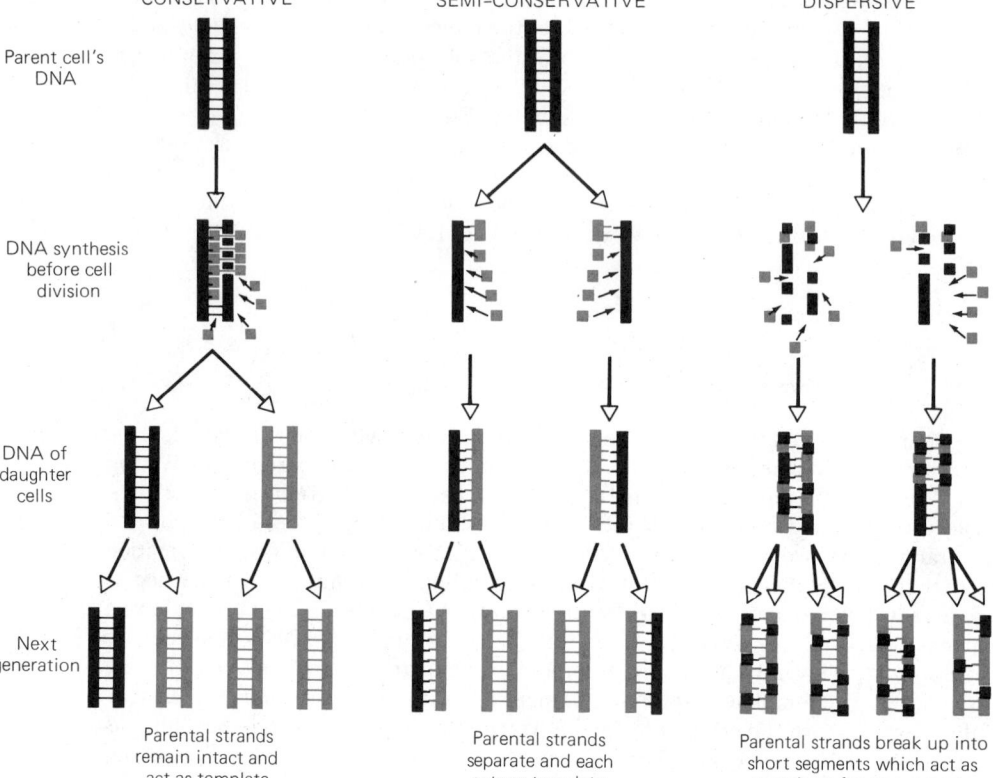

CONSERVATIVE · SEMI-CONSERVATIVE · DISPERSIVE

Parent cell's DNA

DNA synthesis before cell division

DNA of daughter cells

Next generation

Parental strands remain intact and act as template for new double helix.

Parental strands separate and each acts as template for its new complement.

Parental strands break up into short segments which act as templates for new strands. Daughter helices have varying amounts of parental nucleotides.

dium containing ¹⁴N, and then removed cells after one, two, and three generations had passed. Next, the DNA from the cells in each generation was purified, and it was then ready to be analyzed to find out the distribution of ¹⁵N and ¹⁴N.

To do this, Meselson and Stahl placed each batch of isolated DNA on the surface of a salt solution in its own tube, and spun all the tubes in a centrifuge (see Figure 4–10). As the tubes spun, the highest-density DNA migrated downward through the salt solution the farthest; thus DNA containing ¹⁵N migrated toward the bottom farther than ¹⁴N DNA. After spinning the tubes, the investigators could actually see bands of DNA of different density by holding the tubes up to the light. Figure 9–7 shows what patterns of bands would be expected with each hypothesis of DNA replication.

Meselson and Stahl found that in the parental generation, grown in ¹⁵N, all the DNA contained only ¹⁵N. However, when these bacteria were grown in ¹⁴N and allowed to divide to produce one new generation, the DNA of their daughter cells contained both ¹⁴N and ¹⁵N, and settled in a position between that of pure ¹⁴N DNA and pure ¹⁵N DNA. This finding ruled out the possibility that replication is conservative, for in this case, two separate bands, one of ¹⁴N DNA (newly synthesized) and one of ¹⁵N DNA (parental) would be expected. Thus the first of the three hypotheses was incompatible with the experimental evidence.

The evidence from the next generation of bacteria permitted Meselson and Stahl to distinguish between the two remaining hypotheses. When the bacteria originally grown in ¹⁵N had been grown for two generations in a medium containing only ¹⁴N, the DNA isolated from them formed two distinct bands. About half of the DNA was hybrid ¹⁴N-¹⁵N DNA, as in the first generation, and the other half contained only ¹⁴N DNA. This result is incompatible with dispersive replication, which would produce one diffuse band lying between the two actually found. However, this pattern was compatible with semi-conservative replication. Thus Meselson and Stahl concluded that, of their three hypotheses, only semi-conservative replication was supported by the experimental evidence.

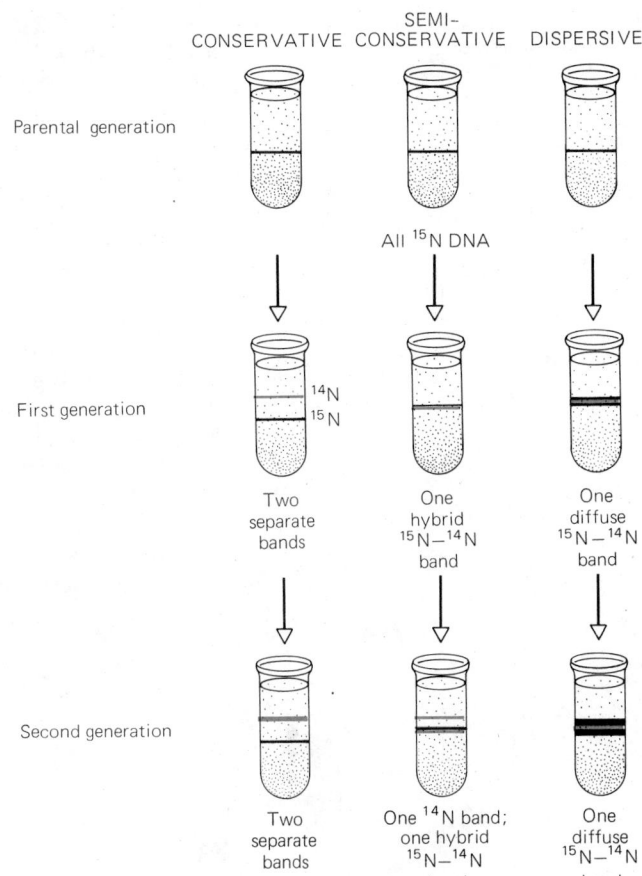

Figure 9-7 Banding patterns expected according to each hypothesis of DNA replication described in Figure 9-6. Meselson and Stahl found the pattern expected for semiconservative replication when they isolated DNA from successive generations of *Escherichia coli.*

Meselson and Stahl worked with bacteria, which are prokaryotes; their DNA lacks the histone proteins completed with the DNA of eukaryotes. Are their results also true for eukaryotes? In 1957, the English biologist J. H. Taylor, using a technique that gives less clear-cut results than the technique used by Meselson and Stahl, had found evidence that the chromosomes of bean seedling root cells (eukaryotes) also replicate in a semi-conservative fashion. This seems to be true for all organisms.

DNA replication begins with the separation of the two strands along the weak hydrogen bonds linking the paired bases. An enzyme then moves along the chain, attaching nucleotides to form the sugar phosphate backbone of a new strand.

Enzymes of DNA Replication

A great deal of research has been devoted to identifying the enzymes that catalyze DNA replication, but we still know very little about this process.

In 1956, Arthur Kornberg and his colleagues at Stanford described the properties of an enzyme now known as **DNA polymerase I**, isolated from *Escherichia coli*. It is now known that DNA polymerase I repairs damaged segments of DNA and also removes bases that have been incorrectly paired during replication.

In 1972, Thomas Kornberg and Malcolm Gefter discovered an enzyme they called **DNA polymerase III**. This seems to be the enzyme primarily responsible for DNA replication, at least in *Escherichia coli*.

In a replicating cell, that DNA which has been formed most recently occurs in short pieces known as **Okazaki fragments** (after their discoverer). Apparently, DNA is replicated to form Okazaki fragments, which are then quickly joined together by an enzyme known as **DNA ligase**. The discovery of a protein known as "swivelase," which unwinds the double helix, has helped to solve the problem of how DNA can uncoil fast enough to permit DNA synthesis to proceed as rapidly as it does.

It is clear that the replication of DNA demands the integrated action of a number of enzymes, but we do not yet know enough to construct a comprehensive model of the process, and almost nothing is known about how DNA replication is regulated.

9-D Structure of Eukaryotic Chromosomes

The genetic material of a prokaryote is one double helix of DNA with its ends joined to each other to form a circle (Figure 9–8). Researchers have found it much harder to work out the precise structure of the many linear chromosomes in the nucleus of a eukaryotic cell. Except during cell division, the chromosomes exist as a mass of fine threads known as **chromatin**, which contains DNA and basic (alkaline) proteins in roughly equal amounts. In chromatin, the DNA double helix with its proteins is supercoiled and looped many times.

We now know that each eukaryotic chromosome contains a single DNA molecule extending from one end of the chromosome to the other, but folded up like a concertina. In 1974, biochemical analysis and electron micrographs showed that the DNA molecule is wound around proteins called **histones**, forming a string of particles called **nucleosomes** (Figure 9–9). One possible role of histones is to protect the

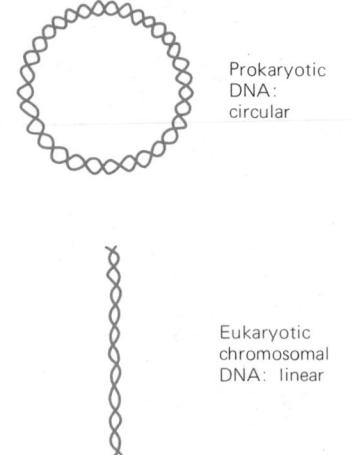

Figure 9-8 DNA is circular in prokaryotes, linear in eukaryotes.

Prokaryotic
DNA:
circular

Eukaryotic
chromosomal
DNA: linear

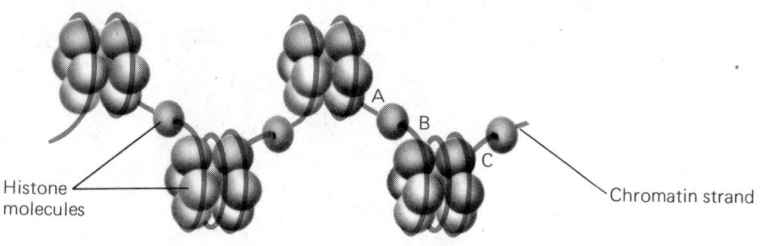

Histone
molecules

Chromatin strand

Figure 9-9 Nucleosomes. Closeup view of part of a eukaryotic chromosome. The chromatin strand is made up of a DNA double helix with its associated protein. Histones are globular proteins. The distance from A to B is about 60 base pairs on the DNA; from A to C is about 200 base pairs.

DNA from being broken down by enzymes in the cell nucleus that attack free DNA. Histones are not the only proteins found in a chromosome, and it is clear that much still remains to be learned about the roles of chromosomal proteins.

In 1977, researchers found methods of determining the sequence of nucleotides in DNA molecules. This is important because the sequence of nucleotides ultimately determines the sequence of amino acids in proteins, as we shall see in Chapter 10. The nucleotide sequences of lengths of DNA from many viruses, bacteria, and eukaryotes have already been determined, and many more DNAs are under examination.

Repetitious DNA

For a number of years, biologists have known that some nucleotide sequences in eukaryotic DNA are repetitions of sequences found elsewhere—on the same or on another chromosome. Our new-found ability to determine the order of nucleotides in DNA has turned up many more of these repeating sequences. In some cases, the reason for a repeat is, at least intuitively, obvious. For instance, every cell contains hundreds of copies of the genes (lengths of DNA) needed to synthesize ribosomes (the organelles of protein synthesis) and histones. These are things that the cell needs occasionally in large amounts (for instance, at the time of cell division), and it is reasonable to suppose that having multiple copies of this genetic information speeds up the synthesis of ribosomes or of the histones needed for new chromosomes. The usefulness of most repetitive DNA, however, is more obscure. In the fruit fly *Drosophila*, for instance, the brief nucleotide sequence A—G—A—A—G is repeated about 100,000 times in the middle of one chromosome! This highly repetitive area is not genetic information used to direct the synthesis of proteins; it may be somehow important to the structure of the chromosome or to the movement of the chromosome during cell division (Chapter 11). However, speculation about the function of repetitive DNA is really mere guesswork at this stage of our knowledge.

9-E Bacterial Restriction Enzymes and Recombinant DNA

In 1978 Werner Arber, Hamilton Smith, and Daniel Nathans won the Nobel Prize for their experiments on bacterial restriction enzymes, enzymes that have revolutionized molecular biology by permitting biologists to analyze genes.

Restriction enzymes are natural defenses of bacteria against invading phages. The enzymes break up the invading viral DNA into useless fragments. Each of the more than 150 restriction enzymes now known attacks double-stranded DNA at a point in the middle of a particular nucleotide sequence (Figure 9–10). (Bacterial DNA is protected from attack by its own restriction enzymes by specific defenders called modification enzymes.) The specificity of restriction enzymes was the key to methods that now permit us to determine the sequence of nucleotides in DNA directly, methods that won the Nobel Prize for Frederick Sanger and Walter Gilbert in 1980, a scant three years after they published their techniques. Before 1977, the nucleotide sequence of DNA could be determined only indirectly, by examining the sequence of amino acids in a protein coded for by the DNA. The disadvantage of such an indirect method is that it produces information only about lengths of DNA that actually dictate the amino acid sequence of a protein. We now know that every organism contains a lot of DNA that does not code for proteins (see Chapter 10).

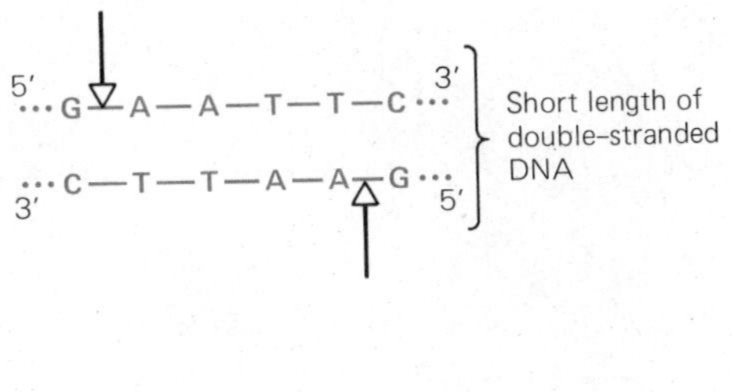

Figure 9-10 The action of one restriction enzyme, called *Eco R1* from *Escherichia coli. Eco R1* cleaves double-stranded DNA between G and A (arrows) wherever the sequence A-T-T-C lies on the 3′ side of a G-A sequence.

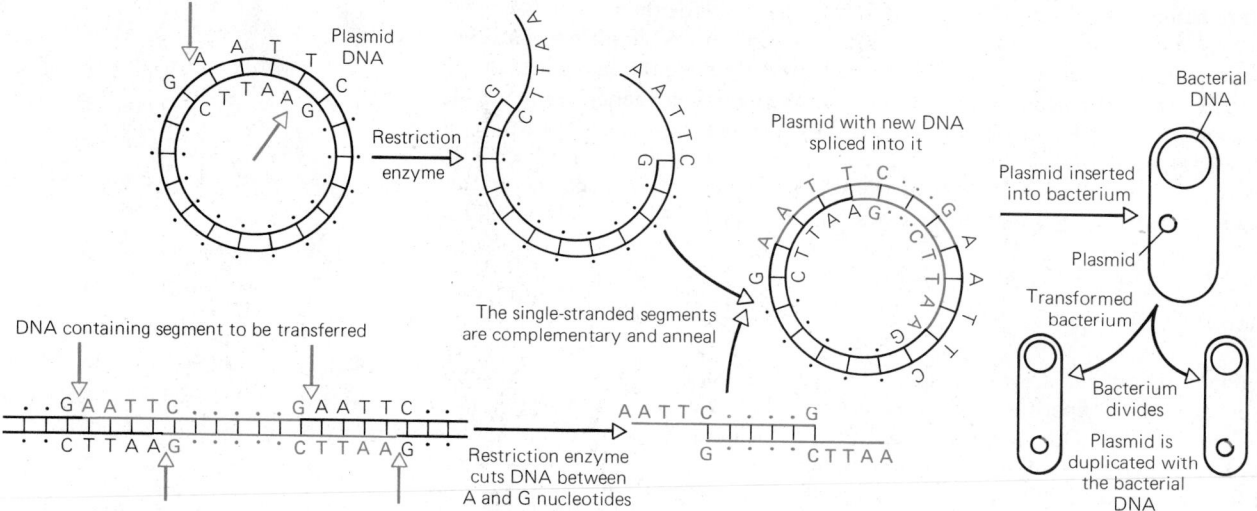

Figure 9-11 How recombinant DNA is made and cloned.

Labels in figure:

Plasmid DNA

Restriction enzyme

Plasmid with new DNA spliced into it

The single-stranded segments are complementary and anneal

DNA containing segment to be transferred

Restriction enzyme cuts DNA between A and G nucleotides

Bacterial DNA

Plasmid inserted into bacterium

Plasmid

Transformed bacterium

Bacterium divides

Plasmid is duplicated with the bacterial DNA

Another important use of restriction enzymes is in producing recombinant DNA. **Recombinant DNA** is made by combining DNA from more than one source. For example, a particular gene can be removed from a human or other animal cell and introduced into a bacterium, usually *Escherichia coli.* The newly introduced foreign gene will be treated as part of the bacterial DNA; it will be replicated with the bacterial DNA before the cell reproduces, and it may be used to produce the protein for which it codes. Because bacteria reproduce at enormous rates, this method permits biologists to **clone** (make many identical copies of) any particular gene for analysis, and also to make a great deal of its protein product comparatively inexpensively.

The technique for cloning recombinant DNA is shown in Figure 9–11. Restriction enzymes are used to cut the gene to be cloned (with some DNA on either side of it) out of the chromosome where it normally occurs. This is more easily said than done; in most cases we don't even know which chromosome contains a particular gene, much less where the gene lies in that chromosome. It is necessary to prepare a number of DNA fragments, insert them into bacteria, and then screen the bacteria to find the ones with the desired gene.

Restriction enzymes are also used to open up the DNA of a carrier that will transport the gene into its *Escherichia coli* host cell. The carrier is either a phage or a **plasmid,** a small, circular DNA molecule that exists outside the main DNA molecule in some bacteria (Figure 9–12). Many restriction enzymes leave short lengths of

Figure 9-12 A plasmid from a bacterium. (Biophoto Associates)

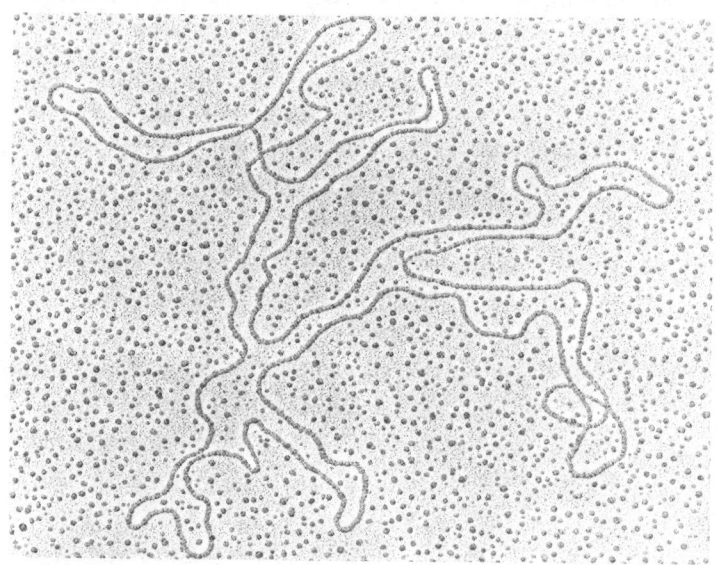

single-stranded DNA at the ends of the pieces they cut. These "sticky ends" will bond, by base-pairing, to other single-stranded DNA, and so will splice the gene to be cloned into the plasmid when the two are mixed. Next, plasmids carrying the new gene are mixed with *E. coli* cells that have been made more permeable than usual by osmotic shock.* Some of the plasmids enter bacteria, where they replicate with the growing population of *E. coli.*

Recombinant DNA techniques can potentially provide us with many useful proteins; some may even be on the market by the time you read this book. Most commercial interest as we write centers on interferon, a protein used to fight viral infections, and on insulin, a protein hormone needed by some diabetics. Both are expensive to extract and purify from animal tissue, and they would be much cheaper if produced by recombination. However, it has proved difficult to produce the pure proteins, probably because genes are sometimes altered during their transfer into bacteria. It should be possible to overcome such technical difficulties.

The ability to move genes from one organism to another opens up vistas of many sorts of **genetic engineering**, creating genetically novel organisms or even providing genes that may be missing in a genetically defective individual. The possibility of introducing new genes into human beings presents serious ethical problems which must be addressed sooner or later.

---

*Osmotic shock   A sudden change in osmotic potential of the medium surrounding a cell.

The evidence that DNA is the genetic material in all organisms from bacteria to oak trees may be summarized thus:

SUMMARY

1. DNA is the substance that transfers genetic information from one cell to another in the process of bacterial transformation.

2. When a phage takes over the genetic machinery of a bacterium, only its genetic material, DNA, enters the bacterium; its protein coat remains outside the bacterial cell wall.

3. All the body cells of the individuals belonging to one species contain the same quantity of DNA. Reproductive cells of the same species contain half this amount. No other substance shows this pattern.

4. The DNA of all members of any species has the same percentages of the four nitrogenous bases.

A DNA strand is made up of a backbone of deoxyribose sugars alternating with phosphate groups. Each deoxyribose sugar is linked to one of four nitrogenous bases, adenine, guanine, cytosine, or thymine. Each DNA molecule consists of two parallel strands of nucleotides oriented in opposite directions. The bases in these nucleotide strands are hydrogen-bonded to their complements on the opposite strand, forming adenine-thymine and guanine-cytosine pairs. The two strands are twisted to form a double helix. A prokaryotic cell's genetic material consists of a DNA double helix with its ends joined to form a circle; a eukaryotic cell nucleus contains several chromosomes, each consisting of a linear molecule of DNA combined with proteins and supercoiled into many loops. Eukaryotic chromosomes contain many repetitive DNA sequences; the functions of most of these repeated sequences are not yet known.

DNA replicates semi-conservatively; the double helix unwinds and each strand serves as a template for the formation of a complementary strand of DNA. Several enzymes are responsible for DNA replication. The complete sequence of events is not yet fully understood, but DNA polymerase III probably forms new DNA as Okazaki fragments, which are then linked by DNA ligase to form the new strand. DNA polymerase I is thought to repair breaks and to correct mistakes in the molecule.

Bacterial restriction enzymes, discovered in the 1970s, are used to make recombinant DNA. By producing eukaryotic proteins in prokaryotic cells, we may be able to provide less expensive treatments for some diseases. The technology of genetic engineering holds out many hopes but also poses many ethical problems.

OBJECTIVES

From your study of this chapter, you should be able to:

1. Describe and explain the evidence that DNA is the genetic material, using these studies as evidence: (a) bacterial transformation; (b) infection of bacteria by bacteriophages; (c) the quantity of DNA in body cells and reproductive cells of a species; (d) comparison of the base composition of DNA in cells from members of the same and different species.

2. Describe the structure of a nucleotide.

3. Describe the structure of a molecule of DNA, and explain why the number of adenine bases in the molecule equals the number of thymine bases and the number of guanine bases equals the number of cytosine bases.

4. Explain the meaning of the terms 3' end and 5' end in relation to a DNA molecule.

5. Describe the experiments of Meselson and Stahl and explain how they provided evidence that DNA replicates in a semi-conservative fashion.

6. Compare and contrast the structures of a eukaryotic chromosome and prokaryotic DNA.

7. Briefly describe how recombinant DNA is made, and state the medical importance of this technology.

SELF-QUIZ

1. DNA is believed to be the genetic material because:
   a. all the body cells of an individual seem to have identical amounts and compositions of DNA, while reproductive cells have half the amount of DNA found in body cells.
   b. the proteins are the same from cell to cell in an individual, but the DNA differs; thus the DNA must be the material that makes different tissues different.
   c. DNA is the largest type of macromolecule found in living organisms.
   d. DNA is found in the cell nucleus.

2. Draw the complementary strand for this DNA template:

$$3' \text{ sugar—A}$$
$$|$$
$$P$$
$$|$$
$$\text{sugar—C}$$
$$|$$
$$P$$
$$|$$
$$\text{sugar—T}$$
$$|$$
$$5' \quad P$$

3. In a DNA molecule, the sugars:
   a. bond covalently to phosphate groups
   b. bond covalently to nitrogenous bases
   c. bond to nitrogenous bases by hydrogen bonds
   d. bond to both phosphate groups and nitrogenous bases by covalent bonds
   e. bond to phosphate groups by ionic bonds and to nitrogenous bases by hydrogen bonds

4. In the first generation of Meselson and Stahl's experiment, the results showed a hybrid band of DNA containing both ^{14}N and ^{15}N. Which of the following is the best interpretation of these results?
   a. the results are consistent with semi-conservative replication
   b. the results support both semi-conservative and dispersive replication
   c. the results are not consistent with conservative replication
   d. neither dispersive nor conservative replication can take place

5. If replication were conservative, which of the following banding patterns would Meselson and Stahl have found in the fourth generation? (Note: concentration refers to concentration of DNA.)
   a. a high concentration band containing only ^{15}N DNA and a high concentration band containing only ^{14}N DNA.
   b. a low concentration hybrid band containing both ^{14}N and ^{15}N DNA and a high concentration band containing only ^{14}N DNA
   c. a high concentration band containing only ^{14}N DNA
   d. a high concentration band containing only ^{14}N DNA and a low concentration band containing only ^{15}N DNA

6. List two similarities and two differences between prokaryotic DNA and a eukaryotic chromosome.

7. Which of the following would not be used in preparing recombinant DNA?
   a. plasmids
   b. phages
   c. DNA polymerase III
   d. restriction enzymes
   e. DNA from two different sources

QUESTIONS FOR DISCUSSION

1. Why is it necessary for eggs and sperm or pollen to contain only half the amount of genetic material found in the other cells of the body?

2. Why is the constancy of DNA content from cell to cell in an organism considered to be evidence that DNA is the genetic material? Is it necessary for cells of an organism to contain identical genetic information? Is it possible for an organism to have different genetic information in different cells of the body?

3. Given the liver of a freshly killed beef steer, how would you determine the amount of DNA per beef liver cell (as opposed to per gram of tissue)?

4. What is the biological importance of the fact that the sugar-phosphate backbones of the DNA double helix are held together by covalent bonds, and that the cross-bridges between the two strands are held together by hydrogen bonds?

5. Various errors can be made in the replication of DNA. Portions of a chromosome may be copied more than once, or may be copied in such a way that they end up in an abnormal position in the chromosome (e.g., ABDECF instead of ABCDEF, where A, B, C, etc., represent long sequences of DNA). In some other errors, portions of a chromosome are omitted entirely (e.g., ABDEF). Using the description of chromosome structure and replication presented in this chapter, can you explain how such accidents might happen?

6. X-rays and ultraviolet rays damage DNA. *Escherichia coli* cells that lack DNA polymerase I are especially sensitive to these rays. Why do you think this is so?

7. Can you think of any dangers and benefits, other than those listed in the chapter, that might result from research on recombinant DNA?

REFERENCES AND FURTHER READING

Abelson, J. "A revolution in biology." *Science* 209:1319, 1980. Abelson introduces an issue of *Science* completely devoted to the discoveries made possible by methods for sequencing DNA. Not easy reading since this field is developing a jargon all its own.

Avers, C. J. *Genetics*. New York: D. Van Nostrand and Company, 1980. A genetics text with a good description of the evidence that DNA is the genetic material.

Avery, O. T., C. M. MacLeod, and M. McCarty. "Studies on the chemical nature of the substance inducing transformation of pneumococcal types." *Journal of Experimental Medicine* 79:137, 1944. The paper that convinced many people that DNA was the genetic material.

Cairns, J. "The chromosome of *Escherichia coli.*" *Cold Spring Harbor Symposium on Quantitative Biology* 28:43, 1963. The description of the circular DNA molecule of a prokaryote and how it replicates.

Kornberg, A. "Biologic synthesis of deoxyribonucleic acid."

Science 131:1503, 1960. An early study on the enzymes of DNA replication.

Maxam, A. M., and W. Gilbert. "A new method for sequencing DNA." *Proceedings of the National Academy of Sciences* (*U.S.*) 74:560, 1977. The most frequently used method for determining the sequence of nucleotides in DNA. This paper joins Avery *et al.* as one of the few hundred papers that are definitely "classics" in biology.

Meselson, M., and F. W. Stahl. "The replication of DNA in *E. coli.*" *Proceedings of the National Academy of Sciences* (*U.S.*) 44:671, 1958. The classic paper demonstrating semiconservative replication in DNA.

Watson, J. D., and F. H. C. Crick. "Molecular structure of nucleic acids. A structure of deoxyribose nucleic acid." *Nature* 171:737, 1953. The Nobel Prize–winning paper on the structure of DNA.

Watson, J. D. *The Double Helix.* New York: Atheneum, 1968. A personal story of the discovery of structure DNA.

CHAPTER 10

PROTEIN SYNTHESIS

In the 1940s scientists began to suspect that the genetic material of organisms determines the structure of their proteins. George Beadle and Edward Tatum worked with **mutations**, inheritable alterations in the genetic material, in the pink bread mold *Neurospora*. Each mutation they studied changed the fungus's ability to synthesize an enzyme needed in one metabolic pathway or another. All proteins, not just enzymes, are synthesized under genetic control.

The discovery that deoxyribonucleic acid (DNA) is the genetic material rapidly generated the suggestion that the linear order of nucleotides in a DNA molecule somehow determines the order of amino acids in a protein and therefore determines the protein's structure and function (Section 3-E). Since 1953, numerous experiments have given support to this hypothesis by showing how the nucleotide sequence of a segment of DNA is translated into the amino acid sequence of a polypeptide* or protein.* First, DNA directs the synthesis of specific ribonucleic acid (RNA)

---

* Polypeptide   A polymer composed of many amino acid monomers.
* Protein   A functional unit made up of one or more polypeptides.

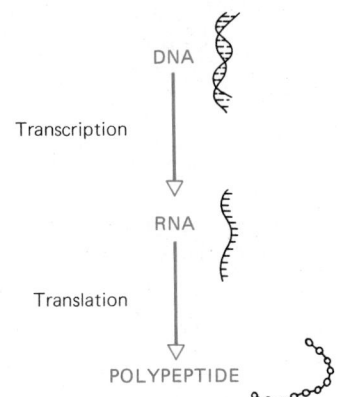

DNA

Transcription

RNA

Translation

POLYPEPTIDE

Figure 10-1 Information flows from the sequence of nucleotides in DNA, to the sequence of nucleotides in RNA, to the sequence of amino acids in a polypeptide or protein. (In some viruses, the genetic material is RNA, in which case the information flow is RNA→DNA→RNA→protein.)

molecules; the nucleotide sequence in one type of RNA molecule, in turn, determines the order in which amino acids will be joined together to form a polypeptide. This transfer of information from DNA to RNA to polypeptides and proteins may be thought of as **gene expression** (Figure 10–1).

Important as genes are, it has always been rather difficult to define a gene, partly because our knowledge and understanding of the term keeps changing. A **gene** is a length of DNA with a particular function; for purposes of this chapter, that function is carrying the information needed to synthesize one kind of RNA or one polypeptide, but genes with other functions exist. For instance, there are lengths of DNA whose function is to control the activity of other genes, and these are often called regulator genes. Perhaps the best advice we can give is not to look for a precise definition of a gene. Like "truth," the word "gene" describes a concept that is fuzzy around the edges.

10-A RNA

RNA, like DNA, comes in long unbranched macromolecules made up of nucleotides. Each nucleotide is made up of a sugar molecule, a nitrogenous base, and a phosphate group (Figure 10-2). The phosphate group of the nucleotide bonds to the 3′ carbon on the sugar of the next nucleotide to form the sugar-phosphate backbone of a nucleic acid (see Figure 9–3).

There are several differences between RNA and DNA. The sugar in RNA is ribose; the sugar in DNA is deoxyribose, which contains one less oxygen atom per molecule than ribose; hence the names ribonucleic acid and deoxyribonucleic acid (Figure 10–3). Another difference between RNA and DNA is that RNA consists of a

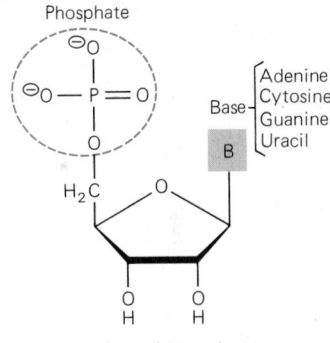

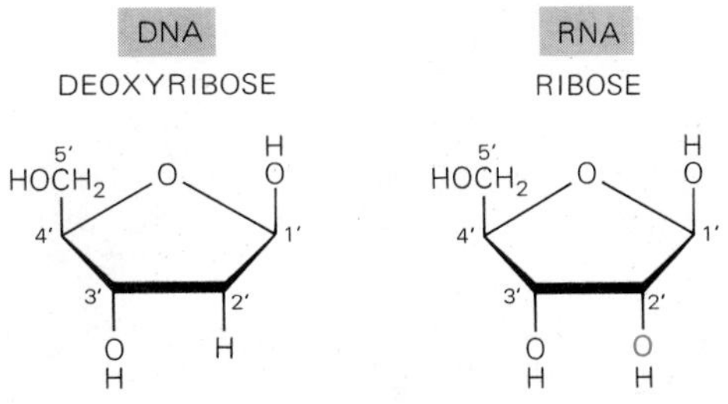

Phosphate

Base {Adenine Cytosine Guanine Uracil

Sugar (Ribose)

Figure 10-2 A ribonucleotide, the basic unit of RNA.

Figure 10-3 Deoxyribose, the sugar found in DNA, lacks the oxygen found on the 2′ carbon of ribose, the sugar found in RNA.

DNA
DEOXYRIBOSE

RNA
RIBOSE

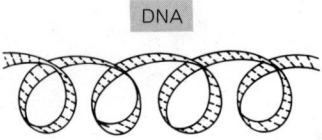

DNA

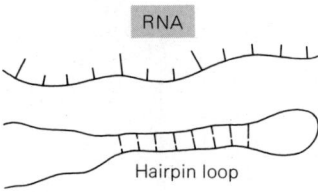

RNA

Hairpin loop

Figure 10-4 Most DNA is double-stranded and wound into a double helix. RNA is always single-stranded.

single strand of nucleotides, although it can form double-stranded sections by folding back on itself in hairpin loops. In contrast to RNA, DNA is usually double-stranded (Figure 10–4), consisting of two complementary chains of nucleotides. The nitrogenous bases found in DNA are adenine, thymine, guanine, and cytosine. RNA also contains adenine, guanine, and cytosine, but uracil is present instead of thymine (Figure 10–5).

There are three main types of RNA in a cell. **Messenger RNA (mRNA)** carries the genetic information—which determines the sequence of amino acids in a protein—from the DNA to the ribosomes, the cell's protein-making machinery. **Transfer RNA (tRNA)** carries amino acids to the ribosomes, where the amino acids are joined together to form a polypeptide. **Ribosomal RNA (rRNA)** is a major component of ribosomes, but its exact role in protein synthesis is unknown.

All RNA is synthesized on a DNA template by a process similar to the synthesis of a new strand of DNA during DNA replication (see Section 9-C). RNA synthesis is known as **transcription** (=''written across'') because it rewrites the genetic message coded in DNA, in the form of an RNA molecule.

The conversion of the genetic information carried by an mRNA molecule to the amino acid sequence of a polypeptide is known as **translation**. Before considering how the various types of RNA interact during translation, it is necessary to understand how genetic information is coded in the DNA molecules that determine RNA structure.

10-B The Genetic Code

DNA, RNA, and polypeptides are all linear, unbranched molecules, and it is reasonable to expect that a particular sequence of nucleotides in an mRNA molecule should be translated into a particular sequence of amino acids in a polypeptide. Because nucleic acids contain four different nucleotides, the genetic ''language'' must have a four-letter ''alphabet.'' Furthermore, to code for the 20 different amino

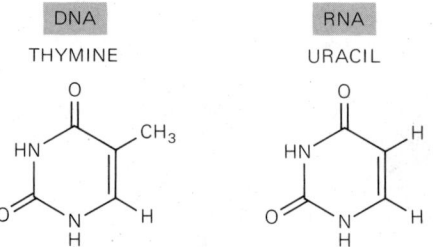

DNA

THYMINE

RNA

URACIL

Figure 10-5 In RNA, the base thymine found in DNA is replaced by the base uracil; either can pair with adenine.

acids found in proteins, more than one nucleotide must make up the code "word" for each amino acid. If code words were two nucleotides long, there would be $4^2 = 16$ different code words, not enough for 20 amino acids (Figure 10-6). The four nucleotides arranged in triplets, however, produce $4^3 = 64$ different code words, more than enough to produce a unique code word for each amino acid. The smallest theoretical size for a code word in DNA is, therefore, three nucleotides.

Francis Crick and others tested this triplet code theory by adding different numbers of nucleotides into the DNA of the bacterium *Escherichia coli*. They reasoned as follows: If the genetic message of DNA is transcribed into mRNA, and later translated into protein starting from one end of the mRNA, then introducing just one or two nucleotides in the middle of the gene will change the entire message after that point into something completely different; for example, if one or two nucleotides containing guanine (G) are added, the message

CAT—CAT—CAT···

might become

CAG—TCA—TCA—T···   or   CAG—GTC—ATC—AT···

However, introducing three nucleotides into the middle of the gene should merely create a short disruption, after which the message will read as in the original version:

CAT—GGG—CAT—CAT   or   CAG—GGT—CAT—CAT

or even

CGG—GAT—CAT—CAT

Experiments bore out this prediction; adding three extra nucleotides into a gene's DNA allowed the bacterium to produce a slightly altered protein, and the bacterium usually survived. Additions of one, two, or four nucleotides changed all subsequent code words and usually killed the bacterium; any proteins produced were so different that they could not perform the functions necessary to the cell's life.

THE FOUR CODE LETTERS
A
C
G
U

SIXTEEN DOUBLETS FROM THE FOUR CODE LETTERS

AA	AC	AG	AU
CA	CC	CG	CU
GA	GC	GG	GU
UA	UC	UG	UU

Figure 10-6 The four different kinds of nucleotides in RNA can be arranged in pairs to form 16 different possible combinations.

"We've cracked the code. It says 'meet me in Red Square on the 15th. Boris'." (Rosemary Smith)

TABLE 10-1 CODONS FOUND IN MESSENGER RNA†
(Codons read in the 5′ → 3′ direction)

SECOND BASE

The abbreviations for the amino acids are:

Ala Alanine
Arg Arginine
Asn Asparagine
Asp Aspartic acid

Cys Cysteine
Gln Glutamine
Glu Glutamic acid
Gly Glycine

His Histidine
Ile Isoleucine
Leu Leucine
Lys Lysine

Met Methionine
Phe Phenylalanine
Pro Proline
Ser Serine

Thr Threonine
Trp Tryptophan
Tyr Tyrosine
Val Valine

FIRST BASE	U	C	A	G	THIRD BASE
U	UUU Phe	UCU Ser	UAU Tyr	UGU Cys	U
	UUC Phe	UCC Ser	UAC Tyr	UGC Cys	C
	UUA Leu	UCA Ser	UAA *Stop*	UGA *Stop*	A
	UUG Leu	UCG Ser	UAG *Stop*	UGG Trp	G
C	CUU Leu	CCU Pro	CAU His	CGU Arg	U
	CUC Leu	CCC Pro	CAC His	CGC Arg	C
	CUA Leu	CCA Pro	CAA Gln	CGA Arg	A
	CUG Leu	CCG Pro	CAG Gln	CGG Arg	G
A	AUU Ile	ACU Thr	AAU Asn	AGU Ser	U
	AUC Ile	ACC Thr	AAC Asn	AGC Ser	C
	AUA Ile	ACA Thr	AAA Lys	AGA Arg	A
	AUG Met	ACG Thr	AAG Lys	AGG Arg	G
G	GUU Val	GCU Ala	GAU Asp	GGU Gly	U
	GUC Val	GCC Ala	GAC Asp	GGC Gly	C
	GUA Val	GCA Ala	GAA Glu	GGA Gly	A
	GUG Val	GCG Ala	GAG Glu	GGG Gly	G

† To use the table, find the row marked with the first base of the codon at the left, and go across this row until you are in the column headed by the second base. Then find the third base marked at the far right of the table. The three *Stop* codons signal positions where the ribosome stops reading and terminates the polypeptide chain. The codon AUG initiates synthesis of a polypeptide.

By 1960, there was considerable evidence for a triplet code, but deciding which sequence of three nucleotides coded for which amino acid appeared overwhelmingly difficult. Finally, biochemists learned to prepare artificial RNAs (polyribonucleotides) with known base sequences. These could be placed into solutions containing ribosomes, amino acids, and other substances needed for protein synthesis and used to produce polypeptides.

In 1961, Marshall Nirenberg and Heinrich Matthaei reported that polyuridylic acid (poly U), a string of ribonucleotides containing only uracil bases, coded for the formation of a polypeptide chain containing only the amino acid phenylalanine. Nirenberg and Matthaei reasoned that the mRNA code word UUU must stand for phenylalanine. (The DNA code would be the base-pair complement of this, AAA.) By means of these and other, somewhat more complicated, experiments, all the amino acid code words were worked out by 1965. These code words, or **codons**, are shown in Table 10-1.

Several features of the genetic code should be noted from this table.

1. The codons shown are the code words found in messenger RNA, not in DNA, where the code triplets will be the complements of those shown.

2. The code contains no punctuation or spaces such as might signal the beginning or end of a codon. In other words, the code must be read from a particular starting point or the whole sequence will be read incorrectly. For instance, the RNA sequence UCUAGAGCUA will produce the amino acid sequence Ser—Arg—Ala if read from left to right. If the reading of the RNA sequence starts at the second nucleotide (C), however, instead of at the beginning, it will produce the completely different amino acid sequence Leu—Glu—Leu. Because the code has no punctuation between codons, an initiation codon, which says in effect "start here," is essential to the correct

reading of a messenger RNA molecule. The initiation codon is the sequence AUG.

3. Inspection of Table 10–1 reveals that the code is **degenerate**; that is, there is more than one codon for most of the amino acids. The degeneracy of the code is biologically useful. For one thing, the deleterious effects of a mutation—an alteration in the sequence of nucleotides—are minimized. If the code were not degenerate, 20 codons would code for amino acids, and 44 would code for nothing (that is, they would act as *Stop* codons). Most mutations would, therefore, lead to *Stop* codons, which would cause premature termination of the polypeptide. Mutations that shorten a polypeptide chain usually lead to inactive proteins, whereas a substitution of one amino acid for another may be harmless.

4. The third base in the codon is often less specific than the first two. In most cases, the degeneracy of the code involves only the third base in the codon. For instance, all of the codons for the amino acid proline have CC as the first two bases.

5. Three of the 64 triplets do not code for any amino acids: UAG, UAA, and UGA are *Stop* signals, which terminate a polypeptide chain.

The genetic code is nearly universal. The same codons code for the same amino acids in all viruses, bacteria, plants, animals, and fungi that have been examined. This is compelling evidence that all organisms on earth today have evolved from a common ancestor. The major groups of organisms have had separate evolutionary histories for hundreds of millions of years, indicating that the code must have been established shortly after life originated, and have continued almost unchanged for the billions of years since. The only known exceptions to the universality of the genetic code are mitochondria. These organelles contain their own DNA and ribosomes, and in the late 1970s it was discovered that some codons mean something different in mitochondria than they do in the rest of the living world—including the cytoplasm surrounding the mitochondria!

Overlapping Genes

Early studies indicated that genes did not overlap one another; in other words, no gene began or ended within the sequence of DNA nucleotides that made up another gene. It was assumed that all genes in all organisms were non-overlapping. In the mid-1970s, however, overlapping genes were discovered in some phages (viruses that infect bacteria), and there is evidence that they also occur in some bacteria and in some viruses that infect animals.

The virus Qβ, which infects *Escherichia coli*, has two genes that start at the same place on the DNA. Usually, transcription begins at this site and ends at a termination signal 400 nucleotides away. But about 3% of the time, the termination signal is missed and transcription continues for another 800 nucleotides until it reaches a double *Stop* signal. Production of the large protein that results is not a "mistake," however; both the long and short polypeptides are vital to the virus. In 1977, Frederick Sanger and his associates published the entire nucleotide sequence of the DNA of the virus φX174, which also infects *E. coli*. (This was the first genetic material to have its nucleotide sequence completely analyzed.) Sanger showed that φX174 contains at least four overlapping genes. The sequences for two genes each contain the sequence for another gene; in each case, the second gene is translated in a shifted reading frame so that a completely different sequence of amino acids is found in the resulting protein.

There is highly suggestive evidence that genes also overlap in bacteria. For instance, the termination signal of one gene in *Escherichia coli* overlaps the initiation signal of an adjacent gene, as also occurs in φX174.

Overlapping genes permit viruses to pack more genetic information into a given length of DNA or RNA than would otherwise be possible. If overlapping genes occur in bacteria (there is no evidence as yet that they exist in eukaryotes), it will overthrow several tenets of molecular biology. For instance, when investigators calculate the number of genes in an organism and predict the effects of mutations and carcinogens upon cells, they assume that genes do not overlap. Where genes do overlap, a single mutation could affect more than one gene.

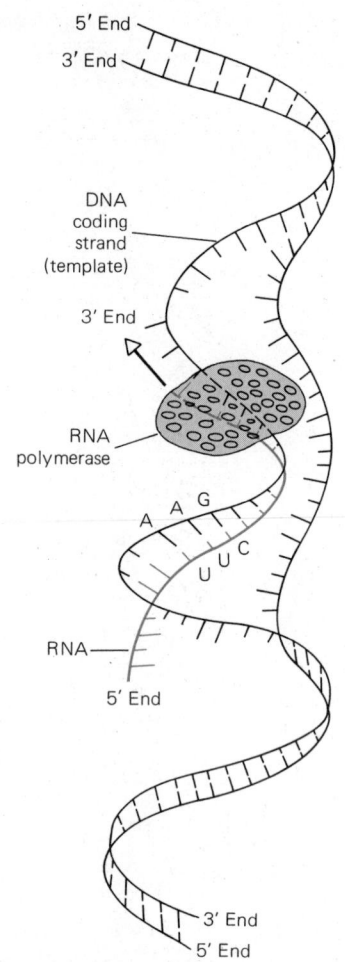

5′ End
3′ End

DNA
coding
strand
(template)

3′ End

RNA
polymerase

A A G

A   C
U U C

RNA

5′ End

3′ End
5′ End

Figure 10-7 Transcription of an RNA molecule from the coding strand of its DNA template.

10-C Transcription of DNA into RNA

One strand of the DNA double helix, the "sense" strand, codes for a series of amino acids. The other DNA strand is the base-pair complement of the first and must, therefore, code for something quite different, presumably nonsense. As we might expect, only the sense strand is transcribed. The other exists, as far as we know, only to provide a template for DNA replication. It is not always the same strand that is transcribed. In some viruses and in some mitochondria and chloroplasts (which contain their own DNA), different parts of both strands code for proteins; this also seems to be true in prokaryotes and in eukaryote nuclei.

During transcription, a length of the double helix of DNA uncoils, and the enzyme **RNA polymerase** attaches to the 3′ end* of the DNA coding strand (Figure 10–7). RNA polymerase pairs individual ribonucleotides to their complementary bases on the DNA strand, and then joins them to form a single strand of RNA.

RNA polymerase attaches to DNA and starts RNA synthesis at a **promoter region** of the gene. The first length of DNA transcribed is a leader, up to 200 nucleotides long, that will never be translated into protein. The polymerase molecule trundles along the DNA molecule, producing a complementary RNA strand as it goes, until it reaches a termination signal, which, in some cases, may be the sequence TTATTT (this is not yet certain). Shortly after this signal, the polymerase leaves the DNA and releases the RNA molecule.

Prokaryotic RNA is ready for translation as soon as it is transcribed; ribosomes may even attach to the beginning of the mRNA and start translating it into protein while the end of the molecule is still being synthesized on the DNA template. However, the mRNA of eukaryotes must be further processed in the nucleus before it enters the cytoplasm and participates in protein synthesis, as we shall see below.

Patchwork Genes

In 1962 Henry Harris pointed out that most of the mRNA produced in the nucleus was destroyed there and never reached the cytoplasm. An unexpected discovery a decade later explained this state of affairs, but raised more questions than it answered (as is so often the case with scientific experiments). Messenger RNA isolated just after transcription was much longer than the same mRNA isolated as it was being translated into a polypeptide; it appeared that the mRNA molecule was shortened before it was translated. Because mRNA is transcribed from a gene, the gene must also be longer than we would expect from looking at the finished polypeptide, with parts that are never translated into protein.

This theory was verified after methods for analyzing the nucleotide sequence of DNA were developed in 1977. A few specific genes, their RNA, and their protein products have now been completely analyzed. It turns out that many eukaryotic genes contain one or more **intervening sequences—introns**, in the jargon of the trade—lengths of DNA that are not translated into amino acids.

Introns are found only in eukaryote genes, but not in all of them; for example, histone* genes are continuous, with no introns. Some genes that contain introns can function normally without them, but others cannot even be transcribed if the intron is deleted.

What, then, is the function of introns? The many answers so far proposed range from protecting RNA against degradation in the nucleus to conserving a supply of unused DNA that is tapped during times of rapid evolution.

Where a gene contains introns, the entire gene is transcribed into pre-mRNA, and then the introns are spliced out, producing a mature mRNA molecule that can

---

* 3′ end, 5′ end The 3′ end of a nucleotide strand is the one at which the 3′ carbon of the sugar of the last nucleotide is not bound to the phosphate of another nucleotide. At the 5′ end, the phosphate bound to the 5′ carbon of the sugar is not bound to another nucleotide (see Figure 9–3).

* Histone A protein with a basic pH, found associated with DNA in eukaryotic chromosomes (Figure 9–9).

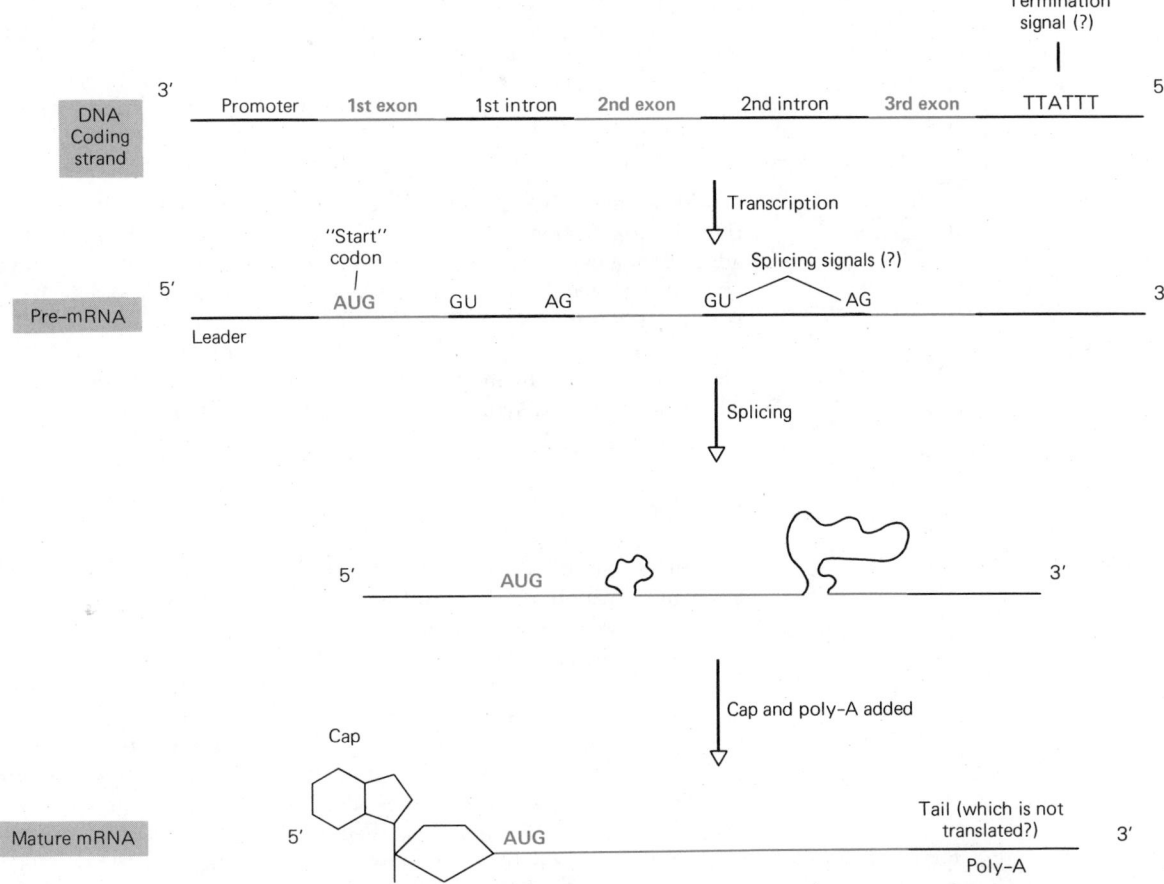

Figure 10-8 Synthesis and maturation of mRNA transcribed from the mouse β globin gene. Only the nucleotides which make up **exons** (colored) are expressed as protein. (The termination signal indicated is probably the signal for RNA polymerase to leave the DNA and not the signal for the end of the protein chain during translation because there is a length of RNA to the left of this signal in the mature mRNA which is apparently not translated and whose function is unknown.)

be translated into protein. We even know (we think) the signals for splicing, although the enzymes involved have yet to be discovered. Figure 10–8 shows the splices needed to mature the mRNA that translates into mouse β-globin, one of the two different kinds of polypeptides in the oxygen-carrying protein hemoglobin. Both the introns removed from this mRNA have the nucleotide sequence GU at their 5′ ends and AG at the 3′ end; these splicing sequences have been found in all the dozen or so messengers completely analyzed to date.

TABLE 10–2 **COMPARISON OF PROKARYOTIC AND EUKARYOTIC mRNA**

EUKARYOTES *(and many of their viruses)†*	PROKARYOTES *(and many phages)†*
Up to 90% of the mRNA is removed as introns during splicing	No splicing; no introns
5′ end of mRNA is capped after transcription	No cap
String of adenines (poly-A) is added at 3′ end during maturation	No poly-A tail
5′ end of mRNA has a "leader" that does not code for protein	Non-coding leader present
3′ end of mRNA has a "trailer" that does not code for protein	Non-coding trailer present

† Why do the viruses resemble the cells they attack? Probably because viral genes are transcribed and translated by the *host's* RNA polymerase, ribosomes, etc., so they must be compatible with the host's system.

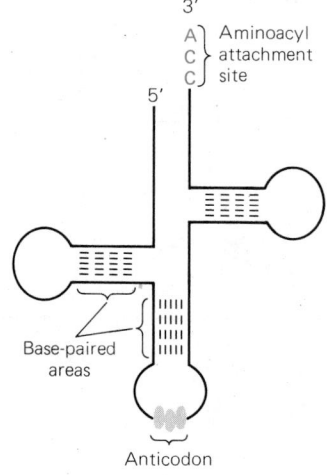

Figure 10-9 The structure of tRNA molecule. In the real, three-dimensional molecule, this structure is twisted and the side arms are close together; further base pairing occurs between complementary bases on the two side loops.

Messenger RNA obviously leads a much more adventurous life than we once thought. In addition to having its introns spliced out, some mRNA is changed by the addition of a cap (consisting of the unusual nucleoside 7-methylguanosine) at the 5′ end of the molecule, and a tail of several adenine nucleotides at the 3′ end, before it is translated into protein. The functions of the cap and tail are still unknown. One suggestion is that the cap interacts with a ribosome to start translation at the correct place; the poly-A tail might attach the mRNA-ribosome complex to a membrane, where most protein synthesis occurs. But these are pure speculation until there is more evidence.

Pseudogenes

Introns are not the only examples of DNA that is never translated into proteins. Even more peculiar are **pseudogenes**, genes that apparently are not even transcribed. Many ordinary, functional genes that produce proteins lie near almost identical pseudogenes. For instance, a pseudogene called α_4 has been found in mice. It is just like the normal α-globin gene except that a change of one nucleotide to another has eliminated the normal protein termination signal (*Stop* codon). The mRNA and the globin molecule this gene would produce if it were transcribed and translated have been searched for but not found, evidence that the gene is not expressed at all.

The discovery of pseudogenes accounts for at least some of the repetitive DNA that is common in chromosomes (Section 9-D). The function of pseudogenes is the subject of furious debate. The most popular theory is that pseudogenes are evolutionary surplus capacity, which could easily mutate into active genes, slightly different from existing genes and possibly having advantageous effects (see the Essay "Proteins as Evolutionary Puzzle Pieces" at the end of this chapter).

10-D Transfer RNA

We turn now to the second kind of RNA that participates in protein synthesis. Transfer RNA (tRNA) molecules are transcribed from **tRNA genes** on the DNA. Their function is to transport amino acids to the mRNA-ribosome complex, where the amino acids will be joined together to form the polypeptide chain. For each amino acid, there is a specific kind of tRNA molecule that will recognize and transport it.

All tRNA molecules for which nucleotide sequences have been determined show the same general shape, usually referred to as a cloverleaf (Figure 10–9). The shape results from base-pairing in some areas of the molecule, which causes characteristic loops. The most important parts of the tRNA molecule are the aminoacyl attachment site and the anticodon.

The **aminoacyl attachment site** is the site at which the amino acid attaches to the tRNA molecule. The site consists of the three bases CCA, at the 3′ end of the tRNA molecule. The attachment of an amino acid to tRNA is called **acylation**, and the tRNA with an amino acid attached to it is called **aminoacyl tRNA** (Figure 10–10). Acylation is carried out by specific enzymes, each of which recognizes the unique shape of one kind of tRNA and of the corresponding amino acid.

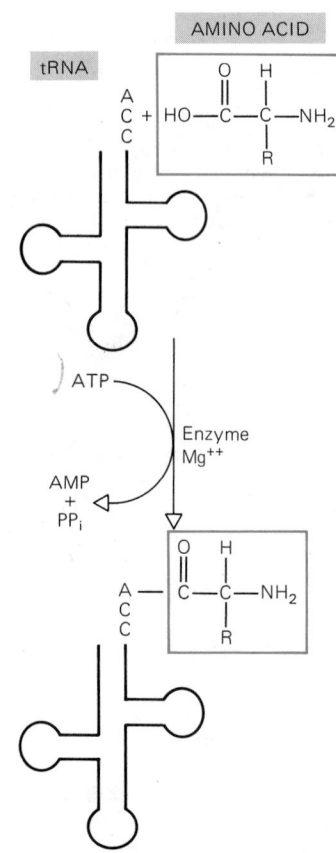

Figure 10-10 Acylation, the attachment of an amino acid to a tRNA molecule. ATP provides energy for the reaction; in this case, ATP is broken down to AMP and a two-phosphate pyrophosphate fragment (PP_i) instead of the more usual ADP and P_i.

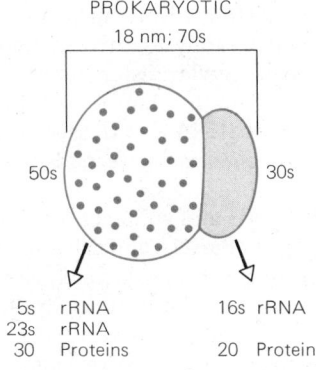

PROKARYOTIC
18 nm; 70s

50s   30s

5s   rRNA   16s rRNA
23s   rRNA
30   Proteins   20   Proteins

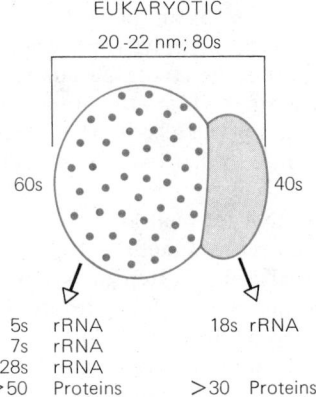

EUKARYOTIC
20-22 nm; 80s

60s   40s

5s   rRNA   18s rRNA
7s   rRNA
28s   rRNA
>50   Proteins   >30 Proteins

Figure 10-11 Structures of prokaryotic and eukaryotic ribosomes. A ribosome is composed of two subunits, one large and one small. ("s" is a unit of measurement of the rate at which a particle migrates to the bottom of a tube during centrifugation; s values for fragments do not necessarily add up to the s value for the intact subunit or ribosome.)

The **anticodon** of a tRNA molecule base-pairs with the appropriate mRNA codon at the mRNA-ribosome complex. This temporarily binds the tRNA to the mRNA, allowing the amino acid carried by the tRNA to be incorporated into the polypeptide in its proper place.

10-E Ribosomal RNA and Ribosomes

Like other RNAs, ribosomal RNA is transcribed on a DNA template. A functional ribosome is composed of two subunits, one large and the other smaller; each subunit is made up of ribosomal RNA and protein (Figure 10–11). Ribosomes found in the cytoplasm of eukaryotic cells are larger than those found in prokaryotic cells or inside mitochondria and chloroplasts.

When the ribosomes are not involved in protein synthesis, the subunits are separate. When protein synthesis begins, messenger RNA binds to a small subunit, and a large subunit then joins to form a functioning mRNA-ribosome complex (Figure 10–12). Despite their enormous importance, we still know very little about ribosomes and how they work.

10-F Protein Synthesis

With DNA, mRNA, tRNA, the ribosome, amino acids, and the necessary enzymes, everything needed for protein synthesis is present. To summarize: DNA carries the genetic information for the sequence of nucleotides in each of the three types of RNA and for the sequence of amino acids in a protein; mRNA carries the information from a gene to the ribosome, and each tRNA molecule carries a molecule of its particular amino acid to the mRNA-ribosome complex.

The first stage in protein synthesis occurs when the two subunits of a ribosome bind to an mRNA molecule. A ribosome has two sites where codons are translated. The initiation codon, AUG, on the mRNA molecule binds to the first of these sites on the ribosome (Figure 10–12). AUG codes for the amino acid methionine. Since the AUG codon means "start here," any nucleotides that precede it in the mRNA molecule are ignored by the ribosome.

Now the anticodon of another tRNA binds to the mRNA codon lined up with the second site of the ribosome. This second tRNA carries the second amino acid that will be incorporated into the new polypeptide chain.

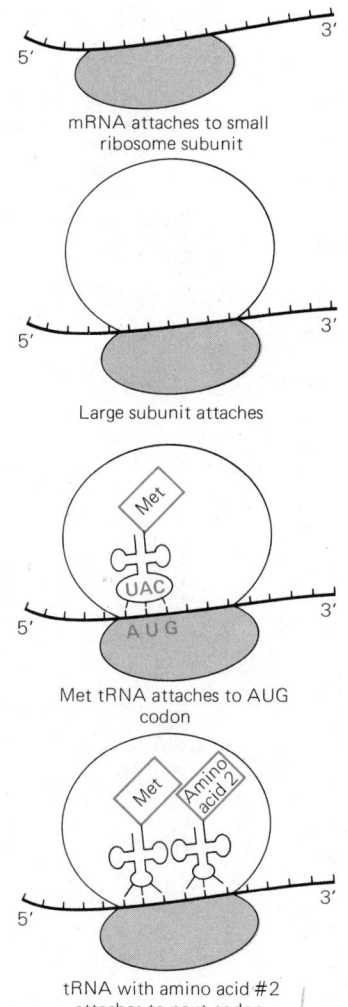

mRNA attaches to small ribosome subunit

Large subunit attaches

Met tRNA attaches to AUG codon

tRNA with amino acid #2 attaches to next codon

Figure 10-12 How protein synthesis begins.

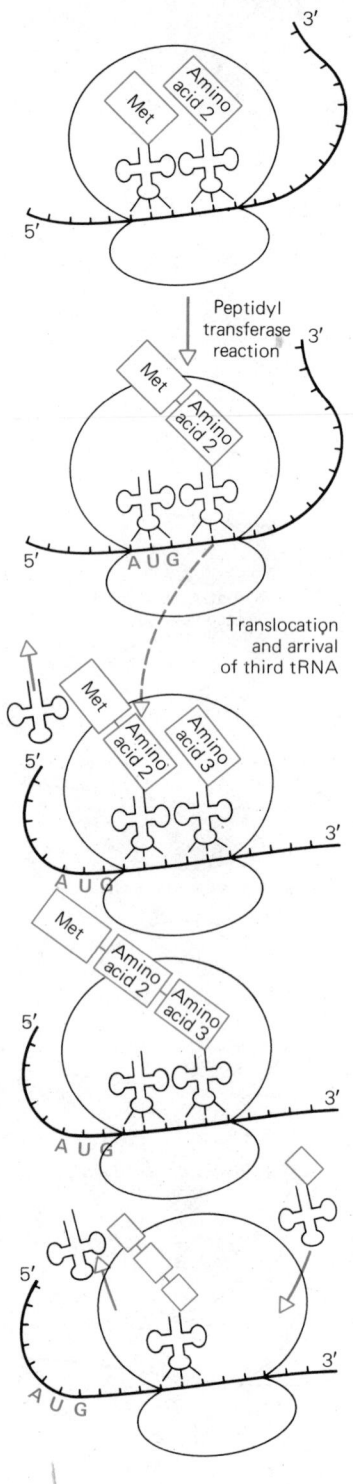

Peptidyl
transferase
reaction

Translocation
and arrival
of third tRNA

The next step, the **peptidyl transferase reaction**, results in peptide bond* formation. The peptidyl transferase enzyme, which catalyzes the reaction, is an integral part of the large ribosomal subunit. In the peptidyl transferase reaction, the amino acid at the first attachment site on the ribosome (in Figure 10–13, methionine) forms a peptide bond with the amino acid whose tRNA is at the second site on the ribosome, so that both amino acids are now attached to the second tRNA (Figure 10–13).

The tRNA for methionine now leaves the ribosome and will eventually pick up another molecule of methionine. Meanwhile, the second tRNA and the mRNA move along the ribosome, a step known as **translocation**; this brings the next codon onto the ribosome, where the appropriate tRNA attaches by its anticodon, bringing the third amino acid into position. The growing polypeptide chain is attached to the newly arrived amino acid on this third tRNA, and the sequence repeats until the ribosome reaches a *Stop* codon on the mRNA. At this point, protein synthesis stops, and the newly formed peptide chain is released from the ribosome (Figure 10–14).

A single mRNA molecule may be translated into protein more than once. Each molecule typically has several to over 100 ribosomes attached to it that transcribe its message as they move along. One mRNA with many ribosomes attached to it forms a cluster known as a **polyribosome** or **polysome** (Figure 10–15).

---

* Peptide bond Covalent bond joining the carboxyl carbon of one amino acid to the α-amino nitrogen of the next (see Figure 3–16).

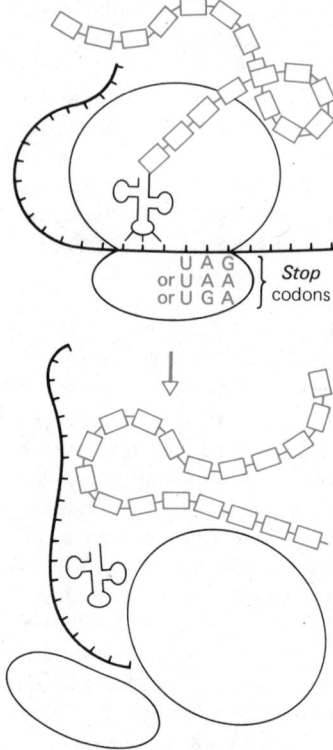

Figure 10-14 When a "stop" codon is reached, the polypeptide is released from the last tRNA and the mRNA is released. The ribosomal subunits separate.

10-G   Antibiotics and Protein Synthesis

A number of antibiotics (antibiological agents) act by preventing protein synthesis. Some are specific in that they block protein synthesis in prokaryotes but not in eukaryotes. These agents are useful antibacterial drugs. Other antibiotics block protein synthesis in all cells, but can be used to kill parasite rather than host cells because the parasite normally carries out protein synthesis faster than the host. The tetracyclines block binding of the aminoacyl-tRNA to the attachment site on a prokaryotic ribosome. Puromycin also binds to a ribosome and becomes linked to the growing peptide chain. Since puromycin cannot translocate on the ribosome, it blocks further formation of the polypeptide at this point; streptomycin also blocks translocation, but only in prokaryotes. Cycloheximide blocks protein synthesis only on the ribosomes of eukaryotes, and so can be used to treat fungal infections.

10-H   Mutations

Mutations are inheritable changes in the genetic material; they occur when the position of a nucleotide or of a segment of several nucleotides in the DNA chain is altered, when nucleotides or segments of DNA are added or removed, or when one nucleotide is changed to another. Mutations may result from mistakes in DNA replication, or they may be caused by any one of a number of **mutagenic agents**, which may act by breaking the DNA molecule or by changing the molecular structure of a nucleotide, thereby changing its pairing behavior. Since the structure of DNA is responsible for the sequence of amino acids in proteins, mutations may cause changes in the protein for which a segment of DNA codes.

One type of simple mutation is the **point mutation**, in which one nucleotide is substituted for another. **Frame-shift mutations** are more drastic than point mutations because they involve either insertions or deletions of one or more nucleotides in the DNA. A frame-shift mutation usually changes the entire sequence of amino acids produced beyond the mutant section of DNA.

Mutations are usually deleterious in their effects because the amino acid sequence of a protein has been determined by millenia of natural selection and cannot be altered much without interfering with the delicate machinery of a cell.

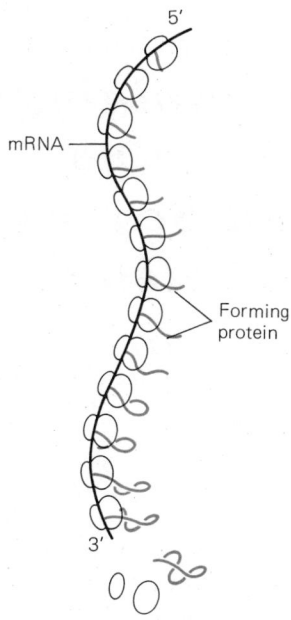

Figure 10-15   A polyribosome, or polysome, consists of an mRNA molecule being translated by several ribosomes simultaneously.

SUMMARY

Many genes code for the sequences of amino acids in the polypeptides produced in a cell. A cell's proteins, in turn, control its structure and metabolism.

RNA is transcribed from the cell's DNA, and thus the nucleotide sequence of RNA is complementary to the nucleotide sequence in DNA. The three main types of RNA in a cell are messenger RNA, whose base sequence is translated into the sequence of amino acids in a polypeptide chain; transfer RNA, which carries amino acids to the ribosome for protein synthesis and brings them into their proper position in the polypeptide chain; and ribosomal RNA, which makes up part of the structure of ribosomes.

Messenger RNA in prokaryotes can be translated by ribosomes as soon as it is synthesized. Eukaryote mRNA, on the other hand, must be processed before it can be translated. Introns (of unknown function) are spliced out of the mRNA molecule, and a cap and tail are added to produce a mature mRNA molecule.

A sequence of three nucleotides in mRNA codes for each amino acid. The code is degenerate in that most amino acids are coded for by more than one codon. The code has no "punctuation marks" except for codons that code for the beginning and end of the polypeptide chain.

Mutations are changes in DNA that are passed on when the DNA replicates; they are also passed to RNA transcribed from the DNA, and changes in mRNA may result in changes in the protein produced.

Essay: Determining the Amino Acid Sequence of Proteins

In 1953 Frederick Sanger determined the amino acid sequences in the polypeptide chains of insulin (see Figure 3–17). This was the first protein structure to be completely analyzed. Since that time the amino acid sequences and the secondary and tertiary structures of a large number of proteins have been worked out.

If a protein contains more than one polypeptide chain, the first step in protein analysis is to separate and purify the individual chains. Next, a sample of each chain is completely hydrolyzed and its amino acid composition determined.

Since the 20 amino acids commonly found in proteins all vary somewhat in their R group structure, they differ in their solubility in different solvents and in the amount of dissociation of H⁺ (and OH⁻) from the R groups in solutions of various pH. Likewise, short peptide chains will differ from one another in these properties, according to their amino acid composition. These differences are the basis for separating amino acids and peptides by the method of chromatography.*

---

*Chromatography The separation of a mixture of substances into its components based on differences in their electrical charges and sizes.

The most common type of chromatography used in protein analysis is **column chromatography**. The column is a large glass tube filled with plastic resin beads that are electrically charged. A sample of the substance to be analyzed is poured into the top of the column, and the various amino acids or peptides in the sample move down the column at different rates, depending on their size and their electric attraction to the resin beads. The column is washed with a solvent "chaser," which can carry the sample molecules toward the base of the column. The molecules that move fastest arrive at the bottom first; those that move slowly arrive later. The vessel that collects the sample at the bottom of the column is changed frequently, so that at the end of the procedure, there are a number of small collecting vessels containing **fractions** that have been collected at various different times and thus contain various different types of molecules. Each fraction will contain only one or a very few of the types of molecules from the original mixture.

After the overall amino acid composition of the polypeptide chain is known, its N-terminal and C-terminal amino acids are determined (Figure 10–A). The N-terminal amino acid has a free —NH₂ not attached to another amino acid by a peptide bond; likewise the C-terminal has a free —COOH that is not peptide bonded (see Figure 3–16). Various reagents will combine with the free amino group; subsequent treatment with acid then frees the amino acid with the reagent still attached to its amino nitrogen. The investigator can isolate and identify the amino acid by chromatography.

Next, the amino acid at the C-terminal end of the polypeptide chain is determined in a similar manner.

Another sample of the polypeptide is subjected to hydrolysis by an enzyme that cleaves only peptide bonds next to certain amino acids. This results in a number of shorter peptides, which must be separated from each other, again by chromatography. Each batch of pure short peptides is divided into samples. One sample is analyzed to determine what amino acids are present; it is first hydrolyzed into its constituent amino acids by acid treatment or by other means; then the amino acid mixture is placed into an amino acid analyzer, a machine that subjects the mixture to column chromatography, collects the fractions as they come off, and analyzes them to determine what types, and how much of each type, of amino acids are present. Another sample of each short peptide is run through an amino acid sequencer. This is a machine that breaks off amino acids from the N-terminal end of each peptide, one by one, collects them, and analyzes them to determine what they are. Both the amino acid analyzer and the amino acid sequencer carry out their tedious repetitive tasks automatically, in a matter of 2 to 4 hours.

The analysis of these short peptides yields several "puzzle pieces" that, when properly fitted together, give the amino acid sequence of the entire polypeptide. To find the proper order, another set of pieces must be obtained by subjecting another sample of the original polypeptide to the action of a different enzyme, and repeating the purification and sequence analysis of the second set of

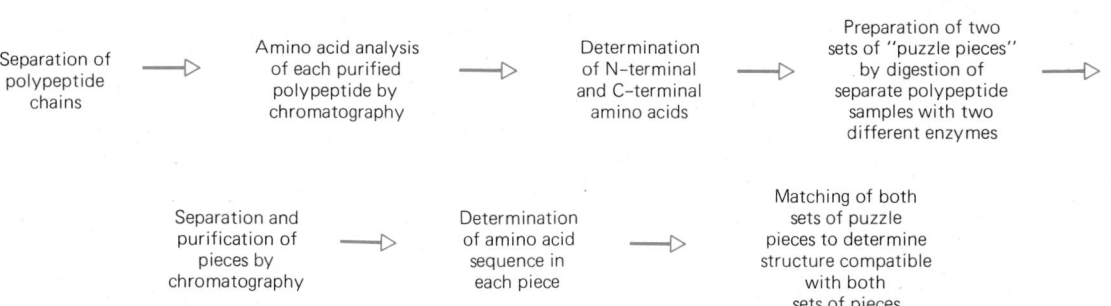

Figure 10-A Summary of the procedure for determining the sequences of amino acids in proteins.

short peptides so obtained. For example, in analyzing the amino acid sequence of the human polypeptide hormone adrenocorticotropic hormone, the first enzyme used to attack the purified hormone and break it into shorter peptides might be trypsin, a digestive enzyme that hydrolyzes peptide bonds in which either lysine (Lys) or arginine (Arg) is at the carboxyl side of the bond. The peptides obtained are shown in Figure 10–B(a).

Previous analysis would have revealed that the N-terminal amino acid of the protein is serine (Ser) and the C-terminal amino acid is phenylalanine (Phe). The first long peptide listed must be at the C-terminal end of the sequence and the second must be at the N-terminal end, but there are still three short peptides and three single amino acids that must be placed in the proper order.

Another sample of the hormone would then be treated with another digestive enzyme, for instance, with chymotrypsin, which attacks peptides so that it frees the C-ends of the amino acids phenylalanine (Phe), tryptophan (Trp), or tyrosine (Tyr). This would produce the peptides shown in Figure 10–B(b). We can now write these pieces out on lengths of paper and move them around until we get two sequences exactly alike, one above the other (Figure 10–B[c]).

This hormone is 39 amino acids long, rather short as naturally occurring peptides go. It was the third peptide to be analyzed (after the two peptides of insulin).

Such work in sequencing hormones is very useful, since it gives the information needed for test-tube production of these molecules, which can be supplied to people who have hormone deficiencies.

In 1977, Sanger used procedures with a similar pattern of logic to analyze the entire genetic material of bacteriophage ϕX174. The discovery of bacterial restriction enzymes, which snip DNA selectively at points surrounded by certain patterns of nucleotides, made such procedures possible (Section 9-E). Many DNA sequences have now been analyzed, providing much of the information about gene structure described in this chapter.

(a) Peptides obtained by trypsin digestion of adrenocorticotropic hormone:

Val—Tyr—Pro—Asp—Ala—Gly—Glu—Asp—Gln—Ser—Ala—Glu—Ala—Phe—Pro—Leu—Glu—Phe
Ser—Tyr—Ser—Met—Glu—His—Phe—Arg
Trp—Gly—Lys
Pro—Val—Gly—Lys
Pro—Val—Lys
Lys $\left.\right\}$
Arg $\left.\right\}$ Single amino acids
Arg $\left.\right\}$

(b) Peptides obtained by chymotrypsin digestion of adrenocorticotropic hormone:

Gly—Lys—Pro—Val—Gly—Lys—Lys—Arg—Arg—Pro—Val—Lys—Val—Tyr
Pro—Asp—Ala—Gly—Glu—Asp—Gln—Ser—Ala—Glu—Ala—Phe
Ser—Met—Glu—His—Phe
Pro—Leu—Glu—Phe
Ser—Tyr
Arg—Trp

(c) Using two sets of peptides (a) and (b) to reconstruct the original sequence
From the first (trypsin) sequencing, we know that

| Ser—Tyr—Ser—Met—Glu—His—Phe—Arg |

comes first. So from our second batch

| Ser—Tyr |

and

| Ser—Met—Glu—His—Phe |

must be arranged like this.
Our first sequencing tells us that we must now look for a peptide beginning with Arg in our second sequence; there is only one, so

| Arg—Trp |

must come next, and we go back to the first set of pieces looking for an N-terminal Trp, which puts

| Trp—Gly—Lys |

as the next sequence, and so on until we have completed the sequence:

Ser—Tyr—Ser—Met—Glu—His—Phe—Arg—Trp—Gly—Lys—Pro—Val—
Gly—Lys—Lys—Arg—Arg—Pro—Val—Lys—Val—Tyr—Pro—Asp—Ala—
Gly—Glu—Asp—Gln—Ser—Ala—Glu—Ala—Phe—Pro—Leu—Glu—Phe.

Figure 10-B Procedure for using two sets of puzzle pieces to determine the amino acid sequence of adrenocorticotropic hormone. Pieces resulting from trypsin digestion are boxed in black, those from chymotrypsin digestion in color.

Essay: Proteins as Evolutionary Puzzle Pieces

The discovery of the genetic code and the development of techniques for determining the sequences of the amino acids in a polypeptide opened an exciting new field, the use of proteins found in modern organisms as "living biochemical fossils." We are all familiar with evolutionary trees for living organisms constructed by comparing these organisms with various fossil forms, and by determining the age of rocks in which the fossils are found. Analysis of protein structure and comparison of proteins from different organisms provide other lines of evidence that can be used to confirm evolutionary trees constructed from fossil evidence, and can sometimes help to shed light on points in the evolutionary tree where the fossil record does not provide good evidence about relationships. The amino acid sequence of a protein molecule can be used, just as the structure of a tooth or a leg bone is, to trace the probable ancestry of living organisms.

To construct an evolutionary tree that encompasses most living organisms, a protein that is common to a variety of organisms must be studied. Cytochrome c, a vital part of the electron transport chain, has been studied in this way. All aerobic organisms can be examined and compared to each other with regard to the structure of this one protein molecule.

A second thing that makes cytochrome c a good choice for evolutionary analysis is that it contains comparatively few amino acids—from 103 to 112 in various organisms studied.

The first step in constructing a protein-structure evolutionary tree is to isolate the protein from a number of different types of organisms and to determine its amino acid sequence. Next, these structures are fed into a computer. The computer compares the amino acid sequences of the various proteins with one another, and tabulates their similarities and differences. The more nearly alike are two species' amino acid sequences, the more closely related are the species likely to be.

If the computer is also supplied with the genetic code, it can construct the "missing links" between related species; that is, it can indicate the probable structure of the ancestral cytochrome c possessed by a species that was the common ancestor of two related species. For example, suppose one species has the amino acid isoleucine in a particular position in its cytochrome c, and another species has proline in that position. If we assume that one of these amino acids was present in the ancestral protein, then two adjacent point mutations must have been required to get from the DNA of one species to that of the other. If, however, we assume that both species came from a common ancestor whose cytochrome c had leucine at that position, then each species could have arrived at its present amino acid by only one point mutation (Figure 10–C). The latter case is more probable. The computer, of course, could also be programmed to detect insertion and deletion mutations; it is often found that two amino acid sequences line up better if it is assumed that one or a few amino acids have been added or deleted during the course of evolution.

One interesting finding in comparative protein studies is that some positions in each particular type of protein molecule are occupied by the same amino acid in every species for which that protein has been analyzed. Is this coincidence? It seems unlikely. These **invariant amino acids** may be part of the protein's active site, or play a role in maintaining the protein molecule's tertiary structure. Changing one of the invariant amino acids would result in a protein with altered function. An individual with such a protein would be at a disadvantage compared to its normal neighbors in the population; hence variations in the invariant amino acids are selected against. Variations in some of the other amino acids in the chain do not seem to make as much difference, although they may well be responsible for the "fine tuning" of a protein that makes it function optimally in a species' particular environment.

The relationships among vertebrates have been studied by comparing the structures of their hemoglo-

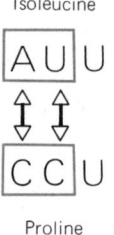

Isoleucine

Proline

(a) Two adjacent point mutations

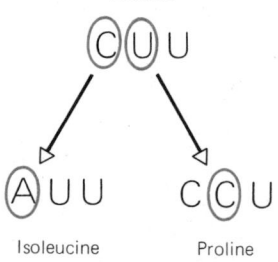

Leucine

Isoleucine   Proline

(b) Two separate point mutations from common ancestor

Figure 10-C Two ways of picturing the evolutionary relationships between proteins having the amino acids isoleucine and proline at a particular position; mRNA codons, rather than the DNA that actually mutates, are shown so you can compare the code with Table 10-1. (a) One amino acid could have been substituted for the other as a result of two adjacent point mutations. (b) Each amino acid could have been substituted for leucine in an ancestral form as a result of two separate point mutations, one in each evolutionary line.

bins. Hemoglobin is a protein that carries oxygen in the blood; it consists of four polypeptide chains: two α-chains and two β-chains. The β-chains have been analyzed in a number of vertebrates, and as described above for cytochrome c, this has allowed some comparisons of relatedness and some guesses as to the nature of the ancestral protein. These studies have revealed that gorilla and human β-chains differ by only one amino acid, while pigs differ from humans by about 17 amino acids, and horses and humans differ at 26 positions.

The α-chains of hemoglobin are very similar to the β-chains, and in fact humans and other vertebrates possess a number of other molecules that are all very similar to the β-chain of hemoglobin. One of these is **myoglobin**, a single-polypeptide protein in skeletal muscle. Myoglobin takes up oxygen from hemoglobin in the blood and holds it until it is

needed by the muscle. There are also different types of hemoglobin chains that are made only during the fetal life of vertebrates; ε-chains are made during early embryonic development and are later replaced by γ-chains. At a late stage in development, the fetus begins to make β-chains, and its γ-chains gradually disappear from its system. (α-chains are produced throughout fetal and adult life.) The fetal types of hemoglobin have a greater affinity for molecular oxygen than that of adult hemoglobin; this permits the fetus to obtain oxygen from the mother's circulatory system. Hence, fetal hemoglobin is an adaptation permitting the fetus to obtain an adequate oxygen supply without direct access to air.

How did so many different types of molecules with such similar structures come into being? Most likely at some point or points in the evolutionary past, ancestral genes were replicated in more than one copy. As

we have seen before (Sections 9-D and 10-C), there is often more than one copy of some genes in the nuclei of modern organisms. Suppose that some copies of the gene underwent mutations that resulted in proteins with slightly different, but advantageous, functions. Individuals that had two or more proteins that were similar to each other, but that were each adapted to performing slightly different functions, would be selected for over individuals that were trying to make one protein serve all these functions by itself. In other words, multiple copies of genes could undergo **adaptive radiation**, changing from one generalized but useful protein to several different, specialized proteins, each one tailored for its particular function, and all together conferring selective advantage on the individual that possessed them over the individuals that were making one unspecialized protein do all the jobs.

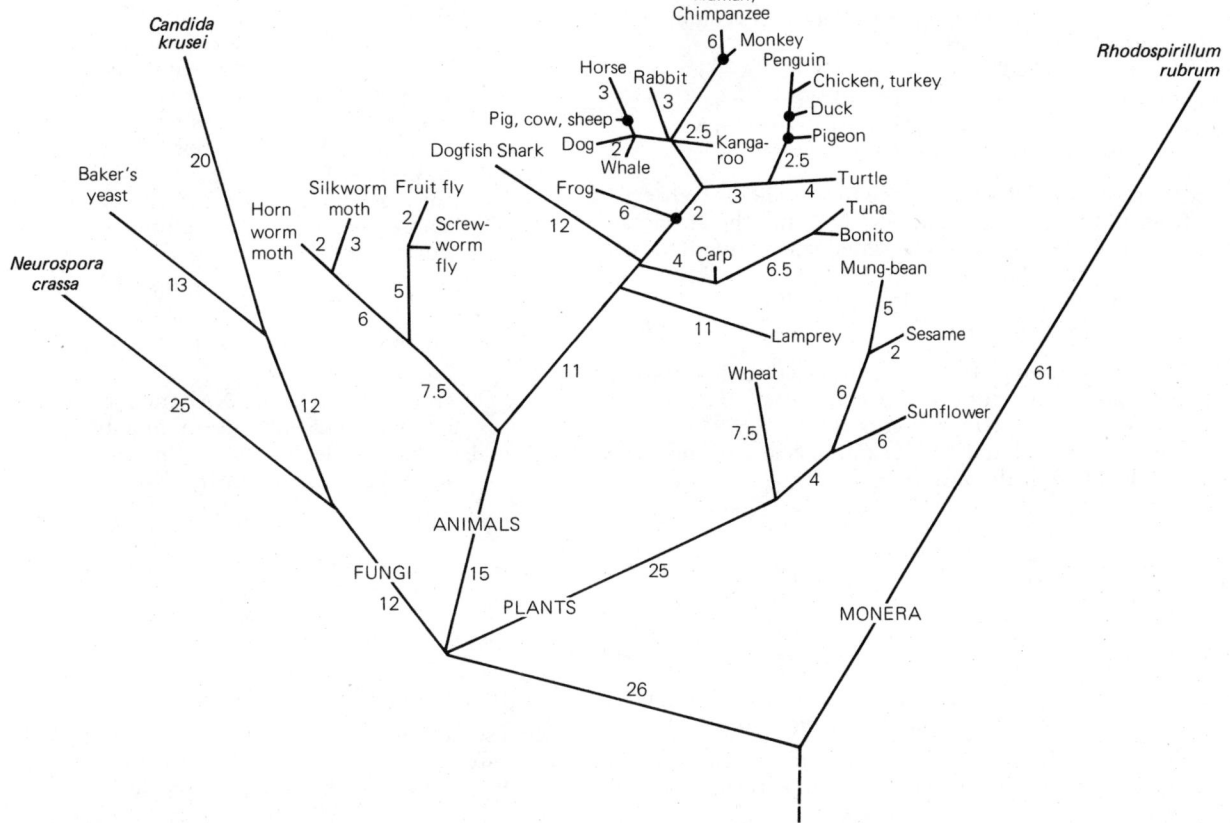

Figure 10-D An evolutionary tree constructed by giving a computer the amino acid sequence of the cytochrome c molecule of each organism. Numbers beside the branches indicate the number of amino acids changed per 100 links between forks. Below each branch point is the ancestral cytochrome c common to all organisms shown above it. (Redrawn from the *Atlas of Protein Sequence and Structure*, Vol. 5, 1972.)

OBJECTIVES

From your study of this chapter, you should be able to:

1. List three differences between DNA and RNA.

2. Describe the genetic code and explain why it must be a triplet code.

3. Given a DNA coding strand and a table of codons, determine the complementary mRNA strand, the codons and anticodons that would be involved in peptide formation from that mRNA sequence, and the amino acid sequence that would be translated.

4. Describe the role of DNA, mRNA, tRNA, ribosomes, and amino acids in protein synthesis.

5. Sketch a transfer RNA molecule and indicate where the anticodon and aminoacyl attachment sites are.

6. List the steps involved in protein synthesis at the ribosome level.

7. Explain the difference between point and frame-shift mutations, in both their effect on DNA structure and their effect on polypeptides produced from the affected DNA code.

SELF-QUIZ

1. Using the base-pairing rules, fill in the bases to be found on the RNA strands transcribed from the following DNA strands:
    a. DNA A—G—G—C—C—T—G—C—T—T—A
       RNA _____
    b. DNA T—G—G—C—A—G—C—T—A—C
       RNA _____
    c. DNA T—T—T—A—C—G—C—A—C—C
       RNA _____

2. Write out the amino acid sequences that would be translated when the following mRNA molecules combine with a ribosome:
    a. A—U—G—C—A—U—A—G—A—A—G
       —G—C—C—U—A—U—U—G—U—A
    b. C—A—U—G—U—U—U—C—U—U—A
       —A—A—G—G—U—C—G—U—U

3. Write out the mRNA sequence that would be transcribed from the following strand of DNA, and the amino acid sequence that would be translated when the mRNA combines with a ribosome:
       T—A—C—A—A—G—T—A—C—T—T
       —G—T—T—T—C—T—T

4. Suppose the two guanosine (G) nucleotides in question 3 are changed to cytosine (C) nucleotides.
    a. What kind of mutation is this?
    b. How would it affect the amino acid sequence translated from the mRNA?

    c. What kind of mutation would result from removal of the G nucleotides from the DNA in question 3?
    d. How would this mutation affect the amino acid sequence translated from the mRNA?

5. According to current ideas concerning protein synthesis:
    a. transfer RNA molecules specific for particular amino acids are synthesized along a messenger RNA template in the cytoplasm
    b. amino acids line up with their mRNA codons on the ribosome and are then linked together by transfer RNA
    c. enzymes that catalyze protein-synthesizing reactions in the cytoplasm are transcribed from activator genes
    d. transfer RNA molecules transport mRNA from the nucleus to the ribosomes
    e. messenger RNA, synthesized on a DNA template in the nucleus, provides information that determines the sequence in which amino acids will be linked during translation

6. List three differences between the structures of DNA and RNA.

7. Transfer RNA is synthesized:
    a. on a DNA template
    b. from a messenger RNA template on a ribosome
    c. on ribosomes without a template
    d. in the nucleolus by the interaction of messenger RNA and chromosomal DNA

QUESTIONS FOR DISCUSSION

1. Why is it important for each type of tRNA to have its own type of enzyme to bind it to an amino acid?

2. Suppose a bacterial cell contained a mutation that changed one of the nucleotides in an anticodon of tRNA.

How might this mutation affect protein synthesis?

3. Pseudogenes look like slightly altered, inactive duplicates of genes that code for proteins. How might they have arisen during the course of evolution?

REFERENCES AND FURTHER READING

(The books and issue of *Science* listed in Chapter 9 also contain information about protein synthesis.)

Clark, B. F., and K. A. Marcker. "How proteins start." *Scientific American*, January 1968.

Crick, F. H. C. "The genetic code." *Scientific American*, October 1962.

Crick, F. H. C. "The genetic code III." *Scientific American*, October 1966.

Crick, F. H. C. "Codon-anticodon pairing: The wobble hypothesis." *Journal of Molecular Biology* 19:548, 1966.

Dayhoff, M. O., C. M. Park, and P. J. McLaughlin. "Building a phylogenetic tree: cytochrome c." In: *Atlas of Protein Sequence and Structure*, Vol. 5, p. 8. ed. M. O. Dayhoff. Washington D.C.: National Biomedical Research Foundation, 1972. Techniques and problems in reconstructing evolutionary events using a computer to compare protein sequences.

Frisch, L., ed. "The genetic code." *Cold Spring Harbor Symposium on Quantitative Biology* 31, 1966. A collection of papers covering experimental evidence about the characteristics of the code and the interactions of molecules during translation.

Nirenberg, M. W. "The genetic code II." *Scientific American*, March 1963.

Nirenberg, M. W., and J. H. Matthaei. "The dependence of cell-free protein synthesis in *E. coli* upon naturally occurring or synthetic polyribonucleotides." *Proceedings of the National Academy of Sciences (U.S.)* 47:1588, 1961. The classic paper on poly-U and mRNA.

Sanger, F. *et al.* "Nucleotide sequence of bacteriophage φX174 DNA." *Nature* 265:687, 1977. The sequence of nucleotides in all the DNA in an organism (if you consider a virus an organism!) has been analyzed for the first time.

Smith, M. *et al.* "DNA sequence at the C terminal of the overlapping genes A and B in the bacteriophage φX174." *Nature* 265:705, 1977. The first unequivocal demonstration that the nucleotide sequences of different genes may overlap.

Wade, N. *The Ultimate Experiment: Man-Made Evolution.* New York: Walker and Co., 1977. The story of the development of recombinant DNA techniques and of the controversies that have surrounded this work.

CHAPTER 11

CELL REPRODUCTION

E very cell has a distinct life history, lasting from the time it is formed by division of an existing cell until it either dies or divides to give rise to two daughter cells. Some types of cells are incapable of dividing and must eventually die. Examples of such cells are nerve and skeletal muscle cells in animals and some cells in the transport systems of higher plants. Most other cells are capable of dividing, and their lives end in division unless they die before they get the chance to divide. A bacterium, for example, is capable of dividing under favorable conditions, but it may die from starvation, antibiotic poisoning, or some other cause first.

Before a eukaryotic cell can divide, its nucleus must go through a complex division that precisely distributes a complete set of chromosomes to each daughter nucleus. This ensures that each daughter cell will inherit a nucleus containing all the genetic information it needs to function. In this chapter we shall consider the two main types of nuclear division, mitosis and meiosis. **Mitosis** produces daughter nuclei with the same number of chromosomes as the original nucleus, whereas **meiosis** produces daughter nuclei with only half the original number of chromosomes. Although mitosis and meiosis technically refer only to division of the nucleus, the words are frequently used to describe division of the entire cell, since division of the cytoplasm usually follows nuclear division. Mitosis and meiosis occur only in eukaryotes. In a prokaryotic cell, the genetic material separates into the daughter cells at division by a mechanism that involves neither mitosis nor meiosis.

Mitosis occurs when a single-celled organism reproduces by dividing into two identical cells, and when a multicellular organism grows, or repairs or replaces damaged or lost parts. For example, when the dead outer cells of the skin slough off, they are replaced by other cells produced by mitotic divisions in the underlying tissue. An embryo develops by mitotic division of the cells descended from the **zygote**, or fertilized egg.

Meiosis is most familiar as the nuclear division by which the **gametes** (sex cells: sperm and eggs) of animals are produced. Meiosis is vital to any organism that reproduces sexually. Without meiosis, the gametes would contain as many chromosomes as the other cells of the parent, and the zygote formed at fertilization would contain twice as many; thus the number of chromosomes would double in each generation.

Before considering meiosis and mitosis we must learn something about chromosomes and how they are organized in cells.

11-A   Chromosomes

Replication

The genetic information of a cell is contained in the DNA of its chromosomes. A **chromosome** is a length of DNA, with its associated proteins.

Before a nucleus can undergo mitosis or meiosis, its chromosomes must duplicate themselves, or **replicate**. In replication, the two strands of the DNA molecule separate, and each serves as the template for the formation of another strand (see Section 9-C). Each chromosome is furnished with a single **centromere**, an area that holds together the two daughter double helices produced when a chromosome replicates (Figure 11–1). After replication, the two identical double-stranded molecules of DNA, with their associated proteins, are called **sister chromatids** as long as they are still attached to each other by the centromere.

Haploid and Diploid Chromosome Numbers

A human male has 46 chromosomes in most of his cells (Figure 11–2). These can be arranged in pairs of similar-looking chromosomes according to their length

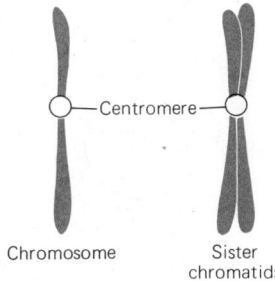

Figure 11-1   A chromosome and the two sister chromatids formed when it replicates.

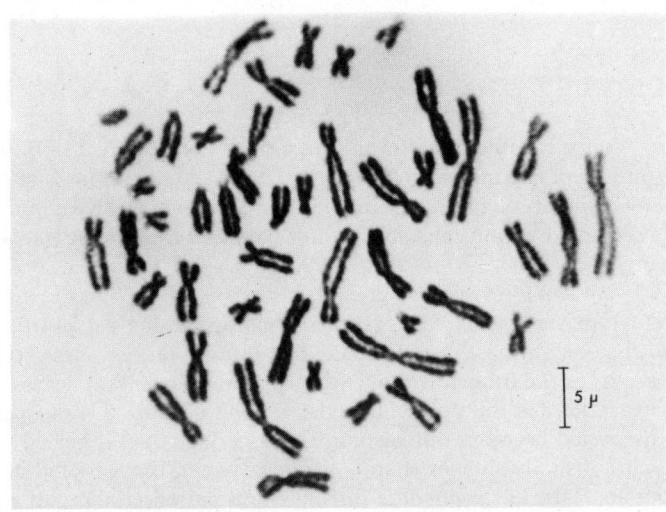

(a)

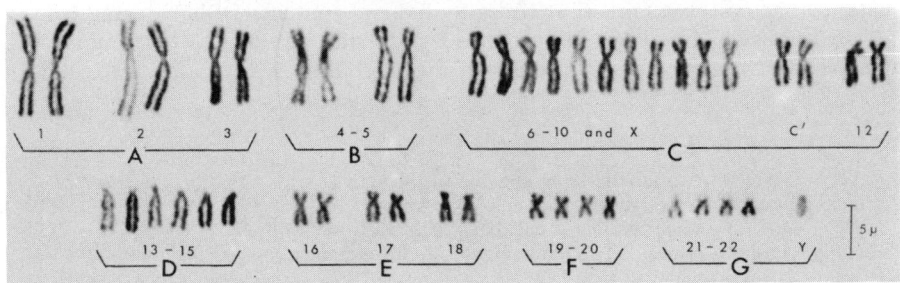

(b)

Figure 11-2   The genetic material of a normal human male. (a) 46 sets of sister chromatids from a body cell. (b) A karyotype, made by cutting out the sets of chromatids from the photograph in part (a) and arranging them in a conventional order. (Carolina Biological Supply Company)

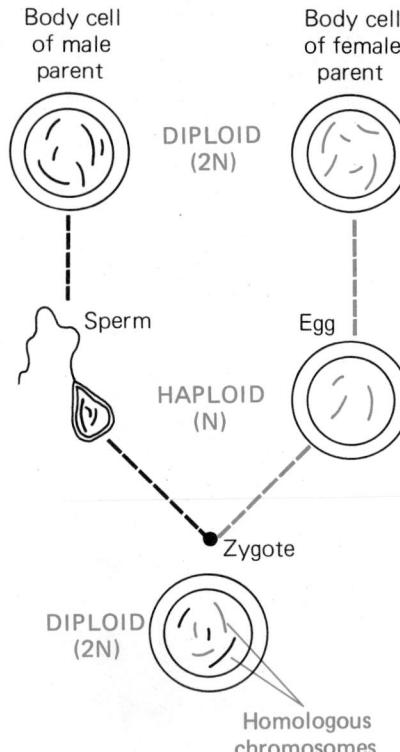

Body cell of male parent

Body cell of female parent

DIPLOID (2N)

Sperm

Egg

HAPLOID (N)

Zygote

DIPLOID (2N)

Homologous chromosomes

Figure 11-3 Haploid cells (egg and sperm) and diploid cells (body cells and zygote) as they occur in the life history of most higher animals. Three pairs of homologous chromosomes, of different lengths, are shown.

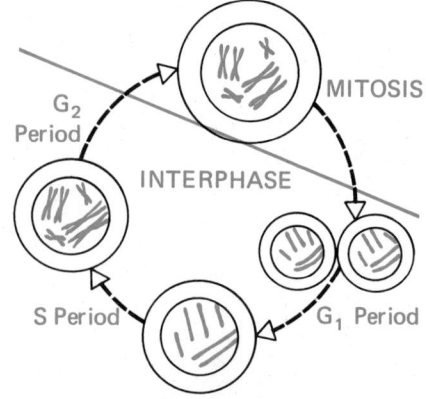

G₂ Period

MITOSIS

INTERPHASE

S Period

G₁ Period

Figure 11-4 A typical cell cycle, Mitosis, or nuclear division, is followed by cell division. The two gap periods (G₁ and G₂) are separated by the synthesis (S) period during which the chromosomes are replicated.

and the position of the centromere. Twenty-two of these pairs contain look-alike chromosomes, but the twenty-third pair is odd; it consists of unlike chromosomes, called X and Y. These chromosomes determine their owner's sex. A human female has two X chromosomes, but no Y chromosome, in her twenty-third pair. In each pair of human chromosomes in both sexes (except the twenty-third pair in males) the two **homologous chromosomes** of the pair appear identical, and in fact both members of the pair contain the same kind of genetic information at the same position on the chromosome. For example, human chromosome #1 contains genes for the enzyme salivary amylase* and for the Rh blood factor.

The possession of pairs of chromosomes is important in the life history of eukaryotic organisms. A human egg contains one member of each pair of chromosomes found in the other cells in the mother's body. A sperm contains one member of each pair of chromosomes found in the father's body cells. When the egg and sperm join at fertilization, the new individual receives one member of each pair of chromosomes from its mother, and one member of each pair from its father, for a total of twenty-three complete pairs of chromosomes.

A cell that contains pairs of homologous chromosomes is said to be **diploid**; the diploid number in humans is 46, or 23 pairs. A cell that contains unpaired chromosomes is said to be **haploid**; an egg or sperm is haploid, and in humans the haploid number is 23 (Figure 11-3). The haploid number of chromosomes in a species is generally designated as **N**; thus N = 23 in human beings, and **2N** (diploid) = 46. N varies from species to species. N = 10 in corn plants and N = 4 in the fruit fly, *Drosophila*.

Some organisms are **tetraploid**, or 4N, with four homologous chromosomes of each type; triploid (3N), hexaploid (6N) and octaploid (8N) organisms are also found, particularly among the plants.

In most animals the zygote and the cells that arise from it by mitotic division contain the diploid number of chromosomes. Some cells in the ovaries or testes eventually undergo the process of meiosis, which produces haploid nuclei. Only the egg and sperm cells have haploid nuclei. Cells that can undergo meiosis are known as **germ cells**; the rest of the body's cells are called body or **somatic** (soma = body) **cells**.

The somatic cells of many lower organisms, such as some protists,* many algae, fungi, and some invertebrate* animals, are normally haploid.

11-B   The Cell Cycle

In cells that are capable of dividing, the **cell cycle** (Figure 11-4) is the period from the beginning of one mitosis to the beginning of the next. The length of the cell cycle varies considerably. The cells in some early embryos divide once every 20 minutes, while most dividing cells of the adult have cell cycles that last from several hours to many days.

The cell cycle has four distinct phases. Mitosis is the period during which the nucleus and cytoplasm divide. The rest of the cell cycle, known as **interphase**, is divided into the remaining three periods. The middle one of these, the **S** period (synthesis period), is the time of DNA synthesis, when the cell replicates its chromosomes in preparation for the next cell division. The **G₁** period (first gap period) is defined as the period between mitosis and the S period; the **G₂** period (second gap period), which is usually shorter than G₁, falls between the S period and the next mitosis. Since its DNA has replicated during the S period, a nominally diploid somatic cell actually contains four times its haploid content of DNA during G₂.

Synthesis of DNA, RNA and proteins is regulated very precisely during the cell cycle. The DNA and histones* present in chromosomes are synthesized only during

---

* Salivary amylase   A starch-digesting enzyme in the saliva.
* Protists   One-celled eukaryotic organisms, e.g., *Amoeba*.
* Invertebrate   An animal without a backbone, e.g., earthworm, fly.
* Histones   Proteins of basic pH, characteristic of chromosomes.

the S phase. In cells with short cell cycles, nuclear DNA is occupied by the enzymes replicating DNA for most of the S phase, so it is not surprising that little RNA transcription occurs during this time; there is little room on the DNA molecules for RNA polymerase.

In cells with longer S phases, the slower rate of DNA synthesis must permit RNA polymerase to occupy the chromosomes at the same time as the enzymes of DNA replication, because RNA synthesis has been found in the S phase as well as during G_1 and G_2 during long cell cycles. Neither RNA nor DNA is synthesized during mitosis. Cytoplasmic protein synthesis, however, is not directly affected by events in the nucleus and occurs throughout the cell cycle in somatic cells.

11-C Factors That Affect the Cell Cycle

We know very little about the factors that control the length of the cell cycle or of its component phases. Because cancerous cells (Section 34-M) divide more frequently than related normal cells, understanding the control of cell division is a pressing medical problem. Research is beginning to produce some tentative answers.

The obvious factors that affect all biochemical reactions also affect cell division. Experiments with the rate of cell division in the bacterium *Escherichia coli*, for instance, have shown that division is fastest when an ample supply of nutrients is available and when the temperature and pH of the environment fall within particular ranges. The genes a cell contains also affect how often it divides: cells from different animals cultured outside the body divide at characteristic rates. These general factors, however, cannot be the only ones controlling the rate of division, because the cells of a particular tissue divide at different rates at different times. There are obviously specific factors in the body that control cell division. Those that have been investigated control either the time when mitosis starts or the rate or timing of DNA synthesis (the S phase) within the cell cycle.

Chalones

In 1960, W. B. Laurence and W. S. Bullough studied how wounding the skin stimulates the cell division that heals the wound. They discovered factors called chalones (pronounced KAY-loans; according to Bullough, the name comes from a Greek word meaning "to slack off the main sheet of a sloop to slow the vessel down"). **Chalones** are substances, mostly peptides and glycoproteins (proteins with sugars attached), that are secreted into the extracellular fluid by healthy cells and that inhibit mitosis in the types of cells that secrete them. Each is specific to the tissue that produces it. It appears that damaged cells stop making chalones; the level of chalones in the extracellular fluid falls, reducing the inhibition and permitting the cells of that tissue to divide and heal the wound. The cells divide until a normal level of chalones builds up again. So far, we have no idea how chalones inhibit mitosis, except that they do *not* act by inhibiting DNA synthesis, which proceeds normally in their presence.

There is now evidence that normal tissues also contain substances that induce mitosis; in other words, their action is the opposite of chalone action. Interactions between these two substances could control rates of cell division in tissues, but we have yet to discover how these factors act and what causes them to be produced.

Cyclic AMP and Calcium Ions

Cyclic adenosine monophosphate (**cyclic AMP** for short) (Figure 11–5) is a nucleotide with many roles as a sort of junior hormone acting within a cell, instead of carrying a message between cells as do more familiar hormones such as insulin. Rapidly dividing cells contain low levels of cyclic AMP, whereas cells that are not dividing contain more cyclic AMP. The concentration of cyclic GMP (Figure 11–5) also varies with the rate of cell division in a tissue, and so does the concentra-

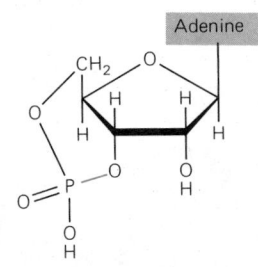

ADENOSINE MONOPHOSPHATE (AMP)

CYCLIC AMP (cAMP)

Figure 11-5 In cyclic AMP the phosphate and sugar are bonded at two places rather than one. Cyclic GMP would have guanine instead of adenine in the position indicated by the colored box.

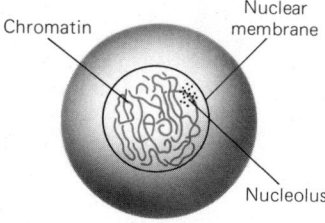

INTERPHASE

Chromatin

Nuclear membrane

Nucleolus

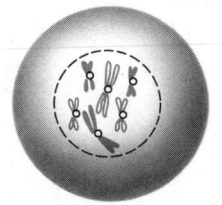

PROPHASE

Sister chromatids visible as chromatin condenses. Nucleolus and nuclear membrane disappear

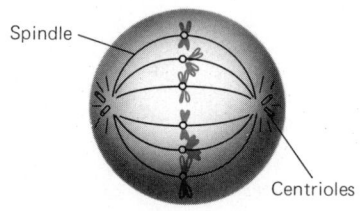

METAPHASE

Spindle

Centrioles

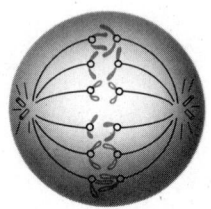

ANAPHASE

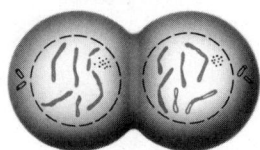

TELOPHASE AND CYTOKINESIS

Figure 11-6 The phases of mitosis. See the text for a description of each phase.

tion of calcium ions. The correlation between cell division and concentrations of these substances is so precise that it seems they must be related, but we do not know whether levels of cyclic nucleotides and calcium cause and inhibit cell division or whether both effects are caused by something else. Levels of these substances are also related to the presence of some hormones that affect cell division.

Drugs

Cell division is affected not only by factors normally found within the body but also by substances used as drugs or present in the environment. Some of these substances slow or stop cell division. Chemotherapy, chemical treatment of cancer, depends upon the fact that some drugs distinguish between cancerous and normal cells in that they block cell division or kill cells only when they are dividing, and many cancer cells divide rapidly. For instance, the nitrogen mustards specifically prevent mitosis, whereas antibiotics such as chloramphenicol, streptomycin, and the tetracyclines prevent cell division indirectly by blocking protein synthesis and so prolonging the G_1 phase of the cell cycle.

11-D Mitosis

Mitosis is the series of events that causes the equal distribution of chromosomes to two daughter nuclei after the chromosome duplication that occurred in the S phase of the cell cycle. Mitosis is a continuous process, but to make it easier to describe, it is divided into four stages according to the appearance of the chromosomes under the light microscope.

When the cell is in interphase, between divisions, its chromosomes are spread out in a tangled mass known as **chromatin**. Mitosis can first be recognized during **prophase**, when the chromatin sorts itself out and condenses into tidy chromosomes, visible under the light microscope as pairs of sister chromatids (Figure 11-6). Condensation is an impressive process. It is comparable to taking a thin strand some 200 m long and coiling it into a cylinder about 1 mm wide by 8 mm long. During prophase, the nucleolus, which is the site where ribosomes are synthesized, usually disappears, and the material that makes up the bulk of the nucleolus becomes scattered through the nucleus. In most eukaryotes (except for some protists and fungi) the nuclear membrane also disappears at the end of prophase. It is thought that the membrane becomes part of cytoplasmic membrane systems and is reformed from them after mitosis. The mitotic spindle (see below) may begin to form during prophase.

Metaphase, the second phase in mitosis, can be recognized by the completion of the mitotic spindle. The **mitotic spindle** (see Figure 11-6) is composed of microtubules (see Section 4-B) built up from subunits that are normally dispersed throughout the cell. In animal and some lower plant cells the ends, or **poles**, of the spindle are occupied by the centrioles (see Section 4-B). Structures in the centromere called **kinetochores** attach each pair of chromatids to microtubules near the midline of the spindle. The resulting configuration, with all of the chromatids lined up at the imaginary **metaphase plate** at the equator of the cell, is characteristic of cells in metaphase.

The third phase of mitosis is **anaphase**. Each centromere splits into two, allowing the sister chromatids to separate and thereby to become chromosomes, which move to opposite poles of the spindle. Interaction between microtubules and chromatids is necessary for this process, which constitutes one of the most precise movements exhibited by a cell. It is reminiscent of the movements of the contractile proteins that cause a muscle fiber to contract, but how it occurs is unknown. In anaphase, each chromosome ends up at one pole of the mitotic spindle, with its sister at the opposite end, so that each pole now has a complete set of chromosomes, the basis for a new nucleus.

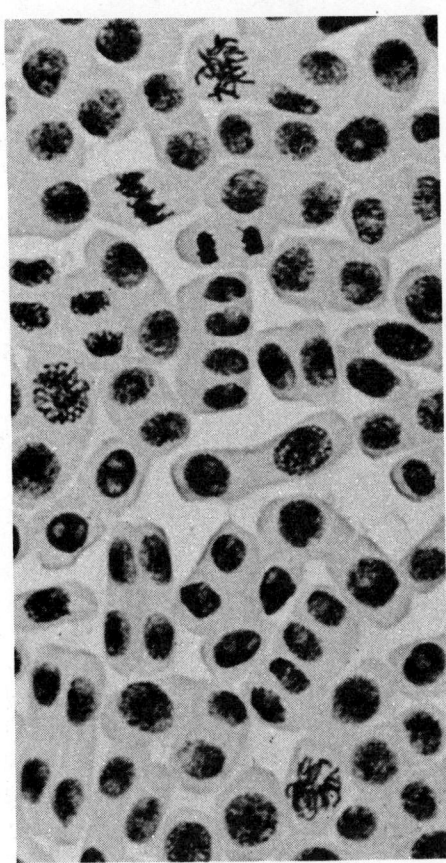

Figure 11-7 Dividing cells from the squashed tip of an onion root. Try to identify the various phases of mitosis, using Figure 11-6 for guidance. (Carolina Biological Supply Company)

Colchicine is a chemical that blocks cell division in eukaryotes, and is sometimes used to produce tetraploid plants from diploid plants. Colchicine binds with a microtubule protein and prevents the mitotic spindle from functioning. Thus, the chromatids cannot segregate into two groups by moving along the spindle, and the nucleus becomes tetraploid. If part of a diploid plant treated in this way is cultured or permitted to form seeds, a tetraploid plant can be grown. Tetraploid plants are often larger and more vigorous than their diploid ancestors; many cultivated vegetables and flowers are tetraploids that have arisen either naturally or by deliberate treatment with colchicine. (Vinblastine and vincristine are chemicals with effects similar to those of colchicine.)

The last stage of mitosis is **telophase**, recognizable as the time at which the chromosomes that have moved to the poles of a mitotic spindle uncoil to become a mass of tangled chromatin once again, a nuclear membrane reforms around each group of chromosomes, and a nucleolus becomes visible in each daughter nucleus. Mitosis is now complete. Telophase is usually accompanied by **cytokinesis**, the process whereby the plasma membrane pinches inward, separating the two daughter nuclei and the two halves of the cytoplasm to form two new cells. In plants, the middle lamella* between the two daughter cells is laid down during telophase, and each daughter cell then builds its own cell wall on its side of the middle lamella.

The events of mitosis ensure that the two sets of daughter chromosomes synthesized during DNA replication separate precisely into the daughter nuclei at cell division, so that each daughter cell receives one copy of each chromosome. Cells of any ploidy (haploid, diploid, tetraploid, etc.) can undergo mitosis.

---

*Middle lamella The common partition between the cell walls of two adjacent plant cells.

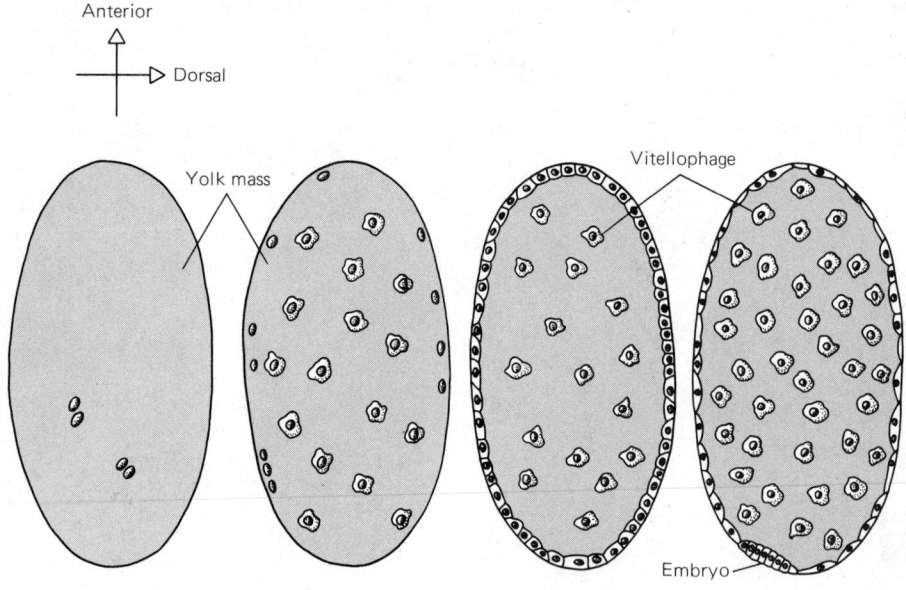

Anterior

Dorsal

Yolk mass

Vitellophage

Embryo

(a) The zygote nucleus has divided twice, to form four nuclei.

(b) The nuclei become enveloped in cell membranes to form cells which continue to divide.

(c) Some of the newly formed cells are vitellophages, which digest yolk, converting it into nutrition for other cells; others form a layer around the outside of the egg.

(d) Some of the cells give rise to the embryo; others form a membrane that protects the embryo.

Figure 11-8 Early development of an insect (a bristletail).

11-E   Syncytia and Coenocytes

Although in most tissues mitosis produces two separate cells—each with its own nucleus, cytoplasm, and plasma membrane—this does not always occur. In certain tissues, nuclear division is not followed by cytokinesis; the result, called a **syncytium** in animals, or a **coenocyte** in plants, is a cell-like unit that contains more than one nucleus within a single plasma membrane.

One example of a syncytium occurs in the embryos of insects (Figure 11–8). After an insect egg has been fertilized, the diploid zygote nucleus undergoes several mitotic divisions that are not followed by cytokinesis. Only after the developing egg contains a number of nuclei (sometimes several thousand) are the first internal plasma membranes formed.

A similar situation sometimes occurs in the endosperm tissue that surrounds and nourishes the embryo of a flowering plant. An extreme case is the coconut; its "milk" is a nutrient fluid filled with nuclei. Eventually the nuclei become surrounded with membranes and cell walls, to form the coconut "meat." Many algae and fungi are also coenocytic, usually with haploid rather than diploid nuclei.

11-F   Meiosis

Meiosis is the process that makes haploid nuclei from diploid nuclei. The simplest way in which this could happen would be by omission of the S phase of the cell cycle, permitting half the chromosomes of a diploid cell to move into each new nucleus during nuclear division and so producing two haploid nuclei. This does not happen. Presumably because meiosis is a form of nuclear division that has evolved from mitosis, the S phase occurs before meiosis as well as before mitosis, and a cell enters meiosis with four times its haploid DNA content. However, because a nucleus cannot divide into more than two daughter nuclei during any one division, two divisions are needed during meiosis to reduce the DNA content of each cell to haploid. Meiosis is long and complicated and frequently takes days to complete instead of hours or minutes as does mitosis.

MEIOSIS I   Homologous chromosomes separate:

MEIOSIS II   Each nucleus divides again, producing haploid nuclei. Division of centromeres permits sister chromatids to separate.

PROPHASE I

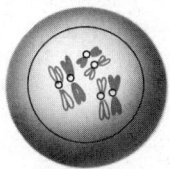

Chromosomes condense and pair with their homologues to form tetrads (groups of four). Tetrad formation is necessary for the later step of separating members of each homologous pair so that each daughter nucleus receives one member of each pair.

PROPHASE II

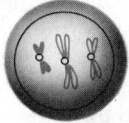 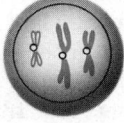

In those organisms (particularly plants) with a period of interkinesis between the two mitotic divisions, the chromosomes must condense again before the second division.

METAPHASE I

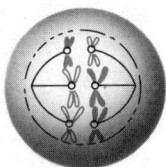

The tetrads move to the equator. The kinetochores of homologous sets of sister chromatids attach to the spindle fibers on opposite sides of the equator as the nuclear envelope begins to disappear.

METAPHASE II

Metaphase II occurs in all organisms that undergo meiosis. Spindles form again and chromatids move to the equator, where their kinetochores attach to the spindle fibers. The centromeres then divide, just as they do in metaphase of mitosis.

ANAPHASE I

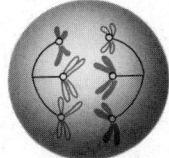

Each set of sister chromatids is pulled toward a pole of the spindle, away from its homologue. The sister chromatids travel as a pair and do not separate until later in meiosis.

ANAPHASE II

The sister chromatids separate to opposite poles.

TELOPHASE I

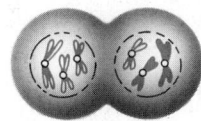

Telophase I differs in different organisms. In some, the nuclear envelope reappears and cytokinesis occurs. In others, the nuclear envelope remains absent and metaphase II starts immediately.

TELOPHASE II

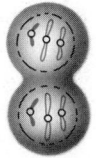 

Four haploid nuclei are formed, each with one member of each pair of chromosomes from the original cell that entered meiosis. The nuclear envelopes reappear and cytokinesis occurs.

Figure 11-9 Meiosis in the formation of male gametes in an animal.

The two series of divisions in meiosis are unimaginatively called meiosis I and II. In meiosis I, all the sets of sister chromatids line up with their homologues, move to the equator, and then separate, moving to opposite poles of the cell. The precise choreography of meiosis I ensures that each daughter nucleus receives one member of each homologous chromosome pair. Unlike the events of mitosis, the sister chromatids travel as a pair and remain joined at the centromere during this division. During meiosis II, each nucleus formed in meiosis I divides again. This time the centromeres divide, releasing the sister chromatids as separate chromosomes, and the resulting nuclei are haploid. The events of meiosis are detailed more fully in Figure 11–9, which you should read carefully.

In animals, meiosis takes place in the ovaries and testes and results in the production of eggs and sperm, respectively. In plants, meiosis occurs in structures called **sporangia** and results in the production of reproductive cells called **spores**; these spores must go through further stages in development before they give rise to eggs and sperm by the process of mitosis. Haploid organisms undergo meiosis following the formation of a diploid zygote during sexual reproduction; each zygote thus gives rise to four new haploid individuals (see Chapter 25).

Besides maintaining the same chromosome number from generation to generation, meiosis allows the genetic material to combine in new ways, a process referred to as **genetic recombination**. One way that genes recombine results from the separation of homologous chromosomes during meiosis I. There is an equal chance that any one chromosome will end up in a daughter cell with either member of any other pair of homologous chromosomes. Thus, if an organism inherited chromosomes ABC from its mother and abc from its father, it is equally likely to produce the gametes containing ABC, AbC, Abc, ABc, aBC, abC, aBc, and abc. This mixes up the chromosomes obtained from the parents and produces a wide variety of combinations in the gametes.

Genetic recombination also occurs during meiosis by the transfer of portions of DNA from one chromosome to its homologue. During prophase I, homologous sets of sister chromatids undergo **synapsis**, in which they line up side by side, attached by their centromeres in a precise alignment, forming a **tetrad** (tetra = four, for the four chromatids involved) (see Figure 11–9). During synapsis, chromatids may attach to each other and exchange segments of genetic material, an event known as **crossing over** (see Section 13-L).

Crossing over is a relatively frequent event. During synapsis the 23 sets of human chromatids will have about 50 areas of crossing over, or **chiasmata** (singular: **chiasma**), among them. Each pair of sister chromatids will be involved in at least one crossover per generation, on average. Crossing over occurs during meiotic prophase I but not during prophase II, and rarely during mitosis. The recombination of genetic material is one of the most important results of meiosis.

SUMMARY

A cell exists from the time it forms by division of another cell until it dies or divides. Cell divisions are of two kinds: mitotic division, in which a cell gives rise to two daughter cells with the same number of chromosomes as the parent, and meiotic division, in which the number of chromosomes in the four daughter cells is half the number in the parent cell.

The period from one nuclear division to the next is known as the cell cycle; it can be divided into mitosis, G_1, the S phase (when DNA synthesis occurs), and G_2. The length of the cell cycle varies from as little as 15 minutes in some early embryos to several days or weeks. Synthesis of DNA is not continuous throughout the cycle, but little is known about how this synthesis, or the length of different phases of the cell cycle, is controlled. Chalones secreted by cells inhibit their subsequent division, but the mechanism is unknown.

Mitosis is a nuclear division in which precise events ensure that the daughter nuclei inherit chromosomes identical to those of the parent nucleus. During prophase of mitosis, the chromatids condense and become visible under the light microscope, and the nucleolus and nuclear membrane disappear. In metaphase, the sister chromatids attach by their kinetochores to the microtubules of the mitotic spindle. During anaphase, each centromere splits into two, releasing the sister chromatids from one another and allowing them to travel to the opposite poles of the spindle. During telophase, the chromosomes at each pole form a nucleus as the nuclear membrane and nucleolus reform, and the chromosomes unravel from their condensed form. Telophase is usually accompanied by cytokinesis, the division of the cytoplasm and cell membrane to form two cells. Mitosis without cytokinesis forms a syncytium or a coenocyte, which contains many nuclei within one membrane.

Meiosis is the nuclear process that ensures production of four haploid nuclei from a diploid nucleus. DNA replication occurs before meiosis as well as before mitosis; this necessitates two meiotic divisions. In the first meiotic division, homologous chromosomes separate during anaphase. Not until the second division do the centromeres divide, permitting sister chromatids to move into different nuclei. During prophase of the first meiotic division, chromatids may exchange genetic information with chromatids of the homologous chromosome. Cytokinesis usually follows each meiotic division.

Meiosis halves the number of chromosomes in a cell in such a way that each daughter nucleus receives one member of each pair of homologous chromosomes. Thus meiosis prevents the chromosome number from doubling in each generation of organisms that reproduce sexually. Meiosis also results in genetic recombination, both by forming new chromosome combinations and by synapsis and crossing over, in which homologous chromosomes exchange genetic information. Mitosis, on the other hand, guarantees that every somatic cell division produces cells with identical chromosome complements.

OBJECTIVES

From your study of this chapter, you should be able to:

1. Define the following terms and be able to use them in context: haploid, diploid, tetraploid, somatic cell, germ cell, fertilization, zygote, interphase, sister chromatids, homologous chromosomes, centromere, mitotic spindle, cytokinesis, chiasma, synapsis.

2. Define and describe the four stages in the cell cycle.

3. State how you would recognize a cell in prophase, metaphase, anaphase, or telophase of mitosis.

4. State why substances that interfere with microtubule function or formation interfere with cell division; name one such substance and state how it is useful to plant breeders.

5. Define syncytium or coenocyte and give one example of how such a structure can be formed.

6. Describe the respective functions of mitotic and meiotic nuclear divisions in the life history of an organism.

7. Compare and contrast what happens to the chromosomes during mitosis with what happens to them in meiosis.

8. Describe synapsis and the formation of chiasmata, and explain their biological importance.

SELF-QUIZ

1. A cell cycle is:
    a. the time from the formation of a cell until its death
    b. the series of events that takes place from the formation of a cell until it divides again
    c. the sequence of events that assures each daughter cell of a set of chromosomes identical with that of its parent cell (mitosis)
    d. the growth of a cell until it is large enough to divide again

2. For the species depicted in Figure 11–3, what is the value of N? of 2N?

3. A diploid somatic cell:
    a. cannot undergo division again
    b. can undergo mitosis but not meiosis
    c. can undergo mitosis or meiosis
    d. can undergo meiosis but not mitosis

4. A cell is in metaphase if:
    a. its chromosomes are visible as distinct threadlike structures
    b. the nuclear membrane is not visible
    c. the chromosomes are lined up on the equator of the cell
    d. the chromosomes are separated into two distinct groups attached to the spindle
    e. the chromosomes are found in two compact groups in two small patches of cytoplasm that are in the process of separating into two distinct cells

5. A cell in prophase of mitosis can be distinguished from a cell in prophase I of meiosis by:
    a. the presence of only half as many chromosomes in the meiotic cell
    b. the formation of tetrads in the meiotic cell
    c. the presence of twice as many chromosomes in the meiotic cell

6. The function of mitotic cell division in the life history of an organism is:
    a. reproduction of identical individuals if the organism is unicellular
    b. growth of an individual if the organism is multicellular
    c. repair of injured tissue
    d. all of the above

7. Substances that interfere with microtubule function interfere with cell division because:
    a. microtubules must be distributed equally to the new cells
    b. microtubules are involved in the precise separation of the chromosomes, which ensures that a complete set of chromosomes gets into each daughter cell
    c. without microtubules, cytokinesis cannot take place, and a syncytium is formed
    d. microtubules are essential to the disappearance of the nuclear membrane, and without them the chromosomes have to stay too close together within the nuclear membrane to be able to separate into two new nuclei

Essay: Radiation and Cell Division

Events in the twentieth century have made all of us aware that radiation is potentially dangerous to living organisms, including ourselves. Cell division is one of the many things that radiation can affect.

From a biological point of view, the most important radiation is **ionizing radiation**—x-rays, γ-rays, and particles such as neutrons, α-particles (helium nuclei), and the like. These are given off by the decay of radioactive elements in the earth's crust and in space. Ionizing radiation has enough energy to ionize substances it strikes, by knocking off electrons. For instance, ionized water, H_2O^+, is very reactive and can cause peculiar reactions in the cell.

In large enough doses, ionizing radiation causes so much disruption that it can kill a cell. In much smaller doses, the most important effect of the radiation is to cause breaks in DNA molecules. If a DNA molecule is sufficiently damaged, it cannot replicate, and this prevents cell division. On the other hand, the damage may show up as mutations that can be replicated and passed on; some mutations of this type in somatic cells are believed to cause cancers, which are characterized by rapid cell division (Section 34-M).

The ability of ionizing radiation in appropriate doses to block cell division is used to treat cancer. Cells are most sensitive to radiation damage just before mitosis. Because cancer cells divide more often than the normal cells that surround them, bombarding an organ with radiation kills or blocks division in many more cancerous than normal cells.

Ultraviolet radiation (**uv**), part of the electromagnetic spectrum (Figure 11–A), is emitted by the sun and by "black light" bulbs. Nucleic acids absorb uv and can be permanently damaged by it. Ultraviolet can kill cells, and in fact it is used to kill bacteria on laboratory equipment that cannot be sterilized by heat or solvents. Less drastically, uv can cause mutations, including mutations that make cells divide more rapidly, as happens in skin cancer, or make them stop dividing. Slowing of cell division is the reason light-colored skin exposed to much sunlight ages so rapidly; damaged cells are not replaced as fast as they otherwise would be. Darker-skinned people are less prone to premature skin aging and skin cancer because the dark pigment, melanin, in their skin absorbs ultraviolet rays and prevents them from penetrating to the DNA of living cells. Ultraviolet radiation does not cause cancer in deeper organs of the body because, unlike x-rays and γ-rays, it is rapidly absorbed by water in living tissue and so does not penetrate beyond the skin.

Both ionizing radiation (from decay of radioactive elements in rocks and in our own bodies) and ultraviolet radiation (from the sun) are part of the natural environment. It has been estimated that about 50% of the ionizing radiation to which the "average" American has been exposed since 1940 is natural, or background, radiation from rocks, cosmic radiation, and the like. "Civilization" exposes us to additional radiation. About 2% of our exposure comes from the fallout from nuclear weapons testing and use; less than 0.2% comes from nuclear power stations and their waste; about 40% comes from medical and dental x-rays; and about 2% from such domestic sources as color televisions and smoke detectors. Most experts think we would be well advised to reduce our exposure to medical x-rays.

It is, however, exceedingly difficult to say how much radiation beyond the inevitable background level should be considered a health hazard. First, it is hard to estimate the background level of radiation; both the kinds and the amounts of radiation a person receives vary with geography, occupation, personal habits, and so forth. Second, the degree of damage resulting from exposure to a particular dose of radiation varies depending on which particular tissues and molecules are struck by the radiation, something we cannot predict.

A third variable is the ability of affected tissue to respond to the damage. All cells contain enzymes that can repair either breaks in DNA or mis-pairing caused by radiation damage to particular nucleotides in the molecule. (Interestingly, some enzymes that can repair damage caused by ultraviolet radiation are activated by light, the very agent that causes the damage!) This repair is not perfect, however, and this is why radiation can sometimes cause cancers and other disorders.

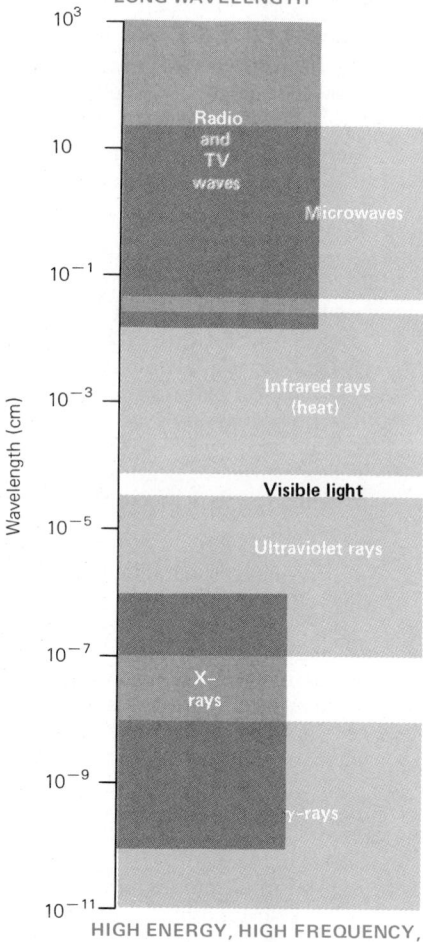

Figure 11-A The electromagnetic spectrum.

QUESTIONS FOR DISCUSSION

1. Tetraploid plants are frequently larger and have larger fruits and flowers than their diploid relatives. What might account for the fact that octaploid plants (8N) tend to be tiny and scrawny, and produce few offspring?

2. Banana plants are hexaploid (6N). Can you explain why they are unable to reproduce sexually? How do you suppose they do reproduce?

3. Since the genetic information is carried equally by egg and sperm, what do you suppose to be the selective advantage of the inequality of size that has evolved between the tiny, mobile sperm and the large immobile egg?

REFERENCES AND FURTHER READING

Biological Effects of Ionizing Radiation Report, *The Effects on Populations of Exposure to Low Levels of Ionizing Radiation.* Washington, D.C.: National Academy of Sciences–National Research Council, 1972. The comprehensive report on human exposure to natural and technologically produced radiation referred to in the essay in this chapter.

Giese, A. C. *Cell Physiology*, 5th ed. Philadelphia: W. B. Saunders Co., 1979. Contains a good discussion of chalones and of the effects of radiation on cell division.

Goodenough, U. *Genetics*, 2d ed. New York: Holt, Rinehart and Winston, 1978. Discusses cell reproduction at the chromosomal and molecular levels, including research evidence and unanswered questions.

Suzuki, D. T., and A. J. F. Griffiths. *An Introduction to Genetic Analysis.* San Francisco: W. H. Freeman, 1976. A genetics textbook with an excellent discussion of cell division.

CHAPTER 12

CELL DEVELOPMENT AND AGING

Most of this book is about the anatomical, physiological, and behavioral adaptations of organisms and their component cells and tissues. We have seen that most of the information that determines these adaptations is coded in an organism's DNA. There is, however, a big, unexplained gap between the inherited information with which an organism is endowed when it starts life as a spore or fertilized egg and the walking, talking, photosynthetic, or flowering adult. How does the information in a spore or **zygote** (fertilized egg) produce the very different adult organism? The answer to this question plainly lies somewhere in the study of development.

In this chapter, we consider development on a cellular and molecular level. As development unfolds, genetically identical cells descended from the original spore or zygote **differentiate**, that is, become different from one another, as different genes are switched on and off in different cells. A gene may be thought of as being **turned on**, or **expressed**, when its DNA is being transcribed into RNA and the RNA translated into protein (if it is messenger RNA; genes for transfer or ribosomal RNA are expressed as the RNA itself [Sections 10-D, E, and F]). Genes that are not being transcribed can be thought of as **turned off**.

We can view development as a precisely coordinated series of events. Each gene is switched on in its turn, and then something resulting from the expression of that gene, or some environmental influence, switches on the next gene in the program.

Many complications enter into this simple-sounding picture, however. For instance, a gene may be transcribed into RNA, and RNA translated into protein, which has no apparent effect on the cell until days or weeks later. Such peculiarities of timing make differentiation very difficult to study. Second, cells in different parts of an organism develop very differently, and the factors responsible for these position effects must be identified.

Differentiation during development overlaps with another developmental process, aging. Aging, too, has effects at the cellular level, and we shall conclude our study of cell development by considering these changes.

12-A Genetic Information in Zygote and Adult

Several lines of evidence have convinced biologists that, with very few exceptions, each cell of an adult organism contains exactly the same genetic information as the zygote from which the cells descended.

The Evidence in Plants

The most direct evidence for chromosomal identity of the zygote and the adult comes from experiments performed by Frederick C. Steward on carrots. He and his coworkers cultured carrot root cells in an artificial medium and found that whole carrot plants developed from single suspended cells (Figure 12–1). Clearly, in this

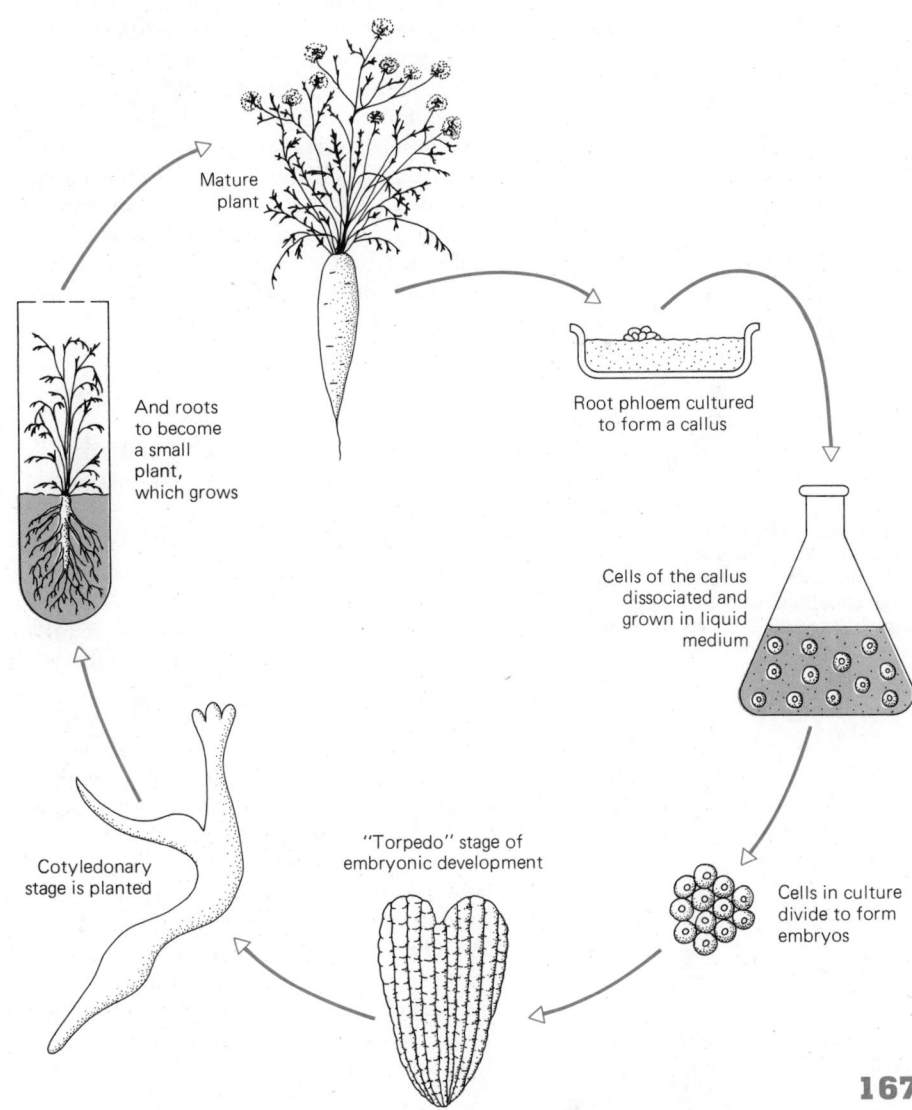

Mature plant

And roots to become a small plant, which grows

Root phloem cultured to form a callus

Cells of the callus dissociated and grown in liquid medium

Cells in culture divide to form embryos

Cotyledonary stage is planted

"Torpedo" stage of embryonic development

Figure 12-1 Stages in the development of a mature carrot plant from cells isolated from the phloem (food-transporting tissue) of a root.

167

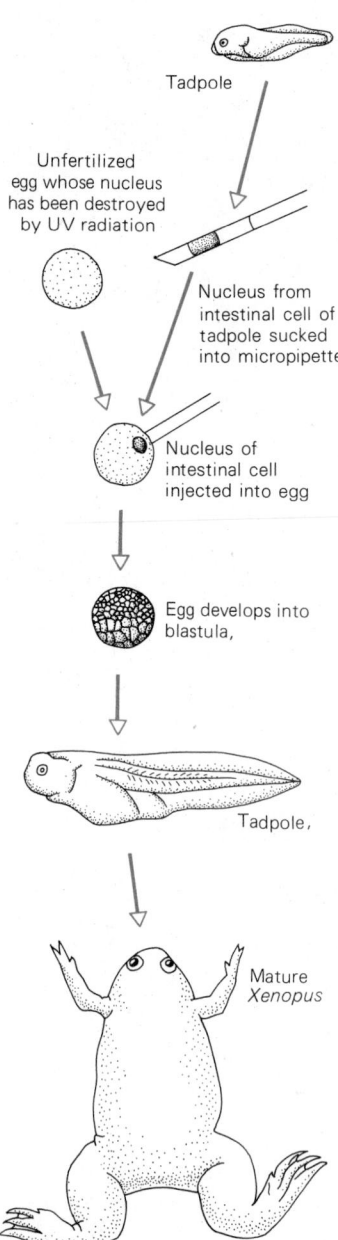

Tadpole

Unfertilized egg whose nucleus has been destroyed by UV radiation

Nucleus from intestinal cell of tadpole sucked into micropipette

Nucleus of intestinal cell injected into egg

Egg develops into blastula,

Tadpole,

Mature *Xenopus*

Figure 12-2 Nuclear transplantation experiment in which a nucleus from a tadpole intestine cell (chosen because it is large enough to be manageable) is injected into an egg and supports the development of a mature frog.

case, a differentiated root cell contained all the genetic information carrots are heir to, and had lost no genes or genetic information during its own differentiation. It still contained all the genetic information necessary for development of all the different cells in an adult carrot. Many other plant cells have the same ability.

The Evidence in Animals

No one has yet grown an entire animal from a single differentiated cell, but experiments implying that this is possible have been done with frogs and mice. These are called **nuclear transplantation** experiments. They were designed to determine whether any genetic information present in the zygote is missing from the nucleus of a differentiated cell. Information would be lost if a gene physically disappeared or became irreversibly inactivated during differentiation. The technique is to remove the nucleus from an egg, replace it with the nucleus of a differentiated cell, and see if this nucleus will take over the role of the zygote nucleus and support development of another complete organism (Figure 12–2). If a nucleus from the gut of a tadpole is transplanted into an egg in this way, the artificial zygote so formed develops normally into a sexually mature adult frog. The conclusion must be that the nuclei of at least some differentiated cells have not lost, during their differentiation, any of the genetic information that is present in the zygote. However, this experiment does not work with all types of differentiated nuclei, and the case has not been made so clearly for animals as for plants.

The most sweeping possible conclusion that we could draw from these experiments is that all the genes present in the zygote of a plant or animal are also present in all the embryonic and adult cells that arise by mitotic divisions of that zygote. Some genes may be inactive, in that they are not transcribed into RNA or translated into protein at any given time, but each is capable of being activated by the appropriate stimulus. It is a trifle reckless, however, to draw this tempting conclusion in our present state of ignorance about how genes are turned on and off; it is safer to say that in most organisms there is no evidence that genetic information is permanently lost from a cell during its differentiation. If this is the case, it seems logical to assume that differentiation of a particular type of cell requires the selective expression of genes; presumably different genes would be active at various times in the differentiation of different cells. In the next section we shall examine the evidence that such differential gene activity does occur.

12-B Evidence of Differential Gene Activity in Prokaryotes

In the bacterium *Escherichia coli*, there is clear evidence that different genes are transcribed at different times under the influence of various regulatory substances inside and outside the cell. The work of François Jacob and Jacques Monod, published in 1961, has led to a general understanding of the genetic control of protein synthesis in prokaryotes. Prokaryotes have a number of **regulatory genes**, which code for proteins whose primary function is to regulate the activity of other genes. An **i** (short for **inhibitory**) **gene** codes for the amino acid sequence of a **repressor protein**. This repressor protein binds to an **operator** region of the DNA that is linked to a **structural gene**, a gene coding for non-regulatory protein. When its operator is combined with the repressor, the structural gene cannot be transcribed. An **operon** is an operator gene plus the structural gene(s) that it controls. The structural gene(s) of an operon can be transcribed only when a specific **inducer** substance combines with the repressor protein and prevents it from binding to and inhibiting the operator gene (Figure 12–3).

Food molecules are the inducers that have been most studied in prokaryotes. A bacterium can use various organic molecules for food, but it does not always make all the enzymes it needs to break down all possible foods. When a new kind of food molecule enters the cell, it may act as an inducer for the synthesis of the enzymes that process it. Jacob and Monod's original work was undertaken to discover how

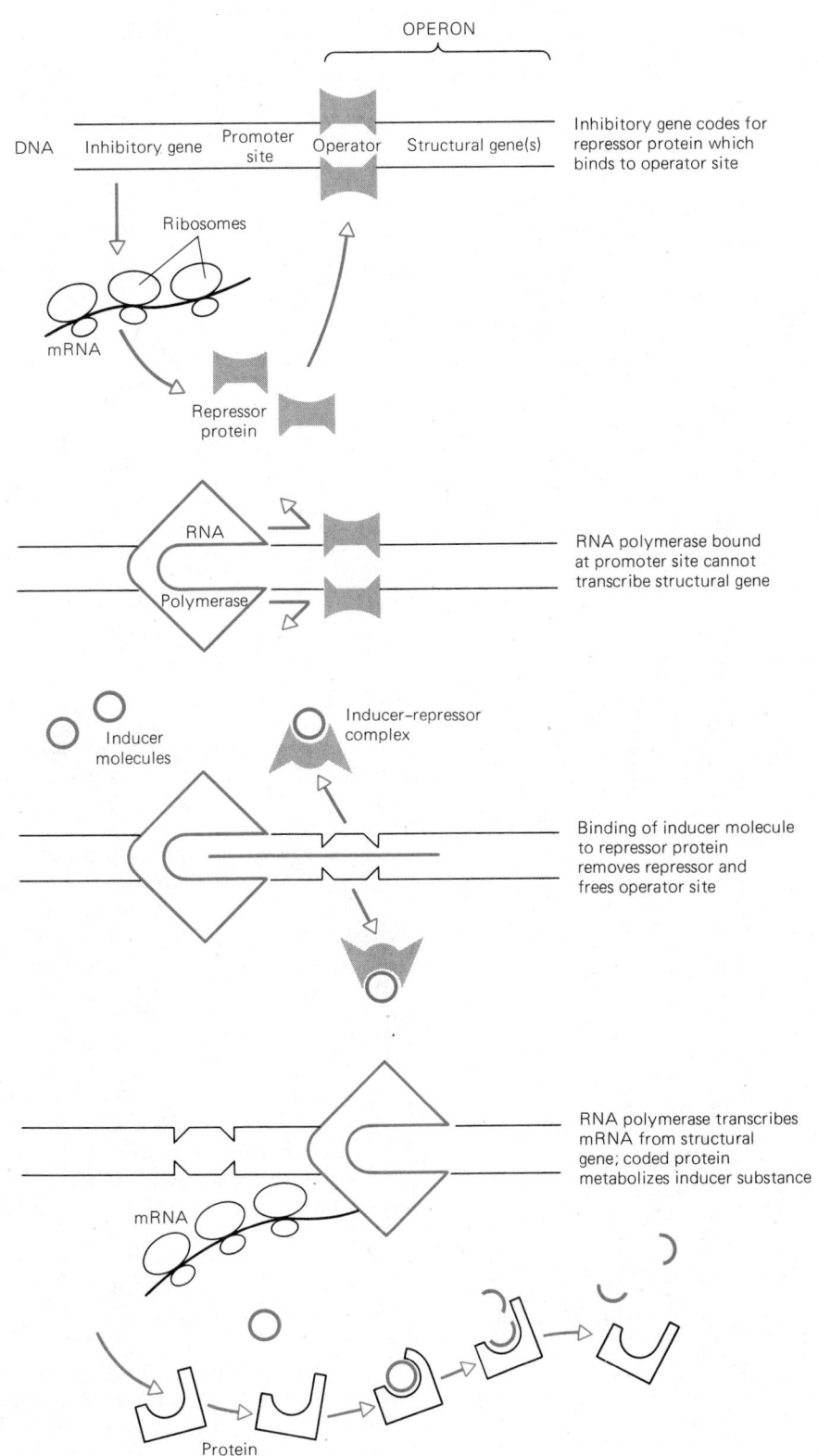

OPERON

DNA — Inhibitory gene — Promoter site — Operator — Structural gene(s)

Inhibitory gene codes for repressor protein which binds to operator site

Ribosomes

mRNA

Repressor protein

RNA Polymerase

RNA polymerase bound at promoter site cannot transcribe structural gene

Inducer molecules

Inducer–repressor complex

Binding of inducer molecule to repressor protein removes repressor and frees operator site

mRNA

RNA polymerase transcribes mRNA from structural gene; coded protein metabolizes inducer substance

Protein

Figure 12-3 Model of gene regulation by induction. Presence of the substrate for the induced protein (an enzyme) leads to transcription of mRNA coding for that protein.

lactose and related sugars induce the formation of the enzyme that hydrolyzes these sugars in *Escherichia coli*.

Bacteria also have other systems that inhibit, instead of inducing, the production of particular proteins when the proteins are not needed.

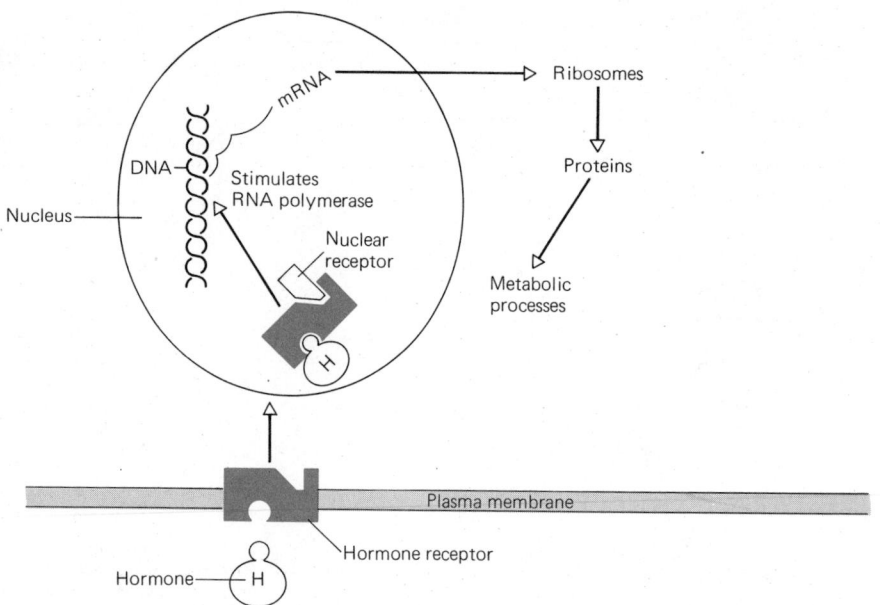

Figure 12-4 The action of steroid hormones that alter transcription.

Sequential Gene Activity

Embryonic development has long been viewed as a complex program of events in which each step in cell differentiation triggers the next, and any event depends upon what has gone before. Such a program can also be seen in some bacterial systems.

Some species of bacteria produce a distinctive structure called an **endospore** within the cell. The endospore has no metabolic activity and is resistant to agents such as heat, drying, and freezing, which would kill a vegetative (growing) cell; the endospore thus permits the bacterium to survive hard times.

An endospore differentiates inside a vegetative cell by a very specific series of steps. For instance, a mutant that lacks the gene necessary to produce one of the enzymes needed for spore formation is known. It carries out the spore-forming reactions until the stage at which that enzyme is needed, but none of the enzymes characteristic of later sporulation is produced. In other words, the transcription of a particular gene is necessary, either directly or indirectly, before other genes can become active. It has also been shown that the RNA polymerase of the vegetative cell is altered during the course of sporulation. It is thought that this enzyme is modified so that it becomes able to transcribe the genes for sporulation and unable to transcribe the genes necessary for further growth of the vegetative cell.

12-C Gene Transcription in Eukaryote Differentiation

Much less is known about the control of gene activity in eukaryotes than in prokaryotes. There is, however, plenty of evidence that different genes are active in different cells and at different stages in development; in some cases we know what is responsible for these differences.

Induction

Like prokaryotic cells, eukaryotic cells can be induced to form certain enzymes by the presence of their respective substrates. Alcohol in the blood, for example, induces an increase in the level of the enzyme alcohol dehydrogenase in the liver, and various amino acids induce the formation of enzymes that break them down.

Induction occurs readily in liver and intestinal cells in animals, but it has never been found in muscle, brain, or other tissues. Induction is much slower in eukaryotes than in prokaryotes. An amino acid may take hours or days to induce an enzyme in eukaryotes, whereas induction takes a matter of minutes in bacteria.

Steroid sex hormones (such as estrogen and progesterone) are needed for the development of sexual maturity in vertebrates. We now know that steroid hormones cause the transcription of particular genes. The hormones bind to receptor proteins. These receptors then carry the hormone into the cell's nucleus, where it stimulates the activity of specific genes (Figure 12–4). Not all the body's cells are sensitive to a particular hormone. Testosterone, for instance, affects the reproductive organs and the cells responsible for beard growth, but not gut or kidney cells. This is because only the sensitive cells bear testosterone receptors. We do not yet know how some cells acquire receptors while others do not.

Giant Chromosomes

The chromosomes of those eukaryotic cells that are not engaged in cell division usually exist in the form of a mass of chromatin in which no structural details can be distinguished. In a few cases, however, the chromosomes are large enough to provide information on gene activity. W. Beerman and U. Clever, in particular, have studied the chromosomes of two insects, the midge *Chironomus* and the larvae of the fruit fly *Drosophila*. Some tissues of these flies—such as the salivary glands, intestines, and Malpighian tubules (excretory organs)—grow by an increase in cell size rather than by an increase in the number of cells. Apparently, the giant cells of these tissues require more genetic material than do normal cells, for as these cells grow their chromosomes replicate up to 10 times (until they are 1024-ploid). Sister chromatids remain attached to each other, and also unwind until they produce **polytene chromosomes** (Figure 12–5). Such a chromosome is a multistranded structure, like a rope, and may be 100 times thicker and 10 times longer than the same chromosome from a normal somatic* cell when it is condensed at mitosis. Each polytene chromosome has a banded structure, whose pattern is faithfully reproduced in all

---

*Somatic cell   A body cell (as opposed to a germ cell, which may give rise to sperm, spores, or an egg).

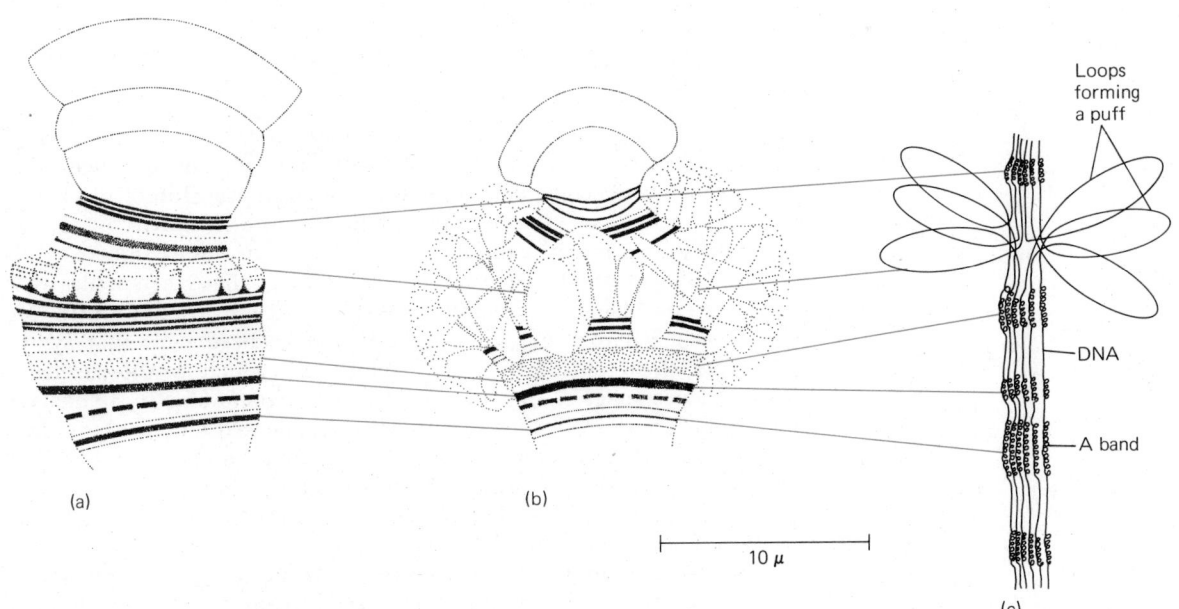

Figure 12-5 A giant polytene chromosome from the salivary gland of the midge *Chironomus*. (a) One end of the chromosome to show the banded pattern. (b) The same chromosome with one band expanded into a puff. (c) Probable structure of the chromosome. Only a few DNA strands are shown.

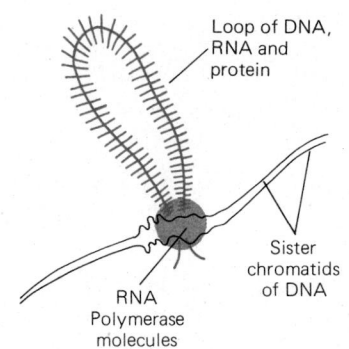

Figure 12-6 Diagram of part of a lamp-brush chromosome. Formation of a loop is visible evidence that genes in that particular part of the chromosome are being expressed.

Loop of DNA, RNA and protein

Sister chromatids of DNA

RNA Polymerase molecules

homologous chromosomes of the same species. The bands that appear dark under the light microscope contain a higher density of DNA and of the chromosomal proteins known as histones than do the regions between the bands. Sometimes the bands appear as "puffs" in which some unraveling of the DNA has occurred.

Puffs are regions of the chromosomes where RNA transcription is occurring. If different genes are expressed in different cells and at different times in development, and puffs represent active genes, we would predict that different bands of the same chromosome would be puffed in different tissues at different times.

Beerman and Clever have described several examples of differential gene activity. For instance, in the salivary glands of *Chironomus pallidivittatus*, four cells produce granules while the rest of the cells in the gland do not. The cells that produce the granules have a puff at one end of one chromosome; the cells that produce no granules have no puff in this position. This is circumstantial evidence that the puff in question represents gene activity necessary for production of the granules.

Differential gene activity during development has also been studied by watching the effects of the insect hormone **ecdysone** on chromosome puffing. All insects develop by going through a sequence of molts, in which they secrete a new outer skin and discard the old one. Ecdysone is necessary for molting to occur. If ecdysone is injected into an insect larva at a time when it is not due to molt, puffs appear and disappear on its polytene chromosomes, in a specific sequence identical with that found during normal molting. In other words, the effect of the hormone is to activate transcription of specific genes in a specific sequence, as one would expect from our model of how differentiation occurs.

Another type of chromosome that has been helpful in the study of differential gene activity is the **lampbrush chromosome** found in the oocytes of many animals (Figure 12-6). In these chromosomes, as in polytene chromosomes, the DNA molecule partially unravels at sites where RNA transcription occurs, and only part of the chromosome is active in RNA synthesis at any one time. Sites of RNA synthesis change with time, as we would expect.

Some of the clearest evidence for an orderly sequence of gene activity in eukaryotic development comes from the work of Norman Wessells and William Rutter on differentiation of the pancreas in mice. They have shown that the substances characteristic of pancreatic cells appear in a precise order during embryonic development, and that inhibiting RNA synthesis will prevent differentiation.

12-D Sequential Gene Activity in Development

Studies of plant and animal chromosomes suggest various ways in which genes may become active and inactive during embryonic development. One way in which a gene may become inactive is through **condensation** of its chromatin. The chromosomes of a cell that is not dividing exist as **chromatin** of various kinds: areas of **heterochromatin** can be distinguished from **euchromatin**, which stains more faintly with the Feulgen reagent, a substance that reacts with DNA (Figure 12-7). One type of heterochromatin is **condensed heterochromatin**. More of this condensed heterochromatin is found in specialized cells that synthesize very few types of proteins than in embryonic cells, which synthesize many proteins. The evidence suggests that when areas of a chromosome in a eukaryotic cell are not being transcribed, they form inactive condensed heterochromatin.

Facultative heterochromatin is found in chromosomes that are inactive in RNA synthesis in the cells of one sex but are euchromatic, or active, in the other sex. This is an example of many genes turned on or off in different animals, and several cases are known.

One of the best studied examples of facultative heterochromatin is found in women and in all other female placental mammals.* A woman inherits one X chro-

---

* Placental mammals Mammals like humans, mice, and horses, in which embryonic development occurs in the uterus and there is no marsupium ("pouch").

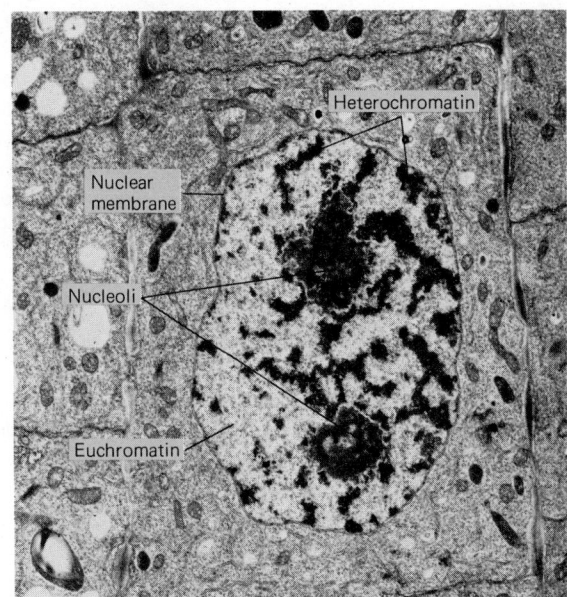

Figure 12-7 Heterochromatin and euchromatin. (Biophoto Associates)

mosome from her mother and one from her father, and one of these is inactive in every cell. (A man inherits only one X chromosome, and it is active in all his cells, as we shall see in Chapter 14.) The process of turning off one of a female mammal's X chromosomes so that it condenses is known as X-inactivation or Lyonization, after Mary Lyon, who studied this phenomenon in mice. The turned-off X chromosome is called a **Barr body**. Every cell of a female usually contains a Barr body, which can be seen under the microscope (Figure 12–8), permitting one to distinguish male from female cells in placental mammals. This test may be used to verify the sex of people entering the Olympics or other sports competitions.

At the beginning of embryonic development, both X chromosomes in a female mammal are active; Lyonization does not occur until after a number of embryonic cell divisions. Furthermore, in the germ cells that will give rise to the ovaries and eggs, both X chromosomes remain active throughout life. But in the other cells, which X chromosome condenses seems to depend on chance, and can also differ from cell to cell in the body. Once Lyonization has occurred in a cell, all the cells (the **clone** of cells) arising by mitosis from that cell will have the same X chromosome condensed as a Barr body. (In calico cats, the two X chromosomes bear genes for different coat color, and these genes are expressed in different parts of the body, depending upon which X chromosome is Lyonized.) The function of Lyonization seems to be to ensure that only one X chromosome is active in female mammals. Strong evidence for this theory comes from finding that the occasional female fetus

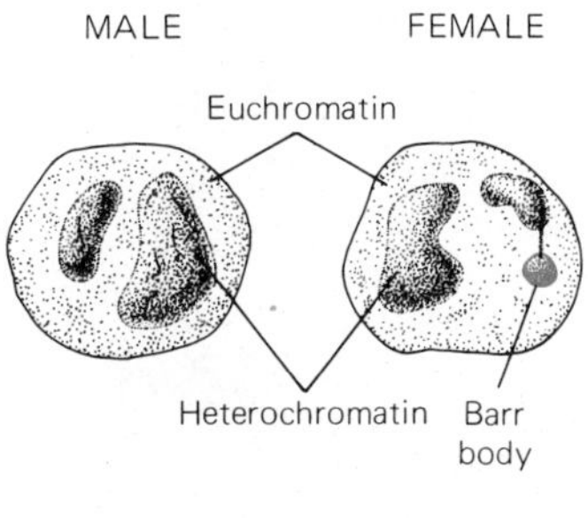

MALE   FEMALE

Euchromatin

Heterochromatin   Barr body

Figure 12-8 Nuclei from white blood cells of a human male and female to show the Barr body found in the cells of all female placental mammals. The Barr body is a highly condensed X chromosome.

that bears three or four X chromosomes, instead of the usual two, still has only one active X chromosome per cell, while the other two or three X chromosomes appear as Barr bodies. There must, presumably, be a selective advantage to having only one genetically active X chromosome, but we do not know what it is. Lyonization, however, is clearly a good example of the fact that certain genes, and in the case of the female placental mammal an entire chromosome, may become completely inactive during cell differentiation.

12-E Effects of Environment on Differentiation

The environment in which a cell finds itself can have a profound effect on its differentiation and can determine which genes will be switched on and which ones will be inactive. In the case of bacterial sporulation (Section 12-B), the stimulus that activates the genes necessary to formation of an endospore is lack of a necessary resource (such as water, oxygen, or food). Some rather more dramatic examples can be found in plants. For instance, many plant seeds (e.g., lettuce) require light for germination. Apparently light is an environmental stimulus that is necessary to switch on the genes involved in germination. As another example, blue light is necessary before a young fern plant will start to grow upward from the embryo (Figure 12–9). Experiments have shown that changes in RNA synthesis, and therefore in gene activity, are necessary to this alteration in the growth pattern, and that blue light brings about these changes.

Considerable research has been devoted to discovering the factors that control differentiation of vascular tissue (the system that transports water and nutrients) in higher plant embryos. Certain plant hormones are vital to this differentiation, as are various environmental factors. Concentrated sugar solutions, for instance, can induce formation of phloem tissue, which normally transports sugar. Phloem tends to form close to the surface of the plant, probably in response to oxygen and carbon dioxide concentrations.

12-F Regeneration and Replacement of Lost Organs

Many organisms, particularly plants and lower animals, have the ability to replace, or **regenerate**, tissues and organs that have been cut off. Regeneration is of interest to students of differentiation in that it offers an example of how a specific stimulus (removing an organ) can cause orderly differentiation. Lack of the organ causes the remaining cells to divide and differentiate in very specific ways so as to replace the organ.

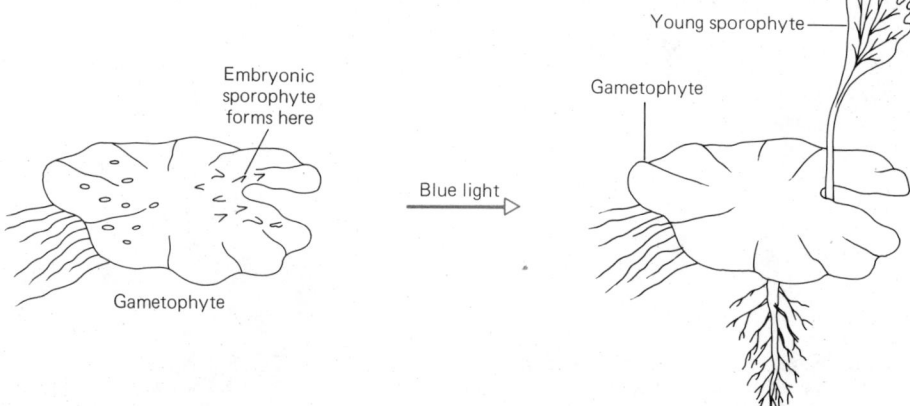

Figure 12-9 Blue light (with a wavelength of 400–500 nm) must fall on a fern gametophyte before the embryonic sporophyte will grow. (See Chapter 26 for fern life history.)

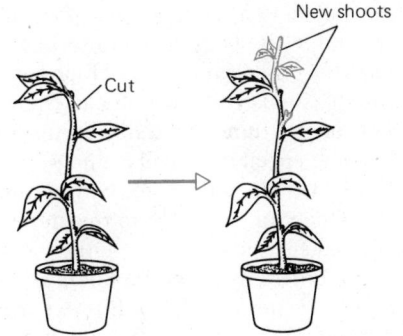

Figure 12-10 Most plants replace lost organs. Here the growing tip of a plant is cut off and two new tips grow up from dormant buds to take its place.

In plants, a lost organ is seldom regenerated, but it is nearly always replaced. If the growing tip of a *Coleus* or any other flowering plant with a single main stem is cut or pinched off, tiny, dormant buds further down the stem start to grow, by cell division and enlargement. These buds then differentiate into stems, leaves, and flowers (Figure 12–10). It is now known that the cell division and gene activity involved in this differentiation are controlled by hormones.

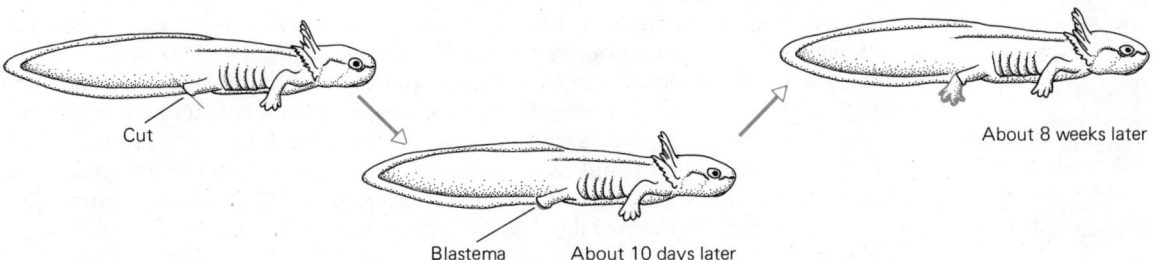

Figure 12-11 Regeneration of a severed limb in a larval salamander. *Ambystoma.*

Most animals, including humans, can regenerate parts of some organs. The human liver, for instance, is capable of regenerating parts that have been removed. However, other animals have more impressive powers of regeneration. There have been many studies of regeneration of lost legs in amphibians such as newts and salamanders (Figure 12–11). A planarian worm can regenerate its head and most other parts of the body (Figure 12–12). In most animals, regeneration of a lost organ starts with formation, at the site of the injury, of a **blastema**—a bud containing a mass of cells that look very like one another. The regenerating organ forms as the blastemal cells divide and differentiate into the various cell and tissue types (e.g., bone, muscle, blood vessel) that make up the new limb. Two major questions are raised by such regeneration: (1) where do the cells of the blastema come from? and (2) what causes them to differentiate in their various ways? There seem to be two possible sources for blastema cells. Either they are cells left over from embryonic

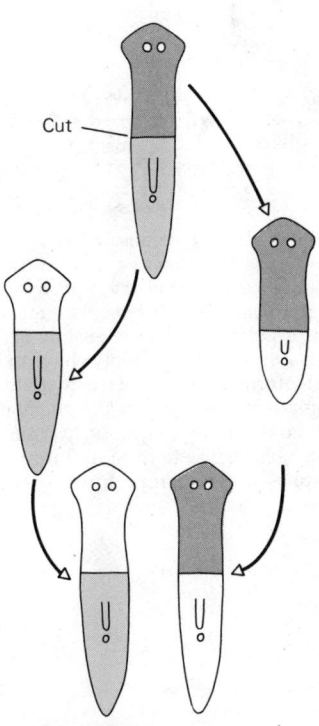

Figure 12-12 Regeneration of a planarian. If the worm is cut in half, the posterior half will regenerate a new front end, and the anterior half will regenerate a new rear end.

**CHAPTER 12
CELL DEVELOPMENT
AND AGING**

175

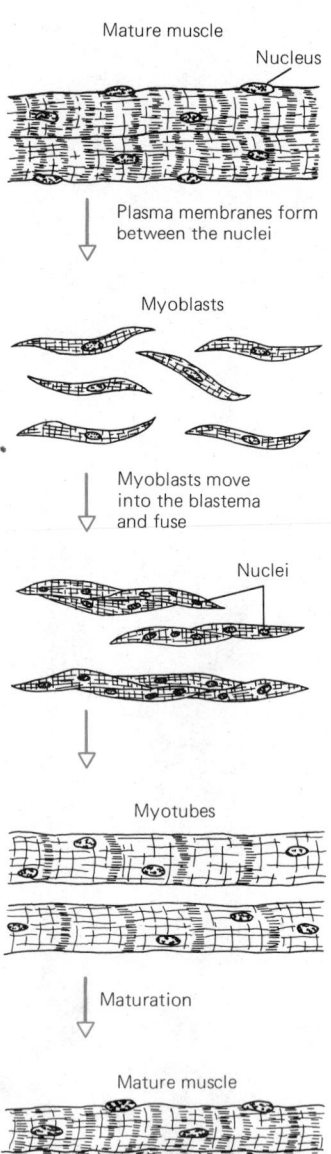

Mature muscle

Nucleus

Plasma membranes form
between the nuclei

Myoblasts

Myoblasts move
into the blastema
and fuse

Nuclei

Myotubes

Maturation

Mature muscle

Figure 12-13 New muscle from old during limb regeneration in a newt. Muscle in the undamaged part of the limb lays down plasma membranes between its nuclei to form myoblasts which are exactly like the embryonic myoblasts from which muscle usually forms. The myoblasts migrate into the blastema, where plasma membranes between them break down once again, and the myoblasts fuse to form multinucleated myotubes which mature into new muscle in the regenerating limb.

development that have not yet differentiated into specialized cells, but are kept around in the body ready to form a blastema, or else they are cells formerly specialized for other functions, which are stimulated by removal of the organ to change into blastema cells, which can then redifferentiate into the cell types needed in the new organ. In newts, it has been shown that at least some blastemal cells arise from existing specialized cells. The syncytium* of a skeletal muscle fiber may break up into blastema cells, each with a single nucleus (Figure 12–13).

One vital stimulus to regeneration—the presence of nervous tissue—has been studied in some detail. Neither a planarian nor an amphibian can regenerate an organ unless many nerve fibers reach the blastema. For instance, frogs cannot normally regenerate a limb. But experiments have shown that if additional nerves are grafted into the limb of a frog, that limb will regenerate if it is later severed. Nerves are said to have a **trophic** effect on regeneration. They secrete some unknown substance that is necessary if regeneration is to occur (Figure 12–14). It is still not known what influences in the blastema cause some of its cells to differentiate into muscle cells, for example, and others into the bone of the regenerating limb.

12-G   Nucleocytoplasmic Interactions in Differentiation

Implicit in most discussion of differentiation is the idea that the nucleus and cytoplasm interact with one another during the course of differentiation. It is clear that the nucleus influences the cytoplasm: RNA transcription in the nucleus controls which enzymes and structural proteins will be present in the cytoplasm. The influence of the cytoplasm on the nucleus is less obvious.

British embryologist John Gurdon has shown that if a nucleus is transplanted from an intestinal cell of a tadpole into a frog egg, the nucleus immediately starts transcribing the genes necessary for early embryonic development (Section 12-A). The only stimulus that could have produced this change is the egg cytoplasm into which the nucleus has been transplanted. It has now been shown convincingly that the first stimuli to determine differential gene activity in the nuclei of an embryo are substances present in different parts of the egg cytoplasm. Studies in which parts of the egg cytoplasm have been removed, or in which nuclei are exposed to parts of the cytoplasm they would not normally encounter, show that specific, unknown messengers in the cytoplasm enter the nucleus and determine how it will differentiate.

---

* Syncytium   A "cell" that contains more than one nucleus within one "cell" membrane.

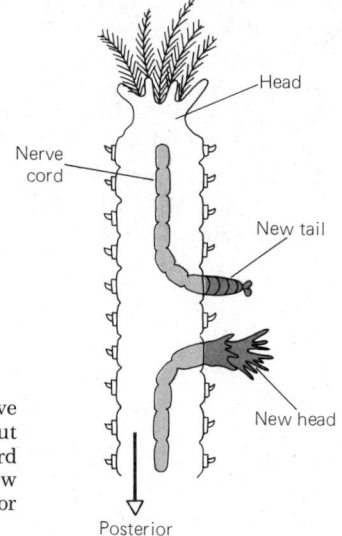

Head

Nerve
cord

New tail

New head

Posterior

Figure 12-14 The effect of nerves on regeneration. The nerve cord of this polychaete worm has been severed and the cut ends moved so they touch the body wall. The part of the cord on the posterior side of the cut induces formation of a new head from the body wall, whereas the part of the cord anterior to the cut induces formation of a new tail.

Different animals are born or hatch at different stages of development. A mouse, for instance, is born hairless, blind, and incapable of regulating its body temperature. The foal of a horse, on the other hand, can run around and use all of its senses within an hour of birth. Further development occurs after birth, in that most animals do not become sexually mature until later in their lives. In addition, some animals undergo **metamorphosis**, a change in body form, before they are sexually mature. Both metamorphosis and sexual maturation involve the same processes of genetic differentiation and growth as have occurred in the embryo.

Metamorphosis is a more-or-less abrupt alteration in an animal's anatomy and physiology as it changes from a larva to an adult. The term is used to cover a number of rather different phenomena, but all of them are controlled by hormones. The best-known examples are those involving rapid change, as when an amphibian tadpole metamorphoses into a frog or a pupa into a butterfly. In many animals, however, the change takes place much more slowly. (Where there is more than one abrupt change during the life history, as there is in butterflies, flies, and beetles, usually only the last change is called metamorphosis.)

The biological significance of metamorphosis is that it permits the young stage of an animal to have a way of life very different from that of the adult. The change from one form to another will be abrupt when the intermediate phase between the two stages is adapted to neither way of life.

Different functions may be served by the different stages in the life history of an animal. For instance, in most sessile* invertebrates the adult is the feeding stage and the larva the distributive stage, whereas in many insects, the adult is the distributive stage and the larva feeds. In all cases, the post-metamorphic adult stage is the one that eventually becomes sexually mature.

Amphibian Metamorphosis

The existence of metamorphosis in amphibians reflects the fact that they are not fully adapted to a terrestrial life. As adults, many amphibians can live on land and exploit the food supply provided by flying insects. As embryos, however, they must live in water, or in such specialized places as pouches on the female's back, because the eggs cannot stand exposure to air. Nearly all amphibians that lay their eggs in water have a swimming larva that stays in the water after hatching and feeds on aquatic algae. For these animals, growing up involves changing from an aquatic herbivore into a terrestrial carnivore, and this requires many dramatic alterations in the animal's anatomy and physiology. The main changes are listed in Table 12-1.

---

* Sessile "Sitting"; attached to a surface, or moving only slowly from place to place.

TABLE 12-1 CHANGES THAT OCCUR DURING FROG METAMORPHOSIS

FUNCTION	AQUATIC LARVA	TERRESTRIAL ADULT
Locomotion	Tail with fins	Legs, no tail
Respiration	External gills	Lungs and skin
	Hemoglobin-oxygen reaction independent of pH	Hemoglobin has different amino acid sequence
		Oxygen loading is pH-dependent
Feeding	Sucking mouth	Big mouth with jaws, sticky tongue
Digestion	Long coiled intestine for digesting algae	Digestive tract short; insect food easily digested
Sensory	Small eyes; lateral line organ (row of pressure-sensitive pits along side of body)	Huge eyes, with different visual pigment; ears
Excretion	Ammonia (NH_3)	Urea ($H_2N-CO-NH_2$)

All of these changes, from the formation of leg bones to the synthesis of a hemoglobin with a different amino acid sequence, are ultimately the result of changes in gene activity that occur during metamorphosis. What makes previously inactive genes start transcribing RNA? Hormonal changes initiate transcription of genes for adult characteristics, although in most cases we still do not know what changes in the animal or in its environment alter hormone secretion and trigger metamorphosis.

It has long been known that removing the thyroid gland from a tadpole will prevent it from going through metamorphosis. If the gland is reimplanted, the animal undergoes metamorphosis. Many experiments have confirmed that the hormone **thyroxin**, secreted by the thyroid gland, must be present for metamorphosis to occur. Thyroxin will also act directly on tissues that change during metamorphosis. If the tail is removed from a tadpole and placed in a bath containing thyroxin, the white blood cells in the tail will digest it so that it gets smaller, just as they would in the intact tadpole. In the control experiment, where the bath contains no thyroxin, the tail is unchanged.

A further interesting point is that thyroxin can affect only those tissues that are in an appropriate state to react to it. If thyroxin is injected into a fairly mature tadpole, the tadpole will metamorphose prematurely, but the same injection will have no effect on a very young tadpole. A tissue cannot react to a hormone until it has reached a particular stage in its own development. As an interesting corollary, in *Necturus maculosus*, a neotenous* salamander that never undergoes metamorphosis but becomes sexually mature as a larva, lack of metamorphosis is not due to the absence of thyroxin. Instead, the tissues of *Necturus* will not react to the hormone at any stage in its life, no matter how much of it is injected into the animal.

(The mechanism responsible for neoteny varies. In another neotenous salamander, the axolotl, injections of appropriate amounts of thyroxin will cause the animal to lose its gills and tail and crawl up onto land. Here the tissues are still sensitive to thyroxin, but the animal does not produce the hormone in amounts sufficient to bring about metamorphosis.)

Molting and Metamorphosis in Insects

Major changes from larval to adult form occur in many insects. Such alterations in an insect's structure can occur only when it molts its exoskeleton and grows a new one. Molting and metamorphosis are therefore intimately related, although the function of molting is primarily to permit an increase in size and the function of metamorphosis is to produce a change in body shape.

Experiments performed in the 1920s showed that if a ligature was tied tightly around the body of an insect before a critical stage in its development, only the front half of the body underwent metamorphosis. This was interpreted to mean that the process was controlled by a hormone produced in the front of the body.

In the 1930s and 1940s, the English insect physiologist Vincent Wigglesworth studied this problem in some detail, working with the blood-sucking bug *Rhodnius prolixus* (Figure 12–15). Wigglesworth found that the molting hormone, ecdysone, was produced by glands in the thorax, just behind the head. The stimulus for the secretion of ecdysone turned out to be distension of the abdomen. Thus, in real life, a molt will occur shortly after a large meal of blood, which distends the abdomen and also provides enough food to last until the molt is completed and feeding is again possible.

Although *Rhodnius* undergoes five molts as it develops, metamorphosis occurs only during the last molt. Why is this? In an ingenious experiment, Wigglesworth diluted the blood of an early nymph (immature stage) by adding blood from an adult and found that the early nymph underwent premature partial metamorphosis. He interpreted this to mean that the blood of an early nymph normally contains a hormone that inhibits metamorphosis during the early molts but disappears at the fifth molt so that metamorphosis then takes place. By diluting the early nymph's

---

*Neotenous   Used of an animal species in which reproductive maturity occurs in an otherwise larval animal.

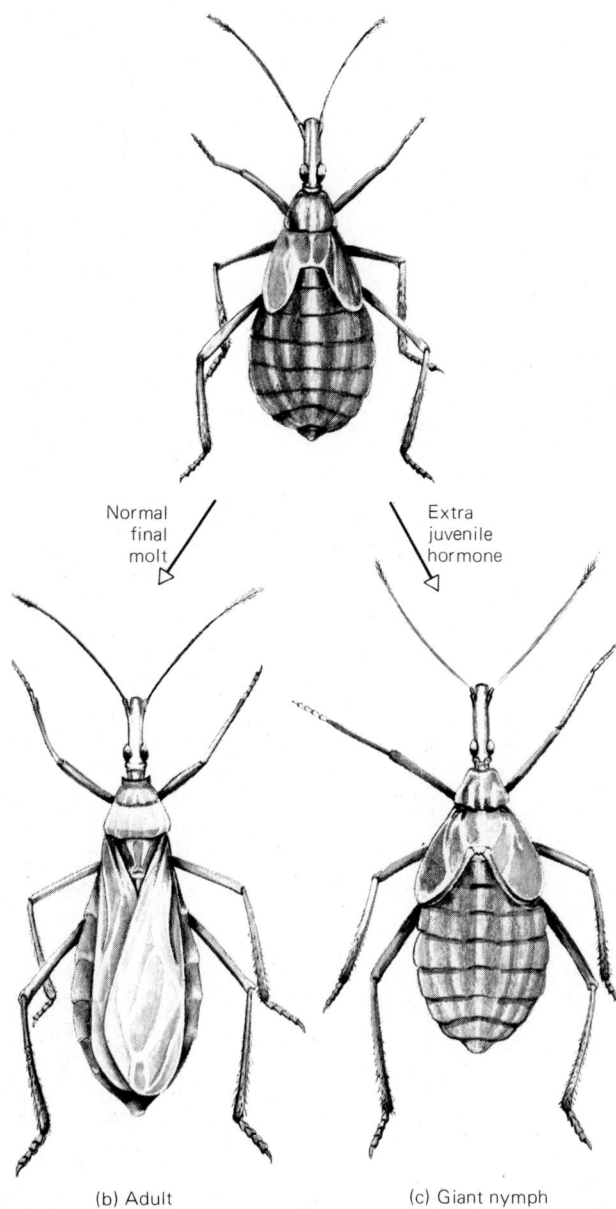

(a) Last nymphal stage

Normal final molt

Extra juvenile hormone

Figure 12-15 Metamorphosis in *Rhodnius prolixus*.

(b) Adult

(c) Giant nymph

blood, he had lowered the concentration of this hormone (**juvenile hormone**) until it could no longer suppress metamorphosis completely. This interpretation has been confirmed by later experiments.

Ecdysone and juvenile hormone are found in many insects. In the absence of juvenile hormone, ecdysone promotes transcription of genes that code for adult characters. In the presence of juvenile hormone, the effect of ecdysone is suppressed, and genes for characteristics of the immature stage are active.

12-I Aging

"As soon as we were born," the *Apocrypha* points out, "we began to draw to our end," for development and aging merge imperceptibly into each other. Why and how do we age and eventually die?

Aging is the sum of changes that accumulate with time and make an organism more likely to die; it begins even before an organism apparently completes development at sexual maturity. Slower healing, for instance, is one of the cellular signs of aging, and even a human teenager's bones heal less rapidly than those of a young child.

Aging and death are genetically programmed. Species have characteristic life-spans that can be altered by selection. An elephant dies of old age when it is about 200 years old, a human at about 80 to 100 years, and some insects at a few days or weeks. Aging assures death at this characteristic age if disease or predators do not kill the animal or plant first. The selective advantage of death produced by aging is much debated. Probably aging eliminates older individuals, leaving food, shelter, and other limited resources to younger individuals, who will usually bear many of the same genes, and whose reproduction is more important evolutionarily.

We do not know how aging occurs; there are at least 20 different theories. Although we can describe many processes characteristic of aging, it has proved impossible to decide which of these is cause and which effect. For instance, the immune system, which defends the body against disease, becomes less efficient with age. Is this why disease becomes more common with age, or vice versa?

As people grow older, their bodies cope less effectively with stress or disease. Because the ability to survive changing conditions depends largely on the immune, nervous, and hormonal systems, researchers have concentrated on these systems, using the hypothesis that degeneration here results in the loss of adaptability seen in the rest of the body: slower healing, hardening connective tissue, brittle bones, and so forth.

The suggestion is often made that aging starts after a particular cell line has divided a certain number of times. However, when this hypothesis has been tested in detail for any particular cell type, it has usually been disproved. When cells removed from the bodies of very old animals are kept in appropriate environments, they may live and divide without any signs of aging for generations beyond the animal's own time of death. This suggests that the environment of a cell determines whether it ages or not. In this case, is there a particular set of cells in the body that actually does age spontaneously and so create an environment in which all the rest of the body ages? We do not know.

One popular theory holds that aging is a cellular process resulting from the accumulation of mutations in all the body's cells. This will not explain aging in all animals, however. During their lifetimes, mammals accumulate proportionately more mutations in the DNA of their cells than those acquired by insects (allowing for their much shorter lives). Similarly, cells die throughout our lives, but this alone cannot explain aging. Human beings who lose cells through disease or accident have survived with many more cells missing from vital organs such as the liver, kidney, brain, and lungs than those normally destroyed during aging.

Searches for a single cause of aging have all ended in failure. It seems likely that the "aging genes" do not control one single system but instead control a multiplicity of minor degenerations. Because the body's systems all interact with one another, minor deficiencies anywhere in the body can accumulate to produce aging of the body as a whole and of its individual systems.

12-J Conclusions

This discussion of development and aging will probably be unsatisfying to the reader with a tidy mind, since it provides a better example than any examined so far of how much we do not know in biology. It should also be stimulating, however, to realize how much scope there is for imaginative new work and new workers in this, as in so many other areas.

What general conclusions can be drawn from the discussion in this chapter?

1. The most important feature that distinguishes one type of specialized cell from another is which genes are expressed in each cell and, therefore, the types of proteins each cell produces.

2. With very few exceptions, as each cell arises by mitotic cell division, it inherits DNA that is identical with that found in every other cell of the organism and in the zygote or spore from which all of the cells have arisen.

3. These first two points, as well as evidence from other sources, lead us to the conclusion that various genes in a cell are switched on and off in specific sequences during differentiation. Genes are not necessarily switched off permanently; they can often be reactivated by appropriate stimuli.

4. Various stimuli in an organism's environment and in its cells control which genes are active. Substances in the egg cytoplasm are important in gene regulation early in development. Unknown trophic factors in nerves are necessary in regeneration, and many hormones are known to control gene activity in plants and animals during embryonic differentiation and in adult life.

SUMMARY

Considerable evidence suggests that, in most animals and plants, all somatic cells of an individual adult organism, as well as the zygote from which they originated, possess the same genetic information. For instance, a mature plant can be grown from a single cell of an adult plant, and the nucleus from a tadpole gut cell can replace the zygote nucleus in supporting development of a mature frog.

During embryonic development, originally identical cells differentiate as each takes on its specialized function in the body. Cells differ from one another in that they contain different proteins that result from the expression of different genes. Differentiation may, therefore, be looked at as the switching on and switching off of different genes in precise sequence.

Evidence that specific genes are switched on and off as a cell differentiates comes from studies such as those on sporulation in bacteria, polytene chromosomes in some flies, lampbrush chromosomes in amphibian oocytes, protein synthesis in

Essay: Human "Clones"

A **clone** is a group of genetically identical individuals produced asexually from each other or from a common parent. The simplest example is a group of bacteria descended, by repeated cell division, from one original bacterium. The bacteria in the clone all inherit the genes of the parental cell.

Discussion of the moral problems raised by the possibility of cloning human beings has been stimulated by experiments similar to those described in Section 12-A. If a nucleus from a tadpole intestine cell is introduced into an egg with its own nucleus removed, the egg may develop into a normal, fertile, frog. Injecting a nucleus from the same individual into each of many egg cells, and raising the resulting individuals to maturity, would produce a clone of frogs.

The members of the clone would be genetically identical because each would contain only the genes contained in the body cells of one individual. Research on test tube babies shows that human eggs can now be induced to develop normally outside the mother's body, at least for a few days. This suggests that the kind of nuclear transplantation successfully performed on frogs might one day be possible with human beings.

People produced by transplantation would all have the genetic characteristics of whoever donated the nuclei. It is frightening to think of a large population of Adolf Hitlers or even of Joan of Arcs. (Remember, however, that so many nongenetic influences affect human development that the members of a clone would be no more alike than are identical twins, who are often noticeably dissimilar.)

The possibility that human beings may one day be able to control the genetic makeup of future populations by producing clones raises moral problems that society has never before encountered. The technical barriers to nuclear transplantation in humans are still so immense, however, that we shall probably not be confronted by the reality in our lifetimes.

However distant human cloning may be as a technique, many people feel that we should be wise to confront the moral problem before we are confronted with the technical reality. For this reason, research on human embryos has been restricted and even completely banned in the United States during much of the 1970s. The ban has now been lifted somewhat to permit experiments with fertilization outside the body that has permitted some previously infertile women to become pregnant. We now have the infertile and their supporters urging that experimentation be permitted, while those troubled by the moral implications urge legislative caution. What do you think? As a voter for the next fifty years, you will probably have to decide.

181

the pancreas of mouse embryos, and the formation of Barr bodies and condensed heterochromatin in mammals.

Various signals inside and outside an organism can cause the changes in gene activity that constitute differentiation. The shape of some plants and the time at which they germinate are determined by light; the genes needed for molting in insects are switched on by the hormone ecdysone; differentiation of phloem in flowering plants is stimulated by sugar. Regeneration of lost limbs and tissues is a useful model system in which redifferentiation can be studied; trophic substances secreted by nerves are necessary to differentiation in this system.

One of the most important influences on gene activity in early embryonic development is the egg cytoplasm, which controls many activities of any nucleus that lies within it. Different areas of the egg cytoplasm cause differential gene activity in nuclei in different parts of the egg.

Aging is the accumulation, with time, of changes that make an organism more likely to die. Since no one theory explains aging, it seems likely that these changes occur in many different, unrelated systems of the body.

OBJECTIVES

From your study of this chapter, you should be able to:

1. Use the following terms in context: differentiation, gene transcription, zygote, chromosome puffs, lampbrush chromosome, blastema, euchromatin, heterochromatin, Barr body, Lyonization, bacterial endospore, sporulation, trophic effect of a nerve.

2. List two systems other than embryonic development of a plant or animal in which cells undergo differentiation.

3. State how one type of specialized cell differs from another type.

4. Describe two kinds of evidence that all the cells of an adult organism contain the same genetic material.

5. Describe the structure of a polytene chromosome; state where such chromosomes are found and why they are useful in studies of gene activity.

6. Describe the process of Lyonization and its effect on gene activity in the cells of a female mammal.

7. Give one plant and one animal example of a situation in which a hormone is a stimulus to differentiation.

8. Give two examples of situations in which environmental stimuli outside the organism influence differentiation.

9. Describe the importance of the egg cytoplasm to differentiation in an embryo.

10. Explain the selective advantage to animals of going through metamorphosis at some stage of the life history; describe how metamorphosis takes place and what factors govern it.

11. Define aging and describe two theories of how it occurs.

SELF-QUIZ

1. During differentiation, cells with the same DNA:
   a. develop similarly
   b. divide at equal rates
   c. contain different genes
   d. transcribe different genes

2. Which of the following does *not* undergo transcription of its genes?
   a. Barr body
   b. polytene chromosome
   c. chromosome puff
   d. euchromatin

3. Nuclear transplantation experiments in frogs have demonstrated that:
   a. gut nuclei of tadpoles contain all the genetic information necessary for the development of normal adults
   b. liver nuclei lack some genes necessary for full development
   c. cell nuclei contain an excess of genetic information

   d. normal development occurs only when certain hormones are added to the nutrient medium

4. All of the following statements are true. Which of them *does not* constitute evidence that all cells of an organism contain the same genetic information?
   a. There are no known examples of the addition of new genetic information during development.
   b. Chromosomes disappear from some cells during development of *Ascaris*, a parasitic roundworm.
   c. Chromosomes look the same in all cells.
   d. Whole carrot plants can be grown from adult carrot root cells.

5. Under normal conditions, the stimulus that initiates metamorphosis is:
   a. concentration of certain hormones
   b. state of nutrition
   c. age of the larva
   d. change of season
   e. decrease in the population of adult members of the species in the vicinity

QUESTIONS FOR DISCUSSION

1. Human embryos with only one X chromosome and no Y (male-determining) chromosome do not develop ovaries and are born sterile. How could you explain this?

2. Ecdysone stimulates the same puffing sequence in some tissues that do not directly affect molting as well as in those which do. Does this disprove the theory that ecdysone triggers molting by causing the transcription of specific genes?

3. Normally, the lens of a frog's eye begins to differentiate from the outer layer (ectoderm) of the embryo when part of the brain touches the ectoderm. If the brain is transplanted to the tail of a frog embryo, a lens will form where the brain touches the tail ectoderm. If a piece of plastic is placed between the growing brain and the ectoderm, the lens will not develop. What do these experiments tell you about the differentiation of the lens?

4. A kidney cell divides more slowly than a fertilized egg. If the egg nucleus is removed and replaced by a kidney cell nucleus, will the resulting hybrid "zygote" divide faster or slower or at the same rate as it did before, and why?

5. Can you think of any reason why the human liver will regenerate whereas brain, muscle, or fingers will not? (Note: the authors cannot.)

6. During the development of the parasitic roundworm *Ascaris*, some cells lose entire chromosomes, and therefore lose genetic information that was present in the zygote. A similar phenomenon accompanies the development of some gall midges. (a) What might be the selective advantage to these organisms of losing these chromosomes? (b) How do you suppose the next generation receives a complete set of genetic information?

REFERENCES AND FURTHER READING

Beerman, W. ed. *Developmental Studies on Giant Chromosomes.* New York: Springer-Verlag, 1972. A description of the polytene chromosomes of insects and other giant chromosomes and the contribution that their study has made to our understanding of gene action during differentiation; for the advanced student.

Grant, P. *Biology of Developing Systems.* New York: Holt, Rinehart and Winston, 1978. A textbook on development with good sections on differentiation and on aging.

Gurdon, J. B. "Transplanted nuclei and cell differentiation." *Scientific American,* December 1968. The technique of nuclear transplantation and a discussion of how far cell differentiation is reversible.

Harris H. *Nucleus and Cytoplasm,* 3d ed. Oxford: Clarendon Press, 1974. A discussion of the interactions of nucleus and cytoplasm, including genes that occur in the cytoplasm and cytoplasmic control of nuclear activity; for the advanced student.

Jacob, F., and J. Monod. "Genetic regulatory mechanisms in the synthesis of proteins." *Journal of Molecular Biology* 3:318, 1961. The classic paper on regulation and messengers in *Escherichia coli.*

Lyon, M. F. "Possible mechanisms of X-chromosome inactivation." *Nature, New Biology* 232:229, 1971. A discussion of Lyonization by the woman who discovered it.

CHAPTER
13

MENDELIAN GENETICS

We have considered genetic material at a molecular level and have seen that genes are segments of DNA molecules, some of which code for the structure of proteins. We have also examined the genetic material of eukaryotes, the chromosomes, at a cellular level, and considered how they are transferred from one cell to another during cell division. In the process of meiosis,* members of homologous chromosome pairs separate from one another, so that the gametes (reproductive cells) receive one member of each pair of homologous chromosomes. After fertilization, the newly formed zygote contains a complete set of chromosomes, half from the male parent and half from the female.

In this chapter we consider genetic characteristics as the traits we observe in individual organisms and study the patterns of inheritance of genetic traits as they pass from parent to offspring. We shall see that these patterns reflect the behavior of chromosomes at the cellular level during meiosis and fertilization.

---

* Meiosis A series of nuclear divisions that reduces the amount of genetic material in each resulting daughter nucleus to half that in the original nucleus.

13-A Mendel's Methods

The study of genetics is based on the work of Gregor Mendel, a monk, and later the abbot, at the monastery of Brünn in what is now Czechoslovakia. Mendel's work was reported at a meeting of the Brünn Society for the Study of Natural Science in 1865, and was published the following year.

Although Mendel's paper presented a completely new and beautifully documented theory of inheritance, it did not lead immediately to a spate of genetic research. The influential scientists studying inheritance at the time reveled in a maze of complex theories, and the few people who read Mendel's paper dismissed his results as trivial because they could be explained by a simple model. Hence Mendel's work received little attention until after his death; it was rediscovered in 1900 almost simultaneously by Hugo deVries in the Netherlands, Carl Correns in Germany, and Erich Tschermak in Austria. In the meantime, chromosomes had been named and their behavior during mitosis and meiosis observed and described; scientists reading Mendel's theory in 1900 could see that the results of Mendel's breeding experiments dovetailed nicely with the movements of chromosomes during meiosis. Once this connection was established, the science of genetics entered a period of productive research.

Mendel's most important conclusion was that inherited variations are transmitted as discrete, unchanging units. This is far from obvious. Careful observers, including Darwin, had always been more impressed by the continuity of variations (consider human height or skin color) than by their distinctness. For this reason, Darwin, and virtually all other biologists before 1900, believed that the inheritable characteristics of parents blended in their offspring. This blending theory, however, was at odds with the theory of evolution by natural selection, which required that inherited variations be maintained in a population from generation to generation. In time, these variations would disappear if they blended with each other in each generation. Unfortunately, Darwin never knew that Mendel's work would have solved this problem for him.

There are several reasons why Mendel succeeded in unraveling the laws of inheritance where other investigators had failed (Table 13–1). First, Mendel was familiar with mathematics and with probability theory, and so he knew the importance of obtaining a large number of offspring in his experiments and of analyzing his data mathematically. In addition, Mendel began by studying just one genetic character at a time, and he followed each character through many generations.

Mendel's results were possible only because he worked with genetically simple characters. This was good judgment, not luck. Before he began his experiments, Mendel spent several years choosing which species of organism and which inherited characters to study. He finally chose garden peas, *Pisum sativum*. Peas were available in many different varieties, each of which bred true to type. Thus tall pea plants of one variety always produced tall offspring, and a variety of white-flowered plants always produced white-flowered offspring. By crossing two different varieties of peas, Mendel could trace the inheritance of particular traits that were different in the two original varieties.

Another characteristic that made peas ideal for Mendel's purposes was their unusual flower structure; a modified petal, the keel, completely surrounds the re-

---

TABLE 13–1 SUMMARY OF REASONS FOR MENDEL'S SUCCESS

1. Chose pea plants, which normally self-fertilize (simple to use).
2. Used pure-breeding varieties.
3. Chose traits with distinctly contrasting forms.
4. Followed one characteristic at a time.
5. Followed characteristic for many generations.
6. Obtained large numbers of offspring.
7. Analyzed results using statistical methods.
8. Luck; happened to choose 7 traits lying on 6 of the 7 chromosome pairs of pea plants (see Section 13-H); the two traits on the same chromosome were quite far apart.

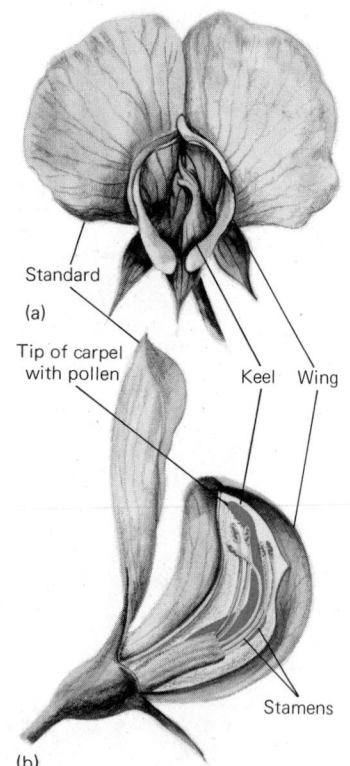

Figure 13-1 Flower of garden pea, the experimental organism used by Mendel in his genetic studies. (a) External view of flower; the petals are modified to form the "standard," "wings," and "keel." (b) Side view of flower, with position of reproductive parts (stamens and carpel) indicated.

productive parts, separating them from the outside world (Figure 13–1). Normally the flower **self-pollinates**; that is, pollen from the male flower parts (**stamens**) is transferred to the female part (**carpel**) of the same flower. Mendel could permit self-pollination or could artificially **cross-pollinate**, taking pollen from flowers of one variety and placing it on the carpels of flowers of another variety. In this way Mendel always knew the male as well as the female parent of any pea seeds he obtained in his studies.

When he had chosen peas, Mendel spent two years performing breeding experiments with a number of different varieties carrying different genetic characteristics. He chose seven characteristics, each of which occurs in two distinct forms, for prolonged study (Figure 13–2).

13-B   A Simple Breeding Experiment

One of the characters Mendel chose to study was flower color. Mendel crossed a pure-breeding strain of red-flowered pea plants with a pure-breeding strain of white-flowered plants. In order to do this, Mendel had to open up a red flower and pluck off a stamen. Then, he had to open a white flower and transfer some of the pollen from the red flower's stamen to the carpel of the white flower. In order to be sure that none of the seeds produced by the white-flowered plant came from self-fertilization, he had to open the white flowers while they were still buds and amputate the stamens so that no pollen would be produced. Mendel also performed the opposite cross, using pollen from white flowers to pollinate red flowers. In order to obtain a large number of offspring for statistical analysis, Mendel had to hand-pollinate many dozens of flowers.

Mendel collected the seeds resulting from these crosses and planted them the following year. When the plants matured, they all produced red flowers (Figure 13–3). He allowed these red flowers to self-pollinate, and from them he collected over 900 seeds. Most of these seeds grew up to be red-flowered plants, but about a quarter of them were white-flowered.

These results show the importance of Mendel's methods. Had he stopped following his plants after one generation, he would have concluded that the white flower characteristic had been obliterated by the red, and had he obtained only a few

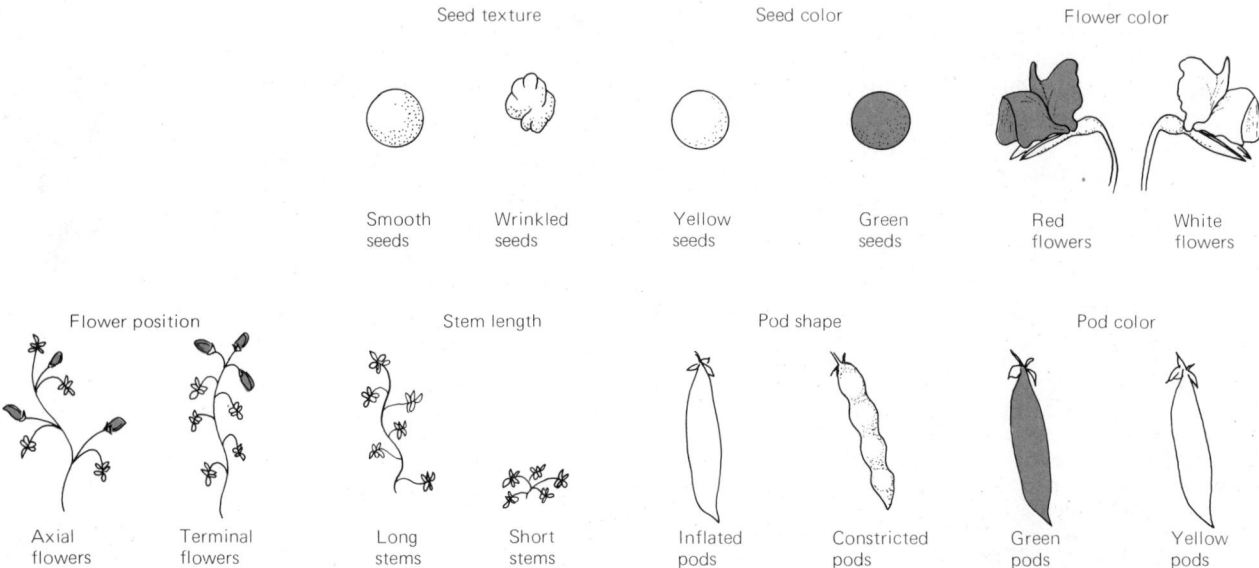

Figure 13-2 Mendel studied seven different traits of pea plants, each of which appeared in two different forms. In each pair, the dominant form (discussed in Section 13-D) is the one on the left.

offspring for each cross, he might have dismissed the reappearance of white flowers as a minor occurrence. However, a statistically solid result of 25% white flowers where none had appeared in the previous generation could not be ignored.

13-C Gene Pairs

Mendel realized that his results could be explained if each plant contained two inherited factors that governed each trait. These factors acted as particles that could pass from parent to offspring and determine the offspring's inherited characters. Mendel postulated that a plant received two factors for each trait, one from each parent. Mendel's "factors" later came to be called genes.

A trait can occur in two or more different forms; for example, flower color may be either red or white. Thus the genes that govern the trait must come in alternative forms, called **alleles**. In our flower-color example, there are two alleles of the pea flower-color gene, the red-flower allele and the white-flower allele. Since genes occur in pairs, a pea plant may have two alleles for red, or two alleles for white, or it may contain one allele for red and one for white. An individual with two of the same allele is said to be **homozygous** for that allele; thus, plants with two alleles for red or two for white are homozygous for flower color. An individual with two different alleles for a trait is said to be **heterozygous**; thus plants with one red and one white allele are heterozygous with respect to flower color.

In Mendel's crosses, the original **parental**, or **P₁**, generation came from pure-breeding stock; thus all his red-flowered plants were homozygous for red flowers, and the white-flowered plants were homozygous for white flowers. Each member of the **first filial**, or **F₁**, generation must have received one allele for red flower color from the red-flowered parent and one allele for white flower color from the white-flowered parent. The F₁ generation was thus heterozygous with respect to flower color.

13-D Law of Dominance

Mendel observed that all of the F₁ generation bore red flowers. What had happened to the white alleles? Since self-crossing of the F₁ plants produced both red- and white-flowered plants, the alleles for white must have been carried in a hidden form in the F₁ plants. This result led Mendel to another conclusion, often called the **law of dominance**: in a heterozygous condition, one allele of a gene may express itself and mask the presence of the other allele; in this case the allele for red flower color masks the allele for white. The allele that expresses itself is called the **dominant** allele, while the masked allele is called the **recessive** allele. Homozygous red-flowered plants cannot be distinguished from heterozygous red-flowered plants just by looking at them. The recessive allele can be detected only if it is present in the homozygous condition, with another recessive allele.

Geneticists often use a shorthand, in which genes are designated by letters of the alphabet; dominant alleles are assigned a capital letter and recessive alleles are designated by the lower case. Thus, the flower-color gene pairs of the P₁ generation of peas can be represented as *RR* for the red-flowered parent and *rr* for the white-flowered parent; the heterozygous F₁ plants would then be *Rr*.

Genotype and Phenotype

As a result of the phenomenon of dominance, it is impossible to tell the **genotype**, or genetic makeup, of an individual that shows the dominant trait merely by inspection. The **phenotype** of an organism is the expression of its genes; the phenotype can be observed in some way, perhaps visually, as in the red flower color of peas, or chemically, as in the tests used to determine the blood type of individuals whose blood looks identical, and so forth. An individual with a dominant phenotype may have a genotype that is either homozygous dominant or heterozygous; an individual with a recessive phenotype, however, must have a **homozygous recessive** genotype.

PARENTAL (P₁) GENERATION

Red   White

FIRST FILIAL (F₁) GENERATION

All red

SECOND FILIAL (F₂) GENERATION

3/4 Red   1/4 White

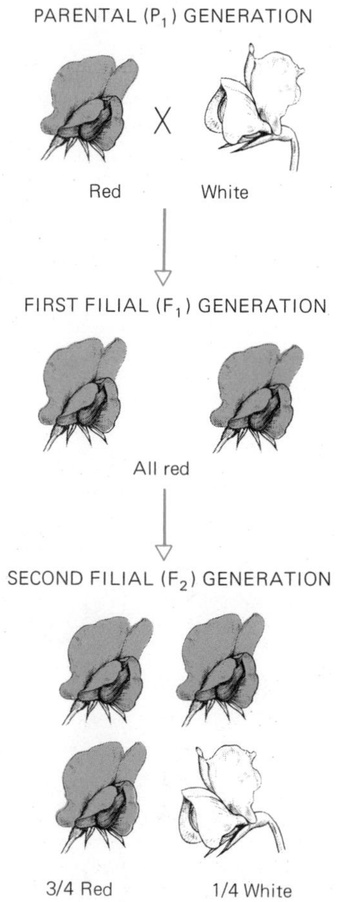

Figure 13-3 Diagram of a cross between pure-breeding red-flowered and white-flowered pea plants (P₁). The offspring (F₁) were then permitted to self-fertilize to produce the plants of the F₂ generation.

An individual's genotype is fixed at the time of fertilization; its phenotype, however, results from the interaction of all of its genes with one another and with factors in the environment. For example, the phenotype "dwarf" in a plant might arise because the plant possesses "dwarf" genes, or because the plant is poorly nourished, even though it has "tall" genes.

13-E   Law of Segregation

In our study of meiosis in Chapter 11, we saw that the members of each pair of homologous chomosomes separate into different nuclei during the formation of haploid nuclei. The steps of meiosis were unknown when Mendel did his work, but Mendel recognized that paired genes must separate from each other during gamete formation; this reasoning is now known as Mendel's **law of segregation**. Mendel saw that the now-single genes could combine in new pairing arrangements at fertilization.

In the flower-color cross, the red-flowered parent RR forms gametes containing a single R gene; the white-flowered parent, rr, forms r gametes (Figure 13–4). At fertilization, the F_1 offspring inherit a complete pair of flower-color genes, R from the red-flowered parent and r from the white-flowered one. The F_1 plants will be red-flowered since R, the dominant allele, will express itself as red flower color and mask the recessive allele. When the F_1 generation reproduces, R and r segregate into separate cells; half the gametes carry the allele for red flowers (R) and half carry the allele for white (r).

When the **second filial**, or F_2, **generation** is formed, either by cross-fertilization or by self-fertilization of the F_1 plants, any egg has an equal chance of being fertilized by any pollen nucleus, so that the members of the F_2 generation have an equal chance of receiving either allele from either parent.

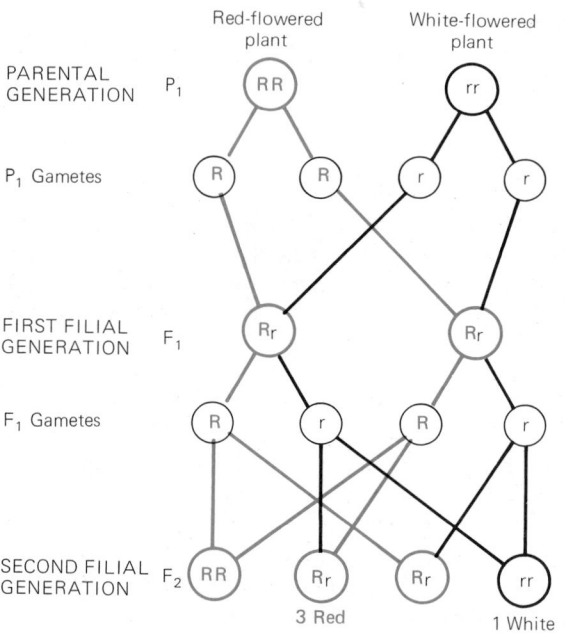

Figure 13-4   Genotypes of gametes and offspring when a plant homozygous for red flower color is crossed with one homozygous for white flower color.

Punnett Squares

When the genotypes of parents used in a cross are known, a convenient way of predicting the genotypes of their offspring is to draw a **Punnett square** (named after the geneticist Reginald Crundall Punnett). To construct a Punnett square, the genes present in the gametes of one parent are written along the top, and those from the other parent are listed down the side (Figure 13–5). The combinations produced by filling in all of the boxes show the possible genotypes of the F_2 individuals and the ratio in which they will occur; in this case, the F_2 generation will include RR, Rr and rr individuals in the ratio of $1:2:1$. If we know that one allele is dominant to the other, we can also predict the phenotypes of the F_2 generation. In this case, since R is dominant to r, the F_2 generation will be expected to contain three times as many red-flowered (RR and Rr) individuals as white-flowered (rr) individuals.

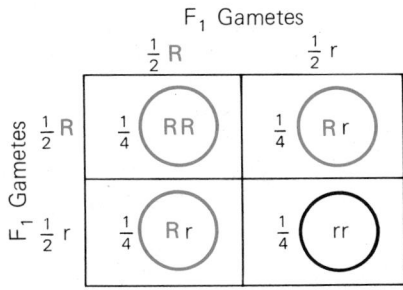

Figure 13-5 A Punnett square diagram of the $F_1 \times F_1$ cross shown in Figure 13-4.

13-F Monohybrid Cross

The cross we have been considering is a **monohybrid cross**, in which only one trait of the parents (flower color, in this case) is of interest. The ratio 3 dominant phenotypes:1 recessive phenotype is expected of F_2 offspring from a monohybrid cross where the alleles exhibit a dominant-recessive relationship. The genotype ratios are 1 homozygous dominant:2 heterozygous:1 homozygous recessive.

Table 13–2 shows the actual numbers of individuals of each phenotype that Mendel found in monohybrid crosses with the seven different pairs of alleles he studied.

13-G Test Cross

If RR and Rr plants both have the same red-flowered phenotype, how is it possible to determine the genotype of a particular red-flowered plant? The usual method is to perform a cross with a plant of known genotype, and observe the phenotypes of the offspring.

In a **test cross**, an organism of dominant phenotype and unknown genotype is crossed with one that is homozygous recessive for the trait in question. A white-flowered pea plant, in our example, has the homozygous recessive genotype, rr, and must pass on an r allele to each of its offspring. If the red-flowered plant of unknown genotype were actually heterozygous (Rr), half the offspring of the test cross would

TABLE 13–2 **RESULTS OF MENDEL'S MONOHYBRID CROSSES INVOLVING SEVEN PAIRS OF TRAITS**

PARENTAL CHARACTERS[1]	F_1	F_2	F_2 RATIO[2]
<u>Smooth</u> × wrinkled seeds	all smooth	5474 smooth : 1850 wrinkled	2.96:1
<u>Red</u> × white flowers	all red	705 red : 224 white	3.15:1
<u>Yellow</u> × green seeds	all yellow	6022 yellow : 2001 green	3.01:1
<u>Inflated</u> × constricted pods	all inflated	882 inflated : 299 constricted	2.95:1
<u>Green</u> × yellow pods	all green	428 green : 152 yellow	2.82:1
<u>Axial</u> × terminal flowers	all axial	651 axial : 207 terminal	3.14:1
<u>Long</u> × short stems	all long	787 long : 277 short	2.84:1

[1]All plants of the parental generation were pure-breeding. The dominant allele of each pair is underlined.

[2]The ratios of number of plants of the dominant phenotype to number of the recessive phenotype in the F_2 generation approximate the expected ratio of $3:1$.

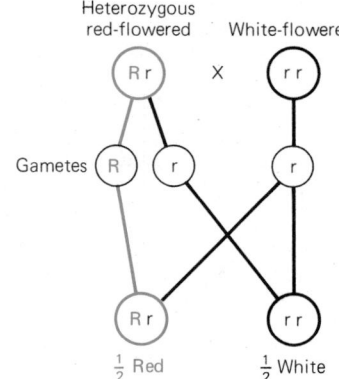

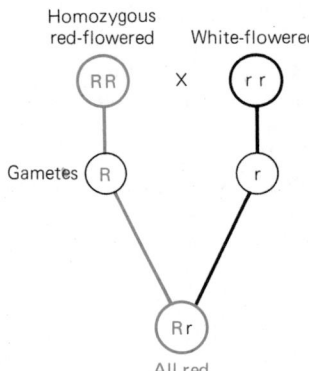

Figure 13-6 Test cross. A red-flowered plant of dominant phenotype but unknown genotype is crossed with a white-flowered (homozygous recessive) plant. By examining the offspring, it should be possible to determine the genotype of the red-flowered plant.

be expected to be white-flowered and half red-flowered (Figure 13–6). On the other hand, if the red-flowered parent were homozygous (RR), it could pass only the R allele to its offspring and they would all be expected to be heterozygous, with red flowers (see Figure 13–6). If any white-flowered progeny (offspring) are obtained from a test cross of a red-flowered plant of unknown genotype, the red-flowered parent must be heterozygous, because the offspring must have obtained the r allele from each parent. If all the progeny have red flowers, however, it is not absolutely certain that the red-flowered parent was homozygous (RR); it is possible, though unlikely, for a plant to have all red-flowered offspring even if its genotype is Rr.

The ratios shown in diagrams of genetic crosses represent expected proportions of offspring of different genotypes. The fact that each female gamete has a 50:50 chance of containing the r allele does not guarantee that half, or indeed that any, of the female gametes will contain the recessive allele. Nor does the fact that half the pollen grains carry the recessive allele guarantee that half the pollen grains involved in fertilization will carry the recessive allele. In practice, there are statistical methods for deciding the probability that all of a given number of offspring will have red flowers, even though the parent is heterozygous. As more progeny, all red-flowered, are obtained from such a cross, it becomes more and more unlikely that the individual of dominant phenotype is a heterozygote. Thus it is possible to be reasonably sure that a given plant is not heterozygous if a certain number of offspring, all red-flowered, are obtained from a cross.

13-H Independent Assortment of Genes

In addition to his monohybrid crosses, Mendel performed **dihybrid crosses** of plants that differed in two pairs of contrasting alleles. In one of these crosses, Mendel crossed plants homozygous for seeds that were both smooth and yellow with plants homozygous for wrinkled and green seeds. All the F_1 offspring were smooth and yellow, showing that these alleles are dominant. When the F_1 plants grown from these seeds were self-fertilized, they produced:

315 smooth yellow seeds

101 wrinkled yellow seeds

108 smooth green seeds

 32 wrinkled green seeds

The ratio of phenotypes in the F_2 is about 9:3:3:1. Mendel proposed that these data were best explained by assuming the **independent assortment** of the genes governing seed color and seed texture during reproduction. Each gene was inherited as if it existed alone and showed no relationship to the inheritance of any other gene. Note that the ratios of smooth to wrinkled and of yellow to green are both 3:1. (We now know that independent assortment is found only for genes borne on different chromosomes.)

In this example, let S = smooth, s = wrinkled, Y = yellow and y = green. The P_1 plants must have had genotypes SSYY (smooth yellow) and ssyy (wrinkled green) and produced gametes SY and sy, respectively. All members of the F_1 genera-

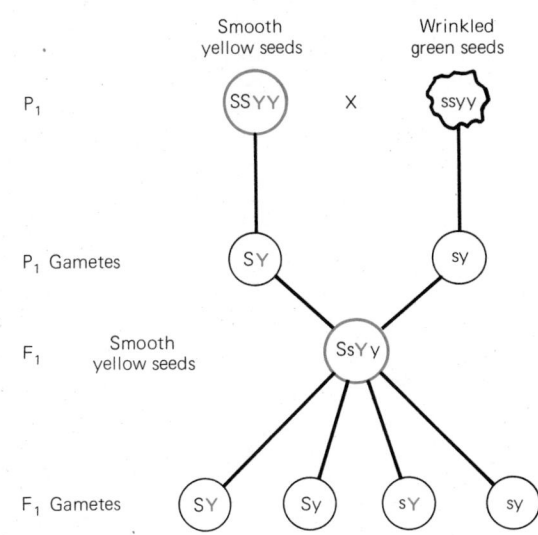

Figure 13-7 Independent assortment of members of two gene pairs during gamete formation in a dihybrid individual. The dihybrid (F_1) individuals produce four different types of gametes in equal proportions.

tion have the genotype $SsYy$. Self-fertilization of the F_1 plants, however, produces a more complex situation. According to Mendel's "law of independent assortment," the members of each gene pair are inherited independently of the members of other gene pairs; thus each F_1 plant produces four kinds of gametes, SY, sy, Sy and sY, in equal proportions (Figure 13-7). To find all possible genetic combinations in the offspring, these gametes can be written along the sides of a Punnett square. Since any female gamete can be fertilized by any male gamete, there are a total of nine possible genotypes, falling into four phenotypes, in the F_2 generation (Figure 13-8).

Table 13-3 shows the genotype and phenotype ratios for this dihybrid cross. A 9:3:3:1 phenotype ratio in the F_2 generation is characteristic of a dihybrid cross. The genotypic ratio will be 1:2:2:1:4:1:2:2:1.

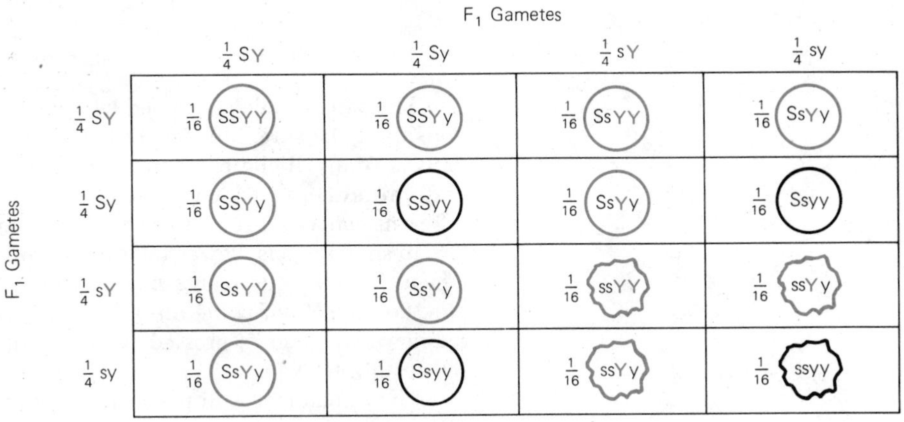

Figure 13-8 Punnett square for the dihybrid cross in Figure 13-7. F_1 gametes are arranged along the top and side of the square; combinations in boxes represent possible genotypes and phenotypes of F_2 individuals.

	F_1 Gametes			
	$\frac{1}{4}$ SY	$\frac{1}{4}$ Sy	$\frac{1}{4}$ sY	$\frac{1}{4}$ sy
$\frac{1}{4}$ SY	$\frac{1}{16}$ SSYY	$\frac{1}{16}$ SSYy	$\frac{1}{16}$ SsYY	$\frac{1}{16}$ SsYy
$\frac{1}{4}$ Sy	$\frac{1}{16}$ SSYy	$\frac{1}{16}$ SSyy	$\frac{1}{16}$ SsYy	$\frac{1}{16}$ Ssyy
$\frac{1}{4}$ sY	$\frac{1}{16}$ SsYY	$\frac{1}{16}$ SsYy	$\frac{1}{16}$ ssYY	$\frac{1}{16}$ ssYy
$\frac{1}{4}$ sy	$\frac{1}{16}$ SsYy	$\frac{1}{16}$ Ssyy	$\frac{1}{16}$ ssYy	$\frac{1}{16}$ ssyy

TABLE 13-3 **PHENOTYPES AND GENOTYPES, AND THEIR RATIOS, RESULTING FROM A SELF-CROSS OF $SsYy$ INDIVIDUALS**

PHENOTYPE	RATIO	GENOTYPE	
smooth, yellow	$\frac{9}{16}$	$\frac{1}{16}SSYY : \frac{2}{16}SSYy : \frac{2}{16}SsYY : \frac{4}{16}SsYy$	or $S\_Y\_$†
smooth, green	$\frac{3}{16}$	$\frac{1}{16}SSyy : \frac{2}{16}Ssyy$	or S_yy
wrinkled, yellow	$\frac{3}{16}$	$\frac{2}{16}ssYy : \frac{1}{16}ssYY$	or $ssY\_$
wrinkled, green	$\frac{1}{16}$	$\frac{1}{16}ssyy$	**must be** $ssyy$

† The notation __ indicates that either allele may be present; e.g., S_yy represents either $SSyy$ or $Ssyy$, both of which will produce a smooth green phenotype.

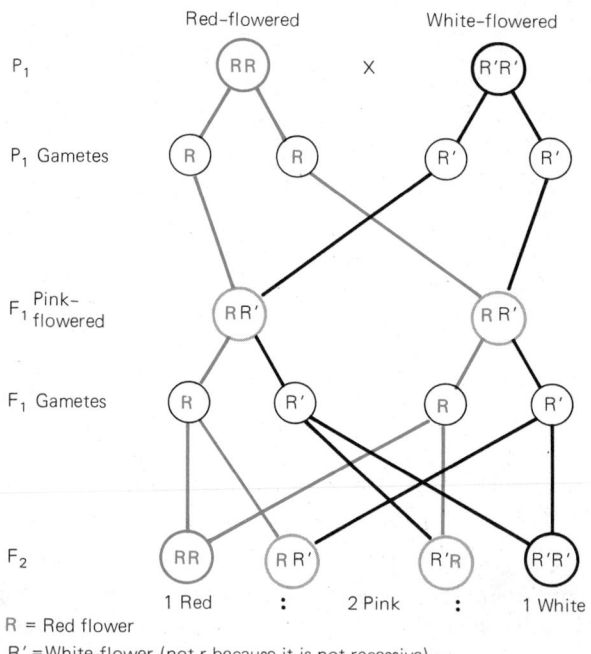

R = Red flower
R' = White flower (not r because it is not recessive)

Figure 13-9 Diagram of a cross involving incomplete dominance in snapdragons.

13-I   Incomplete Dominance

The pairs of alleles studied by Mendel all exhibited a dominant-recessive relationship; to be sure, Mendel chose his pairs because they behaved as distinct alternatives. Many allelic pairs are now known to exhibit **incomplete dominance** or **codominance**, in which the heterozygote has a different phenotype (as well as a different genotype) from that of either homozygote.

In snapdragons, flower color is controlled by alleles that show incomplete dominance. Homozygous plants have either red or white flowers; heterozygotes have pink flowers. What is the ratio of phenotypes expected in the F_2 generation after a red-flowered plant is crossed with a white-flowered plant and the F_1 plants are self-fertilized?

The F_1 plants are all pink, and they are all heterozygous for the alleles for red and white flower color. Half their gametes will contain the allele for red flowers, and half will contain the allele for white flowers. The F_2 phenotype and genotype ratios will both be $1:2:1$, just the same as the F_2 *genotype* ratios for any other monohybrid cross (Figure 13–9). Since the heterozygote produces pink flowers, the expected phenotype ratio for the F_2 plants is 1 red-flowered:2 pink-flowered:1 white-flowered.

13-J   Meiosis and Patterns of Inheritance

Mendel's observations of how genes are passed from parent to offspring correspond precisely with the behavior of chromosomes, the physical structures of inheritance, during meiosis and fertilization.

Diploid eukaryotic cells contain pairs of **homologous chromosomes**; each

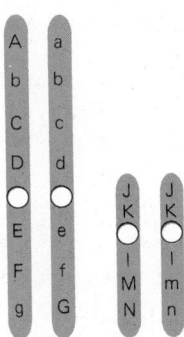

Figure 13-10   Two pairs of homologous chromosomes. Homologues are usually of the same length, have their centromeres (clear circles) at the same positions, and bear alleles for the same traits at the same loci. For example, the pair on the left bears loci A, B, C, D, E, and F; this individual is heterozygous at the A locus (*Aa*), homozygous recessive at the B locus, and so forth.

chromosome is the same length as its homologue; it has its centromere* at the same position, and it bears the same loci. A **locus** is a position on a chromosome where a gene for a particular trait is located. The locus may be occupied by any allele for the trait (Figure 13–10). For example, in pea plants, one chromosome may have the *R* allele at its flower color locus, and its homologue would have either the *R* or the *r* allele at the same locus.

Consider a cell containing a pair of homologous chromosomes bearing the A locus, which is occupied by the *A* allele on one chromosome and by the *a* allele on its homologue (Figure 13–11). As the chromosomes proceed through meiosis, Mendel's law of segregation holds true: *A* and *a* are separated into different gametes. First the two homologous sets of sister chromatids* line up together, forming a **tetrad** (a foursome, with four sister chromatids altogether), and the tetrad then moves to the equator of the meiotic spindle. Here the homologues are pulled apart, so that one set of sister chromatids goes to each pole, and two nuclei are formed, each with one of the two alleles in it. (At the second meiotic division, the sister chromatids separate from one another.)

---

* Centromere   The area where replicated strands (chromatids) of a chromosome remain joined until nuclear division.

* Chromatid   One of the two replicated strands of a chromosome still held together at the centromere.

HOMOLOGOUS CHROMOSOMES
HETEROZYGOUS AT THE "A" LOCUS

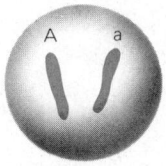

METAPHASE I

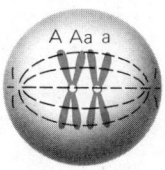

METAPHASE II

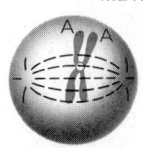 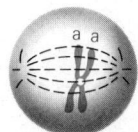

PROPHASE I

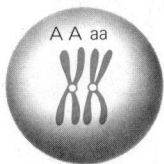

TELOPHASE I

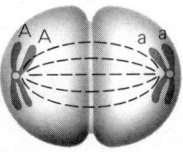

TELOPHASE II

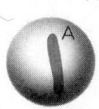 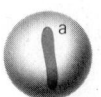

Figure 13-11   The law of segregation reflects the events of meiosis; homologous chromosomes are separated so that each member of the pair ends up in a separate gamete.

To see how the chromosomes follow the law of independent assortment, we next add two more pairs of chromosomes, carrying loci called B and C. At metaphase I of meiosis, each of the three tetrads can line up with either homologue on either side of the equator, and so all eight possible combinations of the A, B, and C chromosomes are equally likely (Figure 13–12).

13-K Linkage Groups

Evidence from genetic studies shows that genes or loci are arranged in linear order, with hundreds of them on each chromosome. The law of independent assortment states that genes segregate independently, but this holds true only if the genes in question are located on different chromosomes. What happens when the two loci are on the same chromosome?

Since the chromosome moves as a unit during meiosis, any loci on the same chromosome are expected to stay together throughout the process rather than assort

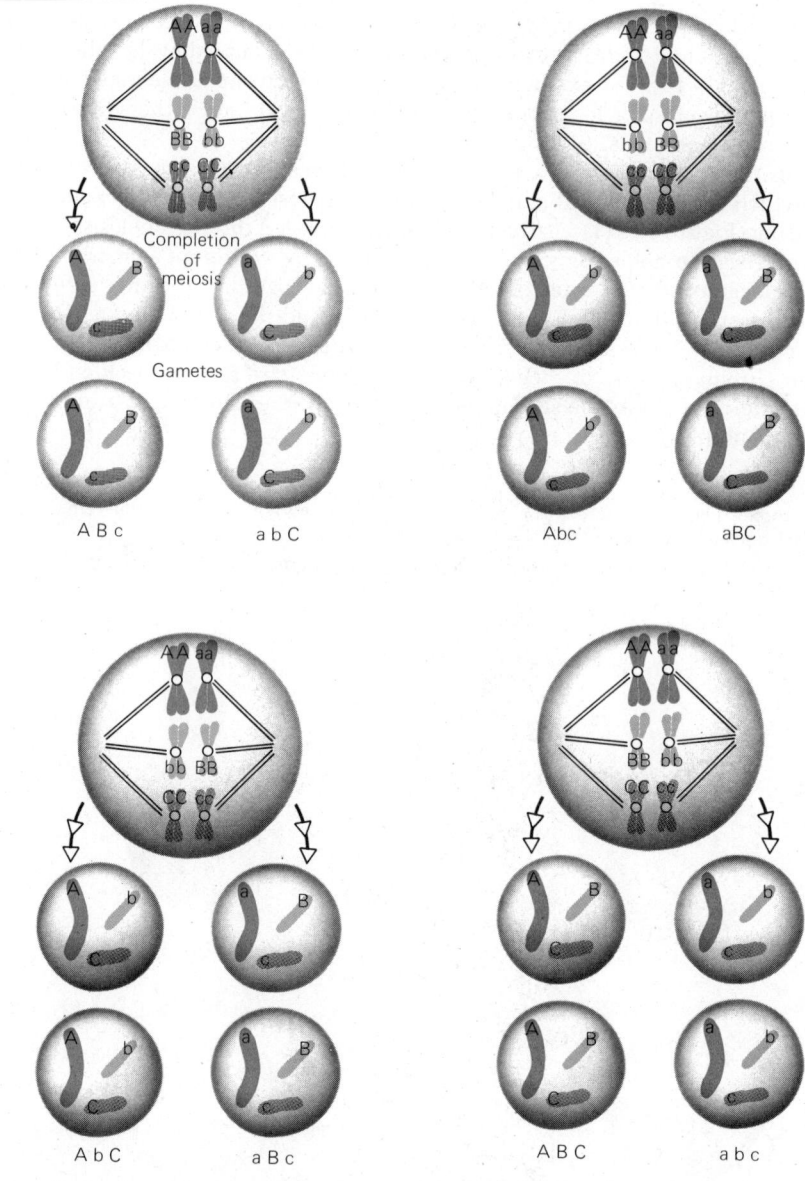

Figure 13-12 Independent assortment is due to various possible alignments of chromosomes during metaphase I of meiosis. The chromosomes of a trihybrid can line up in any of four ways, producing eight possible combinations in the gametes.

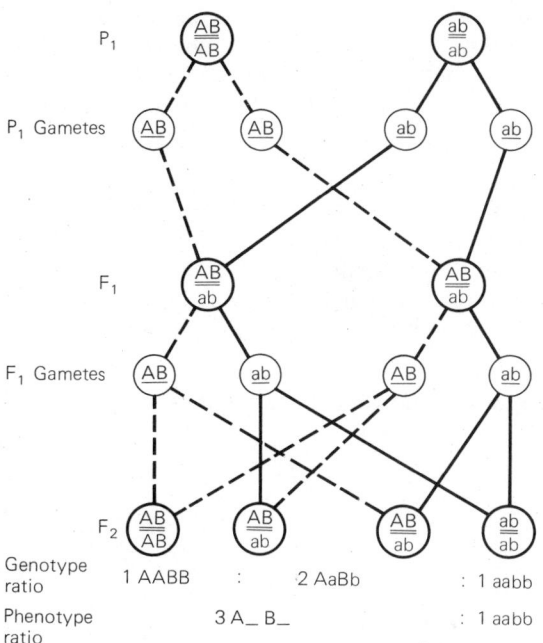

Figure 13-13 Dihybrid cross involving linked loci. Since the A and B loci are on the same chromosome, the alleles on each chromosome move as a unit rather than assorting independently. The genotype and phenotype ratios in the F_2 are the same as for a monohybrid cross rather than the unlinked dihybrid cross.

independently of one another. In other words, they behave as though they were linked, and so they are called **linked loci**. This is a common situation because there are so many genes compared with the number of chromosomes in any organism. In fact, Mendel was quite lucky that his seven pairs of traits were located on six of the seven different chromosome pairs of pea plants; the two loci that were linked were rather far apart on the chromosome (the significance of this will be clear shortly).

We can draw two homologous chromosomes that are heterozygous at two loci, A and B, like this:

$$\frac{\textbf{A} \qquad \textbf{B}}{\textbf{a} \qquad \textbf{b}} \Longrightarrow \text{chromosomes}$$

However, it takes too much time and space to make this drawing; geneticists use a

shorthand method to show linked loci, like this: $\dfrac{AB}{ab}$. In this representation, each

chromosome is shown as a line, and the alleles located on that chromosome are the letters directly above or below the line.

In a cross between parents *AABB* and *aabb* in which loci A and B are linked, the only possible gamete from the homozygous dominant parent is $\underline{AB}$ because *A* and *B* are linked; similarly, the other parent can form only gametes containing $\underline{ab}$ (Figure 13–13). The F_1 individuals are heterozygous, but the gametes they form must be like those of the P_1 individuals because *a* is still linked to *b* and *A* is still linked to *B*; these pairs will not assort independently. The genotypic ratios in the F_2 generation in this cross will be 1 homozygous dominant:2 heterozygous:1 homozygous recessive, and the phenotypic ratios will be 3 dominant:1 recessive; these are the same ratios expected of a monohybrid cross. In a sense, this is a monohybrid cross, although it involves two gene pairs; the fact that the genes are linked makes them assort like one gene pair.

The phenotypic ratio from a cross involving linked loci differs from that expected if the loci were not linked. This deviation from the expected results is the clue showing that two gene pairs are linked. The term **linkage group** refers to all the loci with inheritance patterns that show they are linked to each other, so a linkage group is really all the loci on one chromosome.

13-L Crossing Over

In a cross involving linked genes in the fruit fly *Drosophila*, a male homozygous for the recessive, linked alleles purple eye (*p*) and black body (*b*) was crossed with a female heterozygous at both loci. This mating yielded the following offspring:

151 **wild-type** (showing normal phenotype for both characters) $\dfrac{PB}{pb}$

recombinants
{
8 purple eyes and normal body color $\dfrac{pB}{pb}$

10 normal eyes and black body color $\dfrac{Pb}{pb}$
}

<u>131</u> purple eyes and black body color $\dfrac{pb}{pb}$

Total 300 offspring

The male fly must have had the genotype $\dfrac{pb}{pb}$. The female might have been $\dfrac{PB}{pb}$ or $\dfrac{Pb}{pB}$. In fact, since most of her offspring (the wild-types and the homozygous recessives) inherited either the chromosome containing <u>PB</u> or the chromosome containing <u>pb</u> from her, it seems clear that the first of these two alternatives is the correct one. However, the offspring from this cross are not entirely what was expected from a cross involving linked genes. Although most of the offspring show the expected phenotypes, 18 of the offspring have inherited chromosomes with gene combinations (<u>Pb</u> and <u>pB</u>) that did not exist in either parent. These offspring are known as **recombinants** since they have inherited chromosomes in which the genes of the parents have recombined in new ways. Somehow, the gene for purple eye has moved onto the same chromosome as the gene for normal wings. This occurs by a process known as **crossing over**.

Crossing over occurs during meiosis. When the homologous chromosomes are lined up in the tetrad stage of meiosis, two chromatids, one from each member of the pair, may cross each other and break off, with the broken portions rejoining to the opposite chromatid (Figure 13–14). When the chromatids finally separate into independent chromosomes, some of the chromosomes are **recombinants**, bearing

Figure 13-14 Crossing over occurs during the tetrad stage of meiosis. Chromatids may cross, break off, and rejoin onto another chromatid. Eventually this produces four chromosomes with different gene combinations.

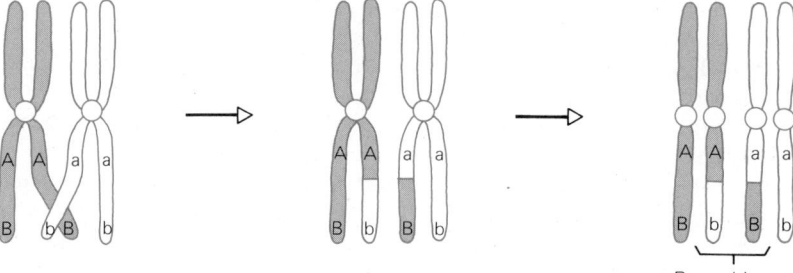

INFORMATION CODING AND TRANSFER

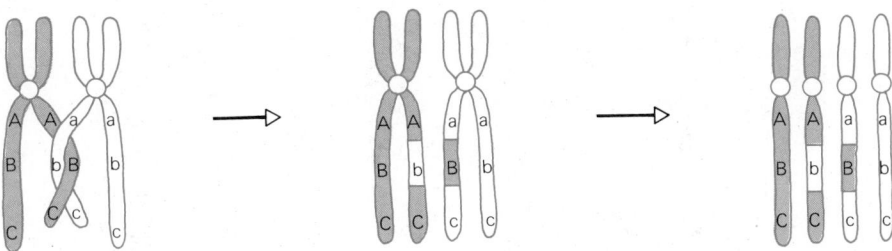

Figure 13-15 Double crossover. Two chromatids in the tetrad cross over twice, exchanging segments between the A and C loci.

new combinations of alleles. Research with the fungus *Neurospora* has shown that all four chromatids may sometimes cross over and exchange genetic material.

The precise mechanism of crossing over is not fully understood, but it must involve some way for the chromatids to break at exactly corresponding points, so that they do not gain or lose genetic material. Note that for each crossover event that occurs, two complementary recombinant homologous chromosomes are produced. For example, when crossover occurs between <u>AB</u> and <u>ab</u>, both the chromosomes <u>Ab</u> and <u>aB</u> are produced. This is easy to see, but something people often forget to allow for when doing genetics problems.

Crossing over has a tremendous evolutionary significance. It permits rearrangement of the various alleles between members of homologous chromosome pairs. If a new, favorable allele should appear on a particular chromosome, and there is already another favorable allele at a different locus on the homologous chromosome, crossing over should produce some chromosomes that carry both favorable alleles. Offspring that receive this combination will be favored by natural selection and will perpetuate the combination. At the same time, offspring receiving the chromosome with the combination of less favorable alleles will be selected against, and the frequency of these alleles in the population will diminish.

The frequency of crossing over is stated as a percentage known as the **crossover value.**

$$\text{Crossover value} = \frac{\text{number of recombinants}}{\text{total number of offspring}} \times 100\%$$

For our *Drosophila* example:

$$\text{Crossover value} = \frac{18}{300} \times 100\% = 6\%$$

The crossover value obtained from looking at the offspring of a cross is often slightly low. This is because double crossovers may occur between the two loci, so that they end up on the same chromosome even though a segment of the chromosome between them has been exchanged. Consider Figure 13–15; if the B and b alleles were not present, you can see that the double crossover between A and C would be undetectable. Offspring from a cross involving these chromosomes would contain <u>AC</u> and <u>ac</u>, the parental combinations. These offspring would be counted as parental rather than recombinant, and so the double crossover would have been counted as no crossover, giving a lower count of crossover events than the number that actually occurred. When you follow all three loci at once, however, you can detect the double crossover because you find some offspring with the combinations <u>AbC</u> and <u>aBc</u>, which are different from the parental combinations.

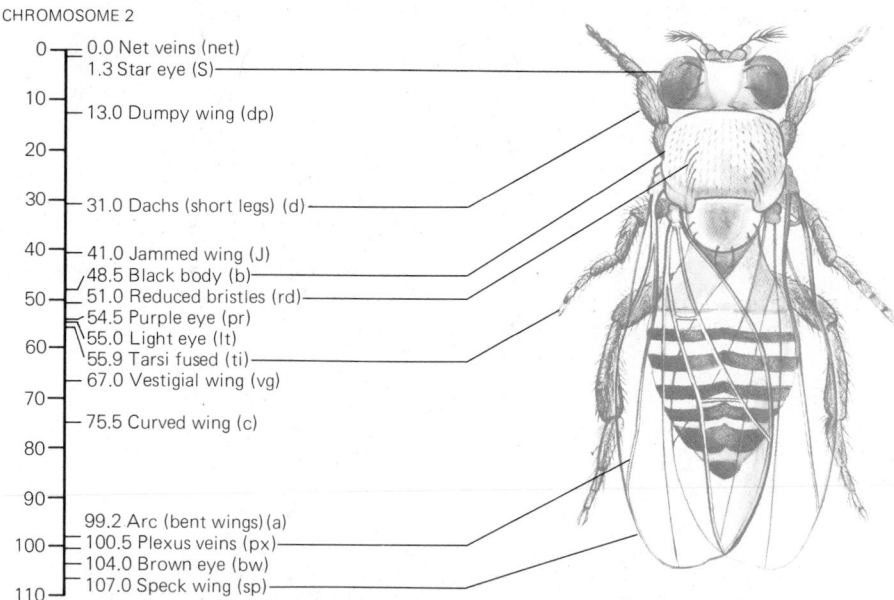

CHROMOSOME 2

0 — 0.0 Net veins (net)
1.3 Star eye (S)

10 — 13.0 Dumpy wing (dp)

20

30 — 31.0 Dachs (short legs) (d)

40 — 41.0 Jammed wing (J)
48.5 Black body (b)

50 — 51.0 Reduced bristles (rd)
54.5 Purple eye (pr)
55.0 Light eye (lt)

60 — 55.9 Tarsi fused (ti)
67.0 Vestigial wing (vg)

70

75.5 Curved wing (c)

80

90

99.2 Arc (bent wings) (a)

100 — 100.5 Plexus veins (px)
104.0 Brown eye (bw)

110 — 107.0 Speck wing (sp)

Figure 13-16 A map of some of the loci known to be present on chromosome 2 of *Drosophila melanogaster*. The map shows mutant alleles identified at each locus. Note that two of the mutations, star eye and jammed wing, are dominant, but that most of the mutations are recessive. The loci for black body and purple eye used as examples in the text are on this chromosome.

Chromosome Mapping

Genes are arranged on chromosomes in linear order, so it is possible, by performing appropriate crosses with linked loci, to make chromosome **maps** showing the order of loci on the chromosome (Figure 13–16). The crossover value between linked loci gives a relative measure of their distance from each other. A 1% crossover value between two loci is arbitrarily set equal to one **map unit** between them. Thus two linked loci are said to be 10 map units apart if experiments show a 10% crossover value between them. If the loci are very close together there will be very few crossovers and a short map distance between them; if the loci are far apart, there will be a greater chance of a crossover's occurring somewhere between them and a larger **map distance** between them. This method of mapping gives an inaccurate map distance between genes that are a long way apart because multiple crossovers between them will lower the number of observed recombinants and give an artificially short map distance between the loci. The most accurate maps are those for which experiments have been done to find the crossover values between as many different loci on the chromosome as possible.

13-M   Probability Theory

The Punnett square is cumbersome to use for a dihybrid cross, and it becomes extremely confusing for following crosses involving three or more gene pairs. It is often more convenient to calculate the probability of the various offspring types directly instead of drawing all of them and then counting them up to determine their various probabilities.

The probability (*P*) of a particular outcome of an event is the fraction of times that outcome is expected to occur in the total of all events. For example, in a cross *Rr* × *Rr*, the probability of any one offspring's having red flowers is three-fourths and the probability of any one offspring's having white flowers is one-fourth. The ratio actually obtained will be close to the predicted probability only when a large number of outcomes is observed (i.e., when a large number of offspring is obtained).

The probability of one kind of outcome, such as red flowers, is defined as:

$$P = \frac{\text{total number of one kind of event}}{\text{total number of all events}} = \frac{\text{number of red flowers}}{\text{number of red + white flowers}}$$

Rules of probability:

1. If an event is certain to occur, its probability is one ($P = 1$). If it will not occur, its probability is zero. If an event might occur, its probability must be somewhere between zero and one.

2. If the probability of an event's occurring is the fraction "q," then the probability that the event will *not* occur is $(1 - q)$. This quantity is often given its own letter, $p = (1 - q)$.

3. The probabilities of all possible outcomes of an event must add up to 1; that is, $p + q = 1$.

4. The probability that events which are independent of each other will both or all occur is the product of their separate probabilities ($P = P_1 \times P_2 \times P_3 \ldots$).

The easiest way to grasp how these rules work is to apply them to some problems.

Q. What is the probability that a tossed coin will turn up heads?

A. $P = \frac{1}{2}$ since heads is one of two equally possible results (definition of probability).

Q. What is the probability that it will not turn up heads?

A. $P = \frac{1}{2}$; that is, $(1 - \frac{1}{2})$, since $\frac{1}{2}$ is the probability of heads (rule 2).

Q. When a coin is tossed, what is the probability that it will land with either heads or tails up?

A. $P = (\frac{1}{2} + \frac{1}{2}) = 1$ since the only possible result is heads or tails (rule 3).

Q. If the probability that a customer will order a hamburger is $\frac{8}{10}$, the probability the hamburger served has been sitting under the heat lamp drying out for a half-hour is $\frac{7}{10}$, and the probability that the ketchup dispenser is empty is $\frac{1}{4}$, what is the probability that the customer will eat a dried-out hamburger without ketchup?

A. (Rule 4.) Since these events are independent of each other, the three probabilities must be multiplied: $P = \frac{8}{10} \times \frac{7}{10} \times \frac{1}{4} = \frac{56}{400} = 0.14$.

Q. In a self-cross of a pea plant heterozygous for both seed color and seed texture $SsYy$, what is the probability of an offspring's being smooth and green (S_yy)?

A. Since the loci are unlinked, the chance of obtaining a smooth seed ($S\_$) is $\frac{3}{4}$ and the chance of obtaining a green seed (yy) is $\frac{1}{4}$. Since these outcomes are independent of one another, they must be multiplied to yield the probability that the seed will be both smooth and green: $\frac{3}{4} \times \frac{1}{4} = \frac{3}{16}$. (Compare with the Punnett square, Figure 13–8.)

To determine the probabilities of a number of events, as is often called for in genetics problems, it is more convenient to use algebra. For example, in determining the possible outcomes of three coins tossed into the air at once, the long way would be to write all the possible combinations of heads and tails:

HHH	all heads $= \frac{1}{8}$	TTT	all tails $= \frac{1}{8}$
HHT		TTH	
HTH	2 heads, 1 tail $= \frac{3}{8}$	THT	2 tails, 1 head $= \frac{3}{8}$
THH		HTT	

The probabilities of these four combinations can be obtained algebraically from the binomial expansion: let p = probability of heads and q = probability of tails; $p = q = \frac{1}{2}$. For two coins, the probability of each combination can be found by squaring the binomial $(p + q)$: $(p + q)^2 = p^2 + 2pq + q^2$. For three coins, the binomial is cubed, and so forth: $(p + q)^3 = p^3 + 3p^2q + 3pq^2 + q^3$. p^3 represents the probability of three heads in a row, p^2q represents the probability of two heads and a tail, etc. Adding together the coefficients $(1 + 3 + 3 + 1) = 8$ gives the total possible outcomes; the coefficient for each term gives the relative probability of that type of outcome. Thus, p^3, with a coefficient of one out of eight possible outcomes, gives the probability of three heads $= \frac{1}{8}$. Similarly, the probability of two heads and a tail is $3p^2q$, out of eight possible outcomes, or $\frac{3}{8}$.

Q. In a cross $Rr \times rr$, what is the probability that the three peas in a pod will grow into two red-flowered plants and one white-flowered plant?

A. The cross will produce Rr and rr plants in an expected $50:50$ ratio. From the binomial expansion, $\frac{3}{8}$ of three-seeded pods would be expected to produce 2 red-flowered : 1 white-flowered.

SUMMARY The modern science of genetics, the study of how genes governing characteristics are inherited, is based on the work of Gregor Mendel. Mendel's success in elucidating the rules of inheritance was largely due to his shrewd choice of an experimental organism, the garden pea, and his use of mathematical analysis to construct a theoretical model that would explain the results he obtained through breeding hundreds of pea plants. Mendel's conclusions may be summarized in modern terms in this way:

1. Genetically based traits are determined by discrete units, or genes, which are passed from parent to offspring during reproduction.

2. A plant or animal contains pairs of genes that determine its genetic characteristics.

3. During meiosis, the two members of each gene pair separate from one another and pass into different cells.

4. At fertilization, each offspring receives a pair of genes for each characteristic, one member of each pair from the gamete of each parent.

5. The genes from each parent remain distinct in the offspring and may reappear in the phenotype of later generations even if they are masked by the phenomenon of dominance in some individuals in intervening generations.

6. During meiosis, the genes of one pair assort independently of genes of other pairs, so long as their loci are on different chromosomes.

The behavior of genetically determined traits in breeding experiments is paralleled by the behavior of the chromosomes during meiotic cell division. This parallelism provides part of the evidence that genes are borne on chromosomes. Loci on the same chromosome are inherited together except when they are separated by the phenomenon of crossing over at the beginning of meiotic cell division.

The laws of probability can be used to forecast the likelihood that the offspring of given parents will inherit a particular set of alleles.

OBJECTIVES

From your study of this chapter, you should be able to:

1. Define and use the following terms: parental (P_1) generation, first filial (F_1) generation, second filial (F_2) generation, dominant, recessive, alleles, homozygous, heterozygous, monohybrid cross, dihybrid cross, segregation, independent assortment, incomplete dominance, homologous chromosomes, locus, linkage groups, crossing over, crossover values, map distance.

2. Give reasons for Mendel's success in elucidating the

laws governing the inheritance of genetic characters, where others had failed.

3. Define and compare the terms phenotype and genotype and their relationship to the terms dominant and recessive.

4. Use a Punnett square to illustrate a monohybrid or independently assorting dihybrid cross, and work out the genotypic and phenotypic ratios expected from such crosses.

5. Explain what is meant by a test cross, and discuss its significance as a genetic tool. Design a test cross to determine the genotype of an organism with a dominant phenotype.

6. Correlate the pattern of inheritance of genetic characteristics in breeding experiments with the behavior of the chromosomes during meiosis and fertilization.

7. Given data from problem situations, identify linkage phenomena and calculate map distances between linked loci.

8. Explain the biological significance of tetrad formation and crossing over during meiosis.

9. Use the rules of probability to solve genetics problems such as those at the end of this chapter.

10. In your own words, state the rules of inheritance that were Mendel's most important contribution to genetics.

SELF-QUIZ

The following problems will test your understanding of the ideas in this chapter.

1. Mendel found that the allele for tallness in peas (*T*) is dominant to its allele for shortness (*t*). What offspring phenotypes would be expected from the following crosses, and in what ratio?
   a. heterozygote self-fertilized
   b. homozygous tall × heterozygote
   c. heterozygote × homozygous short

2. The allele for axial flowers in peas is dominant to the allele for flowers borne terminally. What phenotypic ratios would you expect among the offspring of a cross between a known heterozygous axial-flowered plant and one whose flowers were terminal?

3. Two *Drosophila* (fruit flies) with normal wings are crossed. Among 123 progeny, 88 have normal wings and 35 have "dumpy" wings.
   a. What inheritance pattern is shown by the normal and dumpy alleles?
   b. What were the genotypes of the two parents?

4. If a dumpy-winged female (from problem 3) is crossed with her father, how many normal-winged flies will be expected among 80 offspring?

5. A number of plant species have a recessive allele for albinism; homozygous albino (white) individuals are unable to synthesize chlorophyll. If a tobacco plant heterozygous for albinism is allowed to self-pollinate and 500 of its seeds germinate:
   a. how many of these offspring will be expected to have the same genotype as the parent plant?
   b. how many seedlings will be expected to be white?

6. Sniffles, a male mouse with a colored coat, was mated with Esmeralda, an alluring albino. The resulting litter of six young all had colored fur. The next time around, Esmeralda was mated with Whiskers, who was the same color as Sniffles. Some of Esmeralda's next litter were white.
   a. What are the probable genotypes of Sniffles, Whiskers, and Esmeralda?
   b. If a male of the first litter were mated with a colored female of the second litter, what phenotypic ratio might be expected among the offspring?
   c. What would the expected results be if a male from the first litter were mated with an albino female from the second litter?

7. A kennel owner has a magnificent Irish setter, which he wants to hire out for stud. He knows that one of its ancestors was Erin-go-braugh, who carried a recessive allele for atrophy of the retina. In its homozygous state, this gene produces blindness. Before he can charge a stud fee, he must check to make sure his dog does not carry this allele. How can he go about this?

8. In *Drosophila*, the allele for dachs (short-legged, *d*) is recessive to its allele for normal leg length (*D*) and the allele for hairy body (*h*) is recessive to its allele for normal body (*H*). Make a Punnett square for each of the following crosses:
   a. *DdHh* × *Ddhh*
   b. *DDHh* × *Ddhh*
   c. *DdHh* × *ddhh*
   d. What proportion of the offspring from cross (b) would be expected to show the normal phenotype for both traits?

9. A peony plant with straight stamens and red petals was crossed with another plant having straight stamens and streaky petals. The seeds were collected and germinated, and the following offspring were obtained:
   62 straight stamens, red petals
   59 straight stamens, streaky petals
   18 incurved stamens, red petals
   22 incurved stamens, streaky petals
   a. Which allele in each pair (straight vs. incurved stamens, red vs. streaky petals) is dominant?
   b. What were the genotypes of the parental plants?
   c. What further crosses would you make in order to get a definite answer for part a?

10. In tomato plants, the gene for purple stems (*A*) is dominant to its allele for green stems (*a*) and the gene for red fruit (*R*) is dominant to its allele for yellow fruit (*r*). If two tomato plants heterozygous for both traits are crossed, state what proportion of the offspring are expected to have:
    a. purple stems and yellow fruits
    b. green stems and red fruits
    c. purple stems and red fruits

11. If 640 seeds resulting from the cross in question 10 are collected and planted, determine how many are expected to grow into plants with:
    a. red fruit
    b. green stems
    c. both green stems and yellow fruits

12. If one of the parents from question 10 is crossed with a green-stemmed plant heterozygous for red fruits, what proportion of the offspring would you expect to have each of the following character combinations?
    a. purple stems and yellow fruits
    b. green stems and yellow fruits
    c. green stems and red fruits

13. Pooh had a colony of tiggers whose stripes went across the body. His American pen-pal, Yogi, sent him a tigger whose stripes ran lengthwise. When Pooh crossed it with one of his own animals, he obtained plaid tiggers. Interbreeding among the plaid tiggers produced litters of a majority of plaid members, but some crosswise- and lengthwise-striped animals were also produced. Diagram the crosses made by Pooh, showing the genotypes of the tiggers that account for the coat patterns observed.

14. In cattle, the gene for straight coat (S) is dominant to its allele for curly coat (s). The gene pairs for red (RR) or white (R'R') coat color show an absence of dominance; heterozygotes have a roan coat (RR') (red lightened by intermixed white hairs).
    a. If a curly red cow is mated to a homozygous straight white bull, what will the genotype and phenotype of the calf be?
    b. If the calf is mated to a roan animal with curly hair, what are the possible offspring phenotypes?

15. A farmer has three groups of cows: white ones in the clover patch, red ones in the alfalfa field, and roan in the cornfield. He has a roan bull, Ferdinand, who services the cows in all three fields. (Refer to question 14 for more information.)
    a. What color calves should he expect in each field, and in what proportions?
    b. Ferdinand dies from a bee sting and the farmer decides to make his herd of cows exclusively roan coat in memory of his beloved bull. He sells all the red and white cows, and vows to sell any red or white calves born subsequently. What color bull should he buy to replace Ferdinand, if he wants to sell as many calves as possible?

16. The allele for pea comb (P) in chickens is dominant to the allele for single comb (p), but the alleles for black (B) and white (B') feather color show incomplete dominance, BB' individuals having "blue" feathers. If birds heterozygous for both pairs of genes are mated, determine what proportion of the offspring are expected to be:
    a. single-combed
    b. blue-feathered
    c. white-feathered
    d. white-feathered and pea-combed
    e. blue-feathered and single-combed

17. In a plant heterozygous for two pairs of genes (AaBb), state the chance that a pollen grain it produces will carry
    a. an A allele
    b. an a allele and a b allele
    c. an a allele and a B allele
    d. a B allele or a b allele

18. If the plant in question 17 self-pollinates, figure the probability that a seed will contain
    a. two a alleles
    b. an A allele and an a allele
    c. two a alleles and two B alleles
    d. all four alleles (AaBb)

19. Mr. and Ms. Miller have two sons. What is the probability that their third child will be a boy?

20. The man in question 7 has mated his Irish setter to two bitches known to be heterozygous for the allele for retinal atrophy. Between the two litters, nine pups are obtained, none of which shows retinal atrophy. How certain is the owner that his dog lacks the retinal atrophy gene?

21. In *Drosophila*, the allele for miniature wing (m) is recessive to the allele for normal wing (M), and the gene for vermilion eye (v) is recessive to the allele for normal eye (V). A female heterozygous for vermilion eye and miniature wing was mated to a vermilion-eyed, miniature-winged male. The following offspring were collected:
    - 140 normal wing, normal eyes
    - 3 normal wing, vermilion eyes
    - 6 miniature wing, normal eyes
    - 151 miniature wing, vermilion eyes
    a. What were the linkage groups of the female parent?
    b. What is the crossover value between v and m?

22. A female *Drosophila* heterozygous for the recessive alleles sable body (s) and miniature wing (see question 21) was mated with a sable-bodied, miniature-winged male, and the following progeny were obtained:
    - 250 normal body, normal wings
    - 15 normal body, miniature wings
    - 20 sable body, normal wings
    - 215 sable body, miniature wings
    a. Diagram the linkage groups of the female parent.
    b. Draw the relative positions of the loci of the v, s, and m alleles (using also your answer to problem 21).
    c. What further cross must be made in order to answer part (b) conclusively?
    d. How could the three loci have been mapped in one experiment?

23. In *Drosophila*, the gene for red eye is dominant to its allele for purple eye, and the gene for long wings is dominant to its allele for dumpy wings. A fly heterozygous for both traits is crossed with a fly having purple eyes and dumpy wings. The F_1 are:
    - 609 red eyes, normal wings
    - 614 red eyes, dumpy wings
    - 622 purple eyes, normal wings
    - 616 purple eyes, dumpy wings

    Would you expect that the two loci involved are on the same chromosome or different chromosomes?

24. A young woman had a brother who died in infancy of a rare genetic disease. The disease is caused by a recessive gene found in the heterozygous condition in about 0.0001 of the population. It is lethal only in the homozygous condition.
    a. What are her chances of having a child with this disease?
    b. This woman is thinking of marrying her first cousin. What are her chances of having a child with the disease if she goes through with this marriage?

25. Lois Lane Kent's husband has passed on to their daughter three strange traits: x-ray vision, sensitivity to the mineral kryptonite, and muscles like steel. A book that

Mr. Kent was given by his mother states that the genes for these three traits are dominant and that the vision and muscle loci are 16 map units apart on the same chromosome. The kryptonite locus is on another chromosome. Assuming that an earthman will someday marry her daughter, Ms. Kent is worried about the chances that her grandchildren will inherit the genes that have made their mother such a difficult child to raise. What is the probability that Ms. Kent's grandchildren will have:

a. all three traits?

b. x-ray vision and sensitivity to kryptonite but normal muscles?

26. In sweet peas, flower color is determined by two pairs of genes. Plants with at least one dominant allele in both pairs ($P\_\_C\_\_$) have purple flowers; lack of either P or C produces white flowers. If a plant heterozygous for both loci self-pollinates, what will be the proportions of purple flowers and of white flowers in the offspring?

QUESTION FOR DISCUSSION

Explain why genes linked at a map distance of 50 units or more behave as if they are on different chromosomes. How could you tell that they are on the same chromosome?

REFERENCES AND FURTHER READING

Harrison, D. *Problems in Genetics with Notes and Examples*. Reading, MA: Addison-Wesley, 1970. A useful review and practice book.

Mendel, G. J. *Experiments in Plant Hybridisation*. Edinburgh, Scotland: Oliver and Boyd, 1965. An English translation of Mendel's original paper, together with comments and a biography of Mendel by others.

Srb, A. M., R. D. Owen, and R. S. Edgar. *General Genetics*. San Francisco: W. H. Freeman, 1965. A standard text concentrating on Mendelian genetics.

CHAPTER
14

INHERITANCE PATTERNS AND GENE EXPRESSION

Genes, the units of hereditary information, are lengths of chromosomal DNA coding for the sequence of amino acids in polypeptides or proteins. In Chapter 3 we saw that proteins must fold up properly to work, and that their folding depends on having the correct amino acids in the correct order. A mutation, or change of structure, in the DNA of a gene may prevent the formation of a protein altogether, or it may cause the cell to make an altered or inactive protein. The seriousness of a mutation depends on how much it affects the protein and on how essential the protein is to the organism.

In the twentieth century, it has been realized that many disorders of humans and other organisms are direct results of the phenotypic expression of particular genes. In recent years much research in medicine and genetics has been devoted to understanding how novel forms of a gene arise, how they are inherited, how they are expressed, and how they may be affected by factors in the organism's environment.

In this chapter we consider how new genes arise, how unusual results in breeding experiments can provide information about particular genes, and some of the different factors that control phenotypic expression of genes.

14-A Mutations

A **mutation** is an inheritable change in the genetic material (see Section 10-H). Since a mutant gene frequently codes for a protein different from that encoded by the original gene, a mutant organism may have a novel phenotype that does not resemble that of either parent. Phenotypes like "purple eye" and "black body" in the fruit fly *Drosophila* (Chapter 13) are the result of mutations that geneticists perpetuate in the laboratory as useful tools in genetic studies.

Mutations occur unpredictably; we cannot foretell when any one gene will undergo mutation. In humans, rates of detected mutations vary from 1 in 10^6 to 25 in 10^5 gametes, depending on the gene. Mutations are also recurrent. DNA can be changed in only a limited number of ways, and any particular change or mutation will occur randomly from time to time in a population. A certain low rate of mutation is advantageous to a population because mutation is the ultimate source of all genetic novelty; too frequent mutation, however, may destroy a population by producing characteristics that are unfavorable to survival. Breeding experiments have shown that the rate of mutation in a population of *Escherichia coli* or *Drosophila* can be altered by artificial selection.

At the molecular level, a mutation consists of an alteration, deletion, or insertion of one or more DNA nucleotides (see Section 10-B). Whether the mutation causes insertion or deletion of an amino acid, formation of a *Stop* codon, or replacement of one amino acid by another, however, the delicate evolutionary engineering of a protein is much more likely to be harmed than helped by the change. In somatic cells, body cells, mutations may cause damage including cancer, but mutations in germ cells (reproductive cells) are the only ones that will be passed on to future generations, unless the mutation is maintained by asexual propagation.

Mutagens are agents that cause mutation by altering the structure of DNA; cosmic rays, x-rays, ultraviolet radiation, and various chemicals are among the best-known mutagens.

14-B Lethal Alleles

In some cases, it is easy to understand on a molecular level why one gene is dominant and its allele recessive. If an allele codes for no protein, or for an inactive protein, it will plainly be recessive to an allele that codes for an active or functional version of the protein in question. In the heterozygous condition, the protein encoded by the normal allele will produce the characteristic expression of the allele, while the inactive mutant allele produces nothing. Thus the phenotype of the heterozygous individual will appear normal, and the normal allele will be dominant. In an individual homozygous for the inactive allele, no protein will be created to produce the normal expression of the trait; the recessive phenotype will be the absence of the normal trait (Figure 14–1). Thus a white flower may result from "absence of

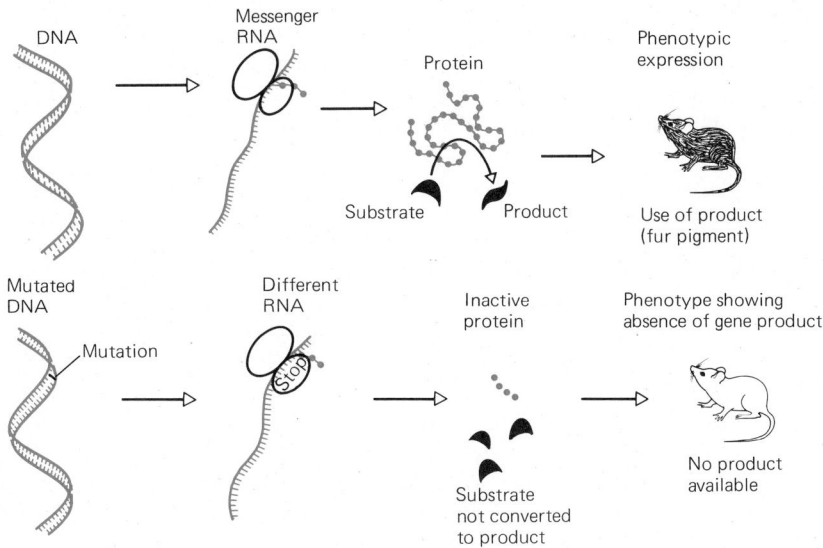

Figure 14-1 How some mutations may result in recessive alleles. The DNA in this example codes for an enzyme needed to make pigment for the fur. The mutant makes an incomplete enzyme, and so the animal does not produce the pigment. An albino animal results.

Kerry

Dexter

Figure 14-2 A lethal allele in Kerry cattle results in short-legged "Dexter" animals in the heterozygous condition; calves homozygous for this allele are severely deformed and die soon after birth.

color" and a dwarf plant from "absence of growth substance."

When an organism is homozygous for an allele that produces no protein or an inactive protein, its phenotype will depend on the protein's importance to the body. If the protein is essential to life, the organism will die, and the allele that fails to encode a functional protein is known as a **lethal** allele. Recessive lethal alleles are much more common than dominant lethal alleles, since a dominant lethal allele will be automatically eliminated from a population—by the death of the affected individual—every time it occurs as a mutation. Recessive lethal alleles, however, will usually occur heterozygously, and so they will be masked by functional alleles; they will be eliminated by selection only when they occur in homozygotes. It has been calculated that each human being carries about 30 lethal recessive alleles in the heterozygous condition. This is higher than the figure for many other organisms and may be the reason that marriages between close relatives produce a higher proportion of abnormal offspring in humans than in most other species.

If just one copy of a "normal" allele does not produce enough of its protein for normal body functioning, the allele will show incomplete dominance. In these cases the heterozygote has a different phenotype from either homozygote. An example of this is found in Kerry cattle. Calves homozygous for the lethal allele are born with tiny legs and internal abnormalities that rapidly result in death. Heterozygotes for this allele are known as Dexter cattle; they have shorter legs than normal (Figure 14–2).

A similar example in humans is the lethal allele that causes brachydactyly, or shortening of the middle bone in the fingers, in heterozygotes; this makes the fingers appear to have only two bones instead of three. In homozygotes, this allele results in abnormal skeletal development. Homozygous babies lack fingers and show other skeletal defects that cause death in infancy.

In a cross between two brachydactylic people, one out of every four children would be expected to be homozygous for the lethal allele and die during infancy; half would be expected to be heterozygous and show brachydactyly; and one-fourth would be expected to be normal (Figure 14–3). This 1:2:1 ratio is typical of lethal alleles when the normal allele is incompletely dominant.

Figure 14-3 A lethal allele in humans. Note the characteristic genotype and phenotype ratios for crosses of (heterozygous × heterozygous) and (normal × heterozygous).

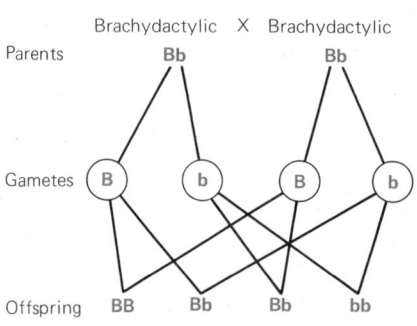

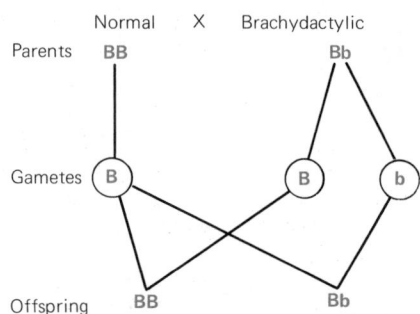

INFORMATION CODING AND TRANSFER

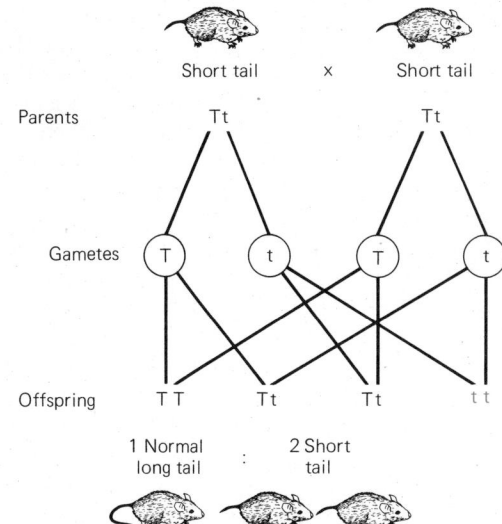

Figure 14-4 A cross involving a lethal allele that shows a 2:1 ratio among the offspring. The short-tailed allele in mice causes death of the homozygous recessive embryo at an early stage of development; it never appears among offspring born to short-tailed parents. One-third of the progeny are normal long-tailed and two-thirds are short-tailed.

Some lethal alleles are mutations of genes that code for proteins so essential that without them the embryo does not develop normally. If a mammalian embryo dies at a very early stage it may be resorbed by the uterus and a 2:1 ratio may be observed in the remaining offspring: two heterozygotes to one normal homozygous offspring (Figure 14–4). In mice, for instance, the short-tail allele (*t*) in the homozygous condition causes early embryonic death. The embryo is then resorbed. Examination of such embryos taken from the uterus early in pregnancy reveals that they lack a backbone and the tissue that forms the muscles, kidneys, etc. Heterozygotes (*Tt*) have shorter tails than normal mice (*TT*).

Sickle Cell Anemia

Perhaps the most famous essentially lethal allele in humans is the one that codes for the hemoglobin of sickle cell anemia. Hemoglobin is a protein found in red blood cells; it is made up of four polypeptide chains, two called alpha and two called beta chains. The sickle allele results from a point mutation (a change in just one nucleotide base) in the normal beta-chain allele; in sickle cell hemoglobin, the beta chain has valine instead of glutamic acid as its sixth amino acid (see Table 10–1). Because of this difference, the hemoglobin molecules stack together when there is not much oxygen in the blood; this causes the red blood cells to assume a "sickle" shape (Figure 14–5). The sickled cells tend to become stuck in the capillaries, the smallest blood vessels, which are just wide enough to permit normal-shaped red blood cells to pass in single file; this cuts off the flow of blood to the areas served by the blocked capillaries. In addition, the sickled cells are frail and break down easily, leaving the victim with fewer red blood cells than normal. The symptoms of sickle cell anemia include tiredness, headaches, muscle cramps, and irritability, all attributable to poor circulation.

The allele for sickle cell anemia shows codominance, in that both it and the normal allele are expressed in heterozygotes; these people have some normal beta chains and some abnormal beta chains. Without special blood tests, they usually do not know that they are carriers of the sickle allele. People who are homozygous for the sickle allele are more severely affected because all their beta chains are abnormal.

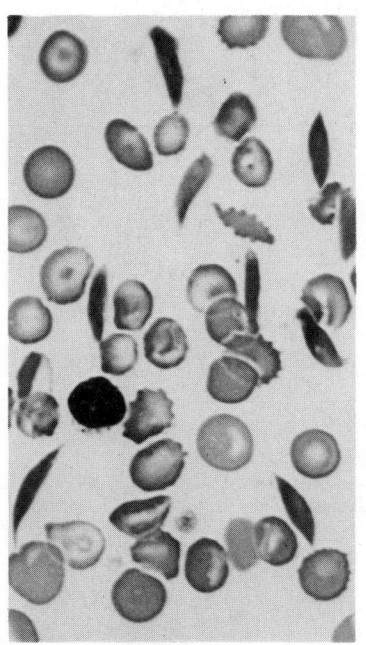

Figure 14-5 Red blood cells from a person with sickle-cell anemia. When the blood is low in oxygen, the cells assume a long, thin, "sickle" shape rather than the usual round, flattened disk form. (Carolina Biological Supply Company)

About 90% of the people who are homozygous for the sickle allele die at an early age. This should be a strong selection pressure that would keep the allele at a very low frequency, the inevitable fate of lethal alleles with no redeeming qualities (see Section 15-A). Yet 20 to 40% of the people in large areas of tropical Africa are heterozygous for the sickle allele. This would require either a fantastically high rate of mutation or a selective advantage for the heterozygote high enough to counterbalance the disadvantage of the recessive homozygote.

In 1953 it was shown that high frequencies of the sickle allele correspond with areas where there is also a high mortality rate from subtertian malaria caused by the protist (unicellular eukaryote) *Plasmodium falciparum*. Possessing at least one copy of the sickle allele lowers a person's chances of developing this form of malaria. Cells containing sickle hemoglobin sickle much faster when they are infected with the parasite. When a cell sickles, it is eaten by one of the body's scavenger white blood cells, which clean up damaged and dead cells. Thus the infected sickle cells are cleared quickly from the body, before the parasites develop to the stage where they can cause disease. In malaria-infested regions, therefore, it is advantageous to be heterozygous for the sickle allele, which protects against a common deadly disease, even though the sickle allele is usually lethal in the homozygous state.

Italian doctors have suggested the same explanation for the high frequency of thalassemia (another recessive hemoglobin abnormality) in districts of Italy with high incidences of malaria.

An individual heterozygous for a genetic condition is referred to as a **carrier**, and an individual homozygous for the condition is called an **affected individual**. (In the case of sickle cell anemia some publications refer to the heterozygote as "having" sickle cell anemia. This is unfortunate since it suggests that the carrier is less fit than the normal homozygote, which is not usually the case.)

Tay-Sachs Disease

Tay-Sachs disease—a metabolic disorder that results in deterioration of the brain and death by about the age of four—is also the result of a recessive lethal allele. In the homozygous condition, this allele causes absence of the enzyme hexosaminidase, which metabolizes a lipid in the brain. Thus, rather than being processed, the lipid accumulates in the nerve cells of the brain and destroys their normal function. One in 30 people of East European Jewish extraction is a carrier for this disease; however, about one-third of the Tay-Sachs cases in the United States are among non-Jewish people.

Cystic Fibrosis

The most common lethal allele in the Caucasian population of the United States causes cystic fibrosis, a disorder characterized by secretion of a thick mucus that can clog the lungs and the ducts of the pancreas and liver. Victims usually die of respiratory infections before their twentieth birthday. Research in 1979 indicated that the culprit might be an abnormal form of the enzyme NADH dehydrogenase, part of the mitochondrial electron transport chain, but this has not been confirmed.

14-C   Inborn Errors of Metabolism

The alleles responsible for Tay-Sachs disease and cystic fibrosis affect enzymes that are part of the body's biochemical processes, or metabolism; genetic disorders resulting from the expression of such alleles are called "inborn errors of metabolism." Recessive alleles that code for errors of metabolism are not necessarily lethal in the homozygote. Phenylketonuria (PKU) and albinism, for instance, are both human hereditary disorders in which an enzyme either is missing or does not function in homozygous recessive individuals (Figure 14–6), but since the metabolic

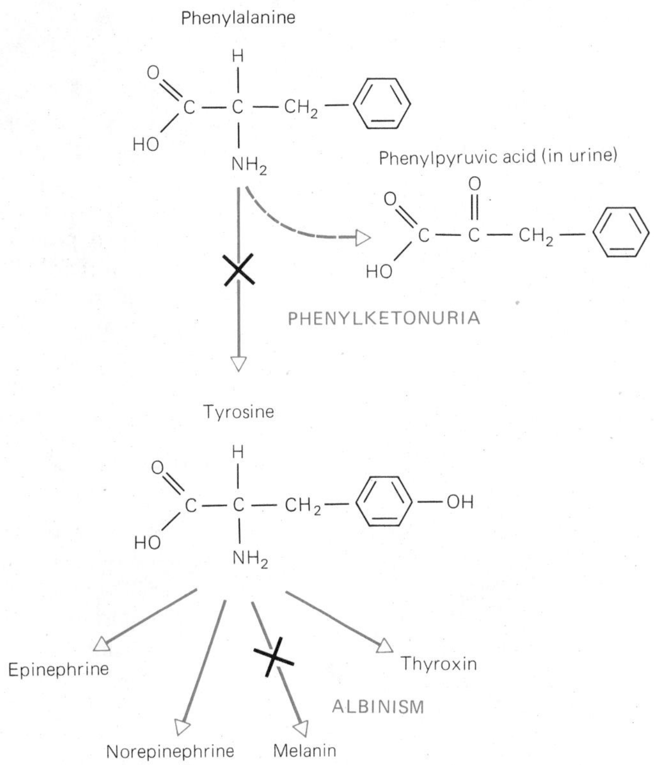

Figure 14-6 The metabolic blocks in phenylketonuria and albinism.

pathways involved are not vital (at least in modern society), the responsible alleles are not lethal.

Victims of PKU are homozygous recessives who lack the enzyme that converts phenylalanine to tyrosine. Phenylalanine is converted instead to phenylpyruvic acid, which accumulates to toxic levels in the bloodstream and eventually is excreted in the urine, giving it a characteristic odor. However, the kidneys do not excrete phenylpyruvic acid fast enough to prevent its damaging various organs, especially the brain, and victims of PKU become mentally retarded if untreated. This disorder can now be controlled, and many states require that newborns be given a blood test for PKU. Children with PKU can be fed a diet low in phenylalanine, which prevents all symptoms of the disease. These children adopt a normal diet in adulthood, after all brain development is complete.

If a woman homozygous for PKU becomes pregnant, although her child may or may not actually have PKU, it will be retarded. High phenylalanine levels in the uterus will cause mental retardation or microcephaly (small head). The PKU-affected woman who wishes to have normal children must go back to a low-phenylalanine diet during pregnancy. Whether or not her children are affected with PKU, they will all be carriers of the disease, since they will each inherit one copy of the recessive allele from her. This poses a significant problem in genetic counseling, since treatment of PKU is enlarging the number of carriers in the population as well as effectively creating a new disease.

Individuals homozygous for the recessive albinism allele lack the enzyme that converts tyrosine into melanin, the dark pigment that gives color to skin, hair, and eyes. True albinos have white hair and very light skin and eyes. You may wonder whether victims of PKU are also albino, since they do not make the tyrosine to turn into melanin. The answer is no, because tyrosine may be obtained in the diet as well as from conversion of phenylalanine. However, people affected with PKU usually have light coloring. A person could be both PKU-affected and albino if he or she were homozygous for both pairs of alleles.

Up to this point, each locus* on a chromosome that we have met was occupied by either of two existing alleles. Most genes, however, are about a thousand nucleotides long, and a mutation anywhere in this long string of nucleotides can change the polypeptide produced. As a result, mutations in different parts of the gene in different individuals in a population may give rise to a large number of different alleles, each producing a different phenotype, which may occupy any one locus.

The human ABO blood group system is an example of such **multiple alleles**. The blood group locus may be occupied by any one of three alleles, I^A, I^B, or i. I^A codes for an enzyme that attaches acetylgalactosamine (a derivative of the sugar galactose) to a protein on the surface of the red blood cells, forming antigen* A; I^B codes for an enzyme that attaches galactose instead, forming antigen B; and no enzyme is produced as a result of the activities of the i allele. Even though there are three alternative alleles in the system, only two can be found in any one person (Table 14–1). I^A and I^B exhibit codominance toward each other, but both are dominant to i.

The ABO alleles are medically important in that they determine which blood groups should not be mixed when giving blood transfusions. Although the ABO and rhesus (Rh) blood groups (see Section 33-D) are the best known and the most medically important, more than 20 loci coding for various blood proteins are known in human beings.

The genetics of ABO blood groups sometimes provides evidence for deciding questions of parentage, such as in paternity suits or in cases when it is suspected that babies have been switched in a hospital nursery. This method can never be used to decide that a particular person definitely is the father or mother of a particular child, but it will sometimes rule out a particular person as the parent. For instance, a man with blood type AB could not have fathered a child with type O blood. On the other hand, a man with an appropriate blood type is not necessarily the father of a particular baby. There are many men in the world whose blood types are such that they could have fathered the baby, and thus the baby's parentage can never be proved conclusively on blood group evidence.

Another example of multiple alleles in humans is the MHC antigens, proteins found on the surfaces of all human cells. There are four loci, with six to 19 different possible alleles apiece, coding for these proteins. This results in such a large number of possible combinations that each person (except identical twins) has what amounts to a unique "chemical fingerprint" on the plasma membranes of every cell

---

* Locus   The position on a chromosome that is occupied by an allele for a particular gene; e.g., the hemoglobin beta-chain locus might be occupied by the allele for normal or sickle beta chains.

* Antigen   A substance that is attacked by an antibody (a protein involved in the body's defense against foreign substances and disease-causing organisms) when the antigen is introduced into an individual that does not normally produce that antigen.

TABLE 14–1   **GENOTYPES AND SERUM AGGLUTININS† IN HUMAN ABO BLOOD GROUPS**

BLOOD GROUP	GENOTYPE	SERUM AGGLUTININ
A	$I^A I^A$ or $I^A i$	anti-B
B	$I^B I^B$ or $I^B i$	anti-A
AB	$I^A I^B$	none
O	ii	anti-A and anti-B

†Serum agglutinin   An antibody found in the liquid part of the blood (i.e., not inside blood cells); it causes foreign antigens (such as those on the red blood cells of a person of a different blood type) to clump together.

PARENTAL
GENERATION

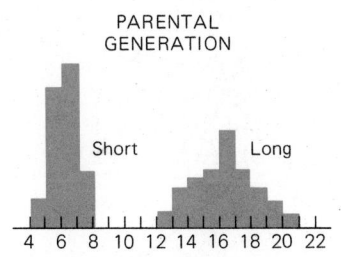

Short   Long

4 6 8 10 12 14 16 18 20 22

F₁ GENERATION

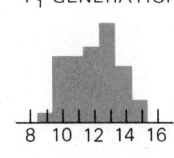

8 10 12 14 16

Figure 14-7 Ear length in corn is controlled by many genes. Crosses between short-eared and long-eared parental strains yielded F₁ plants which bore ears of intermediate lengths; the F₂ generation showed a wide spread in the lengths of ears, although few individuals were found at the extremes of ear lengths seen in the original parental strains. Horizontal axes, length of ears in cm; vertical axes, percentages of individuals with each ear length in each generation. (Redrawn, with permission, from Sturtevant and Beadle, *An Introduction to Genetics*. Philadelphia: W. B. Saunders, 1940, p. 265; paperback edition, New York: Dover Publications, Inc.)

F₂ GENERATION

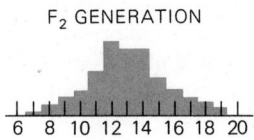

6 8 10 12 14 16 18 20

of the body. These are the proteins that trigger rejection of grafts of tissue from other individuals (Chapter 34). In 1980, for the first time, the proteins were used to determine paternity. The children in question were identical twins born to a rape victim; the twins' MHC antigens matched those of the victim and her husband and were distinctly different from those of the alleged rapist.

14-E   Multiple Loci (Polygenic Characters)

Multiple alleles govern a phenotypic trait when a single locus can be occupied by any one of a number of different alleles. **Multiple loci** govern a phenotypic trait when alleles at several different loci all influence the expression of the trait. (The multiple loci may be on the same homologous chromosome pair or on nonhomologous chromosomes.)

Skin color in humans is **polygenic**, controlled by multiple loci. There is some debate over how many different loci are involved in the production of the pigment melanin, but in any event, a very light-skinned person has recessive, non-melanin-producing alleles at all these loci, while a very dark person has all these loci occupied by alleles that code for melanin-producing enzymes. The effect of the alleles in this case is thought to be additive: the more melanin alleles present, the more melanin is produced.

Polygenic characters are difficult to study because it is hard to disentangle the effects of many genes on a single phenotypic character. Because they are so common, polygenic characters, like human height and the size of tomatoes, misled early observers into a belief in "blending inheritance." Polygenic characters frequently do appear to blend in the offspring because the offspring may show a wide range of phenotypes (Figure 14–7). A further difficulty is that it is frequently nearly impossible to disentangle environmental and genetic influences on a polygenic character. Human height, human intelligence, and the size of an ear of corn are strongly influenced by environmental factors (e.g., nutrition), as well as by the multiple loci that determine the possible range of variation in the phenotype.

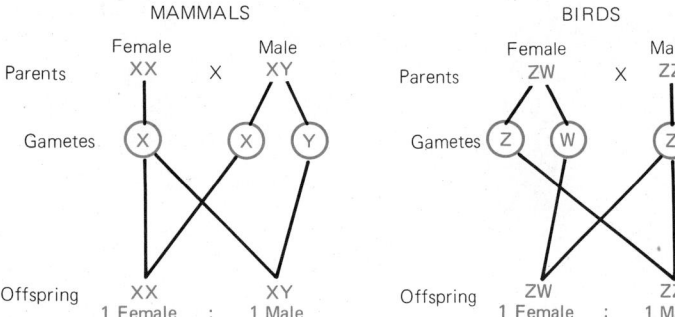

Figure 14-8 Systems of sex determination in mammals and birds. In mammals, females have a pair of X chromosomes; males have one X and one Y chromosome. In birds, males have a pair of like chromosomes, called Z to emphasize that the male/female homozygote/heterozygote system is reversed from the situation in mammals; female birds have a single Z chromosome and a W chromosome, which is sometimes so small as to require special means of detection.

14-F Sex Determination

Sex is one characteristic of an organism that can have dramatic effects upon how a gene is expressed. In most familiar organisms, sex is genetically determined, and about half the individuals are male and half female.

The simplest possible cross that will produce half males and half females (that is, a 1:1 ratio in the offspring) is one between a heterozygote and a homozygote. In mammals and birds, one sex is heterozygous and the other homozygous for a complete chromosome pair, called the **sex chromosomes** (Figure 14-8). In mammals, males are heterozygous for the pair of sex chromosomes, having one X and one Y chromosome; in birds the female is the heterozygote. The sex chromosomes in humans are illustrated in Figure 14-9. Moths, butterflies, reptiles, and some fish and amphibians are like birds in having ZZ males and ZW (or ZO, one unpaired sex chromosome) females. Many higher plants and insects are like mammals in having XX females and XY males.

Possession of particular sex chromosomes causes a zygote to develop into a male or a female via the embryonic process of **sex differentiation**. Although an obvious conclusion would be that sex chromosomes carry the genes for the characters that determine sex differentiation, this is not the case. Wherever they have been studied, the genes for production of male and female characters, such as the sex hormones, are located on chromosomes other than the sex chromosomes.

Several lines of evidence indicate that in most groups of organisms, individuals have genes for both sexes. Under normal circumstances, hormones from the ovary or testis maintain the genetically determined sex. Genetic sex-determining mechanisms act by tipping the sexual balance one way or the other. The work of Calvin Bridges with *Drosophila* showed how such a mechanism can work. Female *Drosophila* have two sex chromosomes (XX); males are XY. Bridges showed that the number of **autosomes**, those chromosomes that are not sex chromosomes, influenced sex in *Drosophila*. For instance, two sets of autosomes and one X chromosome produce a male; two sets of autosomes and two X chromosomes, a female. Bridges concluded that X chromosomes carry mainly genes that cause femaleness, whereas

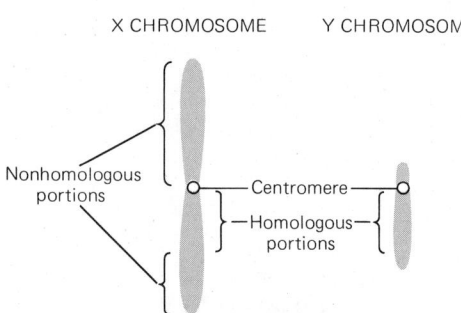

Figure 14-9 Human X and Y chromosomes have homologous portions, containing the same loci; the X chromosome also has two known nonhomologous portions, with no corresponding loci on the Y chromosome. There is tentative evidence for a few nonhomologous loci on the Y chromosome, but if they exist their position is unknown.

SEX CHROMOSOMES	PHENOTYPE†
XX	Normal female
XY	Normal male
—	—
XXX	Female; fertile or sterile
X (Turner's syndrome)	Female; sterile, ovaries rudimentary or absent
XXY (Klinefelter's syndrome)	Male; possible mental retardation
XXXY	Male
XYY	Male; tall, acne-prone; impaired fertility

† Defects in various genes involved in hormone production can alter the phenotype normally exhibited by a particular sex chromosome combination.

autosomes tend to cause maleness. The Y chromosome is of no importance in sex determination in *Drosophila*. Under normal circumstances, the extra X chromosome of a female *Drosophila* (XX) ensures that she will be female.

In humans, by contrast, the Y chromosome seems to be somehow important in determining maleness. People with a Y chromosome are phenotypically male, but people who lack a Y chromosome are phenotypically female (Table 14-2).

Many plants bear both male and female parts on the same plant; all individuals are sexually identical and no sex chromosomes are found.

Various unusual methods of sex determination have been worked out. In the honey bee and some other Hymenoptera,* females are diploid but males are haploid, developing from unfertilized eggs.

In many invertebrates, environmental factors are crucial to sex determination. In animals that cannot move far to find a mate, this may be useful. For instance, in the marine worm *Bonellia viridis*, the developing larva swimming around in the sea belongs to neither sex. Eventually it drifts to the bottom and becomes an adult. If it settles down alone, it will develop into a relatively large female, but if it lands near an existing female, it is attracted to her, and she produces a chemical that causes the larva to develop into a microscopic male. The male migrates into the female's excretory organ and lives there as a parasite.

The snail-like slipper shell, *Crepidula*, lives in stacks of individuals. Young individuals are males, which turn into females as they grow larger and older. The male reproductive tract degenerates, and a female tract develops in its place. Chemicals appear to influence sex determination in this situation as well; if a stack consists entirely of males, some of them will turn into females. Both these systems guarantee that when two or three are gathered together, some will be male and some female.

14-G Sex Linkage

Like autosomes, the sex chromosomes carry various genes, but because these genes are on the sex chromosomes, their inheritance will be related to sex. In mammals, portions of the X chromosome are homologous with part of the Y chromosome, but the X chromosome also has nonhomologous portions, which have no matching loci on the Y chromosome (Figure 14-9). Because genes located on the homologous portions of the sex chromosomes are paired, they behave like normal autosomal loci, and only detailed chromosome mapping will reveal that these loci are on the sex chromosomes.

Alleles located on the nonhomologous portion of sex chromosomes are said to be **sex-linked**. Mammals have many sex-linked characters on the X chromosome, which has large nonhomologous portions. Sex-linked characters show a distinctive pattern of inheritance that depends on the sex of the parents and offspring. In a male mammal, a single recessive allele located on the nonhomologous portion of the

---

*Hymenoptera The group of insects that includes bees, wasps, ants, etc.

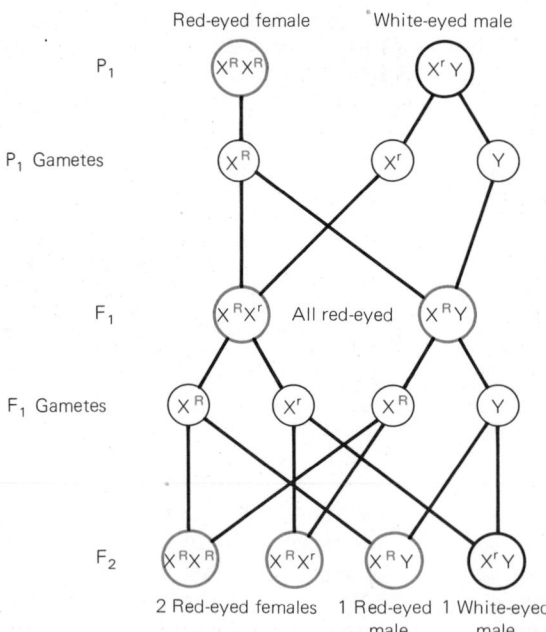

Figure 14-10 Sex-linked traits. In *Drosophila*, the alleles for red eyes (dominant) and white eyes (recessive) are carried on the X chromosome. In a cross between a homozygous red-eyed female and a white-eyed male, all the white-eyed members of the F$_2$ generation are males. The F$_1$ males do not carry the allele for white eyes.

X chromosome will be expressed in the male's phenotype because there is no homologous locus on his Y chromosome to mask it. Therefore, it is possible for a single recessive allele to express itself in the male, while a female must be homozygous recessive in order to show the recessive phenotype. Recessive sex-linked phenotypes, therefore, are more common in men than in women.

An example of a sex-linked character is red eye, which is dominant to white eye in *Drosophila*. If a homozygous red-eyed female is crossed with a white-eyed male, and the F$_1$ offspring mate with each other, the F$_2$ generation exhibits the expected 3 red:1 white ratio, but the eye color is not divided equally between the sexes; all the white-eyed offspring are male (Figure 14–10). This is the distinctive characteristic of sex-linked traits: they usually appear with far greater frequency in the sex with heterozygous sex chromosomes. White-eyed female *Drosophila* are in fact possible, and you should be able to work out how to produce them.

Hemophilia and red-green color blindness are well-known sex-linked traits in humans. In hemophiliac males, one of the factors that cause blood to clot is largely inactive; a victim may bleed to death after even a slight cut. The condition can now be controlled (but not cured) by injections of "clotting factor." Until this development, hemophiliac males nearly always died before they reached reproductive age. With the development of clotting factor, hemophiliac women, previously unknown, will begin to occur in the population. Queen Victoria was the world's most famous carrier of hemophilia. Her hemophiliac son, Leopold, Duke of Albany, and two carrier daughters, Princesses Alice and Beatrice, spread the allele fairly widely through the royal families of Europe, including Prussia, Russia, and Spain. Fortunately, no modern monarchs have inherited the allele.

A familiar example of sex linkage is a locus that influences coat color in cats. As we might expect from the great variety in color and pattern in cats, coat color is controlled by many loci. In the following discussion, assume we are working with a cat whose autosomal genes dictate a black coat (the cat is homozygous recessive for the black allele). Such a cat will not necessarily be black; this is where the sex-linked alleles come in. If the X chromosome of a male (or both X chromosomes of a female) carries the allele for orange coat, the cat produces an enzyme that diverts molecules from the melanin-producing pathway and produces a lighter-colored pigment instead. Males bearing the non-orange allele, and females homozygous for this allele, do not produce this enzyme, and so will be black. Heterozygous females are tortoiseshell (black and orange patches) because of the inactivation of one X chromosome to form a Barr body (Section 12-D) in each cell of female mammals. Barr bodies form randomly from either of the X chromosomes, so that some patches of fur express the black allele and some the orange. Tortoiseshell males are occasionally found, but they are usually sterile; they have the abnormal sex chromosome constitution XXY.

14-H Sex-influenced Genes

A **sex-influenced gene** is one whose expression depends on the sex of the individual. Both sexes may have the gene, but it will be expressed only in the presence of the necessary sex hormones. For instance, women have the genes for production of a beard, testis, or scrotum, but these genes are not expressed in the absence of appropriate concentrations of male sex hormones. Similarly, a bull may carry genes for high milk production, but he will not give milk (he would, however, be useful as a sire for a dairy herd).

In humans, one kind of baldness, called "male pattern baldness," has a genetic origin; the genes are autosomal. A man will become bald if he has only one allele for baldness; in other words, in men the baldness allele acts as a dominant, because male sex hormones somehow stimulate expression of the baldness allele. However, in women this allele acts as a recessive, owing to the influence of the female's hormones, so that a female must have two alleles for baldness before she loses her hair.

Gout is another genetic trait whose expression is influenced by sex. In gout, painful deposits of uric acid salts accumulate in the tissues, particularly in the joints of the big toe. The gene for gout is expressed much less in the presence of female than of male sex hormones.

Gout figures largely as a reason for the temper tantrums of irascible old men in Victorian literature. Avoiding red wine was supposed to alleviate the condition, but this treatment tried the tempers of many sufferers still further. Fortunately, modern medicine can completely prevent the painful uric acid deposits in people with the allele for gout.

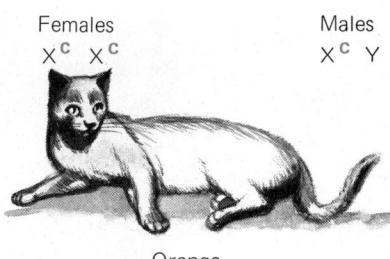

Females $X^C X^C$ Males $X^C Y$

Orange

$X^C X^{C'}$

Tortoiseshell

$X^{C'} X^{C'}$ $X^{C'} Y$

Black

Figure 14-11 Effect of the sex-linked orange allele on the expression of black coat color coded by autosomal genes in cats. In the presence of the orange allele (colored C) the fur is orange; the non-orange allele permits expression of genes for black coat color. Patches of black and orange fur in the heterozygous female reflect formation of Barr bodies.

Figure 14-12 Male pattern baldness, a sex-influenced trait whose expression is stimulated by the presence of male sex hormones.

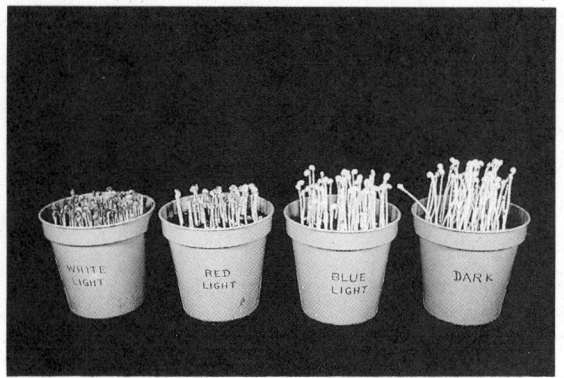

Figure 14-13 The characteristics of living things are determined by the genes (DNA) they contain, and by the environments to which they are exposed. These seedlings all contain similar genes but they look different because each pot was grown in a different environment (different light). (Bio-photo Associates)

14-I Factors That Affect Gene Expression

All of an individual's genes determine its genetic potential: what might be. What actually happens, however, is another matter. The transcription and translation of genes are influenced by the genetic environment (that is, the other genes present) either by way of the enzymes produced by these other genes, or by way of substances made by these enzymes. The influence of enzyme-produced sex hormones on gene expression, as we have just seen, is one example.

Factors impinging from the external environment also play an important role in development. In the last decade or two there have been several true "horror stories" in which drugs taken by pregnant women have caused improper development of the fetus, or cancers (loss of control of cell division), later in the baby's life. A good diet is necessary if a person is to reach the height made possible by his or her genes; in many countries, the modern generation towers above its parents as a result of improved nutrition. Exposure to light makes a person's skin darker (or redder!) and stimulates proper development of germinating seedlings.

A Himalayan rabbit is normally light-colored with black feet, ears, nose, and tail. What makes the fur different colors on different parts of the body? The answer appears to be that the extremities of the animal are cooler than its body, and low temperature causes the growth of black fur. If an area of a Himalayan rabbit's back is shaved and then cooled artificially, the fur in that area will grow black too (Figure 14–14). It is not known precisely how this occurs. Does cold influence transcription of the genes, activity of the enzymes involved in making black pigment, or incorporation of pigment into the hairs?

Figure 14-14 Expression of coat color genes in the Himalayan rabbit is related to skin temperature. Black fur grows on areas of the body with low temperature (below 33°C).

14-J Nondisjunction and Translocation

Occasionally, chromosomes behave abnormally during meiosis. **Translocation** is the attachment of all or part of one chromosome to another; **nondisjunction** is the failure of chromosomes to separate properly. In such a case, the gametes receive one chromosome (or part of a chromosome) too many or too few. A gamete with too many chromosomes may survive to produce an individual with an extra chromosome in every body cell. Such individuals may die as embryos or may be mentally deficient, but a surprising number show no apparent abnormality. A study in a maternity ward in Bradfield, England, showed that 22 of a thousand apparently normal newborn babies had chromosome abnormalities great enough to be visible under the microscope.

One of the best-known human chromosomal abnormalities is Down's syndrome, also known as mongolism. Down's syndrome is usually a result of nondisjunction of chromosome 21 or 22 (see Figure 11–2). People with this syndrome show a number of characteristics, which usually include mental retardation, a "mongoloid" eye fold, stocky build, and a sunny temperament. Down's syndrome and a number of other chromosomal abnormalities are more likely to occur among children born to older than to younger women. Since women are born with all the egg cells they ever produce, this may be a result of aging in egg cells.

Abnormalities produced by nondisjunction or translocation of the sex chromosomes are common; some of these abnormalities produce sterility or mental retardation (Table 14–2). In Table 14–2, note that the Y chromosome produces the male phenotype; people with only X chromosomes are phenotypically female.

The technique of **amniocentesis** now permits detection of abnormal chromosomes and chromosome numbers in a 16-week-old fetus. In this technique, a needle is inserted through the mother's abdominal wall and uterus into the amnion, the sac of fluid surrounding the fetus. Cells that have sloughed into the fluid from the fetus can be collected and examined for chromosomal abnormalities and for a number of metabolic disorders. However, 16 weeks is too late, in some areas, for an abortion to be performed if the fetus is found to be abnormal. New techniques may soon be available for earlier detection of fetal defects.

SUMMARY

Genes express themselves by coding for the sequence of amino acids in proteins. These may be structural proteins, blood proteins, enzymes, etc. Mutations that change the sequence of nucleotides at a genetic locus may result in new alleles that change the coding for amino acid sequences. As a result, the protein normally encoded by the locus may not be produced, or may be produced in a form that is inactive or less active than the normal protein. If the altered locus codes for a protein essential to life, the inability to produce the protein may result in death of the homozygote as a fetus or soon after birth. Changes in the production or function of less essential proteins may still have drastic effects, as exemplified by sickle cell anemia or phenylketonuria.

Many phenotypic characteristics are polygenic; for example, the skin color of humans is thought to be determined by the number of alleles coding for enzymes involved in the production of pigment at a number of different loci.

Sex is determined most simply by one sex being homozygous and the other heterozygous for an entire chromosome pair. This is found in the XX-XY and ZW-ZZ methods of sex determination common to many plants and animals. Sex-linked characters are borne on the sex chromosomes. Sex-influenced characters are the result of autosomal genes whose expression depends on the balance of sex hormones in the individual.

The expression of a gene in a phenotype depends not only upon interactions with other genes but also upon interaction with the internal and external environment. The expression of a gene may depend upon diet, light, temperature, hormones, and other environmental variables.

From your study of this chapter, you should be able to:

1. Explain how changes in the nucleotide sequence of a gene may affect the protein encoded by the gene, and how this is related to phenotypic expression of mutant alleles.

2. Given appropriate data, recognize the 1:2:1 and 2:1 ratios characteristic of lethal alleles and demonstrate knowledge of the inheritance patterns expected from parents carrying lethal alleles by working out crosses correctly.

3. State the possible genotypes of people of blood types A, B, AB, and O, and use your knowledge of these genotypes to solve problems.

4. Explain what is meant by the term multiple alleles and give or recognize examples; explain how this differs from multiple loci and give or recognize examples of multiple loci.

5. State the pattern of sex determination (sex chromosome complement of each sex) for birds, mammals, and *Drosophila*, and use this information in working out sex-linkage problems.

6. Demonstrate your knowledge of the inheritance patterns of sex-linked characteristics by answering relevant problems correctly; recognize the phenomenon of sex linkage when presented with data showing these patterns.

7. Explain the difference between sex-linked and sex-influenced characteristics, and give examples of each.

8. List at least five factors that may affect the expression of a particular gene in an organism.

9. Describe the inheritance pattern found in the human genetic disorders hemophilia, red-green color blindness, sickle cell anemia, Tay-Sachs disease, and phenylketonuria.

10. Describe nondisjunction and translocation of chromosomes.

SELF-QUIZ

1. In the homozygous condition, a recessive lethal allele in cattle produces "amputated" calves with malformations of the limbs, skull, and internal organs. These calves die soon after birth.
   a. What proportion of the normal offspring from a cross of two heterozygotes would be expected to be carriers for this trait?
   b. How could a farmer eliminate this trait from his herd if he finds that some "amputated" calves have been born to his cows?

2. Review the information on brachydactyly, Section 14-B.
   a. If two brachydactylic people marry, what are their chances of having a child with normal fingers?
   b. If a brachydactylic person marries a normal person, what phenotypic ratios can be expected among their offspring?

3. A geneticist studying the various loci that govern coat color in mice is trying to develop true-breeding strains of each possible coat color. He carries out several generations of matings among mice with yellow coats and always obtains some offspring with other colors of coats.
   a. What does this indicate about the genotype of yellow mice?
   b. The geneticist tallies up his results over several generations and finds that he has obtained a total of 184 yellow mice and 95 of other colors. What does this suggest about the nature of the yellow allele?
   c. Why did the geneticist never obtain a homozygous yellow mouse?
   d. How could he prove what became of the homozygous yellow offspring?

4. A farmer orders two dozen baby chicks. When they arrive, he notices that some of the chicks have legs noticeably shorter than normal. The farmer crosses some of these short-legged birds with normal-legged ones, and also makes crosses between the short-legged birds. The following offspring are obtained:

   normal × short: 47 short legs,
                      43 normal leg length
   short × short: 39 short legs,
                      19 normal leg length

   a. What should the farmer conclude about the short-legged genotype?
   b. What happens to the offspring homozygous for the short-leg allele?

5. Below is a pedigree of ABO blood groups for several generations of humans. Circles represent females; squares represent males. Marriages are represented by horizontal lines directly connecting two people, and children are connected to their parents by a vertical line down from the marriage line. For example, (b) and (c) are married to each other, and (d) is one of their two sons. Give the possible genotype(s) for each individual marked with a letter.

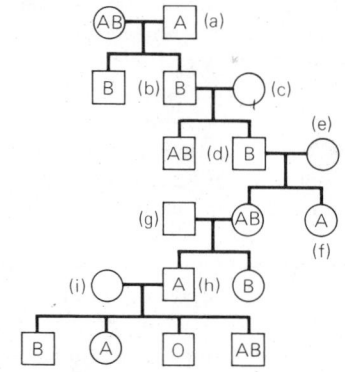

6. Ms. Smith and Ms. Jones gave birth to baby boys (named John and Tom, respectively) on the same day in a large city hospital. After Ms. Smith took her baby home, she began to suspect that it was Ms. Jones's baby, and that the hospital had somehow mixed the infants up. Blood tests revealed that Mr. Smith had blood type O, MN, and Rh$^+$; Ms. Smith had blood type B, N, Rh$^+$; and John Smith had blood type B, M, Rh$^-$. Mr. Jones had blood type A, M, Rh$^+$; Ms. Jones had blood type AB, MN, Rh$^+$; and Tom Jones had blood type O, MN, Rh$^+$. The Rh$^+$ allele is dominant to the Rh$^-$ allele; the M and N alleles show incomplete dominance. Had a mixup occurred?

7. In rabbits, normal coat color (C) is dominant to chinchilla (c^{ch}), which is dominant to Himalayan (c^h), which is dominant to albino (c). What offspring are expected from the following crosses, and in what ratios?
   a. $Cc^h \times c^{ch}c^h$
   b. $c^{ch}c \times c^hc$
   c. $c^{ch}c \times c^{ch}c$

8. In chickens a sex-linked dominant allele causes a feather pattern known as "barred." If a barred hen is mated with a non-barred rooster, what will be the feather pattern and sex of the offspring?

9. What are the expected genotypic ratios among the children of a woman whose father was a hemophiliac, and whose husband is normal?

10. Under what circumstances is it possible for both a father and his son to be hemophiliacs?

11. Red-green color blindness in humans is a sex-linked recessive trait.
    a. If about one in every 12 human males shows this trait, how common is it among human females?
    b. In a large family in which all the daughters have normal vision and all the sons are color-blind, what are the probable genotypes of the parents?
    c. If a normal-sighted woman whose father was color-blind marries a color-blind man, what is the probability that their son will be color-blind?
    d. What is the probability that the couple in (c) will have a color-blind daughter?

12. If a species of mammal has some members which carry a sex-linked lethal trait that causes early death and resorption of the embryo, what sex ratio would be expected among the offspring of a female carrier and a normal male?

13. It is often said that men inherit baldness from their maternal grandfathers via their mothers. In light of what you have learned about this trait, is this a valid statement? Explain.

QUESTIONS FOR DISCUSSION

1. Why do early lethal genes in humans cause miscarriage (expulsion of the embryo in the first 3 months), whereas in mice the defective embryo is usually resorbed?

2. Until the advent of modern technology, hemophiliac men usually died before they reached reproductive age. Nowadays they can be provided with "clotting factor," a serum extract that permits them to lead normal lives and live to have children. The treatment costs about $25,000 a year per person. Can society or should society insist that such men be sterilized so that they cannot perpetuate their disease if taxpayers pay the bill for their medication?

3. One problem with genetic counseling is that people who learn they are carriers for genetic diseases such as hemophilia, Tay-Sachs, sickle cell anemia or phenylketonuria may consider this a terrible stigma. Men have even been known to deny paternity of their children and divorce their wives for infidelity when told that the child had inherited a deleterious recessive gene from each parent. What kinds of arguments and counseling would you use, if you were a genetics counselor, in an attempt to induce a healthier and more productive response to such a discovery?

4. People who are carriers for sickle cell anemia have blood of lower oxygen capacity than most people, so they are probably exposed to greater than usual risks if they become divers, jet pilots, or mountaineers. Otherwise they have no physical handicaps but have nevertheless sometimes been denied access to various professions as a result of the common ignorance and prejudice against genetic disorders. Since this is the case, nationwide screening for sickle cell carriers which has been proposed by some black leaders and health officials may well do more social harm than good. Is it not better for carriers to remain in ignorance of genetic conditions about which nothing can be done? If not, why not?

5. In what way is the treatment of PKU effectively "creating a new disease" (Section 14-C)?

REFERENCES AND FURTHER READING

Brady, R. O. "Hereditary fat-metabolism diseases." *Scientific American*, August 1973. Recounts how researchers discovered the basis of several hereditary diseases and describes treatment attempts.

Bridges, C. B. "Sex in relation to chromosomes and genes." *American Naturalist* 59:12, 1925. The classic paper on sex determination.

Friedmann, T. "Prenatal diagnosis of genetic disease." *Scientific American*, November 1971. Thoughtful article explaining the technology of amniocentesis and pointing out the ethical problems it presents.

Robinson, R. *Genetics for Cat Breeders*, 2d ed. New York: Pergamon Press, 1977. A nontechnical book of interest to cat lovers.

Roth, E. F., Jr., *et al.* "Sickling rates of human AS red cells infected in vitro with *Plasmodium falciparum* malaria." *Science* 202:650–652, 1978.

Saxen, L., and J. Rapola. *Congenital Defects.* New York: Holt, Rinehart and Winston, 1969. A discussion of genetic and other birth defects in humans and other animals.

Sayers, Dorothy L. *Have His Carcass.* London: Harcourt, Brace Jovanovich, 1932. A mystery novel about a human genetic trait.

Stern, C. *Principles of Human Genetics*, 3rd ed. San Francisco: W. H. Freeman, 1973. An excellent human genetics text.

Sutton, H. E. *An Introduction to Human Genetics*, 2d ed. New York: Holt, Rinehart and Winston, 1975. A survey of human genetics with emphasis on genetic theory and processes.

Winchester, A. M. *Heredity, Evolution and Humankind.* St. Paul: West, 1976. Discusses genes, their evolution and expression; contains a good section on human genetics.

PART THREE

EVOLUTION

CHAPTER 15

POPULATION GENETICS

In Chapters 13 and 14 we considered how genes are distributed and expressed in individuals and in the offspring of particular matings. Now we turn to the broader picture, the genetics of populations. The difference between a population of ancient dinosaurs and a population of their descendants, modern birds, lies in the different genes of members of the two populations.

All of the genes in a population (or more strictly, all of the alleles* at all of the loci* of all individuals) are collectively known as the population's **gene pool**. According to one definition, **evolution** is the change in frequencies of alleles in a gene pool from one generation to the next. So, if we can discover how the gene pool of a population changes with time, we shall understand how evolution occurs.

For a population to evolve, it must have genetic variability: if all members of the population were genetically identical, all their offspring would also be identical and the population would not change with time. An important way of discovering

---

*Alleles The lengths of DNA that carry the information for contrasting forms of the same trait (e.g., blue eyes and brown eyes).
*Locus (plural: loci) Position on a chromosome.

why a real population *does* change with time is to construct a model of a population that *does not* change genetically from one generation to the next and then see how a real population differs from the model. We start this chapter with such a model, and then go on to consider some of the factors that cause real populations to deviate from the model and so to evolve over time.

15-A The Hardy-Weinberg Law

Just as Mendel's laws predict the genetic outcome of the mating between a pair of diploid* individuals, so the **Hardy-Weinberg Law** predicts the genetic result of random mating in an idealized population of sexually reproducing diploid organisms that is not undergoing evolution.

Consider a population whose gene pool contains two alternate alleles, *A* and *a*, either of which can occupy one particular locus. Every member of the population must have one of three possible genotypes:* *AA*, *Aa*, or *aa*. The Hardy-Weinberg Law states that the frequencies of the two alleles (their proportions relative to each other in the population's gene pool) will remain unchanged in the next generation, and through all successive generations, if the population meets all of the following conditions:

1. **No mating preferences.** The population must reproduce sexually, and mating must be random with respect to genotype (so that, for instance, an *AA* female does not prefer *aa* to *AA* or *Aa* males when she mates).

2. **No mutation.** The alleles in question must not mutate (or, if they do, the rate of mutation of *A* to *a* must equal the mutation rate from *a* to *A*).

3. **Isolation.** There must be no exchange of genes (gene flow) between the population and any other population.

4. **Large size.** The population must be very large, because the law is based on statistical probabilities. Random sampling errors are more likely to occur in small populations (see Section 15-B).

5. **No selection pressure.** There must be no natural selection with respect to the alleles in question (i.e., no genotype has a reproductive advantage over the others).

Under these conditions, *A* and *a* will remain in the population indefinitely at the same frequencies. You will find the proof of the Hardy-Weinberg Law on page 224.

The Hardy-Weinberg Law expresses the fact that sexual reproduction, with its reshuffling of genes, is not by itself enough to cause evolution. Evolution is a change in gene (that is, allele) frequencies from one generation to the next, and under the conditions of the Hardy-Weinberg Law there is no such change.

The most useful application of the law is that, since it states the conditions under which evolution will *not* occur, it also states that if these conditions are not met, evolution is likely to occur. In other words, evolution is likely to occur if mating is not random, if the population is small, if there is gene flow between populations, or if natural selection occurs.

A very practical application of the Hardy-Weinberg equation is the prediction of how many people in any generation of a human population are carriers for a particular recessive allele.* If we know the number of babies born annually with diseases such as sickle cell anemia or phenylketonuria (PKU), which are expressed only in the homozygous recessive* condition, we can estimate how many people in the parental population are heterozygotes for the gene and therefore carriers of the

---

* Diploid Having paired chromosomes in the body cells.
* Genotype Combination of alleles in an individual.
* Recessive allele Allele not expressed in the heterozygote.
* Homozygous recessive Having two recessive alleles, one at each locus for the gene.

Proof of the Hardy-Weinberg Law

Given: A population of sexually reproducing diploid organisms in which there is one gene locus that can be occupied by either of the alleles A and a. There is no mating preference nor selection pressure based on these alleles, and no net mutation of one to the other; the population is very large and is isolated from genetic exchanges with other populations.

To prove: The frequencies of alleles A and a will not change from one generation to the next in this population.

Let p = the frequency of allele A in the population (the proportion of all alleles that are A).

Let q = the frequency of allele a.

Since all loci for this gene in all members of the population must be occupied by either the A allele or the a allele, $p + q = 1$. Thus if a occurs at 20% of the loci (q = frequency of a = 0.2), the other 80% of the loci must be occupied by A (p = frequency of A = 0.8), and the two frequencies together (0.2 + 0.8) equal one.

Now let us consider what happens when individuals reproduce. The frequencies of the two alleles in the gametes produced by the population are the same as the frequencies of the alleles in the population. AA homozygotes,* of either sex, produce only A gametes; aa homozygotes produce only a gametes; Aa heterozygotes* produce equal numbers of both types of gamete. What are the chances of an A sperm's fertilizing an A egg? Since the frequency of A gametes is p, the frequency of A sperms' fertilizing A eggs is $p \times p = p^2$. Similarly, the frequency of $a \times a$ fertilizations will be q^2. The frequency of fertilizations between A and a gametes will be $(p \times q) + (q \times p)$ (since A sperm, of frequency p, will encounter a eggs, of frequency q, with a frequency of $p \times q$, and vice versa). Therefore, the expected frequencies of the three genotypes in the next generation is:

$$\begin{aligned} AA \quad &p^2 \\ Aa \quad &2pq \\ aa \quad &q^2 \end{aligned}$$

Because the frequencies of the three genotypes must add up to 1, $p^2 + 2pq + q^2 = 1$ (Figure 15–1).

Now that we know the expected frequencies of the three genotypes in the new generation, we can check the Hardy-Weinberg prediction that the allele frequencies will not have changed (the frequencies of A and a should still be p and q, respectively). The frequency of A is p^2 (from the AA homozygotes) $+\frac{1}{2}(2pq)$ (from the heterozygotes; half of their alleles are A, and half are a). Since $q = 1 - p$, from our earlier definition, the frequency of A in our new generation is $p^2 + \frac{1}{2}(2pq) = p^2 + pq = p^2 + p(1 - p) = p^2 + p - p^2 = p$. A similar calculation shows that the frequency of the a allele in this new generation must be q. In other words, as predicted by the Hardy-Weinberg Law, the frequencies of A and a have not changed.

To see how this might work in practice, we will work through a simple example. Suppose we have a population made up of 600 AA males, 400 aa males, 600 AA females, and 400 aa females. If the population meets the conditions of the Hardy-Weinberg equation, what are the proportions of the three genotypes in the next generation, and what are the new allele frequencies of A and a? Out of every 1000 gametes produced by each sex, 600 will be A gametes (since 600 of the 1000 members of each sex are AA). The allele frequency of A among the gametes of each sex (and also in the starting population as a whole) is therefore $600/1000 = 0.6 = p$. Likewise, there will be 400 a gametes. Therefore the allele frequency of $a = 400/1000 = 0.4 = q$. Note that $p + q = 0.6 + 0.4 = 1$. The frequencies of the three genotypes in the next generation can therefore be calculated as follows:

$$\begin{aligned} \text{Frequency of } AA = p^2 \quad &= 0.6 \times 0.6 = 0.36 \\ \text{Frequency of } Aa = 2pq \quad &= 2 \times 0.6 \times 0.4 = 0.48 \\ \text{Frequency of } aa = q^2 \quad &= 0.4 \times 0.4 = 0.16 \end{aligned}$$

(Note that $p^2 + 2pq + q^2 = 0.36 + 0.48 + 0.16 = 1$.)

---

p= F(A) [Frequency of A]
q= F(a) [Frequency of a]

Allele frequencies	$p + q = 1$
Squaring both sides of the equation gives	$(p + q)^2 = (1)^2$
Genotype frequencies	$p^2 + 2pq + q^2 = 1$

Why do we square $(p + q)$ to find the genotype frequencies? Since p is the proportion of A alleles in the population, it is also the proportion of A gametes produced by each sex; likewise, q is the proportion of a gametes. By arranging these on a Punnett square, we obtain the genotype frequencies:

	pA	qa
pA	p^2 AA	pq Aa
qa	pq Aa	q^2 aa

Figure 15-1 Summary of the relationships between allele frequencies (p, q) and genotype frequencies (p^2, $2pq$, q^2).

---

* Homozygotes Individuals having alleles for the same form of the gene at both loci for that gene.
* Heterozygotes Individuals with two different alleles of a gene, one at each of the loci for that gene.

If we assume that this new generation contains 1000 individuals, then the most probable numbers of the three genotypes are 360 *AA* individuals, 480 heterozygotes and 160 *aa* individuals. Note that the genotype frequencies have changed from the starting population. Has this also happened to the allele frequencies? Noting that 1000 individuals contain 2000 alleles, we can calculate the allele frequency for the new generation as follows:

$$\text{Frequency of allele } A = \frac{(2 \times 360) + 480}{2000} = 1200/2000 = 0.6$$

$$\text{Frequency of allele } a = \frac{(2 \times 160) + 480}{2000} = 800/2000 = 0.4$$

The allele frequencies have therefore not changed. Nor would they change if this process were repeated for an infinite number of successive generations. How about the genotype frequencies? These changed from our initial population to the next generation, but the genotype frequencies given by $p^2 + 2pq + q^2$ are in equilibrium with each other, and they will not change over endless successive generations. This can be checked easily by calculating the most probable allele frequencies among the gametes from the second generation. For A, this is (out of 2000 gametes for the whole population) $2 \times 360 + 480 = 1200$. Therefore $p = 0.6$ and, by similar calculation, $q = 0.4$ (Figure 15-2). These, of course, are identical to the allele frequencies of gametes in our original population, and so will give rise to the same proportions of genotypes for generation after generation. If you were to change the allele frequencies in some way, the new allele frequencies would be reproduced indefinitely; after initial adjustment to the proportions determined by the Hardy-Weinberg equilibrium ($p^2 + 2pq + q^2$), the genotype frequencies would also be reproduced indefinitely.

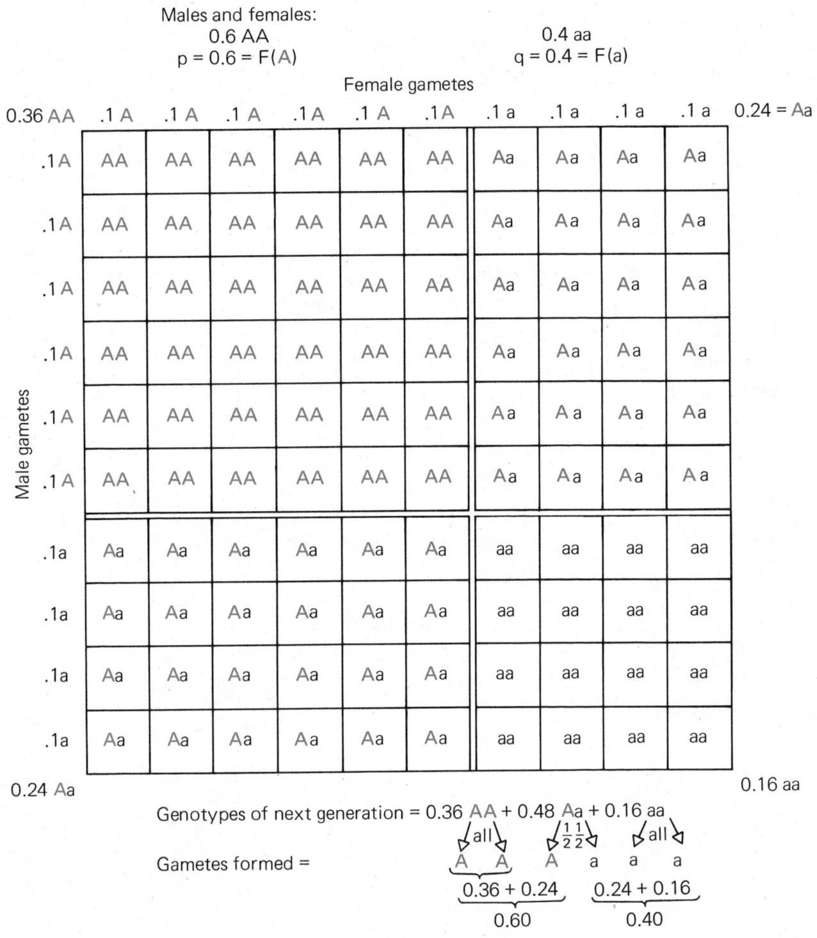

Figure 15-2 Maintenance of the Hardy-Weinberg equilibrium. Beginning with allele frequencies of 0.6 A and 0.4 a in both sexes (top), and in the gametes (sides of Punnett square), the next generation consists of $0.36\,AA\,(=p^2) + 0.48\,Aa\,(=2pq) + 0.16\,aa\,(=q^2)$. The allele frequencies have remained 0.6 A and 0.4 a, and the Punnett square to produce the next generation will look identical to the one shown here.

225

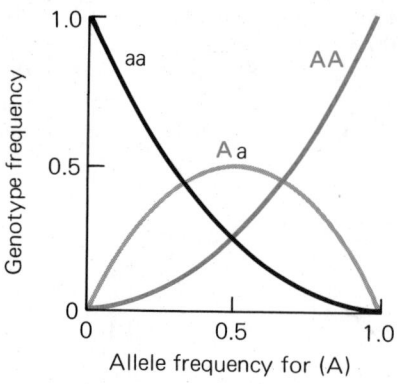

Figure 15-3 The relationship between genotype frequency and allele frequency in a diploid population. The proportion of homozygotes (*AA*, *aa*) for an allele in the population equals the square of the frequency of that allele.

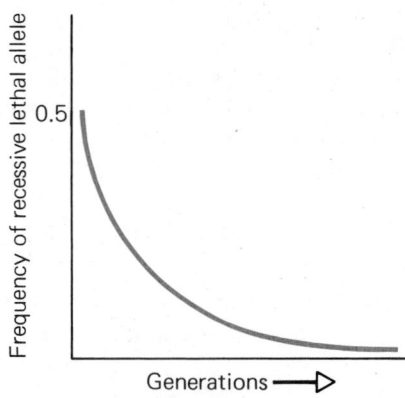

Figure 15-4 The frequency of a recessive lethal allele in a population of diploid organisms. Although the allele will be eliminated by selection wherever it occurs in the homozygous condition, heterozygous carriers of the allele will persist in the population indefinitely.

disease. This value should be recalculated in each generation, since there is selective pressure against these particular homozygous recessive genotypes, and hence the alleles do not meet Hardy-Weinberg conditions. (Recently it has become possible to remove the selective pressure against PKU, and thus the frequency of this trait is likely to stop decreasing in the population.)

The Hardy-Weinberg Law and Recessive Alleles

The Hardy-Weinberg Law has an important implication for lethal recessive alleles* in a population. A lethal dominant allele* will be removed from the population immediately every time it arises by mutation, since it will be expressed in the phenotype* of every individual that carries it, and cause its carrier to die. The allele will be maintained only by recurrent mutation and will therefore be extremely rare. This is not the case for a lethal recessive allele. The lethal phenotype will be expressed only when it occurs as the homozygous recessive; these individuals will always die. But the Hardy-Weinberg equation indicates that many more heterozygous than homozygous recessive individuals will be formed in each generation. The lower the frequency of the allele in the population, the lower the frequency of the recessive homozygote relative to the heterozygote (Figure 15–3). Thus, however rare the homozygous recessive becomes, there are still many heterozygous carriers in the population (Figure 15–4).

15-B Causes of Evolution

The case of the lethal recessive allele illustrates one factor that causes gene frequencies to deviate from the Hardy-Weinberg equilibrium: the selective disadvantage of an allele.

Before considering selection in more detail, we will discuss other factors that produce evolution by causing a change in gene frequency from one generation to the next. These are all the conditions that must *not* be present if a population is to obey the Hardy-Weinberg Law.

---

* Lethal allele Allele whose expression causes death of the individual that carries it.
* Dominant allele Allele expressed in the heterozygote.
* Phenotype The sum total of expression of an organism's genes as influenced by the presence of other genes and by the environment. Not all the genes in the genotype are expressed in the phenotype; presence of genes that are not expressed cannot be detected without special biochemical techniques.

Mutation

A **mutation** is a random inheritable change in the genetic material. Mutation is a recurrent event because a given length of DNA can undergo only a limited number of chemical changes; given enough time, each of these will appear again and again. Thus, in a hypothetical population in which all individuals are homozygous AA, the allele a will eventually be reintroduced by mutation.

Mutation is the ultimate source of genetic variation in a population, but it contributes little to the variation in any one population because it is such a rare event. Any given mutation appears once in between 10^4 and 10^9 individuals. The enormous range in mutation rates is a result of the fact that different loci within a gene pool mutate at different rates.

Most new mutations that arise in a population are disadvantageous and will be selected against. The frequency of such mutant alleles will be maintained by an equilibrium between net mutation rate (e.g., the rate at which A mutates to a minus the rate at which a mutates back to A) and selection.

Gene Flow

A lost allele may be reintroduced into the population by mutation, but it is much more likely to be reintroduced by immigration of individuals (or pollen or sperm) carrying the allele from a neighboring population, resulting in **gene flow** between the two populations (Figure 15-5).

A population such as a group of *Drosophila* in a bottle in the laboratory has a completely closed gene pool (barring escapes), but the gene pools of most populations can exchange genes with the gene pools of nearby populations. Animals may migrate from one area to another, joining and contributing their genes to another gene pool; a tornado may disperse seeds far beyond the bounds of the local population to which the parent plants belonged. Gene flow between populations is therefore the rule and is generally second only to selection as a cause of evolution in local populations.

Comparison of neighboring populations of a single species usually shows that they are not identical. Furthermore, the greater the distance between the populations, the greater the differences between them. Maple trees in Vermont differ from maple trees in Michigan more than they differ from those in New Hampshire. This phenomenon, in which a single character shows a regular gradient of variation across a geographical area, is called a **cline** (Figure 15-6). For instance, song sparrows on the West Coast of North America show a cline of increasing size and decreasing contrast in markings from California to Alaska.

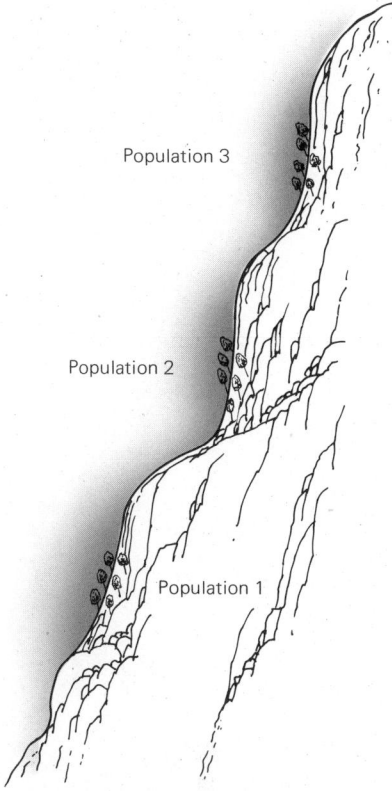

Figure 15-5 Gene flow between populations at different heights on a mountain. The intensity of color indicates the density of pollen grains at particular locations. Some pollen grains will blow away and carry their genes from one population to the next, but more pollen grains will fertilize other flowers within their own populations.

Figure 15-6 A cline. *Achillea* plants collected from different populations in California and Nevada and then grown under uniform conditions reveal clear differences, which must be genetic.

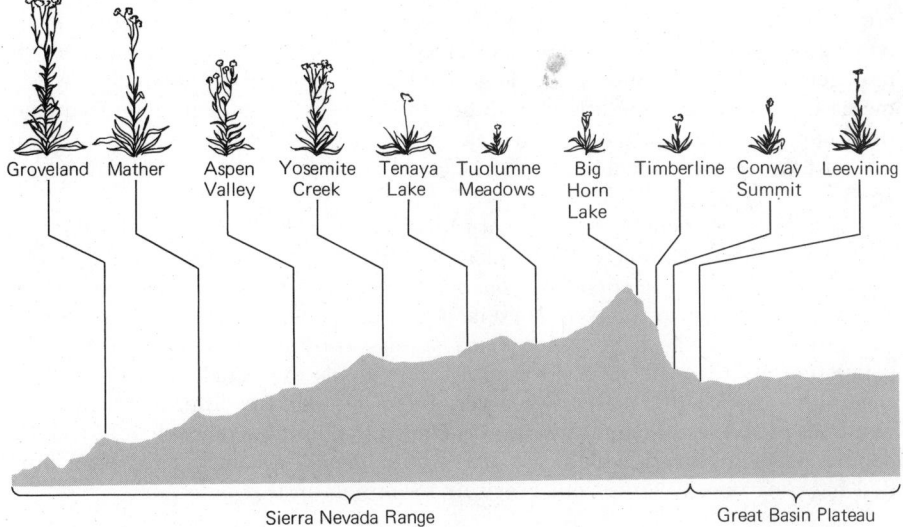

Groveland Mather Aspen Valley Yosemite Creek Tenaya Lake Tuolumne Meadows Big Horn Lake Timberline Conway Summit Leevining

Sierra Nevada Range   Great Basin Plateau

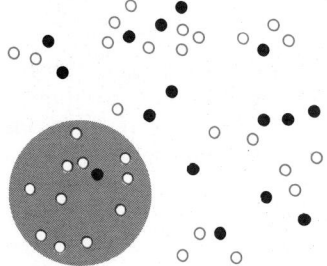

Figure 15-7 Genetic drift in a hypothetical population. The dots represent individuals in a population. Colored individuals carry a rare allele. Suppose that in any one year only 25% of the population breeds—those in the colored circle. By chance only one of the individuals carrying the allele breeds this year, and so the allele is much rarer in the next generation.

There are two main reasons why characters within a species may show clinal variation of this sort. First, if gene flow occurs between adjacent populations, the gene pools of populations that are close to one another will share more alleles than the gene pools of populations that are widely separated. Second, environmental features like climate often vary along gradients. Because these environmental factors act as selective pressures, the phenotypic characters that are best adapted to such pressures will also vary in a gradient.

Gene flow between populations tends to increase the similarity between all the populations of a species. Natural selection has the opposite effect; it tends to make every population uniquely specialized for its particular habitat. Clines are one possible outcome of these two conflicting forces.

Genetic Drift

The Hardy-Weinberg Law holds true only for large populations because the law depends on probability, and the laws of probability apply more accurately to a large sample than to a small one. **Genetic drift** is a change in the proportions of alleles in the gene pool of a population caused by random events.

There is a certain statistical randomness in the passing of alleles from one generation to the next in any population. Sampling of alleles for the next generation can be compared with sampling marbles from a bag. If 900 blue marbles and 100 red marbles (that is, 90% blue and 10% red) are put into a bag, and ten marbles are then withdrawn from the bag, there is a 35% ($0.9^{10} = 0.349$) chance that there will be no red marbles among the ten. By analogy, since organisms form many more gametes than those actually used in fertilization, there is always a chance that an allele will be eliminated from the next generation merely by random sampling error. The chance of losing an allele decreases as population size increases, since large samples are more likely than small samples to approximate the true frequencies of the various characteristics in the entire population. (The more marbles we draw from the bag, the better our chances of getting at least a few red ones.)

Consider a population of five individuals, in which only two breed. The chances are quite good that any particular allele is represented in only one member of the population. If this individual does not breed, the allele will not be present in the next generation; if the individual does breed, the frequency of the allele may well increase in the next generation. In either case, since a change in gene frequency from one generation to the next has taken place, evolution has occurred. Sewall Wright first recognized that genetic drift could cause evolution in this way (Figure 15–7).

The extent of genetic drift in natural populations has long been a subject of controversy. Drift is very hard to measure. Its effect depends largely upon the strength of natural selection, and on population size. Natural selection opposes drift, because it causes nonrandom changes in the gene pool of a population from one generation to the next. Since many species are known in which a breeding population may consist of fewer than 50 individuals, one might be left with the general impression that drift is often very important. On the other hand, there are hundreds of species in which selection is known to eliminate more than 95% of the population each year, in which case drift would have little effect. The actual situation has never been worked out precisely for any natural population.

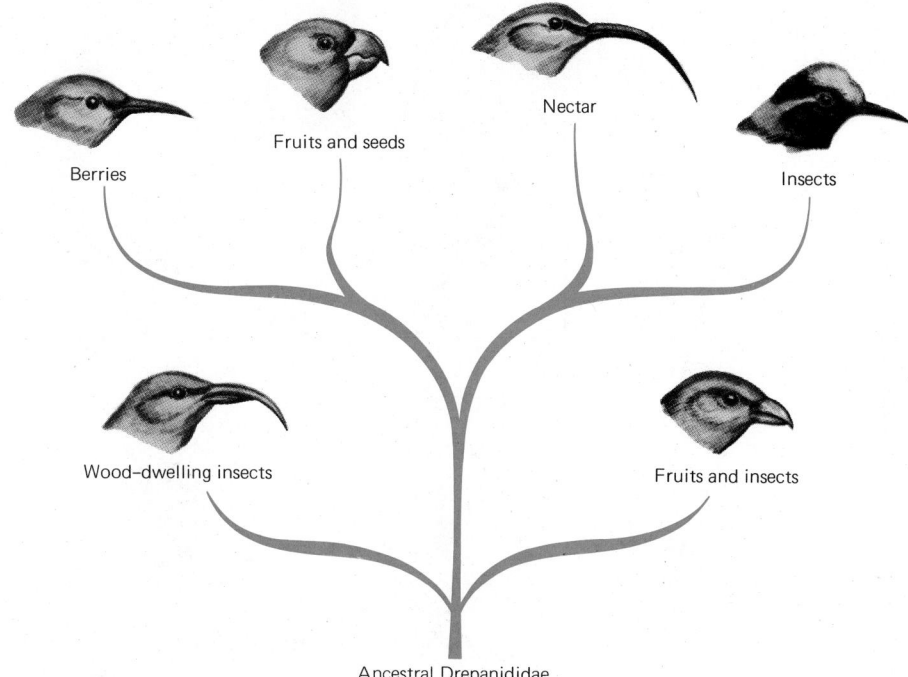

Berries

Fruits and seeds

Nectar

Insects

Wood–dwelling insects

Fruits and insects

Ancestral Drepanididae

Figure 15-8 The probable family tree of Hawaiian honeycreepers. It is likely that small ancestral populations became isolated on different islands and evolved into different species. Each population must have started with a different assortment of genes (founder effect), which were subjected to different selective pressures depending upon the available food; differences in beak size and shape resulted.

One profound effect of genetic drift on evolution is exemplified by the **founder effect**, the change in allele frequencies produced by random sampling when a new population arises from one or a few colonizing individuals. When a few individuals leave a population, the chances are good that they will not contain all the genes that were present in the parent population. If these few individuals become the founders of a new population elsewhere, the gene pool of the daughter population will be different from that of the parental population. Since any new population is in a different environment and will experience different selective pressures from those exerted on the parent population, it is, in practice, usually impossible to tell how much of the genetic difference between the two populations is a result of the founder effect and how much is a result of different selective pressures in the two environments. The founder effect will have a great influence on a population of, say, plants that populate an island from one original seed, or animals such as domestic hamsters, which have all descended from one original pregnant wild female.

Gene fixation is the establishment of one allele in a population as the result of elimination of its alternate allele(s). The locus at which the allele occurs will be homozygous in all individuals after fixation (barring mutation). Fixation may result from genetic drift or from selection against the eliminated allele.

Natural Selection

Natural selection is the nonrandom differential reproduction of genotypes from one generation to the next. Every population consists of a range of phenotypes. When a selective force, such as predation or competition, is at work, some of these phenotypes are more likely to survive and reproduce than others. If the phenotypic characteristics selected for are at least partly under genetic control—as they usually are—then the genes responsible for the favored phenotypes will be represented at higher frequencies in the next generation. In the simplest case of a single pair of alleles, if possession of one allele confers even a slight, but consistent, reproductive

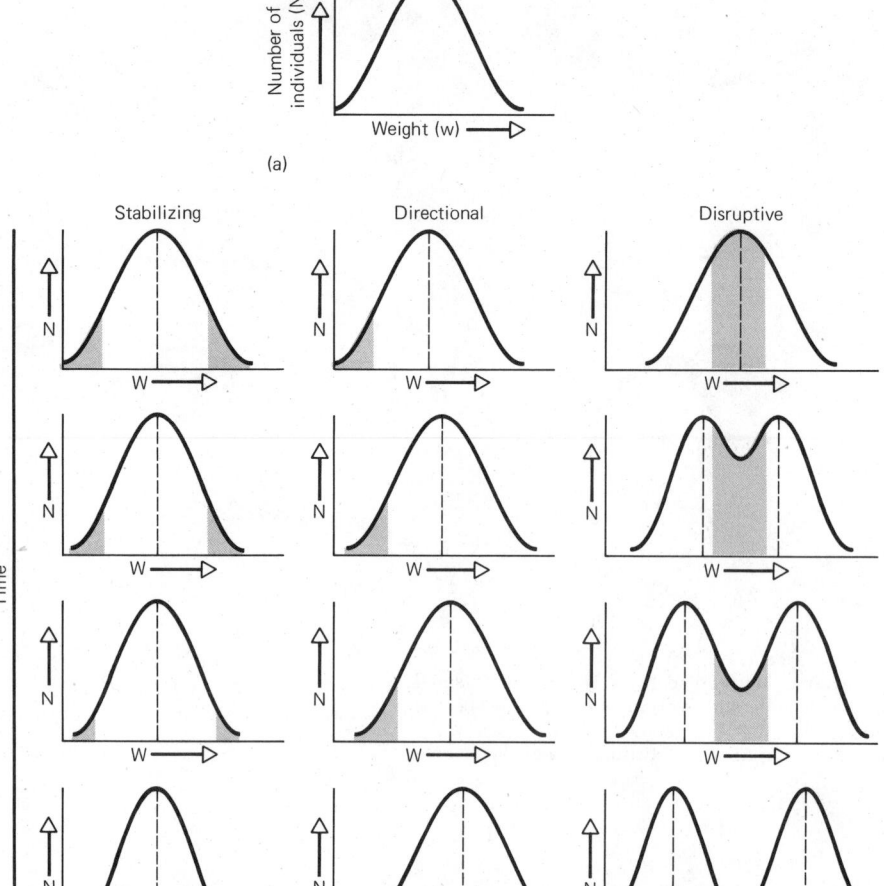

Figure 15-9 Effect of different types of selection on a polygenic character, such as weight, in a population. (a) shows the population before selection, with a normal distribution of individuals of different weights. In (b) the individuals eliminated by selection in each generation are shaded in color. Stabilizing selection eliminates very heavy or very light individuals, leading to a population with less weight variation than in (a). Directional selection in this case eliminates light individuals, leading to a population in which the median* weight (dotted lines) is higher than in (a). Disruptive selection (rarer) eliminates individuals of median weight, producing two distinct populations with different median weights.

advantage, its frequency in the population will steadily increase from one generation to the next, at the expense of the less favored alternate allele. It is said to have a greater **fitness** than the less favored allele. In this way, selection changes the gene frequencies of a population, shifting them away from Hardy-Weinberg equilibrium. Natural selection is by far the most important and potent evolutionary force, and its action has been emphasized throughout this book.

The two most common forms of natural selection are **stabilizing selection**, in which selection tends to favor average phenotypes over extremes in either direction, and **directional selection**, in which the selective force favors the phenotypes at one extreme over the other. The action of these kinds of selection on a normal distribution of phenotypic variation, such as might arise from a character (like human height) under polygenic control (control by many genes), is shown in Figure 15–9. In a population of seeds, if seeds of average size have a better chance of germinating and of growing than seeds that are unusually large or small, and if seed size is a heritable characteristic, the next generation will contain a lower proportion of unusually large or small seeds as a result of stabilizing selection. On the other hand, if birds or other predators tend to eat the larger seeds, but ignore the smaller ones, then they will exert directional selection in favor of smaller seeds.

Disruptive selection takes place when the extremes of a phenotypic range are favored relative to intermediate phenotypes (Figure 15–9). It might happen to our seeds, for example, if a particular kind of beetle specialized in feeding only on seeds of intermediate size, ignoring the very small and very large seeds.

---

*Median   The value that falls between the lowest and the highest 50% of individual measurements.

Mating that is nonrandom with respect to genotype can also bring about evolutionary change. If females consistently choose to mate with males of genotype *AA*, for example, they exert directional selection in favor of the *AA* genotype; if they favor both homozygotes (*AA* and *aa*) relative to heterozygotes (*Aa*), they exert disruptive selection, and if they favor heterozygotes relative to either homozygote, they act as agents of stabilizing selection. Examples are known from studies on *Drosophila* where females prefer to mate with males heterozygous for certain genes to mating with either homozygote. Nonrandom mating resulting from such **sexual selection** can have some bizarre results, such as the tails of peacocks. However, sexual selection is really only a form of natural selection because it gives one genotype a reproductive advantage over another.

15-C Coadapted Gene Complexes

Although for simplicity's sake we often speak of natural selection and the genetics of populations in terms of a pair of alleles at a single locus, genes rarely exert their effects in such a simple manner. Each gene may influence many phenotypic characters, and most phenotypic characters are influenced by many genes. Population geneticists have found dramatic examples of interactions that can occur between the genes present together in the same gene pool and in the same genome (the sum of all the genes of one individual).

If individuals from opposite ends of a cline are mated (something that would not normally occur in nature), it often becomes clear that they are very different from one another genetically, even though their phenotypes may appear quite similar. Sometimes, in fact, the genetic differences are so great that two such individuals cannot breed successfully.

In one such experiment, members of different populations of the swallowtail butterfly *Papilio dardanus* in Africa were mated. In this butterfly, members of different populations, and even of the two sexes, may look very different from one another, since the females mimic a variety of other butterflies (see Chapter 51). When members of adjacent populations mate, each offspring looks like one of the parents, as we would expect on genetic grounds. When individuals from two geographically distant populations are mated, however, the offspring look like nothing seen on earth before—displaying a mishmash of the parental characteristics and of color patterns that neither parent possesses (Figure 15–10).

Figure 15-10 The parents are *Papilio dardanus hippocoonides*. All their offspring inherit the *hippocoonides* gene and would normally resemble the parents. However these parents are from populations that are hundreds of miles apart, and their offspring contain a mixture of modifier genes that have not evolved together. Although all the offspring receive the *hippocoonides* gene, it is completely dominant only in offspring (c). (Note that inheritance of the "tail" is also peculiar in the offspring.)

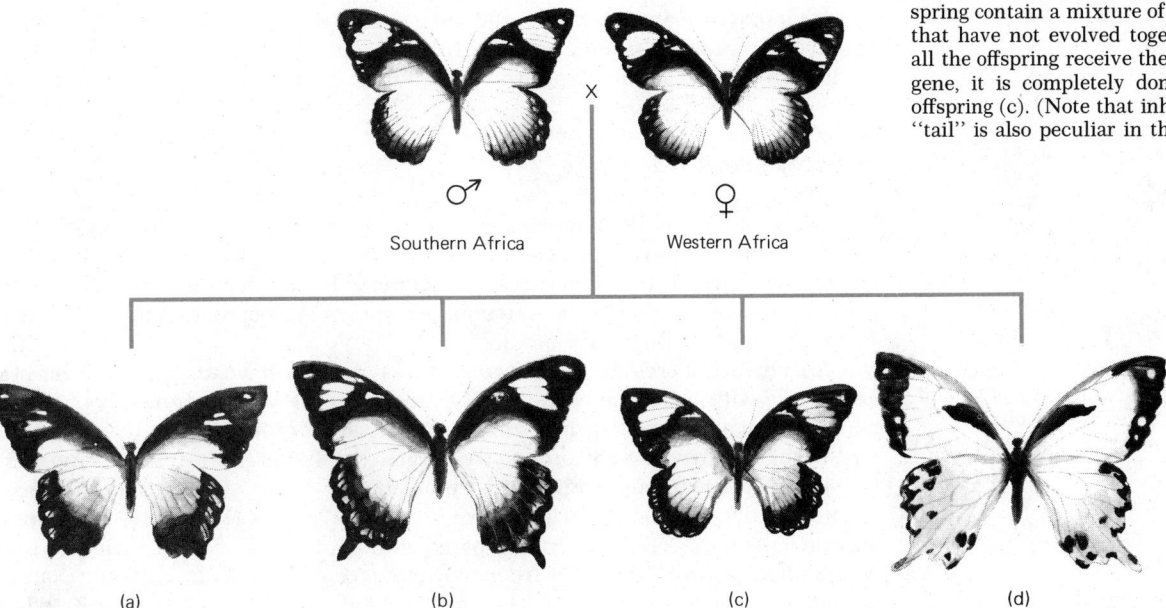

Southern Africa ♂ X ♀ Western Africa

(a)   (b)   (c)   (d)

This example makes the point that when various genes occur together over long periods of time in members of the same or nearby populations connected by gene flow, they become adapted to one another in various ways—that is, a group of genes becomes a **coadapted gene complex**. When genes that are members of the same coadapted gene complex come together in an individual, they interact to produce offspring that are well-adapted to their environment. Genes from distant areas are not adapted to each other in this way; when they come together in the offspring of a mating, they may produce individuals that are phenotypically bizarre. This observation raises the question: what sorts of genetic mechanisms permit genes to interact more harmoniously with one another when they have been subject to evolution within the same gene pool than when they have not?

Evolution of Dominance and Recessiveness

Genes may become adapted to one another by changes in the degree of dominance of one allele over another. We tend to think of dominance as an unchanging characteristic of an allele. Sometimes this is clearly the case; a mutant that results in a cell's producing no protein must be recessive to its normal allele, which causes production of the protein. Several types of observations, however, show that dominance need not be a property of the allele itself. For example, there is an allele that produces **melanism** (dark color) rather than the normal light gray color in the peppered moth in Britain (see Section 16-C). This allele once showed incomplete dominance; in moth collections from the nineteenth century, moths that were presumably heterozygous for the melanism allele are a lighter shade than presumed homozygotes. Today, however, heterozygotes are indistinguishable from homozygotes. In other words, the dominance of the allele for melanism over the nonmelanic allele has increased. Similarly, artificial selection in laboratory populations of *Drosophila* has shown that the degree of dominance or recessiveness of an allele can be modified by selective breeding.

If the degree of dominance can sometimes be altered by selection, what causes dominance? Experiments on the lesser yellow-underwing moth, *Triphaena comes*, provide one answer. A melanic allele is dominant in populations of this moth on islands in Scotland. However, E. B. Ford showed that when moths were crossed with members of populations from different islands, the melanic allele was not dominant in the offspring. In other words, the allele is dominant only in the complex of genes in which it normally occurs, where certain other genes, called its **modifiers**, are also present. If it is advantageous for an allele to become more dominant or less dominant, selection will ensure that the modifiers that affect its dominance will be inherited with it. Thus we would expect selection to favor mutations that move the modifiers onto the same chromosome or part of a chromosome as the modified gene, producing genetic **linkage**. This will reduce the chances of the modifiers' being separated from the gene during gamete formation.

Heterozygote Advantage

The evolution of dominance and recessiveness can lead to **heterozygote advantage**, the selective advantage of heterozygotes over homozygotes at a given locus. The prevalence of heterozygote advantage is suggested by the fact that the heterozygote for an allele is very often more common in a population than the Hardy-Weinberg equilibrium would predict.

A **hybrid** is a cross between two genetic types that have evolved, or been bred artificially, with little or no gene flow between them. **Hybrid vigor** is a phenomenon in which a hybrid offspring is more fit than either of its parents; hybrid vigor is a result of heterozygote advantage. If two genetically different, inbred (and therefore largely homozygous) strains of corn or mice or dogs are crossed, the offspring are nearly always superior in many ways to either parent (Figure 15–11). This is why mongrel dogs are so healthy compared with "pure-bred" dogs of various breeds, which often suffer from a high frequency of congenital ailments. Another example of heterozygote advantage is the case of sickle cell anemia (see Section 14-B), where

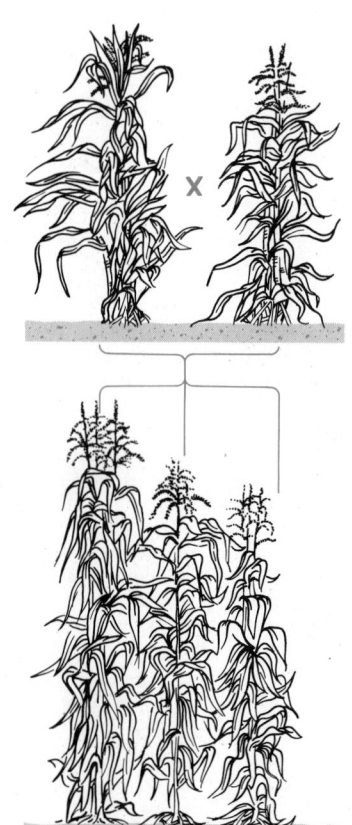

Figure 15-11 Heterozygote superiority in corn. The parents are members of two different inbred strains. Some of their offspring are shown beneath them—taller and more productive than their parents.

the heterozygote is at a selective advantage to either homozygote in areas with a high incidence of malaria.

Heterozygote advantage can arise in several possible ways. First, it may result from the fact that dominance and recessiveness can be altered by selection, and that the products of most genes have more than one effect on the phenotype. Dominance and recessiveness are concepts that apply only to the heterozygote. You cannot tell whether gene B is dominant or recessive by looking at a BB individual, in which all the effects of that gene will be expressed. However, in a heterozygote, which contains two different alleles, there will be strong selection for the beneficial effects of each allele to become dominant and the deleterious effects to become recessive, so that only the advantageous effects of each allele are expressed in the phenotype. After this selective pressure has existed for some time, the heterozygote will be at an advantage to either homozygote because its phenotype will express all the beneficial and none of the detrimental effects of both alleles. In an investigation of the multiple effects of alleles responsible for red or brown testis color in the moth *Ephestia*, it was found that the advantageous effects of both alleles were dominant or nearly dominant, while their disadvantageous effects were recessive.

Second, heterozygote advantage will occur where the presence of two different alleles is favored over the presence of only one of them, for example, where the two alleles control production of different proteins and having both is better than having only one of them. Most organisms have **isozymes**, enzymes that catalyze the same reaction but have different properties, such as different affinities for their substrates and different rates of catalyzing the reaction; different isozymes may work better at different temperatures or substrate concentrations. If isozymes are controlled, as they often are, by different alleles at the same locus, a heterozygote will have two isozymes for a particular reaction, while a homozygote will have only one. Often each isozyme is more useful than the other at a specific stage in the life history or part of the body or at a particular season of the year. Thus a heterozygote is at an advantage over a wider span of conditions than either homozygote.

The importance of heterozygosity in maintaining genetic variation has selected for many adaptations that prevent inbreeding in species whose members live in small breeding populations. Almost all young lions (and some lionesses), for example, leave their pride (social group) of origin, wander, and eventually join another pride. Incest taboos are virtually universal in human societies.

Heterozygote advantage is one phenomenon that maintains genetic variability in a population in spite of natural selection. Where the heterozygote is at an advantage over either homozygote, neither allele will be eliminated by selection, and variation at the locus will continue.

15-D Genetic Load

Because heterozygote advantage maintains recessive genes in the population, it has the disadvantage of producing **genetic load** in a population. Genetic load is the cost to a population of having inferior or lethal homozygotes segregating out in each generation: if there is heterozygote superiority at a great many loci that segregate independently, almost every individual will be an inferior homozygote for some phenotypic character. Since selection does not act on the population as a whole, there can be no selection against the genetic load borne by the population.

Genetic load may be considerably lower than some calculations suggest because alleles have evolved together as part of coadapted gene complexes in such a way that the deleterious effects of each gene are minimized in the resulting phenotypes. Nevertheless, we know of enough dramatic examples of genetic load, such as the segregation of recessive lethals at the sickle cell locus, to suspect that less extreme examples are widespread.

15-E Measurement of Genetic Variation

There is considerable controversy as to what determines the degree of genetic variation found in natural populations. Forces that tend to increase and to decrease

TABLE 15-1 FACTORS THAT INCREASE AND DECREASE GENETIC VARIATION IN A NATURAL POPULATION

FACTOR	EFFECT
Increasing Variation	
Mutation	Introduces variation.
Sexual reproduction	Genetic recombination occurs at gamete formation and at fertilization. New variations may occur in next generation. (However, it does not affect gene frequencies.)
Disruptive selection	Favors phenotypic extremes relative to intermediate phenotypes.
Polymorphism	Retains more than one genetic form of a character in the population.
Gene flow (immigration)	May introduce new genes or genetic combinations.
Population increase	Occurs because selective pressures relaxed. Means more variation in breeding population.
Decreasing Variation	
Stabilizing selection and directional selection	Only individuals with specific genotypes make major contributions to the gene pool of the next generation.
Genetic drift	May randomly change the proportions of genes in the gene pool. Some alleles may fluctuate to extinction and others to fixation.
Emigration	May remove genes or genetic combinations.
Population decrease	Usually associated with increase of selective pressures. Means less variation in breeding population. (Also, loss of variation by genetic drift is more likely in small populations.)

variation act on all populations (Table 15-1), and the amount of variation found at any one time is presumably a result of all of these factors.

Early studies measured variation of readily observable phenotypic characters. For instance, E. B. Ford studied variation in the number of spots on the wings of butterflies and moths, establishing that this character was genetically determined. Since genes often have multiple effects, the appearance of the organism may not be the most important effect of the genes under study. For instance, larval mortality from viruses may differ strikingly between butterflies that have different numbers of wing spots as adults. In this case, viral infection of the larvae rather than, say, predation due to visibility of adult wing pattern may well be the selective pressure determining the frequency of these genes in the population.

Electrophoretic separation* and analysis of proteins is a more recent method of measuring genetic variation. Since proteins are a closer reflection of an organism's genes than are its gross morphological (structural or anatomical) characters, such analysis may give a better estimate of genetic variation. Like the study of morphological characters, electrophoretic analysis of proteins tends to underestimate variation. For instance, enzymes with different isoelectric points* may behave identically at a particular pH; the variability in proteins made up of numerous polypeptides is almost always underestimated, and insoluble structural proteins are usually not analyzed at all. Thus, the figures in the next section, on the extent of variation in natural populations, are almost certainly low estimates.

15-F Variation in Natural Populations

When we study Mendelian genetics, we consider only **polymorphic loci**, loci that can be occupied by two or more alternate alleles. These loci may occur in either the heterozygous or the homozygous condition. In view of the many advantages that

---

*Electrophoretic separation A technique that separates substances by using the fact that they move at different rates (depending on their size and electric charge) when subjected to an electric current.

*Isoelectric point The pH at which the charged groups on a molecule sum to electrical neutrality; charged groups combine with or dissociate from H^+ or OH^- groups in the solution depending on the pH.

Figure 15-12   A British robin. (Biophoto Associates)

accrue to heterozygous individuals, it is surprising to find that population geneticists estimate that 50 to 80% of loci in typical individuals are homozygous in natural populations, and that only 20 to 50% are heterozygous. One reason for high levels of homozygosity might be that certain alleles, formerly present, have been lost by selection or genetic drift at some point in a population's evolutionary past and have not yet been replaced by mutation or gene flow. Another, of course, is that one allele may be consistently superior to any alternatives that have arisen by mutation.

Genetic Polymorphism

Genetic polymorphism is the occurrence in the same place, at the same time, of two or more genetic variants (alleles or larger segments of chromosomes) in such proportions that the rarest of them cannot be maintained by recurrent mutation alone. **Transient polymorphism** occurs while one allele, selectively advantageous compared with its alternate alleles, steadily replaces these other alleles, eventually reaching fixation. More common is **balanced polymorphism**, in which alternate alleles (which may be groups of alleles at linked loci, chromosome segments, or entire chromosomes) are maintained in a population indefinitely. Disruptive selection, the opposing effects of selection and gene flow, and heterozygote advantage can all lead to balanced polymorphism. So can regular variations in the environment such that one genotype is favored at one time and another at another. For instance, in Britain, some robins migrate south every winter, and some stay home. If the winter in Britain is severe, more of the robins that migrate will survive. If the winter is mild, the survival rate is higher among robins that stay at home; they do not face the perilous and exhausting journey of migration. It is obvious that the selection pressures for migrating or staying at home are not constant from year to year; thus both behavior patterns are maintained in the robin population.

Most individuals are either male or female, and both **morphs**, or forms, are almost always present in the same population at the same time. **Sex** is a polymorphism according to the definition because the frequency of the rarer sex is so high, in most organisms, that it cannot be maintained merely by mutation.

The human ABO blood group seems to be another case of balanced polymorphism, in which each allele at the multi-allelic ABO locus is associated with different susceptibilities to diseases. For instance, it seems that people with blood group A are more susceptible to stomach cancer, pernicious anemia,* and diabetes than are people with other blood groups; people of blood group O are more liable to duodenal and gastric ulcers,* smallpox, and pituitary adenoma.* Opposing selective forces are probably responsible for maintaining the different alleles in the population. It is not clear whether heterozygote advantage or disruptive selection is also involved.

Different genetic forms of the sugar maple are also maintained in the population by opposing selective forces. Many more seedlings of the "southern" than of the "northern" form survived the severe drought of 1954 in Ohio, but the "northern" form survives cold winters better, so both types are maintained in the Ohio population.

Frequency-dependent selection, in which phenotypes have different selective advantages depending on their frequency in a population, can also lead to a balanced polymorphism. Many kinds of generalized predators are known to form "search images" for particular prey species or particular phenotypes of a species; while hunting with a particular search image, they will tend to concentrate on one phenotype to a greater extent than that expected from its frequency in the population.

Birds that fed on the peppered moth (Section 15-D) showed this behavior. When both melanic and normal morphs of the moth were observed in places where both forms were more or less equally visible, individual birds were found to feed only on one morph until it became quite rare relative to the other. The birds would then switch their search image and start to feed only on the other morph, ignoring the phenotype that they had eaten more frequently at first. Such frequency-dependent selection tends to favor the rarer over the more common phenotype at any one time, and can thus favor a balanced polymorphism.

In a study in Uganda, snails with cryptic (camouflaging) coloration were eaten by predators more often than conspicuously patterned morphs of the same species, apparently because predators formed search images for the cryptic morph, which was much more abundant. When a phenotype is rare, it may be at an advantage if it is very different from the more common morphs.

A Butterfly Population Explosion

When a natural population grows rapidly in size, it invariably becomes more genetically diverse. E. B. Ford studied the marsh fritillary butterfly, *Euphydryas aurinia*, for 15 years, including a four-year period when an increase in the numbers of the butterfly was observed. During the period of increase, scarcely two of the butterflies caught looked alike; some were so deformed that they could not fly. Both before this population explosion, when the butterflies were very rare, and afterwards, when they were quite common, genetic variability in the population was low, and essentially all the butterflies caught looked identical, although the phenotype before the population explosion was recognizably distinct from the phenotype afterwards.

What conditions would permit a population explosion and at the same time give rise to great variation in the adult population? We would expect population explosions to occur when some selective pressure that has kept the population small is relaxed, so that fewer members die and more reproduce. Perhaps the food supply has suddenly increased, a gene conferring resistance to disease has spread through the population, or a major predator has disappeared. The result in all cases is the same: with selection reduced, many more individuals live to maturity. In addition, many of these are individuals whose phenotypes would ordinarily have been selected against; thus they are genetically different from the usual adult population. Some of them, like butterflies that cannot fly, contain deleterious genetic combina-

---

* Pernicious anemia   Anemia due to degeneration of the stomach lining, depriving the body of the ability to make a factor needed to absorb vitamin B_{12}, which in turn is used in red blood cell formation.
* Duodenal and gastric ulcers   Lesions in the lining of the duodenum (section of the small intestine next to the stomach) and stomach.
* Pituitary adenoma   Tumor in the pituitary gland.

tions, which will probably be selected against and will not be passed on to the next generation. The overall conclusion, however, must be that whether the population is large or small, genetic variation is greater at times when selective pressures are relaxed. At times of stringent selective pressure, a greater proportion of individuals with deviations from the best-adapted genotypes will die or fail to reproduce.

15-G What Does Selection Act On?

In the preceding sections we have discussed effects produced when selection acts on a single gene. This raises the apparently trivial but actually difficult question: what is the unit upon which selection acts?

We usually talk as if selection acts on the phenotypes of individual organisms. At other times, as when we say that selection rapidly eliminates a dominant lethal allele from the population, we speak as if selection acts on individual alleles; there are also situations where selection clearly acts on the genotypes of groups of individuals. The clue to this apparent paradox probably lies in the fact that all three statements are true. Selection acts on segments of DNA, but these are of various sizes in different situations. A lethal gene, however small a segment of DNA, is strongly selected against, and any other genes in the same genome will die with it. Similarly, a polygene (many loci all influencing a single phenotypic character), which may occupy most of a chromosome, or a coadapted gene complex occupying several chromosomes, may have such a strong selective advantage or disadvantage that it decides the fate of the whole genome. The fates of a whole family of related individuals may be determined by the selective advantage or disadvantage of a gene that they all share, so that selection acts on them as if they were one large genome.

15-H Selectively Neutral Mutations

We have discussed many ways in which polymorphisms may be maintained by natural selection. Most biologists would probably agree that these are sufficient to account for the degree of polymorphism observed in natural populations. Others disagree, and in recent years have proposed that natural levels of polymorphism can be explained only by the additional presence of selectively neutral alleles.

A **selectively neutral allele** is one with no selective advantage or disadvantage relative to alternate alleles. If selectively neutral alleles exist, they could certainly contribute to polymorphism. Once such an allele had arisen by mutation, its progress through the population's gene pool would be a **random walk**, depending not on any properties of its own but upon whether it happened to be in one of the genotypes that contributed to the next generation. While a neutral allele was engaged in its random walk, heterozygotes for the allele would be present in the population and polymorphism would result. Calculations show that, on the average, neutral alleles would drift in the gene pool for long periods and contribute strongly to the overall heterozygosity of the population.

Whether there is any such thing as a neutral allele or not is a question that is unlikely to be answered for some time. The best evidence for selective neutrality comes from studies on the evolutionary changes undergone by proteins like cytochrome c, which is found in almost all organisms. A comparison of evolutionary history with the amino acid sequence of cytochrome c in different species shows that some amino acids have been substituted for others at a steady rate in the course of evolution. This constant rate of substitution is unlikely (so it is argued) to result from selection because selection would be more likely to result in uneven rates of evolution. It seems clear that this regular rate of amino acid substitution *could* be produced by selective neutrality, but it is not clear that the action of natural selection can be ruled out because we know very little about the selective forces acting on most of these proteins. That a gene could be selectively neutral for even a year, let alone millions of years, seems unlikely. Every gene that has ever been studied affects other genes, and every amino acid residue is likely to have some effect on the tertiary structure, membrane binding, enzymic activity, isoelectric point, or temperature or pH optimum of the protein in which it occurs.

It might nevertheless be possible for an allele whose selective disadvantage is very small to be maintained in a population through linkage to other alleles at different loci that have large selective advantages. The contribution of selectively neutral or selectively trivial genes to natural polymorphisms remains very much an open question.

SUMMARY

Although natural selection acts on individuals, only populations can evolve. This is because the population is the smallest unit with a gene pool in which the frequencies of alleles can change, and thus evolve.

The Hardy-Weinberg equilibrium describes the frequencies of different genotypes in a population and states that the proportions of different alleles and genotypes will remain the same as long as: (a) mating is random with respect to genotype; (b) there is no net mutation; (c) there is no gene flow; (d) the population is large; and (e) there is no selection for or against the traits being considered. Under these conditions, evolution will not occur. The Hardy-Weinberg equilibrium can, therefore, be used to give a measure of the forces causing evolution when it does occur.

Populations of the same species in different geographical areas tend to form a cline, in which adjacent populations differ from one another along a gradient of some environmental factor. Adaptations to local areas increase this difference; gene flow between adjacent populations decreases it.

Genetic drift may be an important cause of evolution in small populations that are not subject to strong selective pressures.

The gene pool of a population evolves so that each genotype becomes a coadapted gene complex. Genes are often expressed differently when they are inserted into the gene complex of another population. Dominance and recessiveness of alleles may be a result of modifier genes that have evolved in the same coadapted gene complex. Heterozygote advantage may be a result of the evolution of dominance and recessiveness. Heterozygote advantage also contributes substantially to a population's genetic load.

Factors that may increase the variation within the gene pool of a population include mutation, heterozygote advantage, polymorphism maintained by heterozygote advantage or by other factors, disruptive selection, gene flow between populations, sexual reproduction, and relaxation of a selective pressure.

Factors that may decrease variation in a gene pool include gene fixation and the selective pressures that adapt a population precisely to local conditions.

Selection may be thought of as acting on alleles, polygenes, complexes of coadapted genes, individuals, or even groups of related individuals.

OBJECTIVES

From your study of this chapter, you should be able to:

1. Define and use in context: gene pool, evolution, cline, gene flow, genetic drift, founder effect, coadapted gene complex, heterozygote advantage.

2. State, and explain the significance of, the Hardy-Weinberg Law.

3. State and explain the five situations in which the Hardy-Weinberg Law is invalid.

4. Given the necessary data, use the Hardy-Weinberg equation to determine the frequencies of two alleles, and of the genotypes they produce, in a population.

5. Explain why selection is unlikely ever to eliminate a lethal recessive allele from a large population of diploid organisms.

6. Explain why genetic drift is most likely to cause evolution in a small population.

7. Differentiate between stabilizing, directional, and disruptive selection, and give or recognize examples of selective pressures that may cause each.

8. Explain how dominance can evolve.

9. Give one reason for heterozygote advantage and explain the relationship between heterozygote advantage and genetic load.

10. Define polymorphism and discuss three ways it can be maintained.

11. Explain why a rapidly expanding population is more variable genetically than one whose size is constant.

12. List three factors that tend to increase and three that tend to decrease genetic variation in a population, and state how each acts.

1. The Hardy-Weinberg Law states that when two alleles are considered:
   a. dominant alleles will always be more frequent in the population than recessive alleles
   b. heterozygotes will be twice as likely as homozygotes in a population
   c. members of adjacent populations are not able to interbreed with one another
   d. large populations are well-adapted to their environment
   e. allelic frequencies cannot change in a large population unless one is at a selective advantage over the other or mating is nonrandom

2. The Hardy-Weinberg Law allows us to predict that:
   a. sexual reproduction is necessary for evolution
   b. sexual reproduction may be a cause of evolution
   c. sexual reproduction plays no role in evolution
   d. sexual reproduction will cause evolution if individuals prefer mates with one genotype over those with other genotypes

3. a. In certain parts of Africa, people with one normal and one sickle hemoglobin allele enjoy heterozygote advantage over either homozygote (see Sickle Cell Anemia, Section 14-B). Thus the population is experiencing (directional, disruptive, stabilizing) selection.
   b. Suppose that mosquito control measures completely eliminate the threat of malaria from an area where it was once prevalent. If you followed the allele frequencies for the next 30 generations, what changes would you expect to see in the frequencies of the normal and sickle alleles once selection against the homozygous normal individuals is removed?
   c. If, at the time malaria is eliminated, the adult population consists of 10% homozygous normal individuals and 90% heterozygous carriers, and if the people marry without regard to this trait, what will be the percentages of the three genotypes—homozygous normal, heterozygous, and homozygous sickle—in the children born in the next generation?

4. Selection will not eliminate a lethal recessive allele from a large population of diploid organisms because:
   a. there will always be some heterozygote carriers for the allele
   b. gene fixation will occur in the population
   c. heterozygotes are at a selective advantage
   d. the allele will have some good effects, and these will become dominant
   e. the rate of mutation to the lethal allele is higher in a larger population

5. Genetic drift is most important in which situation?
   a. gene flow in a cline
   b. polymorphism
   c. small populations
   d. breeding of flying insects
   e. relaxation of selective pressures

6. Genetic drift is more likely to cause evolution in a small population because:
   a. mating is nonrandom in small populations
   b. random events are more apt to happen to small populations
   c. there is no natural selection in small populations
   d. deviations from statistical averages are more likely to be seen in small populations than in large ones

7. True or False. In order for dominance to evolve, there must be selection for genes at loci other than the one whose dominance is evolving.

8. Which of the following is *not* a partial explanation for heterozygote advantage?
   a. The Hardy-Weinberg equilibrium states that there will always be twice as many heterozygotes as homozygotes.
   b. Two alleles may express themselves in slightly different ways that are adapted to different parts of the life history.
   c. Two different alleles may each operate best under different environmental conditions to which individuals are apt to be exposed.
   d. The presence of one allele may alleviate the detrimental consequences caused by the presence of the other.

9. In each of the following situations, tell whether genetic variability would increase, decrease, or remain the same.
   a. increased mutation rate
   b. decreased natural selection
   c. increased variability in the environment
   d. sexual reproduction

10. A rapidly growing population experiences increasing genetic variability because:
    a. there is a greater chance for polymorphism in large populations
    b. the mutation rate is higher for large populations
    c. there is less selective pressure
    d. a large population can support more genetic experimentation
    e. heterozygote superiority is selectively favored

QUESTIONS FOR DISCUSSION

1. Phenylketonuria (PKU), a genetic disorder due to homozygosity for a certain recessive allele, occurs in about 1 person in 15,000. Roughly what proportion of the population carries the disease?

2. In the United States, about 45% of the population has blood type O, 41% type A, 10% type B, and 4% type AB. Use the information about the genotype(s) of each blood group in Table 14-1, and the Hardy-Weinberg Law, to determine the frequencies of the three alleles involved in the ABO blood group system.

3. What is the effect on the human gene pool of the discovery of expensive medical treatments that permit individuals with previously lethal phenotypes to live? Should society permit research to find ways to treat more such conditions?

4. Biologists believe that primitive eukaryotic organisms were haploid (with unpaired chromosomes) and that diploidy (existence of paired homologous chromosomes) evolved later. What might have been the selective pressure that has resulted in the widespread occurrence of diploidy in organisms today?

5. Can there be selection against a genetic trait, such as Huntington's chorea, which is expressed in the phenotype only after the individual has passed reproductive age?

6. Incest taboos are common in human societies, as are their equivalent in other animals. What function do you think they serve? What arguments against your proposed function can you suggest?

7. Why are mutations almost always deleterious?

8. What characteristics would you expect to see in the genetic makeup of a species that has the ability to colonize a variety of habitats that may become available?

9. According to the Hardy-Weinberg Law, allele frequencies will not change when there is no selective pressure for or against them. How could selectively neutral mutations become prevalent enough to give rise to a polymorphism at that locus in the population?

REFERENCES AND FURTHER READING

Dawkins, R. *The Selfish Gene.* New York: Oxford University Press, 1976. The racy story of natural selection from the gene's point of view; written as science fiction, with some provocative thoughts on such topics as genesmanship and the battle of the sexes; excellent discussion of the evolution of altruism.

Ford, E. B. *Ecological Genetics*, 2d ed. New York: John Wiley & Sons, 1965. A readable book by one of the pioneers in this field.

Kelly, M. G., and J. C. McGrath. *Biology: Evolution and Adaptation to the Environment.* Boston: Houghton Mifflin, 1975. A textbook with a good section on population genetics.

Kimura, M. "The neutral theory of molecular evolution." *Scientific American*, November 1979.

Warren, K. B. ed. Genetics: The Nature and Causes of Genetic Variability in Populations. *Cold Spring Harbor Symposium on Quantitative Biology*. Vol. 20, 1955. A collection of papers on various aspects of population genetics.

Wills, C. "Genetic load." *Scientific American*, March 1970. Experimental and computer-simulation studies on the expression of mutant alleles and their distribution through a population.

CHAPTER 16

NATURAL SELECTION AND EVOLUTION

The theory of **evolution** states that today's organisms have arisen by descent and modification from more ancient forms of life. Evolution can also be defined in more modern terms as a change in the **gene pool** (the total of all the genes present in a population) of a population from one generation to the next. For instance, most biologists believe that human beings evolved from now-extinct animals much like chimpanzees, and that this happened through the erratic accumulation of changes in the gene pool through thousands of generations.

It is hard for us to appreciate today how revolutionary this theory seemed when it was first put forward. Evolution was a shocking idea because we have a tendency, as old as language itself, to think in terms of unchanging types. Plato, for instance, believed that God created eternal, unchangeable prototypes of Horse, Dog, and even of inanimate objects like tables. Real horses, dogs, and tables were merely imperfect versions of their ideal types. The idea that humans, although imperfect, are made in the image of God reflects the same way of thinking.

The theory of evolution says that this way of thought will not do for living organisms: the members of a species, such as salmon, are not deviants from a single ideal salmon, but rather are a group of organisms sharing a gene pool and displaying some of the many variations that the collections of salmon genes can produce. In other words, there is no such thing as a genetically "normal" salmon or human being or oak tree; they are all different versions of each other. This was a difficult idea in the nineteenth century, and for some people, it apparently still is.

Because a gene pool belongs to a whole population, only populations can evolve. Each individual's genetic makeup is fixed at the time of fertilization (barring mutation to another fixed genetic makeup), and the individual is a temporary vessel for some of its population's genes. For a population to evolve, its individuals must contain the raw materials (genetic variability) to produce new and different forms, which may become different species when enough changes have accumulated.

One mechanism that produces evolution is **natural selection**, which results in the differential survival and reproduction of individuals with different genotypes in each generation. In this chapter, we shall first present the logical argument that natural selection must occur, and then go on to study the history of evolution and some examples that scientists have investigated.

16-A   Evolution by Means of Natural Selection

The theory of evolution by means of natural selection is based on three observations: first, as we can see by comparing one cat or human being with another, the members of a species differ from one another. Second, some (though not all) of the differences between organisms are inherited. (Some differences between organisms are not inherited because they are caused by different environments.) Third, more organisms are born than live to grow up and reproduce: many organisms die as embryos or seeds, as saplings, nestlings, or larvae.

The logical conclusion from these three observations is that certain hereditary characteristics of an organism will increase its chances of living to grow up and reproduce, over the chances of organisms with other characteristics. To take an extreme example, if part of your hereditary variability is that you were born with a severe inherited disease of the liver, you have much less chance of living to grow up and reproduce than somebody who was born without this disease. And only if you reproduce can you pass on your inherited characteristics to members of the next generation.

Inherited characteristics that improve an organism's chances of living and reproducing will be more common in the next generation than those that decrease its chances of reproducing. Various combinations of genes will be naturally selected for or against, from one generation to the next, depending on how they affect reproductive potential. For natural selection to cause a change in a population from one generation to the next (that is, to cause evolution), it is not necessary that all genes affect survival and reproduction; the same result occurs if there are just some genes that make an individual more likely to grow up and reproduce.
To summarize:

1. Individuals of a species vary in any one generation.
2. Some variations are genetic.
3. More individuals are produced than live to grow up and reproduce.
4. Individuals with some genes are more likely to survive than those with others.

Conclusion: From the above four premises it follows that those hereditary traits that make their owners more likely to grow up and reproduce will become increasingly common in a population from one generation to the next.

16-B   History of the Theory of Evolution

The theory of evolution as we know it today was put forward in a joint presentation of the views of Charles Darwin and Alfred Russel Wallace before the Linnaean Society of London in July 1858. Before that time, most biologists accepted the view of such authorities as the ancient Greek philosopher and natural historian Aristotle and the Biblical Book of Genesis, and believed that species of organisms had been immutably created in their present form. From time to time philosophers who thought that the living world changed over the centuries made known their views, but by the mid-seventeenth century most of the Western world took the words of Genesis literally and believed that animals and plants had been created in their present forms during the six days of the Creation.

From about 1750 on, however, the evidence that species changed over the ages accumulated steadily, and many people became increasingly convinced of the reality of evolution. Georges-Louis Le Clerc de Buffon and Jean Baptiste de Lamarck were two influential French biologists who believed that species develop progressively and change in a changing environment. In England, the geologists James Hutton, William Smith, and Charles Lyell studied rock strata and the fossils of extinct animals and became convinced that different organisms had lived at different times. So we see that Darwin and Wallace were not the first to suspect the existence of evolution. Their names are linked with the concept because they proposed the theory of natural selection as the mechanism by which evolution occurs. We are always more likely to believe in a process when people explain how it happens than if they merely assert that it does.

Although they never met each other, Darwin and Wallace both came to the same conclusion about the mechanism of evolution as a result of remarkably similar experiences. Through perseverance and influential friends, Darwin had himself appointed naturalist on *H.M.S. Beagle*, a British naval ship embarking on a five-year mapping and collecting expedition. Before he sailed, in 1831, he became enormously interested in Lyell's book *Principles of Geology.* Lyell contended that the world was an ancient arena in which rock formations slowly appeared, changed, and disappeared; similarly, species of animals and plants formed, changed, and became extinct. Darwin's observations of the flora and fauna of South America and the Galapagos Islands (Figure 16–2), and Wallace's observations in the Malay Archi-

Figure 16-1 Charles Lyell. Darwin read Lyell's treatise on geology during the voyage of the Beagle and was initially more interested in geological than in biological changes. Finally, he wrote to Lyell, "At last gleams of light have come, and I am almost convinced that species are not (it is like confessing a murder) immutable." (Smithsonian Institution)

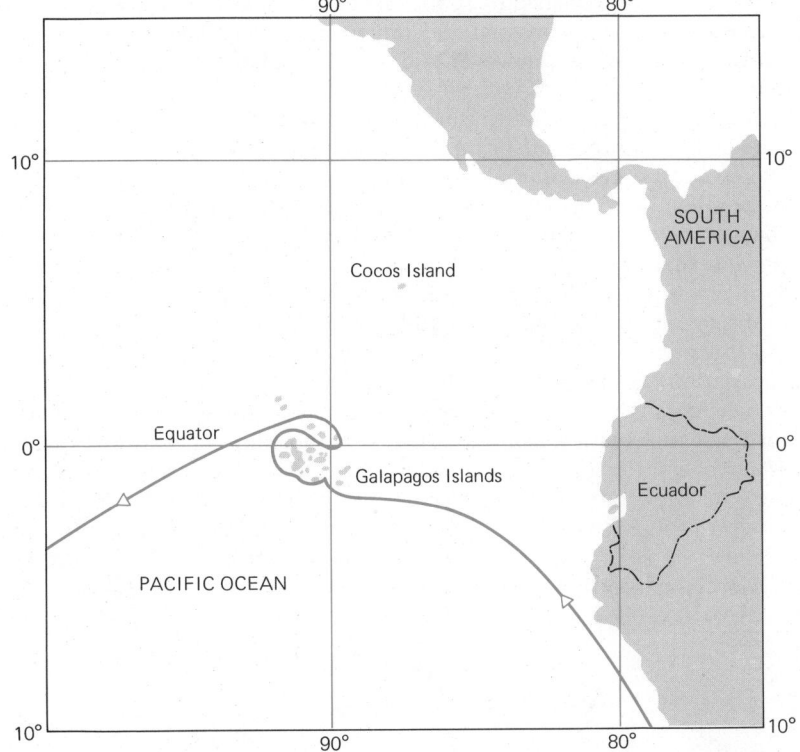

Figure 16-2 Position of the Galapagos Islands with part of the Beagle's route (colored). The effects of evolution are exaggerated on these islands because they are isolated from the mainland and from each other. The few plants and animals that have reached the Galapagos have evolved differently on each island.

**CHAPTER 16
NATURAL SELECTION
AND EVOLUTION** **243**

GROUND FINCH

WOODPECKER FINCH
(A tree finch)

WARBLER-TYPE FINCH

Figure 16-3 Some of the 14 species of Darwin's finches. The woodpecker finch is the most remarkable of Darwin's finches. It has evolved a long bill like a woodpecker's, but not a long tongue; instead it uses twigs to probe for insects in cracks it has enlarged in a tree trunk.

pelago, supported Lyell's theory. Both men became convinced that Lyell was right.

In 1845, Darwin published *The Voyage of the Beagle*, an exciting and readable account of his travels. In it he commented on his observations of the many species of finches unique to the Galapagos Islands: "Seeing this gradation and diversity of structure in one small, intimately related group of birds, one might really fancy that from an original paucity of birds in this archipelago, one species had been taken and modified for different ends" (Figure 16-3). In the same book he showed that he already held the clue to how evolutionary change was brought about, a mechanism he was not to put forward for public discussion for more than 10 years. He said "some check is constantly preventing the too rapid increase of every organized being left in a state of nature. The supply of food, on an average, remains constant; yet the tendency in every animal to increase by propagation is geometrical."

Darwin realized that some force, which he came to call "natural selection," constantly checks the potential population explosions of organisms. He owed this realization to an essay by Thomas Robert Malthus, a clergyman and economist, which Darwin read in 1838. Malthus's *Essay on the Principle of Population* was published in 1789; it argued that every population outgrows its food supply and is eventually reduced by starvation, disease, and war. Populations tend to increase geometrically, while food supplies at best increase only arithmetically. Darwin later made famous Malthus's phrase, "the struggle for existence," which they both saw as an inevitable result of the discrepancy between rapid population growth and limited food supply.

Upon his return to England in 1837, Darwin settled down to a lifetime of writing and thought. In 1858, he received a manuscript from Wallace, of whose work he had no idea, in which he read his own painstakingly documented theory of evolution by means of natural selection. Wallace, meditating on his own experiences and on Malthus's essay, had conceived the idea of natural selection and written his paper in three days. Darwin passed Wallace's paper to Lyell and to the botanist Hooker, who persuaded Darwin to let them present a version of his theory and Wallace's paper at an 1858 meeting of the Linnaean Society. Darwin then worked feverishly to finish his definitive work, *The Origin of Species by Means of Natural Selection*, which was published in 1859. The book sparked immense controversy, a fitting tribute to the most original and important biology book ever written. Not until the twentieth century, however, did most biologists fully accept the idea that evolution occurs by means of natural selection.

One widely held objection to natural selection was the belief that the characters of parents blended in their offspring. If this occurred, the inherited variation in a population would decrease in each generation until all the individuals in the population were alike, leaving natural selection with very little to act upon. We now know that genes do not blend, but Darwin, and most other nineteenth-century biologists, thought that they did. Although Gregor Mendel had published the basic laws of modern genetics (see Chapter 13), his work remained completely unknown until it was rediscovered and publicized in 1901. Mendel's work revealed that genes are inherited independently of one another and do not merge in the offspring; thus, genetic variation is preserved in each generation.

Another major barrier to the acceptance of natural selection was that Darwin illustrated selection with examples of selective breeding of domestic plants and animals. Breeders, for instance, had selected cattle for increased milk production (Holstein), for high-butterfat milk (Jersey), or for meat (Hereford) (Figure 16-4). Darwin never provided a convincing demonstration that selection actually occurs in nature. The example of selection in a wild population described in the next section was not worked out until a hundred years later.

16-C Industrial Melanism

A hundred years after Darwin and Wallace's views of natural selection were introduced, Bernard Kettlewell provided an example of selection in the wild, using the peppered moth, *Biston betularia*, which lives in all parts of England. In collections of moths and butterflies from the last century, *B. betularia* appears as a pale

Figure 16-4 Selection by human beings. These different breeds of cows have been produced by selective breeding for particular characteristics. (Biophoto Associates)

gray, mottled moth. Occasional specimens in nineteenth-century collections are a different variety, *B. betularia carbonaria*, which is a dark mottled charcoal color instead of the usual pale gray (Figure 16–5). We now know that whether this moth is the normal pale form or the dark *carbonaria* form is genetically determined. In the mid-1800s, the dark variety was very rare, but near the industrial city of Manchester, by the end of the nineteenth century, *carbonaria* had become the more common form of the moth, constituting about 90% of the total population. This change in populations of organisms over time is, in itself, evolution.

By the 1930s it was realized that many other species of moths were also becoming darker and darker near industrial cities, although the pale forms remained more common in the countryside. The geneticist E. B. Ford proposed that natural selection had brought about the change to a darker form in industrial areas (a phenomenon known as **industrial melanism**).

Moths are generally nocturnal. They fly, feed, and mate at night; during the day they rest on tree trunks or other surfaces, depending upon camouflage to protect them from predators. Ford proposed that before industrial pollution, the typical, pale gray form of *Biston betularia* had been very well camouflaged against pale,

(a)

(b)

Figure 16-5 The two forms of the peppered moth. (a) On a lichen-covered tree trunk in an unpolluted area. The gray form is so well-camouflaged that it is almost invisible, below and to the right of the black moth. (b) On a soot-covered tree trunk. (Bernard Kettlewell)

TABLE 16-1 NUMBERS OF GRAY AND BLACK PEPPERED MOTHS RECAPTURED AFTER THE RELEASE OF MARKED INDIVIDUALS IN TWO AREAS

LOCATION		GRAY	BLACK
Dorset (unpolluted) 1953	No. released	469	473
	No. recaptured (%)	62 (13.2)	30 (6.3)
Birmingham (polluted) 1953	No. released	137	447
	No. recaptured (%)	18 (13.1)	123 (27.5)
Birmingham (polluted) 1955	No. released	64	154
	No. recaptured (%)	16 (25.0)	82 (53.3)

lichen-covered tree trunks. In polluted areas, however, where industrial smoke had killed the lichens and blackened the tree trunks, the typical form stood out in contrast to these backgrounds and was eaten by predators in much greater numbers than the dark, or **melanic**, form. The most likely predators were birds, which are known to hunt by sight, and against whom camouflage, or lack of it, would be important.

In the 1950s Kettlewell decided to test Ford's hypothesis experimentally. He bred large numbers of the dark and light forms of the moth, marked them, and released them. After their release, he recaptured as many of the moths as possible on subsequent days. If equal numbers of both forms were released and equal numbers fell victim to predators after their release, equal numbers of the two sorts should be recaptured (as long as the two forms behaved in such a way that they were equally likely to be recaptured). Any variation from a 50:50 ratio of the percentage of the two forms recaptured would indicate that one form had suffered a higher mortality than the other after their release. This **mark-release-recapture** technique is often used in ecological experiments.

Kettlewell prepared two batches of moths, each with equal numbers of light and dark forms. He released one batch in a rural, unpolluted area in Dorset, and the other near the polluted industrial city of Birmingham. Not only did he recapture as many as possible of the moths released in both locations, but also he hid in a blind and watched the moths after he had released them. From the blind he saw birds eating moths that were motionless on tree trunks. On one occasion, equal numbers of typical and melanic moths were released in unpolluted Dorset, and watchers in the blind recorded their capture by birds. Of 190 moths eaten by birds, 164 were *carbonaria* and 26 were typical.

The numerical results of Kettlewell's experiments are given in Table 16–1. It is clear that in Birmingham, a polluted industrial area, many more of the dark moths than of the light-colored moths will live long enough to reproduce. The next generation of moths will therefore contain many more dark moths than light ones, since color is inherited. The moths evolve such that the dark form becomes more common in populations in polluted areas than the light form. In other words, distribution of the genotypes of the population changes with time, and that is evolution.

This evolutionary change is brought about by a natural selective force. The selective force in this particular instance is that birds prey upon the moths in such a way that they differentially kill a higher percentage of the light-colored than of the dark-colored moths in industrial areas. Natural selection over many generations of the peppered moth has produced populations of the moth that are well-adapted to survive in their environments, populations whose characteristics change as the environment changes.

Based on the concept of natural selection, we can predict that if pollution is reduced in industrial areas, melanic moths will become rarer and typical forms more frequent in these areas. In fact the Clean Air Act of 1952 has reduced pollution in England, and collections of *Biston betularia* from Manchester in the succeeding 20 years reveal a dramatic decrease in the ratio of dark to light individuals in the moth population. The ability to predict events in this way is the most impressive evidence that can be produced for a scientific theory.

TABLE 16-2 SURVIVAL IN SWISS STARLINGS IN RELATION TO NUMBER OF EGGS LAID[1]

BROOD SIZE (NUMBER OF EGGS IN NEST)	NUMBER OF YOUNG MARKED	RECOVERIES PER 100 BIRDS MARKED[2]
1	65	0
2	328	1.8
3	1278	2.0
4	3956	2.1
5	6175	2.1
6	3156	1.7
7	651	1.5
8	120	0.8
9, 10	28	0

[1] The number of eggs laid during one nesting period is genetically regulated and, like other genetic variations, is acted upon by natural selection. Lack marked all the nestlings in all the nests he could find, and then recaptured them months later when they had left the nest.

[2] The only recoveries scored are those for birds over three months old when they were recaptured.
Source: Lack, *Ecol.* 2, 1948.

16-D   Genetic Contribution to Future Generations

The phrase "survival of the fittest," often used in discussions of evolution, suggests that natural selection selects mainly for survival; it does not. Rather, it selects for the contribution of genes to future generations. Survival is important, in that if an individual does not survive for long it cannot reproduce, but even reproduction is not a guarantee of evolutionary success.

Consider Table 16-2, which shows how many young starlings survived for three months after hatching. The female starlings that seemed to be reproducing most efficiently—those laying nine or ten eggs in one brood—could actually be doomed to evolutionary failure and strongly selected against because hardly any of their young survived. However, females laying four or five eggs per brood had a higher number of offspring surviving for at least three months after they hatched.

David Lack showed that young birds from the larger broods weighed less than those from smaller broods, presumably because the parents could provide adequate food for no more than five or six nestlings. Lack of adequate food was probably a major cause of the high mortality of young from larger broods. Table 16-2 also shows that the most common brood sizes produce the nestlings with lowest mortality rates, as we would predict from the action of natural selection. It seems reasonable to suppose that in years when there is more (or less) food available to the birds than in the year studied in this example, selection would favor birds with broods larger (or smaller) than the average.

Plainly, the reproductive success of a starling is not fully told by the story of one brood. Selection optimizes reproductive success over a lifetime, and the adaptations that produce this success are myriad.

Selection has led to reproductive strategies that are very different from species to species. For example, a herring lays three million eggs at a time; most of them die, but a few survive to carry on the parents' genes. At the other extreme, humans lavish 20 years of parental care on each offspring. The result of the two strategies is the same: selection ensures that the members of future generations will be the descendants of the most reproductively successful members of the present generation.

16-E   Adaptations

When we say that selection causes organisms to become adapted to their environments, we should note that "environment" in this evolutionary context encompasses all of the factors, other than the organism's genes, that can affect whether it

Figure 16-6 An English oak wood in May. The tree in the foreground has been completely defoliated by insects; those in the background are leafing out normally. (Paul Feeny)

lives to reproduce or not. This will include its environment as an embryo, a juvenile, and an adult. Let us, for example, consider an acorn. Whether it successfully resists the selective pressure of its environment depends on the speed and normality of its embryonic development and germination, whether bacteria or fungi infect it as a seed or seedling and destroy it at this stage, whether as a seedling it has enough stored food for rapid growth, whether it escapes being eaten, whether the soil in which it grows can support a large plant, and whether, as a small tree, it avoids death by disease, trampling, or browsing. To make things more complicated, selective pressures may frequently be contradictory. For instance, a hot summer is in one way advantageous to our hypothetical acorn because the acorn can grow faster at higher temperatures, but hot weather also increases the chance that the soil will dry out too much before the root system goes deep enough to obtain a reliable supply of water.

In this section we consider several sets of adaptations that are especially illuminating because biologists have worked out the selective forces involved in their evolution.

Oak Trees and Caterpillars

Many plants synthesize compounds that make the nutrients the plants contain less available to herbivores (plant-eating animals). The relationship between oak trees and herbivorous insects illustrates such a system.

In some years, the caterpillars of the winter moth and other insects are so abundant that they completely defoliate an oak tree (eat all of its leaves) as the leaves open in the spring (Figure 16–6). Such defoliation occurs only early in the summer and never after about mid-June. This is puzzling in light of another observation: many caterpillars hatch early in the spring, before the oak leaves are out, and these caterpillars starve to death. Why have caterpillars not evolved so that they hatch later, when the leaves are fully opened? It looks as if there must be very strong selective pressure for early hatching, to eat the oak leaves while they are as young as possible. What could this selective pressure be?

Do the leaves become less nutritious later in the summer? Paul Feeny showed that the sugar content of the leaves, at least, did not reach a peak until mid-June. He found, however, that when he reared caterpillars in the laboratory, those fed young (early May) leaves gained weight faster than those fed older (late May) leaves. Furthermore, when leaves from either late or early May were ground up and mixed into an agar diet (which ends up with a texture rather like firm gelatin), the caterpillars grew equally well. It thus appeared that the toughness of the late May leaves (removed by grinding the leaves) was one of the selective pressures for early feeding on oak leaves.

Toughness, however, does not seem to be the whole reason for the scarcity of insects on mature oak leaves. Another reason is that mature oak leaves contain only half as much nitrogen per unit weight as do fresh young leaves. What is more, the leaves contain an increasing quantity of tannin as the season progresses. Tannin is the substance that makes tea brown and that is used to tan leather. It complexes protein into large indigestible clumps. Experiments showed that tannins from mature oak leaves could convert leaf proteins into a form that proteolytic enzymes can digest only very slowly. Feeny reared winter moth larvae on artificial diets containing increasing amounts of tannin. As the tannin content of the diet was increased, the larvae gained weight more slowly.

Thus, as the summer progresses, oak leaves become less and less suitable food for winter moth larvae because of their increasing toughness, their lower concentration of nitrogen, and their increasing tannin content, and probably also because of their lower water content. The selective pressure for oak-feeding insects to feed early in the spring is almost certainly that young oak leaves are much more tender and nutritious than older leaves.

How do oak leaves manage to survive the 2 to 3 week period each spring when they are so vulnerable to insects? In some years they don't; the initial leaf crop is entirely eaten and must be replaced by new leaves from dormant buds, depleting the tree's energy reserves. On the average, however, the young oak leaves escape serious

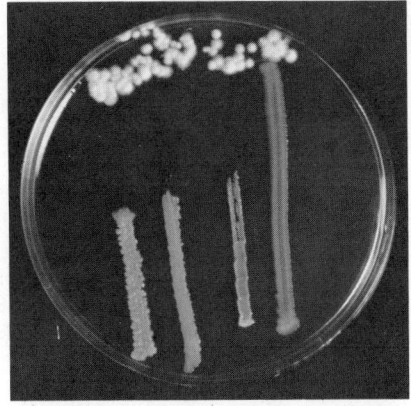

Figure 16-7 Drug resistance in bacteria. Four varieties of bacteria are growing in vertical lines in this dish. The fungus *Penicillium* growing across the top of the dish has produced penicillin, an antibiotic, which has spread out through the dish. Three of the bacterial varieties have been killed as the antibiotic reached them; the fourth (far right) is antibiotic-resistant and it continues to grow. (Biophoto Associates).

damage because their emergence is not exactly predictable, and most young caterpillars hatch too soon or too late to catch the leaves at the right stage.

How can the insects, in turn, adapt to such unpredictability? The females of the winter moth and several other early-feeding species are wingless; thus the tree on which they lay their eggs is likely to be the tree they themselves fed on as caterpillars. Since different oak trees vary by up to two weeks in the time at which they produce their first leaves, a female's progeny probably have a greater chance of hatching at the right time the following spring if she lays the eggs on her own tree rather than if she were to fly around and lay them on randomly selected oak trees.

Resistance to Pesticides and Antibiotics

Several dramatic examples of natural selection in action today are provided by the evolution of resistance to pesticides and antibiotics.

A scale insect feeds on citrus trees in California. In the early 1900s, growers sprayed the trees with cyanide gas, and this killed the scale. But in 1914 some of the insects survived the spraying. The cyanide did not kill them because they possessed a single gene, newly apparent in the population, that permitted them to break down cyanide into harmless compounds. As spraying continued, more insects with the new gene than without it survived to reproduce, and they passed on the gene to their offspring. The frequency of the new gene in the population increased until the whole population was resistant to the spray. Because scale insects, like many other insects, have more than one generation a year, they evolve quickly. To combat the evolution of resistance, growers are encouraged to spray only when necessary and to use different chemicals in different months or years.

Precisely the same thing happens with antibiotics used to kill bacteria that cause human disease. When a bacterial population meets a particular drug, bacteria susceptible to that drug are killed. Bacteria with genes conferring resistance to the drug multiply rapidly once competing bacteria have been removed, and they soon become widespread. Some strains of the bacterium that causes the venereal disease gonorrhea can no longer be killed by any known drug. Since antibiotics and disease-causing bacteria frequently meet in hospitals, it is not surprising that some hospitals harbor drug-resistant bacteria. In many countries, women are now encouraged to give birth at home whenever possible, because mother and infant are safer from bacterial infection at home than in the hospital.

Most countries have outlawed the use of antibiotics in cattle feed. Cattle fatten faster if fed antibiotics, but they also become large-scale breeding grounds for antibiotic-resistant bacteria. Antibiotics are still added to cattle feed in the United States, and drug-resistant bacteria in cattle are becoming increasingly common.

These examples of the evolution of adaptations illustrate merely a few of the less obvious selective pressures that are always acting on all organisms and the adaptations that have evolved in response to them, adaptations that appear ingenious but are really the result of natural selection among randomly produced variations of genes.

Figure 16-8 Jean Baptiste de Lamarck
(Smithsonian Institution)

16-F Lamarckism

Having spoken of evolution and how it is brought about by natural selection, we should mention briefly another theory of evolution. Named Lamarckism after the eighteenth-century biologist who was its main proponent, it holds that characters acquired by one generation can be inherited by the next. Lamarck suggested that giraffes had evolved their long necks because they stretched up to reach the leaves of trees, thereby stretching their necks in their lifetimes, and then passed this character on to their offspring.

Because no one yet knew how genes were inherited, Darwin did not refute this theory when he put forward natural selection as the mechanism of evolution. He did not understand that individuals inherited discrete genes, and thought that the inheritance of acquired characters might have played a minor role in evolution. When, about 1900, Mendel's work on genetics was rediscovered and expanded, such a view became untenable. Nevertheless, the belief persisted into this century. For instance, people cut off rats' tails for generation after generation to see if rats that had lost their tails ever produced tailless offspring, which they did not.

This is not to say that nothing an organism does in its lifetime can affect its offspring's genotype. Use of drugs that destroy chromosomes, or exposure to high levels of radioactivity, may alter the genes passed on to the offspring. However, it is clear that, with a few possible exceptions, nothing an organism does will make its offspring inherit the same characteristic that it has acquired.

The political importance of Lamarckism is an interesting story. The idea that acquired characteristics could be inherited was taken up by T. D. Lysenko, a Soviet agriculturalist. He said that he could breed better varieties of grains by giving the parent plants better conditions. The attraction of this idea to a revolutionary ideology is obvious. It meant that improving the conditions in which people live, such as their health and education, would eventually produce an improved race of people with better genetic potential for health and intelligence. In the 1930s this appeared a much more attractive ideology to a revolutionary society than did the writings of contemporary Western biologists who were saying that the only way to "improve" the human race was by eugenic programs: encouraging healthier, more intelligent people to reproduce, and discouraging the reproduction of defective members of society. So, for ideological reasons, Lysenko rose to the top in Soviet science, and many biologists who opposed his ideas lost their jobs and even their lives. By the 1950s, however, other Soviet scientists had disproved many of Lysenko's theories, and he was finally discredited. But Lamarckism, as a theory, has had an important impact on society and is undoubtedly quite widely believed to this day.

SUMMARY

The theory of evolution asserts that species are not unchangeable, but arise by descent and modification from pre-existing forms. Members of any species of organism differ from one another, and some of their differences are inherited. Natural selection is the differential reproduction of genetically different individuals in each generation; it leads to evolution, a change in the gene pool of a population from one generation to the next.

The theory of evolution by means of natural selection was put forward in 1858 by Darwin and Wallace. Their thinking was stimulated by the writings of Lyell and Malthus, and by observations they made during their own travels. The fact that selection causes the evolution of wild populations was not convincingly shown, however, until the twentieth century. Kettlewell showed that predation by birds was the selective force that led to the evolution of melanic populations of the peppered moth in polluted areas of England.

The anatomical, physiological, and behavioral traits that survive the process of natural selection may be thought of as adaptations that fit an organism to live and reproduce in its particular environment. Adaptations are many and various. The only consistent effect of selection is that it ensures that the genetic contribution of a "successful" individual to future generations is as large as possible.

In another theory of the origin of existing species, which is incompatible with the theory of natural selection but is of historical and political importance, Lamarck proposed that species change with time because they inherit the characters acquired by their parents.

OBJECTIVES

From your study of this chapter, you should be able to:

1. State the three observations leading to the conclusion that evolution occurs under the influence of natural selection.

2. Define evolution.

3. Define natural selection. State what it means to be evolutionarily successful, and for an adaptation to be selected for.

4. Describe the roles of Darwin and Wallace in formulating the theory of evolution and the three main influences that led them to their conclusions.

5. State why Darwin and Wallace's theory of evolution was more convincing than the descriptions of evolution put forward by their predecessors.

6. Describe Kettlewell's experiments with *Biston betularia*, and explain how they demonstrated that industrial melanism in this moth has evolved under the influence of natural selection.

7. Describe Lack's experiments with Swiss starlings, and state what these experiments tell us about the effect of natural selection.

8. Describe Lamarck's theory of how evolution occurs, and explain why Lysenko believed in this theory.

9. Explain why selection does not produce a population of identical, perfectly adapted organisms.

SELF-QUIZ

1. In light of the definition of evolution, which of the following is *not* capable of evolving?
   a. a population of deer
   b. the color of a population of moths
   c. your biology teacher
   d. a population of cattle
   e. the millions of bacteria in your large intestine

2. Which of the following did Kettlewell conclude from his studies on industrial melanism in moths?
   a. a dark moth lays more eggs than a light moth in industrial areas
   b. dark moths are more resistant to pollution than are light moths
   c. pollution caused some moths to become darker than others
   d. dark moths are more likely to survive in polluted area than are light moths
   e. birds prefer the taste of dark moths to the taste of light moths

3. Which bird is most evolutionarily successful?
   a. lays 9 eggs, 8 hatch and 2 reproduce
   b. lays 2 eggs, 2 hatch and 2 reproduce
   c. lays 5 eggs, 5 hatch and 3 reproduce
   d. lays 9 eggs, 9 hatch and 2 reproduce
   e. lays 7 eggs, 5 hatch and 4 reproduce

4. Suppose that you have a pack of 50 assorted dogs. You select the largest male and the largest female from the group, mate them, and sterilize the other members of the pack. Assuming that food supplies remain adequate, you should expect that, in the next generation of dogs:
   a. the young dogs will be, on the average, larger than their two parents
   b. the young dogs will be, on the average, larger than the older members of the pack
   c. the young dogs will be the same average size as the older dogs
   d. all of the young dogs will be larger than the older dogs

QUESTIONS FOR DISCUSSION

For questions 1 to 5, consider Table 16-2.

1. What do you suppose is the disadvantage to a starling of laying a very small clutch of eggs?

2. Suppose the environment changed so that only half as much food was available to the starlings. Would you expect a gradual change in the most frequent brood size? How would this change be brought about?

3. Which female starlings will leave more young per head in the population and hence make the greatest contribution to the genes of the next generation?

4. From what brood size do the greatest number of young survive?

5. Is this also the most frequent family size (assume that Lack marked every bird he could find)?

6. Are all causes of death natural selection? When organisms die in an earthquake, have they been selected against?

7. The embryologist Charles H. Waddington treated fly larvae with heat shock. As a result of this treatment, some of the adult flies showed the abnormal condition "crossveinless" (some of their wing veins were missing). After many generations of this treatment, he let a generation of flies develop without heat treatment and many of them were also crossveinless. Does this experiment provide convincing proof of Lamarckism? If not, what other explanation can you suggest, and what experiments would you perform to test your suggestion?

8. Is human evolution subject to the same pressures as the evolution of other species? Why or why not?

9. Is there any time in its life history when an organism is not subject to selective pressure? Are gametes subject to selective pressure? Are eggs? embryos? Is there selective

Essay: Charles Darwin

(Most of this essay is in Darwin's own words.† Our additions are in italics.)

Charles Darwin was born in 1809 into the sort of upper-middle-class English society that Jane Austen brought to life in her novels.

My father sent me to Edinburgh University where I stayed for two years. I became convinced that my father would leave me property enough to subsist on with some comfort; my belief was sufficient to check any strenuous effort to learn medicine. The instruction at Edinburgh was altogether by lectures, and these were intolerably dull. Dr. Duncan's lectures on Materia Medica at 8 o'clock on a winter's morning are something fearful to remember.

It has proved one of the greatest evils of my life that I was not urged to practice dissection, for I should soon have got over my disgust, and the practice would have been invaluable for all my future work.

---

*Excerpts are from The Autobiography of Charles Darwin (Nora Barlow, Ed.). London: Collins, 1958

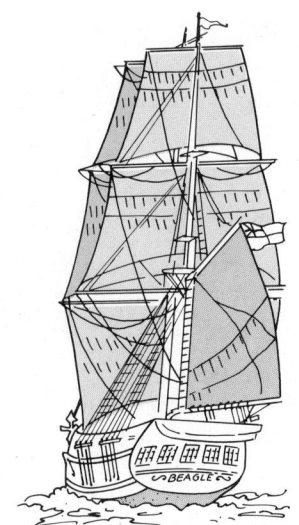

Figure 16-A H.M.S. Beagle

Darwin dropped out of medical school and went to Cambridge to study theology.

From my passion for shooting and hunting, I got into a sporting set. We used often to dine together in the evening and we sometimes drank too much, with jolly singing and playing at cards afterwards. I know that I ought to feel ashamed of days and evenings thus spent, but as some of my friends were very pleasant, and we were all in the highest spirits, I cannot help looking back on these times with much pleasure.

But no pursuit at Cambridge gave me so much pleasure as collecting beetles. No poet ever felt more delight at seeing his first poem published than I did at seeing the magic words "Captured by C. Darwin, Esq." *on the label for an insect.*

In 1831 on returning home I found a letter informing me that Captain Fitz-Roy was willing to give up part of his own cabin to any young man who would go with him, without pay, as naturalist to the voyage of the *Beagle*. Afterwards, I heard that I had run a very narrow risk of being rejected because of the shape of my nose! Fitz-Roy was an ardent disciple of Lavater, and was convinced that he could judge of a man's character by the outline of his features.

The voyage of the *Beagle* has been by far the most important event of my life. As far as I can judge, I worked to the utmost during the voyage from the mere pleasure of investigation. But I was also ambitious to take a fair place among scientific men.

During the voyage, I had been deeply impressed by discovering great fossil animals like existing armadillos. It was evident that such facts as these could only be explained on the supposition that species gradually became modified. It was equally evident that neither the surrounding conditions, nor the will of the organisms could account for the innumerable cases in which organisms of every kind are beautifully adapted to their habitats. I soon perceived that selection was the keystone to man's success in making useful races of animals and plants. But how selection could be applied to organisms

living in a state of nature remained for some time a mystery to me. In October 1838, I happened to read for amusement Malthus on population and, being well prepared to appreciate the struggle for existence which everywhere goes on, it at once struck me that under these circumstances favourable variations would tend to be preserved and unfavourable ones destroyed. The result of this would be formation of a new species.

This was in 1838. It was nearly 20 years before Darwin wrote down his theory, although in the meantime he wrote books and articles on a myriad of other biological subjects. Darwin offers no explanation for his long delay except to say: I had at last got a theory by which to work; but I was so anxious to avoid prejudice, that I determined not for some time to write even the briefest sketch of it.

Early in 1856, Lyell advised me to write out my views pretty fully and I began at once to do so. But my plans were overthrown, for early in the summer of 1858 Mr. Wallace sent me an essay, and this essay contained exactly the same theory as mine. [*Wallace had written his essay in 3 days!*]

Lyell urged that Wallace's essay and the abstract of Darwin's manuscript be published together. I was at first very unwilling to consent, as I thought Mr. Wallace might consider my doing so unjustifiable, for I did not then know how generous and noble was his disposition. Nevertheless, our joint production excited very little attention and the only notice I can remember was by Professor Haughton, whose verdict was that all that was new in them was false, and all that was old was true. This shows how necessary it is that any new view should be explained at considerable length in order to arouse public attention.

Finally Darwin described evolution and natural selection in The Origin of Species, *published in 1859. This version of the theory, full of details and examples, immediately attracted attention and the book became a best-seller. It sparked an immense controversy and public debate, which has continued ever since. In particular, it bothered, and con-*

(a)

(b)

Figure 16-B Fathers of the modern theory of evolution. (a) Charles Darwin and (b) Alfred Russel Wallace. (Biophoto Associates, Linnaean Society, London)

tinues to offend, some Christians. This is for two reasons. First, the theory of evolution contradicts a literal interpretation of the biblical story that the earth and its organisms were created in seven days. This offends fundamentalists who believe in the literal truth of the Bible. Second, the theory superficially leads to a rather deterministic view of life, since neither human nor divine agency is needed for evolution to occur. Darwin himself became trapped in this second point of view. His mind rejected any reality that could not immediately be tested by observation or experiment.

Toward the end of his life he wrote: "I gradually came to disbelieve in Christianity as a divine revelation." His wife, Emma Wedgewood, whose mind was more flexible,

fought him relentlessly on this point. She wrote to him, "May not the habit in scientific pursuits of believing nothing till it is proved, influence your mind too much in other things which cannot be proved in the same way, and which if true are likely to be above our comprehension?" She pointed out to Darwin that bringing scientific method to bear on religious beliefs is a pointless exercise because the two are philosophically distinct realms. Many people before and after Darwin have, however, confused them.

*   *   *

By and large, biologists accepted the theory of natural selection with open arms, since it explained so much and made sense of a million

unexplained facts. There have always been those who resisted the appeal of the theory of evolution and every now and then declare "Darwin was wrong," in the hope of some profitable publicity, usually revealing that they have never read, and do not understand, the theory. Of course modern evolutionists have refined many aspects of Darwin's work, but there can be no question that the theory of evolution has been the single most fruitful piece of biological thinking ever produced.

CHAPTER 16
NATURAL SELECTION
AND EVOLUTION

pressure on young animals that are fed and protected by their parents?

10. Some insects lay eggs on more than one species of larval food plant. There is some evidence that a female is more likely to lay her eggs on the plant species on which she grew as a larva than on any other kind of plant. Is this an example of Lamarckian inheritance? Why?

11. Some people have suggested that we should breed crop plants to have built-in chemical defenses against insect pests. In your opinion, how well would this work?

12. What is the adaptive advantage to a plant of a contact irritant (such as the oil on poison ivy leaves that makes a rash on the skin of passing animals)?

REFERENCES AND FURTHER READING

Bishop, J. A., and L. M. Cook. "Moths, melanism, and clean air." *Scientific American*, January 1975.

Cook, L. M. *et al.* "Increasing frequency of the typical form of the peppered moth in Manchester." *Nature* 227:1155, 1970. These two articles describe the effect of efforts to reduce air pollution on the ratio of melanic to typical morphs in the peppered moth population.

Darwin, C. *On the Origin of Species*. New York: Cambridge University Press, 1975, and Cambridge, MA.: Harvard University Press, 1964, are recent reprintings.

Dobzhansky, T. *et al. Evolution*. San Francisco: W. H. Freeman, 1977. An excellent advanced text by four distinguished evolutionists.

Eiseley, L. C. *Darwin's Century; Evolution and the Men who Discovered It*. Garden City, NY: Doubleday, 1958. An interesting account of the history of the theory of evolution.

Eiseley, L. C. "Charles Darwin." *Scientific American*, February 1956. An abbreviated biography of Darwin and discussion of his work.

Eiseley, L. C. "Alfred Russel Wallace." *Scientific American*, February 1959. A fascinating account of the life of Wallace, how he came to believe in the evolution of species by natural selection, and his views on human evolution.

Feeny, P. P. "Seasonal changes in oak leaf tannins and nutrients as a cause of spring feeding by winter moth caterpillars." *Ecology* 51:565, 1970.

Johnson, C. *Introduction to Natural Selection*. Baltimore: University Park Press, 1976. An excellent modern textbook for those who want to understand natural selection in a little more depth than found in this book.

Mayr, E. "Evolution." *Scientific American*, September 1978. A prominent student of evolution recounts the history of evolutionary theory and explains current ideas about how it works. This article introduces an entire issue devoted to the theme of evolution.

Moorehead, A. *Darwin and the Beagle*. New York: Harper and Row, 1969. A short, beautifully illustrated account of Darwin's travels based on Darwin's diaries.

CHAPTER

17

SPECIATION

The most dramatic event in evolution is **speciation**, the formation of a new species. The history of life is written in the fossil record, where a new species appears, exists for 5 to 10 million years, and then disappears, sometimes leaving one or more descendant species to replace it. Judging from the fossil record, most new species take thousands of years to form, so it is not surprising that we do not see new species forming around us every day. What we know about speciation comes mainly from circumstantial evidence: from the fossil record, from population genetics, and from studies of closely related species.

In this chapter we examine the evidence on how species are formed and conclude that a population (part of an existing species) can evolve into a new species only if it is isolated genetically from the rest of the species. This is an old thought; Charles Darwin realized that new species sometimes formed when populations were isolated on islands. But only in the twentieth century have we realized that there are many different sorts of genetic isolation, some of which do not involve physical barriers like stretches of ocean but can occur in our own backyards.

Before considering how species arise, we must address the somewhat controversial subject of what precisely is meant by the term species.

A **species** is usually defined as a group of organisms capable of breeding with one another and unable to breed with the members of other species. Thus, even though many species may live side by side, they are **reproductively isolated** from one another. Although this idea of a species is a very simple concept, we shall see that it is much harder to pin down in practice. This definition of species was developed largely from the study of vertebrates. Its major disadvantage is that it applies only to organisms that always reproduce sexually. Because there are so many species that usually produce asexually (many prokaryotes, invertebrates, and plants) the definition is of limited use.

17-A   Morphological Species

It is impossible to give a rigorous definition of a species that fits all occasions. The idea that organisms can be neatly divided into species is a very useful one, but we shall see that it does not reflect reality very accurately.

The eighteenth-century naturalist Carolus Linnaeus developed the first working definition of a species. Linnaeus's classification of plants and animals was based on observable morphological (structural, anatomical) differences between organisms. If two organisms were sufficiently different, they were considered to belong to different species. In fact, this is how most organisms are classified today.

The morphological definition of a species has a number of difficulties. Chief among these is the problem of recognizing that individuals at different stages of the life history (for example, caterpillars, chrysalids, and butterflies) all belong to the same species. Another drawback is possible misinterpretation of great differences in size and shape, which may occur among individuals within the same species as a result of environmental effects on growth (as in most plants), or as a result of minor genetic differences. For example, approximately 600 species of snails in the genus *Cerion* have been described from the Caribbean islands, on the basis of differences in size, shape, and color pattern. It has now been shown that these so-called species can interbreed and that their morphological differences probably result from relatively slight differences in genes that regulate growth. In fact, there may be only two species of *Cerion!*

Nevertheless morphological methods are the most practical means of classifying organisms. A biologist who finds a new type of insect, for example, cannot usually perform extensive tests of its reproductive ability but must decide if it belongs to a new species by examining its appearance.

Using morphological descriptions of species has several advantages. Morphological characters are convenient to work with. They also allow for clear communication; with the morphological concept, a person can write a precise definition of a particular species and transmit this definition to other workers. The morphological characters separating the various species are also the natural characters to use in constructing a dichotomous key (Figure 17–1), one of the most important tools of a field biologist. The last major advantage of the morphological species concept is that it permits a **type specimen** of the species to be preserved in a museum for future reference.

17-B   Biological Species

Domestic dogs belong to the species *Canis familiaris*. However, the decision to classify anything from a Pekingese to a Great Dane in the same species was almost certainly not based on morphological characteristics. Domestic dogs are considered to belong to one species because a member of any breed of dog can (usually) breed and produce viable offspring with a member of any other breed. This **biological species** concept, based on the ability of organisms to breed with one another, dates back to Darwin, who pointed out that the members of different species do not breed with one another.

To use the key, begin at the first pair of statements and decide which one pertains to the specimen you are trying to identify. Each member of the pair indicates either a class or the number of the next pair of statements to consult. Continue until you arrive at a statement that gives the class of the specimen.

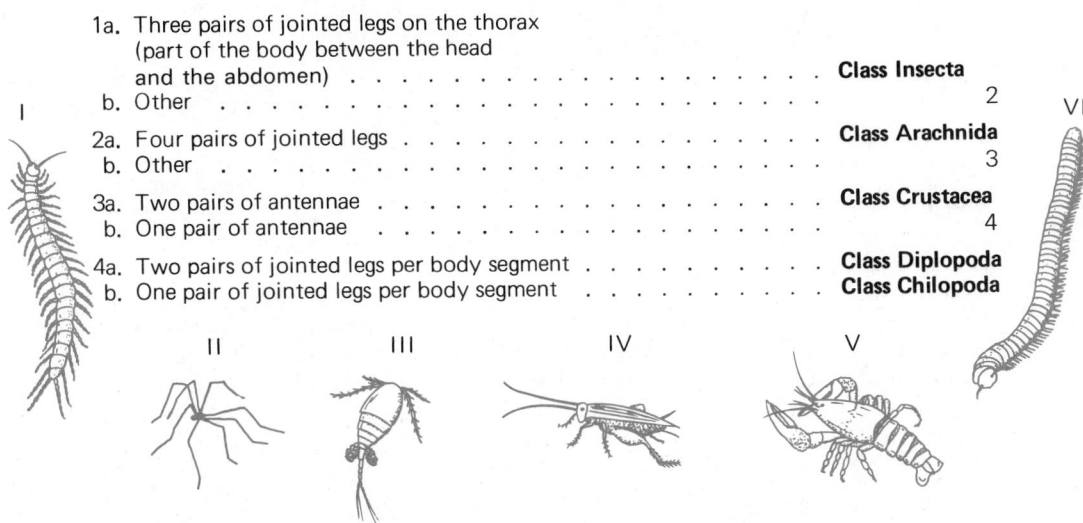

1a. Three pairs of jointed legs on the thorax (part of the body between the head and the abdomen) . . . . **Class Insecta**
 b. Other 2

2a. Four pairs of jointed legs . . . . **Class Arachnida**
 b. Other 3

3a. Two pairs of antennae . . . . **Class Crustacea**
 b. One pair of antennae 4

4a. Two pairs of jointed legs per body segment . . . . **Class Diplopoda**
 b. One pair of jointed legs per body segment . . . . **Class Chilopoda**

Figure 17-1 A dichotomous key, so called because there are two alternative choices at each step. To see how it works, follow the instructions, and identify the classes of the specimens drawn around the edges of the key. For answers, see the Chapter 17 Self-Quiz answers at the back of the book.

Modern genetics has shown the importance of sexual reproduction to evolution. It is sexual recombination of genes that creates the genetic variability found in most populations of organisms. The maternal and paternal chromosome sets must be "compatible" in some way for mitosis and meiosis to occur. If two organisms are genetically very different, their gametes are incompatible, and breeding between the two is impossible because the chromosomes will not line up together or separate properly at mitosis and meiosis. This is why matings between members of different species, such as that between a mare and a jackass, often produce sterile offspring (in this case, a mule)—if they produce any offspring at all. The inability to breed provides a natural means of determining if two organisms are sufficiently distinct genetically to be classified as separate species. Conversely, the ability to interbreed and produce fertile progeny is an indication of compatible genotypes. Two such organisms are closely enough related to be considered as members of the same species.

Species have been defined by two of the twentieth century's most eminent students of evolution in the following ways:

(1) A species is that stage of the evolutionary process at which the once actually or potentially interbreeding array of forms becomes segregated into two or more separate arrays which are physiologically incapable of interbreeding.

—Theodosius Dobzhansky

(2) Species are groups of actually or potentially interbreeding natural populations, which are reproductively isolated from other such groups.

—Ernst Mayr

Notice that both definitions suggest fertility within and sterility between populations. Basing the definition of a species upon these criteria gives the biological

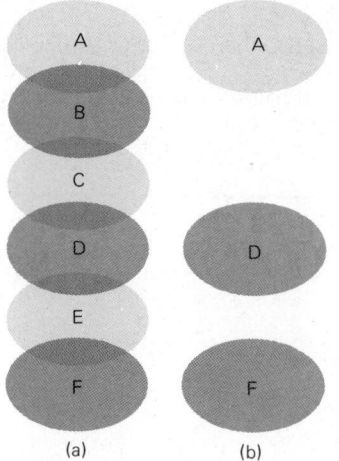

Figure 17-2 A cline series. (a) All populations are considered one species as long as continuous, interbreeding populations link them. If intermediate populations B, C, and E became extinct, then populations A, D, and F would be considered separate species.

species concept advantages and disadvantages opposite those of the morphological species concept. The biological definition is testable, at least in theory. If organisms will breed when in contact under natural conditions and the progeny are viable and fertile, then the organisms belong to the same species.

Another implication of the biological species concept is that the individual is not the basic unit of a species, as it is in the morphological species concept. The smallest unit in which breeding occurs is a population, which can evolve, whereas an individual cannot. A population is cohesive in that its members share a common gene pool.

Considering the population as the basic unit of a species means that as long as a reproductive link can be shown between two individuals, they belong to the same species even if they cannot breed together (Figure 17–2). Thus, individuals from populations at opposite ends of a cline* may be so different genetically that they cannot breed with one another. They are nevertheless members of one species because they are linked by interbreeding populations so that there is, at least in theory, some exchange of genes between them. The European cherry fruit fly, for instance, exists in a cline of populations that stretches across Europe. Members of adjacent populations occasionally interbreed, and so there is gene flow between the populations. When flies from eastern and western Europe are brought together to breed, however, the eggs produced fail to develop; the genetic differences between flies at the two ends of the cline are so great that they cannot interbreed even though they belong to the same species.

The biological species concept, while it is intellectually satisfying, does have some inherent weaknesses not found with the morphological species concept. The definition cannot be applied to organisms that always reproduce asexually or to those known only as fossils. The main stumbling block, though, is the requirement that in order for a group of organisms to constitute a species, all breeding members of that group must have the "potential" to interbreed "naturally." Many populations of organisms considered to be of the same species are totally **allopatric** (allos = other; patria = homeland); that is, they inhabit different areas, and their members never meet. To pass the biological species test, they must interbreed and produce viable progeny when they become **sympatric** (sym = same; hence, living in the same place).

It would seem a simple matter to bring members of the two populations together and see if breeding occurs. However, there have been several instances in which members of different species, living sympatrically and yet reproductively isolated in the wild, have actually bred with one another in captivity. Presumably, confining animals may alter their normal behavior so much that they will breed in the zoo although they never do so in the wild; lions and tigers are an example. Similarly, pollen distribution in plants may be affected by greenhouse conditions so that, for instance, species of primroses that never interbreed in the wild will do so in

---

*Cline A geographical gradient in a character with gradual, continuous changes from one population to the next (Section 15-B).

the laboratory. Such results make us less certain that we are in fact looking at members of the same species if we succeed in breeding members of allopatric populations under artificial conditions. Another problem with breeding experiments is that they are sometimes impractical. (How would you arrange to test northern and southern populations of blue whales? These are believed to be totally allopatric, because their food occurs only in cold waters; the northern and southern hemisphere populations are separated by the vast, warm equatorial zone.)

Even after researchers have tested organisms to see whether or not they belong to the same species, the species must still be defined or described in order to communicate with other biologists. This operational definition must, of necessity, be made in morphological terms, because morphology is the only preservable characteristic of an organism.

17-C Isolating Mechanisms

Under natural conditions, sympatric populations of different species are prevented from breeding with members of other species by various isolating mechanisms. Isolating mechanisms can be divided into two general categories, prezygotic and postzygotic, depending on whether they come into play before or after fertilization.

Prezygotic Isolation

Prezygotic isolating mechanisms are those that take effect before the zygote is formed, that is, before fertilization. Prezygotic means are by far the most common isolating mechanisms. They may be divided into four main categories:

1. **Habitat differences**. Two species may inhabit different areas so that they never encounter one another. A cow cannot attempt to mate with a whale because of the habitat difference. A less farfetched example is provided by the populations of fieldmice and chipmunks that occupy adjacent areas of meadow and woodland. They simply never encounter one another.
2. **Temporal differences in breeding times**. Two species may occupy the same habitat but reproduce at different times. Many species of frogs divide the breeding season among themselves in this way (Figure 17–3). One species may breed in May, but a very closely related species may not breed until June. The temporal difference may also be on a daily basis. Many species of ants have a particular range of hours during which mating flights occur. The more closely related the species, the further apart the timing of their mating flights, reducing the chance of mating between the two species. Similarly, in organisms with external fertilization (fertilization of eggs already shed outside the body), related species release gametes at different times, so that only the "right" sperm and eggs encounter one another.

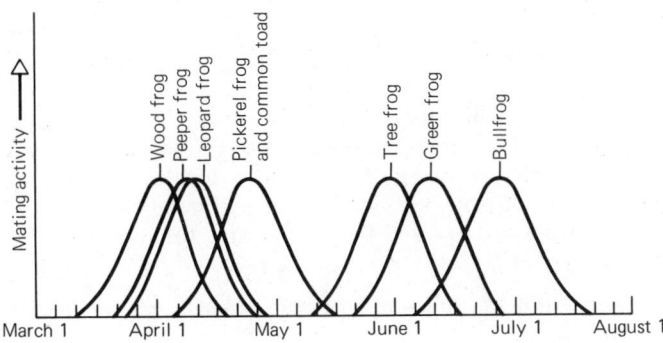

Figure 17-3 Prezygotic reproductive isolation results from mating at different times of year for various frogs and toads in upstate New York. Some of these species can breed with each other in the laboratory, but in the wild this does not occur. Two pairs of species shown have simultaneous mating seasons, but prefer different kinds of breeding habitat. Thus, peepers breed in woodland ponds and shallow water, while leopard frogs breed in swamps, and pickerel frogs breed in streams and ponds, whereas toads mate in ditches or puddles. (Bruce Wallace, Adrian M. Srb, *Adaptation*, 2nd ed., © 1964, p. 82. Reprinted by permission of Prentice-Hall, Inc., Englewood Cliffs, N.J.)

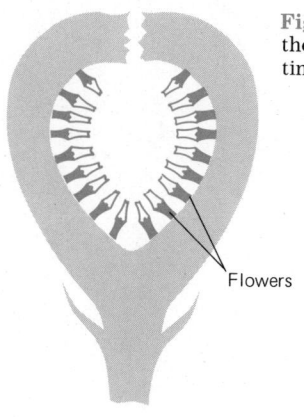

Figure 17-4 Mechanical means of limiting pollination in the fig. The head of flowers is so constructed that only the tiny wasp *Blastophaga psenis* can pollinate it.

Flowers

3. **Mechanical differences**. Organisms that have internal fertilization (including the land plants) may prevent interspecific breeding by mechanical means. For example, the *Calimyra* fig can be pollinated only by a certain species of wasp; other pollinating insects are too big to enter the flowers. The fig flowers (Figure 17–4) are enclosed, so that wind-blown pollen cannot reach them. Thus, if the wasp is not brought along when the fig is transplanted to a new area, the fig will not bear fruit there.

Similarly, the genitals of insects work on a kind of lock and key system. If the male and female are of different species, the genitals will not fit together well, and breeding will be prevented. Indeed, genital anatomy is so specific that it is often used as part of a morphological definition of an insect species.

4. **Behavioral specificity**. Another means of sexual isolation in animals is courtship behavior. There are numerous examples of chain displays in animal courtship. Each step is a prerequisite for the next, and the entire system is species-specific (Figure 17–5). Only members of the same species have the behavioral repertoire to complete the chain correctly so that mating can occur.

Postzygotic Isolation

Postzygotic isolating mechanisms prevent successful reproduction after fertilization has occurred. They are of three main types:

1. **Hybrid* inviability**. A zygote may be formed but not develop properly; the embryo or juvenile dies before reaching sexual maturity. This is usually due to genetic incompatibility, which prevents normal mitosis after fusion of the gamete nuclei. The chromosomes fail to attach to the spindle properly, or the centromeres fail to separate. As a result, some daughter cells do not receive a complete set of chromosomes, and they cannot survive. The more similar the maternal and paternal genes, the longer the embryo may live.

2. **Hybrid sterility**. The chromosomes may be compatible enough so that normal mitosis occurs and the individual develops. Meiosis, however, is much more sensitive to chromosomal imbalance than mitosis. Here the chromosomes may segregate abnormally, resulting in inviable gametes, or meiosis may not occur at all. Abnormal development of the gonads is another common event. This mechanism of species isolation will produce sterility of the hybrid offspring.

3. **Hybrid breakdown**. The last, and least effective, method of species isolation is hybrid breakdown. This occurs when the hybrid offspring are fertile, but produce many infertile or inviable offspring.

Selection for Prezygotic Isolation

In many of the isolating mechanisms discussed above, gametes are wasted; this happens when gametes are released but are somehow prevented from reaching each other, and in hybrid inviability, hybrid sterility, and hybrid breakdown. Wasting gametes is a useless expenditure of time and energy and will be heavily selected against. As a result, most species are isolated by mechanisms that waste as few

---

*Hybrid   Offspring of a cross between individuals of genetically different breeding lines or species.

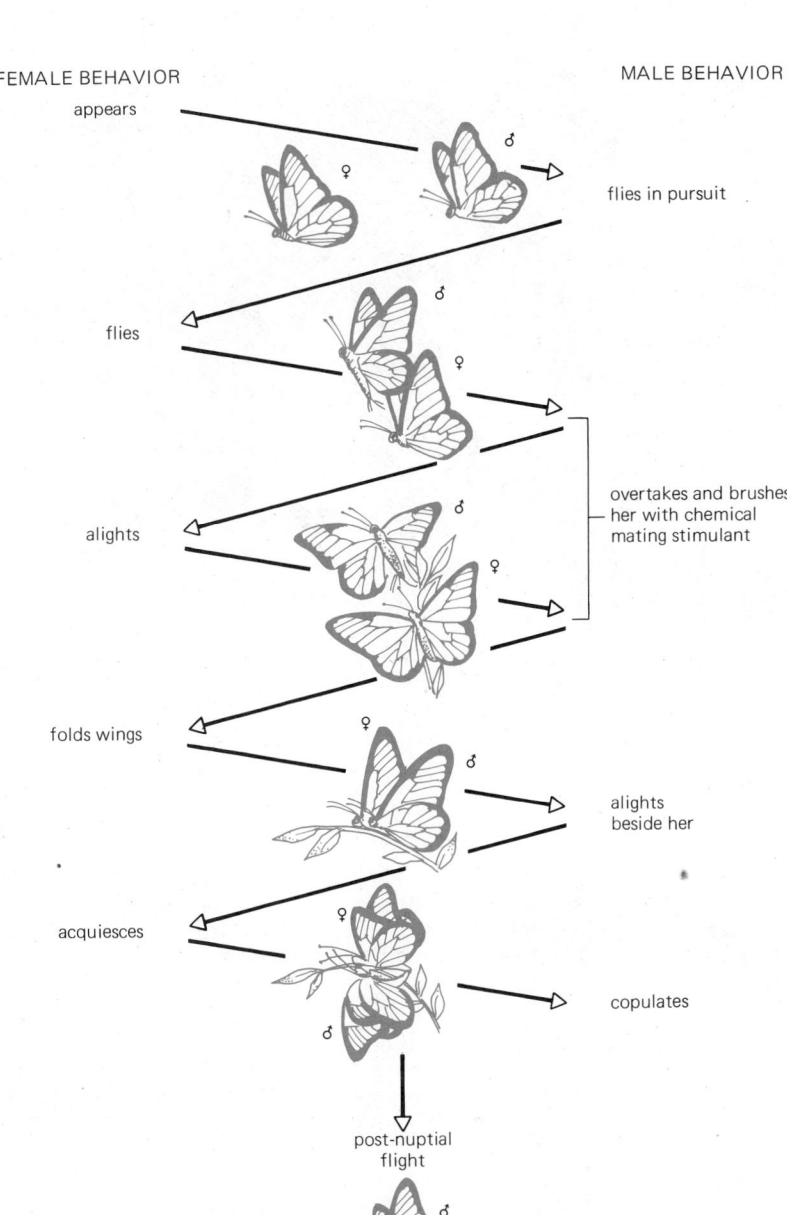

FEMALE BEHAVIOR   MALE BEHAVIOR

appears

                                                       flies in pursuit

flies

                                                       overtakes and brushes
alights   her with chemical
                                                       mating stimulant

folds wings

                                                       alights
                                                       beside her

acquiesces

                                                       copulates

post-nuptial
flight

Figure 17-5 Chain display during courtship in the queen butterfly. Before copulation can occur, male and female must each
behave in a way that stimulates the other to perform the next
step in the chain. (After Brower, L. P., *et al.*, *Zoologica* 50:1,
1965. Copyright © 1965 by the New York Zoological Society.
Reprinted by permission.)

gametes as possible. When two species isolated by mechanisms that waste gametes
occur in the same area, there will be strong selection in favor of individuals with
less wasteful isolating mechanisms. Because prezygotic isolating mechanisms are all
less wasteful of gametes than postzygotic mechanisms, selection will favor prezygotic means of reproductive isolation.

Function of Isolating Mechanisms

This discussion of reproductive isolation assumes that it is a good thing for
related species in the same area not to breed with one another. Why should this be
so? What is the selective advantage to a population of being genetically isolated from
other populations?

Figure 17-6 (a) A St. Kilda wren with an insect in her bill. She is more speckled on head, wings and tail than the mainland wren. (b) A wren from the Scottish mainland bringing insects to her nest. (Biophoto Associates)

As we saw in Chapter 15, each species has a complex of coadapted genes, which work smoothly with one another and which suit individuals to live in the local environment and exploit some of its resources more effectively than other species present. It is selectively advantageous for individuals whose genes make them reproductively successful in a certain area and way of life to become genetically isolated from other individuals whose genes are differently specialized, because the hybrid offspring are often less effective at pursuing either way of life and hence are less likely to leave offspring in their turn.

17-D Speciation

Speciation is the formation of one or more new species from an existing species by the buildup of reproductive isolation between them. For reproductive isolation to occur, gene flow between populations must be reduced to a level at which "foreign" genes entering one population from the other by hybridization can be eliminated by natural selection.

Allopatric Speciation

Allopatric speciation occurs when a population, formerly continuous in range, is split into two or more geographically separated populations. This can occur by subdivision of the original population, as when a glacier cuts across a species' range, or it can result when a small number of individuals colonizes a new habitat that is geographically separate from the original range. For instance, on St. Kilda, an island off the coast of Scotland, there is a species of bird found nowhere else: the St. Kilda wren. This wren closely resembles the species of wren found on the Scottish mainland and has almost certainly evolved from a population of Scottish wrens that migrated to St. Kilda (Figure 17–6).

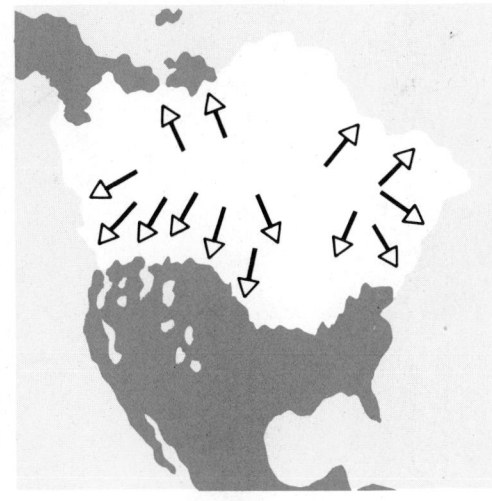

(a)   (b)

Figure 17-7 (a) A glacier. This is Taylor Glacier in Antarctica, which flows slowly into Lake Bonney, whose frozen waters are visible at the bottom right of the picture. (b) Map showing the southernmost extent of the Wisconsin ice sheet, last of the four ice sheets that pushed south into the United States during the four Pleistocene glaciations. (a, U.S. Navy)

Since allopatric populations of a species live in different geographical areas, they do not interbreed. Furthermore, because different environments present different selective pressures, allopatric populations invariably come to differ genetically from one another. The divergence of their gene pools will eventually produce two separate species if the differences become so great that members of the two populations can no longer interbreed. At this stage, it may be impossible to tell, just by looking at them, whether the two populations have evolved into different species or not. The test of speciation comes when the geographical barriers break down and the two populations become sympatric. If interbreeding does not then occur, or at least does not result in widespread hybridization, then the two populations are considered "good" species.

Allopatric speciation is probably the most common method by which new species of animals are formed, and some new species of other kinds of organisms also undoubtedly arise in this way.

PLEISTOCENE GLACIATIONS

The effects of the great glaciations of the Pleistocene epoch on the flora and fauna of North America provide a spectacular example of allopatric speciation on a massive scale. Over the past million years, and ending only a few thousand years ago, four major glaciations covered most of Canada, the northernmost United States, and the western mountains (as well as Northern Europe) with ice, often thousands of feet deep (Figure 17–7).

Each glacial advance pinched off many animals and plants into isolated populations. During the glaciation, these isolated populations diverged extensively from each other. Each glacial retreat permitted some of these isolated, and now different, populations to come back into contact again. In some cases reproductive isolation had proceeded far enough to produce two or more "good" species from one species present before the glaciation; in many other cases, reproductive isolation was not complete, and indeed we can still see various levels of hybridization between adjacent western and eastern populations of many birds, mammals, and insects, each presumably representing populations that survived in different refuges during the last major glaciation.

The wood warblers of North America have undergone impressive allopatric speciation; two dozen or more species can be found hunting insects in the foliage of even a single forest in parts of North America. The ancestral black-throated green warbler was probably distributed across the continent prior to the Pleistocene. At each glacial advance, its range was pushed down to the southeast forest refuge, pinching off a population in the west that subsequently formed a new species, repro-

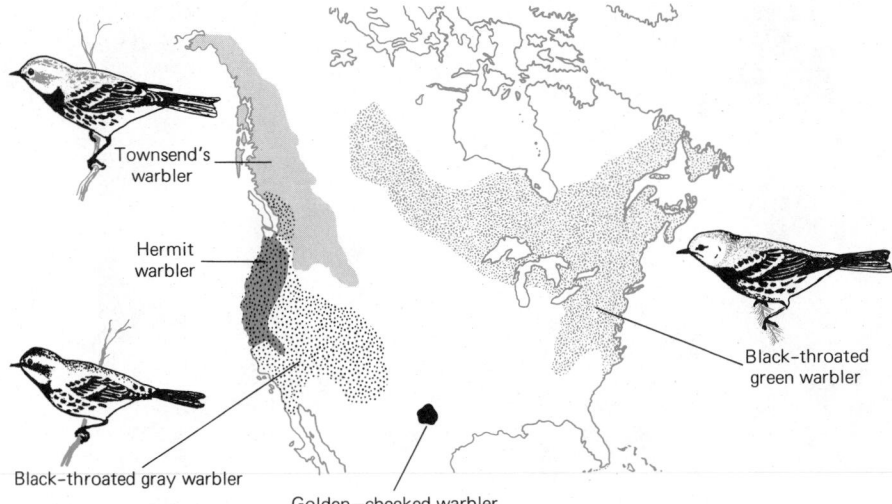

Townsend's
warbler

Hermit
warbler

Black-throated
green warbler

Black-throated gray warbler

Golden—cheeked warbler

Figure 17-8 Breeding areas of members of the black-throated green warbler group. Note that there is very little overlap between the breeding ranges. These five species probably all evolved from a single ancestral species. (From Lancaster, *The Living Bird*, Vol. 3, 1964. The Cornell Laboratory of Ornithology)

ductively isolated by the time of the following interglacial period (Figure 17–8). By the time of the next glacial advance, the black-throated green warbler had again extended across most of the continent, and was again pinched into two populations by the next glacial advance, and so on. Similar multiplication of species from several other ancestral eastern warblers gave rise to a total of 46 species of forest-adapted wood warblers in North America today, and probably to many species that have since become extinct.

Sympatric Speciation

Many organisms are capable of undergoing **sympatric speciation**, the production of new species within a single population. Two genetic mechanisms that can produce instantaneous speciation are fairly widespread in plants. These two mechanisms, autopolyploidy and amphiploidy, are both instances of **polyploidy**, or multiplication of the normal chromosome number. An **autopolyploid** arises when more than two haploid chromosome sets of a single species go to form the zygote. An **amphiploid** is a polyploid with diploid chromosome sets from two different species. Probably more than a third of all plant species have arisen by polyploidy, although the phenomenon is rare in conifers. Nearly all domesticated species and varieties of plants with larger fruit and flowers than those of their wild ancestors are polyploid (Figure 17–9).

Figure 17-9 Flowers of a wild *Primula* (top) and of a garden polyanthus (bottom), a tetraploid variety bred from the wild plant. (Biophoto Associates)

Polyploidy in plants is possible because plants can often be self-fertilized, and most can also reproduce vegetatively. If they could not, the polyploid individual would usually be doomed to extinction because it could not find a genetically compatible mate. Speciation by polyploidy probably also occurs occasionally in sexually reproducing animals. Some species of grasshoppers and primitive fish differ from one another in ploidy and may have arisen in this way. Many animal species that are permanently parthenogenetic (reproducing by means of unfertilized eggs) are certainly polyploid.

A new species may also form by introgression. **Introgression** is the incorporation of the genes of one species into the gene complex of another as a result of successful hybridization. Among plant species, where cross-fertilization between species is much more common than in animals, many examples of introgression are known. If a population of one species receives a large transplant of genes from another species by introgressive hybridization, it might change so radically in a short space of time that it would be considered a new species.

It has been shown experimentally that sympatric speciation can theoretically occur in animals. If a population is polymorphic such that two morphs (forms) are advantageous but hybrids between them are at a selective disadvantage, selection against the hybrids can set up reproductive isolation within the population. Aubrey Manning demonstrated this with a laboratory population of *Drosophila*, using an artificial polymorphism in number of bristles, a character governed by many different genes. In each generation, Manning removed individuals with intermediate numbers of bristles, and only those with high and low numbers were allowed to breed. In just 13 generations, there was complete reproductive isolation between the two groups; females would mate only with males who had a similar bristle number, so that all of their offspring resembled their parents and no hybrids were formed. (Bristle number is not a character normally involved in mate choice but is sometimes linked genetically to characters that are.)

Can sympatric speciation occur in this way within natural populations of animals? In many groups of animals it undoubtedly cannot; in others it is probably the predominant method of speciation. For instance, adults of the apple maggot, *Rhagoletis pomonella*, are attracted to apple trees, which they find by visual and chemical cues. Around 1960, apple maggots colonized the cherry trees in Wisconsin, where apple and cherry orchards grow side by side. This switch probably resulted from a relatively slight genetic change affecting the chemistry by which the flies find their food plant. The populations on apple and cherry trees are now distinct. The cherry race of *R. pomonella* now emerges in the early summer, at the peak ripening time for cherries but before apples ripen. This produces some reproductive isolation from the parent apple race, which continues to emerge in late summer.

These two races, populations with distinct genetically based host preferences, are not yet "good" species since they continue to hybridize to some extent. Gene flow between them has been reduced so greatly, however, that it can only be a matter of time before complete reproductive isolation evolves (if neither population becomes extinct first). The apple race of *Rhagoletis pomonella* itself first appeared on apples in the Hudson Valley in 1864, presumably originating from another race that attacks the fruits of wild hawthorn trees.

The apple maggot example illustrates how a high degree of reproductive isolation can arise very rapidly, even in the middle of a species' range, as a result of very little genetic change. An important feature of this case is that the flies mate on or near the food plant where the female later lays her eggs. This means that a male and a female with the mutation for cherry preference will meet and mate near a cherry tree. At least some of their offspring will probably inherit the trait, and it will not be swamped in the gene pool of the parent population. This sort of mating system is common in insects that are parasites of plants or animals, and it is probable that a substantial number of them arose sympatrically.

Sympatric speciation in vertebrates could arise through a single, fairly major genetic change in a male that has a harem of females or that sires a large litter from one female. Suppose that early in his embryonic development a cell destined to form the male's sperm undergoes mutation; this mutation will be carried by his sperm when he becomes an adult. Further suppose that the mutation will cause his offspring to behave in a way that makes them unlikely to mate with individuals who do not have the mutation. Because the male's body cells did not undergo the mutation, he himself breeds with females lacking the mutation, and he may pass the mutation to many offspring. When they reach adulthood, these offspring could mate with one another, establishing a reproductively isolated population (a new species) in one generation. The genetic change in such a case might be one affecting the season in which the offspring breed, or the enzymes or sense organs determining their food preference. The fossil record shows that 27 species of antelope existed together, eating different kinds of plants, in one period of 6 million years. Since antelope have a harem type of society, it is quite possible that these species could have evolved according to this model.

Parapatric Speciation

Parapatric speciation occurs whenever species evolve as contiguous populations in a continuous cline. It differs from allopatric speciation in that there is never any physical barrier to gene flow, and yet it is not really sympatric speciation because it occurs only at the edge of the parent species' range. A series of reproductively isolated snail populations of the genus *Partula* occurs on vertical overhanging cliffs around the island of Moorea in the Tahiti group. Some hybridization occurs between adjacent populations (which are really species) but the changes in color pattern are very sharp, probably reflecting sharp discontinuity between coadapted gene complexes (see Section 15-C). The environment along such clines may change very gradually, but sharp genotype changes can result because there comes a limit beyond which an existing coadapted gene complex cannot adapt further without a complete reordering of the genome.

This kind of speciation is likely to be found only in plants and animals of low dispersal ability, where adaptations to new conditions on the periphery of the species' range are less likely to be swamped by gene flow from the parent population. Once a new population is established, selection against hybrids is likely to occur because their genes are not compatible with those of the parent population.

Selection Against Hybrids

When two previously separated populations of related organisms come together, they may breed with one another. This situation has various possible outcomes. First, if the two are already "good" species, they will not interbreed at all; second, they may interbreed freely so that they merge into one big population and all genetic distinction between them disappears. A third possibility falls between these two: members of the two populations may interbreed, but if the hybrid offspring between them are at a selective disadvantage compared with the members of either parent population, selection will favor an increase in the reproductive isolation between the two populations.

An example of this third situation occurred in the 1970s in southwestern Arizona. Because of changes in agricultural practices, two populations of yellow sulfur butterflies came together for the first time. Members of the two populations were visibly different and had been considered as two different species; however, when they met, they bred with each other and produced visibly distinct hybrid offspring, as well as offspring that resembled members of both parent populations. Over a span of several years, the hybrids became less and less common, and finally they disappeared almost completely. Butterflies were now mating only with members of their own populations, and two good species had formed.

The actual mechanism that brought about reproductive isolation is known in this case. Males of either population tried to mate with any and all females. When the populations first became sympatric, most females would mate with the first male who came along. In later years, however, more and more females would mate only with males from their own population; when the "wrong" sort of male tried to copulate with her, a female would flip her abdomen away. Whether a female would mate or not depended on the visual appearance of the male. She rejected males whose appearance showed that they were either hybrids or from the other population. It appeared that the hybrid offspring were somehow at a selective disadvantage compared with their "purebred" relatives and that they left fewer offspring. In this situation, any female constituted in such a way that she would mate only with a male of her own population would leave more descendants than her less discriminating sisters. If the mother's sexual preference was genetically determined, at least some of her daughters would be likely to inherit it and behave in the same way. With heavy selection against hybrid matings, genes that permitted a female to avoid mating except with a member of her own population spread rapidly through each population. We do not know how often reproductive isolation arises in this way, since this is probably the only example of its type that anyone has ever been lucky enough to observe in action.

17-E The Frequency of Speciation

It now appears that the paleontological evidence for the creation of new species by gradual evolutionary change is weak. Instead, the fossil record reveals long periods of very little change in a species, interrupted by the disappearance of that species and the sudden appearance of related, morphologically distinct species. The evolution of the horse is sometimes portrayed as a gradual process, with body size steadily increasing and the toes becoming reduced over evolutionary time. In fact, though, speciation probably occurred very many times during the evolution of horses, leading to both species with smaller bodies and species with larger bodies. Since larger-bodied species survived longer than smaller-bodied species, what we see in the fossil record is the appearance of a gradual increase in body size.

Gradual evolution is probably of much less significance in speciation than are the kinds of allopatric and sympatric speciation discussed earlier. What is more, speciation almost certainly occurs much more rapidly and more frequently than used to be thought. The rapid formation of some 600 species of *Drosophila* in the Hawaiian islands, of more than 200 species of cichlid* fish in some African lakes, and of 239 species of gammarid* Crustacea in Lake Baikal (U.S.S.R.) provides dramatic examples of what can happen.

The more specialized a species, the more liable it is to rapid extinction. An animal that eats and breeds only on a particular species of plant, for instance, would be wiped out by the disappearance of that plant. The fossil record shows that species exist for an average of 2 to 10 million years, but the actual range is enormous, from species that probably survive only a few generations to extraordinary examples like some blue-green bacteria that have survived, apparently unchanged, for hundreds of millions of years. (The essay on continental drift at the end of this chapter contains some examples of species that remained unchanged during their isolation on different continents.)

---

* Cichlid   Freshwater fish resembling a sunfish.
* Gammarid   Belonging to a family of Crustacea whose members swim on their sides.

Essay: Continental Drift

Geological changes produce new barriers between populations and change climates; these conditions promote species formation. The gradual drifting apart of the continents was the most dramatic geological change of them all.

Biologists believed in continental drift while geologists still scoffed at the notion. The biologists' belief was based on the distribution of species. Lungfishes, for instance, are restricted to freshwater habitats and are not noted for their ability to disperse long distances; yet the three living species, while clearly related, occur in South America, South Africa, and Australia, respectively. Are we really to believe that ancestral lungfishes crossed major oceans or thousands of miles of inhospitable temperate and subarctic terrain? Flightless birds—the ostrich in Africa, the rhea in South America, and the emu in New Zealand—share closely related species of flightless

insect parasites in their feathers. How did this strange situation arise? The distribution of some more ancient groups of organisms is even more puzzling. Fossils of *Glossopteris*, a genus of Permian plants, occur in South America, Africa, India, and Australia (Figure 17–A); although it is conceivable that the pollen or seeds of these plants could have been dispersed over vast ocean distances by winds, this is hardly a compelling explanation for their distribution. To account for these and many other distributions, especially among fossil animals and plants, theories involving land bridges between the continents developed. These theories were then replaced by the belief that the continents themselves had moved.

During the 1960s the earth sciences underwent a revolution with the general acceptance of the theory of **plate tectonics**, a theory that not only accounts satisfactorily for most of the major geological features on the surface of the globe, but also resolves some formerly puzzling problems in the distribution of animals and plants. The part of plate tectonic theory that concerns us here is the thesis that the outermost layer of the earth is made up of about 15 rigid

plates, which are pushed past, over, and under one another. The plates move because they are "floating" on the mantle of molten rock inside the earth.

Geological evidence, from patterns of rock formations and patterns of magnetism in rocks, strongly supports the fossil evidence suggesting that in Permian and Triassic times, 200 to 250 million years ago, the southern continents (South America, Africa, Antarctica, and Australia), along with India, were united as one large land mass called Gondwanaland (Figure 17–B). North America, Europe, and Asia were meanwhile united as a northern land mass called Laurasia. Moreover, these two supercontinents were themselves united for a time to form a single world land mass, named Pangaea by Alfred Wegener in the 1920s. The rest of the earth was covered by ocean—the ancestor of today's Pacific. The Atlantic, Indian and Antarctic oceans began as rifts in Pangaea when the plates beneath it separated.

Wegener's theory of **continental drift** is now generally accepted. It proposes that the continents have reached their present positions by

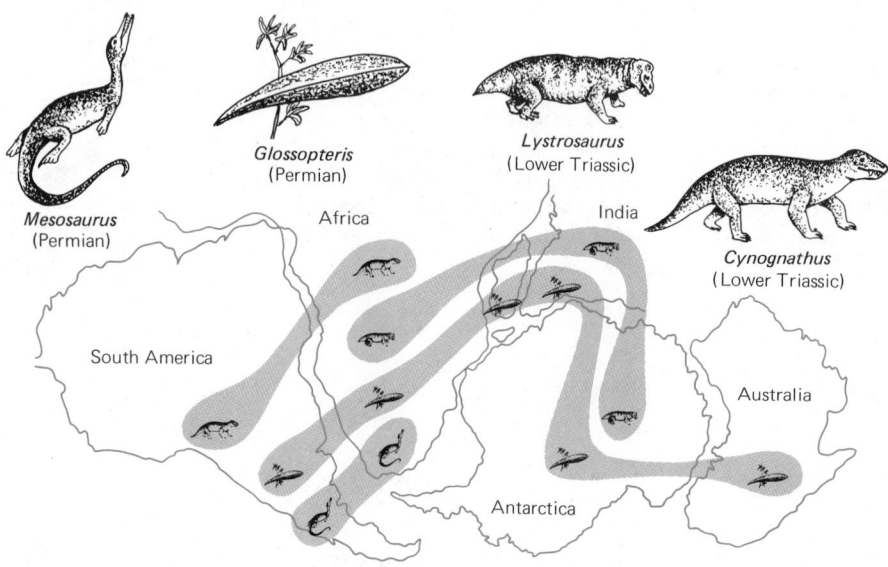

Figure 17-A The distributions of four fossil species found on more than one of the southern continents, which suggested that the southern continents were once joined. (You can find the Permian and Triassic on the endpaper of this book.) (From *Wandering Lands and Animals* by E.H. Colbert. Copyright © 1973 by Edwin H. Colbert. Reprinted by permission of E. P. Dutton)

moving apart from an original union during the Permian and Triassic periods. What happened before the Permian is less clear, although the evidence strongly suggests that the continents were also separated earlier, in the Paleozoic era.

Gondwanaland and Laurasia separated from each other during the Jurassic period; by then the evolution of the dinosaurs was well under way, coniferous trees had existed for millions of years, and the first mammals and birds had evolved. Before Gondwanaland began to split up into the present-day southern continents and India, the dinosaurs and conifers had become supreme, ancestral mammals were well established, and flowering plants had already evolved.

After separation of the continents, the evolution of each group of organisms was able to proceed in different ways. Marsupial (pouched) mammals speciated in Australia and South America, while placental mammals dominated the other land masses. India drifted north and collided with Asia during the Oligocene, the force of the collision giving rise to the Himalaya Mountains. At about the same time, Laurasia was splitting apart. By this time bats, carnivores, primates, ungulates, rodents, and many other orders of mammals had all evolved in Laurasia; thus it is not surprising that North American, Asian, and European mammals are related to each other more closely than any of them are related to the mammals of South America or Australia.

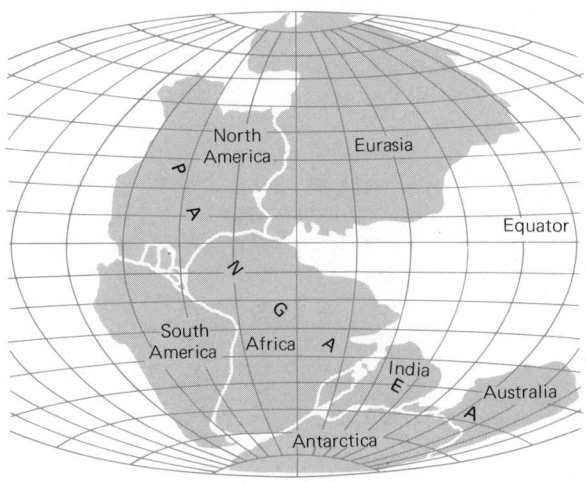

(a) Pangaea: Late Paleozoic Era, 230 million years ago.

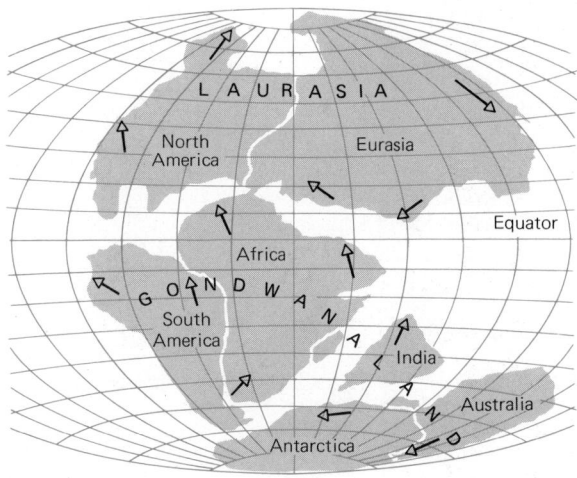

(b) Laurasia and Gondwanaland: Mesozoic Era, 180 million years ago.

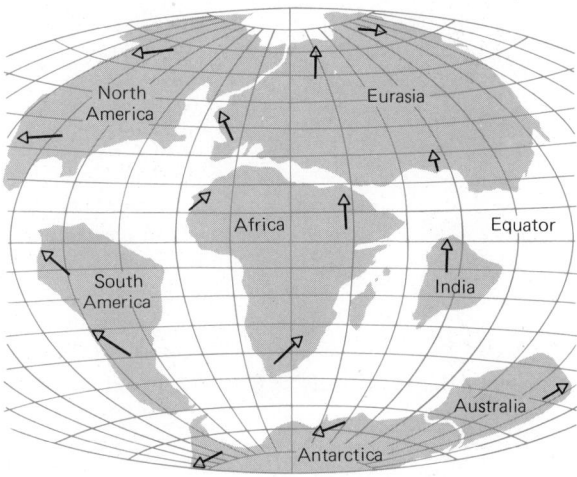

(c) Modern continents (with minor exceptions) had formed by the end of the Mesozoic, 110 million years ago.

Figure 17-B Continental drift (indicated by arrows) produced the modern continents from a single land mass—Pangaea—that existed in the Paleozoic era.

SUMMARY

A species is a group of interbreeding organisms that is reproductively isolated from other such groups. This reproductive isolation protects a coadapted gene pool that gives rise to a limited number of phenotypes; it is clearly legitimate to classify organisms by these phenotypes (i.e., morphologically) because they reflect something that is real in reproductive terms: the coadapted genes of an interbreeding group of organisms. Such morphological classification is generally more practical than classification by ability or inability to interbreed. Organisms do not fit tidily into morphological or reproductive definitions of species.

Reproductive isolation between species may be due to prezygotic or postzygotic mechanisms. Of the two, prezygotic mechanisms are more advantageous because they prevent the waste of gametes and energy that are expended before postzygotic mechanisms can come into play.

Allopatric speciation occurs when two populations of a species become separated so that there is no gene flow between them. Each evolves under the influence of local selective pressures, and they may become so different that they are considered different species. Whether they actually are "good" species or not will be determined only when the two populations again become sympatric. Sympatry will have one of three possible outcomes: introgressive hybridization may occur and the two populations become one; the hybrids may be at a selective disadvantage so that reproductive isolation between the two will increase; or they may already be separate species and not interbreed at all.

Sympatric speciation usually occurs as a result of polyploidy. It might also, theoretically, arise in a polymorphic population in which hybrids between the morphs were at a selective disadvantage, but no such situation has even been conclusively demonstrated in nature. Mutation in one or a few genes may also suffice to create reproductive isolation, and hence speciation, by creating different morphs in sympatric populations.

Parapatric speciation may occur as two adjacent populations develop reproductive isolation from each other. A sharp boundary or contrast in some environmental factor may act as the selective pressure favoring isolation of the two morphs, each of which is better adapted to environmental conditions on its own side of the "boundary."

In light of recent evidence, the traditional concept of speciation as a result of gradual and continuous change in the gene pool is being replaced. It now seems likely that rather abrupt changes in a few loci or chromosomes are rapidly followed by selection for reproductive isolation between "new" and "old" morphs of a population, resulting in two or more "good" species.

OBJECTIVES

From your study of this chapter, you should be able to:

1. List the advantages and disadvantages of basing classification on morphological characters.

2. Give a biological definition of a species.

3. List the advantages and disadvantages of the biological species concept.

4. List and describe seven categories of reproductive isolation; classify each as prezygotic or postzygotic.

5. Distinguish between allopatric, sympatric, and parapatric modes of speciation, and describe how they work. List or recognize examples of each type.

6. List and explain three possible outcomes when two separate populations become sympatric.

SELF-QUIZ

1. Tell which species concept, morphological or biological, should be applied to decide how many species exist in each situation described below.

   a. Fossils that appear identical are found in rocks on the two coasts of North America.

   b. Grublike white larvae are found in a nest of ants.

   c. A colony of ants contains individuals of many different body structures.

   d. You find two parthenogenetic female aphids. (Parthenogenetic organisms reproduce via unfertilized eggs.)

   e. The Rongovian government will not permit the export of the rare red-faced blooper for breeding experiments with a similar population in the Yukon; however, Jack Anderson has managed to obtain (from undisclosed sources) photographs of several specimens, which he has turned over to scientists.

   f. Two morphologically distinct populations of yellow sulfur butterflies interbred and produced distinctly different hybrid individuals.

2. True or False. Amphiploidy can occur only if there are no prezygotic isolating mechanisms between the populations involved.

For each situation below, pick the type of isolating mechanism exhibited and tell whether it is prezygotic or postzygotic.

___ 3. One of two species of crickets in an area mates at dusk, the other at dawn.

___ 4. Female yellow sulfur butterflies avoid mating with the wrong sort of male.

___ 5. Crossing bears from two different populations produces cubs that never produce any offspring as adults.

___ 6. Two species of sparrows that inhabit lowland valleys live on opposite sides of the Rocky Mountains.

   a. behavioral
   b. habitat
   c. hybrid breakdown
   d. hybrid inviability
   e. hybrid sterility
   f. mechanical
   g. temporal

QUESTIONS FOR DISCUSSION

1. What, in biological terms, is a "race" of people or any other animal?

2. Female *Drosophila* select their mates on the basis of variations in the courtship dance that the males perform. In Manning's experiment (see Section 17-D), bristle number was chosen as a convenient characteristic by which people could select different flies. At the beginning of the experiment, genes governing bristle number were occasionally linked with (i.e., on the same chromosome as) genes determining the pattern of the courtship dance. At the end of the experiment this situation had changed. What would you predict had happened to the linkage between genes controlling bristle number and those controlling the dance?

3. You are asked to catalog the species of insects living in a remote area that has never been visited by biologists before. What criteria would you use to determine whether individuals you collect are in the same or different species? What problems might you encounter in using this procedure, and how would you resolve these problems?

4. A visitor to an island notices that several plant species on the island are somewhat different from those on the mainland. What differences might there be, and how would they have come about?

REFERENCES AND FURTHER READING

Bush, L. "Modes of animal speciation." *Annual Review of Ecology and Systematics* 6:339, 1975.

Colbert, E. H. *Wandering Lands and Animals.* New York: E. P. Dutton, 1973. A very readable introduction to continental drift.

Dobzhansky, T. *Genetics and the Evolutionary Process.* New York: Columbia University Press, 1970. An advanced book on genetics, evolution, and speciation by one of the great evolutionists.

Gould, S. J. *Ontogeny and Phylogeny.* Cambridge, MA: Harvard University Press, 1977. Summarizes the modern view of what the fossil record has to tell us about speciation.

Lack, D. *Darwin's Finches.* New York: Cambridge University Press, 1947. (The *Scientific American* article, "Darwin's Finches," April 1953, is a brief summary of this book.)

Mayr, E. *Animal Species and Evolution.* Cambridge, MA: Belknap Press, 1966. The basic reference on allopatric speciation in animals.

Stebbins, G. L. *Processes of Organic Evolution*, 2d ed. Englewood Cliffs, NJ: Prentice-Hall, 1971. A good, fairly advanced book on evolution and speciation.

Tauber, C. A., and M. J. Tauber. "A genetic model for sympatric speciation through habitat diversification and seasonal isolation." *Nature* 268:702, 1977. Modern work suggests that sympatric speciation is much more common than was first suspected. This paper discusses some of the ways in which it can occur.

White, M. *Modes of Speciation.* San Francisco: W. H. Freeman, 1977. An advanced book on speciation in animals and plants, covering just about every aspect of the subject.

CHAPTER 18

EVOLUTION AND REPRODUCTION

We have seen that an individual's evolutionary success is measured in terms of the proportion of its genes present in future generations of the population. In evolutionary terms, it is useless for an individual to have genes that allow it to find food, avoid predators, survive harsh weather, and carry on the business of living, unless it also has genes that make it reproduce.

The actual anatomy, physiology, and behavior involved in reproduction are covered in other chapters. This chapter considers different patterns of reproduction found in the living world and the kinds of selective pressures that have encouraged the proliferation of the responsible genes. Most of the chapter is devoted to sexual reproduction, a rather complicated process with many logistical drawbacks. What advantages of sexual reproduction more than compensate for its disadvantages and make it so widespread among organisms? How did sexual reproduction evolve in the first place? Beyond these questions lies the question of why an organism has one type of sexual system rather than another—the variations are almost endless.

In all of this discussion we must keep in mind the selective pressures acting on genes that govern reproduction; successful reproduction automatically selects for the genes bringing that method of reproduction about. We shall see that different selective pressures bring about the variations in reproductive patterns we see around us today.

18-A Is Sex Necessary?

At first glance it seems that sex *is* necessary. Most animals, and many higher plants, rely on sexual reproduction to perpetuate their genes. On the other hand, sexual reproduction is much less common among lower plants, bacteria, fungi, and unicellular eukaryotic organisms.

Although it is widespread, sexual reproduction wastes more energy than does asexual reproduction. If sperm, pollen, or eggs are released into the water or air to find each other by chance, millions of them are lost and so the energy spent to make them is wasted. In animals with internal fertilization, fewer eggs and sperm are lost, but the animal has to spend a lot of time and energy finding and courting a mate. Further energy is wasted because sexually produced offspring are so small at first that many of them are killed by predation, inability to find food or water, and so forth.

By contrast, when an organism reproduces asexually, only one or a few offspring usually form at a time, and the offspring often remain attached to the parent until they reach a much larger size. Many unicellular organisms reproduce asexually simply by dividing into two identical, smaller cells. **Budding** of smaller individuals, which eventually detach from the parent, and **parthenogenesis**—development of an unfertilized egg into a new individual—are common forms of asexual reproduction in animals. Many plants also produce new individuals, which remain attached to the parent until they are quite large. These forms of asexual reproduction require more energy per offspring than is needed for sexual reproduction, but each asexually produced offspring receives a better start in life than does a sexually produced organism and has a good chance of surviving. Thus, asexual reproduction is less apt to waste energy than is sexual reproduction.

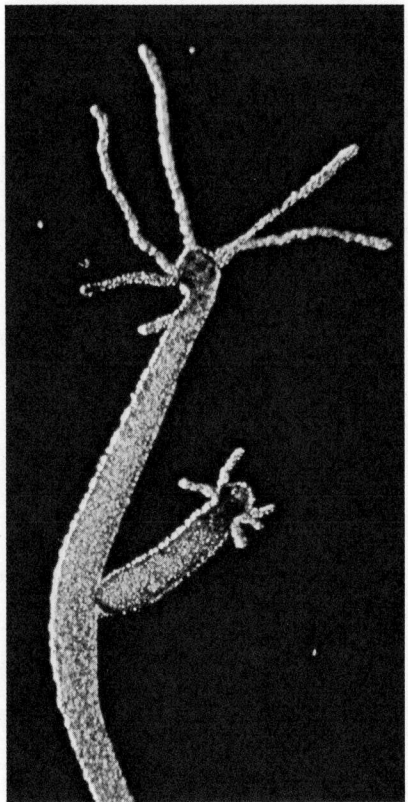

Figure 18-1 *Hydra,* a commonly studied invertebrate animal, produces new individuals asexually by budding. When the bud is large enough, it breaks off and starts an independent life. (Biophoto Associates)

As a result, an asexually reproducing organism like a strawberry plant can, on the average, produce many more surviving offspring in a season than can a similar plant that reproduces sexually. Why, then, do so many organisms use so much energy and waste so many cells reproducing sexually? Sexual reproduction must have tremendous adaptive value, or it could not have survived and become so widespread.

18-B Selection for Sexual Reproduction

The main biological difference between sexual and asexual reproduction is that sexual reproduction involves **genetic recombination**, and so it produces more genetic variability in a population. An organism that reproduces asexually passes on all its own genes to all its offspring, and therefore they are all alike genetically—that is, the original organism plus its offspring form a **clone**. The only genetic variation that can arise in such a population comes from **mutations**.

Sexual organisms can also undergo mutation, but their main—often enormous—variability arises during meiosis and fertilization. Each of us inherits only half our mother's genes and half our father's genes, and these genes were shuffled during meiosis. This is why we show many genetic differences from our parents and our brothers and sisters, and why we are even more different from people who are not close relatives.

Some organisms can reproduce both with and without sex. When we study such organisms, we find that there are times when the genetic variation produced by sexual reproduction is advantageous and other times when it is not. Many unicellular organisms, algae, and small invertebrates reproduce asexually during the summer and then reproduce sexually when the temperature declines and days become shorter in the fall. Experiments have shown that many of these organisms change from asexual to sexual reproduction because of changing environmental conditions. Depriving them of food, heat, light, or oxygen will often effect the switch.

Figure 18-2 Different varieties of domestic fowl show the enormous genetic variation that is possible within a species that reproduces sexually. (Biophoto Associates)

It makes sense that environmental conditions should cause the switch from no sex to sex in this way. Under this system, an organism that is well adapted to its environment can reproduce numerous, equally well-adapted copies of itself while conditions remain constant during the summer. When conditions then change in the fall, the organism reproduces sexually, creating many genetically different offspring. Because they are different, there is a good chance that some of them will survive the changing conditions.

Many species of bacteria and other "lower" organisms occur all over the world in a variety of habitats, and some of these species have existed virtually unchanged for more than 500 million years; it is hard to imagine an environmental change so catastrophic and widespread as to threaten them with extinction. Most higher plants and animals, on the other hand, are specialized and can live only in the few places that supply their particular needs. Individual species survive for only a few million years (which is not long in evolutionary terms).

A species with only asexual reproduction (dandelions are a good example) may do very well for a while, but is ultimately doomed to extinction much more surely than is a species that reproduces sexually. A change in the environment that kills dandelions will kill *all* dandelions in the area because they are all very similar genetically. Organisms need the genetic variability produced by sexual reproduction if they are to form new species adapted to new environments. If they do not form new species, the whole group may become extinct. We think of the ruling reptiles (including dinosaurs) as a large group of higher animals that became extinct. But in fact the ruling reptiles' descendants, the birds, are alive and well and living all over the world. Without sexual reproduction, the reptiles could not have left these very different, successful descendants.

Sexual reproduction, then, is inefficient for an organism that lives all over the world and is tolerant of environmental change. More specialized species live in constant danger of dying out, leaving no descendants, if they do not have the genetic diversity that sexual reproduction provides.

Figure 18-3 Dandelions: widespread and successful plants that reproduce asexually. (Biophoto Associates)

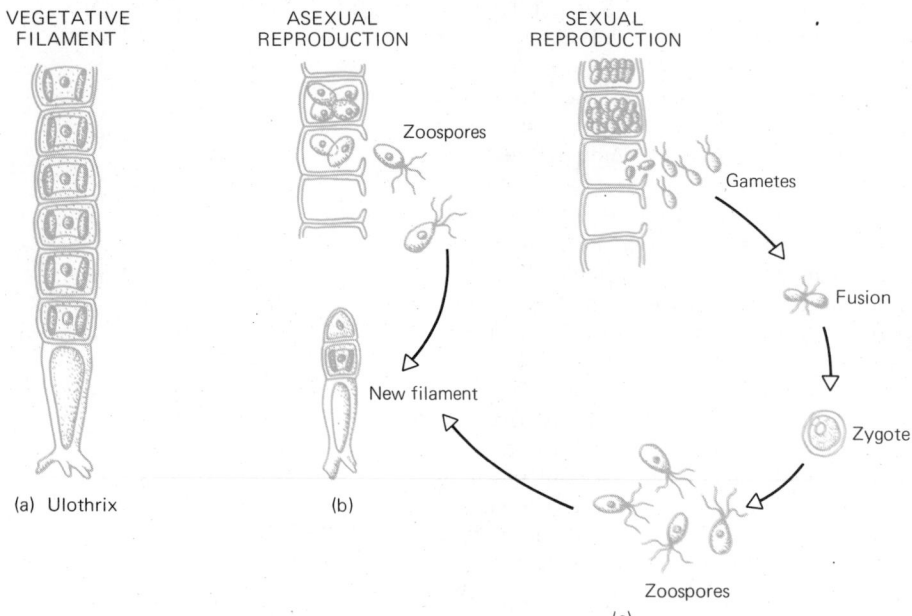

VEGETATIVE
FILAMENT

ASEXUAL
REPRODUCTION

SEXUAL
REPRODUCTION

Zoospores

Gametes

Fusion

New filament

Zygote

(a) Ulothrix

(b)

Zoospores

(c)

Figure 18-4 Life history of *Ulothrix*. In asexual reproduction (b), each zoospore germinates into a filament with the same genetic makeup as the parent filament. In sexual reproduction (c), gametes fuse to form a diploid zygote, which undergoes meiosis to form four zoospores of new genetic types.

18-C   Origin of Sexual Reproduction in Eukaryotes

Sexual reproduction involving meiosis and fertilization at some stage in the life history occurs only in eukaryotes. Genetic recombination in prokaryotes is known, but the amount of genetic information transferred from one individual to another is not constant, and we do not know how common such recombination is in nature (see Section 22-B).

The first sexual reproduction among eukaryotes was probably similar to that seen even today in organisms like *Ulothrix*. This is an alga consisting of a filament of haploid (having unpaired chromosomes) cells produced by division of one original cell. During the growing season, *Ulothrix* reproduces asexually; some cells in the filament undergo a number of internal divisions, producing several small flagellated cells called **zoospores** (zoo = animal, due to the fact that the spores swim by means of their flagella; spore = reproductive cell). The zoospores are released by the rupture of the parent cell, and they swim off to suitable locations, settle, and divide, starting new filaments (Figure 18–4).

Sexual reproduction probably evolved as a modification of this program. In sexual reproduction, some of the cells of the filament divide more times than for asexual reproduction, and the resulting flagellated cells are very tiny. These are released and swim about in an apparently aimless fashion. There may be thousands swarming in the water at once, because all the *Ulothrix* filaments in an area release these tiny cells at the same time. When two of them collide, they may stick together and then fuse to become one. This is clearly a sexual event, and it forms a diploid (having pairs of homologous chromosomes) zygote. (One theory on the origin of fertilization holds that it occurred as an accidental fusion of two undersized spores, which turned out to be advantageous. The characteristics of the first gametes that caused them to fuse were thus perpetuated by natural selection. It is less easy to imagine how the other half of sexual reproduction, meiosis, originated.)

The *Ulothrix* zygote secretes a protective cover around itself and becomes dormant. When it resumes activity, it immediately undergoes meiosis and produces four new haploid zoospores, each of which settles down and produces a new filament by mitosis. Note that *Ulothrix*, like many other primitive plants, is diploid only in the zygote stage; the vegetative filaments and zoospores are haploid. Thus each

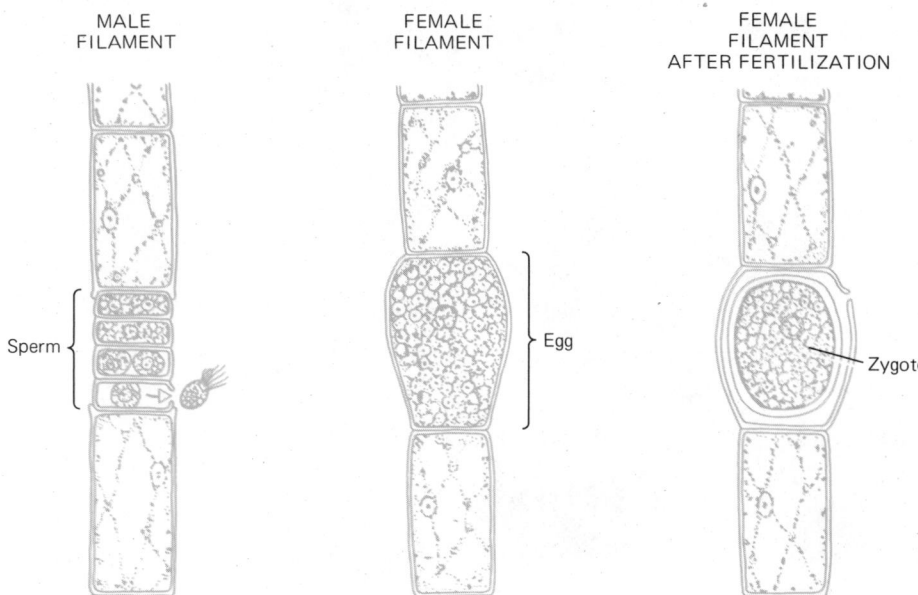

MALE
FILAMENT

FEMALE
FILAMENT

FEMALE
FILAMENT
AFTER FERTILIZATION

Sperm

Egg

Zygote

Figure 18-5 Sexual reproduction in *Oedogonium*. Small, flagellated gametes (sperm) from male filaments swim to large, immobile gametes (eggs) retained in the female filaments.

fertilization/meiosis sequence results in four new genetic combinations, assuming that both independent assortment of chromosomes and crossing-over between chromosomes occur during meiosis (Sections 13-H and 13-L).

In *Ulothrix* we see a very simple form of sexual reproduction. The gametes cannot be distinguished from one another by sight, although some workers claim that only gametes from different filaments will fuse with one another. Sexual reproduction of this type, in which the gametes appear identical, is called **isogamy** (iso = same; gamy = gametes) and is found in various relatively simple organisms.

An evolutionary advance over isogamy is found in *Oedogonium*, another filamentous alga, which is very similar to *Ulothrix* in form and habitat. In *Oedogonium*, single cells in some filaments become eggs by acquiring nutrients from neighboring cells. In other filaments, cells undergo several internal divisions, forming many flagellated, swimming sperm. The sperm are released from the parent cell and swim to the egg-bearing filaments, attracted by a chemical exuded from the egg cells. The first sperm to arrive penetrates and fertilizes the egg; the egg membrane becomes resistant to the entry of any more sperm, and the newly formed zygote secretes a protective wall around itself and becomes dormant (Figure 18–5). As in *Ulothrix*, when the zygote resumes activity, it undergoes meiosis to form four zoospores that are genetically different from either parent, and each zoospore forms a new filament.

This type of sexual reproduction is called **oogamy** (oo = egg); there is a large, nonmotile gamete provided with much food, and a smaller gamete that must move or be moved to the larger gamete. By convention, we call the larger gamete **female** and the smaller one **male**.

The Origin of Oogamy

The earliest form of sexual reproduction was probably isogamy. In the course of evolution, however, oogamy has become the most widespread system of sexual reproduction. All higher plants and all animals that reproduce sexually are oogamous. How did oogamy originate?

It seems likely that some early alga started to produce larger gametes, which contained more stored food than usual. These fused with the usual small gametes from other plants. Even today there are algae that reproduce in this way; the zygote

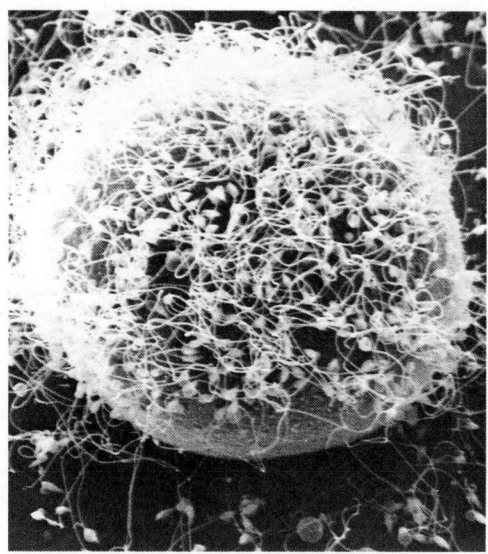

Figure 18-6 The egg of a sea urchin surrounded by a tangle of sperm with spherical heads and threadlike tails. (Biophoto Associates)

is formed when a large flagellated gamete fuses with a smaller one (fusion of unlike gametes is called **anisogamy** [an = not; anisogamy includes oogamy]). The selective advantage of such a situation was undoubtedly that spores formed from the resulting zygotes contained more stored food, so that the filaments they produced were larger when they began to grow, grew faster, and were superior to the filaments produced by smaller zygotes in the competition for light and nutrients. Eventually the larger gamete became so laden with nutrients that a flagellum could not move it, and it sat where it was until a motile gamete found it.

In higher plants and animals, the egg contains not only stored food but also the information necessary to carry out the early stages of embryonic development; this information is in the form of already-made messenger RNA (Section 36-G). Thus the major advantage of oogamy is that the embryo can be larger and more differentiated before it has to provide its own food. Higher animals could probably never have evolved without an egg containing stored food and genetic information for embryonic development. It is hard to imagine even a worm developing if the tiny embryo had to form a mouth and feed itself when it contained only two or four cells.

Thus, an egg that contains food and information, and is therefore too large to be motile, is of enormous selective advantage. But a nonmotile egg is no use without a motile sperm that can reach it. So sperm and egg vitally complement each other and have evolved together.

18-D Evolution of Mechanisms That Ensure Fertilization

When an organism reproduces asexually, it is "doing its own thing," independent of other individuals. (The fact that many organisms in a population are reproducing asexually at the same time reflects environmental conditions providing them all with enough energy to put into reproduction.) In sexual reproduction, however, production of each new individual involves the union of two kinds of cells, and the production and distribution of these cells must be coordinated for successful reproduction to occur. One common coordinating mechanism is the existence of breeding seasons, with members of a population all coming into breeding condition in response to some environmental cue such as temperature or duration of daylight (see Figure 17–3). Another is mating behavior, including chemical secretions by some members of the population that stimulate onset of reproductive condition in the other members; examples are found among the locusts of Africa, which respond to chemicals secreted by the first male to reach sexual maturity, and many fungi.

In most familiar species, there are two sexes, and they are determined by a genetic mechanism that produces half males and half females. Thus any individual has a 50% chance of finding that its nearest neighbor is of the same sex as itself, and the two will be unable to breed together. When there are only a few individuals in a population, or when individuals cannot move far to find suitable mates, there is selection for less familiar forms of sexual reproduction. We have already seen one form: in both *Ulothrix* and *Oedogonium*, environmental factors caused all individuals in the population to produce gametes at the same time; motile gametes were released, and found mates either in the water or still associated with the (female) parent. Release of gametes into the water, by one or both sexes, occurs in many other algae, in some marine (ocean-dwelling) worms, and in sea urchins.

A second kind of adaptation that eliminates the problem of finding a mate is seen in many invertebrates, in which sex is determined, not genetically, but by environmental factors. An individual settling in a new habitat develops as a female, and a new individual joining her responds to her chemical secretions by becoming a male (Section 14-F).

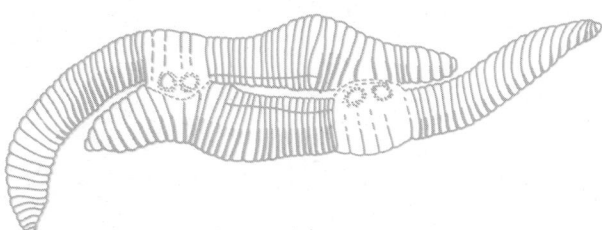

Figure 18-7 Mutual copulation between two earthworms.

Still another pattern found among organisms that cannot move around to find mates is the existence of reproductive organs of both sexes in the same individual. In animals this is called **hermaphrodism**, and it is common in earthworms, leeches, and many kinds of snails. In addition to being assured that any other member of the species it encounters is a potential mate, hermaphrodism also gives each animal a chance to be both a mother and a father! Plants with reproductive organs of both sexes are more commonly said to be **monoecious** (mono = one; oekion = house). Hermaphroditic animals and monoecious plants may be able to self-fertilize, but there is often selective pressure to cross with other individuals.

18-E   Evolutionary Roles of Male and Female

In the rest of this chapter we shall discuss the familiar systems in higher animals with separate male and female sexes, and see their effects upon social structure. We begin by considering differences in the role of male and female in sexual reproduction and go on to consider some of the evolutionary consequences of these roles in different animals.

The female is, in a sense, a limiting resource for the sexual reproduction of a species. At the simplest level, because eggs are bigger than sperm, they take more energy to produce. A male's evolutionary success is usually limited, not by the number of sperm he can produce, but by his ability to deliver his sperm to as many eggs as possible. The female's evolutionary success is limited by the number of her eggs that survive to become part of the breeding population. For some females this means laying as many eggs as her energy budget will permit and letting them raise themselves, as in the case of a herring (a fish), which may lay a million eggs a year. At the opposite extreme, raising offspring to maturity may mean gestating them in her own body and caring for and feeding them for years afterwards, as in the case of many mammals. Which program is favored by evolution depends on the ecology of the species.

Because a female's parental investment in future offspring is usually greater than the male's, the selective pressures acting on a female may actually conflict

Figure 18-8 *Drosophila* courtship. Here the male (right) vibrates his wings to produce his genetically determined "song," to which the female (left) may or may not respond. (Biophoto Associates by courtesy of Dr. W. L. Burnet)

with those acting on a male. While it may be advantageous for a male to copulate with as many females as possible in order to raise his chances of fathering surviving offspring, it is apt to be advantageous for a female to be much more choosy. She produces fewer eggs and so has fewer second chances if her first mate is genetically unfit.

Under this selective pressure, it is not surprising that females of all species of animals studied show discrimination in their choice of mates. A simple example comes from a laboratory experiment with fruit flies (*Drosophila*). Of the females, only 4% failed to reproduce, whereas 24% of the males failed to copulate even once. These celibate males courted females just as vigorously as did successful males, but no female ever accepted them. In other words, females do discriminate between different potential mates. The female who discriminates and copulates only with genetically fit males will be at a selective advantage. On the other hand, it may be to a male's advantage to appear genetically fit even when he is not, because females may then be deceived into mating with him. This has been envisioned as an evolutionary battle of the sexes, with skilled salesmanship among the males and an equally well-developed sales resistance and discrimination among the females.

Sexual Differences

A female's reproductive success is not usually limited by her ability to find mates. It is more likely to be limited by her inability to rear her young. A male who demonstrates that he can contribute to raising offspring will be attractive to females. For instance, among birds, the male of choice is often the holder and defender of a territory that provides food and shelter needed by the female and her young. Defending a territory is also of direct selective advantage to the male, for the territory he holds promotes the welfare of his offspring. Once a female finds a good territory, she need not bother about preliminaries but may mate with the resident male without further courtship. In such a situation, a male's ability to hold and defend a territory will determine his breeding success. Males of other species may compete for control over valued resources other than territories, because possession makes them irresistibly attractive to females. For instance, we know that a man with wealth and status attracts some women no matter how unattractive he is physically.

The different sexual roles of male and female may lead to their having different appearances, a phenomenon known as **sexual dimorphism** (di = two, morph = form). For example, female birds are much more likely than males to have drab colors. Because females are vulnerable as they sit on their eggs, it is advantageous for them to be inconspicuous. This camouflage works; when the male is the more conspicuous sex, mortality is invariably higher among males than among females. When defending a territory, a male may flaunt vivid coloration or unusual, exaggerated postures, making him more visible not only to other males who might think of invading but also to females, who may notice what a nice territory he has. An interesting variation on males' use of color to advertise their valuable property to females is found among some bowerbirds and weaverbirds. The male African village weaverbird, for instance, is dull-colored, but he builds a colorful nest and jumps up and down beside it saying, in effect, not "look at me" but "look at the gorgeous nest I have built for you." If no females are attracted, and if the color of the nest starts to fade, he will tear it to pieces and build another one.

Figure 18-9 A sexually dimorphic species: male and female Barrow's goldeneye. (William Camp)

(a)   (b)

Figure 18-10 Male weapons and decorations. (a) A red deer's antlers are used in fights with other males and identify him to females. (b) A peacock displays his tail, used in courtship displays. (Biophoto Associates)

Another type of sexual dimorphism is found in the possession of weapons by the male. Large antlers or horns in many hoofed animals, long tusks in boars, and the enormous size of male seals all give a male an edge in combat against other males for mates or breeding territories. Hence there is selective pressure for the males, but not the females, to possess these traits.

Strong sexual dimorphism of this type is rare in **monogamous** species (species whose members take just one mate for a whole breeding season). If the sex ratio is more or less equal, a male is almost bound to find a mate if he stays around long enough. Selective pressure on the monogamous male is mainly to stay alive; he is not involved in much competition with other males and tries to stay out of the way of predators. So he is not under selective pressure to evolve the huge horns, elongated canines, and other offensive weapons that are the hallmarks of a male destined either for glory or, much more likely, for early death. In monogamous species such as beavers or geese, the male shows as much discrimination in his choice of mate as the female. Like the female, he has only one chance to choose a mate, and it will be just as advantageous for him as for the female to choose a mate who will produce healthy, successful offspring.

Mating Systems

The courtship behavior of any species ensures accurate recognition so that there is little chance of a female's copulating with a member of the wrong species. In **polygamous** species, those in which each animal may mate with more than one other, males frequently have vast sex drive and little discrimination. They will court almost anything vaguely appropriate, and the females must recognize and pick out a male. The male's appearance, physique, and courtship behavior assist in this. As a corollary, in monogamous animals, which mate with only one other individual, the sex drive in both sexes is about equal, the sexes are often indistinguishable in appearance and behavior, and courtship is mutual.

Figure 18-11 Sexually monomorphic, monogamous animals: Fischer's lovebirds. (William Dilger)

Figure 18-12 A typical polygynous species: a male impala (with horns) and his harem. (Biophoto Associates, N.H.P.A.)

(d)

POLYGAMY. Polygamy may be divided into **polyandry**, in which one female mates with more than one male, and the much more commonly found **polygyny**, in which one male mates with many females. Polygyny may evolve where a female gets a better share of some limited resource by joining a mated pair or a male and his harem than by mating with an unmated male. The resource in question might be limited food or better protection from predators in a group than in a twosome, or it might even be peace and quiet and a helping hand to raise the young. An example of such a situation is known from studies of Japanese hamadryas baboons. A particular male was quite capable of defending his territory against all comers, but he was so belligerent that there was constant turmoil in his troop, and females always abandoned him after a short stay. They left to settle down with a troop where there might be less food but where they had a chance of giving birth and raising their young in peace and quiet.

Another selective force for harems with only one male may be that females do not have to share food and other resources with males whom they do not need for copulation. In such situations, the females help the resident male to defend their territory. On the other hand, where it's "all hands to the barricades" to fight off predators, nonreproductive males may be welcome members of the troop. Both kinds of social organization occur among various monkeys and apes.

All other factors being equal, polygyny is a more favorable system for the male of a species (or at least for the successful males, who will be the only ones that reproduce). However, polygyny is possible only where the female does not need the male's full-time help in bringing up the young. Its presence or absence, therefore, will usually depend on the advantages or disadvantages to the female.

MONOGAMY. Under what circumstances does monogamy evolve? When the combined energy of both parents is needed to raise the young, there will be selective pressure on males to exhibit monogamous behavior. For example, monogamy (more or less) is the most common form of human sexuality (although both types of polygamy arise under various circumstances). We can infer that this is so because the human infant is so demanding to raise: males who stayed and helped their mates to raise the children left more offspring in the next generation than those whose mates were left to raise the young alone.

Figure 18-13 These Barbary rock apes, like most monkeys and apes, live in permanent, polygamous troops. (Biophoto Associates)

ECOLOGY AND MATING SYSTEMS. An important factor in determining the mating system of a species is its ecology, since such factors as the distribution of food, water, nesting sites, and shelter in the environment determine the distribution and social behavior of individuals.

Where food is scarce and found in small, isolated pockets, individuals tend to be solitary and come together only for a short time, in the breeding season. (Bears, badgers, and moose behave like this.) Couples may come together only to mate, or they may form short-lived pairs or colonies while they raise the young. On the other hand, it seems that most monkeys and apes can live in troops because their diet is mainly plentiful plant food; the all-important (in evolutionary terms) females and young can eat well and still enjoy the protection of living in groups. As we would expect from this, in situations where food is occasionally inadequate, groups containing only one male are the rule and there will be considerable competition among males to enter a troop, since this is the only way they can breed.

18-F Selfishness and Altruism

We have seen that a species' ecology and evolutionary history is often reflected in its mating system. We can go a step further and see that the mating system has effects of its own that limit, and may even dictate, the evolution of particular adaptations by the species. For one thing, the mating system has a profound effect upon how closely members of a species are related to one another. For instance, in a polygynous herd of horses where only one male mates, all of one year's offspring will be more closely related than they would be in a monogamous group where the year's offspring have different fathers. The degree to which members of a species are related can have some unusual evolutionary results. One of these is selection for **altruistic behavior**, behavior whereby an individual increases the welfare of another individual at the expense of its own welfare.

At first sight, it is difficult to explain the evolution of altruism. Altruistic behavior favors the reproductive success, not of the altruistic individual, but of another member of the species. If natural selection acted only on the individual and its genotype, it could not possibly produce such behavior.

However, genes last much longer than do the individuals that serve as their temporary carriers, and selection can be thought of as acting, not on individuals, but on alleles, polygenes, or coadapted gene complexes. As a result, an allele may spread through a population at the expense of a particular individual that carries the allele. In particular, an allele that favors altruistic behavior toward close relatives improves the chances that the allele will survive in the population because there is a good chance that closely related individuals will also contain replicas of the allele. To take an extreme example, if an individual dies to save ten close relatives, one copy of the "kin-altruism" allele is lost but a large number is saved. **Kin selection** is the kind of natural selection that can favor the survival of altruism alleles.

How closely related must organisms be for kin selection to operate? In 1964 William Hamilton showed how the necessary degree of kinship could be calculated by considering the **index of relatedness** of two individuals, that is, the probability that they share a particular allele. An individual in a sexually reproducing species has inherited any particular allele from either its mother or its father; since sperm and eggs each contain 50% of the parent's genes, the chance that the allele is also present in a brother or sister is 50%. The degree of relatedness of brothers and sisters is therefore $\frac{1}{2}$. Note that this is the *average* degree of relatedness; by the luck of the meiotic draw, an allele may be present in more or fewer brothers or sisters. The relatedness between a parent and its offspring is exactly $\frac{1}{2}$, since half the parent's genetic material is inherited by the offspring.

Indices of relatedness can be calculated for other relatives. For instance, in a monogamous species, the chance that a grandparent, uncle, aunt, nephew, or niece also has a particular allele of yours is $\frac{1}{4}$. The index of relatedness with a first cousin or great grandchild is $\frac{1}{8}$, with a second cousin,* $\frac{1}{32}$, and with an identical twin 1. Third cousins* are much less special; the index of relatedness is only $\frac{1}{128}$, probably not much greater than the chances of sharing the allele with any other individual in the whole population.

We would therefore expect the degree of altruistic behavior to tail off towards relatives who are less close. Specifically, an allele that prompted an individual to lay down his or her life for a relative would have to save more than two brothers, sisters, or children, more than four uncles and aunts, or more than eight first cousins, and so on, to be selectively advantageous. Otherwise it would not survive, on the average, in enough other bodies to compensate for its loss in the altruistic individual.

Organisms, of course, do not calculate their degree of relatedness to others before embarking on a course of action. Selection favors alleles that act as if such calculations had been done. And in many situations, altruism is futile—the individual benefited may not contain the altruistic gene. It is the long-term statistical pattern that counts: By promoting the appropriate level of altruism towards other individuals, alleles are statistically likely to survive. Altruistic suicide is not particularly common. What counts more often is the statistical risk of death. Even a third cousin may be worth saving if the risk to the altruist is small. In fact, the most common acts of altruism pose little risk to the altruist but contribute markedly to the welfare of related individuals.

---

* Second cousins Children of first cousins are second cousins to each other.
* Third cousins Children of second cousins are third cousins to each other.

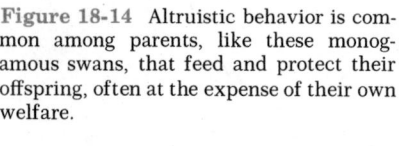

Figure 18-14 Altruistic behavior is common among parents, like these monogamous swans, that feed and protect their offspring, often at the expense of their own welfare.

How do organisms know who their relatives are? In small groups, such as herds of ungulates (hoofed mammals) or troops of monkeys, and especially in polygynous groups, there is a good chance that all individuals are related; alleles that promote altruism towards any individual of the same group might spread through such a population, perhaps explaining why altruistic behavior is so common within groups of monkeys, lions, wolves, and dolphins. Altruism between brothers and sisters is not so marked as between parents and their offspring in most animals; the reason may be partly that the parent/child relationship is highly predictable. There is little probability that the relationship is not really what it seems to be. In many species, a mother can be more sure that her offspring are her own than can a father. Fathers are vulnerable to deception and therefore may be expected to put less effort into caring for young than do mothers. Similarly, maternal grandmothers are more sure of their grandchildren than are paternal grandmothers; a grandmother can be sure of her daughter's children, but her son might have been deceived into rearing young that do not contain his (and therefore her) genes.

Reproduction is costly; females in particular invest considerable amounts of energy, nutrients, and time in offspring. What determines exactly how much, on the average, the evolutionarily successful female spends on any particular offspring? Does she devote equal amounts of time and resources to each? R. L. Trivers has made it possible to examine such questions by comparing **parental investment**, which is defined as any investment by the parent in an individual offspring that increases the offspring's chance of surviving (and hence of its future reproductive success) at the cost of the parent's ability to invest in other offspring. If a hawk, for example, feeds a rabbit to one of two nestlings, the major part of the parental investment that this rabbit represents is the increased probability that the other nestling will die because it did not receive the rabbit. Another part of the parental investment is the decreased ability of the female to lay other eggs and rear other young because of the time and energy she invested in catching the rabbit.

Clearly, parental investment must not be spread too thinly over many offspring, because then few or none of them may receive enough food to survive. Equally clearly, devoting too much investment to too few offspring will be selected against because more prolific females will leave more progeny in the next generation.

How is parental investment distributed among existing offspring, all of whom are equally related to the mother? Because the parental investment required to raise a runt* is greater than that required to finish raising a larger elder brother or sister,

---

*Runt   The smallest member of a brood or litter. In some birds, the parents begin to incubate the eggs the day before the last one is laid; thus, this one hatches the day after the others, and is a day smaller and less developed.

Figure 18-15 Red deer stags fighting for leadership of a herd of females, a prerequisite for reproduction. The energy he puts into fighting is part of a stag's parental investment in his offspring. (Biophoto Associates, N.H.P.A.)

ALTRUISM

The altruist reduces her own fitness but increases that of her sister so much that the genes they share are more common in the next generation.

SELFISHNESS

 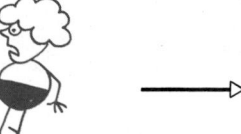  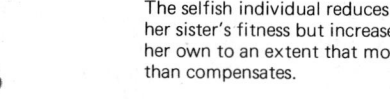

The selfish individual reduces her sister's fitness but increases her own to an extent that more than compensates.

SPITE

Our heroine (with hatchet) lowers the fitness of an unrelated (unshaded) competitor while damaging her own position; however, the reduction of competition benefits her sister to an extent that more than compensates.

Figure 18-16 Conditions for the evolution of altruism, selfishness, and spite. The family is represented by an individual and her sister. The sister in purple is the actor; on average, her sister shares half her genes (purple half of body). The hatchet indicates behavior that harms another. The plate of food symbolizes any valuable resource. (Adapted by permission from *Sociobiology: The New Synthesis* by E. O. Wilson, Harvard University Press, 1975)

it would clearly pay a parent to concentrate on saving the elder offspring first, and even feed the runt to the others, if food shortage is acute. On the other hand, if food were more abundant than normal, it would be better to feed the runt first, because the others are better able to survive a short period without attention. Weaning* will presumably take place when the parental investment involved in further attention to existing children would be more advantageously spent on raising new offspring.

Altruism between siblings (brothers and sisters) is somewhat different from that between parents and offspring because the age differences are less and because the certainty of the relationship is less secure. Nevertheless, siblings share, on the average, 50% of their genes, and altruism would be expected. Should an offspring always try to grab food for itself, or should it share? Since an offspring is 100% related to itself, and only about 50% related to a sibling, the offspring "should" try to grab food for itself unless this action costs its siblings at least twice the amount by which grabbing would benefit itself (see Figure 18–16). This might occur, for example, when the sibling is much younger and would be placed in great jeopardy by the elder sibling's grabbing all the food.

The Social Insects

Colonies of social insects have long intrigued biologists because of their very high degree of specialization and cooperation and because the majority of individuals in a colony are sterile—they will never leave any offspring. This is altruism on a grand scale. Colonies of wasps, ants, termites, and social bees, especially honeybees, are indeed impressive societies. Food is shared communally, and information is exchanged in elaborate ways, including a great range of different chemical signals and the elaborate "dances" by which worker honeybees indicate the direction and distance of a food source to their nestmates. Workers are fearless in the defense of a colony; many sacrifice their lives on its behalf.

A colony of social insects is a huge family, sometimes numbering several million individuals, all descended from the same mother—the reproductive female, or

*Weaning   Withdrawing the mother's milk (more generally, withdrawing parental care).

Figure 18-17 Altruistic behavior in honeybees. (a) A worker foraging for nectar to contribute to the hive's communal honeycombs. (b) A queen (center) surrounded by workers who raise her offspring. (a, Carolina Biological Supply Company; b, Biophoto Associates)

queen. The workers, which seldom or never reproduce, can sometimes be divided into several distinct "castes," each with a particular set of duties to perform. In the more advanced insect societies, queens and their mates do nothing except reproduce; they are completely taken care of by workers, who feed them, defend the colony, and rear the young in special nurseries within the colony (Figure 18–17). In some of the ants and termites, the queen becomes nothing but a gigantic egg factory, incapable of moving. She dwarfs the workers who feed her, clean her, and remove her eggs to the nurseries. Every now and again, depending on ecological conditions, winged reproductive males and females are formed; they leave the colony in a swarm, and the young queens mate and then form new colonies of their own.

Among the ants, bees, and wasps (all in the insect order Hymenoptera), workers are infertile females; in the termites (order Isoptera), workers also include infertile males. How do the workers gain an advantage great enough to have selected for the evolution of such altruistic, even Kamikaze, individuals? In ants, bees, and wasps, the answer lies in a curious aspect of sex determination. The single mature queen in the colony receives enough sperm during her mating flight to last her for the rest of her life of perhaps 10 years or more. She uses these sperm to fertilize the eggs as she lays them, producing female workers whose subsequent differentiation into different castes is determined by the conditions of their upbringing and particularly by the kind of food they eat. But not all of the queen's eggs are fertilized; the unfertilized ones turn into males, which thus have no father. All they have is a single (haploid) set of chromosomes derived from their mother. Thus the sperm cells from a reproductive male all contain the same haploid set of chromosomes, and his degree of relatedness to his mother is 100%. His mother, on the other hand, is diploid and her son receives only 50% of her genes so, strange as it may seem, *her* relatedness to *him* is only 50%. Full sisters share the same father (whose sperm are genetically identical), so that if a particular allele present in one worker was obtained from her father, there is a 100% chance (barring mutation) that the allele will be present in each of her sisters. However, there is only a 50% chance that any allele a worker obtained from her mother will also be present in any one sister. In other words, the relatedness between sisters averages $\frac{3}{4}$ (not $\frac{1}{2}$ as in normal sexual animals). This gives rise to the curious circumstance that a female hymenopteran is related more closely to her sisters than she is to her mother! She can therefore further the interests of her genes more by farming her mother as a sister-making machine than she can by reproducing. More accurately, an allele that promotes making sisters will replicate more rapidly than will an allele for making offspring. This selects for sterility of workers, which has evolved independently at least 11 times in the Hymenoptera and once elsewhere (the termites), an extraordinary example of the complex effects that can be produced by a particular sexual system.

Essay: Human Sexual Systems: Some Speculations

Studies on mate selection in modern human societies show that people do not choose their mates at random from the population; rather, they tend to marry according to certain characteristics, choosing people of similar educational background, race, religion, and social status. Marriage customs are governed by all sorts of rules and traditions. The most widespread of these are the taboos and laws against incest—mating with a genetically related individual.

A tendency against incest occurs in most species. For instance, female pigeons discriminate against males of the same genetic type as themselves. In fruit flies, females who do not discriminate against their male relatives leave only a quarter as many offspring as those that do. Incest taboos are more extensive in human than in most other animal societies. One reason may be that, as studies have shown, human mother–son, father–daughter, and sister–brother matings have a very high chance of producing abnormal offspring. Humans seem to carry a relatively large proportion of the harmful genes that invariably show up in inbred stock.

Another example of nonrandom mate choice in humans comes from a nineteenth-century study on marriages of members of the British peerage. The study was undertaken to find out why the families of great men died out much more frequently than did families of the British population in general. The study showed that a considerable proportion of peers married heiresses. An heiress, almost by definition, comes from a small family, since she would not be much of an heiress if her parents' wealth were divided among 10 children. Heiresses who married peers were found to perform reproductively almost exactly as their mothers had before them. If an heiress were one of two children, there was a very high chance that she would also have two children. (In this case, mate selection by the peer and the heiress had the result of coupling relative infertility with high achievement of money and status.)

Modern *Homo sapiens* has a tendency toward monogamy but has presumably evolved from a polygynous ancestor. The evolution of human monogamy was probably made almost inevitable by the evolution of an infant that takes so much energy and time to raise. Humans mature more slowly than most other animals, and it seems likely that, throughout much of human history, both parents have had to do their share to raise human offspring to sexual maturity with consistent success. It is as much in the interest of the man as of the woman that his offspring reach maturity, and so the man, too, would be better off monogamous unless he could provide for the children of more than one wife.

There was probably another selective pressure, unique to humans, tending to make early humans monogamous: the possession of material goods, which was of vast importance to human evolutionary success. It would be strongly advantageous to bequeath objects like clothes or a cave to your genetic children. It is possible that before people acquired material possessions, selection favored polygyny for male and monogamy for female strategy, so that the two sexes were at evolutionary war with one another. From the moment the first man invented his first tool, however, woman had him in an impossible bind. (This is not to suggest that men rather than women invented tools, merely that only the man's tools matter to this argument.) Man's bind at this point was that woman could obscure a man's paternity, so that he wouldn't know which were his children or who should inherit his axes. What happened, in essence, was that woman promised man sexual fidelity and, therefore, identifiable offspring to leave his axes to, in exchange for a permanent commitment on his part to helping her raise the children. It did not matter to primitive woman whether he mated with other women, except where doing so led to injury at the hands of jealous husbands or to neglect of his obligations to her and their children—hence the origin of the double standard.

This highly speculative theory of the origin of human monogamy would account for another interesting fact. In societies that condemn adultery among their members, a woman's adultery is considered a much greater offense than a man's. This may well be because female adultery threatens a man's ability to identify his own children. If he cannot identify her children as his own, a man is wasting his time, in evolutionary terms, in helping a woman raise the children.

(Rosemary Smith)

All species have a certain amount of genetic variation among their members as a result of mutation. The amount of such variation is slight, and members of species that reproduce asexually are, therefore, genetically very similar to one another. Members of sexual species are much more variable because genes are reshuffled during meiosis, and form new combinations at fertilization. Asexual reproduction wastes less energy than sexual reproduction and is common in species that occur in many places in the world and are adaptable to changing conditions. For more localized and specialized species, the energy wasted in sexual reproduction is worthwhile because at least some of the genetically different individuals in a sexually reproducing species can usually survive and evolve in changed conditions.

Sexual reproduction probably originated as the accidental fusion of two asexual reproductive cells. The advantage of having one gamete stuffed with a food supply for the new individual selected for the evolution of oogamy, the most common form of sexual reproduction today. Adaptations such as specific mating seasons, environmentally determined sex, and hermaphrodism help to overcome the difficulties encountered when two individuals must act in concert to achieve reproduction.

The sexual system of a species is determined by the amount of energy each sex puts into producing and rearing offspring and by ecological factors such as the distribution of food, prevalence of predators, etc. Because eggs take more energy to produce than do sperm, females are usually more selective than males in their choice of mate. Sexual dimorphism may arise when the selective pressures on the two sexes conflict, or when males and females have different roles in reproduction. Sexual dimorphism tends to be greater in polygamous species. In monogamous species, the members of both sexes must choose their mates with discrimination, and the roles and behavior of the two sexes are more similar.

An allele that makes an individual perform altruistic behaviors detrimental to itself can be selected for when the behavior enhances the survival of other copies of the allele in the individual's relatives, allowing that allele to spread through the population. Altruistic behavior is most likely to arise in species in which closely related individuals spend much time together.

SUMMARY

OBJECTIVES

From your study of this chapter, you should be able to:

1. Define isogamy, oogamy, hermaphrodism, polygamy, polyandry, polygyny, monogamy.
2. State the advantages and disadvantages of sexual reproduction contrasted with asexual reproduction.
3. Discuss the advantages enjoyed by organisms that combine sexual and asexual reproduction in the life history.
4. Give at least one advantage of oogamy and of hermaphrodism.
5. Give some reasons why selective pressures acting on a female may be different from those acting on a male, and describe some resulting differences between the two sexes in a species.
6. Give evidence for the theory that female choice almost invariably exists in mating systems.
7. Describe what is meant by altruistic behavior and kin selection, and explain how alleles for altruistic behavior may spread through a population.

SELF-QUIZ

1. An advantage of sexual reproduction over asexual reproduction is that it:
   a. increases the mutation rate
   b. increases genetic variability in a population
   c. produces larger offspring
   d. reduces the risk of death during development for the offspring
   e. gives organisms something to do on Saturday night
2. Some organisms reproduce asexually when environmental conditions are (favorable, unfavorable) and sexually when conditions are (favorable, unfavorable) for growth.

3. The main advantage of oogamy is that it:
   a. has separate male and female sexes
   b. provides for the nourishment of the growing zygote
   c. assures cross-fertilization
   d. involves two nonmotile gametes
4. It is generally true that males are an abundant resource, and that to be evolutionarily successful they should try to mate with as many females as possible. One possible exception to this generalization might occur when:
   a. there are many more females than males
   b. there are about equal numbers of males and females

289

c. the father's care is required to raise the offspring

d. there are many predators

e. the male holds a territory against other males

5. Studies on female choice of mates in *Drosophila* showed that:

    a. more males than females reproduce

    b. females who reproduce leave more offspring per individual than do males who reproduce

    c. males who do not reproduce fail because they never court females

    d. males who do not reproduce fail because females will not accept them

    e. females who do not reproduce fail because males do not court them

6. When food is distributed in such a way that an animal must spend a large part of its day wandering from one place to another to find enough to eat, what type of mating system would you expect it to have?

    a. monogamy

    b. polyandry

    c. polygamy

    d. polygyny

7. A human male who remains a bachelor is most likely to enhance his evolutionary success by altruistic behavior towards the children of

    a. his sister

    b. his brother

    c. his mother

    d. his grandmother

    e. his niece

QUESTIONS FOR DISCUSSION

1. Does the desirability of genetic variations imply that monogamy in humans is a counter-evolutionary tendency? Why? (Hint:compare the average human family size with the number of possible genetically different eggs or sperm that a person could produce; for humans, with 23 pairs of gene-bearing chromosomes, each person could produce 2^{23} different kinds of reproductive cells.)

2. What are the selective advantages of internal over external fertilization?

3. List as many possible selective advantages as you can think of for courtship rituals.

4. Do humans have courtship rituals? Is courtship in humans mutual or is it carried out predominantly by one sex? Why?

5. What selective pressures might be responsible for the evolution of polyandry?

6. Female walruses must bear their young on land, but suitable stretches of beach are scarce. Similarly, nest sites for gulls are scarce. Both have crowded breeding grounds. Walruses eat molluscs (mussels, clams, etc.) whereas gulls will eat almost anything—including the egg next door. Explain what mating system you would expect each of these animals to show.

7. What is the advantage of hermaphrodism over separation of the sexes? What selective pressures might have led to evolution of species with separate sexes? Why are there never more than two different sexes in a species?

8. We all know people who don't reciprocate a favor when it is their turn. What might be some possible outcomes if such a nonaltruistic allele were present in a population together with an altruistic allele?

9. Explain sibling rivalry for parental favors and parental impartiality towards offspring in terms of degree of relatedness.

10. Do you think that altruistic behavior in humans is genetically controlled?

11. Trivers and Hare weighed the fertile males and females produced in colonies of 20 species of ants and found that the investment in females was three times the investment in males (by weight). Can you explain how this situation may have been selected for?

12. How might the invention of birth-control methods and labor-saving household appliances affect the monogamous mating system of humans (see Essay)?

REFERENCES AND FURTHER READING

Campbell, B., ed. *Sexual Selection and the Descent of Man, 1871–1971.* Chicago: Aldine, 1972. A collection of essays by some prominent students of the topic.

Morris, D. *The Naked Ape.* A male chauvinist's approach to the evolution of human sexual systems. Highly readable.

Morgan, E. *The Descent of Woman.* New York: Stein and Day, 1972. A feminist's reply to Morris.

Thornhill, R. "Sexual selection in the black-tipped hanging-fly." *Scientific American,* June 1980. Observations of bizarre mating habits in one species of insects support Darwin's theory of sexual selection.

Wiley, R. H., Jr. "The lek mating system of the sage grouse." *Scientific American,* May 1978. A case study of one of the less familiar forms of polygyny.

Wilson, E. O. *Sociobiology: The New Synthesis.* Cambridge, MA: Belknap Press, 1975. The classic work on the subject of sociobiology (the biology and evolution of social behavior). It has stirred controversy because its author—famous for his studies of insect societies—treats humans just like other animals when considering the origin of behavior patterns. Good discussion of altruism.

CHAPTER
19
ORIGIN OF LIFE

If we could trace the ancestry of living organisms, we should find a long line of cells going back billions of years. Each cell has come from the division of a previously existing cell . . . but where did the first cell come from? Most scientists today believe that cells arose as the result of hundreds of millions of years of chance chemical events, some of which were selected for and preserved in the course of evolution. In this chapter we shall consider the evidence supporting the theory that living things have arisen, step by step, from nonliving chemicals.

19-A The Death of Spontaneous Generation

For many centuries people believed that organisms appeared from nonliving matter by a process known as **spontaneous generation**. The Greek philosopher Aristotle wrote that frogs and insects were generated from moist earth. People frequently observed that stagnant watering troughs produced worms, and spoiled meat produced maggots.

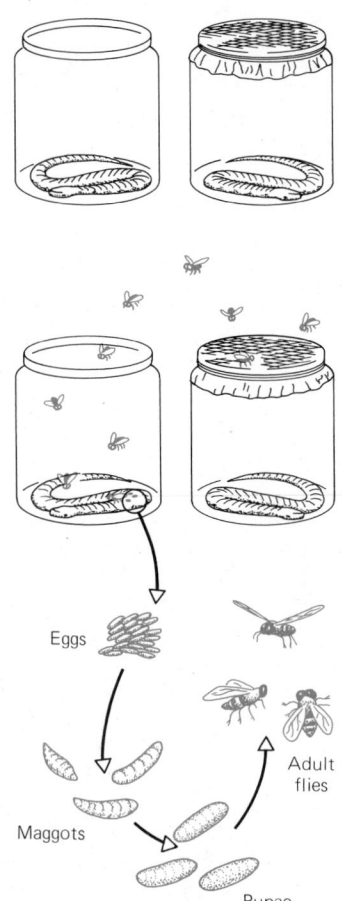

Figure 19-1 Redi's experiment. Flies entered the open jars and laid eggs on the dead animals; the eggs developed into maggots, which became pupae and eventually emerged as adult flies. Flies laid eggs on the muslin of the covered jars, but no maggots appeared on the dead animals.

The Italian Francesco Redi is credited with dealing the first important blow to the theory of spontaneous generation in 1668. Redi placed dead snakes and eels in a number of glass jars, and covered some of the jars with fine muslin (Figure 19–1). Flies soon arrived and laid eggs on the dead animals in the open jars; the eggs hatched into maggots. No maggots appeared in the covered jars, although flies settled on the muslin and laid eggs there. Before long, similar experiments convinced most people that macroscopic organisms came from the eggs or seeds of their parents.

Meanwhile, in the late 1600s, the Dutch lens grinder Anton van Leeuwenhoek built a microscope and discovered that tiny organisms, invisible to the naked eye, lived in ponds, ditches, soil, and many other places—including his own body. He proposed that they, too, arose by reproduction of others of their kind.

The Abbé Lazzaro Spallanzani agreed with Leeuwenhoek. Spallanzani did a series of experiments in which he prepared glass flasks of nutrient broth, corked them, sealed the corks, heated the flasks, and watched to see if any microorganisms appeared in the broth. Most of the time they did not, but microorganisms appeared soon after he broke the seal and allowed air to seep in. People with opposing viewpoints did not accept these experiments as disproving spontaneous generation. They declared that heating the flasks had rendered the air unfit to support life, or had destroyed "vital molecules" that float around in the air until they enter matter, giving it life. Since Spallanzani's sealed, heated flasks sometimes came out teeming with microorganisms, the issue remained in doubt. (We now know that some bacteria make resting spores that survive heat treatment and grow afterward—accounting for Spallanzani's erratic results.)

The theory of spontaneous generation of microorganisms was finally laid to rest in the nineteenth century by the experiments of Louis Pasteur in France and John Tyndall in England. They demonstrated that bacterial cells are present in air, and that if air is purified before it passes into a flask of heat-sterilized nutrient broth, no bacteria will develop in the broth. Pasteur drew the necks of glass flasks into S-shaped curves (Figure 19–2). Air could enter a flask freely, but it did not travel quickly enough to carry bacteria along with it. Any cells in the air were trapped at the bottom of the curve. Tyndall purified the air entering his flasks by passing it through a flame or through absorbent cotton. These processes, too, removed bacteria from the air and kept the broth clear. Thus by the late 1870s most scientists agreed that all living organisms, of whatever size, came from reproduction of previously existing organisms.

19-B Conditions for the Origin of Life

Louis Pasteur is often credited with disproving the theory of spontaneous generation, more than a century ago. But Pasteur himself once remarked that his fruitless 20-year search for spontaneous generation did not lead him to believe it was impossible. What Pasteur showed was that life did not arise in his flasks under the conditions he used (sterilized nutrient medium, clean air) in the amount of time he waited. He did not show that life could *never* arise from nonliving matter under *any* set of conditions.

The theory of the origin of life presented in this chapter differs from the old concept of spontaneous generation mainly in the length of time envisaged as necessary for the process. Most scientists now believe that the origin of life was an inevitable stage in the evolution of matter, and that it has probably happened time and again in many parts of the universe, wherever conditions were appropriate.

What conditions must have existed for life to begin? There are three basic requirements: appropriate chemicals, including water, various inorganic ions, and organic substances; absence of molecular oxygen (O_2); and lots and lots of time. Water is abundant on the earth, and inorganic molecules are easy to obtain by dissolving rocks or volcanic gases in water. Leaving the problem of obtaining organic molecules for later, let us assume that they are available and consider the importance of the other two conditions.

A Reducing Atmosphere

Absence of oxygen gas was a requirement for the origin of life because oxygen would have combined with organic molecules and destroyed them so quickly that they would not have had a chance to interact and form more complex structures. The oxidizing atmosphere of the earth today probably precludes spontaneous generation. (Another reason that new life does not arise from nonliving matter today is that any organic molecules that were formed would quickly be absorbed by some existing organism.)

Shortly after 1950, it was proposed that the early earth had a **reducing atmosphere**, so called because the reducing agent hydrogen was believed to have been the most common element. This reasoning came from the enormous abundance of hydrogen compared with other elements in the solar system, especially in the gases that formed the sun and planets; even today, the atmospheres of Jupiter and Saturn are primarily hydrogen and hydrogen compounds of oxygen (H_2O, water) and nitrogen (NH_3, ammonia). These gases, plus the reduced form of carbon (CH_4, methane) were proposed as the constituents of the earth's atmosphere when life was evolving.

By the late 1970s, however, geological studies had produced convincing evidence that such highly reducing conditions probably did not exist beyond the earliest stages of the earth's formation, and the atmosphere during the time when life was evolving was only mildly reducing. Hydrogen gas (H_2) is so light that most of it must have escaped into space very early, and the earth is so near the sun that ammonia would have been decomposed by light, leaving hydrogen gas, which again would have escaped, and nitrogen gas (N_2). Moreover, carbon may have been partly or even predominantly in the form of carbon monoxide and carbon dioxide, though this is more open to question. The important point of both the older and the newer views, however, is that there was no free oxygen (O_2); any oxygen formed combined quickly with iron or other metals. Most of the oxygen gas in the atmosphere today is believed to be a by-product of photosynthesis.

Time

The other important requirement for the origin of life is plenty of time. The events necessary for the beginnings of life were extremely unlikely. Given enough time, however, the occurrence of even very improbable events is inevitable. For example, suppose the probability that an event will occur in a year is one in a thousand. The probability that it will not occur is then 0.999. The probability that it will not happen for two years is $(0.999)^2$, for three years $(0.999)^3$, etc.

Table 19–1 shows that there is a very small probability that the event will not happen at least once in 8128 years, or conversely, a very high probability (0.9997) that it will happen. The events involved in the origin of life were very much less probable than 1/1000, but there was much more time available. Geological evidence indicates that the earth formed about 4.6 billion years ago, and the earliest known fossils of prokaryotic cells are found in rocks formed about 1.1 billion years later. So, as unlikely as living systems are, they had such a long time to arise that their evolution was probably inevitable.

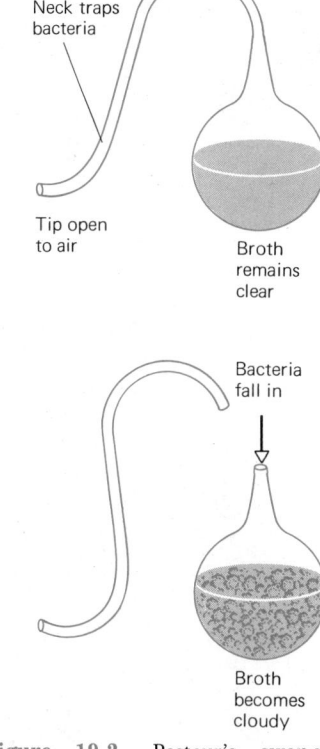

Figure 19-2 Pasteur's swan-necked flasks. Air could enter the flask freely through the open tip, but it did not travel quickly enough through the curved neck to carry bacteria along with it. Any cells present were trapped at the bottom of the curve, while air could continue on into the flask. Only if the neck was broken off could bacteria enter the flask and putrify the nutrient broth.

TABLE 19–1	**PROBABILITY THAT AN EVENT WILL NOT HAPPEN**	
Given this probability:	in 1 year	0.999
then:	in 2 years	0.998
	in 3 years	0.997
	in 4 years	0.996
	in 1024 years	0.359
	in 2048 years	0.129
	in 4096 years	0.017
	in 8128 years	0.000276

19-C The Search for the Beginning of Life

We can never know precisely how life originated. However, it is possible to look for evidence as to how it could have happened. In 1924 a Russian, Alexander Oparin, published a theory of how life could have arisen from simple molecules on the early earth. An Englishman, J. B. S. Haldane, published a paper in 1929 that said essentially the same thing. (He had never seen Oparin's paper, which had not been translated from the Russian.) Research since then has largely borne out the predictions made by Oparin and Haldane. Scientists have simulated prebiotic (before life existed) conditions in their laboratories; surprisingly, the nonliving systems formed in these artificial environments exhibit many properties that we consider characteristic of life.

Abiotic* Production of Organic Monomers

The twentieth-century quest for the origin of life received a major impetus in 1953 when Stanley Miller made a laboratory simulation of prebiotic conditions as they were then envisioned—the highly reducing atmosphere. Miller circulated a mixture of reducing atmosphere gases past an electrical discharge (representing lightning storms, which would have been an energy source promoting chemical reactions in the primitive atmosphere) and collected the products in a flask of water, representing the sea (Figure 19–3). After a week, the solution in the flask contained many organic compounds of low molecular weight (Figure 19–4).

The amino acids found by Miller were of special interest. The enormous variety of proteins and the amazing properties of enzymes were widely appreciated; indeed, proteins were regarded as *the* class of substances necessary for life. Thus Miller's demonstration that amino acids would form under conditions thought to have existed on the prebiotic earth generated much excitement.

Many investigators have repeated Miller's experiment (which is so simple that it frequently appears at high school science fairs) or variations of it, using different proportions of the starting gases, and sometimes including hydrogen sulfide (H_2S), carbon dioxide, or inorganic salts. Energy sources used, besides electric sparks,

---

* Abiotic   Without life: occurring neither within living cells nor under their influence.

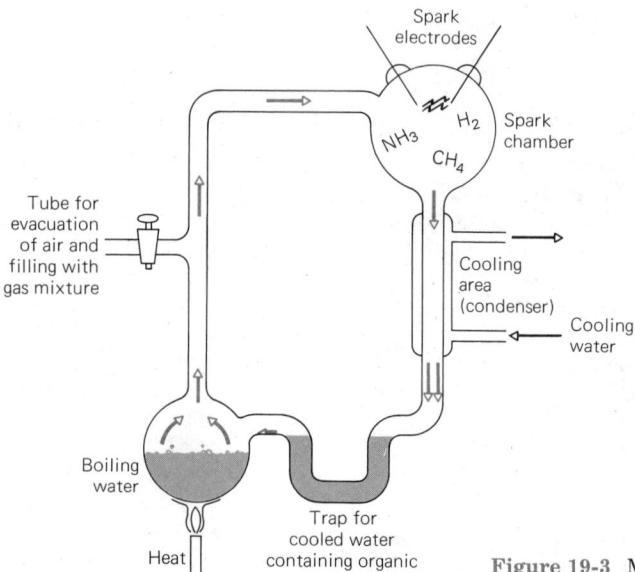

Figure 19-3 Miller's apparatus for simulating prebiotic conditions. The "atmosphere" was a mixture of hydrogen gas (H_2), methane (CH_4), and ammonia (NH_3) in a glass chamber the size of a soccer ball. Sparks from electrodes represented lightning, a source of energy. The "sea" was boiled to produce water vapor, which could provide oxygen to the reaction and also carry organic molecules back to the sea in "rainfall."

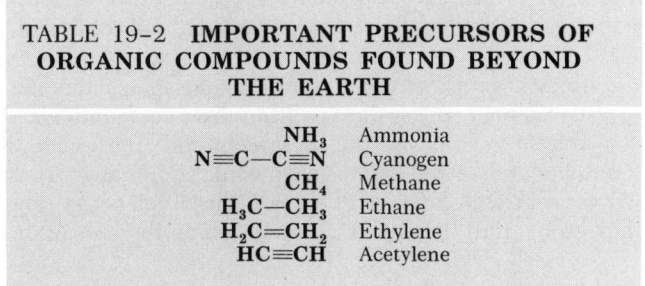

AMINO ACIDS

Figure 19-4 Some of the organic compounds found by Miller in his simulation experiment. Compounds shaded in gray are not found in modern organisms.

Formic acid

Acetic acid

Propionic acid

Succinic acid

Glycolic acid

Lactic acid

Urea

Sarcosine

Glycine

α-Alanine

β-Alanine

Aspartic acid

Glutamic acid

AMINO ACIDS

include heat, bright sunlight, ultraviolet light, electron beams, and particles emitted by the decay of radioactive materials, all possible sources of energy on the prebiotic earth. A long list of organic substances can now be added to the ones Miller obtained in his simulations.

Experiments using the more recent concept of an only mildly reducing atmosphere give lower yields; thus, it would take longer to accumulate enough monomers for the next steps to occur.

In addition to laboratory experiments, astronomy and geology also provide evidence that organic monomers can form without the agency of living organisms. Several important small molecules that are raw materials or intermediates in the synthesis of organic monomers are found beyond the earth and its atmosphere —in stars, dust clouds, space, and the atmospheres of other planets (Table 19–2). Meteorites, chunks of material that fall from space, contain a wide variety of organic compounds, some quite complex. Organic compounds are even now formed outside living organisms on earth, although at a very low rate; hot metallic carbides extruded by volcanic activity form hydrocarbons when they come into contact with water.

TABLE 19-2 **IMPORTANT PRECURSORS OF ORGANIC COMPOUNDS FOUND BEYOND THE EARTH**

NH_3	Ammonia
$N{\equiv}C{-}C{\equiv}N$	Cyanogen
CH_4	Methane
$H_3C{-}CH_3$	Ethane
$H_2C{=}CH_2$	Ethylene
$HC{\equiv}CH$	Acetylene

All these lines of evidence support the contention that organic compounds could have formed on the early earth by the action of available forms of energy. Without oxygen to destroy them or organisms to absorb them, these compounds would have accumulated in the oceans until, as Haldane put it, the sea had the composition of a "hot, dilute soup."

19-D   Formation of Polymers

The next step in chemical evolution was the polymerization of organic compounds into macromolecules. In a solution of organic compounds, the condensation of monomers into polymers takes place at a much slower rate than that of the opposite reaction, the hydrolysis of polymers into their constituent monomers. However, a number of workers have created laboratory conditions that shift the equilibrium of the reaction

$$\text{monomers} \; \underset{+H_2O}{\overset{-H_2O}{\rightleftharpoons}} \; \text{polymer}$$

in favor of polymerization.

Sidney Fox found that heating a dry mixture of amino acids produced long-chain **proteinoids**, protein-like molecules with molecular weights of over 10,000. At the temperatures he used (about 60°C), water released by the condensation of amino acids evaporates quickly, preventing the hydrolysis of the proteinoids back into their constituent amino acids. Some researchers have suggested that such events might have happened in hot volcanic areas, or that tidal pools might have trapped amino acids, which would have polymerized as the water evaporated. High tides or rainstorms might have washed the resulting polymers into the sea.

It may be that none of these special conditions was necessary for formation of organic polymers. It has been found that experiments like Miller's produce short peptides (amino acid polymers) directly by reactions between hydrogen cyanide polymers and water, rather than by the polymerization of amino acids.

Short peptide chains are fairly unstable and tend to hydrolyze into amino acids in solution. Longer chains, however, remain intact, stabilized by interactions between different parts of the molecule (similar to secondary and tertiary structure in proteins).

Some proteinoids exhibit enzyme-like properties; for example, they can catalyze some chemical reactions, and their catalytic properties can be destroyed by overheating and by chemicals that inhibit enzymes. The conclusion must be that molecules similar to enzymes, which are so vital to life, could have been produced on earth before living organisms existed.

19-E   Formation of Aggregates

When two or more types of polymer are shaken together in water, **coacervate droplets** form. They resemble the oil droplets that form when oil is shaken in water, but they are much more stable. If fats are present, they coat the droplets with a "skin" reminiscent of a lipid cell membrane. Oparin proposed that coacervate droplets were the ancestors of the first living organisms, and studied their properties extensively.

Coacervate droplets prepared from certain materials accumulate small molecules in comparatively high concentrations from their surroundings. Droplets prepared from the protein gelatin and the polysaccharide gum arabic, for example, concentrate a number of dyes and amino acids, while simple sugars and nucleotides remain equally concentrated inside and outside the droplets. As droplets take up substances, they grow, until they reach a size at which they break up easily when agitated.

Oparin studied the condensation of nucleotides to form short polynucleotides under prebiotic conditions. When they reached a certain size, the polynucleotides

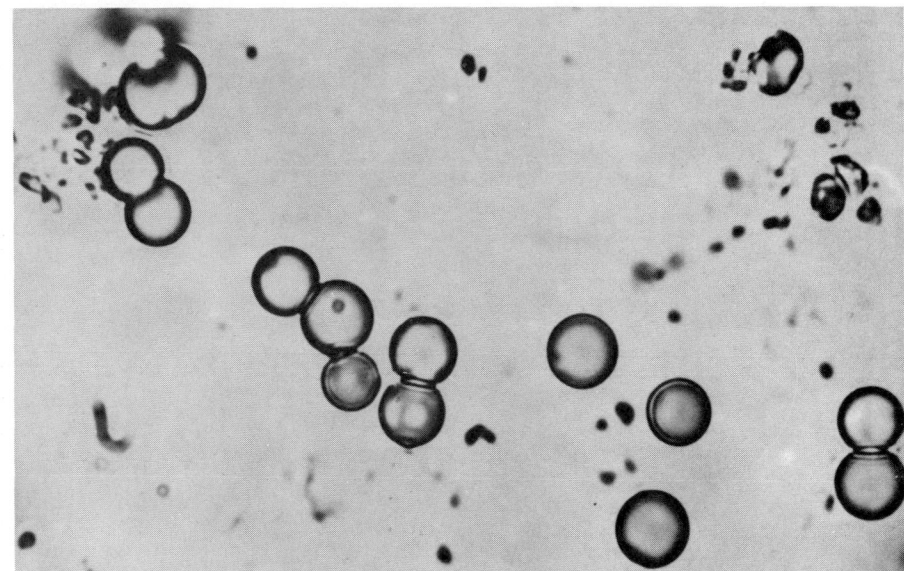

Figure 19-5 Proteinoid microspheres, each 1 to 2 μm in diameter. (Sidney Fox)

formed coacervates. Furthermore, once coacervates formed there was a marked shift in the equilibrium of the reaction

$$\text{nucleotides} \; \underset{+H_2O}{\overset{-H_2O}{\rightleftharpoons}} \; \text{polynucleotides}$$

in favor of enhanced polymerization. Thus coacervates stabilize their constituent polymers.

Fox found that when he added water to proteinoids, they formed **microspheres** (Figure 19–5). Since microspheres are denser than water, they sink and the water shields them from destruction by ultraviolet light. Fox believes that microspheres are better candidates than coacervate droplets for prebiotic systems. Microspheres are more stable than coacervate droplets and of uniform size (about 2 μm in diameter). Microspheres have osmotic properties, swelling when placed in hypotonic* solutions and shrinking when placed in hypertonic* solutions; they also accumulate monomers from the surrounding medium, catalyze reactions in their interiors, and bud off new microspheres as they grow. They have a double layer of proteins as a boundary between the interior and the external environment (Figure 19–6). The protein boundary is selectively permeable, and admits polynucleotides (the theoretical precursors of nucleic acids) very readily. In addition, microspheres have internal structure: watery areas, lipid-like areas, and boundary-layer areas provide sites for distinct chemical activities, just as in modern cells. In fact the interior of a microsphere looks so much like that of simple bacteria that even experts confuse them in electron micrographs.

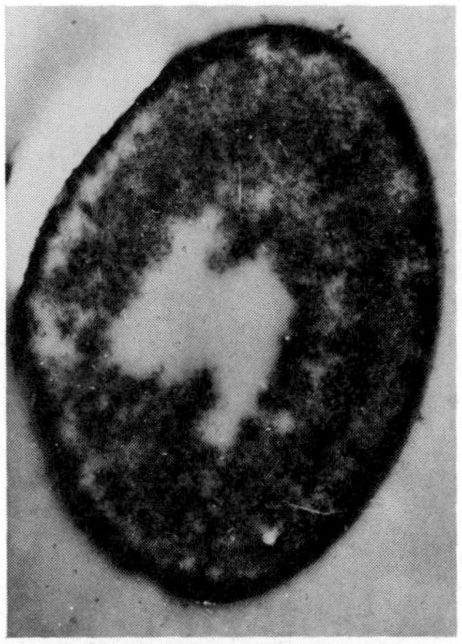

Figure 19-6 Electron micrograph of a proteinoid microsphere. Note the clearly defined boundary layer and the separation of contents into two distinct areas. (Sidney Fox)

19-F   Beginnings of Metabolism

Coacervate droplets and microspheres show **emergent properties**, which reflect the principle that "the whole is greater than the sum of its parts." This is true on both the structural and the functional levels.

Oparin prepared coacervate droplets that accumulated the dye methylene red in comparatively high concentrations from the surrounding solution. When he added NADH to the solution, the droplets took it up as well, and the dye in the droplets was reduced by the transfer of hydrogens from NADH, forming reduced dye and NAD$^+$

---

*Hypotonic Having a lower concentration of dissolved substances than another solution used for comparison; in this case, the solution inside the microspheres.
*Hypertonic Having a higher concentration of dissolved substances.

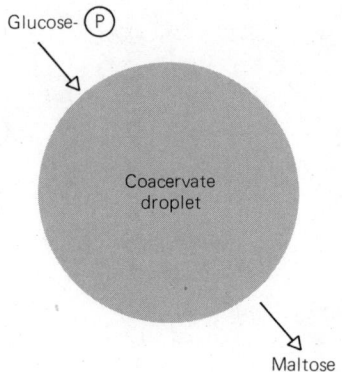

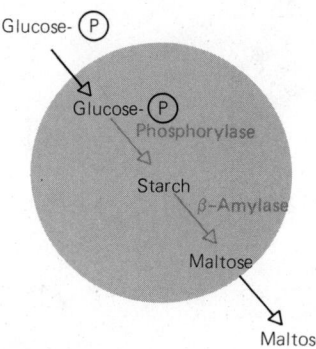

Figure 19-7 A two-step "metabolic pathway" observed by Oparin. Coacervate droplets absorb glucose phosphate from the surrounding solution; enzymes incorporated by the coacervate polymerize the glucose phosphate into starch, and digest the starch to maltose. The appearance of maltose in the surroundings indicates the completion of this simple "metabolic pathway."

as end products. This reaction occurred much faster inside the droplets than it would have in a solution of the dye and NADH alone. The droplets would also take up the enzyme NADH dehydrogenase from the solution, and they would then reduce the dye even faster (an example of another simple pathway is shown in Figure 19–7).

Microspheres can also accumulate substances from their surroundings, and catalyze chemical reactions internally. These activities of coacervates and microspheres may be interpreted as a rudimentary form of metabolism.

Development of Metabolic Pathways

The simple metabolism of a prebiotic aggregate is a far cry from a living cell with its many thousands of reactions. Oparin wrote: "The path followed by nature from the original systems of protobionts* to the most primitive bacteria and algae[1] was not in the least shorter or simpler than the path from the amoeba to man."[2] Along the way, the protobiont must have added more and more chemical reactions to its metabolism; as their numbers increased, these reactions had to be coordinated so that substrates were channeled into the metabolic pathways where they were needed; the catalysts for each reaction had to become more efficient, and the mechanisms for protein synthesis and the replication of genetic material had to evolve, allowing the protobiont and its offspring to make many copies of the metabolic enzymes and other proteins.

Prebiotic systems arrived at the threshold of life by the process of **chemical selection**, a term analogous to natural selection, but applied to nonliving systems. At first, chemical selection probably selected for mere longevity; the most stable aggregations would outlast their less stable neighbors. When droplets broke up, their components might recombine with debris from other droplets, and any favorable new combinations would survive.

As more and more long-lived aggregates accumulated, those with particular patterns of activity probably started to outlive their fellows. Any reaction that contributed to the stability of a system would confer a selective advantage on it. At first, such systems would have been so rare that the substrates they needed for their stabilizing reactions would have been abundant in the surrounding organic soup. However, as these systems grew and fragmented, there would have been more and more similar systems competing for increasingly scarce raw materials. Any system that acquired a catalyst that could convert another available molecule to the molecule it needed would have an advantage over its competitors. Suppose, for example, that the crucial molecule A was in short supply because of intense competition among aggregates taking it up faster than it was being produced by lightning storms, volcanoes, and so forth. Any aggregate that acquired the means of converting an

---

* Protobionts   Early aggregates, forerunners of living organisms.
[1] By "algae," Oparin probably meant the blue-green bacteria, photosynthetic prokaryotes, which are seldom called algae anymore.
[2] This statement displays "poetic license" and should not be taken to imply that an amoeba was a direct ancestor of humans.

abundant molecule, B, into the scarce A, would have an advantage: this would happen until there were so many systems using the conversion B → A that B also became scarce. At this point, a system that evolved a catalyst for the reaction C → B would survive, while B-users would disappear. This sort of process would have been repeated, with chemical pathways becoming longer and longer. There would also have been tremendous selective pressure for prebiotic systems to acquire increasingly effective catalysts for each metabolic reaction. Those systems that could catalyze the most reactions the fastest would have accumulated molecules from the broth and grown and reproduced faster than their competitors.

Origin of Energy Metabolism

One of the first requirements of a metabolic system is the ability to trap and use energy to drive various reactions. Recently biologists have realized that a membrane-bound hydrogen ion (H^+) pump is a universal feature of organisms, and some have proposed that such a pump may have provided the energy needed for primitive aggregates or cells to accumulate certain small molecules and ions from the external medium (Section 6-D). Such a pump may have been driven by light, an abundant source of energy that drives relatively simple pumps in some kinds of bacteria today. Elaboration of these pumps eventually produced the modern electron transport systems of aerobic respiration and oxygen-generating photosynthesis.

An older and more conventional view is that energy metabolism was originally based on ATP. Adenine is the simplest of the nucleotide bases; it can be formed by polymerization of five molecules of hydrogen cyanide and modification of some of the resulting bonds by reaction with water (Figure 19–8). Addition of a ribose molecule and three phosphate groups at the proper positions would result in ATP. The other nucleoside triphosphates, GTP, CTP, UTP, and TTP, might have been just as useful as energy donors (they do donate energy to various metabolic reactions in modern organisms), but they were probably not formed abiotically in as large amounts as ATP.

Thus ATP was probably a relatively common and available molecule that hydrolyzed easily to yield energy. Recent evidence shows that ATP promotes some kinds of reactions in prebiotic chemical simulations. As the supply of ATP available from the external soup ran low, systems that could use the energy of their pump-generated H^+ gradients, or the energy released by the breakdown of other compounds, to resynthesize ATP from ADP and inorganic phosphate would have been at a selective advantage.

The need for an ATP-regenerating system is believed to have selected for the series of anaerobic reactions known as fermentation. This process is undoubtedly extremely ancient; fermentation of glucose to pyruvate seems to be universal in modern organisms, many of which appear to be little changed from fossils found in Cambrian and even Precambrian (before 600 million years ago) rocks. The tricarboxylic acid cycle and oxidative phosphorylation found in aerobic organisms are highly efficient means of extracting further energy from the products of fermentation. This aerobic phase of respiration is believed to be a relative latecomer in the evolution of metabolism.

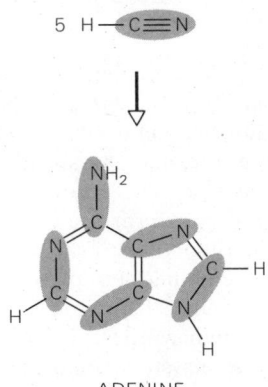

Figure 19-8 Adenine is basically a polymer of five molecules of hydrogen cyanide, although the chemical mechanism for its formation may be more complicated than indicated here.

ADENINE

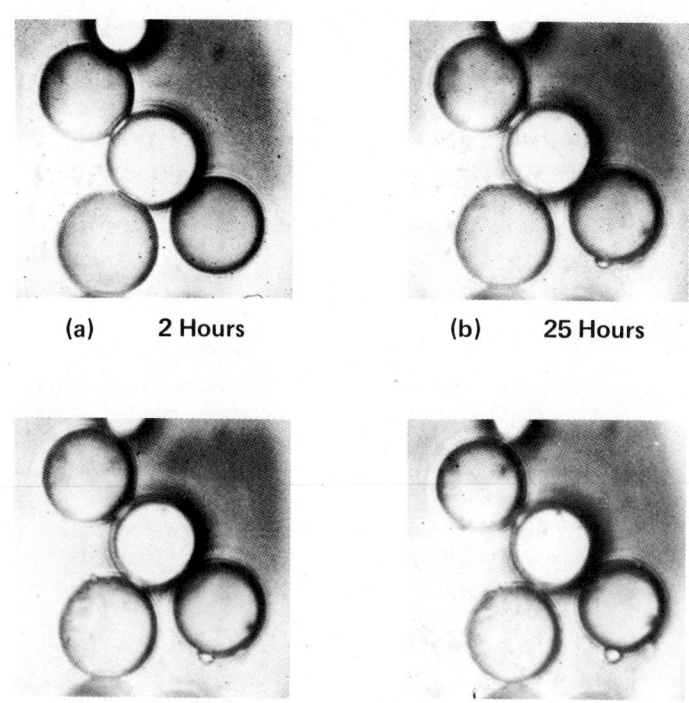

(a)   2 Hours

(b)   25 Hours

(c)   42 Hours

(d)   68 Hours

Figure 19-9 Budding in a proteinoid microsphere. Note growth of the bud on the microsphere at the lower right. (Sidney Fox)

19-G   The Beginnings of Biological Information

Self-organized aggregates of macromolecules have many structural and functional similarities to modern cells. However, they cannot be considered "living" because they cannot reproduce. True, microspheres produce buds that eventually become independent (Figure 19–9), but there is always the danger that the bud, the parent, or both will end up lacking some molecule crucial to the rudimentary metabolism that made the parent "successful."

In the reproduction of modern cells, each offspring receives a set of the information it needs to make all of the parent's proteins; the offspring can then make identical copies of these proteins as required. A cell's DNA contains this biological information; the sequence of DNA nucleotides determines the order of nucleotides in messenger, transfer, and ribosomal RNA, and the messenger RNA nucleotide sequence, in turn, dictates the order of amino acids in the cell's proteins (Chapter 10). Thus, biologists diagram information flow in modern organisms as:

$$DNA \longrightarrow RNA \longrightarrow protein$$

This system is remarkable for the degree of division of labor among the three kinds of substances. DNA is very stable, and thus it can serve as a file copy of the information needed to make RNAs. It can be used to make as many copies of RNA as needed, and can spend the rest of its time being replicated in preparation for cell division. This frees RNA to spend all of its time in protein synthesis. Proteins, in turn, perform the actual work of the cell's metabolism—regulating exchange of substances with the environment, catalyzing the reactions of metabolism, and even assembling new proteins and nucleic acids.

Once biologists had worked out the mechanism of information transfer and protein synthesis in modern cells, they began to examine how such a complex interactive system could have originated. If we assume that DNA came first, as in the information flow diagram above, we meet the problem that nucleic acids lack any noticeable catalytic activity. Left to themselves, nucleotides polymerize very slowly and do not even replicate existing nucleic acid strands faithfully. It seems unlikely that any DNA formed abiotically could have made much evolutionary progress on its own.

An alternative to the DNA-first theory is that the system of biological information flow evolved backwards from the present-day sequence, just as we believe metabolic pathways must have:

$$\text{protein} \longrightarrow \text{RNA} \longrightarrow \text{DNA}$$

At first this theory seemed equally unlikely; although proteins do have catalytic ability, they cannot copy themselves directly. It was assumed that proteins lack any information content; if the 20 common amino acids found in modern proteins had assembled at random into proteinoids, they would have formed too many different kinds of molecules, with too many different possible combinations into aggregates, for chemical selection to have sorted them out as quickly as we think evolution must have occurred. Numerous experiments have now shown, however, that amino acids do have informational properties, in that the proteinoids they form have nonrandom sequences. Certain amino acids are more likely than other equally common ones to become parts of proteinoids, and likewise amino acids are more likely to join in some configurations than in others. (For example, if two kinds of amino acids, A and B, are allowed to form dipeptides under prebiotic conditions, the sequence AB may occur almost exclusively, whereas by random reactions we would expect AB, BA, AA, and BB all to occur with equal frequency.) Only a small fraction of the theoretically possible proteinoids probably formed on the early earth, and so chemical selection had a more manageable task.

It now seems likely that biological information did evolve from the protein rather than the DNA end, but this does not mean that a full-fledged protein system evolved and then added RNA and DNA in turn. Proteinoids and nucleic acids doubtless interacted and evolved together, with chemical selection perpetuating neither favorable proteinoids nor favorable polynucleotides alone, but rather those aggregates in which the two groups of compounds became more and more precisely associated.

Recent experiments have shown a possible first step in this association. Microspheres contain areas with different microenvironments—hydrophilic, hydrophobic, and so forth. It turns out that amino acids and the polynucleotides containing their modern-day anticodons both have affinities for the same kinds of microenvironments, and different amino acid/anticodon combinations tend to congregate in different microenvironments. This suggests that the genetic code did not arise randomly, but itself has an informational basis.

We can also find evidence that the first informational polynucleotides were probably RNAs. The three-dimensional structure of nucleotides is such that amino acids become associated with the 2'-hydroxyl group of the sugar (Figure 19–10). The deoxyribose sugar of DNA lacks this 2'-hydroxyl group and thus cannot serve as a "handle" to hold amino acids in a precise sequence until they polymerize. In the metabolism of modern cells, deoxyribonucleotides are produced by removing the hydroxyl group from ribonucleotides; this suggests that ribonucleotides may be the evolutionary forerunners of deoxyribonucleotides. Furthermore, some viruses have an enzyme that uses RNA as a template for DNA synthesis, rather than the other way around. Overall, the evidence suggests that DNA was added to the informational store at a later stage in evolution than RNA.

Possessing even a rudimentary system of protein synthesis would have been of tremendous advantage to a prebiotic system because it would have allowed more or less exact duplication of proteinoids that catalyzed favorable reactions. At first, replication of nucleic acids was probably quite unfaithful, and the frequent changes in nucleotide sequences would have given rise to new amino acid sequences in polypeptides, some of which would have proved to be better catalysts than self-ordered proteinoids.

RIBONUCLEOTIDE

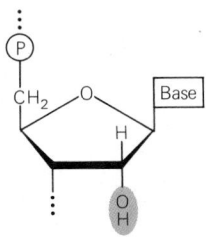

DEOXYRIBONUCLEOTIDE

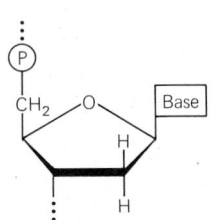

Figure 19-10 A ribonucleotide and a deoxyribonucleotide, showing the hydroxyl group of a ribonucleotide that associates with amino acids (colored area).

TABLE 19-3 SOME IMPORTANT MILESTONES IN THE ORIGIN OF LIFE

TIME, BILLIONS OF YEARS AGO	EVENT
4.6	Earth originates
4.3	Conditions on earth stabilize
3.8	Ocean mineral content similar to today's; atmosphere like today's but without O_2; carbon as CO_2 (Isua formation rocks, Iceland)
3.5	Earliest known stromatolites (mats of prokaryotic cells) (Australia)
2.3	Oxygen-generating photosynthesis
2.0	Great diversity of bacteria (Gunflint chert, Ontario) Atmospheric oxygen reaches 1% Aerobic respiration appears
1.45	Eukaryotic cells; sexual reproduction allows more rapid evolution
0.7	Soft-bodied animals (jellyfish, worms)
0.6	Hard animal skeletons (Cambrian period begins)

Subsequent events are outlined on the endpaper of this book

19-H   Heterotrophs and Autotrophs

With the advent of reliable reproduction of genetic information, the story of the origin of life can be considered complete. The appearance of the first true organisms marked the end of the era of chemical selection and the beginning of the era of natural selection. Survival was no longer enough; competition among cells required evolution of more efficient means of acquiring energy and converting it into offspring.

There was probably a long period when all organisms were **heterotrophs**; that is, they grew and multiplied at the expense of organic molecules made outside their own bodies. At first, these molecules were provided by the surrounding dilute soup. Organisms must have had some sort of metabolism, based on protein catalysts, for converting the food molecules they acquired to meet their own needs.

As the nutrients in the primordial soup became scarcer, competition for them grew more fierce. Some organisms may have taken to devouring their neighbors to obtain nutrients; others evolved means of acquiring energy from other sources. They became **autotrophs**, able to manufacture food from inorganic molecules.

The evolution of autotrophy was a very important step in the evolution of life because it freed the living world from dependence on the slow production of food by abiotic means, and greatly increased the amount of life the earth could support.

We tend to think of autotrophy as the ability to carry on photosynthesis, but there were other, less successful evolutionary "experiments" in autotrophy, which are still found among some bacteria. There are many species of **chemosynthetic** bacteria, which obtain energy from a variety of inorganic chemical oxidations and fix* carbon dioxide even in the absence of light.

Examining the metabolic capabilities of various groups of bacteria allows us to outline how the process of photosynthesis may have developed by gradual stages. Probably the first step was the use of light energy to create an electrochemical gradient of H^+ across a cell membrane that could power the synthesis of ATP. This reaction is part of modern photosynthesis (see Chapter 8).

The use of light energy to release hydrogen, which was used in turn to reduce carbon dioxide, probably evolved later. The purple non-sulfur bacteria, Athiorhodaceae, use light energy to transfer hydrogen from the end products of glycolysis

---

* Fix   To incorporate into a less volatile compound; in this case, to incorporate into a larger organic molecule.

to carbon dioxide. This process allows the bacteria to use the glycolytic waste products as energy sources.

Other primitive photosynthetic organisms use other hydrogen donors to fix carbon dioxide. Purple and green sulfur bacteria (Thiorhodaceae) use hydrogen gas and hydrogen sulfide (H_2S) as hydrogen donors. The use of water as a hydrogen donor probably came quite late; it is difficult to extract hydrogen from water. However, because water is much more readily available than other hydrogen donors, organisms that can use water as a hydrogen source have a considerable selective advantage. The only bacteria that use water as a hydrogen donor for photosynthesis are the blue-green bacteria, some of which may have been the ancestors of the higher groups of photosynthetic organisms.

19-I   Aerobic Respiration

Photosynthesis that uses water as a hydrogen donor also releases oxygen in the form of O_2. Free oxygen is a threat to life because it destroys the flavins, which are important coenzymes; the appearance of oxygen in the atmosphere must have presented a grave environmental crisis. However, some organisms survived, probably by being able to protect their flavins from oxidation by combining their organic wastes with oxygen, thus ridding themselves of both at once. Bioluminescent organisms today use such a process to produce light through the oxidation of organic compounds.

The extremely important metabolic pathway of aerobic respiration, which uses oxygen in the breakdown of organic molecules to carbon dioxide and water, probably began like this. Glycolysis was a common source of organic wastes. Aerobic respiration may have evolved by interposing reactions between the end products of glycolysis and their combination with oxygen.

The yield of ATP from aerobic respiration is much higher than the ATP yield of anaerobic fermentation. This tremendous bonus in the amount of energy available from food molecules allowed organisms to grow and reproduce at a much more rapid rate than they could with only an anaerobic system. It also allowed them to use part of their energy budgets to "experiment" with new enzymes and new structures that were energetically demanding but gave them superior competitive abilities and allowed them to outstrip their anaerobic neighbors. In the modern world, anaerobic organisms are restricted to habitats that lack sufficient oxygen to support aerobic life.

19-J   Origin of Cellular Organelles

The next big jump in evolution was probably the rise of eukaryotic cells containing membrane-bound organelles. An intriguing theory advanced in the early part of the twentieth century was that mitochondria and chloroplasts originated as bacteria and blue-green bacteria, respectively, that came to live within other cells (see Section 22-J).

Eukaryotes differ from prokaryotes in that eukaryotes can undergo mitosis and meiosis (see Chapter 11). Mitosis allows the nucleus to duplicate its chromosomes precisely and to distribute them equally to two daughter nuclei. Meiosis is necessary to sexual reproduction, which involves fertilization.

The ability to reproduce sexually was an extremely important development because it allowed for genetic variation by the reshuffling of existing genes, in addition to the genetic variation produced by mutation to form new genes. Genetic variability is the raw material for natural selection, and so for evolution. The fossil record indicates that eukaryotes evolved much faster than prokaryotes, diversifying into organisms with a variety of sizes, shapes, and lifestyles. Although eukaryotes have dominated the earth for at least 700 million years, prokaryotes have also survived and flourished; many of them are believed to have changed very little in more than a billion years.

Organisms often appear to be at the mercy of their environments. The environment sets limits and organisms must succeed within these limits or die. In fact, however, organisms can and do influence and change their environments in many ways.

What changes have organisms made on the face of the earth? Heterotrophs have consumed all the nourishment from the primordial soup; water-splitting autotrophs have added molecular oxygen to the air and thus promoted selection for the evolution of aerobic respiration. Accumulation of oxygen produced by photosynthetic organisms has built up an ozone layer high in the atmosphere. Ozone (O_3) is formed from molecular oxygen by the action of ultraviolet light (uv); this ozone acts as a filter and prevents much of the sun's ultraviolet radiation from reaching the surface of the earth. Ultraviolet light is very destructive to proteins and nucleic acids. The first organisms were protected by the water surrounding them, which acts as a shield against ultraviolet light, and they probably also had mechanisms for repairing damage caused by uv, just as all organisms do today. Nevertheless, uv was probably one of several environmental factors that held early organisms back from moving out of the water and colonizing the land. Vast resources of sunlight and minerals were available on land, and the first organisms to make the move found little competition. Terrestrial plants evolved and flourished, and were soon joined by terrestrial animals. The trees and grasses that now clothe much of the land have also added oxygen to the atmosphere, changed the patterns of water flow from the land to the seas, and speeded the formation of soil from rock. Thus organism and environment have molded one another during the history of life on our planet.

SUMMARY

Until the nineteenth century, people believed that living organisms were constantly being formed from nonliving substances when conditions were favorable. Pasteur and Tyndall destroyed the belief that organisms are formed spontaneously on earth under present-day conditions. However, most scientists today believe that life originally arose spontaneously.

The conditions under which life began, however, were very different from those on earth today. The primitive earth probably had a mildly reducing atmosphere, composed of the gases in today's atmosphere with the exception of oxygen. Such an atmosphere would have been conducive to the formation and stabilization of organic compounds, which gradually polymerized, formed macromolecular aggregates, and evolved systems of metabolism, information transfer, and reproduction, eventually becoming living organisms.

Some important events during the early history of life were the evolution of photosynthesis and aerobic respiration, and the acquisition of intracellular organelles. Organisms then changed their environment from an earth of barren water and rock under a somewhat reducing atmosphere to one of teeming oceans and verdant landscapes in an oxidizing atmosphere. Each environmental change caused by organisms exerted selective pressures to adapt to the new environment, which in turn changed the environment even more. Thus living organisms and their environment have shaped each other during the evolution of life on earth.

Five questions about the origin of life have tentatively been answered:

1. How did enzymes arise without previous enzymes to make them?

2. How did cells come into existence without cells to make them?

3. How did membranes originate?

4. How did informational macromolecules arise before the complex genetic code?

5. How did reproduction begin?

OBJECTIVES

From your study of this chapter, you should be able to:

1. Recognize what is meant by the terms spontaneous generation, information (as applied to biological molecules), and emergent properties.

2. Give two reasons why it is unlikely that spontaneous generation could occur under the conditions that exist on earth today.

3. Compare the environment in which life is believed to have arisen with the present environment on earth.

4. Discuss the importance of time in the theory of the origin of life presented in this chapter.

5. Trace the steps by which life may have originated on earth, from the formation of organic monomers through the rise of eukaryotic organisms.

6. State what evidence we have that these steps were possible, describing the experimental work of Miller, Fox, and Oparin.

7. List (a) kinds of places and (b) sources of energy, on the prebiotic earth, that could have been important in the formation of organic molecules and prebiotic systems.

8. Describe the order and steps by which anaerobic respiration, aerobic respiration, and photosynthesis may have evolved.

9. Compare fermentation and aerobic respiration and describe their significance for the origin and evolution of life.

10. State the significance of the evolution of the first autotrophs for the evolution of early life.

11. Describe changes in the environment that resulted from the presence of living organisms, and explain how the evolution of modern organisms was dependent on the change to modern environmental conditions.

SELF-QUIZ

1. Mitochondria contain particles made up of about 100 polypeptide molecules of several different types. These particles catalyze the production of acetyl CoA and carbon dioxide from pyruvate. When disassembled into the component polypeptides, no enzymatic activity is seen. These particles are examples of:
   a. spontaneous generation
   b. emergent properties of systems
   c. the origin of life
   d. informational macromolecules
   e. proteinoid microspheres

2. Number the following structures and processes in the order in which they are believed to have evolved:
   aerobic respiration
   proteinoids
   water-splitting photosynthesis
   organic monomers
   acquisition of intracellular organelles
   fermentation

3. In Stanley Miller's classic experiment:
   a. nucleic acids were formed
   b. ultraviolet radiation was used
   c. oxygen was one of the starting ingredients
   d. water was strictly excluded from the system
   e. amino acids were formed

4. List two changes in the environment that resulted from evolution of the process of water-splitting photosynthesis.

5. Which of the following is *not* true of the microspheres studied by Sidney Fox?
   a. They are formed by addition of water to proteinoids.
   b. They are of uniform size, about 2 μm in diameter.
   c. Microspheres grow and bud off new microspheres.
   d. They are bounded by a layer of lipid.
   e. The boundary of a microsphere shows selective permeability.

QUESTIONS FOR DISCUSSION

1. Are the laboratory simulations that were presented in this chapter (e.g., Miller's reducing atmosphere simulation, Oparin's coacervate droplets, and Fox's microspheres) experiments or observations? Are there any controls?

2. Suppose that a scientist claimed to have produced life from nonliving materials under laboratory conditions. What criteria must such an "organism" meet before *you* agreed that it was truly living?

3. Figure 19–4 shows some compounds that were formed in Miller's simulation of the early earth but are not found in living organisms. Can you explain this?

4. Sidney Fox and his colleagues have pointed out that the five questions listed at the end of the chapter summary can all be answered on the basis of the properties of proteinoids and the microspheres they form when placed in water. Can you describe these properties?

5. What is the evidence that aerobic respiration evolved before the eukaryotic condition?

REFERENCES AND FURTHER READING

Dickerson, R. E. "Chemical evolution and the origin of life." *Scientific American*, September 1978.

Dickerson, R. E. "Cytochrome *c* and the evolution of energy metabolism." *Scientific American*, March 1980. Builds an evolutionary tree for kinds of photosynthesis and respiration in prokaryotes.

Farley, J. *The Spontaneous Generation Controversy from Descartes to Oparin*. Baltimore: Johns Hopkins University Press, 1977. An historical account of philosophical and scientific controversy over the origin of life; full of fascinating tidbits of information and interesting ideas about the development of scientific thought.

Fox, S. W. *et al.* "Chemical origins of cells." *Chemical and Engineering News* 48:80–94, 1970. This and the next article are fairly technical, requiring a good chemical background.

Fox, S. W. "Chemical origins of cells: 2." *Chemical and Engineering News* 49:46–53, 1971.

Kerr, R. A. "Origin of life: new ingredients suggested." *Science* 210:42, 1980.

Oparin, A. I. *Genesis and Evolutionary Development of Life*. Translated by Eleanor Maass. New York: Academic Press, 1968. A fairly comprehensive theory of the origin of various aspects of life.

Orgel, L. E. *The Origins of Life*. New York: John Wiley and Sons, 1973. A discussion of chemical selection and the origin of life by one of Miller's associates.

Ponnamperuma, C. *The Origins of Life*. New York: E. P. Dutton, 1972. Excellent illustrations and photographs.

Schopf, J. W. "The evolution of the earliest cells." *Scientific American*, September 1978. A prominent student of early microfossils explains the evidence for current beliefs about the early history of life.

Wald, G. "The origin of life." *Scientific American*, August 1954. Interesting discussion of the role of time in probability.

Weber, A. L., and J. C. Lacey, Jr. "Genetic code correlations: amino acids and their anticodon nucleotides." *Journal of Molecular Evolution* 11:199, 1978.

CHAPTER 20

CLASSIFICATION OF ORGANISMS

In the sixteenth and seventeenth centuries, when Europeans began to travel to the far corners of the world, they brought back tales and actual specimens of thousands of exotic plants and animals that were unknown in their homelands. At about the same time, early microscopists found a myriad of tiny organisms, whose existence had not been suspected, living around them almost everywhere they looked. The number of different kinds of organisms soon became almost overwhelming.

The Swedish botanist Carolus Linnaeus is credited with inventing the first orderly system for classifying organisms. Linnaeus based his classification on morphology (structure, anatomy); for example, two species of trees with similar leaves and bark fell close together in his scheme. After Darwin's theory of evolution gained acceptance, biologists decided that classifying organisms according to their evolutionary relationships would be more natural. Fortunately, the morphological characters used by Linnaeus usually reflect evolutionary relationships; hence biologists have been able to retain most of the names and groupings of Linnaeus's thorough and meticulous system.

Our system of naming and classifying organisms is undoubtedly inadequate to

Figure 20-1 Carolus Linnaeus, who described himself thus: "Brown-eyed, nimble, hasty, did everything promptly." Linnaeus classified plants by their sexual parts with group names such as *Polyandria*, meaning "twenty or more males in bed with the same female." This emphasis on sex shocked some of his contemporaries. The Bishop of Carlisle wrote: "To tell you that nothing could equal the gross prurience of Linnaeus's mind is perfectly needless," and Goethe worried about the embarrassment chaste young people might suffer when reading botany textbooks. (Biophoto Associates, National Portrait Gallery, London)

the task. Biologists believe that there are some 10 million different species of organisms in the world, of which only about 15% have been described. (Today we are destroying natural habitats so rapidly, by pollution, our population explosion, and the destruction of forests, rivers, and fields, that most of the remaining species will be extinct before they are ever described!) Until better systems of classification become widespread, however, we must make do with what we have. We need scientific names for organisms so that we can communicate with one another and know when we are talking about the same species, and we need some sort of classification scheme to make sense of the vast diversity of organisms all around us.

In this chapter we shall consider how living things are classified and introduce the five-kingdom system of classification that we use in this book.

20-A   Species

The basic unit for classifying organisms is the species. In the case of organisms that reproduce sexually (and many do not), a **species** is a group of organisms that actually or potentially interbreed with one another to produce fertile offspring and that do not interbreed with members of other species under natural conditions. Interbreeding may be either direct or via populations that link various individuals. For instance, all the dogs in the world belong to one species; a dog in Australia could undoubtedly breed with a dog in Jamaica even though it might never have the chance. You may argue that a Chihuahua is so small that it could not breed with a Doberman pinscher for physical reasons, and this is undoubtedly true, but in this case the two belong to the same species according to the second part of the definition because they are linked by interbreeding populations: the Chihuahua could mate with a toy poodle, which could mate with a Labrador, which could mate with a Doberman.

20-B   Binomial Nomenclature

Linnaeus proposed that every species be given a unique Latin **binomial**, that is, a two-word name, to set it apart from every other type of organism. The first word in this binomial designates the genus to which the species belongs; a **genus** (plural, **genera**; adjective, **generic**) contains one species or a group of similar species. The second word in the binomial denotes the species itself. For example, the binomial for the grizzly bear is *Ursus horribilis* (horrible bear). This name indicates that the grizzly has been placed in the genus *Ursus*, and within that genus, has been given the specific, or species, epithet *horribilis*. The complete binomial *Ursus horribilis* is

(a)

(b)

Figure 20-2 A taxonomic accident: (a) *Pieris japonica*, a common garden plant, was given the same generic name as the cabbage white butterfly, *Pieris rapae* (b), and its relatives. (Biophoto Associates)

needed to distinguish the species of bear known as the grizzly from other species in the genus *Ursus*, such as *Ursus arctos*, the Alaskan brown bear.

The name of a genus can be used only for one species or for a small group of related species. (However, because plants, animals, and microorganisms are named by different international commissions on nomenclature, the same genus name has sometimes been used more than once; thus *Pieris* is a genus of butterflies and of garden shrubs!) Specific epithets, however, are adjectives, and the same one may be combined with different generic names and used for a number of unrelated organisms; for example, *Erythronium americanum*, the trout lily; *Euarctos americanus*, the American black bear; *Hepatica americana*, a relative of the buttercup; and *Coccyzus americanus*, the yellow-billed cuckoo, all bear adjectives meaning "American" as the second word in their binomials.

20-C Taxonomy

Taxonomy is the classification of organisms. The classification system consists of a hierarchy of categories. The most inclusive categories are the **kingdoms**; the other main categories, in descending order, are **phylum**, **class**, **order**, **family**, **genus**, and **species**. (You can remember this order by memorizing the sentence "King Phil came over for Gene's special.")

A **taxon** (plural, **taxa**) is a group of organisms defined by the classification scheme, such as a particular species or class. For example, Ursidae (a family) is a taxon including the species in the genus *Ursus* as well as those in other genera of bears, such as the polar bear, *Thalarctos maritimus*, and the American black bear, *Euarctos americanus*. Some taxa contain only one group at the next lower level; for example, there are numerous families that contain only one genus and one species.

Table 20–1 gives the classification for human beings, and here you can see the seven important levels of the hierarchical classification. Table 20–2 gives guidelines for the use of binomials and names of taxa.

Linnaeus named and classified all of the plants and animals known to him in his massive books *Systema Naturae* and *Species Plantarum*. Linnaeus used morphology as his criterion for classification because he believed in divine creation and because of his respect for the work of such Greek philosophers as Plato, who wrote that every object has an intrinsically defined being or "essence." Thus, according to Plato, although there are many different sorts of tables, the "essence of table" is a flat surface supported by legs. A biologist who believed in divine creation thought about classification in the same terms. God had created the perfect dog (or horse, or dandelion), and although various imperfect versions of what we call a dog exist, from the medley of dogs in the world we should be able to discern the "essence of dog," the morphology of the original, perfect dog.

TABLE 20-1 CLASSIFICATION OF *Homo sapiens*

CATEGORY	TAXON	CHARACTERISTICS
Kingdom	Animalia	Heterotrophic, multicellular organisms lacking cell walls and possessing a motile stage in the life history.
Phylum	Chordata	Animals with a dorsal, hollow nerve tube, notochord, and pharyngeal gill slits at some stage of the life history.
Class	Mammalia	Chordates with only one bone in the lower jaw, hair or fur, young nourished by milk from mammary glands of female parent.
Order	Primates	Originally arboreal (tree-living) mammals with flattened fingers and nails, vision the most important sense, and poor sense of smell.
Family	Hominidae	Primates with bipedal locomotion, flat faces, binocular color vision.
Genus	*Homo**	Hominid with large brain, speech, long childhood.
Species	*Homo sapiens**	High forehead, body hair reduced, prominent chin.

* Homo Latin: man.
* Sapiens Latin: knowing, wise.

Linnaeus believed that a species could be described by listing the morphological characteristics of a "perfect" member of the species. This led to a practice that is still widespread today, that of naming **type specimens**. When a new species is described, the author takes a typical specimen, declares it to be the type specimen of the species, and preserves it. The type specimen then becomes the definition of the species. If later workers need to know whether they are really working on the same species of daisies that the original author described, they compare their daisies with the type specimen. This is essential to research, since without it two workers may argue forever about why they each find that their own local daisies respond very differently to an experimental situation, never realizing that their results differ because they are working on different species.

In the case of microorganisms, instead of preserving one or a few dead specimens, it is now possible to preserve clones of genetically identical individuals alive by freezing them in liquid nitrogen ($-196°C$). Researchers can then obtain portions

TABLE 20-2 CONVENTIONS FOR USING BINOMIALS AND NAMES OF HIGHER TAXA

Capitalize:
1. Genus, but usually *not* species
2. Latin names of taxa above genus level, but *not* their Englisn counterparts

Examples
Ursus horribilis
Mammalia, mammals
Hominidae, hominids

Italicize or underline:
Genus and species (binomial), but *not* Latin or English names above genus level

Homo sapiens
Hominidae

Spell out:
1. The generic name the first time you use it in each paragraph
2. The specific epithet every time you use it

Abbreviate:
1. The generic name to its first letter at the second and subsequent mentions in the same paragraph
2. When you know the genus but not the species of the organism(s) you are discussing (in this case, always spell out the genus name, even if you have already mentioned it previously in the same paragraph)

U. horribilis

Ursus sp. (one, unknown, species)
Ursus spp. (more than one species of *Ursus*)

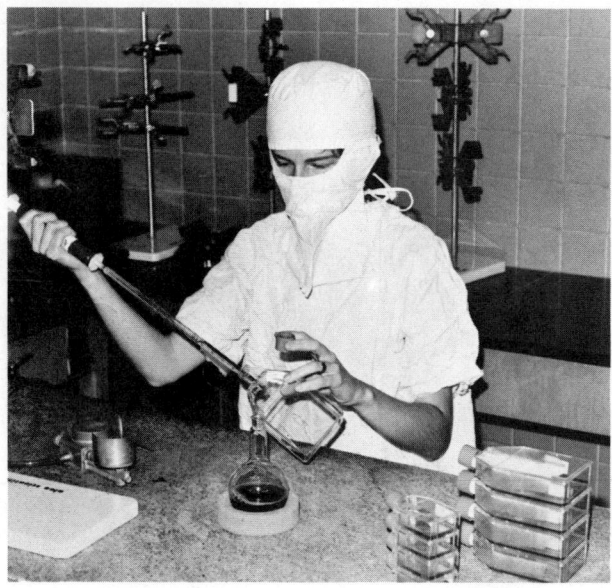

Figure 20-3 Preserving strains of microorganisms. Using sterile techniques to exclude other organisms, a worker transfers samples of a culture to fresh medium. (John and Marian McGrath)

of a culture of a particular species, and even of particular genetic strains within a species, for experimentation or comparison (Figure 20–3).

Although thousands of the species and taxa names introduced by Linnaeus are in use today, general acceptance of the theory of evolution has led biologists to classify organisms in a way that is, at least theoretically, different from that used by Linnaeus. The theory of evolution states that species arise from other species and change with time. This ongoing process of evolution makes a "species" almost impossible to define. Organisms change all the time, as they exchange genes with one another and are subjected to different selective pressures at different times and in different places (see Chapters 15, 16, 17).

The evolutionary history of a species is its **phylogeny**, and modern taxonomists try to classify organisms according to their phylogenetic (sometimes written "phyletic") relationships. This amounts to drawing an evolutionary family tree for an organism. In many cases it leads to the same result as Linnaeus's morphological classification, since organisms that have evolved from a common ancestor are likely to be more similar than those that have not. This is not always the case, however; the Australian marsupial* wolf looks, and in many ways behaves, very much like the North American timber wolf. But a study of the evolutionary history of each leads to the remarkable conclusion that the Australian wolf is much more closely related to the very different-looking kangaroo than it is to the timber wolf.

Although a phylogenetic taxonomy sounds straightforward, it leaves many problems. One difficulty is that it is very hard to define a species. Members of the same species are supposed to interbreed, but how then does one classify all the plants, bacteria, and animals that reproduce asexually and never or hardly ever "breed"? Another difficulty in phylogenetic taxonomy is deciding how closely organisms must be related in order to fall within the same taxon. Are all the species of the dog family phylogenetically close enough to one another to be lumped into the same genus, and if not, how many genera should we use? How alike must different members of a family be? There is never a simple answer to such questions. In the end, biologists usually come back to using morphological criteria, even though these present some problems of their own.

Molecular geneticists have also begun to analyze the detailed genetic makeup of organisms and to measure the extent of differences between them. It is now possible, for example, to estimate the number of mutations (inheritable changes in the DNA) that has occurred in two species since they diverged from a common ancestor.

---

*Marsupials Group of mammals whose young undergo part of their development in a pouch.

Similarly, biologists can determine the amino acid sequence of a protein, such as cytochrome c or hemoglobin, that occurs in most members of a particular group of organisms. Baboon hemoglobin and human hemoglobin differ by fewer amino acids than do human and horse hemoglobin, and so it is reasonable to conclude that humans and horses have had longer separate evolutionary histories, and are less closely related, than are humans and baboons. In general (and reassuringly) the new biochemical techniques of taxonomy confirm the phylogenetic relationships worked out on morphological grounds during the last hundred years (see Essay, "Proteins As Evolutionary Puzzle Pieces," Chapter 10).

20-D   The Five Kingdoms

The most inclusive taxa are the kingdoms. Linnaeus's system of classification had two kingdoms, the plants (Plantae) and animals (Animalia). This seemed reasonable in his day, since the macroscopic (visible to the unaided eye) land plants and animals were the most familiar organisms, and they were clearly very different. Plants did not move around; they did not eat, but seemed to need only water in order to grow. Animals were **motile**; that is, they could move from place to place. Animals had to eat plants or each other in order to stay alive. On a microscopic level, plants could be seen to have cell walls, which animal cells lacked. Fungi seemed to be aberrant plants, since they had cell walls and root-like structures that absorbed food from living or dead organisms, but lacked the green pigments of the other "plants."

However, discovery and examination of a wide variety of microscopic organisms since the time of Linnaeus have revealed many organisms that do not fit neatly into either the plant or the animal camp. Some organisms, such as *Euglena*, seemed to fit both descriptions. *Euglena* (Figure 23–4) has a rather stiff covering, not as thick as a plant cell wall but certainly affording more protection than a cell membrane. *Euglena* also has chloroplasts and can carry on photosynthesis when exposed to light. However, it also has a flagellum with which it can swim—an animal-like characteristic—and it can engulf other organisms and digest them as food, just like an animal. Bacteria presented another problem, since they have cell walls but often also possess flagella that make them motile; most cannot make their own food, but some can carry on photosynthesis. As more organisms were discovered and studied, the division into plant and animal kingdoms seemed to become more and more artificial and the zone between the two became more and more confusing.

Recent attempts to revise biological classification at the kingdom level have been many and varied, and prominent books and articles often use conflicting schemes of classification. For the sake of consistency, this book uses one system of classification, based on ecologist Robert Whittaker's system of five kingdoms.

The first organisms on earth probably belonged to the kingdom Monera, composed of the organisms with prokaryotic* cells. The other four kingdoms, made up of organisms with eukaryotic* cells, are separated according to two main criteria: degree of complexity and mode of nutrition. The kingdom Protista is composed of **unicellular** (one-celled), eukaryotic organisms. Eukaryotes with **multicellular** (many-celled) body structures are placed into one of the other three kingdoms, kingdom Plantae, kingdom Animalia, and kingdom Fungi—which are separated from one another largely on the basis of nutrition. Most of the members of the kingdom Plantae (plants) are photosynthetic (and secondarily absorptive; plants obtain water and minerals by absorbing them from their surroundings). Most members of the kingdom Animalia (animals) are **ingestive**; that is, they engulf or swallow their food and digest it internally. The Fungi are absorptive; they absorb organic molecules from outside their bodies directly through their exterior cell membranes. It seems likely that each kingdom contains members that evolved the characteristics typical of that kingdom independently; thus each kingdom probably contains members from several evolutionary lines. We shall consider each kingdom in turn.

---

* Prokaryotic Lacking a nuclear membrane separating the DNA from the cytoplasm, and without other membrane-bound organelles.
* Eukaryotic With a nuclear membrane surrounding the genetic material, and with various membrane-bound organelles in the cytoplasm.

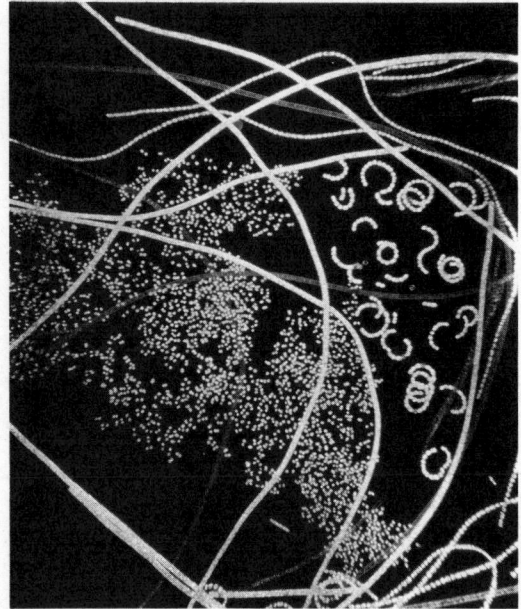

Figure 20-4 Monera: these tiny filaments, necklaces, and clumps of dots are various species of blue-green bacteria from the surface of a pond. (Biophoto Associates)

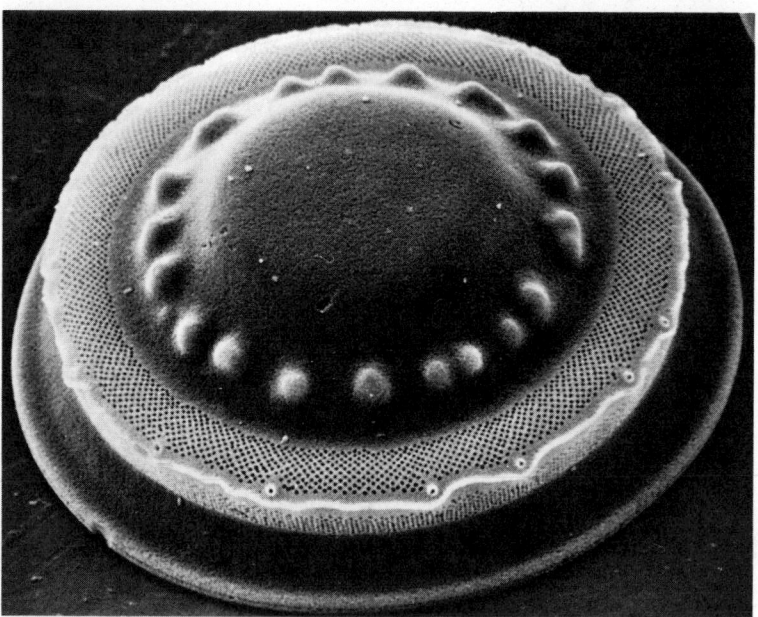

Figure 20-5 Protista: the shell of a diatom, a protist found floating in ponds or lakes, magnified 3,000 times. (Biophoto Associates)

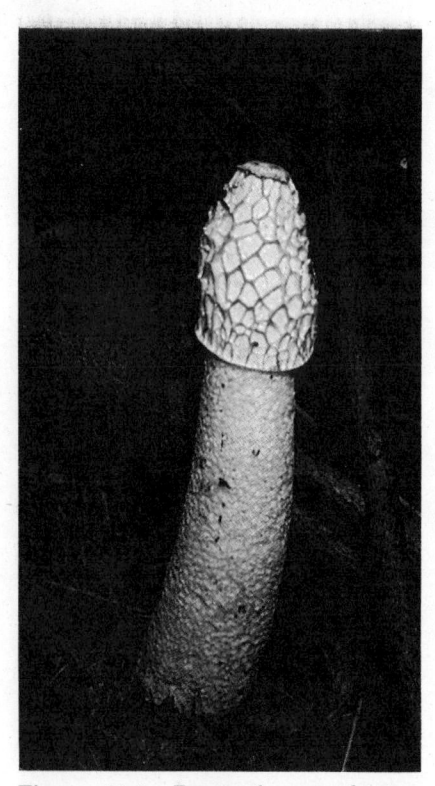

Figure 20-6 Fungi: the reproductive structure of the aptly named *Phallus impudicus*. (Biophoto Associates)

Kingdom Monera

All prokaryotic organisms, that is, the bacteria and the blue-green bacteria, are placed in the kingdom Monera. Prokaryotes lack nuclear membranes and mitochondria, chloroplasts, and other membrane-bound organelles found in eukaryotic cells. Their DNA is present as one circular double helix without the associated proteins found in eukaryotes, and they divide and reproduce without the nuclear divisions of meiosis or mitosis found in eukaryotes (see Chapter 11). Prokaryotes also differ from eukaryotes biochemically in such things as the materials found in their cell walls, the size and composition of their ribosomes, and in some of their metabolic pathways.

Kingdom Protista

The Protista are the one-celled eukaryotic organisms. As eukaryotes, they have nuclear membranes and linear chromosomes (with associated proteins) that can go through the processes of meiosis and mitosis. The first eukaryotes were undoubtedly protists, some of which gave rise to the fungi, animals, and plants that dominate life on earth today. Modern protists show modes of life that closely parallel those of higher eukaryotes, some being photosynthetic, some ingestive, and some absorptive, some motile and some nonmotile, some with cell walls and some without; all combinations of these characteristics are represented in various members of this group.

Kingdom Fungi

The fungi are often classified as plants because they are nonmotile, and because they have an external wall that resembles the cell wall of plant cells in many ways. Fungi, however, cannot make their own food; they must absorb food from a living or nonliving organic source. In many cases, they excrete digestive enzymes that digest food outside their bodies before they can absorb it. Whittaker felt that fungi should be separated from plants because they are completely absorptive, whereas most plants are only secondarily absorptive (that is, plants absorb water and minerals and carbon dioxide, but make their own organic molecules). In addition, fungi and green plants differ in cell wall composition, body plan, and reproduction.

Figure 20-7 Plantae: a water lily, member of a group of plants that live with their roots in the mud and their leaves at the surface of shallow bodies of fresh water.

Kingdom Plantae (Plants)

All members of the plant kingdom are eukaryotes and have cell walls that contain cellulose. Most contain chlorophyll and carry on photosynthesis inside chloroplasts, although a few species have lost their chlorophyll and obtain all of their nutrients by absorption. The plant kingdom includes the multicellular algae as well as all the familiar multicellular land plants—the mosses, ferns, grasses, shrubs, and trees.

Kingdom Animalia (Animals)

Animals are multicellular, eukaryotic, heterotrophic* organisms that obtain food mainly by ingestion. Most animals can move, and this permits them to acquire food from their environment. All but the simplest animals produce gametes (eggs and sperm) in multicellular organs, and the fertilized eggs develop into multicellular embryos.

---

* Heterotrophic Dependent on other organisms for organic (food) molecules.

Figure 20-8 Animalia: the tentacles belong to sea anemones, members of a group of simple animals; the eye at the left belongs to a fish, member of a group of more complex animals. The fish eats the anemones' leftovers and is protected from predators by the anemones' stinging tentacles. (U.S. Navy)

Difficulties with the Five-Kingdom System

The five-kingdom system, like all other classificatory schemes that have been suggested, poses many problems. There are always organisms that do not fit neatly into the pigeonholes provided by scientists.

The division between eukaryotes and prokaryotes is primitive and fundamental. A separate kingdom for the prokaryotic Monera seems appropriate; the four eukaryotic kingdoms, however, are unsatisfactory for two main reasons. First, the kingdom Protista contains a very heterogeneous collection of phyla. The aim of modern taxonomy is that each grouping of organisms should contain organisms related more closely to each other than to organisms in other groups. The kingdom Protista violates this rule by containing, for instance, organisms that are related more closely to some plants than to any other protistans. Second, many different protistan lines gave rise to multicellular groups, and so none of the other kingdoms has a single common ancestor. For instance, different groups of fungi clearly arose from different protistan groups rather than from a common fungal ancestor. Furthermore, all the members of the animal kingdom, for instance, are not related by a common ancestor as they should be under an ideal phyletic classification; they are merely allied by certain similarities and by their multicellularity. Despite the inadequacies of the four eukaryote kingdoms, these groupings are widely used today, not because they are natural but because they are convenient.

The difficulties inherent in any system of biological classification are due to evolution's being a continuous process. For the sake of convenience we separate what is really a continuous series of organisms showing gradual evolutionary changes into boxes whose dimensions we define. There are bound to be organisms that do not fit into the boxes.

20-E   The Fossil Record

The attempt to classify organisms according to phylogeny relies heavily on our knowledge of life in the past, gleaned from the fossil record (Table 20–3). A **fossil** is any preserved evidence of life long past: a body or part of a body, an impression of the surface of the body such as a footprint, organic molecules like oil and coal, which are remains of organisms, or even a **coprolite**, which is preserved excrement.

How is a fossil formed? Usually, when an organism dies, scavengers and decay organisms rapidly destroy it. Occasionally, however, a body may come to rest in an acid bog, or be buried under a layer of mud that cuts off oxygen, conditions that retard decay and may permit the body to be preserved.

Very few fossils retain their original organic matter. Usually, the organic components are replaced over the course of millions of years by inorganic minerals. For example, suppose an organism died and sank to the bottom of the sea near the coast. Mud washed down a nearby river might cover the fossil-to-be, shutting out oxygen so that the body decayed very slowly. As successive layers of mud piled up, the pressure might have converted the fossil bed to rock. Meanwhile, the fossil might have decayed slowly, leaving behind its imprint in the rock. Alternatively, the soft parts might have rotted, leaving the hard parts. The original minerals of hard parts such as shell or bone would have been replaced, ion by ion, by more stable chemicals dissolved in the water that seeped through the mud and rock.

Rocks laid down by the settling and packing of small particles are called **sedimentary rocks**; examples are shale, sandstone, and limestone. The history of a sedimentary deposit is not over when it hardens into rock. It may be buried still further, or it may be caught between other masses of rock and crushed and crumpled as the earth's crust contracts and shifts. Under intense heat and pressure, sedimentary rocks undergo internal rearrangements and become extremely hard. The end products of this process are called **metamorphic rocks**, because they have been greatly changed. Examples are slate (from shale) and marble (from limestone). In the metamorphic process, fossils are usually pushed, pulled, heated, and flattened, until they are utterly destroyed. Because many of the oldest sedimentary

TABLE 20-3 THE FOSSIL RECORD

ERA	PERIOD		LIFE	YEARS AGO (Millions)
Cenozoic	Quaternary	Recent	Humans; glaciations force migrations and extinction of many terrestrial organisms.	
		Pleistocene		
				2.5
	Tertiary	Pliocene	Most modern mammals and flowering plants evolve.	
		Miocene		
		Oligocene		
		Eocene		
		Paleocene		
				65
Mesozoic	Cretaceous		First flowering plants. Most dinosaurs disappear.	
				135
	Jurassic		Dinosaurs dominant. First birds evolve from reptiles.	
				190
	Triassic		Gymnosperms* and reptiles dominant terrestrial forms. First mammals.	
				225
Paleozoic	Permian		First gymnosperms.	
				280
	Carboniferous	Pennsylvanian	Swampy forests of ferns, etc. First reptiles.* Insects on land.	
		Mississippian		
				345
	Devonian		First land vertebrates.*	
				395
	Silurian		First land plants.	
				430
	Ordovician		First vertebrates: fish.	
				500
	Cambrian		Many algae, all invertebrate* phyla.	
				600
Precambrian	Precambrian		Few fossils: Monera, fungi, soft-bodied invertebrates.	
				4500

deposits have undergone metamorphic processes, much of the most ancient fossil record has been destroyed.

Only a minute proportion of the animals and plants alive at any one time will be preserved by fossilization. Animals with hard skeletons have a much better chance of being preserved in the fossil record than those without them. Thus vertebrates are well represented by fossils whereas worms are very poorly represented. Organisms living in areas where continued sedimentation occurs are more likely to be

---

* Gymnosperms   Plants that bear seeds but no flowers; e.g., conifers (pine, fir, etc.).
* Reptiles   Vertebrates with dry skin, scales covering body, eggs laid on land.
* Vertebrates   Animals with backbones, e.g., fish, dog.
* Invertebrates   Animals without backbones, e.g., worm, fly.

(a)

(b)

Figure 20-9 Fossils. (a) A fern that lived nearly 150 million years ago. (b) A fish from rocks in Scotland laid down more than 100 million years ago. (Biophoto Associates)

preserved than those in other areas. Sedimentation is most widespread on the sea floor, and it is also fairly continuous in slow-moving rivers and in lakes and swamps. Sedimentation on land is a rarer phenomenon, although flood deposits or drifting sands may produce sedimentation. A terrestrial (land) animal that undergoes death by drowning or whose body is swept down a river after death is more likely to be fossilized than one that never ventures near a large body of water. Forests in particular produce few fossils, largely because the acidity of the leaf litter on a forest floor is likely to dissolve the skeleton away rather quickly. All of these factors make any collection of fossils subject to a very large "sampling error."

We sometimes see fairly complete descriptions not only of the structure but also of the life and habits of an extinct organism. How do we obtain this information just from the animal's fossilized parts? First, certain features of an animal's skeleton, such as the placement and size of the limbs, provide information about its habitat and its mode and speed of locomotion. The size and shape of the teeth may give evidence as to the animal's probable diet. Such things as a heap of bones in the position of the animal's stomach may tell us what it ate. A group of minute animal skeletons in the region of the pelvic girdle of a larger one strongly suggests that the young developed inside their mothers' bodies. The type of deposit in which a fossil is found, its relationship to other rock formations in the vicinity, and the other fossils present can also reveal much about the habitat of the fossilized organism: whether it was from the mountains or the plains, from a freshwater or marine environment.

Early Fossils

Until quite recently, the earliest known fossils were from the Cambrian period of geological time, which began about 600 million years ago. Most major groups of organisms had evolved by the Cambrian period, and their origins were veiled in mystery because most Precambrian rocks are deeply buried, or are metamorphic, or have been extensively eroded.

Recently, however, scientists have developed techniques for examining thin sections of rock microscopically, and this has revealed many fossils of unicellular organisms even in Precambrian deposits. Bacteria-like microfossils have been reported from the Fig Tree chert (a kind of rock) in South Africa, which is over 3 billion years old, and from the underlying Onverwacht shale, thought to be 3.4 to 3.6 billion years old. The Bulawayan limestone in South Africa contains fossils believed to be photosynthetic blue-green bacteria from 3 billion years ago, and the 1-billion-year-old Bitter Springs formation in the Northern Territory of Australia has revealed prokaryotes (and possibly eukaryotes) of about 30 different species. Some of the presumed blue-green bacteria in this assemblage are structurally indistinguishable from forms alive today! The oldest definite fossils are stromatolites (mats of bacteria), about 3.5 billion years old, from an area called North Pole in western Australia.

Further microscopic and chemical analysis of ancient rocks can be expected to provide a more complete picture of the early history of life on earth. Perhaps some of the evolutionary mysteries that now puzzle taxonomists will be cleared up by new fossil evidence.

SUMMARY

Taxonomy is the branch of biology concerned with relationships between organisms and with their classification. The basic unit of classification is the species; each species is given a unique Latin binomial, consisting of the genus name and species epithet.

Species are grouped into progressively more inclusive taxa. The main levels in the taxonomic hierarchy, from most to least inclusive, are: kingdom, phylum, class, order, family, genus, and species. A taxon in each higher level contains one or more taxa of the next lower level.

In theory, living things are classified by phylogenetic relationships. In practice, morphology is the criterion usually used. Other features of organisms, such as physiology, biochemistry, and behavior, are also used.

This book uses the taxonomic system that divides organisms into five kingdoms: Monera, the prokaryotes; Protista, the unicellular eukaryotes; Plantae, the plants; Fungi, the fungi; and Animalia, the animals. This classification is based largely on the mode of nutrition and cellular organization of organisms. Several evolutionary lines of organisms are almost certainly grouped in each kingdom.

The fossil record is important in determining phylogenetic relationships among organisms. Unfortunately, the fossil record is subject to sampling error due to differences in habits, habitat, amount of hard tissue, and size among various organisms of the past. Many groups of organisms have undoubtedly been lost without a trace because their members lived and died under conditions unsuitable for fossilization. Recent advances in techniques for studying microfossils and the biochemistry of fossils promise to extend our knowledge of early life on earth.

OBJECTIVES

From your study of this chapter, you should be able to:

1. Give the meaning of the following words and use them correctly: taxonomy, taxon, morphology, type specimen, phylogeny, species.

2. Describe Linnaeus's contribution to taxonomy and the basis for his classification.

3. State the theoretical basis for modern biological classification.

4. Write the Latin binomial of an organism correctly.

5. List the seven main hierarchical levels into which organisms are placed.

6. List the five kingdoms of organisms used in this book, and state the criteria used to assign species to each kingdom.

7. List or recognize conditions that favor fossilization and conditions that make it unlikely that an organism will be fossilized.

SELF-QUIZ

1. The Latin binomial for the common dog is properly written:
   a. canis familiaris
   b. Canis Familiaris
   c. Canis familiaris
   d. *Canis familiaris*
   e. *canis familiaris*

2. You are given a microscope slide on which is mounted some biological material. On examining it, you observe that there are numerous individual cells containing chloroplasts and swimming around rapidly. You should conclude that this material belongs in the kingdom:
   a. Plantae   d. Protista
   b. Fungi   e. Monera
   c. Animalia

3. Fossilization is favored by:
   a. lack of hard, abrasive projections
   b. absence of oxygen
   c. an abundance of nutrients
   d. warm temperatures
   e. strong sunlight

4. Modern classification is based on:
   a. taxonomy
   b. phylogeny
   c. morphology
   d. fossils
   e. autotrophy

QUESTIONS FOR DISCUSSION

1. Although modern taxonomy tries to classify organisms on the basis of their phylogenetic relationships, in practice most organisms are classified according to their morphology. Why is this so?

2. Why is the classification that Linnaeus produced without benefit of the theory of evolution so similar to that of modern evolutionary biologists?

3. Why is it so difficult for biologists to agree on a classificatory scheme that a dozen different classifications may be in use at one time?

4. If you set up a new scheme for classifying organisms, what criteria would you use to divide them into kingdoms (nutrition? cell structure? size? habitat?) and why?

REFERENCES AND FURTHER READING

Blackwelder, R. E. *Taxonomy: A Text and Reference Book.* New York: John Wiley and Sons, 1967. A massive advanced book on taxonomy; useful for reference.

Gurin, J. "In the beginning." *Science 80,* July-August, 1980. A short, lively account of recent finds of ancient fossil microorganisms.

International Commission on Zoological Nomenclature. International Code of Zoological Nomenclature adopted by the XV International Congress of Zoology. London: International Trust for Zoological Nomenclature, 1964. The work of the commission that is officially responsible for resolving disputes over the names of animals and animal classification.

Knoll, A. H., and E. S. Barghoorn. "Precambrian eukaryotic organisms: A reassessment of the evidence." *Science* 190:52, 1975. Argues that artifacts of preservation may have been mistaken for cell nuclei.

Leedale, G. F. "How many are the kingdoms of organisms?" *Taxon* 23:261, 1974. Discusses the difficulties with various methods of dividing organisms into kingdoms.

Morris, S. C., and H. B. Whittington. "The animals of the Burgess shale." *Scientific American,* July 1979. An account of an unusually rich fossil find.

Savory, T. *Animal Taxonomy.* London: Herremann Educational, 1970. A very readable introduction to taxonomy.

Schopf, J. W., and D. Z. Oehler. "How old are the eukaryotes?" *Science* 193:47, 1976. A counterargument to the article by Knoll and Barghoorn, based on size and patterns of cell clusters.

Whittaker, R. H. "New concepts of kingdoms of organisms." *Science* 163:150, 1969. The proposal of a five-kingdom system, with discussion of the rationale for such a system and the problem of drawing boundaries.

CHAPTER

21

VIRUSES

Viruses are tiny particles composed largely of nucleic acid and protein. They lack many of the features of living cells.

Viruses were discovered in the late nineteenth century, when it was found that the infectious agents that cause smallpox and tobacco mosaic were too small to be seen with the light microscope and would pass through filters that stopped all known bacteria. It was also discovered that viruses reproduced only inside living cells. Once inside a host cell, the virus takes over the host's metabolic machinery, and the host cell then produces viral (of virus) proteins and nucleic acids. Because they lack the metabolic machinery to reproduce their genetic material independently of a host cell, however, viruses are not considered real living organisms and do not belong in the five kingdoms. In 1935 tobacco mosaic virus was crystallized, emphasizing another difference of viruses from living cells, since only relatively simple chemical compounds can be crystallized, not complex cells. Nevertheless, the crystallized virus was capable of infecting cells.

Most of our knowledge of viruses comes from work on **bacteriophages**, often simply called **phages**; these are viruses that infect only bacteria. Phages have long

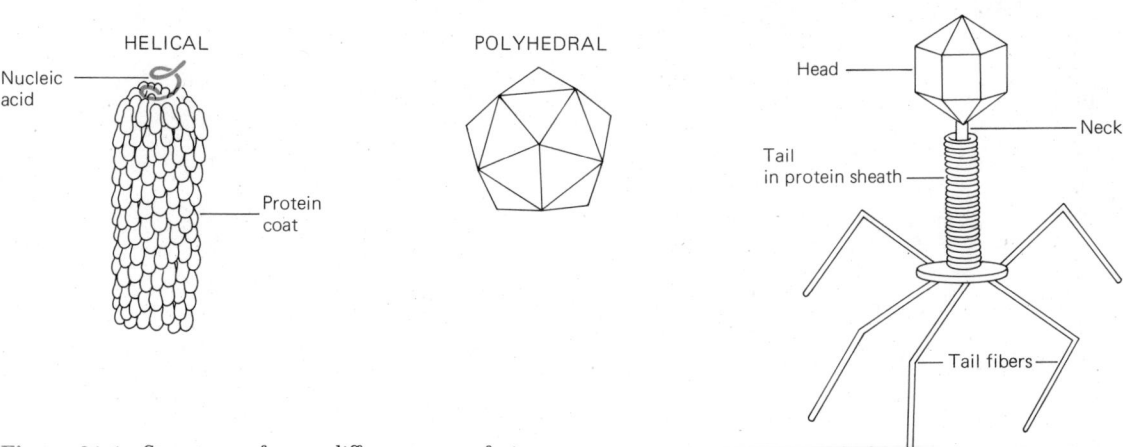

HELICAL

Nucleic acid

Protein coat

POLYHEDRAL

T-EVEN

Head

Neck

Tail in protein sheath

Tail fibers

Figure 21-1 Structures of some different types of viruses.

been produced by the millions in cultures of bacteria, whereas until recently it was more difficult to culture living animal cells and the more complex viruses that infect them.

21-A   Structure of a Virus

A virus particle consists of a molecule of RNA or of DNA (but not both), which constitutes all its genetic material, or **genome**, and a surrounding protein coat. In some viruses, the protein coat is surrounded by an outer envelope of carbohydrate and lipid. Viruses lack almost everything else normally found in a cell. They have no ribosomes nor any of the enzymes necessary for protein synthesis, and have none of the enzymes needed for energy production. Viruses usually contain only those enzymes necessary for them to invade a cell and replicate their own nucleic acids.

Virus particles are generally either polyhedral or helical, or a combination of both (Figure 21–1). Their shape is a result of the organization of the subunits that

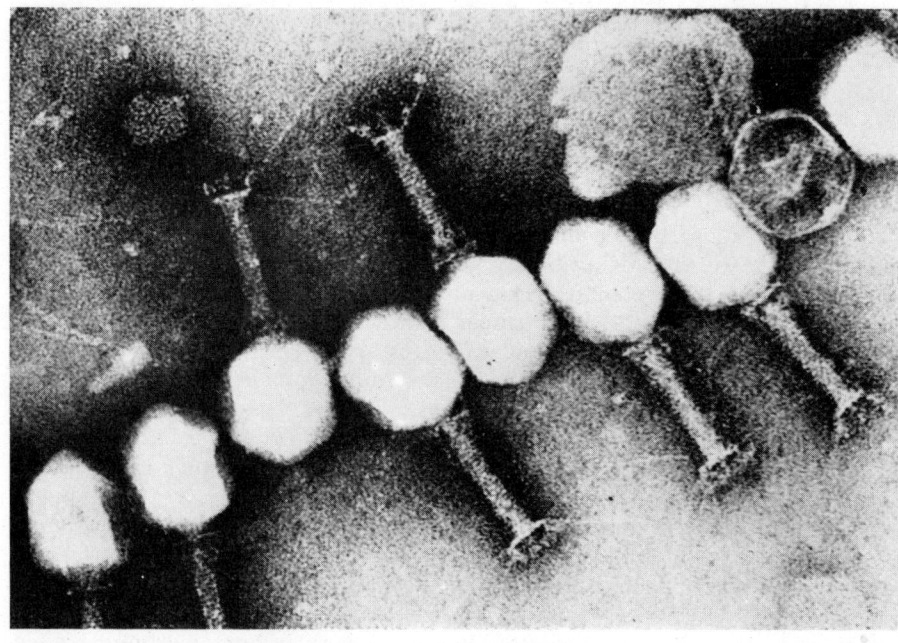

THE DIVERSITY OF ORGANISMS

Figure 21-2 Electron micrograph of bacteriophage T4, which infects *Escherichia coli.* (Carolina Biological Supply Company)

make up their protein coat. T-even phages,* which infect the bacterium *Escherichia coli*, have a rather more complicated structure (see Figures 21–1, 21–2).

21-B   The Lytic Cycle

Viruses show two main types of life history. A **lytic cycle** occurs when a virus invades a cell, reproduces, and is dispersed when the cell breaks, or **lyses**.

When a T-even phage collides with an *Escherichia coli*, it attaches to the bacterium by its tail fibers, which bond to specific receptors on the *E. coli* cell wall (Figure 21–3). An enzyme (phage lysozyme) in the phage tail then breaks down part of the bacterial cell wall. The tail sheath of the phage contracts, injecting the viral DNA into the cell. Soon after infection, the DNA of the host bacterium is destroyed. All messenger RNA synthesized from then on is transcribed from the phage DNA. The metabolic machinery of the host continues to function, but since all the messenger RNA present is transcribed from the viral DNA, the bacterial ribosomes produce phage protein coat. ATP produced by the bacterium supplies the energy used in the synthesis of phage DNA and proteins. The phage DNA codes for a number of enzymes and other proteins, such as the enzymes that break down the host DNA and assemble the completed phage protein into a coat. When as many as several thousand copies of the phage DNA and protein coat have been produced inside the bacterial cell, the two components are assembled into whole virus particles in the process known as maturation (Figure 21–4). Finally, a phage-encoded lysozyme digests the bacterial cell wall from within, causing the cell to lyse, or burst, and to release the complete phage particles. A cell invaded by a lytic virus is almost invariably killed by it.

---

*T-even phages   There is a "T" series of phages of *Escherichia coli*: T1, T2, etc.; the T-even phages are those with even numbers (T2, T4, T6).

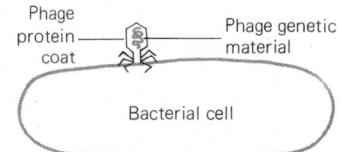

Phage attaches to bacterial
cell wall

Phage injects genetic
material into bacterium

Phage takes over bacterial
metabolism, causing synthesis
of new phage proteins and
nucleic acid

Phage protein and nucleic
acid assemble into complete
phages

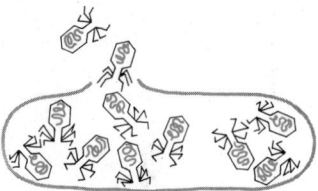

Host cell wall lyses,
releasing phage

Figure 21-3   Reproductive cycle of a lytic phage.

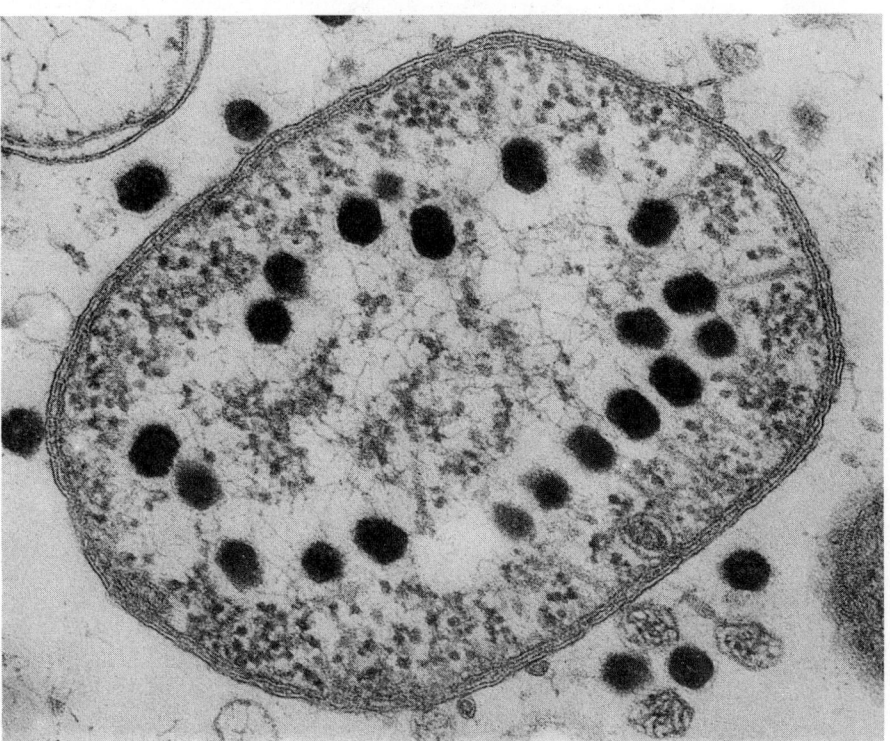

Figure 21-4   Electron micrograph of a cell of the bacterium *Escherichia coli* infected with bacteriophage T2. The black hexagons inside the cell are new phages produced after the phage genetic material was injected into the bacterium. Note the empty phage coats attached to the cell wall. The thin fibrous material in the cell is DNA. (L. D. Simon. *Virology* 38:287, 1969)

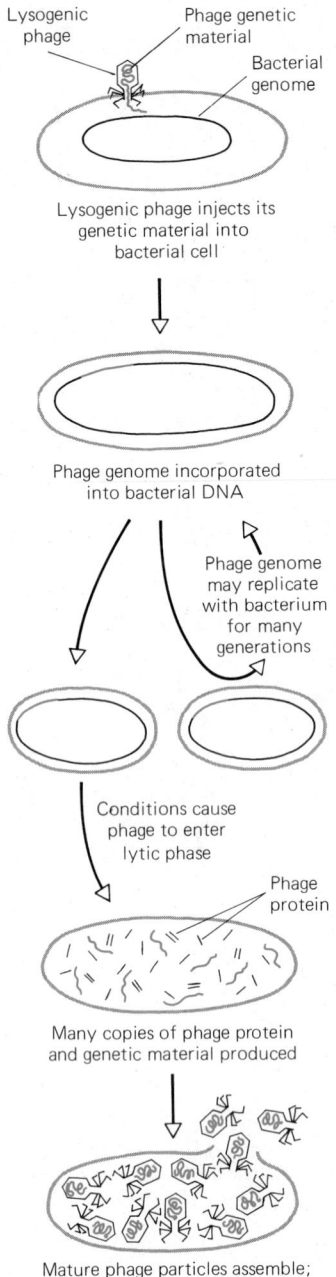

Lysogenic phage

Phage genetic material

Bacterial genome

Lysogenic phage injects its genetic material into bacterial cell

Phage genome incorporated into bacterial DNA

Phage genome may replicate with bacterium for many generations

Conditions cause phage to enter lytic phase

Phage protein

Many copies of phage protein and genetic material produced

Mature phage particles assemble; released when phage enzyme lyses bacterial cell wall

Figure 21-5 Reproductive cycle of a lysogenic, or temperate, phage. The host may survive for many generations with the phage genetic material incorporated into the host genome, until some condition triggers the phage to become lytic.

21-C The Lysogenic Cycle

The lytic cycle described above is that of a virulent (extremely damaging), lytic bacteriophage, which can only replicate and cause cell lysis. Other phages, known as **temperate** phages, may either replicate and lyse the cell they invade, or may instead enter a dormant phase in which the phage DNA is joined to that of the host cell (Figure 21–5) and replicated with it over many cell generations. A host cell containing such a temperate phage is called a **lysogenic** cell. Certain external stimuli can cause a lysogenic cell's phage DNA to enter the lytic cycle, releasing intact phages.

Some lysogenic bacterial cells are of importance to human health. For instance *Corynebacterium diphtheriae*, the bacterium that causes diphtheria, synthesizes the toxin responsible for the disease only when it contains a particular phage. Similarly, the streptococcal bacterium that causes scarlet fever produces symptoms only when it is lysogenic, and the same may be true of the bacterium *Clostridium botulinum*, which causes botulism. An important consequence of the lysogenic relationship is that phages released when a cell lyses may carry with them a portion of the bacterial DNA. A phage may then introduce such bacterial DNA into a new bacterium that it infects. This process is known as **transduction** ("carrying across"), and it produces genetic recombination in the new host bacterium. The ability of some viruses to carry DNA from one cell to another is used in recombinant DNA experiments (see Section 9-E) as well as in studies on bacterial genetics.

21-D RNA Viruses

Viruses that contain RNA instead of DNA as their genetic material have novel mechanisms of replication. A virus whose genetic material is single-stranded RNA usually replicates by first synthesizing a second RNA strand, whose base sequence is

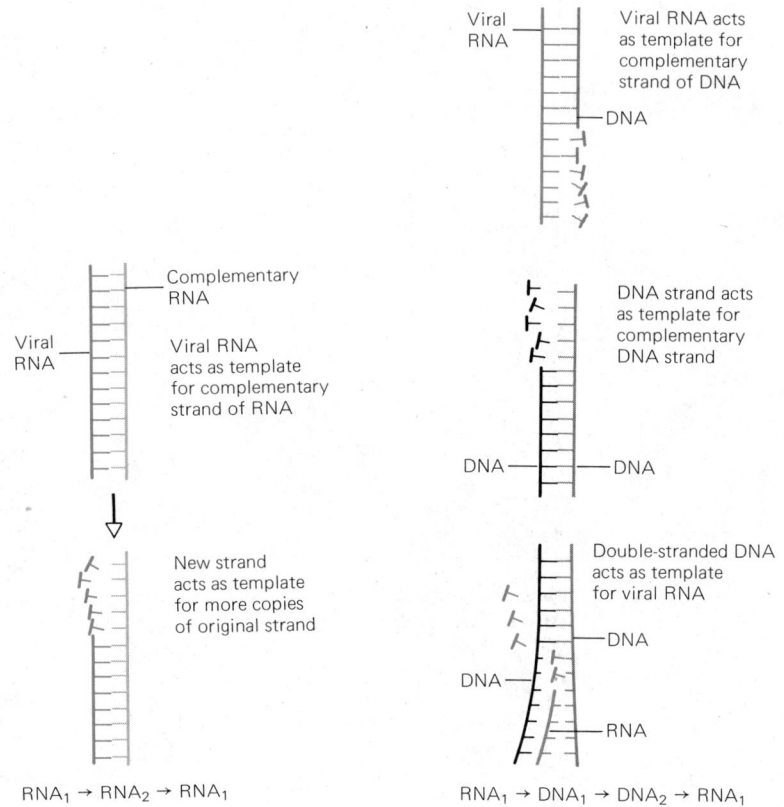

Viral RNA

Complementary RNA

Viral RNA acts as template for complementary strand of RNA

New strand acts as template for more copies of original strand

$RNA_1 \rightarrow RNA_2 \rightarrow RNA_1$

Viral RNA

Viral RNA acts as template for complementary strand of DNA

DNA

DNA strand acts as template for complementary DNA strand

DNA

DNA

Double-stranded DNA acts as template for viral RNA

DNA

DNA

RNA

$RNA_1 \rightarrow DNA_1 \rightarrow DNA_2 \rightarrow RNA_1$

Figure 21-6 Two different ways in which the genome of an RNA virus may replicate.

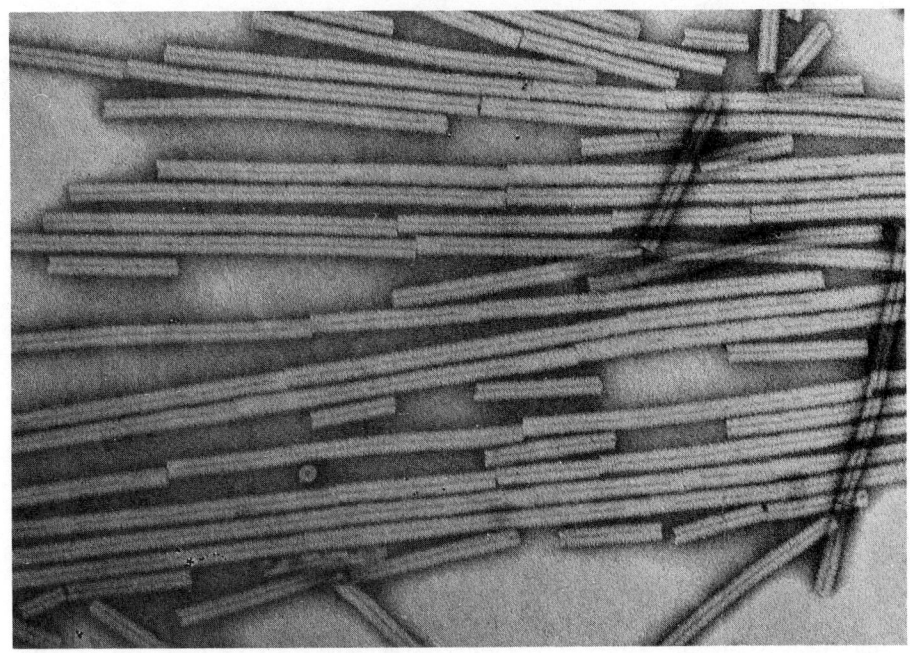

Figure 21-7 Tobacco mosaic virus, the first virus shown to have RNA as its genetic material. (Biophoto Associates)

complementary to that of the first strand. This second strand then acts as a template for new copies of the viral genome identical to the original viral RNA strand (Figure 21-6).

Other viruses with genomes of single-stranded RNA first cause an infected cell to synthesize the unusual enzyme **RNA-directed DNA polymerase** (also known as **reverse transcriptase**), whose discovery in the late 1960s caused a stir and even disbelief in scientific circles. Up until that time, biologists believed that biological information must invariably flow from DNA to RNA by the transcription of RNA on a DNA template. However, this polymerase catalyzes synthesis of a DNA strand complementary to the virus's RNA genome, and this DNA then acts as the template for its complement, thus forming double-stranded DNA. Multiple copies of the virus's RNA are produced on the DNA template.

The polio virus has a single-stranded RNA genome that acts directly as messenger RNA when it attaches to the host cell's ribosomes. It then serves as the template for replication of the viral RNA.

21-E Virus Diseases of Plants and Animals

Viruses are fairly specific in the cells they can infect. Thus polio viruses can attack only the cells of humans and a few other primates.* This specificity is probably because the virus must interact with specific receptor sites on the host cell surface.

Some plant diseases known to be caused by viruses are tobacco mosaic, tobacco necrosis, alfalfa mosaic, rice dwarf, and wound tumor. Plant viruses may be spread by the wind or by insects. It is usually impossible to cure virus infections in plants. Farmers try to prevent the spread of viral infections by practicing soil hygiene and by burning infected plants. Breeding programs are devoted to developing virus-resistant strains of important crop plants.

Viruses that infect animals include those causing rabies, chickenpox, polio, and influenza, as well as many known to produce various forms of cancer. *Herpesvirus hominis* is an example of a virus that causes eukaryotic cells to become lysogenic. It may cause human cancers, although this is not certain. Herpes viruses cause cold sores, a venereal infection, mononucleosis, and probably hepatitis. Even after the

---

*Primates The order of mammals containing humans, monkeys, apes, lemurs, and so on.

infection has cleared up, a person who has once been attacked by the virus will always be prone to develop cold sores at times of stress or general ill health. It appears that the virus is present in a lysogenic form and that external stimuli can cause it to enter the lytic cycle at any time.

Except when they cause cancer, most viruses cause disease by disturbing the metabolism of an infected cell and eventually destroying it. The virus reproduces in the dying cell, and its progeny invade neighboring cells. Because viruses are fairly specific in the cells they attack, a particular virus attacks only the cells of specific tissues. The common cold virus attacks the cells lining the human respiratory tract; the scrapie virus attacks the nervous system of sheep. Viruses that cause cancer are different in that they alter the infected cell but do not kill it.

Viruses and Cancer

"Cancer" is a lay term for a variety of abnormal growths, more properly called **tumors**, some of which are caused by viruses. The most noteworthy characteristic of tumor viruses is that they **transform** cells (a process not related to bacterial transformation described in Chapter 9) by altering the cellular DNA. Transformed cells often undergo drastic changes in morphology and metabolism such that they become unresponsive to the normal controls over cell division. The viral DNA or RNA remains and multiplies with the host cells as the tumor grows. In a cell transformed by polyoma virus, for instance, the viral genome is joined to the host genome as it is in lysogenic bacteria.

For some time it was difficult to imagine how the genome of an RNA virus could become joined to the host DNA, as seems to occur with all **oncogenic** (tumor-causing) viruses and with other temperate viruses. It is now clear that RNA viruses with the code for reverse transcriptase use this enzyme to recode their RNA genomes as DNA, which can attach to the host cell genome. At present it appears that only oncogenic viruses code for this enzyme (although some oncogenic viruses have a DNA genome and do not need it). This offers some hope that destruction of reverse transcriptase may be a specific way to destroy oncogenic RNA viruses.

Various RNA and DNA viruses clearly cause cancers in other animals; Rous sarcoma virus (in domestic fowl) and polyoma virus are examples. It therefore seems quite likely that some human cancers could be caused by viruses. At the time of writing, no viruses have definitely been shown to cause cancers in humans. However, one kind of herpes virus, the Epstein-Barr virus, which probably infects almost all human beings, appears to be intimately related to Burkitt's lymphoma, a cancer of the lymph nodes below the jaw, and to nasopharyngeal carcinoma (a tumor in the space behind the nose); possibly these cancers develop when some environmental factor stimulates a change in the virus's activity. In addition, virus particles virtually identical to those that cause mouse mammary (breast) tumors have been found in the milk of women whose families have a high incidence of breast cancer. Many researchers are now investigating the possibility that some viruses cause human cancer, because methods of prevention or cure often cannot be developed until the causes of cancer are known. (Cancer is discussed in more detail in Chapter 34.)

21-F Viroids

In the 1960s and 1970s, several kinds of infectious particles even smaller than viruses were isolated. These **viroids** consist of short, single strands of RNA without any protective coat of protein or other materials. Diseases known to be caused by viroids include spindle tuber of potatoes, exocortis of citrus trees, chlorotic mottle and stunting of chrysanthemums, and several other diseases of important crop plants. Viroids are also suspected of causing some animal diseases, including some rare human nervous system diseases.

SUMMARY

Viruses resemble living things in that they contain nucleic acids and protein. They differ from living organisms in that many of them can be crystallized and in that they lack the metabolic machinery to synthesize proteins and replicate nucleic acids. After the viral genome is injected into a specific host cell, it takes over the cell's metabolic machinery, causing it to transcribe and replicate the viral nucleic acid and synthesize the viral coat. Cells reproduce by dividing into two, but hundreds of viruses may be produced in one host cell. Viruses are responsible for many diseases, including some tumors, in plants and animals.

OBJECTIVES

From your study of this chapter, you should be able to:

1. Discuss similarities and differences between viruses and living organisms (see also Section 1-E).
2. Describe the structure of a virus.
3. Describe lytic and lysogenic cycles.
4. Explain how the genome of an RNA virus is replicated.

SELF-QUIZ

1. One reason that viruses are considered nonliving is that:
   a. they lack replicable nucleic acids
   b. their nucleic acids do not code for proteins
   c. they cannot make their own food molecules
   d. they cannot carry out their own reproduction
   e. they do not undergo genetic change (mutation) and so do not become adapted to changes in their environment

2. A virus always contains:
   a. DNA, RNA, and proteins
   b. nucleic acids and lipids
   c. proteins and nucleic acids
   d. DNA and proteins
   e. nucleic acids, proteins, lipids, and carbohydrates

3. Examine each statement below, and tell whether it is true of lytic viruses, lysogenic viruses, or both.
   a. Many virus particles may be released from each host cell.
   b. The host's DNA is destroyed by the virus.
   c. Part of the host's DNA may be carried to a new host cell by the virus.
   d. The viral DNA may be incorporated into the host's genome.
   e. Viral proteins are synthesized on host ribosomes.

4. True or False. The genomes of some RNA viruses are replicated by the formation of RNA templates, whereas the genomes of others are replicated by the formation of DNA templates; in both cases, the viral RNA genome is then transcribed from the template.

QUESTIONS FOR DISCUSSION

1. If viruses cause cancer in humans, it will be very difficult to produce convincing evidence that they do so. Why?
2. The geneticist Robert L. Sinsheimer has said that one reason experiments with recombinant DNA might be dangerous is that a phage could be used to carry a gene for a bacterial restriction enzyme (which destroys forms of DNA that are not specifically protected against it) into a eukaryotic cell. What might be the deleterious effects of such a transduction?
3. In what ways do viruses resemble living organisms, and what criteria of "life" do they lack?

REFERENCES AND FURTHER READING

Butler, P. J. G., and A. Klug. "The assembly of a virus." *Scientific American*, November 1978. How tobacco mosaic virus particles self-assemble.

Campbell, A. M. "How viruses insert their DNA into the DNA of the host cell." *Scientific American*, June 1976. Describes experimental techniques and evidence for incorporation of lysogenic viral DNA into host cell DNA.

Diener, T. O. "Viroids." *Scientific American*, January 1981. How the existence of viroids was established.

Gallo, R. C. "Reverse transcriptase, the DNA polymerase of oncogenic RNA viruses." *Nature* 234:194, 1971. The original description of transcription of DNA from RNA in cancer-causing viruses.

Goodenough, *U. Genetics*, 2d ed. New York: Holt, Rinehart and Winston, 1978. Presents recent information on transcription, mapping, and recombination of viral genomes.

Henle, W., G. Henle, and E. T. Lennette. "The Epstein-Barr virus." *Scientific American*, July 1979.

Horne, R. W. "The structure of viruses." *Scientific American*, January 1963.

Kaplan, M. M., and H. Koprowski. "Rabies." *Scientific American*, January 1980. History, transmission, study, and modern treatment of a dread viral disease.

Kellenberger, E. "The genetic control of the shape of a virus." *Scientific American*, December 1966.

Luria, S. E. "The recognition of DNA in bacteria." *Scientific American*, January 1970. Bacteria fight back by destroying certain types of viral DNA.

Smith, K. M. *The Biology of Viruses*. New York: Oxford University Press, 1965. An excellent introduction to viruses and viral disease.

Temin, H. M. "RNA-directed DNA synthesis." *Scientific American*, January 1972. The story of how RNA viruses replicate their DNA.

Wood, W. B., and R. S. Edgar. "Building a bacterial virus." *Scientific American*, June 1967.

CHAPTER

22

MONERA

AEGJ

The kingdom Monera contains all of the prokaryotic organisms; the two phyla we shall study are the bacteria and the cyanobacteria (also called blue-green bacteria or simply blue-greens). (The cyanobacteria were formerly called blue-green algae, but many biologists have decided that the term ''alga'' is used for too many different organisms and should be replaced with more precise terminology wherever possible.) Most of the Monera are unicellular organisms, although they frequently occur as filaments (strings) or colonies of independent cells.

The prokaryotes have always been of great importance to human societies. Bacteria cause devastating diseases, spoil food, and kill crops; cyanobacteria are responsible for forms of water pollution that may cause illness or death to fish, livestock, and humans. On the other hand, if all of the Monera on earth disappeared tomorrow, all of the plants and animals would soon die, for Monera produce the ammonium and nitrates needed by higher plants for protein synthesis and play a major role in recycling nutrients.

Because of their tiny size, these important organisms were almost unknown until Louis Pasteur's extensive studies of bacteria in the nineteenth century. In the

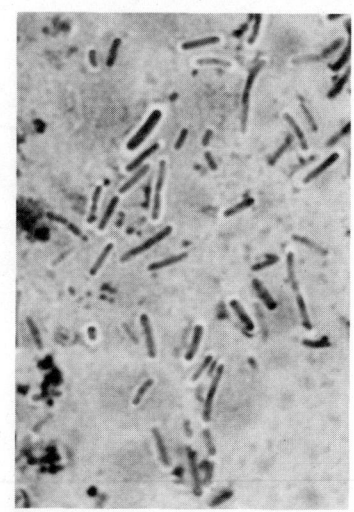

Figure 22-1 Cells of *Klebsiella pneumoniae,* showing capsules. This organism causes a serious, though infrequent, form of pneumonia; excessive consumption of alcohol may lower resistance to these infections. (Carolina Biological Supply Company)

Mesosome

Ribosomes

Cell wall

Capsule

Nuclear area

Plasma membrane

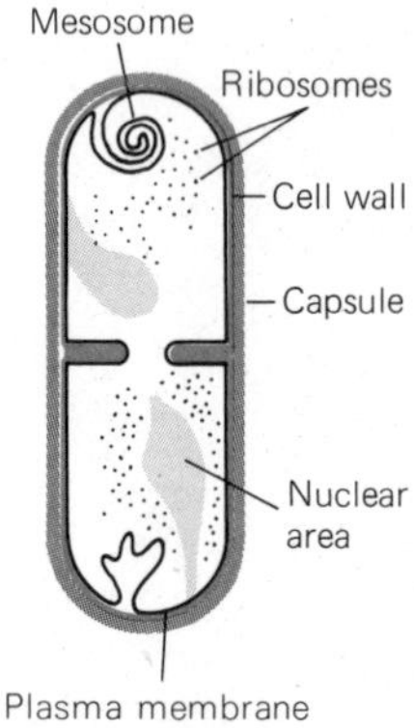

Figure 22-2 Diagram of a heterotrophic bacterium undergoing cell division.

twentieth century, widespread use of the electron microscope and the development of an array of biochemical techniques have led to a vast expansion in our knowledge and understanding of these diminutive organisms.

22-A Prokaryotic Cells

Prokaryotic cells differ from eukaryotic cells in three main ways: (1) the structure and chemical composition of their cell wall; (2) the absence of membrane-bound organelles; and (3) the organization of their genetic material.

The prokaryotic cell wall is unique in that it is made up of a mucocomplex substance, a polymer of glucose derivatives attached to various amino acids. The walls of some bacteria also have outer layers of lipopolysaccharide, a polymer composed of lipid and sugar monomers; these lipopolysaccharides are of interest because of their toxic properties. Many diseases caused by bacteria can be duplicated by injecting only the purified lipopolysaccharide of the cell wall into a host organism.

Besides protecting the cell from injury, the cell wall prevents cells in hypotonic* media from bursting as they gain water by osmosis (see Section 5-F). If bacteria are cultured in isotonic* solutions, however, they will survive even if their cell walls are digested away by enzymes. Such bacteria do not divide, indicating that the cell wall is necessary to reproduction.

Some bacteria surround themselves with a **capsule** of polysaccharide or polypeptide outside the cell wall (Figure 22–1). *Streptococcus pneumoniae* causes bacterial pneumonia only when it is encapsulated. Apparently, the body's phagocytic cells, which engulf and destroy bacteria, can destroy the naked but not the encapsulated form of this bacterium. The cell walls of cyanobacteria are usually surrounded by a sheath of gelatinous material secreted by the cells.

In nature, prokaryotic cells produce a dense feltlike mat of polysaccharides (sugar polymers) that enables them to stick to surfaces—soil particles, rocks in stream beds, or human teeth or gums, for example.

Prokaryotic cells range in size from about 1 to 10 μm; a few eukaryotic cells are smaller than about 7 μm in diameter, but most are much larger. It may be that the size of prokaryotic cells is restricted by their lack of cellular organelles. The only organelles found consistently in all prokaryotes are ribosomes, on which messenger RNA is translated into proteins. However, prokaryotic ribosomes are smaller than the ribosomes in the cytoplasm of eukaryotes.

Many biochemical reactions can take place only on the surface of a membrane, and some prokaryotes lack all membranes except for their plasma membrane. However, it would be wrong to give the impression that no prokaryotes contain internal membranes. In some bacteria, the plasma membrane develops many loops and indentations that increase its surface area, forming a **mesosome** (Figure 22-2); this is probably important to formation of a new cell wall during cell division, and to the separation of DNA prior to cell division. Furthermore, photosynthetic prokaryotes may have internal membranous structures containing photosynthetic pigments and enzymes (see Figure 8–2). Some aerobic* bacteria also have internal membrane systems housing respiratory enzymes.

There are no simple rules concerning the metabolism of prokaryotes. Some are aerobic and some anaerobic,* and some use nitrate (NO_3^-) or sulfate (SO_4^{2-}) instead of oxygen as electron acceptors in respiration. Most cyanobacteria are photosynthetic, and some bacteria carry out primitive forms of photosynthesis using hydrogen gas, sulfur compounds, or organic molecules instead of water as sources of hydrogen. This varied group of organisms contains members capable of adaptation

---

* Hypotonic Having a lower osmotic pressure than that of the solution to which it is compared (in this case, the solution inside the cell).

* Isotonic Having an equal osmotic pressure.

* Aerobic Requiring molecular oxygen for cellular respiration.

* Anaerobic Not requiring molecular oxygen for the breakdown of food to release energy.

THE DIVERSITY OF ORGANISMS

to living almost anywhere and to using many different energy sources. As an example of this rapid metabolic evolution, there are bacteria that can survive on a mixture of aviation fuel and aluminum; this is surely a recent adaptation to modern life!

22-B Reproduction in Prokaryotes

Prokaryotic DNA is complexed with different proteins from those in a eukaryotic chromosome. (Some people feel that a prokaryote's DNA should therefore not be called a "chromosome.") A prokaryote's genetic material is contained in an unpaired, circular DNA molecule having only about one-thousandth the DNA found in even the smallest eukaryotic cell. A well-nourished prokaryotic cell replicates its DNA well in advance of cell division, and may contain perhaps two or three copies at once if the rate of cell division does not keep up with the rate of replication. The DNA is attached to the plasma membrane, and a cell containing at least two complete copies of the DNA divides into two equal or unequal halves, with at least one DNA molecule in each. Mitosis and meiosis do not occur in prokaryotes.

Nothing resembling the familiar eukaryotic process of sexual reproduction by fusion of gametes occurs in prokaryotes. However, if we consider that the essential function of sexual reproduction is to transfer genetic information between individuals, then some prokaryotes do have sexual systems: many kinds of bacteria sometimes undergo sexual processes wherein one individual donates DNA to another. Subsequently, the recipient bacterium has some of its DNA replaced by DNA from the donor cell, or else DNA that is diploid* in places. No form of sexual reproduction has been found in any cyanobacteria.

Genetic recombination in bacteria was first shown by Joshua Lederberg and Edward Tatum. They cultured two strains of *Escherichia coli;* one strain could not synthesize three amino acids needed for growth, while the other strain could not synthesize two other needed amino acids, nor the vitamin biotin. Each strain could grow only when supplied with the substances it could not synthesize. After growing the two strains together in a nutrient-rich broth for a while, Lederberg and Tatum plated some of the cell mixture on agar* that contained sugars and salts but no amino acids or vitamins. A small fraction of the cells grew and reproduced; they must have been able to make all of the substances they needed, and were therefore **recombinants**, containing genes from both original strains.

There are three mechanisms by which genes can be transferred between bacteria:

1. **Transformation**, in which DNA released by a broken cell is taken up, sometimes in surprisingly large pieces, by a second bacterium.

2. **Conjugation**, in which two cells of different "mating types"—the equivalent of sexes—come close together and are joined by a protein bridge, or **pilus**, through which fragments of DNA pass from one cell to the other.

3. **Transduction**, in which bacterial DNA is carried from one bacterial cell to another inside a bacterial virus, or bacteriophage.

In all three processes, only fragments of the donor DNA are transferred. Once inside the recipient, the transferred DNA fragment lines up against the homologous* portion of the recipient's DNA. Enzymes then break up and remove that part of the recipient's DNA, substituting the newly arrived piece in its place (Figure 22–3). Recombination is precise in that there is no increase or decrease in the number of nucleotides in the recipient's DNA, only a one-for-one substitution of donor for recipient nucleotides.

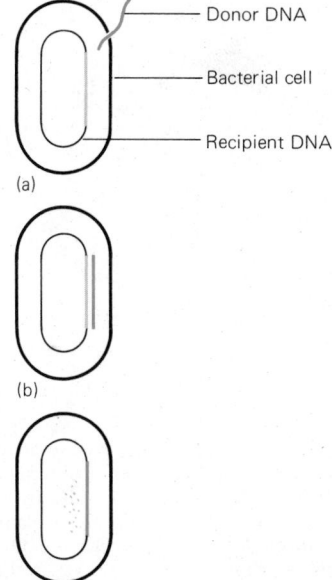

Figure 22-3 Genetic recombination in a bacterium. (a) Donor DNA enters recipient cell. (b) Donor DNA lines up next to homologous portion of recipient cell's DNA. (c) Donor DNA becomes incorporated into recipient DNA strand, and the original portion of the recipient DNA is destroyed by enzymes.

---

* Diploid Having paired molecules of DNA that contain the same kind of genetic information.
* Agar A medium of a gelatin-like consistency used for growing microorganisms in laboratory culture.
* Homologous Containing the same kinds of genetic information.

It is not clear whether these processes of recombination take place regularly in nature, or whether they occur because of special conditions found only in the laboratory. Each mechanism is known to take place in only a handful of bacterial species, and the first two are unknown in blue-greens.

In most eukaryotes, sexual reproduction brings about genetic recombination, the raw material for natural selection. How have prokaryotes, with little or no sexual reproduction, had such a long and successful evolutionary history?

Mutation* is the only source of genetic variability in prokaryotes, but nevertheless they evolve rapidly, partly because they reproduce with such speed. (A bacterium can produce millions of offspring in a week under ideal conditions.) If the number of individuals is large enough, mutation can produce so many genetic variants that genetic recombination would be superfluous. In other words, in prokaryotes, a large population coupled with a high rate of population growth can produce a great deal of genetic variation among individuals.

Prokaryotes reproduce by cell division and by fragmentation of colonies. Many produce resting **spores** that contain stored food and are resistant to heat and drying. Under favorable conditions, a new cell germinates from the spore.

22-C   Cyanobacteria

Nearly all cyanobacteria are photosynthetic, with chlorophyll *a* being the only type of chlorophyll present. Cyanobacteria also contain the phycobilin pigment phycocyanin, which is blue, and often the red phycobilin, phycoerythrin, as well. Various carotenoids contribute yellow and red pigmentation. The members of this group show more specialization of membranes, and of local areas within membranes devoted to particular functions, than that in other prokaryotes. Many cyanobacteria contain gas bubbles that enable them to float near the surface of the water, where sunlight is plentiful. Many also have an outer sheath of gelatinous material that may contain pigments of various colors; the sheath may also contain toxins (poisons) that make them inedible to many herbivores. No cyanobacteria possess cilia or flagella, but some species exhibit a mysterious gliding movement.

Cyanobacteria may form filamentous or clustered colonies (Figure 22–4). Some of the filamentous forms show division of labor within a colony: besides the ordinary vegetative cells, which carry on photosynthesis, a filament may contain spore-pro-

---

* Mutation Inheritable change in the DNA carrying an organism's genetic information.

(a)

(b)

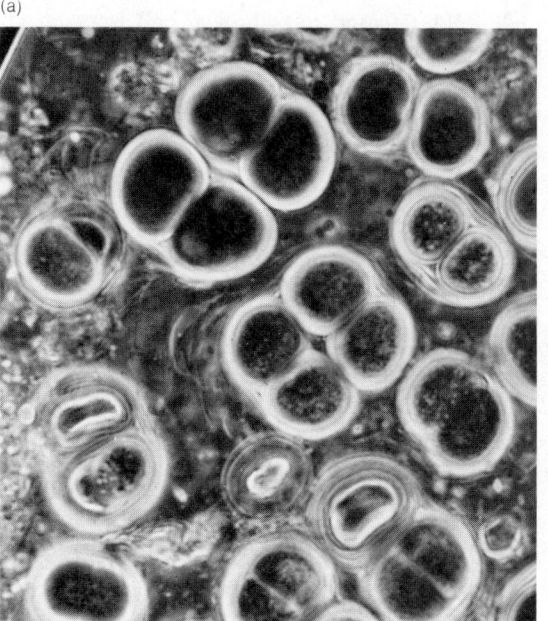
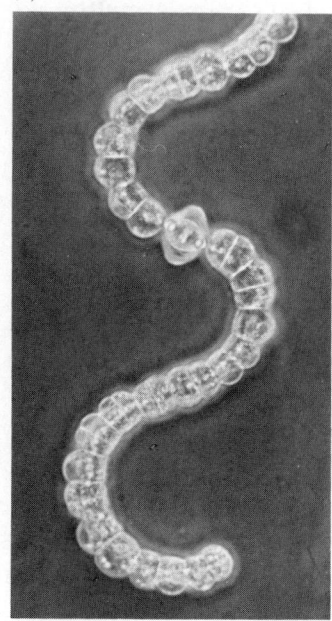

Figure 22-4 Representative cyanobacteria. (a) *Gloeocapsa,* a spherical form that lives in small clusters surrounded by gelatinous sheaths. (b) *Anabaena,* a filamentous form. Most of the cells are photosynthetic; the larger, diamond-shaped cell is a nonphotosynthetic heterocyst, which fixes nitrogen. (Biophoto Associates)

ducing cells, other cells specialized for attachment to the substrate, and **hetero-cysts**. Heterocysts are the cells that carry out **nitrogen fixation**, the reduction of nitrogen gas (N_2) to ammonia (NH_3) (see Section 22-F).

Some cyanobacteria tolerate extremes of temperature, pH, and salinity that would kill photosynthetic eukaryotes. Some are not photosynthetic but heterotrophic, obtaining all or some of their food from outside the body. Some are anaerobic and some aerobic. All in all, they are a very versatile group, with different species adapted to a wide range of different environments.

Ecological Importance of Cyanobacteria

Cyanobacteria are found in almost every moist environment, in the sea, in fresh water, and on land. Nearly all species are photosynthetic, and they must live near the surface, where they can obtain light. They are important members of the **phytoplankton** (= plants floating) in the sea and fresh water. They produce oxygen as a by-product of photosynthesis, and some provide food for heterotrophs.

Population explosions of cyanobacteria are one of the more unpleasant effects of water pollution because many forms are toxic and slimy. During the 1940s and 1950s, Lake Washington, near Seattle, was subject to thick, smelly **blooms** (population explosions) of cyanobacteria of the genus *Oscillatoria*, which thrived on the nutrients—especially phosphates and nitrates—in the sewage discharged into the lake. Such blooms are invariably accompanied by a decline in the fish population; as blue-greens die, they provide food for an enormous array of aerobic bacteria, which use all the oxygen in the water, and any fish that are not killed by blue-green toxins suffocate. Fish cannot control populations of these blue-greens by eating them because their toxins and gelatinous sheaths make the cyanobacteria inedible. During the 1960s, public outcry about the condition of Lake Washington led to the installation of improved sewage facilities, and the lake has since become steadily clearer and regained its fish population.

22-D Bacteria

Bacteria are the most numerous organisms on earth; billions of them may be found in a handful of soil. They are also found in nearly every habitat, from the sea floor to icebergs and hot springs. Their wide distribution reflects the variety of metabolic adaptations that enable various bacteria to live in environments of unusual chemical and thermal characteristics. The immense biological success of bacteria undoubtedly results from these metabolic capabilities, coupled with small size, rapid reproductive rate, and the ability to form resistant spores and so survive adverse conditions. These characteristics permit bacteria to live in many habitats that are here today and gone tomorrow. A drop of rain on a leaf will evaporate in less than a day, but in that time a bacterium may have divided several times and produced spores, which will blow away when the drop dries.

Classification of Bacteria

The classification of bacteria uses criteria very different from those applied to other organisms, for although the aim of taxonomy is to classify organisms by their phylogenetic (evolutionary) relationships, it is much easier to apply this ideal to large organisms with distinctly different forms than to microorganisms that may all look alike. Another problem is that it is impossible to define species of bacteria by the usual criterion of interbreeding populations, since bacteria reproduce asexually. Thus no one pretends that bacteria are classified by their evolutionary relationships except insofar as similar organisms are likely to be related. Futhermore, because there is so much metabolic diversity among the bacteria, their biochemistry is not as useful for classificatory purposes as is the more conservative biochemistry of most higher organisms. Cell structure is the criterion used to divide bacteria into classes.

Here we shall discuss only a few of the most important groups of bacteria.

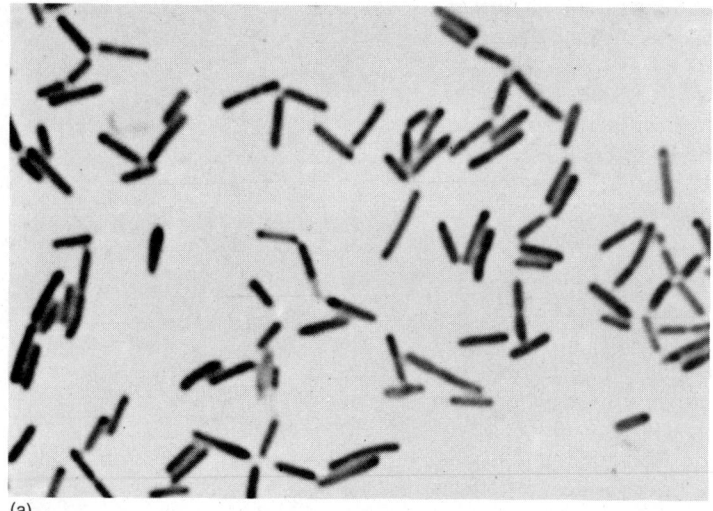

(a)

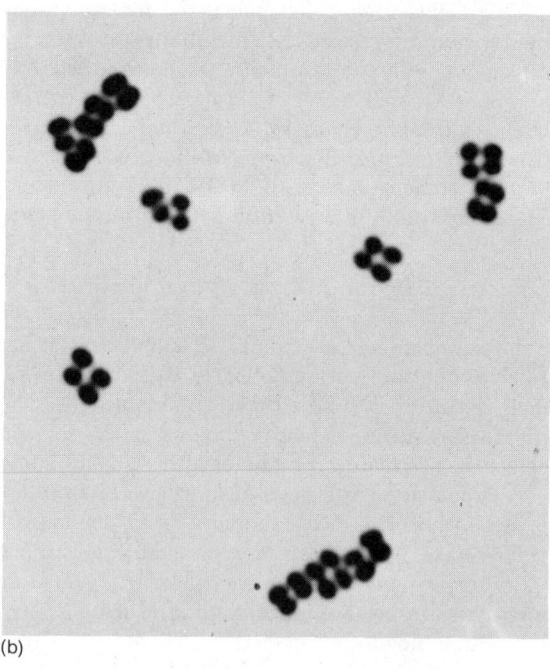

(b)

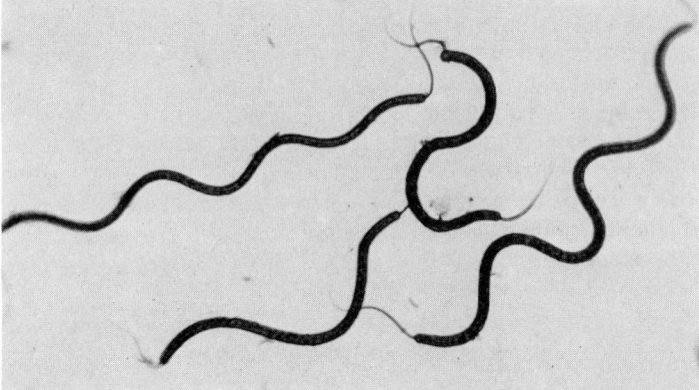

(c)

Figure 22-5 Eubacteria come in three shapes. (a) Typical bacilli, or rods. (b) Cocci, or spheres. *Gaffkya tetragena* occurs as clumps of four cells, living on the mucous membranes that line the respiratory tract. (c) Spirilla, or corkscrew shapes. *Spirillum volutans* is a large spirillum with a flagellum at each end. (Carolina Biological Supply Company)

EUBACTERIA. Eubacteria are a heterogeneous collection containing some of the best-known bacteria. All have rigid cell walls, and some move by the lashing of very thin flagella, composed of the protein flagellin rather than microtubules, which form the flagella of eukaryotes. Eubacteria reproduce by **binary fission**, in which one cell divides into two identical daughter cells. After division, daughter cells may stick to each other to form chains or clusters, although no division of labor occurs between the cells.

Eubacteria are commonly described by their shape. Rod-shaped bacteria such as *Escherichia coli* are called **bacilli** (singular: **bacillus**). Spherical bacteria are known as **cocci** (singular: **coccus**), and spiral bacteria are called **spirilla** (singular: **spirillum**) (Figure 22–5).

The **Gram staining** procedure, named for the Danish microbiologist Hans Christian Gram, has long been used to characterize bacteria. **Gram-positive** cells take up the violet stain used in this procedure; when they are then washed with alcohol, the pores in their cell walls close and the alcohol does not enter the cells and wash the stain out. In contrast, **Gram-negative cells**, which have a layer of lipopolysaccharide covering the cell wall, lose the violet stain when washed with alcohol, and when a red dye is then placed on the cells, the Gram-negative cells take it up. The distinction between Gram-positive and Gram-negative types of bacteria is medically important: Gram-positive bacteria are more readily destroyed by most **antibiotics**, substances that destroy microorganisms, than are Gram-negative bacteria. Gram-positive bacteria are also more susceptible to the action of **lysozyme**, an enzyme in nasal secretions and other body fluids that destroys their cell walls. This enzyme was discovered by Alexander Fleming when he noticed that bacteria died after he had sneezed on them.

MYXOBACTERIA. Myxobacteria are mobile bacteria that glide along with movements that appear similar to those of amoebas. They are common in soil and in decomposing organic matter. Some form reproductive structures, called fruiting bodies, that look a little like mushrooms (Figure 22–6). This is one of the most complex methods of reproduction known among prokaryotes; the fruiting bodies release **cysts**, which are like the spores of other bacteria in that they are resistant to adverse environmental conditions and germinate when conditions are favorable. Most myxobacteria are **saprobes**, feeding on nonliving organic matter; they secrete enzymes that break down foodstuffs outside the cell, and they then absorb the resulting simple organic molecules. They may feed on other bacteria or on the cellulose of bark and decaying wood.

SPIROCHETES. The spirochetes are curved or spiral bacteria that move by means of an axial filament (see Figure 22–6). Most of them are long and thin. The genus *Spirochaeta*, probably observed by van Leeuwenhoek as early as 1683, has members up to 500 µm long. Many can survive only under highly anaerobic conditions, and others are parasites in warm-blooded animals. *Treponema pallidum*, which causes syphilis, is a parasitic spirochete.

RICKETTSIAE. Rickettsiae are tiny, rod-like bacteria about 1 µm long (see Figure 22–6). Most of them are Gram-negative. Rickettsiae are parasites that usually live inside cells, and can frequently be detected inside the cells of arthropods such as ticks or insects. Insect bites may transmit these bacteria to mammals. Although most rickettsiae are relatively harmless to their hosts, the one that causes Rocky Mountain spotted fever has gained some notoriety for the group.

ACTINOMYCOTA. The Actinomycota, commonly called actinomycetes, resemble fungi in that they produce branching, multicellular filaments (see Figure 22–6). Most of them are saprobes, and some are anaerobic.

Various actinomycetes are important ecologically. Some break down lipids and waxes in dead plants and animals, and some species are pathogenic (disease-causing). *Mycobacterium tuberculosis* causes tuberculosis. Leprosy is caused by *Mycobacterium leprae*, a remarkably slow-growing bacterium that divides only once every twelve days; this explains why it takes so long for the disease to appear once a victim is exposed to it. Various actinomycetes, such as species of *Actinomyces*, are normal members of the bacterial population found in the human intestine and respiratory passages.

Many actinomycetes, particularly members of the genus *Streptomyces*, are the source of valuable antibiotics, substances that kill microorganisms. These antibiotics were discovered during a systematic search for antibacterial agents that would kill Gram-negative bacteria. A number of actinomycetes in the soil were screened during the search. The discovery of the first known actinomycete antibiotic, streptomycin, was announced in 1943. Among the other antibiotics isolated from actinomycetes are tetracycline, chloramphenicol, erythromycin, neomycin, and nystatin. Many antibiotics are produced by growing the bacteria in huge tanks; others can now be synthesized artificially.

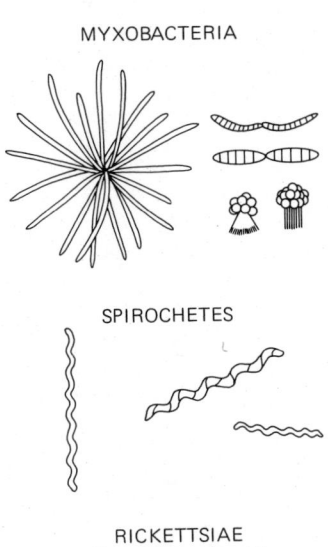

MYXOBACTERIA

SPIROCHETES

RICKETTSIAE
(in a eukaryotic cell)

Cell nucleus   Bacterium

ACTINOMYCOTA

Figure 22-6 Some members of the groups of bacteria discussed in Section 22-D.

22-E Autotrophic Bacteria

Autotrophs are organisms that can use inorganic molecules to synthesize the food molecules they need for growth and energy production. The best-known autotrophs are green plants, which produce food molecules from carbon dioxide and water through the use of light energy. There are also some bacteria, from several different groups, that can carry out photosynthesis. Bacterial photosynthesis differs in several ways from that of plants and cyanobacteria. First, photosynthetic bacteria do not use water as the hydrogen donor; hence they produce no oxygen during photosynthesis. Hydrogen donor molecules used by various photosynthetic bacteria include hydrogen sulfide, fatty acids, alcohols, and other substances. Secondly, bacterial chlorophylls differ from those of plants and cyanobacteria in absorbing light most strongly in the near-infrared portion of the spectrum, rather than the visible range. They can thus photosynthesize in what we would consider darkness. Furthermore, many photosynthetic bacteria are anaerobic. Thus some photosynthetic bacteria, such as the "sulfur bacteria," are ideally suited to life in the mud at the bottom of relatively stagnant bodies of water. This mud usually contains no oxygen, since aerobic bacteria near the surface use the little oxygen dissolved in the water, and most of the visible light has been filtered out by the water and algae in the surface layers, with only the longer wavelengths reaching the bottom. Furthermore, anaerobic mud usually contains hydrogen sulfide (H_2S), which smells like rotten eggs, produced from the breakdown of plant protein by sulfide bacteria. The photosynthetic sulfur bacteria use this hydrogen sulfide (instead of water) as a source of electrons for photosynthesis:

$$CO_2 + 2H_2S \longrightarrow (CH_2O)_n + H_2O + 2S$$

Under normal conditions, the sulfur bacteria in a pond or a lake use up the hydrogen sulfide as soon as it is produced by sulfide bacteria. If sewage is dumped into a body of water, however, the sulfide bacteria grow rapidly on nutrients in the sewage and cloud the water. Under such conditions, the photosynthetic sulfur bacteria will grow poorly because they receive even less light than usual, and hydrogen sulfide will rise to the surface, producing the characteristic stench of a sewage-polluted lake.

Chemosynthesis is a form of autotrophic nutrition found only in various bacteria. In chemosynthesis, inorganic chemical reactions, involving the oxidation of ammonia, nitrates, sulfides, or iron, are used instead of light as the energy source for the synthesis of organic from inorganic compounds.

22-F   Nitrogen Fixation

Nitrogen fixation is the reduction of gaseous nitrogen (N_2) from the air to ammonia (NH_3). This is a vital ecological process; all green plants need nitrogen, in the form of ammonium ions or nitrates, in order to synthesize amino acids for their proteins, but most inorganic nitrogen exists as gaseous N_2 in the air. The nitrogen molecule contains triple covalent bonds ($N{\equiv}N$), making it very stable. The only organisms that can reduce these bonds are some bacteria and those cyanobacteria with heterocysts.

Nitrogen fixation is a complex series of reduction reactions requiring a powerful reducing agent, electrons from photosynthesis or respiration, ATP, and the enzyme **nitrogenase**. Because this enzyme is readily inactivated by oxygen, it cannot function in a cell that gives off oxygen during photosynthesis. Some photosynthetic cyanobacteria fix nitrogen inside their heterocysts—nonphotosynthetic cells that do not produce oxygen and can therefore house the nitrogen-fixing machinery.

The nitrogen-fixing, photosynthetic cyanobacteria are extremely self-sufficient; they need only carbon dioxide and some inorganic substances to meet their few metabolic needs. They may colonize new habitats, such as rocks newly exposed at the surface of the earth by earthquakes or volcanoes, before anything else can live there.

Nitrogen-fixing bacteria have been intensively studied because of their importance to agriculture. It has been known for centuries, for instance, that growing **legumes** (e.g., peas, beans, clover, alfalfa, vetch) as part of crop rotation improves the fertility of the soil. This is because bacteria of the genus *Rhizobium* live in nodules on the roots of legumes (Figure 22–7). These bacteria use sugars produced by the legume's photosynthesis and supply the plant with ammonium. Plant breeders and microbiologists have cooperated in attempts to produce *Rhizobium* associations with plants other than legumes; this would both reduce the dependence of agriculture on expensive nitrogen fertilizers and improve the nutritional quality of plant protein. These attempts have not yet succeeded, but they have produced use-

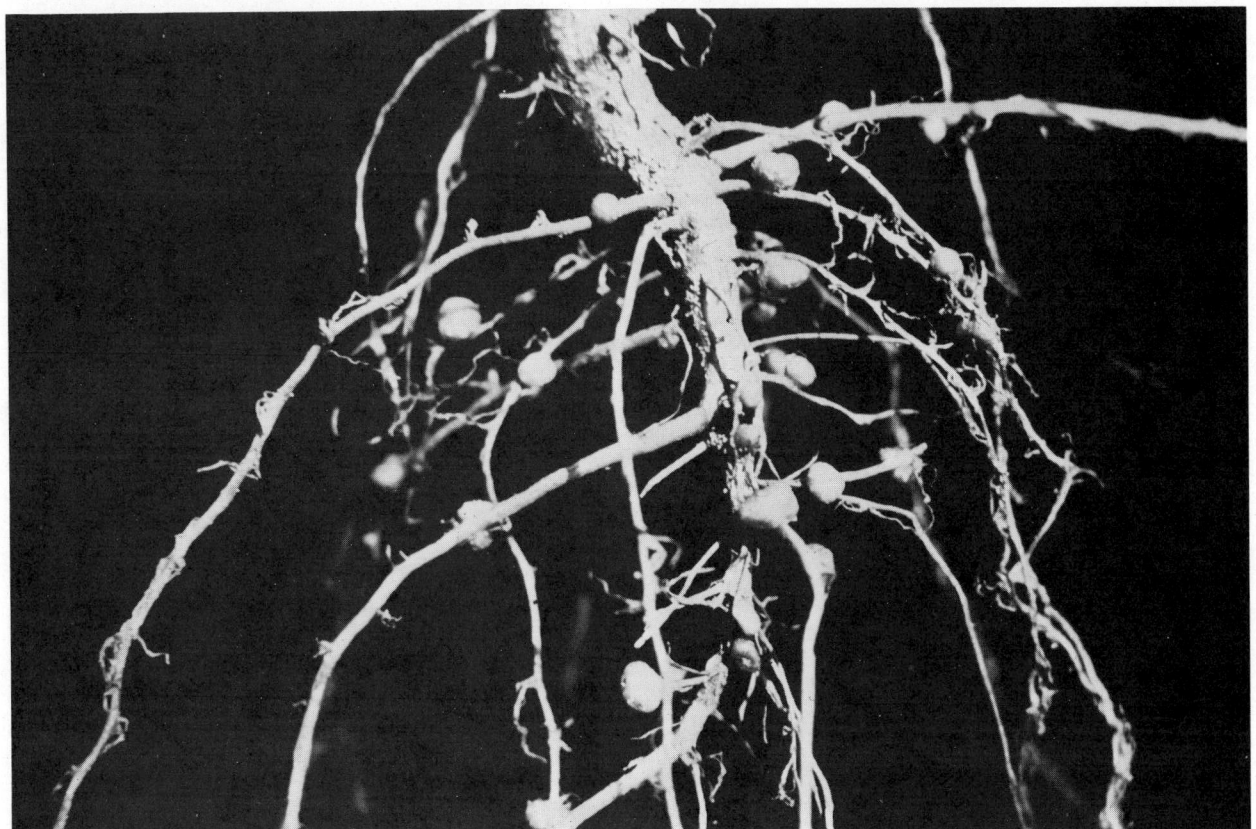

Figure 22-7 The roots of a cowpea plant. The swellings are root nodules that house nitrogen-fixing *Rhizobium* bacteria. (Carolina Biological Supply Company)

ful information on the genetics and biochemistry of nitrogen fixation and on the relationship between *Rhizobium* and its host plant.

Nitrogen fixed into the form of ammonia or ammonium can be oxidized by **nitrifying bacteria** to nitrites (NO_2^-) and nitrates (NO_3^-). These organisms are chemosynthetic, using the energy released by oxidation of the nitrogenous compounds to drive their metabolic processes (Figure 22–8).

22-G   Heterotrophic Bacteria

Most bacteria are heterotrophic saprobes, which means they absorb small organic molecules as nutrients. In order to do this, they must usually secrete enzymes that release the small molecules by digesting large organic molecules outside the cell. Heterotrophic bacteria and fungi are thus the most important decomposing organisms; they break down dead plants and animals into smaller organic compounds. Eventually, these decomposers convert the carbon in dead organisms into carbon dioxide, which can then be used by green plants in photosynthesis. Without decomposers, carbon and many other compounds would be irretrievably locked up in the bodies of dead organisms, and life would eventually cease.

Other heterotrophic bacteria are **parasitic**, receiving nutrition from the living body of another organism. Some parasitic bacteria cause disease; others, however, are vital to the lives of their hosts.

The **flora** of an animal includes the bacteria that normally live on or in it; these may be saprobes, living on substances such as dead, sloughed skin cells, or parasites absorbing nutrients from the living tissues. The bacterial flora of human beings includes *Staphylococcus epidermidis* on the skin, *Staphylococcus aureus* in the nostrils, and *Bacteroides fragilis* and *Escherichia coli* in the intestine.

NITROGEN FIXATION:

$N_2 \rightarrow 2N$; (−160 Kcal/Mole)

$2N + 3H_2 \rightarrow 2NH_3$ (13 Kcal/Mole)

(Azotobacter, Beijerinckia, Clostridium, Rhizobium, Cyanobacteria)

NITRIFICATION:

$NH_4^+ + 1\frac{1}{2} O_2 \rightarrow NO_2^- + H_2O + 2H^+$
Nitrite

65 Kcal/Mole
(Nitrosomonas)

$NO_2^- + \frac{1}{2} O_2 \rightarrow NO_3^-$
Nitrate

17 Kcal/Mole
(Nitrobacter)

Figure 22-8 Nitrogen fixation, the reduction of gaseous nitrogen (N_2) to ammonia or ammonium (NH_3, NH_4^+), can be carried out by various members of the bacterial genera named here and by some cyanobacteria. Other bacteria carry out nitrification of ammonia or ammonium, ultimately converting nitrogen into the form of nitrates. Nitrifying bacteria are chemosynthetic, using the energy released by the oxidation of nitrogenous compounds to drive their metabolism.

The normal flora of an animal may benefit the host by aiding digestion, by providing nutrients, and by competing with pathogenic bacteria: microbes, including pathogens, grow better without competition from other microbes for food and space. Some bacteria also produce antibiotics that actively inhibit the growth of other organisms. The result of bacterial competition is that normal, established populations of bacteria usually prevent invading pathogenic species from establishing themselves. Foreign bacteria invade the skin, gut, or mucous membranes more readily if the resident bacteria have been removed by antibiotics, douching, and other means. Beyond this, both mice and men receive vitamins from their gut flora and are more susceptible to bacterial and nutrient-deficiency disease if they lose their vitamin-producing inhabitants. And many animals that can digest cellulose (e.g., cows and colobus monkeys) owe this ability to specific bacteria in their stomachs.

On the other hand, members of an animal's flora may cause disease if they become established in tissues where they are not normally found. *Escherichia coli* causes cystitis, or inflammation of the bladder, and *Staphylococcus aureus* can cause serious infections if it gets into a wound; this is one reason why surgeons wear masks.

22-H   Pathogenic Bacteria

A **pathogen** is an organism capable of producing disease. Pathogenicity is not an unchangeable property; as we have seen, many bacteria cause disease under some conditions but not others. Although bacteria have been known since 1676, it was not until the nineteenth century that they were recognized as a major cause of disease.

Some pathogenic bacteria destroy the cells of their host, but most cause disease because they produce **toxins**, poisonous substances that damage the host's metabolism. Bacterial toxins may be divided into two classes: exotoxins and endotoxins. **Endotoxins** ("inside toxins") are lipopolysaccharides in the cell walls of Gram-negative bacteria; they produce fever and damage to the circulatory system. **Exotoxins** ("outside toxins") are much less common than endotoxins. Exotoxins are proteins, and they are secreted from the bacterium into the surrounding medium; they may be carried around the host's body in the bloodstream. Exotoxins are produced by the bacteria that cause diphtheria, tetanus (lockjaw), and botulism, among others. Since exotoxins are proteins, they have different structures and may have very different effects on the body. For instance, the exotoxins of *Vibrio cholerae* (which causes cholera) and *Shigella dysenteriae* (dysentery) cause severe diarrhea, whereas the exotoxin of *Yersinia pestis* causes the headaches and boils characteristic of the plague.

Control of Bacterial Disease

Preventing the spread of pathogenic bacteria depends on understanding how these organisms normally disperse. Diseases like diphtheria, scarlet fever, whooping cough, and tuberculosis are caused by airborne bacteria, usually released in cough or sneeze droplets. Thus the traditional method of preventing the spread of these diseases is to quarantine anyone who catches them. The incidence of tuberculosis, once a common disease, decreased dramatically after the passage of ordinances against spitting in public places. Nowadays immunization (see Section 34-G) is an effective method of preventing many bacterial diseases.

The most important preventive measures against bacterial diseases are hygiene and sanitation. In the last century the Hungarian physician Semmelweiss and the American doctor Oliver Wendell Holmes almost simultaneously found that mortality of women after childbirth could be reduced about tenfold if midwives and doctors washed their hands and their instruments between patients; in addition, the English surgeon Joseph Lister developed aseptic surgical techniques that reduced bacterial infections in patients following surgery. Keeping food and water reasonably free of bacteria saves many more lives today than do antibiotics. Cleanliness is particularly important in protecting infants from respiratory and intestinal infections. Indeed, the increase in average life expectancy in most countries in the last 40 years can be ascribed almost entirely to better hygiene, which has decreased infant mortality.

Heating to 60°C for 30 minutes destroys exotoxins, which are proteins, and kills most bacterial cells. For this reason pasteurization (heating) has proved a very easy and effective means of protection against botulism in canned food, brucellosis and tuberculosis transmitted in milk, and dysentery caused by drinking water contaminated with human sewage. In addition, antibiotics have saved thousands of lives. It is a truism that more lives were lost to disease than to enemy action during all wars until about 1940. During World War II, for the first time, antibiotics reduced the mortality from diseases contracted under unsanitary field conditions and from infected wounds.

Drug Resistance in Bacteria

When antibiotics were first introduced in the 1940s, they were hailed as miracle drugs that would soon eradicate most infectious diseases. However, as most students of evolution would have predicted, bacteria that happened to possess mutations making them resistant to antibiotics were heavily favored by natural selection. When a bacterial population is exposed to a particular drug, susceptible cells die and resistant ones multiply rapidly, since their competition has been removed. Resistant strains soon become widespread. For example, some strains of *Neisseria gonorrhoeae*, which causes gonorrhea, can no longer be killed by any known drug.

Some of the bacterial mutations that confer drug resistance are well understood. Some bacteria, for instance, make penicillinases, enzymes that destroy penicillin. Others have enzymes that add a molecular group to the drug, covering the site where the drug would normally bind to the bacterium.

Bacteria can also become resistant to drugs by transfer of genes that confer resistance from one cell to another. In 1959 a group of Japanese scientists discovered that every drug-sensitive member of a population of *Shigella* had acquired resistance within an hour after some drug-resistant *Shigella* were added to the culture. Sometimes, the genes involved in drug resistance exist as an **episome**, a segment of DNA that, because it is not attached to the bacterial chromosome, can be transferred between individuals of the same or different species. Fortunately, most common disease-producing bacteria do not seem to undergo this process of **infectious drug resistance transfer.**

22-I   Bacteria and Food

All food contains bacteria. Although milk is sterile when it leaves a healthy cow, it contains several types of bacteria by the time it reaches the table. *Streptococcus lactis* and species of *Lactobacillus* ferment* the milk sugar, lactose, and produce lactic acid. The production of lactic acid lowers the pH of the milk until the milk proteins coagulate. Pasteurization reduces the population of these bacteria and retards spoilage. On the other hand, *S. lactis* is necessary to produce many cheeses, including cottage cheese, and species of *Lactobacillus* are necessary to produce yogurt. Bacterial fermentation is also important in the manufacture of a number of other foods, such as sauerkraut and pickles. Vinegar is produced by allowing bacteria such as *Acetobacter aceti* to oxidize ethyl alcohol in apple cider or wine to acetic acid.

"Food poisoning" comes from toxins produced by bacteria growing in food. The common bacteria *Staphylococcus aureus* and *Clostridium perfringens* both produce toxins when they grow on suitable foodstuffs. These toxins, however, are seldom lethal. On the other hand, the common anaerobic soil bacterium *Clostridium botu-*

---

*Ferment   Release energy by anaerobic breakdown.

linum produces the toxin that causes botulism, one of the most dreaded, but fortunately rare, forms of food poisoning. This toxin interferes with nerve activity, causing paralysis and, if the breathing muscles are paralyzed, death. The toxin is easily destroyed by heat, but the bacterium also produces endospores that must be heated for a long time to kill them. Since *Clostridium* can grow only under strictly anaerobic conditions, fresh and frozen foods are safe from botulism, but canned foods provide an anaerobic environment where the spores can grow and produce toxin. Commercial canned goods are usually safe from *C. botulinum* unless the equipment used for heat-sterilization is defective. Most cases of botulism are due to inadequate heating of home-canned goods. The most effective safeguard against botulism is to can only acid foods such as fruit and pickles, since *Clostridium* cannot survive at a low pH; vegetables such as beans and peas do not contain enough acid to kill the bacteria.

Food contaminated by populations of *Salmonella* bacteria may also cause illness in people who eat it; pork, poultry, and eggs are common sources of *Salmonella* infection. The main symptom of this infection is a bout of diarrhea shortly after eating the contaminated food. In this case, it is the living bacteria, not toxins, that cause the trouble.

22-J   Origin of Mitochondria and Chloroplasts

An intriguing theory that plastids and mitochondria evolved from free-living prokaryotes was proposed at the turn of the century and has been championed by Lynn Margulis during recent years. According to the theory, these prokaryotes were engulfed by some other prokaryotic cell, and took up residence to the mutual benefit of both parties. During the course of evolution, the engulfing cell underwent other changes and eventually became a eukaryotic cell. The evidence in favor of this theory is as follows:

1. Some prokaryotes can grow inside eukaryotic cells today.
2. Plastids and mitochondria are about the size of prokaryotic cells.
3. Like prokaryotes, plastids and mitochondria contain DNA that is circular and not complexed with proteins.
4. Plastids and mitochondria synthesize at least some of their own proteins. Their ribosomes are the same size as prokaryote ribosomes, and chloroplast ribosomes can be hybridized with those of prokaryotes. **Hybridization** of ribosomes involves combining small subunits from one source (e.g., chloroplasts) with large subunits from another (e.g., bacteria). The hybridization is considered successful if the hybrid ribosomes can carry out protein synthesis *in vitro.** Ribosomal subunits from the cytoplasm of eukaryotes will not hybridize with the ribosomes of prokaryotes or of eukaryotic organelles.
5. Many antibiotics block protein synthesis in prokaryotes, and in the mitochondria and plastids of eukaryotic cells, but not in the cytoplasmic ribosomes of eukaryotic cells.
6. When a prokaryote synthesizes a protein, the polypeptide chain always starts with formylmethionine (f-met). This amino acid plays no part in protein synthesis in eukaryotes. Recent work with yeasts (eukaryotes) has shown that when yeast mitochondria synthesize protein, the chain always starts with f-met, but when a protein is synthesized in the cytoplasm of these yeasts, f-met is absent.
7. Eukaryote cells deprived of their mitochondria or plastids cannot replace them. Apparently these organelles arise only by division of existing plastids or mitochondria.

In vitro   Latin: "in glass" (e.g., test tubes), as opposed to *in vivo,* in the living organism.

Those who disagree with the theory that plastids and mitochondria originated as prokaryotes engulfed by another prokaryote argue as follows:

1. Although the protein-synthesizing machinery of mitochondria is like that of prokaryotes, the two are not identical.

2. Like nuclear chromosomes, and unlike prokaryote DNA, mitochondrial genes sometimes contain intervening sequences of DNA (Section 10-C).

3. Mitochondria and plastids require proteins coded by genes on the nuclear chromosomes and synthesized on cytoplasmic ribosomes (mitochondria make only about 10% of the proteins they need; the rest is imported from the cytoplasm). If these organelles were originally prokaryotes, some of their genes must have moved into the cell nucleus, which seems unlikely.

4. Other membranous structures, such as nuclear membranes, the Golgi apparatus, etc., do not appear to be descended from engulfed prokaryotes; presumably they arose by proliferation of membranes that then engulfed certain other structures in prokaryotes. Mitochondria and plastids could have originated in the same way.

5. It is not surprising that the DNA and ribosomes of mitochondria and plastids resemble those of prokaryotes. After all, prokaryotes were probably the ancestors of eukaryotes. Cytoplasmic ribosomes and nuclear DNA have merely evolved differences from their prokaryote ancestors faster than have mitochondria and plastids.

This controversy can never be settled conclusively. Even advocates of the theory that mitochondria and plastids originated as free-living prokaryotes disagree about how they got into the first eukaryotic cell. Some feel that they were parasites; others argue that the first eukaryotic cell formed by the fusion of several prokaryotic cells. It is also possible that organelles originated as prey that were resistant to digestion. In any event, this controversy has stimulated a lot of interesting work on mitochondria and plastids and the division of labor in eukaryotic cells.

SUMMARY

The kingdom Monera contains all of the prokaryotic organisms, i.e., the cyanobacteria (blue-greens) and the bacteria. Prokaryotic cells differ from eukaryotic cells in having a cell wall of mucocomplex substance and circular DNA molecules with few proteins attached, and in lacking membrane-bound organelles. The internal membranes in some prokaryotic cells include the mesosomes of some bacteria, and systems of membranes containing the enzymes of photosynthesis or respiration. Reproduction among prokaryotes is usually by binary fission. No form of sexual reproduction has ever been observed in cyanobacteria, but some bacteria exchange genetic information as a variable amount of DNA passes from a donor to a recipient bacterium. Many prokaryotes form spores under adverse conditions. Their small size, metabolic diversity, and rapid reproduction probably account for the evolutionary success of prokaryotes.

Cyanobacteria are nearly all photosynthetic; they are distinguished from photosynthetic bacteria by possessing chlorophyll a and phycobilins, and by using water as an electron donor and releasing molecular oxygen as a by-product of photosynthesis. Cyanobacteria have a gelatinous outer sheath. Some occur as colonies or filaments in which different cells have different functions. Cyanobacteria are often members of "blooms" produced by an excess of nutrients in bodies of fresh water.

Bacteria are the most numerous and ubiquitous organisms on earth. Eubacteria are divided into bacilli (rods), cocci (spheres), and spirilla (curves or spirals). The cell wall is covered with lipopolysaccharide in Gram-negative bacteria. The actinomycetes are common soil organisms; many life-saving antibiotics were first isolated from members of this group.

Prokaryotes exhibit a wide diversity of metabolic capabilities. Many bacteria are autotrophs, either photosynthetic or chemosynthetic. Nitrogen-fixing bacteria and cyanobacteria convert gaseous nitrogen into a form that can be used by plants to make amino acids. Most bacteria are heterotrophic saprobes or parasites. Saprobic bacteria are vital ecologically as decomposers that break down organic material and permit its constituents to be recycled. Some bacteria are part of the normal microbial flora of animals, and some are pathogenic, causing disease by the production of toxins.

Immunization, improved hygiene, and antibiotics have reduced human and animal deaths from bacterial disease. One disadvantage of antibiotics is that their use selects for mutant bacteria resistant to the antibiotic in question.

Some bacteria are important in the production of foods such as yogurt, cheese, and vinegar. Food poisoning is caused by ingesting exotoxins or pathogenic bacteria in food.

Some workers believe that mitochondria and plastids arose when some prokaryotes took up residence inside others.

OBJECTIVES

From your study of this chapter, you should be able to:

1. Use the following words and terms correctly: capsule, mesosome, spore, heterocyst, autotroph, heterotroph, photosynthesis, chemosynthesis, nitrogen fixation, decomposer organism, saprobe, parasite, pathogen.

2. Describe the distinguishing characteristics of members of the kingdom Monera, and of the phylum Cyanobacteria and the phylum Bacteria.

3. List and describe the four sources of genetic change known among prokaryotes, and discuss the contributions of these sources and of reproductive rate to the evolutionary success of prokaryotes.

4. Explain what is meant by a "bloom" of cyanobacteria; list some factors that can cause such blooms, and describe measures to prevent them.

5. List the three main shapes of Eubacteria and give their Latin names.

6. State how photosynthesis in bacteria differs from that of cyanobacteria and eukaryotes.

7. Describe the metabolism of a sulfur bacterium and the conditions that encourage or discourage the growth of these bacteria.

8. State what the "normal flora" of an animal is, and list three reasons why it is important to the animal.

9. Summarize the ecological roles and importance of prokaryotes.

10. Describe the difference between bacterial endotoxins and exotoxins, and explain how each causes disease.

11. State what "drug resistance" of a bacterial population is, and explain how it arises and how it differs from "infectious drug resistance transfer."

12. Give some examples of the use of bacteria in food production.

13. Give evidence for and against the theory that mitochondria and plastids arose from free-living prokaryotes.

SELF-QUIZ

1. An organism should be placed in the kingdom Monera if:
   a. it consists of a single cell
   b. it has a cell wall
   c. it is surrounded by a capsule
   d. it lacks a nuclear membrane separating its genetic material from the cytoplasm
   e. it causes diseases

2. Prokaryotic cells differ from eukaryotic cells in:
   a. structure of genetic material
   b. possessing cell walls
   c. possessing vacuoles
   d. possessing ribosomes
   e. all of the above

3. Newly started rice paddies produce poor crops until they have established a flourishing population of cyanobacteria. This is probably because:
   a. the rice needs nitrogen fixed by the cyanobacteria
   b. the rice cannot compete with weeds, which are poisoned by toxins produced by the cyanobacteria

   c. the cyanobacteria use up surplus nutrients from sewage in the rice paddies
   d. the cyanobacteria lack proteins on their DNA
   e. rice plants do better with a protective coating of cyanobacteria

4. When you must take antibiotics for a long time, a doctor will frequently prescribe a combination of antibiotics instead of only one. Why?
   a. The combination of drugs aids in singling out the pathogenic bacterium without affecting others.
   b. Some of the antibiotics may encourage the development of non-pathogenic bacteria, but these can be combatted by careful choice of a second antibiotic.
   c. There is a very small chance that single bacteria resistant to both antibiotics are present.
   d. You may be allergic to one of the antibiotics but not to them all.
   e. It takes longer for a bacterial cell to develop resistance to two drugs to which it is exposed.

5. True or False. A bloom of photosynthetic cyanobacteria can lead to shortage of oxygen in a lake or pond.

6. True or False. Many bacteria protect their hosts from pathogens by producing antibiotics.

7. True or False. All food containing bacteria is unsafe for consumption and should be thrown out.

QUESTIONS FOR DISCUSSION

1. What is the adaptive value to an organism of producing a toxin? What is the disadvantage of toxin production?

2. Why are fossil prokaryotes more difficult to find and study than fossils of other organisms?

3. Many prokaryotes can synthesize all the molecules they need when provided with a nutrient medium containing a supply of inorganic salts and one type of small organic molecule (e.g., a monosaccharide or fatty acid) as an organic carbon source. How can prokaryotes be so self-sufficient if their DNA is so small compared to even one of the many chromosomes found in eukaryotic cells?

4. Defend one of the positions on the origin of mitochondria and chloroplasts by setting forth plausible arguments against each point made by the opposition.

REFERENCES AND FURTHER READING

Braude, A. I. "Bacterial endotoxins." *Scientific American*, March 1964. Discusses the nature of endotoxins and their role in disease.

Costerton, J. W., G. G. Geesey, and K.-J. Cheng. "How bacteria stick." *Scientific American*, January 1978.

Delwiche, C. C. "The nitrogen cycle." *Scientific American*, September 1970. Various bacteria play a role in the many transformations of inorganic nitrogen.

Fogg, G. E. *et al. The Blue-green Algae*. New York: Academic Press, 1973. A compendious and well-written text on cyanobacteria including discussion of their economic importance.

Gibor, A. "Mitochondria and chloroplasts: Reproduction, development and heredity." In: *Topics in the Study of Life:* *The Bio Source Book*. New York: Harper and Row, 1971, pp. 48–54. Presents and discusses evidence pro and con the theory of the symbiotic origin of organelles.

Nester, E. W., C. E. Roberts, N. N. Pearsall, and B. J. McCarthy. *Microbiology*, 2d ed. New York: Holt, Rinehart and Winston, 1978. Wide coverage of aspects of the bacteria and other microorganisms, with much recent information and a useful cross-referencing system.

Stanier, R. Y., E. A. Adelberg, and J. L. Ingraham. *The Microbial World*, 4th ed. Englewood Cliffs, NJ: Prentice-Hall, 1976. An excellent text covering the discovery, structure, physiology, and ecology of bacteria and other microorganisms, including several chapters on symbiotic relationships, parasitism, disease, and disease responses of the host.

23

PROTISTA AND THE ORIGIN OF MULTICELLULARITY

The unicellular eukaryotes are classified in the kingdom Protista. The evolution of the first eukaryotic cells from prokaryotes, about 1.45 billion years ago, was a tremendous advance. The development by eukaryotes of specialized organelles surrounded by membranes permitted cells to become larger. Eukaryotic cells have a number of linear chromosomes, with a greater total amount of genetic information than prokaryotic cells; the evolution of mitosis and meiosis in eukaryotes guaranteed an orderly sorting of these chromosomes into the cells formed during cell division. Through crossing-over and independent assortment of chromosomes during meiosis, sexual reproduction, which is much more common among eukaryotes than among prokaryotes, permitted an increase in the genetic variability in each generation of a eukaryotic organism.

Members of the kingdom Protista have radiated into many different habitats and ways of life. Some are photosynthetic; modern unicellular **algae**, with chloroplasts, large cell vacuoles, and cell walls, undoubtedly resemble the ancestors of the plant kingdom. Other protists, known as **protozoans**, ingest their food and lack chloroplasts and cell walls; they are probably much like the ancestors of animals.

Many protists have become highly specialized for their own ways of life, with organelles much more complicated than those found in the cells of plants and animals.

In this chapter we shall consider some of the specializations found in the kingdom Protista. Then we shall consider some of the selective pressures that may have operated in the evolution of the multicellular fungi, plants, and animals, the subjects of the remaining chapters in this section of the book.

23-A Kingdom Protista (Protozoa and Unicellular Algae)

Protists are found wherever there is water: in the sea, in fresh water, in moist soil, or in the bodies of animals. They are an enormously varied group. Photosynthetic forms float or swim on the surface of a pond or ocean; amoebas crawl over the mud at the bottom, and various **sessile** forms grow attached to objects such as rocks and pondweed. Furthermore, although most protists live as solitary cells, many form **colonies** of similar, but independent, cells.

Ideally, the classification of organisms reflects their evolutionary relationships. Most of the evolution of the protists, however, is obscure. If eukaryotic cells originated only once in evolutionary history, then the Protista are all descended from a common ancestor. But it is just as likely that modern protists are the descendants of many early pioneers of eukaryotic structure. The relationships among the various protistan groups will probably never be fully established, although studies of protistan biochemistry and genetics are providing clues.

The Protista may be divided into two main groups: heterotrophs and autotrophs. However, since evolution has no respect for the boundaries of artificial systems of classification invented by humans, borderline species, which cannot be classified satisfactorily, are inevitable in any biological classification; the border between heterotrophic and autotrophic protists is just as fuzzy as the lines separating protists from plants, animals, and fungi. Before the protists were separated into a kingdom by themselves, botanists considered the autotrophic protists to be plants (unicellular algae), and zoologists considered the heterotrophs to be animals (protozoans). Such a division is still useful in many ways, for it seems that animals evolved from heterotrophic protists and plants from autotrophic protists. Many other protists, however, have no known multicellular relatives.

Although most unicellular eukaryotes are classified as Protista, some appear to be so closely related to a particular evolutionary line among the plants or fungi that they are classified in one or the other of these kingdoms, with their multicellular

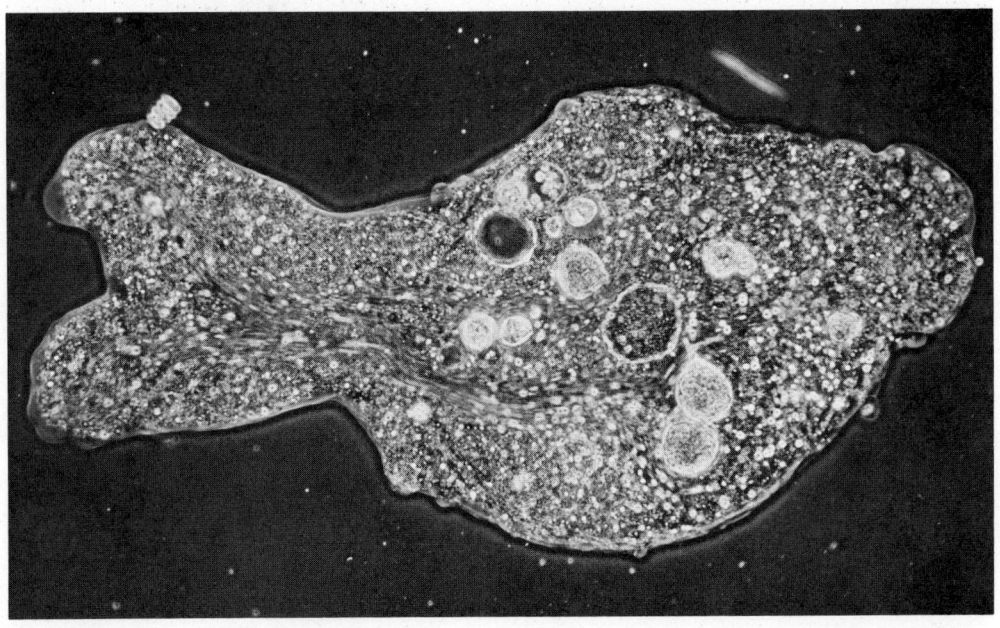

Figure 23-1 *Amoeba proteus*, a common freshwater protist. (Biophoto Associates)

relatives. It is also difficult to decide how to classify organisms made up of many cells that show little division of labor between them. Is such an organism best regarded as a protist colony or as an undifferentiated plant or animal? Again, it will generally be classified with its closest identifiable relatives.

23-B Physiology of Protists

The Protista are such a heterogeneous lot that very few generalizations can be made about their physiology (how they work). Some have a cellulose cell wall, a gelatinous sheath or a shell that provides protection and support and may also prevent the cell's bursting when it takes in water by osmosis from a hypotonic* environment. Many protists that lack cell walls have **contractile vacuoles** that collect excess water and expel it from the cell. Contractile vacuoles are prominent in freshwater protists but are also found in some marine forms.

Oxygen and carbon dioxide enter and leave protists by diffusion across the plasma membrane. Although nearly all protists can respire aerobically, many of them are also capable of living indefinitely using anaerobic fermentation when oxygen is not available.

One of the major secrets of their success is that, under adverse conditions, many protists can enclose themselves in **cysts**. A cyst is resistant to desiccation (drying out) and temperature fluctuation; this permits its owner to survive until favorable conditions return. Some desert protists emerge from their cysts to feed and reproduce only for the few hours each year when water is available. Cysts may also permit protists to be dispersed worldwide; an encysted cell may be carried from one continent to another in dried mud on the foot of a migrating duck, in the fur of a shipboard rat, or in human clothing.

Some protists move by means of cilia, flagella, or **pseudopodia**—flowing extensions of the cytoplasm; other protists have no means of locomotion.

How does an organism use a flagellum to propel itself? Wavelike beats, or undulations, pass from one end of the flagellum to the other. The beating acts like a propeller, pulling the cell through the water if it is anterior (in the front, or leading, end) and pushing if it is posterior (at the rear end) (Figure 23–2). If there is nothing to stop it, the cell will move forward in a spiral path, rotating as it goes. Some flagellated cells do not rotate because they are held in position against the bottom of a pond or against some other surface.

Cilia have the same structure as that of flagella but are shorter and generally more numerous. Ciliary locomotion can be compared to rowing a boat, with each cilium bending on its return stroke so that it offers the least possible resistance to the water.

Pseudopodial or amoeboid locomotion is more complex and less well-understood than flagellary or ciliary locomotion. In pseudopodial locomotion, an amoeba extends a pseudopod ("false foot") from its body, and the rest of the body flows into the forward extrusion.

---

* Hypotonic Having a lower osmotic pressure than that of a reference solution; in this case, the interior of the cell.

Figure 23-2 Modes of locomotion found among protists.

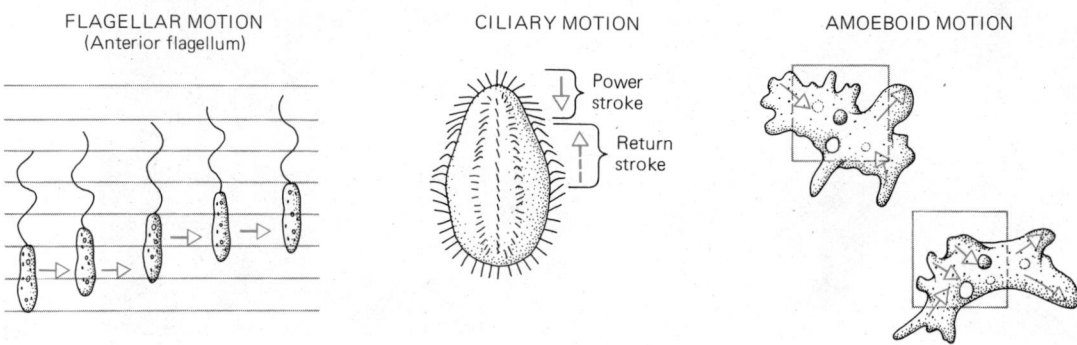

FLAGELLAR MOTION (Anterior flagellum)

CILIARY MOTION
Power stroke
Return stroke

AMOEBOID MOTION

Protists feed in various ways. Heterotrophic protists ingest food by endocytosis (Figure 5–9) or absorb small organic molecules from their environment. Many protists have more than one type of nutrition: some species can switch from photosynthesis to endocytosis to absorption, as conditions dictate.

Most protists have sensory mechanisms that permit them to detect **stimuli** such as touch, light, and chemicals. Light detectors, or **eyespots**, consist of photosensitive pigments contained in small organelles. Many protists can detect objects around them that touch their cilia or flagella. They also detect chemicals in their environment, presumably via changes these substances produce in protein molecules in their plasma membranes. Protists respond to stimuli with appropriate behaviors, usually moving toward or away from a stimulus.

Protists usually reproduce by division of the cytoplasm after a mitotic division of the nucleus. Most can also undergo one of a variety of types of sexual reproduction. Like most organisms that can reproduce either sexually or asexually, protists tend to reproduce asexually when conditions favor feeding and growth, and sexually when they do not. There is a clear adaptive advantage to this system: the offspring of asexual reproduction are genetically identical to one another (except for possible mutations), and a protist that is well adapted to the prevailing conditions can produce hundreds of equally well-adapted offspring if it divides asexually. When the food supply decreases or the temperature changes, however, sexual reproduction permits the production of genetically diverse offspring, increasing the chances that at least some of them will be able to survive in the altered environment.

In considering some of the more important phyla of the Protista, we will examine three phyla of unicellular algae first, and then four phyla of protozoa.

TABLE 23-1 THE KINGDOM PROTISTA

UNICELLULAR ALGAE (Mainly Autotrophs)

Phylum Pyrrophyta (~1000 species)	Dinoflagellates. Pectin and cellulose cell walls; mostly marine*; two flagella; chlorophylls *a* and *c* and carotenoids; e.g., *Gonyaulax, Noctiluca, Gymnodinium, Ceratium.*
Phylum Euglenophyta (~450 species)	*Euglena* and its relatives. No cell wall; mostly freshwater; 1–3 anterior flagella; chlorophylls *a* and *b* and carotenoids; e.g., *Peranema.*
Phylum Chrysophyta (~5800 species)	Diatoms and their allies. Cellulose and pectin cell walls, containing silica in diatoms; marine or freshwater; diatoms lack flagella, allied forms may have one or more flagella; chlorophylls *a* and *c* and carotenoid pigments; e.g., *Navicula, Chaetoceros.*

PROTOZOANS (Mainly Heterotrophs)

Phylum Zoomastigina	Flagellates without chloroplasts. No cell wall; freshwater or parasitic; one or more flagella; e.g., *Trypanosoma, Trichomonas, Lophomonas.*
Phylum Sarcodina (~40,000 species)	Amoebas, foraminiferans, heliozoans and radiolarians. No cell wall but some secrete shells containing silica, calcium carbonate, etc.; marine and freshwater; pseudopodia, no flagella; e.g., *Arcella, Amoeba, Difflugia, Globigerina, Actinosphaerium.*
Phylum Sporozoa	Parasitic protists. Flagella or pseudopodia; complex life history; e.g., *Toxoplasma, Plasmodium, Monocystis, Gregarina.*
Phylum Ciliophora (~8000 species)	Ciliates. Two types of nuclei, cilia for locomotion and food collection; no cell wall; freshwater and marine, some parasites; e.g., *Paramecium, Tetrahymena, Vorticella, Euplotes, Stentor.*

* Marine Living in the sea.

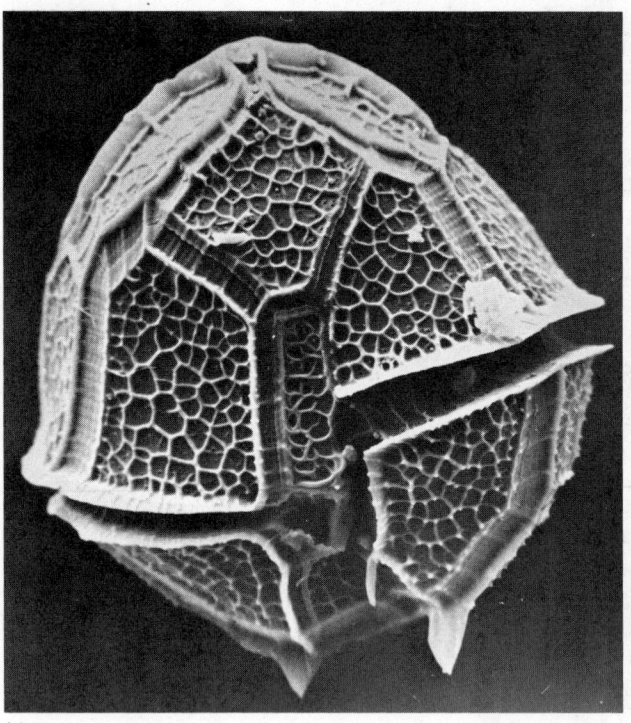

Figure 23-3 Scanning electron micrographs of planktonic protists. (a) a dinoflagellate, magnified more than 2,000 times. (b) Half of the pillbox-like shell of the diatom *Aulacodiscus.* (Biophoto Associates)

23-C Phytoplankton

Many photosynthetic protists and other small photosynthetic organisms (cyanobacteria and members of some of the groups of multicellular algae) are members of the **phytoplankton** (phyto = plant; plankton = wanderer)—the autotrophic organisms that float near the surfaces of oceans, lakes, and ponds. Phytoplankton are ecologically important as the ultimate source of food for aquatic animals. They also produce an estimated 30 to 50% of the oxygen in the atmosphere.

Phytoplankton need light, but because light is absorbed rapidly as it passes through water, phytoplankton can survive only if they remain near the surface of the water. Some phytoplankton retain their favorable position by using one or more flagella to swim upward. Others have projections that act as water wings, increasing their surface resistance and enabling them to stay afloat. Still others store their extra food as oil rather than starch; oil makes the cell buoyant so that it floats near the surface. Many organisms have more than one of these adaptations.

When oil-storing forms of phytoplankton die and fall to the bottom of a body of water, they may be covered by sediments and subjected to enormous pressures. Such phytoplankton, of hundreds of millions of years ago, were the source of the fossil fuel oil we use today.

Phytoplankton also have seasonal cycles. The number of photosynthetic organisms that a body of water can support is usually limited either by the light intensity (and temperature) or by the concentration of nutrients (especially nitrates and phosphates). In winter in temperate* climates, the phytoplankton population is small because there is not much light. In summer in temperate climates, and throughout the year in the tropics, nutrients are limiting. As day length increases in the temperate zones in the spring, an enormous burst of reproduction occurs among the phytoplankton. All the dissolved nutrients are absorbed and then carried to the bottom as planktonic organisms die and sink. Thus, despite plentiful light, the phytoplankton

*Temperate Having distinct summer and winter seasons.

population declines in midsummer because few nutrients are available. Later in the year, water from the deeper layers of the sea wells up, bringing nutrients back to the surface. Much of the ocean is sparsely populated—a watery desert—due to the scarcity of nutrients needed by the phytoplankton, which in turn are necessary as food for animal life. Phytoplankton density is greatest in those areas of the ocean where currents bring nutrient-rich waters up from the deep throughout the year.

23-D   Phylum Pyrrophyta (Dinoflagellates)

The Pyrrophyta, or dinoflagellates (see Figure 23–3a), are primitive in that their chromosomes contain few proteins and are attached to the nuclear membrane. Although they rarely reproduce sexually, they reproduce readily by cell division, and make up one of the two main groups of marine phytoplankton. Ancient dinoflagellates formed large oil deposits. Some members of the group are colorless, but most contain chlorophylls *a* and *c* and carry on photosynthesis.

Dinoflagellates have two flagella, one wrapped around the middle of the cell and the other projecting from the rear. Some are naked, while others are covered with cellulose "armor plating." Several species, including the well-known *Noctiluca*, are bioluminescent, that is, able to give off light (like a firefly). Alister Hardy described an encounter with these organisms in the English Channel:

I looked over the side to see a small shoal of fish, most likely mackerel, lit up by each individual being covered by a coat of fire; they were being chased this way and that by some much larger fish similarly aflame. On putting over a tow-net, which came up brilliantly illuminated, the sea was seen to be full of very small *Peridinium*-like dinoflagellates of the genus *Goniaulax*.

Gonyaulax (as it is now spelled) has another claim to fame. Some species produce a nerve poison that can kill humans. During population explosions of *Gonyaulax*, shellfish that have eaten large quantities of this dinoflagellate become unfit for human consumption. Since *Gonyaulax* contains a red pigment, such population explosions may color the water red, a fitting warning signal commonly known as a "red tide."

23-E   Phylum Euglenophyta

The Euglenophyta are named after the familiar genus *Euglena* (Figure 23–4). Most members of this group live in fresh water; they are especially abundant in polluted habitats. These protists have one or more flagella. They lack cell walls but may have elastic, transparent **pellicles** just beneath their plasma membranes. Their chloroplasts contain chlorophylls *a* and *b* and carotenoids. The Euglenophyta reproduce asexually by dividing longitudinally into two halves; sexual reproduction is unknown.

Members of the phylum Euglenophyta illustrate the difficulty of classifying everything as heterotrophic or autotrophic. If *Euglena*, for instance, is raised in the dark, where it cannot carry on photosynthesis, it loses its green color and becomes a heterotroph, ingesting food through a **gullet**.

The Euglenophyta are commonly considered to be the protists most resembling the ancestors of both plants and animals. The reasons for this are partly negative: members of most other protist phyla are highly specialized and plainly represent the end results of long periods of evolutionary adaptation to their own ways of life. A more positive reason is that animals and most lower plants have flagella, at least in their sperm, making it reasonable to look for their origins among the relatively unspecialized flagellates such as the Euglenophyta.

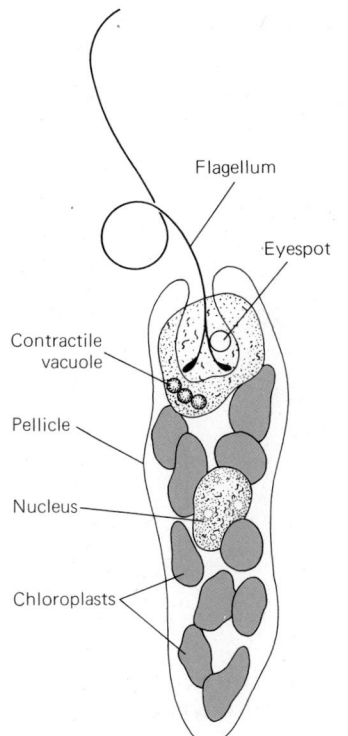

Flagellum

Eyespot

Contractile vacuole

Pellicle

Nucleus

Chloroplasts

Figure 23-4 *Euglena* (Euglenophyta). The light-sensitive eyespot permits the cell to move towards available light, using its flagellum for locomotion. The contractile vacuole expels excess water taken in by osmosis.

23-F Phylum Chrysophyta (Diatoms and Golden Algae)

Chrysophytes include the diatoms (Figure 23–3b) and the "golden algae," most of which are photosynthetic. These protists are found in the sea, in fresh water, and in wet spots on rocks, plants, and wood. Diatoms and pyrrophytes are the two main groups of marine phytoplankton. In unpolluted fresh water, diatoms (with green algae, Chapter 25) are the most important group of phytoplankton.

Diatoms have a rigid cell wall impregnated with pectin and with silica (SiO_2), which constitutes an important part of both glass and sand. Often the cell wall has an intricate pattern of pits or ridges. Sexual reproduction is common among the chrysophytes, although asexual reproduction predominates in this group.

The siliceous walls of diatoms are very resistant to decay. As a result, they have accumulated on the ocean floor in enormous numbers. Such an area of the sea floor may later be raised above sea level by geological activity. Deposits of "diatomaceous earth" are mined for commercial use as a fine abrasive in silver polishes and toothpaste and as the packing in air and water filters.

The chrysophytes contain chlorophylls a and c and the accessory pigment fucoxanthin, a carotenoid that gives them their characteristic yellow-brown color. They store much of their surplus food as oil and, with other phytoplankton, were important in the formation of petroleum deposits.

23-G Phylum Zoomastigina (Zooflagellates)

The phylum Zoomastigina contains heterotrophic flagellates. Some are free-living whereas others are parasites (Figure 23–5) or **symbionts**, organisms that live in close association with members of another species. The parasitic forms often have complex life histories involving two hosts. The phylum Zoomastigina includes the symbionts in the guts of termites and wood roaches; these flagellates engulf and digest the wood eaten by their insect hosts. Because the insects cannot make their own wood-digesting enzymes, they are completely dependent on their symbionts.

Of great medical and veterinary importance are parasitic zooflagellates of the genus *Trypanosoma*. Trypanosomes are found living in all groups of vertebrates, but most species parasitize mammals. Human diseases caused by trypanosome infection include sleeping sickness (Africa) and Chagas' disease (South and Central America). Nagana is a trypanosome disease of cattle that makes cattle farming impossible in much of sub-Saharan Africa. Both sleeping sickness and nagana are transmitted to their mammalian hosts by bites of the tsetse fly, the host during part of the trypanosome life history.

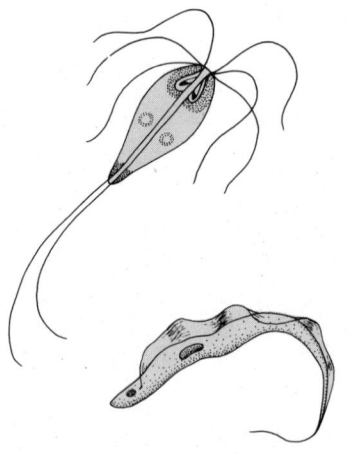

Figure 23-5 Parasitic flagellates of the phylum Zoomastigina.

23-H Phylum Sarcodina

The Sarcodina are protists that move and engulf their prey with pseudopodia. *Amoeba proteus*, a freshwater species, is one of the most thoroughly studied. It is easy to keep in captivity on a diet of bacteria, and is large enough to be used in studies of amoeboid locomotion and endocytosis. It is also used in experiments on nucleus-cytoplasm interactions, since its nucleus can be removed or transplanted by microsurgical techniques to find out how the cytoplasm reacts.

Because *Amoeba proteus* is such a familiar creature, sarcodines are often thought of as naked and amoeboid. In fact this is far from the case. Many amoebas have shells, and all members of the other three classes in the phylum, the foraminiferans, heliozoans, and radiolarians, also have shells (Figure 23–6).

Most **foraminiferans** inhabit the warmer oceans of the world and secrete a **calcareous** ($CaCO_3$) shell full of holes through which they poke long thin pseudopodia. The pseudopodia branch and join outside the shell to form a net that traps and digests the organism's prey.

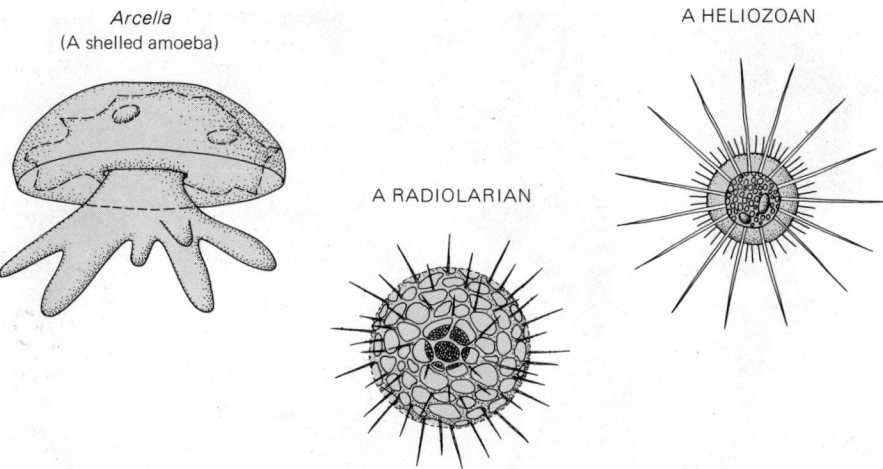

Arcella
(A shelled amoeba)

A HELIOZOAN

A RADIOLARIAN

Figure 23-6   Members of the protistan phylum Sarcodina (not drawn to scale).

When foraminiferans die, their shells sink to the bottom. Millions of years' worth of this foraminiferan debris has formed chalk rocks or limestone, such as the famous white cliffs of Dover in England. Foraminiferan fossils are also common in deposits where oil has accumulated from prehistoric phytoplankton. As an oil well is drilled, the bit passes through successive species of such Foraminifera that evolved at different times. By identifying the foraminiferan species in a particular layer, a geologist can estimate the age of the rock and decide whether oil deposits should be sought above or below that layer.

Radiolarians, an exclusively marine group of sarcodines, secrete elaborate and beautiful exoskeletons that usually contain silica. Like foraminiferans, they extrude pseudopodia through holes in the shell, but unlike foraminiferans, they draw their prey inside the shell for digestion. Many radiolarians and foraminiferans are colonial or multinucleate, containing many nuclei within a common plasma membrane; some reach a size of several millimetres in diameter. They are included in the kingdom Protista because their nearest relatives are clearly the single-celled radiolarians and foraminiferans.

Heliozoans, or "sun animals," live mainly in fresh water. They may be free or stalked. Some are naked, but some have skeletons of silica or foreign particles, such as sand grains, embedded in a gelatinous test (protective covering). Like the foraminiferans and radiolarians, they extrude long thin pseudopods, which they use to capture food.

23-I   Phylum Sporozoa

All of the sporozoans are parasites, usually with the complicated life histories characteristic of this way of life; some need two different host species to complete the life history. Some have flagellated gametes and some have pseudopods. The presence of both modes of locomotion indicates that this is an artificial phylum, composed of members of two evolutionary lines rather than one. The ancestors of the sporozoans were probably flagellates and sarcodines that took up parasitic existences and underwent **convergent evolution**, that is, evolved many similar adaptations to their similar ways of life. Most feed by absorbing small organic molecules from their hosts.

Sporozoans are the cause of coccidiosis, which can devastate flocks of chickens and herds of calves, and of malaria in humans. The sporozoan *Toxoplasma* may well be the most common human parasite and is also found in birds. It causes the disease toxoplasmosis, whose symptoms may be so slight as to pass unnoticed, and which is sometimes diagnosed as a form of arthritis.

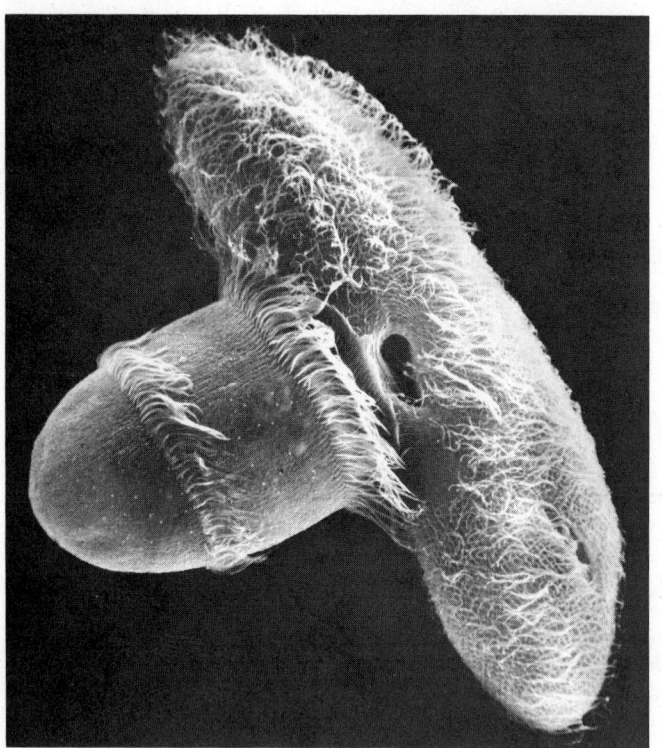 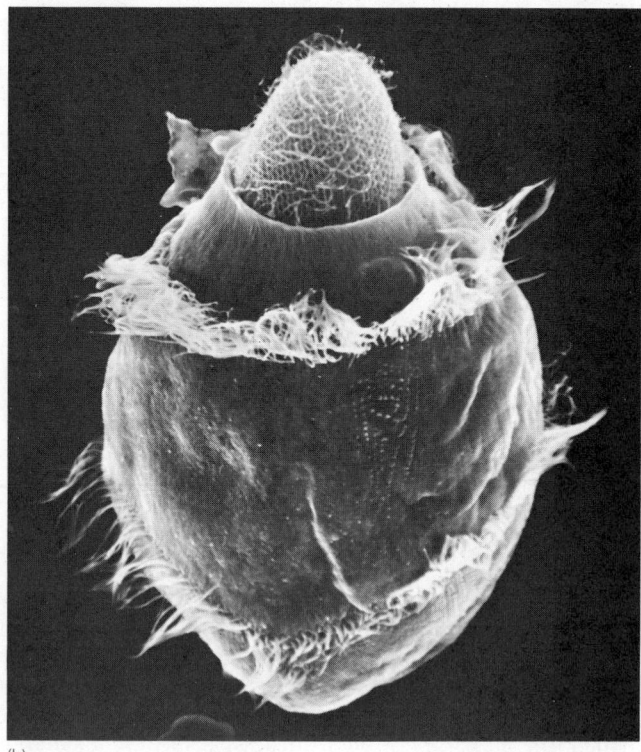

Figure 23-7 Ciliate eats ciliate. (a) A *Didinium* (at left) attacks a *Paramecium* that looks much too big for it. The cavity halfway along the *Paramecium* is its gullet. (b) The *Didinium* has stretched enormously and engulfed all but the tip of the *Paramecium*. (Biophoto Associates, courtesy of Drs. G. Antipa and E. Small)

23-J   Phylum Ciliophora (Ciliates)

The ciliates are extremely complex, highly organized, heterotrophic protists with cilia either all over the body or in specialized areas of the cell surface (Figure 23–7). The body wall contains a pellicle and often numerous **trichocysts**, thread-like organelles that can be discharged to the outside. Some trichocysts have barbed tips, and some eject poison; trichocysts serve for anchorage, defense, or prey capture.

Most ciliates prey on bacteria, small animals, or fellow protists; some eat organic particles from the surrounding water, some are symbionts, and a few are parasites. Specialized cilia around the mouth region sweep food into a gullet. Food taken in here enters a food vacuole, which fuses with a lysosome containing digestive enzymes. The products of digestion are absorbed into the surrounding cytoplasm, and undigested remains are discharged at a specific site on the cell surface. Likewise, the contractile vacuoles, which discharge excess water, have specific sites where they expel their contents.

A peculiar characteristic of the ciliates is that they possess more than one nucleus: each cell has one or more small **micronuclei**, each containing one copy of the genetic material, and a large **macronucleus**, which contains up to 500 times as much DNA as that of the micronucleus. The relationship between the micronucleus and macronucleus is not fully understood, but it seems that the micronucleus controls sexual reproduction and heredity, while the macronucleus controls cell growth and metabolism.

Paramecium is the most familiar genus of freshwater ciliates (Figure 23–8). Some paramecia are hosts for other organisms. For example, *Chlorella*, a unicellular green alga, sometimes lives in the cytoplasm of *Paramecium bursaria*. This arrangement presumably aids both parties; the alga gains protection and the ciliate has captive photosynthesis. Other strains of paramecia contain **Kappa particles**, composed of nucleic acid. These paramecia are known as killers because they produce a toxin lethal to paramecia that do not contain the particles. **Lambda particles**, found in other strains of paramecia, have been grown outside their ciliate hosts and have been identified as Gram-negative bacteria.

23-K   On to Multicellularity

Once upon a time all organisms on earth were unicellular. Wherever such life was abundant, there must have been strong selective pressure for increased size. A large organism could eat more of its neighbors and be eaten by fewer. A single cell cannot just become larger and larger; eventually the center of the cell is too far from the outside to obtain the substances it needs from its environment fast enough. As cell size increases, the ratio of surface area to volume decreases, and a larger volume of cytoplasm is serviced by a relatively decreasing area of cell surface (Figure 23–9). Because the exchange of food, oxygen, and cell wastes takes place through the surface, the decrease in the surface-to-volume ratio means that each unit volume of cytoplasm receives a smaller amount of supplies per unit of time. Thus the size of single cells is limited, and an increase in size must be accomplished by an increase in cell number, so that each cell has a surface-to-volume ratio sufficient to sustain its metabolism.

Some protists are colonial, with a number of cells joined together to form a larger organism. Each cell is small and able to exchange substances with its environment in an efficient manner, but the cells are still more or less identical. True multicellularity implies division of labor among the cells of an organism such that different cells carry out different tasks.

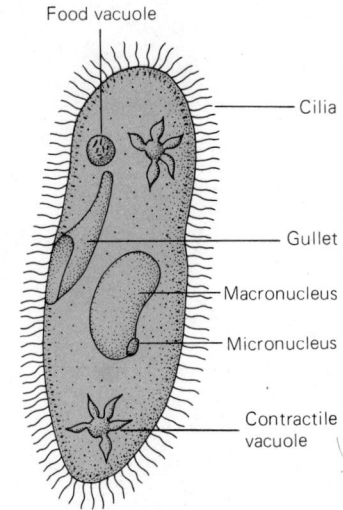

Figure 23-8 *Paramecium*, a ciliate. The body is completely covered with cilia. Cilia also sweep food particles into the gullet. The cytoplasm at the end of the gullet engulfs food into vacuoles by phagocytosis. Food is digested by enzymes in the food vacuoles as they move around in the cytoplasm.

Cube side length	Surface area (SA)	Volume (V)	SA/V
1 cm	6 cm²	1 cm³	6:1
2 cm	24 cm²	8 cm³	3:1
3 cm	54 cm²	27 cm³	2:1

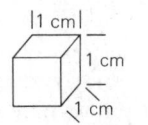

Surface area = 1 cm X 1 cm X 6 = 6 cm²
Volume = 1 cm X 1 cm X 1 cm = 1 cm³

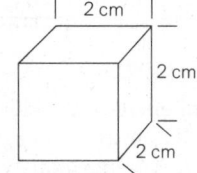

Surface area = 2 cm X 2 cm X 6 = 24 cm²
Volume = 2 cm X 2 cm X 2 cm = 8 cm³

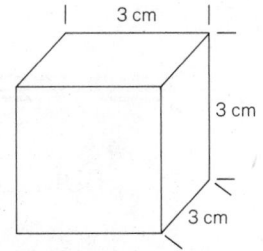

Surface area = 3 cm X 3 cm X 6 = 54 cm²
Volume = 3 cm X 3 cm X 3 cm = 27 cm³

Figure 23-9 Surface-to-volume ratios. Surface area increases with the square of the linear dimension of the object, whereas volume increases with the cube of the linear dimension. If the objects are of the same shape, the larger one will have less surface area per unit volume. That is, the surface-area-to-volume ratio (SA/V) decreases as an object increases in size while retaining the same shape.

**CHAPTER 23
PROTISTA AND THE
ORIGIN OF
MULTICELLULARITY**

TABLE 23-2 BASIC FUNCTIONS THAT MUST BE CARRIED OUT BY EVERY ORGANISM

Feeding or making food	Dispersal (locomotion, scattering seeds or larvae)
Gas exchange	Support and protection
Waste removal	Coordination of all functions (nerves, hormones, etc.)
Internal transport of food, gas, etc.	
Sensing environmental stimuli	Reproduction

Usually each cell can carry out its own basic functions, but in multicellular organisms each cell is also specialized to help carry out one of these functions for the entire body.

Every cell and every organism must carry out certain basic life functions (Table 23-2). The story of the kingdoms Fungi, Plantae, and Animalia can be viewed as the evolution of increasingly specialized groups of cells that contribute to the existence of the individual organism. **Tissues** are groups of similar cells that are specialized to perform a particular function. **Organs** are made up of several different types of tissues whose interaction results in the performance of a specific function.

The first sign of division of labor is usually the specialization of certain cells for reproduction, while others acquire energy by feeding or photosynthesis. Other early specializations include structures for protection and anchorage.

The simplest multicellular organisms are small, with no cell very far from the watery environment that provides food and removes body wastes. As organisms became larger, the inner cells were further and further from the environment. Organisms that evolved means of transporting substances from the environment to these cells, and vice versa, were able to increase in size, whereas those without such systems had to remain small. Finally, a large-bodied organism needs mechanisms that coordinate the activities of the various parts into a working system. Coordination exists in even the smallest of single cells, but it is perhaps most impressive in the complex systems of sense organs, nerves, hormones, and muscles found among the higher animals.

We tend to look upon increases in size and complexity as "progress" and to despise yesteryear's smallness and simplicity. However, unicellular organisms do have some advantages over larger creatures. A single cell, for instance, can live in a tiny space and needs only a little food before it is ready to reproduce. This allows unicellular organisms to exploit many habitats that are not available to larger forms. The ubiquity and diversity of bacteria and protists today attests to their evolutionary success.

SUMMARY

The origin of eukaryotic cells led to the evolution of a wide array of unicellular heterotrophic and autotrophic organisms that are classified in the kingdom Protista. Protists may have an enormously complex organization within a single cell.

Autotrophic protists are found either in the phytoplankton or growing on moist surfaces. Heterotrophic protists may be free-living predators on other protists, bacteria, and small multicellular organisms, or may be symbionts or parasites. Some parasitic protists have had enormous impact on human health and economy.

Selective pressure for large size probably led to the evolution of multicellular fungi, plants, and animals from protistan ancestors. In protists, division of labor occurs within the cytoplasm and organelles of a single cell. In multicellular plants and animals, division of labor among cells and tissues is added to the division of labor within a single cell.

OBJECTIVES

From your study of this chapter, you should be able to:

1. Give the definitive characteristics of members of the kingdom Protista.

2. Describe the three main types of locomotion found among protists.

3. Explain what phytoplankton are, their ecological importance, and what factors control the fluctuation in their numbers.

4. List distinguishing characteristics of members of the following groups: Chrysophyta, Pyrrophyta, Euglenophyta, Sarcodina, Zoomastigina, Ciliophora.

5. Place each of the following in the correct phylum: dinoflagellate, *Gonyaulax*, *Euglena*, diatom, *Amoeba*, *Paramecium*.

SELF-QUIZ

1. An organism should be placed in the kingdom Protista if it is _____, _____, and is not clearly related to members of the kingdom Fungi, Plantae, or Animalia.

2. For each description below, choose the mode of locomotion described:

   _____ a. Cytoplasm flows out into temporary extrusions from the cell body
   _____ b. Oar-like beating propels the cell through the water
   _____ c. Propeller-like undulations pull or push the cell through the water

   i. cilia
   ii. flagella
   iii. pseudopodia

3. Populations of phytoplankton in temperate climates are controlled by scarcity of _____ in the summer and by scarcity of _____ in the winter.

4. A dinoflagellate would have which of the following sets of characteristics?

   a. a mouth, contractile vacuole, and mitochondria
   b. a macronucleus, a micronucleus, and a chloroplast
   c. chloroplasts, one circular chromosome, and cilia
   d. a contractile vacuole, pseudopods, and cilia
   e. chloroplasts, mitochondria, and flagella

QUESTIONS FOR DISCUSSION

1. Can you think of selective pressures other than the one discussed in this chapter that might have selected for evolution of multicellular organisms?

2. How do the protists you studied in this chapter carry out each of the functions listed in Table 23–2?

REFERENCES AND FURTHER READING

Barnes, R. D. *Invertebrate Zoology*, 4th ed. Philadelphia: Saunders, 1980. An excellent text and reference book.

Carson, R. L. *The Sea Around Us*. New York: Oxford University Press, 1951. Contains a delightful discussion of the importance of algal photosynthesis in the sea.

Curtis, H. *The Marvellous Animals*. Garden City, NY: Natural History Press, 1968. A delightfully written introduction to the kingdom Protista.

Hardy, A. *The Open Sea*. Part I, The World of Plankton. Boston: Houghton Mifflin, 1965. An anecdotal account of planktonic protists and invertebrates; wonderfully illustrated.

CHAPTER

24

FUNGI

The kingdom Fungi contains eukaryotic, multicellular organisms adapted to absorbing food molecules from their surroundings. As decomposers, the fungi are vitally important to members of the plant and animal kingdoms. When a dead leaf drifts to the forest floor or an animal dies of disease, fungal and bacterial spores floating in the air have already settled on it. These spores quickly germinate and begin to break down the dead organism, releasing small organic molecules that they can use as food, as well as minerals that may be absorbed by the decomposer or by nearby plants.

Fungi are also important in the economy of human societies. Some fungi cause tremendous losses of food and crops every year, but fungi also make important contributions to the production of foods and medicines. In this chapter we shall study the fungal way of life, the classification of fungi, and the importance of fungi to humans.

Nutrition

Fungi cannot produce their own food by photosynthesis as do green plants, nor do they have mouths to ingest organic matter as do animals. They absorb food through their cell walls and plasma membranes. Fungi may be **saprobes**, absorbing food from dead organic matter, or **parasites**, absorbing food from the living bodies of other organisms.

Fungi secrete digestive enzymes, which hydrolyze the organic matter around them into small organic molecules and minerals that the fungus can absorb. Some of the substances produced by such enzyme action are not absorbed by the fungus, and these, together with the waste products of the fungus's metabolism, enrich the area around the fungus. Humans use this characteristic by deliberately introducing fungi to increase the nutrient content of certain foods.

Body Plan

The absorptive lifestyle of fungi is intimately associated with two important characteristics: production of spores and mycelial growth. A **spore** is a tiny, usually haploid, cell that disperses the fungus to new habitats, usually by floating through the air. The production of many tiny spores increases the chance that at least a few will fall onto a suitable food source. When this happens, the spore germinates, starts absorbing food, and grows into a threadlike **hypha**. The hypha grows rapidly and branches until it resembles a tangled mass of threads. The body of a fungus, made up of many hyphae, is called a **mycelium** (Figure 24–1). The mycelium is well suited to absorbing food. Its high surface-to-volume ratio permits the surface exposed to the external food source to absorb enough food to nourish the enclosed volume of cytoplasm. A hypha releases chemicals that cause other hyphae to grow away from it. As a result, the fungus spreads out through its food source, and competition between hyphae is reduced. Parasitic fungi absorb nutrients from the body fluids of the host, and parasites of plants may produce specialized hyphae called **haustoria** (singular: **haustorium**) that penetrate a plant's cell wall and lie against the plasma membrane, where they can absorb food.

Some hyphae are **coenocytic**, with many nuclei lying in the same cytoplasm. Others are divided by **septa** (singular: **septum**) into compartments containing one or more nuclei. A **dikaryotic hypha**, or **dikaryon**, contains two (haploid) nuclei between each pair of adjacent septa; it is usually designated as N + N to distinguish it from haploid (N) and diploid (2N) cells.

The rigid cell walls of a fungus are composed of cellulose, other polysaccharides, and chitin. Cellulose, a polymer of glucose, is also found in plant cell walls. The other polysaccharides present in the fungal cell wall are glucose polymers, but with the glucose units arranged in different ways from that in cellulose. **Chitin** is a polymer composed of monomers of a nitrogen-containing glucose amine. (Oddly enough, the other group to which chitin is important is the arthropods—the insects, spiders, crabs, and their kin—animals whose external skeletons contain chitin.)

Reproduction

Fungi may reproduce vegetatively* when mycelia break up into fragments, each of which grows into a new individual. Spores may be formed asexually or as a result of sexual processes. Spores are often produced on structures that hold them away from the food source, up in the air where they can be blown into a wide range of possible habitats.

---

*Vegetative reproduction   Reproduction by growth of an individual's body or fragments of its body; reproduction without production of gametes or spores.

SPORE

HYPHA

Hypha

MYCELIUM

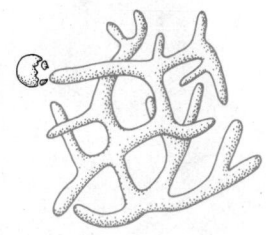

HAUSTORIA

Haustorium   Fungal hypha

Plant cell

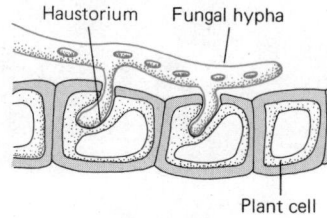

COENOCYTIC HYPHA

Nuclei

SEPTATE HYPHA

Nuclei

SEPTA in hypha

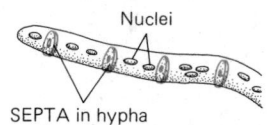

DIKARYON

Nuclei

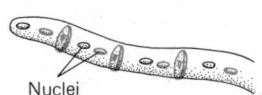

Figure 24-1   Some basic features of fungal anatomy.

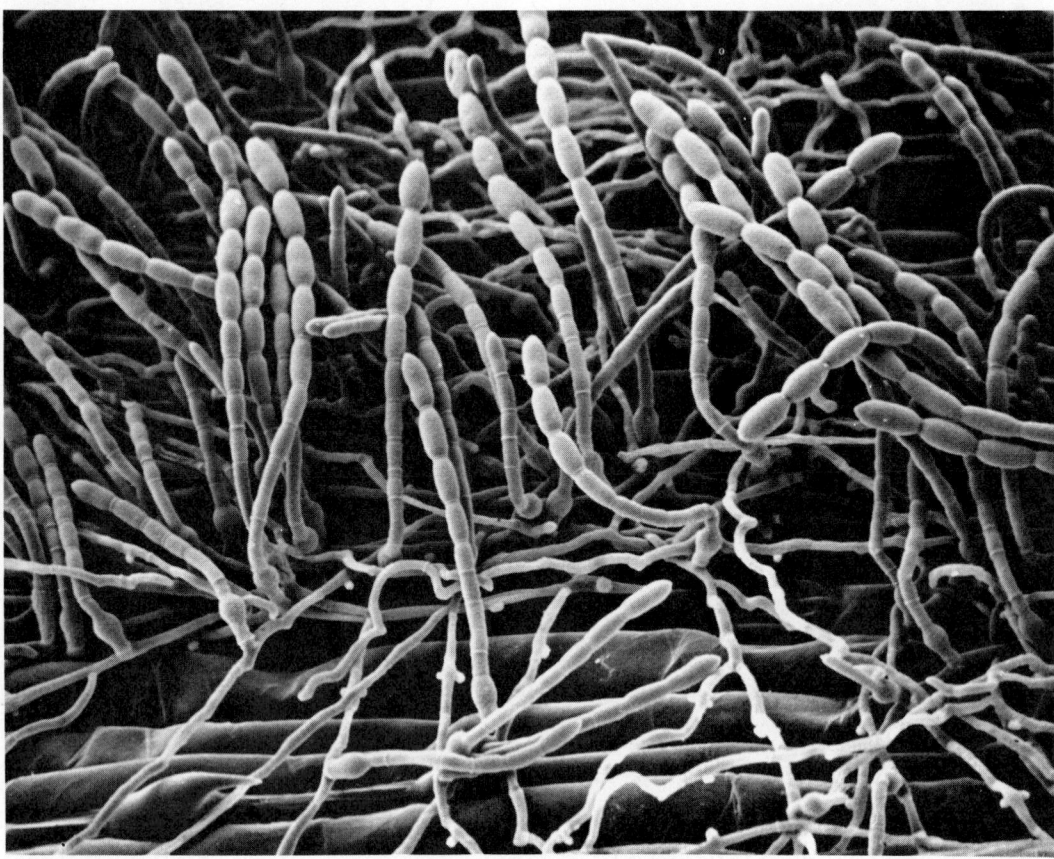

Figure 24-2 Mycelia of powdery mildew on the surface of a barley plant. The tips of the vertical, aerial hyphae are pinching in and forming spores. (Biophoto Associates)

The parts of a fungus we normally see are reproductive structures. The above-ground parts of mushrooms and cup fungi are **fruiting bodies**, large, complex structures composed of many hyphae. Fruiting bodies disperse spores produced by sexual processes; the mycelia grow as tangles of microscopic threads hidden in the food source.

In some species of fungi, sexual reproduction takes place between mycelia of different **mating types**, the equivalent of sexes in other organisms; mating types are usually designated as + and −. Other species of fungi have no separate mating types, and sexual reproduction may take place between different hyphae of the same mycelium. Sexual reproduction often involves the secretion of hormones by one or both partners, with hyphae growing toward one another along the concentration gradient of the hormone until contact is made.

24-B Classification of Fungi

The fungi probably arose along several different lines of evolution from protistan ancestors. The two most important types of organisms classified in this kingdom are the true fungi and the slime molds. Most of the remarks in this chapter refer to true fungi, except in Section 24-I, which specifically discusses the slime molds.

Fungi were originally classified in the plant kingdom. Botanists use the term **division** instead of phylum in the taxonomic hierarchy, and this usage is retained in fungal classification.

The true fungi, or Eumycophyta, are divided into classes on the basis of their characteristic sexual reproductive structures (or lack thereof), which is one of the few consistent characteristics in each class. Eumycophytes almost always have cell walls, and they reproduce by means of a wide variety of types of spores.

TABLE 24-1   DIVISIONS AND CLASSES OF THE KINGDOM FUNGI

Division Eumycophyta	True fungi. All classes contain saprobes and parasites with cell walls and (usually) hyphae. Unless otherwise indicated, fungi are terrestrial and reproduce both sexually and asexually by means of spores.
Class Oomycetes	"Water molds" and their terrestrial relatives. Unwalled flagellated spores and walled airborne spores; coenocytic hyphae; no chitin in cell walls. E.g., *Phytophthora infestans* (late blight of potatoes), *Plasmopara viticola* (downy mildew of grapes), *Albugo candida* (white rust on cabbage and other plants), *Peronospora parasitica* (downy mildew on crops), *Saprolegnia* (mold on dead insects, fish, frogs, etc., in water).
Class Zygomycetes	Sexual reproduction by zygospores; coenocytic hyphae; chitinous cell walls. E.g., *Rhizopus* (the black bread mold), *Entomophthora muscae* (parasite on the housefly), *Pilobolus* (a dung fungus).
Class Ascomycetes	"Sac fungi." Sexual reproduction by ascospores formed in asci; some unicellular (yeasts); hyphae of multicellular forms divided by perforated septa; short dikaryotic stage before sexual reproduction. E.g., *Neurospora crassa* (the pink bread mold), *Saccharomyces cerevisiae* (bread and wine yeast), *Claviceps purpurea* (ergot disease of grasses), *Sordaria* (a dung fungus), *Peziza* (a cup fungus), *Endothecia parasitica* (the chestnut blight fungus), *Aspergillus* (a common mold on foods; one species used to ferment beans for *shoyu*, soy sauce); *Penicillium* (various species used for production of penicillin and in the production of Roquefort and Camembert cheeses).
Class Basidiomycetes	"Club fungi." Sexual reproduction by basidiospores borne by basidia. Septate hyphae; long dikaryotic stage in life history. Mushrooms and toadstools, bracket fungi, rusts, and smuts. E.g., mushrooms (poisonous *Amanita* species as well as *Agaricus campestris*, the mushroom sold in grocery stores); *Polyporus* (bracket fungi); *Lycoperdon* (puffballs); *Phallus impudicus* (stinkhorn); *Puccinia graminis* (wheat rust); *Ustilago* (smuts of corn, oat, wheat, etc.)
Class Deuteromycetes	"Fungi imperfecti," not known to reproduce sexually. E.g., *Microsporum gypseum* (ringworm of dogs), *Epidermophyton floccosum* (athlete's foot), *Botrytis cinerea* (spear rot of asparagus), *Fusarium culmorum* (root rot of wheat).
Division Myxomycophyta	Slime molds. Differ from true fungi in having a mobile amoeboid stage in the life history.
Class Myxomycetes	Acellular slime molds. Feed as coenocytic plasmodia.
Class Acrasiomycetes	Cellular slime molds. Feeding amoebas congregate into a pseudoplasmodium before forming asexual fruiting body.

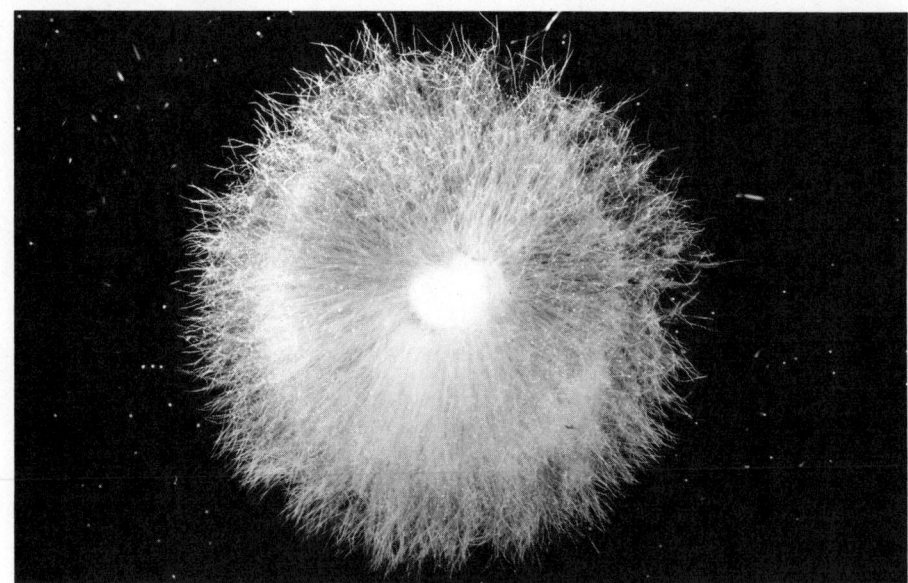

Figure 24-3 The mycelium of the oomycete *Achyla* growing on a hemp seed in a dish of water. (Carolina Biological Supply Company)

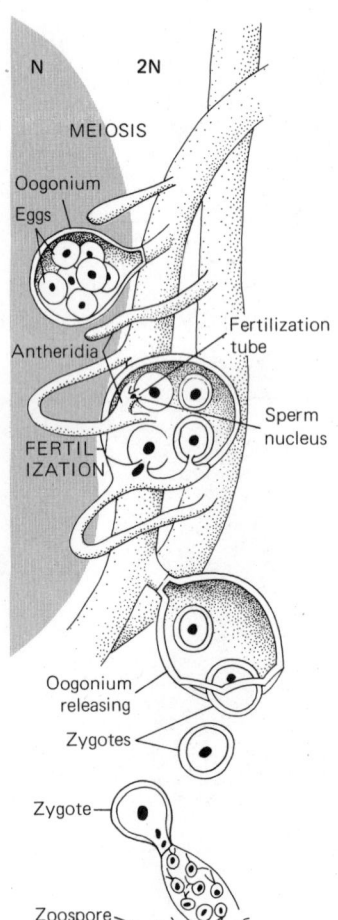

Class Oomycetes

Despite their common name of "water molds," there are many terrestrial oomycetes, including some very damaging plant parasites. The distinguishing characteristic of the oomycetes is the production of flagellated spores, which can swim to new food sources, a primitive* trait not found in more advanced fungi. Many oomycetes also produce airborne spores. The hyphae are coenocytic. The cell wall contains cellulose but no chitin, and the flagellated spores have no cell walls.

"Oomycetes," literally translated, means "egg fungi" and refers to the method of sexual reproduction (as do all the names of the classes of Eumycophyta). As the fungus uses up its food source, enlarged tips (called **oogonia**) form on some of the hyphae. Cytoplasm flows into these tips, and one to several large eggs form, each consisting of a nucleus surrounded by dense, nutrient-rich cytoplasm. Meanwhile, other hyphae nearby form slender, hooked tips, which grow out to the oogonia. When a hooked male (**antheridial**) hypha contacts an oogonium, it grows into it and releases haploid nuclei, which fuse with the nuclei of the eggs to form zygotes. After a period of dormancy, a zygote germinates into a short hypha that forms a sporangium and releases flagellated spores, which swim about until they find a suitable place to grow and form mycelia (Figure 24–4).

---

* Primitive Believed to have arisen early in the course of evolution (not to be confused with "simple").

Figure 24-4 Sexual reproduction in the oomycete *Saprolegnia*, a common saprobe on organic matter in the water. Haploid stages in the life history (N) are shown against a colored background. This form has hyphae with diploid (2N) nuclei.

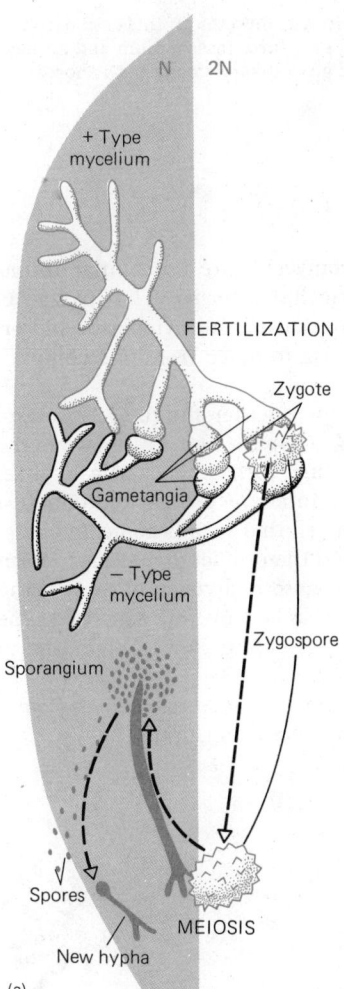

+ Type
mycelium

FERTILIZATION

Zygote

Gametangia

− Type
mycelium

Zygospore

Sporangium

Spores

New hypha

MEIOSIS

(a)

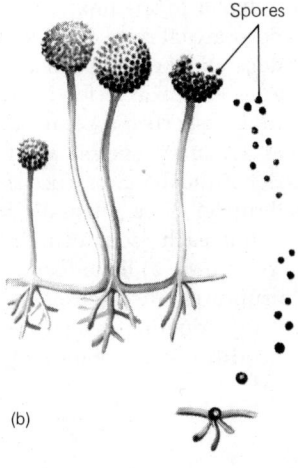

Spores

(b)

Figure 24-5 (a) Sexual reproduction in *Rhizopus*, the black bread mold, requires mycelia of opposite mating types (called + and −). During most of its life history, *Rhizopus* is haploid (N) (shown against colored background). (b) Asexual reproduction in *Rhizopus*.

Class Zygomycetes

Zygomycetes are terrestrial fungi with coenocytic hyphae and cell walls composed mainly of chitin. Their sexual reproduction is characterized by formation of zygospores (Figure 24–5a).

In a common zygomycete, *Rhizopus* (black bread mold), sexual reproduction requires members of two different mating types. When hyphae of opposite mating types meet, each forms a **gametangium** (gamete-producing structure) with a cross wall that closes it off from the rest of the mycelium. The two gametangia then fuse, and some of the nuclei pair and fuse to form diploid zygote nuclei. The fused gametangia become covered with a thick protective wall containing many identical zygote nuclei; this resistant **zygospore** enters a period of dormancy. Meiosis occurs when the zygospore germinates; it then sends up an aerial hypha, which forms a **sporangium** and releases many haploid spores (Figure 24–5a).

Asexual reproduction is much more common than sexual reproduction in *Rhizopus*. Sporangia form atop aerial hyphae, and the spores can be carried away by air currents (Figure 24–5b).

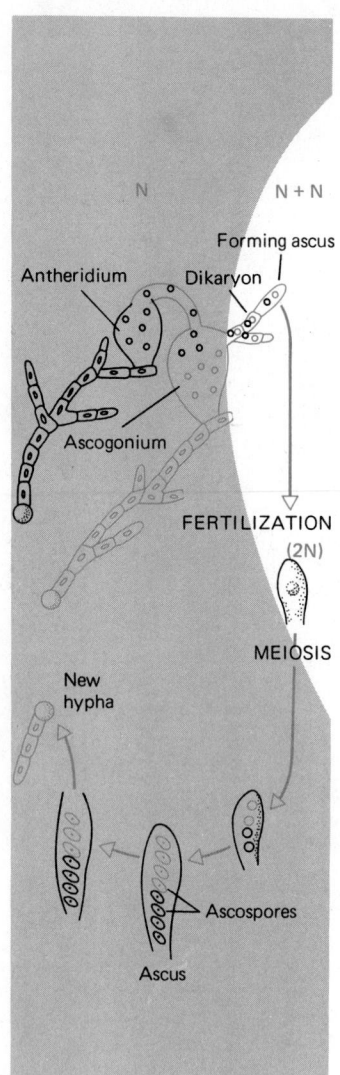

N   N + N

Forming ascus

Antheridium   Dikaryon

Ascogonium

FERTILIZATION

(2N)

MEIOSIS

New
hypha

Ascospores

Ascus

Figure 24-6 Generalized scheme of sexual reproduction in Ascomycetes, in this case with two mating types (black and colored). After fusion of enlarged structures (ascogonium and antheridium) containing many nuclei, haploid nuclei pair up and grow dikaryotically for a short time. Asci form on the dikaryotic hyphae.

Class Ascomycetes

Among the more than 30,000 species of Ascomycetes are unicellular forms, called yeasts, as well as a great variety of multicellular forms. The hyphae of multicellular forms are divided by perforated septa that permit cytoplasm and organelles such as nuclei, ribosomes, and mitochondria to move from one compartment of the hypha to the next.

Just before sexual reproduction, two hyphae grow together and their cytoplasm fuses, but their nuclei remain separate so that the new hyphae that develop from the fused structure are dikaryotic. These hyphae, together with other sterile hyphae, form a fruiting structure typical of the species. Here characteristic reproductive cells, or **asci** (singular: **ascus**) will form. Within a cell that is to become an ascus, the two nuclei of the dikaryon fuse (fertilization) and then undergo meiosis to form four haploid nuclei. This is usually followed by one mitotic division of each of the four nuclei; thus each ascus usually contains eight haploid nuclei, which become enclosed in their own walls to form eight **ascospores** (Figure 24-6). When the tip of an ascus ruptures, the ascospores are liberated.

Asexual reproduction is also common among the ascomycetes. Many ascomycetes form **conidia**, which pinch off asexual **conidiospores** (Figure 24-7).

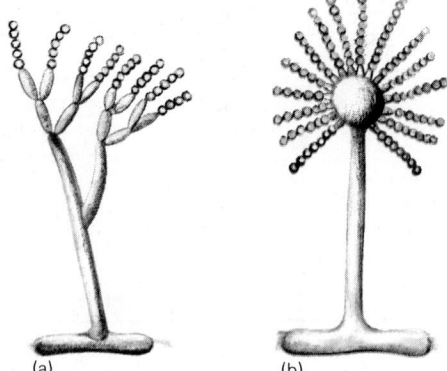

Figure 24-7 Conidiospores pinch off from the tips of aerial hyphae in the asexual reproduction of *Penicillium* (a) and *Aspergillus* (b).

(a)   (b)

Class Basidiomycetes

Basidiomycetes (e.g., mushrooms, bracket fungi, and smuts), like ascomycetes, have perforated septa that divide their hyphae into compartments. Here too, a haploid spore germinates to produce a hypha, which will fuse with another hypha of the same species (and of opposite mating type if the basidiomycete is a species with different mating types). Thereafter the mycelium is dikaryotic (N + N); the organism usually spends most of its life history in this stage.

The sexual reproductive structure, the **basidium** (plural: **basidia**) or club, is produced at the tip of a dikaryotic hypha. Within each basidium, the two haploid nuclei fuse, forming the only diploid nucleus in the life history. Meiosis follows immediately, forming four haploid nuclei. These move to the outer edge of the basidium, and here the cell wall forms fingerlike extensions into which the nuclei move, forming **basidiospores**. Behind the nuclei the extensions close, separating the basidiospores from the rest of the fungus, and forming a delicate stalk, which is easily broken. Once released, the basidiospores are wafted away on the slightest air current (Figure 24-8).

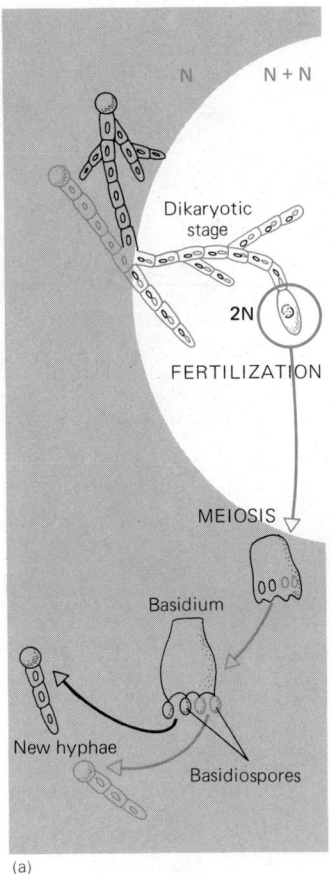

(a)

(b)

Figure 24-8 (a) Generalized diagram of sexual reproduction in Basidiomycetes. The life history is characterized by an extended time in the dikaryotic (N + N) stage. (b) Basidia and basidiospores in the fruiting body of a mushroom. (Biophoto Associates)

The most familiar basidiomycetes are the mushrooms. In these fungi, a well-fed mycelium will form an underground mass of hyphae that differentiates into a bulbous base, a stalk, and a knoblike cap. Some morning after an autumn rain we awake to find that the hyphae composing the stalk have swelled with moisture and elongated, carrying the cap above ground. The cap opens like a belated umbrella after the rain has ceased, and numerous basidia along the edges of the gills or pores beneath the cap prepare to loose their spores. The gills or pores on the underside of a mushroom increase the surface area where spores can be formed and discharged (Figure 24–9).

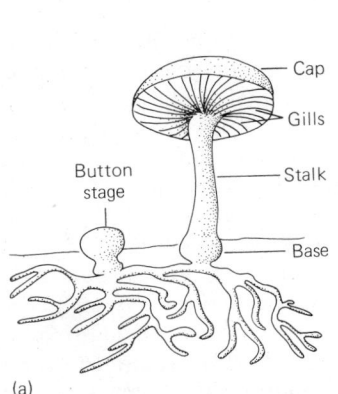

(a)

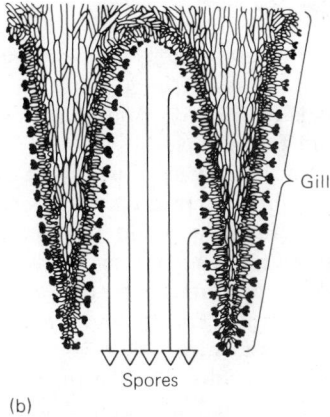

(b)

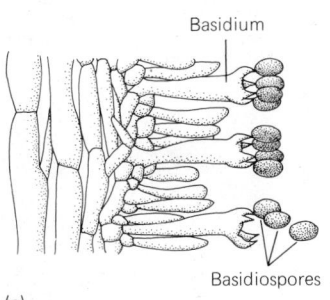

(c)

Figure 24-9 (a) A mushroom is the fruiting body of a basidiomycete. The vegetative, feeding mycelium remains underground. (b) Basidia with basidiospores form all along the edges of the gills. (c) Closeup of basidia and basidiospores on gill. (Adapted from Wilson, C. L., et al., *Botany*, 5th ed. © 1971 Holt, Rinehart and Winston.)

Figure 24-10   A bracket fungus, the fruiting body of a basidiomycete.

Bracket fungi are also fruiting bodies of basidiomycetes; they are often found on rotting wood (Figure 24–10).

Class Deuteromycetes

The class Deuteromycetes contains a rummage-sale collection of fungi that cannot be assigned to any other class because their sexual reproduction (if they have any) has never been observed, and so they cannot be classified by sexual reproductive structure. Because they do not show sexual reproduction, these forms are called "imperfect fungi." The group is undoubtedly polyphyletic (containing forms descended along several different evolutionary lines). The organisms that cause ringworm and athlete's foot are members of this group. Other deuteromycetes cause important diseases of crops, including strawberry leaf blight and bitter rot of grapes.

24-D   Fungi as Decomposers

All fungi are heterotrophic, so they must obtain their food from outside sources. The thousands of species of fungi have adapted to a variety of lifestyles. One of their ecological roles is to break down organic matter: dead animals and animal wastes, dead plants, and fallen leaves.

Fungi are not the only organisms that consume dead organisms. Large animals such as crows, vultures, coyotes, and hyenas consume large quantities of carrion (dead animals); many insects and other invertebrates also feed on dead plants and animals, but fungi and bacteria are the most important decomposers.

Fungi and bacteria are adapted to different conditions, and they share food sources in various ways. For instance, some fungi can grow in relatively high concentrations of salts and sugars that would kill most bacteria. Many fungi also tolerate extreme acidity; acid foods like pickles and jams (fruit is acid) are safe from attack by bacteria but not by fungi. Their ability to absorb water from damp air permits fungi (unlike bacteria) to grow in environments where there is no liquid water. On the other hand, many bacteria tolerate anaerobic conditions better than do fungi. Although fungi such as yeasts can carry on anaerobic fermentation, no fungus can grow and reproduce without oxygen.

Unlike parasitic fungi, which cause many diseases, most saprobic fungi are beneficial to humans, and even those that form molds on food are usually harmless. Some saprobic fungi cause considerable inconvenience, however, by consuming such unlikely substances as leather, hair, wax, cork, and polyvinyl plastics. Although their action in the town dump is convenient, fungi are a nuisance when they break down telephone insulators, clothes, shoes, books, or rafters. During their long years of global supremacy, the Spanish and British navies lost more ships to wood-rotting fungi than to enemy action. It was quite common for the bottom of a ship to fall out in mid-ocean!

24-E Fungal Diseases

Some fungi seem to have evolved from saprobes, living on dead matter, via intermediate forms living on almost-dead matter such as injured tissues, to parasites, which absorb nutrients from living organisms.

Hundreds more fungal diseases are known for plants than for animals. Ordinarily, the skin and mucous membranes provide a healthy animal with considerable natural defense against penetration by fungal hyphae; a sick or injured animal is more susceptible to fungal infection. Many animals also have a natural **fungicide** (fungus-killer) in the form of the fatty acids and their salts, secreted in sweat.

Fungi can easily enter a higher plant via its stomata, pores in the surface of a leaf or stem that permit the exchange of gases between the air and the interior of the plant. Air spaces within the leaves permit the fungus to obtain all the oxygen it needs and still reach the food in the plant's cells. Because plant cells must have access to air for photosynthesis and respiration, a plant contains more air spaces than an animal and so provides a more hospitable environment for fungi. When fungi do invade animals, they attack areas exposed to the air, such as the skin, lungs, and mucous membranes.

Fungal spores that land on the surface of a leaf will germinate and grow into stomata or injured areas (Figure 24–11). Once inside, the hypha grows between the cells packed loosely about the air spaces. The hypha can absorb nutrients from the fluid between cells, or it can invade leaf cells by sending haustoria in next to their cytoplasm. Finally the well-fed mycelium is ready to reproduce. It may grow hyphae out through the stomata, or it may form a clump of sporangia under the outer epidermal layer of cells; these sporangia grow and finally rupture the epidermis of the host, releasing the spores into the air (Figure 24–12).

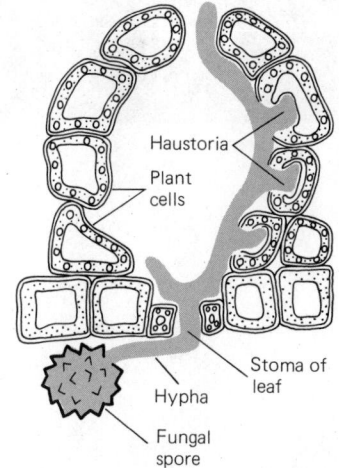

Figure 24-11 Fungi may enter plants through their stomata and grow through interior air spaces. Here, the fungus (color) has grown through a stoma of a leaf and sent haustoria into some of the internal photosynthetic cells.

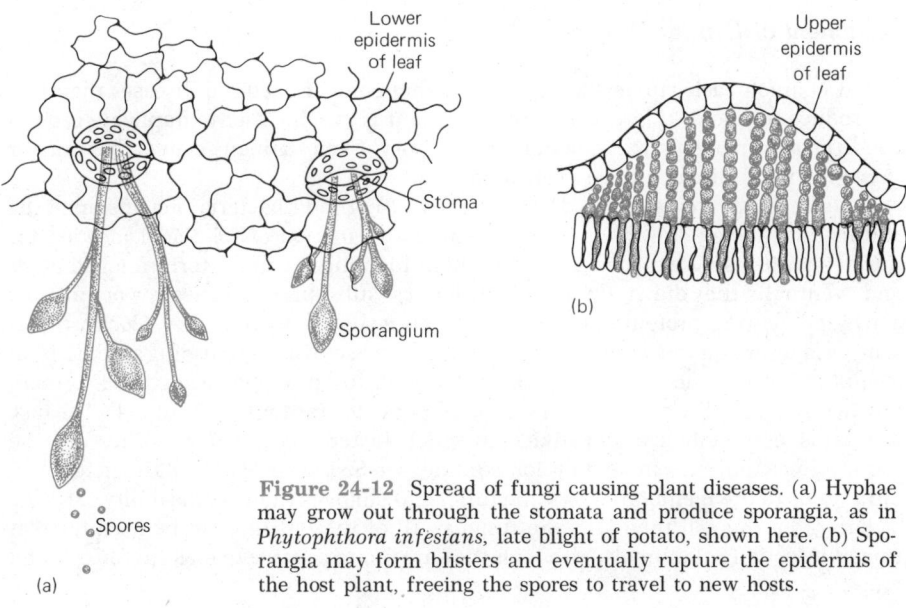

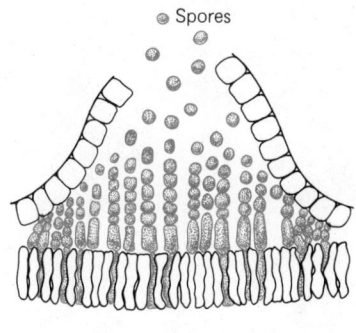

Figure 24-12 Spread of fungi causing plant diseases. (a) Hyphae may grow out through the stomata and produce sporangia, as in *Phytophthora infestans*, late blight of potato, shown here. (b) Sporangia may form blisters and eventually rupture the epidermis of the host plant, freeing the spores to travel to new hosts.

Figure 24-13 Powdery mildew of lilac. This mild disease thrives on lilac leaves without causing much apparent damage to the bush.

One of the most historically important fungal plant parasites is *Phytophthora infestans*, the oomycete that causes late blight in potatoes; it was responsible for the Irish potato famine. The potato, a native of South America, was introduced into Europe in the sixteenth century. Potatoes require little labor and produce high yields of one of the most nutritionally perfect plant foods; by the nineteenth century, they were almost the only crop grown in Ireland. However, **monocultures** (plantings of only one crop) are always particularly susceptible to the rapid spread of disease, and in 1845 and 1846, late blight destroyed essentially the whole Irish potato crop, leading to a devastating famine. Between 1845 and 1851, a million people died in Ireland and a million and a half emigrated, mainly to the United States and Canada. The European potato shortage stimulated the agricultural economy of North America, where grain for export has been produced in increasing quantities ever since.

Fungal diseases of plants are still among the greatest threats to human welfare. Hundreds of millions of dollars are spent each year to protect crops by treatment with fungicides, and to develop new fungicides and new strains of crops resistant to fungal attack.

Treatment of Fungal Diseases

It is much easier to develop chemicals that combat bacterial diseases than it is to produce fungicides. This is mainly because it is much easier to duplicate growth conditions for parasitic bacteria in the laboratory. Even so, many agents that cure or prevent fungal diseases have been produced.

Many drugs are useful in destroying both fungi and bacteria. For instance the sulfonamides (e.g., sulfanilamide) prevent the formation of folic acid, needed for purine synthesis. Organisms treated with sulfonamides cannot form nucleic acids and eventually they die. A number of antibiotics (substances that destroy organisms) act by preventing protein synthesis. Some are specific, in that they block protein synthesis in prokaryotes but not in eukaryotes; these drugs are useful antibacterial agents, but are of no use against the eukaryotic fungi. Other drugs block protein synthesis in all cells, but can be used to kill parasites rather than host cells because the parasites metabolize and divide so much faster. Among these drugs are the tetracyclines, puromycin, and cycloheximide (see Section 10-G). The fungicide nystatin binds to the membranes of some fungi and changes their permeability; leakage of ions from the cell then kills the fungus. All of these drugs can be used to treat fungal infections of animals, although they are too expensive to be used for treating crops.

Besides these antibiotics, many more general antifungal agents are known. Sodium propionate is a fatty acid salt used by bakeries to inhibit growth of mold in bread. It is also used to treat fungal infections of the skin, as are sulfur, benzoic acid, salicylic acid (aspirin), gentian violet, and potassium permanganate.

Prevention of Fungal Diseases

Some fungal diseases of animals can be controlled by prolonged treatment with proper medication. However, once a fungus becomes established in a plant or an animal, or in wood, paper, or leather goods around the house, it is virtually impossible to eradicate: measures that prevent spores from germinating are the most effective defenses against fungi.

The discovery of chemical sprays that kill fungal spores in crops was a milestone in the battle between humans and fungi for food. In the 1860s most of the grape vines in France were torn up and grafted onto rootstocks of American vines to protect them from root-feeding North American aphids,* which had almost destroyed the French wine industry when they were accidentally introduced into that country. Unfortunately, the American vines also imported *Plasmopara viticola*, downy mildew of grapes. This disease almost wiped out the French grape harvest in wet years during the 1870s.

A preventive measure against downy mildew was discovered in 1882 by Pierre Millardet of the University of Bordeaux. A local vineyard owner was spraying his grapes with a vile-looking mixture of copper sulfate and lime to discourage passersby from eating the grapes; Millardet noticed that these vines were free of mildew. He made up a mixture of copper hydroxide and lime, which he called **Bordeaux mixture**, the first and one of the most successful fungicides ever produced by human ingenuity. Germinating fungal spores produce just enough acid to dissolve the very insoluble copper hydroxide. The copper kills the spore, and the host plant is not affected by the remaining undissolved copper. Bordeaux mixture is also effective against germinating spores of late blight of potatoes and many other fungi that parasitize plants.

---

* Aphids Small soft-bodied insects that suck the juices of plants.

Figure 24-14 An American elm killed by Dutch elm disease, caused by a fungus. The disease is steadily spreading and eradicating virtually every American elm in its path. Attempts at control are aimed at killing the bark beetle that spreads the disease. The healthy-looking American elm in the background had to be felled within a year after this shot was taken.

Symbiosis ("together living") is an intimate relationship between members of two different species. If the relationship is thought to be beneficial to both species, it is called **mutualism**, as in the case of lichens (discussed below) or of the bacteria that break down cellulose in the stomach of a cow. If the relationship is beneficial to one species and has no known influence one way or the other on the second species, it is called **commensalism**. **Parasitism** is a symbiosis in which one species benefits and the other is harmed. In fact, these subdivisions of symbiosis are seldom useful because they are defined in terms of harm and benefit, which are usually difficult to determine. For instance, even a parasite may benefit its host by making it resistant to attack by a second parasite, and it is impossible to prove that a relationship has no effect on one species, as the definition of commensalism requires.

Mycorrhizae

A number of fungal species, particularly some common forest mushrooms (basidiomycetes), may exist as free-living forms or in a symbiosis with a plant root; such an association is called a **mycorrhiza** ("fungus root"). The fungus benefits the plant by absorbing minerals from the soil and passing them on to the plant. When pine trees were introduced to Australia and Puerto Rico, they grew very poorly until supplied with soil containing the appropriate mycorrhizal fungi, after which they grew rapidly. It is now known that plants of most families must form mycorrhizae with the proper fungi for best growth. It is not yet clear what benefits the fungus derives from the relationship.

Lichens

A **lichen** is composed of an ascomycete or basidiomycete fungus and photosynthetic cyanobacteria (Chapter 22) or green algae (Chapter 25), living symbiotically. The photosynthetic members of lichens are sometimes found living by themselves, but the fungi are usually not found growing alone in the wild, although they can be grown alone in the laboratory. The fungus in a lichen obtains organic compounds from the alga, but it is unclear how the alga benefits from the association. Possibly the alga obtains water and minerals from the fungus. The association between fungus and alga certainly extends the area in which each species can grow; a lichen may be found in areas where neither its fungus nor its alga can grow alone.

A lichen lives an unharried life with a very low metabolic rate and slow rate of growth. Lichens are extremely resistant to drought and cold. They are the most important autotrophs of areas in the **tundra**, the vegetation belt that stretches around the north pole and that also occurs at high altitudes in mountainous regions (see Chapter 48). The plants of the tundra are usually sparse and stunted, but lichens can survive in this cold, windswept environment where there is hardly any soil. Lichens absorb minerals from the air and so can grow on tombstones and rocky islands in the ocean where nothing else can survive. Scientists have studied the size and growth rate of lichens found on the mysterious stone heads of Easter Island to determine how long the heads have been there.

Lichens are particularly sensitive to air pollution. The death of lichens in an area invariably indicates an increase in air pollution. Conversely, the return of lichens to an area is a sign that a pollution-control program is working.

Most lichens reproduce by **fragmentation**, in which small pieces break off and blow away. In some, the fungus forms spores that blow away and find the appropriate alga only by luck.

Lichens are important in many ways. They are usually the first organisms to colonize a bare rock, and they probably contribute, although very slowly, to breaking down rock into soil. They are also the main food of most tundra animals.

Many lichens have crystalline deposits of unusual organic acids on their surfaces. Their function to the lichens is unknown, but it has been proposed that they

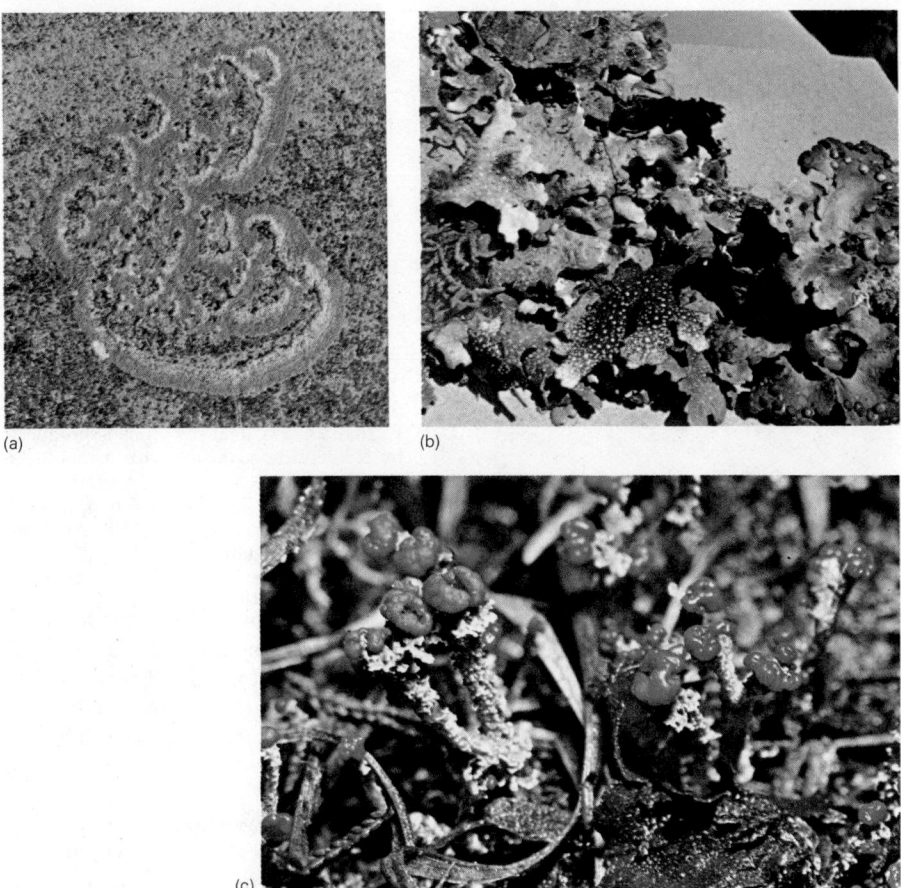

(a)

(b)

(c)

Figure 24-15 Three main body forms of lichens. (a) Crustose forms resemble blotches of thick paint. (b) Foliose forms resemble leaves. (c) A fruticose, or shrubby, form called "British soldiers." (b,c Biophoto Associates, N.H.P.A.)

protect the cells from high light intensities. Many of these chemicals have long been prized as beautiful dyes. Harris tweed is produced with lichen dyes, as are fabrics from the Orkney and Shetland Islands north of Scotland.

24-G   Fungi for Food

Human beings have long known that fermented foods usually keep better than the food from which they are made. This was a compelling motive for producing wine, beer, cheeses, sauerkraut, fermented sausages, and yogurt before refrigerators were invented. Fermented foods are also often more nutritious, flavorful, and digestible than their raw materials.

Beer is made by fermenting germinated grain (malt), usually barley, flavored with hops. During germination, the plant embryo breaks down its starchy food supply to monosaccharides, which are then fermented by yeast.

The yeast *Saccharomyces cerevisiae* makes alcohol in the production of wine and beer. A different strain of the same yeast is used in bread-making. Here, *Saccharomyces* does not produce alcohol because it grows under aerobic conditions, but its respiration gives off carbon dioxide, which is trapped in the dough as bubbles that give a light texture to the bread.

English Stilton and French Roquefort, Brie, and Camembert are all cheeses that get their special flavors from specific ascomycetes introduced as part of the production process.

Figure 24-16 Real soy sauce is produced by fermenting soybeans with a mold, *Aspergillus oryzae*, shown completely covering the beans. The beans are packed in huge earthenware crocks with 20% salt brine. (Keith Steinkraus)

Although most modern soy sauce (*shoyu*) is produced by hydrolyzing soybeans with hydrochloric acid, the best *shoyu* is still made by fermenting boiled soybeans and wheat with the ascomycete *Aspergillus oryzae* for about a year (Figure 24–16). The Chinese invented *shoyu* thousands of years ago. It added flavor and vital amino acids, produced by the fungus and by bacteria, to a low-protein diet of rice. The use of fungi to increase the nutrient content of foods has been very important in many countries.

Some fungi that produce diseases are the same as those used in food production. An interesting example is *Botrytis cinerea*, which causes spear rot in asparagus but is greatly valued in some wine-producing areas. Grapes infected with this "noble rot" ferment to produce Sauternes and Barsac in France, and Beerenauslesen in Germany and in the eastern United States.

Growing the grocery store mushroom, a basidiomycete, is a million-dollar industry in many parts of the world. The morels and edible truffles are ascomycetes (Figure 24–17). In France, truffle hounds and pigs are trained to hunt the underground truffles by smell; scientists discovered how to "farm" these valuable fungi (selling at $400 per pound in New York in 1981) only in the late 1970s, and it may be several years before the crop shows good yields, because the fungi are grown as mycorrhizae on the roots of tree seedlings.

Figure 24-17 A morel, an edible fruiting body of an ascomycete. (Biophoto Associates, N.H.P.A.)

(a)

(b)

Figure 24-18 Poisonous fungi. (a) An ergot in a head of rye. (b) *Amanita muscaria* or fly agaric. (a, Carolina Biological Supply Company; b, Biophoto Associates, N.H.P.A.)

24-H Chemical Defenses of Fungi

The ascomycete *Claviceps purpurea* infects the flowers of rye and other cereals and produces a structure called an **ergot** where a seed would normally be found in the head of grain (Figure 24–18). The ergot produces numerous extremely toxic substances. Humans may be poisoned by ergots when they eat bread made from infected rye, and ergotism, also called St. Anthony's Fire, caused the death of thousands in medieval Europe. Ergot also supplied the chemicals from which lysergic acid diethylamide (LSD) was first synthesized.

The toxins produced by the *Claviceps* ergot, and the notorious poisons of some mushrooms, protect these fungi from predators and parasites. Every year dozens of people die from eating poisonous mushrooms. In Europe and North America, the usual culprits are some members of the genus *Amanita*. Most professional **mycologists** (people who study fungi) consider it so hard to distinguish edible from poisonous mushrooms that they will never eat a mushroom collected from anywhere but a grocery store. Still, amateur mushroom hunters continue to collect and eat mushrooms.

More than thirty years ago, Alexander Fleming noticed that his bacterial cultures had been killed by the ascomycete *Penicillium notatum*. Further study of this phenomenon resulted in the discovery of the world's most widely used and effective antibiotic, **penicillin**. The antibiotic presumably serves the fungus that produces it by reducing competition from bacteria; humans find it very useful for the same purpose. Various fungi are also used to synthesize vitamins, amino acids, enzymes, sterols (precursors of steroid hormones), and other organic compounds on a commercial scale. Even the substances produced by ergots are valuable drugs. Although they are deadly in high concentrations, small quantities are used to induce labor, control bleeding, ease migraine headache, and treat high blood pressure and varicose veins.

24-I Division Myxomycophyta (Slime Molds)

Whereas the Eumycophyta, or true fungi, almost always have rigid cell walls, members of the Myxomycophyta spend part of their lives as mobile, amoeba-like organisms with soft cell walls that make them look more like animals than fungi. At this stage in the life history they also engulf organic material and bacteria, which we think of as characteristic of animal and protist cells, rather than absorbing food as the true fungi do.

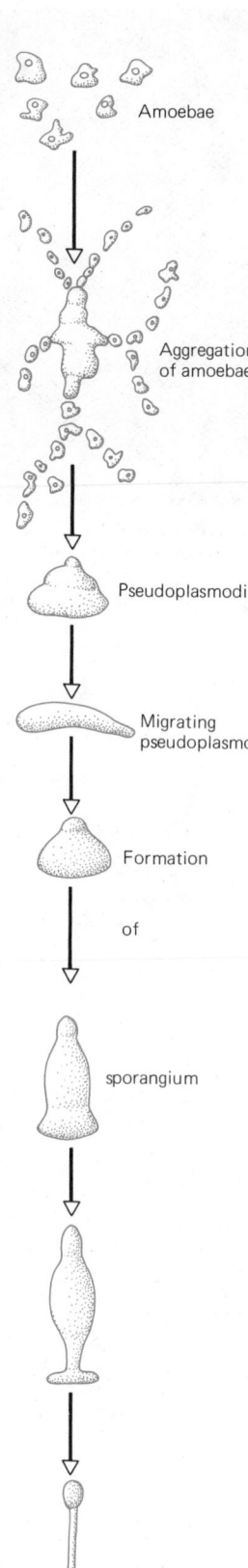

Amoebae

Aggregation
of amoebae

Pseudoplasmodium

Migrating
pseudoplasmodium

Formation

of

sporangium

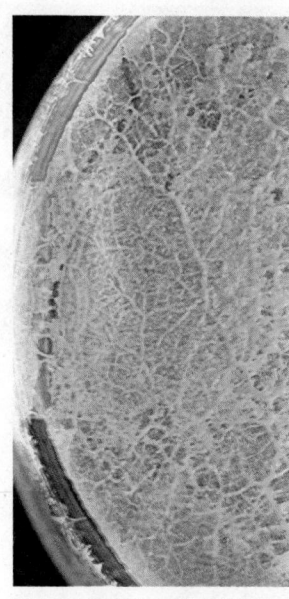

Figure 24-19 The plasmodium of the acellular slime mold *Physarum* is coenocytic, like a giant amoeba with many nuclei.

Figure 24-20 Developing fruiting bodies of *Arcyria ferruginea*, a myxomycete. (Biophoto Associates)

The amoeba-like feeding stage of an acellular slime mold (class Myxomycetes) is a coenocytic structure called a **plasmodium** (Figure 24–19). It usually forms by the fusion of many small plasmodia. The plasmodium streams around in soil, wood, dung, or decayed vegetation, engulfing bacteria or particles of food. It bears little resemblance to a fungus until conditions become unfavorable for feeding and growth. Then the plasmodium turns into a fungus-like fruiting body with cell walls and produces spores sexually (Figure 24–20).

In their feeding stage, cellular slime molds (class Acrasiomycetes) exist as uninucleate **myxamoebas**, almost indistinguishable from true amoebas. When their food supply is exhausted, some of the myxamoebas secrete a nucleotide, cyclic AMP, which attracts other myxamoebas. The myxamoebas crawl along the cyclic AMP concentration gradient and form a slug-like clump, or **pseudoplasmodium**. The pseudoplasmodium may crawl around for a while; then it stops and forms a fruiting body containing spores with cellulose cell walls (Figure 24–21). After they are released, the spores germinate and form new myxamoebas. Sexual reproduction has never been reported in the cellular slime molds.

Figure 24-21 Aggregation and formation of sporangium in a cellular slime mold, *Dictyostelium discoideum*.

SUMMARY

Members of the kingdom Fungi are eukaryotic saprobes and parasites that obtain their nutrients by absorption and have a cell wall in at least some stages of the life history. Spores are involved in both sexual and asexual reproduction.

On the basis of sexual reproductive structures, the division Eumycophyta can be divided into four main classes: Oomycetes, Zygomycetes, Ascomycetes, and Basidiomycetes. A fifth class, Deuteromycetes, consists of all the true fungi for which no sexual reproduction is known.

The most important ecological role of fungi is as decomposer organisms, which break down dead plants and animals and absorb the resulting small food molecules. Inevitably, however, some of these molecules escape the fungi, and minerals released by fungal breakdown are important to green plants, which absorb and reuse these vital nutrients.

Fungi are also important as parasites, causing diseases of both plants and animals. Most fungal diseases attack plants. Once established, a fungal disease of a plant is often impossible to combat, and fungal diseases of animals can be cured only with much difficulty. Thus prevention of infection is important in combating fungal diseases.

Lichens are symbiotic associations between fungi and green algae or cyanobacteria. Lichens are important food producers in cold or barren areas; they may also help to form soil from bare rock. Other fungi form associations known as mycorrhizae with the roots of higher plants and benefit these plants by increasing their supply of minerals.

Many fungi produce toxic compounds that serve to deter predators or reduce competition from bacteria or other fungi.

Fungi are economically harmful because they destroy seed, standing crops, and harvested food, as well as clothing, homes, and other possessions. Fungi also cause disease and death to human beings, livestock, and pets. On the positive side of the ledger, fungi are useful in the production of many edible fruiting bodies, of fermented foods, and of drugs, antibiotics, and various organic chemicals.

The Myxomycophyta are the slime molds, which live either as coenocytic plasmodia or as unicellular amoebas. They feed by engulfing food particles, and they lack rigid cell walls. These characteristics suggest affinities with the protozoan amoebas; slime molds are classified as fungi because they form fruiting bodies with walled spores, a habit strongly suggesting fungal alliances.

OBJECTIVES

From your study of this chapter, you should be able to:

1. Describe why the activity of fungi as decomposers is vital to life on earth.

2. Describe the major characteristics of fungi, and tell how fungi differ from plants and animals.

3. Define the terms saprobe, parasite, hypha, mycelium, dikaryon, haustorium, spore, sporangium, ascus, basidium, fruiting body, symbiosis.

4. Describe both sexual and asexual life histories of a fungus such as *Rhizopus*, the black bread mold.

5. List the divisions and classes of the fungi with the distinguishing characters of each.

6. Explain why there are many more fungal diseases of plants than of animals.

7. List or recognize ways in which fungi are economically beneficial and ways in which they are economically harmful.

8. List at least three ways in which fungi may spread.

9. Discuss methods of controlling fungi.

10. State what lichens and mycorrhizae are, and explain their ecological roles.

SELF-QUIZ

1. Commercial mushrooms are grown in soil enriched with horse manure. These mushrooms are:
   a. autotrophic
   b. parasitic
   c. saprobic
   d. chemosynthetic

2. The organism below that would have haustoria is:
   a. *Phytophthora infestans*, which causes late blight of potatoes
   b. black bread mold (*Rhizopus*)
   c. yeast
   d. commercial mushroom

3. The taxonomy of the true fungi is based on:
   a. life history
   b. sexual reproductive structures
   c. mode of nutrition
   d. complexity of vegetative structures

4. Which is *not* a means of limiting the growth of fungi?
   a. spraying with compounds containing sulfur
   b. refrigeration
   c. dehydration
   d. scrupulous cleanliness
   e. humidification

5. Fungal pathogens invade a living host by:
   a. digesting away the epidermal layer with their powerful enzymes
   b. growing in through a break in the epidermis
   c. secreting hormone-like substances that cause the host's cells to accept them as part of the body
   d. growing hyphae under a cell of the host and prying it up as a lever would

6. The bracket fungi found on trees are:
   a. fruiting bodies of mycelia growing hidden in the tree trunk
   b. mycelia absorbing nutrients from the exposed surface of the wood
   c. sporangia
   d. lichens

7. In the mycorrhizal association between a pine tree and a fungus:
   a. the fungus eventually depletes the tree's mineral supply
   b. the fungus secretes toxic materials that inhibit the growth of nearby trees
   c. the fungus absorbs nutrients from the soil
   d. the fungus converts nitrogen into a form the tree can use

8. Fungi are probably most widely spread by:
   a. airborne spores
   b. ingestion and subsequent deposition in the feces of animals
   c. migration of insects with spores or bits of hyphae stuck to them
   d. fragmentation of vegetative mycelia
   e. water currents

9. A fungus with cellulose cell walls, coenocytic hyphae, and unwalled, flagellated spores belongs to the class:
   a. Oomycetes
   b. Zygomycetes
   c. Ascomycetes
   d. Basidiomycetes
   e. Deuteromycetes
   f. Acrasiomycetes

QUESTIONS FOR DISCUSSION

1. How is the anatomy of fungi related to their way of life?

2. In what ways is a dikaryon similar to a diploid organism? How do the two differ?

3. How can disasters such as the Irish potato famine be prevented?

4. When plants are moved from their original habitat into a new one, a disease that was a minor nuisance in the home country may become a major disaster. This was the case with the late blight of potatoes, downy mildew on grapes, and white pine blister rust. What possible reasons can you think of for this?

5. Why do you think monocultures of an agricultural crop are more susceptible to disease than mixed plantings?

6. Fresh water heavily polluted with industrial wastes contains few fungi, compared to unpolluted waters. How might this affect the life in a lake or stream?

7. Is it safe to conclude that *Penicillium* secretions aid the fungus by reducing competition from bacteria in nature because they do so in the laboratory? What experiments could you do to investigate this question?

REFERENCES AND FURTHER READING

Alexander, M. *Microbial Ecology*. New York: John Wiley and Sons, 1971. Deals with roles of microbes in a variety of habitats.

Alexopoulous, C. J. *Introductory Mycology*, 2d ed. New York: John Wiley and Sons, 1962. A clear and thorough textbook on fungi and their allies.

Hansen, J. "Let them eat truffles." *Science 80*, December 1980. About the modern truffle industry.

Jensen, W. A., and F. B. Salisbury. *Botany: An Ecological Approach*. Belmont, CA: Wadsworth, 1972. The chapters on fungi show how intriguing these organisms can be.

CHAPTER 25

LOWER PLANTS

The plant kingdom consists of multicellular photosynthetic organisms, along with those nonphotosynthetic organisms and unicellular organisms that are clearly close relatives of the photosynthetic multicellular forms. Since most plants are photosynthetic and can make their own food molecules from inorganic substances, they are more self-reliant than animals, and many selective pressures acting on plants are different from those affecting animals. The photosynthetic way of life places a high selective premium on the ability to obtain light. The multicellularity of plants allowed the evolution of specialized nonphotosynthetic structures that hold the photosynthetic structures in positions where they receive a reliable supply of sunlight. In addition, there is often a premium on rapid growth; reaching a large size quickly assures a plant of its place in the sun, and shades out its competitors.

Plant cells have stiff cell walls, which provide some protection against herbivorous animals and prevent the plant cell from absorbing so much water that it bursts. Most multicellular plants cannot move about; they are too large to swim with flagella, and stiff cell walls are incompatible with the evolution of muscle tissue.

375

The fact that plants move very little makes it doubly important for a plant to grow where it can obtain sufficient sunlight. Many plants have evolved mechanisms that increase the likelihood that their offspring start life in favorable habitats. Nonmotility also means that plants must have special adaptations if they are to undergo sexual reproduction. Since plants themselves cannot come together to mate, other mechanisms must bring their gametes together. Many plants have only asexual, or vegetative, reproduction and do not encounter this problem (see Section 46-I).

The possession of photosynthetic machinery and cell walls, the lack of motility, and the problems of obtaining light and of sexual reproduction are common to most members of the plant kingdom. During the course of evolution, plants have produced adaptations that now permit them to exploit just about every habitat with sufficient light, moisture, and mineral salts to support the growth of photosynthetic organisms.

Most people are more familiar with land plants than with seaweeds, the multicellular algae that float in quiet waters or cling to rocks and coral reefs in coastal waters or shallow seas. In recent years, however, the surge of interest in the sea and its resources has stimulated new work by **phycologists**, people who study algae.

In this chapter we shall examine the three groups of multicellular algae and consider some algal life histories.

25-A The Multicellular Algae

Algae are photosynthetic organisms with relatively unspecialized bodies. The groups of algae that have at least some multicellular members are placed in three **divisions** (the equivalent of phyla in the animal kingdom) of the plant kingdom: division Rhodophyta (red algae), division Phaeophyta (brown algae), and division Chlorophyta (green algae).

The three groups of multicellular algae are believed to have arisen from three different groups of unicellular ancestors (Figure 25–1). Two of the three divisions, the red and brown algae, are almost exclusively marine, but contain a few freshwater forms. The third group, the green algae, is well represented in both marine and freshwater environments, and also contains forms that live in moist terrestrial (land) habitats. This group is believed to have given rise to the land plants, which we shall study in the next chapter. The division Chlorophyta embraces many more different species than either of the other divisions, reflecting the adaptation of the green algae to a wide variety of habitats. The red and brown algae, the two lines that have remained almost exclusively in the sea, include species that have attained great size and complexity.

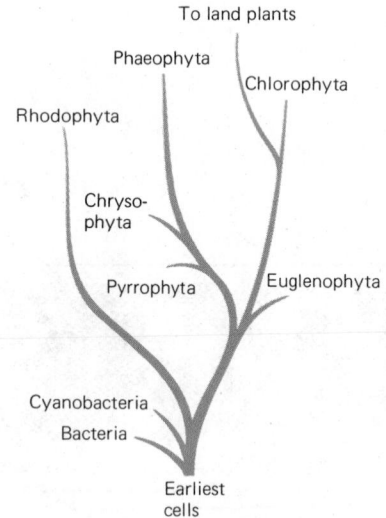

Figure 25-1 A possible evolutionary scheme for the relationships of the multicellular algae and other photosynthetic organisms.

25-B Criteria for Classification

The multicellular algae are classified according to a number of criteria. Morphology was the criterion used by Linnaeus. However, each group of multicellular algae exhibits a wide range of morphology, and biochemical criteria have proved to be more useful in classifying algae. Like the cyanobacteria and the photosynthetic protistans, all multicellular algae contain chlorophyll *a*, but the other forms of chlorophyll (*b*, *c*, *d*) and the other types of pigments present are used to divide the algae into their main groups. Indeed, the names of the algal divisions are based on the characteristic colors of their members. Not all green algae are green, or red algae red, but the majority of members in each group do exhibit the color of their division name, because a particular accessory pigment predominates in each group. These accessory pigments capture light of various wavelengths not absorbed by chlorophyll *a*, and pass this trapped light energy to chlorophyll *a*, which uses it in photosynthesis.

Plants store various organic compounds that they manufacture during photosynthesis. The form in which food is stored (oils, starch, etc.) is another criterion used in algal classification.

TABLE 25-1 DIVISIONS OF MULTICELLULAR ALGAE

Division Rhodophyta (~4000 species)	Red algae. Chlorophyll *a*, phycobilins, carotenoids; store modified starch; chloroplast membranes not stacked; no flagellated cells in life history; unicellular to multicellular forms; most marine, some freshwater, a few terrestrial. E.g., *Chondrus crispus* (Irish moss), *Polysiphonia, Porphyridium.*
Division Phaeophyta (~1500 species)	Brown algae. Chlorophylls *a* and *c*, carotenoids; store laminarin; all multicellular, some very large and elaborate; almost all marine. E.g., *Fucus, Ascophyllum, Nereocystis, Laminaria.*
Division Chlorophyta (~7000 species)	Green algae. Chlorophylls *a* and *b*, carotenoids; store starch; unicellular to multicellular but not very large or elaborate; mostly freshwater, many marine, some terrestrial. E.g., *Chlorella, Chlamydomonas, Spirogyra, Ulothrix, Oedogonium, Ulva.*

The type of chlorophyll, the accessory pigments present, and the form of food storage molecules do not change rapidly during evolution. These molecules are intimately connected with the very existence of the plant, and even a small change in the structure of a molecule may affect its function so drastically that the plant cannot survive. Such biochemical characteristics are said to be **conservative**, since evolution has not altered them rapidly.

Characteristics such as morphology and reproductive patterns tend to be more **plastic**, or able to change without harm to the plant, and variations in structure and reproduction have been selected for when they confer an adaptive advantage on plants living in different habitats. These characteristics are often not good indications of evolutionary relationships. Nevertheless, they were used by early botanists to classify the multicellular algae, for lack of better criteria.

Modern techniques have led to the reshuffling of some algae into groups to which they had not previously been assigned. Many that were once classified as separate organisms are now known to be different forms of the same species, which grow into very different-looking plants depending on the chemistry, depth, temperature, or movement of the water in which they live.

25-C Division Rhodophyta

The division Rhodophyta, or red algae, contains single-celled forms as well as forms that grow as filaments, branching structures, and broad flat plates or ruffles (Figure 25-2). Some of the more complex red algae reach sizes of up to a metre long,

(a)

(b)

Figure 25-2 Members of the division Rhodophyta. (a) *Odonthalia* grows on submerged rocks; (b) *Plumaria* grows on rocks in the intertidal zone. (Biophoto Associates)

but most are small and delicate. All red algae live attached to some surface: a rock, coral reef, animal shell, or even a larger alga. Larger forms attach to the surface by means of a specialized part of the body called a **holdfast**; other forms cement themselves to the surface and spread over it to form a crust.

The red algae are believed by some to have descended from the blue-green bacteria via an evolutionary branch that diverged quite early (see Figure 25–1). One reason for this belief is similarity of pigments: both groups probably use chlorophyll *a* as their only chlorophyll, and in addition have two types of phycobilin accessory pigments: phycocyanin and phycoerythrin. But whereas the blue phycocyanin predominates in blue-greens (giving them a blue-green hue from the combination of phycocyanin with chlorophyll), phycoerythrin predominates in the red algae, masking the green chlorophyll, and giving most members of this group their characteristic red (pink to red-black) appearance. The function of phycobilins has been debated but it has now been shown that light energy absorbed by phycobilins is passed on to chlorophyll and used in photosynthesis. Like all other organisms that carry on oxygen-producing photosynthesis, red algae also contain carotenoid accessory pigments.

The cell structure of red algae also suggests affinities with the blue-greens: in the chloroplasts of red algae, the photosynthetic membranes are not arranged in stacks as they are in the chloroplasts of other multicellular plants; their arrangement is reminiscent of the arrangement of membranes in the cells of blue-greens. Furthermore, like the blue-greens, the red algae do not have any flagellated cells, not even flagellated sperm cells, which are found in all other groups of algae and in some land plants. There is no evidence that red algae have ever had flagella, unlike higher land plants, which have clearly lost the ability to form flagella during the course of evolution.

The red algae include more marine species than the brown or green algae, although there are more total species of green algae. Red algae may be found far up the shore in the splash zone, where the ground is never covered by water even at high tide. At the opposite extreme, red algae occur at depths of almost 100 metres in clear seas, and are probably the only organisms able to carry on photosynthesis at such depths.

Red algae are prominent on tropical coral reefs, where they play an important part in reef-building. Until recently, the construction of reefs was attributed solely to coral animals, which secrete calcium carbonate (lime) tubes around themselves (see Section 27-C). However, many algae, especially red algae, extract calcium carbonate from the sea and use it to attach themselves firmly to their substrate, at the same time contributing to the structure of the reef. Reef-building corals do not occur below about 70 metres. This may be due partly to the fact that algae, which cement the reef and attract small animals that the corals prey upon, are absent below this depth.

25-D Division Phaeophyta

The division Phaeophyta, or brown algae, contains the biggest and most complex algae. This group contains no unicellular or colonial forms; its members range in size from microscopic filaments to the Pacific kelps that may reach lengths of 70 metres.

Some scientists believe that the Phaeophyta are related to the unicellular Chrysophyta (Section 23-F), because the two groups show biochemical similarities. Both contain chlorophyll *c* as well as the ubiquitous chlorophyll *a*, and both have similar carotenoid pigments such as fucoxanthin. There are no phycobilin pigments in either group. Phaeophyta store their food reserves as **laminarin**, while the storage product of chrysophytes is a similar compound called **modified laminarin**.

Brown algae are especially noticeable in cool, shallow waters along the seacoast in temperate and subpolar areas (see Figure 25–4). Most forms grow attached to a solid substrate. There are a few small freshwater forms.

The larger members of the Phaeophyta are interesting because of their complex tissue differentiation and their adaptations to life along the coast, where they may

Reproductive structures

Air bladder

Blade

Stipe

Holdfast

Figure 25-3 The brown alga *Fucus* shows the considerable differentiation of parts found in many of the large brown algae.

THE DIVERSITY OF ORGANISMS

be pounded by the surf and then left high and dry when the tide goes out. The brown alga *Fucus* is a common and familiar seaweed of the intertidal zone* (Figure 25–3). This plant reaches lengths of a foot or more. One end of the body is specialized as a holdfast, which anchors the plant to the substrate. When the tide is in, the rest of the plant floats near the surface of the water, close to the light. The free ends of the plant are buoyed up by **air bladders**, filled with gas. Numerous swellings on these free ends contain chambers where the reproductive cells are formed. *Fucus* is also covered with a gelatinous material, which reduces evaporation and keeps the plant from drying completely when it is out of the water at low tide. A dark pigment shields the living tissue from bright sunlight until the tide returns.

The *Fucus* plant body shows an appreciable differentiation of tissue. The outside of the plant is encased in a layer of tightly packed cells. These secrete the protective coating and also contain chloroplasts, as is appropriate, since these are the cells that receive the most light. Inside this layer is a layer of larger, more loosely spaced cells with fewer chloroplasts. In the center of the plant body is a column of tough filaments that hold the plant together. Strands of mucilage in this region hold water, and release it gradually to replenish water lost from the outer layers when the plant is exposed to air.

As the plant grows, the tips get longer and longer, and the soft outer layers of the lower part of the plant may be worn away. The tough center fibers remain and form the **stipe**, or stalk, which connects the expanded, photosynthetic **blade** to the holdfast. The cells of the stipe transport food produced by the blade to the holdfast, which is usually deeper in the water and shaded so that it cannot carry on photosynthesis to feed itself. Some of the larger kelps, such as *Laminaria* (Figure 25–5b), actually have cells resembling the phloem tissue that transports food from the leaves to the roots of land plants. The brown algae and the land plants are on separate evolutionary lines, and their possession of similar food-transporting tissues is an example of **convergent evolution**, that is, evolution of similar adaptations in organisms that are not related but are exposed to similar selective pressures.

Laminaria and *Nereocystis* are kelps that are much larger than *Fucus*; thus they can live anchored in deeper water and still have parts that float at the surface, near the sunlight. Their holdfasts keep the kelps in position so that they are not washed ashore nor swept out to sea, where they might sink too far down to carry on photosynthesis.

---

*Intertidal zone   Area of shore that is covered by water at high tide but exposed to the air at low tide.

(a)

(b)

Figure 25-4 Members of the division Phaeophyta. (a) A form that grows as a mass on rocks or other algae. (b) *Fucus* plants with swollen reproductive structures at their tips fill most of this picture. Note the air bladders near the forks in the plants at the top right. The more slender plants at the bottom left are *Ascophyllum*, another brown alga. (Biophoto Associates)

(a)

(b)

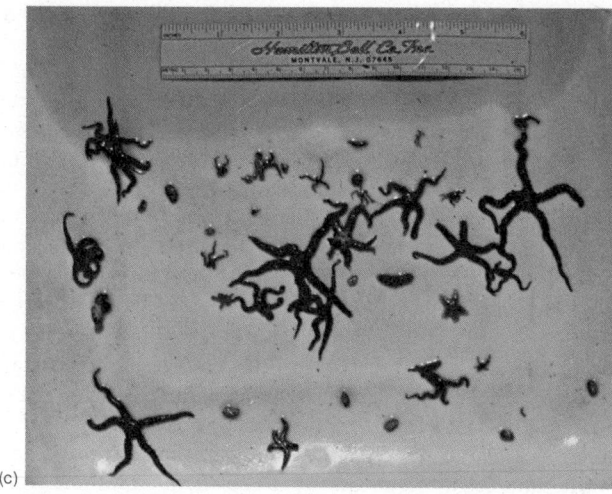
(c)

Figure 25-5 Some adaptations of brown algae. (a) Air bladders provide buoyancy and hold the photosynthetic blades up near the surface of the water where more light is available. What might be the adaptive advantage of the spiny structures lining the edges of the blades? (b) The holdfast anchors the alga firmly to the rocks and prevents its being washed ashore or swept out to sea. (c) These animals were all found sheltering in the holdfast of the plant pictured in part (b). (a, Steven Webster)

The brown algae provide a retreat for many small bottom-dwelling animals (Figure 25–5c). They may also provide a surface for the attachment of other algae that grow as epiphytes—plants attached to the surface of larger plants but not harming them.

25-E   Importance of Seaweeds

Multicellular algae provide food for herbivores such as snails and sea urchins, which may consume a large proportion of the algal biomass (the total amount of material present in living bodies).

The brown algae include many forms that are economically important. Farmers living near seashores have long used certain species of brown algae as fertilizer for

Figure 25-6 Nori (*Porphyra*), a red alga, growing on nets at a Japanese seaweed farm. The nets are positioned to expose the alga at low tide, as in its natural intertidal habitat. The dried algae are a common food in the orient. (Biophoto Associates)

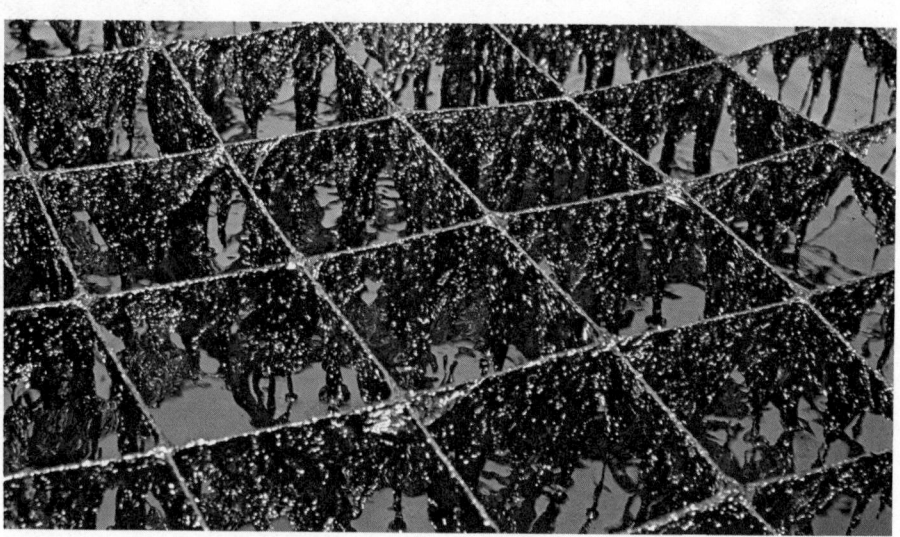

crops and as fodder supplements for livestock. Algae tend to accumulate the important nutrients nitrogen, potassium, and iodine in great quantities. Recently, some developing countries have founded industries to dry algae as food for livestock. Adding small quantities of dried algae to the fodder of cattle and poultry has yielded dramatic improvements in fertility and disease resistance. Some forms of brown and red algae are also important food plants for humans in the Orient and along the coasts of Europe. Some kelps are processed to extract a component of their cell walls called **alginate**. This substance is an ingredient in about half the ice cream produced in the United States; it gives ice cream a smooth texture and helps prevent the formation of ice crystals. Alginate is also used in various drugs, cosmetics, and dental impression materials.

The red alga known as Irish moss produces **carrageenan**, a substance used in puddings, candies, and ice cream. **Agar**, also an extract of a red alga, is used as a base for nutrient media in the laboratory culture of microorganisms.

25-F Division Chlorophyta

The Chlorophyta, or green algae, are believed to be on the main line of evolution to the land plants. Like the land plants, they contain chlorophyll *b* as well as chlorophyll *a*. They also contain many carotenoids, and lack phycobilins. The division Chlorophyta is the largest algal group in number of known species; these species have great diversity of forms and occur in a variety of habitats. Most are quite small and relatively simple compared with the larger brown algae; there is no tissue differentiation, although there may be cells specialized as holdfasts or for reproduction. Some of the Chlorophyta are single-celled; others are simple or branched filaments of cells, with or without holdfasts; some form hollow balls of cells, while some are broad, flat sheets (Figure 25–7). Many species of green algae are coenocytic, with many nuclei inside the same membrane.

Marine intertidal and shallow-water habitats are populated largely by members of the Rhodophyta and Phaeophyta, but some green algae grow in these areas too. Some species of green algae secrete a calcareous cement that helps them attach to sandy shores, where the particles shift so readily that most other types of algae cannot establish their holdfasts. Some species of green algae also contribute limy material to coral reefs and lagoons. Most Chlorophyta, however, are found in fresh water, both still and running, and some live on moist rocks, soil, and tree trunks in terrestrial habitats.

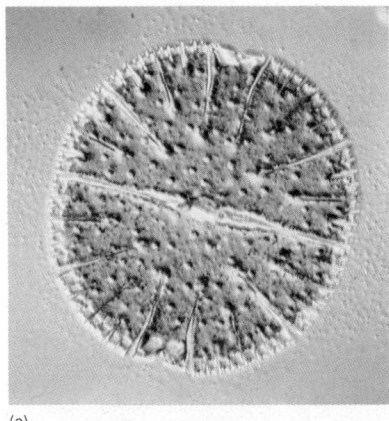
(a)

(b)

(c)

Figure 25-7 Members of the division Chlorophyta. (a) A desmid, a single-celled form with two mirror-image halves. (b) A calcium-secreting marine form. (c) *Ulva*, or sea lettuce, a marine intertidal form. (a, Biophoto Associates; b, Steven Webster; c, J. M. Kingsbury)

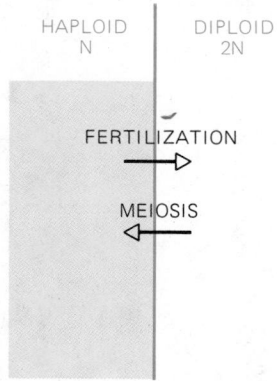

FERTILIZATION

MEIOSIS

Figure 25-8 Basic format used in plant life history diagrams in this book. Stages of the life history with haploid chromosome numbers are shown on a colored background; stages with diploid chromosome numbers are shown on white. Fertilization occurs whenever the life history line crosses from color to white (haploid to diploid); at meiosis, the line crosses from the white to the colored area (diploid to haploid).

25-G Life Histories

The life histories of green algae illustrate an important evolutionary trend that we shall follow with some care in the evolution of higher plants. Each chlorophyte life history mentioned here has parallels among the Rhodophyta and Phaeophyta.

In our study of life histories we shall use a standard format (Figure 25-8), which can be explained in terms of the human life history (Figure 25-9). An organism's chromosomes carry its hereditary material. Human beings have 46 chromosomes, or 23 pairs per cell (the **diploid** number for humans). When gametes (eggs and sperm) are produced by meiosis, each gamete formed receives one member of each pair of chromosomes, for a total of 23 chromosomes (the **haploid** number for humans). When a sperm fertilizes an egg, a zygote with the diploid number of chromosomes is formed. Mitosis of the zygote gives rise to the diploid body of a human being.

Thus, **meiosis** is the process that reduces the number of chromosomes from the diploid to the haploid number; **fertilization** is the process that returns the chromosome content to the diploid number.

In their general outline, the life histories of the brown alga *Fucus* and of many other algae follow the pattern described for humans. A completely opposite type of life history may also be found in many algae, for example, in *Ulothrix*. *Ulothrix* is a

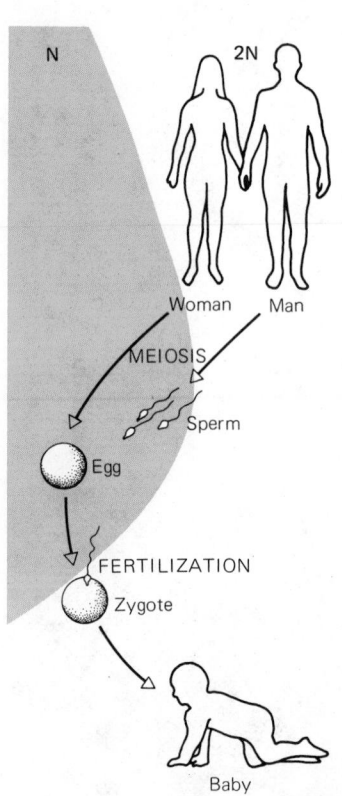

Figure 25-9 Human life history. Human beings are diploid; the gametes (eggs and sperm) are the only haploid cells in the human life history.

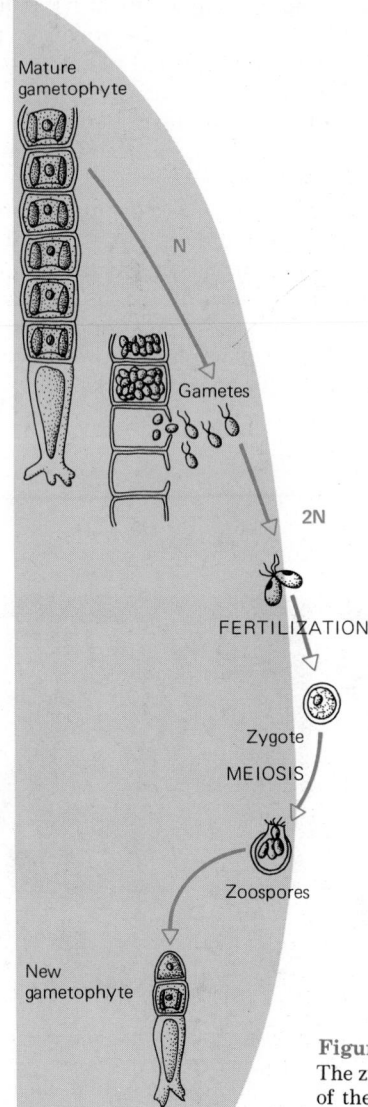

Figure 25-10 Life history of *Ulothrix*. The zygote is the only diploid cell; the cells of the filament, the gametes, and the zoospores are all haploid.

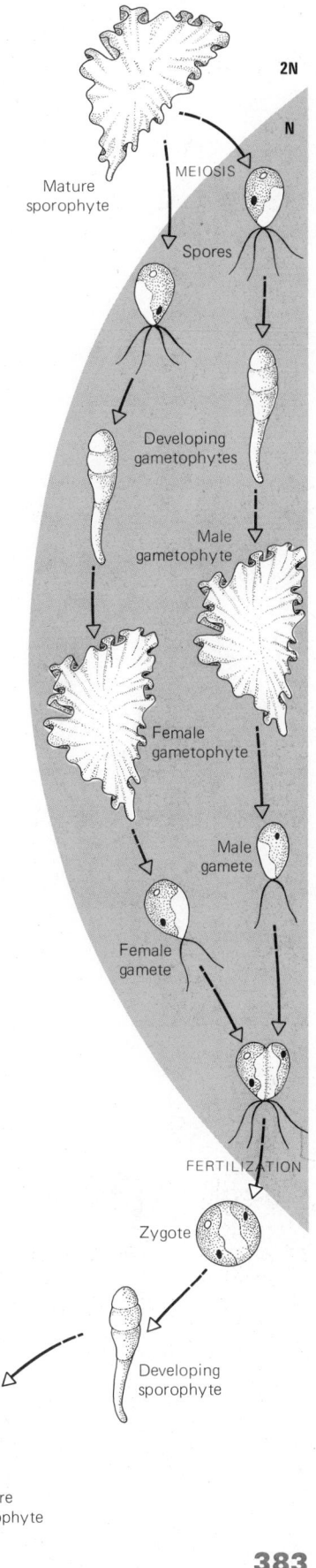

Figure 25-11 Life history of *Ulva*. In this green alga, the multicellular diploid plant (sporophyte) and the multicellular haploid plant (gametophyte) appear alike to the unaided eye. In many other species there are distinct morphological differences between sporophyte and gametophyte.

filamentous green alga that lives attached to rocks in streams. The cells of the filament are haploid, and the zygote is the only diploid cell in the life history (Figure 25-10). Before sexual reproduction, the cells of a *Ulothrix* filament divide many times by mitosis to form tiny flagellated gametes, which are released when the cell walls rupture. The gametes swim away, pair up, and fuse, forming diploid zygotes. Each zygote then secretes a very thick, resistant wall about itself and enters a dormant period. When the zygote breaks dormancy, it undergoes meiosis, producing four haploid cells, called **zoospores** because they have flagella and can swim like flagellated protists. Eventually each zoospore settles down and divides to form a new haploid filament. In life histories of this type, the sole diploid cell, the zygote, is a resting stage and does not carry on the normal life activities of the species; it must undergo meiosis to form haploid cells that carry on photosynthesis and growth.

Alternation of Generations

The life histories of most plants are intermediate between the two types described above, for humans and for *Ulothrix*; such life histories have a multicellular haploid stage, which alternates with a multicellular diploid stage, a situation known as **alternation of generations**.

In these plants, as in *Ulothrix*, meiosis does not give rise to gametes; rather, it produces **spores**, haploid cells that do not undergo fertilization. These spores divide by mitosis to produce multicellular haploid plants. Eventually, the haploid plants produce gametes. Since the plants are already haploid, however, the gametes are produced by mitosis, rather than by meiosis. The haploid gametes then undergo fertilization and produce diploid zygotes.

A life history with both a multicellular diploid form and a multicellular haploid form is found in the green alga *Ulva*, or sea lettuce, which grows near the tide line. The multicellular diploid body (Figure 25-11) is known as the **sporophyte** (sporo = spore; phyte = plant), because it produces spores. Each haploid spore grows into a multicellular haploid plant, which is called a **gametophyte** because it produces gametes. Fusion of two haploid gametes produces a diploid zygote, which grows into a multicellular diploid sporophyte.

The sporophyte and gametophyte of *Ulva* happen to be identical in appearance. (They can be told apart only by counting the number of chromosomes in a cell.) More often, sporophyte and gametophyte of the same species are strikingly different. Careful study of members of the three divisions of multicellular algae has shown that some forms originally classified as different species and even genera are really just sporophytes and gametophytes of the same species.

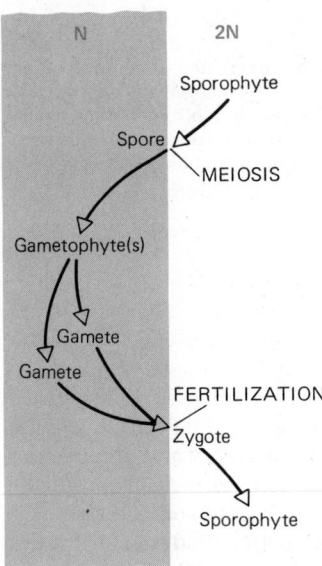

Figure 25-12 General scheme for alternation of generations. The diploid sporophyte produces haploid spores by meiosis; each spore grows into a haploid gametophyte (male, female, or bisexual, depending on species). Gametophytes produce gametes, which unite during fertilization to form the diploid zygote; the zygote grows into a new sporophyte.

The phrase "alternation of generations" indicates that the sporophyte generation is the parent of the gametophyte generation, which is the parent of the next sporophyte generation, and so forth (Figure 25–12). These generations are linked by the alternation of meiosis with fertilization:

$$\text{sporophyte} \xrightarrow{\text{(meiosis)}} \text{spore} \to \text{gametophyte} \to \text{gamete} \xrightarrow{\text{(fertilization)}} \text{zygote} \to \text{sporophyte}$$

Sexual versus Vegetative Reproduction

Sexual reproduction in many algae is initiated by some unfavorable condition in the environment, such as lack of nutrients, unfavorable temperature, or shortening periods of daylight. The selective advantage of this response is that sexual reproduction results in recombination of genetic material from different individuals. The offspring of sexual reproduction thus differ from one another more than the offspring of vegetative reproduction, producing at least some individuals that will probably survive better under the changed circumstances.

Following sexual reproduction, the secretion of a heavy, resistant wall around the zygote protects it from freezing, drying, or similar unfavorable conditions. When conditions improve, the zygote breaks dormancy and carries out the next step in the life history: either germination to form a new sporophyte or meiosis to produce the first cells of new gametophytes. Furthermore the zygote, being small, can be encased in a protective coat with little energy expenditure and tends to escape the notice of predators. Thus it is ideal for its role of carrying the species through hard times.

Many algae also have means of asexual or vegetative reproduction, in which cells of the plant body produce genetically identical dispersal cells, often with flagella, that can swim off and start new plant bodies (Figure 25–13). This permits successful genetic combinations to be multiplied many times and distribute themselves widely when conditions are favorable for their growth.

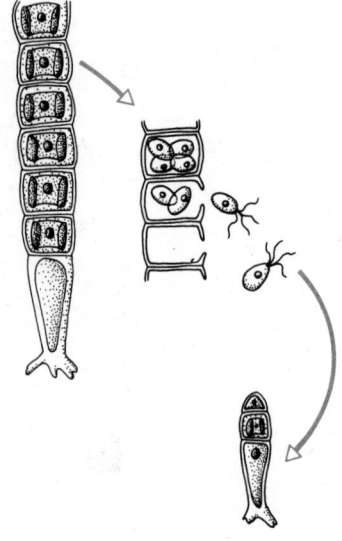

Figure 25-13 Vegetative reproduction in *Ulothrix*. Vegetative cells of the filament divide by mitosis and give rise to flagellated zoospores, larger than gametes. Each zoospore is capable of settling down and producing a new filament.

SUMMARY

The three main groups of multicellular algae included in the plant kingdom probably arose independently from different groups of photosynthetic protists. These lower plants have not evolved the adaptations that enable higher plants to live on land; hence the multicellular algae are largely restricted to marine and freshwater habitats. Furthermore, because they require light for photosynthesis, they live only in shallow water where sufficient light is available. Holdfasts or gelatinous secretions attach most forms to rocks or other substrates, and air bladders in some forms allow the photosynthetic parts of the plant to float near the surface, where there is adequate light. Plants in the intertidal zone are subject to desiccation when they are exposed to heat and air at low tide. A gelatinous coating helps these plants to retain moisture, and dark pigments screen the cells from the sun's rays.

Basic differences in biochemistry and physiology seem to have favored different groups of algae in different habitats. The red algae tend to live high up the shore or in the deepest layers of water that are penetrated by sunlight. Brown algae dominate shallow waters, particularly in colder seas. Green algae are present in most marine environments but are usually less conspicuous there, whereas many species flourish in freshwater habitats.

Multicellular algae provide food for various forms of marine life. Their ability to accumulate essential minerals from seawater has also made them useful as food for livestock and humans. In addition, they produce various economically useful substances.

Alternation of diploid and haploid generations, with similar or different body structure, is characteristic of plants. Sexual reproduction can often be seen as an adaptation that carries the species through hard times via the production of a resistant zygote. Vegetative reproduction can spread favorable genetic combinations in good times.

OBJECTIVES

From your study of this chapter, you should be able to:

1. List the characteristics that set members of the plant kingdom apart from members of the other kingdoms.

2. List or recognize the distinguishing characteristics of members of the divisions Rhodophyta, Phaeophyta, and Chlorophyta.

3. State how each of the following is of adaptive advantage to the algae that possess it: gelatinous secretion, accessory photosynthetic pigment, holdfast, air bladder, sexual reproduction, asexual reproduction, zoospores.

4. Define the following terms, and be able to use them: spore, gamete, zygote, fertilization, zoospore, holdfast, stipe, blade, air bladder.

5. Briefly explain alternation of generations, stating which stages in the life history of a plant are diploid and which are haploid.

SELF-QUIZ

Give the name of the algal division whose members fit each description below.

_____ 1. May reach several metres in length; no unicellular forms known.

_____ 2. Contain chlorophylls *a* and *b*; store food as starch.

_____ 3. Contain chlorophylls *a* and *c*; store food as laminarin.

_____ 4. Chloroplasts contain chlorophyll *a* and phycobilins; internal chloroplast membranes occur singly.

For each adaptation of algae on the left, pick its function from the list on the right. The functions may be used one or several times or not at all.

Adaptation	*Function*
_____ 5. Accessory photosynthetic pigment	a. maintains a favorable location
_____ 6. Gelatinous secretion	b. obtains more nutrients
_____ 7. Holdfast	c. protects from desiccation
_____ 8. Air bladder	d. obtains more energy

(*Self-quiz continues on next page.*)

9. A haploid reproductive cell that may divide and give rise to a new plant is called a(n):
   a. spore
   b. gamete
   c. zygote
   d. embryo

10. Which of the following stages in the life history of a plant are diploid?

   a. gamete
   b. zygote
   c. gametophyte
   d. sporophyte
   e. spore

11. Beginning with the zygote stage, arrange the terms in question 10 in the order found in a plant's life history.

QUESTIONS FOR DISCUSSION

1. Nonmotile plants must be able to get gametes together for fertilization. How do *Ulothrix* and *Ulva* accomplish this?

2. Red algae do not have flagellated cells; how do you suppose their gametes might get together for sexual reproduction?

3. What modifications in the structure and life histories of aquatic algae would be necessary for them to live permanently on land?

REFERENCES AND FURTHER READING

Boney, A. D. *A Biology of Marine Algae.* New York: Hutchinson Educational, Ltd., 1966. Covers algal ecology, major groups, and some experimental work.

Dawson, E. Y. *Marine Botany: An Introduction.* New York: Holt, Rinehart and Winston, 1966. Covers marine algae, fungi, and grasses, economic importance of algae, and history of the study of algae.

Jackson, D. F., Ed. *Algae and Man.* New York: Plenum Press, 1964. Several articles pertain to public health and medical aspects of algae.

Jensen, W. A., and F. B. Salisbury. *Botany: An Ecological Approach.* Belmont, CA: Wadsworth, 1972. Chapters 17 and 18 treat algae in their ecological groupings rather than by taxonomy; an interesting change.

Nisizawa, K., Ed. *Proceedings of the Seventh International Seaweed Symposium.* Tokyo: University of Tokyo Press, 1971. Articles covering many aspects of ecology, economics, biology, and distribution of algae.

CHAPTER

26

HIGHER PLANTS

Some of the green algae were probably the ancestors of the land plants. Consider a green alga growing on the mud at the edge of a lake. The plant is in constant danger from extreme conditions. If the water recedes, the plant may lose too much water to the air and die. On the other hand, a heavy rain may sweep the plant into an unsuitable habitat, or bury it under mud and silt. Under such conditions, several adaptations will be selected for: anchorage structures that keep the plant in one place, a large size that keeps it from being buried completely, and a surface coating that reduces water loss. Some division of labor between cells would also be advantageous because the anchoring cells would be buried and could no longer carry on photosynthesis. If our hypothetical plant were to evolve such adaptations, it could survive better not only at the edge of the lake, but also higher up on the shore, completely out of the water. The ancestors of land plants were almost certainly algae that evolved such adaptations.

In this chapter, we shall examine the problems of life on land and the adaptations of land plants that permit them to cope with these problems. We shall then examine some of the groups of living land plants, considering the adaptations that equip each group for life in increasingly rigorous terrestrial environments, and noting several evolutionary trends among the land plants.

TABLE 26-1 COMPARISON OF WATER AND LAND AS HABITATS FOR PLANTS

	WATER	LAND
Water	Close to each cell	Under land surface; evaporates quickly above surface
Minerals	Close to each cell	On or under land surface
Gases	Dissolved at low concentrations	Plentiful in the air
Support	Provides buoyancy, support	Much less support for parts in air
Light	Cuts out some wavelengths, and lowers intensity	More light available
Temperature	Little fluctuation, slow change	Changes more rapid, wider extremes
Reproduction	Motile gametes swim	Water seldom available for swimming gametes
Dispersal	Water carries offspring to new locations	Water seldom available to carry offspring to new locations

26-A Life on Land

The main challenges facing a plant are obtaining light, water, carbon dioxide, oxygen, and minerals for metabolic processes, and bringing the gametes together for fertilization.

A land environment offers many advantages to plants that can function out of water (Table 26–1). More light is available on land than in water because the water itself absorbs much of the sun's energy. Carbon dioxide for photosynthesis and oxygen for respiration are present in much higher concentrations in air than in water. In addition, the first plants to colonize land did not have to compete with other plants for these resources, nor were they subject to predation, because terrestrial animals had not yet evolved.

The one big disadvantage of moving out of water is that water becomes hard to obtain. For a terrestrial plant, the only reliable source of water lies well below the surface of the soil or in rock crevices, where it is too dark to carry on photosynthesis.

Living on land presents other problems. A plant surrounded by air quickly loses much of its body water by evaporation; thus only plants with a protective water-proof coating have survived on land. Although there is plenty of light on land, the ultraviolet (uv) light in sunshine damages nucleic acids, so plants exposed to much uv need pigments or equivalent protections to prevent damage to their genetic material. Also, because air is so much less dense than water, it gives virtually no support to the plant body, and land plants either have remained small or have evolved supporting structures. Temperature changes are wider and more rapid in air and soil than in water; land plants have adaptations that permit them to survive these fluctuations. Finally, many aquatic algae rely on water for reproduction; in abundant water sperm can swim to eggs and young plants can disperse to new locations. On land, plants carry out these functions without a constant supply of water, either by taking advantage of rain or dew to carry the reproductive cells or by providing these cells with a waterproof coating before they travel through the air.

Plants could not move onto land successfully until they evolved the adaptations needed for survival there. It is not clear whether land plants arose from one or from several species that successfully adapted to the new environment. There is good chemical evidence that the ancestral species belonged to the green algae, or Chlorophyta. Like the green algae, land plants possess chlorophylls a and b, and β-carotene, as their photosynthetic pigments. Both groups also store most of their food reserves in the form of starch.

Adaptations to Life on Land

In a terrestrial habitat, the resources a plant needs are segregated; water and minerals lie below the surface of the soil, while light and air are above it. The division of labor in the bodies of land plants reflects this division of resources.

TABLE 26-2 ADAPTATIONS OF LAND PLANTS TO TERRESTRIAL ENVIRONMENTS

PROBLEM	ADAPTATION
1. Obtaining water and mineral nutrients when they no longer surround the entire plant	Rhizoids or roots
2. Transporting water within the plant	Xylem
3. Transporting food from sites of manufacture to sites of use	Phloem
4. Preventing evaporation from surfaces exposed to air	Cuticle
5. Obtaining gases for photosynthesis and respiration	Stomata
6. Obtaining sunlight for photosynthesis	Leaves
7. Supporting body in medium lacking buoyancy	Xylem
8. Coordinating plant growth and plant response to changes in environment	Hormones
9. Getting gametes together without reliable supply of water for sperm	Pollen
10. Dispersing new individuals to suitable locations	Airborne spores, then seeds

Underground structures serve as anchors and absorb water and minerals. The photosynthetic structures above ground produce enough food for all the cells in the plant (Table 26–2).

Eventually, some plants evolved **vascular tissue**, which transports substances between various parts of the plant. There are two types of vascular tissue: **phloem**, which conducts organic materials, mainly food, from sites of manufacture to sites of use or storage, and **xylem**, which transports mainly water and minerals from the roots to the stems and leaves. Xylem also provides support for the plant body. A plant with xylem may be compared to a building whose plumbing pipes double as supporting columns.

Above the ground level, the plant's surface is covered by a waxy **cuticle**, which is essentially impermeable to water and reduces the evaporation of water from the plant into the air. However, the cuticle is also impermeable to carbon dioxide and oxygen, which a plant must exchange with the air. Portions of the plant above ground can exchange gases through tiny pores called **stomata** (singular: **stoma**). The stomata are surrounded by pairs of **guard cells**, which can regulate the size of the opening (Figure 26–1). Some water is inevitably lost through the stomata, but much less than if evaporation proceeded freely from the entire above-ground surface of the plant.

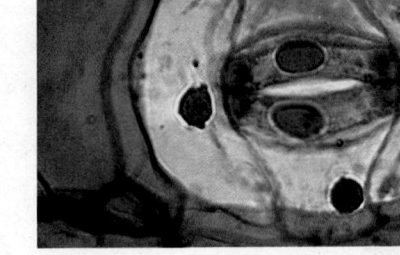

Figure 26-1 A stoma ("mouth") between a pair of lip-like guard cells on the underside of a *Tradescantia* leaf. The guard cells are pink. Cell nuclei are red. (Carolina Biological Supply Company)

Land plants have also developed a variety of hormones, which coordinate the activities of the plant and its response to environmental cues. Some of the large brown algae called kelps have hormones, but hormones play a much more important role in land plants. Hormones make roots grow down into the soil, and stems turn up towards the light. Hormones also initiate reproduction, dormancy, and the breaking of dormancy (see Chapter 47).

TABLE 26-3 **CHARACTERISTICS OF EMBRYOPHYTES**

1. Usually contain chlorophylls *a* and *b*, and β-carotene (a few specialized as parasites and saprobes); usually store food as starch
2. Exhibit alternation of generations; sexual generation (gametophyte) alternates with asexual generation (sporophyte)
3. Macroscopic (can be seen with unaided eye)
4. Distinct tissues (groups of cells specialized to perform particular functions)
5. Multicellular reproductive structures (sexual and asexual)
6. Embryo retained within protective jacket of sterile cells in parent

Reproduction on Land

Land plants exhibit **alternation of generations** (see Section 25-G), in which the haploid **gametophyte** generation produces gametes and the diploid **sporophyte** produces reproductive cells called spores.

Flagellated, swimming spores and sperm are of limited use on land. Furthermore, reproductive cells and young individuals are extremely susceptible to desiccation because a small structure has a great deal of surface area exposed to evaporation, compared to the small volume of water that it can hold. All land plants have multicellular reproductive structures, where developing reproductive cells—sperm, eggs, and spores—are protected, at least for a time, by a surrounding jacket of sterile cells. Land plants are sometimes called **embryophytes** because the zygote remains within the parent plant as it develops into an embryo (Table 26–3). This term is more appropriate than "land plants," because some embryophytes have become re-adapted to aquatic habitats, and some algae, especially green algae, have become adapted to life on shaded rocks or tree trunks well out of the water.

In this chapter we shall follow a few important trends in the evolution of the life histories of land plants:

1. The life histories of land plants have changed from those of the earliest land plants, in which the gametophyte is dominant and the sporophyte dependent on it, through forms in which gametophyte and sporophyte are independent, to the condition in "higher" plants where the sporophyte is dominant and the gametophyte is dependent on it.

2. In the "lower" land plants, spores are the main means of dispersal; in higher plants, there is a new structure, the seed, which is well adapted for establishing a new individual in a new location.

3. Lower land plants produce flagellated sperm that must swim to the eggs in a liquid medium; in higher plants pollen grains are carried to female reproductive structures by the wind or by animals; after reaching the female parts, pollen produces sperm nuclei (Figure 26–2).

Figure 26-2 (a) The eggs of lower land plants are fertilized by flagellated, swimming sperm, such as this fern sperm, which has many flagella. (b) Higher land plants, such as pines, have evolved pollen, which can be carried through the air. Once arrived near the female gametophyte, the pollen grain grows a pollen tube, through which the sperm nucleus migrates to reach the egg. (Carolina Biological Supply Company)

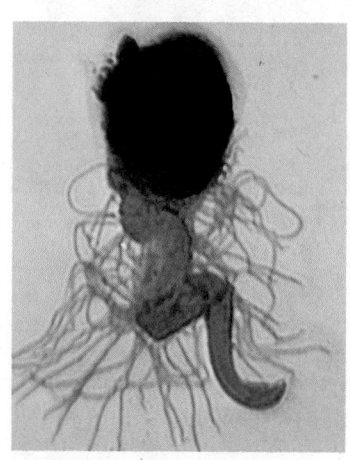

(a)

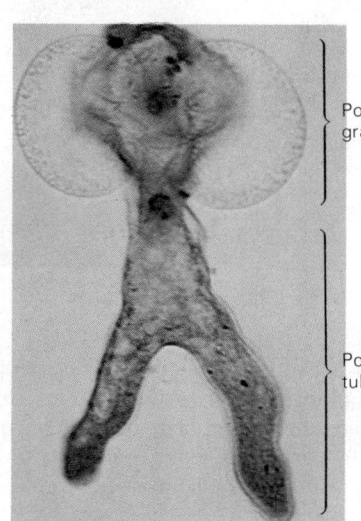

Pollen grain

Pollen tube

(b)

| | COMMON | ESTIMATED NUMBER OF |
GROUP	NAME	LIVING SPECIES
TABLE 26-4 **SUMMARY OF THE EMBRYOPHYTES** (The classification system used here is only one of many that have been proposed [or used].)		
Bryophyta	Liverworts, hornworts, mosses	23,600
Tracheophyta	Vascular plants	
Lycopsida	Club mosses	1200
Sphenopsida	Horsetails	40
Pteropsida		
Filicineae	Ferns	10,000
Gymnospermae	Gymnosperms	665
Angiospermae	Flowering plants	285,000

Table 26-4 and Figure 26-3 show the various groups of embryophytes we shall meet. In this book we consider Bryophyta and Tracheophyta as divisions, but some taxonomic schemes consider the next lower level or the second lower level shown in Table 26-4 as divisions.

26-B   Division Bryophyta

The division Bryophyta includes the mosses, liverworts, and hornworts. The familiar low-growing moss plants are gametophytes. It is tempting to describe a moss gametophyte as composed of diminutive roots, a tiny, slender stem, and minute leaves, but botanists would censure such a description because the terms "root," "stem," and "leaf" are correctly applied only to organs that contain vascular tissue. Bryophytes lack vascular tissue, so we must resort to other terminology in describing their structure.

The rootlike organs of bryophytes are called **rhizoids**. These slender, colorless projections anchor the plant and absorb water and minerals. The moss gametophyte has a short main axis, with flat, green projections that carry on photosynthesis (Figure 26-4). Generally these leaflike structures are only one or two cells thick, and so they are very susceptible to drying out. This, added to the lack of vascular tissue, makes it advantageous for the plant to remain small so that the slow movement of water from the rhizoids keeps pace with the needs of the top of the plant.

Moss gametophytes have other adaptations that allow them to grow on land despite their lack of vascular tissue. They grow very close together; this crowding forces the moss plants to hold each other up, and also creates thousands of tiny spaces that hold water like a sponge. Thus the mode of growth increases the water supply available to the colony. However, because water evaporates quickly when exposed to direct sunlight, mosses can usually survive only in the shade. Their photosynthetic pigments and enzymes can function in dim light.

Some species of moss have bisexual gametophytes, containing both male and female organs; others have separate male and female plants. Sperm shed from male

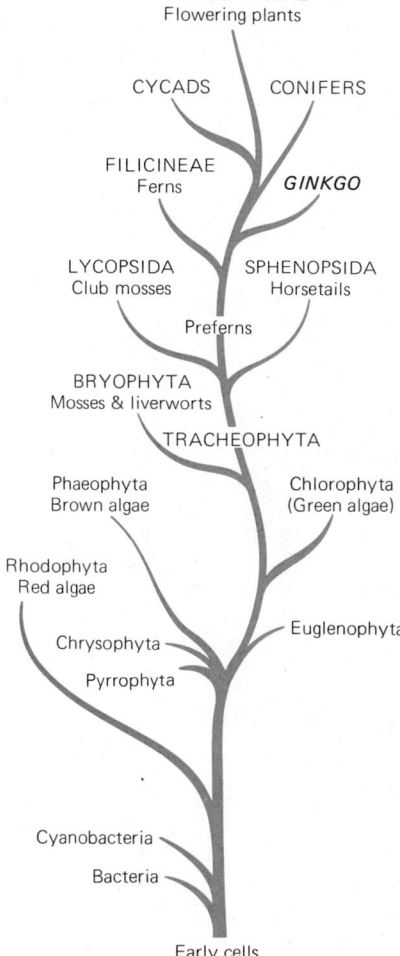

Figure 26-3   Possible evolutionary relationships in the plant kingdom, including relationships to autotrophic organisms in the kingdoms Monera and Protista.

Figure 26-4   Gametophytes (green) and sporophytes (brown) of a moss. Each sporophyte has grown from a zygote formed in the female organ of a gametophyte. (Biophoto Associates)

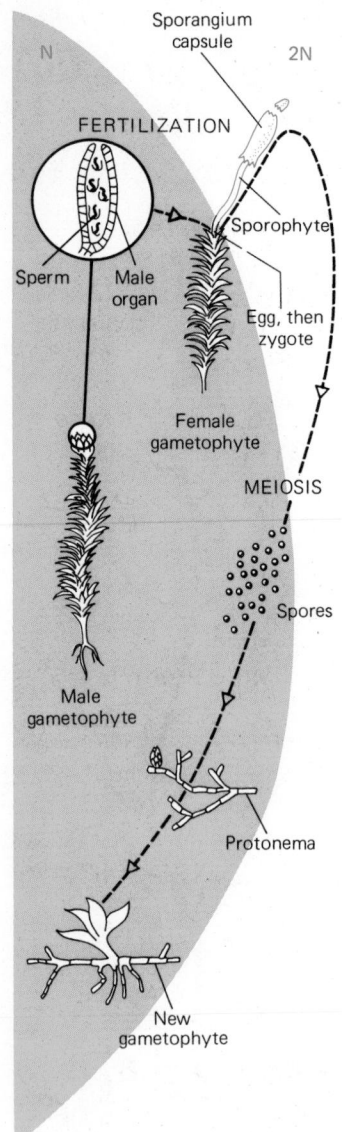

FERTILIZATION

Sporangium capsule

N   2N

Sporophyte

Sperm   Male organ

Egg, then zygote

Female gametophyte

MEIOSIS

Spores

Male gametophyte

Protonema

New gametophyte

organs swim through a film of moisture to the top of the female plant (or branch), where an egg has developed. A sperm fertilizes an egg, forming a diploid zygote. The zygote, still at the top of the female gametophyte, grows into a sporophyte, a long stalk with a capsule on top. The sporophyte may carry on photosynthesis, but it relies on the female gametophyte for water and minerals, and probably for some of its food. Some moss sporophytes have what appear to be rudimentary conducting cells, which transport food and water.

Cells inside the capsule of the sporophyte undergo meiosis, producing haploid spores. When the spores are mature, the capsule bursts and flings them into the air. If a spore lands in a suitable place, it germinates to form a **protonema**, a filament that grows along the surface of the ground. Buds arising from the protonema develop into a clump of new gametophytes (Figure 26-5).

At first glance, liverworts seem very different from mosses because liverwort gametophytes are flat and ribbonlike. Liverwort gametophytes typically have a **dichotomous** pattern of branching; that is, the growing tip of the plant forks into two, and the new fork forks again, and so forth. Then, as the older parts of the plant die, the new forks become separate individuals. Another means of vegetative reproduction in some liverworts is by **gemmae**, tiny balls of vegetative cells that pinch off from the parent plant. They are produced inside cunning little cups that grow on the upper surface of the plant (Figure 26-6). When raindrops splash into the cups, the gemmae bounce out; if they come to rest in a suitable spot, they can grow into new plants. Sexual reproduction is similar to that of mosses.

In summary, the Bryophyta are very small plants, and they are confined to moist, usually shady habitats by their lack of true vascular tissue and by their requirement for water for swimming sperm during sexual reproduction. These restrictions in size and habitat do not mean that bryophytes are unsuccessful; they would not be here if they had not succeeded in their own particular way. In fact, there are about 23,600 species of living bryophytes.

Bryophytes serve an important ecological function. Because they are small and have no roots, they can exist on bare rock, and they help to break the rock surface into particles. Organic material is added as the bryophytes die, so that eventually soil forms and larger plants with roots can grow (Figure 26-7).

Figure 26-5 Life history of a moss.

Figure 26-6 Liverworts with many gemmae cups. (William Camp)

Figure 26-7 A bed of moss growing on this rock has provided a water-retaining, nutrient-rich layer in which rooted plants, in this case ferns, can grow. (William Camp)

TABLE 26–5 DIFFERENCES BETWEEN BRYOPHYTA AND TRACHEOPHYTA

BRYOPHYTA	TRACHEOPHYTA
No vascular tissue	Vascular tissue present in sporophytes
No true roots, stems, leaves; rootlike rhizoids anchor plant and absorb water	Most modern forms have roots, stems, and leaves in sporophyte generation
Not more than a few centimetres tall	Usually more than a few centimetres high
Sporophyte dependent on gametophyte generation	Sporophyte independent, self-sufficient
Gametophyte generation dominant	
Gametophyte independent, self-sufficient	Sporophyte generation dominant
	Gametophyte independent ("lower" vascular plants) or dependent on sporophyte (gymnosperms and angiosperms)
Water needed for swimming sperm during sexual reproduction	Water needed for swimming sperm except in angiosperms and higher gymnosperms
Lack of vascular tissue restricts to moist environments	Wide range of habitats

26-C Division Tracheophyta: The Vascular Plants

The Tracheophyta, or vascular plants, include the most advanced and complex forms of plant life. There are many groups of vascular plants; each has adaptations that represent major evolutionary jumps, conferring tremendous advantages over previously existing groups of plants. The vascular tissue characteristic of the sporophyte generation was the first such evolutionary leap.

By definition, since they have vascular tissue, the organs of tracheophyte sporophytes can correctly be called **stems**, **roots**, and **leaves**. During the course of evolution, the sporophyte has become the larger and more conspicuous, or **dominant**, generation; at the same time, the gametophyte generation, which usually lacks vascular tissue, has become smaller and less conspicuous.

Several groups of lower vascular plants shed light on the stages of evolution that may have led to the plants of more advanced groups. These primitive groups were once abundant, but more advanced plants have crowded them out in most habitats, and only a few representatives survive today.

The fossil record indicates that early vascular plants had no roots or leaves. The sporophyte consisted of an underground stem, called a **rhizome** (not to be confused with a rhizoid), which gave rise to aerial stems growing to heights of several inches above the ground.

26-D Lycopsida (Club Mosses and Ground Pines)

After the evolution of a vascular sporophyte, the next evolutionary advance among the tracheophytes was the acquisition of tiny leaves and roots. Sporophytes in the subdivision Lycopsida, the club mosses and their relatives, have small, slender roots, usually growing from a rhizome; many tiny leaves, with one **vein** (strand of vascular tissue) apiece, cover the aerial stems. As in bryophytes, the sporophytes bear **sporangia** (singular: **sporangium**), the spore-producing structures, on the upper portions of the stems (Figure 26–8). This position assures that spores will disperse as widely as possible, increasing the chances that some of them will land in places where they can grow.

Figure 26-8 The sporangia of this club moss are arranged in clusters, or strobili, held aloft on stems resembling branched candelabra.

26-E Sphenopsida (Horsetails or Scouring Rushes)

The sphenopsids are nicknamed "horsetails" because some of them resemble horses' tails, or "scouring rushes" because people used them to scrub out dirty pots and pans before the invention of steel wool. Sphenopsids were ideal for this because their bodies have a tough coating of silica, a very abrasive material.

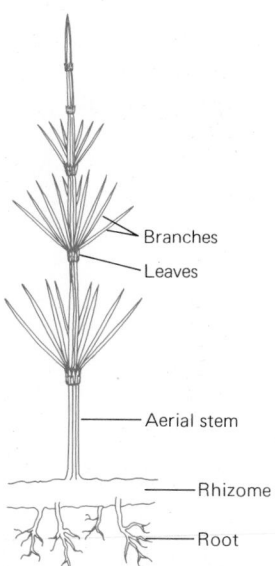

Figure 26-9 *Equisetum* has well-developed underground stems, or rhizomes, and aerial stems, in this case with branches. The roots are thin and the leaves are tiny.

(a)   (b)

Figure 26-10 Horsetails. (a) A vegetative shoot of *Equisetum telmateia*, with thin green branches (the purple flowers belong to other kinds of plants). (b) Fertile shoots of *E. arvense;* the cone-like strobili at the tips of the shoots bear many sporangia. (Biophoto Associates, N.H.P.A.)

In sphenopsid plants, tiny roots, as well as aerial stems, arise from the rhizome (Figure 26–9). Tiny leaves, which are so small that they are often overlooked, are arranged around the "joints" of the aerial stems. Some members of the group have slender branches that resemble the needles of young pine trees, and indeed, these plants are sometimes mistaken for young pines. The sporangia grow in clusters at the top of the stem. In some species sporangia are borne at the tops of the green vegetative shoots, while in others they are produced by specialized nongreen **fertile shoots** (Figure 26–10).

Both lycopsids and sphenopsids can reproduce vegetatively by growing new shoots, either from a rhizome or from a running stem that trails above the surface of the ground. Their life histories are comparable to those of ferns, which we shall consider shortly.

A respectable number of lycopsids, about 1200 species, is still with us, but only one genus, *Equisetum*, with about 40 species, has survived to represent the sphenopsids. Fossil evidence shows that both these groups were once represented by fairly large trees, ranging up to 50 metres tall in the case of lycopsid trees and up to 30 metres for some of the sphenopsid trees. The species that have survived in these two groups are quite small and probably always have been small. In other words, when their larger relatives died out, they were already adapted to the conditions in which they now grow. Members of these groups are not very conspicuous, although they are really quite common and easy to find if you know what to look for.

26-F   Pteropsida: The "Large Leaved" Vascular Plants

The remaining vascular plants show a distinct evolutionary advance, in the possession of leaves with many veins, unlike the tiny, single-veined, scalelike leaves of lycopsids and sphenopsids.

26-G   Filicineae: Ferns

The most primitive class in the subdivision Pteropsida is the Filicineae, or ferns. The leaves of ferns, called **fronds**, are usually highly subdivided. Fronds usually arise from a rhizome, which has many small roots. Ferns may reproduce

(a)

(b)

(c)

Figure 26-11 Ferns. (a) A royal fern (*Osmunda regalis*) in a northern forest; the two brown fertile fronds in the center bear many sporangia. (b) Sporangia clustered on the back of a green frond of a polypody fern (*Polypodium vulgare*). (c) A tropical tree fern. (b, Biophoto Associates, N.H.P.A.)

vegetatively by extension of the rhizome underground, with new sets of fronds produced above ground at intervals.

Sporangia form either on the undersides of the green vegetative fronds (where worried owners sometimes mistake them for some sort of disease) or on separate nongreen **fertile fronds** (Figure 26–11).

Tropical tree ferns may be quite tall, but most temperate species are seldom more than a metre high. Many ferns died out at the same time that lycopsid and sphenopsid trees became extinct, but the survivors have evolved extensively since then, giving rise to many new fern species.

Life History of Ferns

Cells within the sporangia of a fern sporophyte undergo meiosis, forming haploid spores, which are shed into the air and eventually drift to the ground. A spore grows into a small, green photosynthetic gametophyte (Figure 26–12). The gametophyte absorbs water via rhizoids like those of a bryophyte gametophyte. Most fern

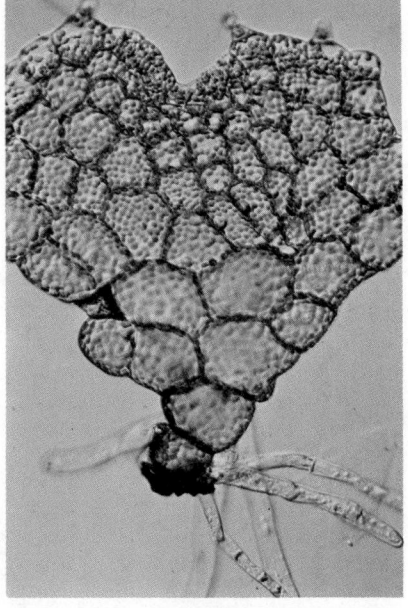

Figure 26-12 A fern gametophyte. Note the green photosynthetic cells and the colorless rhizoids. (Biophoto Associates)

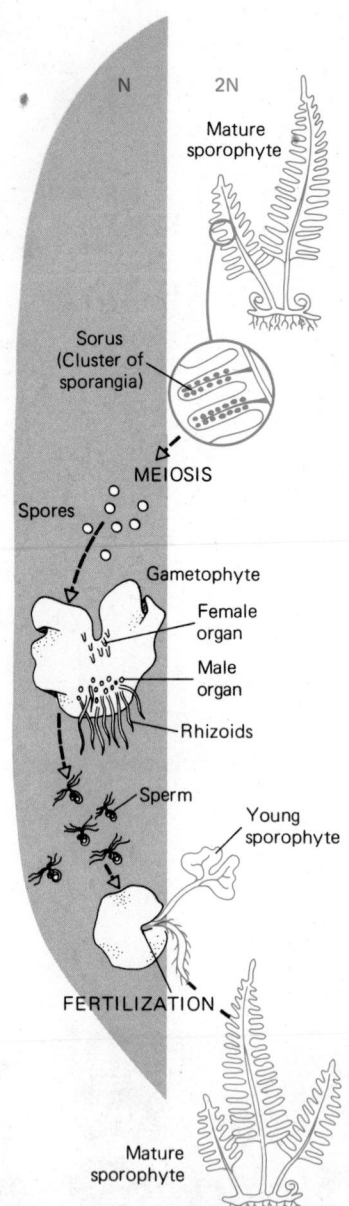

N 2N

Mature sporophyte

Sorus (Cluster of sporangia)

MEIOSIS

Spores

Gametophyte

Female organ

Male organ

Rhizoids

Sperm

Young sporophyte

FERTILIZATION

Mature sporophyte

Figure 26-13 Life history of a fern. The sporophyte and gametophyte are independent of each other; the sporophyte is much larger than the gametophyte, which is a few millimetres across. The male gamete is a swimming sperm.

gametophytes produce both male and female reproductive structures. However, the eggs and sperm mature at different times, ensuring that the eggs of one gametophyte usually will be fertilized by sperm from another. Fern sperm have flagella, and must have moisture on the surface of the soil to swim through. The zygote remains with the (female) gametophyte while it develops into a sporophyte embryo. The young sporophyte pushes a leaf up into the air and a root down into the soil. For a short while the sporophyte depends on the gametophyte for nourishment, but it soon establishes itself as an independent plant (Figure 26–13).

The life history of the ferns is basically similar to that of lycopsids and sphenopsids. Because of this similarity, lycopsids and sphenopsids are sometimes called **fern allies**. In these three groups, the gametophyte and sporophyte live independently of each other, and in this respect the life history is intermediate between those of bryophytes and higher vascular plants.

Gametophytes of sphenopsids and some lycopsids are green like those of ferns; in other lycopsids, however, the gametophytes are brown or colorless **saprobes**, living on dead organic matter in the soil. Interestingly, the gametophytes of lycopsids and of some ferns live symbiotically with a fungus that invades their tissues.

The possession of vascular tissue gives members of these groups many advantages over the bryophytes in the struggle to live on land. Their roots can penetrate the soil and take up water from it, while their leaves (and any aerial stems) can thrust into the air and compete for sunlight. However, sexual reproduction still depends on water for flagellated sperm to swim in (Figure 26–2a). Thus these plants grow mainly in areas where water is frequently available. The sporophytes are **perennial**; that is, they live for many years. So all is not lost if conditions are unfavorable for sexual reproduction one year. In addition, these plants do not depend on sexual reproduction alone to produce new individuals. Many species show vigorous vegetative growth, producing large clumps of individuals.

TABLE 26-6 FERNS AND "FERN ALLIES"

GENERAL CHARACTERISTICS

1. Both gametophyte and sporophyte independent.
2. Sperm require free water in environment in order to swim to egg.

SUBDIVISION LYCOPSIDA

1. Many fossil forms, about 1200 living species.
2. Some fossil forms large trees; living forms small, low-growing, less than ⅓ metre high.
3. Body with underground or above-ground trailing stems as well as erect stems; roots present, numerous small leaves cover stems and branches.
4. Sporangia borne near tops of stems, on leaves or modified leaves.
5. Gametophytes autotrophic or saprobic; with symbiotic fungus.

SUBDIVISION SPHENOPSIDA

1. Many fossil forms; one genus of living plants, with 30–40 species.
2. Some fossil forms large trees; living forms small, maximum less than a metre high.
3. Creeping underground stem (rhizome) gives rise to hollow erect stems, with tiny leaves; some with slender branches; roots present.
4. Sporangia form in clusters at top of stems; on vegetative stems in some species, on nongreen fertile stems in others.
5. Gametophytes autotrophic.

SUBDIVISION PTEROPSIDA

Class Filicineae
1. Many fossil forms; about 10,000 living species.
2. Some large trees (tropical); many smaller forms.
3. Most have underground stem (rhizome), roots, and large, many-veined leaves (fronds).
4. Sporangia on underside of fronds or on separate nongreen fertile fronds.
5. Gametophytes usually photosynthetic.

26-H Gymnosperms

The redwoods of the coastal mountains in California, Canadian pines and hemlocks, cypress standing knee-deep in Southern swamps, and palmlike cycads of the tropics are all members of the various groups lumped together under the term gymnosperms. The largest group of gymnosperms is the conifers, but there are other, less familiar groups. Nearly all gymnosperms are trees, but there are also some shrubs and even woody vines. Although fewer than 700 species of this once dominant group survive today, they cover large areas of the earth's surface. What is probably the oldest tree in the world, a 4900-year-old bristlecone pine (*Pinus longaeva*) in the mountains of eastern Nevada, is a gymnosperm; so are the tallest tree, a redwood (*Sequoia sempervirens*) over 100 metres high, and the tree with the greatest bulk, a giant sequoia (*Sequoiadendron giganteum*) nicknamed "General Sherman," which is over 80 metres tall, 20 metres around at its base, and 3500 to 4000 years old.

Gymnosperms show many striking advances over the plants discussed so far. These new characteristics have allowed many gymnosperms to adapt to life in very dry places, where mosses and most ferns could not survive.

One of the main features contributing to gymnosperm success was the evolution of a new reproductive structure, the **seed**. The word "gymnosperm" literally means "naked seed," a reference to the absence of an ovary, which surrounds the seeds of the most advanced group of plants, the angiosperms. Although they are "naked," gymnosperm seeds are not unprotected, as you know if you have ever pried a seed out of a pine cone (Figure 26–15a).

Seeds are much better adapted for dispersal than are the spores that disperse the more primitive land plants. (Note, however, that seeds are not equivalent to spores in the life histories of these two types of plants.) Instead of the single haploid cell of a spore, seeds contain many cells, organized into three main parts: the outer, protective **seed coat**, the multicellular **embryo** of the new sporophyte generation, and a **food supply** that the new plant uses while its roots and leaves develop from their embryonic rudiments. The seed is an extraordinary evolutionary invention for the survival and perpetuation of terrestrial plants.

But the seed is not the only reproductive modification in this group. In gymnosperms the sporophyte forms two kinds of spores (small **microspores** and larger **megaspores**), which give rise to separate male and female gametophytes. Such

Figure 26-14 Mature giant sequoias are the largest of living plants, dwarfing the six-foot man in the picture.

(a)

(b)

Figure 26-15 (a) Seed cone of a red pine (*Pinus resinosa*). The seeds have thin, papery "wings" which aid in their dispersal by the wind. Two pale, rounded seeds can be seen near the base of a scale at the lower left. (b) Staminate, or pollen, cones of Austrian pine (*P. nigra*). Pollen cones are small, soft, and short-lived. (a, William Camp; b, Biophoto Associates)

(a)

(b)

Figure 26-16 Cycads. (a) A plant with new fronds forming in the center. (b) Mature female cones, with one cone opened to show its red seeds. (Biophoto Associates)

heterospory occurs in a few lycopsids and ferns, but becomes universal in the gymnosperms.

Gymnosperm gametophytes are very tiny, much smaller than the gametophytes of mosses or ferns. The gametophytes develop within the body of the sporophyte parent, which provides nourishment, moisture, and protection. Thus, a gymnosperm gametophyte can survive anywhere the sporophyte can.

A further reproductive adaptation is that gymnosperms do not need free water for fertilization. Instead, a new male reproductive structure, the **pollen grain**, is transported by the wind. Pollen grains are specialized immature male gametophytes, which arise from microspores, are released into the air, and complete their development when they reach the female gametophyte. The pollen of some gymnosperms does produce flagellated sperm, but these sperm are released right at the female gametophyte and swim in fluids secreted by the plant.

Last, gymnosperms have woody tissues, made up of xylem. New wood is added each year; thus the plant grows in diameter as well as in height. Wood strengthens the stem and allows the plant to grow tall and compete for sunlight.

These adaptations have made the gymnosperms a very successful group of plants for about 345 million years. Although individual species of gymnosperms died out during two major periods of extinction in the geological past, many have survived. We shall mention four kinds of living gymnosperms: the cycads, *Ginkgo*, *Ephedra*, and the conifers.

The **cycads** are found growing mainly in tropical or semitropical regions. They look rather like palm trees except that they have cones (Figure 26–16). Cycads have separate microsporangiate ("male") and megasporangiate ("female") plants, and the sperm are flagellated. The living cycads are mere remnants of a group that was once much more abundant. Cycads were a dominant group in the flora 300 million years ago, and there is evidence that some dinosaurs ate cycad leaves and seeds.

Another interesting, and more familiar, gymnosperm is *Ginkgo biloba* (Figure 26–17), the sole surviving species of a group that was once quite large. *Ginkgo* is a very popular tree for street plantings in many American cities, since it is remarkably resistant to urban smog. *Ginkgo* trees are either male or female, and the pollen releases flagellated sperm. *Ginkgo* trees planted in cities and parks are usually males

(a)

(b)

Figure 26-17 *Ginkgo.* (a) A male specimen. (b) A branch of a female tree, showing a seed and leaves.

Figure 26-18 *Ephedra viridis,* also known as Mormon or Mexican tea, growing on the southern rim of the Grand Canyon. The leaves of this desert-adapted gymnosperm are much reduced; the green stems carry on photosynthesis. (Paul Feeny)

because the females produce foul-smelling seeds.

Ephedra, a wiry shrub that grows in the deserts of the southwestern United States, is a member of another gymnosperm group (Figure 26–18).

The most familiar gymnosperms are the **conifers**, or cone-bearers. Most have cone-like reproductive structures and needle-like or scale-like leaves, with little surface area and thick cuticle. Pines, firs, cedars, yews, hemlocks, junipers, larches, spruces, and cypresses are all conifers. Many conifers do well in poor or shallow soil, where nutrients are scarce and where there are often long cold or dry spells. This is probably one reason why the conifers are so common today.

Life History of a Pine

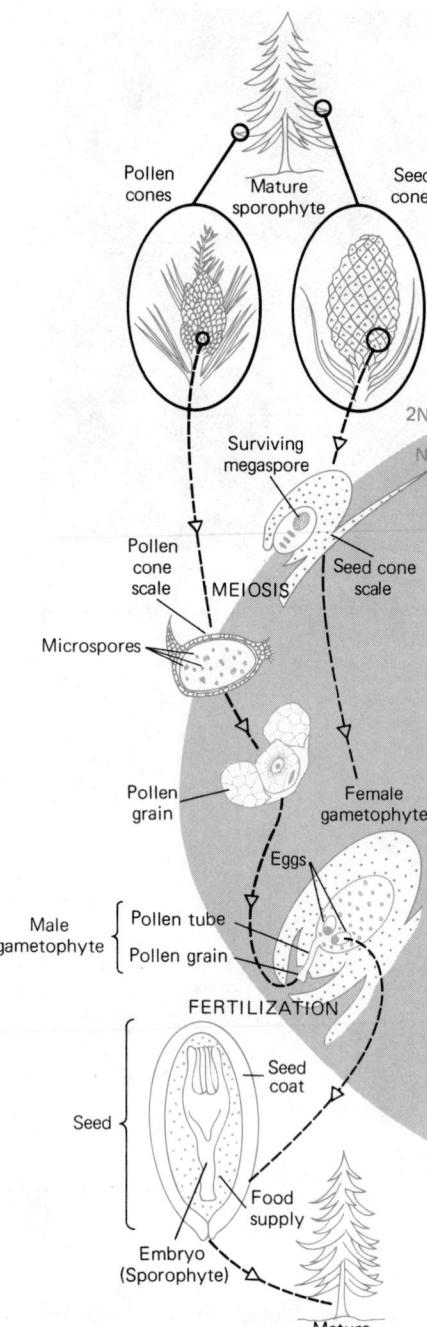

Let us consider the life history of a typical conifer, a pine (Figure 26–19). A mature pine tree produces two types of cones: the seed cone, and the staminate, or pollen, cone. Cells in megasporangia in the seed cone form megaspores by meiosis. Three of the four megaspores resulting from meiosis disintegrate; the fourth develops into a female gametophyte.

Meanwhile, the staminate cones produce smaller microspores (see Figure 26–15b). The microspores begin to develop into male gametophytes, but before they complete their development they are released as pollen grains. In the center of the pollen grain are a few haploid nuclei, and at the sides are flattened, winglike structures, which permit the pollen grain to be dispersed by the wind.

When pollen is shed, the scales of the seed cone are spread apart, and the pollen falls down among them, landing near the female gametophytes at the bases of the seed cone scales. After this **pollination** (the receiving of pollen), the seed cone keeps growing and the scales close for a time. Each pollen grain sends a tiny **pollen tube** toward the female gametophyte (see Figure 26–2b). At the same time, the female gametophyte continues to develop and eventually produces several eggs.

Fertilization occurs when a sperm nucleus leaves the pollen tube and fuses with an egg nucleus. Several zygotes are produced in each female gametophyte, and the embryos compete with one another until one of them crowds out all the others. Meanwhile, the female gametophyte becomes a food supply by absorbing food from the parent sporophyte, and the protective seed coat develops. When the seeds are ready to be shed, the scales of the seed cone open once more and the seeds fall or blow out and spin through the air on their thin, papery wings.

All of this takes a surprisingly long time. During its first spring, a seed cone forms and opens its scales for pollination. Male and female gametophytes take about a year to mature in pines, and fertilization takes place in the second spring of the seed cone's existence. (This points up the frequently overlooked fact that pollination and fertilization are two separate events, and the terms cannot properly be used interchangeably.) The cone's second summer is occupied by the development of the embryo and the absorption of its food supply from the parent sporophyte; not until late summer do the scales open once again to release the seeds. Many conifers reproduce faster than pines, and their seed cones release seeds in the same year that they first appear.

26-I Angiosperms: The Flowering Plants

Spectacular numbers of angiosperms exist today: an estimated 285,000 species, or six times the number of species of all other plant groups combined. Flowering plants are also impressively varied, ranging from duckweeds to dogwoods, onions to oak trees. They range in size from tiny *Wolffia*, about one millimetre long, to towering *Eucalyptus* trees that vie with the redwoods for botanical height records. Representatives of the angiosperms grow in deserts, on the tops of mountains, and in polar regions, salt marshes, lakes, and streams; their flowers borrow every hue of the rainbow.

Angiosperms are crucial to the existence and economy of human beings (Table 26–7). We may occasionally nibble pine seeds between meals, swill spruce beer, or sauté a batch of fern fiddleheads (fronds not yet opened up) as an exotic spring side-dish, but almost all of our plant food comes from angiosperms, as does most of the food for domestic animals. And, although gymnosperms provide most utility lumber, angiosperm trees are used for objects that require wood of particular beauty or strength. Cotton and linen are obtained from angiosperm plants, and their fibers can be dyed with pigments extracted from roots and berries of other angiosperms. Flowering plants give us medicines and drugs, teas and spices. Finally, we spend millions of dollars each year buying live plants and the wherewithal to care for them. All in all, the angiosperms are a most important group.

Figure 26-19 Life history of pine. Microspores are produced by pollen cones; megaspores are produced by seed cones. Pollen grains, which develop from microspores, are carried by the wind. Gametophytes develop protected by the sporophyte plant (the pollen tube is the mature male gametophyte). The seed, consisting of an embryonic sporophyte plus its food supply inside a protective seed coat, serves as the dispersal unit in the life history.

(a)

(b)

(c)

(d)

Figure 26-20 The variety of flowering plants. The beaver-tail cactus (a) receives little water in its desert home; at the opposite extreme, water lilies (b) live surrounded by water. (c) Wispy Spanish moss (really an angiosperm) festoons the branches of a sturdy oak tree. (d) The snow plant, a resident of mountains in the western U.S., cannot photosynthesize, but is thought to live as a saprophyte on dead matter. (a,b, Biophoto Associates, N.H.P.A.; d, William Camp)

TABLE 26-7 ECONOMIC IMPORTANCE OF LAND PLANTS

Bryophytes	Sphagnum moss: Used as fuel (peat) in Ireland, Scotland, etc.; used as mulch and planting medium in gardening and nursery industries
Lycopsids	Formerly used as Christmas greens, the ground pine is now rare and protected in most areas
Ferns	Foliage used in florist industry; plants sold for house and garden
Gymnosperms	Lumber: Douglas fir, hemlock, spruce, various pine species, cedar, redwood
	Turpentine: Distilled from pine trees
	Pulp for paper: Various conifers
	Christmas trees: Spruces, pines, firs, eastern red cedar
	Landscape plants: Spruces, junipers, yews, cedars, cypress, hemlock, pines, cycads, *Ginkgo*
	Gin: Sometimes prepared by redistilling spirits with juniper "berries" for flavoring
Angiosperms	Food: Fruits, berries, seeds, nuts, grains, stalks, leaves, roots, tubers; extraction of juices, syrups, and fats
	Clothing: Cotton, linen
	Lumber: Oak, maple, ash, birch, poplar, walnut, cherry, pecan
	Fuel: Wood, charcoal
	Landscaping: Grass, oak, maple, magnolia, birch, and many other trees; flowering shrubs; annual and perennial herbaceous flowers
	Beverages: Coffee; tea; fermentation of many angiosperm species to make beer, wine, and liquors
	Drugs and medicines: Tobacco; aspirin (originally derived from willow bark); morphine, opium; marijuana; atropine (from *Belladonna* plant); digitalis (from foxglove); various tonics, from sassafras, dandelion, coltsfoot, etc.

Evolution of the Angiosperms

The present geological age is sometimes called the age of mammals and angiosperms because these are the dominant forms of animal and plant life on earth. Where did the angiosperms come from? Botanists are not really sure. The fossil evidence so far does not link the angiosperms clearly to any other group. Angiosperms seem to have sprung up quite suddenly and radiated rapidly into a variety of different forms. Fossils of angiosperm plants and their pollen are found in abundance in rock laid down in the late part of the Cretaceous period (around 70 million years ago). During this time there was much drying and cooling around the world. The dinosaurs, most of the cycads, and some conifers became extinct, while mammals and angiosperms started to dominate the living world.

As yet, no one has produced a widely accepted answer for the spectacular evolutionary success of the angiosperms, although several theories have been proposed. One theory holds that angiosperms have simply outcompeted more primitive plants, thanks to their possession of more efficient vascular tissue. This permits many angiosperms to have relatively large leaves, since evaporation of water into the air can be compensated by fresh delivery from the roots. With a large surface area for photosynthesis, an angiosperm can produce food and grow quickly. In addition, angiosperm seedlings can often survive with less light than can gymnosperm seedlings.

The speed at which some angiosperms grow and reproduce may account for the rapidity of their evolution, as well as their phenomenal success in taking over vast areas once covered by coniferous forests, such as the middle of the eastern coast of the United States. Only in parts of the South, with nutrient-poor, sandy soil, and frequent fires, and in the Canadian north, with its shallow, rocky soil and long, frigid winters, do the conifers still hold sway. Some botanists theorize that this is due to the evergreen habit of most conifers, which enables them to take advantage of every beam of sunshine.

At least part of the reason for angiosperm success is their adaptation to life on an earth that is populated not only by other plants but also by an abundance of animal life. When previous groups of land plants were evolving, terrestrial animals were few, but by the time the angiosperms evolved, land animals were already well-established and diversified.

Perhaps the first relationship between flowering plants and animals that springs to mind is the pollination of many angiosperms by animals, especially in-

(a)

(b)

Figure 26-21 Angiosperm flowers. (a) The colorful flowers of a hardhead thistle attract animal pollinators. (b) The flowers of a grass are wind-pollinated. Pollen is released from the dangling yellow stamens into the breeze. The small, feathery white structures are the female flower parts, which must receive pollen from passing air currents. (Biophoto Associates, N.H.P.A.)

sects. Flowers pollinated by animals tend to be large and showy, attracting the animals visually, and they often have an olfactory attractant, or odor, as well as sweet nectar that the animal drinks. Each type of flower has features that adapt it to particular pollinators. Angiosperms may also rely on animals to disperse their seeds. Indeed, it has been proposed that widespread dispersal of seeds by birds was a crucial factor in the evolution of so many species of angiosperms. Some species of angiosperms still rely on the wind for pollination, and these generally have small, inconspicuous flowers, such as those of oak or maple trees and of the grasses (Figure 26-21b). Many angiosperm seeds are also dispersed by the wind.

Reproduction in Angiosperms

The most distinctive feature of angiosperms is the flower. The parts of a flower are modified, highly specialized, leaves. The two outermost sets of flower parts are the **sepals** and the **petals**. Although petals and sepals are sterile, they are often modified into large, brilliantly colored structures that attract pollinators, and thus play a crucial role in reproduction. The **stamens** are the "male" structures, which produce pollen. A "female" structure, or **carpel**, consists of three parts: the **stigma**, often covered with a sticky substance to which pollen grains adhere, the **style**, and the **ovary** (Figure 26-22).

The ovary encloses one or more **ovules**, in which the tiny female gametophyte develops. After fertilization, the ovule matures and becomes the seed coat surrounding the embryonic plant. The term "angiosperm" literally means "hidden seed;" the name comes from the fact that the seed is hidden within a fruit, which develops from the ovary. To a botanist, a **fruit** is the matured ovary surrounding the seed. Fruits are not necessarily juicy and delicious; the pods of peas and milkweed are fruits, and so are pumpkins and peanut shells.

Like gymnosperms, angiosperms produce two kinds of spores: megaspores produced in the ovules develop into female gametophytes, and microspores produced by the stamens develop into pollen.

The female gametophyte of an angiosperm is even smaller and simpler in structure than that of a gymnosperm. In most flowering plants, the female gametophyte contains several haploid nuclei, among them the **egg nucleus** and two **central nuclei**, which in some species fuse to form a single central nucleus (Figure 26-23).

The pollen grain, as in gymnosperms, is an immature male gametophyte, which is transported to the stigma by the wind or by an animal. Here it produces a pollen tube, which grows down the style and into the ovary, where it releases two haploid sperm nuclei. One of these fertilizes the egg nucleus, to produce a diploid zygote nucleus. The other fertilizes the central nucleus (or nuclei), forming a triploid (3N) **endosperm nucleus**, which divides and gives rise to a triploid nutritive tissue, the **endosperm**. As the zygote develops into an embryo, the endosperm absorbs food from the parent sporophyte, and the layers of the ovule wall develop into the seed coat. The ovary wall develops into the fruit.

Double fertilization, of the egg nucleus and of the central nucleus, is the unique feature of the angiosperm life history; its selective advantage is not clear.

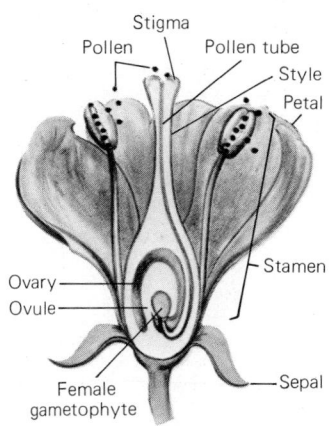

Figure 26-22 Parts of a flower. Pollen produced by the stamens lands on the sticky stigma of the carpel and grows down the style to the female gametophyte inside the ovule.

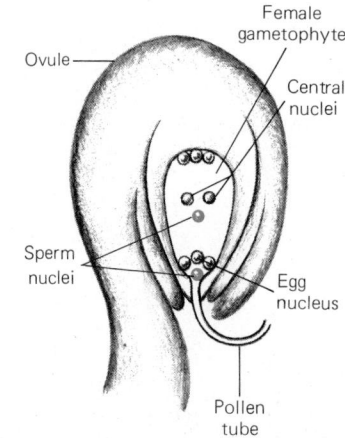

Figure 26-23 Double fertilization in angiosperms. One sperm nucleus fertilizes the egg, forming a zygote (diploid); the other sperm nucleus fertilizes the two haploid central nuclei (or a diploid central nucleus if these have already fused), forming a triploid endosperm nucleus.

SUMMARY

Land offers plants advantages that are not found in water: more sunlight, more carbon dioxide, and more oxygen. However, land plants must have adaptations that allow them to cope with the problems of a terrestrial existence: lack of water, evaporation of the water that is available, and lack of support. Rhizoids or roots allow land plants to obtain water from the soil. Tracheophytes have evolved vascular tissue, which transports water taken in by the roots, supports the stems and leaves in the air, and transports food from the photosynthetic parts to the roots. A waxy cuticle retards evaporation from the leaves and stems, and stomata allow gases to enter the leaves with minimal water loss. Leaves expose large surface areas for photosynthesis, and hormones coordinate the activities of different parts of the plant with one another and with cues from the environment. In the most advanced land plants, pollen grains and pollen tubes permit sexual reproduction without the need for water in which sperm can swim, and seeds supply food and protection to young individuals, increasing their chances of survival.

The adaptations of plants to life on land show several evolutionary trends:

1. The sporophyte has replaced the gametophyte as the dominant stage in the life history. During this shift, the size of the sporophyte progressively increased, while that of the gametophyte decreased.

2. The increase in size of the sporophyte was accompanied by an increase in strength and efficiency of its vascular tissue.

3. As tissues became more and more specialized, new organs were added progressively:

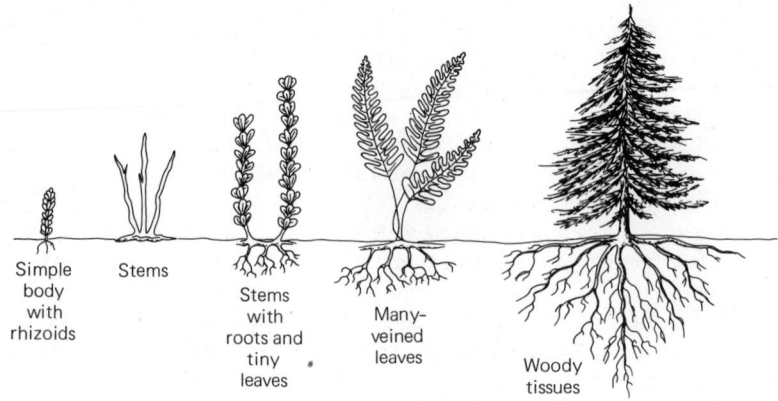

Simple body with rhizoids

Stems

Stems with roots and tiny leaves

Many-veined leaves

Woody tissues

4. The male gametophyte evolved into a waterproof pollen grain, which traveled to the female gametophyte before releasing the sperm. Sperm no longer faced a long, dangerous swim to the egg.

5. The female gametophyte of all land plants retains the egg and protects the zygote as it develops into an embryo; in higher land plants, the female gametophyte itself is retained on or in the sporophyte parent, which contributes food and protective coatings to the seed, a new dispersal structure containing the embryo of the next sporophyte generation.

6. In the higher vascular plants, the spore, a single haploid reproductive cell that is the dispersal stage of lower land plants, has become differentiated into megaspores, which give rise to female gametophytes, and microspores, which give rise to pollen grains (male gametophytes). Instead of being shed, the megaspores remain in the sporophyte, which protects the female gametophyte and plays a major role in the production of the seed. Seeds replaced spores as the dispersal stage in the life history.

7. As plant structure and reproduction became more and more independent of water and moisture in the environment, plants underwent adaptive radiation and spread into more and more different habitats.

OBJECTIVES

From your study of this chapter, you should be able to:

1. Define the following words briefly, and use them correctly: vascular tissue, cuticle, stomata, gametophyte, gamete, egg, sperm, sporophyte, spore, sporangium, meiosis, fertilization, zygote, embryo, rhizoid, rhizome, stem, root, leaf, pollen, pollination, seed, stamen, carpel, fruit.

2. Contrast water and land environments as habitats for plant life.

3. List the problems faced by plants living on land that are not faced by plants living in water, and for each problem name the adaptation(s) of land plants that allow them to survive in the face of the problem.

4. List the characteristics of embryophytes and explain why embryophytes are believed to be descended from green algae.

5. List the adaptations of bryophytes that allow them to live on land, and the characteristics that restrict them to very moist land environments.

6. Trace the evolutionary advances shown by bryophytes, lycopsids (club mosses) and sphenopsids (horsetails), ferns, gymnosperms, and angiosperms.

7. Compare and contrast any of the groups mentioned in objective #6 with respect to structure, reproduction, and life history.

8. Place plants correctly into one of the groups listed in objective #6.

9. (a) Explain what alternation of generations is; (b) trace the shift from dominance of the gametophyte to dominance of the sporophyte during evolution of land plants, giving examples; and (c) discuss the selective value of this shift.

SELF-QUIZ

1. During the course of evolution, sporophytes became dominant over gametophytes due to what adaptation?
   a. airborne pollen
   b. seeds
   c. vascular tissue
   d. diploidy
   e. stomata

2. A seed encloses an individual of the next sporophyte generation in the form of a(n):
   a. spore
   b. gamete
   c. zygote
   d. embryo
   e. pollen grain

3. An adaptation of land plants that reduces evaporation of water from the body surface into the air is:
   a. roots
   b. rhizoids
   c. stomata
   d. cuticle
   e. wood

4. Mosses are adapted to live in a land environment in that:
   a. they have means of vegetative reproduction
   b. they have alternation of generations
   c. they have dependent sporophytes
   d. they hold water near their bodies
   e. they are no more than a few inches tall

5. An advance shown by gymnosperms that was not found in any of the more primitive groups of plants was:
   a. growth exceeding a few feet high
   b. dominance of the sporophyte over the gametophyte generation
   c. protection of the gametophyte within the sporophyte body
   d. presence of true leaves
   e. production of reproductive structures at tips of the plant

QUESTIONS FOR DISCUSSION

1. Parental care of the young was an adaptation that played a major part in the evolutionary success of birds and mammals. What parallels can you find in the plant kingdom?

2. Most conifers in temperate climates keep their leaves through the winter. Most angiosperm trees in the same environment are deciduous; that is, they drop their leaves each fall and produce a new set each spring. What are the advantages and disadvantages of being evergreen? Of being deciduous?

3. Why are there so many species of angiosperms? (You may be interested in reading Philip Regal's theory, listed in the references; do you agree with his argument?)

4. What energy expenditures must a plant make if it is pollinated by animals? What energy savings does the plant gain by having animal pollination? What is the adaptive advantage to a plant of being animal-pollinated rather than wind-pollinated?

REFERENCES AND FURTHER READING

Banks, H. *Evolution and Plants of the Past*. Belmont, CA: Wadsworth, 1970. A lively presentation of paleobotany.

Coulter, M. C., revised by H. J. Dittmer. *The Story of the Plant Kingdom*, 3d ed. Chicago: University of Chicago Press, 1964. An interesting and highly readable evolutionary approach to the study of plants.

Jensen, W. A., and F. B. Salisbury. *Botany: An Ecological Approach*. Belmont, CA: Wadsworth, 1972. Interesting examples, an ecological/evolutionary approach, and appropriate experimental results make enjoyable reading.

Regal, P. J. "Ecology and evolution of flowering plant dominance." *Science* 196:622, 1977. A theory that angiosperms speciated so rapidly because of particular interactions with animals.

Scagel, R. F. *et al. Plant Diversity: An Evolutionary Approach*. Belmont, CA: Wadsworth, 1969. Breadth and depth appropriate for the serious student.

Weier, T. E., C. R. Stocking, and M. G. Barbour. *Botany: An Introduction to Plant Biology*, 5th ed. New York: John Wiley and Sons, Inc., 1974. A time-tested botany text; especially well illustrated.

Wilson, C. L., W. E. Loomis, and T. A. Steeves. *Botany*, 5th ed. New York: Holt, Rinehart and Winston, 1971. Chapters 23 to 28 present a clear introduction to plant groups and their evolution.

CHAPTER 27

THE LOWER INVERTEBRATES

In this chapter we begin our survey of the kingdom Animalia. Animals are multicellular heterotrophs that ingest their food—eat now, digest later—rather than digesting food externally and then absorbing it, as the fungi do. Animals produce gametes (eggs and sperm) in multicellular structures; after fertilization, the zygote develops into an embryo. Depending on who is counting, the animal kingdom contains up to 29 phyla. Although we will refrain from introducing all of them, it will still take four chapters to present this kingdom. One subphylum of the phylum Chordata contains all the vertebrates, the animals with backbones. The rest of the chordates, and the members of all other animal phyla, are commonly called **invertebrates,** animals without backbones.

There are few clues to the origin and early evolution of the animals. The scanty fossil record of the Precambrian era, which ended 600 million years ago, shows soft-bodied animals such as jellyfish and worms. The abundant and diverse fossils of the Cambrian era include representatives of most animal phyla, but there is little evidence of their evolutionary history.

Any reconstruction of early animal evolution relies heavily on comparisons of the structure and embryonic development of living forms. We consider some simple body plans to be **primitive**, that is, to have arisen early in evolution, and assume that similar forms gave rise to the more complex structures found in some **higher** (later-evolved) forms. We can study animals of some early evolutionary lines directly because they became specialized for ways of life that have changed little in hundreds of millions of years, and their modern descendants resemble these ancient ancestors quite closely.

One striking attribute of animals is the high degree of tissue, organ, and organ system complexity in the more advanced forms. The ingestive mode of life must

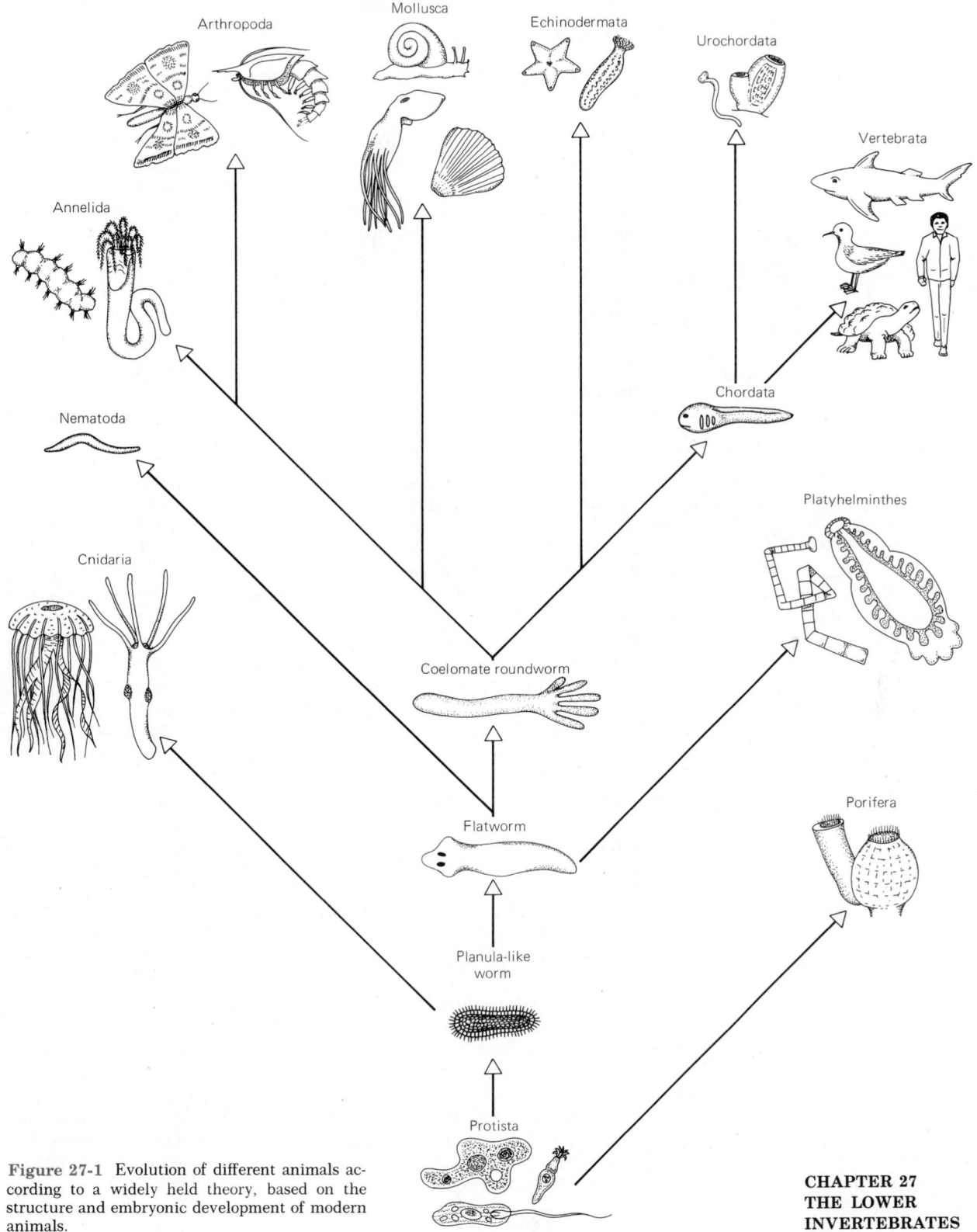

Figure 27-1 Evolution of different animals according to a widely held theory, based on the structure and embryonic development of modern animals.

often exert selective pressure for large size, superior strength, and the ability to detect and obtain food with precision. This favors the evolution of muscular, sensory, and nervous systems—features unique to animals.

The structure of most animals boils down to a double-walled tube. The inner tube is the digestive system, which processes ingested food; the cells lining the tube are specialized to deal with food—to digest it, absorb it, and push it through the body. The tissues of the outer tube, on the outside of the animal's body, will inevitably be specialized for dealing with the outside world. Here we find protective structures such as skin, shells, horns, spines, and slime glands, as well as the sense organs and the nervous system, which tell an animal what is going on in the world around it. Between these inner and outer body layers are packed the other organs of the body: the muscles, blood vessels, reproductive organs, and so forth. These are concerned with internal functioning and have no direct contact with the outside world or the gut.

In this chapter we shall study the so-called lower invertebrates—animals whose body forms we believe originated quite early in evolution. In these forms we can trace a possible sequence of events in the evolution of the basic animal body plan. Before we consider the animals themselves, however, we shall look at some of the environmental pressures that must have helped to shape animal evolution.

27-A   Animal and Environment

Most early animals (and plants too) evolved in the sea, which is a much more stable and hospitable environment for life than either fresh water or land. For instance, the salt composition of the sea is very similar to that of a cell; most marine animals therefore have no osmotic problems of water gain or loss. The temperature of the sea changes slowly, within fairly narrow limits, and the small animals, plants, and protists in the **plankton** (the total of all floating organisms) provide a constantly renewed source of food. In contrast, fresh water usually contains fewer nutrients and so supports less life. It is also invariably hypotonic to living cells; a freshwater animal must constantly expend energy to retain its salts and expel the excess water that enters its body by osmosis. Land is an even more difficult environment, in that water is often in short supply, making death by dehydration a constant danger for many land animals. Relatively few animals have invaded fresh water successfully, and only two groups, the arthropods (spiders, insects, and their kin) and the higher vertebrates (reptiles, birds, and mammals), have really conquered the problems of life on land.

Despite its advantages, life in the sea poses certain difficulties. The surface of the sea, which contains most of the food, is constantly tossing about. Fish, which are vertebrates, swim well enough to maintain their positions despite water movement, but few invertebrates are powerful enough swimmers to do this. Invertebrates have evolved two sorts of adaptations that permit them to cope with the problem of being swept away by the sea. One is to remain small enough to float around as members of the **zooplankton**, or floating animals, close to their food supply of phytoplankton and smaller zooplankton. The alternative is for an animal to be **sessile**, firmly anchored to a stable object. Even invertebrates like crabs and worms, which are not sessile, lead a fairly sedentary life, creeping along over rocks and seaweed or burying themselves in sand.

Most sessile animals are filter feeders. They use cilia or muscles to set up water currents past (or through) their bodies and filter out or seize any food that appears in the current. Other sessile animals hang out a net of tentacles or mucus and trap passing food particles.

A sessile organism needs protection from mobile predators. It may have active protection such as stingers, or passive protection such as a thick shell or a coating of toxic mucus.

Sexual reproduction poses problems for sessile animals, since they cannot move around and encounter a mate. Most overcome this by behavioral adaptations that permit all the members of a species to release their sperm and eggs at the same time in response to environmental cues such as a particular water temperature or a full moon.

TABLE 27–1	**SOME LOWER INVERTEBRATE PHYLA**
Phylum Porifera (~10,000 species)	Sponges. Sessile animals with specialized cells but no tissues or organs; no anterior end, mouth, or digestive cavity; body often of indefinite form, organized around water canals; skeleton of spicules or spongin; most are marine.
Phylum Cnidaria (~10,000 species)	Coelenterates. Free-swimming or sessile; inner and outer tissue layers well-developed; mouth surrounded by tentacles bearing nematocysts; gastrovascular cavity; no anus; solitary or colonial; radial symmetry†; most are marine.
Phylum Platyhelminthes (~12,700 species)	Flatworms. Free-living or parasitic; bilateral symmetry†; three well-developed layers with organs; gut has mouth but no anus; most are marine or freshwater.
Phylum Nematoda (~12,000 species)	Roundworms. Free-living or parasitic; gut with mouth and anus; fluid-filled body cavity; marine, freshwater, and damp soil habitats.
Phylum Rotifera (~1500 species)	Rotifers. Sessile or free-swimming; anterior end with a ring of cilia around mouth; most are microscopic; almost all live in fresh water.

†Radial and bilateral symmetry are explained in Figure 27–13.

It is also noteworthy that many sessile animals can reproduce asexually by fission (dividing into, usually, two parts), or by budding (production of a smaller individual attached to its parent); the bud may later become independent, or may remain attached as part of a colony. Such asexual reproduction avoids the problem of synchronizing gamete release.

An adaptation of most marine invertebrates to their relatively sedentary life is the production of a tiny, **motile** (able to move about) larval stage in the life history. A **larva** is the juvenile stage of an animal. It usually differs from the adult in its structure and habitat, as well as in its food supply. The larval stage may function in dispersal, since it is small enough to live in the plankton and be swept away to colonize new habitats. Another advantage of a larval stage is that potential competition for food between the planktonic larva and its parents is reduced.

27-B   Phylum Porifera (Sponges)

Sponges are simple, multicellular animals (Figure 27–2). Although they show some cell specialization, they have no tissues and very little interdependence among their cells. This lack of interdependence was shown in 1907 by the embryologist H. V. Wilson, who discovered that a living sponge may be pushed through fine silk so

Figure 27-2 Sponges. (a) A solitary sponge, with a worm crawling over the edge of the hole where water exits. (b) A sponge colony in the Sea of Cortez; all the individuals push water out through a common opening (center). (Steven Webster)

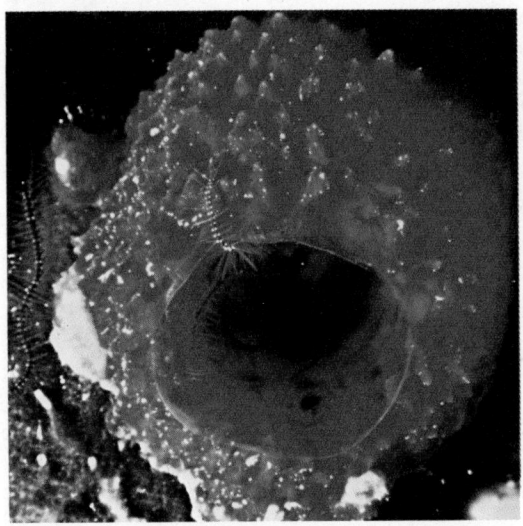

(a)

(b)

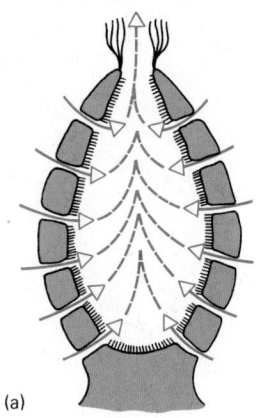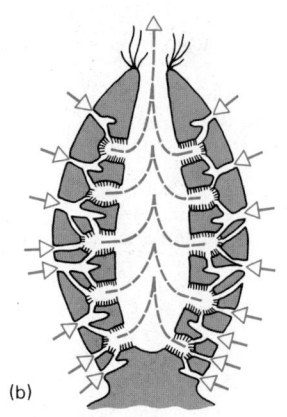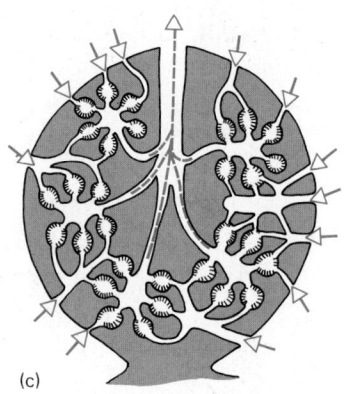

(a)   (b)   (c)

Figure 27-3 Three basic types of arrangement of the body wall and canals in sponges. Black fringe represents flagella of choanocytes (Figure 27-4), whose beating draws water into the sponge's body through numerous tiny openings on the sides. Water leaves the body through the opening at the top. Flagellated cells may line the main cavity of the body (a), or the body wall may be folded in such a way that the flagellated cells line pouches connected to more complicated canal systems (b, c).

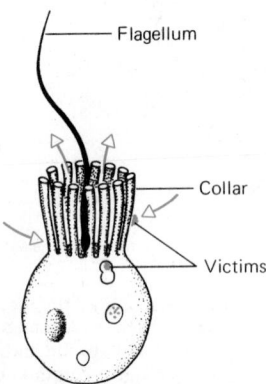

Figure 27-4 A choanocyte ("collar cell"), found only in sponges and some protozoa. Lashing of the flagellum draws a current through the collar, a ring of projections (microvilli) which trap food particles. Food is ingested by endocytosis and digested intracellularly.

that it is broken up into individual cells and cell debris. If this mixture is then allowed to stand, the cells start crawling around and aggregate into larger masses. Finally, after about three weeks, a functional sponge is re-formed. Such treatment would kill any other adult animal.

Sponges are sessile and move very little; the ancient Greeks thought they were plants. In the eighteenth century, however, people discovered that sponges have flagella that set up water currents through the body, and sponges were, therefore, reclassified as animals. Sponges are **filter feeders**; food particles (including small organisms) brought into the body by water currents are ingested by special **choanocytes**, which are unique to sponges (Figure 27-4); the choanocytes pass food on to amoeba-like cells, which distribute it to other cells. Waste is ejected into the water current, and gas is exchanged by diffusion across the cell surfaces.

The size of sponges varies from a few millimetres to the size of a bathtub. Their structure is maintained by a skeleton of a fibrous protein, **spongin** (as in the once-familiar household sponge), or of **spicules** made of calcium carbonate or silica. Sponges are classified according to the chemical composition and structure of these skeletal structures.

Like most other animals, sponges produce flagellated sperm and immobile eggs. These may be carried into the sea by water currents leaving the sponge, or fertilization may take place in the interior cavity of the sponge. The zygote undergoes mitosis to form the embryo, a hollow ball of cells, which usually develops cilia and becomes a free-swimming larva. The larva lives among the plankton for a while and then settles to the bottom, attaches, and forms a young sponge.

Although most sponges are marine, there are a few freshwater forms. Because they are sessile, sponges can live only in clear water, which is free of debris that would bury them or clog their pores.

27-C Phylum Cnidaria (Coelenterates)

The phylum Cnidaria contains the sea anemones, corals, and jellyfish, some of the most beautiful of all animals. The inner and outer layers of the body are well developed, but the middle layer, or **mesoglea** (= "middle glue"), ranges in composition from a noncellular membrane to masses of jellylike, nonliving substance containing wandering, amoeba-like cells. The gut is not a complete tube, with mouth and anus, but is rather a blind sac, with one opening serving to ingest food and expel indigestible matter.

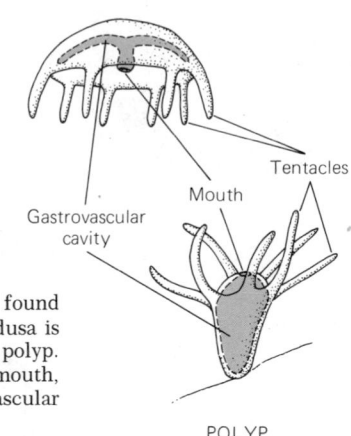

Figure 27-5 A siphonophore, a large colonial hydrozoan cnidarian, which swims by contractions of its many bells. It traps prey, which include quite large fish, with nematocysts on its trailing tentacles. (J. M. King, courtesy of Alice Alldredge)

MEDUSA

Tentacles

Mouth

Gastrovascular cavity

POLYP

Figure 27-6 The two body forms found in cnidarians. The free-floating medusa is an inverted version of the sessile polyp. Both have tentacles surrounding the mouth, which leads into a blind gastrovascular cavity.

There are two basic, and essentially similiar, body forms in the Cnidaria: **polyp** and **medusa** (Figure 27–6). The polyp is sessile, and the medusa is free-swimming, moving by weak contractions of its bell.

Most cnidarians are carnivores. They do not chase their food but are "sit and wait" predators that trap their prey. Special structures called **nematocysts** (Figure 27-7) are a unique characteristic of Cnidaria and are most concentrated on the tentacles, which surround the mouth. When a nematocyst is touched, it everts a threadlike structure. The thread may twist about bristles on the body of the prey, entangling it, or may secrete a sticky or a paralytic substance. The tentacles then pull the prey into the mouth. Digestion is started in the **gastrovascular cavity** (gastro = stomach; vascular = vessel), which functions in both digestion and internal transport. Small particles of food are engulfed from the cavity by amoeba-like cells, which complete digestion and distribute food to other cells. Any undigested remnants of the prey are ejected through the mouth opening.

Most cnidarians are marine, but there are a few freshwater forms.

Class Hydrozoa

The most primitive class of Cnidaria, the Hydrozoa, is thought to have contained the ancestors of both the other classes.

The polyp and medusa body forms may be combined in many different ways in the Hydrozoa. Usually polyps, either alone or in colonies, are the feeding stage in the

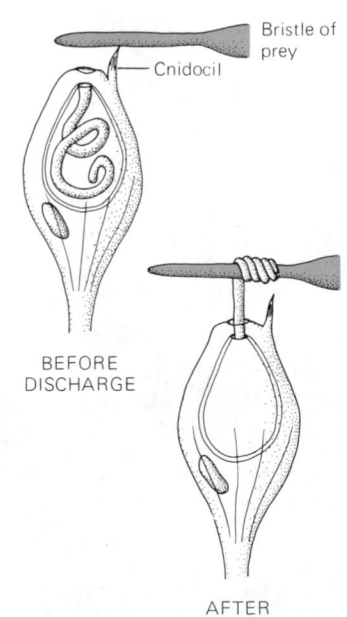

Bristle of prey

Cnidocil

BEFORE DISCHARGE

AFTER DISCHARGE

Figure 27-7 Nematocysts, specialized structures found in Cnidaria. The cnidocil acts as a trigger. Touching the cnidocil causes the nematocyst to eject its thread, which may entangle the prey (as shown) or secrete a sticky or poisonous substance that immobilizes the prey.

TABLE 27-2 **THE CLASSES OF THE PHYLUM CNIDARIA**

Class Hydrozoa (~3700 species)	Most have both polyp and medusa in the life history; polyp stage often colonial; most marine, a few freshwater; e.g., Portuguese man-of-war, *Hydra*, some corals, *Obelia*.
Class Anthozoa (~6100 species)	Medusa stage completely absent; sessile, often colonial; all marine; e.g., sea anemones, most corals, sea fans, sea pansies.
Class Scyphozoa (~200 species)	Polyp stage reduced; free-swimming; large medusas (up to 2 m diameter); all marine; e.g., jellyfish, sea wasp, sea nettle.

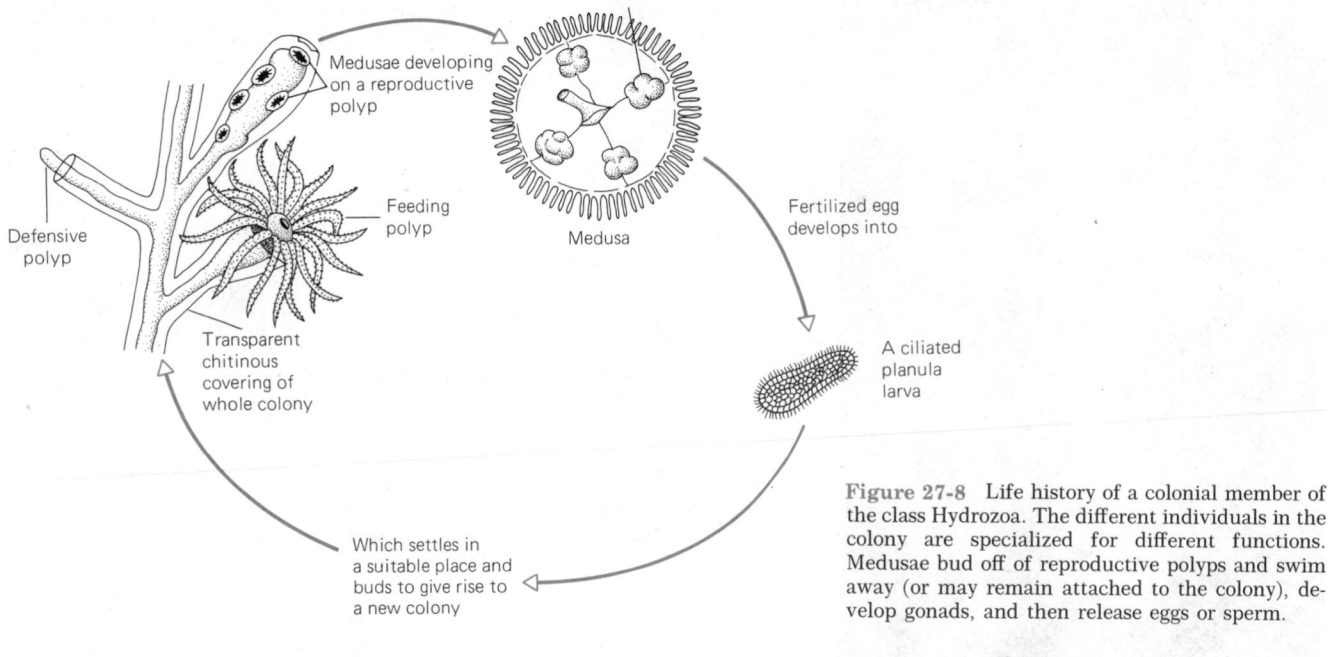

Gonad

Medusae developing
on a reproductive
polyp

Feeding
polyp

Defensive
polyp

Transparent
chitinous
covering of
whole colony

Medusa

Fertilized egg
develops into

A ciliated
planula
larva

Which settles in
a suitable place and
buds to give rise to
a new colony

Figure 27-8 Life history of a colonial member of the class Hydrozoa. The different individuals in the colony are specialized for different functions. Medusae bud off of reproductive polyps and swim away (or may remain attached to the colony), develop gonads, and then release eggs or sperm.

life history and reproduce asexually, giving rise to other polyps or to medusas. The medusas swim off and then reproduce sexually, giving rise to a new polyp generation (Figure 27–8).

Hydra, a freshwater genus that has no medusa in its life history, is the most commonly studied hydrozoan. This is unfortunate because it is not at all typical of the class.

Class Scyphozoa (Jellyfish)

Members of the cnidarian class Scyphozoa have a greatly reduced polyp stage; the most prominent stage in the life history is the planktonic medusa. Some species of Scyphozoa can swim quite actively by contraction of muscle fibers arranged around the bell. The almost transparent body of these animals makes them nearly invisible to prey, predator, and unwary swimmer alike (Figure 27–10).

Figure 27-9 Hydrozoan polyps feeding at night. (Steven Webster)

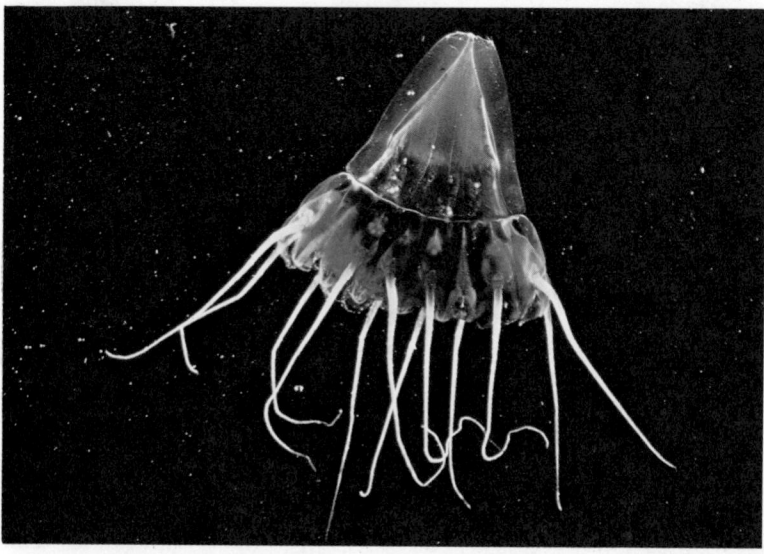

Figure 27-10 A scyphozoan medusa. (Bruce Robison)

Figure 27-11 The tiny polyps of a coral (class Anthozoa). (Steven Webster)

Figure 27-12 A sea anemone, member of the class Anthozoa. The tentacles trap food and pass it to the mouth (center). (Biophoto Associates)

RADIAL SYMMETRY

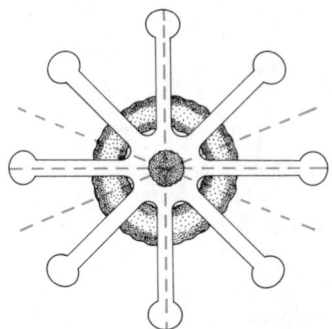

BILATERAL SYMMETRY

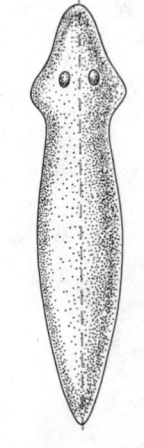

Class Anthozoa (Sea Anemones and Corals)

The third class of Cnidaria is the Anthozoa. All adult anthozoans have the polyp body form. The medusa stage has apparently been lost during evolution. A tiny ciliated larva is the distributive phase in the life history. Most corals are tiny colonial anthozoans with a calcareous skeleton common to the whole colony (Figure 27-11). They have played important roles in the formation of rock layers and islands in the warmer seas of the world. Sea anemones are larger anthozoans attached to rocks and other surfaces around the ocean shore, mainly between high and low water marks (Figure 27–12).

27-D   Body Symmetry

The bodies of cnidarians are **radially symmetrical**; that is, the animal can be divided into mirror-image halves by several different planes passing through its long axis (Figure 27–13). Many of the larvae of other invertebrates, as well as some protists, also have this type of body symmetry. Radial symmetry permits an animal to detect the approach of food or danger from any side, an advantage to a sessile or passively drifting animal. However, animals with well-developed muscular systems can move and obtain food more efficiently if they are **bilaterally symmetrical**, with only one plane that divides the body into two mirror-image halves. Bilateral symmetry allows an animal to have a more streamlined shape and to concentrate the power of its muscles and appendages into producing motion along a specific axis.

Accompanying the trend toward bilateral symmetry in animals is the evolution of **cephalization** (kephale = head)—the concentration of sensory and nervous tissue at the "head" end of the body, where it can monitor the area that the animal is entering and adjust the direction of motion before the rest of the body follows.

Figure 27-13 Radial and bilateral symmetry. In radial symmetry, a cut made in several planes perpendicular to the page would cut the animal into mirror-image halves. A bilaterally symmetrical animal has only one plane of symmetry; the colored line shows the only plane through which the animal could be cut to yield mirror-image halves.

TABLE 27-3	**THE CLASSES OF THE PHYLUM PLATYHELMINTHES**
Class Turbellaria (~3000 species)	Free-living flatworms. Predators and scavengers; body surface ciliated; most marine, some freshwater, a few terrestrial.
Class Trematoda (~6000 species)	Flukes. Ectoparasites and endoparasites, almost always of vertebrates; two suckers attach to host; life history with or without intermediate host.
Class Cestoda (~1500 species)	Tapeworms. Endoparasites of vertebrates; scolex attaches to host; proglottids produce eggs and break off after fertilization; no head or digestive system; life history with one or more intermediate hosts.

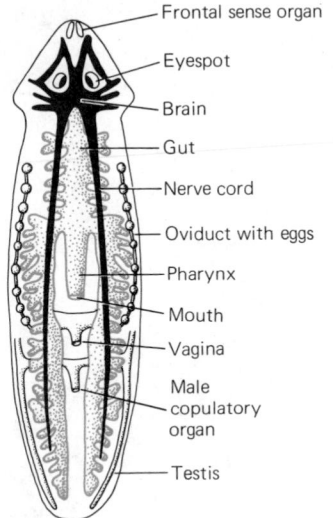

Figure 27-14   The internal anatomy of a free-living platyhelminth worm.

(Labels: Frontal sense organ, Eyespot, Brain, Gut, Nerve cord, Oviduct with eggs, Pharynx, Mouth, Vagina, Male copulatory organ, Testis)

27-E   Phylum Platyhelminthes (Flatworms)

The phylum Platyhelminthes is the most primitive group of animals that show bilateral symmetry, the beginnings of cephalization, and the development of true organs instead of the rather simple tissues found in members of the Cnidaria (Figure 27–14). Flatworms also have a well-developed middle layer, containing their **gonads** (reproductive organs), an excretory system, and distinct layers of muscles.

Flatworms retain some primitive characteristics: the mouth is still the only opening into the digestive tract (there is no mouth at all in the tapeworms), and there is still no specialized circulatory system or blood to carry substances around the body. The gut and excretory organs extend throughout the body and carry out this function. Gas is exchanged across the general surface of the flattened body.

The phylum Platyhelminthes includes three classes of worms. Two of these, the Trematoda and Cestoda, contain only parasitic forms.

Class Turbellaria

Members of the class Turbellaria are free-living flatworms with cilia distributed over the entire body surface. They are often recognizable by eyespots on the top of the head; most are marine but a few, like the planarian *Dugesia* (often studied in the laboratory), live in fresh water. A few species are terrestrial, but these spend most of their time in moist areas, under logs and leaf mold, for example. Turbellarians feed on smaller organisms or on dead organic matter (Figure 27–16). Small bits of meat are used as bait to attract turbellarians for collection. The ventral surface (underside) of the body secretes a slimy mucus that serves as a track for the worm to glide along, propelled by the beating of its cilia.

Many turbellarians reproduce asexually by budding or by transverse fission. Their sexual reproduction is highly variable, but in general they are **hermaphroditic**; that is, they have both male and female organs in the same individual. Sexual reproduction occurs by mutual copulation between two individuals, with internal fertilization. The embryos of some species hatch directly from the eggs as miniature adults. In other species, a ciliated planktonic larva is formed.

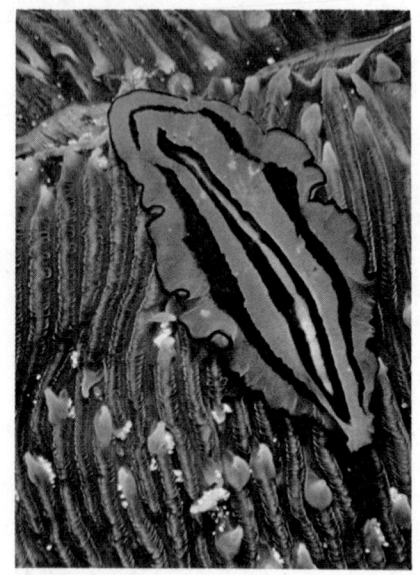

Figure 27-15   A turbellarian crawling across coral. (Biophoto Associates, N.H.P.A.)

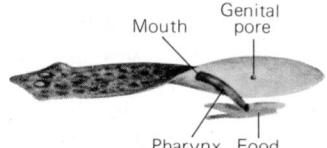

(Labels: Mouth, Genital pore, Pharynx, Food)

Figure 27-16   A turbellarian feeding. The pharynx is everted through the mouth and food is sucked into the digestive system. Undigested food is expelled through the mouth.

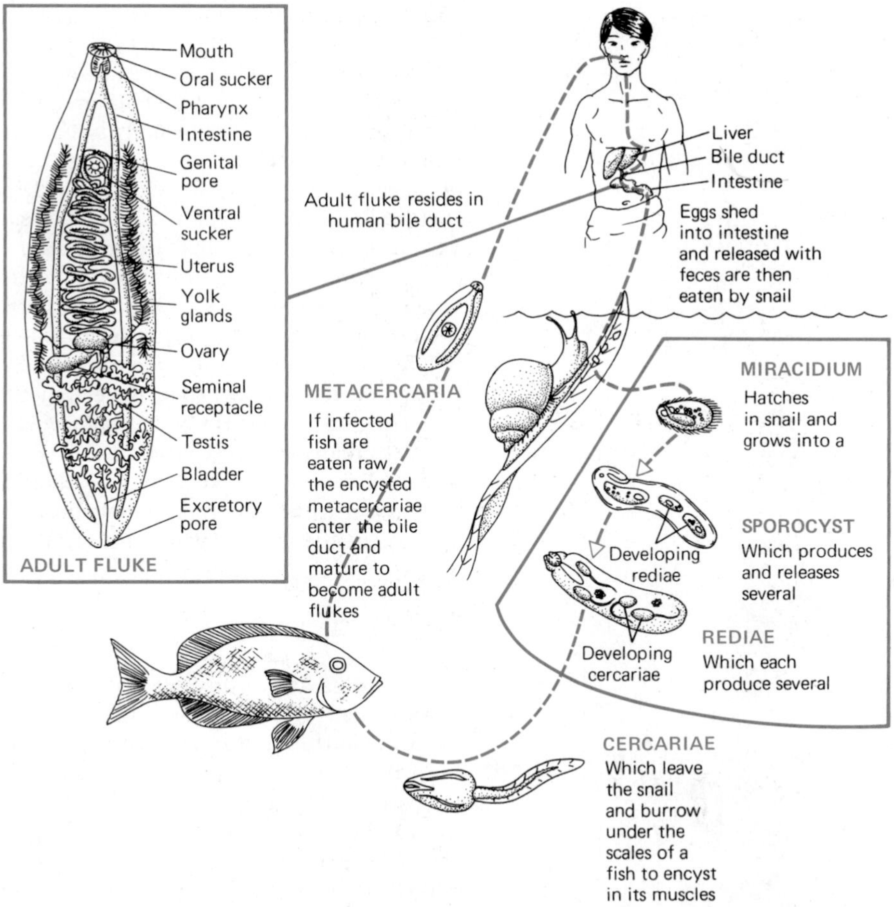
Mouth
Oral sucker
Pharynx
Intestine
Genital pore
Ventral sucker
Uterus
Yolk glands
Ovary
Seminal receptacle
Testis
Bladder
Excretory pore
ADULT FLUKE
Adult fluke resides in human bile duct
Liver
Bile duct
Intestine
Eggs shed into intestine and released with feces are then eaten by snail
MIRACIDIUM
Hatches in snail and grows into a
SPOROCYST
Which produces and releases several
Developing rediae
REDIAE
Which each produce several
Developing cercariae
CERCARIAE
Which leave the snail and burrow under the scales of a fish to encyst in its muscles
METACERCARIA
If infected fish are eaten raw, the encysted metacercariae enter the bile duct and mature to become adult flukes

Figure 27-17 Life history of the Chinese liver fluke, *Opisthorchis sinensis.* Eggs eaten by a suitable aquatic snail may reproduce asexually and make several hundred offspring each. The motile cercaria discharged from the snail finds a fish host, burrows through its skin, and forms a cyst in the muscle. An encysted metacercaria becomes active after the fish muscle is eaten by a suitable host; it makes its way to the bile duct, where it matures to adulthood.

Class Trematoda (Flukes)

Members of the class Trematoda, the flukes, are flatworms living as ectoparasites or endoparasites (external or internal parasites). Most of them are from one to a few centimetres long. They usually have adhesive organs or suckers near the mouth at the anterior end of the body, with a second sucker on the ventral surface. The muscular **pharynx** pulls host tissue or body fluids in through the mouth and passes it into the two branches of the intestine. Most of the body is occupied by reproductive organs. Most adult flukes are hermaphroditic, and they may produce half a million eggs during a lifetime. Many flukes have complex life histories involving several different hosts (Figure 27–17). The host of the adult is called the primary or definitive host; hosts of larval stages are called intermediate hosts.

Blood flukes of the genus *Schistosoma* infect about 200 million people in over 70 nations in the tropics. These flukes cause enlargement of the liver and spleen, disorders of the urinary tract, bloating of the abdomen, and wasting of the arms and legs; there is as yet no known cure. This problem is growing as the building of dams creates more habitat for the aquatic snails that serve as intermediate hosts. Fluke eggs enter the water in human feces, and cercariae leave their snail hosts and burrow through the skin of people standing in the water. As a general rule, it is a good idea to avoid swimming in fresh water in the tropics.

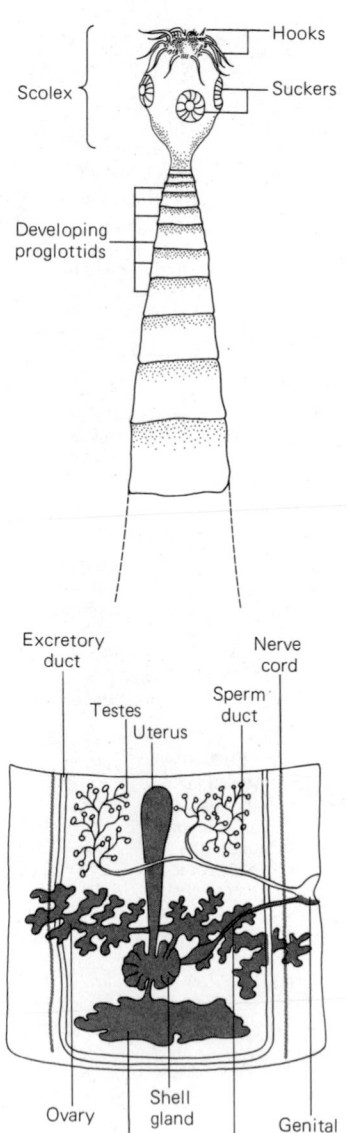

Figure 27-18 Anatomy of the pork tapeworm, *Taenia solium*. The hooks and suckers of the scolex attach to the wall of the host's digestive tract. Proglottids develop just behind the scolex. A mature proglottid (below) is little more than a reproductive system containing both male and female organs, which may fertilize one another or mate with another proglottid. Ripe proglottids detach from the end of the worm and are shed in the host's feces.

Class Cestoda (Tapeworms)

Tapeworms are highly specialized parasites found in the intestines of probably every species of vertebrate. Their anatomy is unlike that of the Turbellaria or flukes. The anterior of the body usually has a **scolex**, bearing a circle of hooks or suckers, which attach to the lining of the host's intestine. A tapeworm has no head, mouth, or digestive system. It absorbs digested food from the host's gut through its body surface. Immediately behind the scolex is a growing region that constantly produces new body sections, called **proglottids**, by budding. A proglottid contains little except reproductive organs. It mates with itself or with an adjacent proglottid and, when full of fertilized eggs, breaks away from the rest of the worm and passes out of the host with the feces. If the fertilized egg is then eaten by another host, the egg hatches and develops. Many cestodes must invade several different host species in succession to complete their development.

The pork tapeworm *Taenia solium* is one of the more common tapeworms that infect humans in the United States (Figure 27–18). People can avoid becoming hosts to this worm by cooking pork thoroughly, so as to kill any worm cysts in the pork muscle. Fish are intermediate hosts for many species of tapeworms that infect humans in the Orient, where raw fish is a popular dish.

27-F Parasitism

Parasites are organisms that extract their food from living hosts. **Ectoparasites** attach to the outside of the host; examples are ticks, fleas, and some flukes. **Endoparasites** live inside the host's body. Many phyla of animals include parasites; two of the most important parasitic groups are the flatworms, which we just discussed, and the roundworms, covered in the next section.

Finding food is often difficult for parasites because appropriate hosts may be few and far between. Many parasitic worms compensate by producing hundreds of thousands or millions of eggs, the large numbers ensuring that at least a few find a host; most of the young starve to death, unable to find a suitable host. In some fluke species, larvae that successfully find hosts reproduce asexually, further increasing the chances of some individuals' surviving to adulthood. Parasitic worm life histories often involve intermediate hosts, and there may be a cyst stage that can survive for a long time, until it meets an appropriate host.

Internal parasites must have certain anatomical and physiological adaptations to their way of life. For example, roundworms have a protective cuticle covering the outside of the body; this prevents them from being digested by their hosts.

A parasite's food is pretty much pre-digested by its host. Intestinal parasites are surrounded by digested food, and other parasites need only suck in nutritious blood. There is not much for the parasite's digestive system to do, and most parasites have reduced digestive systems. The energy freed by this savings is devoted to expansion of the reproductive system. In fact, tapeworms have no digestive system whatever; they absorb all their food across the body wall, and their bodies are little more than "egg factories."

An interesting complication of a parasitic way of life is that a parasite is unsuccessful if it does too much damage to the host. From the point of view of the parasite, the ideal host-parasite relationship is one in which the host remains alive long enough to permit the parasite to reproduce many times. There is thus strong selection for those parasites that do not kill their hosts.

The worst outbreaks of diseases occur when parasites—be they viruses, bacteria, protists, fungi, or animals—first come into contact with a particular population of hosts. History is full of examples of parasites, such as the plague or syphilis bacteria, that wiped out huge proportions of new-found host populations, but this also, of course, wiped out most of the parasites. Such an initial evolutionary encounter selects heavily for those hosts with defense mechanisms against the parasite, and for those parasites that are less virulent, until, after a while, the parasite will do less damage to its new-found host population than it did initially.

TABLE 27-4 SOME PARASITIC WORMS COMMON IN HUMANS

NAME	SYMPTOMS	MEANS OF INFECTION
Platyhelminthes		
Chinese liver fluke	None in mild cases; destruction of liver, bile stones, and clogging of liver ducts in severe cases	Eating raw fish
Blood fluke (schistosomes)	Enlargement of liver and spleen Urinary disorders Bloated abdomen, wasted arms and legs	Drinking or wading barefoot in water containing infected person's urine. Infects about 200 million people in 70 nations. Not found where there are modern sewage disposal systems.
Swimmer's itch	Itching after exposure of skin to infested water	Burrowing of fluke larvae of species that cannot successfully infect humans
Bladder worm (*Echinococcus*) (immature stage of a worm that lives in dogs when adult)	Cysts up to the size of an orange; symptoms depend on part of body invaded	Infected dogs licking people's hands or faces or contaminating drinking water
Pork, beef, and fish tapeworms	Immature worms: cysts Adult worms may cause diarrhea, loss of weight, perforation of intestine	Eating undercooked meat containing worm cysts
Nematoda		
Pinworm	Anal itching	Females lay eggs around anal opening: hands may transfer eggs to mouth, maintaining infection in same person. Physical contact may also transfer to other people
Hookworm	Anemia, lethargy	Young worms burrow through skin (bare feet) from moist soil and grass contaminated by feces of infected humans

27-G   Phylum Nematoda (Roundworms)

Most nematodes, or roundworms, are so small that they attract little notice. However, they are important inhabitants of marine and freshwater mud and of the soil; any handful of soil contains thousands of tiny white roundworms. Some roundworms are unpleasant parasites.

A nematode contains nearly all of the kinds of organs and tissues that one finds in even the most complex animal: a digestive tract with mouth and anus, muscles, nerves, excretory organs, and reproductive organs (Figure 27–19). The only major organ system that is lacking is a circulatory system, but nematodes do have a **pseudocoelom**, a body cavity between the digestive tract and the other organs, which is filled with fluid that may carry food molecules and dissolved gases from one place to another. A tough cuticle covers the outside of the body.

About 12,000 species of nematodes are known, and many thousands more species are undoubtedly still to be described. About 50 species parasitize humans; the most harmful of these are the hookworm, the intestinal roundworm *Ascaris lumbricoides*, the guinea worm, the filaria worm, and *Trichinella spiralis*, which causes trichinosis.

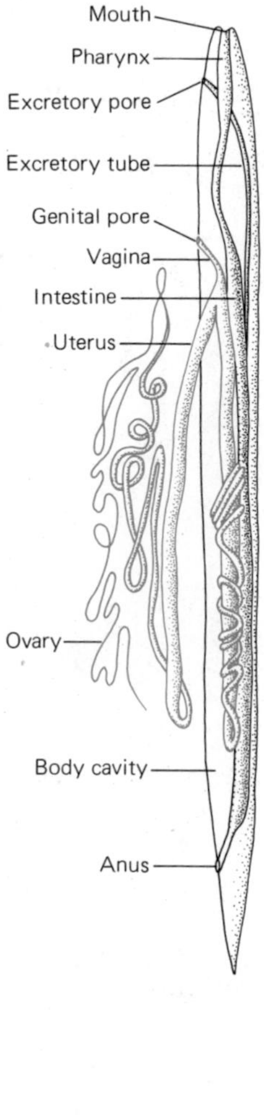

Mouth
Pharynx
Excretory pore
Excretory tube
Genital pore
Vagina
Intestine
Uterus
Ovary
Body cavity
Anus

Figure 27-19 Internal anatomy of a female *Ascaris,* a parasitic roundworm. Advances over platyhelminths include a gut with an anus as well as a mouth, and a fluid-filled body cavity. As in most parasites, reproductive structures (color) are prominent.

417

27-H   Phylum Rotifera

Rotifers are among the lesser known invertebrates, but they are very likely to be found in a jar of water taken from a pond or stream, from a roof gutter, or from water trapped in a clump of moss or in a cemetery urn. Many rotifers are no larger than some protists, but they are nevertheless multicellular and possess all of the organs listed above for the nematodes; in fact the two phyla are believed to be closely related.

Rotifers are also known as "wheel animals" because their mouths are surrounded by a crown of cilia whose beating looks like a wheel turning. The cilia create a feeding current, which pulls small organisms and organic particles into the animal's mouth. Jaws in the pharynx grind the food and pass it on to the stomach and intestine. Although small males are produced occasionally, most rotifers are females, and most produce only one, comparatively large, egg at a time (Figure 27–20).

A peculiarity of rotifers is that the adults' cells never divide. Mitosis ceases following embryonic development, and the adult cannot grow or repair itself. Not surprisingly, rotifers are used in research designed to determine what causes cells to undergo mitosis.

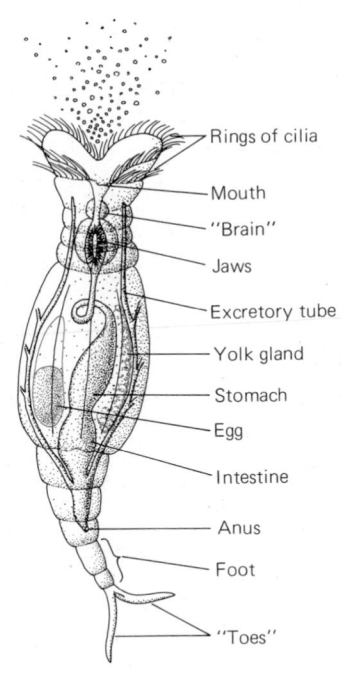

Rings of cilia

Mouth

"Brain"

Jaws

Excretory tube

Yolk gland

Stomach

Egg

Intestine

Anus

Foot

"Toes"

Figure 27-20   Anatomy of a rotifer. Internal organs are easily seen through the transparent body wall of many forms. The "toes" are often provided with glands that secrete "glue" by which the rotifer can anchor itself to the substratum. Most rotifers are females, and their offspring are usually more females (note the single large egg).

SUMMARY

The so-called lower invertebrates introduced in this chapter are believed to have evolved in the sea, where most forms still live. These animals are either planktonic, or bottom-dwelling with planktonic larvae. Some have adapted to life in fresh water, in damp terrestrial habitats, or in the watery interior of a host's body.

The sponges, phylum Porifera, exhibit a primitive level of organization, with specialized cells but no tissues or organs. They are sessile filter feeders, using their flagellated choanocytes to draw water into the body and strain out food particles. Their cells show marked independence of one another.

Members of the phylum Cnidaria have two well-developed tissue layers separated by a mesoglea, but they lack most organs. Both sessile polyps and swimming medusas have the same body plan—a ring of tentacles bearing nematocysts surrounds the mouth, which leads into a blind sac, the gastrovascular cavity. *Hydra*, jellyfish, corals, and sea anemones belong to this group.

The Platyhelminthes, or flatworms, show several more advanced features, including bilateral symmetry, cephalization, and three well-developed tissue layers; there are several organ systems, but not a circulatory system. The digestive system has only one opening, the mouth. Free-living flatworms and parasitic flukes and tapeworms belong to this phylum.

The Nematoda (roundworms) and the Rotifera ("wheel animals") possess complete digestive tracts, with both mouth and anus, and many other organ systems. There is a fluid-filled body cavity that may serve for transport, but no circulatory system as such.

OBJECTIVES

From your study of this chapter, you should be able to:

1. Use the following terms correctly: invertebrate, primitive, plankton, zooplankton, sessile, motile, larva, filter feeder, hermaphroditic, parasite, ectoparasite, endoparasite.
2. List the characteristics of the kingdom Animalia and of the phyla Porifera, Cnidaria, Platyhelminthes, Nematoda, and Rotifera.
3. When presented with any member of the phyla in objective 3, name the phylum to which it belongs.
4. Explain the adaptive advantages of the following characteristics: radial symmetry, bilateral symmetry, cephalization, mobile stage of the life history, sessile stage of the life history.

SELF-QUIZ

1. Sea anemones are members of the phylum _____ .
2. Sponges are members of the phylum _____ .
3. An animal that must move a great deal will experience selective pressures favoring (bilateral, radial) symmetry.

Matching: For each group listed below, pick out *all* the characteristics listed which pertain to that group.

	Group		*Characteristic*
_____	4. Rotifera	a.	gut with mouth and anus
_____	5. Platyhelminthes	b.	sessile members
_____	6. Nematoda	c.	parasitic members
_____	7. Cnidaria	d.	excretory system
_____	8. Porifera	e.	nematocysts
		f.	no gut present
		g.	choanocytes
		h.	bilateral symmetry

QUESTIONS FOR DISCUSSION

1. Most freshwater invertebrates have lost the larval stage found in their marine relatives' life histories. Can you explain an adaptive advantage of this situation?
2. How do the organisms covered in this chapter get along without special organs for gas exchange and internal transport?
3. What is the adaptive advantage to a sponge of the complex series of canals and feeding chambers shown in part (c) of Figure 27-3 compared to the simple structure in part (a)?
4. What selective pressures have made it advantageous for Chinese liver flukes to produce hundreds of eggs at a time and for rotifers to produce only one or a few at once?

REFERENCES AND FURTHER READING

Barnes, R. D. *Invertebrate Zoology*, 4th ed. Philadelphia: Saunders, 1980. An excellent text and reference book covering the invertebrates except insects.

Buchsbaum, R. *Animals Without Backbones*, 2d ed. Chicago: University of Chicago Press, 1976. A classic elementary textbook on invertebrates.

Goreau, T. F., N. I. Goreau, and T. J. Goreau. "Corals and coral reefs." *Scientific American*, August 1979. Biology, growth, and ecology of reef-building corals.

CHAPTER

SOME HIGHER INVERTEBRATES

The platyhelminths and all higher animals develop from embryos with three major layers of cells: the **endoderm**, which comes to form the lining of the gut; the **ectoderm**, which comes to form the skin, sense organs, and nervous system; and the **mesoderm**, which forms between the other two layers and gives rise to the muscles, blood vessels, reproductive organs, kidneys, glands, and so forth (Figure 28–1). In the annelid worms, described in this chapter, a space between layers of the embryonic mesoderm develops into a **coelom**, a fluid-filled body cavity. All annelids and the animals that evolved after them (molluscs, arthropods, echinoderms, and chordates) are basically coelomate, although the coelom has secondarily disappeared in the members of some phyla.

The coelomate animals may be divided into two evolutionary lines, largely on the basis of their early embryonic development, which often provides evidence of phylogenetic relationships. One line contains those animals known as **protostomes**, including three major invertebrate phyla, the annelids, molluscs, and arthropods, which we shall consider in this chapter. Higher invertebrates of the phyla Echinodermata and Chordata belong to the other, or **deuterostome**, evolutionary line, leading to the vertebrates; they are considered in Chapter 29.

28-A The Coelom

The origin of the coelom was—for several reasons—one of the most important steps in animal evolution. First, the coelom separates the muscles of the gut from the muscles of the body wall; as a result, the gut can move independently of the body wall, and movement of food down the gut is not dependent on locomotory movements. Second, the fluid that fills the coelom may act as a simple circulatory system, transporting waste, food, and gases around the body. More importantly, the coelom provides space where a true circulatory system with blood vessels can develop and function without being squeezed by other organs. A heart can pump only in such a free space; a constant blood flow would be impossible if the heart were squashed by other organs every time the animal moved a muscle. Without a circulatory system, every molecule of oxygen an animal uses must diffuse across the body surface to each cell that needs it. Because the surface-to-volume ratio of an organism decreases as body size increases, such a system will support respiration only in a very small, flat, or sluggish animal. Similar considerations apply to the transport of food and waste to and from the body cells. A circulatory system immediately overcomes these limitations. Possession of a circulatory system, made possible by the evolution of a coelom, permitted the increases in size and metabolic rate seen in many higher animals.

A coelom also provides space in which the gonads (reproductive organs) can expand, allowing gametes (eggs and sperm) to accumulate over a long period of time. Most coelomates have brief, annual breeding seasons during which they produce large numbers of offspring at the time most favorable for their development.

28-B Phylum Annelida

Annelids, or segmented worms, differ from the platyhelminth flatworms and the nematode roundworms in several ways. First, annelids possess a coelom. In addition, the body of a typical annelid is **metamerically segmented** (Figure 28–2), or divided into repeated sections by partitions called **septa** (singular: **septum**). Most annelids also differ from other worms in having a **closed circulatory system**, consisting of blood vessels with muscular walls. Some of the vessels are enlarged and especially muscular, and function as "hearts," pumping blood through the vessels.

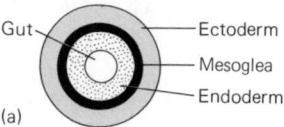

TWO LAYERS, NO COELOM
(e.g. CNIDARIA)

Gut — Ectoderm — Mesoglea — Endoderm

(a)

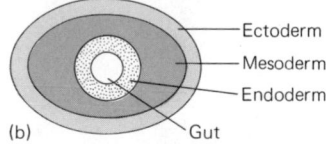

THREE LAYERS, NO COELOM
(e.g. PLATYHELMINTHES)

Ectoderm — Mesoderm — Endoderm

(b) — Gut

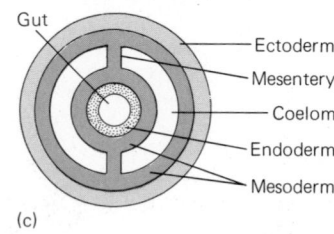

THREE LAYERS, COELOM
(e.g. ANNELIDA)

Gut — Ectoderm — Mesentery — Coelom — Endoderm — Mesoderm

(c)

Figure 28-1 Body plans of members of some animal phyla in cross section. (The function of the mesenteries is to suspend organs in the coelom.) Cnidaria (a) and Platyhelminthes (b) were discussed in Chapter 27.

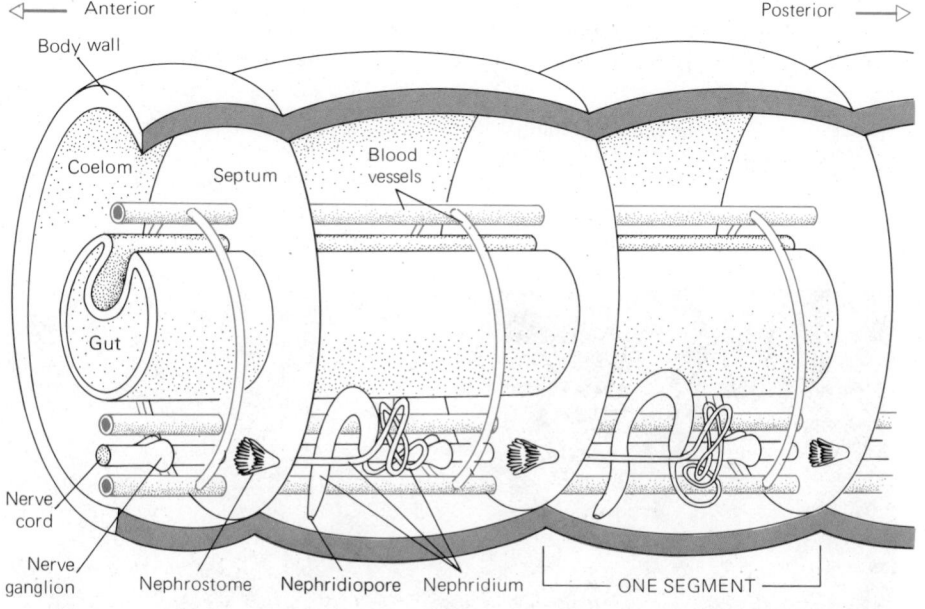

Figure 28-2 The metameric, or segmented, structure characteristic of annelids is illustrated by this diagram of part of an earthworm. Organs such as nerve ganglia, nephridia (excretory organs), etc., are repeated in most metameres of the body.

TABLE 28-1 THE PHYLUM ANNELIDA AND ITS MAJOR CLASSES

Phylum Annelida	Segmented worms; coelomate, with circulatory system (except Hirudinea); setae in some; gut with mouth and anus; respiration by skin or gills.
Class Polychaeta (~3500 species)	Mostly marine, segmented worms; well-developed head; each segment usually bears parapodia with setae; tube-dwelling and free-crawling; development via trochophore larva; e.g., *Aphrodite* (sea mouse), *Sabella* (fanworm), *Nereis* (ragworm), *Arenicola* (lugworm).
Class Oligochaeta (~2400 species)	Terrestrial and freshwater segmented worms; parapodia absent, setae present; head reduced; e.g., *Lumbricus* (earthworm), *Megascolides* (giant Australian worm), *Tubifex* (often used to feed pet fish).
Class Hirudinea (~300 species)	Leeches. Body usually flattened, segmentation reduced, coelom reduced; setae absent; anterior* and posterior* suckers; ectoparasites, predators, and scavengers; e.g., *Hirudo medicinalis* (medicinal leech).

Annelids use gills or the general body surface for exchanging gases with their environment. Their excretory organs are **nephridia**, tube-like organs leading from the coelom to the exterior. The nephridia excrete nitrogenous wastes (and some water and salts). The cells lining the nephridial tube absorb specific substances from the fluid in the tube, and in this way they regulate the composition of the body fluid.

Class Polychaeta (Bristle Worms)

Of the three annelid classes, the polychaetes probably resemble the ancestral annelids most closely. The polychaetes are almost all marine; most live in mud and sand at the bottom of shallow coastal waters, although a few species are planktonic. Most are free-living, but a number build tubes or live in permanent burrows. Some polychaetes have **gills**, extensions of the body surface that increase the surface area available for gas exchange. The polychaetes have external fertilization and nearly all develop via a planktonic **trochophore** larva (Figure 28–3).

A free-living polychaete is readily identified by the pair of fleshy, paddle-like appendages, called **parapodia**, on each segment (Figure 28–4). These parapodia usually bear stiffened **setae** (singular: **seta**), bristles composed of chitin.* *Aphrodite* is a polychaete commonly called a "sea mouse" because the entire dorsal (upper) surface is covered with a mat of fine setae that looks like fur.

---

*Anterior   At the front end of an animal.
*Posterior   At the rear end of an animal.
*Chitin   A stiff polysaccharide, which also makes up the outer covering of arthropods (e.g., insects).

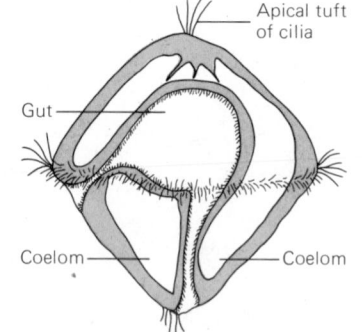

Figure 28-3 The planktonic trochophore larva of a polychaete (annelid) worm. Many molluscs (see Section 28-C) have a similar larva.

Labels: Apical tuft of cilia; Gut; Coelom; Coelom

Figure 28-4 A nereid worm, a polychaete, wriggling across a coral. Note the segmented body and the numerous setae, borne on short parapodia, on each segment. (Steven Webster)

Figure 28-5 A serpulid worm, a polychaete that lives in a tube. Its feathery, spiral arms filter food out of the water. (Steven Webster)

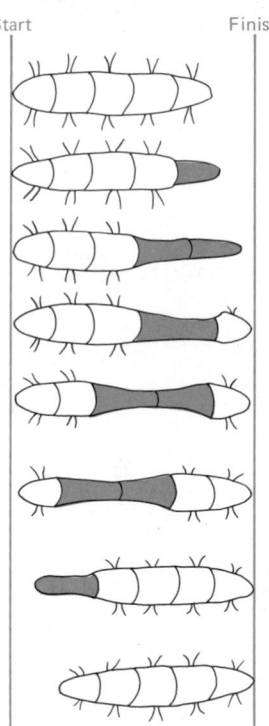

The burrowing polychaetes, such as the lugworm *Arenicola*, much prized as bait, have reduced parapodia. They burrow by expanding some segments of the body to grip the sides of the burrow with their setae. The segments forward of the anchored point become long and thin so that they can be pushed through the mud or sand. The front segments then expand to grip the side of the burrow, while the segments behind relax their hold and are pulled forward. This type of movement can occur because the annelid body is divided into segments, each with a compartment of fluid-filled coelom (see Figure 28-6). If the coelom were not partitioned by septa, muscular contraction in the middle of the body would push the coelomic fluid both forward and backward, and it would be impossible to expand individual segments to grip the burrow while other segments were contracted. The coelom and its fluid thus act as a water-filled, or **hydrostatic**, skeleton, rather like a bag filled with water. Because the fluid cannot be compressed, it provides a firm base against which the muscles can work. Furthermore, since this skeleton is fluid rather than solid, it can change shape as the muscles contract against it, permitting great flexibility. Anyone who has dug for lugworms will testify that this system permits them to burrow rapidly and efficiently.

The habit of living in tubes has evolved separately in a number of polychaete families (Figure 28-5). Tube-dwelling polychaetes build their tubes by cementing sand grains and bits of shell together with mucus and other secretions. Their parapodia are usually reduced or absent, and special feeding arms are developed from the head. The tube-dwellers are carnivores, feeding on other animals, or filter-feeders, straining small organisms or food particles out of the surrounding water.

Figure 28-6 An earthworm crawling, using its hydrostatic skeleton. Each segment acts as a closed sac of fluid surrounded by two sets of muscles, which can contract alternately so as to shorten and lengthen the segment. When the segment is short and wide, its setae (black bristles) are extended, pushing against the substrate and keeping the worm from slipping backwards. Each segment is extended in turn (color); at the end of the sequence the worm has moved forward.

Class Oligochaeta (Earthworms and Their Relatives)

Oligochaetes are freshwater, terrestrial, and marine annelids, most of which burrow in mud or soil; others live among water weeds. Some are less than half a millimetre long, while the giant Australian earthworm may exceed 3 m in length. Like polychaetes, oligochaetes are metameric and have setae, but they never have parapodia. The oligochaete head is reduced, probably because of the burrowing habits of most of these worms. Like burrowing polychaetes, oligochaetes move by muscular contractions working against a hydrostatic skeleton: each part of the body in turn is extended, anchored and contracted (Figure 28-6).

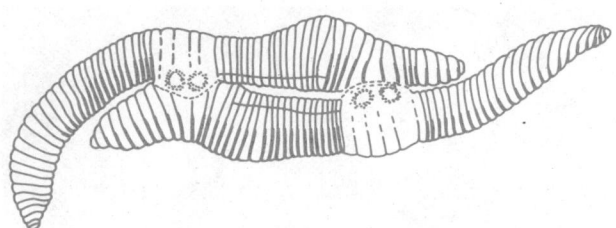

Figure 28-7 Earthworms copulating.

Most oligochaetes feed on dead organic matter, particularly plant material. Earthworms and other species ingest soil as they burrow, digesting any organic matter on the soil particles and voiding the inorganic particles as worm casts, which can often be found in a garden. Unlike the polychaetes, in which the sexes are separate, the oligochaetes are **hermaphroditic**; that is, they contain reproductive organs of both sexes. During copulation, two animals touch each other and each passes sperm to the other (Figure 28–7). Fertilized individuals later lay eggs in a cocoon.

Class Hirudinea (Leeches)

There are only about 300 species of leeches, making this the smallest annelid class. Most live in fresh water, but there are some marine species and a few species that are found in moist habitats on land. Many leeches are external parasites (**ecto-parasites**) that suck the blood of their victims. Fish, turtles, amphibians,* water birds, and snails are all preferred hosts of one species or another. Not all leeches are ectoparasites, however; many are predators, eating other animals, or scavengers, eating dead organic matter.

Leeches vary in length from about 0.5 to 20 cm. The body is often flattened. There are usually 34 segments, but these may look from the exterior as if they are subdivided; inside the body, the segmentation is much reduced.

There is a sucker at each end of the body. Some leeches move inchworm-fashion by attaching the posterior sucker to a surface, extending the body forward, attaching the anterior sucker, and pulling the posterior up to join the front end. Others are swift and graceful swimmers, throwing their long bodies up and down into a series of curves.

In the leeches, the coelom is reduced by the presence of blocks of muscle used in locomotion and in feeding, and is not used as a hydrostatic skeleton. A series of spaces or **sinuses** replace the coelom and circulatory system. Gas exchange is usually over the general body surface. Like oligochaetes, leeches are hermaphroditic. Leeches often stay with their eggs, arching the body over them and moving rhythmically up and down, thus keeping the eggs ventilated with fresh oxygen-bearing water. The young hatch as little leeches, not as larvae; they often fasten onto the parent until they have had their first meal.

Only a few species of leeches have the rasping "teeth" needed to break through the tough skin of a mammal. Species that do have these teeth also secrete a local anesthetic that keeps the wound from coming to the attention of the victim, and an anticoagulant that keeps the blood from clotting. Their powerful sucking muscles can remove blood quickly, and their distensible gut and body wall allow them to swell to many times their normal size as they gorge themselves. Opportunities to feed may be infrequent, and leeches can survive for months between meals.

In the last century, Western doctors often relied on the medicinal leech, *Hirudo medicinalis*, to bleed patients. Indeed, the doctors themselves were commonly known as "leeches," and every doctor kept an aquarium with leeches that could be popped into a handy container in his little black bag when he went out on housecalls.

---

* Amphibians Frogs, salamanders, newts, and their relatives.

TABLE 28-2 THE PHYLUM MOLLUSCA AND ITS MAJOR CLASSES

Phylum Mollusca	Bilaterally symmetrical with segmentation reduced; body covered by a mantle, which may secrete a shell; head and muscular foot usually present; gas exchange by gills or lining of the mantle cavity; excretion by nephridia; circulatory system with heart; development via trochophore-like larva or direct.*
Class Amphineura (~650 species)	Marine; shell of 8 plates; one foot, used for locomotion; head reduced; e.g., *Chaetopleura* (Atlantic chiton), *Chiton*.
Class Bivalvia (~11,000 species)	Marine and freshwater; flattened shell with 2 valves; head reduced; paired gills; most filter feeders; mantle forms siphons; e.g., *Tridacna* (giant clam), *Teredo* (shipworm), *Cardium* (cockle), *Pecten* (scallop), *Mytilus* (mussel), *Ostrea*, *Spondylus* (oyster), *Mya*, *Ensis* (razor clam), *Pinna* (pen shell).
Class Gastropoda (~75,000 species)	Marine, freshwater, or terrestrial; asymmetric body with a usually coiled shell; shell reduced or absent in some; foot for locomotion; radula present; e.g., *Patella* (limpet), *Haliotis* (abalone), *Littorina* (periwinkle), *Strombus* (conch), *Buccinum* (whelk), *Aplysia* (sea hare), *Aeolis* (sea slug), *Lymnaea* (pond snail), *Planorbis* (ramshorn snail), *Helix* (escargot), *Limax* (slug).
Class Cephalopoda (~650 species)	Marine; head surrounded by prehensile* tentacles, usually with suckers; shell external, internal, or absent; mouth with or without radula; locomotion by jet propulsion using siphon made from mantle; large eyes; direct development; e.g., *Nautilus*, *Octopus*, *Argonauta* (paper nautilus), *Sepia* (cuttlefish), *Loligo* (squid), *Architeuthis* (giant squid).

28-C   Phylum Mollusca

The phylum Mollusca contains chitons, snails, octopuses, and the "shell-fish"—clams, oysters, limpets, scallops, and the like. A prominent theory holds that molluscs and annelids share a common ancestor, which was partly segmented when the two lines diverged; most members of the molluscan line lost their tendency toward segmentation, whereas other molluscs and the annelids evolved more pronounced segmentation. The molluscan body consists basically of a muscular **head-foot**, with the body on top surmounted by a **mantle**, a flattened piece of tissue, which usually secretes a calcareous* shell. Between the mantle and the body lies a **mantle cavity**, used for gas exchange, and in some species, for feeding and locomotion (Figure 28–8). As in annelids, the excretory organs are nephridia.

---

*Direct   Egg hatches as a miniature version of the adult, with no distinct larval stage.
*Prehensile   Grasping.
*Calcareous   Composed of or containing calcium carbonate ($CaCO_3$).

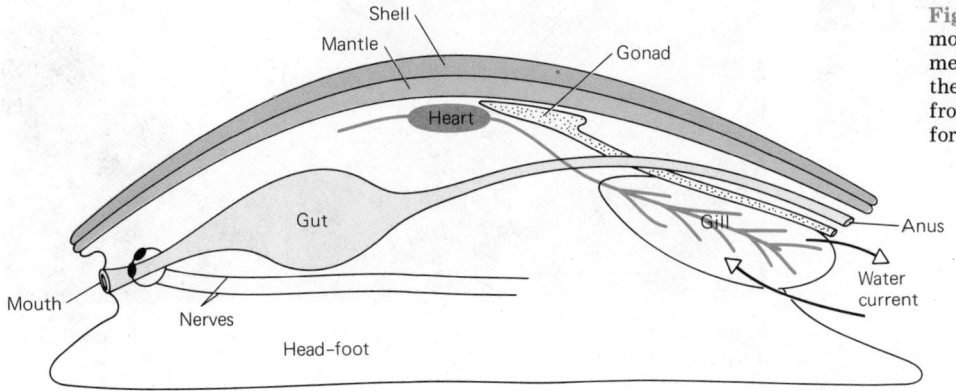

Figure 28-8 A hypothetical ancestral mollusc, showing the general anatomy of members of this phylum. The muscles of the lower body form a continuous mass from which both the head and foot are formed.

More than 100,000 species of molluscs have been described, making this one of the largest animal phyla, second only to the Arthropoda. Although most molluscs are marine, members of the phylum live in nearly every kind of habitat. Molluscs rank third in number of species adapted to terrestrial life, after arthropods and vertebrates. The molluscan species with shells are restricted to areas that provide enough calcium carbonate to make these shells, but naked molluscs, such as slugs, can survive without lime.

Class Amphineura (Chitons and Their Relatives)

The class Amphineura contains the chitons—oval, flattened creatures, with a tough mantle and eight crosswise skeletal plates (Figure 28–9). They are found in the intertidal zone and shallow water along the coast, firmly attached to rocks by their muscular feet. They move around very slowly, scraping algae off the rocks with a **radula**, which resembles a tongue made of sandpaper and which is continually replaced; the radula is an admirable algae-scraper. Respiration is by gills, which lie protected in the mantle chamber. When exposed to air at low tide, chitons, like many other molluscs, pull themselves firmly down onto a rock; the mantle covers the rest of the body tightly like a suction cup, and prevents desiccation (drying out). Eggs are discharged directly into the sea, where they are fertilized and develop by way of a trochophore larva.

Class Gastropoda (e.g., Snails, Slugs, Limpets)

In terms of number of species, the class Gastropoda is the largest molluscan class, containing many familiar forms such as snails and slugs. A single shell, which may be spirally coiled or a flattened cone, is characteristic of the class (Figure 28–10). Shelled gastropods start life by secreting a tiny shell and then secrete more shell around the opening, enlarging the shell as they grow. The shell is reduced or

Figure 28-9 A chiton grazing algae as it creeps over a rock on the seashore. Note the shell, composed of eight plates. (Steven Webster)

Figure 28-10 A terrestrial gastropod. Note the head-foot and the coiled shell. (Biophoto Associates, N.H.P.A.)

Figure 28-11 A nudibranch, a marine gastropod without a shell. This form can swim by undulating its body. (Biophoto Associates, N.H.P.A.)

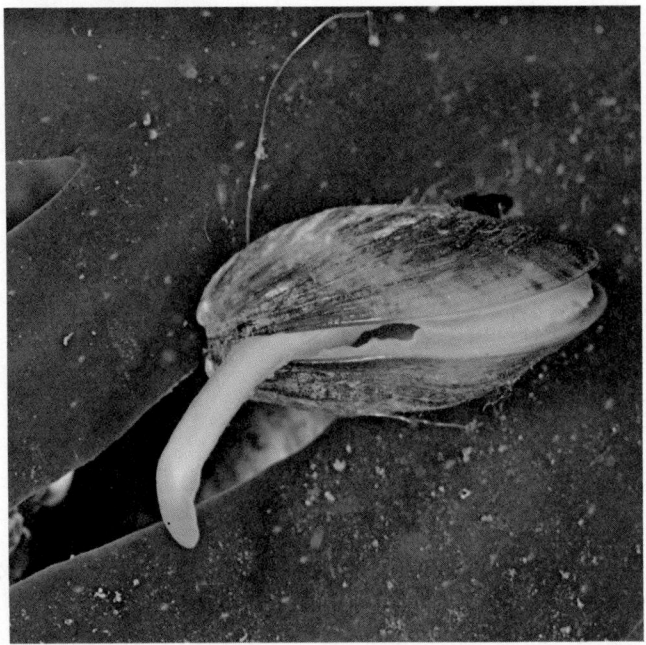

Figure 28-12 A marine bivalve. Note the edges of the mantle visible within the shell, and the long foot protruding out toward the left. (Biophoto Associates)

absent in families like the nudibranchs (marine slugs) and in terrestrial slugs. All gastropods have well-defined heads, usually with eyes and tentacles, and most gastropods have elongated flattened feet by which they creep around; the class is so diverse, however, that its members have few other features in common.

Most gastropods have gills inside the mantle cavity. However, in some forms the mantle cavity acts as a lung, permitting the animal to obtain oxygen from air instead of from water. Some gastropods have both gills and lungs.

Like chitons, many intertidal gastropods, such as limpets and periwinkles, always return to the same "home" on a rock when the tide is out. Here they pull their shell tightly down against the groove they have worn in the rock, holding on by suction of the foot. Land-living slugs and snails have anatomical and behavioral adaptations that prevent excessive water loss. The entrance to the mantle cavity, which must be kept moist, is small, and snails can pull their shells down onto the ground around them in the heat of the day. Slugs, having no shell, are more susceptible to desiccation than snails, and tend to restrict their activities to cool, moist places and times of the day.

Some of the world's loveliest animals are nudibranchs—commonly known by the unlovely name of sea slugs. The mantle, shell, and gills of these gastropods have disappeared, leaving a naked body that is often brilliantly colored and covered by numerous projections that increase the surface area for gas exchange (Figure 28–11).

Class Bivalvia (Bivalved Molluscs)

The bivalves are a large group with world-wide distribution, but they are not as varied structurally as are the gastropods; they are mainly marine molluscs in which the body is flattened between the two valves of a hinged shell (Figure 28–12). Most bivalves can protect the body by withdrawing it inside the closed shell, but some are too large to fit completely into their shells. The edge of the mantle is drawn out to form two siphons, one for water entering the mantle cavity and one for water leaving. Most bivalves are ciliary filter feeders; cilia on the gills beat, drawing a water current across the gills, and the cilia strain out food particles. Food becomes trapped

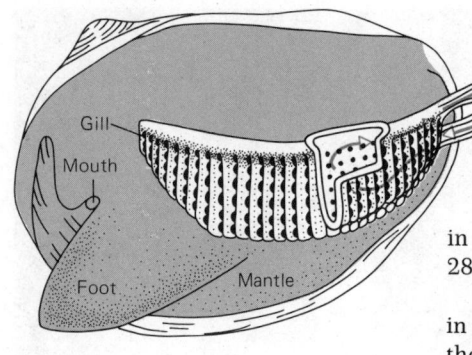

Gill
Mouth
Foot
Mantle
Exhalant siphon
Inhalant siphon

Figure 28-13 The gills of a clam are built like large, flat sieves, which strain food out of the incoming water current and pass it, trapped in strands of mucus, to the mouth.

in strands of mucus, and the gill cilia move these strands to the mouth (Figure 28–13).

As one might expect of filter feeders, many bivalves live sedentary lives, buried in sand or mud in the case of the clams or attached to the substratum by threads in the case of mussels. A few species are more mobile and can swim for brief periods by clapping the valves of their shells together; others use the muscular foot to burrow or to creep around. Many bivalves are important as "shellfish" for human food, and a number of species are specially cultivated for this purpose.

Class Cephalopoda (Squids, Octopuses, Cuttlefish, and Nautiluses)

All members of the class Cephalopoda are marine. The cephalopods are the most specialized of molluscs, and among the most advanced of all invertebrates. Behavioral studies have shown that octopuses are quite intelligent and can solve many problems. The fossil record shows that cephalopods evolved from ancestors with straight or curved shells, but today, the only living cephalopods with external shells are those of the genus *Nautilus*, which have curled shells filled with air.

Cephalopods appear to be quite closely related to gastropods, but the body has been rearranged. The mouth is now in the middle of the foot, whose edges are drawn out to form tentacles lined with rows of suckers (Figure 28–14). All cephalopods swim by jerky jet propulsion. They take water into the mantle cavity and expel it forcibly through a siphon formed from the edge of the mantle, which the animal can point in various directions to determine which way it will move. Most cephalopods also have a large ink sac that permits them to eject into the sea a cloud of brown or black ink containing the black pigment melanin. Ink from the genus *Sepia* was used for years to make sepia-colored writing ink. Ejecting ink seems to be an escape behavior: a cephalopod when attacked will eject a cloud of ink and shoot off in another direction. Perhaps the predator confuses the ink cloud with the cephalopod.

The large eyes of cephalopods are their main sense organs. The eyes are quite similar to those of vertebrates and look remarkably like those of a sleepy human being. Close behind the eyes lies a brain that is surprisingly large for an invertebrate.

All cephalopods are carnivorous, tearing their fish or invertebrate prey apart with beaklike jaws. While the squids, cuttlefish, and nautiluses are active swimmers, octopuses have taken up a more sedentary mode of life and usually lie near the entrances of their dens in rock crevices, waiting for prey. Some octopods can inject poison into their prey through their beaks. Their poison is evidently a nerve or muscle toxin, since death results from paralysis of the breathing muscles.

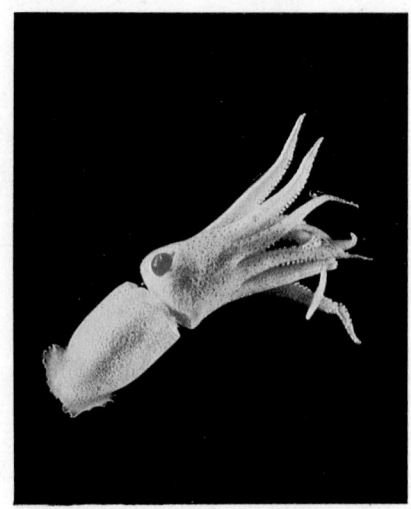

Figure 28-14 A juvenile squid swimming in the plankton. Squids swim by expelling water from a siphon made from the mantle. The tentacles are derived from the foot. (Langdon Quetin)

Like most molluscs, cephalopods have gills in the mantle cavity, but unlike other molluscs, they have a closed circulatory system, which transports food and oxygen to the cells rapidly and so supports their very un-mollusc-like high rate of activity. The cephalopods also have pigment cells called **chromatophores** in their skin and can change color rapidly by expanding and contracting different-colored chromatophores. The sexes are separate, and fertilization is internal. During copulation, the male reaches a tentacle into his own mantle cavity, where he picks up packages of sperm called spermatophores, and then places them into the female's mantle cavity. The female then lays one or a string of fertilized eggs, which either become attached to the substratum or remain free-floating, depending on the species. Female octopods look after their eggs once they are laid, removing debris and squirting water from their siphons over the nest until the eggs hatch as miniature versions of the adults.

Cephalopods were once caught for fish bait on the Grand Banks off Newfoundland; they are also prized as food in various parts of the world.

Essay: Giant Squid

Giant cephalopods appear as sea monsters in stories from many nations. Beasts similar to octopuses appear in a number of ancient Greek and Roman legends, but the most familiar mythical cephalopod is the "kraken" of Norwegian folklore. Scandinavians from the sixteenth century onwards described encounters in which "one of these Sea-Monsters will drown easily many great ships provided with many strong Mariners."

For years scientists were uncertain whether the kraken was fact or fiction. In the original 1735 version of his *Systema Naturae*, a catalogue of all the known animals, Linnaeus included the kraken under the name *Sepia microcosmos*, plainly believing that it was a giant squid. By the time the second edition was published, Linnaeus had apparently become convinced that this animal existed only in the minds of imaginative sailors and omitted it from his list. Wild and exaggerated stories always seem to have accompanied descriptions of the kraken, and for about 200 years most people believed that there was no such animal.

In 1861, however, a French naval vessel encountered a kraken near the Canary Islands, and her crew's description of the incident could not be dismissed as just another tall story. The animal lay on the surface long enough for them to make a detailed drawing, after which they harpooned and lassoed the beast, bringing a portion of it—which weighed 20 kg—on board the boat. The kraken had eight arms, each 2 m long, and two longer tentacles. It could open its beak to almost half a metre, and its weight was estimated at more than 1000 kg, or about one ton.

It is said that this incident prompted Jules Verne to describe the battle between a kraken and a submarine in his *Twenty Thousand Leagues Under the Sea*. Finally, at the end of the nineteenth century, some specimens and parts of giant squid were collected by fishermen, mainly near Newfoundland.

One species of giant squid is called *Architeuthis harveyii* after the Reverend Moses Harvey, an amateur naturalist, who lived in St. John's, Newfoundland, and recognized the importance of the fishermen's catches. In 1873 he acquired part of a tentacle that was 6 m long and had been cut off by a fisherman in a small boat who had been attacked by a squid. Finally, Harvey became the proud possessor of an intact *Architeuthis*. He wrote:

I stood on the shore of Logy Bay . . . and . . . thought of how I would astonish the savants, and confound the naturalists, and startle the world at large. I speedily completed a bargain with the fishermen, whom I astonished by offering 10 dols. to deliver the beast carefully at my house.

Next day, to my great satisfaction, a cart arrived at my door almost filled with the hideous, corpse-like creature, which I speedily stowed away in an outbuilding, in a huge vat filled with the strongest brine . . . A stream of daily visitors came to gaze in shuddering horror at the dead giant.

Harvey found that the tentacles were 8 m long and that the animal was 10 m long overall. Larger specimens have been reliably reported, but few have been captured intact or photographed to show the whole body, as this one was.

Harvey's specimens went to A. E. Verril, Professor of Zoology at Yale, who in 1879 published all the material he had accumulated on giant squid. He named two species of *Architeuthis*, and other people have described another 10. There is not, however, nearly enough information for us to be sure how many species really exist. *Architeuthis* is plainly a deep sea creature with poor swimming ability; it comes to the surface only under exceptional circumstances. Few people who sight giant squid are equipped to capture and preserve them for proper study.

Giant squid are undoubtedly the largest and heaviest invertebrates ever described. The largest apparently authentic measurements come from a specimen stranded on a New Zealand beach in 1888. This animal was nearly 20 m in overall length, with tentacles of 16 m and a body nearly 2 m long. Members of other species with bodies almost twice as long have been measured, but they all had shorter tentacles. It has been estimated that *Architeuthis* may grow until the body is more than 4 m long. An animal of this size has suckers about 3 cm in diameter, but whales have been found scarred with the marks of suckers nearly 20 cm across. Such suckers presumably belonged to a larger squid than has ever been seen.

The best-known predator of giant squid is the toothed cachalot, or sperm whale, which may grow up to 20 m in length and would be considerably heavier than the largest reported squid. Captured sperm whales are often described as bearing the marks of titanic battles with giant squid: sucker scars and gouges from a huge beak on their skins, tentacles and beaks in their stomachs. Moby Dick was a sperm whale, and the attacks of these voracious carnivores have occasionally sunk quite large ships.

A naturalist aboard the whale-ship *Cachalot* in about 1875 left us a dramatic account of a fight between a sperm whale and a giant squid:

I was leaning over the rail, gazing steadily at the surface of the sea, where the . . . tropical moon made a broad path . . . There was a violent commotion in the sea . . . and . . . a very large sperm whale was locked in deadly conflict with a cuttlefish, or squid, almost as large as himself, whose interminable tentacles seemed to enlace the whole of his great body. The head of the whale especially seemed a perfect net-work of writhing arms—naturally, I suppose, for it appeared as if the whale had the tail part of the mollusc in his jaws, and, in a business-like, methodical way, was sawing through it.

By the side of the columnar black head of the whale appeared the head of the great squid, as awful an object as one could well imagine even in a fevered dream. The eyes were remarkable for their size and blackness, . . . at least a foot in diameter.

All round the combatants were numerous sharks, like jackals round a lion, ready to share the feast, and apparently assisting in the destruction of the huge cephalopod.

429

TABLE 28-3 THE PHYLUM ARTHROPODA AND ITS MAJOR CLASSES

Phylum Arthropoda	Segmented animals with a jointed exoskeleton containing chitin; jointed appendages; body cavity a hemocoel; respiration through body surfaces or by gills or tracheae; early embryo syncytial* in many species; marine, freshwater, and terrestrial forms.
Class Arachnida (~57,000 species)	Body having 1 or 2 main parts; 6 pairs of appendages (chelicerae, pedipalps, 4 pairs of walking legs); mostly terrestrial; e.g., spiders, scorpions, ticks, mites, harvestmen.
Class Crustacea (~25,000 species)	Body of 2 or 3 parts; antennae, chewing mouthparts, 3 or more pairs of legs; development usually via nauplius larva; skeleton usually impregnated with $CaCO_3$; mostly marine; e.g., shrimps, krill, lobsters, crabs, barnacles, cladocerans, ostracods, copepods.
Class Insecta (~700,000 species)	Body divided into head, thorax and abdomen; antennae; mouthparts modified for chewing, sucking, or lapping; usually with 2 pairs of wings and 3 pairs of legs; breathing mainly by tracheae; excretion by Malpighian tubules; mostly terrestrial. (For examples, see Table 28-4)
Class Diplopoda (~7000 species)	Body with distinct head bearing antennae and chewing mouthparts; most segments of body grouped in pairs covered by a single skeletal plate, with each apparent segment bearing 2 pairs of walking legs and 2 pairs of spiracles; breathing by tracheal system; terrestrial, scavengers on dead vegetable matter, or herbivores.* The millipedes. (Figure 28-15)
Class Chilopoda (~2,000 species)	Body with distinct head bearing large antennae and chewing mouthparts; appendages of first body segment modified as poison claws; remaining segments bearing a pair of walking legs each; terrestrial in damp areas including houses; predaceous on insects. The centipedes. (Figure 28-16)

28-D Phylum Arthropoda

The Arthropoda is the largest phylum of animals. It is impossible to do justice to this enormous group without writing a complete book on the subject. Here we can do no more than list the main groups of arthropods and mention a few of the more remarkable animals found in this phylum.

The term arthropod, literally translated, means "jointed foot," an important characteristic of the arthropods. Their major evolutionary step forward was the

---

* Syncytial Not divided into individual cells; having many nuclei enclosed within one membrane.
* Herbivore An animal that feeds on plants or parts of plants.

Figure 28-15 A millipede, member of the arthropod class Diplopoda (see Table 28-3). (Biophoto Associates)

Figure 28-16 A centipede, member of the arthropod class Chilopoda (see Table 28-3). (Biophoto Associates)

Figure 28-17 This crab shows the hard exoskeleton and jointed appendages characteristic of the phylum Arthropoda. (Biophoto Associates)

development of a tough external **cuticle**, which covers the body and acts as an external skeleton, or **exoskeleton**. This is composed of layers of protein complexed with the strong, flexible polysaccharide **chitin**. The exoskeleton consists of a set of plates, which cover not only the body but also the jointed appendages. Hinges between adjacent plates permit mobility, while the tough plates protect against attack and injury as well as water loss.

Arthropod survival in terrestrial habitats often depends upon the efficiency of their waterproofing. Arthropods are one of the two animal groups (the other is the higher vertebrates of the phylum Chordata) with many members that have adapted to life on land. Since Roman times, people in Africa and Europe have mixed fine dust with stored grain to keep it free of insects. The dust abrades the soft cuticle between segments, destroying the insect's waterproofing until it dries up and dies.

One result of possessing an exoskeleton that cannot expand is that arthropods must **molt**, or shed their cuticles, in order to grow. In many arthropods, successive molts take the animal through a series of larval stages. Some larval arthropods look more and more like the adult with each molt, as new adult structures are added. Other larval arthropods may be very different from their adults and have a totally different way of life, as in the case of caterpillars, which are larval stages of butterflies and moths.

Arthropods probably evolved from annelid ancestors, since the basic primitive body plan of the arthropods is an elongated, segmented body. A hard exoskeleton protects the soft tissues and provides a surface for the attachment of muscles that move the jointed appendages (Figure 28–17). There is one pair of jointed appendages per body segment in this ancestral plan. Each appendage starts as a short stump near the body, and it then splits into two sections, one toward the midline of the body and the other toward the outside. This primitive type of appendage became variously modified during the course of evolution; frequently one or the other of the two sections was lost, but some appendages, such as the antennae of the lobster, retain both parts. Segmentation of the body also became modified during the course of evolution, and many modern arthropods show little sign of segmentation. In many groups of arthropods, the segments became grouped to form distinct parts of the body, such as the head, thorax, and abdomen in the familiar housefly. The body cavity of arthropods is known as a **hemocoel** (hemo = blood; coel = cavity). The body fluid, or **hemolymph**, also serves as the blood. Arthropods have an **open circulatory system**, in which the hemolymph flows through channels within and between the body organs, rather than being contained within discrete blood vessels.

Class Arachnida (Spiders, Scorpions, Ticks, Mites)

The most familiar arachnids are predators or ectoparasites, although many mites feed on plants or dead organic matter. Arachnids have six pairs of jointed appendages. The first pair, or **chelicerae**, are adapted for feeding by piercing their prey or host and sucking out its body fluids. The chelicerae often have associated poison glands that inject a substance that anesthetizes or kills the prey. The **pedipalps**, the next pair of appendages, act as sensory organs that detect touch and chemicals and help to hold the food in place while an arachnid is eating. In male spiders, the pedipalps aid in sperm transfer to the female during mating. The

Figure 28-18 Dinner in the Namib desert: a scorpion, a member of the class Arachnida, killing a grasshopper, a member of the class Insecta. (Biophoto Associates, N.H.P.A.)

Figure 28-19 A garden spider feeding on an insect it has caught in its web. (Carolina Biological Supply Company)

pedipalps of scorpions end in a large pincer. The other four pairs of appendages in the arachnids are the walking legs.

Spiders are the best known arachnids. Their 32,000 species are distributed all over the world. All have spinnerets with which they spin silk, and which may be used to make the webs that trap their prey and the cocoons that protect their eggs. The silk is a fine, elastic protein. It is extruded from the spinnerets as a liquid but hardens as it meets the air. All spiders are carnivorous, and usually inject a paralytic poison into their prey (Figure 28–19).

The arachnids most important to humans are the ticks and mites. Ectoparasites may transmit disease; for example, ticks may carry Rocky Mountain spotted fever, and chiggers (larval mites) may carry Asian scrub typhus. Chiggers and the adult mites that cause mange and scabies may cause severe irritation of the skin. The red spider mites that infest and kill house plants have many relatives that are serious agricultural pests. The vast majority of mites, however, live around us unnoticed because of their tiny size—a species that reaches a millimetre or two is enormous as mites go. These masters of miniaturization live in an incredible variety of habitats and feed on living or dead organisms of all descriptions.

Class Crustacea (e.g., Crabs, Wood Lice)

The class Crustacea is a large and diverse group of arthropods that includes familiar animals such as lobsters, shrimps, crabs, crayfish, barnacles, wood lice, pill bugs, and water fleas. In general, crustaceans are aquatic arthropods with two pairs of antennae, three pairs of feeding appendages formed for chewing, and legs that are often forked. However, a number of species of crustaceans do not conform to this description.

Various species of Crustacea are the most prominent animals in plankton (Figure 28–20). Indeed, the crustaceans known as copepods are so abundant in plankton

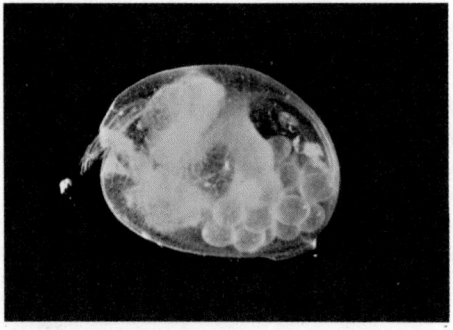

Figure 28-20 An ostracod, a tiny planktonic crustacean. The pink spheres are eggs. (Bruce Robison)

Figure 28-21 A lobster, a decapod crustacean with a cuticle reinforced by calcium carbonate. (Biophoto Associates)

that there may well be more copepods than any other group of multicellular animals in the world—including the insects. Planktonic crustaceans are the main food of many vertebrates and include the "krill" upon which the biggest whales feed.

The decapod (10-footed) crustaceans include the larger bottom-living and swimming forms such as shrimps, lobsters, crabs, hermit crabs, and crayfish. Most have a cuticle reinforced with calcium carbonate, making it harder and more brittle than that of most other arthropods (Figure 28–21).

Barnacles are marine crustaceans that are highly modified for a sessile way of life. Essentially, barnacles are attached to the substrate by their heads, and shovel food into their mouths with their thin feathery legs (Figure 28–22). The body is protected by calcareous plates, which may or may not be cemented directly to the substratum. Barnacle eggs hatch into planktonic larvae, which attach to any solid substrate and grow very rapidly. A layer of barnacles ruins the streamlining of any ship, from a nuclear submarine to a wooden sailboat, and scraping barnacles from the bottom of a boat has been a chore for sailors ever since prehistoric times.

Class Insecta

Insects are, without doubt, the most successful invertebrate group in terrestrial habitats. It has been estimated that the insects on earth weigh 12 times as much as the humans, and that there are 300 million insects for every person alive. Nearly a million insect species have been described, more than the number of all other animal species put together. Insects, bats, and birds are the only living animal groups with members that can fly.

Insects range in size from tiny beetles only 0.1 mm long to tropical moths with a wingspan of 30 cm. Some extinct insects were even larger. Nearly all insects are terrestrial; aquatic forms clearly had land-dwelling forebears.

Figure 28-22 A barnacle. These odd crustaceans attach to a solid surface by their heads and secrete calcareous shells around themselves. They use their legs to shovel food into their mouths. (Steven Webster)

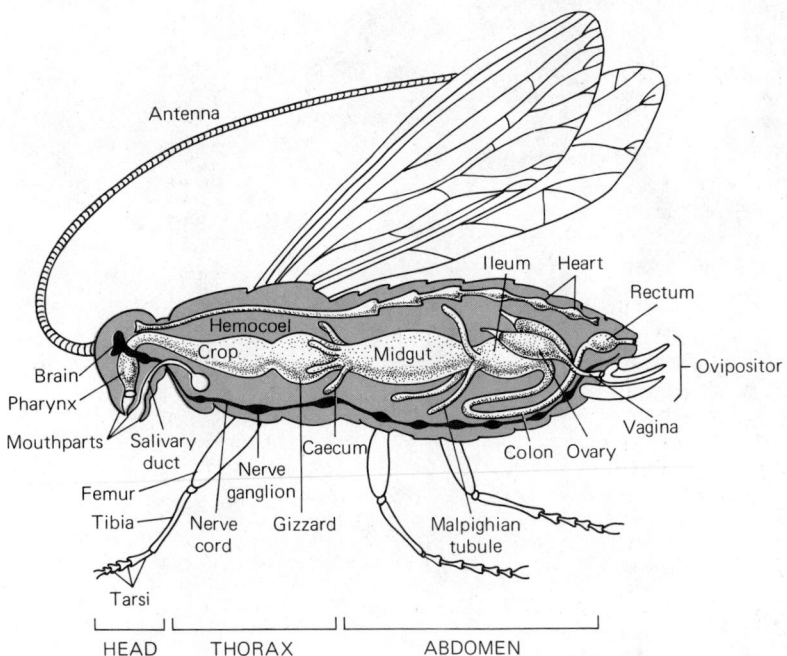

Antenna

Ileum   Heart

Rectum

Hemocoel

Crop   Midgut

Brain

Pharynx

Ovipositor

Vagina

Mouthparts   Salivary   Caecum   Colon   Ovary
duct

Femur   Nerve
ganglion

Tibia   Nerve   Gizzard   Malpighian
cord   tubule

Tarsi

HEAD   THORAX   ABDOMEN

Figure 28-23 The anatomy of an insect. Note that the circulatory system is open; the heart pumps fluid into the hemocoel, the general body cavity. The nerve cord is solid and ventral. The Malpighian tubules are the main excretory organs; they empty into the gut. Fertilization is internal, and the female uses her ovipositor to place the eggs in appropriate positions when she lays them.

The insect body is usually divided into head, thorax, and abdomen. The thorax of adult insects characteristically bears three pairs of walking legs and frequently one or two pairs of wings. The head bears one pair of antennae, specialized mouthparts, and, usually, compound eyes (Figure 28–23).

The adaptations of insects to terrestrial life are varied and marvellous. The cuticle not only provides support on land, but also forms a waterproof covering that minimizes water loss. Insects breathe by means of **tracheae**, air-filled tubes that admit air from the outside world via tiny pores called **spiracles** and carry it by way of a series of finer and finer branches throughout the body. The small size of the spiracles reduces evaporation; in many species, the spiracles can be closed. Insects have evolved sense organs that are effective on land, including eyes that permit color vision in some species, receptors on the antennae, and mouthparts that detect vibrations, touch, airborne chemicals, and the chemical composition of food. Wings develop as outgrowths from the cuticle, strengthened by rib-like structures called **veins**. Insects have specialized excretory structures, the **Malpighian tubules**, which ensure that a minimum of water is excreted from the body. Like all arthropods, insects have an open circulatory system; hemolymph fills the body cavity. Since insects are basically terrestrial, they have internal fertilization. The eggs are laid with a waterproof covering, which protects them from dehydration. Females use their sense organs to select a place to lay their eggs where the young will find food when they hatch. The young undergo a series of molts, and the higher insects pass through larval stages (e.g., maggots) that are completely different from the adult in appearance and way of life.

The French physiologist Claude Bernard pointed out that for an animal to be free to move into new habitats, it must be capable of regulating its internal environment in the face of external conditions that are inhospitable to life. Most invertebrates are restricted to moist or aquatic habitats because they cannot maintain the

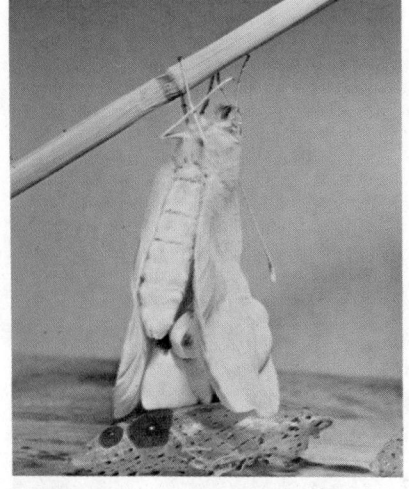

Figure 28-24 A butterfly, a member of the order Lepidoptera, emerging from its chrysalis and starting to spread its wings. A life history with complete metamorphosis (egg → caterpillar → pupa (chrysalis) → adult) is typical of several insect orders. (Biophoto Associates)

TABLE 28-4 **THE MAJOR ORDERS OF THE CLASS INSECTA**

Ametabola†
Development: Egg → Immature → Adult
Mayflies. Wings membranous, forewings large and triangular; mouthparts chewing in naiads, vestigial* in wingless, with chewing mouthparts and simple eyes. Primitive insects descended from ancestors that were never winged.

Order Thysanura
(~700 species)
Bristletails; e.g., silverfish, firebrats. Soft, many-segmented bodies; long antennae; usually 2 or 3 long appendages at posterior tip of abdomen.

Order Collembola
(~2000 species)
Springtails. Most have specialized spring-like pair of appendages at tip of abdomen that fit into pair of hooklike appendages nearer front of abdomen.

Paurometabola†
Development: Egg → Nymph → Adult
The nymphs are small but wingless versions of the adults; at each molt, the nymph becomes larger and the wing pads increase in size; the last molt produces a (usually) winged adult. Nymphs live in the same habitat as adults, and have the same mode of life.

Order Orthoptera
(often subdivided into several different orders)
(~23,000 species)
Grasshoppers, crickets, mantids, walking sticks, roaches, etc. Forewings straight and leathery, hindwings membranous and folded under forewings at rest; chewing mouthparts.

Order Isoptera
(~1800 species)
Termites. Wings membranous, 2 pairs alike, or absent; chewing mouthparts; social.

Order Dermaptera
(~1100 species)
Earwigs. Forewings short or absent, hindwings highly folded; chewing mouthparts; pair of pincer-like appendages at tip of abdomen.

Order Anoplura
(~200 species)
Sucking lice; e.g., head and body lice, pubic louse. Wingless; piercing and sucking mouthparts; stout claws.

Order Hemiptera
(~40,000 species)
True bugs (the only insects properly called bugs); e.g., chinch bugs, stink bugs, plant bugs, water boatmen, water striders, assassin bugs, bed bugs. Forewings leathery at base and membranous at tip, hindwings membranous; mouthparts form jointed, piercing and sucking beak.

Order Homoptera
(~20,000 species)
Cicadas, aphids, scale insects, leaf hoppers, spittle insects. Wings membranous and held rooflike over the abdomen, or absent; mouthparts form sucking beak.

Hemimetabola†
Development: Egg → Naiad → Adult
The naiads are aquatic with gills; at its last molt the naiad emerges from the water and becomes an aerial adult. Thus naiads and adults have different modes of life.

Order Ephemeroptera
(~1500 species)
The immature stages are miniature versions of the adults; size increases with each molt. Adults are adults; 2 or 3 long processes at tip of abdomen.

Order Odonata
(~5000 species)
Dragonflies and damselflies. Wings membranous, long and narrow, chewing mouthparts.

Holometabola†
Development: Egg → Larva → Pupa → Adult
The larvae lack compound eyes; they bear no resemblance to the adult, and usually have a different mode of life (and often a different habitat); the pupa is an immobile, non-feeding stage during which the larva metamorphoses to the (usually) winged adult form.

Order Neuroptera
(~4600 species)
Lacewings, ant-lions, aphid-lions. Wings membranous, 2 pairs similar; mouthparts sucking in larvae, chewing in adults.

Order Lepidoptera
(~110,000 species)
Butterflies, moths, skippers. Wings very large, covered with scales; mouthparts chewing in larvae (caterpillars), long coiled sucking proboscis in adults.

Order Coleoptera
(~280,000 species)
Beetles. Forewings hard, hindwings membranous and folded under forewings; chewing mouthparts.

Order Trichoptera
(~4500 species)
Caddis flies (adults) and caddis worms (larvae). Wings moth-like, covered with small hairs; mouthparts chewing in aquatic larvae, vestigial in adults.

Order Diptera
(~85,000 species)
True flies; e.g., mosquitoes, gnats, midges, crane flies, houseflies, horseflies. Only one pair of wings, hindwings reduced to knoblike halteres; mouthparts variable in larvae, piercing and sucking or vestigial in adults.

Order Siphonaptera
(~1100 species)
Fleas. Adults wingless; mouthparts chewing in larvae, sucking in adults.

Order Hymenoptera
(~100,000 species)
Wasps, bees, ants. Wings membranous when present; mouthparts both sucking and chewing; females with stinger or piercing ovipositor at tip of abdomen; some forms social.

†Orders grouped on the basis of similar life histories but not necessarily related phylogenetically.
*Vestigial   Much reduced in size and nonfunctional.

Figure 28-25 A green lacewing in flight. This beneficial insect devours aphids, insect pests that suck the juices of plants. (Biophoto Associates)

salt compositions of their bodies in fresh water or conserve body water when exposed to air. Insects have won their freedom by evolving the ability to tolerate fluctuations in the salt and water content of some of their body fluids, fluctuations that would be fatal to most other organisms.

An obvious question is why most of the insects living today are so small. The mechanics of an exoskeleton and the fact that flapping flight requires less energy of a lighter animal undoubtedly impose some theoretical upper limit on the size of a flying insect. Nevertheless, modern insects do not reach this limit, and some extinct insects were much larger. The answer, then, must lie in the fact that insects, by their size, occupy habitats where vertebrates, the other main group of terrestrial animals, cannot compete with them. Leaf miners are insects that live within a living leaf. Other insects live inside a single seed or inside the egg of another insect. In these tiny habitats they face little competition, and with their small size they need little food to grow to maturity. The evolutionary tendency to small size has been an enormously successful one for the insects.

HUMANS AND INSECTS   Insects perform many roles that are vital to human life. Without bees and other insects, for instance, many flowering plants would never be pollinated. Numerous beetles, ants, and flies are important decomposers, breaking down the dead bodies of plants and animals. Nevertheless, people have devoted more time to killing insects than to praising them. It is perhaps unduly gloomy to conclude that we are losing the battle against insects, who will one day inherit the earth, but those who believe this have good reason for their opinion.

Insects attack human beings directly with bites and stings. Much more important, blood-sucking insects transmit many diseases. Malaria, river blindness, and sleeping sickness carried in this way blind and kill several million people a year. Insects also transmit plant diseases, such as Dutch elm disease and many viral diseases of crop plants. Insects probably do more damage indirectly, however, by competing with people for food and other crops. It is estimated that in 1975 the gypsy moth, tussock moth, southern pine beetle, and spruce budworm destroyed enough forest trees to build nearly a million houses in the United States alone. In the United States, insects destroy some 10% of all crops grown. Crops in tropical climates suffer from even greater insect damage than do crops grown in more temperate areas, since insects grow and reproduce faster in the tropics. In Kenya, officials estimate that insects destroy 75% of the nation's crops. A locust swarm in Africa may be 30 m deep along a front 1500 m long, and will consume every fragment of plant material in its path, leaving hundreds of square miles of country devastated.

Pesticides have not proved to be the answer to the insect problem. This is partly because so many pesticides are almost as dangerous to people as they are to insects, and partly because pesticides act as selective pressures for the evolution of resistant strains of insects, which evolve too fast for the very expensive pesticide research to

Figure 28-26 A Colorado potato beetle, a destructive agricultural pest. The order Coleoptera (beetles) contains the greatest number of insect species. (Paul Feeny)

keep up. The idea that any insect species can be eliminated by insecticides without substantial damage to human beings is probably a dream. Consequently, **entomologists** (scientists who study insects) are now directing more of their efforts to the more realistic goal of using a combination of chemical and biological methods to control damaging outbreaks of particular species of insects. Biological controls include sterilizing males and releasing them (used for species in which females will mate only once), breeding pest-resistant plants, and introducing specific predators and parasites of pest insects. Yet despite the fact that human beings have waged war on insects since the two have existed together, human efforts have apparently not succeeded in exterminating a single species of unwanted insects.

SUMMARY

A coelom is a fluid-filled body cavity that originates as a space between layers of mesoderm during embryonic development. The coelom permitted the evolution of more efficient modes of digestion, circulation, and reproduction and thus led to an enormous burst of adaptive radiation that resulted in the spectacular variety of coelomate animals.

The phylum Annelida includes the classes Polychaeta (bristle worms), which contains mostly marine forms; Oligochaeta (earthworms and the like), which contains mostly freshwater or terrestrial forms; and Hirudinea (leeches), which contains mostly freshwater forms. Many polychaetes and oligochaetes are segmented and move by means of a hydrostatic skeleton acting together with their setae. Leeches lose their segmentation during development, and move by means of powerful body muscles.

Most members of the phylum Mollusca are unsegmented marine animals. A mantle covers most of the body and usually secretes a calcareous shell; gas exchange occurs through gills in the mantle cavity or through the lining of the mantle cavity itself. Molluscs also have a muscular foot, which is used for locomotion or burrowing; cephalopods use their tentacles, derived from the foot, for gripping prey. The main molluscan classes are the Amphineura, or chitons; the Gastropoda, or snails and slugs; the Bivalvia, or shellfish; and the Cephalopoda, the nautiluses, squids, and octopuses. Many of the gastropods are well adapted to life in terrestrial habitats.

The phylum Arthropoda contains more known species and a larger weight and number of individuals than are found in any other animal phylum. Arthropods are highly successful in marine, freshwater, and terrestrial habitats. They have segmented bodies and a varied array of jointed appendages used for feeding, locomotion, fighting, and sensing stimuli in the environment. The entire body of an arthropod is covered with an exoskeleton, which is flexible at the joints; the exoskeleton protects the body from injury and desiccation. The crustaceans are mainly marine, but there are many freshwater forms and a few that are terrestrial. The classes Arachnida and Insecta contain mainly animals adapted to life in a variety of terrestrial environments.

OBJECTIVES

From your study of this chapter you should be able to:

1. State what a coelom is and discuss its evolutionary importance to animals.
2. List the distinguishing characteristics of annelids, molluscs, and arthropods.
3. Discuss reasons for the success of the annelids, molluscs, and arthropods.
4. When presented with an annelid, mollusc, or arthropod, name the phylum to which it belongs.
5. Name, describe, and give an example of each of the three classes of annelids.
6. Name, describe, and give examples of four classes of molluscs.
7. Name, describe, and give examples of each of three major classes of arthropods.
8. Describe the adaptations of insects to a terrestrial way of life.

SELF-QUIZ

1. The possession of a mantle, a trochophore-like larva and a calcareous exoskeleton is characteristic of the:
    a. Annelida
    b. Cnidaria
    c. Crustacea
    d. Mollusca
    e. Arthropoda

2. Clams are placed in the class:
    a. Bivalvia
    b. Trematoda
    c. Gastropoda
    d. Cephalopoda
    e. Amphineura

Match (choose all correct lettered items for each description):

____ 3. Part of the insect respiratory system

____ 4. Excretory organ of an insect

____ 5. Locomotory apparatus of a polychaete

____ 6. Tongue-like rasping organ of some molluscs

____ 7. Feeding apparatus of a spider

a. chelicerae
b. hydrostatic skeleton
c. Malpighian tubule
d. mantle
e. muscular foot
f. radula
g. setae
h. spiracle
i. suckers
j. trachea
k. walking legs

8. The evolutionary importance of a coelom is that:
    a. it permitted animals to have a circulatory system and other internal organs which move
    b. it permitted animals to move onto land with an internal storage place for extra body fluid
    c. it provided the possibility of evolving a hard, protective exoskeleton
    d. it allowed organisms to have excretory systems
    e. it paved the way for evolution of locomotory appendages

9. Which of the following sets of animals all belong to the same class?
    a. crab, scorpion, lobster
    b. shrimp, barnacle, pill bug
    c. leech, tick, flea
    d. snail, slug, scallop
    e. clam, nautilus, oyster

Match:

____ 10. Black widow spider

____ 11. Medicinal leech

____ 12. Chiton

____ 13. Earthworm

____ 14. Octopus

____ 15. Barnacle

a. Amphineura
b. Arachnida
c. Bivalvia
d. Cephalopoda
e. Crustacea
f. Gastropoda
g. Hirudinea
h. Insecta
i. Oligochaeta
j. Polychaeta

QUESTIONS FOR DISCUSSION

1. In several species of leeches, individuals fix their spermatophores (sperm packets) onto the external body wall of another leech. When many individuals are kept in a jar together, the largest individuals receive more spermatophores than do smaller individuals. What is the adaptive advantage of this behavior?

2. Why is the exoskeleton of an arthropod so much more effective as a waterproofing device for a terrestrial animal than is the calcareous exoskeleton (shell) of a mollusc? (Think of the difference between the shell of a lobster and the shell of a snail.)

3. It is thought that the main limit to the size of insects is the tracheal system. Terrestrial vertebrates with lungs can grow to much larger sizes. Why is this?

4. Describe briefly how members of the phyla Annelida, Mollusca, and Arthropoda feed, reproduce, and breathe. Be sure to note differences among members of the various classes (listed with Self-Quiz problems 10 through 15) in each phylum.

REFERENCES AND FURTHER READING

(See also references in Chapter 27)

Barrington, E. J. *Invertebrate Structure and Function.* Boston: Houghton Mifflin, 1967. A modern text emphasizing structure/function relationships.

Borror, D. J. *et al. An Introduction to the Study of Insects,* 4th ed. New York: Holt, Rinehart and Winston, 1976. A good introductory entomology text.

Evans, H. E. *Life on a Little-Known Planet.* New York: E. P. Dutton, 1968. An entertaining and enlightening account of our insect neighbors by a leading scientist.

Lane, F. W. *Kingdom of the Octopus.* New York: Sheridan House, 1965. This fascinating book on cephalopods contains Harvey's account of the acquisition of *Architeuthis.*

Morton, J. E. *Molluscs,* 4th ed. London: Hutchinson, 1967. A comprehensive introduction to molluscan biology.

CHAPTER 29

THE ORIGIN OF VERTEBRATES

Humans, crocodiles, birds, and frogs are members of the subphylum Vertebrata of the phylum Chordata. Although we have a remarkable fossil record of the evolution of vertebrates into a wide range of forms living in a variety of habitats, details of the origin of the earliest chordates are lost, probably forever, owing to the poor fossil record from the Precambrian and Cambrian geological periods. In this chapter we shall consider the major modern invertebrate groups most closely related to the vertebrates—the echinoderms and the invertebrate chordates—and draw what deductions we can about how the earliest chordates arose and what they were like.

29-A Phylum Echinodermata

Sea stars, brittle stars, sea cucumbers, sea lilies, sea urchins, and sand dollars belong to the invertebrate phylum Echinodermata. Although one would hardly guess it, the echinoderms are close relatives of the chordates. The evidence for this relationship comes from a study of the early embryonic development of inverte-

TABLE 29-1 **THE PHYLUM ECHINODERMATA AND ITS CLASSES**

Phylum Echinodermata	Marine; adults with pentaradial symmetry, a calcareous skeleton, and a water vascular system with tube feet.
Class Crinoidea	Stalked or free; branched arms with feeding grooves; ciliated tube feet used for feeding and respiration.
Class Asteroidea	Free-moving with arms merging into disc; suctional tube feet used for locomotion and feeding.
Class Ophiuroidea	Free-moving with thin arms marked off from disc and used for locomotion; tube feet used as sensory organs and for feeding.
Class Echinoidea	Free-moving with skeleton of fused plates; no arms; suctional tube feet used for locomotion and respiration.
Class Holothuroidea	Free-moving with mouth at one end of elongated body; skeleton reduced; some tube feet used for locomotion, others modified into tentacles around mouth.

brates. The first series of cell divisions of the zygote, and later the formation of the gut and coelom, are quite similar for echinoderms and chordates. These patterns of echinoderm and chordate development are very different from the patterns shown by the annelids, molluscs, and arthropods.

The name Echinodermata means "spiny-skinned" and refers to the most distinctive characteristic of the group: the **calcareous** ($CaCO_3$) spines and plates that form a skeleton just under the skin in all members of the phylum (Figure 29–1). Adult echinoderms are also unique in their **pentaradial symmetry**: the body is divided into five parts around a central disc, where the mouth lies (see Figures 29–4 and 29–5). In some ways, their physiology is simply organized, for these animals have no head and no excretory system. A distinctive feature is their **water vascular system**, a series of fluid-filled vessels that may be used in feeding and locomotion. Alterations of pressure in this system permit an echinoderm to extend and retract its **tube feet** or podia (Figure 29–2). Tube feet are interesting organs found only in echinoderms. Each foot is hollow and has a suction cup on the end; it operates through changes in hydraulic (fluid) pressure in the water vascular system. Some echinoderms use their many small tube feet for their slow locomotion, and predatory forms may use the feet to catch prey.

All echinoderms are marine. They are primarily bottom-dwelling forms, and most can move about, although very slowly. Sea lilies are the only extant group of echinoderms with sessile members.

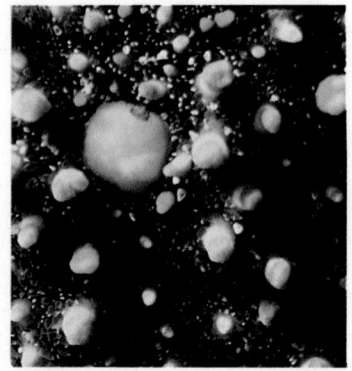

Figure 29-1 Echinoderms are named for their spiny skins. This is part of the surface of a sea star (class Asteroidea) to show its calcareous spines. (Steven Webster)

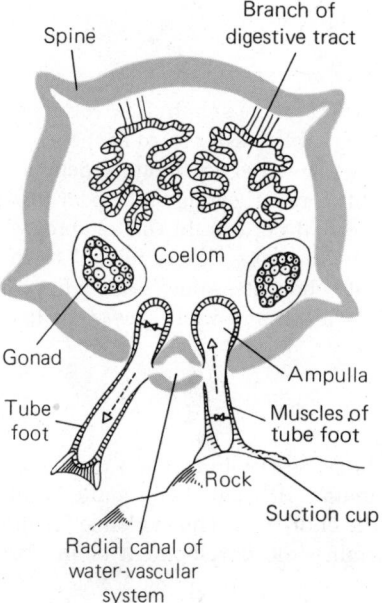

Figure 29-2 Diagrammatic cross-section through the arm of a sea star (asteroid). The arm is surrounded by calcareous plates (colored). When muscles in the ampulla contract, fluid is forced into the tube foot and it extends. When muscles in the wall of the tube foot contract, they force fluid back into the ampulla and radial canal. This lowers the fluid pressure in the tube foot and it shortens. The low pressure between the sucker of the foot and the substrate causes the foot to stick like a suction cup.

Figure 29-3 The feathery (yellow) arms of a crinoid reach out over the surface of a coral. (Steven Webster)

Class Crinoidea (Sea Lilies and Feather Stars)

Some crinoids are sessile, attached to the substratum by a stalk, and some are mobile, but all use their branched, feathery arms to trap food particles in a mucous web (Figure 29–3). Food is then carried to the mouth by ciliated tube feet. The tube feet also provide surfaces for gas exchange.

Class Asteroidea (Sea Stars)

The familiar asteroids, or sea stars (Figure 29–4), are flattened, with a central disc from which radiate five or more arms. The mouth is located on the underside of the central disc. Asteroids move mainly by their tube feet, which are confined to the underside of the body. The asteroid skeleton is organized as a series of plates that permit a certain amount of movement in the arms. Most sea stars are carnivorous and feed on snails, crustaceans, bivalves, polychaetes, other echinoderms, and even fish, using their tube feet to grip their prey. The "crown of thorns" sea star feeds on cnidarian polyps and is notorious for the damage it does to such coral reefs as the Great Barrier Reef of Australia.

Sea stars are famous for their ability to regenerate a whole body from one arm that is still attached to part of the central disc. At times misguided fishermen have tried to destroy asteroids preying on mussel and oyster beds by hacking them into pieces, a practice that merely contributes to a population explosion of the echinoderms.

Figure 29-4 This asteroid (sea star), from a coral reef near Mozambique, displays the pentaradial symmetry characteristic of adult echinoderms. (Biophoto Associates, N.H.P.A.)

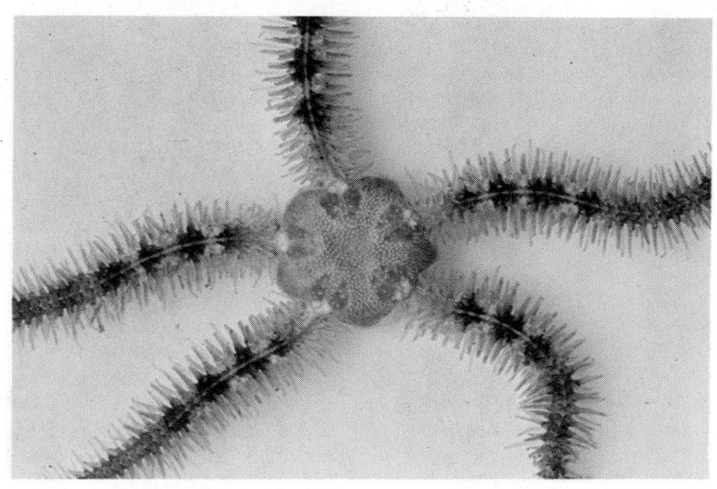

Figure 29-5   An ophiuroid. (Biophoto Associates)

Class Ophiuroidea (Brittle Stars, Serpent Stars)

Ophiuroids look much like asteroids except that their arms are sharply marked off from the central disc (Figure 29–5). Ophiuroids also move very differently from asteroids, by wriggling their jointed arms rather than using their tube feet. The tube feet are used mainly as sense organs and for passing food along to the mouth. Many ophiuroids are covered with spines, which may give the arms a feathery appearance. Some ophiuroids can actually use the arms to swim, which is unusual among echinoderms.

Most ophiuroids eat organic detritus from the sea floor and whatever small organisms they encounter. They have a very simple gut with a saclike stomach and no intestine or anus.

Class Echinoidea (Sea Urchins and Sand Dollars)

Echinoids are tubby little echinoderms very different in appearance from the graceful crinoids and ophiuroids. They live mouth-down on the bottom of the sea, protected from intrusion by brittle calcareous spines, which easily penetrate soft flesh and are very hard to remove (Figure 29–6). The calcareous skeletal plates are fused into a sphere or, in the case of sand dollars, an envelope pierced only by holes for the mouth, anus, and tube feet. The mouth contains a chewing apparatus with a ring of large triangular teeth (Figure 29–7). Echinoids can chew almost any sort of

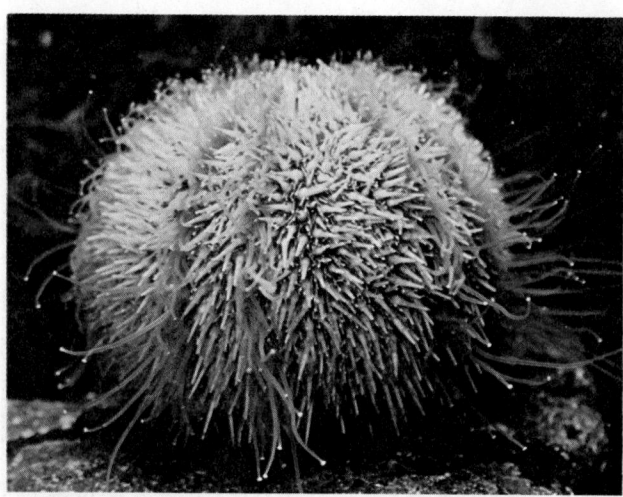

Figure 29-6 A sea urchin of the rocky shore. This species is sometimes eaten as human food. (Biophoto Associates)

Figure 29-7   Teeth protrude from the mouth of a sea urchin. (Steven Webster)

(a)

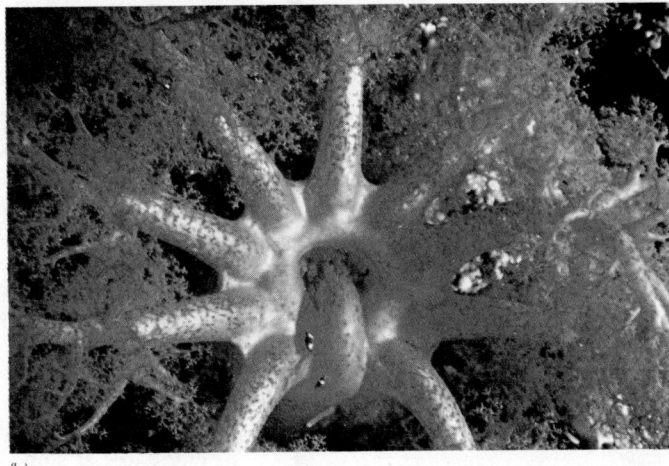
(b)

Figure 29-8 Holothuroids. (a) A sea cucumber crawling along. (b) An individual
of another species wiping food and mucus
off a tentacle into its mouth. (a, Biophoto
Associates, N.H.P.A.; b, Steven Webster)

organic material they encounter. They walk either on their spines, which are hinged
to the rest of the skeleton by a sort of universal joint, or on their tube feet. The tube
feet are also used for gas exchange.

Class Holothuroidea (Sea Cucumbers)

Holothurians look like large flabby sausages as they wash gently to and fro, or
lie partly buried, on the sea floor (Figure 29–8). One would never guess that they are
echinoderms, because their calcareous skeleton is reduced to microscopic plates
scattered through the skin. Like echinoids, holothurians lack arms. They are elongated, with the mouth at one end so that they are really lying on their sides. Holothurians can move slowly, either by using their tube feet, or by wriggling with
caterpillar-like movements of the body. Some are quite competent burrowers. The
mouth is surrounded by modified tube feet known as tentacles. In many species the
tentacles secrete a net of mucus, which is used to catch small planktonic organisms
for food. Other holothurians feed like earthworms, engulfing sand and mud from
which they digest the organic matter before voiding the inorganic particles through
the anus at the other end of the body.

29-B Characteristics of Chordates

The phylum Chordata contains all those animals that have a stiff, rod-like **notochord** at some stage in their lives. The notochord serves as an internal skeleton
and prevents the body from shortening when the body muscles contract. In the
vertebrates the notochord is surrounded or replaced by a column of vertebrae during
embryonic development. This does not occur in the invertebrate chordates, which
make up two subphyla.

TABLE 29-2 **CHARACTERISTICS OF THE PHYLUM CHORDATA**

1. Notochord at some stage of life.
2. Dorsal tubular nerve cord. The brain develops from a swelling of the
   anterior end of the tube.
3. Pharyngeal gill slits at some stage in the life history.
4. Segmentation of at least part of the body in most groups.
5. Post-anal tail at some stage in the life history.
6. Ventral heart.
7. Endoskeleton.

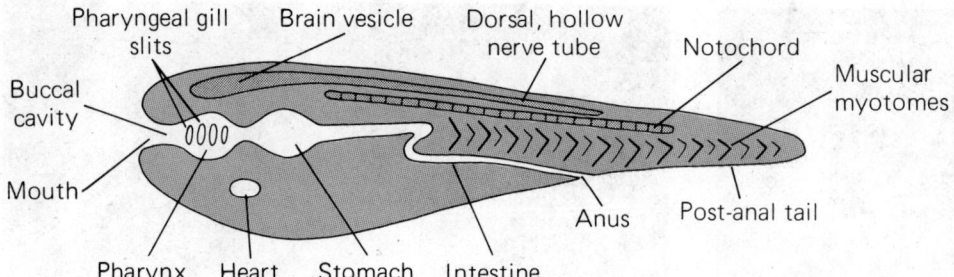

Figure 29-9 Diagram of a generalized, primitive chordate.

Whether they are vertebrates or not, all chordates have a number of important features in common (Figure 29-9). The notochord is the first of these. In addition, all chordates at some time in their lives have **pharyngeal gill slits**, which lead from the **pharynx**, the throat cavity between the mouth and the stomach, to the exterior. Chordates also have a nerve cord in the form of a hollow tube located on the dorsal side of the body. This **hollow dorsal nerve cord** distinguishes the chordates from most invertebrates (e.g., annelids, molluscs, arthropods, echinoderms), whose main nerve cord is solid and ventrally placed (see Figure 29-10 for an explanation of terms describing anatomical positions).

Their notochord, pharyngeal gill slits, and hollow dorsal nerve cord are enough to distinguish chordates from members of all other phyla. But chordates have a number of other common characteristics as well; most are more or less segmented, and they have an endoskeleton (internal skeleton). In addition, all chordates have a tail that extends behind the anus at some stage in their lives (in nonchordates the anus is usually, although not always, at the end of the tail).

This collection of chordate characteristics comprises a particularly successful set of adaptations. The combination of an internal notochord with segmented blocks of muscle, the **myotomes** (Figure 29-9), allowed early chordates to swim quickly and efficiently by a side-to-side wiggle of the body, permitting them to swim much farther and faster than other animals. These primitive "fish" swam forward, taking in food and water through their mouths and letting the extra water escape through the gill slits. Sense organs in the head detected where the animal was going and found food. The brain and nervous system in this area became well developed.

Before attempting to reconstruct the origin of the vertebrates from their invertebrate ancestors, we shall discuss some animals that may be similar to the "missing links" in this story, the two groups of invertebrate chordates.

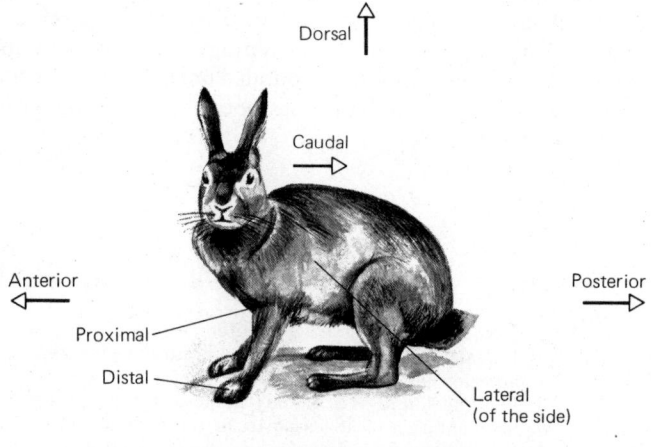

Figure 29-10 The relative positions of parts of an animal's body are described by these anatomical terms. Caudal means toward the tail. The proximal part of a limb is the part nearer to the point of attachment to the main part of the body, and the distal part is farther from the point of attachment.

29-C Subphylum Urochordata (Tunicates)

Urochordates, the sea squirts and their relatives, are an entirely marine group. As adults, these animals look nothing like any other chordate. They are classified as chordates because they have a tadpole-like larval stage with a notochord and a hollow dorsal nerve cord. However, the notochord disappears during metamorphosis to the adult form.

One class of tunicates contains sessile animals, the sea squirts, which are quite common around sea coasts; the other two classes contain planktonic forms that are rarely seen because, although common, they are fragile and transparent.

Class Ascidiacea (Sea Squirts)

Ascidians are sessile urochordates with the odd property of secreting a test (protective outer covering) of the polysaccharide cellulose, which is found nowhere else in the animal kingdom. Ascidians are filter feeders: they take in water through the mouth, filter it through a complicated basket derived from the pharyngeal gill slits of the larva, and push it out through a pore (Figure 29–11). Many ascidians live in colonies, which may share a common mouth and a common cellulose test (Figure 29–12).

The ascidians have a motile, tadpole-like larva, which looks almost exactly like our drawing of a generalized chordate (see Figure 29–9). The main difference is that not all ascidian tadpoles have mouths. The larval period is very brief, and the larva does not feed. When it is released from its egg case, the tadpole swims to the surface with efficient fish-like wriggles, using the action of its myotomes against its notochord. It then drifts a short distance in the plankton, turns, and swims down to the bottom again. Here it searches out a suitable surface on a rock or a dock piling and attaches itself by suckers on the tip of its nose. Then it metamorphoses into an adult. The tail is resorbed into the body, the tiny pharyngeal gill slits subdivide to form a pharyngeal basket, and the adult ascidian takes shape. Since the ascidian tadpole is the most primitive known chordate, it looks as if the typical chordate characteristics evolved, not as adaptations of an adult to its way of life, but in a larva.

The vertebrates could easily have evolved from an animal closely related to the tunicate tadpole. The trouble with this theory is that the tadpole metamorphoses into an adult that looks nothing like any vertebrate. The tadpole never reproduces, so how could it have become an ancestor? A solution to this problem was suggested in 1928 by British zoologist Walter Garstang, and is now widely accepted. Garstang proposed that vertebrates evolved from a tadpole-like creature that failed to metamorphose but developed functional gonads and reproduced while still a larva. This suggestion is not as far-fetched as it may look at first glance: a number of living animals develop some degree of sexual maturity while they are larvae (Section 30-E) and there is good evidence that groups other than the vertebrates have also originated in precisely this way.

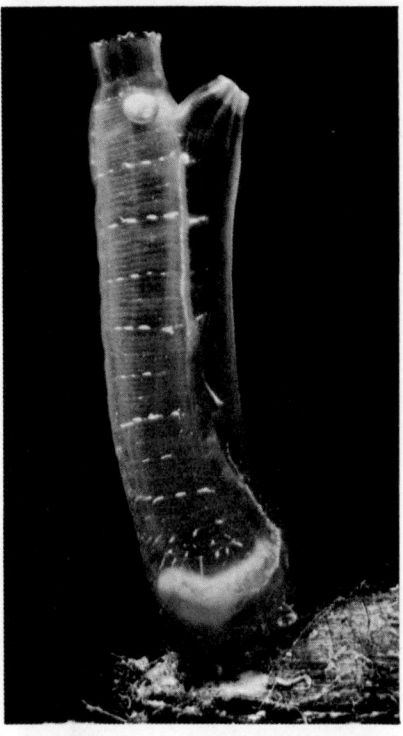

Figure 29-11 A solitary sea squirt. The "basket" filling the body is pharyngeal gill slits arranged to filter food out of the water. The pink object at the base is the stomach; water leaves the body via the transparent tube at the right of the body. (Biophoto Associates)

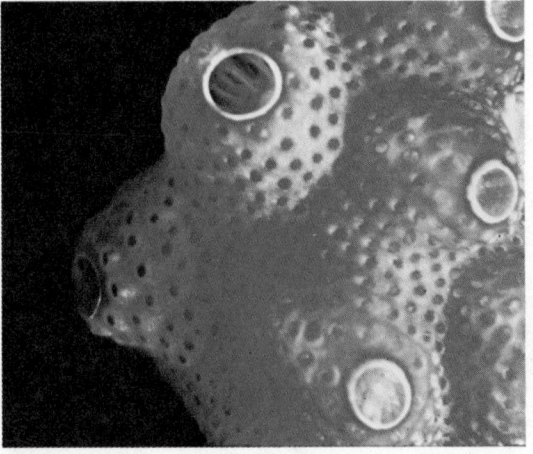

Figure 29-12 A colonial ascidian. Water enters and leaves the colony by way of pores (surrounded by white rings) which supply more than one individual. (Steven Webster)

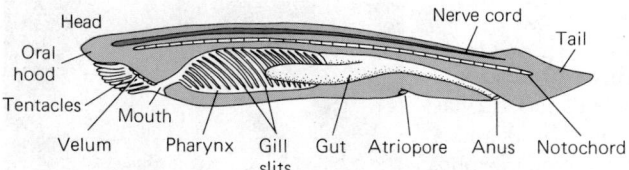

Head
Oral hood
Tentacles
Velum
Mouth
Pharynx
Gill slits
Gut
Atriopore
Anus
Notochord
Nerve cord
Tail

Figure 29-13 Diagram of an amphioxus. The chordate characters that can be seen are a notochord, dorsal nerve cord, tail, and pharyngeal gill slits.

29-D Subphylum Cephalochordata

The subphylum Cephalochordata contains only amphioxus, the lancelet. Only two genera, *Asymmetron* and *Branchiostoma*, and 29 species are recognized. Amphioxus looks like a simple fish with too many gill slits, or like a urochordate tadpole with a vastly expanded gill system, which is probably how it evolved. The adults burrow in sand in shallow tropical and warm temperate oceans. They are fish-like animals but swim poorly because most of the body is occupied by the enormous pharyngeal gill basket, leaving little room for swimming muscles. An amphioxus lives buried in the sand with only its head end protruding. The **oral hood**, which projects beyond the entrance to the mouth, bears tentacles with chemical and tactile (touch) sensory cells (Figure 29–13). Under the hood, the entrance to the mouth itself is protected by a further ring of sensory tentacles, the **velum**.

An amphioxus feeds, like a urochordate, by using cilia to pull a current of water into its mouth. Any food in the current is trapped in a mucous net secreted by an organ called the **endostyle**. The mucus passes down the pharynx into the alimentary canal, where the food is digested, and the water current is pushed out through the gill slits. The gill slits do not lead from the pharynx directly to the exterior as they do in the familiar bony fish, but open into a large **atrium**, which in turn opens to the exterior via a pore, the atriopore.

Swimming movements are accomplished by V-shaped myotomes. When the muscles of one side of the body contract, the animal bends to that side. The notochord consists of a series of discs flattened against each other. The whole cord is encased in a fibrous sheath that prevents it from dislocating.

29-E Vertebrate Characteristics

Vertebrates differ from other chordates in having a vertebral column, which replaces the notochord to a greater or lesser extent. Various other characters are found in most, but not all, vertebrates. A tail that extends back beyond the anus is common. All vertebrates have some sort of a liver, endocrine organs,* and kidneys that differ completely from the various excretory organs in invertebrates. A ventral heart, closed blood vessels, and some degree of segmentation also occur in all vertebrates. **Cephalization** is pronounced; that is, sense organs and nerves are concentrated at the front end of the body so that vertebrates have very obvious heads.

Primitively, vertebrates were aquatic animals that moved as sharks do today, by sinusoidal swimming with the segmented myotomes pulling against a dorsal notochord or vertebral column and hence pushing the paddlelike tail against the water. Early fish fossils lacked movable jaws, and it is likely that the original feeding method was a type of filter feeding like that of amphioxus; food was probably filtered out of a water stream, which entered through the mouth, passed over the gills, and exited through the gill slits. (Gills are used for gas exchange in modern fish, but feeding was probably their original function.)

29-F Class Agnatha

The most primitive vertebrate group is the class Agnatha (a = without; gnathos = jaw). The earliest vertebrate fossils are ostracoderms, found in rock laid down in the Ordovician to Devonian periods (see the end paper of this book). The

*Endocrine organs Organs composed of ductless glands that secrete hormones.

ostracoderms were fish-like agnathans, in many cases covered with heavy bony plates, which increased their chances of fossilization. Their modern relatives are cyclostomes (= round mouths), the lampreys and hagfishes. These animals not only show us the living remnants of the most primitive type of vertebrate organization, but also present a set of peculiar specializations. Their odd habits may explain why these creatures survive today, while most of their close relatives have been extinct for hundreds of millions of years.

Adult lampreys and hagfishes are long, narrow, more-or-less cylindrical creatures without paired fins and without jaws (Figure 29–14). The adult lamprey is a semi-parasite, and hagfishes are scavengers.

Lampreys are found in temperate regions throughout the world, while hagfishes inhabit temperate and subtropical waters of the Atlantic and Pacific. The adults of most species of lampreys are marine. They have a sucking mouth and a rasping tongue covered with teeth, which are used to break the skin and suck the blood of bony fish. They do not usually kill their prey. The sucker is also used to attach the adult to stones when it is at rest. The larval form, called an **ammocoete**, lives for as long as seven years as an amphioxus-like filter feeder buried in the mud of a river or stream, whereas the adult usually lives for only a few years and dies after migrating upriver to spawn. Many lampreys spend most of their lives in fresh water and are sometimes the scourge of those few freshwater fisheries that have not been destroyed by man-made water pollution.

The skeleton of a lamprey consists of the notochord, which persists throughout life, and of various small **cartilages**, parts of the skeleton made up of cartilage (familiar cartilages in the human body include the tip of the nose and the external ear). The heads of cyclostomes are greatly modified from the usual vertebrate pattern because of their unique feeding habits.

Hagfishes are exclusively marine; much less is known about them than about lampreys. They have no larval stage. Hagfishes apparently attack only dead, diseased, or disabled fish and also feed on various invertebrates. They do not suck and rasp at their prey as do lampreys, but can bite using cartilage rods on either side of the mouth.

The mouth and pharynx of the ammocoete larva of a lamprey greatly resemble those of an amphioxus. There is, however, one difference between the two organisms that is of great evolutionary importance. An ammocoete, like an amphioxus, filters food from a current of water that flows in through the mouth and out through the gill slits. In an amphioxus this current is propelled by cilia, mainly on the gills. This is also undoubtedly the way in which the hypothetical tadpole ancestor of these animals fed. In the ammocoete, however, the gills are not ciliary but muscular, and it is muscular action in the walls of the gill pouches that propels the feeding current. Since muscular movement can propel water much faster than ciliary movement, the feeding method of an ammocoete allows more food to be gathered in a given period of time. As a result, an ammocoete can grow to a greater size on filtered food alone than an amphioxus can.

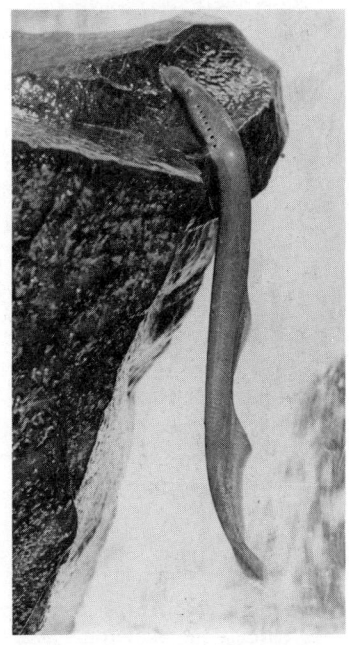

Figure 29-14 An adult lamprey, *Entosphenus tridentatus,* attached to a rock by its sucker. Note the external gill slits and the unpaired fin. (Carolina Biological Supply Company)

SUMMARY

The echinoderms are close relatives of the chordates. This relationship is deduced from the embryology of members of the two groups, and is not evident in the adults. Adult echinoderms are pentaradially symmetrical; the body is usually covered with a thin epidermis overlying a calcareous endoskeleton. Tube feet and a water vascular system are unique features of this group. The echinoderms are all marine, living as filter feeders, carnivores, grazers, or scavengers.

The phylum Chordata contains those animals with a notochord, a hollow dorsal nerve cord, and pharyngeal gill slits. Most chordates living today are vertebrates—chordates in which the notochord is surrounded or replaced by a vertebral column of bone or cartilage. Urochordates and cephalochordates are the only living invertebrate chordates.

The ancestors of chordates were filter feeders. They drew in water containing plankton through the mouth, extracted the food, and pushed the surplus water out through the pharyngeal gill slits. Feeding and not gas exchange is almost certainly the primitive function of gills. With increasing size and larger food, greater speed to catch the food became selectively advantageous. The notochord is the primitive vertebrate skeleton. With myotomes attached to it at the sides of the body, it permitted fast swimming by throwing the body sideways in S-shaped folds; this was probably the most important factor in the success of early vertebrates. Such sinusoidal locomotion can be seen today in such animals as the lamprey and sharks. With fast locomotion, development of the head region with sense organs out in front became more important.

OBJECTIVES

From your study of this chapter you should be able to:

1. List characters that would distinguish members of the following taxa: Echinodermata, Chordata, Urochordata, Cephalochordata, Vertebrata, Agnatha.

2. Describe the habitat and mode of life of members of the groups listed in objective #1.

3. When presented with any echinoderm, name the class to which it belongs.

4. Describe the evolutionary advances of early chordates and vertebrates over their ancestors, and explain the adaptive value of these new characteristics.

SELF-QUIZ

1. The members of the Echinodermata *do not* show:
   a. calcareous skeleton
   b. pentaradial symmetry
   c. tube feet
   d. metameric segmentation
   e. slow locomotion

2. Filter feeding is found in which of the following? (Choose all correct answers.)
   a. adult urochordates
   b. adult cephalochordates
   c. adult agnathans
   d. larval agnathans

3. A long, cylindrical animal that has tube feet is a member of the:
   a. Chordata
   b. Ophiuroidea

   c. Urochordata
   d. Echinoidea
   e. Holothuroidea

4. You would be most likely to find an adult urochordate:
   a. in a mountain stream
   b. in a large river such as the Mississippi
   c. preying on clams
   d. along Cape Cod, attached to the piling of a dock

5. Which of the following chordate characteristics contributes *least* to its efficiency of locomotion?
   a. myotomes
   b. pharyngeal gill slits
   c. notochord
   d. post-anal tail
   e. streamlined body shape

QUESTIONS FOR DISCUSSION

1. An early theory held that vertebrates might have originated from annelids (turn an annelid worm upside down and it looks quite like a fish). What similarities between the two groups led to such a theory?

2. Why is filter feeding such a common way of life among invertebrate animals?

3. Cephalization is pronounced in the vertebrates but not in the cephalochordates, urochordates, or echinoderms. What differences in selective pressures may have caused this difference in degree of cephalization?

REFERENCES AND FURTHER READING

Alldredge, A. "Appendicularians." *Scientific American,* July 1976. An account of an interesting and little-known group of urochordates.

Barrington, E. J. W. *The Biology of the Hemichordates and Protochordata.* Edinburgh: Oliver and Boyd, Ltd., 1965. A thorough discussion of the lower chordates and some of their invertebrate relatives.

Jensen, D. "The hagfish." *Scientific American,* February 1966.

Nichols, D. *Echinoderms,* 4th ed. London: Hutchinson, 1969. An authoritative and readable book on the echinoderms.

CHAPTER
30

VERTEBRATE ANATOMY AND EVOLUTION

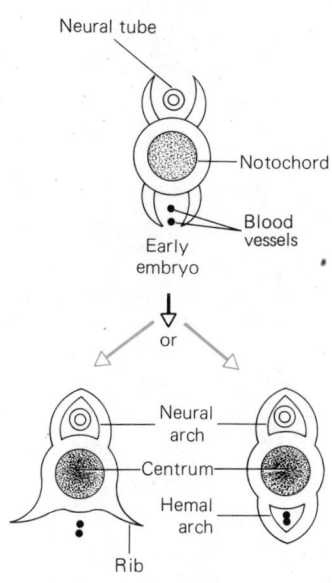

The vertebrate animals make up a subphylum of the phylum Chordata—animals with a notochord, pharyngeal gill slits, and a dorsal hollow nerve cord. During vertebrate evolution the notochord became encased in bony or cartilaginous vertebrae, which form the axis of the skeleton (Figure 30–1). This permitted rapid, efficient locomotion by providing a strong but flexible rod for the segmental myotomes (muscles) to pull against in the body's sinusoidal swimming movements.

An animal that can move fast can become carnivorous, feeding on other animals, and a carnivore can spend much less time actually eating than a herbivore because energy and nutrients are more densely packed in animal tissue than in plant material. As we shall see, most of the evolutionary advances made by the vertebrates have been associated with a carnivorous way of life. Vertebrates are among the most

Figure 30-1 Embryonic development of a vertebra. In evolution, and in embryonic development, the centrum begins as a bony or cartilaginous covering that surrounds, and eventually replaces, the notochord. An arch above the centrum protects the neural tube, which will become the spinal cord. In addition, a vertebra may have a hemal arch, which protects the large blood vessels below the centrum. Ribs and processes for muscle attachment may be present on one or both arches. Together, the vertebrae form a flexible but incompressible column and give vertebrates their speedy locomotion.

TABLE 30-1 THE CLASSES OF THE SUBPHYLUM VERTEBRATA

Class Agnatha	Jawless fishes; gill openings separate; skeleton cartilaginous; notochord persists throughout life; marine and freshwater. The lampreys and hagfishes.
Class Chondrichthyes	Cartilaginous fishes; cartilaginous skeletons, jaws, notochord replaced by vertebrae in the adult; gill openings separate; paired pectoral and pelvic fins; tail fin usually asymmetrical. The sharks, skates, and rays.
Class Osteichthyes	Bony fishes; bony skeletons and jaws; gill openings all covered by a single operculum; paired pectoral and pelvic fins; tail fin usually symmetrical; many have a swimbladder; marine and freshwater; e.g., herring, salmon, sturgeon, eels, sea horse, electric eel.
Class Amphibia	Tetrapods that lay eggs without an amnion or shell; respiration via lungs and skin; scales absent. The urodeles (e.g., salamanders and newts), anurans (frogs and toads), and caecilians.
Class Reptilia	Tetrapods with amniotic eggs and scaly skin. The squamata (snakes and lizards), chelonians (turtles, tortoises, and terrapins), and crocodilians (e.g., crocodiles, caimans, and alligators).
Class Aves	Birds. Tetrapod vertebrates with feathers; high body temperature; bipedal, most species have more than one mode of locomotion; forelimbs usually modified to form wings; e.g., sparrows, penguins, ostriches.
Class Mammalia	Tetrapods with young nourished by milk from mammary glands of females; most viviparous and covered with hair; only vertebrates with only one bone in each side of lower jaw. The monotremes (echidna and duck-billed platypus), marsupials (e.g., opossum, kangaroo), and placental mammals (e.g., humans, bats, whales, rodents, dogs, cattle, elephants, horses).

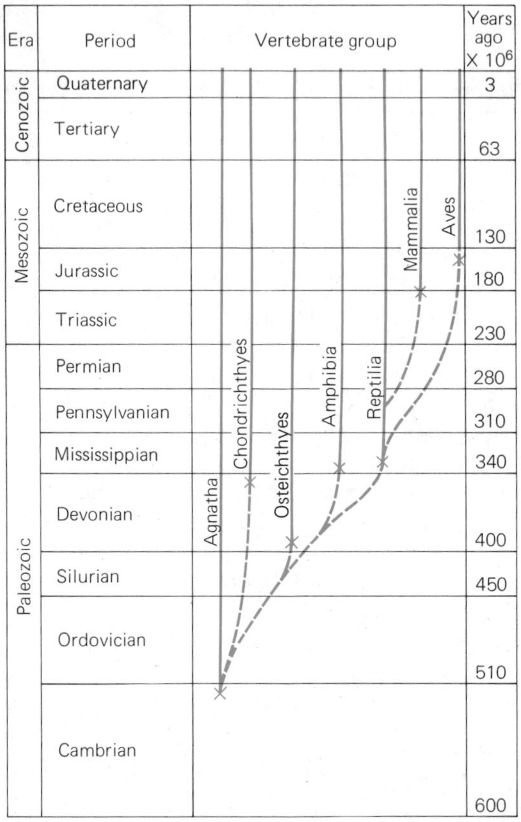

Figure 30-2 Chart of the fossil record and probable family tree of the classes of vertebrates. The crosses indicate when members of each group first appear in the fossil record.

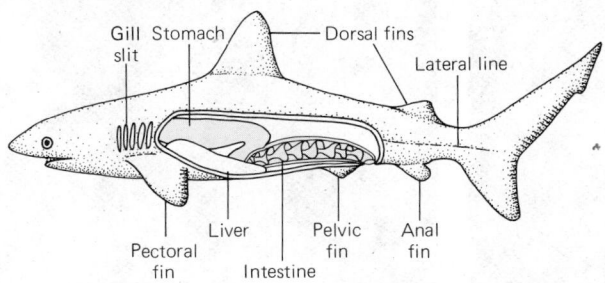

ELASMOBRANCH

Gill slit · Stomach — Dorsal fins
Lateral line
Liver · Pelvic fin · Anal fin
Pectoral fin · Intestine

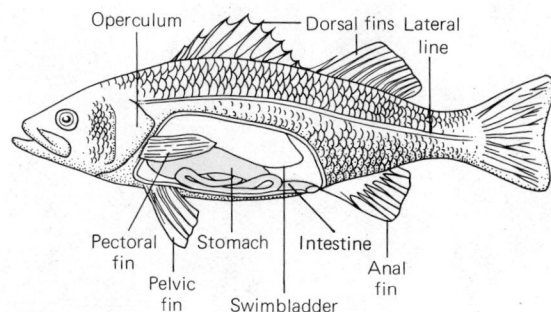

TELEOST

Operculum — Dorsal fins · Lateral line
Pectoral fin · Stomach · Intestine · Anal fin
Pelvic fin · Swimbladder

Figure 30-3 Comparison of a shark and a bony fish. The body cavity occupies much less of the body than do the swimming muscles. The large, fat-filled liver of the shark and the swimbladder of the bony fish both provide buoyancy. A bony fish has a long, thin intestine, which provides a large surface area for absorption of digested food into the body. The shark's intestine is shorter and contains a spiral valve, which increases its internal surface. The gill slits are separate in the shark but covered by a common operculum in the bony fish. Note the different positions of the fins in the two types.

numerous and diverse animals; they have radiated over the globe, with species at home in the sea, in fresh water, on land, and in the air.

In this chapter we shall examine the classes of vertebrates and some of the evolutionary advances that account for their success.

30-A Evolution of Fishes

Available evidence suggests that the agnathan fishes described in Chapter 29 gave rise to two large and successful groups of fishes: the **Chondrichthyes**, or cartilaginous fishes (most of which belong to a group called **elasmobranchs**), and the **Osteichthyes**, or bony fishes. These two groups have made two major evolutionary advances over their agnathan ancestors. First, part of the gill skeleton has moved forward and has evolved into jaws, permitting the Chondrichthyes and Osteichthyes to bite and chew their food instead of sucking or filtering it as agnathans do. Second, both groups have paired fins: **pectoral** fins at the front and **pelvic** fins at the back, as well as unpaired **dorsal**, **anal**, and **tail** fins (Figure 30–3). Paired fins allow a fish to balance and maneuver in ways that are not possible without them. Paired fins are also important phylogenetically because they gave rise eventually to the paired forelimbs and hindlimbs of terrestrial vertebrates.

30-B Class Chondrichthyes
(Cartilaginous Fish: Sharks and Rays)

Of the hundreds of species of Chondrichthyes known from Paleozoic fossils (see Figure 30–2), only about 250 species survive today; these are the sharks, dogfish, rays, and skates. They have skeletons composed entirely of cartilage, and almost all are marine. The sharks are notorious carnivores and scavengers; the flattened skates and rays, which live on the sea floor, feed mostly on invertebrates. Although they are not quite so complicated anatomically as the bony fish, the Chondrichthyes are a successful group well adapted to their role as predators of the seas.

The teeth of a shark are enlarged versions of the **denticles** found all over its skin (Figure 30–4). Sharks are not picky eaters. One killed in the Adriatic Sea had in its stomach two raincoats, part of a horse, an automobile license plate, and a length of rope. The stomach of a small dogfish, such as you might dissect in the laboratory, is more likely to contain crustaceans and fish. Digestion takes place mainly in the stomach, while absorption occurs in the intestine, which, unlike that of most vertebrates, is short and fat. The surface area for absorption is increased, however, by a **spiral valve**, which runs down the inside of the intestine. The liver of a shark is enormous in proportion to its body size. This is because sharks store lipids in their livers, and by altering the lipid content of the liver, they can slowly increase or

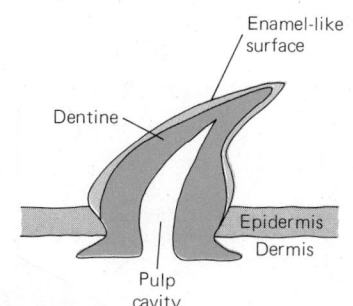

Enamel-like surface
Dentine
Epidermis
Dermis
Pulp cavity

Figure 30-4 The placoid scale of a shark. These scales, also known as dermal denticles, are embedded in the dermis of the skin all over the body and also make up the teeth in the jaws and roof of the mouth.

CHAPTER 30
VERTEBRATE ANATOMY
AND EVOLUTION

451

Figure 30-5 A white tip reef shark. Note the separate gill slits, asymmetrical tail, and triangular dorsal fin. (Biophoto Associates, N.H.P.A.)

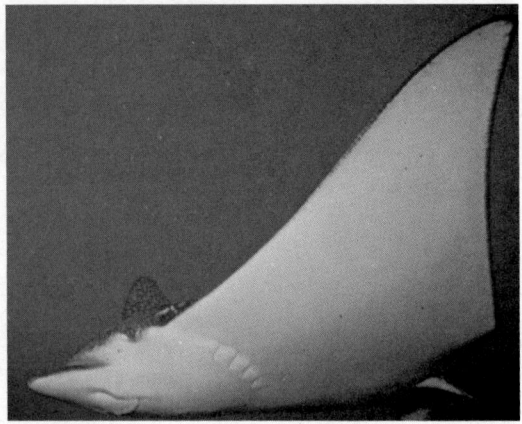

Figure 30-6 An eagle ray. Note the row of gill slits on the ventral surface of the body. (Biophoto Associates, N.H.P.A.)

decrease the buoyancy of the whole body, reducing the effort needed to stay at a particular depth in the water.

Sharks swim in primitive vertebrate fashion, by sinuous waves of the body, using their segmental muscles and jointed backbones. Sharks (and bony fish) that spend all of their time swimming actively depend on their constant forward motion to pass oxygen-rich water over their gills, where gas exchange occurs. If these active swimmers are restrained, they "drown" because they lack the pumping muscles in the head that allow less active fish to stay in one place and draw a water current across the gills.

Sharks have internal fertilization. Females produce a few, large eggs each, with a great deal of yolk to supply food for the embryo. The male uses his claspers to pass sperm into the female's body, where the eggs are fertilized. In some species the large eggs are covered with a leathery coat before being deposited; in most species, however, the mother retains the embryos in the **oviduct**, the tube leading from the ovary toward the exterior of the body, until they are fully developed baby sharks. Many of these sharks are **ovoviviparous**—that is, the embryo receives all its nourishment from the yolk of its egg; a few, however, are **viviparous**—nourished by the mother, with close contact between the blood vessels of the oviduct and those that grow out from the embryo over the egg yolk. These embryos receive nourishment from the mother's bloodstream. The embryos of the sand tiger, a shark commonly kept in public aquariums, obtain nourishment by eating their fellow embryos as well as additional eggs coming from the ovary! This prenatal cannibalism results in the birth of only two offspring at a time, one from each oviduct.

The pressure and chemical sense organs of sharks are highly efficient. Sharks have large **olfactory organs** in their heads, which permit them to detect chemicals (notoriously blood) in the water. In addition, all kinds of fish possess **lateral line systems**, composed of pressure receptors arranged along the sides of the body, which permit them to detect vibrations or sound waves in the water. In the head of a shark, the pits of the lateral line organ are modified to form the **ampullae of Lorenzini**, long canals filled with mucus. These are very sensitive electroreceptors. Any animal generates electric fields, and experiments have shown that a dogfish shark can use its electroreceptors to detect an electrical potential of as little as 2 microvolts emitted by a worm that the shark can neither see nor smell. Once the olfactory and lateral line organs have guided the shark to the vicinity of its prey, the eyes guide the actual attack on a victim.

The two largest sharks (like the largest whales) are not predaceous but gentle filter feeders. One of these, the whale shark, filters more than a million litres of water an hour. It lives mainly on plankton.

The pectoral fins of skates and rays are greatly enlarged and are used for locomotion (Figure 30-6). Some rays have poison spines, which they use to defend themselves, on the back or tail. The electric ray repels intruders with an organ that can produce quite a powerful electric shock.

Although the number of species of Chondrichthyes has declined since the Permian period, the bony fish, or Osteichthyes—the second major class of fish that evolved from agnathan ancestors—are still evolving new species. Bony fish anatomy and physiology are so versatile that this group has given rise to a much greater diversity of types than has any other aquatic group. Most modern bony fish are members of the superorder **Teleostei**. (A super order is a group containing several orders.)

The diversity among teleosts is almost as bewildering as that of the insects. Among the teleosts are filter feeders like herrings; parrot-fishes, which crunch up coral; insectivores (insect-eaters) like trout; and predaceous carnivores like piranhas, barracuda, and blennies. Teleosts come in all shapes and sizes. Boxfish are practically spherical, moray eels are huge and snake-like, a sea horse looks like a horse's head on a monkey's tail, and a stonefish looks like an irregular lump of rock.

Teleosts have undergone impressive adaptive radiation—the formation of new species, adapted to different ways of life, from common ancestral species. This radiation has resulted largely from a few evolutionary advances. One of the most important was the development of a **swimbladder**, found in many bony fish. The swimbladder is a gas-filled sac formed as an outgrowth of the pharynx. By altering the gas pressure in the bladder, a fish can alter its buoyancy so that it can remain at any depth in the water with no muscular effort.

Most of the push in teleost locomotion comes from the tail; the paired fins are usually used for fine control. The pelvic fins are usually further forward and higher on the body than those of a shark (see Figure 30–3). The gills of bony fish do not open separately to the exterior, as do those of cartilaginous fish, but are instead covered by a common **operculum**. Water for respiration moves in through the mouth and out through the gills as a result of the movement of muscles in the head and at the base of the operculum.

A striking feature of bony fish is that they occur both in fresh water and in the sea, and many, like trout, salmon, and eels, can move from one to the other. Life in fresh water poses problems of osmoregulation, regulation of the body's water content, because water tends to enter the fish by osmosis through the thin gill membranes. Teleosts have two main adaptations that permit them to live with this problem. Their kidneys, the main organs of osmoregulation, excrete very dilute urine when the fish is in fresh water. In addition, freshwater fish have special cells in the gills that can absorb salts from the water by active transport.

Teleost reproduction is as varied as everything else about this group. Unlike the Chondrichthyes, most of the teleosts have external fertilization, and many lay large numbers of small eggs (as many as 3 million for herring) (Figure 30–7). Some build nests and care for the eggs; others do not. In pipefishes and sea horses, the males have brood pouches, where they incubate the young. Other species have internal fertilization, and the young develop within the female's body. In some deep-sea fishes the male is reduced to a tiny parasite that lives permanently attached to the female.

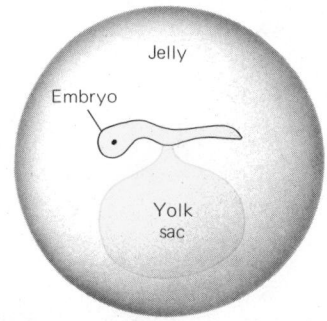

Figure 30-7 A teleost egg. The embryo obtains its food from the yolk sac and is protected and buoyed up by the jelly coat. The egg lacks the membranes that protect the embryos of terrestrial vertebrates. The egg of an amphibian resembles that of a teleost.

Figure 30-8 A bony fish, class Osteichthyes. Note the placement of the pelvic fins far forward on the body compared to those of the shark in Figure 30-5. (Biophoto Associates, N.H.P.A.)

Figure 30-9 Bony fish come in varied shapes and colors, as shown by this imperial butterfly fish. (Biophoto Associates, N.H.P.A.)

Many teleosts have sharp, protective spines on their dorsal fins or opercula. In some species, like the stonefish, rockfish, and scorpion fish, these spines are connected to poison glands and can inject nerve or muscle poisons powerful enough to kill human beings or large fish. Several species of puffer fish protect their eggs with a coating of jelly containing **tetrodotoxin**, a nerve poison so potent that less than a microgram will kill a human being. (This makes eating puffer fish, considered a delicacy in Japan, a bit like Russian roulette. Chefs must be specially licensed to prepare the fish; nevertheless, fatal accidents still occur.)

Members of four different families of teleosts have evolved the ability to produce electric discharges. *Electrophorus*, the electric eel of the Amazon, can generate up to 550 volts, which it uses for offense and defense. More commonly, teleosts use electric organs and electroreceptors for navigation (see the essay, Chapter 40).

Many deep-sea fishes are luminescent. Some have light-producing cells; others have organs containing luminous bacteria. Light flashes are probably used to signal the opposite sex and to startle attackers. In the deep-sea angler fish, the luminous tip of a fin is used as a lure that attracts prey. In addition, many fish can change color by using tiny muscles in the skin to alter the size of different chromatophores (color cells).

At times when the water has a low oxygen content, some fish breathe air. A goldfish in its bowl will gulp air at the surface when its water is low in oxygen. Some fish have special organs that can absorb gaseous oxygen, but most use the swimbladder, which is connected to the pharynx in some teleosts. A swimbladder used to breathe air in this way is very like the lung of a terrestrial vertebrate, and it is not difficult to imagine that some air-breathing fishes, using their pectoral and pelvic fins to move about on land, were the ancestors of the first terrestrial vertebrates. In fact, there are some modern fish that do leave the water and crawl about on land.

30-D The Move to Land

Many selective pressures probably contributed to the evolution of vertebrates that could live, for at least part of the time, on land. In the Devonian period, the seas were teeming with carnivorous fish, and any fish that could remove itself or its eggs onto land would lower its mortality rate impressively. Plant life on land was well established, and terrestrial insects were evolving and multiplying rapidly. A vertebrate that ate plants or insects, and that could survive on land, would have had little competition for a burgeoning food supply. In addition, air contains more oxygen— about 20%—whereas water seldom contains more than 4% oxygen. However, any fish that survives on land for even a short time must possess adaptations to existing surrounded by air, which has very different properties from those of water.

Air is less dense than water and provides less support to the body. The pectoral and pelvic fins and the backbone of a fish that moved from the water to the land thus had to have sturdy bones and strong muscles to permit the fish to support itself on land. Furthermore, gills cannot be used to breathe on land. The surface tension of water makes the feathery gill filaments stick together when the fish comes out of water into the air; a swimbladder/lung or body surface that keeps its shape in air is necessary.

Dehydration is probably the biggest problem faced by an aquatic animal moving to land. Plainly the body surface must be waterproofed, but this is not enough. The respiratory surface must be kept moist because gases can pass across plasma membranes only in solution. Water will also be lost in the urine and feces. Lungs and swimbladders lose less water by evaporation than does an external respiratory surface. Fish that can survive on land for long periods, like the lungfish, have an internal respiratory surface. Their kidneys can also produce a concentrated urine, and the walls of the digestive tract can absorb water into the body, thus reducing the amount lost in the feces.

In addition, sound, light, and chemicals travel differently in air than in water. The sense organs of an aquatic animal will work on land, but they are less than ideal.

The fossil record shows that there evolved in the Carboniferous era, about 300 million years ago, a number of animals that looked much like modern lungfish. These creatures had fish-shaped bodies, short stubby legs, and no gills. They were the first land vertebrates, ancestors of modern amphibians and reptiles.

TABLE 30-2 ORDERS OF AMPHIBIA

ORDER	NUMBER OF SPECIES	HABITATS	EXAMPLES
Urodela	320	Freshwater and moist terrestrial throughout world	Salamanders, mud puppies, newts
Anura	2000	Freshwater, moist terrestrial; some desert and brackish water species	Frogs, toads
Apoda	55	Moist tropical terrestrial	Caecilians (look like worms)

30-E Class Amphibia (Frogs, Salamanders, and Their Relatives)

Amphibians, reptiles, birds, and mammals are known as **tetrapods** ("four feet"). They are the land vertebrates. Only reptiles, birds, and mammals are fully emancipated from the water, however; most amphibians must return to water to reproduce, and their larval tadpole stage is fully aquatic. Although adult amphibians of many species are terrestrial, most must still live in moist environments. Even the desert toads spend most of their time in burrows where the humidity is high. Few amphibians are able to cope with the osmotic complications of life in the sea. Their body fluids, like those of most other vertebrates, have a higher (less negative) osmotic potential than that of sea water. However, there is a tropical species that lives in the brackish water of mangrove swamps.

The two largest amphibian orders are the Urodela (e.g., newts, salamanders, hell-bender, mud puppy) and the Anura (frogs and toads). Urodeles are the more generalized and show the transition from fish to tetrapod more clearly. The limbs of a urodele contain small bones and muscles, like those found at the base of the pectoral and pelvic fins of lungfish. In addition they have pectoral and pelvic limb girdles (shoulder and hip girdles), which eventually evolved until they formed a strut between the backbone and limbs in all higher vertebrates. The anurans have very specialized skeletons that give them their ability to jump long distances. The backbone is vastly shortened, with many vertebrae fused together, and the limb girdles are firmly attached to the backbone. Caecilians belong to a third, small order, the Apoda, which means "without feet." They are legless, wormlike, tropical animals adapted to burrowing in leaf litter on the forest floor.

Figure 30-11 A Central American tree frog, member of the amphibian order Anura. The tympanic membrane can be seen just behind its eye. (Mark Rausher)

Figure 30-10 A fire salamander of the amphibian order Urodela. Compare this with the rather similar lizard, which is a reptile (Fig. 30-14). The salamander's skin is moist, but the lizard's is covered with dry scales. (Biophoto Associates, N.H.P.A.)

Amphibians have a soft glandular skin, which is used for gas exchange in most species, despite the fact that most amphibians also have small lungs. (The aquatic larvae have gills.) Only the caecilians have scales. The sense organs of amphibians are also adapted to life on land. In particular, the anterior end of the lateral line organ (the pressure receptors of fish) has evolved into simple ears, which respond to sound (pressure) waves in air. Amphibians were also the first vertebrates to develop true tongues. In most frogs and toads the tongue is long and sticky and can be shot out rapidly to catch flies.

Another amphibian adaptation to terrestrial life is the production of **vasopressin** (antidiuretic hormone), a hormone that increases the amount of water resorbed into the body from the urine.

Despite these adaptations to life on land, amphibians are not emancipated from water to the same extent as reptiles, birds, and mammals. This is most obvious in reproduction. Amphibian eggs are without the membranes and shell that protect the eggs of higher tetrapods from desiccation. Many lay their eggs in fresh water, but species in all the major groups have evolved the ability to lay their eggs elsewhere. Some salamanders, frogs, and caecilians lay their eggs in damp places on land. In the frog *Pipa dorsalis*, the eggs develop in pouches on the mother's back. A male frog of the genus *Rhinoderma* carries the developing young in his vocal pouch, and the female *Rheobatrachus* carries the tadpoles in her stomach. In a number of species of amphibians, the young develop in the female's oviduct and are born as miniature adults.

While the tadpoles of frogs and toads are herbivorous, salamander larvae are carnivorous and look much more like miniature adults than anuran larvae do. Some salamanders, such as one species of *Ambystoma*, never undergo metamorphosis but breed although they look like larvae even when they become breeding adults (a phenomenon known as paedogenesis).

Amphibians are visible in large numbers only during the breeding season, when they congregate around ponds and streams. At mating sites, male frogs and toads produce mating calls that permit the female to recognize and approach a male of her own species. For the rest of the year they are secretive and silent, and little is known about their behavior. Many amphibians **hibernate** on land or under water during the winter; metabolism slows down and the animal survives on energy stored in the body. Since they are poorly protected from desiccation, many amphibians also **aestivate**, or go into hiding during summer, emerging only at night when it is cooler and more humid, or on wet days.

Most amphibians are protected from predators by poisons in the mucus on their skin (and in the jelly surrounding their eggs). The disgust of a cat or dog that attacks a toad is dramatic evidence that the defense works. (Frog legs are skinned before they are served as human food.) Some tropical species are extremely poisonous and advertise themselves to potential predators by spectacular fluorescent green and red coloration. Colombian Indians tip poison arrows with virulent nerve poisons such as **batrachotoxin**, which they obtain by heating the frogs *Dendrobates* and *Phyllobates* over a fire; this induces the frogs to secrete quantities of poisonous mucus. A milligram or less of batrachotoxin will kill a human being.

30-F Class Reptilia

The Mesozoic era is sometimes known as the "age of reptiles" because reptiles of all shapes and sizes are the predominant animals found in marine, freshwater, and terrestrial fossil beds laid down during this time. With the insects, reptiles dominated animal life on land for about 200 million years and are still very much with us today.

Whether reptiles descended from early amphibians, or from a group of lungfish different from the ancestors of the amphibians, is a hotly debated question. Whatever the answer, it is clear that reptiles are better adapted to life on land than are amphibians. Their main advantage is the evolution of an egg that is protected from dehydration and so can be laid on land. Their major disadvantage compared with their mammalian and avian (bird) descendants is that they have less ability to maintain a high body temperature independently of the environment.

THE DIVERSITY OF ORGANISMS

The **cleidoic** (closed) or **amniotic** egg (Figure 30–12), which reptiles, birds, and some primitive mammals lay, has a number of features that adapt it to life on land. In the amniotic egg, the developing embryo is surrounded by a membrane, the **amnion**, which encloses the **amniotic fluid**. This fluid protects the embryo from dehydration and from being jolted around. Two membranous sacs are attached to the embryo. One is the **yolk sac** (found also in fish embryos), containing yolk which the embryo uses for food. The other is the **allantois**, which stores the embryo's nitrogenous waste until hatching and is found only in reptiles, birds, and mammals. Blood vessels grow out from the embryo through the membranes of the yolk sac and allantois until they come close to the surface of the egg, where they take in oxygen from the environment and release carbon dioxide. The embryo, amnion, yolk sac, and allantois are all surrounded by a membrane called the **chorion**, which controls the overall permeability of the egg. The egg is permeable to oxygen and carbon dioxide and impermeable to water. The chorion in turn is surrounded by the outer **egg shell**; the reptilian egg shell is leathery and contains some calcium carbonate.

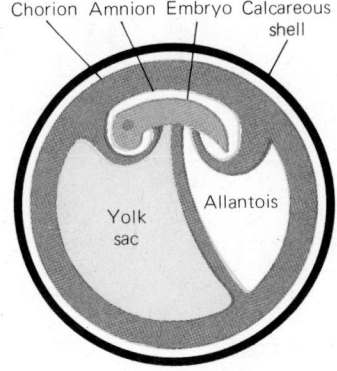

Figure 30-12 The amniotic egg of a reptile or bird.

Since embryonic development proceeds faster at higher temperatures, it is selectively advantageous to keep the eggs warm. Reptiles have numerous adaptations that accomplish this. Female pythons coil around the eggs and brood them for three months until they hatch. Many snakes, especially in colder climates, retain the embryos inside the mother's body. Reptiles may also bury their eggs in warm sand or in rotting vegetation, which is warmer than the surrounding environment.

Because the egg is laid in a leathery shell or the young develop within the mother's body, reptiles have internal fertilization, and the male has a penis (or even two).

The skin of a reptile is dry and scaly. The scales are made up of **keratin**, a protein that is also present in feathers, horns, and hair. In crocodiles and tortoises the scales are replaced by the skin as they wear away. In lizards and snakes, the scales are shed several times a year during a molt.

The first reptiles were small carnivores that looked rather like small dogs. Their limbs were stronger and were tucked further under their bodies than those of amphibians, enabling them to move faster; their jaws were more firmly attached to the skull, permitting them to subdue and eat larger prey; and their skins were waterproofed and scaly, minimizing the loss of water across the outer body surface. Without a moist skin, reptiles had to breathe entirely with their lungs.

The adaptive radiation of the early reptiles is a fascinating story. There were reptiles that flew, walked, and swam. The dog-like ancestors of the mammals became fast-running carnivorous **quadrupeds** (animals walking on four legs); many members of the group that gave rise to the birds had a tendency to **bipedalism** (walking on two legs) and had reduced forelimbs (Figure 30–13). Some reptiles were

Figure 30-13 *Camptosaurus*, a birdlike dinosaur. Compare this with the skeleton of a modern bird (Fig. 30-21). The main resemblance is that both are bipedal. The dinosaur had a heavier but weaker skeleton; the rib cage was not reinforced as it is in a bird, and the vertebrae and limb girdles were separate. Some of the muscles that pulled the hind limb back were attached to the tail; thus the tail is heavy and ribbed. This feature is found in nearly all reptiles. In mammals and birds, the hind leg muscles attach only to the pelvic girdle, and the tail skeleton is much smaller or even absent.

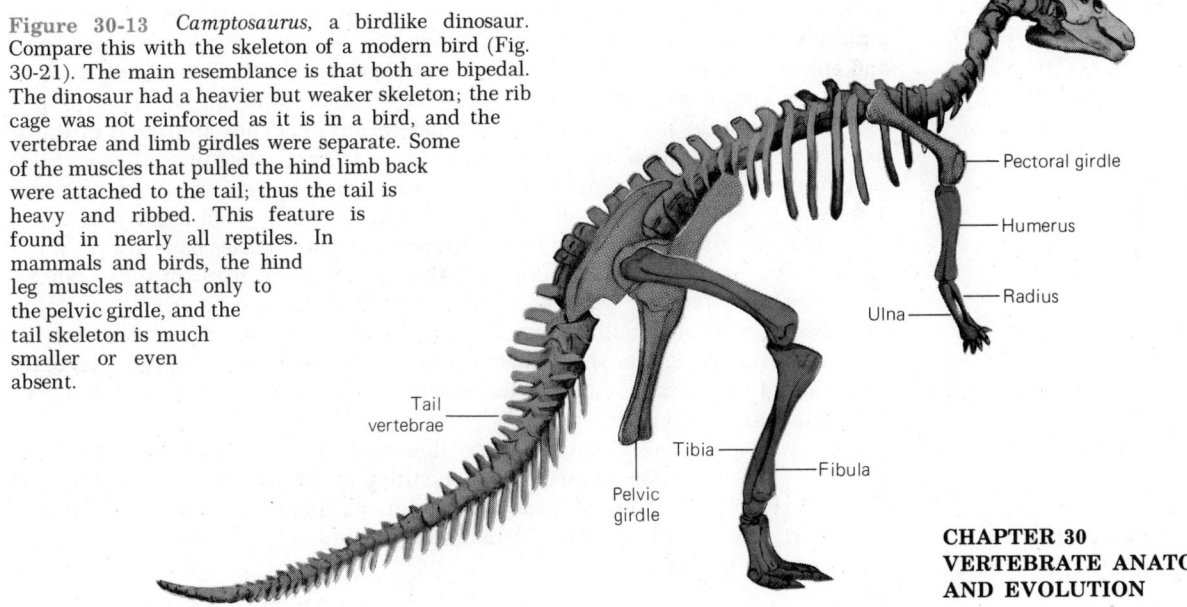

**CHAPTER 30
VERTEBRATE ANATOMY
AND EVOLUTION**

insectivorous, some voracious carnivores, and some placid herbivores of enormous size. (Although the huge brontosaurs are often depicted wallowing in ancient swamps, recent comparisons with modern-day animals support the idea that they were actually dry-land animals, grazing in treetops as the equivalents of giraffes and elephants.)

There is still some mystery about the extinction of many of the reptiles, for large numbers of them, including all the dinosaurs and all the flying reptiles, disappeared from the fossil record during a short time at the end of the Cretaceous period. But it seems that the birds and mammals, with their constant high metabolic rate, finally out-competed the reptiles for most of the world's resources, leaving only the relatively few specialized reptiles that are with us today.

It is incorrect to call reptiles cold-blooded. They maintain a body temperature considerably higher than that of their surroundings, but they do not usually generate most of their heat by their metabolism, as birds and mammals do. The reptiles' adaptations to permit high body temperature are mainly behavioral. Many reptiles must lie in the sun for a period before they become warm enough to be active. Some extinct reptiles had large sails on their backs, and it is believed that they controlled their temperature by altering the angle of the sail to the sun. (Many butterflies do this today with their wings.)

Modern Reptiles

Of the vast array of prehistoric reptiles, representatives of only three orders have survived to the present day.

TABLE 30-3 **ORDERS OF MODERN REPTILES**

ORDER	HABITATS	EXAMPLES
Chelonia	Freshwater, marine, terrestrial	Turtles and tortoises
Squamata	Mainly terrestrial; some freshwater and marine	Lizards, snakes, geckos, iguanas
Crocodilia	Mainly freshwater; some marine, most partly terrestrial	Crocodiles, alligators, caimans

ORDER SQUAMATA

The Squamata, the lizards and snakes, are the reptiles most like their prehistoric ancestors. Lizards are particularly easy to mistake for salamanders, which are typical amphibians; they differ from salamanders, however, in the greater strength and efficiency of their skeletons, in the possession of dry scaly skin and of claws on their toes, and in the fact that their eggs are adapted to life on land.

Lizards are the most widespread of modern reptiles. They are found in jungle treetops, grasslands, deserts, rivers, and sea coasts. They range in size from geckos weighing one or two grams to the 100-kilogram Komodo dragon of Indonesia, which is reliably reported to kill small pigs on occasion. The Malaysian flying dragon is a lizard that can glide from tree to tree. A number of lizard species have lost their legs in a way that parallels the evolution of snakes. The marine iguana of the Galapagos is even at home for long periods in the sea, where it feeds on marine algae. Most chameleons, with their rolling, turreted eyes, are insectivorous, as are a majority of the smaller lizards. Larger species will eat mammals and birds when they can catch them, and a few species, like the basilisk (whose gaze, in legends, is death), eat vegetable matter and fruit.

Many people think of snakes when the word "reptile" is mentioned. A few dangerous snakes, and ignorance and superstition about many harmless ones, have been responsible for giving all reptiles a bad name. In fact snakes are not typical reptiles; they are highly specialized anatomically, and very few species are dangerous to human beings.

Figure 30-14 A lizard, showing the dry scales and clawed toes characteristic of reptiles. (Biophoto Associates)

Figure 30-15 A gecko, an insect-eating lizard whose sucker-like toes enable it to hang upside down. (Biophoto Associates)

The evidence suggests that snakes evolved from lizards and probably lost their legs originally as an adaptation to burrowing. Most species move by using their body musculature to throw the body into curves; scales on the ventral surface, or the curves of the body itself, provide traction. The group includes expert swimmers, burrowers, and climbers. The backbone is greatly elongated, and most of the vertebrae bear long, flexible ribs that hold the body in shape.

The jaws of a snake are exceptionally mobile and loosely attached to the rest of the skull. The two halves of the lower jaw are joined in front only by elastic tissue, and the mouth can be enlarged to enormous size, permitting a snake to engulf prey much larger than the diameter of the head. The front of the windpipe can be protruded into the throat so that breathing is not obstructed as prey passes slowly down the throat. The teeth tend to curve backwards, preventing the prey from popping back out of the mouth. Egg-eating snakes have a special tooth in the top of the mouth that is used to break the shell of birds' eggs. In some snake species two or more teeth are specialized as hollow fangs connected to muscular salivary glands that produce venom and pump it through the fangs into the prey. Some snake venoms contain nerve toxins and toxins that paralyze muscles. Others contain agents that lyse blood cells.

(a)

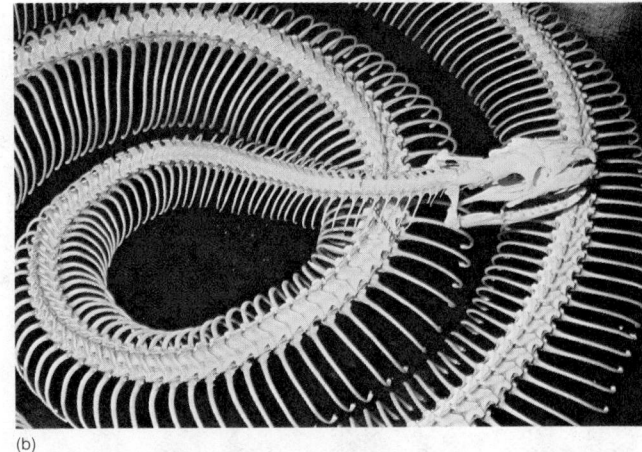

(b)

Figure 30-16 (a) A snake hatching from its leathery egg. (b) The skeleton of a snake is mostly vertebrae with their attached ribs. (Biophoto Associates, N.H.P.A.)

A snake's tongue is not poisonous, as is popularly supposed; it flickers as it carries chemicals from the air or ground to the organs of Jacobson—sense organs that detect chemicals—located in the roof of the mouth. Snakes have well-developed ears, which respond mainly to vibrations of the ground detected through the lower jaw. Pit vipers and some boas also have heat-detecting organs on the head, which allow them to strike warm-blooded prey accurately on dark nights or in deep burrows.

The most primitive snakes retain traces of their presumed lizard ancestry, such as hindlimb girdles and rather immobile jaws. Most of these snakes belong to the New World boas and the Old World pythons, including the world's biggest snakes. The longest are the awe-inspiring South American anacondas, which may reach nearly 10 metres in length.

ORDER CHELONIA

Turtles, terrapins, and tortoises belong to the order Chelonia, one of the most ancient reptilian groups. They are clearly very specialized and have survived because the protection afforded by their bony or leathery shell has been highly effective. Most are herbivorous, and there are species adapted to life on land, in fresh water, and in the sea. Some can withdraw their legs and heads completely into their shells. Many marine species are famous for their annual migrations to the beaches where they lay their eggs.

ORDER CROCODILIA

The order Crocodilia contains the closest living relatives of the extinct ruling reptiles, and of their descendants, the birds. Among the ruling reptiles, known as the Saurischia, dinosaurs had a strong tendency towards bipedalism, and crocodiles, although they are quadrupeds, plainly belong to this line of evolution because their hind limbs are longer than their forelimbs. Three groups of crocodilians survive today: the crocodiles of Africa, Asia, and America; the alligators of the southern United States and China and the caimans of Central America; and the gavial of Southeast Asia. All spend much of their time in water, and have a special arrangement of their nostrils that permits them to breathe while the rest of the body is submerged. All are carnivorous. Females lay their eggs in nests and guard the nest

Figure 30-17 A turtle, member of the reptilian order Chelonia, sunning herself. (Biophoto Associates, N.H.P.A.)

Figure 30-18 A crocodile, member of the reptilian order Crocodilia. (Biophoto Associates)

and newly hatched young from their cannibalistic relatives and other predators. Fear of crocodilians and desire for their skins to make shoes and handbags have brought most species close to extinction; however, efforts to protect alligators in the southern United States have been spectacularly successful, and controlled hunting is now permitted in some areas.

30-G Class Aves (the Birds)

Birds can be simply defined as the only organisms with feathers. Birds and mammals both originated from different, early groups of reptiles.

The earliest known bird is *Archaeopteryx*, represented by several specimens from the Jurassic limestone of Bavaria, laid down 150 million years ago. *Archaeopteryx* was a bird because it had feathers (and a wishbone), but in all other ways it looked much more like one of the small, bipedal dinosaurs from which the birds presumably originated. *Archaeopteryx* had a long tail, with separate vertebrae (modern birds have a greatly reduced tail skeleton), teeth (birds have a beak or bill), and small wings with claws on the ends of the toes (the alterations of the forelimb skeleton that turned them from legs into wings were not complete) (Figure 30–19). This bird probably could not fly by flapping its wings, nor could it fold its wings close to the body as modern birds do.

The origin of bird flight has been a subject of much debate. One theory holds that birds probably used their forelimbs at first for stabilizing themselves when jumping from branch to branch, and later as parachutes (as used by flying lizards, flying frogs, and flying squirrels). According to another theory, birds were bipedal insectivores, running along the ground catching insects. The broadening and flattening of the forelimbs would have allowed such animals to make sustained jumps into the air and so increased their insect-catching prowess. Still another theory holds that wings evolved as "insect nets," used to scoop flying prey toward the mouth. Each theory proposes that the wings and their associated bones and muscles eventually became capable of true, flapping flight.

Feathers probably evolved from reptilian scales, although there is no trace of this evolution. Feathers are one of the main reasons for bird success. They are among the strongest known materials, by weight; they are flexible, and they are also excellent insulation (Figure 30–20). Feathers form a flying surface that is a distinct

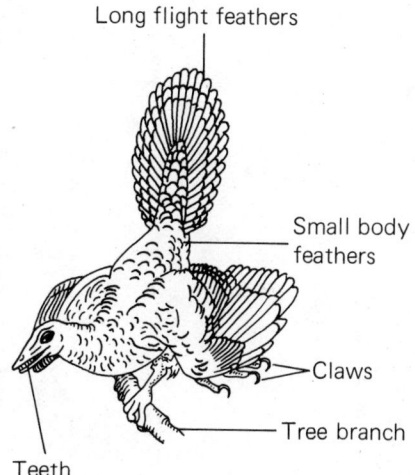

Figure 30-19 *Archaeopteryx*, the first known bird. Note the heavy tail and the claws on the digits in the wing, neither of which is present in a modern bird.

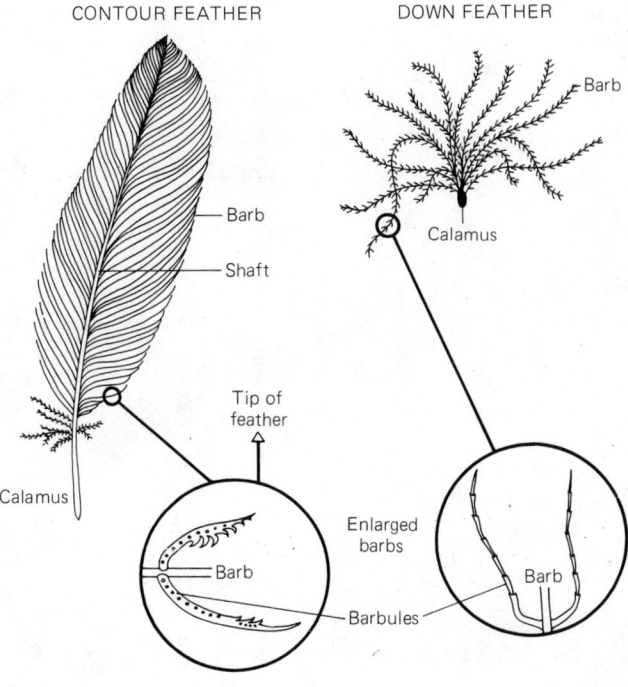

Figure 30-20 Feathers that perform different functions. The contour feathers provide a smooth surface for flight and streamlining. When the bird preens, it draws the barbs through its bill from base to tip. As a result, the hooked barbules on one barb hook onto the ridged barbules on the next one, linking the barbs in a smooth but flexible surface. Down feathers are the last word in insulation. Their long, unhooked barbules provide a mass of spaces where air is trapped and warmed by the body. They are the first feathers to develop on young birds, and they underlie the contour feathers on parts of the body of the adult. Contour feathers have a little bit of down at the base; thus they supply insulation as well as a smooth body surface.

461

improvement over the **patagium**, of skin and connective tissue, used by bats and the extinct flying reptiles. The trouble with a patagium is that a tear may put the entire wing out of order. With feathers, however, birds can suffer considerable damage to the wings without losing the ability to fly. A patagium also invariably stretches from forelimb to hindlimb. Birds use only the forelimbs for flight; the legs are free to be used for running or swimming, and nearly all birds have two different types of locomotion (e.g., flying and swimming, running and flying). Feathers are molted periodically; colorful feathers are used in courtship displays and other types of social signaling.

Birds are **endothermic**, using metabolic heat to maintain their body temperature at a high level. The down feathers, and a layer of fat just under the skin, insulate the body against changes in temperature. Endothermy permits birds (and mammals) to be active whatever the external temperature, which is not possible for reptiles or amphibians.

The strongly social nature of many birds and their complex behavior patterns are a result of two main factors. First, a bird's brain is quite large and complex, as it must be to control the intricate muscular movements of flying. Second, birds build nests and incubate their eggs; because they have a high body temperature, and therefore must feed more often than most reptiles, both parents often collaborate in nest-building, incubating, and feeding. This has led to extensive interaction between the sexes, and has apparently resulted in complex courtship, monogamy,* or reproduction in large colonies in many bird species.

Birds are anatomically conservative, without the great range of structural modifications found in other vertebrate classes. This is undoubtedly because most birds fly, and flight is structurally demanding. The proportions of a bird's skeleton are very different from those of other tetrapods (Figure 30–21). The center of gravity has been brought forward from its position in most other tetrapods, so that it lies under the wings and above the legs. The teeth are replaced by a light horny bill or beak and a **gizzard**, a grinding chamber in the gut (found also in crocodiles). In addition, most of the bones in birds' skeletons are fused to one another, strengthening the air frame, and only the legs, neck, bill, and wings are separately moveable. The tail vertebrae are fused together into a tiny bump behind the pelvic girdle, and the tail is made up of feathers that can be moved by muscles attached to the reduced tail skeleton. Bird bones tend to be hollow and reinforced by internal girder-like structures similar to those in an airplane wing; the whole skeleton is both light and

---

* Monogamy   The mating of one male with one female, either for life or for the duration of one breeding season.

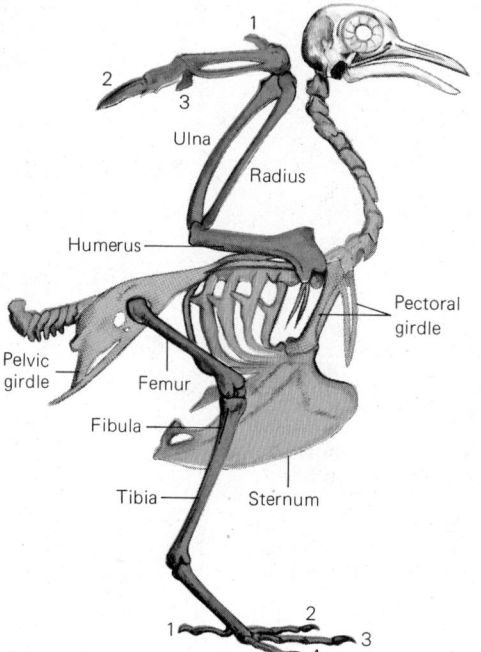

Figure 30-21 Skeleton of a modern bird (a pigeon). Note the major features that distinguish it from the skeleton of a reptile or mammal. The main part of the body is very light but strong. The vertebrae between the pectoral and pelvic girdles are fused to make a rigid backbone. Overlapping processes on the ribs, and their fusion to the sternum, make the rib cage rigid. The pectoral girdle reaches all the way back to the pelvic girdle, and the girdles are fused to the backbone. All these features give the skeleton the strength to withstand the shock of landing, while leaving the head and neck, and the tail with its reduced skeleton, free to move. The digits are numbered to show what has happened to them as the wing evolved. Reptiles have four digits on each limb. One of these has completely disappeared as the wing evolved from a reptilian limb.

Figure 30-22 Family life is important among the birds. These blue tits share the task of feeding their nestlings. (Biophoto Associates, N.H.P.A.)

strong. The most important flight muscles are the **pectorals**, or breast muscles, which attach the base of the wings to the **keel** of an extended **sternum**, or breastbone. The lungs of a bird have bypass tubes that connect them with fine air sacs that spread throughout the body and even into some of the bones. The air sacs allow heat to be lost rapidly, particularly during flight, when the flight muscles and heart generate large quantities of metabolic heat.

Birds are primarily visual animals, with binocular and color vision almost as good as that of the Primates (the order of mammals that contains monkeys, apes, humans, etc.) And like primates, birds are arboreal (tree-dwelling) animals, which tend to have the good vision needed for landing and jumping in trees.

Birds have internal fertilization, although the male has no penis in most species. The egg must be fertilized before passing through the oviduct, where the calcareous shell is added to prevent the egg's being squashed during incubation.

Bird flight ranges from flapping flight, such as that of sparrows, robins, chickadees, and so forth, to the soaring flight of the hawks, vultures, and albatrosses. Soaring birds ride the thermal currents in the air, like a glider (though more efficiently), and flap their wings infrequently.

Birds that live on land usually eat seeds and fruit (finches, parrots, fowl, pheasants), insects and insect larvae (thrushes, swifts, woodpeckers, wrens, swallows),

Figure 30-23 Puffins, sea-going birds that eat small fish and nest in holes at the tops of cliffs. (Biophoto Associates)

Figure 30-24 A white-headed wood-pecker, a bird that drills holes in dead trees and feeds on wood-boring insects. The nest is a larger hole excavated in an old tree trunk; here father brings food to his nest-lings. (William Camp)

smaller vertebrates (owls, eagles, falcons, hawks), or carrion (vultures and crows). A large number of birds find most of their food, in the form of invertebrates, algae, and fish, in fresh water or in the sea. Some of these water-feeding birds are wading birds, such as the sandpipers and herons, while others are capable of swimming and diving, with feet modified as paddles. Gulls, pelicans, ducks, geese, and cormorants fall into the latter category. The most highly specialized water birds are the penguins, which cannot fly because their wings as well as their feet are modified as paddles. The other birds that cannot fly are typified by ostriches, which rely on their speed to escape predators.

30-H Class Mammalia

Mammals originated from early reptiles some 200 million years ago, during the Triassic period; this was well before the first birds appeared. Early mammalian fossils are scarce, but it seems that the first mammals evolved adaptations that permitted them to avoid competition with reptiles, which dominated the earth at this time. The mammals were small, about the size of a small mouse or shrew. The sizes and shapes of their teeth suggest that they ate mainly insects, probably supplemented with buds, roots, fruit, and so forth. They had large eye sockets, suggesting that they were nocturnal like the large-eyed modern primates, and were well on the way to becoming endothermic. It is quite likely that endothermy was advantageous, partly because it permitted mammals to be active during the cooler night and so avoid competition with reptiles.

The extinction of many of the dominant reptiles over a span of several million years at the end of the Cretaceous period permitted a burst of mammalian evolution; many forms became larger, and by adaptive radiation mammals came to exploit many of the resources formerly monopolized by reptiles.

Two features—perfection of quadrupedal locomotion and new adaptations to carnivory—are the secrets of mammalian success. The adaptations for fast quadrupedal locomotion include a narrow foot track, with the legs slung much further under the body than in reptiles, and development of the limb girdles, which in some fast mammals may become highly mobile. (Part of the pectoral girdle of a cat moves back and forth several inches with each leg movement, lengthening the stride.) Muscles that move the limbs become clustered at the top of the leg, making the lower leg slim and light; this enables the animal to move its legs more quickly. Furthermore, muscles moving the hind leg backward are no longer attached to the tail vertebrae, as they are in reptiles, but to a spur on the back of the pelvic girdle. Thus the tail of mammals is freed from use in locomotion and can be used for other purposes (e.g., as a fly swatter, as a fifth hand, or for balance) (Figure 30–25).

Figure 30-25 The skeleton of a mammal (a cat). The scapula (shoulder blade) is only loosely attached to the backbone, so that it moves back and forth, giving the cat its long stride. Most of the muscles that move the hind limb are attached to the pelvic girdle so that the tail, unlike that of a reptile, is not involved in locomotion. The limbs are slim and the foot track narrow, contributing to the rapid movement characteristic of mammals.

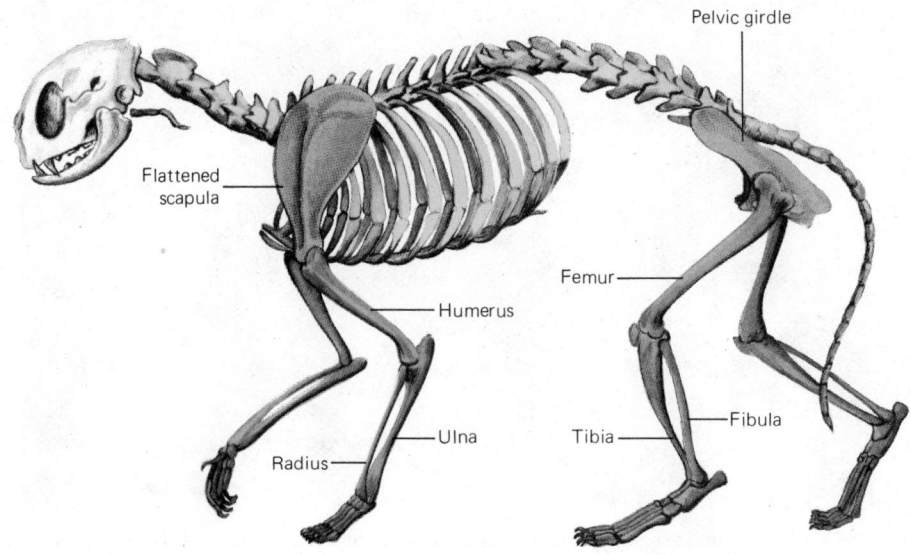

Pelvic girdle

Flattened scapula

Humerus

Femur

Ulna

Tibia

Fibula

Radius

Most mammals are viviparous. The mother's blood vessels lie close to those of the fetus, supplying the fetus with food and oxygen and removing waste. Various degrees of viviparity occur in elasmobranchs, bony fish, and reptiles, but it reaches its highest development in mammals. The egg is tiny, with very little nourishing yolk; the embryo is completely dependent on the mother. Viviparity permits the mammalian mother to remain mobile while incubating embryos that, like bird embryos, must be warm to survive. While not all mammals are viviparous, all female mammals nourish the young after they are born with milk produced in mammary glands.

Mammals, like birds, are endothermic; the body is insulated by the hair or fur and the layer of fat beneath the skin. Endothermy is an enormous advantage to a carnivore, making it ready for action when its prey is not. It has also permitted birds and mammals to exist in extreme temperatures that other land vertebrates cannot survive: Leech's petrel, penguins, and polar bears survive in polar areas; camels and vultures are among the few animals active at noon in the desert.

The evolution of specialized teeth is another advance seen in mammals. Whereas reptiles and fish have teeth that are all roughly the same size and more or less conical, early mammals rapidly evolved different kinds of teeth: chisel-like **incisors** for cutting, pointed **canines** for gripping and tearing, and grindstone-like **molars** for crushing and breaking. During vertebrate evolution, too, the number of separate bones that make up the lower jaw, has been steadily reduced until mammals have just one bone on each side (the mandible). These alterations in the lower jaw and teeth permit a carnivore to grip a struggling victim more firmly, with less chance of dislocating the jaw (Figure 30–26).

Although the earliest mammalian design is that of a fast, quadrupedal carnivore, many modern mammals are neither fast, quadrupedal, nor carnivorous: the success of this group has led to extensive adaptive radiation, during which herbivory, bipedalism, flight, and many other adaptations have evolved.

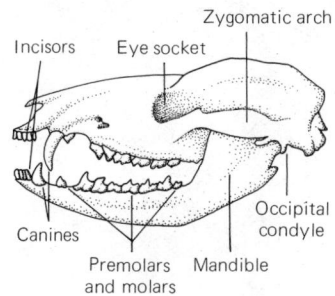

Figure 30-26 The skull of a mammal to show different kinds of teeth. The zygomatic arch protects the eye from being jogged by the jaw muscles. The backbone attaches to the skull at the occipital condyle.

TABLE 30–4 **MAJOR ORDERS OF MAMMALS**

Order Monotremata	Egg-laying mammals: the duck-billed platypus (*Ornithorhynchus*) and the echidna (*Tachyglossus*).
Order Marsupialia	Mammals with marsupial pouches; e.g., opossum, kangaroo, koala bear.
Order Insectivora	Small insect-eating mammals; e.g., moles, shrews.
Order Dermoptera	Flying lemurs.
Order Chiroptera	Bats.
Order Primates	Lemurs, monkeys, apes, humans; e.g., *Lemur, Tarsius* (spectral tarsier), *Cebus, Macacus, Pongo* (chimpanzee), *Pan* (gorilla), *Homo*.
Order Edentata	Sloths, anteaters, armadillos.
Order Lagomorpha	Rabbits, hares, pikas.
Order Rodentia	Rodents: mice, rats, voles, beavers, porcupines, guinea pigs, hamsters, jerboas, squirrels, gophers.
Order Cetacea	Whales, dolphins, porpoises.
Order Carnivora	Dogs, cats, hyenas, bears, wolves, raccoons, pandas, weasels, otters, badgers, skunks, jackals, civets, mongooses, etc.
Order Pinnipedia	Seals, sea lions, walruses.
Order Proboscidea	Elephants
Order Perissodactyla	Odd-toed ungulates; e.g., horses, zebras, asses, tapirs, rhinoceroses.
Order Artiodactyla	Even-toed ungulates; e.g., pigs, hippopotamuses, camels, deer, giraffes, buffaloes, domestic cattle, gazelles, goats, llamas, antelopes, sheep.

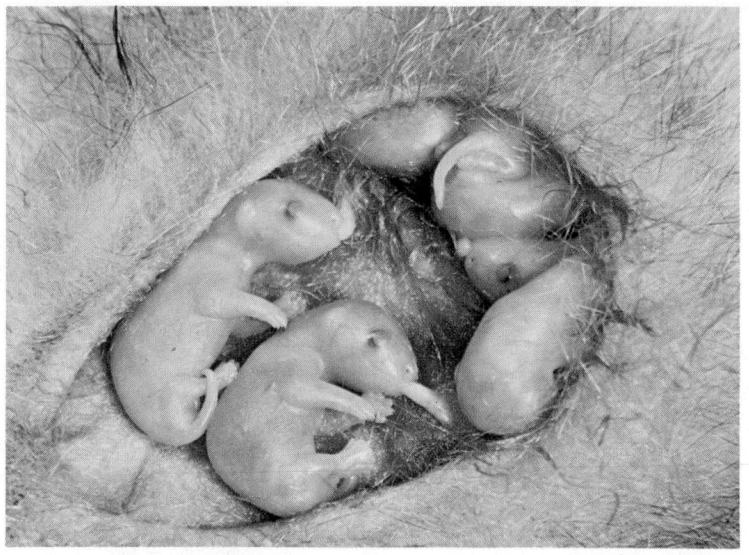

Figure 30-27 The young of the opossum complete their development attached to a teat in the marsupium, or pouch, as shown here. (Carolina Biological Supply Company)

Monotremes

The monotremes are one of the three major groups of mammals. There are only two genera of living monotremes: the duck-billed platypus, *Ornithorhynchus*, and the spiny anteater or echidna, *Tachyglossus*. They are distinguished from other mammals by the fact that they lay eggs, which are very similar to reptilian eggs, although the mothers suckle the hatched young with milk from mammary glands in typical mammalian fashion. The monotremes thus provide evidence that viviparity and suckling did not necessarily evolve at the same time. Monotremes lay their eggs in burrows in the ground, and incubate the eggs and young with the warmth of their bodies.

The duck-billed platypus, found only in Australia, lives rather like a water rat on the edges of streams. It eats aquatic invertebrates with its duck-like bill and swims with its flattened tail. The echidna, found in Australia and New Guinea, digs out ant nests and traps the ants with its sticky tongue.

Marsupials

Marsupials are mammals whose young are born at an early stage of development and finish developing in a pouch, or **marsupium**.

Marsupials are found only in Australia and America, although fossil evidence shows that they once lived in Europe. The most familiar American marsupial is the opossum, which was once restricted to Central America and the adjacent parts of North and South America. In recent years, however, it has expanded its range dramatically in North America and is now a familiar sight in many areas of the United States. We do not know why opossums have become so numerous and widespread. Well-known Australian marsupials include kangaroos and the koala bear.

Most mammals have a placenta, an organ that permits substances to be exchanged between mother and fetus. Most marsupials have no placenta. The eggs are yolky; the embryo feeds on this yolk and on fluid in the uterus* as it develops, depending on the mother only for protection. The embryos are born as little as 8 days after fertilization and crawl from the mother's cloaca* to her pouch. Once they arrive in the pouch, the young attach their lips firmly to a teat, and muscles around the mother's mammary gland pump milk into them while they complete development (Figure 30–27).

---

* Uterus The organ in females of viviparous mammals where the young develop, situated between the oviducts and the exterior of the body.
* Cloaca The vestibule, found in most vertebrates, into which urine, feces, and sperm or eggs are discharged before they leave the body.

Placental Mammals

Most modern mammals are placental. The egg has hardly any yolk, and the first organ to form during embryonic development is a **placenta**, across which the mother supplies the offspring with food and oxygen and removes its wastes. The placenta contains both fetal and maternal tissues, and many blood vessels; thus the mother's blood vessels run very close to those of the fetus, and substances can pass easily between them. The maternal side of the placenta is part of the wall of the uterus, and the fetal placenta forms from the membranes of the chorion and allantois.

The young are born at various stages of development, from the furless, blind babies of a mouse, to a horse foal, which can walk within a few minutes of birth. Many mammalian species have developed complicated social systems, which probably originated from the advantage gained when both parents cared for and fed the growing young.

Modern placental mammals are classified into a number of orders, with adaptations to many different ways of life. Here we will consider the most important orders briefly, starting with the mammals that appear first in the fossil record and most closely resemble their reptilian ancestors.

Hedgehogs, shrews, and moles are members of the order **Insectivora**, small nocturnal mammals that feed on insects and other invertebrates. They have changed very little since the Cretaceous period. Some early insectivores gave rise to the bats and primates.

Bats (order **Chiroptera**) and birds are the only vertebrates with true flapping flight. It is undoubtedly advantageous for bats to avoid competing with the more numerous birds as much as possible. Most birds are active only during the day, and most bats are nocturnal. Some bats have developed echolocation systems (see Essay, Chapter 39), which permit them to fly and to catch food in the dark.

The order **Primates** includes lemurs, monkeys, apes, and human beings. They retain many features of early mammals, with additional adaptations to arboreal life (see the essay at the end of this chapter).

Members of the order **Rodentia**, mammals with gnawing teeth, are the most successful modern mammals apart from *Homo sapiens*. They live in all parts of the world. Three thousand rodent species have been described—as many as all other mammals put together. One striking rodent characteristic is that they have nearly all remained small. Most rodents also reproduce very rapidly. Gophers, rats, mice, moles, lemmings, squirrels, chipmunks, beavers, and porcupines are all successful rodents adapted to various ways of life. The **lagomorphs**, the rabbits and hares, are an order of rodent relatives specialized for jumping.

Figure 30-28 A mouse, member of the mammalian order Rodentia. (Biophoto Associates)

Figure 30-29 A hare, order Lagomorpha, coming out of a "run" —a path it has made beneath the vegetation of a field. (Biophoto Associates)

Figure 30-30 An Atlantic bottle-nosed dolphin, member of the mammalian order Cetacea. (Paul Feeny)

Whales belong to the order **Cetacea**, mammals that have become extremely well adapted to a permanent life in the sea; some of their adaptations will be considered in Chapter 33. Cetaceans can be divided broadly into the toothed and the baleen whales. Toothed whales, including the dolphins and porpoises and the sperm, bottlenose, white, and killer whales, feed mainly on fish and large invertebrates. Baleen whales are filter feeders; surprisingly, they are the largest whales, and include the blue whale, the largest animal that has ever lived. Baleen whales feed on crustaceans in the plankton, which they engulf in huge mouthfuls. They then use their large tongues to push water and plankton against the sieve of whalebone or baleen plates that lines the jaws, pushing out the water and retaining the plankton. Recent research has shown that whales are intelligent, sociable animals, able to communicate with one another by sound. They can also hear echoes of their own voices that have bounced off other objects; this process of echolocation helps them identify food and other objects in the water.

Whales are hunted for substances such as baleen, oil, and meat for pet food. Although they produce nothing that cannot be obtained from other sources, many of the larger whales have been hunted to the brink of extinction. Tuna trawlers drown great numbers of porpoises and dolphins, which travel and feed with schools of tuna. The mammals get tangled in the nets and cannot come up to breathe.

Members of the order **Carnivora**, including the cats, dogs, skunks, and bears, are the most specialized mammalian hunters. Their bodies are the ultimate development in the trend towards quadrupedal carnivory. Their behavior is complicated and involves the ability to learn much of their hunting behavior. Their sense of smell is usually well developed, and carnivores use odors, often produced by anal glands, in many of their social interactions as well as for hunting. Social and family behavior is common and often involves teaching the young, as well as protecting and feeding each other. Both the modern carnivores and the hoofed mammals arose from a common ancestral population in Paleocene times.

(a)

(b)

Figure 30-31 Order Carnivora. (a) Lionesses sharing a kill with their young. (b) A raccoon, a carnivore by classification although it will eat many kinds of food besides meat. (Biophoto Associates)

Figure 30-32 A walrus, distinguished from other pinnipeds by its huge tusks. (Biophoto Associates, N.H.P.A.)

Pinnipeds, the seals, walruses, and sea lions, are undoubtedly descended from carnivorous ancestors. They are less fully adapted to aquatic life than cetaceans, since most pinnipeds must come on land to copulate and to bear their young. Their limbs are less reduced than those of a whale, and they can move about (clumsily) on land, although they may stay at sea for long periods of time. They feed almost exclusively on fish. The main exceptions are walruses, sad-eyed giants that live on the seashore, where they gouge molluscs from the rocks with their enormous canine teeth.

There are three orders of **ungulates**, mammals that walk on the tips of their toes; most of them are hoofed mammals, divided according to whether they walk on an even number of toes (deer, cattle, etc.) or an odd number (horses, rhinoceroses, and tapirs). The only other ungulate group is the elephants, with two modern genera, one in Africa and the other in Asia.

Ungulates are the mammals most highly specialized for a herbivorous diet. Because their teeth are flattened and used to crush and grind tough plant material, they are not well equipped to fight a potential predator. Most ungulates rely on running fast to escape their enemies, and they also tend to feed in herds, where every animal's eyes watch out for danger.

The main defenses of elephants (order **Proboscidea**) are probably their large size and huge tusks. The weight and maneuverability of an elephant's trunk, developed from the nose and upper lip, make it a formidable opponent. In order to sustain their huge bodies, elephants must feed for as much as 18 hours a day.

Figure 30-33 An elephant, order Proboscidea, using its trunk to squirt a drink into its mouth. (Biophoto Associates, N.H.P.A.)

Figure 30-34 Zebras are perissodactyls. Here, some eat while others keep an eye on the photographer. (Biophoto Associates, N.H.P.A.)

Figure 30-35 Artiodactyls. This ewe and her lamb show the fur and nursing behavior characteristic of mammals.

The order **Perissodactyla**, containing the odd-toed ungulates, is a relatively small group; its members range from the tapirs, small, rather pig-like animals that live in forests, to the enormous, portly rhinoceroses, which have a thick horny skin and very little hair. Horses, asses, and zebras of various species are more widespread; they are as well adapted to grazing in herds on grassy plains as many of the even-toed ungulates described below.

Just as the Carnivora can be considered the ultimate in mammalian carnivores, so the order **Artiodactyla**, containing the even-toed ungulates, represents the peak of evolution for herbivores. One reason for their evolutionary success was the development of a **ruminant** digestive system, a series of stomachs in which symbiotic bacteria break down plant cellulose (see Section 31-D). In addition, artiodactyls are keen of sense and fleet of foot. Some of them are also reasonably intelligent. Their adaptive radiation is impressive: they have given rise to such diverse forms as pigs, hippopotamuses, and camels. Cattle, sheep, and deer often have horns or antlers, which probably evolved originally as defensive weapons. No artiodactyl will fight if it can help it; injuries, especially to their delicate legs, are usually fatal and must be avoided. Many artiodactyls, however, give a very good account of themselves with horns and hoofs if they are cornered.

SUMMARY   The agnathan fishes, the first vertebrates, probably gave rise to the bony and cartilaginous fishes, which were the dominant forms of animal life on earth in the later Paleozoic era. The bony fish gave rise to the first terrestrial vertebrates, amphibians, which still had largely aquatic reproduction. With the evolution of the amniotic egg, the adaptive radiation of land vertebrates began. Reptiles—adapted to every way of life, from swimming in the sea to flying through the air—were the dominant animals throughout the Mesozoic era. Early quadrupedal reptiles evolved into dog-like primitive mammals, and bipedal dinosaurs gave rise to birds. With the extinction of many reptile groups towards the end of the Cretaceous, birds and mammals inherited the earth and have radiated widely ever since.

Essay: Human Evolution

All the people on earth today belong to the species *Homo sapiens*, in the mammalian order **Primates**. Primates can be divided into the **prosimians**, or lower primates, which include tarsiers, lemurs, and lorises, and the **anthropoids**, including monkeys, apes, and humans. Different modern primates illustrate various stages in primate evolution, from animals resembling the primitive Cretaceous mammals to the great apes and humans.

In many mammals, such as deer with their antlers and hoofs, the skeleton is highly modified for a particular way of life. The skeletons of primates, however, look very much like those of primitive mammals. The most outstanding primate adaptations are found in the development of the nervous system, with its corresponding range of behaviors. The main evolutionary trend in the nervous system of primates is the enormous expansion of the cerebral hemispheres of the brain, and particularly of the frontal lobes (Figure 30–A).

Evolution of the nervous system is closely tied to the **arboreal**, or tree-dwelling, lifestyle characteristic of modern primates and their ancestors. Arboreal life demands well-developed sense organs and muscular dexterity. Good vision is almost essential to an animal such as a bird or a monkey that must jump and land on a branch.

Early in primate history, a mouse-like prosimian started living in trees. Modern representatives of this group are the tree shrews, which have many features in common with modern rats and mice. The most advanced prosimian is the Indonesian tarsier, an arboreal, nocturnal creature with huge eyes, fully stereoscopic vision, and nails, rather than claws, on all its digits. In addition, the upper lip is free of the gums, as it is in higher primates. This feature is a large part of the reason that higher primates have such mobile and expressive faces. Higher primates use facial expressions for communication. This reflects the shift to vision as the dominant sense, rather than smell, which is used in communication by most other mammals. During primate evolution, the snout has become progressively reduced. This is probably an adaptation that gives the eyes an unimpeded view of the world. The snout is shorter, the jaws of primates have become shorter, and some of the teeth have been lost. This process is still going on, as evidenced by the variability with which people produce wisdom teeth.

Monkeys, apes, and humans are the anthropoid primates. Anthropoids have stereoscopic color vision and rounded heads. Many of them can learn complex behavior patterns, thanks to their relatively large brains.

The upright posture of anthropoids has had an important effect on their evolution. Even quadrupedal monkeys such as baboons sit upright for long periods, and so free their hands to manipulate food, handle their young, and so on (see Figure 30–C). Few primates besides chimpanzees and humans actually walk upright, but some arboreal monkeys spend long periods in a vertical position as they swing through the trees by their arms, a type of locomotion called **brachiation**. It has been argued that the upright posture of humans is a result of our descent from a brachiating ancestor. Opponents of this theory note that the fingers of brachiators are specialized into a quick-release hook, which is ideal for catching and releasing a branch, whereas human fingers, like those of many ground-dwelling primates, are free to manipulate objects and show no signs of a brachiating ancestry.

There are only four genera of apes: gibbon, orangutan, gorilla, and chimpanzee. All live in the Old World, and their characteristics bridge the anatomical and behavioral gap between monkeys and humans.

Monkeys are basically quadrupedal. The proportions of their limbs are similar to those of other mammals, and their rib cages are compressed from side to side, as in most quadrupeds. In contrast, the arms of apes are long in comparison with their hind limbs (selected for by brachiation?), and their rib cages are flattened from front to back. The tail of an ape is reduced to the few fused vertebrae of the **coccyx**; the loss of the tail makes the upright sitting posture more comfortable. The brains of apes are also relatively larger than those of monkeys.

Few areas of research have produced as much vituperation and confusion as the search for the fossils of our ancestors. The fossil evidence of human ancestry is fragmentary; seldom does the most spectacular find consist of more than a few teeth and part of a jawbone. You can imagine the difficulty of trying to deduce the brain size of a fossil species from such evidence, let alone of deciding whether it walked upright or on all fours. Yet such deductions are made all the time. One reason for the scarcity of fossils is that many of the animals in question were forest-dwellers. Few fossils are preserved in forests, since the acidity of the soil and conditions favorable to decomposition tend to destroy corpses quickly.

Much of the muddle, however, stems from human vanity. Research-

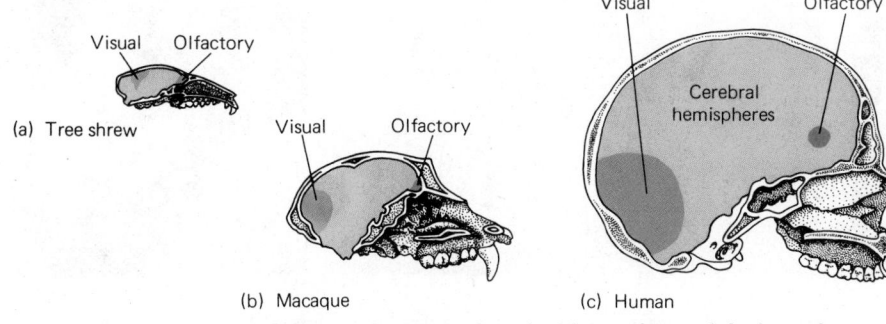

Figure 30-A Expansion of the visual center and cerebral hemispheres of the brain during primate evolution. Macaques are ground-dwelling monkeys, and tree shrews are arboreal.

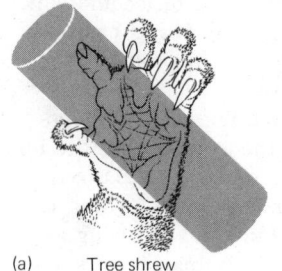
(a)   Tree shrew

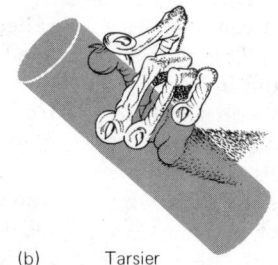
(b)   Tarsier

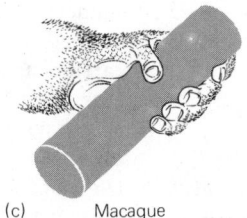
(c)   Macaque

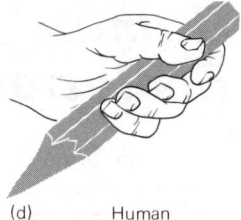
(d)   Human

Figure 30-B Hands of primates. This series of drawings shows the evolutionary trend from relatively immobile digits with claws to the human hand with opposable thumb and with fingertips protected by nails.

ers inevitably hope to discover vital clues to human ancestry. As a result, dozens of fossils that in retrospect are those of monkeys (and even of modern humans!) have been hailed as "the missing link" between apes and humans. In addition, for many years almost every hominid fossil that was found was not just given a new specific name, but was often even assigned to a new genus to make it more memorable.

There are more similarities than differences between apes and **hominids** (members of the human family). In fact, some people question whether humans would have been placed in a family by themselves if taxonomy had been invented by any other species. Nevertheless, the four genera of modern apes have some features that are common to them all but are not found in hominids. For instance, the arms and backbone of modern apes show more extreme adaptations to brachiation than those displayed by hominids. Modern apes

have powerful canine teeth, large incisors, and specialized feet, features not shared by humans. Since our nearest relatives are the African apes, the search for the presumed common ancestor of apes and humans has centered in African fossils. The most likely human ancestor from the rocks of the African Miocene is represented by a group of fossils of apes with a strong tendency toward bipedalism, which may represent the first divergence of the hominids from the African apes. We cannot be sure of this, since the evidence consists of only teeth and fragments of bone. The fossil record yields no further evidence on human evolution for another 15 million years.

Hominid fossils reappear in the early Pleistocene, about 3.5 million years ago, again in Africa. Raymond Dart discovered the first specimen in 1924 in Bechuanaland; he named it *Australopithecus africanus*. Australopithecines had essentially human

teeth, with small incisors and canines, and a posture more upright than that of the modern apes. They were ape-like in having large, heavy jaws, and they had brains that were little larger than those of the modern apes. It seems that these hominids were scavengers and hunters, but there is no convincing evidence that they used tools.

During the Miocene, many forest areas turned into open grasslands. Hominids probably came out of the forests during this period. They may well have roamed the savanna in bands, like modern baboons. Unlike all other primates, however, they rapidly became more nearly bipedal after they reached the open plains.

Although the reason for ground-dwelling bipedalism is hotly debated, the result was clearly the freeing of the hands for catching animals, throwing stones, and other activities. Bipedalism and the change from a largely herbivorous to an omnivorous diet evolved simultaneously, as the

(a)

(b)

(c)

Figure 30-C Primates. (a) A slender loris, a prosimian. Note the big, forward-facing eyes, the reduced nose, and the grasping toes. (b) A baboon, an anthropoid. (c) A chimpanzee, member of the modern-day species most closely related to humans. (Biophoto Associates; a, b, N.H.P.A.)

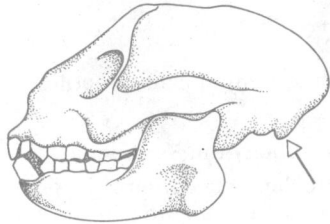

(a) *Archaeolemur,* a lemur

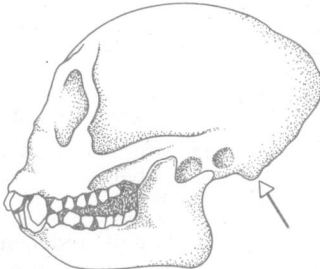

(b) *Cebus,* a new World monkey

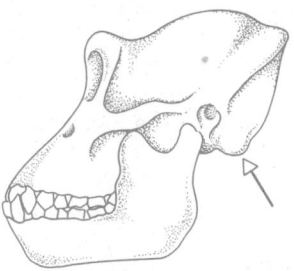

(c) *Gorilla*

Figure 30-D Changes in the skull during primate evolution. The arrow shows the angle at which the head joins the backbone at the foramen magnum. The joint between the head and the neck has moved down under the back of the skull as primates have become more bipedal. Note the decrease in relative length of the jaw and increasingly forward-looking position of the eyes.

piles of animal bones associated with the bipedal skeletons of *Australopithecus* show. We can only speculate that the advantages of cooperation in hunting and group defense may have selected for the development of language for communication.

The first fossils classified in the genus *Homo,* because they are so similar to modern humans, date from the mid-Pleistocene in Africa, about 2 million years ago. These fossils, e.g., those of *Homo erectus* (Figure 30-E), are of a fully bipedal, omnivorous, tool-using hominid. Some *H. erectus* bones are found in caves, suggesting the use of more-or-less permanent home bases. Besides animal bones and stone tools, some of the caves contain heaps of charcoal and charred bones, showing that fire had been domesticated and brought

indoors by this time. Presumably this habit originated in the use of natural fires to keep warm, cook food, or split stones.

This stage in human evolution is correlated with the colonization of other, colder areas by emigrants from Africa. Their anatomical or physiological adaptations alone will not permit human beings to survive winters as cold as those of Central Europe and China; behavioral adaptation or technological expertise is necessary. Plainly the pre-human brain had developed to the point where *Homo erectus* could produce social and technological solutions—such as fire, clothing, stored food, and communal living in caves—to the problems of surviving cold winters.

These solutions probably led to the development of some of the most important features of *Homo sapiens.* Permanent settlements in Eurasia may have been responses to the difficulty of a nomadic life during the cold winters. Cooperative effort permitted the group to trap game animals; during the summer, the main source of food would have been plant materials—seeds, fruits, leaves, nuts, roots, and berries—gathered as they were used. The use of fire opened up a new range of plant foods; cooking can remove volatile toxins and can also soften plants and make them more digestible. A settled hunter-gatherer society such as this must have been the precursor to the development of agriculture. It would demand social organization and communication between individuals that would be a strong selective pressure for the development of language, social rites, laws, and customs. We find these reflected in the decorated tools, pots, and dwellings that began to appear in Eurasia about 20,000 years ago.

The selective pressures for the evolution of that most human feature, a brain that is large relative to body size, are still obscure and widely debated. One of the difficulties with this discussion is that there is no consensus as to how much brain is necessary to how much intelligent behavior. In general, it is clear that the more precise its control over its hand muscles and the more complex its learned behavior, the more nerve cells we would expect to find in the brain of a hominid (or any

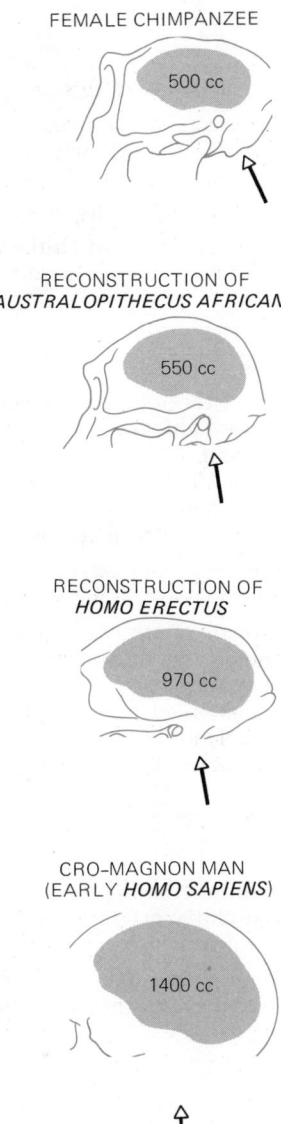

FEMALE CHIMPANZEE

500 cc

RECONSTRUCTION OF *AUSTRALOPITHECUS AFRICANUS*

550 cc

RECONSTRUCTION OF *HOMO ERECTUS*

970 cc

CRO-MAGNON MAN (EARLY *HOMO SAPIENS*)

1400 cc

Figure 30-E Changes in proportions of the skull from great ape to human. Note the increase in size of the brain case, change in position of the foramen magnum (arrow) as hominids became more bipedal, and relative reduction in size of the teeth and jaws as hominids changed from a purely herbivorous to an omnivorous diet.

other animal). We know, however, that large areas of the human brain can be destroyed without altering behavior in any way. There seem to be a lot of reserve nerve cells in the brain which we can normally do without. In short, we do not know enough about the organization and functioning of our own brains, let alone of the brains of ancestral hominids, to speak in any but the most speculative terms about when, and under what selective pressures, particular changes in the nervous system occurred.

473

OBJECTIVES

From your study of this chapter, you should be able to:

1. List characters that would permit you to distinguish members of the following classes: Chondrichthyes, Osteichthyes, Amphibia, Reptilia, Aves, Mammalia.

2. Name the class of any vertebrate presented to you.

3. State the order in which the following groups appeared in the fossil record: invertebrates, agnathans, Chondrichthyes, bony fishes, amphibians, reptiles, birds, mammals.

4. List the classes of the subphylum Vertebrata.

5. List and explain the problems encountered by previously water-dwelling vertebrates in adapting to life on land, and explain adaptations that enable modern terrestrial vertebrates to live on land.

SELF-QUIZ

1. Which of the following is *not* a vertebrate?
   a. an amphioxus
   b. a lamprey
   c. a shark
   d. a kangaroo
   e. a duck-billed platypus

2. What is the most characteristic feature of Aves, found in no other class of living vertebrates?

3. Which pair of animals is most closely related?
   a. horse and rhinoceros
   b. bat and rat
   c. duck-billed platypus and Canada goose
   d. whale and seal
   e. hephalump and elephant

4. The Atlantic salmon belongs to the class _____ .

5. List three problems of terrestrial life that previously aquatic vertebrates had to overcome before they could invade the land.

6. The first major vertebrate class to be totally independent of permanent bodies of water during reproduction was the _____ .

7. A vertebrate that has a body covering of scales, glandless skin, no limbs, and internal fertilization and that suns itself to increase its body temperature in the morning belongs to the class _____ .

8. True or false? Mammals appear in the fossil record before birds.

QUESTIONS FOR DISCUSSION

1. We often talk as if *Homo sapiens* were the most highly evolved and specialized mammal. In what ways is this true, and in what ways is it not true?

2. Many mammals have adapted to a life permanently at sea. Why haven't they gone back to using gills for respiration?

3. What are the advantages of social and parental behavior that have made it profitable for organisms to spend some of their energy budgets in these activities? What are the drawbacks of such types of behavior?

4. Why do so few birds live as grazers on leaves and grass?

REFERENCES AND FURTHER READING

Bellairs, A., and R. Carrington. *The Life of Reptiles*. New York: Universe Books, 1970. An authoritative introduction to the reptiles.

Buffetaut, E. "The evolution of crocodilians." *Scientific American*, October 1979.

Colbert, E. H. *Evolution of the Vertebrates*, 2d ed. New York: John Wiley and Sons, 1969. A well-written account of living and extinct vertebrates.

Marshall, N. B. *The Life of Fishes*. Cleveland: The World Publishing Co., 1966. A readable account of all aspects of fish biology.

Noble, G. K. *The Biology of the Amphibia*. New York: Dover Publications, 1954. Comprehensive biology, physiology, and ecology of amphibia.

Romer, A. S. *The Vertebrate Body*, 5th ed. Philadelphia: W. B. Saunders Co., 1977. A standard text on the comparative anatomy of vertebrates, living and extinct.

Welty, J. C. *The Life of Birds*, 2d ed. Philadelphia: W. B. Saunders Co., 1975. A readable account of all aspects of bird biology.

Young, J. Z. *The Life of Mammals*, 2d ed. Oxford: Clarendon Press, 1975. An excellent, readable introduction to mammalian biology.

PART FIVE

ANIMAL BIOLOGY

CHAPTER 31

ANIMAL NUTRITION AND DIGESTION

Animals are heterotrophs; they cannot make their own food from inorganic substances but must ingest organic molecules from the environment. Animals can be broadly divided into **herbivores**, which eat plants; **carnivores**, which eat animals; and **omnivores**, which eat both. Animals with each mode of nutrition have digestive systems suited to handling and digesting the type of food eaten.

Feeding is a necessary evil. The longer an animal spends feeding, the less time it has for activities that are more selectively advantageous, such as reproduction and raising its offspring. Thus, there is strong selective pressure for an animal to feed as rapidly as possible. It was long assumed that an animal's feeding habits were determined largely by its energy needs, but recent work makes it clear that many animals have more difficulty in obtaining specific nutrients than in acquiring sufficient energy.

The food that an animal ingests is **digested**, or broken down into small organic molecules, which can be absorbed from the gut. In this chapter we shall consider what food an animal needs, how animals obtain food, and the role of the digestive system and liver in processing food so that it can be used by the body.

(a)

(b)

(c)

(d)

Figure 31-1 Feeding methods. (a) A kingfisher with a fish it has caught. (b) A blowfly sucks up liquid food. (c) Lions. (d) A mare with her foal. (Biophoto Associates)

31-A Nutrients

The nutrients that any animal must ingest as food may be divided, for convenience, into **macronutrients**—nutrients that are needed in large quantities—and **micronutrients**—nutrients that are required in lesser amounts.

Macronutrients

The macronutrients are fats, carbohydrates, and proteins. All three can serve as energy sources because they can be broken down into molecules that are respired to produce ATP (see Chapter 7). The amount of energy available from a given amount of a macronutrient is commonly measured as its **caloric value**, the number of Calories of heat that it will produce when fully oxidized (Table 31-1). All three classes of macronutrients can also provide carbon atoms to form organic polymers. In addition, proteins supply amino acids for building the body's own proteins.

Macronutrients can be stored until the body needs them for energy. Carbohydrates are stored as the polysaccharide glycogen, in muscle and liver, and fats are stored as fat. Proteins cannot be stored, but excess protein is broken down to amino acids, which are deaminated* and then processed as fats or carbohydrates.

---

* Deaminated Having the amino (—NH$_2$) group removed.

TABLE 31-1 **THE CALORIC VALUES OF MACRONUTRIENTS**	
MACRONUTRIENT	*CALORIES† (Kcal) PER GM*
Protein	4.4
Fat	9.3
Carbohydrates	4.1

† A calorie is the amount of heat needed to raise the temperature of one gram of water by 1°C. The "Calories" in food are actually kilocalories, which may be indicated by a capital C.

1 kilocalorie (Kcal) = 1 Calorie = 1000 calories.

PROBLEMS WITH MACRONUTRIENTS

The most common dietary problem for people in industrialized countries (especially the United States) is obesity: if more calories are ingested in the food than are used up, the excess is stored as fat. Overeating and lack of physical activity are the usual causes of obesity, although it is clear from the profitable trade in reducing drugs and gadgets that millions of people would rather not believe this.

A much more common problem for most animals is protein deficiency. This usually occurs, not because the diet contains too little total protein, but because it does not contain enough of the essential amino acids. All animals can convert some amino acids into others, but the **essential amino acids** must be supplied in the diet because an animal cannot synthesize them from other amino acids. Essential amino acids may be used as they are or may serve as essential precursors of other amino acids. The essential amino acids are the same for most animals.

One of the best-known protein deficiency diseases is **kwashiorkor**, found in African populations where the diet consists primarily of cornmeal. Such a diet contains very little of the essential amino acid tryptophan. Victims of kwashiorkor, particularly growing children who need much protein, are lethargic, have edema,* and fail to grow normally. Another protein deficiency disease has been produced by makers of baby formula who enlarged their market through advertisements persuading women in developing countries that bottle-feeding instead of breast-feeding was the chic, Western thing to do. Millions of women in these countries began buying baby formula, despite the fact that breast milk is cheaper and better for babies. Many of these women, who don't have enough money even to feed the adults in the family, who can't read instructions, and who live in unhygienic conditions, end up feeding their babies diluted, contaminated baby formula. Dilute milk does not contain enough essential amino acids to nourish a baby properly, and nutritionists ascribe millions of infant deaths to this advertising campaign.

Fat deficiencies can also cause nutritional diseases. Certain fatty acids are essential in the diet; they are constituents of cell membranes and of some hormones. People who live on diets of fish, rice, or fruit, which are all very low in fat, develop a craving for fats and treat them as a delicacy.

Micronutrients

Micronutrients are the substances an organism must have in its diet in small quantities because it cannot synthesize them for itself or because it cannot synthesize them as fast as it needs them. Micronutrients can be divided into **vitamins**, which are organic compounds, and **minerals**, which are the inorganic micronutrients. Various deficiency diseases result from shortage of certain vitamins and minerals in the diet.

VITAMINS

The vitamin needs of various animals differ. For instance, many animals can make their own ascorbic acid from other molecules, and so vitamin C is not necessary in their diets, as it is in the human diet. Vitamins needed in the human diet are generally divided into two categories: water-soluble and fat-soluble. **Water-soluble vitamins** (Table 31–2) are coenzymes needed in metabolism. The water-soluble vitamins are easily excreted by the kidney. The **fat-soluble vitamins** (Table 31–3) are fewer in number than the water-soluble vitamins and have various poorly understood functions. Fat-soluble vitamins that are not needed immediately may be stored in fatty tissue. Because these vitamins are not soluble in water, the body's enzymes must process them before they can be excreted by the kidneys. As a result, some of them can accumulate to toxic levels if eaten in amounts larger than the body can use. Some modern "diets" recommend dangerous doses of certain fat-soluble vitamins. (In 1978, for instance, two people on high-vitamin diets died of vitamin A poisoning.) The key to avoiding such dangers is common sense and moderation. Almost anything can be poisonous if consumed in sufficient quantity.

* Edema   Swelling due to retention of excessive amounts of fluid in parts of the body.

TABLE 31-2 FUNCTIONS AND DEFICIENCY SYMPTOMS OF WATER-SOLUBLE VITAMINS

VITAMIN	METABOLIC* ROLE	DEFICIENCY SYMPTOMS
Thiamine (B_1)	Coenzyme in carbohydrate metabolism.	Beriberi, loss of appetite, indigestion, fatigue, weakening of the heart and blood vessel walls, edema.
Riboflavin (B_2)	Coenzyme in protein and carbohydrate metabolism, as part of FAD.*	Inflammation and breakdown of skin, swollen tongue, eye irritation.
Niacin	Coenzyme in energy metabolism; part of NAD^+* and $NADP^+$.*	Pellagra, fatigue, neuritis,* skin eruptions.
Pyridoxine (B_6)	Coenzyme in many phases of amino acid metabolism.	Anemia,* irritability, convulsions, neuritis.
Pantothenic acid	Coenzyme in acetylation* of CoA.*	Similar to other B vitamins.
Biotin	Coenzyme in decarboxylation* and deamination.	Rare; minute amounts required.
Folic acid	Coenzyme in formation of nucleotides and heme.*	Some types of anemia.
Cobalamin (B_{12})	Coenzyme in formation of proteins and nucleic acids.	Pernicious anemia.
Ascorbic acid (C)	Helps build intercellular cement for bone, teeth, cartilage, connective fibers in skin and blood vessels; maintains resistance to infection, frees iron to make hemoglobin.	Scurvy, anemia, slow wound healing.

TABLE 31-3 FUNCTIONS AND DEFICIENCY SYMPTOMS OF FAT-SOLUBLE VITAMINS†

VITAMIN	PHYSIOLOGICAL ROLE	DEFICIENCY SYMPTOMS
A (retinol)	Part of visual pigments in eye.	Night blindness. Drying and damage of skin and mucous membranes.
D (calciferol)	Increases absorption of calcium and phosphorus and their deposition in bones.	Rickets in children.
E (tocopherol)	Protects red blood cells, vitamin A, and unsaturated fatty acids from oxidation; important in muscle maintenance.	Hemolysis* of red blood cells; anemia, sterility (in rats).
K (menadione)	Needed in the synthesis of prothrombin, which is necessary for blood clotting.	Hemorrhage* in the newborn, who lacks the intestinal bacteria that synthesize vitamin K.

†Vitamins A, D, and K are toxic in large amounts.

* FAD, NAD^+, $NADP^+$, CoA   Coenzymes in cellular respiration (see Chapter 7).

* Neuritis   Inflammation of nerves.

* Anemia   A deficiency of hemoglobin or of red blood cells, resulting in a deficiency of oxygen supply to the tissues.

* Acetylation   Addition of acetyl $\left(CH_3-C {\overset{O}{\underset{}{\lessgtr}}} \right)$ group.

* Decarboxylation   Removal of carboxyl (—COOH) group from a molecule, in the form of CO_2.

* Heme   Iron-containing group in hemoglobin and some cytochromes (see Fig. 32–15).

* Metabolic   Having to do with *metabolism*, the total of all the chemical reactions that an organism carries out as part of its life processes.

* Hemolysis   The bursting of blood cells.

* Hemorrhage   Bleeding, e.g., from a ruptured blood vessel.

Essay: Diet and Cardiovascular Disease

Cardiovascular diseases, those that affect the heart and blood system, cause about half the deaths in the United States. They include **hypertension** (high blood pressure) and **atherosclerosis** (hardening and blockage of the arteries). Cardiovascular diseases cause death in many ways, such as strokes and heart attacks (including "coronaries," blockage of the arteries that supply blood to the heart muscles). Hypertension can be controlled by drugs; dietary measures such as losing weight and reducing the intake of sodium are also helpful. Reducing sodium intake may be difficult because meat contains quite a lot of sodium, and many commercially prepared foods contain sodium in added salt, preservatives, and monosodium glutamate (MSG). In one study, a hamburger, a cheese pizza, and a meat pie each contained more sodium than a portion of French fries liberally sprinkled with table salt!

In atherosclerosis, the artery walls develop deposits of lipids, including cholesterol, that reduce the internal diameter of the blood vessel and also make its walls less elastic. For many years, doctors have believed that the cholesterol level in the blood was correlated with atherosclerosis. Several studies have now shown that blood cholesterol levels have little or no correlation either with the amount of cholesterol in the diet or with death from cardiovascular disease.

New studies suggest that susceptibility to heart attacks is more closely correlated with levels of substances in the blood called HDLs and LDLs than with cholesterol. Cholesterol does not travel free in the blood but is combined with lipid-protein molecules. Most of the cholesterol is carried by low-density lipoproteins (LDLs); some is carried by high-density lipoproteins (HDLs). Studies of American whites and blacks and of Israeli, Japanese, and Hawaiian men (that is, of a varied collection of people) have shown that the risk of heart attack is greater the lower the HDL concentration and the higher the LDL concentration in the blood. Researchers have also found that some people who have permanently high levels of HDL or low LDL concentrations because of their genetic makeup apparently never die of atherosclerosis.

The factors correlated with high HDL levels are those long known to be associated with a low risk of heart attack—being female (a woman's HDL level is about 10% higher than a man's), being slim, exercising, not smoking, and consuming moderate amounts of alcohol (rather than consuming none or large amounts). However, where people have attempted to alter these factors, for instance by losing weight or by taking up running, there is no convincing evidence that they have reduced their risk of heart attack. You will notice that these studies talk about correlations, not about cause and effect. This is because cardiovascular disease takes a long time to develop and because few experiments are possible on human subjects, which makes scientific investigation—and determination of cause and effect—of such diseases difficult. Even an obvious correlation can be misleading. For instance, runners have a low incidence of cardiovascular disease. Is this because people with high HDL levels are predisposed to run or does running raise HDL levels? Runners are also more likely to consume alcohol and less likely to smoke than are other members of the general population. Is this cause or effect or neither? We do not know.

While research on this complex subject is still inconclusive, it is probably fair to say that the evidence to date shows that, unless you are obese, no major alteration in diet will have much effect on your chances of dying from a heart attack.

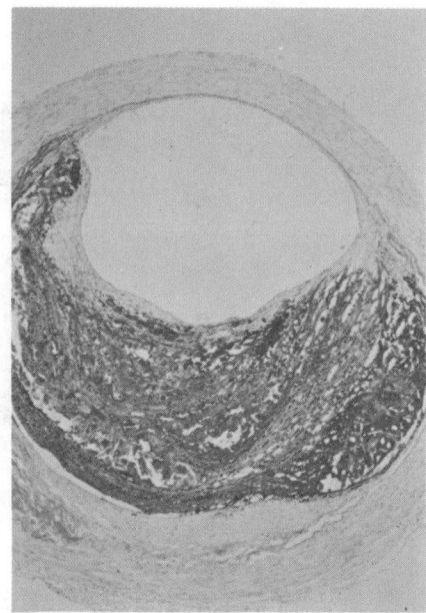

Figure 31-A Atherosclerotic deposit (red) in an artery (gray). (Biophoto Associates)

TABLE 31-4 PHYSIOLOGICAL ROLES AND SOURCES OF THE IMPORTANT MINERALS

MINERAL	PHYSIOLOGICAL ROLES	FOOD SOURCES
Sodium (Na)	Major extracellular fluid cation* Osmotic (water) balance Buffering (acid-base balance) Absorption of glucose into cells Transmission of electro-chemical impulse in muscles and nerves	Table salt (NaCl), milk, meat, eggs, baking soda, baking powder, carrots, beets, spinach, celery
Potassium (K)	Major intracellular fluid cation Buffering Regulation of nerve and muscle function Glycogen formation Protein synthesis	Whole grains, meat, legumes, fruits, vegetables
Calcium (Ca)	Component of bones and teeth Blood clotting Muscle contraction Nerve impulse transmission Plasma membrane permeability Enzyme activation (ATPases)	Milk, cheese, green leafy vegetables, whole grains, egg yolk, legumes, nuts
Phosphorus (P)	Bone formation Phosphorylation* of glucose, glycerol, fatty acids; aids in their absorption and transport Energy metabolism (enzymes, ATP) Buffer system Component of DNA and RNA	Milk, cheese, meat, egg yolk, whole grains, legumes, nuts
Magnesium (Mg)	Constituent of bones and teeth Activator and coenzyme in carbohydrate and protein metabolism	Whole grains, nuts, meat, milk, legumes
Chlorine (Cl)	Major extracellular fluid anion* Buffering Water balance Hydrochloric acid in stomach	Table salt
Sulfur (S)	Part of proteins Activates enzymes High-energy sulfur bonds in energy metabolism Detoxification reactions	Meat, eggs, cheese, milk, nuts, legumes

MINERALS

We need some minerals in relatively large amounts. Sodium and potassium, for instance, are vital to the working of every nerve and muscle in the body (Table 31–4). Large quantities of these minerals (particularly sodium) are excreted in the urine every day; sodium excretion is also a vital part of sweating, which is necessary to the regulation of body temperature in some mammals. Calcium is necessary to muscular activity and, with phosphorus, is needed in large amounts for bone formation. Other minerals are known as **trace minerals** (Table 31–5). Some of them are needed in tiny amounts for the activities of enzymes in various metabolic pathways. The functions of other trace minerals are unknown or poorly understood.

---

* Cation A positively charged ion.
* Phosphorylation Addition of a phosphate group.
* Anion A negatively charged ion.

TABLE 31-5 **PHYSIOLOGICAL ROLES AND SOURCES OF TRACE MINERALS**

MINERAL	PHYSIOLOGICAL ROLES	FOOD SOURCES
Iron (Fe)	Component of the heme group found in hemoglobin, myoglobin,* and cytochromes	Liver, meats, egg yolk, whole grains, enriched bread and cereal, dark green vegetables, legumes, nuts
Copper (Cu)	Associated with iron in hemoglobin synthesis and the absorption and transport of iron Present in cytochromes and in red blood cells Involved in bone formation, maintenance of nervous tissue	Liver, meat, seafood, whole grains, legumes, nuts
Iodine (I)	Component of thyroid hormone, which regulates cellular respiration	Iodized salt, seafoods
Manganese (Mn)	Ions necessary in urea formation, protein metabolism, glycolysis, and tricarboxylic acid cycle	Cereals, soybeans, legumes, nuts, tea, coffee
Cobalt (Co)	Constituent of vitamin B_{12}, essential for red blood cell formation	Vitamin B_{12} in meat
Zinc (Zn)	Essential enzyme constituent of carbonic anhydrase (buffering function), carboxypeptidase, lactic dehydrogenase Needed for storage of insulin Seems to be required for normal senses of smell and taste	Widely distributed; liver, seafood
Molybdenum (Mo)	Constituent of specific enzymes involved in the conversion of purines to uric acid, aldehyde oxidation	Organ meats, milk, whole grains, leafy vegetables, legumes
Fluorine (F)	Small amount improves resistance to tooth decay	Water, treatment by dentist
Selenium (Se)	Associated with fat metabolism	
Chromium (Cr)	Associated with glucose metabolism	

31-B Digestive Systems

Digestion is the hydrolysis of food macromolecules into monomers, particularly amino acids, glycerol, monosaccharides, and fatty acids. Food can be digested either **intracellularly**, that is, in vacuoles within cells, or **extracellularly**, in a digestive cavity in the body of the animal.

Intracellular digestion predominates in protists and many lower animals such as sponges. Extracellular digestion plays a more prominent role as we progress up the evolutionary scale. Cnidarians, such as *Hydra*, jellyfish, and sea anemones, capture surprisingly large prey animals, and spend a long time digesting them before the next meal, in contrast to the sponges, which feed continuously. For extracellular digestion, the cnidarians have a **gastrovascular** ("food-blood") cavity, which doubles as a transport system (see Figure 27–6). Food is pushed into the gastrovascular cavity by the tentacles. Cells lining the cavity secrete digestive enzymes into the cavity, and digestion begins. As small particles separate from the food, they are picked up by cells lining the cavity and digestion is completed inside these cells.

---

*Myoglobin   Oxygen-storing molecule in muscles.

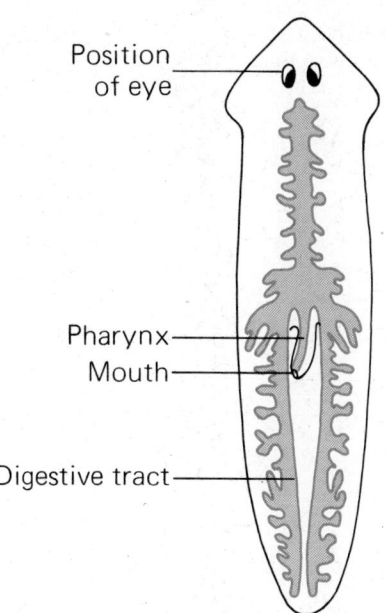

Figure 31-2 A digestive tract with only one opening. This freshwater platyhelminth (flatworm) has a mouth but no anus. The branched gut also acts as a circulatory system, distributing food to all parts of the body.

The free-living platyhelminths have a muscular **pharynx** that allows them to suck up food in an efficient manner. They also have a highly branched intestine, with a large surface area for secretion of digestive enzymes and absorption of food (Figure 31–2). Digestion is completed intracellularly. Both the cnidarians and the platyhelminths must discharge undigested wastes the same way they came in, through the mouth.

In all higher animals, the **digestive tract** is basically a tube with a mouth for ingestion and an anal opening for **egestion**. Such a tubular digestive system allows the food to move in only one direction. Additional food can be taken in while previously eaten food is being digested, and different parts of the tube become specialized and perform different functions (Figure 31–3). The food is broken down step by step, and the nutrients released are absorbed further down the tube. In earthworms and in some arthropods, for example, we find not only a sucking pharynx for food intake but, beyond it, a **crop**, which is a large, thin-walled chamber that stores food, followed by a **gizzard**, which is a muscular, thick-walled chamber, often with hard, tooth-like structures that break up the food into smaller particles as the muscular walls churn. Mechanical breakdown increases the surface area of the food particles, allowing more efficient chemical attack by the digestive enzymes. The intestine is the next part of the digestive tract; the front part of it secretes enzymes, while following portions are increasingly specialized to absorb food molecules. Often there is some intracellular digestion in cells of the intestinal lining as food passes from the intestine to the blood. Many molluscs and arthropods have large digestive glands, where food is digested intracellularly. The digestive tracts of many animals have outpocketings called **caeca** (singular: **caecum**), blind sacs that hold food destined for a longer stay in the intestine (Figure 31–4).

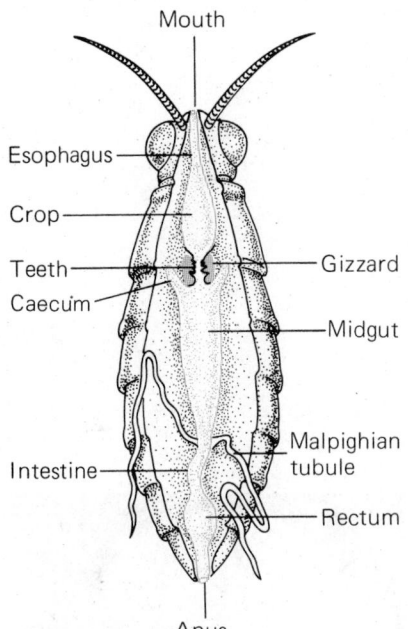

Figure 31-4 An insect's digestive tract. (The Malpighian tubules are excretory structures; they empty into the gut.)

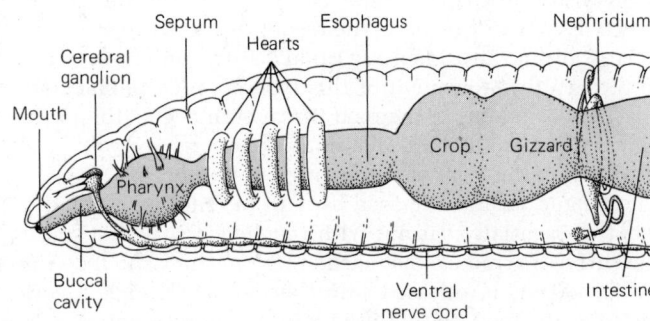

Figure 31-3 Front end of the gut of an earthworm (color).

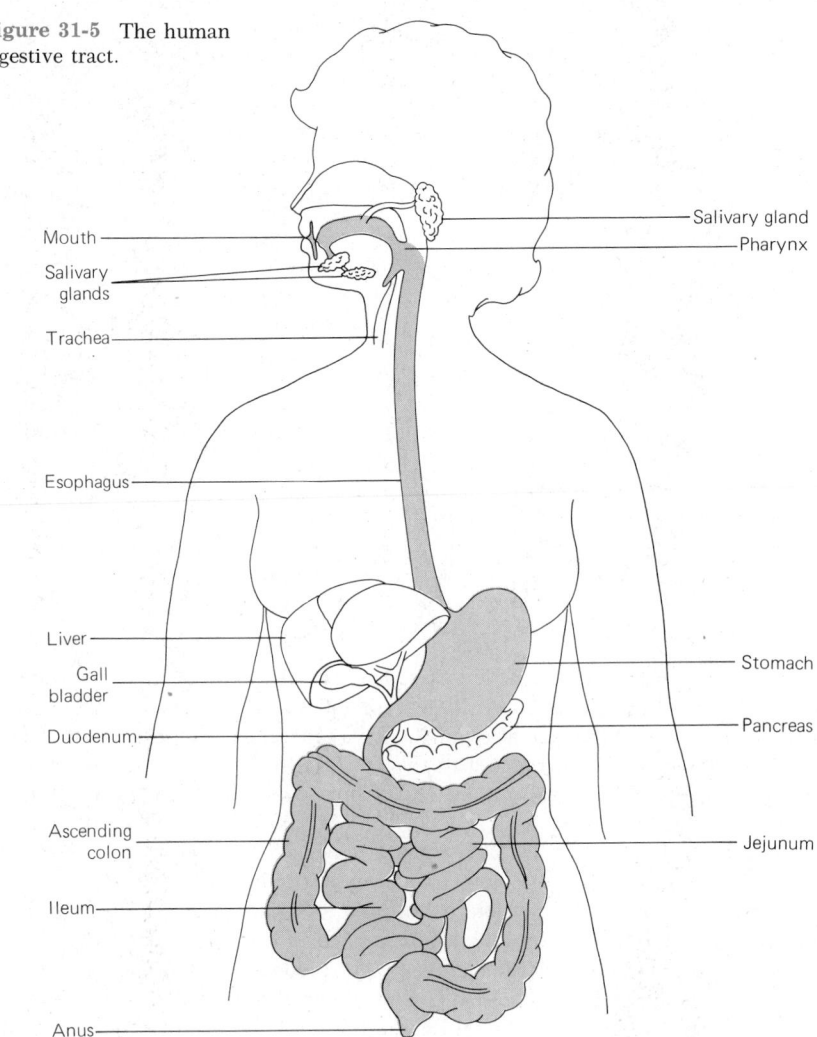

Figure 31-5 The human digestive tract.

Salivary gland
Pharynx
Mouth
Salivary glands
Trachea
Esophagus
Liver
Gall bladder
Duodenum
Ascending colon
Ileum
Anus
Stomach
Pancreas
Jejunum

Figure 31-6 A human skull. (Biophoto Associates)

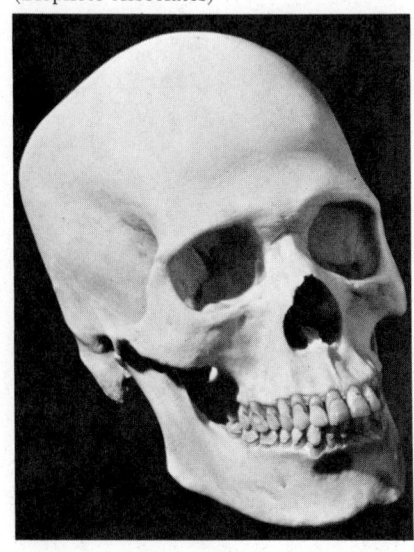

31-C   Human Digestion

The Human Alimentary Canal

The digestive tract of humans may also be called the **gut**, **alimentary canal**, or **gastrointestinal tract** (Figure 31–5). Basically, it has five functions: (1) food intake; (2) food storage and transport; (3) mechanical breakdown and digestion of food; (4) absorption of nutrients; (5) formation and evacuation of feces.

Food enters the human gut by manipulations of the mouth and associated structures such as the lips, tongue, and jaw muscles. Once a bite of food has been swallowed, its journey through the rest of the gut is completed automatically. The muscular walls of the alimentary canal contract in waves, starting from the end closest to the mouth. Movement of this type is called **peristalsis**.

Besides taking in the food, the mouth begins the process of mechanical breakdown, using teeth of various shapes and sizes (Figure 31–6). The **incisors**, chisel-like teeth in the front of the mouth, cut bite-size pieces of food from a larger portion. The tongue pushes the food back to the **molars**, which are the millstones of the mouth, essentially flat but with roughened surfaces, so that when the jaw moves, the surfaces grind against each other, mashing the food into small particles. Meanwhile, **saliva** is released into the mouth through a series of ducts. The saliva moistens the food, so that it will not scratch the delicate mucous membranes of the mouth and digestive tract, and it also contains a starch-digesting enzyme.

As a bite of food is swallowed, it passes through the **pharynx**, gliding over the

TABLE 31-6 SOME MAMMALIAN DIGESTIVE ENZYMES

ORIGIN	ENZYME	ACTION
Salivary glands	α-Amylase	Hydrolyzes starch and glycogen to dextrins (small polysaccharides) and maltose
Stomach	Pepsin	An endopeptidase, hydrolyzing proteins into polypeptides
Pancreas	Lipase	Hydrolyzes triglycerides into fatty acids, glycerides and glycerol
	α-Amylase	As in saliva
	α-Glucosidase } β-Galactosidase	Hydrolyze polysaccharides to monosaccharides
	Trypsin } Chymotrypsin	Endopeptidases
	Carboxypeptidase A } Carboxypeptidase B	Exopeptidases, hydrolyzing peptide bonds at the C-terminal (carboxyl) end of a peptide chain
	Aminopeptidases	Exopeptidases, hydrolyzing peptide bonds at the N-terminal (amino) end of the peptide chain
	Ribonuclease	Hydrolyzes RNA into nucleotides
	Deoxyribonuclease	Hydrolyzes DNA into nucleotides

epiglottis, a sort of trap door over the entrance of the trachea,* and into the **esophagus**. A wave of peristalsis passes down the walls of the esophagus, pushing the lump of food down to the stomach.

The **stomach** serves several functions. It is the widest part of the gut, in keeping with its function as a holding chamber that collects food and releases it into the intestine in small servings. Having a stomach enables us to take infrequent meals and spend the intervening hours in various pursuits. The muscular walls of the stomach churn the contents around, until they have the consistency of thick soup. Food at this stage is called **chyme**. Glands in the stomach wall also release acid and digestive enzymes that start digestion of the protein in our food. The chyme is released in spurts into the **duodenum**, which makes up the first part of the **small intestine**. A circular muscle called a **sphincter** closes the stomach off from the intestine except when it relaxes briefly to permit a small amount of chyme to squirt through, propelled by muscular contractions.

In the intestine, more digestive enzymes are added to the chyme. Some are secreted by the lining of the intestine itself, and some by the pancreas, which empties its digestive enzymes into the duodenum by way of a duct. Bile from the gall bladder enters through another duct (see Section 31-G). Digestion continues in the small intestine, and nutrients are absorbed through its lining. The remainder of the food passes into the **colon** or **large intestine**, where millions of bacteria live and work. The large intestine absorbs water, minerals, and vitamin K, which is produced by intestinal bacteria, and pushes the remaining **fecal matter** into the rectum, where it is held until it is voided. **Defecation**, or expulsion of the feces from the body, depends on contraction of the walls of the rectum and relaxation of the **anal sphincter**, another circular muscle at the very end of the digestive tract.

Human Digestive Enzymes

The digestive enzymes are **hydrolases**, enzymes that break substances down by the addition of water into connecting bonds. **Glycosidases** hydrolyze the glycosidic bonds that join monosaccharides together in carbohydrates; **lipases** hydrolyze lipids; and **proteases** hydrolyze peptide bonds in proteins (Table 31-6). **Exopeptidases** are proteases that cleave peptide bonds and free the terminal amino acid in a chain, whereas **endopeptidases** cleave peptide bonds in the interior of a peptide chain (compare Table 31-6 and Figure 31-7). Endopeptidases

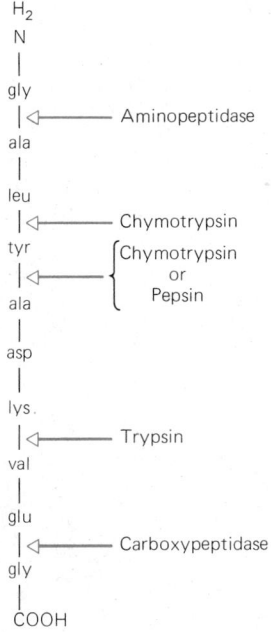

Figure 31-7 Different mammalian proteolytic enzymes break the amino acid chain of a protein in different places. Here the formula of a peptide indicates where some different digestive enzymes cleave the chain.

---

*Trachea   Windpipe; tube that conducts air from the pharynx to the lungs.

begin protein digestion by breaking a long polypeptide into shorter chains; exopeptidases then attack the end of a chain and break off amino acids one by one.

Digestive enzymes are produced by two types of glands: those with ducts and those without. The glands with ducts are the **pancreas** and **salivary glands**, which form during embryonic development as outpouchings of the gut and remain connected to it by their ducts, which carry their enzyme secretions into the canal. In addition, ductless, enzyme-secreting **tubular glands** line most of the length of the gut. (The gut is also lined by millions of **mucous glands** producing **mucus**, which lubricates the food and prevents it from injuring the gut by friction.)

Each day the three pairs of salivary glands discharge more than one litre of saliva into their ducts, which lead to the mouth. Saliva is mainly mucus, but it also contains an α-amylase, a digestive enzyme that hydrolyzes starch into smaller saccharides.

The fluid in the stomach contains hydrochloric acid (HCl), which gives it the remarkably low pH of less than 2.0. Some cells in the stomach lining secrete hydrogen ions, and others chloride ions, to produce this solution. Still other cells in the stomach lining secrete **pepsin**, a name encompassing a number of proteases found only in vertebrates; these enzymes will work only at a very low pH. One of the most important actions of pepsin is to digest **collagen**, a major constituent of fibrous tissue in meat, which is soluble at the low pH found in the stomach.

It is remarkable that the stomach does not digest itself, because the hydrochloric acid and pepsin it secretes can digest most flesh. The stomach protects itself by secreting acid- and enzyme-proof mucus, which coats the stomach wall. Even so, life for a stomach cell is short; the stomach surface loses about half a million cells a minute, and all its cells are replaced every three days. Ulcers occur when the stomach does not secrete enough mucus to protect itself. Similarly, the esophagus and duodenum on either side of the stomach (Figure 31–5) can become ulcerated if they are overexposed to stomach acid. People are more prone to ulcers if they live for some time under stressful conditions that they cannot control, but why this psychological condition should cause the stomach to malfunction is not known. The usual cure for ulcers is merely to relieve the stress and then rest the stomach by eating small meals of foods that do not aggravate the ulcer.

Most of the digestive enzymes are produced in the pancreas and empty into the small intestine close to the stomach. Because the pancreatic enzymes work best at a pH range of 7 to 8, the first role of the pancreas is to secrete a concentrated solution of sodium bicarbonate, which neutralizes the hydrochloric acid entering the intestine from the stomach.

Pancreatic juice contains enzymes that work on all three major food types: proteins, carbohydrates, and fats (see Table 31–6). Proteolytic enzymes are synthesized by the pancreas in inactive forms, which do not digest the pancreas. (Occasionally this system breaks down, and the pancreas will then be completely digested within a few hours by its own enzymes, in the condition known as acute pancreatitis, which is usually fatal.) The inactive enzymes become active forms after they reach the intestine. The protease trypsin is released by cleaving off part of an inactive molecule, trypsinogen. This is done by the enzyme **enterokinase**, secreted by the intestine. The other pancreatic enzymes are activated by the action of trypsin itself.

The secretion of digestive enzymes into the gut is regulated by nervous action and by hormones (Figure 31–8). Consequently the enzymes are present only when needed and not at other times.

Absorption of Food

About 10 litres of fluid must be absorbed from the digestive tract into the blood every day. Of this, about 1.5 litres consists of fluid we have drunk, and about 8.5 litres consists of the fluid secreted into the gut as digestive enzymes and mucus. Of these 10 litres, about 9.5 litres are absorbed into the small intestine and a small amount in the colon.

In the stomach, only lipid-soluble substances like alcohol and a few drugs are absorbed. The intestine is highly adapted for absorption; it is lined by folds and by

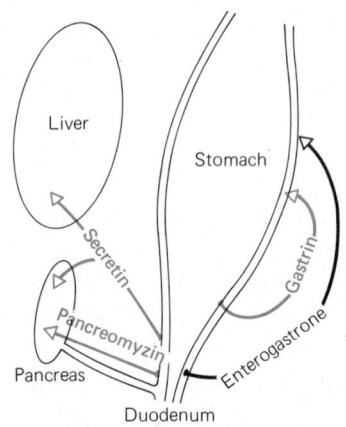

Figure 31-8 The hormonal control of some digestive secretions in humans. Colored arrows indicate hormones that stimulate their target organs; black arrows indicate inhibitory hormones. Pancreomyzin is secreted by the duodenal mucosa when chyme enters the duodenum, and it stimulates the secretion of digestive enzymes by the pancreas. Enterogastrone inhibits peristalsis and the secretion of hydrochloric acid by the stomach.

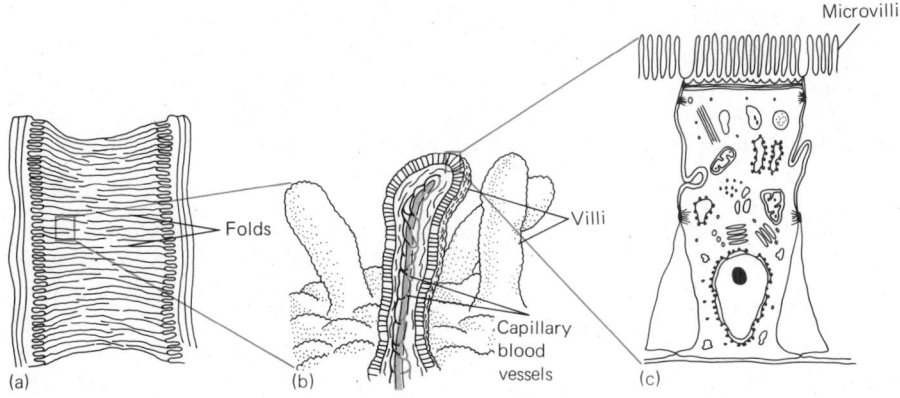
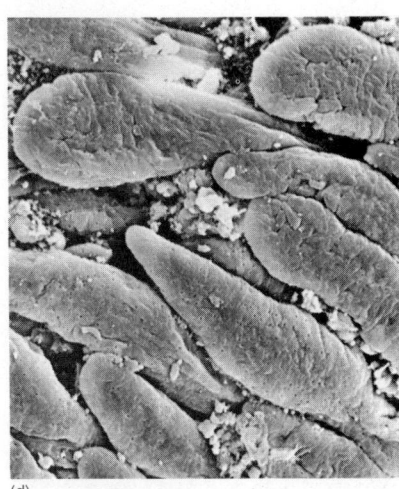

Microvilli

Folds

Villi

Capillary
blood
vessels

(a) (b) (c)

(d)

Figure 31-9 The small intestine has a large surface area across which food can be absorbed. (a) The lining of the intestine is highly folded. (b) Finger-like projections called villi line the intestine. The villi contain capillaries which transport food absorbed from the gut. (c) The plasma membranes of cells covering the villi are folded into microvilli. The absorptive area of the human small intestine has been estimated to be about the size of a tennis court. (d) Scanning electron micrograph of the small intestine of a mouse, magnified 550 times. (Biophoto Associates)

a "brush border" of finger-like villi, which in turn are covered with microvilli each about 1 nm long (Figure 31–9). These folds on folds enormously increase the internal surface area of the intestine. The cells lining the intestine contain large quantities of digestive enzymes, which apparently digest food molecules while they are actually passing through the cells.

We do not yet understand how monosaccharides are absorbed from the gut. Sodium is moved by energy-requiring active transport and presumably supplies the energy for glucose transport in some unknown manner, since glucose transport stops whenever sodium transport is blocked, even though ATP is still available. Amino acid absorption is also an active process. Water is absorbed by sodium transport: sodium is actively transported out of the intestine into the fluid surrounding the cells (the extracellular fluid) outside the gut. Negatively charged chloride ions follow passively, attracted by the positively charged sodium ions. The accumulation of sodium and chloride ions lowers the osmotic potential outside the intestine compared with that inside, and water moves out by osmosis.

Fats are absorbed mainly as fatty acids and monoglycerides.* What seems to happen is that these lipid-soluble molecules diffuse across the membranes of cells lining the intestine. Once inside the cells they are converted mainly into triglycerides, which are then voided from the cell by exocytosis* as tiny fat droplets, about 0.5 nm in diameter, called **chylomicrons**. Chylomicrons contain small amounts of cholesterol as well as triglycerides, and are coated by a layer of protein, which makes them water-soluble. Chylomicrons travel by way of the extracellular fluid and the lymphatic system and enter the blood vessels near the heart. A small quantity of fatty acids is absorbed directly into the bloodstream near the intestine without being converted into triglycerides.

Sodium, other ions, and water are the main substances absorbed in the large intestine. This absorption ensures that only about 100 ml of water and small amounts of inorganic ions are lost in the feces every day. The feces are about three-fourths water and one-fourth solid matter. Of the solid matter, about 30% is made up of bacteria (from the intestine where they live), 15% is inorganic matter, 3% is protein, 20% is fat, and 30% is undigested roughage.

---

*Monoglycerides Molecules composed of a glycerol molecule covalently bonded to one fatty acid molecule.

*Exocytosis Expulsion from the cell by fusion of a membranous sac with the plasma membrane, followed by opening of the sac to the cell's exterior.

31-D Feeding and Digestion in Herbivores

Most of the aquatic plant life on earth consists of tiny floating plants. It is, therefore, not surprising that most aquatic herbivores are filter feeders, straining these minute plants out of large volumes of water. The filtering system usually traps the plants in a mucous secretion, which is then moved to the gut by ciliary action (Figure 31–10).

Animals that eat terrestrial plants have special nutritional problems. Aquatic plants enjoy the support of the surrounding water, and so have little supporting tissue. Terrestrial plants, however, have more elaborate cell walls, which contain a higher proportion of hard-to-digest polysaccharides, notably cellulose. Breaking cell walls and digesting cellulose are major problems for most terrestrial herbivores.

Most herbivorous insects cannot digest plant cell walls. Their mouthparts are adapted to breaking or piercing cell walls so that they can feed on the cytoplasm. One of the reasons herbivorous insects such as locusts and grasshoppers are so destructive of crops is that they utilize only a small fraction of the food they eat. Cell walls and starch grains generally appear in the feces unchanged. There are some exceptions, such as termites, which feed on wood; their guts are populated by flagellates (protists) that secrete wood-digesting enzymes.

Herbivorous mammals have greatly enlarged molars that enable them to grind plant cell walls (see Figure 31–11); however, their most common nutritional adaptation is a collection of symbiotic* microorganisms housed in a specialized portion of the digestive tract. These microorganisms include cellulose-fermenting bacteria, which can use polysaccharides as food under anaerobic conditions. The bacteria synthesize an array of polysaccharide-splitting glycosidases. Since the food must be exposed to the bacteria for fairly long periods of time, it must pass through the gut slowly. Consequently, herbivores tend to have longer intestines than do omnivores or carnivores of similar sizes. (Some bacterial fermentation of food probably occurs in the intestines of all terrestrial vertebrates, including omnivores such as pigs, rats, and humans.)

Many vertebrate herbivores have a caecum, a sac set off to one side of the gut at the junction of the small and large intestines, and used as a fermentation chamber. (The human appendix is probably a vestigial* caecum, which has become reduced in size as we evolved towards a less herbivorous diet.) In rats and horses, the caecum and colon are enlarged to house fermentation.

Digestion aided by symbiotic bacteria has reached its greatest complexity in the ruminants. **Ruminants** are mammals in which the stomach has an alkaline pH and is divided into fermentation chambers. Most ruminants are artiodactyls, the even-toed ungulates including sheep, cattle, and deer. Ruminant digestion has also evolved in unrelated animals, including some marsupials (pouched mammals), colobus monkeys, and sloths.

The digestive tract of a ruminant such as a sheep may be taken as an example

---

* Symbiotic Living in close association.
* Vestigial Small and nonfunctional.

Figure 31-10 A sabellid, a marine annelid worm, feeds by filtering food out of the water with its tentacles. (Biophoto Associates)

Figure 31-11 The skull of a mammalian carnivore and the skull of a herbivore to show the characteristic difference in the teeth.

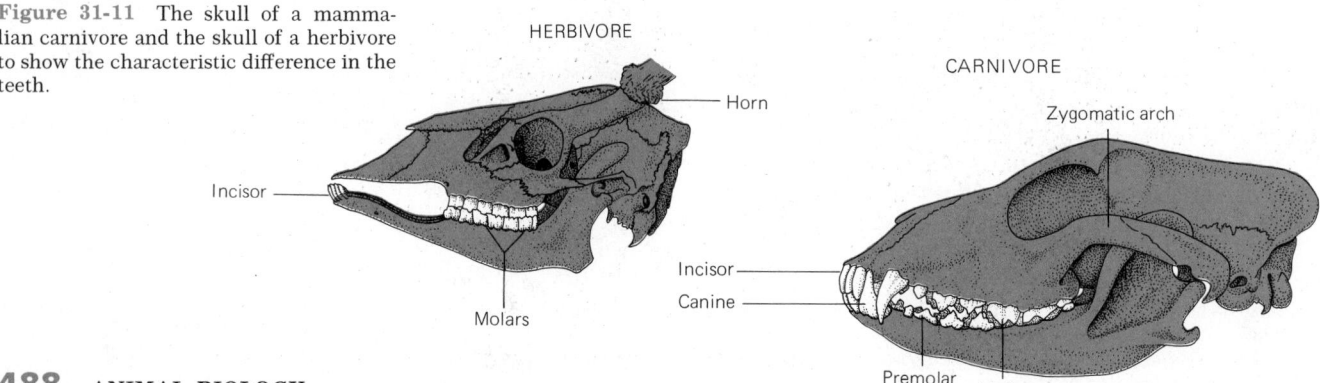

HERBIVORE

Horn

Incisor

Molars

CARNIVORE

Zygomatic arch

Incisor

Canine

Premolar Molar

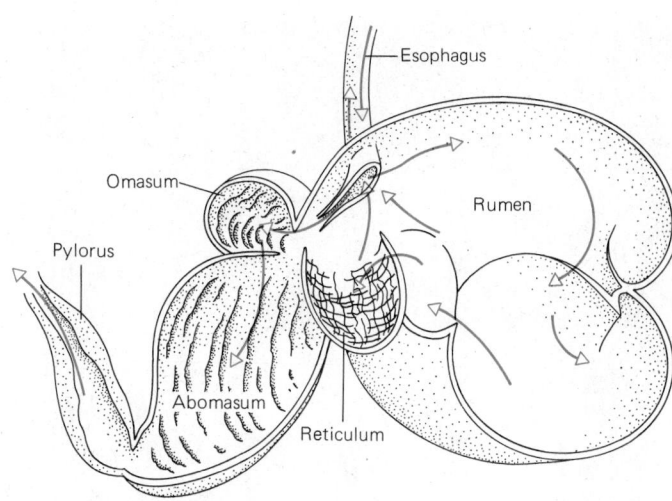

Figure 31-12 The stomach of a sheep with the wall cut away to show its four compartments. Colored arrows show the several directions in which the food travels.

(Figure 31–12). In such an animal, the salivary glands secrete about 10 to 15 litres of alkaline saliva per day; this saliva maintains a pH of about 8.5 in the stomach. The stomach is divided into four chambers. Food descends first into the **rumen** and the **reticulum**. In these two chambers the food is mixed to a pulp and fermented by anaerobic bacteria and protozoans. Sugars released by carbohydrate breakdown are used by the microorganisms; the host's share is fatty acids produced by the microorganisms during fermentation. The fatty acids are absorbed directly by the sheep's rumen. The contents of the rumen are then regurgitated into the mouth as "cud," which is chewed some more. On its second descent the food bypasses the rumen and the reticulum and enters the **omasum**, where it is mechanically churned. Finally it enters the **abomasum**, which corresponds to the stomach of other mammals. Here the ruminant recaptures many of the nutrients that the symbiotic microorganisms have used up by digesting the microorganisms themselves.

There are many advantages to using symbionts for digestion. Some microorganisms can synthesize amino acids using urea and ammonia; animal enzymes cannot do this. Thus microorganisms are valuable when the diet is low in protein. In addition, symbionts synthesize many vitamins, especially those of the B group, which can be used by the host. Herbivores such as baboons, which do not have ruminant digestion, have to eat meat occasionally (grasshoppers, snakes, baby monkeys) to replenish their B vitamins. A ruminant needs few dietary vitamins except vitamin A (which is common in plants) and vitamin D (which is less so). It has been calculated that microbial symbionts use about 6.5% of the calories in ingested food, and some of this is recaptured when the host digests the microorganisms.

31-E Adaptations of Mammalian Carnivores

Herbivores eat food that is hard to digest. Carnivores, on the other hand, experience their main nutritional problems in catching their food in the first place; once it is caught, their worries are pretty well over. Since the food of carnivores consists of other animals, its composition is very similar to that of their own bodies, with little of the waste that results from eating plants with thick cell walls and a high water content.

Carnivorous mammals have teeth adapted to killing their prey and shredding it into bite-size pieces (Figure 31–11). The **canine** teeth are often elongated into formidable fangs that can inflict swift and extensive damage on the prey. The muscles and bones of the skull are powerful, enabling them to subdue their meals without damaging their jaws. The molars are modified so that they resemble the blades of short saws, adapted to shredding meat into chunks that can be swallowed. Extensive chewing is not necessary, since there are no thick cell walls to break. The strong stomach acid and powerful proteolytic enzymes make short work of the food, and the intestine is short compared with that of herbivores and omnivores of the same size.

(a)

(b)

Figure 31-13 Bird bills. (a) The short, incurved bill of a barn owl is ideally suited to ripping animal prey apart. (b) A pelican's bill is huge with a soft, fleshy bag on its undersurface. The bird scoops up a large volume of water and then pushes the water out of the closed bill, retaining any food within the bill. (Biophoto Associates, N.H.P.A.)

31-F Feeding in Birds

A bird's jaws are composed of a **beak** or **bill** made up of bone and keratin, a protein also found in hair and fingernails. Modern birds have no teeth and so cannot grind food in the mouth. In insectivorous and seed-eating birds this function is taken over by a **gizzard**, a muscular sac formed from the hindmost portion of the stomach. It may contain stones that the bird picks up and swallows; the stones are replaced as they wear smooth.

The beak of a bird is modified according to the feeding habits of the species (Figure 31–13). The jaw muscles are particularly strong in carnivorous birds and in birds that eat large, hard seeds. Carnivores may also have remarkably powerful digestive enzymes capable of breaking down even "the bones and the beak" of their prey, although cormorants, hawks, and owls regurgitate bone, fur, and feathers in characteristic pellets.

A special organ found particularly in seed-eating birds is the **crop**, a storage sac that lies between the esophagus and the stomach. Storing food in the esophagus or crop ensures an almost continuous supply of food to the stomach and intestine for digestion. In animals with a high metabolic rate, and therefore with a high rate of food consumption, this reduces the frequency of feeding. Food takes only about an hour to pass through the body of a young bluejay, and a thrush fed blackberries will void the seeds 45 minutes later.

Birds generally utilize their food more efficiently than do mammals. A three-week-old stork converts about 33% of the weight of its diet of fish into stork. This compares with a food utilization efficiency of about 10% in a young mammal.

31-G   Functions of the Mammalian Liver

The liver is a large organ consisting of several lobes; it lies along the stomach and the upper part of the intestine just below the diaphragm. It is riddled with an enormous supply of blood vessels.

The liver is a gland that supplies bile to the intestine. In adult vertebrates, this function is a minor one, but the liver originally arose as a digestive gland in lower chordates. Throughout the liver, a network of tiny tubules collects **bile**—a solution of salts, bilirubin (made when hemoglobin from red blood cells is broken down in the liver), cholesterol, and fatty acids. Bile accumulates in the **gall bladder**, which empties into the small intestine by way of a duct. (Gall stones are formed when cholesterol precipitates out of the bile in the gall bladder.) Bile has two functions in the intestine. First, it acts as a detergent, breaking fat into small globules that can be attacked by digestive enzymes. Second, and more important, bile salts aid in the absorption of lipids from the intestine; removal of the gall bladder sometimes causes difficulty with lipid absorption.

Digested food molecules absorbed into the bloodstream from the intestine pass directly to the liver by way of the **hepatic portal vein**. Before these molecules pass on into the rest of the body, the liver may change their concentration and even their chemical structure. The liver performs a vital role in detoxifying otherwise poisonous substances. In addition, it stores food molecules that reach it from the intestine, converts them biochemically, and releases them back into the blood at a controlled rate. For instance, the liver removes glucose from the blood under the influence of the hormone insulin and stores it as glycogen. When the level of glucose in the blood falls, the liver breaks down glycogen and releases glucose into the blood.

The liver also synthesizes many of the blood proteins (e.g., albumins) and releases them into the blood when they are needed. In addition, the liver destroys old red blood cells and recycles their components, and it also converts nitrogenous wastes into the form of urea for excretion by the kidneys. With the kidneys, the liver is vital in regulating what the blood contains when it reaches all the other organs of the body. Because the liver is the body's major organ for making all these biochemical adjustments, severe liver damage or loss of the liver is rapidly fatal.

31-H   Stored Food and Its Uses

Muscular activity uses up considerably more than half our caloric expenditure. Most of this muscular activity goes to maintain the body's temperature. The contraction of tiny muscles that we don't even notice releases heat and warms the body; this is why we use up more calories in winter, especially in an under-heated house. Most of our other daily calorie usage goes for functions like urine formation and nervous system activity, whose caloric requirements vary little from day to day.

The body's carbohydrate stores (glycogen in the liver and muscles) would supply its energy needs for only about 12 hours if they were used alone. However, a human being of normal weight can usually survive without food for at least six weeks, using fat reserves for energy (Figure 31–14). A hibernating animal, with a lowered temperature and metabolic rate, can survive for months on the fat reserves that it built up by eating ravenously during the autumn.

Figure 31-14 The fate of stored food in a starving human being whose initial body weight was 15% fat. (A human being of average weight takes weeks to die from starvation although death from thirst occurs in a few days.)

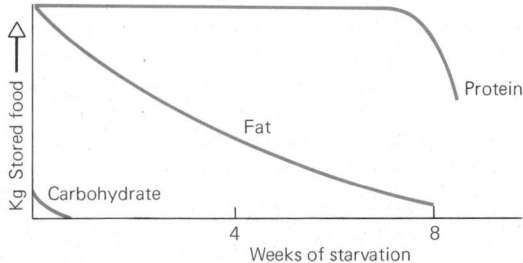

Because a given weight of fat provides about twice as many calories as the same weight of carbohydrate or protein, energy is stored most compactly in the form of fat. Fat is stored in the fat cells of **adipose tissue**, a storage tissue found under the skin, between muscle fibers, in the breasts and buttocks, between folds of the intestines in the abdomen, and elsewhere in the body. Fat is constantly exchanged between the bloodstream and adipose tissue: every molecule of fat in adipose tissue is replaced about every three weeks.

31-I   Regulation of Feeding

Feeding is governed by two kinds of controls: long-term and short-term. Long-term regulation ensures that enough stored food is maintained in the body. It can be altered by the action of hormones that ensure, for instance, that an animal builds up its fat reserves before its hungry young are born or hatched, and before hibernation. Short-term regulation ensures that an animal eats regularly on a day-to-day basis, so that food passes through the gut more or less continuously.

The control of feeding is poorly understood. The brain contains various centers that, when stimulated, start or stop feeding and control the selection of food. There are various hypotheses, but no real answers, as to what stimulates these brain centers in a normal animal.

Habit has a major effect on short-term feeding. People accustomed to three meals a day become hungry and experience muscular contractions ("hunger pangs") of the stomach if they miss one meal. However, people living alone, or working intensely so that they are not reminded of mealtimes, frequently miss meals without feeling hunger. Distension of the stomach inhibits the feeding center in the brain; this is the main reason we stop eating after a large meal. On the other hand, people and animals whose food never reaches the stomach (because a tube has been inserted into the esophagus to divert food to the outside of the body instead of the stomach) also stop eating after awhile. This suggests that the quantity of food that has passed the mouth is monitored in some unknown way.

The "ideal weights" on life insurance tables have been adjusted upward since the 1960s, after studies showed that many people as much as 20% heavier than their "ideal weights" lived longer than those of ideal weight. There is in fact considerable variation in how much body fat an individual carries or should carry. If you have struggled to lose weight over and over and always seem to return to a weight somewhat over your ideal weight, it is very likely that your long-term regulator is "set" to the higher weight and that, as a result, you are no less healthy than your slimmer friends. Those of us in that situation should probably give up fighting the battle of the bulge, exercise to maintain our health, and hope that seventeenth-century fashions in bodies will return in our lifetimes!

Why an animal eats *what* it does is even more complicated and less well understood than why an animal eats as much as it does. For carnivores, things are reasonably simple because carnivores eat food that is highly nutritious and generally nontoxic. Carnivores rapidly learn to avoid food that tastes unpleasant, as you realize if you have ever seen a cat attack a toad.

Herbivores face a bigger problem: many plants contain toxic chemicals, and even more (including many of our common foods) are toxic in large amounts. Furthermore, very few herbivores can eat just one kind of plant because no single plant species contains the complete mix of nutrients that animals need. Herbivores can tell the difference between most species of plants by smell, touch, or other senses. A sheep presented with a field of grasses, most of which are new to it, adds only one new species to its diet at a time and eats very little of that. Presumably, this gives the long-term learning system time to "tell" the sheep whether the new plant is a good thing or not. Using this system, a herbivore can work up to a varied and nutritious diet. Small, slow-moving herbivores, such as caterpillars, eat only one plant in their lives—the one where they hatch from the egg. Special arrangements make up for the nutrient deficiencies of the food plant. For example, the caterpillar of a tiger swallowtail butterfly cannot obtain all the essential amino acids and sodium it requires from its diet of cherry tree leaves. However, the caterpillar hatches with extra supplies of these nutrients, deposited in the egg by its mother, who obtained them from plant nectar as an adult.

Figure 31-15 Salt feeding behavior. These tiger swallowtail butterflies are "puddling." They suck up sodium-rich fluid from sand just as deer and cattle lick up minerals from salt licks. (Paul Feeny)

Specific Appetites

One of the least-understood aspects of feeding is the development of an appetite for a specific substance. Often (but not always) an animal with a long-term deficit of fat in its diet will develop a craving that can be satisfied only by fat. Children with calcium deficiencies have been known to eat the plaster off walls (which is mainly $CaCO_3$). Laborers in the tropics frequently develop salt deficiency because of the volume of sweat they produce; they drink salt water or salty beer and find it delicious, although it tastes repulsive to anyone who is not short of sodium.

Sodium appetite has been the subject of considerable research. Many herbivores experience a constant shortage of sodium because many plants contain very little of this essential animal nutrient. Indeed, the size of an animal population may actually be limited by the availability of sodium. Isle Royale in Lake Superior has a population of about 1200 moose, and measurements of sodium eaten and excreted show that they are using every speck of available sodium; the island could not support any more of these large herbivores.

Historically, salt has been a valuable, and frequently expensive, commodity in the trade of human populations living too far from the sea to obtain salt. In New Guinea to this day an inland village that has a mineral spring will evaporate the water to produce "salt" containing sodium and other ions. This "salt" can be traded for large quantities of food or clay pots.

Animals increase the sodium content of their diet in various ways. Aquatic plants generally contain more sodium than do the nearby land plants, and moose or elk knee-deep in a swamp, grazing on water weeds, are a common sight in the wilds of Canada. Colobus monkeys in India do the same thing, and African elephants will travel many miles from their usual feeding grounds to salt springs. "Salt licks" are areas where the soil contains more sodium than it does elsewhere. These licks are gathering places for many different herbivores in such places as Yellowstone Park and many game parks in Africa. Any provident farmer sets out cakes of salt for horses and cows.

Plainly, specific appetites are valuable because they lead animals to search out and eat food containing nutrients that the body needs. In some way, the deficiency of a particular nutrient changes the reaction of the sense organs or the brain to potential food so that salt, tryptophan, or whatever it is, tastes or looks much more appetizing to an animal that needs that substance, but we have no idea how this happens.

Essay: Scurvy

Scurvy is a disease that is hardly ever seen these days. It plagued sailors on long voyages before the end of the eighteenth century.

From 1740 to 1744, the Englishman George Anson sailed around the world aboard *H.M.S. Centurion.* He captured a fortune in treasure from Spanish possessions on the West Coast of America and from galleons in the Pacific. Richard Walter and Pascoe Thomas, who sailed on the *Centurion*, wrote the story of this voyage during which more than half of the 400 men on board died of scurvy. Here is their account of the disease as the ship battled gales off Cape Horn, the southernmost tip of South America:

And now as it were to add the finishing stroke to our misfortunes, our people began to be universally afflicted with the most terrible, obstinate, and at sea, incurable disease, the scurvy, which quickly made a most dreadful havoc among us, beginning at first to carry off two or three a day, but soon increasing and at last carrying off eight or ten; and as most of the living were very ill of the same distemper, and the little remainder who preserved their healths better, in a manner quite worn out with incessant labour, I have sometimes seen four or five dead bodies sewn up in their hammocks, others not, washing about the decks, for want of help to bury them in the sea.

Its symptoms are inconstant and innumerable, and its progress and effects extremely irregular; for scarcely any two persons have the same complaints and where there hath been found some conformity in the symptoms, the order of their appearance has been totally different. . . . The common appearances are large discoloured spots over the whole surface of the body, swelled legs, putrid gums, and above all, an extraordinary lassitude of the whole body, especially after any exercise, however inconsiderable. And this lassitude at last degenerates into a proneness to swoon on the least exertion of strength, or even the least motion. . . . At other times, the whole body but more especially the legs, were subject to ulcers of the worst kind, attended with rotten bones, and such a luxuriancy of fungous flesh as yielded to no remedy. But a most extraordinary circumstance, and what would be scarcely credible upon any single evidence, is that the scars of wounds which had been for many years healed, were forced open again by this violent distemper: of this, there was a remarkable instance in one of the invalids on board the *Centurion*, who had been wounded about fifty years before at the Battle of the Boyne; for although he was cured soon after, and had continued well for a great number of years past, yet in his being attacked by scurvy, his wounds, in the progress of his disease, broke out afresh, and appeared as if they had never been healed: Nay what is still more astonishing, the callous of a broken bone, which had been completely formed for a long time, was found to be hereby dissolved, and the fracture seemed as it had never been consolidated. Indeed, the effects of this disease were in almost every instance wonderful; for many of our people, though confined to their hammocks, appeared to have no inconsiderable share of health, for they ate and drank heartily, were cheerful, and talked with much seeming vigour, and with a loud strong tone of voice: and yet on their being the least moved, though it was only from one part of the ship to another, and that in their hammocks, they have died before they could well reach the deck: and it was no uncommon thing for those who were able to walk the deck and do some kind of duty, to drop down dead in an instant on any endeavour to act with their utmost vigour, many of our people having perished in this manner during the course of the voyage.[1]

Sad to relate, a cure for scurvy was known at this time. Various people had shown that scurvy was a nutritional deficiency which could be cured by eating raw fruit and vegetables. Since fresh vegetables would not survive long voyages in the days before refrigeration, this was not a very helpful observation, until it was found that acid fruit juices would retain their anti-scurvy properties for long periods. By the 1820s, the British were known as "limeys" because their sailors consumed a daily ration of lime juice, but it was almost a century after Anson's voyage before this habit became widespread.

Scurvy is caused by dietary deficiency of ascorbic acid, or vitamin C, which most vertebrates (but not humans or monkeys) can synthesize for themselves. Ascorbic acid is necessary to the synthesis of the amino acid hydroxyproline, which is required for connective tissue synthesis and repair. Without it, the body's connective tissue slowly breaks down; hence the loosening of teeth in degenerating gums and the breakdown of scar tissue and blood vessels. Sufferers from scurvy usually died as a result of hemorrhage from broken blood vessels.

---

[1] Reprinted with permission from L. A. Wilcox, *Anson's Voyage.* New York: St. Martin's Press, Inc.

H.M.S. Centurion

Because animals are heterotrophs, their diet must contain all of the organic and inorganic substances they need for metabolism, growth, and energy. Animals obtain fats, carbohydrates, and proteins, and vitamins and minerals, in their food.

The function of digestion is to break food down into molecules that can be absorbed into the body from the gut. In vertebrates, digestive enzymes are synthesized in the salivary glands, pancreas, and lining of the digestive tract. Many animals, particularly herbivores, harbor symbiotic microorganisms, which secrete digestive enzymes, in the alimentary canal.

Digested food is absorbed into the blood and extracellular fluid by diffusion and by active transport across the enormous surface area of the intestine.

The liver plays a major role in controlling the fate of newly absorbed food molecules. It stores excess glucose as glycogen, synthesizes many blood proteins, and converts nitrogenous and other wastes into a form that can be excreted by the kidneys. Excess carbohydrate or protein is converted into triglycerides and stored in the fat cells of adipose tissue.

Feeding is regulated by long-term and short-term control mechanisms that are not well understood. These controls ensure that the alimentary canal is efficiently occupied most of the time and that the animal maintains its body reserve of fat without spending unnecessary time feeding. All animals have complicated regulatory systems that control what, as well as how much, they eat.

OBJECTIVES

From your study of this chapter, you should be able to:

1. Name the major classes of macronutrients and micronutrients, and list the general functions of each class.
2. Explain the selective advantages of the following digestive adaptations found in the animal kingdom: extracellular digestion, discontinuous feeding, digestive tract with mouth and anus, crop, gizzard, and caeca.
3. List the parts of the human digestive tract in order, and state what happens to food in each part of the tract.
4. List the organs that secrete digestive enzymes in mammals, and state the type of substrate digested by the enzymes secreted by each organ.
5. List the parts of the gut in which various substances are absorbed, and which types of substances are absorbed in each part.
6. Describe the digestive adaptations of herbivorous and carnivorous mammals and of birds.
7. Discuss how symbiotic microorganisms contribute to the nutrition of their hosts.
8. List the functions of the mammalian liver, and explain the importance of this organ.

SELF-QUIZ

Matching: For each numbered phrase below choose the letter of the correct class of nutrient on the right. More than one letter may be correct; choose all that apply.

____ 1. Inorganic nutrients

____ 2. Macronutrient that cannot be stored in the body

____ 3. May be the source of energy for the body's metabolism

____ 4. Source of material for cell membranes

____ 5. Coenzymes for metabolic enzymes

____ 6. Digested by enzyme in saliva

____ 7. Absorbed in large intestine

a. protein
b. carbohydrate
c. fat
d. water-soluble vitamins
e. fat-soluble vitamins
f. minerals

8. The main advantage of having a digestive tract with a mouth and anus is:
   a. it permits different parts of the gut to become specialized to perform different parts of the digestive process in turn
   b. it permits an animal without teeth to have a means of grinding its food
   c. it permits animals to eat a great deal at once and digest it while doing something else
   d. it permits animals to eat larger organisms as food
   e. it permits animals to eat food in larger chunks

9. In humans, digestion of food is completed in the:
   a. mouth
   b. stomach
   c. small intestine
   d. large intestine
   e. rectum

(Quiz continues on next page)

10. In humans, protein digestion is carried out by enzymes secreted by the:
    a. stomach, pancreas, and salivary glands
    b. liver, salivary glands, pancreas, and small intestine
    c. salivary glands, stomach, pancreas, and small intestine
    d. liver, stomach, pancreas, and small intestine
    e. stomach, small intestine, and pancreas
11. A portion of the stomach that has evolved extremely thickened muscular walls and is quite efficient at grinding hard food is called a(n):
    a. rumen
    b. gizzard
    c. crop
    d. omasum
    e. caecum
12. Which of the following is probably *not* an action of symbiotic microorganisms of the gut?
    a. use of the host's food for their own nutrition
    b. extracellular digestion
    c. aerobic respiration
    d. breakdown of substrates that the host cannot digest
    e. manufacture of vitamins needed by the host animal
13. Which of the following is *not* a function of the mammalian liver?
    a. secretion of digestive enzymes for export to the gut
    b. regulation of blood glucose and amino acid content
    c. production of the nitrogenous waste urea
    d. production of plasma proteins for the blood
    e. destruction of old red blood cells
    f. detoxification of poisonous substances

QUESTIONS FOR DISCUSSION

1. Herbivores can seldom survive by eating only one species of plant (e.g., corn lacks the amino acid tryptophan, many plants contain too little sodium). This is probably no evolutionary accident. What's in it for the plant?

2. Why does it take longer to become hungry after a protein-rich meal than after a meal that is mostly carbohydrate?

3. Trace the fate of a piece of pepper pizza through the human digestive tract. (*Contents of pizza*: crust: carbohydrate and various B vitamins; cheese: protein, fat, calcium, phosphorus; tomato: vitamin C, potassium; pepper: vitamin A, iron, cellulose).

4. Some kinds of stress can upset an animal's normal nutrient balance. For example, infection increases the rate of utilization of vitamin C. How might the organism compensate for this disturbance? Will this invariably change the optimum dietary level?

REFERENCES AND FURTHER READING

Arms, K., P. Feeny, and R. C. Lederhouse. "Sodium: stimulus for puddling behavior by tiger swallowtail butterflies, *Papilio glaucus*." *Science* 185:372, 1974. A description of the experiments that showed that puddling butterflies obtain sodium by this behavior.

Botkin, D. B., *et al.* "Sodium dynamics in a northern ecosystem." *Proceedings of the National Academy of Sciences* (U.S.) 70:2745, 1973. The work that showed that the Isle Royale moose population appears to be limited by the availability of sodium.

Gordon, M. S. *Animal Physiology: Principles and Adaptations*, 2d ed. New York: Macmillan Publishing Co., 1972. An excellent textbook on the comparative physiology of different animals.

Harlan, J. R. "The plants and animals that nourish man." *Scientific American*, September 1976. An interesting discussion of human food plants and animals; where they originated and how they have become tamed or cultivated.

Scrimshaw, N. S., and V. R. Young. "The requirements of human nutrition." *Scientific American*, September 1976. A list of human nutritional requirements and a discussion of factors that affect them; interesting and readable.

Watt, B. K., and A. L. Merrill. *Handbook of the Nutritional Contents of Foods*. New York: Dover Publications, 1975. The United States Department of Agriculture sponsored this collection of data, giving proximate composition and content of vitamins and minerals for raw, processed, and prepared foods.

CHAPTER 32

GAS EXCHANGE IN ANIMALS

From a climber struggling to the top of Mount Everest with an oxygen cylinder, to a shark gliding through the depths of the ocean, every animal is continually exchanging gases with its environment. Small animals can obtain oxygen and give off carbon dioxide directly through plasma membranes on the outer surfaces of their bodies. The evolution of larger animals has been possible, however, only because they have evolved specialized respiratory systems, which permit them to take up oxygen from their environments and eliminate carbon dioxide from the body. This chapter describes some of the arrangements for gas exchange found in animals and some of the theoretical considerations that govern their working.

32-A Ventilation and Respiration

Molecular oxygen in high concentrations is dangerous to living tissues because it oxidizes organic molecules. For this reason, little oxygen can be stored in the body, and oxygen must be obtained continuously from the environment. Our own bodies can survive for weeks without food, and for days without water, but only for minutes without oxygen.

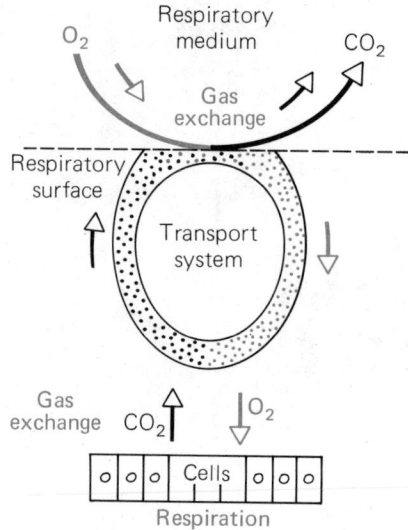

Figure 32-1 Gas exchange occurs between the respiratory surface, the body fluids, and living cells.

Most of the earth's molecular oxygen is in the air. Some of it, however, is dissolved in bodies of water and in the water of the soil. Either water or air may serve as the **respiratory medium** for an animal, that is, as the immediate source of oxygen.

If each cell of an animal is exposed to the respiratory medium, the cells can pick up oxygen directly. In larger animals, however, many cells live deep inside the body and must obtain oxygen from the **extracellular fluid** that surrounds them. In such animals the extracellular fluid is in contact with the blood, which in turn is in contact with the **respiratory surface**, where the blood takes up oxygen from the respiratory medium. Here the blood also releases carbon dioxide, a gaseous end product of respiration, which it has picked up from the body tissues (Figure 32-1). This **gas exchange** should not be confused with cellular respiration.

The respiratory system of an animal consists of the respiratory surface, any passageways that connect the respiratory surface to the outside of the body, and any muscular structures used to ventilate the respiratory surface. **Ventilation** is the process of moving the respiratory medium so that the respiratory surface is constantly exposed to a fresh supply of oxygen. **Breathing** is ventilation of a respiratory surface with a particular medium: air.

Ventilation is necessary because gas exchange depends on diffusion. The greater the concentration gradient* of gases across the respiratory surface, the faster gases will diffuse between the respiratory medium and the blood. Near the respiratory surface oxygen is quickly removed, and carbon dioxide rapidly builds up in the air or water. Both these effects reduce the concentration gradient and slow diffusion across the respiratory surface. Ventilation creates a current that brings a fresh supply of air or water, which renews the oxygen and removes the carbon dioxide at the respiratory surface, permitting gas exchange to occur as rapidly as possible.

32-B   Facilitated Diffusion of Oxygen

Until very recently, it was thought that oxygen moves from the environment into the blood by diffusion. In the early 1970s, however, Barry Burns and Gail Gurtner demonstrated that perhaps 15 to 30% of oxygen transfer from the air to the blood in the lungs of mammals is accomplished by **facilitated diffusion** (Section 5-C). A type of cytochrome* is the oxygen carrier. Facilitated diffusion speeds up diffusion, allowing the blood to pick up oxygen faster than would be possible without the carrier.

The first studies on facilitated diffusion of oxygen were investigations of oxygen exchange in the placentas of pregnant sheep. The **placenta** is an organ in the uterus of a pregnant female mammal where capillaries* from the mother's bloodstream pass close to capillaries of the fetal circulation. Burns and Gurtner found that a drug that specifically inhibits a cytochrome known as $P450$ caused a drop of about 80% in the rate of oxygen transfer from the maternal to the fetal blood.

It has long been known that one of the functions of $P450$ is to oxidize toxic substances, rendering them less harmful to the body. The lungs perform much of the body's detoxification, surpassing the liver, the main organ of detoxification, in the handling of some toxins. The ready supply of oxygen at the lung surface makes it an ideal site for oxidizing foreign substances.

Problems can arise when the body is detoxifying an inordinately large amount of a foreign substance. Tobacco smoke or anesthetic gases in the lung, for instance, may force $P450$ to use up much of the oxygen it carries in detoxifying these substances, instead of transferring it to the blood. When this happens, the body tissues may experience oxygen deficiency.

Pregnant women are especially affected by tobacco, drugs, and anesthetics, for not only do these substances reduce the amount of oxygen transported from the lungs into the mother's bloodstream, but they also restrict the oxygen supply to the

---

*Concentration gradient   The change in concentration over a distance.
*Cytochrome   An electron carrier molecule consisting of a protein and a porphyrin ring (see Figure 32-15) containing a metal ion.
*Capillaries   The smallest blood vessels in a circulatory system; exchange of substances between the blood and the fluid surrounding the blood vessels takes place through the thin walls of capillaries.

fetus. If the cytochrome *P450* in the placenta is tied up in the detoxification of substances arriving in the mother's blood, the fetus loses a large percentage of its oxygen supply. This may account for the observation that nurses who work with anesthetics have a 30% higher rate of miscarriage and birth defects in their infants than the rate observed in women in other occupations, and that women who smoke tend to give birth to small babies.

32-C Problems of Gas Exchange

What problems are involved in moving oxygen from the external medium to the extracellular fluid surrounding a cell? First, the amount of oxygen available from the environment varies. For instance, there is comparatively little oxygen in an aqueous environment, and the warmer or the more salty the water, the less oxygen it contains (Table 32–1). Air, by comparison, is rich in oxygen, and oxygen diffuses half a million times faster in air than in water. It looks, at first glance, as if aquatic animals are at a disadvantage. On the other hand, gas molecules must cross plasma membranes in solution, and so respiratory surfaces must always be wet. Animals that obtain their oxygen from air lose large quantities of precious water by evaporation from their respiratory surfaces.

The rate at which an animal can absorb oxygen depends on the amount of oxygen available and on the area of respiratory surface exposed to the respiratory medium. Small animals can often obtain all the oxygen they need through the general body surface. As the volume of an animal increases, however, its surface-area-to-volume ratio decreases. Larger animals must have some sort of modification in their shape that increases the area of the respiratory surface relative to the volume of the body.

Since carbon dioxide is much more soluble in water than is oxygen, any respiratory system that can supply an animal with enough oxygen is more than capable of disposing of carbon dioxide at a fast enough rate.

A final consideration in gas exchange is how much oxygen an animal needs. The **metabolic rate** is the rate at which an organism uses food by way of cellular respiration; it is usually measured as the volume of oxygen used per unit of body weight per unit of time. Thus:

$$\text{oxygen requirement} = \text{metabolic rate} \times \text{body weight}$$

Homeothermic ("warm-blooded") animals, which maintain constant body temperatures higher than those of the surroundings, have high metabolic rates because they continually oxidize food, producing heat to replace the heat lost across the body surface. Since surface-to-volume ratio increases with decreasing size (see Figure 23–9), small homeothermic animals lose relatively more heat than do large ones and so must use relatively more oxygen (Figure 32–2).

TABLE 32–1 **OXYGEN CONTENT OF SOME RESPIRATORY MEDIA**	
MEDIUM	OXYGEN CONTENT (*ml/litre*)
Sea water at 5°C	6.4
Fresh water at 5°C	9.0
Fresh water at 25°C	5.8
Air	209.5

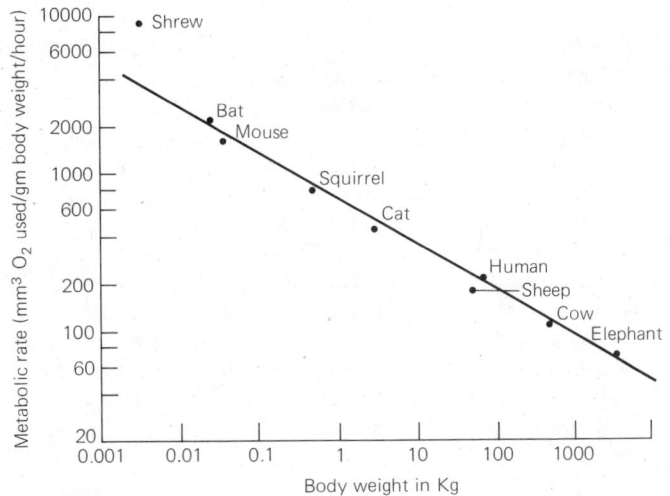

Figure 32-2 The metabolic rates of homeothermic (warm-blooded) animals are inversely proportional to the size of the animal.

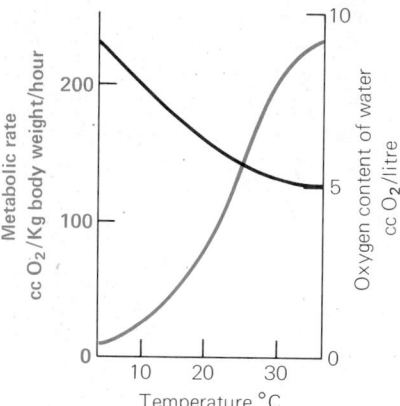

Figure 32-3 Metabolic rates of poikilothermic animals vary with the temperature of the environment (colored line). For an aquatic animal, higher temperatures decrease the amount of oxygen available, since less oxygen dissolves in warm than in cold water (black line).

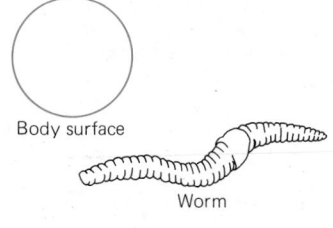

Body surface

Worm

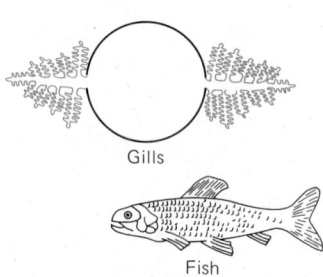

Gills

Fish

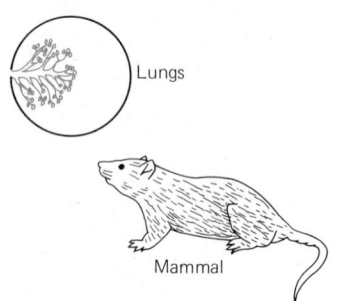

Lungs

Mammal

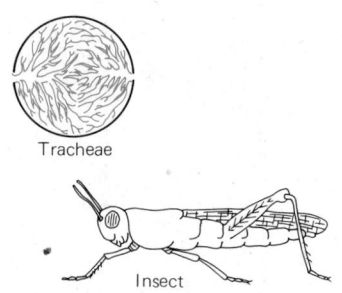

Tracheae

Insect

Figure 32-4 The four main types of respiratory surfaces (color) found in different animals.

Poikilotherms, animals that do not maintain a constant body temperature, may show a drastic change in oxygen consumption with changing environmental temperature (Figure 32–3). The higher the temperature, the faster the chemical reactions of metabolism occur, the higher the animal's metabolic rate, and the more oxygen it needs. An animal that obtains its oxygen supply from water has a dual problem as the temperature rises. It needs more oxygen because of the rise in its metabolic rate, but there is less oxygen available because less oxygen is dissolved in water at higher temperatures (see Table 32–1). A few active fish, such as trout, can obtain enough oxygen only in very cold water. Thus the discharge of waste heat (for instance, from a power plant) into a lake or stream may ruin the trout fishing.

32-D   Respiratory Surfaces and Ventilation

Four main types of respiratory surface are used by animals: the body surface, gills, lungs, and tracheae (Figure 32–4).

The Body Surface

Some animals, such as earthworms, obtain all the oxygen they need across the general body surface. This is possible only if the animal is fairly small, so that it has a high surface-to-volume ratio. In addition, the body surface must be kept moist. Third, the animal's metabolic rate must be low enough so that it does not use oxygen at a great rate. Fourth, the thin moist skin must be protected from injury; it is often covered with a slimy mucus that makes it too slippery to be damaged by sharp objects.

Gills

Animals such as fish and many water-dwelling arthropods, molluscs, and amphibians carry out gas exchange through **gills**, feathery tissue outgrowths that are exposed to the water and that exchange gases with it across the thin gill membranes.

An animal that ventilates its respiratory surfaces with water faces a problem in that water is heavy for the amount of oxygen it contains. Therefore, a large proportion of the animal's energy must be used to push a constant current of water across the gill membranes. It is inefficient to pass water back out of the gill area in the opposite direction from the way it came in: that involves overcoming the inertia of the water in order to stop it, and then pushing it out the other way. Instead, in most aquatic animals, water passes in through one opening and out by another.

The gills of a bony fish are located behind the head (Figure 32–5). Water enters through the mouth, passes from the pharynx across the gills, and leaves via the opening behind the **operculum**, which covers the gills.

Blood in the gills of a fish circulates in a pathway that increases the blood's efficiency in removing oxygen from the water. In such a **countercurrent ex-**

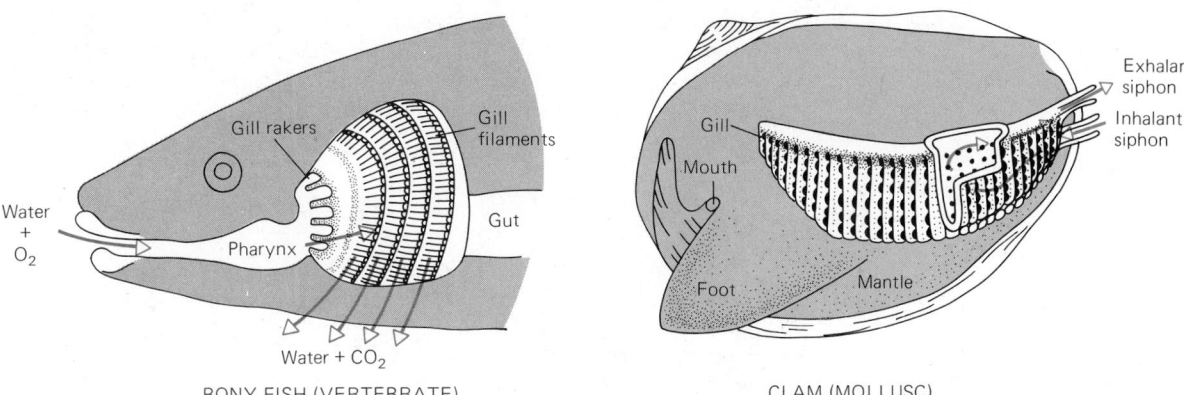

BONY FISH (VERTEBRATE) CLAM (MOLLUSC)

Figure 32-5 A one-way current of water crosses the respiratory surface in gill-breathing animals. In the fish, water enters through the mouth and passes out across the gills. In the clam, the water enters through the inhalant siphon, passes through pores in the gills, and then leaves through the exhalant siphon. The clam's gills are also used for filter feeding; food particles are trapped in mucus on the gills and passed to the mouth.

change system, two fluids exchange substances (or heat) with one another as they flow in opposite directions. The blood capillaries in the gills are arranged so that blood flows opposite to the direction in which water flows past the gills (Figure 32–6). This arrangement maintains an oxygen concentration gradient between the water and the blood at all times. Water entering the gill area encounters blood that has already picked up some oxygen and is about to leave the gills. However, this blood still has the capacity to hold more oxygen, which it picks up from the fresh, oxygen-rich water that it encounters. As the water passes on, it loses more and more of its oxygen to the blood, but it encounters blood that is less and less saturated with oxygen. The water always contains more oxygen than the blood it encounters, and thus the water continues to lose oxygen and the blood continues to gain it (Figure 32–7). The countercurrent mechanism is so efficient that the gills of a fish may remove more than 80% of the oxygen from the water in the respiratory current.

In animals such as many molluscs and the lower chordates (amphioxus, tunicates [Sections 29-C, 29-D]), ventilation of the gills is inextricably tied to feeding; food is filtered out of water drawn into the gill area, and gas exchange occurs at the

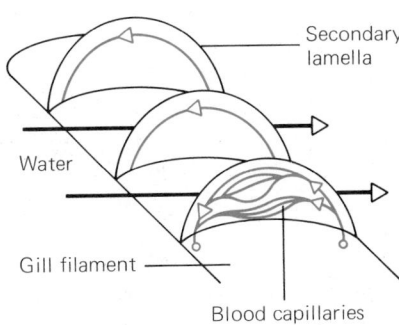

Figure 32-6 Countercurrent flow in the gill of a fish. Water (black arrows) crosses the gills in the opposite direction from the blood (colored). Secondary lamellae are fine extensions of the gill filaments (see Figure 32-5).

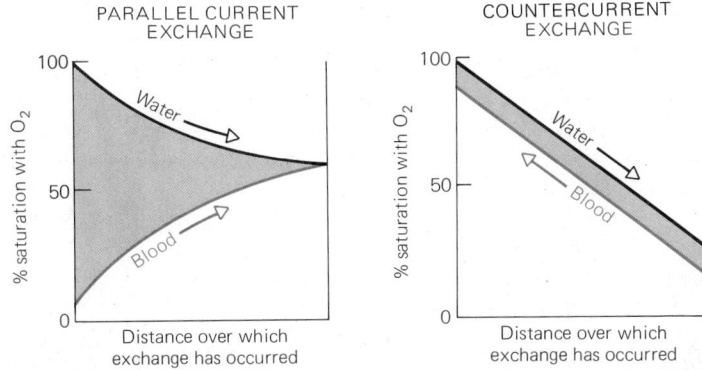

Figure 32-7 The relative efficiencies of oxygen exchange between water and blood by countercurrent and parallel current mechanisms. With parallel currents, the oxygen gradient (color shading) between the two media is steadily reduced and the blood takes up less and less oxygen as the two media travel together. With a countercurrent mechanism, the oxygen gradient between the two media is maintained and the blood may take up as much as 80% of the oxygen originally present in the water.

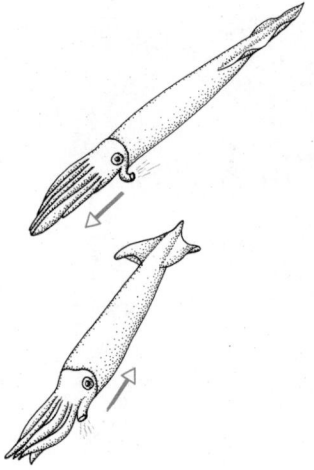

Figure 32-8 The squid can move forward or backward (colored arrows), depending on which way the siphon is aimed as it ejects water from the mantle cavity surrounding its gills.

same time. Some animals use respiratory water currents for locomotion by squeezing water forcibly out of the gill area. Squid, for instance, can eject water from the siphon with considerable force, creating a jet-propulsion stream that moves them rapidly forward or backward depending on which way the siphon is aimed (Figure 32–8).

Lungs

The problems faced by animals with lungs are entirely different from those faced by animals with gills. Air is about a thousand times less dense than water, so that less energy is needed to move a given volume of air than of water to and from the respiratory surface. In addition, a given volume of air contains about 30 times as much oxygen as the same volume of cold water. Furthermore, oxygen molecules diffuse 500,000 times faster in air than in water. Thus it is possible for an animal to ventilate the lungs by a **tidal**, or in-and-out, flow rather than by the one-directional stream usual with gills. This makes it unnecessary to have two respiratory openings on the surface of the body. On the debit side, breathing air presents the problem of water loss from the respiratory surface.

A respiratory surface such as a gill, which is more or less directly exposed to the outside environment, makes an unsatisfactory air-breathing organ because the surface tension of the water covering the gill membranes causes the soft gill processes to stick together, preventing the circulation of air past the surface of the gills. Some fish can survive in the air for the short periods of time needed to migrate from one pool to another, and their gill branches have extra supports that hold the feathery processes of the gills apart from one another.

During the evolution of terrestrial animals, lungs—internal structures less subject to desiccation and collapse than gills—have evolved as respiratory surfaces. Vertebrate lungs originated as outpocketings of the pharynx. They connect to the air via long narrow internal respiratory passageways, which minimize water loss from the respiratory surface of the lungs.

Air enters the body through the nose or mouth. It is more healthful to breathe through the nose because its complex passages warm, moisten, and filter the incoming air. Air passes through the pharynx, a common passageway for both air and food, and enters the **trachea** (windpipe) by way of the **larynx**, also known as the voice box or Adam's apple. The walls of the trachea contain cartilaginous rings, which hold the tube open. The inner surface of the trachea is lined with cilia, which keep the air passages clear by moving foreign particles up into the pharynx, where they can be swallowed. The posterior end of the trachea divides into two **bronchi**, which in turn divide into finer and finer tubes, the **bronchioles** (Figure 32–9). The small-

Figure 32-9 (a) The human respiratory system. (b) The bronchioles end in tiny sac-like alveoli throughout the lungs; gas exchange takes place across the moist surfaces of the alveoli.

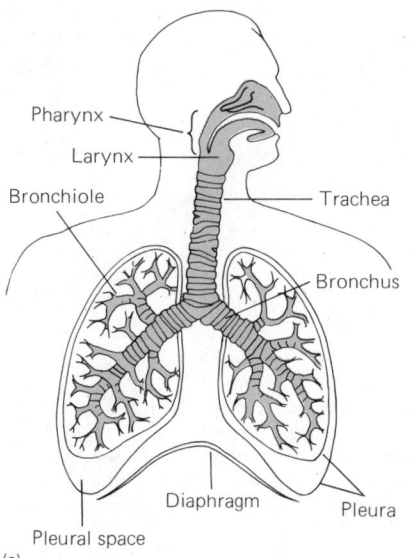

Pharynx

Larynx

Bronchiole

Trachea

Bronchus

Diaphragm

Pleura

Pleural space

(a)

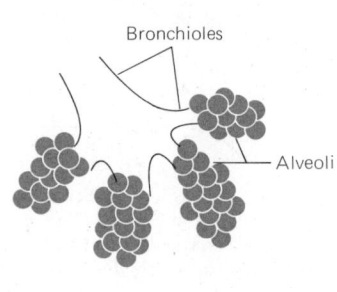

Bronchioles

Alveoli

(b)

Figure 32-10 Positive pressure breathing in a frog. Movements of the floor of the mouth force air into the lungs.

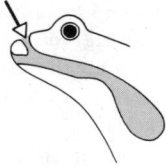 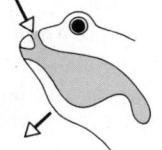 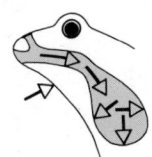 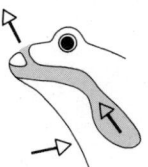

(a) Nostrils open   (b) Floor of mouth lowered   (c) Nostrils closed, floor of mouth raised   (d) Nostrils opened, abdominal muscles contracted

est bronchioles end in a myriad of tiny sacs, the **alveoli**, whose thin walls are the actual respiratory surfaces. A vast network of capillaries surrounds the alveoli; here blood picks up oxygen for transport to the rest of the body.

Much air stays in the alveoli even during maximum exhalation. This residual air keeps the moist walls of the alveoli from sticking together, or collapsing. As another protection against collapse, the surfaces of the alveoli secrete a **surfactant**, a substance that lowers the surface tension, and hence the attraction, of the fluid covering the membranes. Infants born prematurely often have not yet synthesized this surfactant and suffer from respiratory trouble as a result. The thin walls of their alveoli stick together like the sides of a wet plastic bag.

How does an air-breathing vertebrate ventilate its lungs? There are two main ways, positive pressure breathing and negative pressure breathing. **Positive pressure breathing** evolved first, and it bears some relationship to the ventilation movements of a fish. A frog breathes by a positive pressure mechanism (Figure 32-10). At the beginning of a ventilation cycle, the frog opens its nostrils and lowers the floor of its mouth. Enlargement of the mouth cavity creates a partial vacuum inside the mouth, and air enters through the nostrils. (This part of the frog's breathing cycle actually operates on the negative pressure principle, described in the next paragraph.) The frog then closes its nostrils and raises the floor of its mouth, pushing the air through the pharynx and into the lungs. After holding the air in its lungs, the frog reverses the chain of events. It opens its nostrils and contracts its abdominal muscles while relaxing the muscles of its pharynx, and the air flows back out.

Mammals have **negative pressure breathing** (Figure 32-11). They have a respiratory muscle, the **diaphragm**, not found in other vertebrates (some reptiles have an incomplete diaphragm). The diaphragm extends across the bottom of the chest cavity, beneath the lungs, so that the chest cavity is closed off from the abdominal cavity. During inhalation, the muscles between the ribs contract and lift the ribs; at the same time, the diaphragm contracts and so moves lower. These movements increase the volume of the chest cavity and correspondingly reduce its internal pressure. Air then rushes into the nose, down the trachea, and into the lungs, moving down the air pressure gradient. During exhalation, relaxation of the rib muscles and diaphragm decreases the volume of the chest cavity, increasing the

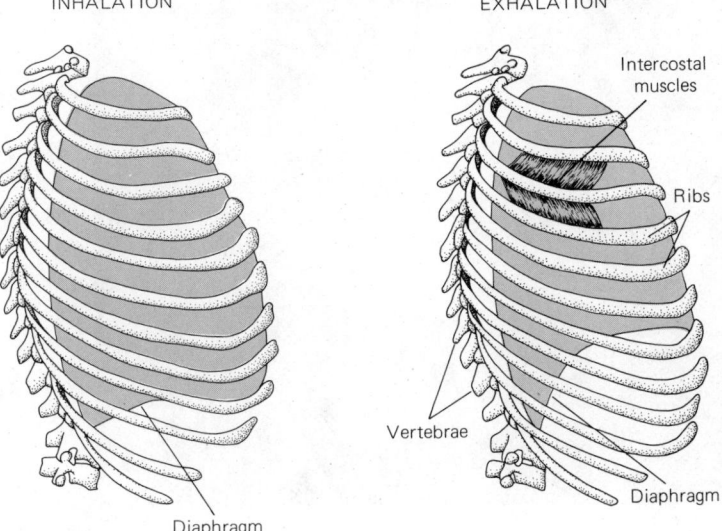

INHALATION   EXHALATION

Intercostal muscles

Ribs

Vertebrae

Diaphragm

Diaphragm

Figure 32-11 Negative pressure breathing. As the ribs are lifted up and out and the diaphragm is lowered, the size of the chest cavity increases and the pressure within it decreases. Air rushes in through the nose, mouth, or both. Exhalation occurs when the breathing muscles relax; it is a passive process.

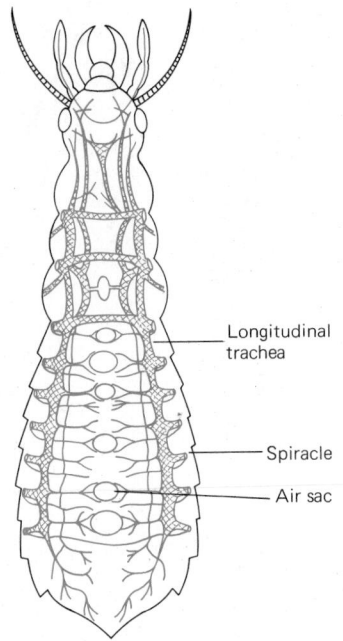

Figure 32-12 The tracheal system of an insect.

Longitudinal trachea

Spiracle

Air sac

pressure inside it and forcing air back out of the lungs. (You can get an idea of how negative pressure breathing works by closing your mouth and holding your nose while you expand your rib cage and lower your diaphragm; you will feel the partial vacuum created. Then remove your fingers from your nose and you can hear and feel the air rushing in as the pressure equalizes.)

What is the advantage of negative rather than positive pressure breathing? First, since air is highly compressible, positive pressure breathing requires a good deal of force to push any appreciable volume of air down a long trachea and into the lungs, an exertion comparable to blowing up a balloon on the end of a long tube, except that the balloon is inside the body rather than outside. This requires strong muscles to compress the air enough to push it down the trachea, and a large mouth to hold enough air at a breath to fill the lungs. Plainly this is an inefficient way to move air, and animals with such systems tend to have fairly low metabolic rates. Negative pressure breathing requires less muscular effort to move a given volume of air. Furthermore, negative pressure breathing allows an animal to eat and breathe at the same time. If it were necessary to push air from the mouth into the lungs, any food in the mouth might also be pushed into the trachea and cause an obstruction. Negative pressure breathing creates a more gentle stream of air, which is less apt to pull food along into the air passages.

Tracheal Systems

Air-breathing vertebrates have lungs. The other major group of land animals, the terrestrial arthropods (insects, centipedes, millipedes, and some spiders), breathe air by means of **tracheae** (singular: **trachea**), a system of air tubes that extend throughout the body. The tracheal system does not rely on the circulatory system to transport oxygen from a gas exchange surface to the body's cells. Instead, the tracheal system consists of a series of fine tubes that start at **spiracles**, tiny openings at the body surface, and branch into all parts of the body, ending close to every cell (Figure 32–12). Chemicals given off by cells that are short of oxygen induce tracheae to grow branches into that area.

How do these arthropods ensure a fresh supply of air in the tracheal system at all times? The smallest insects have only to open their spiracles and let the diffusion of gases do the rest. For larger or more active insects, the distances involved are too great for diffusion to be adequate. The abdominal muscles are alternately contracted and relaxed, pumping air into and out of the tubes and air sacs of the tracheal system.

Figure 32-13 A scanning electron micrograph of one of a caterpillar's spiracles. The hairlike structures prevent dust from entering the trachea. (Biophoto Associates)

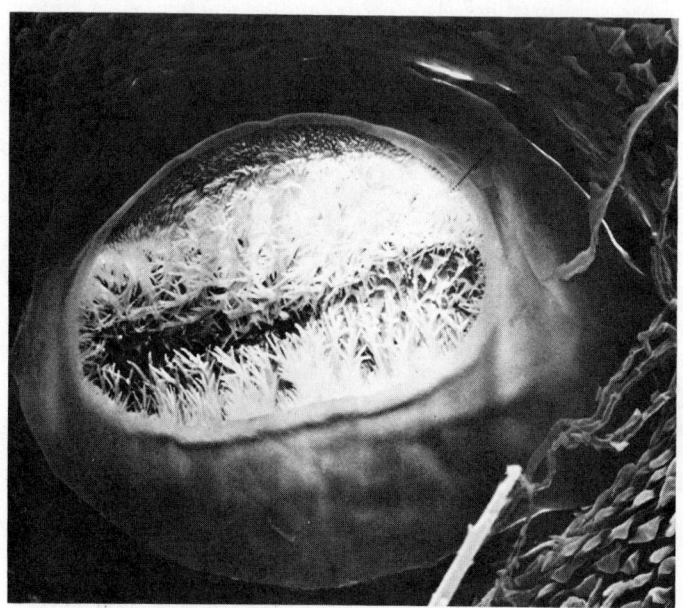

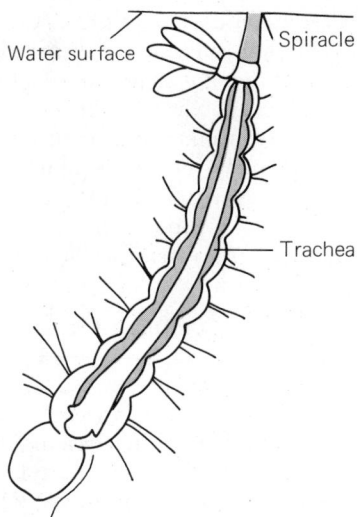

Figure 32-14 Modification of the tracheal system in a mosquito larva.

Most insects are terrestrial, but during the course of evolution some terrestrial insects have invaded freshwater habitats. These insects have adaptations for obtaining gaseous oxygen for their tracheal systems while they are submerged. Some aquatic forms, such as the naiads (immature forms) of mayflies, have gills containing closed endings of the tracheal system. As long as the water in which they live is well aerated, they can obtain oxygen by diffusion across the gill surfaces and through the membranes of the tracheal system. A mosquito larva has a special siphon tube, which encloses the tracheal tubes and has a spiracle opening on the end (Figure 32–14). The larva floats head-down just under the surface of the water, with only its siphon protruding above the surface to obtain air. If it is disturbed, the larva will dive below the surface, but it must return soon for a fresh supply of oxygen. One of the most interesting adaptations of aquatic insects is found in some diving beetles that hold an air bubble under their wings and over the openings of their spiracles. Thus, when they dive, they carry additional air supplies in their own little scuba tanks. As the oxygen in the air bubbles is used up, more oxygen diffuses in from the surrounding water.

32-E Respiratory Pigments

Water cannot carry much dissolved oxygen, especially if it also contains dissolved salts. Since the body fluids of animals also contain salts, it is not surprising that large active animals, which use lots of oxygen, have adaptations that increase the amount of oxygen their blood can carry. The most common adaptation is to possess **respiratory pigments**, large protein molecules that pick up oxygen in areas where the oxygen pressure is high and release it in places where the oxygen pressure is low. The respiratory pigment hemoglobin in the red blood cells of a mammal carries 98% of the oxygen in the blood.

Hemoglobin (Hb) is a general name for a group of oxygen-carrying compounds that have molecular weights ranging from 17,000 to several million. What all of them have in common is a heme group (Figure 32-15), with an iron atom at its center. It is this iron atom that actually binds the oxygen. When oxygen is bound to the iron atom, the pigment is said to be **oxygenated** (not oxidized) and is called **oxyhemoglobin (HbO)**. Oxyhemoglobin appears bright red; in the deoxygenated state hemoglobin is a darker, purplish red. The hemoglobin of vertebrates consists of four polypeptide chains, each with a heme group attached.

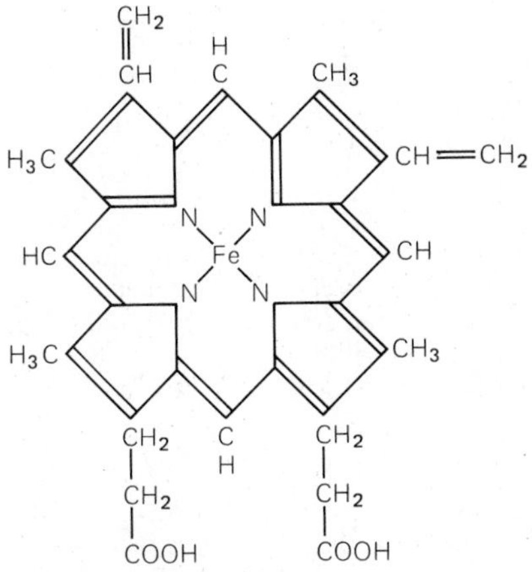

Hemoglobins are found in almost all vertebrates, and all vertebrate hemoglobins are believed to be related, descended from some common ancestral hemoglobin. Hemoglobins also occur in various invertebrates, including earthworms and some

Figure 32-15 The heme group of a hemoglobin molecule. The structure around the iron atom, called a porphyrin ring, is attached to hemoglobin protein chains.

other annelids, and in the larvae of chironomid flies that live in the murky bottoms of polluted ponds. Since these cases seem to be isolated from each other evolutionarily, it is thought that hemoglobins have evolved independently in various forms of life. Hemoglobins probably evolved from cytochromes that contained heme groups; such cytochromes are found in all aerobic organisms.

Another group of respiratory pigments is the **hemocyanins**, copper-containing compounds with molecular weights ranging from 50,000 to 74,000. When oxygen is bound to the copper, the compound appears bluish; deoxygenated hemocyanin is colorless. Many arthropods and molluscs have hemocyanins in their blood.

Oxygen Dissociation Curves

An **oxygen dissociation curve** is a graph that shows how much oxygen a respiratory pigment carries when it has been permitted to reach equilibrium with air or water containing various amounts of oxygen. In general, the higher the partial pressure* of oxygen (horizontal axis), the more nearly saturated with oxygen the pigment will be (vertical axis). Notice that the hemoglobin dissociation curve (Figure 32–16) is S-shaped. The top end of the dissociation curve shows the ability of the hemoglobin to take up oxygen from the environment; the **loading pressure** or **loading tension (T_L)** is the environmental oxygen pressure at which the hemoglobin will reach an equilibrium at which it is saturated (i.e., at which virtually all of its iron atoms carry oxygen). The **unloading pressure** or **unloading tension (T_U)** is arbitrarily defined as the oxygen pressure at which the pigment carries only half as much oxygen as it can hold when saturated.

The steep part of the curve occurs at the oxygen pressures normally present in the fluids deep within the body. At these oxygen pressures, a relatively small drop in oxygen pressure in the tissues results in a large dissociation of oxygen from hemoglobin. For example, human hemoglobin becomes 97.5% loaded with oxygen as it passes through the lungs, where the oxygen pressure is 100 to 110 mm Hg.* The oxygen pressure in the tissue fluids is about 32 mm Hg. At this oxygen pressure, hemoglobin is a little over 60% saturated, so that about 37% of the oxygen is given up to the tissue fluid (see Figure 32–16b). During exercise, the oxygen pressure in

---

*Partial pressure   The portion of air pressure that is attributable to a particular type of gas in the air.

*Hg   Chemical symbol for mercury; air pressure is commonly measured as the height to which it can push a column of mercury. Air at sea level has a pressure of 760 mm Hg = 1 atmosphere.

Figure 32-16   (a) Oxygen-hemoglobin dissociation curve for human hemoglobin. T_L = loading tension; T_U = unloading tension. (b) Human dissociation curves at pH 7.4 (black) and pH 7.2 (colored).

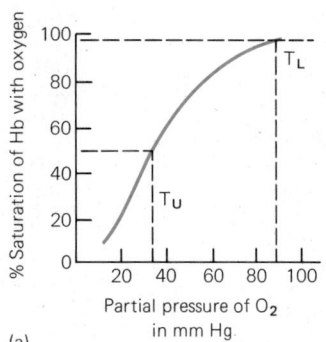

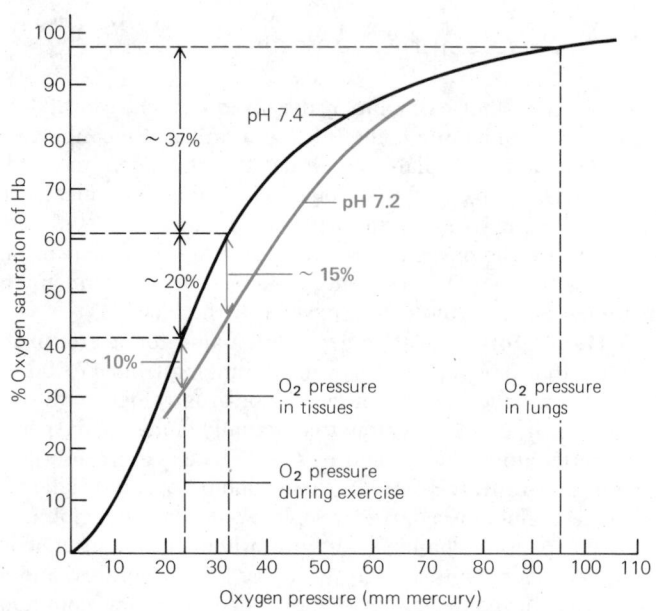

the tissue fluid falls lower than normal, and the hemoglobin releases more of its oxygen than usual.

The extent to which oxyhemoglobin dissociates is determined primarily by the oxygen pressure in the fluid around the red blood cell, but it is also influenced by pH. Oxyhemoglobin dissociates more readily in an acid environment (colored curve in Figure 32–16b). When more oxygen is used, more carbon dioxide is formed and picked up by the blood. Carbon dioxide reacts with water to form carbonic acid, making the blood more acidic and causing hemoglobin to release a greater supply of oxygen to the tissues that need it most. In resting tissues, the amount of carbon dioxide entering the blood changes its pH only a little and may cause the blood to give up only an additional 1 to 2% of its oxygen, but a tissue such as an exercising muscle releases a great deal of carbon dioxide and lactic acid and hence may cause an additional 10% dissociation of oxyhemoglobin (Figure 32–16b). When blood returns to the lungs, release of CO_2 decreases its acidity and facilitates formation of oxyhemoglobin.

Adaptations of Hemoglobin

Different vertebrates have hemoglobins with different properties. For example, the loading pressure for the hemoglobin of a given species is always well below the lower limit of oxygen pressure normally found in its environment. Thus, even at the lower limit of normal environmental oxygen pressure, the hemoglobin will still become saturated with oxygen. On the other hand, the properties of its hemoglobin limit the environments in which a species can live.

The unloading pressure reflects the oxygen pressure of the animal's tissue fluids. A small mammal has a metabolic rate higher than that of a large one, so the small mammal uses oxygen more quickly. The smaller the animal, the further its oxyhemoglobin dissociation curve is shifted to the right (Figure 32–17). In general, the smaller an animal, the higher will be the unloading pressure of its hemoglobin—that is, the more oxygen its hemoglobin will give up for a given drop in oxygen pressure. Is a mouse at a disadvantage in having its hemoglobin loading pressure so far to the right? Does this not reduce the amount of oxygen that can be picked up in the lungs? Actually, the partial pressure of oxygen in the air at sea level is about 150 mm Hg, still well above the loading pressure of mouse hemoglobin.

An animal may have hemoglobin with different properties at different times in its life. Before birth, human fetuses produce several kinds of hemoglobin that are not made by the body of an adult. After birth, fetal hemoglobin is gradually replaced by adult hemoglobin. All of the oxygen a fetus gets comes from its mother's bloodstream. If fetal and adult hemoglobins had the same dissociation curves, the fetus would not be able to pick up very much of the oxygen released by the mother's blood. But a fetus's hemoglobin has a lower loading pressure than that of the mother, permitting it to pick up oxygen at oxygen pressures that cause the mother's hemoglobin to release oxygen (Figure 32–18).

32-F Carbon Dioxide Transport

Carbon dioxide produced during aerobic respiration must be carried by the blood to the lungs, where it is excreted. Some carbon dioxide combines with hemoglobin and other proteins in the blood. Instead of attaching to hemoglobin's iron atoms, as oxygen does, carbon dioxide attaches to amino acids in the polypeptide chains. Oxygenated and deoxygenated forms of hemoglobin have different shapes; deoxygenated hemoglobin takes up CO_2 more readily, whereas oxygenated hemoglobin releases CO_2 more easily. The overall result is to facilitate transport of carbon dioxide at low oxygen concentrations (in the tissues) and cause its release where oxygen levels are high (in the lungs).

Most of the carbon dioxide in the blood is carried as bicarbonate ions dissolved in the blood plasma or in the interior of the red blood cells. Red blood cells also contain the enzyme **carbonic anhydrase**, which catalyzes the formation of car-

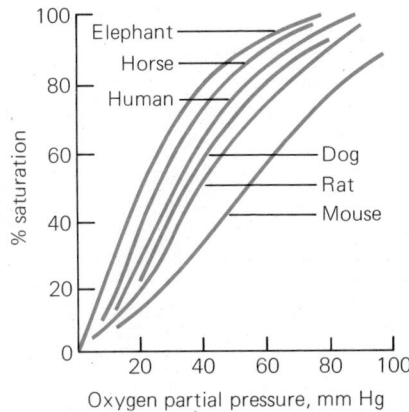

Figure 32-17 The smaller the animal, the farther its hemoglobin loading curve is shifted to the right. This allows small animals to obtain enough oxygen to sustain their high metabolic rates. Their hemoglobin unloads more of its oxygen at any given oxygen pressure than does the hemoglobin of a larger animal.

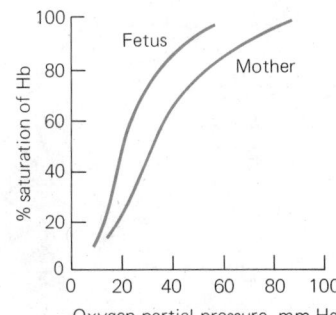

Figure 32-18 In mammals, fetal hemoglobin has a higher affinity for oxygen than does maternal hemoglobin. This allows the fetus to obtain the oxygen it needs from the mother's blood.

EQUATION 1

$$H_2O \;+\; CO_2 \;\xrightarrow[\text{Carbonic anhydrase}]{}\; H_2CO_3 \;\longrightarrow\; H^+ \;+\; HCO_3^-$$

Water   Carbon dioxide (6-7% of blood CO_2)   Carbonic acid   Hydrogen ion   Bicarbonate ion (86-88% of blood CO_2)

Figure 32-19 Transport of carbon dioxide and blood buffering. In Equation 2, the hydrogen ions formed in Equation 1 are taken up by negatively-charged proteins, notably hemoglobin, thus minimizing the change in pH. When the blood arrives in the lungs, these equations reverse (colored arrows) and carbon dioxide gas is released.

EQUATION 2

$$H^+ \;+\; HCO_3^- \;+\; K^+ \;+\; Hb^- \;\longrightarrow\; K^+ \;+\; HCO_3^- \;+\; HHb$$

Hydrogen   Bicarbonate   Potassium   Hemoglobin   Acid hemoglobin

bonic acid from water and carbon dioxide. Once formed, carbonic acid dissociates readily into hydrogen and bicarbonate ions (Figure 32–19, Eq. 1).

If all of the hydrogen ions formed by the action of carbonic anhydrase were allowed to remain in the bloodstream, the blood would be very acidic indeed. But the hydrogen ions are neutralized by another reaction, in which they combine with hemoglobin. In addition to being a respiratory pigment, hemoglobin is a negative ion that becomes associated with positive potassium ions. When the blood becomes more acidic than usual, the potassium ions associated with hemoglobin are displaced by hydrogen ions, and the hemoglobin becomes **acid hemoglobin** (Figure 32–19, Eq. 2), reducing the acidity of the blood. If the blood becomes too basic, acid hemoglobin dissociates, releasing hydrogen ions and the negative hemoglobin ion, which takes up potassium. This renders the blood more acidic, but the equilibrium of the overall reaction lies far in favor of acid hemoglobin.

When the blood returns to the lungs, it gives up some of its carbon dioxide into the air in the lungs. This reduces the concentration of carbon dioxide in the blood, altering the equilibrium between carbon dioxide and carbonic acid. The equilibrium is restored by the dissociation of carbonic acid into water and more carbon dioxide. This, in turn, alters the equilibrium of Equation 1 in Figure 32–19, so that hydrogen and bicarbonate ions combine to form more carbonic acid.

The blood gives up only about 10% of its carbon dioxide as it passes through the lungs. The other 90% is retained, mostly in the form of bicarbonate ions, which act as important blood **buffers**, substances that keep the pH from fluctuating. Therefore, although carbon dioxide is a waste product, its presence is essential in regulating the pH of the blood.

32-G Regulation of Ventilation

The oxygen and carbon dioxide pressures in the blood of vertebrates must be continually adjusted so that they stay within certain so-called physiological limits. This adjustment is made by the rate of breathing. As blood passes through the capillaries of the lungs, the gas in the blood comes into equilibrium with the gas in the alveoli; the higher the breathing rate, the greater will be the percentage of oxygen in the alveolar air, and the lower will be the percentage of carbon dioxide, because the air in the lungs is replaced more completely and more frequently. With slowed breathing the reverse is true. The breathing rate, in turn, is controlled by the oxygen and carbon dioxide content of the blood and cerebrospinal fluid (the fluid around the brain and spinal cord); specialized receptor organs monitor the pressures of these two gases in the body fluids and send nerve signals to the diaphragm and rib muscles, speeding or slowing the rate of breathing (Figure 32–20).

If the carbon dioxide content of the blood drops below a certain critical level, breathing is inhibited. By purposely hyperventilating, that is, taking several deep breaths in rapid succession, you can hold your breath longer; swimmers often do this so that they can swim underwater for a longer time. At each breath, however,

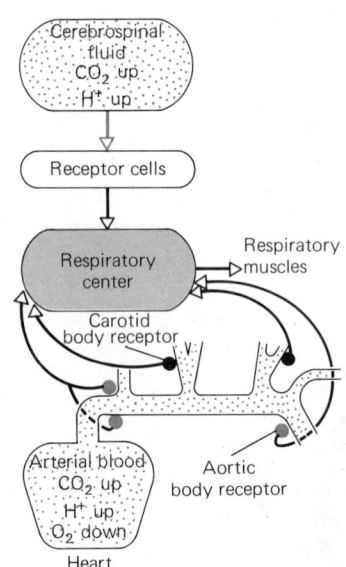

Figure 32-20 Action of CO_2 receptor cells. An increase in the percentage of CO_2 and H^+ in the blood and cerebrospinal fluid is detected by receptors that send the information, via nerves (black lines with arrows), to the respiratory center in the medulla of the brain (see Figure 38-6). The respiratory center sends impulses that stimulate the respiratory muscles, increasing the breathing rate.

Essay: Human Adaptations to High Altitude

During the sixteenth century the Spaniards conquered the Incas of Peru and lived in the high Andes. They were the first to record the adverse effects of the lack of oxygen and low atmospheric pressure at such high altitudes on people accustomed to living at lower elevations.

The fundamental problem of gas exchange at high altitude is that the air pressure is lower than it is at sea level. Thus, although oxygen is present in air in the same proportions at these high altitudes, there is less oxygen available because there is less air. Consequently, the blood picks up less oxygen, and the body experiences **hypoxia**, or "altitude sickness," with symptoms that include headaches, weakness, and breathlessness.

Human beings can adapt to living at high altitudes without hypoxia by two main types of adaptations: those of a native of the lowland living temporarily at a high altitude,

and those found in people who live and die at high elevations. The first type of adaptation was studied particularly when the Olympics were held at the high altitude of Mexico City in 1968; it consists mainly of an increase in the rate of breathing, which takes about three weeks to develop fully. This increases the oxygen pressure in the alveoli and therefore in the blood, but has the disadvantage of lowering the carbon dioxide concentration of the blood, which makes the blood more alkaline. During acclimation, the cells of the body increasingly remove bicarbonate ions from the blood and so restore the pH of the blood to normal.

People who live all their lives at high altitudes do not breathe faster than people who live at sea level, but have a different set of adaptations, which develop during childhood. A child born and raised at high altitudes develops more alveoli and more blood vessels in the lungs than does a child living at lower altitudes.

Natives of both lowlands and highlands living at high altitudes slowly produce more red blood cells, and therefore more hemoglobin. This increases the capacity of the blood to carry oxygen. When it reaches the tissues, however, the blood at high

altitudes carries oxygen at a lower pressure than at low altitudes. Two adaptations in highland natives ensure that enough oxygen reaches the cells. First, more capillaries supply blood to the tissues than in lowland people; this shortens the distance for diffusion of oxygen from the blood to the tissues. Second, the oxygen dissociation curve of a highland native lies to the left of the normal position (Figure 32–A). Thus hemoglobin has a higher affinity for oxygen and picks up and releases oxygen at the relatively lower oxygen pressures present in the alveoli and tissues at high altitudes.

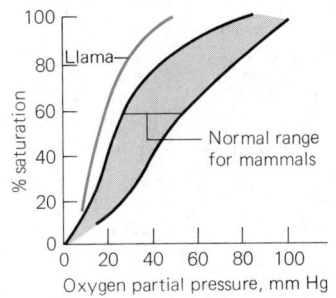

Figure 32-A Adaptation of llama hemoglobin to high altitudes. The llama's hemoglobin dissociation curve is to the left of that of most other mammals, allowing the llama to obtain adequate supplies of oxygen despite the low oxygen pressure at high altitudes.

---

the carbon dioxide content of the blood is lowered, and if it goes too far down, you lose consciousness. This may seem annoying, but it is really an important defense mechanism of the body that prevents you from lowering the carbon dioxide pressure of your blood to a dangerous level. Similarly, if you intentionally hold your breath, the carbon dioxide level in the blood rises, and above a certain level, you again lose consciousness. Once this happens, the breathing reflexes take control again, and cause you to inhale. It is impossible to kill yourself just by holding your breath, although some children have discovered that parents do not know this and can be panicked by this method.

Insects have feedback control over gas exchange in their tracheal systems. Their spiracles open and close in response to carbon dioxide levels in the tissues. Circulation of fresh air is controlled partly by pumping movements of the abdomen and sometimes by movements of the flight muscles.

The main branches of the insect tracheal system are filled with air; only a film of fluid covers the walls of these tubes (Figure 32–21). Farther into the system, though, the finer tubes (tracheoles) contain fluid, since capillary force pulls body fluids into these fine tracheal endings. If an insect becomes very active, carbon dioxide and breakdown products of food molecules accumulate, lowering the osmotic potential of the tissue fluids. In response to the lowered osmotic potential in the tissues, fluid will move from the tracheoles into the tissues, allowing air in the tracheoles to penetrate closer to the cells that need it. Since oxygen moves more quickly through air than through fluid, this mechanism speeds delivery of oxygen to the cells by leaving a shorter distance for it to travel through fluid.

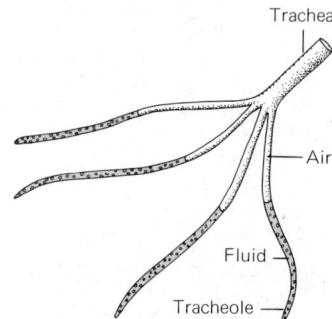

Figure 32-21 The endings of an insect trachea. The level of fluid in a tracheole varies with the insect's metabolic rate.

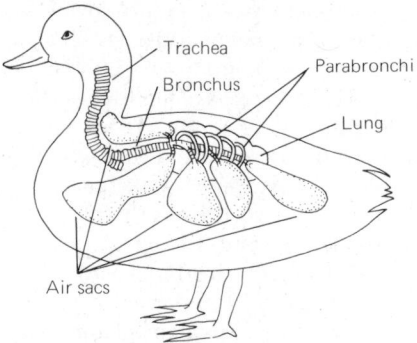

Figure 32-22 Air sacs in a bird. Branches of the bronchus connect to the air sacs. Parabronchi form channels for one-way air flow within the lung; they connect with the bronchus at both ends.

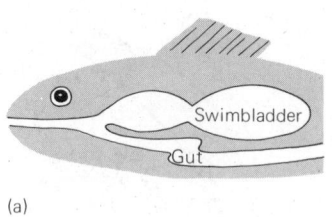

(a)

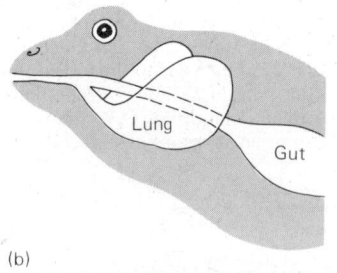

(b)

Figure 32-23 (a) The swimbladder arises as an outpouching from the dorsal part of the pharynx. (b) Lungs arose as outpouchings of the ventral part of the pharynx, although they often come to lie dorsal to the digestive tract.

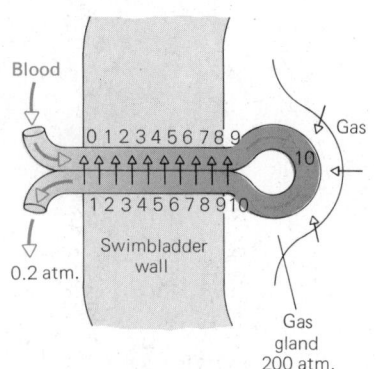

Figure 32-24 Countercurrent exchange in the rete mirabile in the swimbladder. Gas is exchanged between the long incoming and outgoing portions of capillaries, so that the blood in the body is kept in equilibrium with the low gas pressure in the water, even though the blood comes into equilibrium with the much higher gas pressure in the swimbladder as it flows through the walls of the swimbladder. Colored arrows indicate the direction of blood flow; black arrows indicate movement of gases. Black numerals in capillaries represent differences in gas pressure between outgoing and incoming blood.

32-H Air Sacs in Birds

Besides their small, compact lungs, birds also have a number of **air sacs** that extend throughout the body cavity (Figure 32–22). Unlike the lungs, these air sacs have neither alveoli nor an extensive supply of blood capillaries. The air sacs help to lower the density of the bird's body, a useful adaptation to flight, but much more importantly, the air sacs allow the bird to dissipate the tremendous amount of heat generated during flight.

The trachea, bronchi, and other air tubes in birds are so arranged that air may or may not pass through the lungs as it enters the air sacs from the outside; the route taken depends on the oxygen demands of the bird and its need to dissipate heat. It has been estimated that a pigeon in flight uses one-quarter of the air it inhales for gas exchange and three-quarters for cooling the body by evaporating water, mainly across the surfaces of the air sacs.

31-I Swimbladder Physiology

The **swimbladder**, or **air bladder**, of bony fishes is an organ in which gas is stored; in most fishes, its primary function is to provide variable buoyancy by altering the density of the fish's body. The amount of air in the bladder can be regulated so that the fish can float at a particular depth, rise, or sink, with little muscular effort. Nerve endings in the wall of the swimbladder detect the degree of stretching of the wall, and a feedback mechanism adjusts the amount of gas in the bladder.

The swimbladder arises as an outpocketing of the pharynx (Figure 32–23). Some fish still have a connection between the swimbladder and pharynx, and these fish can "spit out air" to increase their density, or swallow bubbles of air to fill the swimbladder. Other fish have no connection between the digestive tract and the swimbladder and so must use the blood to exchange gas with the swimbladder.

A swimbladder with a connection to the pharynx may function as an accessory breathing structure in fish that live in oxygen-depleted water. Such fish come to the surface to gulp air into the swimbladder; oxygen is picked up by the blood passing through the vessels in the walls of the swimbladder.

The **rete mirabile** (= miraculous net) consists of a tremendous number of hairpin-shaped capillaries lying side by side in the wall of the swimbladder. The parts of the capillaries going toward the swimbladder are pressed close to the parts coming back out (Figure 32–24). This arrangement allows for countercurrent exchange, which keeps the volume of gas in the swimbladder constant. The air in the swimbladder of a fish living at depths of 2000 to 3000 metres is under about 200 atmospheres pressure. Blood coming into the swimbladder carries oxygen at about 0.2 atmosphere (the same as the oxygen pressure in the water) and encounters air at 200 atmospheres pressure in the swimbladder. Under these conditions, the blood will pick up huge quantities of gas from the swimbladder; however, none of this extra gas must be taken away if the volume of the swimbladder is to remain constant. The countercurrent exchange system of the rete mirabile provides for this by permitting the outgoing blood to give up its extra gas to the incoming blood. Thus, by the time the incoming blood reaches the swimbladder, it is already in equilibrium with the air there, so that no more air is removed.

All animals whose cells carry out aerobic respiration must obtain oxygen from their environments and expel carbon dioxide. Since carbon dioxide is much more soluble in water than is oxygen, obtaining sufficient oxygen is the more difficult of the two. The amount of oxygen an animal needs depends largely upon its metabolic rate, which increases with activity and, in a warm-blooded animal, with increasing surface-to-volume ratio and decreasing size.

Gas exchange with the environment can take place only across a moist surface. Small or inactive animals can obtain all their oxygen by diffusion across the general body surface. A high surface-to-volume ratio is found in animals whose bodies are small, flattened, or covered with projections. Larger or more active animals usually have part of the body specialized as a respiratory surface and transport gases to and from this surface by way of a circulatory system.

The four main types of respiratory organs are the body surface, gills, lungs, and tracheal systems.

In organisms with lungs, the gas content of the blood is regulated by controlling the rate of breathing and, therefore, the gas composition in the alveoli.

Many animals have respiratory pigments in their blood that vastly increase the blood's oxygen-carrying capacity. The oxygen dissociation curve for a pigment shows the oxygen pressures at which the pigment will pick up and release oxygen. A study of these curves shows that the dissociation properties of pigments are adapted to the oxygen pressures under which they must normally operate.

Carbon dioxide is transported mainly in the form of bicarbonate ion, an important buffer that helps to maintain the pH of the body fluids at a constant level.

The lungs of vertebrates are closely related to the swimbladders of bony fishes. The swimbladder is a gas-filled organ that permits the fish to change its buoyancy; it may also be used as an accessory breathing organ.

OBJECTIVES

From your study of this chapter, you should be able to:

1. Distinguish between the processes of ventilation and respiration.

2. Discuss the advantages and disadvantages of air and water as respiratory media.

3. Describe, in qualitative terms, the relationship between metabolic rate, size (for a homeothermic animal), and environmental temperature.

4. Compare and contrast lungs and gills with respect to structure, function, and their advantages and disadvantages.

5. Explain how the countercurrent exchange mechanism works in the gills of a bony fish; state the importance of this adaptation to the fish.

6. Explain how the positive pressure and negative pressure mechanisms of breathing work, and state the advantages of a negative pressure breathing mechanism.

7. Describe the main differences in structure and function between the tracheal systems of insects and other types of respiratory systems.

8. (a) State the function of respiratory pigments; (b) draw an oxygen loading curve for hemoglobin and label both axes, the loading tension, and the unloading tension, and state their significance to the organism; (c) state the main factor governing whether hemoglobin loads or unloads oxygen; (d) state how pH influences loading and unloading; and (e) predict the change in shape and position of such a curve with a change in the animal's size or environment.

9. Compare and contrast the transport of oxygen and carbon dioxide in the blood.

10. Describe how hemoglobin, other blood proteins, and carbon dioxide buffer the blood.

11. Explain how ventilation movements are regulated in the human body.

12. List and explain two ways ventilation of the tracheal system is enhanced in an active insect.

13. Describe two functions of the swimbladder of bony fishes.

1. A cat's hemoglobin dissociation curve will lie further to the (left/right) than that of a bear.

2. Both hyperventilation and holding of one's breath can cause loss of consciousness. Under *normal circumstances*, why does this happen?
   a. alteration of carbon dioxide levels in the blood
   b. alteration of oxygen levels in the blood
   c. loss of hemoglobin from red blood cells
   d. distress of the lungs
   e. excess dissociation of oxygen from hemoglobin

3. Which of the following is a handicap to using the general body surface as a respiratory surface?
   a. having a low surface-to-volume ratio
   b. plentiful supply of water in the environment
   c. low metabolic activity
   d. thin slimy mucus covering the skin

4. A disadvantage of using air as a respiratory medium is:
   a. it carries less oxygen than water does
   b. it increases the risk of desiccation
   c. oxygen diffuses faster in air than in water
   d. air contains nitrogen as well as oxygen
   e. air pressure changes more than water pressure with changes in temperature

5. The metabolic rate of a poikilothermic animal increases:
   a. with increasing environmental temperature
   b. with decreasing environmental temperature
   c. with increase in size
   d. with decrease in muscular activity
   e. with increase in age

6. The countercurrent exchange mechanism in the gills of fish works by:
   a. running the respiratory medium and blood in the same direction
   b. running the respiratory medium in a direction perpendicular to that of the blood flow
   c. maintaining a gradient such that the respiratory medium always contains a concentration of oxygen higher than that of the blood

   d. maintaining a gradient such that the respiratory medium always contains an oxygen concentration lower than that of the blood
   e. keeping the oxygen concentration in the respiratory medium equal to that in the blood

7. The main difference between the insect tracheal system and most other types of respiratory systems is:
   a. tracheal systems do not rely on the blood to transport oxygen to the tissues
   b. insects do not ventilate their tracheal systems
   c. insects do not dispose of carbon dioxide via their tracheal systems
   d. insects exchange both carbon dioxide and oxygen via their tracheal systems
   e. oxygen need not be in solution to cross the membranes in the tracheal systems of insects

8. The main factor that determines the saturation of hemoglobin with oxygen is:
   a. oxygen concentration in the blood
   b. carbon dioxide concentration in the blood
   c. pH of the blood
   d. hemoglobin concentration in the blood
   e. breathing rate

9. As blood passes through the capillaries around the alveoli of the lungs, in which direction would you expect each of the following equations to go (i.e., toward the left or toward the right)?
   a. $Hb + O_2 \leftrightarrow HbO$
   b. $H^+ + Hb \leftrightarrow HHb$
   c. $CO_2 + H_2O \leftrightarrow H_2CO_3$
   d. $H_2CO_3 \leftrightarrow H^+ + HCO_3^-$

10. Which of the following is the main function of swimbladders in fish?
    a. adjustment of total body density
    b. gas exchange
    c. incubation of the eggs
    d. locomotion by jet-propulsion
    e. detection of chemicals in the water

QUESTIONS FOR DISCUSSION

1. Ice fish are a family of bony fishes that inhabit Antarctic waters. These fish have no hemoglobin in their blood. How do you think they are able to survive without this respiratory pigment that all other adult vertebrates possess? What characteristics would you expect a member of this family to show?

2. Many adult amphibians use gills for breathing. There are many reptiles, birds, and mammals that spend almost all their time in water. Why do you think it is that none of them has evolved so that it retains the embryonic gills and uses them for gas exchange in the adult stage?

3. The evolution of coverings for the gills necessitated arrangements for openings to and from the gill area as well as muscles to draw a current of water across the gills. Yet very few animals have gills without some sort of covering. What is the advantage of a gill covering that has selected for evolution of the covering plus all these accessory arrangements?

4. Do you consider that the respiratory surfaces of your lungs are exposed to the environment? Why or why not?

5. A long-term smoking habit can destroy the cilia in the lining of the air passages. How might this affect a person's health?

6. Draw a hypothetical dissociation curve for frog hemoglobin, and on the same graph draw the curve you would expect for frog tadpole hemoglobin. Explain why you drew your curves as you did. (Tadpoles live in water and metamorphose into air-breathing frogs.)

7. Which would you expect to have larger swimbladders, marine fish or freshwater fish? Why?

8. Flounders, sole, and other fish that live on the bottom of the sea lack swimbladders. How can you account for this?

9. Mackerel and other fish that spend their entire lives swimming actively in the upper layers of the ocean also lack swimbladders. How can you account for this?

REFERENCES AND FURTHER READING

Altman, P., and D. Dittmer, eds. *Respiration and Circulation*. Bethesda, MD: Federation of American Societies for Experimental Biology, 1970. An authoritative book; good chapters on gas exchange, particularly in vertebrates.

Avery, M. E., N-S. Wang, and H. W. Taeusch, Jr. "The lung of the newborn infant." *Scientific American*, April 1973. Describes the physiological changes in the lung just before birth and efforts to speed these changes in premature infants.

Comroe, J. H., Jr. "The lung." *Scientific American*, February 1966. Anatomy and physiology of the human lung and respiratory tract, describing physiological measurement techniques and applications.

Hughes, G. M. *Comparative Physiology of Vertebrate Respiration*. Cambridge, MA: Harvard University Press, 1965. A short work treating various aspects of vertebrate respiration.

CHAPTER 33

ANIMAL TRANSPORT SYSTEMS

The earliest organisms lived in the sea, which provided them with oxygen, carried away carbon dioxide and other wastes, and surrounded them with an environment of relatively constant temperature and chemical composition. The body fluids of a higher animal may be likened to a tiny captive sea, because they perform the same functions inside the animal's body that the sea outside performs for more primitive organisms.

The cells of a multicellular animal live in an environment of **extracellular fluid (ECF)**; cells obtain food and oxygen from the ECF and discharge their wastes into it. If the ECF is to remain a favorable cellular environment, it must constantly receive fresh food and oxygen, and be cleansed of its wastes. These are the roles of the transport system.

Even the smallest animal must have a means of transporting substances within its body. Oxygen must be moved from the environment into the body, and then to the ECF around each cell; food molecules must move from the site of digestion to the ECF; and waste products must be removed from the ECF and expelled into the environment. Various fluid systems, often called **vascular systems**, facilitate such transport in most members of the animal kingdom. A **circulatory system** is a vascular system in which the transport fluid moves rhythmically in a particular direction, usually because it is propelled by a muscular pumping structure.

The evolution of an efficient transport system permitted cells to obtain and use food and oxygen more rapidly, and so to respire faster, than would otherwise have been possible. An animal's metabolic rate* gives an indirect measure of how efficiently its circulatory system supplies oxygen and food for cellular respiration. A higher metabolic rate is advantageous because it permits an animal to move faster. Table 33–1 gives metabolic rates for a variety of animals.

In this chapter we shall start with relatively simple animals and follow the increasing complexity in transport systems of animals farther up the phylogenetic (evolutionary) scale. We shall then consider the mammalian circulatory system in some detail.

In addition to transporting gases, food, and wastes, the circulatory system carries hormones, as well as molecules and cells that help to protect the body from disease (see Chapter 34). Another important, and indeed inevitable, function performed by circulatory systems is the distribution of heat; many animals can control their exchange of heat with the environment by adjustments within the circulatory system. At the end of this chapter we shall consider how animals regulate their body temperatures.

33-A Transport in Invertebrates

Cnidaria

Cnidarians are slow-moving creatures with low metabolic rates (see the figures for the sea anemone and jellyfish, Table 33–1). They obtain most of their oxygen through the body's thin outer layer of cells; food is transported by the gastrovascular cavity, which does double duty as a digestive (gastro) and transport (vascular) system.

In *Hydra* the gastrovascular cavity occupies the center of the body cylinder and extends into each tentacle (Figure 33–1a). In the much larger jellyfish *Aurelia*, the cavity extends into a series of radiating canals, which branch at intervals so that nutrients reach tissues all over the body (Figure 33–1b). The fluid in the canal system is pushed in defined pathways, propelled by cilia lining the canals. Wastes are released across the general body surface.

The metabolic rate of a jellyfish increases when it swims or feeds, but the contraction of its muscle fibers during these activities also speeds the flow of fluid through the canals, automatically speeding delivery of food as the demand increases.

---

* Metabolic rate   The rate of the total of an organism's biochemical reactions; usually measured as the rate of oxygen consumption by respiration, since respiration produces the energy needed for the other biochemical processes.

TABLE 33–1 **METABOLIC RATES OF SOME ANIMALS**†

ANIMAL	O_2 CONSUMPTION ($\mu l/g$ body weight/hr)
Sea anemone	15
Jellyfish	5
Planarian	75
Earthworm	60
Snail	250
Octopus	280
Squid	320
Crab	80
Cockroach	450
Butterfly–resting	600
–flying	100,000
Starfish	400
Sea urchin	15
Sea squirt	5
Goldfish	420
Trout	225
Frog	55
Iguana	60
Parakeet–resting	4,500
–flying	22,000
Mouse–resting	2,000
–running	20,000
Dog	360
Human	200

†At typical environmental temperatures.

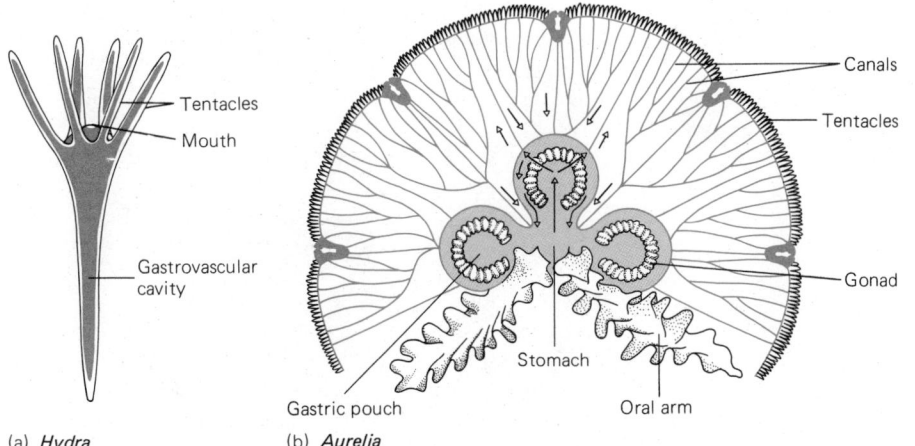

(a) *Hydra*   (b) *Aurelia*

Figure 33-1 Gastrovascular cavities (color) of two cnidarians. Arrows show movement of digested food in *Aurelia*.

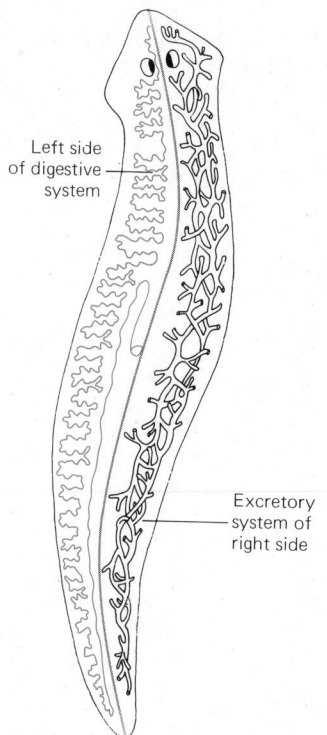

Left side of digestive system

Excretory system of right side

Figure 33-2 In planarians, the digestive system (color) and excretory system (outlined in black) branch throughout the body and perform their separate functions. Only half of each system is shown.

Planaria

Free-living flatworms have higher metabolic rates than cnidarians. However, because they have flattened bodies, they can still rely on the general body surface for gas exchange, and they have no separate circulatory system or blood. Food is distributed by the digestive cavity, which branches and rebranches into all parts of the body, providing a large surface area from which food can be absorbed.

The excretory system of a planarian also branches throughout the body and collects waste substances that must be expelled (Figure 33-2).

Annelida

A major advance of the annelids was the evolution of a fluid-filled body cavity, the coelom, which permitted the various muscular organs to move independently of one another (see Section 28-A). Although circulation in earthworms is still influenced by movement of the body muscles, the beating of the five pairs of muscular hearts (enlarged blood vessels) in the coelomic cavities moves the blood even while the animal is at rest. The earthworm's hearts provide much faster circulation than that produced by the ciliary action of cnidarians.

An earthworm has a simple **closed circulatory system**, in which the blood never leaves the vessels (Figure 33-3). Exchange between the blood and the extracellular fluid bathing the cells takes place across the thin walls of the blood vessels.

Another adaptation contributing to the efficiency of transport in earthworms is the presence of a respiratory pigment similar to the hemoglobin of vertebrates. This allows the blood to transport more oxygen per unit of volume, decreasing the amount of blood that must be moved in order to meet the tissues' oxygen demands.

What is the advantage of having a circulatory system that moves the blood steadily at all times? First, it permits much greater division of labor among organs and tissues. In platyhelminths, which have no circulatory system, the gut and the excretory organs branch throughout the body, serving as their own circulatory systems. In an annelid, by comparison, the gut is simple and unbranched, specialized only for digestion. The separate circulatory system carries the digested food to all of the body cells.

The localization of function in specific organs is one of the major reasons for the physiological efficiency of higher animals. An outstanding example is the clumping of nerve cells to form a brain; information is processed more rapidly and the animal responds faster than it could if messages had to travel around a nerve network scattered throughout the body. However, since nerve cells are easily damaged

Figure 33-3 (a) Arrangement of the major blood vessels in several segments of an earthworm. (b) Some blood vessels found in a single segment. The blood of an earthworm flows in a definite, one-way circuit.

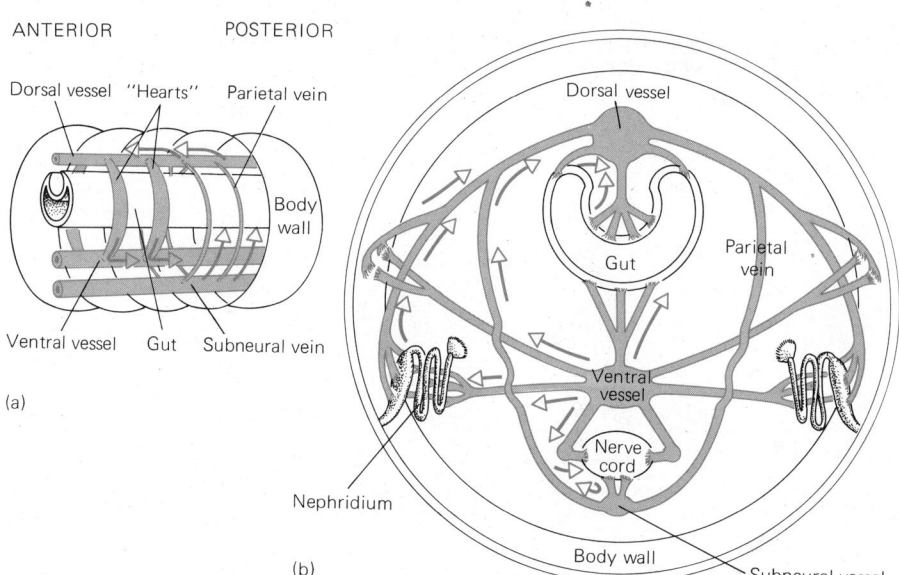

ANTERIOR   POSTERIOR

Dorsal vessel   "Hearts"   Parietal vein

Body wall

Ventral vessel   Gut   Subneural vein

(a)

Dorsal vessel

Gut

Parietal vein

Ventral vessel

Nerve cord

Nephridium

Body wall

Subneural vessel

(b)

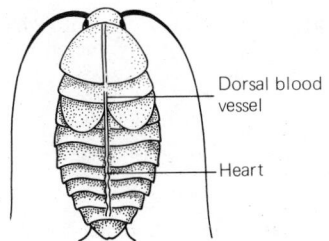

Figure 33-4 The heart of a cockroach can be seen just beneath the exoskeleton of the dorsal body surface.

Dorsal blood vessel

Heart

by temporary shortages of food and oxygen, a brain with many nerve cells is possible only when there is an efficient circulation to supply these needs at a rapid rate.

Insects

The circulatory system of an insect is of peculiar interest. Since many insects are as active as vertebrates (and much more active than most other invertebrates), we would expect to find a very efficient circulatory system, probably consisting of closed vessels with a powerful heart. However, in a dissected insect, even a large one such as a big cockroach, it is difficult to see any trace of a circulatory system. Surprisingly, most insects have **open circulatory systems**, with very few blood vessels. The blood (also called **hemolymph**) is not distinct from the other body fluids, as it is in most animals. Instead, the entire body cavity, called a **hemocoel**, is filled with hemolymph.

The blood of an insect is moved by contractions of a long, thin-walled heart, lying just below the dorsal* surface of the exoskeleton (Figure 33–4), and by movements of the body muscles. The heart pushes blood forward and out through a vessel that usually ends in the head, near the brain. In many insects, auxiliary hearts at the bases of appendages (wings, antennae, legs, and so forth) pump blood into channels that extend out into the appendages.

How can such an open circulatory system work efficiently enough to supply an insect's needs? The secret probably lies in the fact that the hemocoel is not the main system for the transport of oxygen. The high blood pressure and rapid circulation of mammals are necessary because some mammalian organs, such as the brain, need vast amounts of oxygen. However, food, hormones, and waste materials need not be moved with the same level of speed and efficiency. In an insect, oxygen reaches cells by way of the **tracheal system**, a series of air-filled tubes completely separate from the circulation (Figure 32–12), and other substances can travel more slowly via the hemolymph.

In a closed circulatory system, the blood moves in one direction along a defined route. In an open system, such as that of an insect, there is still a predictable pattern of flow, and the fluid in a tissue generally moves in one direction, rather than sloshing back and forth at random. In addition, the lack of blood vessels means that the tissues are bathed directly in blood; there is no vessel wall for food to traverse on its way to the cells. Furthermore, although the heart may provide the main impetus to the blood in a resting insect, the movement of other muscles speeds circulation greatly in an active insect.

33-B Circulation in Vertebrates

All vertebrates have closed circulatory systems, and exchange of substances between the blood and the extracellular fluid occurs only across the thin walls of the **capillaries**, the finest blood vessels. Contractions of a strong, muscular heart exert the pressure needed to force the blood through the fine tubes of the capillaries. Blood travels from the heart to the capillaries through large blood vessels called **arteries** and returns from the capillaries to the heart through **veins**.

---

* Dorsal   Toward the back, or uppermost surface, of an animal.

CHAPTER 33
ANIMAL TRANSPORT
SYSTEMS

517

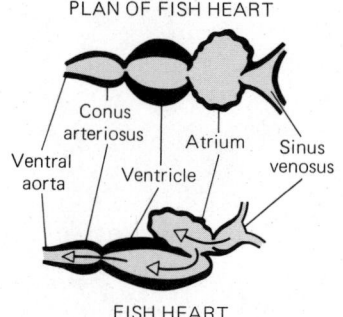

PLAN OF FISH HEART

Conus arteriosus

Ventral aorta

Ventricle

Atrium

Sinus venosus

FISH HEART

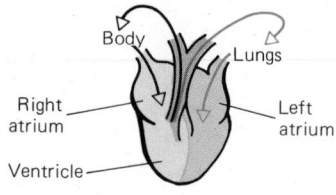

AMPHIBIAN HEART

Body

Lungs

Right atrium

Left atrium

Ventricle

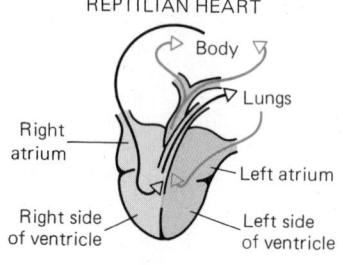

REPTILIAN HEART

Body

Lungs

Right atrium

Left atrium

Right side of ventricle

Left side of ventricle

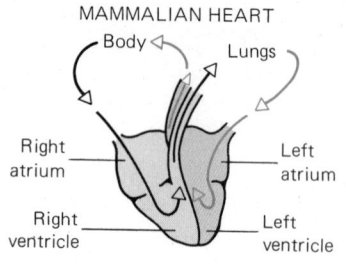

MAMMALIAN HEART

Body

Lungs

Right atrium

Left atrium

Right ventricle

Left ventricle

Figure 33-5 Blood flow through the hearts of vertebrates. (Gray indicates deoxygenated blood; color indicates oxygenated blood.)

Fish

In fishes, the heart consists of a series of four chambers, which collect the blood and then pump it out (Figure 33–5). Blood leaves the heart via the short, muscular **ventral* aorta**, and travels to the gills, where it picks up oxygen. From the gill capillaries, blood flows into the **dorsal aorta**, whose branches distribute blood to the capillaries of all the body organs (Figure 33–6). Blood returns to the heart through the veins.

The type of circulation in which blood passes through the heart only once in a complete circuit around the body is known as a **single circulation**. Such a circulatory system has the advantage that all of the blood going to the body has already been oxygenated in the gills. A disadvantage is that the narrow gill capillaries offer considerable resistance to the passage of blood, so that blood leaving the gills is at a much lower pressure than when it entered. Thus, no matter how hard the heart pumps, the blood traveling in a fish's dorsal aorta is at a relatively low pressure, since it has had to pass through the capillaries in the gills; this slows the rate of delivery of oxygen to the cells and limits the metabolic rate that fish can attain.

Amphibians and Reptiles

In higher vertebrates, the problem of low blood pressure in the body capillaries is overcome by a **double circulation**, which passes blood through the heart twice in a complete circuit. Blood is first pumped from the heart to the lungs, where it is oxygenated. As blood passes through the heart for the second time, its pressure is raised again before it goes out to the capillary beds in the rest of the body. The hearts of birds and mammals are divided into two sides, right and left; each side has an atrium (receiving chamber) and a ventricle (pumping chamber). The right side of the heart receives deoxygenated blood from the body and sends it to the lungs, and oxygenated blood returns to the left side of the heart, which then sends it to the body (Figures 33–5, 33–6).

---

* Ventral   Toward the underside, or belly, of an animal.

Figure 33-6 (left) The single circulation of fish. Blood passes through the heart once during each circuit of the body. (right) In the double circulation of birds and mammals, blood must pass through the heart twice before it returns to the same point.

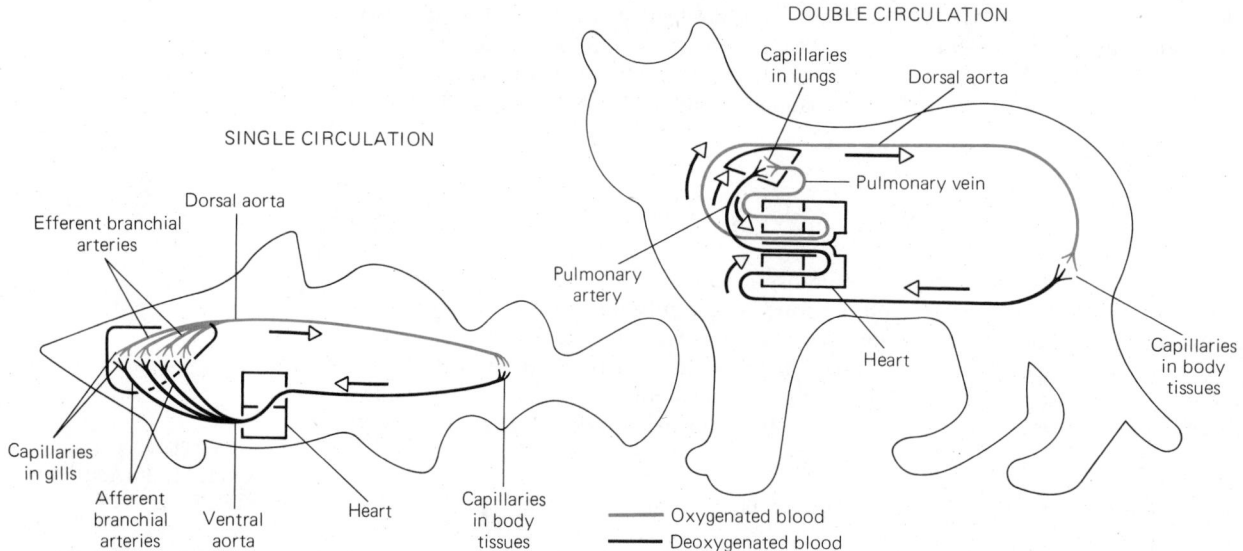

SINGLE CIRCULATION

Dorsal aorta

Efferent branchial arteries

Capillaries in gills

Afferent branchial arteries

Ventral aorta

Heart

Capillaries in body tissues

DOUBLE CIRCULATION

Capillaries in lungs

Dorsal aorta

Pulmonary vein

Pulmonary artery

Heart

Capillaries in body tissues

Oxygenated blood
Deoxygenated blood

The hearts of amphibians and reptiles (except crocodilians) are not fully divided into two halves (Figure 33–5). In amphibians, some blood that has just come from the lungs returns to the lungs instead of passing out to the rest of the body each time the heart contracts. This is not so inefficient as it looks; most amphibians absorb more oxygen through the skin than through the lungs or gills. Thus the **cutaneous** veins returning to the heart from the skin often contain more oxygen than the **pulmonary** veins from the lungs, and there would be little advantage in keeping the blood from these two sources separate.

In reptiles the circulatory system is more advanced. The blood obtains oxygen only in the lungs, and is immediately returned to the heart via the pulmonary vein. There are two atria, but in most reptiles the ventricle is only partially divided. However, physiological tests have shown that the valves in the ventricle work in such a way that there is little mixing of blood from the two sides. Thus, the heart is functionally, but not anatomically, divided. In reptiles, blood travels from the heart to the body via paired dorsal aortae, one on each side of the body.

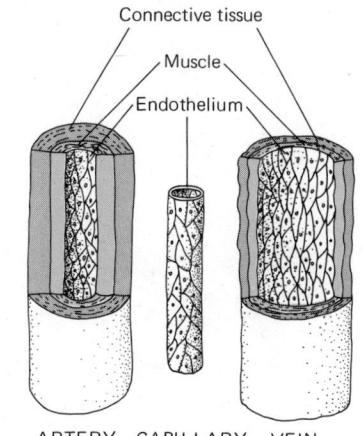

Figure 33-7 Blood vessels (not to scale).

Mammals and Birds

Blood leaving the heart for the body by way of two aortae, as it does in reptiles, loses pressure faster than it would if it were traveling in one large vessel. Thus it is not surprising to find that in mammals and birds, with their very high metabolic rates, one of the pair of dorsal aortae has disappeared. This adaptation apparently arose independently in mammals and birds while they were evolving from two different groups of reptiles. Birds have retained the right, and mammals the left, of the reptilian paired aortae.

Both birds and mammals have double circulations with the ventricles completely separated. This undoubtedly evolved under the influence of two selective pressures. First, keeping oxygenated and deoxygenated blood separate in the heart ensures that blood reaching the body organs from the aorta contains as much oxygen as possible. Second, to animals with a high metabolic rate, it is important for the blood in the aorta to be under considerable pressure. The blood loses pressure as it passes through the capillaries of the respiratory organ. Returning it to the heart after it passes through the lungs permits the heart to raise the pressure again before the blood goes out to the rest of the body. Higher blood pressure means faster circulation; oxygen and food reach the tissues faster, and waste is removed more rapidly.

33-C The Mammalian Circulation

Blood Vessels

Arteries, capillaries, and veins are the pipes through which blood travels to the tissues (Figure 33–7). Arteries are vessels that carry blood away from the heart; their walls are muscular and highly elastic. The arteries branch and rebranch into smaller **arterioles**, which divide even further into capillaries.

Capillaries are so small that red blood cells must pass through them in single file. The capillary walls are only one cell thick, in keeping with their role as the sites where substances pass between the blood and the extracellular fluid. The far ends of capillaries rejoin to form larger vessels called **venules**, and these finally combine to form **veins**, blood vessels leading back to the heart. The walls of veins contain connective tissue and muscle, as do arteries, but veins are much less elastic, and tend to have larger internal diameters.

Another important difference between veins and arteries is that veins contain **valves**, flaps of tissue that help to keep blood flowing in one direction. These valves open under the pressure of blood going toward the heart, and close when it begins to go backward (Figure 33–8).

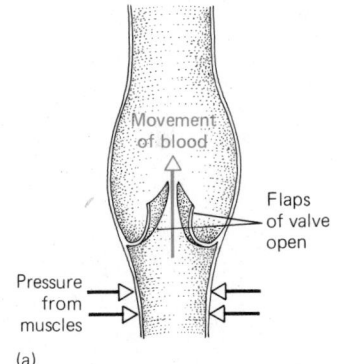

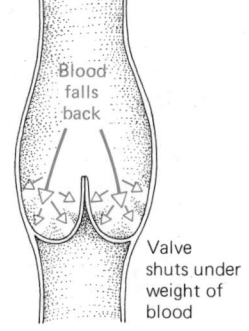

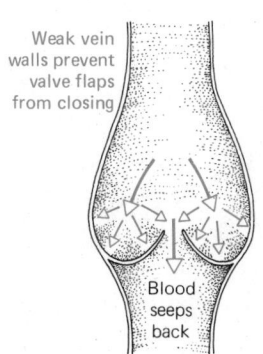

Figure 33-8 Valves. (a) When blood flows toward the heart, the valve opens and allows it to pass. (b) If blood moves in the reverse direction, it fills the cuplike flaps of the valve and presses the edges together, preventing backflow. (c) The walls of a varicose vein are weak and allow blood to collect and distend them so that the edges of the valve flaps cannot meet; blood may then return through the valve, and circulation is impaired.

519

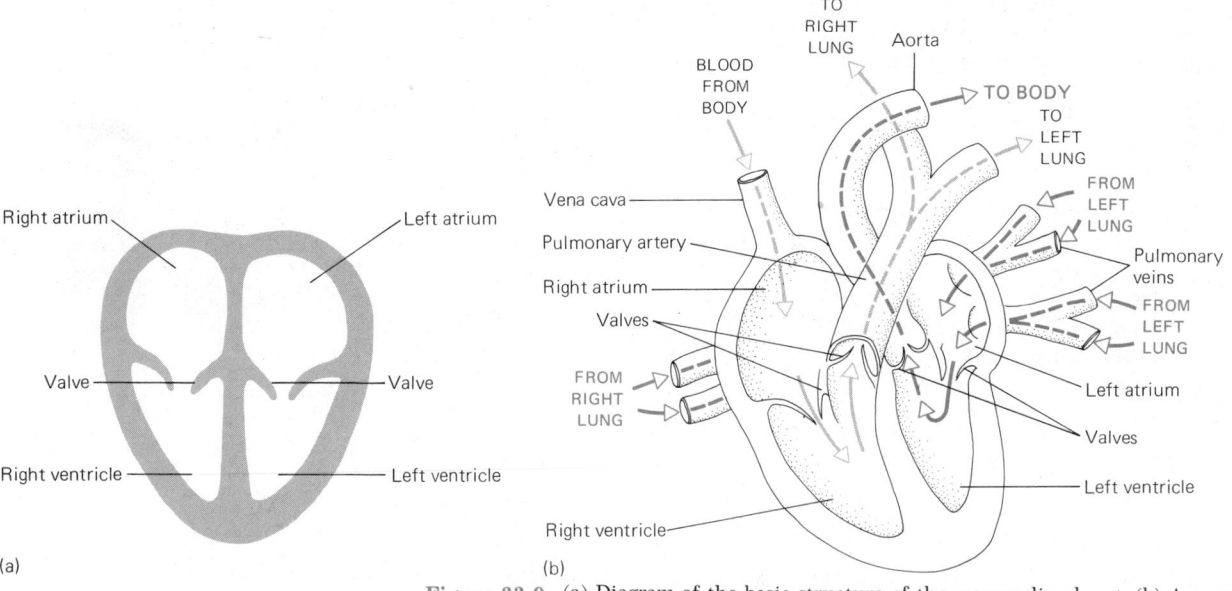

Right atrium

Left atrium

Valve

Valve

Right ventricle

Left ventricle

(a)

TO
RIGHT
LUNG

Aorta

BLOOD
FROM
BODY

TO BODY

TO
LEFT
LUNG

Vena cava

FROM
LEFT
LUNG

Pulmonary artery

Pulmonary
veins

Right atrium

FROM
LEFT
LUNG

Valves

FROM
RIGHT
LUNG

Left atrium

Valves

Left ventricle

Right ventricle

(b)

Figure 33-9 (a) Diagram of the basic structure of the mammalian heart. (b) A more realistic view of the heart, showing the major associated blood vessels. (In both views the owner of the heart is facing us, so that its right is on our left and so forth.)

When the walls of a vein are weakened, blood may collect in the vein and so distend it that the valve flaps cannot meet. Since the valve cannot now prevent blood from flowing backwards, pools of blood collect in the weakened vein. Such **varicose veins** can be very painful if the weakness is in a large vein. **Hemorrhoids** are varicose veins in the walls of the rectum; these veins have been damaged by pressure due to conditions such as constipation or pregnancy.

Like the veins, the heart is provided with valves that direct the flow of blood in a one-way path. Valves between the atria and ventricles prevent backflow of the blood into the atria when the ventricles contract, and valves between the ventricles and arteries prevent blood from falling back into the heart when the ventricles relax after pumping the blood out (see Figure 33–9).

The Circuit of Blood in the Body

Closed blood vessels, with strategically placed valves preventing backflow, ensure that the blood of a vertebrate flows in only one direction and in definite channels. Blood returns to the heart from the body via two large veins, the **venae cavae**, flows through the right atrium, and continues on into the right ventricle (Figure 33–9). From the right ventricle, blood passes through a valve into the **pulmonary artery**, which carries it to the lungs. In the lungs the blood flows through capillaries surrounding the **alveoli**, the air sacs of the lungs. Blood picks up oxygen and loses carbon dioxide across the thin walls of the alveoli and lung capillaries. The blood then flows through venules and veins that eventually join to form the **pulmonary veins**, which carry freshly oxygenated blood back to the heart. This time, though, the blood enters the left atrium and passes through the valve into the left ventricle. When the left ventricle contracts, the oxygenated blood passes through a valve into the **aorta**, the main artery to the body. The wall of the left ventricle is much thicker and more muscular than the wall of the right ventricle; it must push the blood throughout the body, not just on the short journey to the lungs.

The aorta gives rise to many branch arteries, which take blood to the wall of the heart itself **(coronary arteries)**, to the head **(carotid arteries)**, and to the digestive system, kidneys, and all of the other organs (Figure 33–10).

Blood leaving capillaries in the body enters venules and finally veins, all of which empty into the venae cavae before they join the right atrium of the heart.

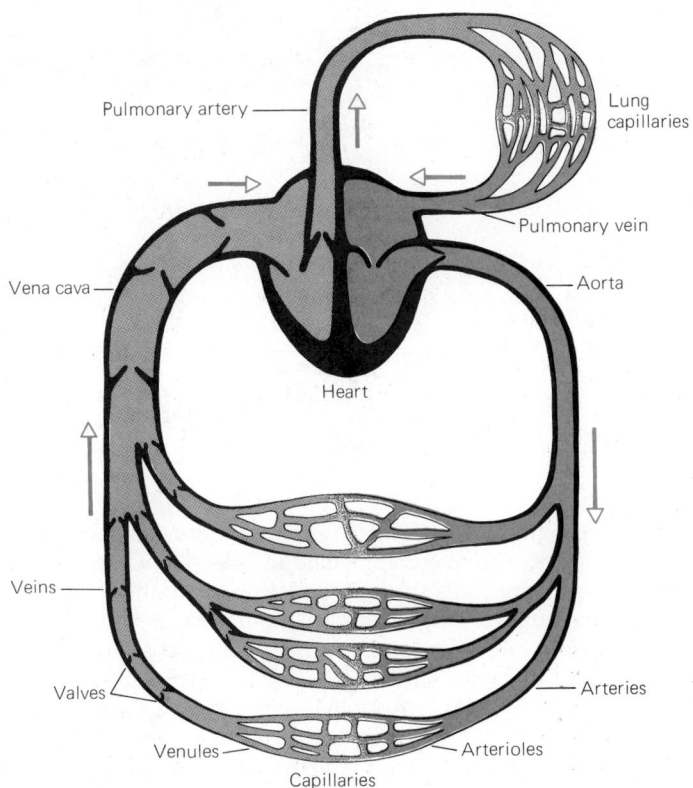

Figure 33-10 Basic scheme of the circulatory pathway in a mammal. The heart and its blood vessels have been somewhat rearranged for clarity. Gray represents deoxygenated blood; color represents oxygenated blood.

The Heart Cycle

The heart beats continuously throughout an animal's life. Indeed, the beating of the heart is so crucial that its cessation has long been taken to indicate death. Each heartbeat is initiated by a "pacemaker," a small mass of tissue known as the **sinoatrial node**, located at the entrance to the right atrium. Control of the heartbeat is discussed in Section 40-B.

When the ventricles contract, they exert considerable pressure on the blood. This point in the heartbeat is known as **systole**, and the pressure of the blood during ventricular contraction is known as the **systolic pressure**. Following contraction, the heart relaxes and blood rushes in from the venae cavae and pulmonary vein, partially filling the ventricles. This part of the heartbeat cycle is known as **diastole**, and the blood pressure at this time is called the **diastolic pressure**.

The brachial artery in the arm, just above the elbow, is usually used for measuring blood pressure. Blood pressure is expressed as the ratio of the systolic pressure over the diastolic pressure (both measured in mm Hg); e.g., 120/75 is considered the "average" blood pressure for a man of 20. Systolic pressure indicates the force with which the left ventricle pushes blood. Diastolic pressure indicates the resistance of the blood vessels; it is useful in diagnosing hardening of the arteries or strain on their walls.

Heart sounds may be heard through a stethoscope as the heart beats. The first sound, often referred to as "lubb," is heard when the ventricles contract. A higher-pitched, shorter, sharper sound, "dup," is made by the snapping shut of the valves between the ventricles and the pulmonary artery and aorta.

Blood Pressure and Circulation

Blood flows through the circulatory system from the region of highest pressure, the contracting ventricles, to the region of lowest pressure, the entrance to the heart. Blood pressure is determined mainly by the heart rate, the volume of blood expelled at each stroke, and the resistance of the blood vessels to the flow of blood; the greater any of these, the higher the blood pressure.

TABLE 33-2	CIRCULATION IN THE HUMAN BODY		
	VOLUME *(cc)*	*VELOCITY* *(cm/sec)*	*PRESSURE* *(mm Hg)*
Aorta	100	40	100
Arteries	325	40–10	100–40
Arterioles	50	10–0.1	40–25
Capillaries	**250**	**less than 0.1**	**25–12**
Venules	300	less than 0.3	12–8
Veins	2200	0.3–5	10–5
Vena cava	300	5–20	2

As the ventricles push blood into the arteries, the artery muscles relax and the elastic walls expand to accommodate the blood flowing through them. As the blood passes, the artery walls contract and exert pressure on the blood. Blood pressure decreases steadily in the arterioles, capillaries, venules, and veins (Table 33-2). Blood pressure falls mainly because the vessel walls exert frictional resistance to the movement of the blood; although the main arteries help the heart to push the blood along, the walls of the other blood vessels offer only resistance to its flow. The narrower the vessel, the greater the resistance, and so the speed of blood flow drops almost to nothing in the capillary beds. The veins are flabby and do not help to push the blood back to the heart. In addition, blood in the veins below the heart must be returned against the pull of gravity. Thus, blood tends to collect in the veins. In humans, the veins normally contain 2200 cc* of blood, and all of the other blood vessels together contain only about 1300 cc.

Blood in the veins obviously does return to the heart sooner or later, propelled mainly by the muscles of the body. When muscles contract, they squeeze against the outsides of veins, forcing blood along. The valves in the veins permit blood to flow only towards the heart; when the muscles again relax, the valves keep the blood from falling back. Since muscular contraction is needed to push blood through the veins, it is more tiring to stand still than to walk for an equal period. Standing allows blood to collect in the veins of the feet and legs. The feet swell with stranded blood, and the body temporarily loses the use of blood that should be distributing oxygen and nutrients to other tissues. Recent studies have shown that students who jiggle their feet are more alert and perform better on long exams than their peers who sit still.

* cc Cubic centimetre; the same volume as one millilitre.

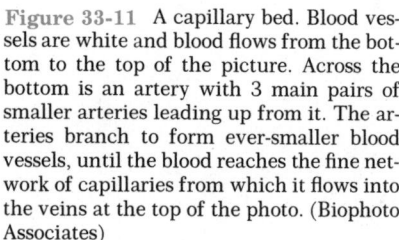

Figure 33-11 A capillary bed. Blood vessels are white and blood flows from the bottom to the top of the picture. Across the bottom is an artery with 3 main pairs of smaller arteries leading up from it. The arteries branch to form ever-smaller blood vessels, until the blood reaches the fine network of capillaries from which it flows into the veins at the top of the photo. (Biophoto Associates)

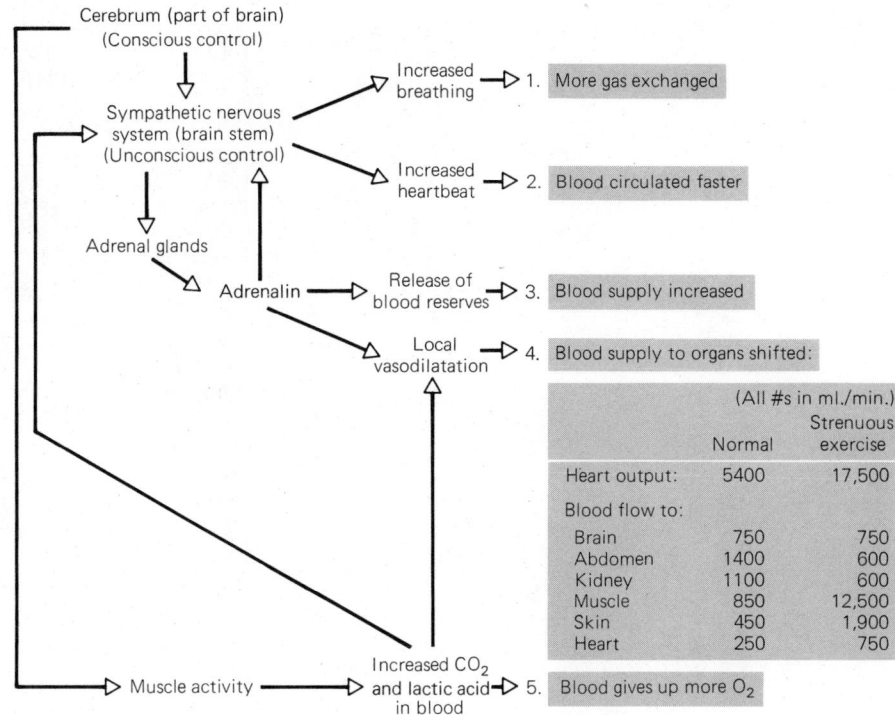

Figure 33-12 Some of the human body's responses to exercise.

Cerebrum (part of brain)
(Conscious control)

Sympathetic nervous system (brain stem)
(Unconscious control)

Increased breathing ⟹ 1. More gas exchanged

Increased heartbeat ⟹ 2. Blood circulated faster

Adrenal glands

Adrenalin ⟹ Release of blood reserves ⟹ 3. Blood supply increased

Local vasodilatation ⟹ 4. Blood supply to organs shifted:

Muscle activity ⟶ Increased CO_2 and lactic acid in blood ⟹ 5. Blood gives up more O_2

	(All #s in ml./min.)	
	Normal	Strenuous exercise
Heart output:	5400	17,500
Blood flow to:		
Brain	750	750
Abdomen	1400	600
Kidney	1100	600
Muscle	850	12,500
Skin	450	1,900
Heart	250	750

In the arteries, arterioles, and capillaries, the blood pressure is not always constant. Muscles in the walls of a blood vessel may contract or relax, changing the vessel's diameter, and this changes the blood pressure by making it harder or easier for the blood to pass through the vessel. If the heartbeat rate is increased, or if the volume of blood pumped per beat is increased, there is a general increase in blood pressure.

The Circulatory System's Adjustments to Exercise

The circulatory system adjusts in various ways to changes in physiological conditions. These adjustments are usually controlled by negative feedback* systems, which ensure that the composition of the extracellular fluid remains almost constant. We shall consider, as an example, some of the circulatory system's responses to vigorous exercise.

As exercise begins, the nervous system sends impulses to the **adrenal glands**, anterior to the kidneys, causing them to release the hormone **adrenalin** into the bloodstream. Adrenalin causes the **spleen** (an organ behind the stomach) and other organs that store blood to discharge some of their reserves into the circulatory system, increasing the volume of blood available. Adrenalin also causes local **vasodilatation**, or widening of the arterioles and capillaries, in the skin, muscles, and heart, increasing the blood supply to these organs (Figure 33–12). In compensation for this increase, the blood supply to the abdomen and kidneys is reduced; adrenalin causes the small vessels in these organs to constrict. This tradeoff of blood supplies helps to maintain the blood pressure. There is not enough blood to fill the whole circulatory system in the dilated state.

Adrenalin also stimulates increases in the breathing rate and in the heartbeat rate. Oxygen is taken in and carbon dioxide is given off faster, and the heart sends the blood through the body faster, speeding delivery of extra oxygen to the muscles and hastening removal of waste products.

---

*Negative feedback Mechanism whereby the change detected in some condition stimulates compensating activity that brings the condition back to its normal range.

TABLE 33-3	**MAIN COMPONENTS**
	OF THE BLOOD

Water	45–54% vv†
Salts	
Sodium	2400 mg/l
Potassium	80 mg/l
Calcium	80 mg/l
Magnesium	28 mg/l
Chloride	2600 mg/l
Bicarbonate	1500 mg/l
Plasma Proteins	7–9% wv††
Blood Cells	40–50% wv
White cells (leukocytes)	$4.7–9.7 \times 10^3/\mu l$
Red cells (erythrocytes)	$3.6–5.5 \times 10^6/\mu l$
Platelets (thrombocytes)	
Substances Transported by Blood	
Sugars	
Amino acids	
Fatty acids, glycerol	
Hormones	
Nitrogenous wastes	
Carbon dioxide	
Oxygen	

† vv means volume per volume; e.g., 12 ml per 100 ml is 12% vv.

†† wv means weight per volume; e.g., 13 g per 100 ml is 13% wv.

During exercise, the muscles produce more carbon dioxide and lactic acid than usual. These substances make the blood more acidic as it passes through the muscles, and an increase in acidity does three things: it makes the blood give up more of its oxygen in the muscles, it increases the dilatation of the blood vessels in the muscles, and it also stimulates the nervous system to increase the secretion of adrenalin and the breathing and heartbeat rates still further.

These are only a few of the interactions involved in the body's adjustment to exercise, but they illustrate the complexity of the physiological mechanisms that adjust the body's vital functions to changes in its activity.

33-D Blood

The familiar red fluid called **blood** is really a tissue made up of a liquid matrix containing several types of cells (Table 33–3). About half the volume of blood is made up of a fluid called **plasma**, and the other half is blood cells. The plasma contains various salts and a great variety of plasma proteins. **Serum** is plasma from which the proteins involved in clotting have been removed.

The blood cells can be divided into three main groups: the **white cells** or **leukocytes**, the **red cells** or **erythrocytes**, and the **platelets** or **thrombocytes** (Figure 33–13). There are many different types of white blood cells; most of them help protect the body from disease. White blood cells will be discussed in Chapter 34.

Erythrocytes are by far the most numerous cells in the blood (3.6 to 5.5 million per microlitre). Their main function is oxygen transport. Mature mammalian red blood cells have no nuclei and very little other internal structure; they contain little but hemoglobin, a protein that binds oxygen. Red cells are produced from populations of nucleated, dividing cells in the bone marrow. Their production is precisely regulated, and if the red cell population falls, the bone marrow immediately steps

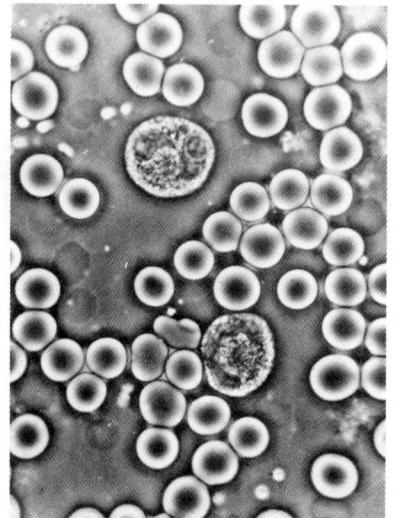

Figure 33-13 A blood smear. The small cells are red blood cells, the big, granular ones are white blood cells, and the faint, transparent circles are red blood cell ghosts—empty plasma membranes. (Biophoto Associates)

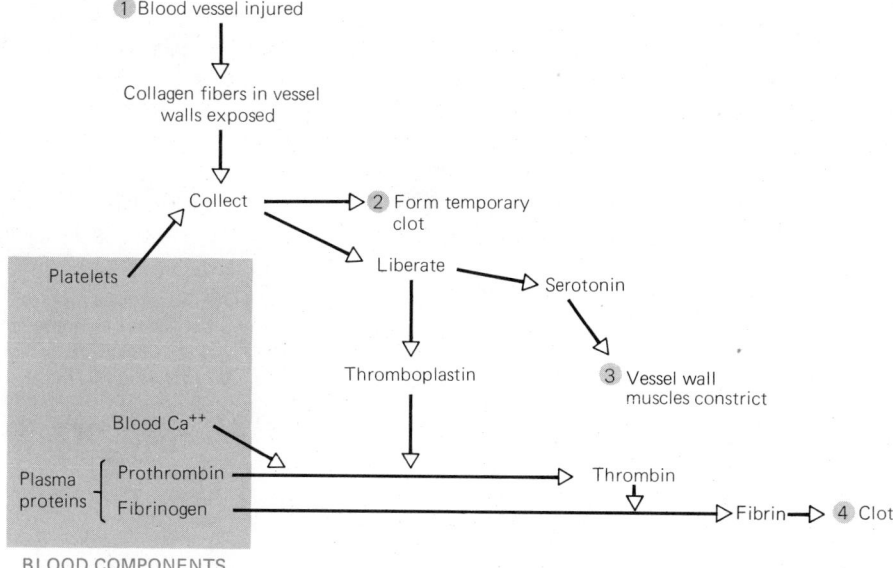

Figure 33-14 A simplified diagram of some reactions involved in the clotting of blood.

up production. Living at high altitudes (see Essay, Chapter 32) and leading a physically active life are two other factors that increase erythrocyte production; both conditions increase the body's oxygen demand. Red blood cells usually survive for about four months after they reach the bloodstream; they then break up and white blood cells destroy their remains by phagocytosis. **Anemia** is a condition in which the blood contains fewer erythrocytes or less hemoglobin than usual as a result of unusually slow production or fast destruction of red cells or hemoglobin. Anemia is a symptom that may be caused by a variety of diseases.

The platelets are important in blood clotting. Platelets are not really cells, but are formed by the pinching off of parts of the cytoplasm of large cells in the bone marrow.

Inflammation and Blood Clotting

Inflammation is the set of changes that a tissue undergoes in response to injury. Damaged tissues release **histamine**, an amino acid derivative. Histamine increases the blood supply to the damaged area and also increases the permeability of the capillaries, so that fluid and clotting proteins from the blood leak into the tissues. A blood clot soon seals off the injured area from the external environment and so delays invasion by bacteria and other foreign substances. White blood cells are chemically attracted to the damaged area. When large numbers of white blood cells engulf bacteria and dying cells, they themselves eventually die. In a serious local inflammation, these dead cells accumulate as **pus**.

Clotting begins when the wall of a blood vessel is broken or damaged; it is an extremely complex process, and our description will be simplified. The injured cells release substances that attract blood platelets to the site. When the platelets come into contact with fibers of the structural protein collagen exposed by the injury, they disintegrate and form a temporary plug for the injured vessel. They also release two substances. The first is **serotonin**, which causes the muscles in the blood vessel wall to contract and constrict the vessel, reducing blood loss. Platelets also release the enzyme **thromboplastin**, which catalyzes the change of one of the plasma proteins, **prothrombin**, into **thrombin**. Thrombin is an enzyme that catalyzes the change of another plasma protein, **fibrinogen**, into **fibrin**. Strands of fibrin form a meshwork around the disintegrated platelets. Still another plasma protein catalyzes the conversion of the loose fibrin meshwork into a tough, hard, permanent plug or **clot**, which seals off the injured part of the blood vessel from the exterior (Figure 33–14).

TABLE 33-4 **FUNCTIONS OF THE LIVER IN REGULATION OF BLOOD COMPOSITION**
Blood glucose $\rightleftharpoons$ Liver glycogen
Amino acids
Uptake for manufacture of blood proteins
Clotting proteins
Albumins
Excess deaminated
Red blood cells and hemoglobin
Broken down in liver
Iron conserved
Porphyrin rings excreted in bile

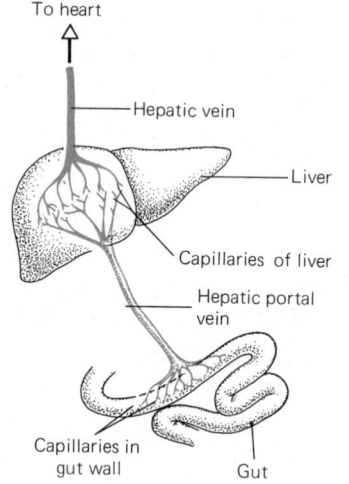

Figure 33-15 The hepatic portal system.

Control of Blood Composition

The contents of the blood (and hence of the extracellular fluid) are regulated within very narrow limits. Various systems monitor the substances in the blood and add or remove them as circumstances dictate. For instance, if the number of red blood cells falls, the resulting oxygen shortage causes kidney cells to secrete the hormone **erythropoietin** into the blood. When this hormone reaches the bone marrow, it stimulates increased production of red blood cells. These cells boost the oxygen-transporting capacity of the blood. As the blood's oxygen level returns to normal, erythropoietin production stops and red blood cell production returns to normal.

The amount of water, salts, and nitrogenous waste in the blood is controlled largely by the kidneys, as discussed in Chapter 35.

The liver is the organ most responsible for regulating the concentration of food molecules in the blood. Blood passing through capillaries in the wall of the intestine picks up food from the digestive tract and carries it directly through the **hepatic portal vein** to capillary beds in the liver (Figure 33–15). (The hepatic portal system is one of the few places in the circulatory system where blood passes through two separate capillary beds before it returns to the heart.) The liver removes excess glucose from the blood and stores it as the polysaccharide **glycogen**. When the level of blood glucose falls too far, the liver breaks down glycogen and releases glucose into the bloodstream.

The liver also has a number of functions in protein metabolism (Table 33–4). If the blood picks up excess amino acids from the digestive tract, the liver breaks them down. The liver also uses amino acids to manufacture plasma proteins, including clotting proteins and albumins. In addition, hemoglobin from old red blood cells is broken down in the liver. The hemoglobin iron atoms are saved and reused; the porphyrin ring (Figure 32–15) is excreted in the bile.

Human Blood Groups

People belong to different blood groups because their genes code for different blood proteins. The most famous blood proteins are those of the ABO and Rh series, but there are many others.

Let us first consider the ABO blood groups. People with type A blood have a glycoprotein* called A on their red blood cell membranes. These people also have a protein, anti-B protein, in their blood plasma. People with type B blood are just the reverse; they have glycoprotein B on their red cell membranes and anti-A protein in

* Glycoprotein Protein bonded to carbohydrate.

Figure 33-16 Agglutination of type A blood cells given in transfusion to a type O person. The recipient's anti-A antibodies attach to the glycoprotein A on the surface of the type A red cells. Clumps of type A red cells and anti-A antibodies precipitate out of the bloodstream.

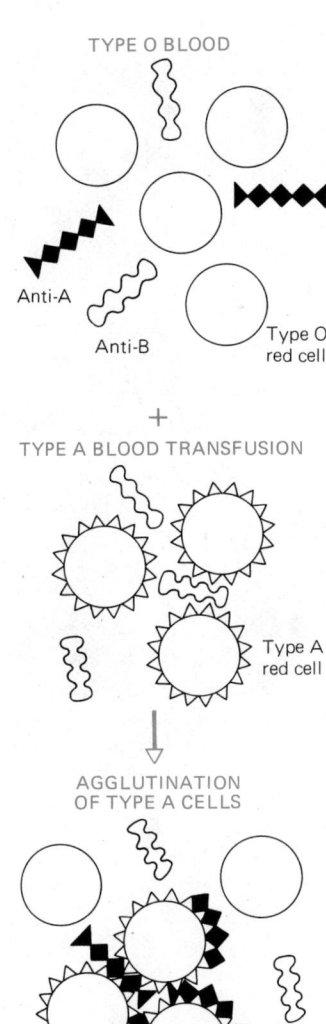

TYPE O BLOOD

Anti-A

Anti-B

Type O red cell

+

TYPE A BLOOD TRANSFUSION

Type A red cell

AGGLUTINATION OF TYPE A CELLS

their blood plasma. Type AB blood has both A and B on the red cells and neither anti-A nor anti-B in the blood plasma, while type O blood has neither glycoprotein on the red cells and both anti-A and anti-B in the plasma (Table 33–5). The A and B red cell glycoproteins are called **antigens**; the anti-A and anti-B proteins are said to be **antibodies** to their respective glycoproteins. The characteristic of an antigen-antibody pair is that they stick to each other (see Chapter 34). For example, plasma containing anti-B antibodies causes type B red blood cells to **agglutinate**, or stick together; this makes the blood cells useless.

For a blood transfusion to be successful, ABO blood groups must be compatible between donor and recipient. If they are not, the blood cells will agglutinate. For instance, when a person with type O blood is given a transfusion of type A blood, the type A red cells encounter the anti-A antibody in the type O plasma. The type A red cells agglutinate (Figure 33–16). The agglutinated red cells can clog blood vessels and stop the circulation to various parts of the body, which can in turn cause severe damage and perhaps death. Before giving a transfusion, hospital personnel generally mix a small sample of donor and recipient blood on a glass slide and make sure that no agglutination occurs.

The Rh blood groups differ from the ABO groups in that the ABO antibodies are always present in the blood, whereas Rh antibodies are produced only when foreign Rh antigens enter the body. **Rh-positive** people have an Rh antigen on their blood cells. People without Rh antigens are **Rh-negative**. If an Rh-negative person is accidentally given a transfusion of Rh-positive blood, anti-Rh antibodies will develop over a 2 to 4 month period in the Rh-negative person. Any Rh-positive blood that subsequently enters this person's circulation will agglutinate.

The Rh blood groups are best known for their role in producing **erythroblastosis fetalis** in newborn babies. In this condition, the red blood cells agglutinate; death usually results unless the infant is given a transfusion to change all of its blood. The condition develops as a result of an Rh-negative mother's carrying an Rh-positive fetus. During birth some of the baby's blood may enter the mother's circulation and she will then produce anti-Rh antibodies to it. If she later carries another Rh-positive fetus, her anti-Rh antibodies diffuse across the placenta into the fetus, where they will cause its blood cells to agglutinate. Since Rh-negative is a recessive trait and Rh-positive people are more common, an Rh-negative woman is quite likely to marry an Rh-positive man and bear Rh-positive babies. Leakage between the fetal and maternal circulations is not inevitable, however, and many Rh-negative women bear several Rh-positive children without problems. Nowadays, an Rh-negative mother may receive injections of anti-Rh antibody after delivery, to remove any Rh factor that may have entered her blood from the blood of the fetus and so keep the mother's body from manufacturing antibodies that could damage future fetuses.

	ANTIGEN ON RED BLOOD CELL MEMBRANES	ANTIBODY IN BLOOD PLASMA
TABLE 33–5 BLOOD PROTEINS IN HUMAN ABO BLOOD GROUPS		
BLOOD TYPE		
A	A	anti-B
B	B	anti-A
AB	A and B	neither
O	neither	anti-A and anti-B

33-E The Lymphatic System

In many ways, the body's capillary beds are the most important parts of the circulatory system, for it is here that the exchange of substances between blood, extracellular fluid, and cells takes place. Most substances, such as glucose and oxygen, leave the blood for the extracellular fluid by diffusing down the concentration gradient between the two fluids. Waste and carbon dioxide return to the blood in the same manner. In addition, water and larger molecules, such as hormones and small proteins, enter and leave the blood either by moving through the spaces between the cells of the capillary walls or by pinocytosis across these cells (see Figure 5–10).

Water leaves the capillaries under the pressure generated as blood is forced through a tube of small diameter. Toward the end of a capillary, so much water has been lost that the proteins left behind lower the blood's osmotic potential, and most of the water returns to the capillary. Overall, however, slightly more fluid leaves than enters the blood in the capillary beds.

This excess fluid is eventually collected and drained away through the **lymphatics**, thin-walled vessels with valves that ensure one-way flow (Figure 33–17). The lymphatics eventually join to form the **thoracic duct** and the **right lymph duct**, which empty into veins near the heart. Often these are the only lymph vessels large enough to be visible. The lymphatics perform several vital functions:

1. They drain excess water from the extracellular fluid back into the circulatory system.

2. They serve as a temporary reservoir for fluids taken into the body. Some of the fluid absorbed from the digestive tract finds it way into the lymphatic system, which releases it gradually, reducing the workload of the kidneys.

3. They carry large molecules, such as large proteins and hormones, to the bloodstream. Such molecules are too large to cross the wall of a capillary and so cannot reach the bloodstream directly.

4. Some food molecules, especially fats, move into the lymph rather than into the blood when they are absorbed from the intestine. The lymphatics form the main route by which such molecules reach the blood.

5. Lymph nodes occur in several areas of the body. These nodes are an important part of the body's defense against disease (see Chapter 34).

33-F Temperature Regulation

The temperature of a living cell determines the rate of its metabolic processes. An organism can grow faster and respond to the environment more rapidly if its cells are kept warm, and the ability of some animals to maintain a constant, relatively high body temperature is believed to be a major reason for their evolutionary success. We usually think of all animals except mammals and birds as "cold-blooded," but in fact, many lower vertebrates and invertebrates also have adaptations that permit them to regulate their body temperatures.

An enzyme functions efficiently only at or near its optimum temperature, which lies in a range spanning no more than about 15°C for each enzyme. This limits the temperature range within which animals can live. Natural bodies of water (excluding hot springs) are hospitable to animal life in that their temperatures always fall between about −2°C and 40°C and change very slowly. Temperatures too high or too low for living cells to survive are common on land, however, and to make matters worse, the temperature on land can change very quickly.

Animals show two basic types of adaptations that permit them to cope with adverse changes in environmental temperature:

1. Some animals tolerate the change, and their tissues are not killed by temperatures higher or lower than the optimum range.

2. An animal may do something that keeps its body temperature within the optimum range even though the temperature outside falls or rises. This is called **thermoregulation**—controlling the temperature of the body.

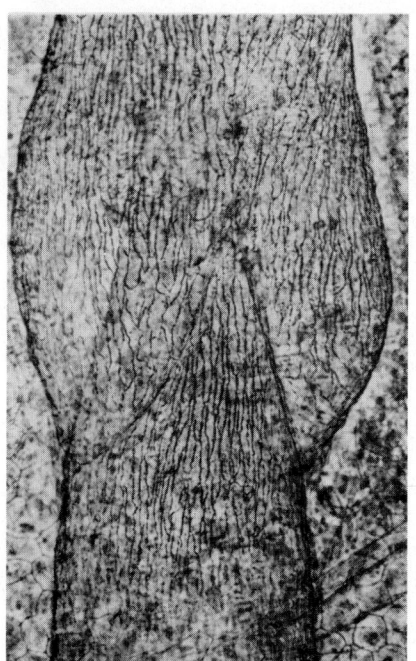

Figure 33-17 Lymph vessel, showing a valve. (Carolina Biological Supply Company)

The unscientific term "warm-blooded" corresponds approximately to the scientific term **homeothermic**, applied to animals that maintain a constant temperature at all times, such as most birds and mammals. "Cold-blooded" corresponds roughly with **poikilothermic**, used to describe an animal whose body temperature is usually the same as that of its environment (Figure 33–18). Aquatic invertebrates are thoroughly poikilothermic; reptiles are much less so. Various animals fill every part of the range between complete homeothermy and complete poikilothermy.

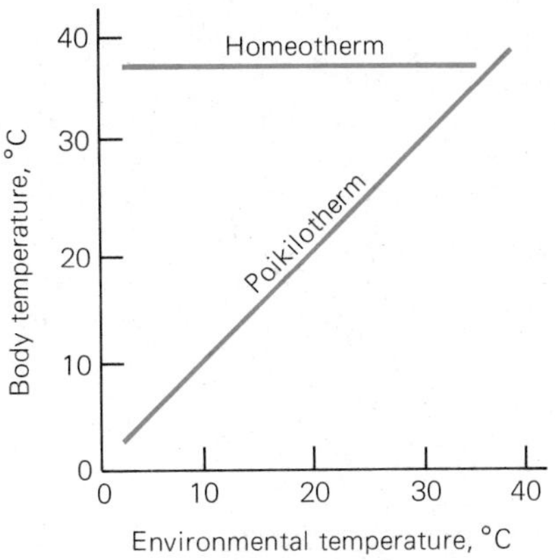

Acclimation

The vast majority of reptiles, amphibians, fish, and invertebrates can tolerate body temperatures in the range of about 0°C to 35°C. However, a rapid change of 30°C will inactivate most enzymes and kill the animal. When the temperature changes slowly, many animals can undergo **acclimation**—adjustment to a change—and remain active. For instance, if a goldfish has been living in water at 30°C, its nerve impulses will be blocked at temperatures below 10°C, but after the same fish has spent some time in water at 15°C, its nerves will function at temperatures down to 1°C; it has become acclimated to the lower temperature.

In some fish and molluscs, and probably in many other animals, acclimation is due to the transcription* of previously inactive genes and synthesis of new enzymes that function at the new temperature. In a number of fish and aquatic invertebrates, this changeover from one set of enzymes to another occurs regularly as the water temperature changes with the seasons.

Endotherms and Ectotherms

Most land animals regulate their body temperatures. **Ectotherms** use behavioral adaptations for thermoregulation; they move into areas where the environmental temperature is appropriate. **Endotherms** depend largely on physiological adaptations to control body temperature. Every animal produces **metabolic heat** from the chemical reactions going on in its body. In ectotherms, this metabolic heat generally escapes into the environment. On the other hand, endotherms, mainly birds and mammals, have a layer of insulation as well as physiological mechanisms that control the rate of metabolic heat loss.

The distinction between ectothermy and endothermy is seldom clear-cut: most animals fall somewhere between the two. Thus, desert rodents, which are endotherms, escape the midday heat by burrowing, a behavioral adaptation. Moths, which are primarily ectotherms, fly poorly unless their wing muscles are at least at 35°C, and they warm the muscles by physiological means, contracting them (fluttering their wings) before they take to the air. Behavioral and physiological thermoregulation, then, are found in both ectotherms and endotherms.

Torpor and Hibernation

Thermoregulation uses considerable energy. There is no advantage in being warm and active when neither food nor mates are available, and many animals, both ectotherms and endotherms, maintain high body temperatures only at certain times of the day or year. At other times, they conserve energy by allowing their body temperatures to fall.

A period of daily **torpor** or inactivity is common in most terrestrial invertebrates, amphibians, and reptiles. A torpid animal's body temperature varies with the temperature of its surroundings. A mountain lizard becomes torpid overnight when the temperature drops, and then raises its temperature just before sunrise (before the external temperature rises); it is ready to go hunting when day breaks.

Daily torpor is also common in small mammals. The loss of heat to the cooler environment is more acute for a smaller mammal because it has a larger ratio of

Figure 33-18 Relation of body temperature to environmental temperature in a homeotherm and a poikilotherm.

---

* Transcription Synthesis of RNA using a DNA template.

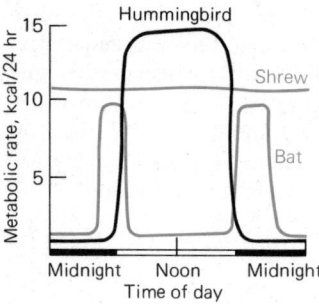

Figure 33-19 Daily fluctuations in metabolic rate of three small "warm-blooded" animals. The hummingbird and bat use less than half as much food per day as a shrew of the same size because they are active only when their food is available. At other times they become torpid, allowing their body temperatures and metabolic rates to fall.

surface area, which loses heat, to body volume, which produces the heat. The smaller the animal, then, the higher its metabolic rate must be to maintain a given temperature, since each cell must generate more heat than would each cell of a larger animal. A small mammal such as a shrew must eat more than its weight in food each day to heat itself. Small mammals use more than 80% of their food calories merely to maintain their body temperatures; a shrew or a bat is never more than a few hours away from starving to death. A period of torpor, when its body temperature falls to the level of the environmental temperature, relieves the constant drain on the animal's energy resources (Figure 33–19).

Many vertebrates hibernate in winter or **estivate** in summer. Metabolic rate and body temperature decrease, and the animal reduces its energy expenditure by avoiding activity at a time of year when food or water is scarce. The temperature of a hibernating mammal becomes equal to the environment, as long as the latter is not too cold. For instance, a hibernating rodent's body temperature does not fall below 10°C.

Behavioral Thermoregulation

A dog lying in the shade on a hot day and bees shivering in their hive on a cold one both exemplify behavior patterns that control body temperature. The dog avoids being further warmed by the radiant heat of the sun (Figure 33–20); the bees produce metabolic heat by muscle movement. The bees also close the entrance to the hive and huddle together, slowing the loss of their metabolic heat.

Behavior that decreases the loss of metabolic heat to the environment can help terrestrial animals keep their bodies above environmental temperatures, but this is

Figure 33-20 Heat exchange between a homeotherm lying in the sun and its environment. Figures in color are temperatures in °C.

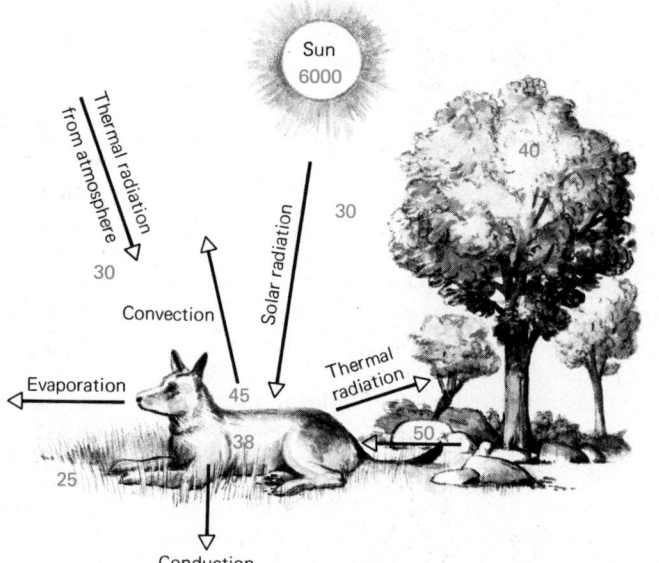

(a)

(b)

Figure 33-21 Thermoregulation. (a) A penguin in the frigid Antarctic keeps its chick warm. (b) A lizard sunbathing on a log in the sun. (a, U.S. Navy; b, Biophoto Associates)

not practical for any gill-breathing animal. As blood in the gills picks up oxygen, it also loses metabolic heat that it absorbed in the interior of the body. The blood in the gills quickly reaches thermal equilibrium with the water outside, and metabolic heat is lost almost as fast as it is generated. Only a few large fish, such as the tuna, can raise their body temperature above that of the surrounding water.

The vast majority of terrestrial ectotherms move into areas that have favorable temperatures. Many earthworms, reptiles, and arthropods retire to burrows during the hottest or coldest parts of the day; this is advantageous because the soil changes temperature more slowly than air does. Millipedes, houseflies, and crabs escape the heat of the sun by avoiding light.

For most animals the need to avoid extreme heat is only an occasional problem. The muscles of most vertebrates and terrestrial arthropods work most efficiently at the surprisingly high temperature range of 35 to 40°C; because temperatures on earth are usually colder than this, most terrestrial animals, rather than avoiding heat, need to capture as much heat as possible from their environments. Living in warmer areas is the simplest way to do this, and in fact there are many more species of poikilotherms of all sorts in the tropics than in temperate or cold climates.

In cold areas, sunbathing is the most common behavioral adaptation to capturing environmental heat; even though the air temperature may be very cold, an animal can often absorb much radiant heat by sitting directly in the sun's rays. Many reptiles sunbathe in positions that expose as much of their body surface area as possible, or extend special folds of skin while sunbathing. Many poikilotherms are also fairly dark in color, and many mammals in colder areas have black-tipped noses, ears, and paws. These dark-colored areas absorb radiant heat more readily; they also lose heat more rapidly, and it is common to find that these animals have behavior patterns—such as curling up to sleep or moving about only when the sun is out—that reduce their heat loss.

The insects that pollinate an Arctic species of poppy, *Papaver radicatum*, sunbathe in the open flower and may raise their temperatures to 35°C, about 20°C above the air temperature. The poppy flower itself turns so that it faces the sun, an adaptation that encourages the visits of the insects.

Physiological Thermoregulation

Birds and mammals maintain high body temperatures, from 35 to 42°C depending on the species. They do this by producing a great deal of metabolic heat and regulating the rate this heat is lost to the environment. When the environment is too

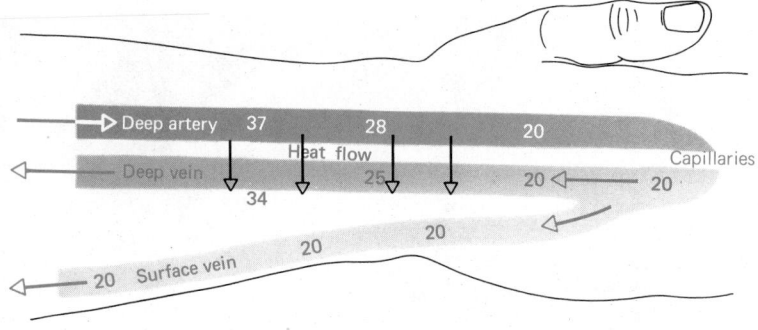

Figure 33-22 Countercurrent heat exchanger between an artery and a vein deep in the human arm. Heat is lost in the capillaries of the hand. As the blood runs through the adjacent deep vein and artery, it exchanges heat so that the arterial blood is cooled by the time it reaches the hand, and venous blood is warmed as it returns to the body. When heat conservation is unnecessary, the heat exchanger can be bypassed, and most of the blood is returned to the body via a surface vein where its temperature does not change. The nervous system controls the degree of constriction of the two veins and so controls heat loss.

Colored arrows = blood flow
Black arrows = heat flow
Numbers = temperature in °C

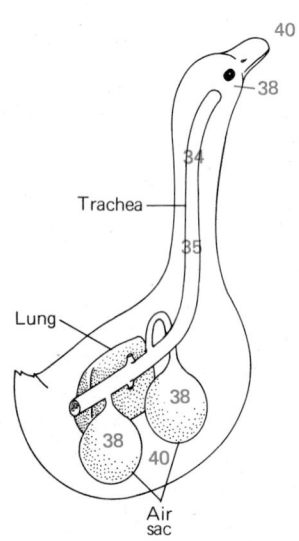

Figure 33-23 Temperature distribution in a bird when both body and air are at 40°C. Birds must dissipate enormous amounts of heat, particularly during flight, by evaporative cooling in the air sacs and trachea. There are many more air sacs than are shown here. Colored numbers = temperature in °C.

cold, for example, the brain induces shivering, muscular movement that produces metabolic heat. When the environment is too warm, the animal becomes less active, and it may rest in the shade or in a burrow.

Layers of fat plus a body covering of fur or feathers insulate the interior of the endotherm's body from the environment and decrease heat exchange; such insulation is not found in ectotherms. Cold-climate endotherms, such as whales, seals, and polar bears, tend to have thicker layers of insulation than their relatives native to more temperate areas. At the other extreme, a camel living in a hot desert is protected from gaining too much heat from the sun by the thick layer of fur and fat insulating its back.

Several physiological mechanisms enable an endotherm to decrease its heat loss when its body temperature falls. Because heat is lost across the body surface, blood coming to the skin from the interior of the body is usually warmer than the skin. When more blood passes through the skin, more heat is lost from the body. When the body is too cool, nervous and hormonal signals constrict the surface blood vessels and decrease the blood flow to the skin, reducing heat loss.

Extremities, such as the ears, nose, and legs, are often slim and streamlined, with little insulating fat. Endotherms stand to lose considerable amounts of heat from the blood in these parts of the body. To take extreme examples of this problem, consider how much heat must be lost through the legs of a bird fishing in an icy mountain stream, or through the paws of an arctic fox trotting across a snowfield. Endotherms cope with this difficulty by means of **countercurrent heat exchangers**: the blood vessels entering and leaving the ear or leg run next to each other, and the blood entering the limb is cooled so that it has little heat to lose to the environment, while the blood returning to the body is warmed almost to body core temperature (Figure 33–22).

When they need to cool themselves, most endotherms radiate heat by increasing the supply of warm blood to the skin and by rearranging their fur or feathers so that more heat escapes. Most of them also employ **evaporative cooling** as a second line of heat control: as water evaporates from the skin or respiratory tract, the heat needed to vaporize the water is removed from the body. Human beings and some other mammals have sweat glands in the skin for evaporative cooling; many carnivores, ungulates, and primates pant to increase evaporation from the respiratory tract. The metabolic heat generated by a flying bird is tremendous, and on its way to and from the lungs, air passes through numerous air sacs, which give an enormous surface for evaporative cooling (Figure 33–23).

Such physiological and behavioral mechanisms enable birds and mammals to maintain a high, constant body temperature. These endotherms can move about when the environmental temperature is so low or so high that other animals cannot be active. It is not surprising that the coldest regions of the world are populated only by mammals and birds. The northernmost part of the Arctic is populated only by birds known as Leech's petrels, and only penguins and seals inhabit the southernmost zone of life around the edges of Antarctica.

Essay: Adaptations of Diving Mammals

Many mammals have become adapted to diving for food or protection. These animals face a variety of problems. First, they must still breathe air, but their trips to the surface may be infrequent (Table 33–6). Second, as an animal dives and then resurfaces, it encounters changes in pressure of about one atmosphere (760 mm Hg) for every 10 metres of depth. At high external pressures the gas in the lungs is compressed and forced into solution in the blood; then, when the animal returns to the surface, pressure is reduced and the dissolved gas, mainly nitrogen, comes out of solution and forms bubbles that may block blood vessels (Caisson disease or "the bends"). Third, water has a high heat capacity, and thus it absorbs heat faster than air, threatening a warm-blooded animal with death from **hypothermia**, or chilling of the body. The adaptations of diving animals that permit them to overcome these problems are exaggerations of features found in other animals.

All diving mammals show **bradycardia**, or slowing of the heart rate, when they dive; for example, a seal's heart rate drops from 150 to 10 beats per minute. Bradycardia is seen in other animals, including humans, when they submerge, and fish show it when removed from water; the advantage of bradycardia is probably that it saves the body energy and oxygen. In addition, blood vessels constrict and reduce the circulation to the kidneys, gut, and so forth. This conserves oxygen for use by the brain, which must not be allowed to experience oxygen deficiency. Furthermore, although circulation to the head is maintained, the respiratory center in the nervous system of a diving animal tolerates relatively high levels of carbon dioxide, and thus it does not stimulate the animal to breathe while submerged. Other body organs carry out fermentation rather than respiration of their food, and most of the carbon dioxide and lactic acid they produce is retained in the tissues; when the animal resurfaces, a surge of metabolic wastes enters the bloodstream.

Many diving mammals can store extra oxygen for use during a dive. They have more red blood cells, which permit the blood to carry more oxygen. Their muscles also contain extra quantities of **myoglobin**, an oxygen-binding pigment related to hemoglobin. Seals carry most of their oxygen store in the blood, whereas whales store more oxygen in myoglobin.

It is perhaps surprising that the lungs do not carry more oxygen than usual during a dive; extra-large lungs would permit an animal to carry down more oxygen. In fact, seals exhale as they submerge, and whales have much less lung volume per unit of weight than do other mammals. These are adaptations that prevent Caisson disease; the less air in the lungs, the less nitrogen will dissolve in the blood during a dive (air is 79% nitrogen). Furthermore, as the animal dives, compression of its lungs forces much of the air into the air passages, whose walls are impermeable to gases. This air cannot dissolve in the blood. An important behavioral adaptation to diving is slow resurfacing, so that the nitrogen dissolved in the blood comes out of solution gradually and returns to the lungs instead of blocking the blood vessels.

Diving mammals reduce heat loss by reducing the flow of blood to the skin during a dive. Their bodies are often shaped with a low surface-to-volume ratio, and the blood vessels to their appendages are arranged in such a way that they can act as countercurrent heat exchangers (see Figure 33–22).

On its return to the surface, a diving mammal must spend several minutes breathing to expel carbon dioxide and replenish its oxygen supply before it can dive again.

Figure 33-A A porpoise comes to the surface to breathe. The blowhole (dorsal nostril) and the horizontal tail fluke are visible. The forelimbs are modified to form flippers and the hindlimbs of the porpoise's terrestrial ancestors have completely disappeared. The dorsal fin encloses several enlarged vertebral processes. (U.S. Navy)

TABLE 33–6 DURATION AND DEPTH OF DIVING FOR SOME MAMMALS		
SPECIES	DURATION (minutes)	DEPTH (metres)
Beaver	15	shallow
Muskrat	12	shallow
Walrus	10	80
Gray seal	20	100
Bottle-nosed whale	120	unknown
Blue whale	49	100
Most people	1	shallow
Trained skin divers	2.5	20

SUMMARY

Most animals have transport systems that move molecules and heat within the body. In many primitive animals, transport of food is carried out by the gastrovascular cavity. Animals with coeloms have true circulatory systems. An open circulatory system has few blood vessels, and blood bathes the cells directly. The closed circulatory systems of many invertebrates and of all vertebrates have blood vessels through which blood is pumped by the heart.

The circulatory systems of vertebrates show an evolutionary trend from single to double circulation. The double circulation of birds and mammals achieves complete separation of oxygenated and deoxygenated blood as well as elevation of the blood pressure in the body's capillaries.

The quick and orderly flow of blood is accomplished by a muscular pump, the heart, and a set of pipes, the blood vessels. Blood flows through the circuit from the region of high pressure, the contracting ventricles of the heart, through the vessels at progressively lower pressure, until it returns to the heart. Valves in the veins and the heart prevent backflow. Blood pressure and blood supply in various parts of the body may be regulated by dilatation and contraction of arterioles and capillaries.

The circulatory system responds to changes in the body's activities so that the body's new needs are met. During exercise, the amount of blood flowing and the rate of flow are increased, and more blood is diverted to the active muscle tissues.

Blood is a tissue consisting of a watery matrix that carries salt, proteins, and blood cells. White blood cells defend the body. Red blood cells contain hemoglobin, a protein that combines with oxygen in the lungs and releases it in the capillaries of the body tissues. The blood platelets are important components in the clotting mechanism of the blood. Clotting helps to plug the vessel walls after injury, preventing loss of vital fluids or entry of pathogenic organisms.

The lymphatic system consists of vessels that collect extracellular fluid, proteins, and digested fats, and empty them into the venous system.

Living cells can function only at certain temperatures. Animals cope with adverse temperature changes either by tolerating them or by making behavioral or physiological responses that help to maintain their bodies within a suitable temperature range. Ectotherms have more behavioral than physiological adaptations for thermoregulation. They may move into hotter or colder areas, or they may sunbathe, burrow, or huddle together to maintain a favorable body temperature. Torpor, hibernation, and estivation are temporary reductions in metabolic rate that allow many animals to reduce the energy they would have to expend on thermoregulation under extreme conditions.

Thermoregulation by physiological mechanisms is most apparent in endotherms. They produce large quantities of metabolic heat and regulate its escape into the environment by such means as insulation, alteration of the blood supply to the skin, regulation of the temperature of blood reaching the extremities, and evaporation of water from the body surface.

OBJECTIVES

From your study of this chapter, you should be able to:

1. Describe the basic transport system in a cnidarian, an earthworm, an insect, a fish, and a mammal.

2. Describe how an open circulatory system such as that of an insect differs from the closed circulatory system of an earthworm or a vertebrate.

3. Describe the double circulation of birds and mammals and list its selective advantages over the single circulation of fishes. Describe the circulation of amphibians and reptiles, and explain how their systems meet the needs of these animals.

4. Describe the structure, and state the main functions, of arteries, veins, capillaries, and the heart.

5. Describe the locations of valves in the circulatory system, and briefly explain their structure and function.

6. Trace the flow of blood through the mammalian circulatory system, indicating the sites at which oxygen, carbon dioxide, and food molecules enter and leave the bloodstream, and using the correct names for the chambers of the heart and for the major arteries and veins labeled in Figures 33–9b and 33–10.

7. Describe adjustments made in the circulation as the body's needs increase during exercise.

8. List the principal substances found in blood.

9. State the principal functions of red blood cells, white blood cells, and platelets.

10. List four ways in which the liver helps to maintain the blood contents at a steady level; state why the hepatic portal system is important to the regulation of the blood's composition.

11. Explain why the ABO blood groups of recipient and donor must be matched for a successful blood transfusion.

12. Describe how and why an Rh-negative mother's blood may damage the blood of a baby born to her.

13. List or recognize four functions of the lymphatic system.

14. Describe the mechanism and routes by which fluid moves between the circulatory and lymphatic systems.

15. Define ectothermy and endothermy and explain the differences between them.

16. Describe how endothermic and ectothermic animals regulate the temperatures of their bodies.

SELF-QUIZ

1. Place an "X" in the boxes to indicate which transport systems exhibit the feature mentioned:

	cnidarian	earth-worm	insect	fish	mammal
food transport					
oxygen transport					
high-pressure fluid picks up food					
muscular circulatory pump(s)					

2. Which of the following is *not* a difference between the circulatory systems of mammals and those of insects?
   a. Substances must cross a vessel wall to reach the cells of mammals, but the blood of insects is in direct contact with their cells.
   b. Blood moves randomly in an insect, but in a definite path in a mammal.
   c. Insects do not transport oxygen in their transport system, whereas mammals do.
   d. Mammals have much higher blood pressure than that of insects.

3. Select *two* advantages of a double circulation over a single circulation.
   a. In the double circulation, all the blood going to the tissues is oxygenated, whereas in the single circulation it is not.

   b. In the double circulation, the blood can transport more types of substances.
   c. In the double circulation, the blood is at higher pressure when it enters the body tissues.
   d. In a double circulation, the blood travels around the body faster.
   e. In a double circulation, there are twice as many blood vessels servicing the body tissues.

4. The greatest amount of oxygen will be lost from the blood while it is traveling through:
   a. the capillaries around the alveoli
   b. the left atrium of the heart
   c. the arteries
   d. the capillaries in the body
   e. the veins

5. If you were asked to dissect an animal so as to reveal a valve, all of the following places would be good to try *except*:
   a. the opening between the right atrium and the right ventricle
   b. the fork where the pulmonary artery splits and one branch goes to each lung
   c. the base of the aorta where it leaves the left ventricle
   d. a vein in the arm
   e. a lymph vessel that empties into the thoracic duct

6. Trace the path of a fat molecule from the time it leaves the intestine until it is deposited in the fatty tissue of the body. Name, in order, all the structures it passes through on the way. (Give the shortest possible route.)

(Quiz continues on next page.)

7. Increasing the adrenalin content of the blood would be expected to decrease the flow of blood to the:
   a. brain
   b. skin
   c. liver
   d. heart
   e. lungs

8. Which sentence below states the main difference between an ectotherm and an endotherm?
   a. An ectotherm obtains its body heat from the environment, whereas an endotherm retains its own internally generated heat.
   b. An ectotherm always has a cooler body temperature than that of an endotherm.
   c. An ectotherm always has the same temperature as the water or air around it, whereas an endotherm always has the same body temperature.
   d. An ectotherm picks up heat from its environment, whereas an endotherm loses heat to its environment.
   e. Ectotherms live in water, whereas endotherms live on land.

9. State whether each phrase below is characteristic of endotherms, ectotherms, or both.
   a. Regulation of body heat is mainly by moving to locations with favorable temperatures.
   b. Heat is generated by the body's metabolism.
   c. An insulating body covering reduces heat exchange with the environment.

QUESTIONS FOR DISCUSSION

1. List some forces, besides contraction of the heart, that may move fluids in the bodies of animals.

2. A sea urchin can live without any apparent inconvenience after its heart has been removed. Explain how this is possible (see Table 33–1).

3. Explain how the open transport system of a butterfly can deliver the oxygen and sugar needed to sustain the high metabolic rate of the flight muscles as the butterfly flits from flower to flower (see Table 33–1).

4. What restrictions in size and activity are imposed on animals that possess an open circulation combined with a tracheal system?

5. Can you think of any reasons why cephalopods (squid, octopus, etc.) are the only molluscs with closed circulatory systems, and why other molluscs (snails, clams, chitons, etc.) manage well with open systems?

6. Birds and mammals have four-chambered hearts, and are homeothermic. In what way might these two characteristics be linked?

7. Arteries usually lie deep in the body, whereas veins lie near the surface. What is the advantage of this arrangement?

8. Why does blood flow to the skin increase so much during exercise?

9. What would happen if a person with type AB blood received a transfusion from a person with type A blood? What would happen if a person with type B blood received a transfusion of O blood?

10. A clot formed within an intact blood vessel is called a thrombus. If part of a thrombus breaks loose, it may be carried to another part of the circulatory system and lodge in another blood vessel, creating a blockage called an embolism. Explain how this might damage the body.

11. Misinformed people often define arteries as blood vessels that contain oxygenated blood, and veins as vessels that contain deoxygenated blood. What is wrong with these definitions?

12. Many mammals hibernate, but birds do not. Why not?

REFERENCES AND FURTHER READING

Carey, F. G. "Fishes with warm bodies." *Scientific American,* February 1973. Tuna and some sharks have countercurrent heat exchangers that reduce heat loss through the gills.

Gardiner, M. S. *Biology of the Invertebrates.* New York: McGraw-Hill, 1972. Contains much detailed information; for the advanced student.

Gordon, M. S. *Animal Physiology: Principles and Adaptations,* 2d ed. New York: Macmillan, 1972. A well-written modern comparative physiology; discusses the changes in physiological systems in different animals.

Heinrich, B., and G. A. Bartholomew, "Temperature control in flying moths." *Scientific American,* June 1972. Relates anatomy and physiology of moths to control of their body temperatures.

Kanwisher, J. "Temperature regulation." In *Comparative Physiology,* L. Goldstein, Ed. New York: Holt, Rinehart and Winston, 1977. A good general discussion of thermoregulation in vertebrates and invertebrates.

Prosser, C. L., Ed. *Comparative Animal Physiology,* 3d ed. Philadelphia: W. B. Saunders, 1973. Several expert authors cover a wide range of physiological topics; extensive reference list at end of each chapter.

Ramsay, J. A. Chapter 2: Circulation. *Physiological Approach to the Lower Animals,* 2d ed. New York: Cambridge Univ. Press, 1968. A brief, but charming, comparative invertebrate physiology.

Whittow, G. C., Ed. *Comparative Physiology of Thermoregulation,* Vol. 1. New York: Academic Press, 1970. A fairly technical discussion of means of temperature control in fishes, amphibians, and reptiles.

Wood, J. E. "The venous system." *Scientific American,* January 1968.

Zucker, M. B. "The functioning of blood platelets." *Scientific American,* June 1980.

CHAPTER 34

DEFENSES AGAINST DISEASE

Few people suffer twice from diseases such as measles, chicken pox, and mumps. Our first encounter with the pathogens (disease-causing organisms) that cause these diseases evokes some response from the body that equips it to dispose of the same kinds of pathogens more effectively when they next invade. A response that is bigger and better the second time around marks this reaction as a product of the immune system. Study of immune responses makes up the science of immunology.

If you have had measles, you are immune to further attacks of measles but not of German measles or of the common cold: the immune response to measles is **specific** for the virus that causes measles. The specificity of an immune response also means the body must be able to distinguish the measles virus as **foreign**, something that does not normally occur in the body. Since the immune response to a bout of measles protects the body from measles viruses encountered later in life, immunity also obviously involves some sort of **memory**: the body remembers that it has previously encountered the virus.

These three characteristics—specificity, recognition of foreignness, and memory—are the basic features of immune responses.

Immune responses that protect against specific invaders are known only in vertebrates, but the vertebrate immune system is only an evolutionary extension of less specific protections against disease that all animals possess. For instance, **phagocytes**, cells that ingest bacteria, dead body cells, and other debris, are found in all animals. Similarly, many body fluids contain substances that kill bacteria, fungi, and viruses. Examples are the acid in the human stomach, a substance in semen, and **lysozyme**, a bactericidal (bacteria-killing) enzyme found in tears, nasal secretions, and saliva. Some of these protections are there all the time; others, like a bactericide in the body fluids of lobsters, are secreted only when the animal becomes infected by a pathogen.

In this chapter, we shall consider two unspecific defense systems—the skin and interferon—in our own bodies, and then study the specific responses of the immune system. We shall concentrate on the mammalian immune system, which has been most intensively studied, but lower vertebrates have similar, though less sophisticated, systems.

Immunology is a complicated and fast-moving subject; research journals print new findings faster than the eye can read, and this leads to a certain amount of confusion and lots of argument. Here we try to simplify the subject and to present only conclusions that will stand the test of time, but past experience suggests that at least a few of the assertions in this chapter will have been disproved by the time you read this book.

34-A Unspecific Defenses

Skin

Skin is the body's main protective organ (Figure 34–1). Its importance can be gauged from the fact that whether a burn victim lives or dies depends almost entirely on how rapidly and completely damaged skin can be replaced by healing, by grafted skin, or by a plastic skin substitute (now being tested in cases where large areas of skin have been destroyed).

The epidermal cells at the surface of the skin are constantly dying and sloughing off. They are replaced by the division of cells in the basal layer; minor damage to the skin is quickly repaired, and the body maintains a constantly renewed mechanical barrier against injury, infection, and water loss.

Figure 34-1 A section through human skin.

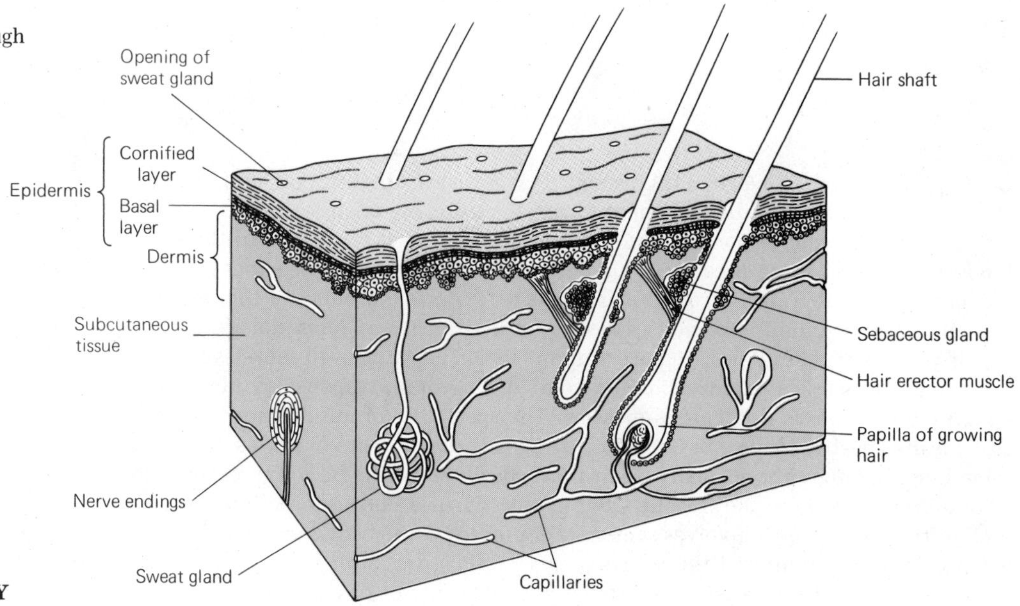

Opening of sweat gland

Hair shaft

Epidermis { Cornified layer

Basal layer

Dermis

Subcutaneous tissue

Sebaceous gland

Hair erector muscle

Papilla of growing hair

Nerve endings

Sweat gland

Capillaries

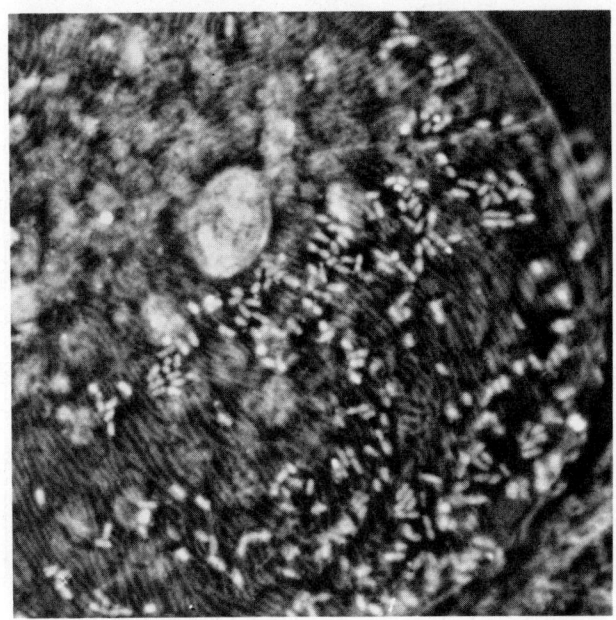

Figure 34-2 Part of an epithelial cell from the mouth with its resident bacteria (white rods). (Biophoto Associates)

The skin also produces chemical defenses: its sebaceous glands produce oil and wax, and its sweat glands produce sweat. These secretions contain lactic acid and fatty acids, which lower the pH enough to kill or inhibit the growth of many fungi and bacteria. The skin is also home to millions of other bacteria. These local residents produce substances that protect their territories—our skin—from invasion by foreign bacteria, many of which might cause disease.

Besides protecting the body against disease and injury, the skin plays vital roles in temperature regulation (Section 33-F), production of cholesterol, and the detection of stimuli such as pressure and temperature (Chapter 39).

Interferon

The immune system protects the body against viruses, as well as against other invaders, frequently destroying them in the bloodstream so that they never reach the cells of vital organs such as the brain or liver. If viruses elude the immune system and invade a body cell, however, the infected cell releases **interferons**, proteins that help to protect neighboring cells from the virus. Interferons work by interfering with viral replication. Most pathogenic viruses have RNA, not DNA, as their genetic material, and the replicating form of viral RNA seems to be the main stimulus for a cell to produce interferon.

Interferon produced in response to one virus gives cells some protection against later invasion by the same or another virus; this is one reason people seldom have flu twice in one winter even if there are several strains of influenza virus around.

Before 1980, interferon was extracted, at vast expense, from human cells; it should be much cheaper to produce now that interferon genes have been spliced into bacteria (see Chapter 9). The first experiments using genetically engineered interferon against human cancers believed to be caused by viruses have been disappointing. This may be because we know so little about the specificity of interferon molecules; we do not know whether a particular virus induces the production of one or of several different kinds of interferons.

The skin and its secretions are a first line of defense against pathogens, and interferons combat any viruses that reach body cells. The immune system, subject of most of this chapter, lies between these two levels; it destroys pathogens in the body fluids, after they have penetrated the skin or mucous membranes but before they have reached most of the body's cells.

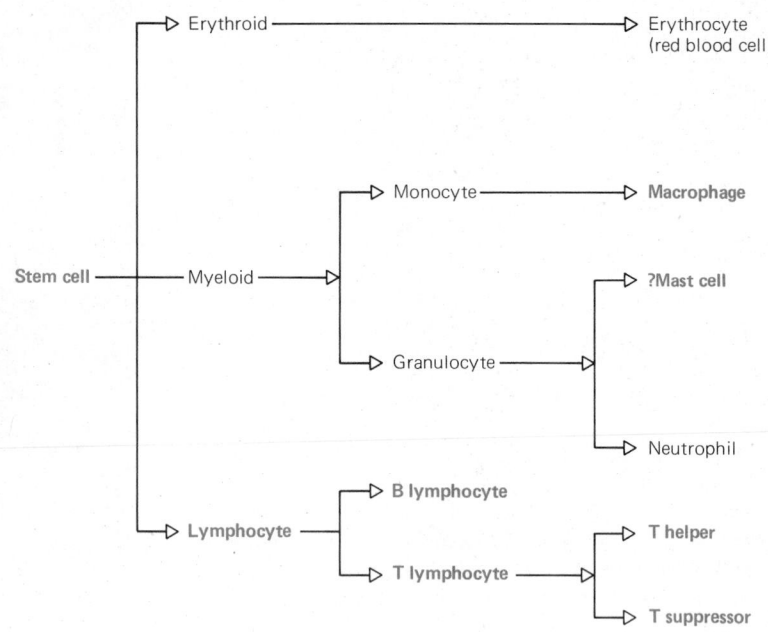

Figure 34-3 Ways in which a hemopoietic (blood-forming) stem cell can differentiate. Each stem cell (on the left) may either divide and give rise to new stem cells or differentiate to produce one of the blood cells listed down the right (or several more types that are not listed). The question mark indicates that researchers are not sure mast cells come from granulocytes. Cell types mentioned in the chapter are in color.

34-B The Immune System

The role of the immune system is to recognize and destroy foreign antigens that invade the body. An **antigen** is any substance that can stimulate the body to mount an immune response against it. The most common antigens are protein-containing substances from another organism. For instance, the glycoproteins on a cell's plasma membrane act as antigens and provoke an immune response when an organ is transplanted from one person into another; bacterial and viral proteins act as antigens if the virus or bacterium invades the bloodstream. Note that a substance can be an antigen for one person but not for another: the glycoproteins on your liver cells are not antigens to your body, but they would act as antigens if injected into another person.

Various white blood cells are the agents responsible for immune responses. These cells are all formed in the **hemopoietic** (blood-cell-forming) tissue of the marrow that fills the center of many bones (Figure 34–3), but they move out to live and work in other parts of the body.

Anatomy of the Immune System

Many of the cellular interactions involved in immune responses take place in **lymphoid tissue**, which occurs in the lymph nodes, in the spleen, and in the main places where pathogens invade the body—the linings of the respiratory, alimentary, genital, and urinary tracts. (The tonsils are lymphoid tissue in the throat.)

In Chapter 33 we saw that fluid filters out of the blood in the body's capillary beds and joins the extracellular fluid, or lymph, that surrounds all cells. The lymph drains slowly into thin-walled lymphatic vessels, which drain back into the blood via the thoracic duct (Figure 34–4). At intervals along the lymphatic vessels, the lymph passes through **lymph nodes**, networks of sinuses lined with white blood cells. The lymph nodes filter foreign matter out of the lymph. The spleen, a spongy organ near the stomach, is a similar filter for the blood.

Lymphocytes

Lymphocytes are white blood cells of crucial importance in all immune responses. Lymphocytes circulate throughout the body, from the bloodstream, through the lymph, and back into the blood. Most of an animal's lymphocytes can be removed experimentally by draining the thoracic duct through a tube. If a rat is treated in this way, it produces hardly any immune response to an antigen. Injecting lymphocytes back into the experimental rat restores the immune response.

Lymphocytes have at least three roles in the immune response: they recognize antigens, they destroy antigens, and they "remember" the reaction in some way.

Lymphocytes may be divided into two major groups, T lymphocytes and B lymphocytes. T lymphocytes (or T cells) take their name from the thymus gland, a lymph gland at the base of the neck. (This gland is familiar to gourmets under another name, "sweetbreads.") After these cells are formed by cell division in the bone marrow, they travel to the thymus gland, where they undergo further differentiation before they can play their normal role in the immune system. If the thymus is removed from a young mammal shortly after birth, the animal grows up without the immune response that normally rejects tissue transplanted (grafted) from another individual.

B lymphocytes take their name from the **bursa of Fabricius**, a lymph gland attached to the intestine of birds. Newly formed lymphocytes become B lymphocytes by undergoing differentiation in the bursa. If a chicken's bursa is removed early in life, the bird loses its ability to mount an immune response to bacteria or viruses in the bloodstream. Mammals have no bursa; the differentiation B lymphocytes undergo in the bursa of a bird is thought to take place in the bone marrow in a mammal.

The experiments with removal of the thymus or bursa make it clear that T and B cells are responsible for different kinds of immune responses. The immune response that combats microorganisms in the blood is called a **humoral response** (humoral means fluid) because it involves proteins that are released into the blood by B cells (it also involves the B cells themselves). The graft-rejection response is called a **cell-mediated response** because the reaction involves only cells—T lymphocytes that have matured in the thymus. The distinction is not so clear-cut as was once thought, however; we shall see that certain T cells are necessary to the humoral response. First, though, we will consider what is known about the cell-mediated response.

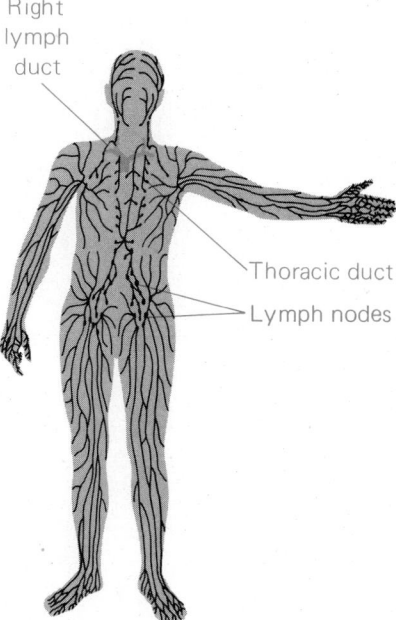

Right lymph duct

Thoracic duct

Lymph nodes

Figure 34-4 Location of the major lymphatic vessels and lymph nodes in the human body.

34-C   Cell-Mediated Responses

The medical goal of replacing diseased and damaged tissue by transplants from other individuals is usually thwarted by cell-mediated responses. Immunologists often study these responses using skin grafts, which are easy to work with and do not harm the recipient. If skin is transplanted from one mouse to another, the graft is quickly invaded by blood vessels, and it looks healthy for several days. Within nine days, however, the blood supply to the graft diminishes, and the graft is infiltrated by T lymphocytes. The graft begins to wither, and within a day or two it sloughs off. A second graft from the same donor to the same host is rejected faster than the first one, implying that the body "remembers" the graft antigens. T lymphocytes appear to be responsible for this memory as well as for graft rejection, although little is known about either phenomenon. It seems that T lymphocytes have receptors on their surfaces that bind the graft antigens, and that this is how the host "recognizes" the graft.

Medical researchers are primarily interested in preventing rejection of grafts in the first place. One way of doing this is to use drugs that suppress the body's immune responses. Such drugs are invariably used to reduce graft rejection after heart or kidney transplant operations. Most such drugs work by suppressing lymphocyte cell division. The other, and safer, way to minimize graft rejection is to match the tissue antigens of donor and recipient as closely as possible. To understand why this is effective, we must learn a little more about the antigens that induce the cell-mediated response.

The Major Histocompatibility Complex

Most of the body's cells bear, on their surfaces, glycoproteins that, among other functions, identify the cells as belonging to a particular tissue in a particular individual; these glycoproteins may act as antigens. In 1980, Baruj Benacerraf, Jean Dausset, and George Snell won the Nobel Prize for their 40 years of work on the mammalian gene complex that determines at least some of these antigens. The gene complex is known as the **major histocompatibility complex (MHC)**. "Histocompatibility" means compatibility between tissues and refers to the fact that organs can be transplanted successfully from one animal into another only if their MHC antigens are compatible. If host and donor have very different MHC antigens, the host's immune system will reject a transplanted organ despite massive doses of immunosuppressant drugs. The success of organ transplants has been much improved since the development of techniques for matching MHC antigens as closely as possible before attempting to transplant an organ.

Animals living beyond the reach of medical care do not exchange organs. Clearly, the role of these antigens in graft recognition seen in the laboratory is not the normal function of MHC antigens and the many other transplantation antigens that also exist. Animals all the way back to annelid worms reject foreign grafts, which suggests that the existence of the responsible antigens has some powerful advantage. Many people think that this advantage is protection of the body against cancer. Cells that become cancerous almost certainly lose some of their MHC antigens, and they definitely develop novel antigens; this enables the immune system to recognize them as abnormal and to destroy many of them before they proliferate. Most cancerous cells are probably killed by the immune system without ever causing noticeable symptoms.

34-D Humoral Immune Responses

Humoral immunity, the body's main defense against pathogenic bacteria and viruses, depends on B lymphocytes. In response to a bacterial or viral antigen, some of the B lymphocytes release **antibodies** that trigger the antigen's destruction and recognize it if it appears again. An antibody is any one of a large class of proteins called **immunoglobulins** found in the blood. (The terms "immunoglobulin" and "antibody" can be used interchangeably.)

An antibody is identified by its ability to combine with a specific antigen. When a solution of antigen is mixed with blood serum containing antibody to that particular antigen, the antigen and antibody molecules stick to each other and precipitate out of solution. An antibody reacts to the antigen's three-dimensional shape and distribution of electrical charges; no covalent bonds or chemical reactions are involved. In this respect, antigen-antibody binding resembles the way an enzyme binds to its substrate (Section 3-E).

The specificity of humoral immune responses depends on antibody-antigen binding: the closer the fit between the molecular shapes of the two molecules, the stronger the binding between them and the greater their specificity for each other

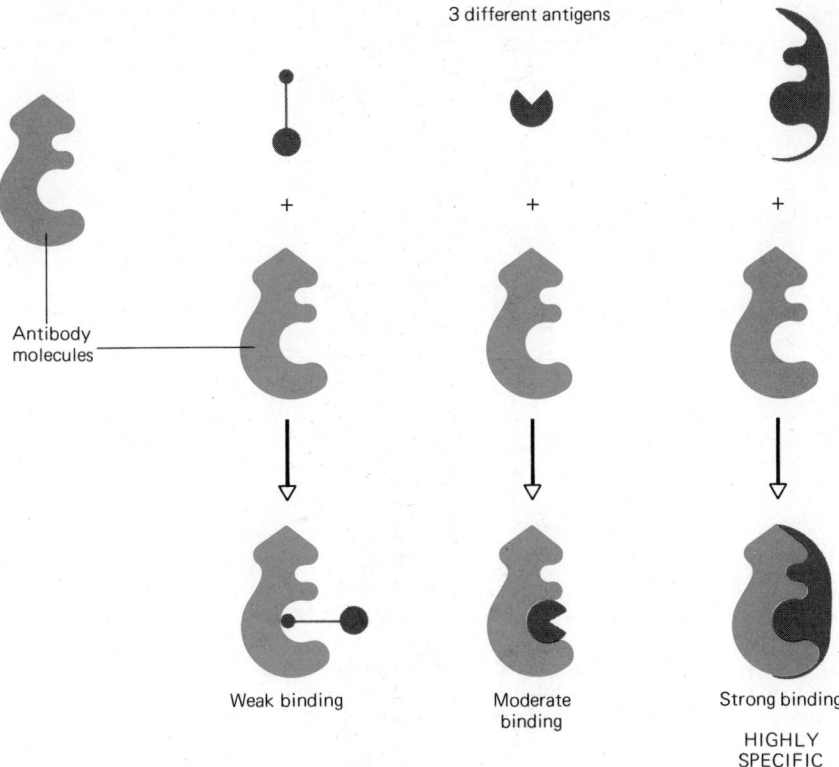

3 different antigens

Antibody molecules

Weak binding

Moderate binding

Strong binding

HIGHLY SPECIFIC

Figure 34-5 Immunological specificity. The particular kind of antibody shown in this diagram (color) binds one of the three antigens shown (gray) much more strongly than it binds the other two; the antibody is said to be specific for the antigen it binds most strongly.

(Figure 34–5). On the other hand, specificity is not absolute; many different antigens may be capable of binding an antibody more or less tightly.

The human body probably contains antibodies to about 100,000 different antigens and produces small amounts of all of them even if it never encounters the corresponding antigens. Immunoglobulins are synthesized by B lymphocytes and released into the bloodstream, where they trigger the first steps of a humoral response.

If we inject an antigen, such as a bacterial toxin, into a rabbit for the first time, one of the thousands of different immunoglobulins in the blood will almost certainly bind this antigen to some extent. The complex of bacterium + immunoglobulin is now recognized, in some unknown way, by a phagocytic **macrophage**, which ingests the whole lot (Figure 34–6). The macrophage frees the bacterial antigen and pushes it out onto the macrophage plasma membrane. In its travels through the body, the macrophage soon reaches a lymph node, where hundreds of B lymphocytes

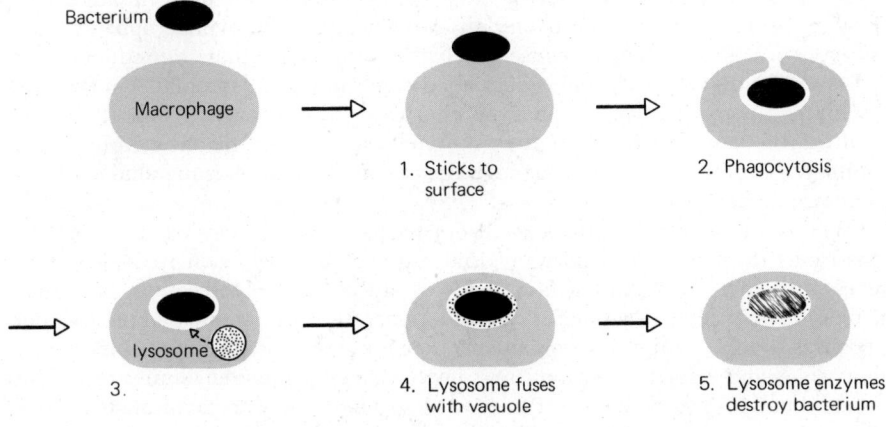

Bacterium

Macrophage

1. Sticks to surface

2. Phagocytosis

lysosome

3.

4. Lysosome fuses with vacuole

5. Lysosome enzymes destroy bacterium

Figure 34-6 Phagocytosis of a bacterium by a macrophage (one of the phagocytes listed in Figure 34-3). A lysosome is a sac of digestive enzymes.

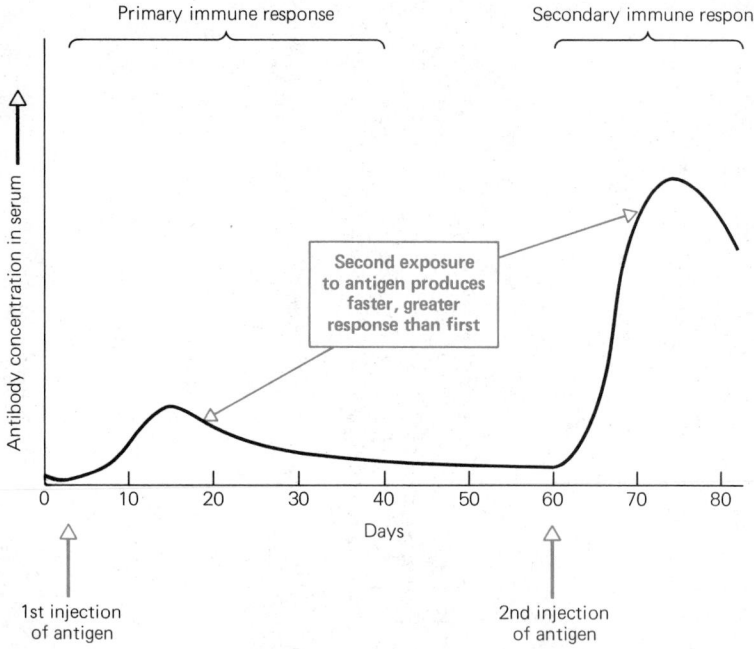

Figure 34-7 Primary and secondary immune responses. The graph shows the amount of antibody to a specific antigen detected in the blood of a rabbit. The second time the antigen is injected, the rabbit produces the specific antibody more rapidly and in greater amounts.

reside. Each lymphocyte bears on its surface a sample of the antibody it synthesizes, and the bacterial antigen on the macrophage's surface will bind to any appropriate antibody. The binding of an antigenic macrophage stimulates a lymphocyte to divide, forming a clone of cells. (T lymphocytes known as "helpers" are vital to clone formation and immunoglobulin synthesis, although how they work is not known.) Some members of this clone leave the lymph node and float off to seed other lymph nodes. All synthesize the immunoglobulin specific to the bacterial antigen, and unlike the original lymphocyte in the lymph node, many of them secrete the antibody. Within a few days of the initial infection, a great deal of that antibody appears in the blood.

34-E Primary and Secondary Immune Responses

The flood of antibody into the bloodstream during a humoral response to the body's first encounter with a bacterial toxin marks the **primary immune response** to the antigen (Figure 34–7). During the primary response, the antigen will eventually disappear from the blood, bound by antibody and eaten by macrophages. Then "suppressor" T lymphocytes cause the lymphocyte clone that is producing antibodies to stop dividing. The clone does not die out, however; it remains in the body, an enlarged population of lymphocytes that react to that particular antigen. As a result, if the same bacterium enters the body again, a **secondary immune response**, faster and more extensive than the primary response, is marshaled to repel boarders (Figure 34–7).

The clone of lymphocytes formed during the primary response to an antigen constitutes the body's "memory" of that antigen. Because each lymphocyte produces only one, or at most a few, types of antibody, the body must build up a memory clone for each antigen it encounters before it has an arsenal of secondary responses to most of the microorganisms it encounters. This is why babies have so many colds and infections in their first few years: they must encounter many antigens and build up many clones of memory cells before they are immune to as many diseases as the average adult.

Figure 34-8 Two macrophages in a scanning electron micrograph of part of the human intestine. (Biophoto Associates)

In the humoral immune response we know that immunological memory consists of an enlarged clone of B lymphocytes sensitized to a particular antigen. In the case of cell-mediated responses, we know less about the memory.

34-F   The Complement Reaction

Both the cell-mediated and the humoral immune responses may sometimes be augmented by further activities of the immune system. Let us consider how one of these, the complement reaction, assists the humoral response.

If it is to remove a microorganism from the body fluids, a phagocytic cell such as a macrophage must be stimulated, first to stick to the microorganism, and second to engulf it. We have seen that some bacteria bear surface antigens that bind immunoglobulin, and that this binding is sufficient to stimulate a macrophage to ingest the whole complex. Other bacteria, however, such as some virulent pneumococci, streptococci, and staphylococci, have capsules that reduce immunoglobulin binding, or produce substances that repel macrophages. The **complement reaction** assists the immunoglobulins to destroy such resistant bacteria.

The complement reaction resembles blood clotting in that one reaction in the series triggers the next, producing a cascade of reactions that amplify the response as it proceeds. Complement reactions can be triggered by the binding of immunoglobulins to a resistant bacterium or by the polysaccharide coat on some bacteria. In the latter case, the reaction produces enzymes that digest the bacterial coat; the damaged bacterium can then be engulfed by a macrophage. This reaction is most familiar when it occurs in a local infection such as a pimple or a boil. The complement reaction attracts and stimulates **mast cells** (see Figure 34–3), which release histamine, causing local inflammation and attracting phagocytes to the site of infection. **Pus** in an infection is the cellular debris produced by this complement-stimulated attack on bacteria by phagocytes.

The complement reactions, then, enhance phagocytosis stimulated by immunoglobulins and induce phagocytosis in some cases where immunoglobulins are ineffective to do so.

34-G   Vaccination

Vaccination against specific diseases produces a primary immune response and thereby creates an immunological memory, ready to trigger a secondary response at the next appearance of the disease antigen. The practice of vaccination, however, began long before people understood how it works. Edward Jenner, an English physician, developed the first vaccine, against smallpox. Jenner noticed that dairy workers who had had the relatively mild disease cowpox seemed to be immune to smallpox. Jenner found that rubbing pus from cowpox sores into scratches in the skin prevented people from coming down with smallpox later. In this case, the antigens of smallpox and cowpox are so similar that the same antibodies work against them. (It is interesting to note that smallpox was not only the first disease to be prevented by vaccination, but also the first disease to be eradicated. The last known outbreaks occurred in the Indian subcontinent and Africa in the late 1970s. Large-scale international vaccination programs greatly reduced the annual number of smallpox cases, but the disease persisted for many years at low levels. The final conquest came after health officials adopted a different strategy: searching out pockets of infection [people were given reward money for each case they reported], quarantining the victims, and inoculating their friends and relations.)

Since Jenner's time, medical researchers have developed vaccines for a number of bacterial and viral diseases, including polio and influenza, which proved especially difficult. "Booster shots" serve to jog the body's immunological memory into producing more antibodies and more cells, ensuring that there are plenty of memory cells available if a diphtheria or whooping cough bacterium should invade.

34-H Passively Acquired Immunity

An animal is said to be passively immune when it contains antibodies that were not synthesized in its own body. A newborn baby is passively immune, temporarily protected from disease by immunoglobulins that reach it from the mother's blood before birth. These maternal antibodies are steadily used up over a period of a few months until the baby's immune system is sufficiently mature to take over.

The breast-fed newborn is also protected by colostrum, a thin fluid produced by the mammary glands after childbirth before the flow of milk begins. Colostrum contains antibodies believed to protect the human infant's digestive tract from infections. Once the normal bacterial inhabitants of the digestive tract become established, they themselves suppress the invasion of dangerous newcomers. Human babies do not absorb antibodies from colostrum into the blood, although the young of some other mammals do.

Passive immunity can also be used medically. There are some antigens so virulent that there is little hope the body's own primary immune response will be able to avert serious damage or death. If by some mischance such an antigen enters the body, the victim can sometimes be protected temporarily by injections of antibodies produced by another animal. These antibodies are usually prepared by giving several small injections of an antigen, such as tetanus or snake venom, to a horse and later collecting samples of the horse's blood, which now contains antibodies to that antigen. The horse's serum can then be stored until it is needed to protect a patient from that specific antigen. Such injections should not be used lightly, however, because the recipient will produce an immune response to the horse proteins in the serum; this might produce a dangerous secondary reaction if the patient were ever again injected with horse serum.

Artificial clones of antibody-producing cells have now been produced and can be maintained in the laboratory. Each of these clones produces a single antibody. Researchers hope that these **monoclonal antibodies** will provide a safe substitute for horse serum when patients require passive immunity.

However they are acquired, the antibodies involved in passive immunity eventually disappear from the recipient's body, and the immunity is lost.

34-I Immunoglobulins

Immunoglobulins are proteins; each molecule is made up of two identical heavy peptide chains, two identical light peptide chains, and, in some cases, short "junction segments," all held together by disulfide links (Figure 34–9). A healthy mammal contains many thousands of different immunoglobulins, each in such small amounts that it is very hard to obtain enough of any one type to analyze its struc-

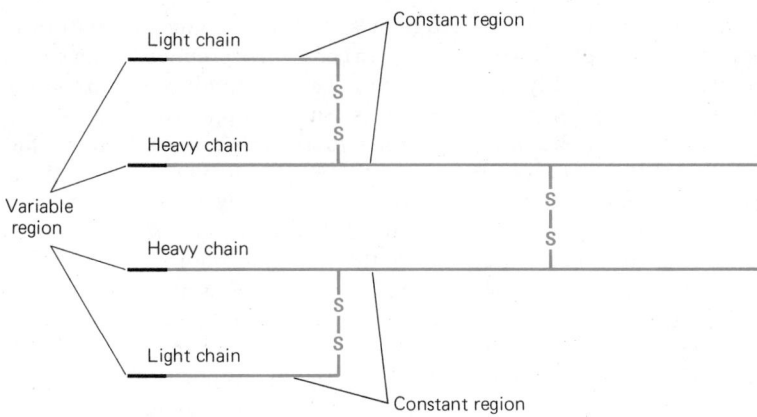

Figure 34-9 Structure of an immunoglobulin molecule. Each consists of two identical light and two identical heavy peptide chains joined by disulfide bonds (S—S). Part of each chain is variable (black) and part is constant (colored).

ture. This problem was first overcome using blood from patients with myeloma, a type of cancer in which an antibody-producing cell divides uncontrollably and produces a vast clone of cells, all synthesizing only one type of immunoglobulin. Now that we can also mass-produce monoclonal antibodies in the laboratory, it is much easier to study antibody structure. Researchers have analyzed the amino acid sequences of many immunoglobulins as well as the nucleotide sequences of the corresponding genes.

The immunoglobulins are divided into five main groups, some with functions that are not yet understood (Table 34–1). The groups are known as G, M, A, D, and E (the letters have no significance, which is most unhelpful). Within any one of these classes, parts of the heavy and light chains are constant from one specific antibody to the next, while other parts of the molecule are variable (Figure 34–9).

It now seems certain that the variable portion of the molecule accounts for the antigen specificity of an immunoglobulin, and the constant region determines its general biological activity. For instance, in the G immunoglobulins, those that combine with bacteria and viruses in the blood, the constant region is responsible for (a) attracting macrophages that engulf and destroy an invading bacterium and (b) enabling the immunoglobulin to cross the placenta, so that maternal antibodies can reach the fetus.

The Genetics of Immunoglobulin Production

The existence of antibodies poses an interesting genetic problem. Antibodies are proteins, and organisms are believed to require one gene for each different kind of protein they produce. An individual mammal has about 30,000 different genes that code for proteins, but it may produce 100,000 different antibodies—in addition to all the other proteins it needs—during its lifetime. How can this be? The answer was not discovered until after 1977, when it became possible to analyze the nucleotide sequence of DNA and RNA. The secret of antibody diversity lies in the fact that antibody genes occasionally move around on the chromosome and from one chromosome to another. So far we have no idea how this happens.

Recall that each antibody is made up of two pairs of peptide chains, and that each chain has a variable and a constant region (Figure 34–9). During early development, antibody genes line up to bring together the DNA segments that code for a variable region and a constant region. In mice, immunologists have analyzed about 250 genes for variable regions and probably 8 genes for constant regions, as well as about 12 joining segments. Combined in various ways in different developing cells, these genes permit the synthesis of about 10,000 possible different antibody molecules from fewer than 300 genes.

GROUP	MAJOR CHARACTERISTICS
G	Combat microorganisms and their toxins in extracellular fluid; can cross the placenta
A	Found in mucous secretions; defend external body surfaces
M	Particularly effective at sticking bacteria together and immobilizing them
D	Found on the surfaces of lymphocytes
E	Effective against parasitic infections; responsible for symptoms of allergy

TABLE 34-1 THE IMMUNOGLOBULIN GROUPS

Other rearrangements occur later, in adult life, during the maturation of a lymphocyte. For example, B lymphocytes are constantly produced from stem cells in the bone marrow; those B lymphocytes with class M immunoglobulin on their surfaces migrate to the spleen and lymph nodes. Such lymphocytes may be stimulated to secrete M immunoglobulin or may undergo a "class switch," after which they will produce antibodies of a completely different group (immunoglobulin G, A, or E). This class switch is also brought about by rearrangement of the lymphocyte DNA; this involves bringing a new gene for the constant region into action, because it is the constant region that determines the immunoglobulin's class.

Much still remains to be discovered about the genetics of immunoglobulin production, but it is obvious that the discovery of "jumping genes," which can be arranged and transcribed in various combinations, explains the ability of a mammal to produce a vast range of antibodies from relatively few antibody-determining genes.

34-J Failures of the Immune System

The immune system is vital in protecting the body from disease; when something goes wrong with it, the consequences are frequently fatal. Occasionally, lymphocytes fail to form during embryonic development. For instance, T lymphocytes fail to develop if the thymus gland is abnormal. Since T helper cells are necessary to formation of the B lymphocyte clones that synthesize large amounts of antigen-specific immunoglobulin, babies with poorly developed thymus glands can produce only small amounts of each circulating antibody. A baby born without T lymphocytes usually dies of the first pathogens it encounters. A few such babies have been saved by keeping them in sterile environments and by transplanting bone marrow cells and thymus tissue, which may permit the individual to make his own antibodies. It is noteworthy that thymus-deficient animals do have B cells and circulating antibodies.

Cases are known in which the B system and antibodies are absent but T cells and T responses are normal. Individuals without B systems usually succumb to bacterial infections early in life.

Autoimmunity is a dangerous condition in which the body develops antibodies to its own antigens. Normally, the body "learns" to recognize its own proteins and other antigens during development and cannot produce antibodies to them (Section 34-L). Occasionally, however, the self-recognition system breaks down. In most cases, we do not know why this happens. However, in some instances it is because the body is stimulated to produce a large amount of an antibody in response to a foreign antigen that is very similar to one of the body's own antigens; in this case, the antibody may then go on to destroy the body's similar protein as well as the foreign antigen. For example, antibodies formed during a bout of rheumatic fever may later cause autoimmune reactions that break down the body's proteins, particularly in the heart. Rheumatoid arthritis and a number of other devastating, although fairly rare, diseases are also thought to be caused by autoimmunity.

34-K Anaphylaxis and Allergies

A second exposure to an antigen usually leads to a secondary reaction that is greater and swifter than the primary reaction; however, it occasionally leads to a reaction so violent as to cause tissue damage. This is known as a **hypersensitivity reaction**.

If an antigen such as egg albumin is injected into a guinea pig, it has no obvious effect. If the injection is repeated three weeks later, the sensitized animal produces the symptoms of general **anaphylaxis**: the muscles of the bronchiole walls contract, constricting the air passages to the lungs, and the capillaries dilate. The animal will probably die unless injected with adrenalin, which counteracts the bronchial constriction. Similar **anaphylactic shock** occurs in human beings who are **allergic** to such things as penicillin or insect stings. About 10% of the population has irritating but less dangerous allergies to substances (such as pollen or the mites in house dust) that never get into the bloodstream.

Group E immunoglobulins are responsible for allergic reactions. Most people produce very little immunoglobulin E, but when these antibodies are produced they bind to mast cells (see Figure 34–3). When an allergy-producing antigen (an **allergen**) reaches such a bound mast cell, the antigen binds to the immunoglobulin and the mast cell self-destructs, releasing histamine, which causes a hypersensitivity reaction. Histamine makes arterioles and capillaries dilate and increases the permeability of capillary walls, so that fluid escapes and swells the tissues.

Presumably the E immunoglobulins did not evolve because allergies were so valuable to afflicted members of the human race. The normal role of these antibodies is almost certainly to protect the body from infection by platyhelminth parasites (tapeworms and flukes). Mast cells are concentrated in the lining of the gut and nasal passages, the areas usually invaded by these parasites. Although flatworm infestations are not a major threat to human health in many parts of the world today, this is largely due to improved hygiene in the last 200 years; such parasites undoubtedly played an important role in human evolution. Nowadays, in the small percentage of people with high levels of immunoglobulin E, the reaction is triggered, not by platyhelminth antigens, but by alternative (allergy-causing) antigens such as hens' eggs, animal dander, and chocolate.

Those who suffer from allergies may derive some consolation from the thought that they are apparently less likely to develop tumors than are other members of the population, although we do not know why this is so.

34-L   Immunological Tolerance

A mature immune system produces an immune response or an allergic reaction to a foreign antigen, but not to any of the thousands of potential antigens in its own body; that is, it distinguishes "self" from "not self." How does the body develop this **tolerance** to its own antigens?

Normally, if one injects a little blood from one cow into another, the recipient's immune system rapidly destroys the introduced blood cells, recognizing them as foreign. However, nonidentical (dizygotic) twin cattle do not produce immune reactions to injections of blood from the other twin. Such cattle grow up with some blood cells from the other twin in their blood; they receive the foreign cells as fetuses when they share the same placenta and their blood systems are partly linked.

This evidence led Sir Francis McFarlane Burnet, one of the fathers of immunology, to suggest that the body learns not to produce antibodies to antigens it encounters as a fetus. The theory was borne out by experiments in which foreign cells and other antigens were injected into young animals before or immediately after birth. The earlier an animal encounters an antigen, the more likely it is to become tolerant to that antigen; that is, it will not produce an immune response if the antigen is introduced into the body again in later life. Tolerance can also occasionally be induced in older animals, but it is much less easy to induce and less predictable than in a fetus or newborn.

Allergies are sometimes treated by attempting to induce tolerance to allergens. Small amounts of the allergen are repeatedly injected, in the hope that the body will cease to react to it. This treatment does not always work; since we know little about how tolerance is produced, we do not know why. Part of the explanation probably lies in the fact that the immune response and the genes that control the immune system vary enormously among individuals.

34-M   Cancer

Tumors are clumps of cells that grow and multiply abnormally. Some tumors are harmless, like the common wart or fibroid cysts of the uterus; on the other hand, a tumor may become **malignant**, growing uncontrollably, destroying nearby tissues, and eventually causing death. A "cancer" is a malignant tumor.

Cancerous cells differ from normal cells in three main ways: they divide more rapidly, they do not stick to each other so firmly as do normal cells, and they **dedifferentiate**: that is, they look as if they have reverted to an early stage in their development. When ciliated cells in the bronchi are transformed into malignant cells, for instance, they lose their cilia, dedifferentiating into formless cells that divide as rapidly as embryonic cells. If tumor cells lose their normal ability to stick tightly to neighboring cells, they may come unstuck and **metastasize**, or travel to other parts of the body, where they may start new tumors.

Nobody knows if there is a common cause for all cancers. Genetic changes probably always occur in cells that become cancerous. This is sometimes a secondary effect rather than the cause of the cancer in the first place, however, because while many **carcinogens** (cancer-causing agents) are also mutagens (causing genetic changes), some are not.

Some forms of cancer are produced by viruses, and there are people who think that all cancers will turn out to be of viral origin. This theory is extremely difficult to study. To produce convincing evidence, one would have to extract a virus from a cancerous cell and show that it transformed a normal cell into a cancerous one. This is often almost impossible because cancer usually takes many years to develop, at least in humans. If a virus is introduced into a cell that becomes cancerous years later, it is always possible that some event in the intervening years, and not the virus itself, caused the cancer.

The viruses that cause influenza and colds are replicated and cause disease when they invade a cell. Other viruses, including some that are known to cause cancer, do not immediately destroy the cells they have entered; they just sit there, often for long periods, doing no damage. The genetic material of the virus becomes part of the genetic material of the cell it has invaded (Figure 34–10); it is duplicated with the cell's own DNA and passed on to the daughter cells at cell division just as if it were one of the cell's own genes. Latent viruses of this sort are very common. We undoubtedly have dozens of them in our bodies. For instance, most adults in northern Europe and the United States contain Epstein-Barr (EB) virus. Such a latent virus may live indefinitely and be passed from parent to child in an egg or sperm. (Transmission of a virus from parent to child may be one reason susceptibility to some kinds of cancer is hereditary.) A familiar latent virus is the herpes virus, which when stimulated by unknown factors leaves its latent state, is reproduced, and causes the cell damage known as a "cold sore." (There is also evidence that herpes may sometimes cause cancer.)

Studies on rats and mice suggest that some carcinogens cause cancer by prodding latent viruses already in the cell into activity. For instance, the EB virus is invariably found in children with Burkitt's lymphoma, a lymph cancer common in parts of Africa. Burkitt's lymphoma is not common in the United States, however, although the EB virus is widespread, which suggests that whatever factor besides the EB virus is necessary to produce the cancer is less common in the United States than in Africa.

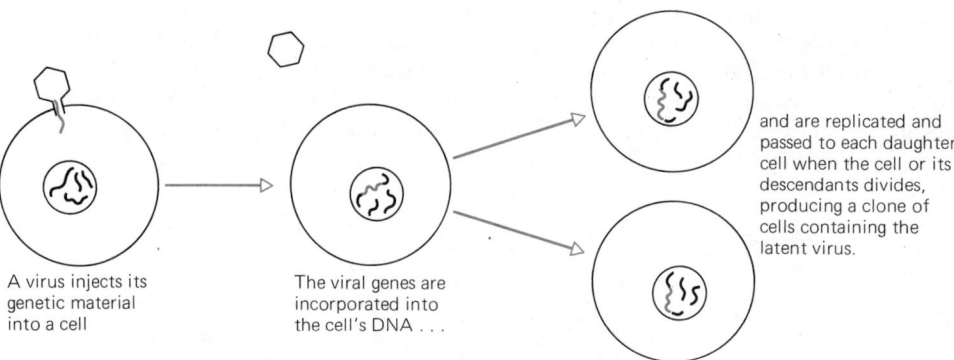

A virus injects its genetic material into a cell

The viral genes are incorporated into the cell's DNA . . .

and are replicated and passed to each daughter cell when the cell or its descendants divides, producing a clone of cells containing the latent virus.

Figure 34-10 A latent virus in a cell.

TABLE 34-2	SOME SUBSTANCES KNOWN TO BE CARCINOGENIC
SUBSTANCE	COMMENTS
Asbestos dust Chromium compounds Some petroleum products	Workers in these industries have high risk of lung cancer
Tobacco smoke	Quit now!
Estrogen	Mammalian hormone carcinogenic in large amounts
X-rays	Many people are exposed to unnecessary medical x-rays
Benzene	Used to be a common solvent in chemical labs

Cancer is the second most common cause of death in the United States and accounts for 20% of all deaths. To put this figure, which scares many people, into perspective, an American has almost as great a chance of dying of an act of homicide as of cancer, and before the age of 45, homicide is a much more probable cause of death (although perhaps that is not a particularly comforting thought!). Deaths from cancer have increased in the twentieth century. This is mainly because cancer is largely a disease of later life, and people are living longer instead of dying of infectious bacterial diseases in early life as they used to. (Life expectancy now is about 80 years, compared with about 45 years in 1900.) Some people, however, think that there has also been an increase in cancer because carcinogens are becoming more common in our environment (Table 34–2).

In recent years, treatments that produce genuine cures have been found for some types of cancer, such as one form of childhood leukemia, and the death rates from other cancers have been reduced. This is partly due to new treatments and partly because people are more aware of the need to seek treatment as soon as they suspect that they may have cancer.

SUMMARY

All animals have unspecific defense mechanisms that protect them from diseases. Examples of unspecific mechanisms include the defense systems provided by skin, interferon, phagocytes, and bactericidal fluids. Vertebrates, and to a lesser extent invertebrates, are also protected by much more specific immune responses.

Immune responses are characterized by specificity, by the recognition of antigens as either part of the body or as foreign, and by "memory" that the body has encountered a particular foreign antigen before. As a result, the second or later reaction to an antigen is faster and more extensive than the first.

T lymphocytes are responsible for the rejection of tissue transplanted from other individuals by the cell-mediated response. The histocompatibility antigens label cell surfaces and determine whether or not a tissue or organ transplant will be rejected by the recipient's immune system.

B lymphocytes are responsible for the humoral immune response triggered by binding of antigens with antibodies in the blood. The body produces hundreds of thousands of different kinds of immunoglobulins that bind antigens with greater or lesser specificity. Exposure of the immune system to an antigen causes a primary immune response, in which the lymphocytes that produce the antibody specific to that antigen are stimulated to divide and to produce and release more antibody. After the invasion has been defeated, an enlarged clone of these lymphocytes remains as the body's "memory" of the antigen, so that a secondary immune response to the antigen is greater and more rapid than the first. Foreign antigens are destroyed by phagocytes such as macrophages, which are stimulated to engulf the antigen only when it is bound by an antibody; in the case of some bacteria this reaction requires the assistance of the complement reactions.

Vaccination stimulates a primary immune response to an antigenic pathogen so that the body responds with an effective secondary response if it later encounters the pathogen.

The vast diversity of immunoglobulins in the body is produced by recombination of relatively few genes, each of which specifies part of an immunoglobulin molecule. The body does not produce antibodies to its own antigens. In some unknown way, the immune system comes to tolerate the body's own antigens during embryonic development.

Anaphylactic shock and allergic reactions occur when an allergen induces mast cells bound to E group immunoglobulins to release histamine.

Cancer is a collection of diseases in which normal mechanisms that control cell division, differentiation, and adhesion do not function. Latent viruses, duplicated when the host cells duplicate, are probably necessary to the formation of many types of cancer. Development of such a cancer is probably triggered, not by the virus itself, but by another infection or by an environmental carcinogen. Many cancers are destroyed by the immune system, which recognizes tumor cells as foreign by the presence of different antigens on their surfaces.

OBJECTIVES

From your study of this chapter, you should be able to:

1. Describe three unspecific defenses of animals against disease.

2. List three characteristics of an immune response and give an example of each.

3. State what each of the following is and describe its role in immune responses: antigen, antibody, histocompatibility antigen, bone marrow, lymph node, thymus gland, bursa of Fabricius, T lymphocyte, B lymphocyte, macrophage, memory cell.

4. Describe what is known of the cell-mediated immune response, and give an example of such a response.

5. Describe the steps by which a bacterium is destroyed during a primary humoral immune response that does not involve the complement reaction.

6. List three roles played by lymphocytes in the immune system.

7. Explain how vaccination protects an animal from catching a serious case of a disease.

8. Describe how an allergic reaction comes about.

9. State how a cancer cell differs from a normal cell.

10. Describe the role of the immune system in cancer therapy.

SELF-QUIZ

Match the following structures with the function each performs:

____ 1. A foreign macromolecule that may endanger the body

____ 2. Site of a filter that removes invaders from the body

____ 3. Long-lived cells that help the body respond quickly to previously encountered antigens

____ 4. Macromolecules that agglutinate foreign molecules in the bloodstream

a. lymph node
b. B lymphocyte
c. thymus gland
d. antibody
e. antigen
f. memory cells

5. The introduction of a bacterial antigen into the body triggers a response specifically against that antigen by:
   a. causing antibody molecules to assume a shape that permits them to bind and agglutinate the antibody
   b. causing mutations in cells so that they produce antibodies to the antigen that caused the mutation
   c. causing cells with the proper antibody to disintegrate and release the antibody
   d. stimulating reproduction of cells that make the antibody to that antigen

6. If most of its lymphocytes were drained out of an animal via the thoracic duct, the animal would *not*:
   a. lose its ability to combat cancer
   b. produce new lymphocytes
   c. lose its ability to reject skin grafts
   d. lose its ability to mount a secondary immune response
   e. produce an increased number of red blood cells

7. Skin can be grafted from one identical human twin to another, time after time, without being rejected because:
   a. immunoglobulins in the blood do not react to antigens on the other twin's cells
   b. the twins have the same MHC genes and antigens, and so the twins' cells do not stimulate a cell-mediated response in each other
   c. the twins have been exposed to each other's MHC antigens as fetuses and, as a result, do not mount immune responses against each other
   d. macrophages cannot destroy the twin's cells because the cells are not bound to antibody
   e. B lymphocytes do not react to the cell surface antigens of an individual with the same MHC genes

8. Vaccination protects the body against catching a disease because:
   a. it provides antibodies synthesized by another animal
   b. it makes the disease organism histocompatible with the body
   c. it produces an enlarged clone of memory cells against that disease
   d. it builds up an immunological tolerance for the disease antigen
   e. it releases large amounts of unspecific defensive secretions

9. Assuming that cancerous cells bear cell surface antigens, which of the following statements about cancer therapy is likely to be true?
   a. Attempts to produce antibodies to tumors will be futile because each tumor contains different cells bearing thousands of different antigens.
   b. The cell-mediated response against MHC antigens cannot destroy a tumor once it has metastasized.
   c. Tumor antigens will stimulate immune responses, and therapy that increases the magnitude of such responses should be effective.
   d. Tumors caused by bacteria are likely to bear viral antigens on their surfaces and so may be attacked by interferon and anti-viral immunoglobulins.
   e. Vaccination by a single cancer antigen that would protect against all kinds of cancer should be theoretically possible.

QUESTIONS FOR DISCUSSION

1. What might be the selective advantage of a baby's being born before its immune system has matured?
2. Studies have shown that multiple sclerosis, poliomyelitis, mononucleosis, and Hodgkin's disease are more common among children and young adults who have few or no siblings, few playmates, uncrowded homes, and well-educated, well-to-do parents. How might these factors affect their immune systems' ability to fend off the viruses responsible for these diseases?

REFERENCES AND FURTHER READING

Golub, E. S. *The Cellular Basis of the Immune Response.* Sunderland, MA: Sinauer Associates, 1977. An excellent little paperback. Concentrates on the cellular immune responses involved in organ transplantation and similar reactions which we have covered only scantily in this chapter.

Gordon, B. L. *Essentials of Immunology*, 2d ed. Philadelphia: F. A. Davis Co., 1974. Immunology with a decidedly medical slant.

Roitt, I. M. *Essential Immunology*, 3d ed. Oxford: Blackwell Scientific Publications, 1977. The best introduction to modern immunology; highly readable.

CHAPTER 35

EXCRETION

A great deal of evidence suggests that life began in the sea. First, most invertebrates and primitive plants are marine (living in salt water). Second, the body fluids of most invertebrates resemble seawater in their salt concentration. The body fluids of vertebrates are also similar to the sea in salt composition, but they are only about one-third as concentrated. The sea forms an almost ideal medium for life, largely because its properties are so constant. It has a stable salt composition and a constant pH. Because of its high thermal capacity, its temperature changes very little and very slowly. Fresh water and land, the other major environments of the earth, lack these properties.

The conditions under which animal cells can live are very restricted. This is illustrated by the remarkable similarity among the body fluids of all animals and by the difficulty of culturing animal tissue outside the body. Most cells cannot survive even seemingly small variations in the composition of the surrounding fluid: the heart stops if there is a small increase in the potassium content of the blood; a slight increase in the magnesium content of the body fluids blocks nerve function.

Every animal must maintain the composition of its extracellular fluid, which surrounds all cells, in the condition suitable for life. This is part of the general problem of **homeostasis**, or "staying the same," which includes regulation of tem-

TABLE 35-1 EXCHANGE OF SUBSTANCES BETWEEN THE HUMAN BODY AND ITS ENVIRONMENT

SUBSTANCE	DAILY INTAKE	DAILY OUTPUT
Water	1 litre in fluids 1 litre in foods 0.35 litre from oxidation of food	1 litre in urine 0.75 litre in perspiration 0.5 litre in expired air 0.1 litre in feces
Solid food	2 kilograms	0.15 kilogram fecal weight
Oxygen and carbon dioxide†	12,450 litres (about a 2.3 m cube)	12,450 litres

† Because one molecule of carbon dioxide is produced for every molecule of oxygen consumed, the volume of oxygen consumed equals the volume of carbon dioxide produced.

perature, of pH, and of the amounts and proportions of salts and water. Homeostasis might be easier if an organism were a self-contained system, but every organism must constantly take in substances from its environment, use them in the chemical reactions of metabolism, and discharge the resulting waste products back into the environment (Table 35-1). With the constant flow of substances into the organism, through it, and out again, the task of maintaining the constant composition of the fluid surrounding its cells is formidable.

35-A   Connections Between Fluid Compartments

Single-celled organisms maintain their internal environments by exchanging materials directly with the external environment. In a multicellular body, however, most cells are not in contact with the environment, but live surrounded by extracellular fluid. This is the most crucial fluid in the body, because its composition determines what substances can enter and leave a cell. One of the functions of the circulatory system is to maintain the correct composition of the extracellular fluid.

Let us follow a hypothetical drop of fluid in its journey through the main fluid-containing compartments of a vertebrate's body. The drop leaves the heart via the aorta, the largest artery, and is pushed into smaller arteries, then arterioles, until it reaches a capillary bed. The walls of capillaries are thin and permeable, and the blood pressure forces some fluid out of the capillary into the extracellular fluid. Most of this fluid is resorbed by the capillary, but our drop remains in the extracellular fluid and eventually finds its way into a lymph vessel. The lymphatic system returns surplus extracellular fluid to the blood by way of ducts that empty into the large veins near the heart; this completes a round trip for our drop of fluid. Alternatively, it could have seeped out of capillaries in the covering of the brain and become part of the **cerebrospinal fluid**, which bathes the brain and spinal cord. This fluid also drains back into the body's veins and returns to the heart.

Because all the fluid compartments of the body are in contact with one another, an animal can regulate the composition of all its body fluids by controlling the content of any one of them. In vertebrates the liver and kidneys are the most important organs that monitor the composition of the blood and thus keep the composition of all the body fluids constant.

35-B   Substances to Be Excreted

The chief wastes produced by the body are carbon dioxide and water from breakdown of organic molecules, and nitrogenous wastes from breakdown of proteins. Carbon dioxide is excreted across the respiratory surfaces of the body. Excretory organs such as kidneys have the two major functions of removing nitrogenous wastes and regulating the body's water content. In addition, excretory organs control the body's content of salts and of substances like spices, drugs, and hormones, which occur in lesser amounts. Onions, garlic, and some other spices have volatile

TABLE 35-2 IMPORTANT WASTE SUBSTANCES PRODUCED BY THE HUMAN BODY, AND SITES AT WHICH THEY ARE EXCRETED

SUBSTANCE EXCRETED	EXCRETING ORGAN(S)
Nitrogenous wastes	Kidneys Skin (small amount in sweat)
Water	Kidneys Skin Lungs
Salts	Kidneys Skin (in sweat)
Carbon dioxide	Lungs
Spices	Lungs Kidneys

components that leave the body through the lungs. Other parts of the same spices are excreted through the kidneys. Penicillin and other drugs are removed from the system primarily via the kidneys. The kidneys, liver, and lungs carry out **detoxification**, altering substances to forms that are not poisonous to the body.

Although the kidneys control the salt composition of the blood, some salts are also excreted through the skin in the sweat, and some leave with the feces. When large amounts of water and salts are lost through the skin as sweat, the urine output of the kidneys diminishes. A human being working in desert conditions may lose sweat at the rate of more than one litre an hour, and this produces a loss of 10 to 30 grams of sodium chloride per day. This heavy loss of salt causes no physiological difficulty at the time, but drinking water after such heavy sweating dilutes the extracellular fluid and leads to muscle cramps—an illustration of the importance of maintaining the composition of the body fluids.

Note that undigested food from the gut does not appear on our list of substances excreted by the body. Food that passes down the digestive tract and out the anus is not excreted in the scientific sense of the term; it is **egested**—that is, it travels through and is expelled from the body without ever passing through a plasma membrane to become a part of the body. The term "excretion" is correctly applied only to substances that must cross plasma membranes to leave the body.

Nitrogenous Wastes

Nitrogenous wastes are produced by the breakdown of proteins in the body. The first step in this process is the breakdown of proteins to amino acids; this occurs in the digestive tract as food proteins are digested, and in the body cells, where pro-

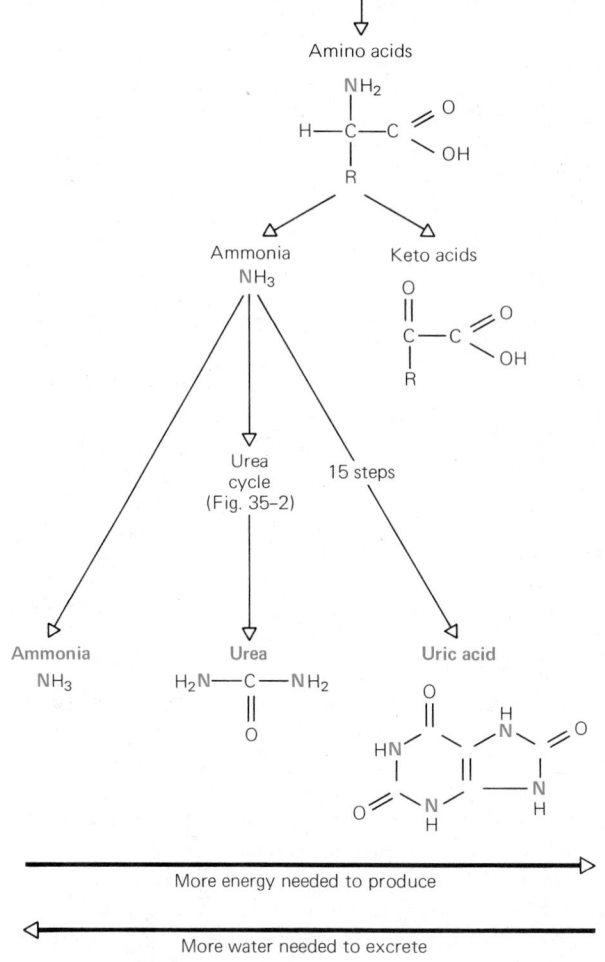

Figure 35-1 Nitrogenous wastes are formed by deamination of excess amino acids resulting from digestion of proteins. The first product of deamination is ammonia; in many animals ammonia is converted to urea by the urea cycle (ornithine cycle). Still other animals convert ammonia to uric acid by an even more lengthy pathway (not shown). Production of urea and uric acid expends energy but conserves water.

teins are constantly made and destroyed. Fats and carbohydrates are stored in the body for future use, but the body cannot store proteins or amino acids. Amino acids in excess of the body's immediate needs are **deaminated**; that is, their amino (—NH$_2$) groups are removed. This leaves keto acids,* which can be used for energy or stored as a carbohydrate or fat. Each —NH$_2$ group removed picks up another hydrogen atom and becomes NH$_3$, ammonia (Figure 35–1).

Ammonia is toxic in any but the most dilute solution. Some animals, especially animals that live in water, dissolve the ammonia in large quantities of water from their environment and excrete this dilute solution. Since ammonia is the first metabolic breakdown product of amino acids, it can be produced with very little energy. Thus it is energetically advantageous to excrete ammonia. However, the fact that excretion of ammonia requires large quantities of water is often disadvantageous, particularly to land-dwelling animals.

Some land-dwelling animals, such as adult amphibians and mammals, are often short of water, and they conserve water by excreting urea instead of ammonia. (Members of the class Chondrichthyes, which are marine, also excrete urea, but theirs is a different case and we will discuss their unique adaptation later.) Such animals convert ammonia to urea by way of the **urea** or **ornithine cycle** (Figure 35–2). This biochemical conversion requires energy, but for an animal with a limited supply of water, it is worth using this energy to produce urea. Since urea is much less toxic than ammonia, it can accumulate in higher concentrations without

---

* Keto acid   A molecule with both a keto $\left(\begin{array}{c} \diagup \\ \diagdown \end{array} C{=}O\right)$ and a carboxyl $\left(-C{\displaystyle \mathop{\lessgtr}^{OH}_{O}}\right)$ group.

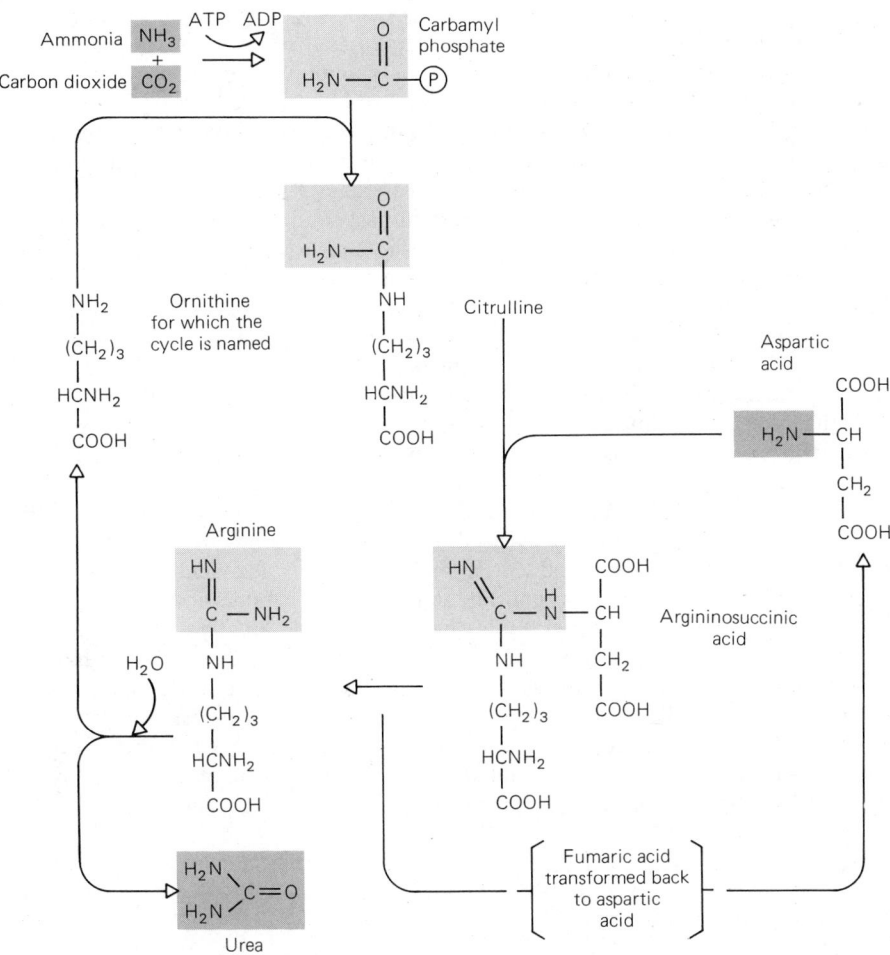

Figure 35-2 The urea or ornithine cycle. Carbon dioxide and a molecule of ammonia are combined and attached to a molecule of ornithine, forming citrulline. Addition of another ammonia, in the form of an amino group, results in a molecule of arginine. Arginine is then broken apart to give a molecule of urea and the original molecule of ornithine, which goes through the cycle again.

TABLE 35-3 ADVANTAGES AND DISADVANTAGES OF NITROGENOUS WASTES IN RELATION TO HABITAT

WASTE	ADVANTAGES	DISADVANTAGES	HABITAT	EXCRETED BY
Ammonia	Produced with little energy.	Toxic in concentrated solution. Must be excreted in lots of water.	Water	Marine and fresh-water invertebrates, bony fish, amphibian larvae.
Urea	Less toxic than ammonia. Less water needed to excrete it.	Requires some energy to produce.	Land / Sea	Adult amphibians, mammals. Chondrichthyes
Uric acid	Very little water is excreted with it.	Requires considerable energy to produce.	Land	Reptiles, birds, insects.

damaging the tissues, and can be excreted in a more concentrated form, requiring less water.

In mammals, most urea is formed by the liver, which takes excess amino acids out of the blood, deaminates them, and incorporates them into urea molecules. The brain and kidneys also form urea, but in lesser amounts.

Certain animals, notably reptiles, birds, and insects, eliminate ammonia by combining it with carbon dioxide and a number of other molecules to form **uric acid**. Uric acid synthesis takes about 15 steps and requires a great deal of energy. This investment of energy may be worthwhile as a water conservation measure, however, since uric acid can be excreted in almost solid form. In fact it is so insoluble that it precipitates spontaneously out of a concentrated solution.

Since the kidneys can handle nitrogenous waste only when it is in solution, birds and reptiles, which excrete uric acid, pass a dilute solution of uric acid from the kidneys into the **cloaca**, a common reservoir at the end of the urinary, digestive, and reproductive tracts. Here most of the water is resorbed and uric acid crystals precipitate and mix with the feces so that the two are voided together. A similar situation is found in insects.

Although most mammals do not produce uric acid, humans, the great apes, and Dalmatian dogs produce it in small quantities (2.5 to 6.4 mg% in the blood). However, in mammals, uric acid is produced by way of purine* metabolism rather than by the protein-breakdown pathway used by birds and reptiles. Human beings with certain metabolic disorders produce larger quantities of uric acid than usual, and this causes the disease known as gout. In gout, uric acid crystals accumulate in some of the joints of the body, especially in the toes, causing great pain.

Lower animals, which live in the sea or in fresh water or other moist habitats, excrete most of their nitrogenous wastes as ammonia. More advanced groups of animals that evolved the metabolic pathways and excretory organs for production and excretion of urea or uric acid were able to move into drier habitats. However, Table 35-4 shows that animals do not always excrete the form of nitrogenous waste predicted by their taxonomy; habitat plays an important part, too. Thus, although

---

* Purine   Member of a group of nitrogenous bases chemically related to purine (e.g., adenine and guanine, found in DNA and RNA).

TABLE 35-4 PROPORTION OF NITROGENOUS WASTE EXCRETED AS AMMONIA, UREA, AND URIC ACID BY VARIOUS ANIMALS

ANIMAL	PERCENTAGE† OF WASTE NITROGEN EXCRETED AS:		
	AMMONIA	UREA	URIC ACID
Protozoans	>98		
Cnidarians	52	4	
Earthworm	72	5	2
Molluscs: Squid	67	2	2
Terrestrial snail	22	17	7
Arthropods: Marine crustacean	87		
Terrestrial insect	0		92
Aquatic insect	90		
Frog: Tadpole	78	20	
Adult	3–38	62–88	<1
Reptiles: Freshwater turtle	15	39	19
Green turtle (marine)	29–51	0–12	1–6
Python	9		89
Birds	3	10	87
Mammals: Mouse	4	84	2

† The figures do not always add up to 100 percent because varying amounts of nitrogen are excreted in the form of nitrogenous substances other than the three mentioned here.

particular enzyme systems may enable an organism to live in a habitat that would otherwise be inhospitable to it, it may not use these enzymes if its habitat makes them superfluous.

For example, not all reptiles excrete uric acid, and some animals that could produce urea do not. Conversely, most fish, although surrounded by water, do excrete a little urea. The lungfish, a peculiar animal able to crawl on land for short periods of time, normally lives in freshwater ponds and excretes ammonia. When the water dries up, the lungfish buries itself in the mud, reduces its metabolism to the bare minimum necessary to maintain a dormant state, and accumulates nitrogenous wastes in the form of urea. When it rains and another pool of water collects, the lungfish excretes a large quantity of urea and then returns to excreting ammonia for as long as the water supply remains plentiful.

Why do reptiles and birds excrete uric acid, whereas mammals turn their nitrogenous wastes into urea? One might expect all three groups to have the same excretory end product, since they have fairly similar habitat ranges as well as similar evolutionary origins from prehistoric reptiles. The production of urea or uric acid correlates with the mode of reproduction in these primarily terrestrial vertebrates. Birds and reptiles lay their eggs on land, and the shells of these eggs are fairly impermeable to water. The water enclosed within its shell is all the water the embryo will have until it hatches. Thus, if the embryo produced ammonia, or even urea, the concentration of nitrogenous wastes might become toxic before the embryo was ready to hatch. Uric acid, however, precipitates from solution and sits as a mass of solid crystals, leaving the water in the egg free for other uses. The body fluids of a mammalian embryo, on the other hand, make contact with its mother's fluids via the placenta, permitting it to rely on her system to dispose of its wastes; in effect, any water available to the mother can also be used by the mammalian embryo. Thus a mammalian embryo does not have to use the extra energy needed to produce uric acid rather than urea. Mammals appear to have lost the metabolic pathway by which birds and reptiles produce uric acid.

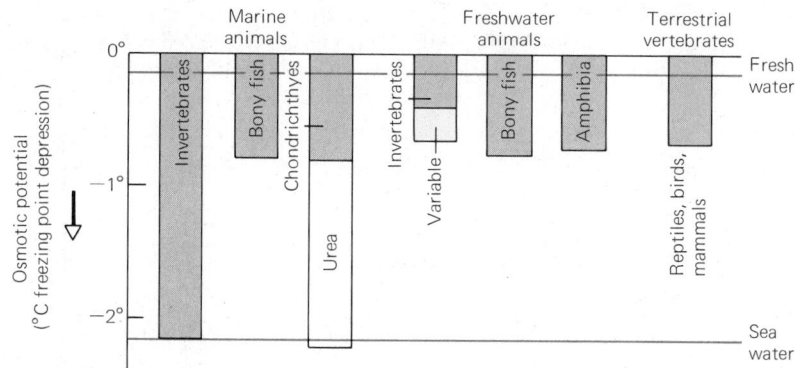

Figure 35-3 Osmotic potential of body fluids in various animal groups compared with osmotic potential of sea water and fresh water.

35-C Osmoregulation in Different Environments

The form of nitrogenous waste excreted by an animal depends on the availability of water, and this, in turn, depends on the water balance between the animal's body fluids and its environment. In general, the body fluids of marine invertebrates are isotonic* with the sea water in which they live. Freshwater invertebrates and most vertebrates have body fluids about one-third as concentrated as those of marine invertebrates; they are hypotonic* to sea water, but are hypertonic* to fresh water (Figure 35–3).

Marine fish and other marine vertebrates must prevent osmotic water loss to their hypertonic environment, and uptake of too many salts by diffusion. Freshwater organisms have just the opposite problems; they must prevent loss of salts by diffusion and water uptake by osmosis. They do this in part by excreting large volumes of dilute urine, but they must also conserve salts while ridding their bodies of nitrogenous wastes. How do vertebrates live with the osmotic problems posed by their environments?

Freshwater fish are covered with a mucous secretion that retards passage of water and salts through the body surface. A freshwater bony fish does not drink water, but it must pass water over its gills to obtain oxygen, and water inevitably enters the body across the permeable gill membranes. Such a fish eliminates water by producing a great deal of very dilute urine, but it loses salts both via the urine and via diffusion through the gill membranes. This is counteracted by the activity of special cells in the gills that take up salt from the environment by active transport (Figure 35–4). Freshwater fish also take in salts as part of their food.

---

*Isotonic Of the same osmotic pressure and osmotic potential.
*Hypotonic Having a lower osmotic pressure (and therefore a higher osmotic potential) (see Section 5-F).
*Hypertonic Having a higher osmotic pressure (and therefore a lower osmotic potential).

Figure 35-4 Osmotic adaptations of bony fish living in fresh water and in salt water. What changes must a salmon undergo as it returns from the sea to spawn in the stream where it hatched?

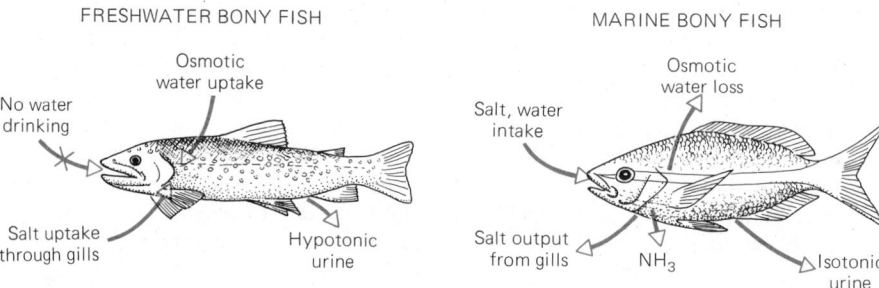

FRESHWATER BONY FISH

Osmotic water uptake

No water drinking

Salt uptake through gills

Hypotonic urine

MARINE BONY FISH

Osmotic water loss

Salt, water intake

Salt output from gills

NH_3

Isotonic urine

A marine bony fish must rid itself of excess salts, and the gill cells of such a fish pump salts out of the body rather than in. The marine fish loses water through the gills because its body fluids are less concentrated than is sea water. It also loses water in its urine, which is approximately isotonic with its body fluids. A marine fish makes up for these losses by drinking sea water and excreting much of the salt through its gills, again by active transport (see Figure 35–4). In a sense, although the marine bony fish is surrounded by water, it actually lives in a physiological desert, because it tends to lose water to its hypertonic environment; the fish has evolved adaptations that conserve water, enabling it to survive in this desert.

The major exceptions to the rule that the body fluids of vertebrates are about one-third as concentrated as sea water are the marine Chondrichthyes (sharks, skates, and rays). These cartilaginous fishes have evolved an interesting and unusual method of coping with their marine environment. Like most other vertebrates, they have body fluids with a salt concentration about one-third as great as that of sea water, but they also produce large quantities of urea and retain much of it in their body fluids (Figures 35–3, 35–5). Their tissues have become adapted to functioning at high levels of urea, which would be toxic to most other organisms. The combination of salts and urea lowers the osmotic potential of the body fluids to slightly below that of sea water, so that these fish actually absorb some water from the sea through their gills by osmosis, and this water can be used for excretion.

Birds that live at sea without access to fresh drinking water acquire water in a manner similar to that of a bony fish; they drink sea water, and a **salt gland** (or **nasal gland**) in the head excretes a very concentrated salt solution. This drips out of the **nares**, or nostrils. Birds also excrete uric acid in very concentrated form; their kidneys and cloaca conserve as much water as possible (Figure 35–6).

Marine mammals, such as whales and porpoises, take in sea water along with their food. Their kidneys can produce a urine several times as concentrated as sea water (Figure 35–7). This is especially important for the carnivorous marine mammals, because their high-protein diet yields much urea to excrete.

The kidneys of some land vertebrates can also produce highly concentrated urine. Laboratory rats can live indefinitely when all they are given to drink is sea water. Sea water is too concentrated to support human life, for although the human

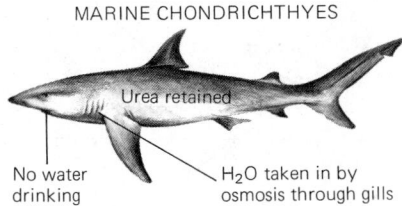

MARINE CHONDRICHTHYES

Urea retained

No water drinking

H$_2$O taken in by osmosis through gills

Figure 35-5 Osmotic adaptations in marine Chondrichthyes.

MARINE BIRDS

Salt water excreted

Drinks salt water

Concentrated urine via cloaca

Figure 35-6 Osmotic adaptations of marine birds.

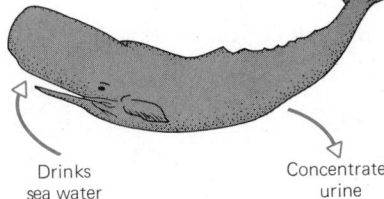

Figure 35-7 Marine mammals drink sea water but excrete the salts in a very concentrated urine, conserving the water.

Drinks sea water

Concentrated urine

kidney can produce urine that is slightly more concentrated than sea water, this is not enough to offset other water losses from the body. Water is lost constantly through the lungs and skin. Furthermore, the high content of magnesium and sulfate in sea water may produce diarrhea and increase water loss by way of the feces. For every swallow of sea water, even more of the precious body water must be used to excrete the salts taken in. Thus humans lost at sea are indeed surrounded by "water, water everywhere, nor any drop to drink."

If our hypothetical mariners had read this far in the chapter, they might recall that bony fish have body fluids only one-third as concentrated as sea water, and think that eating fish would be easier on the kidneys. This is of little help, however, because fish is high in protein that will force the body to produce large quantities of urea, again requiring more water for urine production. It is, however, possible to improve the osmotic situation by drinking the dilute body fluids squeezed from bony fish. In addition, shipwrecked sailors can eat algae, which contain more carbohydrate and less protein than fish. It usually takes only a few days for a human to die of thirst without drinking at all (humans do not starve to death for about two months), but people have survived at sea with no fresh water for more than two months by eating a low-protein diet and drinking any available hypotonic fluids (human urine may be one of these).

35-D   How Excretory Organs Work

Any excretory structure does three things:

1. It collects fluids from somewhere inside the body, usually from the blood or from spaces between organs.
2. It modifies the composition of this fluid by resorbing substances the body needs to retain, or by active transport of waste substances into the excretory product.
3. It provides some means of expelling the excretory product from the body.

During excretion, an organism expends metabolic energy. First, it uses energy in the breakdown of proteins and in the formation of urea or uric acid; second, energy is used in the active transport mechanisms, such as the sodium pump, that help to modify fluids collected from the body into final excretory products. Although human kidneys make up less than 0.5% of the body weight, they use 7.2% of the oxygen consumed by the body. Pumping blood from the heart to the kidneys takes another 2.7% of the body's total oxygen consumption, so that about 10% of the human body's energy is spent just moving blood to the kidneys and cleansing it.

35-E   Excretory Organs of Invertebrates

A freshwater protozoan excretes ammonia through the plasma membrane, but it must also rid itself of excess water that enters by osmosis. Most protozoans do this through **contractile vacuoles**, which accumulate water and then contract, squeezing this water back into the environment (Figure 35-8). Like all other freshwater organisms, protozoans face a scarcity of available salts. Before the contractile vacuole expels its contents, salts are removed by active transport.

Members of the marine phylum Cnidaria live in an isotonic environment and lose most of their wastes by diffusion. Freshwater flatworms (phylum Platyhelminthes), on the other hand, have organized, multicellular excretory systems, which permit them to void excess water (Figure 35-9); body fluids are collected into **flame cells** by the beating of cilia. The fluid then passes through a series of tubules until it reaches an excretory pore at the surface of the body.

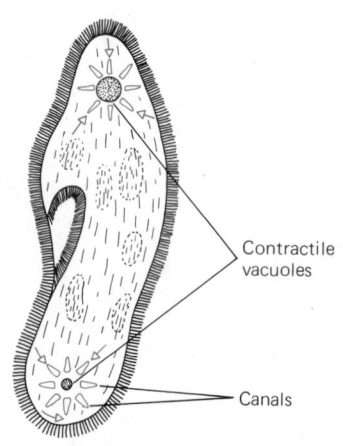

Figure 35-8 Contractile vacuoles in *Paramecium* excrete excess water that has moved in from its hypotonic environment. Fluid is collected through the canals; most of the salts are resorbed before the vacuole contracts and expels the fluid. Every 4–8 minutes the vacuoles eject a volume of water equivalent to the volume of the entire cell.

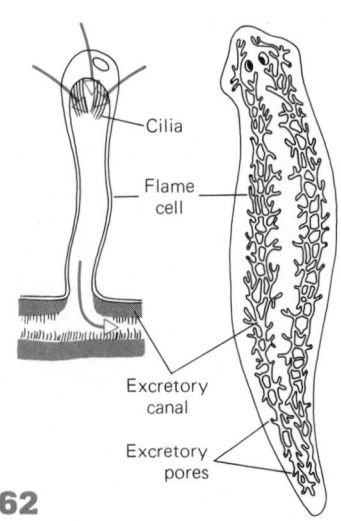

Figure 35-9 Excretory system of a planarian (phylum Platyhelminthes). Flame cells collect the body fluids and pass them down a system of excretory ducts to pores on the surface of the body.

562

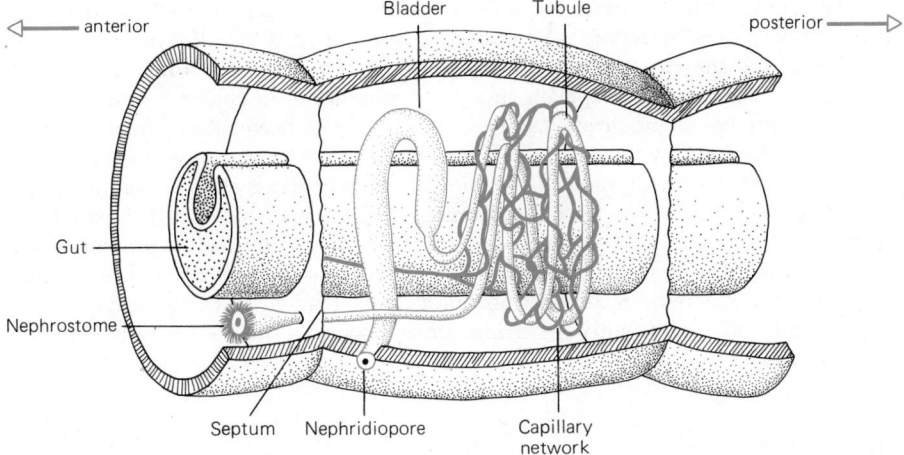

anterior ◁——   Bladder   Tubule   posterior ——▷

Gut

Nephrostome

Septum   Nephridiopore   Capillary network

Figure 35-10 An earthworm nephridium. The nephrostome is a funnel-shaped opening that collects fluid. The remainder of the nephridium lies in the next segment back. Capillaries of the circulatory system intertwined with the tubule resorb needed substances before the fluid reaches the enlarged bladder, from which the urine is discharged through the nephridiopore. Each segment, except a few in the anterior end, contains a pair of nephridia.

The functional unit of excretion in an earthworm is the **nephridium** (Figure 35–10). The nephrostome, a ciliated funnel, opens into the coelom and draws coelomic fluid into the long, thin, coiled tubule of the nephridium. As fluid flows through the nephridium, substances that the body needs are reclaimed and passed into surrounding capillaries of the circulatory system. Fluid that is expelled from the body through the nephridiopore contains water, nitrogenous wastes, and any salts that have not been resorbed.

Like the earthworm, most other invertebrates have nephridia, but insects have evolved a completely different system. They have long, slender **Malpighian tubules**, attached at one or both ends to the gut (Figure 35–11). Nitrogenous wastes from the body fluid are transformed into uric acid, which is then moved down the Malpighian tubule into the gut; cells in the lining of the rectum resorb water from these wastes before they are eliminated as fairly dry fecal pellets.

35-F The Kidney

Every vertebrate has a pair of kidneys. In lower vertebrates, the kidneys are long thin organs extending along either side of the backbone; mammalian kidneys, on the other hand, are very compact, and are the only "kidney-shaped" kidneys. The kidneys lie behind the **peritoneum**, the membrane lining the abdominal cavity.

The functional units of vertebrate kidneys are **nephrons**, which closely resemble the nephridia of earthworms. Each human kidney is made up of about a million nephrons.

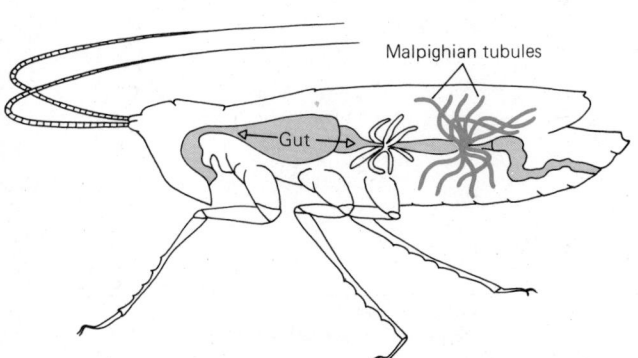

Malpighian tubules

Gut

Figure 35-11 The slender Malpighian tubules of insects discharge uric acid into the gut.

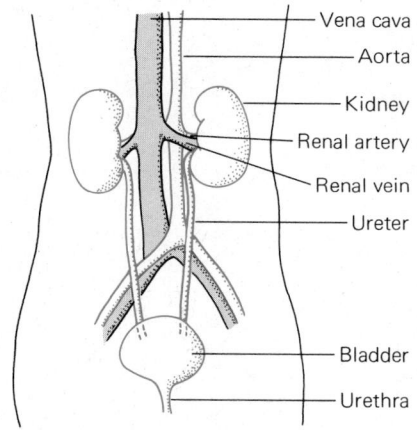

Figure 35-12 The urinary system of a human. The kidneys lie in the small of the back, behind the peritoneum and against the spinal column.

The **renal artery** carries blood from the dorsal aorta into the kidney. Here the artery divides into arterioles and then into capillaries. Fluid filters out of these capillaries into the cup-shaped end of a nephron. The **filtrate** (fluid) then travels through the tubule of the nephron, and the remainder of the blood follows along in capillaries outside the nephron. As the blood and filtrate travel along, the cells of the nephron tubule modify the content of both fluids, **resorbing** some substances from the filtrate and returning them to the blood, and **secreting** other substances from the blood to the urine that is being formed. The purified blood in the capillaries returns to the heart via the **renal vein**, whereas the urine finally passes down a collecting duct, leaves the kidney, flows down a **ureter**, and is stored in the **urinary bladder**. Water may be resorbed from the urinary bladder under appropriate hormonal conditions. Eventually the urine is expelled from the body via the **urethra** (Figure 35–12).

Functions of the Nephron

The human body contains about 5.6 litres of blood; 1.2 to 1.3 litres of blood pass through the kidneys each minute, for a daily total of about 1600 litres. This is nearly one-quarter of all the blood pumped out by the heart, and so a very high proportion of all the blood is passing through the kidneys at any time. There are only about 3 litres of blood plasma in the human body; thus every drop of plasma in the blood passes through the kidneys, where its contents are monitored, checked, and altered, about 560 times in any day, or essentially constantly. About 180 litres of filtrate pass through the nephrons in a day, at the rate of about 125 ml/min. Most of the filtrate is resorbed, so that only about 1 litre of urine is produced daily.

Let us follow the process of urine formation as the urine-to-be passes along a nephron (Figure 35–13). Blood from the renal artery flows through an **afferent arteriole** into the **glomerulus**, a capillary bed lying in the **nephric capsule** (or **Bowman's capsule**) of each nephron. The blood is under pressure and the walls of the capillaries and capsule are permeable, so that much of the fluid from the blood filters into the capsule, leaving behind large proteins and whole cells, which are too large to pass through the filter, along with the rest of the fluid.

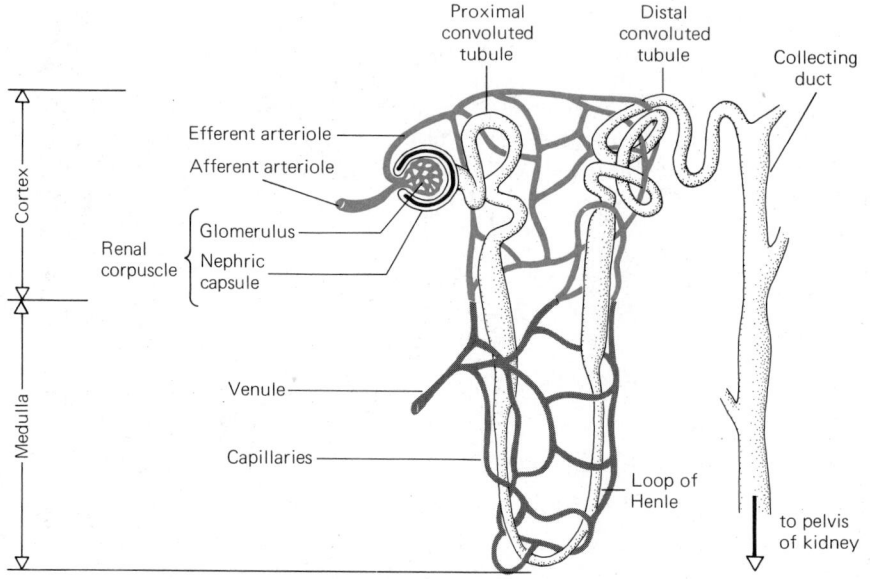

Figure 35-13 A nephron and its associated blood supply. A human has about 2 million nephrons.

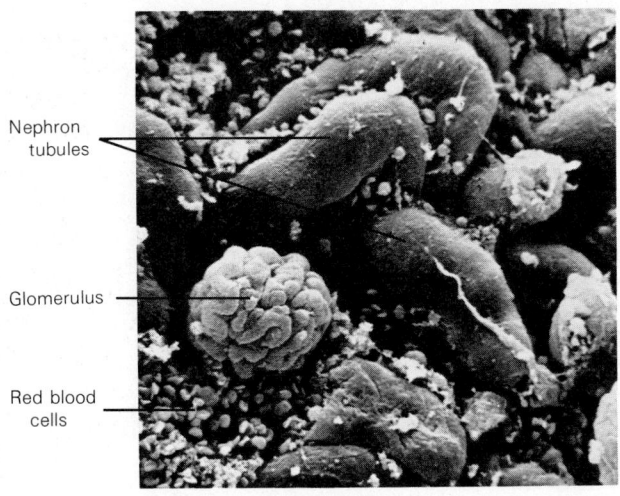

Nephron
tubules

Glomerulus

Red blood
cells

Figure 35-14 Scanning electron micrograph of the kidney cortex. The red blood cells have leaked from blood vessels broken during specimen preparation. (Biophoto Associates)

From the capsule, the glomerular filtrate passes into the tubule of the nephron, which has four main parts: the **proximal convoluted tubule**, the U-shaped **loop of Henle**, the **distal convoluted tubule**, and finally the **collecting duct**. As the filtered fluid passes through the tubule, its composition is altered by the secretion and resorption of various substances.

In the proximal convoluted tubule, a considerable amount of resorption takes place (Figure 35–15). Sodium and glucose are returned to the blood by active transport; water follows these solutes passively by osmosis. In many mammals and birds the loops of Henle are quite long, and their activities (discussed below) permit the production of relatively concentrated urine. From the loop of Henle, the fluid in the nephron passes to the distal convoluted tubule, another area of resorption and the chief area where secretion into the tubule occurs; the main substances secreted are potassium and hydrogen ions, by active transport, and ammonia, by diffusion. Secretion of hydrogen ions is a means of regulating the pH of the blood; when the blood becomes too acidic, more hydrogen ions and fewer potassium ions are secreted. Sodium is resorbed in the distal tubule, being exchanged for hydrogen ions to preserve the balance of electric charges. Various drugs, such as penicillin, are also secreted into the urine in this area. The collecting duct, along with the loop of Henle, plays a vital role in water balance.

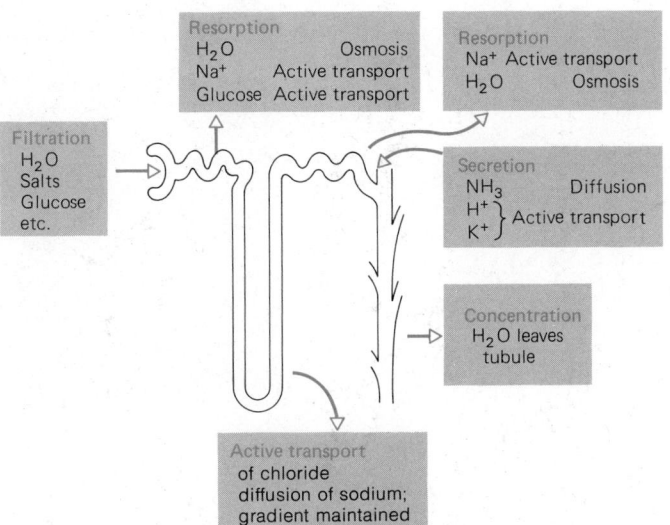

Resorption
H_2O Osmosis
Na^+ Active transport
Glucose   Active transport

Resorption
Na^+ Active transport
H_2O Osmosis

Filtration
H_2O
Salts
Glucose
etc.

Secretion
NH_3 Diffusion
$\left.\begin{matrix} H^+ \\ K^+ \end{matrix}\right\}$ Active transport

Concentration
H_2O leaves
tubule

Active transport
of chloride
diffusion of sodium;
gradient maintained

Figure 35-15 Summary of activities in the different parts of the nephron.

TABLE 35-5 **RELATIVE CONCENTRATIONS OF SUBSTANCES IN THE HUMAN GLOMERULAR FILTRATE AND IN URINE**		
SUBSTANCE	CONCENTRATION IN PLASMA (125 ml/min) IN mEq/l	CONCENTRATION IN URINE (1 ml/min) IN mEq/l
Na^+	142	128
K^+	5	60
Ca^{2+}	4	4.8
Mg^{2+}	3	15
Cl^-	103	134
HCO_3^-	28	14
Creatinine	1.1	196
Glucose	varies	0

THE LOOP OF HENLE

Although water is resorbed from the nephron, there is no known carrier that transports water molecules; a cell moves water across a membrane by using the active transport of solutes to set up a concentration gradient such that water molecules follow the transported substance by osmosis. The loop of Henle sets up an ionic gradient of this type. However, the actual concentration of urine takes place in the collecting ducts. To understand how this works, it is necessary to realize that the loops of Henle and collecting ducts lie intermingled in the **medulla** of the kidney (Figure 35–16). The other parts of the nephron lie outside the medulla in the outer region, or **cortex**, of the kidney.

Chloride in the filtrate is actively transported through some of the cells lining the loop of Henle and into the extracellular fluid outside. The negatively charged

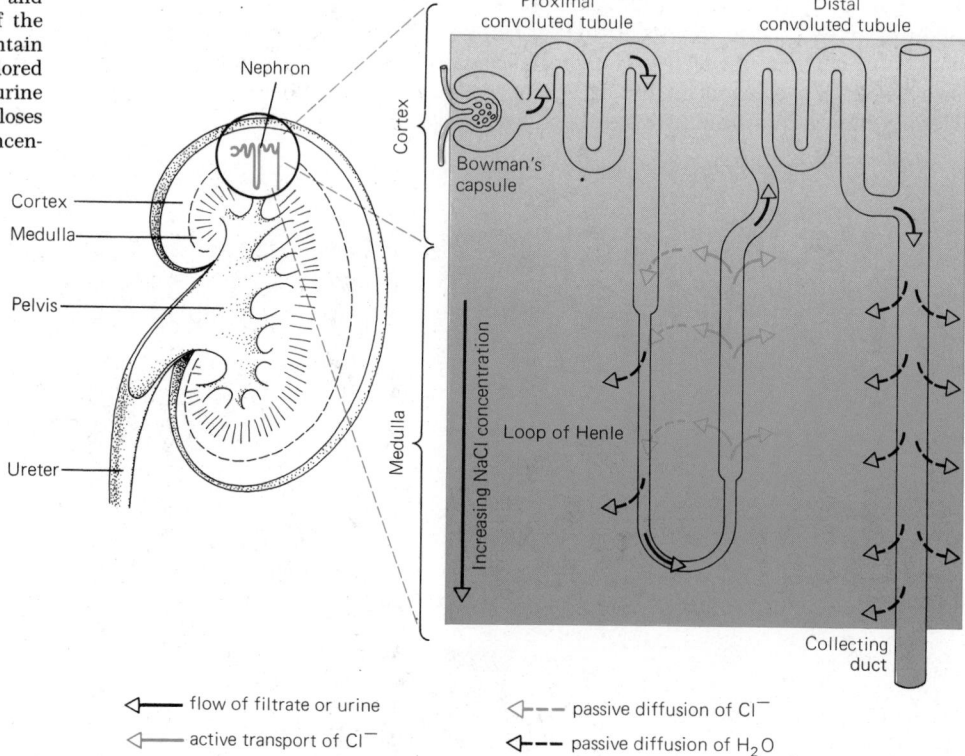

Figure 35-16 The loops of Henle and collecting ducts lie in the medulla of the kidney. Here the loops of Henle maintain a high sodium chloride gradient (colored shading) in the extracellular fluid; as urine passes through the collecting ducts it loses water by osmosis and becomes concentrated.

chloride ions attract positively charged sodium ions, which follow. Once the sodium chloride concentration outside the collecting duct is greater than that inside, water will cross the membrane by osmosis. The chloride "pump," the active transport system for chloride, therefore moves not only chloride ions but also sodium ions and water.

The active transport of chloride out of the loop of Henle does not start until the filtrate reaches the thick section of the ascending part of the loop. Since chloride transport begins nearer the bottom than the top of the loop, more chloride is transported out at the bottom; as the filtrate moves up, there is less and less chloride available to be transported out. Hence there is a gradient of NaCl, with the higher salt concentration near the bottom of the loop (colored shading in Figure 35–16). The ascending part of the loop of Henle is relatively impermeable to water, so that water cannot follow the ions out at this point.

The walls of the descending part of the loop of Henle are quite permeable, and chloride transported out of the ascending part of the tubule can diffuse back into the filtrate in the descending part. As the filtrate once again carries this chloride up the ascending part of the loop, the chloride is again actively transported out of the tubule. By constantly reusing the same chloride, the loop of Henle is able to set up a much higher gradient than would be possible if it used the chloride in the original filtrate only once. The real function of the loop of Henle is to set up a high concentration of sodium chloride in the medulla of the kidney. In some animals, urea is also transported out of the loop of Henle. This has not been shown in humans, but high concentrations of urea do exist in the medulla of the human kidney because urea diffuses readily out of the tubule and then, like sodium chloride, becomes "trapped," diffusing back into the descending loop and being moved around again and again.

With the glomerular fluid in the two parts of its loop flowing in opposite directions, the hairpin-shaped loop of Henle is also a countercurrent exchange system, with sodium, chloride, and urea constantly moving from the ascending to the descending part, so that there is very little difference in solute concentration between adjacent parts of the loop. The loop of Henle is often referred to as a **countercurrent multiplier** because this countercurrent exchange of the recycling solutes permits the osmotic potential in the medulla to be lowered so far.

From the loop of Henle, urine passes through the distal convoluted tubule to the collecting duct. The actual concentration of the urine takes place in the collecting duct. The walls of the collecting duct are permeable to water. As the urine passes down the collecting duct, it re-enters the medulla of the kidney, where the extracellular fluid has a low osmotic potential due to the sodium and chloride transported into it by the loop of Henle. Here water passes out of the urine and into the extracellular fluid by osmosis. The sodium, chloride, and water in the extracellular fluid of the medulla eventually diffuse back into the capillaries or into the lymphatic system.

35-G Regulation of Kidney Function

The function of the kidneys is under the minute-by-minute control of many factors, all of which work together to ensure that the composition of the blood, and therefore of the body's extracellular fluid, remains constant. For instance, the filtration rate remains nearly constant because certain cells in the kidneys can detect changes caused by variations in blood pressure and adjust the constriction of the muscles in the walls of the afferent and efferent arterioles, maintaining the proper blood pressure in the glomeruli. The rate of urine formation, and urine composition, on the other hand, change dramatically under different circumstances. For instance, although drinking a large quantity of water, which dilutes the blood, raises the filtration rate only slightly, the rate of water resorption drops so that the excess water is excreted from the body.

Urine composition and the rate of urine formation are largely regulated by hormones.

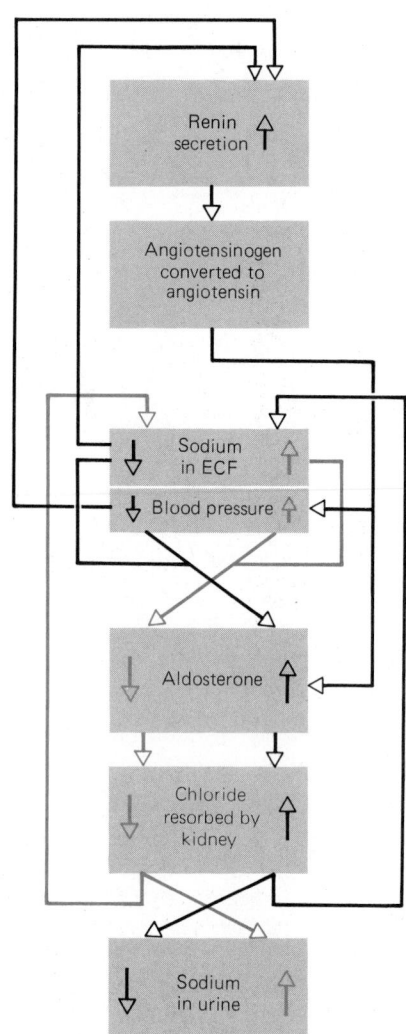

Figure 35-17 Interaction among aldosterone, renin, and angiotensin. Black arrows indicate events that tend to return low sodium levels to normal, and colored arrows indicate adjustments that lower sodium levels when there is too much sodium.

Vasopressin (ADH)

Vasopressin (also known as **antidiuretic hormone, ADH**) is a hormone released from the posterior pituitary gland in the brain. Its presence increases resorption of water from the urine. Loss of water from the body stimulates vasopressin secretion and slows the loss of water via the urine. The body may detect a decrease in its water content in one of two ways: either as a reduction in blood volume (e.g., caused by severe bleeding) or as a decrease (greater negativity) in the osmotic potential of the blood due to loss of water (e.g., from sweating).

Vasopressin appears to work by increasing the permeability of the collecting ducts to water. In the absence of vasopressin, the walls of the ducts are practically impermeable to water and very little water is resorbed from the urine.

Insufficient vasopressin production results in the disease known as diabetes insipidus, characterized by thirst and the production of large quantities of dilute urine. It is much less common than diabetes mellitus, which is characterized by sugar in the urine.

Aldosterone, Renin, and Angiotensin

Sodium accounts for about 90% of all the positive ions outside the cells of the body. Since water moves passively, the volume of water in the extracellular fluid is determined by the amount of salt. Thus the amount of sodium in the body is the single most important factor determining the volume and concentration of the blood and extracellular fluid.

The body's sodium content is controlled by the balance between intake in the diet and loss in the urine. (Sodium is also lost in the feces and in sweat, but less control can be exerted over these.) Vertebrates have developed nervous and hormonal systems that permit them to regulate how much sodium they eat and how much leaves the body. These mechanisms have conferred great selective advantages by permitting survival in a greater range of habitats than would otherwise be possible.

Angiotensin and aldosterone are hormones that together control how much sodium is resorbed from the filtrate in the nephron. **Aldosterone** is one of several steroid hormones secreted by the cortex of the **adrenal gland**, attached to the kidney. Aldosterone promotes resorption of sodium by the distal tubule. The rate of aldosterone secretion, and therefore its concentration in the blood, is determined by the salt content of the blood. A slight decrease in the sodium content of blood reaching the adrenal cortex causes an immediate increase in aldosterone secretion, which leads to resorption of more salt from the filtrate, and hence, to a decrease in the amount of sodium in the urine.

While aldosterone secretion can be controlled directly by the sodium concentration of the blood, it is also controlled indirectly by renin and angiotensin. **Renin** is secreted by the kidneys in response to a decrease in sodium or potassium in the blood, or in response to a reduction in blood pressure. Renin is an enzyme; it converts the glycoprotein angiotensinogen in the plasma into angiotensin. **Angiotensin** causes constriction of blood vessels and an increase in aldosterone secretion, both of which increase the blood pressure either directly or by raising its sodium concentration; an increase in blood sodium concentration decreases loss of water in the urine and therefore maintains blood volume (Figure 35–17).

SUMMARY   Living cells require a relatively constant chemical environment. Cells are continuously altering their immediate environment by taking in substances from it and converting them into new substances, some useful, some wastes which must be removed from the cell and from its immediate environment. Various specialized mechanisms have evolved in the animal kingdom as a means of maintaining a constant internal chemical environment in the face of the continuous flow of materials between the external environment, the extracellular fluid, and the interiors of cells.

The problem of excretion at first seems to be simply the problem of removing nitrogenous wastes before they reach toxic levels; however, the availability of salts and water in the animal's environment imposes restrictions on how this removal can be accomplished, and the animal must have adaptations that allow it to dispose of nitrogenous wastes while at the same time maintaining the proper salt and water (osmotic) balance in its body fluids. When all of these factors are taken into consideration, excretion is seen to require not only collection and disposal of excess substances, but also a means of retaining those that are needed.

The function of the vertebrate kidney is to maintain the composition of all of the body's fluids within the narrow limits necessary if the cells of the body are to function. The basic unit of the kidney is the nephron, a long tube closely associated with a capillary bed of the circulatory system. Blood plasma is filtered, under hydrostatic pressure, into one end of this tube. As the filtrate passes through the nephron, substances needed by the body are resorbed through the cells of the nephron tubule into the extracellular fluid, and then into the capillaries, either by diffusion or by active transport. Unwanted substances are secreted from the blood into the filtrate. After being changed in these ways during its passage through the tubule, the filtrate is collected as urine. Homeostatic mechanisms under hormonal control regulate the amount and composition of the urine produced.

Essay: Adaptations of Mammals to Sodium-Deficient Environments

Many arid and mountainous environments contain little sodium. Although no animal can survive with much less than the usual amount of sodium in its body, most mammals (and some insects) have adaptations that permit them to survive in such areas.

These adaptations may be divided into those that increase the dietary intake of sodium and those that reduce loss of sodium from the body. Herbivores, including reindeer, elephants, gorillas, elk, moose, sheep, and kangaroos, have been seen eating soil, drinking sea water, or consuming other improbable items of diet whose only redeeming value appears to be their high sodium content. Plainly an appetite for salt when the body is deficient in sodium is a useful adaptation, and it is common in mammals.

A number of physiological differences have been recorded between kangaroos, sheep, foxes, cattle, and rabbits in sodium-deficient mountainous areas of Australia and members of the same species in environments on the coast, where there is plenty of sodium. Whenever possible, animals in mountainous areas selected food plants containing sodium, and in addition held sodium loss from the body to a minimum. For instance, the urine of mountain animals contained virtually no sodium, whereas kangaroos and wombats on the coast had up to 300 milliequivalents per litre of sodium in their urine. This difference is undoubtedly due to the higher levels of renin, angiotensin, and aldosterone found in the mountain-dwelling animals. During the day rabbits produce soft feces, which they eat to extract further sodium (and vitamins). The fecal pellets egested by highland rabbits after the second digestion contained practically no sodium.

Sodium in the feces may be reduced by resorption from the gut. Sodium-deficient ruminants* also reduce the sodium loss in their feces by changing the composition of their saliva. Ruminants such as cattle and sheep produce many litres of saliva a day. This usually contains a high concentration of sodium bicarbonate, which produces an alkaline aqueous environment in the stomach for the microbial symbionts that ferment the animal's food. Much of the sodium bicarbonate solution in the stomach is absorbed back into the body as the feces pass down the intestine, but some is lost with the feces. Aldosterone decreases the amount of sodium in the saliva and causes potassium to be secreted in its place. Although the need to secrete large quantities of sodium in the saliva every day seems to impose a sodium-supply problem on ruminants, the effect may, in fact, be rather the other way around. Sheep survive temporary sodium shortages better than do non-ruminants such as humans or foxes. This is probably because the ruminant stomach provides a large store of sodium which can be steadily replaced by potassium and used for more vital functions in the body during times of sodium shortage.

---

* Ruminants   Mammalian herbivores in which the stomach contains fluids of an alkaline (basic) pH and is divided into fermentation chambers housing microorganisms that digest the food.

OBJECTIVES

From your study of this chapter, you should be able to:

1. (a) List four substances excreted by the human body; (b) name the organ(s) that excrete(s) each substance; (c) state the difference between excretion and egestion.

2. (a) State the origin of nitrogenous wastes in animal metabolism; (b) name three common nitrogenous wastes considered in this chapter; (c) state the advantages and disadvantages of each substance as an excretory end product, and relate these advantages and disadvantages to the animal's habitat; (d) name the animal groups commonly associated with each of these excretory end products; (e) use your knowledge of parts (c) and (d) to predict the main nitrogenous waste excreted by an animal.

3. Explain the relationship between nitrogenous waste formation, osmoregulation, energy expenditure, and habitat as factors in the problem of body fluid regulation. Interpret Figures 35–3 and 35–18 (in Self-Quiz) as they relate to this problem.

4. (a) List two problems encountered by a freshwater animal in maintaining homeostasis, and the adaptations of freshwater bony fish that enable them to cope with these problems; (b) list the problems encountered by animals in maintaining homeostasis in a hyperosmotic environment, and the adaptations of (1) marine Chondrichthyes, (2) marine bony fish, (3) marine birds, and (4) marine mammals in meeting these problems.

5. List three essential steps common to the action of excretory organs in animals.

6. Name the excretory structures found in the following forms: protozoans, planarians, earthworms, vertebrates, and insects; briefly explain how each works.

7. Define the terms resorption and secretion, and be able to explain kidney function in terms of these activities.

8. Draw or identify, and give the functions of, the following parts of a nephron and its associated structures: nephric (Bowman's) capsule, glomerulus, renal artery, renal vein, proximal convoluted tubule, distal convoluted tubule, loop of Henle, collecting duct (tubule).

9. Explain how the loop of Henle concentrates the glomerular filtrate.

10. Describe the effects of vasopressin, aldosterone, renin, angiotensin, blood pressure, sweating, and excessive bleeding on the volume and composition of the urine, as these are outlined in this chapter.

SELF-QUIZ

1. The pack rat, a rodent, often goes for long periods without drinking, eats leaves of juicy plants, and moves about in the open only in the evening and at night. From these habits, you can guess that it lives in:
   a. desert areas
   b. the Arctic tundra
   c. the woodlands of the eastern U.S.

2. The main nitrogenous waste substance excreted by the pack rat will probably be:
   a. ammonia
   b. urea
   c. uric acid

3. Study Figure 35–18. The crab *Carcinus* inhabits estuaries (mouths of rivers where they join the sea). Salinity of sea water is about 35 g per litre.
   a. Would the crab expend more energy resting on the bottom during high tide or low tide? (The salinity would be higher during high tide, when ocean water pushes into the mouth of the river.)
   b. What do you think it is using this energy for?
   c. What is the main nitrogenous substance it excretes?

4. An advantage of excreting nitrogenous wastes in the form of uric acid is that:
   a. uric acid can be excreted in almost solid form
   b. the formation of uric acid requires a great deal of energy
   c. uric acid is the first metabolic breakdown product of amino acids
   d. uric acid may be excreted through the lungs
   e. uric acid is highly toxic, so it is important for the animal to get rid of it

5. The main excretory structure in houseflies is the:
   a. Malpighian tubule
   b. flame cell
   c. nephron
   d. contractile vacuole
   e. nephridium

6. Salmon have gills that are more permeable to water than to salts. Salmon hatch in freshwater streams, and then migrate to the ocean. Once they reach the ocean, you would expect the rate of uptake of water into their bodies through the gills to:
   a. increase
   b. decrease
   c. remain the same

7. Urine leaves the kidney via:
   a. the renal vein
   b. the urethra
   c. the bladder
   d. the ureter
   e. the collecting duct

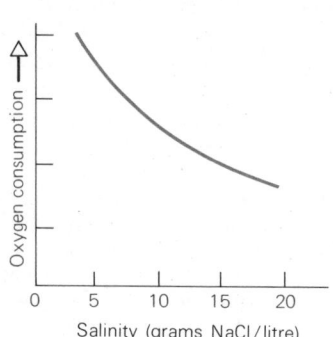

Figure 35-18 Oxygen consumption of an estuarine crab, *Carcinus*, as a function of salinity in the external environment.

8. Match each structure on the left with its function from the list at the right.

___ Loop of Henle

___ Renal artery

___ Proximal convoluted tubule

___ Glomerulus

___ Distal convoluted tubule

   a. carries blood into the kidney
   b. area where a considerable amount of resorption takes place
   c. main area of secretion
   d. filtration of blood
   e. plays a role in concentration of urine

9. Filtration into the kidney tubule is accomplished by means of:
   a. active transport
   b. hydrostatic blood pressure
   c. an osmotic potential gradient
   d. secretion
   e. diffusion

10. Severe dehydration causes a decrease in osmotic potential in the blood. This causes a(n) (increase/decrease) in the amount of urine produced. This change in urine production is caused primarily by:
   a. an increase in the amount of water filtered out of the blood
   b. a decrease in the amount of water filtered out of the blood
   c. an increase in the amount of water resorbed
   d. a decrease in the amount of water resorbed

11. A hormone that causes constriction of the efferent arterioles (leading out of the glomeruli) would cause a(n) (increase/decrease) in the blood pressure in the glomeruli and would cause a(n) (increase/decrease) in the amount of urine produced.

QUESTIONS FOR DISCUSSION

1. Why does an increase in aldosterone secretion increase the volume of extracellular fluid and increase blood pressure?

2. If we divide animals into marine and freshwater invertebrates, marine and freshwater bony fish, cartilaginous fish (class Chondrichthyes), amphibians, and terrestrial vertebrates, which animals are in the following osmotic situations?
   a. approximately in osmotic equilibrium with their environment
   b. must have adaptations to guard against "desiccation" (loss of body water to the environment)
   c. must have adaptations to prevent gain of excess water from the environment
   d. in danger of taking up too many salts from the environment
   e. must have adaptations to prevent loss of salts to the environment

3. Many fish lack a urinary bladder, but urinary bladders are found in amphibians and all higher vertebrates. What is the advantage of having a urinary bladder?

4. Would you expect a dolphin to have longer or shorter loops of Henle than those in your kidneys?

5. Certain reptiles have kidneys with no glomeruli in the nephric capsules. What effect would this have on urine formation? What type of habitat are such animals adapted to?

REFERENCES AND FURTHER READING

Baldwin, E. *An Introduction to Comparative Biochemistry*, 4th ed. New York: Cambridge University Press, 1964. A short and provocative book with emphasis on osmoregulatory problems and their solutions by animals.

Ganong, W. F. Renal Function. *Review of Medical Physiology*, 8th ed. Los Altos, California: Lange Medical Publications, 1977.

Prosser, C. L., Ed. *Comparative Animal Physiology*, 3d ed. Philadelphia: W. B. Saunders Co., 1973. Chapters 1, 2, and 7. For advanced students. Extensive reference list follows each chapter.

Schmidt-Nielsen, K. "Salt glands." *Scientific American*, January 1959. An account of the structure and function of salt glands and their contribution to osmoregulation, particularly in sea birds.

Schmidt-Nielsen, K. "Marine vertebrates—problems of salt and water." *Topics in the Study of Life*. New York: Harper and Row, 1971.

Scoggins, B. A., *et al.* "The physiological and morphological response of mammals to changes in their sodium status." *Memoirs of the Society of Endocrinology* 18:577–601, 1972. An advanced discussion of how hormonal control permits mammals to regulate their sodium content by regulating the amount of sodium excreted by the kidneys and by other mechanisms.

Smith, H. W. *From Fish to Philosopher*. Garden City, New York: Doubleday, 1961. An evolutionary history of internal homeostasis in vertebrates delightfully written by an eminent renal physiologist.

CHAPTER 36

SEXUAL REPRODUCTION AND EMBRYONIC DEVELOPMENT

Almost all species of animals can reproduce sexually although, as we saw in Chapter 18, they don't always do so. An individual produced asexually may develop from an egg, a bud, or various other parts of the body of either parent. Sexual reproduction is monotonous in contrast: the individual's development starts only when two dissimilar gametes fuse at fertilization. An **egg**, the female gamete, contains food for the embryo, and molecules that control early development. As a result of all this luggage, an egg is too large to be motile and must sit around waiting to be found. A motile **sperm** is the necessary complement to the nonmotile egg, and the two have evolved together.

In most animal species the two gametes come from different parents, and the anatomy, behavior, and physiology of the parents must be coordinated so that gametes are produced at the same time and brought together in the same place. Hormones often play an important role in this coordination.

Fusion of the gametes at fertilization produces a single-celled **zygote** that is not equipped to live as its multicellular parents live until it has undergone a period of **embryonic development**. First, it must divide and form many cells. Second, the many cells must differentiate from one another, taking on the different shapes and chemical characteristics that specialize them for their later roles in different parts of the body. Third, the embryonic cells must move around, spread out, or clump together, creating the many different assemblages of cells that give each organ, and the body as a whole, its characteristic form. While all this is going on, the embryo is supplied with food—either from the egg yolk or by exchange of food between the maternal and fetal bloodstreams (as in most mammals), or by a combination of the two.

Because evolutionary success is defined in terms of reproductive success, we have touched on reproductive adaptations in several other chapters: in the discussions of evolution of reproductive patterns and sex roles in Chapter 18, and of reproduction in animals adapted to different habitats and ways of life in Chapters 27 to 30. Reproductive adaptations are many and varied, and rather than try to paint such a broad picture here, we shall confine ourselves in this chapter to the study of the anatomy and physiology of sexual reproduction in vertebrates, emphasizing human reproduction, which involves most of the same hormones and organs as in any other vertebrate. Then we shall turn to selected aspects of embryonic development, with the emphasis now on other vertebrates because experimentation with human embryos is greatly restricted. However, the early development of vertebrates is very similar from one species to another, and we can gain useful information about our own development by studying that of the more convenient frog or chick embryo.

36-A   Human Reproductive Organs

The human reproductive organs can be roughly divided into the internal organs and the external or accessory sex organs. In addition, each sex has **secondary sexual characteristics**, such as enlarged breasts in the human female or a higher metabolic rate in the human male, which are not part of the actual reproductive apparatus.

Female Reproductive Organs

The external sex organs of a woman consist of the **labia majora** and **labia minora**, **clitoris**, **vestibule**, and **hymen**; these organs are known collectively as the **vulva** (Figure 36-1). Note that the urinary and genital openings are separate in the human female. In virgin women, the vaginal orifice is covered by the hymen, a membrane with slits in it through which the menstrual flow passes.

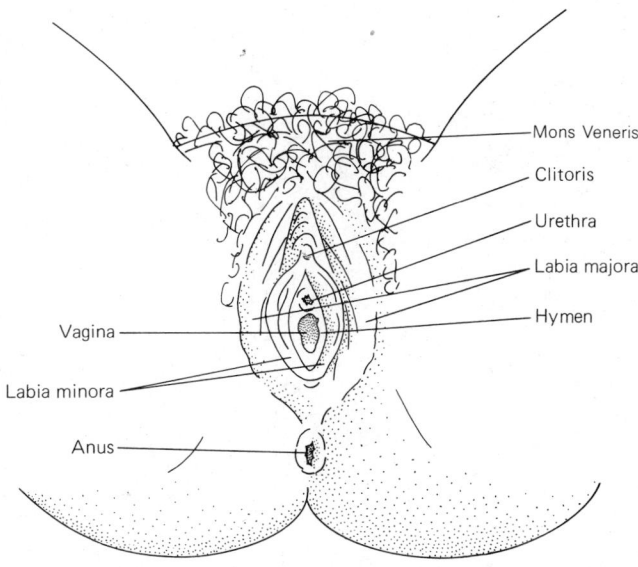

Vagina

Labia minora

Anus

Mons Veneris

Clitoris

Urethra

Labia majora

Hymen

Figure 36-1   Vulva of an adult woman. (The mons veneris is a fatty pad over a bone.)

CHAPTER 36
SEXUAL REPRODUCTION
AND EMBRYONIC
DEVELOPMENT

573

The internal female reproductive organs consist of the ovaries and fallopian tubes (the oviducts), uterus, and vagina (Figure 36-2). The two **ovaries** are the female **gonads**, or gamete-producing organs; the ovaries produce eggs. Girls are born with all the (as yet immature) eggs they will ever have, and so damage to an ovary is potentially more harmful to fertility than is damage to a testis, which continues to produce new sperm throughout a male's adult life. An ovary contains many follicles, each consisting of an immature egg surrounded by nutritive follicle cells. When mature, the follicle ruptures and releases its egg into the coelom. The beating of cilia in the **fallopian tube** draws the egg into the tube and on toward the uterus. Fertilization usually occurs in the fallopian tubes. An unfertilized egg dies and disintegrates about 72 hours after ovulation (release of a mature egg).

The **uterus** is a highly elastic organ whose main function is to contain the developing embryo. The uterus is about the same size and shape as a pear. Its walls contain masses of smooth muscle that expel the baby during childbirth. The external opening of the uterus is the **cervix**, made up largely of the biggest, most powerful sphincter* muscle in the body. Its strength is necessary to hold about 15 pounds of fetus and fluid in the uterus during pregnancy. The cervix protrudes into the upper end of the vagina.

The **vagina** is the receptacle for the penis during copulation, or sexual intercourse, and the pathway to the exterior for a baby during childbirth. In accordance with these functions it has extremely elastic walls. It is muscular, but the vaginal muscles are much weaker and thinner than those of the uterus.

Male Reproductive Organs

In a male, **spermatozoa**, or sperm, are produced in the **testis**, the male gonad. For unknown reasons, the 37°C temperature of the mammalian body is so high that it prevents production of sperm; mammalian testes usually lie in the cooler **scrotum** outside the body cavity. The testes form in the abdominal cavity during development and normally descend through a canal in the abdominal wall into the scrotum before birth. (The presence of this canal between the abdominal cavity and the scrotum is one reason men are so much more prone to hernias* than are women, who have no such weakness in the abdominal wall.)

Sperm develop from cells lining the **seminiferous tubules** of the testis. Muscular action in the seminiferous tubules carries mature sperm to the **epididymis**,

---

* Sphincter   A circular muscle whose contraction closes a tube.
* Hernia   A split in a sheet of muscle such as that of the abdominal wall or diaphragm.

Figure 36-2 The internal reproductive organs of a woman.

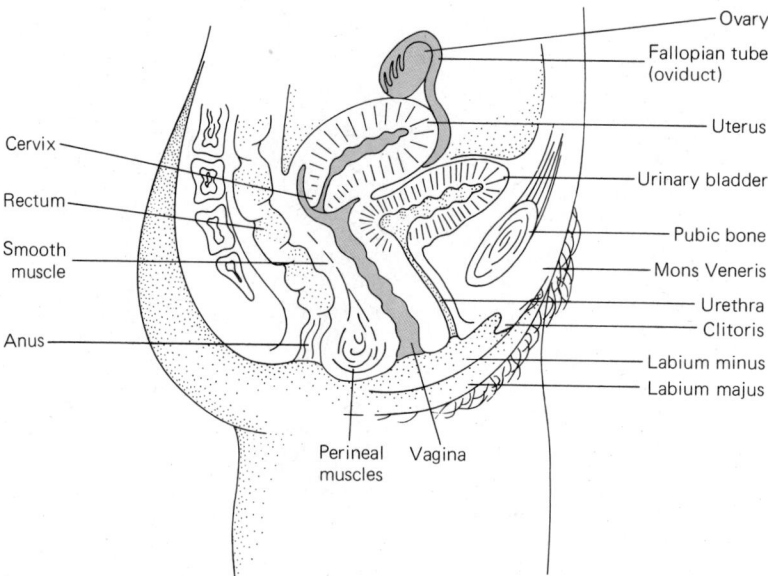

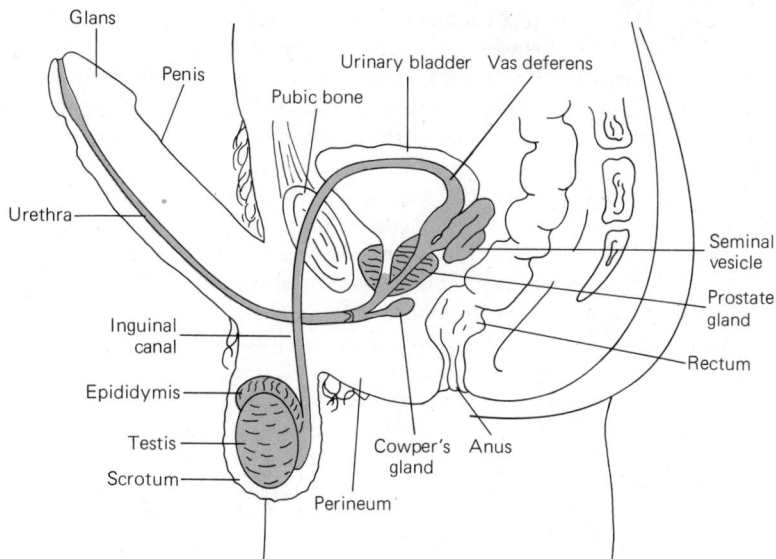

Figure 36-3 Cross section through the pelvic region of an adult human male during erection. Reproductive organs and path of sperm in color.

where they are stored and where their final maturation takes place. During sexual stimulation, the sperm move through the **vas deferens** (plural: **vasa deferentia**) by contraction of its walls, until they reach the **seminal vesicles**. The sperm then move into the urethra, where they are joined by secretions from the seminal vesicles and the **prostate** and **Cowper's glands**. The sperm and their attendant secretions, collectively called **semen**, leave the penis via the urethra (Figure 36–3). The **penis** is an external sex organ that introduces semen into the vagina during sexual intercourse; urine leaves the male's body by the same route, although the two fluids cannot pass through the urethra at the same time.

36-B Physiology of Sexual Intercourse

During **copulation** (often called **sexual intercourse** in humans), the male introduces sperm into the body of the female; this permits fertilization to occur inside the female's body rather than in an external, and possibly unfavorable, environment. Depending on the species, the resulting embryo may develop internally, or the female may lay a fertilized egg. Copulation occurs in many invertebrates from the platyhelminths on up the phylogenetic tree, and in some members of all vertebrate classes except Agnatha. The term copulation is usually reserved for the mating of species in which the male has an organ such as a penis that is used to deposit sperm in the female. Some species can copulate without this; in most species of birds, for instance, the males have no penis, and birds of these species mate by placing their cloacal openings tightly together. This seems odd in view of the fact that most male reptiles have at least one penis (sometimes two), and presumably so did the ancestors of modern birds.

Let us now consider the physiology of human copulation. Before the penis can enter the vagina, it must become at least partially erect under the influence of sexual stimulation. Sexual stimulation can be brought about by any of the senses, and usually most effectively by touch. The external genitals, and especially the glans of the penis, are the most sexually sensitive areas.

The penis consists of three cylinders of spongy tissue that extend into the body (Figure 36–4). The cells within the cylinders contain many hollow bodies, which are usually empty and collapsed. During sexual stimulation, about 10 times the normal volume of blood is carried from the arteries into these hollow bodies. The blood filling the spaces presses against the outsides of the veins, narrowing them so that the flow of blood leaving the penis is restricted; the penis thus enlarges and becomes rigid. Sexual stimulation also increases the muscular contractions that move the sperm.

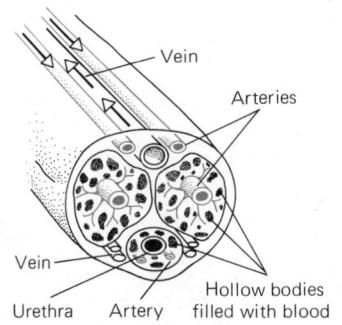

Figure 36-4 Cross section of a human penis.

CHAPTER 36
SEXUAL REPRODUCTION
AND EMBRYONIC
DEVELOPMENT

Sexual stimulation of the female is also a response to many stimuli. The clitoris is the organ most sensitive to touch. In a sexually stimulated female, the vulva becomes swollen because of an increased blood supply, and the walls of the vagina secrete fluid that acts as a lubricant for the entrance of the penis. Lubrication is caused by contraction of muscles around the blood vessels in the wall of the vagina, forcing fluid out of the enlarged vessels.

When the erect penis has been inserted into the vagina, stimulation by the movement of the genitals against each other may result in **orgasm**. Orgasm is characterized by an increase in heart rate and blood pressure, engorging of various tissues with blood, and faster and deeper breathing, which finally result in an explosive burst of muscular contractions. In men, orgasm is accompanied by the release of semen, which is called **ejaculation**. The muscles in the sperm ducts contract in peristaltic waves to eject semen. In women, orgasm is characterized by regular spasms of the muscles surrounding the vagina.

Orgasm is not necessary in either the male or the female for fertilization; a small amount of semen may be released before ejaculation, and this may contain enough sperm to fertilize an egg.

36-C   Hormones and Reproduction

Hormones of the Male

A variety of hormones maintains the reproductive state of the adult male. The most important of these is **testosterone**, which is produced by the testes. Its presence ensures maturation of sperm and also maintains male secondary sexual characteristics, such as beard growth, deep voice, and strong muscles. Men isolated from most sexual stimulation—for instance, on expeditions to Antarctica—produce less testosterone and have slower beard growth than usual.

The Menstrual Cycle

At puberty (10 to 14 years of age), the **pituitary** gland beneath the brain starts a series of hormonal cycles that periodically render a woman fertile (capable of becoming pregnant) until the cycles cease at the menopause, some 30 to 40 years later. These hormonal changes and the effects they produce are called **menstrual cycles** (Figure 36-5).

Figure 36-5 Levels of sex hormones in the bloodstream during the phases of one menstrual cycle in which pregnancy does not occur. Secretion of each hormone is influenced by the level of one or two of the others (see text).

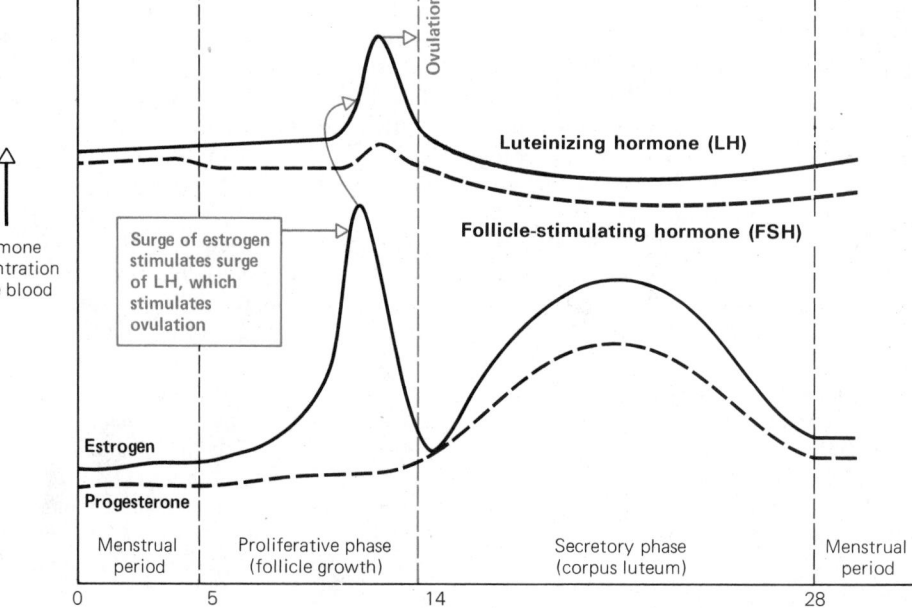

Hormone concentration in the blood

Surge of estrogen stimulates surge of LH, which stimulates ovulation

Ovulation

Luteinizing hormone (LH)

Follicle-stimulating hormone (FSH)

Estrogen

Progesterone

| Menstrual period | Proliferative phase (follicle growth) | Secretory phase (corpus luteum) | Menstrual period |

0   5   14   28

Day of menstrual cycle

Human menstrual cycles are notoriously variable, but the "model" cycle lasts 28 days. At the beginning of the cycle, the pituitary secretes increasing quantities of **follicle-stimulating hormone (FSH)**. FSH causes an ovarian follicle to mature and to produce another hormone, **estrogen**. A surge of estrogen from the maturing follicle stimulates a dramatic increase in secretion of still another hormone, **luteinizing hormone (LH)**, by the pituitary. This LH, together with FSH, brings about the final maturation of the follicle, culminating in rupture of the follicle and release of a mature egg. Still under the influence of LH, the cells of the ruptured follicle grow and form a **corpus luteum**, which secretes more estrogen and also yet another hormone, **progesterone**. Progesterone and estrogen stimulate the **endometrium**, the lining of the uterus, to thicken and to secrete nourishing fluid, in preparation to receive a fertilized egg. The levels of progesterone and estrogen at this time inhibit secretion of LH and FSH from the pituitary. (Note that the estrogen surge stimulates the LH surge, whereas somewhat lower estrogen levels inhibit LH secretion.) The subsequent drop in LH and FSH levels deprives the corpus luteum of the hormones that stimulate it; it begins to degenerate, and its hormone secretion reaches such a low level that the surface of the endometrium dies and sloughs off in a menstrual period. Without the ovarian (corpus luteum) hormones to inhibit it, the pituitary increases secretion of FSH once more, and the cycle repeats. This sequence is different if fertilization and pregnancy occur, as we shall see next.

Pregnancy

Sperm released into the vagina during ejaculation swim through the cervix and uterus into the fallopian tubes, where fertilization usually occurs. (This description of pregnancy applies primarily to humans, but the situation is similar in other mammals.)

The fertilized egg, already undergoing development, moves slowly down into the uterus. About a week after fertilization, it starts to **implant** in the endometrium of the uterus. The most important organ of pregnancy then begins to form. This is the **placenta**, produced partly by the uterine wall and partly by the embryo (Figure 36-6). It immediately starts to secrete **human chorionic gonadotropin (HCG)**, a hormone related to LH. Elevated HCG, like elevated LH, promotes production of progesterone, which inhibits production of FSH and LH by the pituitary gland. This prevents the next ovulation, and thus prevents formation of any new embryo while the first one is developing. If the embryo is abnormal, or dies, secretion of HCG by the placenta stops and the HCG level in the bloodstream falls, permitting the body to **abort** or **miscarry** the pregnancy. It is estimated that as many as three out of five human embryos that implant are abnormal and abort in this manner. If a hormone from the mother's body maintained pregnancy, there would be no way to discard dead and abnormal embryos.

By the same reasoning, it is not surprising that a hormone produced by the developing fetus determines when birth shall occur (at least in sheep, the only

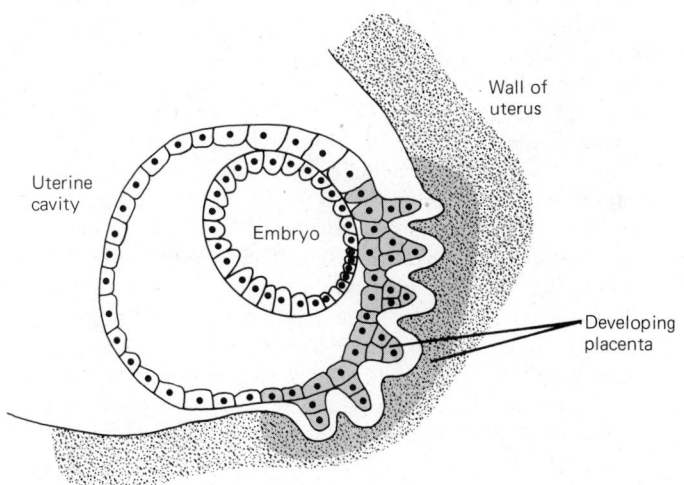

Figure 36-6 A developing mammalian embryo implanting in the wall of the uterus.

Wall of uterus

Uterine cavity

Embryo

Developing placenta

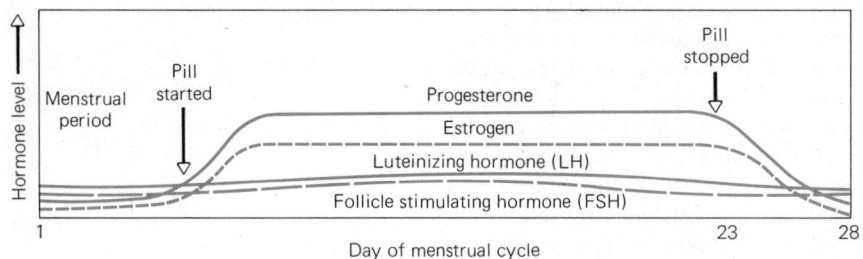

Figure 36-7 Blood levels of reproductive hormones in a woman taking the combination contraceptive pill. (Compare with the curves in Figure 36-5.)

animals on which detailed experiments have been done). The average period of **gestation**, or pregnancy, is 270 days in humans, but this is very variable. It is important that the baby be born when it is mature, and not when the mother feels like it. In sheep, and presumably in humans, the fetus signals that it is mature by secreting hormones from its now-developed adrenal glands. These fetal hormones diffuse across the placenta and build up in the mother's bloodstream until they cause secretion of the hormone **oxytocin** by the mother's pituitary gland. Oxytocin stimulates the muscles of the uterus to contract and cause birth.

Expulsion of the placenta causes the maternal pituitary to secrete a hormone that initiates **lactation**, or milk production, by the mammary glands. Once this has occurred, milk supply is thereafter on a supply-and-demand basis; the more a baby suckles, the more milk is produced about 24 hours later.

36-D Birth Control

Since the dawn of history, human beings and other (particularly social) animals have regulated their reproduction, so that the number of offspring bears some relation to the parents' ability to raise them. Rats living in crowded conditions have lower fertility and a high rate of abortion; a bird may abandon her eggs if her neighbors of the same species have many young already hatched. Carnivorous birds with uncertain food supplies usually feed the weakest nestling last; if food is abundant, this nestling may receive enough to live; if not, its nestmates will receive all the food, and the weakling will be the first to die. Thus valuable energy has not been spent on offspring that are not likely to survive.

There are several modern methods of birth control in use among human beings. They include abortion, sterilization, and **contraception**, which really means preventing fertilization but is also used for birth control methods that prevent implantation of the fertilized egg.

One of the most effective methods of contraception is commonly called "the pill." The "combination" pill is the type most often used. Birth control pills contain synthetic hormones that alter normal hormone levels in the body. The combination pill contains varying proportions of synthetic estrogen and progesterone, hormones that inhibit the release of the pituitary hormones FSH and LH (Figure 36–7). Since FSH and LH promote maturation and ovulation of eggs, ovulation does not occur in a woman taking the pill, and pregnancy is prevented because there is no egg for sperm to fertilize. The decrease in progesterone level when the pill is stopped causes a menstrual period to occur. (The main reason the pills are not used continuously is that most women feel more normal when they have a period every month.)

The medical risks associated with use of the pill are those of pregnancy itself. The elevated progesterone level in the blood produces a hormonal state qualitatively like that of a pregnancy and an increased tendency to such events as excessive blood clotting (although there is no form of birth control or legal abortion that is nearly as dangerous to a woman as pregnancy itself).

Some years ago, people became concerned that the estrogen in contraceptive pills might increase a woman's susceptibility to cancer, because estrogen administered to counteract symptoms of menopause definitely increases a woman's chances of developing uterine cancer. Further research showed that although the "sequential" pill, which is no longer used, might have been suspect, a woman is actually *less* susceptible to various types of cancer after she has been taking modern contraceptive pills for several years.

As effective as hormonal pills in preventing unwanted pregnancies are intra-uterine devices, or IUDs, which are inserted into the uterus. An IUD affects the lining of the uterus in such a way that the fertilized egg cannot implant, but there is considerable controversy about how this occurs.

Another method of birth control is the use of a rubber diaphragm, smeared with spermicidal (sperm-killing) jelly each time it is used. The diaphragm blocks the entrance to the uterus so that sperm cannot leave the vagina. Fertilization is prevented both by the physical obstruction of sperm by the diaphragm and by the spermicidal action of the jelly.

Responsibility for use of the contraceptive methods mentioned so far rests mainly on the woman. The condom is the contraceptive most commonly used by men. A condom is rolled onto the erect penis shortly before intercourse. It catches the semen so that sperm do not enter the female reproductive tract.

The rhythm method of birth control consists of avoiding sexual intercourse during those days of the menstrual cycle when there is an egg present to be fertilized. Since eggs may live for three days and sperm remain viable for up to two days in the female reproductive tract, this means avoiding intercourse two or three days before and two or three days after ovulation. The difficulty with this method is in deciding when ovulation occurs. It usually occurs 14 days before the start of the next menstrual period, but knowing this is of no help in predicting when ovulation will occur. In fact ovulation can occur at any time during the cycle, including during the menstrual period. Various methods, such as keeping a chart of resting rectal temperature, have been developed to increase the reliability of estimating when ovulation is going to occur. The temperature method is also used as a guide to the time of ovulation when a couple is trying to conceive.

Vasectomy and Tubal Ligation

Sterilization is any more-or-less permanent change that prevents an animal from reproducing sexually. Sterilization of either sex is the fastest-growing method of contraception in the world. The most common sterilization operation for men is **vasectomy**, that is, severing and tying off the vasa deferentia. This is a simple operation, usually performed under local anesthesia (Figure 36–8). Afterwards, sperm are still produced but are resorbed into the body, and the fluid ejaculated contains only the secretions of various glands, although its volume is little less than when it also contained sperm. A device that acts as a sort of partial vasectomy is being tested; this is a tiny valve that can be put inside the vas deferens and turned on or off to permit or block the passage of sperm. Vasectomy has gained popularity as more sperm banks become available in several countries. A man about to undergo a vasectomy may have a sample of his sperm frozen and stored in one of these banks. Hundreds of normal children have already been produced from such sperm by **artificial insemination**, that is, introduction of sperm into the vagina by means other than intercourse.

Sterilization of a woman usually involves **tubal ligation**, the cutting and tying off of the fallopian tubes. This operation can also now be performed under local anesthesia. After tubal ligation the ovary continues to function as it did before, but sperm cannot reach the eggs, and thus fertilization cannot occur.

Abortion

Induced abortion is probably the oldest human birth control method known. Abortion was accepted, and fairly common, in the United States and Europe until the early nineteenth century. Before this time, doctors approved of, or at least tolerated, abortion because a woman had up to a one-third chance of dying in childbirth, so that abortions often saved women's lives. After midwives and doctors found that washing their hands and clothes improved their patients' survival rates, childbirth became less dangerous to a woman than a nineteenth-century abortion, and doctors started to oppose abortion. Religious and ethical opposition did not develop until some time later. The situation has changed since World War II now that the danger of abortion to a woman is again less than the risk of a completed pregnancy, and abortion is, once again, legal in many countries.

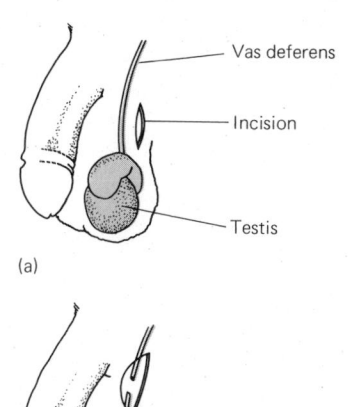

(a)

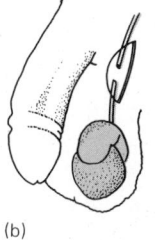

(b)

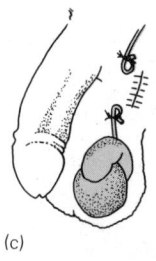

(c)

Figure 36-8 Vasectomy, the surgical procedure in which the vas deferens is severed so that sperm can no longer enter the semen.

There are three main methods of abortion. In **vacuum curettage** and **dilation and curettage** ("D and C") the fetus is sucked or scraped out of the uterus, usually under local anesthesia. ("D and C" is used for diagnosis and treatment of a number of uterine disorders as well as for aborting early pregnancies.) These methods are used early in pregnancy. After the fifteenth week of pregnancy, **saline injection** can be used. An injection of salt solution kills the fetus, and the uterus subsequently expels the fetus and the placenta.

36-E Gamete Formation

We will begin our consideration of the embryonic development of vertebrates with the formation of the **gametes**, the sperm and egg.

Spermatogenesis

The formation of sperm is essentially the same in all animals. Sperm or spermatozoa (singular: spermatozoon) are the male gametes. Since it contains little cytoplasm, a sperm is very small, and in nearly all species it is motile, swimming with the flagellum that forms its tail.

Spermatogenesis is the formation of sperm in the testis. It involves two proc-

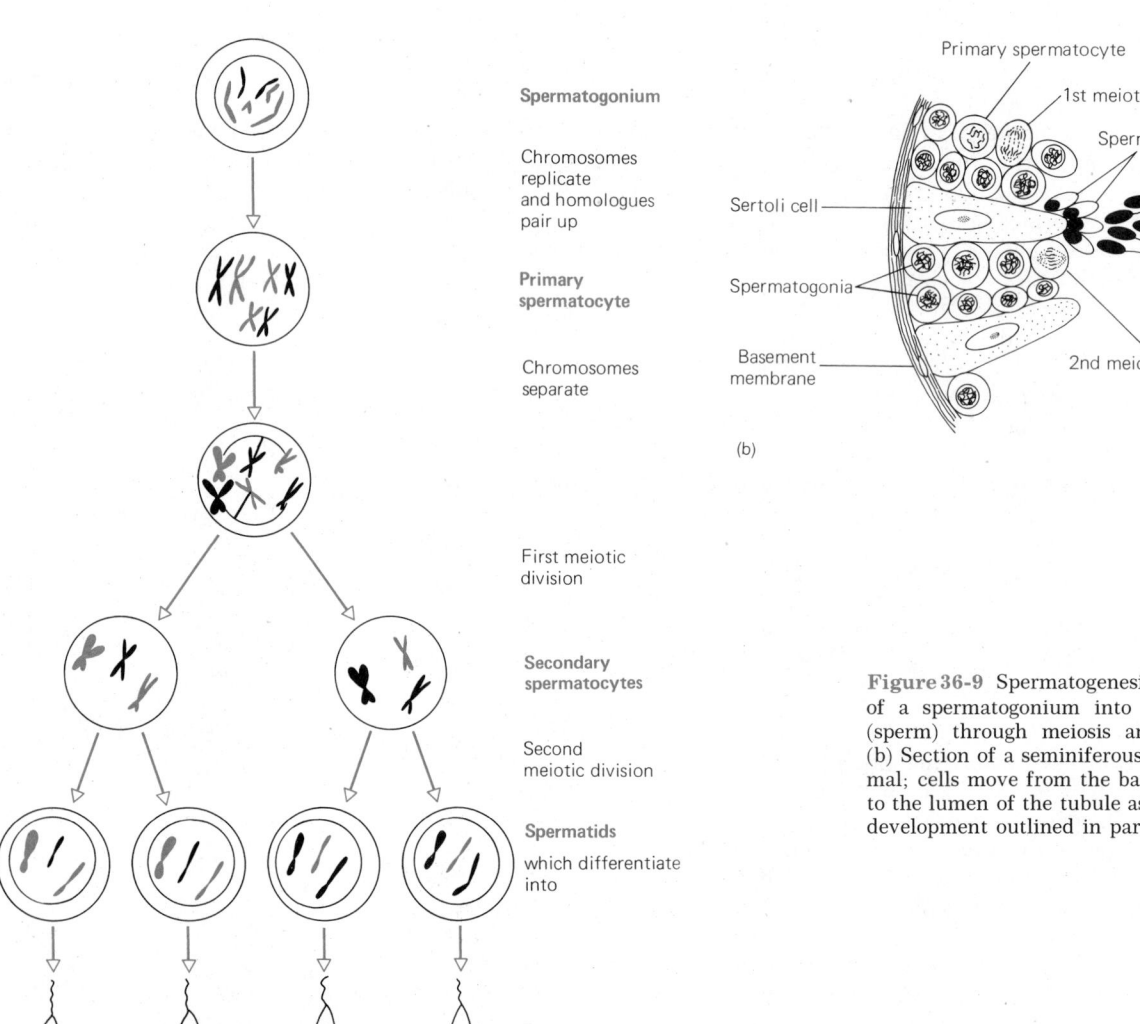

(a)

(b)

Figure 36-9 Spermatogenesis. (a) Development of a spermatogonium into four spermatozoa (sperm) through meiosis and differentiation. (b) Section of a seminiferous tubule of a mammal; cells move from the basement membrane to the lumen of the tubule as they undergo the development outlined in part (a).

esses: meiosis, whereby haploid* sperm nuclei are produced from diploid* nuclei, and differentiation of the developing spermatozoon into an efficient, swimming cell with a head, a tail, and many mitochondria.

The testis is made up of masses of seminiferous tubules, in which the stages of spermatogenesis take place in an orderly sequence. Cells lining each tubule divide, and developing sperm are pushed toward the lumen* of the tubule as they mature (Figure 36–9).

Oogenesis

Oogenesis, the formation of eggs or **ova** (singular: **ovum**) is more complicated than spermatogenesis. Here, meiotic nuclear division is accompanied by unequal division of the cytoplasm, so that one diploid cell produces only one large ovum and a number of smaller cells called **polar bodies**. During the course of evolution, the egg became the main source of stored food, ribosomes, messenger RNA, and other cytoplasmic components that support the early development of the embryo. The biological problem in oogenesis is to synthesize or transport all of the necessary substances into the cytoplasm. This involves the metabolism of the egg itself and of the cells around it in the ovary.

Primary oocytes (Figure 36–10) are diploid cells that form by mitosis in the ovary. A primary oocyte then undergoes the first meiotic division to form a large **secondary oocyte** and a tiny polar body. Depending on the organism, this first polar body may go through the second meiotic division to produce two haploid polar bodies. The secondary oocyte undergoes the second meiotic division to produce a haploid ovum and a haploid polar body. The polar bodies are really just a means of shedding chromosomes from the developing egg cell, and they soon die.

---

* Haploid   Containing the number of chromosomes found in a gamete, which is half the number
   of chromosomes found in a body cell (in most animals).
* Diploid   Containing the number of chromosomes found in a body cell, which in most animals
   equals twice the number found in a gamete.
* Lumen   Hollow area in the center of a tube.

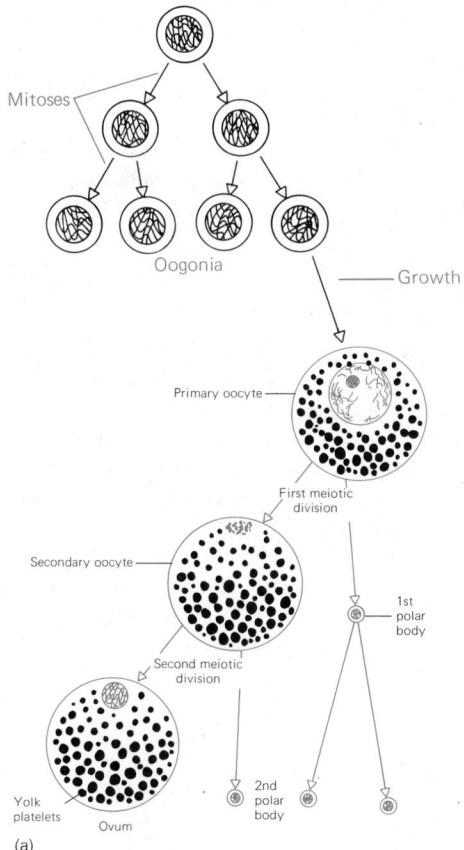

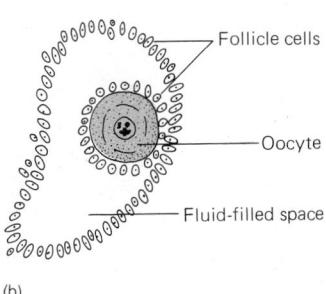

Figure 36-10   (a) Steps in oogenesis in a vertebrate. The oogonia grow into primary oocytes by their own activity and by transfer of material from adjacent follicle cells. (b) A mature follicle in the ovary of a mammal.

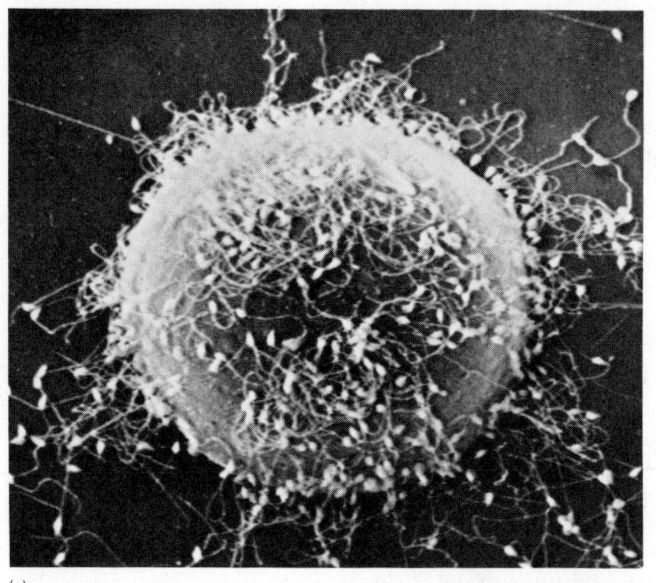

(a)

(b)

Figure 36-11 (a) An egg surrounded by sperm. (b) Close-up of the surface of the egg and several sperm; only one sperm will fertilize the egg. (Biophoto Associates)

Although spermatogenesis always proceeds to completion whether or not there is any chance of the resultant sperms' fertilizing an egg, oogenesis usually stops somewhere, at various stages in different animals. In humans, a secondary oocyte rather than an ovum is released from the ovary. Only if the secondary oocyte is penetrated by a sperm will the second meiotic division occur, producing a polar body and an ovum whose chromosomes can then combine with those from the sperm to produce the zygote nucleus.

36-F   Fertilization

Fertilization is the reaction between a haploid sperm and a haploid egg that produces a diploid **zygote**. It serves two functions in an animal's life history. First, fertilization restores the diploid number of chromosomes, in new combinations that result from crossing over and independent assortment during meiosis (Sections 13-H, 13-L). Second, the sperm's breaking through the egg cell membrane is the stimulus that normally initiates development; pricking the unfertilized egg of many animals with a pin will also induce development.

Fertilization is a complicated case of two cells' recognizing one another. Whether fertilization is **internal**, within the female's body, or **external**, in water surrounding the mating animals, eggs attract sperm by exuding chemicals into the surrounding fluid. When a sperm touches the jelly around the egg, it sticks there, immobilized by a substance that "recognizes" the sperm's plasma membrane (Figure 36–11). Apparently a certain amount of this substance must be neutralized by reacting with sperm before any sperm can penetrate the jelly to reach the egg. A single sperm can probably never fertilize an egg by itself; thousands of sperm usually stick to the jelly coat before fertilization can occur. When enough sperm are stuck to the egg, the sperms' active role ends and the egg takes over. The egg membrane engulfs the head of a sperm and pulls the sperm nucleus into the egg, leaving the rest of the sperm outside (Figure 36–12). As soon as this happens, some sort of electrical and chemical reaction runs throughout the egg cell membrane and makes it impermeable to further sperm.

36-G   Embryonic Development

Embryonic development is a complex process involving cell division, cell differentiation, and cell movement, as the genetic information in the zygote expresses

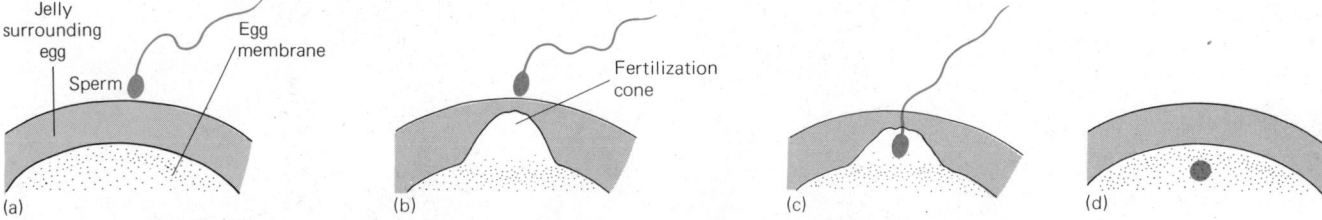

Figure 36-12 Sperm entry into a frog's egg. (a) The sperm sticks to the jelly coat that surrounds the egg. (b) The egg forms a fertilization cone, which (c) engulfs the sperm and then (d) retracts, digesting the sperm tail as it does so.

itself and forms the mature animal. In order to understand just how complex this process is, let us consider three major aspects of development. The first of these is **differentiation** into cell types (see Chapter 12). From one fertilized zygote, which is one type of cell, cells as different as liver, muscle, nerve, and skin cells are produced. These cells differ from one another in that they have synthesized different enzymes and structural proteins. This must be due to the fact that different genes have become active in different cells during development.

The second aspect of embryonic development is **growth**. How and why do cells divide and grow? At various stages in development, depending on the animal, food from the outside reaches the embryo and enables it to increase its overall size. The mammalian embryo is nourished by the placenta; a frog hatches into a tadpole, which eats for itself. Before this point, no overall growth occurs; however, the number of cells in the embryo has increased enormously by cell division.

A third aspect of embryonic development is the **formation of shape**. This consists, on a molecular level, of the rearrangement of proteins and other molecules to form larger structures within cells and, on a larger scale, of the movement and building of cells into specific patterns to form organs. There is little solid evidence on how particular shapes of organs form during development, although acres of paper have been covered with models of how this *might* occur. As we follow the various stages of development, we shall consider some of the more reasonable suggestions.

Embryonic development may be divided into four main stages: cleavage, gastrulation, neurulation, and organogenesis.

Cleavage

The main functions of cleavage are to produce a large number of cells by rapid cell division and to segregate different factors in the zygote cytoplasm into different cells; these factors determine how the various cells develop later. Most cells grow in

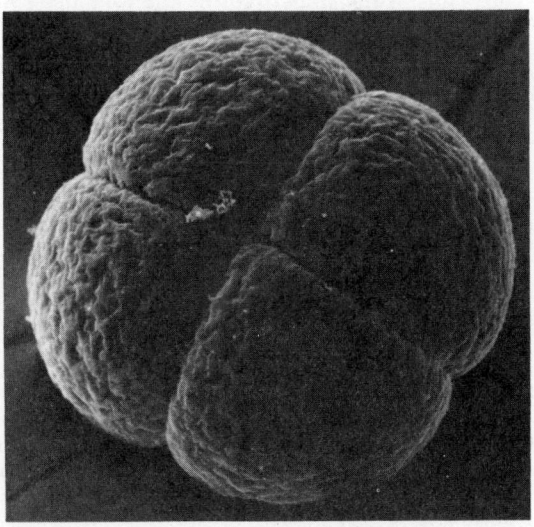

Figure 36-13 Scanning electron micrograph of the four-cell cleavage stage of an ascidian embryo. Ascidians are invertebrate chordates and have regular radial (as opposed to spiral) cleavage. (John Wourms)

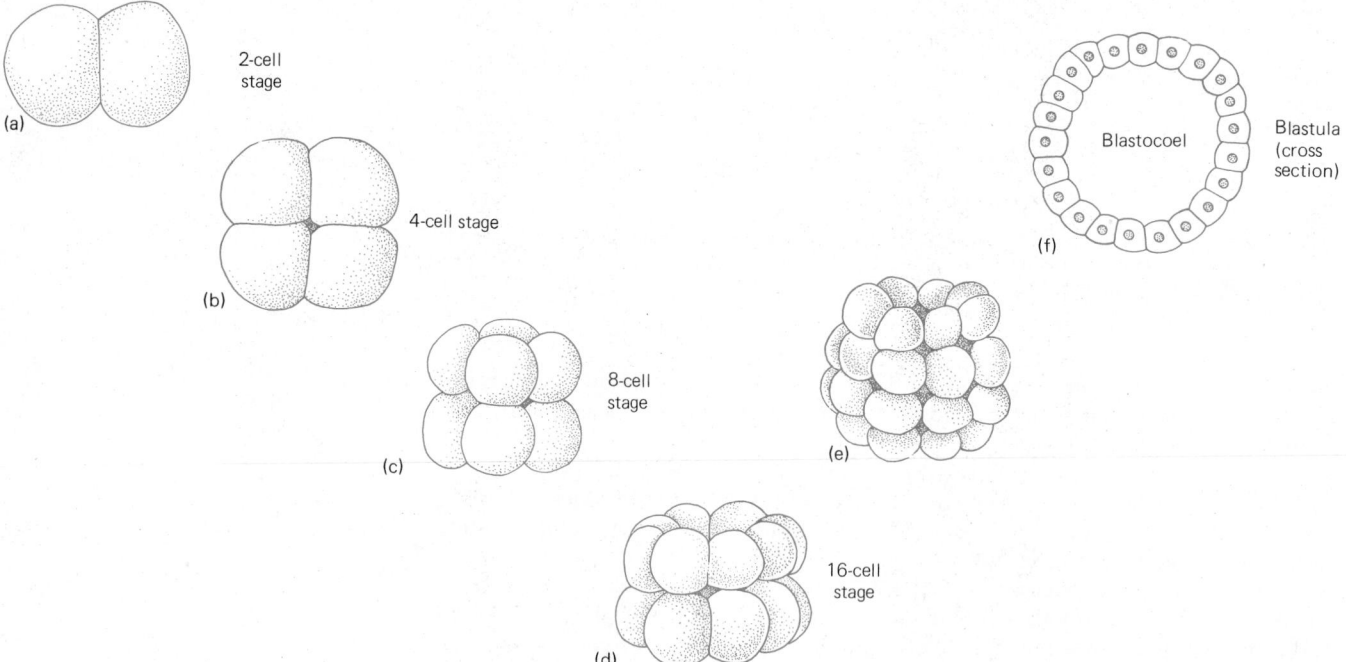

Figure 36-14 Radial cleavage, which occurs in vertebrates and their invertebrate relatives (e.g., echinoderms). This diagram shows radial cleavage in an egg that contains little yolk. (Contrast it with radial cleavage in a yolky bird's egg, Figure 36-15.)

size before they divide, but no cell growth occurs during cleavage; the zygote divides into smaller and smaller cells. The end result of cleavage is an embryo known as a **blastula**, which is hollow in most animals (Figure 36–14).

The precise pattern of cleavage in a given animal depends largely on the amount of yolk in the egg. In animals whose eggs contain little yolk, such as mammals and many invertebrates, the cell divisions of cleavage produce many cells of roughly the same size. Eggs like those of a bird contain such large amounts of yolk that cleavage cannot pass completely through the yolk, and a bird's egg cleaves only in a small area on top of the yolk (Figure 36–15).

CONTROL OF CLEAVAGE In the 1920s studies on the snail *Cepaea* provided evidence that the genes of the mother control cleavage. This snail has a shell that may be coiled either to the right or to the left. A snail has spiral cleavage (Figure 36–16), and the direction of the cleavage spiral predicts the future curve of the adult snail's shell. Thus, if the embyro forms a left-handed spiral during cleavage, it will have a

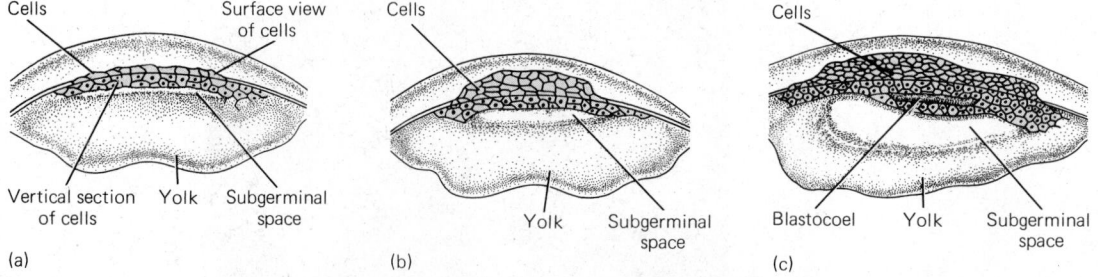

Figure 36-15 Stages in the cleavage of a hen's egg; vertical sections. Cleavage is restricted to a small disk of cytoplasm on top of the yolk. A subgerminal space appears beneath this disk and separates the developing embryo from the yolk.

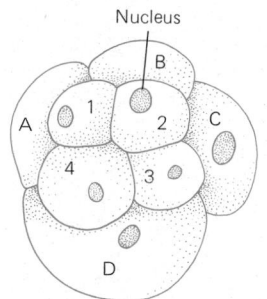

Figure 36-16 Spiral cleavage in a mollusc. The four large cells (A, B, C, D) are formed first and then the four smaller ones on top in turn: 1, 2, 3, then 4, and so on.

left-handed coil to its shell when it is an adult. We usually assume that the embryo's own genes control development, but in this case they do not. The pattern of cleavage (and so of the adult shell) in *Cepaea* is dictated by the organization of the cytoplasm in the egg, which was laid down under the influence of the mother's physiology and directed by her genotype. The genes in the snail zygote have no influence whatsoever.

Further evidence that points to the same conclusion comes from studies in which hybrid embryos have been produced by artificial fertilization between two different species of frogs. Some hybrid embryos grow into normal animals, but most do not. The genes of two different species are usually incompatible and incapable of producing normal embryonic development. In these hybrid embryos, cleavage proceeds normally. The incompatibility of the genes does not interfere with development until later.

A third piece of evidence for maternal control of cleavage is the fact that very little RNA is synthesized during cleavage. In addition, if a fertilized egg is injected with actinomycin D, a drug that inhibits RNA synthesis, cleavage is normal, but the embryo later dies. It seems that gene action, the synthesis of messenger RNA, is not necessary to cleavage, although it becomes vital later during development.

On the other hand, inhibitors of protein synthesis do interfere with cleavage. Protein synthesis is necessary for cleavage, and thus messenger RNA must be present even though very little of it is synthesized during cleavage. Increasing evidence shows that messenger RNA is synthesized and stored in the egg before fertilization, and that this stored RNA controls protein synthesis during cleavage.

Gastrulation

After formation of a blastula, the next stage in embryonic development is **gastrulation**, during which the cells rearrange themselves into three distinct layers. Although cell division continues during gastrulation, the most obvious event is cell movement. A cavity called the **blastopore** forms at the side of the blastula (Figure 36–17). Cells from the surface of the blastula around the blastopore move through

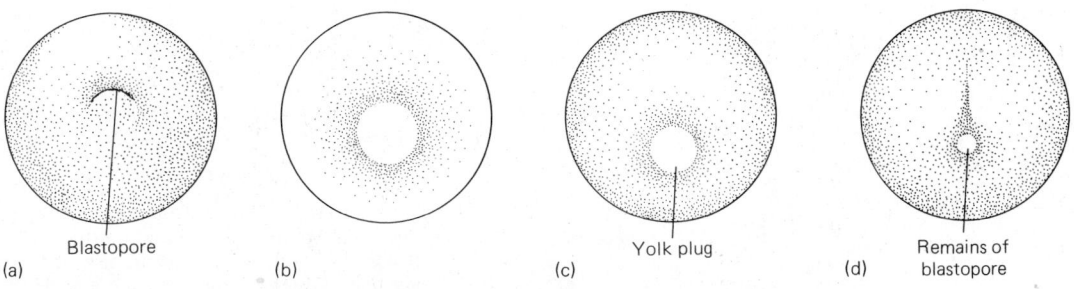

(a)   Blastopore   (b)   (c)   Yolk plug   (d)   Remains of blastopore

Figure 36-17 External view of gastrulation in a frog embryo to show the blastopore as it forms, enlarges, and then becomes smaller.

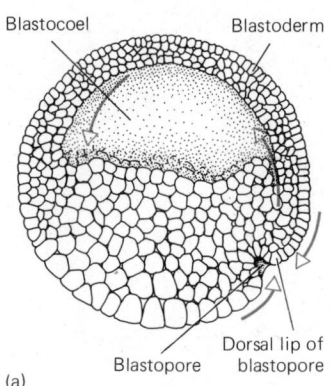

Blastocoel   Blastoderm

Dorsal lip of
blastopore

Blastopore

(a)

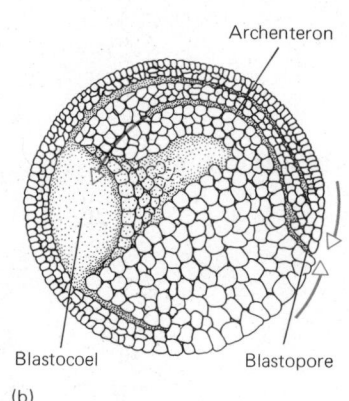

Archenteron

Blastocoel   Blastopore

(b)

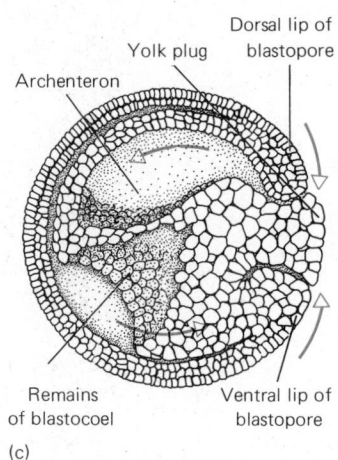

Yolk plug   Dorsal lip of
blastopore

Archenteron

Remains
of blastocoel

Ventral lip of
blastopore

(c)

Figure 36-18 Gastrulation in a frog. Sheets of cells from the outside of the embryo move in through the blastopore, finally obliterating the blastocoel and forming a new cavity, the archenteron, or gastrocoel. Arrows show the directions of cell movement.

the blastopore into the hollow interior (Figure 36–18).

During gastrulation, the single-layered hollow blastula becomes a **gastrula,** with cells arranged in three layers called the **germ layers**, and with a new central cavity, the **gastrocoel** (or **archenteron**), which will eventually form the lumen of the gut (Figure 36–19).

The outermost germ layer of the gastrula is called the **ectoderm**. This will form the **neurectoderm**, which develops into the nervous system, and the **epidermis**, which gives rise to skin, hair, nails, sweat glands, and so forth. The innermost layer of cells is the **endoderm**, next to the gastrocoel. From the endoderm will come the gut lining, the digestive glands, and so forth. Between the ectoderm and the endoderm lie the cells of the third germ layer, the **mesoderm**, which will form the skeleton, muscles, gonads, and allied structures in the adult. A split in the mesoderm at a later stage will form the coelom (see Figure 36–23c).

Complications are introduced by the fact that no organ in the body contains only cells from one germ layer, and the germ layers become considerably intermixed later in development.

In gastrulation, for the first time, the embryo's genes take over much of the control of development. The messenger RNA laid down in the egg has been used up, and transcription of the embryo's genes is reflected in rapid RNA synthesis. From this time on, drugs that inhibit RNA synthesis will kill the embryo.

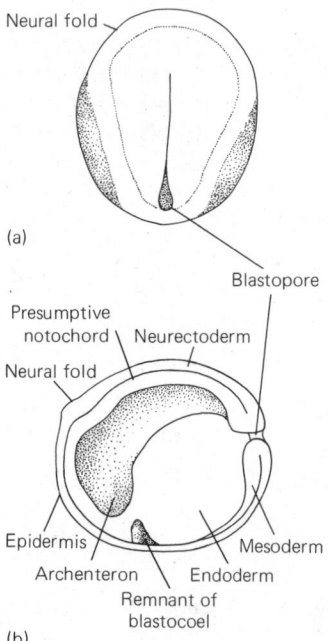

Neural fold

(a)

Blastopore

Presumptive
notochord   Neurectoderm

Neural fold

Epidermis   Mesoderm

Archenteron   Endoderm

Remnant of
blastocoel

(b)

Figure 36-19 The end of gastrulation and beginning of neurulation in a frog. (a) External view; (b) longitudinal section. During neurulation, the embryo elongates. The neural folds rise up and finally fuse together in the midline to form the neural tube (see Figure 36-21).

Gastrulation introduces several new developmental processes, most of which involve interactions between cells. The first is cell movement. Some cells move by themselves; others move as sheets of cells. An animal cell moving by itself looks much like an amoeba: it puts out pseudopodia, which attach to something ahead of the cell and pull the rest of the cell up to the point of attachment. Cells do not move randomly; they follow particular tracks through the embryo, and when they have reached wherever they are going, they stop. Hence cells must "recognize" both the tracks they follow and their destination. If embryonic cells are cultured in a laboratory dish, they will move along lines scratched in the bottom of the dish; it may be that cells in the embryo merely follow railroad tracks of connective tissue as they travel. When they reach their destinations, however, they must recognize their surroundings precisely. This they probably do by responding to the unique "signature" that each type of cell bears on its surface (Section 4-B). The surfaces of the cells at its destination induce the traveling cell to stop moving and to **adhere**, or stick, to the cells on either side of it. Adhesion between cells must be very specific and change with time to account for many of the events in embryonic development. In a sheet of cells, for instance, each cell has only a small area of attachment to the identical cells on either side of it but much larger areas attached to the different cells above and below it. A moving sheet of cells, such as cells moving during gastrulation, has no attachment to the cells beneath it.

Little is known about cell adhesion. Cells can be separated from one another by substances that dissolve the glue-like hyaluronic acid that holds cells together in life, but we have no idea what causes cells to produce this glue at particular times or to produce more glue on one side of the cell than on another. This is an important area of medical research because cancer cells (Section 34-M) come unstuck more readily than normal cells, and do not respond to signals to stop moving and dividing in the same way as normal cells.

Embryonic induction is another complex interaction between cells, in which one cell in an embryo alters the fate of another. This phenomenon is seen for the first time at gastrulation, but it is also common later in development. We shall consider a later example first, since it is more clear-cut.

The lens of the eye is formed from the ectoderm of the head as a result of contact with part of the brain, which grows out to form a bulge called the optic vesicle (Figure 36–20). When the optic vesicle touches part of the ectoderm of the head, it induces that bit of the ectoderm to form the lens. If something like a piece of cellophane is placed between the growing optic vesicle and the ectoderm, the lens never develops. If optic vesicles are transplanted from their normal site in the head to elsewhere in the body, they induce lens tissue to form in any ectoderm that they touch. They cannot, however, induce tissue other than ectoderm to form a lens. Hence the fate of a tissue is determined both by the activity of its own genes and by its environment: genetic activity determines that a cell has become ectoderm rather than mesoderm or endoderm, but unless it touches the optic vesicle, ectoderm will not develop into a lens. The whole process is called induction because contact with the optic vesicle **induces** ectoderm to form lens instead of skin or whatever else it would otherwise have formed.

Primary embryonic induction is the process by which ectoderm, mesoderm, and endoderm are induced to form in the gastrula. This process has been intensively studied, but we still understand little about it.

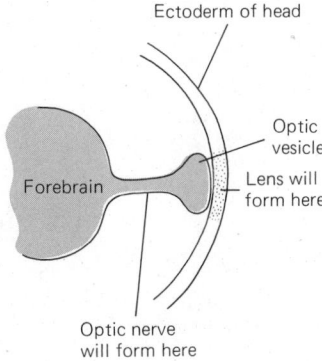

Ectoderm of head

Optic vesicle

Lens will form here

Forebrain

Optic nerve will form here

Figure 36-20 Induction. Contact between the optic vesicle and ectoderm induces formation of the lens of the eye from ectodermal tissue.

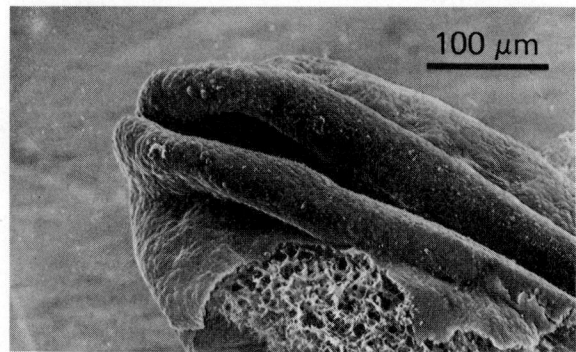

Figure 36-21 Scanning electron micrograph of a chick neurula. The lip-like ridges that will fuse to form the neural tube are moving towards each other at the head of the embryo (left). Figure 36-22 (b) shows a cross section of this stage. (Kathryn Tosney)

Neurulation

After the three germ layers have formed in the gastrula, the neural tube and head begin to form in the process of **neurulation**. The neurectoderm, destined to form the nervous tissue, begins as two parallel folds that finally join and form a tube, the **neural tube**, from which all of the nervous system develops (Figures 36–21 and 36–22).

Figure 36-22 Changes in cell shape during neurulation. Eight cells have been colored. Compare their shapes in parts (a) to (c) from beginning to end of neurulation. The fibers in the colored cells show the position of microfilaments whose contraction produces these changes in shape.

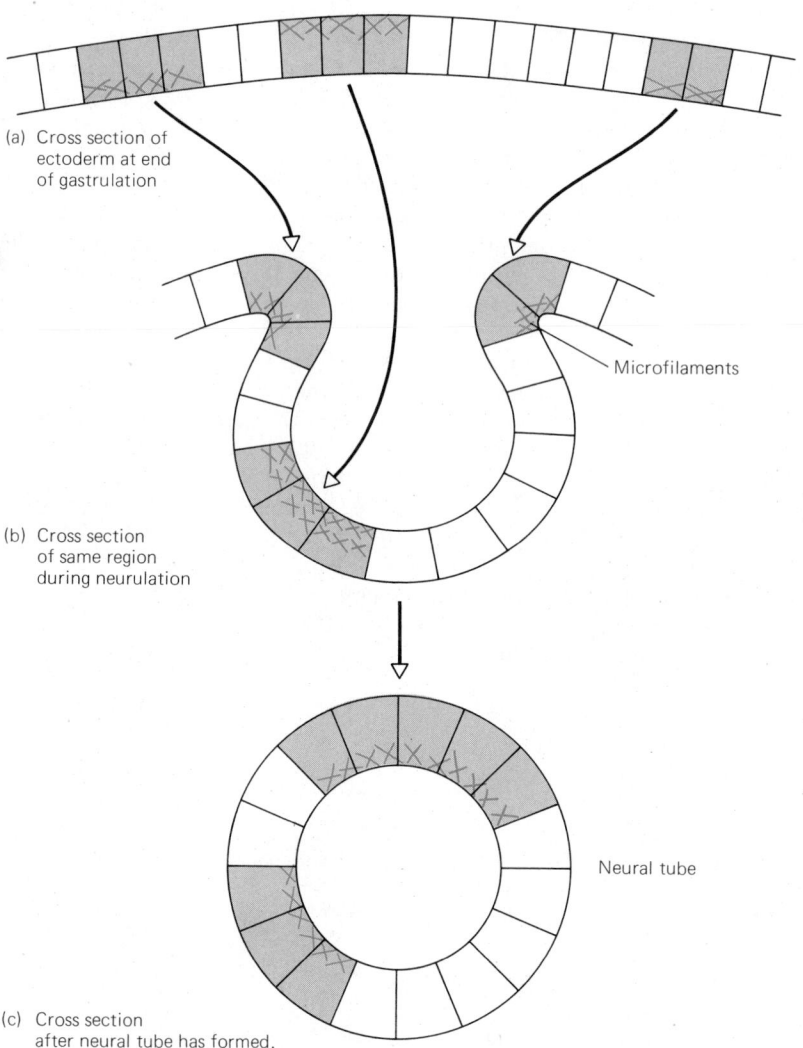

(a) Cross section of ectoderm at end of gastrulation

Microfilaments

(b) Cross section of same region during neurulation

Neural tube

(c) Cross section after neural tube has formed.

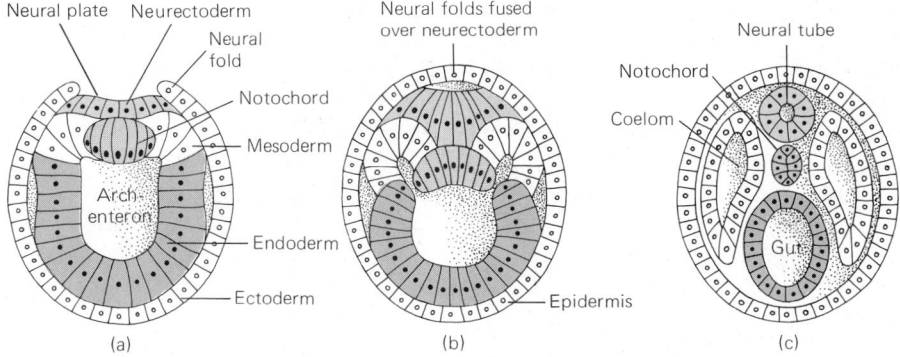

Figure 36-23 Formation of organ rudiments in a chordate with very little yolk in the egg (e.g., an amphioxus).

In neurulation, we see cells changing shape, something that occurs frequently during embryonic development. Consider the change in shape of the cells that will form the wall of the neural tube from the time they are part of a flat sheet of ectoderm until they become pie-shaped cells lining a tube. This change of shape results because contractile fibers called microfilaments within the cell contract, shortening the plasma membrane on what will become the inside of the tube or the inside of any fold (Figure 36–22). The contraction of microfilaments attached to the plasma membrane could account for all changes in cell shape. However, microfilaments are not always visible in electron micrographs of cells that are changing shape, so it is possible that other mechanisms are also sometimes involved.

Organogenesis

After neurulation, all of the organ systems of the body start to form in a complicated series of cell interactions (Figures 36–23 and 36–24). As far as we know, all of the processes essential to organ formation are ones that we have already mentioned: differential gene activity, movement of cells, differential recognition and adhesion among cells, inductive interactions between cells, and changes in cell shape. Here we shall outline the main events in human organogenesis.

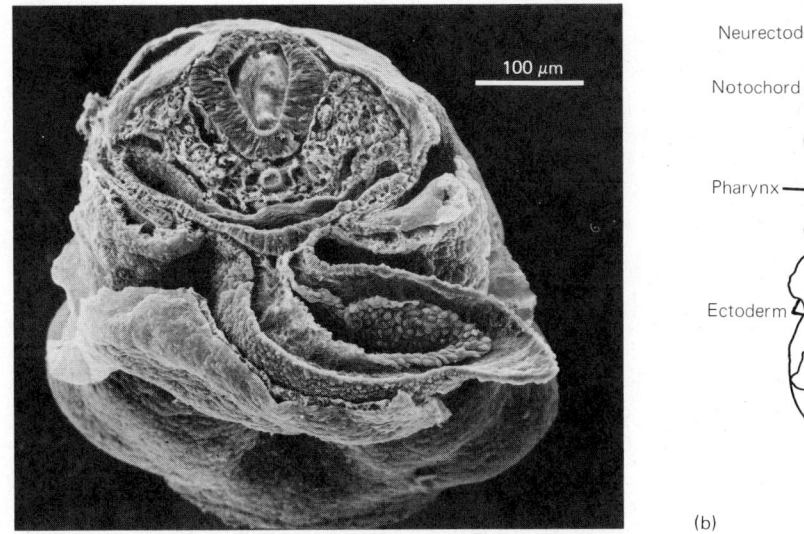
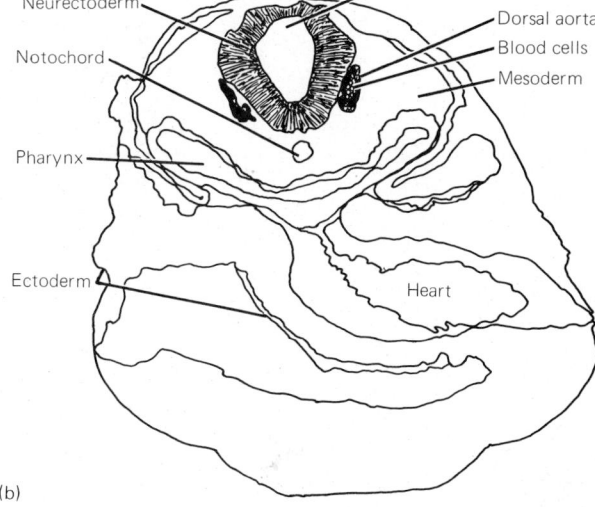

Figure 36-24 (a) Section through the top half of a chick embryo during organogenesis, after 48 hours of incubation. (b) A diagram to show the main features of the embryo. (Kathryn Tosney)

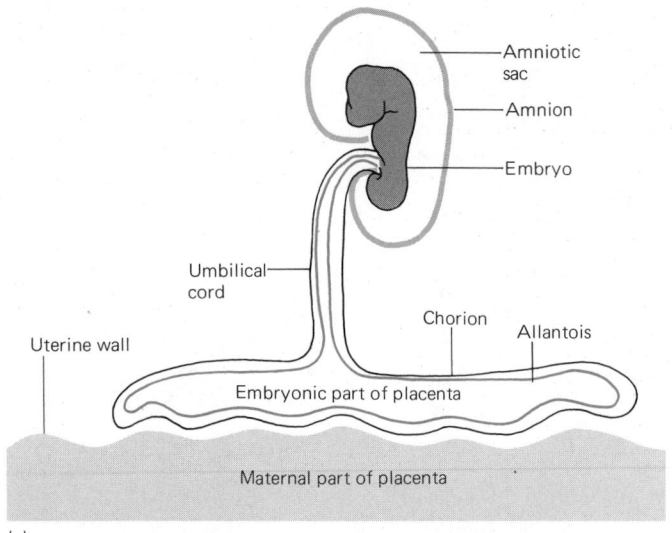

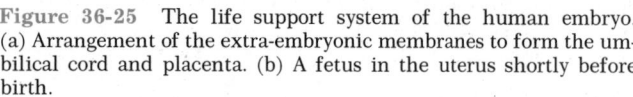

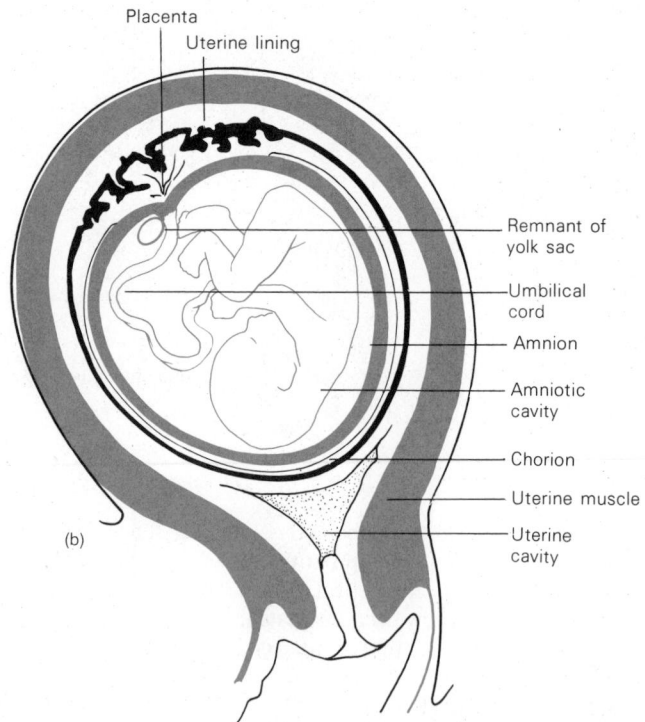

Figure 36-25 The life support system of the human embryo. (a) Arrangement of the extra-embryonic membranes to form the umbilical cord and placenta. (b) A fetus in the uterus shortly before birth.

The human embryo undergoes neurulation during the third week after fertilization. At this stage, the embryo is only about 2 mm long and is almost indistinguishable from the embryo of a frog or a sea urchin, but already it is surrounded by its membranes and attached to the uterus at the placenta (Figure 36–25). The placenta is now tiny, but by the time of birth it will develop into an organ that looks like raw hamburger, about the size of a stack of four dinner plates. The membranes joining the embryo to the placenta develop into a cord, the **umbilicus**, which grows thicker and longer as development proceeds.

By the end of the third week the embryo has entered the stage of organogenesis; in this stage the major organ systems begin to form: the nervous system, gut, and blood vessels. The heart, shaped like a lumpy tube, starts to pulsate. Drugs and diseases are most apt to damage the embryo at these early stages when the organ

TABLE 36–1 **OUTLINE OF HUMAN DEVELOPMENT**	
DAYS AFTER FERTILIZATION (*Approximate; varies a lot*)	*WHAT'S HAPPENING*
0–8	Cleavage
6	Implantation
21	Neurulation
24	Nervous system, gut, and blood vessels start to develop
28–35	Embryo most susceptible to damage by German measles, thalidomide, other drugs, etc.
42	Testes differentiating in male
75	Primary oocytes enter first meiotic division in female
90	All major organ systems formed but tiny
270	Birth (parturition)

systems are forming. After about three months of development, the embryo, although it is still small (about 30 mm long), is more or less fully formed and much less susceptible to malformations. The drug thalidomide, prescribed as a tranquilizer for pregnant women during the 1960s, stunted development of the limbs from the tiny limb buds during the fourth and fifth weeks of development. Similarly, if a pregnant woman has German measles during the fourth through twelfth weeks of pregnancy, the disease may damage the embryo's heart, eyes, ears, or brain, which are developing at that time. There is evidence that alcohol and caffeine (in coffee, tea, and colas) may also damage the developing fetus.

During the third month, the fetus begins to move, and the mother may feel its movements. From this point onward, the most obvious progress is growth in size. There are still changes taking place, however. The nervous system is still immature, and so are the circulatory and respiratory systems; the fetus cannot survive outside the mother. The youngest fetus known to have survived a premature birth was about 20 weeks old, and it required continuous assistance to breathe, feed, and maintain its body temperature for many weeks thereafter.

Birth

In humans, the date of birth averages about 270 days after conception, but there is much variation in the time a baby takes to develop. The process in which uterine contractions expel the baby and the placenta is called **labor**, and can be divided into three main stages. The first stage is **dilation**, which usually lasts from 2 to 20 hours and ends when the cervix of the uterus is fully open or dilated. The second stage, **expulsion**, which lasts from about 2 to 100 minutes, begins with **full crowning**, the appearance of the baby's head in the cervix, and continues while the baby is pushed, head first, down through the vagina into the outside world, where it draws its first breath.

The third, or **placental**, stage begins when the baby is born. The uterus continues to contract while the umbilical cord is clamped, and some minutes after the baby is born the uterus expels the placenta. The umbilical cord can now be severed, and the baby's independent existence begins.

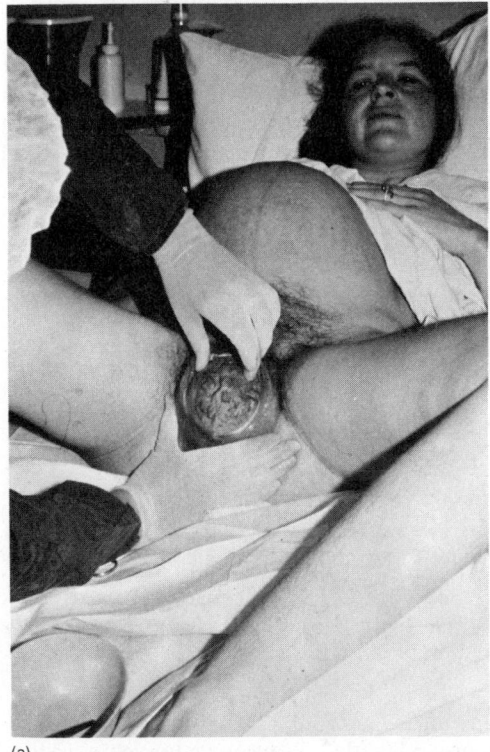
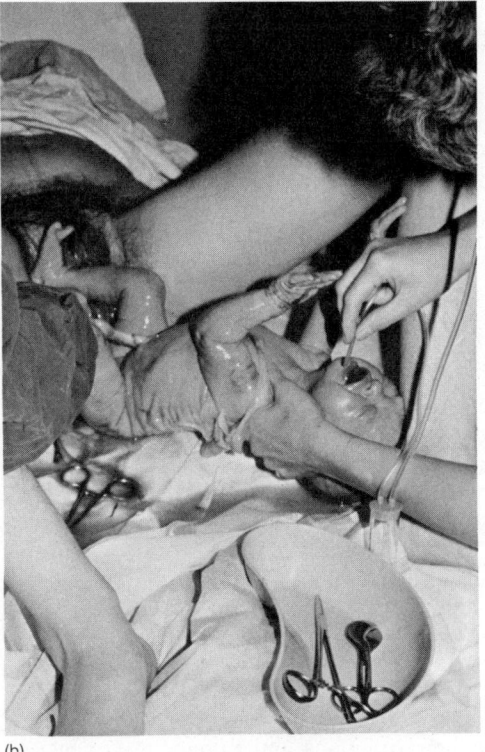

(a)

(b)

Figure 36-26 Human birth. (a) The head emerges. (b) The midwife clears mucus from the newborn's mouth. (Biophoto Associates)

36-H Maturation

The term "embryonic development" is usually used to cover only that period in an animal's life before it leaves the protection of the egg or of the mother's body at hatching or birth. Animals continue to change in various ways after this time. For instance, different mammals are born at different stages of development. A mouse is born hairless, blind, and incapable of regulating its body temperature. The foal of a horse, on the other hand, can run around and use all of its senses within an hour of birth. Further development also occurs after birth, in that most animals do not become sexually mature until later in their lives. In addition, some animals undergo **metamorphosis**, a change in body form, before they are sexually mature (Section 12-H). Both metamorphosis and sexual maturation involve the same processes of genetic differentiation, growth, and morphogenesis as those that occurred in the embryo. Aging is another change that occurs after embryonic development is complete (Section 12-I).

SUMMARY

In the human male, sperm form in the testes and move through ducts where glandular secretions are added. During sexual intercourse, semen is ejaculated into the woman's vagina; sperm swim through the cervix and uterus to the fallopian tubes, where fertilization of an egg released from one of the ovaries occurs. Hormones control the production of eggs and sperm and maintain pregnancy. Various methods of birth control prevent conception or embryonic development by interfering with various stages in the reproductive cycle.

An animal starts life as a zygote, which undergoes embryonic development and eventually becomes a self-sufficient individual. The main cellular and molecular processes governing development are genetic differentiation between cells and the interactions among cells. Development can be broken down for convenience into the stages of cleavage, gastrulation, neurulation, and organogenesis.

The pattern of cytoplasm laid down in the egg during its formation in the ovary determines what happens during cleavage. From gastrulation onwards, the zygote's own genes control development.

Neurulation is followed by organogenesis. In the human embryo, the major organ systems have formed after about three months of development. The fetus cannot survive outside the uterus until about six months after fertilization. Birth occurs at about nine months and involves powerful uterine contractions that expel the fetus and placenta from the uterus through the vagina.

Following embryonic development, maturational changes must take place before an individual is sexually mature.

OBJECTIVES

From your study of this chapter, you should be able to:

1. List the organs in the human male and female reproductive tracts, give their functions, describe how gametes are formed and released, and explain how pregnancy begins.

2. State how "the pill," IUD, diaphragm and jelly, condom, rhythm, induced abortion, and sterilization of males or females operate as birth control methods. State the part of the reproductive process with which each method interferes.

3. Describe the processes of spermatogenesis, oogenesis, and fertilization, and state their functions and importance.

4. Name the four main stages into which embryonic development is divided, and list definitive features of each.

5. Describe evidence that suggests that the pattern of cleavage is determined by factors in the egg cytoplasm and that the genes of the zygote do not become important in development until gastrulation.

6. List or recognize body parts formed from each of the three primary embryonic germ layers (ectoderm, mesoderm, and endoderm).

7. Use the following words and phrases correctly: blastocoel, blastopore, blastula, cleavage, embryonic induction, gastrula, gastrulation, neurula, neurulation, zygote.

SELF-QUIZ

Associate the reproductive organs on the right with the description on the left:

___ 1. Tube for conducting sperm
___ 2. Receptacle for penis
___ 3. Production of seminal secretions
___ 4. Conducts eggs
___ 5. Holds baby in uterus
___ 6. Produces sperm
___ 7. Prepares nutritive lining for embryo

a. cervix
b. fallopian tube
c. Cowper's gland
d. ovary
e. prostate gland
f. testis
g. urethra
h. uterus
i. vagina
j. vas deferens

8. From the list of reproductive organs and passages above, construct a (correct) route for the passage of sperm from the site of production to the site of fertilization.

For each of the following birth control methods, choose the correct means of interference with reproduction:

___ 9. Diaphragm and jelly
___ 10. Vasectomy
___ 11. "The pill"
___ 12. IUD
___ 13. Tubal ligation
___ 14. Induced abortion
___ 15. Condom

a. prevents fertilization of egg
b. prevents embryo implantation
c. prevents completion of embryonic development of implanted embryo
d. prevents ovulation
e. prevents sperm formation
f. prevents release of sperm into seminal fluid

Match the correct stage of development with each of the following characteristics:

___ 16. Pattern depends on amount and distribution of yolk
___ 17. Not affected by injection of RNA synthesis inhibitor
___ 18. Formation of embryonic spinal cord
___ 19. Results in formation of skeleton and muscles from mesoderm
___ 20. Rapid cell division without enlargement
___ 21. Influenced by embryo's genome
___ 22. Produces gastrocoel

C = Cleavage
G = Gastrulation
N = Neurulation
O = Organogenesis

23. Which of the following is *not* evidence that the pattern of cleavage is determined by factors in the egg cytoplasm rather than by the genes of the zygote?
    a. Cleavage is prevented by injection of a protein synthesis inhibitor.
    b. Cleavage is not affected by injection of inhibitors of RNA synthesis into the zygote.
    c. Interspecific hybrid zygotes go through normal cleavage but die at the beginning of gastrulation.
    d. The direction of the first division of cleavage in the snail *Cepaea* is determined by the mother's genotype.

24. The blastopore:
    a. is obliterated during gastrulation
    b. induces formation of the lens of the eye
    c. is the hollow where the embryo implants in the uterus
    d. is the hollow space inside a blastula
    e. is the hole through which some external cells pass to the inside of the embryo

QUESTIONS FOR DISCUSSION

1. Does vasectomy affect male potency?
2. Does abortion affect a woman's subsequent fertility?
3. Can a woman become pregnant the first time she has sexual intercourse?
4. Why do you think the venereal disease gonorrhea is now epidemic in the United States and Western Europe?
5. What is the adaptive advantage of an embryo's being provided with preformed RNA transcribed from its mother's genes?

REFERENCES AND FURTHER READING

Beaconsfield, P., G. Birdwood, and R. Beaconsfield. "The placenta." *Scientific American*, August 1980.

Boolootian, R. A. *Human Reproduction.* New York: John Wiley & Sons, 1971. A very good, popularly written account of human reproduction and sexuality.

Browder, L. W. *Developmental Biology.* Philadelphia: Saunders College, 1980. A straightforward, well-illustrated text.

Bryant, P. J., S. V. Bryant, and V. French. "Biological regeneration and pattern formation." *Scientific American*, July 1977. A description of regeneration experiments in insects and vertebrates and the contribution of this work to an understanding of embryonic differentiation.

Conklin, E. G. "The orientation and cell lineage of the ascidian egg." *Journal of the Academy of Natural Sciences* (Philadelphia) 2:13, 1905. A delightfully written early account, interesting in that it shows the ingenuity of a scientist asking a question that was difficult to answer with the limited techniques available at the time.

Swartz, D. P., and R. L. V. Wiele. *Methods of Conception Control.* Raritan, NJ: Ortho Pharmaceutical Corp., 1971. A straightforward account for the layperson.

Tietze, C., and S. Lewit. "Legal abortion." *Scientific American*, January 1977. An interesting account of the sociological effects of legalizing abortion.

Wolpert, L. *The Development of Pattern and Form in Animals.* New York: Oxford University Press, 1974. A readable, well-illustrated account of how cells interact to form functional structures during development.

CHAPTER 37

NEURONS

Digestion, respiration, circulation, excretion, and reproduction are processes that may be isolated for ease of study, but they are inseparable in a living animal. All must be coordinated in such a way that they work together. We have already seen several examples of coordination, for example, control of appetite and of breathing.

The nervous and endocrine (hormonal) systems of an animal are primarily responsible for integrating its activities. The sense organs receive **stimuli**, such as light, chemicals, or pressure, which carry information about the body's internal and external environments. This information is relayed to the nervous system, which directs appropriate muscular or hormonal adjustments to maintain internal homeostasis or to make appropriate behavioral responses to the external environment (Figure 37–1).

The remaining chapters in this section will deal with the nervous system and with the sense organs, muscles, and glands that work with it in coordinating the activities of a living animal. We will begin with **neurons**, the cells that carry messages in the nervous system. **Sensory neurons** carry information from receptor cells in the sense organs to other neurons in the **central nervous system (CNS)**,

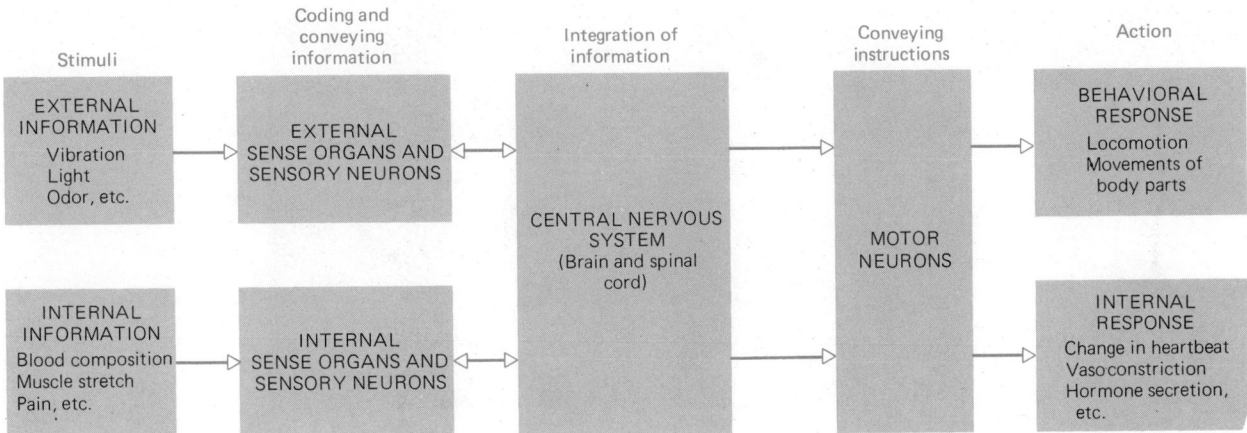

Figure 37-1   The flow of information into, through, and out of the nervous system.

the brain and spinal cord. **Motor neurons** carry messages to the body's **effectors**, its muscles and glands. **Interneurons** are the neurons between sensory and motor neurons; they form a network that processes sensory information and sends out appropriate instructions to the effectors via the motor neurons. Neurons pass information to one another across small spaces, or **synapses**, that separate two neurons; each neuron may synapse with many others so that a single neuron may receive information from, or send information to, many parts of the body. The arrangement and number of neurons and their synaptic connections in an animal's nervous system determine how the animal responds to stimuli and the kinds of behavior it can perform.

37-A   Structure of a Neuron

Every neuron has a **cell body** or **soma**, containing the nucleus and most of the cell's metabolic machinery: ribosomes, mitochondria, Golgi apparatus, and so forth. A neuron also has long, thin extensions, or **processes**, whose number and type vary depending on the type of neuron (Figure 37–2).

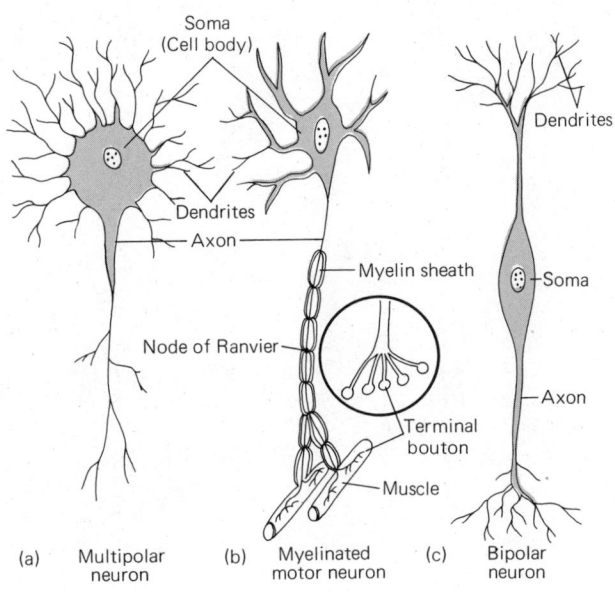

(a)   Multipolar neuron   (b)   Myelinated motor neuron   (c)   Bipolar neuron

Figure 37-2   Structure of different types of neurons. (a) A multipolar neuron has a central cell body, numerous dendrites, and a branched axon. (b) A motor neuron to a vertebrate muscle. This one has an axon surrounded by myelin, insulating material made up of the plasma membranes of Schwann cells (see Figure 37-5). (c) A bipolar neuron has two main branches, on opposite sides of the cell body.

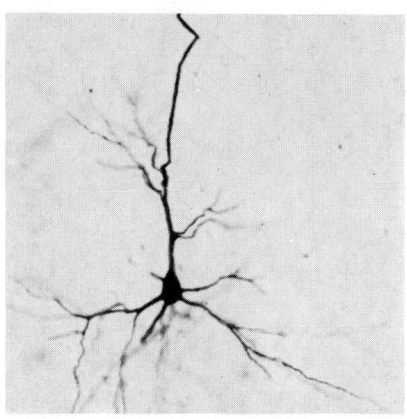
(a)

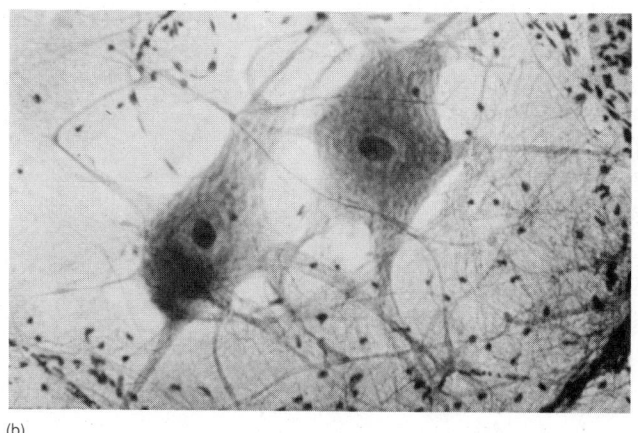
(b)

Figure 37-3 Neurons as they appear under the light microscope. (a) A neuron showing the thick axon with several branches, the dark cell body, and dendrites. (b) Two cell bodies lie in a tangled mass of dendrites and axons. (Carolina Biological Supply Company)

Two types of processes can be distinguished by their functions. **Dendrites** receive information, which may come from other neurons, from other types of cells, or from the external environment. In the vertebrate brain the dendrites of a single cell may receive as many as 100,000 connections from other neurons, although 1000 to 10,000 connections is about average for a human neuron. Not all connections come from different neurons; one neuron may send more than one connection to another.

A neuron also has one process that carries information to the next cell; this is called the **axon**. Although dendrites vary in diameter along their lengths, the axon has a relatively constant diameter. Furthermore, although the dendrites and soma may modify incoming information, an axon transmits an unaltered message to its eventual target.

A neuron may carry a message over a considerable distance; for example, some neurons extend from the spinal cord to the toe of an elephant. The longest individual cells found in any animal are some of its neurons.

37-B Glial Cells

The nervous system contains accessory cells called **glia** or **glial cells** in close association with the neurons (Figure 37–4). There are about 10 times as many glial

Figure 37-4 A glial cell with an axon that is part of a neuron growing in a culture dish in the laboratory. (The leading edge of a moving animal cell usually has the ruffled membrane form shown in this scanning electron micrograph.) (Norman Wessells)

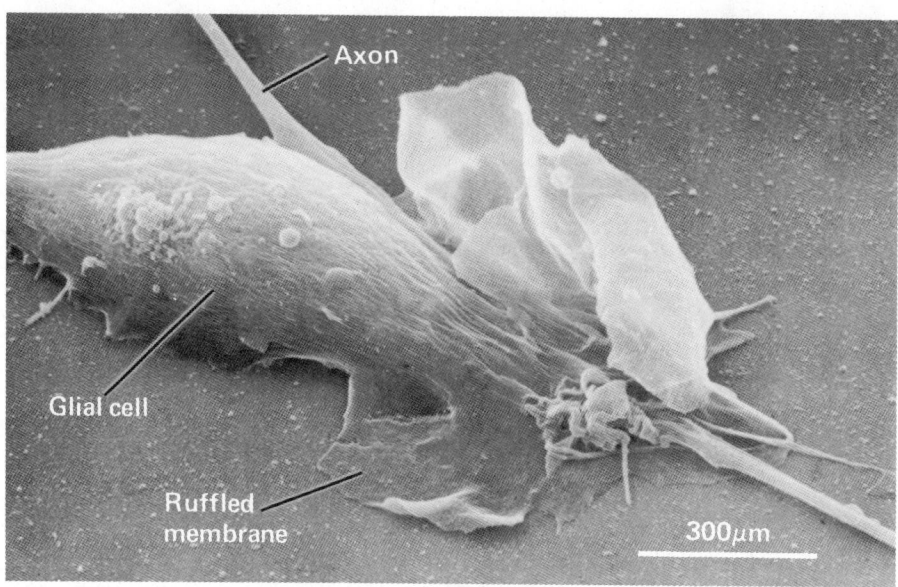

Axon

Glial cell

Ruffled membrane

300μm

cells as neurons in the human nervous system. The functions of many types of glial cells are not well understood, since these cells were largely ignored until quite recently, but it is becoming clear that they play a much more active role in the nervous system than was previously thought. Some glial cells, the **microglia**, are now thought to enter the central nervous system from the blood, and to act as scavenger cells.

Schwann cells are associated with axons that form the nerves which link various parts of the body with the central nervous system. A Schwann cell forms a great excess of plasma membrane, which wraps around the axon several times, forming a structure called the **myelin sheath** (Figure 37–5). The membrane has a high percentage of lipid, and there is very little cytoplasm in the outer layers of the wrapping. Each Schwann cell covers about 1 mm of the axon's length. There are gaps, called **nodes of Ranvier**, between adjacent Schwann cells; at these nodes the naked axon is exposed for a very short distance (see Figure 37–2).

Some axons within the central nervous system also have myelin sheaths; these sheaths are formed by **oligodendroglia**, another kind of glial cell.

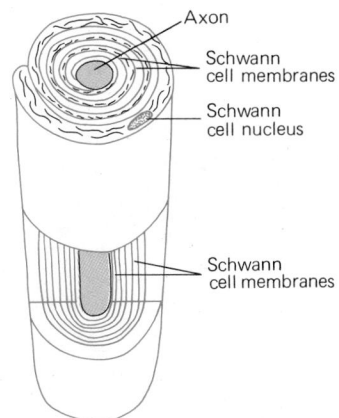

Figure 37-5 Cross section of a myelin sheath formed around an axon by a Schwann cell. The plasma membrane of the Schwann cell forms several layers of wrapping around the axon.

37-C   Resting Membrane Potential

Like most other living cells, neurons have an asymmetric distribution of ions across their plasma membranes, such that the inside of the cell is electrically negative with respect to the surrounding fluid. As a result, there is an electrical **potential difference** across the membrane. A. L. Hodgkin and A. F. Huxley, studying the giant axons of squids (Table 37–1), won the Nobel Prize for showing how this potential difference contributes to the functioning of a neuron.

The ability of a neuron to accept and relay information results from (1) the differences in the ion distribution between the inside and outside of the cell and (2) short-lived variations in the permeability of the plasma membrane to certain ions. These features, combined, permit neurons to produce rapidly moving electrical impulses, which travel down the axon to the synapse.

The asymmetric distribution of ions across the membrane of a resting neuron depends on an active pumping mechanism in the membrane as well as on the permeability properties of the membrane. A **sodium-potassium pump** expends energy, in the form of ATP, to eject sodium ions from the cell and to accumulate potassium ions inside. The continuous working of the pump keeps sodium and potassium from reaching equilibrium on the two sides of the membrane as they diffuse down their concentration gradients.

TABLE 37–1 **CONCENTRATIONS† OF IONS IN THE GIANT AXON OF A SQUID AND IN THE FLUID THAT SURROUNDS IT**			
ION	*SEAWATER*	*SQUID BLOOD*	*SQUID AXON‡*
K^+	10	20	400
Na^+	450	450	50
Cl^-	550	550	100
Ca^{2+}	10	10	0.5
Mg^{2+}	55	55	10
Organic anions	—	—	400

† Concentrations in millimoles.
‡ Squids and many other active invertebrates have neurons with "giant axons" of such large diameter that they are much easier to study than the very fine axons of vertebrate neurons.

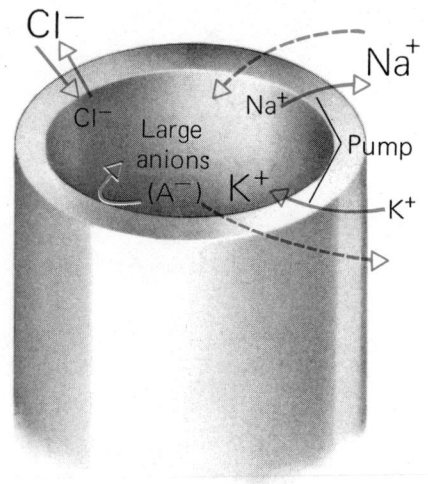

Figure 37-6 Ion distribution in a resting neuron. The sodium-potassium pump keeps sodium more concentrated outside the cell and potassium more concentrated inside. Large protein anions (A⁻) cannot escape from the cell. Chloride tends to remain outside the cell, attracted by the net positive charge there and repelled by the net negative charge inside the cell.

The membrane potential of a resting neuron is determined by the membrane's relative permeabilities to potassium and sodium ions. Potassium flows out more freely than sodium flows in, and therefore a net positive charge accumulates outside the membrane. Since like charges repel one another, this positive charge outside the membrane eventually reaches a level that reduces the outflow of positively charged potassium ions. Eventually the potassium outflow just equals the inward flow of sodium, and no more excess charge accumulates outside the membrane. The net charge imbalance across the membrane thus acts as a force that has diminished the flow of potassium ions. In a typical neuron this force, the **resting potential**, has a value of -70 millivolts; the sign is negative because the inside of the neuron is negatively charged with respect to the outside.

The negative ions distribute themselves passively in response to the membrane potential. The cell's proteins are negatively charged; they are confined to the inside of the cell because the membrane is impermeable to them. The membrane is somewhat permeable to chloride ions, but the tendency of chloride to become equally concentrated on the two sides of the membrane is counteracted by the attraction of the net positive charge outside the cell and the repulsion of the net negative charge inside the cell. Thus most of the chloride ions are outside of the cell (Figure 37-6).

The resting potential has two sources of stored energy: the sodium accumulated on one side of the membrane, and the stockpile of potassium on the other. In the resting neuron, the flow of various ions across the membrane is such that their movement causes no net change in charge, but if this resting flow is disturbed, the ions' electric charges alter the membrane's electrical potential. Proteins in the membrane form permeability channels with "gates" that can open or close. Electrically gated channels are opened by changes in membrane potential, whereas chemically gated channels are opened by the binding of chemicals that change the position of part of the gate-forming protein. Opening more of the sodium or potassium channel gates permits more of the corresponding ions to cross the membrane. Furthermore, these events change the membrane potential in a local area and open electrically gated channels in adjacent areas of the membrane where, in turn, the membrane's electrical potential changes as ions flow across.

Neurons carry information in the form of just such brief changes in membrane permeability to sodium and potassium, changes that are detectable as rapidly moving disturbances in the resting potential of the membrane. Such a change often starts as the result of a stimulus applied to a small part of the membrane (usually the dendrites or soma). This causes a small, localized disturbance in the membrane potential. Some of these disturbances become large enough to spread along the entire neuron.

37-D   Local Changes in Potential

A stimulus received by a neuron causes a small **local response** in the membrane potential, which may **depolarize** the membrane (decrease the potential difference across the membrane) or **hyperpolarize** it (increase the potential difference). A local response in a receptor cell of a sense organ is called a **generator potential**, whereas such a change in a non-receptor cell is called a **postsynaptic potential**. A local change in membrane potential is proportional to the intensity of the stimulus.

Excitatory, or depolarizing, stimulations make the membrane more permeable to sodium for a brief period, and the inrush of positive sodium ions causes the membrane potential to become temporarily less negative (Figure 37-7). Some **inhibitory** stimuli are hyperpolarizing, increasing the permeability of the membrane to potassium, and allowing potassium to flow out; thus the inside of the cell becomes even more negative with respect to the outside than it was at rest. Other inhibitory stimuli do not hyperpolarize the membrane, but rather increase the membrane's permeability to chloride. This permits more chloride ions to follow along when sodium enters the cell, offsetting some of the positive charge brought in by sodium. Thus a larger number of sodium ions (from a stronger stimulus) is needed to depolarize the cell.

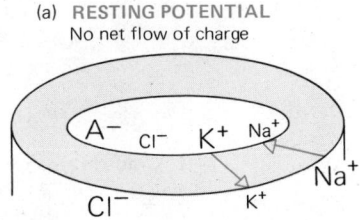

(a) RESTING POTENTIAL
No net flow of charge

(b) DEPOLARIZATION—EXCITATION
Increased Na⁺ inflow

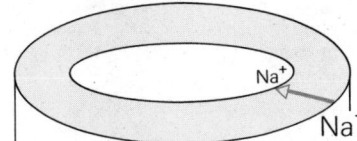

(d) HYPERPOLARIZATION—INHIBITION
Increased K⁺ outflow
Slight net Cl⁻ outflow

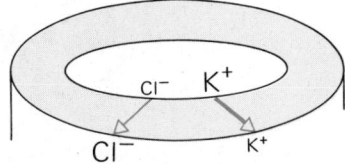

(c) RECOVERY FROM DEPOLARIZATION
Increased K⁺ outflow
Net Cl⁻ inflow

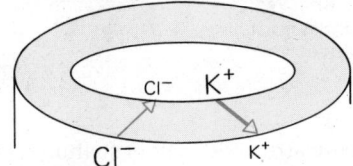

(e) RECOVERY FROM HYPERPOLARIZATION
Decreased K⁺ outflow; normal Na⁺ inflow
Net Cl⁻ outflow

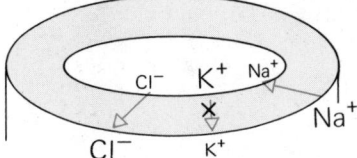

(f) MEMBRANE POTENTIAL CHANGES

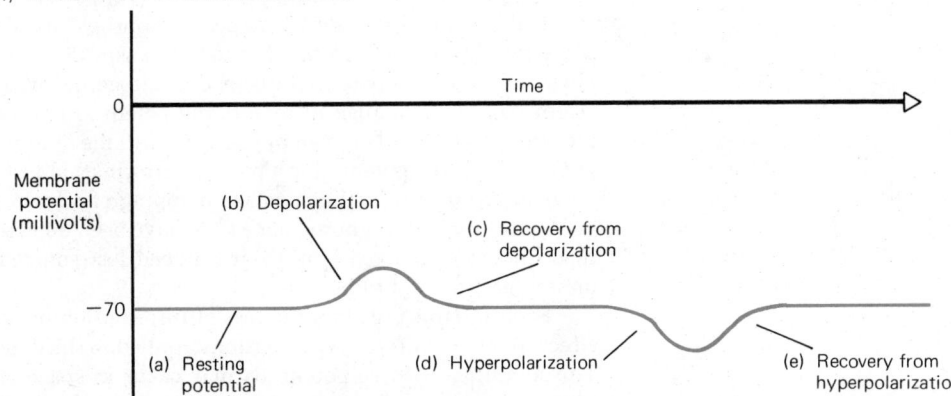

Figure 37-7 Local responses. (a) Resting membrane, showing base level ion distribution and net flow of each kind of ion (A^- = organic anions). In b–e, only changes in ion flow from the resting state are shown. (f) Graph of membrane potential changes during situations a–e.

How does the resting potential return to normal? After an excitatory stimulation, the membrane's increased permeability to sodium is quickly turned off, and is followed by an increase in permeability to potassium; the resulting outflow of potassium restores the membrane potential to its resting value. After an inhibitory stimulation, the membrane's increased negative potential greatly reduces the potassium flow. However, the normal (but small) inward flow of sodium drives the potential back to its resting value, and then the outward flow of potassium resumes its normal

TABLE 37-2 A COMPARISON OF THE PROPERTIES OF LOCAL RESPONSES AND ACTION POTENTIALS

LOCAL RESPONSE	ACTION POTENTIAL (a.p.)
1. Graded response	All-or-none response
2. Caused by an electrical stimulus of any intensity	Caused only by an electrical stimulus at or greater than threshold intensity
3. The extent (amplitude) of the electrical depolarization of the membrane is proportional to the intensity of the stimulus	No matter what the intensity of the stimulus, an a.p. in a given neuron is always of the same amplitude
4. Spreads across the membrane from the point of stimulation, but decays rapidly as it goes (its amplitude decreases)	Passes down the whole length of the axon with no change in its amplitude
5. Lasts for as long as the stimulus exists	Always lasts for the same period of time (i.e., as long as it takes to reach the end of the axon) no matter what the intensity of the stimulus
6. May sum with nearby potentials	No summation
7. No refractory period; another response may be started at any time (will undergo temporal summation if previous response still active)	Refractory period during which no new a.p. can be initiated

rate, equal to the rate of inward movement of sodium. (The sodium-potassium pump is not needed to restore the resting potential on a short-term basis; a neuron may respond to thousands of stimulations even with its pump stopped by chemical inhibitors. However, the pump is necessary to maintain the ion distribution in the long run.)

Local changes in membrane potential propagate along the membrane from the point of stimulation. However, the cytoplasm does offer some resistance to ionic flow, and so a local potential disappears altogether within a short distance from its origin. Since the change in membrane potential is proportional to the intensity of the stimulus, a small change in potential will die out more quickly than a larger one. Such a dissipated potential is a piece of information lost. This may be advantageous, for many stimuli, like the constant dripping of a leaky faucet or the itch of a mosquito bite, are better ignored once they have been detected. The messages from these repetitive or prolonged stimuli are indeed lost somewhere in the nervous system under certain conditions.

Summation prevents the loss of information by adding together the electrical effects of many different stimulations applied to the dendrites and soma of a neuron. The summing of these potentials may occur in space or in time. **Spatial summation** occurs when the potential changes due to two or more simultaneous stimuli overlap in space on the surface of the cell. **Temporal summation** occurs when there are two or more stimulations very close in time in the same area of the membrane, so that the later stimulations add their effect before the first potential has completely dissipated (Figure 37-8).

37-E Action Potentials

Summation of local potentials in a neuron's dendrites and soma may cause an electrical potential to be propagated in the axon. Each axon has a **threshold**, a level of depolarization that must be exceeded if it is to respond to the stimulus. When

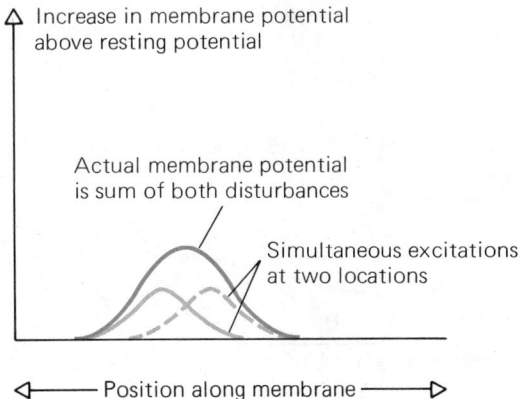

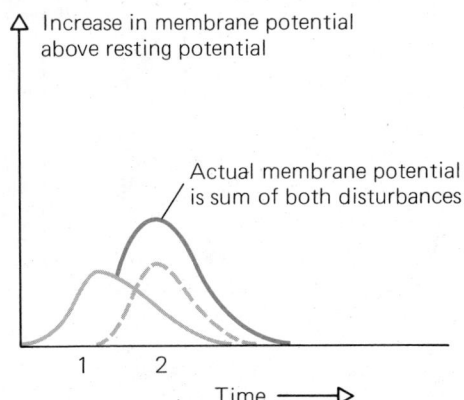

Figure 37-8 Summation of separate excitatory impulses: (a) Spatial summation of two depolarizations given simultaneously at adjacent points of the membrane. (b) Temporal summation of two depolarizations given at same point; pulse given at time 1 has not died out before another is added at time 2.

summation of local potentials reaches this threshold level, the resulting depolarization is **self-propagating**; that is, it causes an above-threshold depolarization in adjacent parts of the membrane and thus spreads down the axon without alteration.

This self-propagating wave of depolarization down the axon is called an **action potential**, **nerve impulse**, or **spike potential** (because of its shape on a recording of its electrical activity; see Figure 37–10). All stimuli above the threshold level cause action potentials of the same size. Since a neuron fires an action potential maximally or not at all, action potentials are termed "**all or none**" responses.

What happens in an action potential? The sodium gates open, and sodium ions rush into the axon. The decreased electrical potential increases the permeability to sodium even further, by opening more sodium gates for longer periods of time; thus inflow of sodium is a self-reinforcing process (Figure 37–9). All of this quickly reverses the membrane potential, with the inside now positive relative to the outside. The sodium ions that have just entered the axon spread to adjacent, resting areas, causing electrically gated channels to open and depolarizing these areas as well. As the cycle continues, a wave of depolarization sweeps along the axon (Figure 37–10). After the action potential has passed each part of the membrane, the sodium gates in that area close and return the sodium permeability to the normal level. Next, the potassium gates open and the permeability to potassium increases; potassium leaves the cell, restoring the membrane's resting potential. The potassium gates close and return the potassium permeability to its resting value.

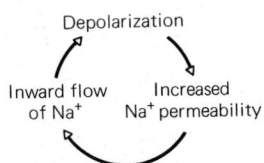

Figure 37-9 Circular interaction between permeability and potential during propagation of the nerve impulse.

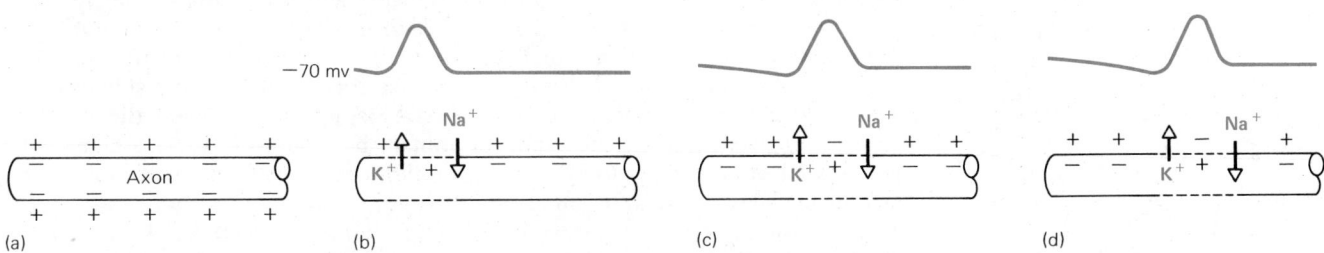

Figure 37-10 Spread of an action potential along an axon. (a) Resting state, with the inside of the membrane negative with respect to the outside. (b, c, d) Successive stages in conduction of the impulse along the axon from left to right. The increased sodium influx to the right of the action potential is caused by ions spreading to that region and depolarizing the membrane. This propagates the action potential by the sequence of events depicted in Figure 37-9.

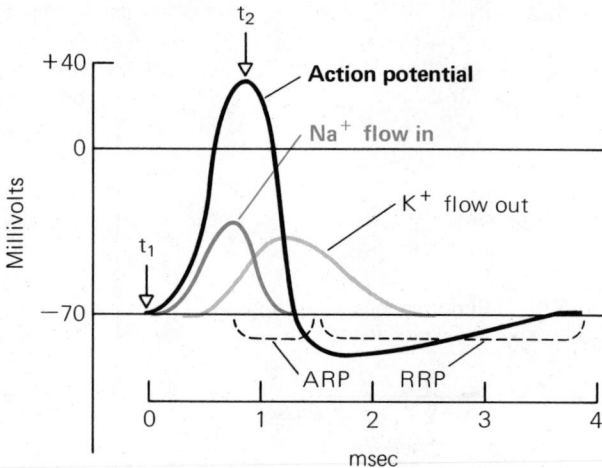

Figure 37-11 Action potential recorded as a nerve impulse sweeps past one point on the axonal membrane. Superimposed on the graph are the flows of sodium and potassium. During the absolute refractory period (ARP), no new impulse can be generated; during the relative refractory period (RRP), a stronger-than-threshold stimulus can initiate another action potential.

A graph of the electrical changes at one point on the axonal membrane as the action potential passes that point can be made by placing one electrode (electric current conductor) inside the membrane and another outside (Figure 37–11). When the electrodes are attached to an oscilloscope, the instrument will record the difference in electric charges at the two electrodes, and thus measure the electrical potential difference across the membrane. A baseline reading of −70 millivolts (see Figure 37–11) means that the inside electrode detects an electrical potential which is 70 millivolts more negative than that detected by the outside electrode. The rising phase of the action potential begins at t_1; this signals depolarization of the membrane as sodium rushes in. The potential difference across the membrane declines to zero, and reverses briefly as the inside becomes about 30 millivolts positive with respect to the outside (at t_2). Now the **repolarization**, or falling phase, begins as potassium rushes out, and the inside of the cell once more becomes negative with respect to the outside. During the first 0.5 to 1.0 millisecond* of this phase, the membrane is in an **absolute refractory period**, when no new action potential can be initiated, no matter how great the stimulus applied. After this comes the longer **relative refractory period**, lasting for 2 to 3 milliseconds. During this time another action potential can be initiated if the stimulus given is stronger than the normal threshold stimulus for the resting axon.

Speed of Propagation

The structure of an axon helps to determine how fast it can transmit nerve impulses. Even the most inefficient axons transmit action potentials at fairly rapid rates. There are two main factors that increase the rate of impulse conduction. First, an increase in diameter of an axon decreases the internal electrical resistance along its length, and so increases the rate at which sodium spreads to either side from its point of entry when it crosses the membrane; this, in turn, increases the speed of impulse propagation along the axon. All animals have axons of various diameters. Those involved in escape reactions in invertebrates are the thickest. Examples are the "giant" fibers (axons) of annelids, arthropods, and molluscs. If you have ever missed a swat at a cockroach, you have been the victim of the activity of such a giant fiber, which runs from the cerci (structures resembling twin tailpipes at the roach's posterior end) to the brain, and informs the brain of any slight air currents that disturb the cerci on your downstroke.

The second adaptation, found mainly in vertebrates, is a myelin sheath. The

*Millisecond   One-thousandth of a second.

layers of fatty wrapping insulate the axon and greatly increase electrical resistance across the membrane. Since current cannot pass through the myelin sheath, it must spread down the axon until it reaches a node of Ranvier, where myelin is absent (Figure 37–12). Here the potential across the membrane reaches threshold quickly and starts another action potential, which travels the short distance to the other side of the node, where the current again races to the next node. The result is an extremely fast form of conduction termed **saltatory** ("jumping") **conduction**, whereby the current generated by the activity at one node spreads with little loss to the next node, which may be as much as 1 to 2 mm away, and so on from node to node. The action potential at the node regenerates the original signal so that it can reach the next node with sufficient intensity to start an action potential there. Although some unmyelinated giant squid axons have been found to conduct action potentials at speeds of nearly 20 metres per second, much smaller mammalian myelinated axons may transmit impulses at up to 100 metres per second.

37-F   Synaptic Transmission

The summation of local potentials in the dendrites and soma of a neuron may initiate an action potential, which is transmitted down the axon. The axon ends at a synapse, an area where the membrane of the axon lies very close to the membrane of the next cell in line, which may be another neuron or part of a muscle or gland. A synapse where an axon ends close to a muscle is called a **neuromuscular junction**. Bernard Katz first worked out how synapses function by studying neuromuscular junctions.

Information travels across most synapses by way of chemical messengers called **neurotransmitters**. Neurotransmitters are found in tiny membranous sacs, or **vesicles**, in the **synaptic knobs**, or **boutons**, enlargements of the end of the axon (Figure 37–13). The presence of many such knobs allows a neuron to pass information to several other cells, or to make more synapses with one cell than with another. The plasma membrane of the synaptic knob is called the **presynaptic membrane**. It lies close to the **postsynaptic membrane**, part of the dendrite or soma of another neuron or part of the surface of a muscle or gland cell. Between the presynaptic and postsynaptic membranes lies a space called the **synaptic cleft**, which is about 20 nm wide.

As an action potential reaches the end of an axon, it causes some of the vesicles to discharge their contents into the synaptic cleft. The transmitter molecules cross the cleft and attach to receptor molecules in the postsynaptic membrane. This opens a chemically gated permeability channel in the postsynaptic membrane, and the resulting flow of ions changes the membrane's electrical potential. Even though the synaptic cleft is very narrow, the process of synaptic transmission is much slower

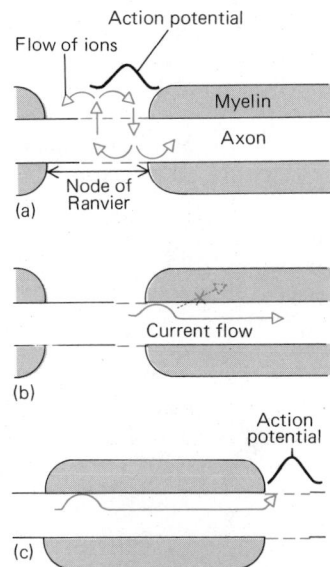

Figure 37-12 Conduction along a myelinated neuron. Colored arrows represent flow of current. (a) The action potential travels across the node of Ranvier only as fast as the ions can cross the membrane and change its potential. (b) In parts of the axon surrounded by myelin, the current cannot cross the membrane and thus spreads down the axon's interior to the next node. This conduction is rapid because it is not limited by the speed of ion flow. (c) When the current reaches the next node of Ranvier, it lowers the membrane potential, thus starting another action potential.

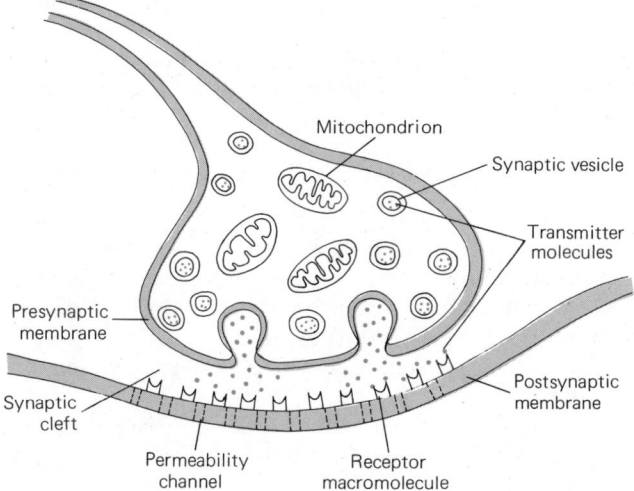

Figure 37-13 Transmission across a synapse. When an action potential arrives, vesicles fuse with the presynaptic membrane and discharge their neurotransmitter. Transmitter molecules cross the synaptic cleft and attach to receptor macromolecules in the postsynaptic membrane. Binding of the transmitter to the receptor opens permeability channels and permits an increased flow of certain ions across the membrane.

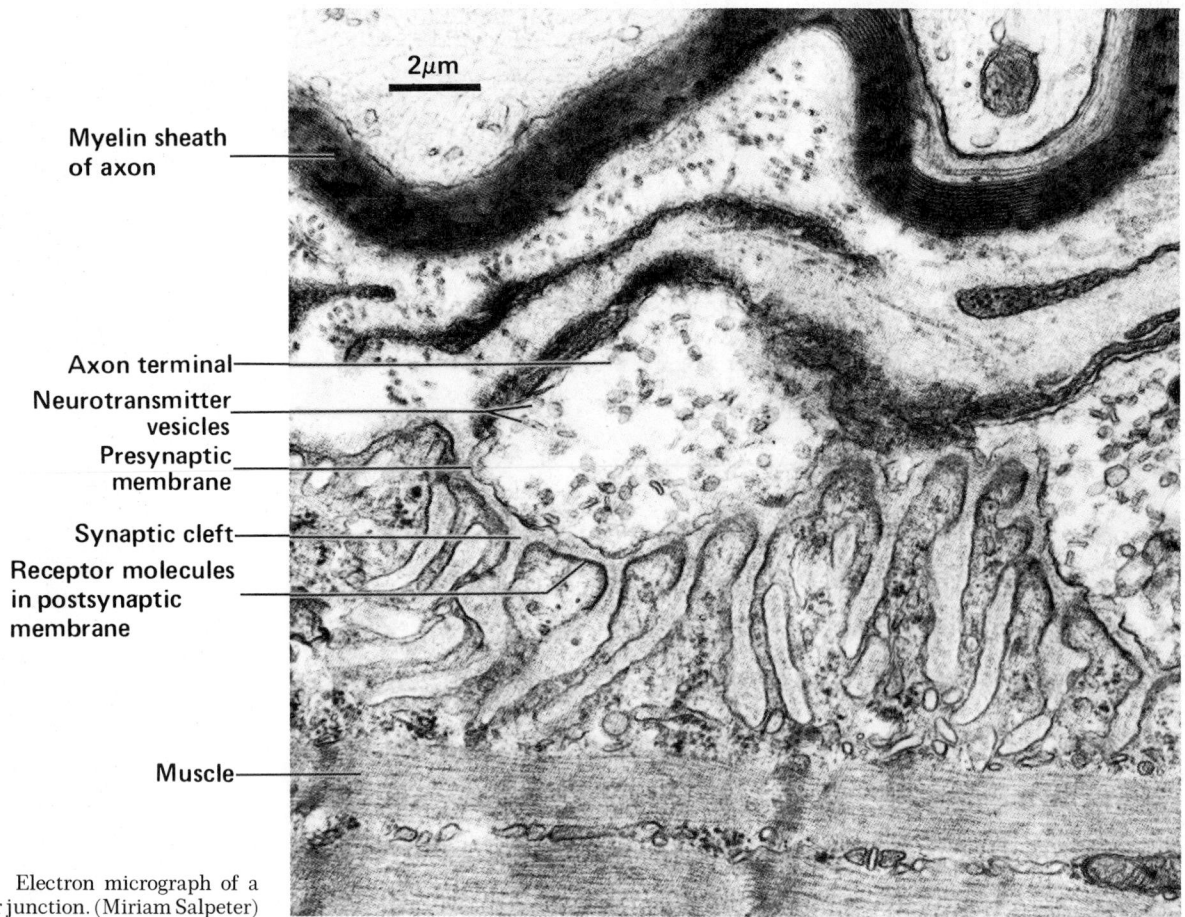

Myelin sheath
of axon

Axon terminal

Neurotransmitter
vesicles

Presynaptic
membrane

Synaptic cleft

Receptor molecules
in postsynaptic
membrane

Muscle

2μm

Figure 37-14 Electron micrograph of a neuromuscular junction. (Miriam Salpeter)

than conduction along the axon. Thus, in general, the more synapses in a neural pathway, the slower is transmission along that pathway.

The arrival of a neurotransmitter that depolarizes the postsynaptic membrane of a neuron causes a local **excitatory postsynaptic potential (EPSP)**. Only if the postsynaptic cell is depolarized enough to exceed its own axon's threshold will it transmit an action potential down its axon to another synapse. Often a postsynaptic cell will not fire an action potential if it receives information from only one presynaptic cell.

Although a neuron will transmit an impulse in both directions, the overall transmission of information through the nervous system occurs in only one direction because of the properties of synapses. Information can cross most types of synapse in only one direction because only the presynaptic membrane can release neurotransmitters, and only the postsynaptic membrane has receptors for these neurotransmitters.

Inhibitory Synapses

Not all synapses are excitatory. At an **inhibitory synapse** the postsynaptic membrane may be hyperpolarized; alternatively, the postsynaptic membrane at an inhibitory synapse may become temporarily more permeable to chloride ions (see Section 37-D). Either event results in an **inhibitory postsynaptic potential (IPSP)**, which increases the amount of depolarization necessary to reach the threshold for an action potential.

The sum of depolarizations and hyperpolarizations caused by excitatory and inhibitory synapses impinging on a cell determines whether or not it will fire an action potential. When more inhibitory synapses are active, more excitatory impulses will be needed to produce a threshold depolarization. Whether a particular synapse is inhibitory or excitatory depends upon the properties of the postsynaptic membrane. The same neurotransmitter can act at both kinds of synapses.

Electrical Synapses

Although the most common synapses are "chemical" synapses, where transmission is by way of a neurotransmitter, more and more "electrical" synapses have been discovered in recent years. Electrical synapses have no neurotransmitter; the membranes of the two cells lie only 2 nm apart, and they form a low-resistance gap junction (Section 4-C). Here current can jump directly between the cells, eliminating the slow process of chemical transmission.

Electrical synapses usually permit two-way transmission of impulses because current can flow in either direction. Since electrical transmission is faster than chemical transmission, why do animals have chemical synapses at all? The reason seems to be that chemical synapses permit more complex and varied modification of information.

37-G   Neurotransmitters

The most common neurotransmitters are shown in Figure 37–15. Acetylcholine and noradrenalin (also known as norepinephrine) act as transmitters in synapses between neurons both inside and outside the central nervous system, while other known transmitters (serotonin, dopamine, glutamate, and γ-aminobutyric acid) have so far been found only inside the central nervous system, at least in vertebrates. There are probably still other transmitters to be discovered. Only one type of transmitter has so far been found at any one synapse.

After they have acted, transmitter molecules must be removed or destroyed. Otherwise, their action would continue indefinitely, and all useful information

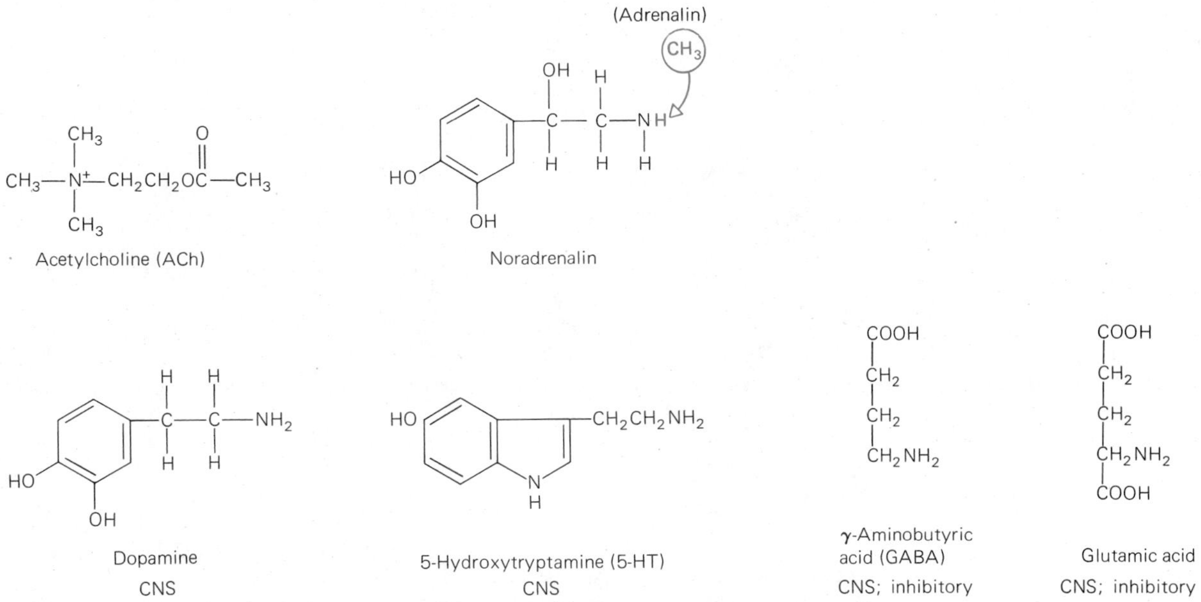

Figure 37-15 Molecular structures of some neurotransmitters. Most are found only in the central nervous system (CNS), but acetylcholine and noradrenalin are also found in the peripheral nervous system (nerves outside the brain and spinal cord). Adrenalin has a methyl (CH₃) group in place of the H shown in color in noradrenalin.

MODE OF ACTION	DRUGS THAT ACT AT ACETYLCHOLINE SYNAPSES	DRUGS THAT ACT AT CATECHOLAMINE* SYNAPSES
Inhibits manufacture of neurotransmitter	Hemicholinium (blocks neuron's uptake of choline to make acetylcholine)	Alphamethylparatyrosine (AMPT) (a sedative) Disulfiram (used as an antioxidant in the rubber industry; used as a drug to treat alcoholics, since violent illness ensues when alcohol is consumed after ingestion of the drug)
Decreases stores of neurotransmitter		Reserpine (sedative or depressant, used for epilepsy or hypertension) Tetrabenazine (blocks storage of neurotransmitter in vesicles)
Decreases level of molecules that deactivate neurotransmitter	Diisopropyl fluorophosphate (insecticide; chemical warfare; applied locally to treat glaucoma) Physostigmine Neostigmine (used in treating myasthenia gravis, where more muscular stimulation is required; antidote for atropine poisoning)	Iproniazid (antidepressant; once used to treat tuberculosis, now only occasionally for psychotic depression) Nialamide (antidepressant)
Enhances release of neurotransmitter		Cocaine Amphetamine (CNS stimulant)
Inhibits release of neurotransmitter	Botulin toxin (botulism poisoning)	Alphahydroxybutyrate
Activates receptor site	Pilocarpine (from a plant; mimics acetylcholine at most sites) Carbachol (mimics acetylcholine at most sites) Muscarine (from fly agaric, a mushroom; at synapses with smooth* and cardiac* muscle, glands, some CNS sites) Nicotine (from tobacco; at synapses with skeletal muscle,* some CNS sites not affected by muscarine)	Noradrenalin, adrenalin Isoproterenol (inhaled as treatment for asthma; stimulates heart, relaxes bronchial tubes, dilates blood vessels to skeletal muscles)
Blocks receptor sites	Atropine (from some plants, e.g., deadly nightshade; with artificial respiration, is antidote for anticholinesterase* poisoning; works at synapses between neurons and effectors) Scopolamine (found in many atropine-producing plants; with morphine, induces "twilight sleep" for childbirth Curare (used in poisoned darts, and in research to immobilize animals while leaving physiology intact; works at skeletal muscles)	Phentolamine (used in behavior research; blocks adrenalin sites responsible for constriction of blood vessels, thus blocking rise in blood pressure) Propranolol (used in research, not clinically) Chlorpromazine (used to treat schizophrenia or bad drug trips) Haloperidol
Inhibits resorption of neurotransmitter (by catecholamine-releasing presynaptic membrane)		Imipramine (antidepressant, drug of choice) Cocaine (CNS stimulant) Amphetamine

† Many familiar drugs, such as caffeine, LSD, morphine, amphetamine, and marijuana, have more generalized effects, and in most cases their exact mode of action is not well understood.

* Catecholamines A group of chemicals containing an amine group; includes adrenalin, noradrenalin, dopamine (see Figure 37–15).
* Smooth muscle The type of muscle that lines the walls of internal organs, e.g., digestive tract, blood vessels.
* Cardiac muscle The type of muscle that makes up the vertebrate heart.
* Skeletal muscle The type of muscle that moves parts of the skeleton, e.g., muscles of legs, back.
* Anticholinesterase Agent that blocks the effect of acetylcholinesterase.

would be lost. Noradrenalin is removed from a synaptic cleft by being resorbed by the presynaptic membrane. Acetylcholine is broken down by the enzyme **acetylcholinesterase**.

Many insecticides and nerve gases, such as the organophosphates, act as inhibitors of acetylcholinesterase. In the presence of these inhibitors, acetylcholine in the synapses keeps stimulating the postsynaptic membranes, and an animal's nervous system soon goes wild, transmitting one nerve impulse after another. These impulses in turn cause contraction of the muscles in uncontrollable spasms and, eventually, death.

Drugs and the Brain

The synapse offers an ideal point for the use of drugs to intervene and regulate the nervous system. Many drugs affect the nervous system, but we understand the actions of only a few (Table 37–3).

Opium, from the seedpod of a poppy, has been used as a drug since classical Greek times, not only because it is the most effective pain-killer ever discovered, but also because of the euphoric state it induces. Opiates were used as pain-killers in the Civil War in the United States, and addiction to opiates has been a social problem in the United States ever since. The search for a non-addictive opiate has been intense, but all the opium derivatives ever produced—including morphine, Demerol, methadone, codeine, and heroin—eventually produce addiction in many people who take them. Opiates bind to postsynaptic receptors in the brain and block the binding of any neurotransmitters that are released. This prevents the transmission of nervous impulses along a tract of nerves by which the body normally "tells" the brain that it is in pain. (Pain is a useful biological reaction, for when the brain is informed that some part of the body is in pain, perhaps from a cut or burn, it causes the body to move away from the source of the damage.)

Recently, substances called **endorphins** have been isolated from various parts of the brain. Their natural function is not fully understood, but it is interesting that, chemically, they are similar to morphine. The existence in the brain of receptors for endorphins may explain why the brain is so sensitive to morphine, opium, and related drugs.

Drugs such as LSD, psilocybin, mescaline, yohimbine, and barbiturates all act on synapses in the brain. Some of them produce hallucinations in ways that are not yet understood. The action of amphetamines (Benzadrine, Dexedrine, and "speed") is more easily understood. Like Sevin and other insecticides, they prevent the normal disappearance of transmitter molecules from synapses, so that the synapses continue to fire long after they would otherwise have stopped. As with the insecticides, overdoses are fatal.

The effects of alcohol on the nervous system are poorly understood. A recent study suggests that alcohol has the short-term effect of enhancing the action of γ-aminobutyric acid, the neurotransmitter that helps relieve anxiety; however, long-term use of alcohol decreases the brain's content of γ-aminobutyric acid, and so makes it necessary to consume increasing amounts of alcohol to remove anxiety. Alcohol also kills neurons faster than they would otherwise die—at the rate of about 10,000 per ounce of alcohol consumed. This probably accounts for the mental deterioration seen in some alcoholics.

37-H Interpretation of Messages

A nerve impulse is an all-or-none phenomenon; it either occurs or it doesn't. However, the stimulus that initiates a nerve impulse may be large or small, and an animal can in fact detect differences in intensity of stimuli. How can a neuron with only two modes—either "on" (transmitting an action potential) or "off" (not firing an action potential)—send information about the magnitude of stimuli?

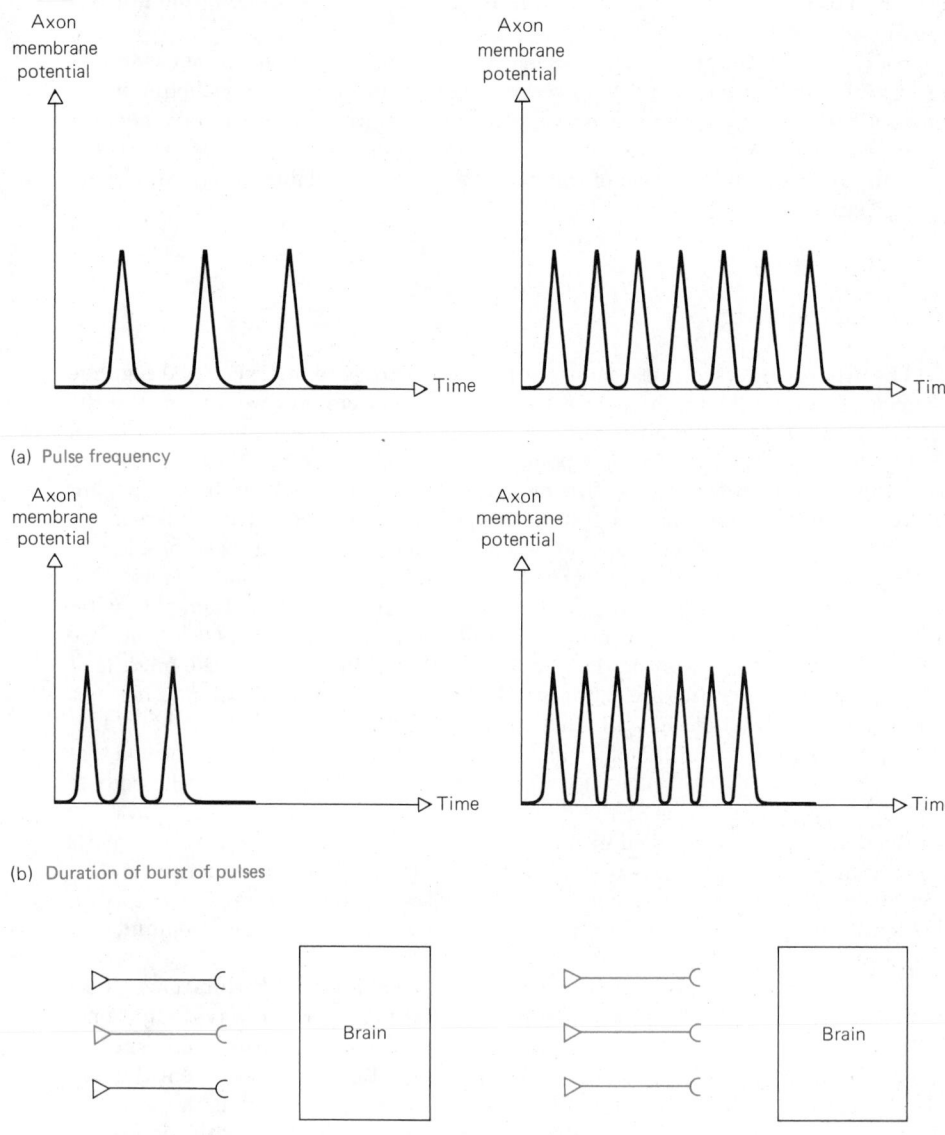

WEAK STIMULUS   STRONG STIMULUS

Axon membrane potential

Time

Axon membrane potential

Time

(a) Pulse frequency

Axon membrane potential

Time

Axon membrane potential

Time

(b) Duration of burst of pulses

Brain   Brain

(c) Number of neurons firing

Figure 37-16 Coding of stimulus intensity using neurons that produce only all-or-nothing responses. A strong stimulus will cause the neuron to (a) fire more rapidly, (b) fire a longer burst of impulses, or (c) cause more receptor cells to fire (color).

The nervous system encodes the strength of a stimulus in three ways (Figure 37–16). First, the intensity of a stimulus may be reflected in the frequency of action potentials. A weak stimulus initiates only a few action potentials per second, whereas a strong stimulus initiates many. The frequency cannot be raised indefinitely in proportion to stimulus intensity, because a neuron cannot fire during the absolute refractory period. However, when the stimulus is greater than threshold intensity, the neuron will fire during its relative refractory period. A weaker stimulus, which cannot induce firing during the relative refractory period, will produce a slower rate of firing.

Stimulus intensity may also be reflected in the length of time occupied by a burst of action potentials. A weak stimulus may give rise to a short burst of pulses in the neuron, a stronger stimulus to a longer burst.

The third way of encoding the stimulus intensity is in the number and kinds of neurons the stimulus activates. The threshold needed to initiate a nerve impulse varies from one neuron to another. Thus, a stimulus of low intensity will cause only a few neurons to fire action potentials, whereas a stimulus of higher intensity will fire all of these neurons as well as others with higher thresholds.

The nervous system receives all of its information in the form of trains of action potentials. How does the brain know whether an incoming stimulus represents a light shining, or a sound, or pressure on the skin, or an odor? Such qualitative information is built into the system in the way the neurons are "wired" to each other. That is, a neuron "knows" that it is responding to light when it is stimulated by a neuron that received its information from light receptor cells. If light receptor cells fire an action potential for any other reason, for instance, because of a hard blow on the head, the brain will "see" a flash of light, even though there was no light received by the eye.

Thus a relatively simple event, the passage of an all-or-none action potential, can be used to transmit information about the intensity and type of stimulus by using the timing of impulses and the "wiring" of neurons as additional sources of information.

37-I Neurosecretion

The transfer and integration of information by means of conducted impulses and synaptic potentials is not the only function of neurons. Many nerve cells produce neurosecretory substances or neurohormones and release these into the blood, where they help coordinate various physiological activities. For instance, the adrenalin in the bloodstream during stress is secreted partly by endocrine glands and partly by the axons of some neurons. (Noradrenalin is essentially the same as adrenalin.)

The secretions in neurosecretory cells are packaged in vesicles called secretory granules, which are much larger than synaptic vesicles. They release their contents in much the same way. In invertebrates, neurohormones regulate such functions as growth, maturation, metamorphosis, heart rate, reproduction, and even color changes. Vertebrates also have neurosecretory cells, notably those in various areas of the hypothalamus in the brain (see Section 41-B).

SUMMARY

Information passes along a neuron in the form of electrical changes across the neuron plasma membrane. When it reaches a synapse, the information is usually transmitted as a chemical that can cross the synaptic cleft to the membrane of the next cell and disturb its electrical balance in turn.

The resting neuron has two sources of stored energy: a high external concentration of Na^+ and a high internal concentration of K^+. This asymmetrical ion distribution is maintained, in the long run, by the sodium-potassium pump. The differential permeability of the membrane to different ions results in a resting potential across the membrane. Changes in this membrane potential occur when a stimulus opens the gates of ion channels through the membrane, allowing certain ions to cross the membrane and depolarize or hyperpolarize that part of the membrane. Small local changes in the membrane potential in the dendrites or soma may sum temporally and spatially until they exceed the threshold needed for the axon to produce an action potential. An action potential travels down an axon faster if the axon has a large diameter or if it is electrically insulated by a myelin sheath.

The intensity of a stimulus is coded largely in terms of the frequency of action potentials, the total length of a burst of action potentials, and the thresholds of neurons that are firing. The type of stimulus received is coded by the specific wiring of the nervous system.

Some nerves have neurosecretory functions in addition to, or instead of, their better-known role as conveyors of electrical impulses.

OBJECTIVES

From your study of this chapter, you should be able to:

1. Describe the basic structure and function of neurons.
2. Define the sodium-potassium ionic pump, and explain its role in maintaining the membrane potential.
3. Contrast the properties of local responses (generator potentials and postsynaptic potentials) with those of action potentials.
4. Draw a graph of the potential changes that occur during an action potential, and relate them to the flow of sodium and potassium ions across the axonal membrane.
5. Describe the myelin sheath, and explain its effect on impulse conduction.
6. Draw a model synapse with its principal components, explain the function of each component, and explain how information is transmitted to the postsynaptic cell.
7. Explain the difference between an excitatory and an inhibitory synapse in terms of effect on the postsynaptic membrane.
8. Explain how information about the intensity and type of a stimulus is transmitted in the nervous system.

SELF-QUIZ

1. Impulses leave a neuron via the:
   a. dendrites
   b. nucleus
   c. myelin sheath
   d. axon
   e. soma

2. Tell whether each of the following is true of a local response or of an action potential.
   a. It has a threshold.
   b. It is an all-or-none response.
   c. It has a refractory period.
   d. It is a graded depolarization.
   e. It sums with other similar electrical responses.

3. The myelin sheath around the axons of some vertebrate neurons:
   a. is rich in lipids because it is formed by the membranes of Schwann cells
   b. is a secretory product of the Schwann cells
   c. is produced inside the axon and extruded out through the membrane
   d. is continuous all along the length of the axon
   e. secretes neurotransmitter substances for release at the synaptic boutons

4. Write a short sentence describing the importance of each of the following components of a synapse:
   a. neurotransmitter substance
   b. neurotransmitter vesicle
   c. receptor molecules
   d. enzymes that destroy neurotransmitter

5. In an inhibitory synapse:
   a. information travels from the postsynaptic to the presynaptic cell
   b. the neurotransmitter used is different from the neurotransmitter used in excitatory synapses
   c. there are no receptor molecules on the postsynaptic membrane
   d. the postsynaptic receptor molecules cause hyperpolarization of the membrane rather than depolarization when combined with the neurotransmitter
   e. the stimulus is transmitted electrically rather than chemically

6. A neuron transmits information about the intensity of a stimulus that it has received by:
   a. changing the size of its action potentials
   b. changing the speed at which its action potentials travel
   c. releasing different types of neurotransmitters
   d. transmitting different numbers of nerve impulses in a given time period
   e. changing the number of boutons that release neurotransmitter

QUESTIONS FOR DISCUSSION

1. Why does the sodium-potassium ionic pump theoretically need energy (from hydrolysis of ATP) if it merely exchanges one positive ion for another?

2. In terms of spatial and temporal summation, what is the advantage of having various complicated, branching dendrites instead of a more regular, spherical surface for synaptic input?

3. After an action potential arrives at the presynaptic membrane, what factors determine whether the postsynaptic cell will fire an action potential?

4. Why don't all the neurons in an animal's nervous system have either giant or myelinated axons? What is the selective advantage of having some neurons with axons that conduct impulses more slowly?

5. Although synapses that occur between the axon of one cell and the dendrites of another have been most intensively studied, many parts of the nervous system have synaptic arrangements in which the presynaptic membrane of a dendrite of one cell releases neurotransmitter that is received by the dendrite or soma of another cell. Furthermore, the parts of the nervous system that process information about vision and smell contain many reciprocal synapses, areas where two synapses occur side by side, with one synapse sending information from cell A to cell B, while the neighboring synapse is arranged to send information from cell B to cell A. Such synapses release neurotransmitter in response to local potential changes that are not large enough to initiate action potentials. What would be the effect of such arrangements on the functions of the neurons involved? How would this affect the overall working of the nervous system?

REFERENCES AND FURTHER READING

Axelrod, J. "Neurotransmitters." *Scientific American*, June 1974. Covers chemistry, manufacture, release, and activity of neurotransmitters, as well as aspects of drugs affecting the nervous system.

Cooke, I., and M. Lipkin, Eds. *Cellular Neurophysiology.* New York: Holt, Rinehart and Winston, 1972. A source book.

Katz, B. *Nerve, Muscle, and Synapse.* New York: McGraw-Hill, 1966. A standard short treatment by the discoverer of the generator potential.

Levitt, R. A. *Psychopharmacology: A Biological Approach.* Washington, DC: Hemisphere Publishing Company; New York: John Wiley and Sons, 1975. The biological effects of drugs.

McLennan, H. *Synaptic Transmission*, 2d ed. Philadelphia: W. B. Saunders, 1970. Detailed account of types and functioning of synapses.

Morell, P., and W. T. Norton. "Myelin." *Scientific American*, May 1980. Origin, properties, and importance of the fatty sheath that surrounds many axons.

Schwartz, J. H. "The transport of substances in nerve cells." *Scientific American*, April 1980.

Shepherd, G. M. "Microcircuits in the nervous system." *Scientific American*, February 1978. Discussion of complex synaptic arrangements that occur at sensory organs.

Stevens, C. F. "The neuron." *Scientific American*, September 1979. A short, modern summary.

CHAPTER 38

THE VERTEBRATE NERVOUS SYSTEM

During the course of animal evolution, nervous systems have become increasingly complex. This is mainly because animals have become larger and more mobile. A rapidly moving animal must have more neurons than a sedentary one. It moves more muscles, and it encounters more objects, from pebbles to predators, to which it must react appropriately if it is to survive. A sluggish parasitic nematode (roundworm) may have as few as 160 neurons, and a correspondingly limited behavioral repertoire; an octopus, with precise control over its eight tentacles and considerable ability to learn new behavior patterns, has more than a billion; a human being has about 10 billion neurons.

Large numbers of neurons are not the only requirement for a complex nervous system; their organization is also crucial. In the simplest nervous systems, the nerve nets of cnidarians, the neurons are scattered throughout the body (Figure 38–1). A cnidarian's reaction to most stimuli is generalized because a cnidarian neuron transmits impulses to all of its neighbors. Furthermore, the reaction of the animal depends on the strength of the original stimulus; only part of the body responds to a weak stimulus, whereas the entire animal responds to a strong one.

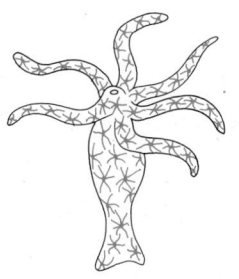

(a) Hydra (Cnidaria)

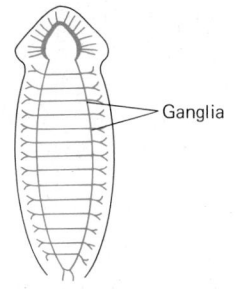

Ganglia

(b) Planarian (Platyhelminthes)

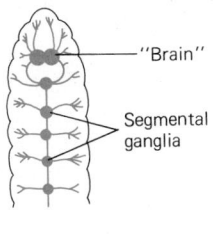

"Brain"

Segmental ganglia

(c) Earthworm (Annelida)

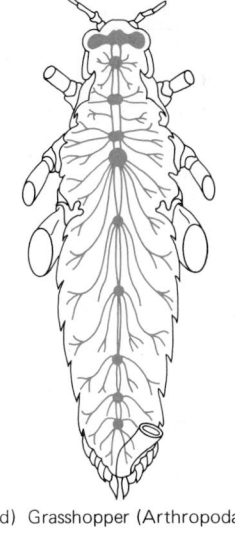

(d) Grasshopper (Arthropoda)

Figure 38-1 Nervous systems of some invertebrates.

The platyhelminths (flatworms) show a significant advance over the cnidarians. The cell bodies of many of the neurons of platyhelminths are clustered together, forming **ganglia** (singular: **ganglion**). Depending on the patterns of synapses within a ganglion, incoming information may be passed to one or to several other neurons. Precise and limited interactions between different neurons can occur in a way that is not possible with a nerve net. Furthermore, different ganglia may become specialized to govern particular parts of the body.

Another important trend in the evolution of nervous systems is correlated with the evolution of bilateral symmetry. Animals with a definite body axis move in a particular direction—forward. This created selective pressure for **cephalization**, or formation of a head. A head bears the major sense organs, such as the eyes and ears, which detect what is happening in the outside world; the animal's leading end can sample the new environment for food or safety as it moves. An animal reacts to the information provided by its sense organs by way of "decision-making" neurons and the neurons that control its muscles. The closer these neurons are to the sense organs, the faster the animal can react. Consequently, nervous tissue has become more and more concentrated in the head as cephalization has increased. The ganglia in the head eventually became large enough to be called a **brain**, an organ that is the main nervous control center of the body.

38-A   Classification of Neurons

Neurons may be classified by their structure, by the neurotransmitter they secrete, by their position in the body, or in many other ways. In considering the functional organization of the nervous system, the most important feature of a neuron is what information it takes where. The progress of information through any nervous system can be divided into three main steps: **sensory** activities, the collection of information; **central processing** of that information in the nervous system; and **motor responses** to the first two steps (Figure 38–2). A **receptor** detects some change in its environment, be it blood chemistry or sound waves in the air, and converts the change into nerve impulses. The impulses travel through **sensory neurons**, which carry information from receptors to the central nervous system.

In vertebrates, the **central nervous system (CNS)** consists of the brain and spinal cord; in invertebrates it usually consists of one or more ganglia. Most of the central nervous system is composed of **interneurons**, which relay messages from one neuron to another. In the process, sensory information from various parts of the body is **integrated**, assuring that the response made will be appropriate for existing conditions. (For example, your response to a plate of pastries depends on how full your stomach is, as well as how delicious the pastries look and smell.) During integration, information may be modified, augmented, or suppressed. From the central nervous system, **motor neurons** relay information to the body's **effectors**, the organs that carry out a response. The most common effectors are glands, which secrete hormones, digestive enzymes, etc., and muscles, which move parts of the body.

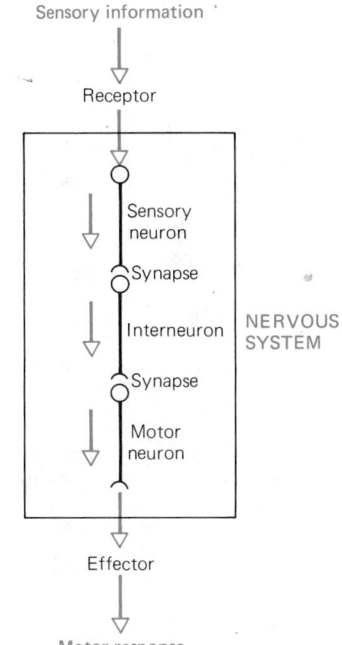

Sensory information

Receptor

NERVOUS SYSTEM

Sensory neuron

Synapse

Interneuron

Synapse

Motor neuron

Effector

Motor response

Figure 38-2 The nervous system processes sensory information and determines an appropriate response.

Central nervous system
  Brain
  Spinal cord

Peripheral nervous system
  Nerves
  Ganglia

Divisions of peripheral nervous
system:
  Somatic (voluntary)
  Autonomic (involuntary
    homeostatic control)

Figure 38-3 The major divisions of the vertebrate nervous system.

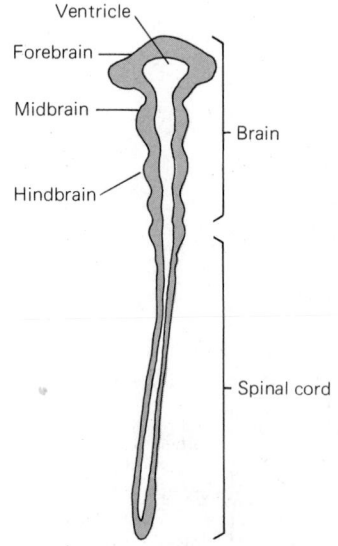

Figure 38-4 The brain and spinal cord as they appear during early development of a vertebrate embryo. The canal in the center will become the ventricles of the brain and the canal in the spinal cord of the adult.

38-B Divisions of the Nervous System

The nervous system of a vertebrate can be divided into the central and peripheral nervous systems. The central nervous system contains many entire neurons as well as the cell bodies of other neurons whose dendrites* and axons* lie in the peripheral nervous system.

The brain is protected by the skull; the spinal cord by the flexible cylinder of the vertebral column. Within these bony coverings, the brain and spinal cord are covered by three layers of membranes called **meninges**. The meninges contain branches of the circulatory system that exude part of the blood plasma to form **cerebrospinal fluid**, which bathes the central nervous system. The fluid, contained within the meninges, also cushions the nervous system from jarring. (Disorders of the central nervous system can sometimes be detected by examining samples of cerebrospinal fluid withdrawn from the spinal cord. Inflammation of the meninges is called **meningitis**, and is often a serious condition because the meninges are so close to the vital tissues of the CNS.)

The **peripheral nervous system** is all of the nervous system that is not part of the CNS; it consists of **nerves**, which are bundles of axons covered with protective sheaths of connective tissue, and some **peripheral ganglia**, which are clusters of neuron cell bodies lying outside the CNS.

The peripheral nervous system has two parts, which serve different functions and can be distinguished by the effectors they control. The **somatic nervous system** contains both motor and sensory neurons; it controls the activity of the muscles attached to the skeleton, which we think of as being under "conscious" control, including the muscles we use to smile, run, sing, or draw. The **autonomic nervous system**, on the other hand, contains only motor neurons, and controls muscles and glands that usually operate without our being aware of them; these effectors control such things as blood pressure and the movement of food in the gut. Another way to think of the difference between the two systems is that the somatic system permits us to react to the outside world, while the autonomic system is concerned mainly with homeostatic mechanisms within the body.

38-C The Vertebrate Brain

The central nervous system of a vertebrate grows from a tube formed during early embryonic development (Figure 38-4). As the embryo develops, the front part of the tube enlarges in a series of uneven bulges, until it is possible to pick out three main parts, unimaginatively called the forebrain, midbrain, and hindbrain, as well as the long, straight spinal cord.

All vertebrates, from fish to mammals, have brains with the same basic structures (Figure 38-5). In the course of evolution, some parts of the brain have changed very little, while others have virtually exploded overnight when viewed on an evolutionary time scale. Some areas of the brain have retained their primitive functions, whereas others have taken on new functions as vertebrates evolved.

Hindbrain

The most obvious part of the hindbrain of a fish is the **medulla**, the enlargement of the spinal cord as it enters the brain (see Figure 38-5). Through it pass many of the sensory and almost all of the motor nerves on their way to or from higher centers. ("Higher centers" are parts of the CNS that integrate a broad range of information.) The medulla also contains many **nuclei**, or clusters of neuron cell bodies (ganglia within the CNS), that receive sensory input and send out motor signals which control such automatic, or **reflex**, functions as breathing, swallowing, vomiting, and the constriction of blood vessels.

---

*Dendrites Extensions of a neuron that receive information from other cells or from the environment.
*Axon Extension of a neuron that transmits nerve impulses to the next cell(s) in line.

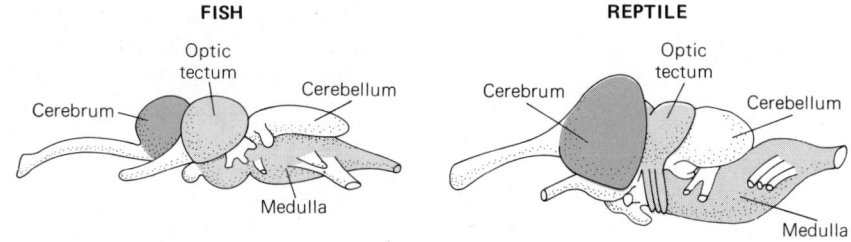

FISH

REPTILE

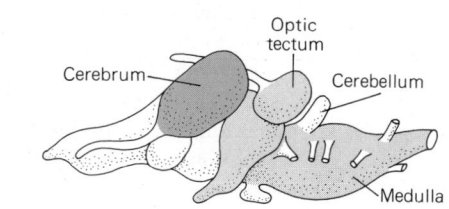

AMPHIBIAN

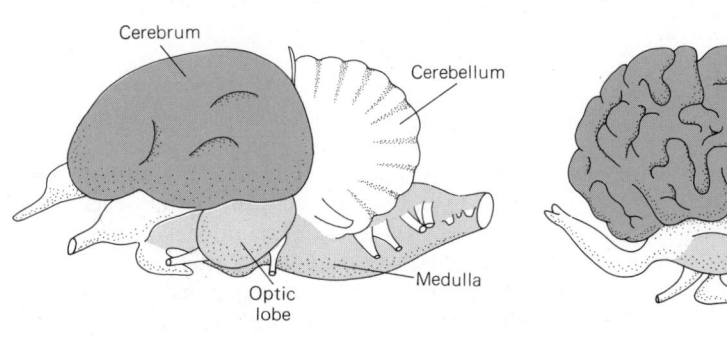

BIRD

MAMMAL

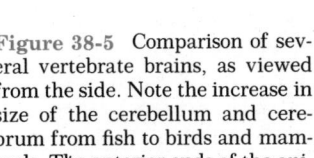

Figure 38-5 Comparison of several vertebrate brains, as viewed from the side. Note the increase in size of the cerebellum and cerebrum from fish to birds and mammals. The anterior ends of the animals are to the left.

TABLE 38-1 FUNCTIONS OF MAJOR PARTS OF THE VERTEBRATE BRAIN

DERIVATION	NAME	FUNCTION
Hindbrain	Medulla	Passage of messages between brain and spinal cord; control of visceral reflexes
	Cerebellum	Coordination of equilibrium and movement
Midbrain	Tectum (in lower vertebrates)	Association
	Anterior colliculi (mammals)	Reflexes of iris and eyelid
	Posterior colliculi (mammals)	Receives sensory information from ear
Forebrain Diencephalon	Thalamus	Relays olfactory messages to midbrain (in fish). Area of sensory integration (in higher vertebrates)
	Hypothalamus	Controls emotional states and drives (pleasure, pain, thirst, sex, rage, etc.)
	Posterior pituitary	Releases hormones (see Section 41-B)
Telencephalon	Olfactory bulb	Receives olfactory information (most important telencephalon area in fish)
	Corpus striatum	Complex behavior patterns (in birds)
	Cerebral hemispheres (cerebral cortex)	Well-developed only in mammals. Sensory and motor association, visual and auditory processing, seat of "intelligence," and, in humans, of ability to use language, both written and spoken

The medulla has changed little during vertebrate evolution compared with the cerebellum, the other major structure of the hindbrain. The **cerebellum** is an outgrowth of the medulla, and during evolution it has acquired much of the central control of equilibrium (balance) and movement.

Midbrain

During evolution the midbrain has changed more in function than in size or structure. In fish and amphibians it is the principal area for **association**, that is, the channeling of sensory input into appropriate motor pathways. In these lower vertebrates, a major part of the midbrain is the **tectum** or **optic lobe**, which receives signals from the **optic nerves**, carrying visual information from the eyes. In some cases, the tectum also receives sensory information from the hindbrain. In mammals, the analysis of vision has moved out of the midbrain and has become a function of part of the forebrain. The midbrain of mammals consists of the **anterior colliculi**, which control reflexes of the iris and eyelids, and the **posterior colliculi**, which analyze and relay information coming in from the ear via the auditory nerve.

Forebrain

The **forebrain** has undergone a great deal of change in vertebrates. It can be divided into two major parts, the diencephalon and the telencephalon. Lying more or less immediately in front of the midbrain, the **diencephalon** contains the thalamus, hypothalamus, and the posterior lobe of the pituitary gland (Figure 38–6). In fish, the **thalamus** relays information from the olfactory (sense of smell) organs to the midbrain, but in other vertebrates it functions as one of the centers in which all sensory information is integrated. Immediately below the thalamus is the **hypothalamus**, a vitally important organ where the nervous and hormonal systems interact (Section 41-B). Evolutionarily, this section of the forebrain is the oldest, and even in lower vertebrates it is a relatively well-developed center for such homeostatic functions as the control of body temperature, osmoregulation,* growth, sexual drive and maturity, thirst, and hunger.

While the diencephalon slowly expanded to handle increased sensory input, the **telencephalon**, the anterior part of the forebrain, grew astoundingly in both size and complexity during vertebrate evolution. In fishes and amphibians it is concerned with little other than olfactory information. In reptiles and birds, however, the most important part of the brain is the **corpus striatum** at the base of the telencephalon. This structure is particularly well-developed in birds, and is responsible for their complex behavior patterns.

Just above the corpus striatum is the **cerebrum**, divided by a fissure into the right and left **cerebral hemispheres**. Beginning with early mammals, there is a progressive increase in the surface area and in overall importance of the **cerebral cortex**, the outer layer of the cerebral hemispheres. The original, deeper layers of the hemispheres, the **hippocampus** and **limbic structures**, remain ventral and are important, in primates* at least, in regulating emotional state and possibly short-term memory. Above them, the cerebral cortex has expanded in all directions, and lies like the cap of a wrinkled mushroom over the rest of the brain. This expansion has greatly increased the surface area of the cerebral cortex, which has correspondingly become highly folded and convoluted in more intelligent animals. The **gray matter** of the cerebrum is composed of thick layers of unmyelinated* cells. Many myelinated fiber tracts are interspersed with the gray matter. These **white matter** tracts consist of bundles of axons connecting with other areas of the brain. One of the largest myelinated tracts, the **corpus callosum**, connects the bases of the cerebral hemispheres with each other; its function is to let the right half of the brain know what the left half is doing.

---

*Osmoregulation   Regulation of salt and water balance.
*Primates   The order of mammals that includes humans, apes, and monkeys.
*Unmyelinated   Lacking the fatty sheath that surrounds some axons (see Section 37-B).

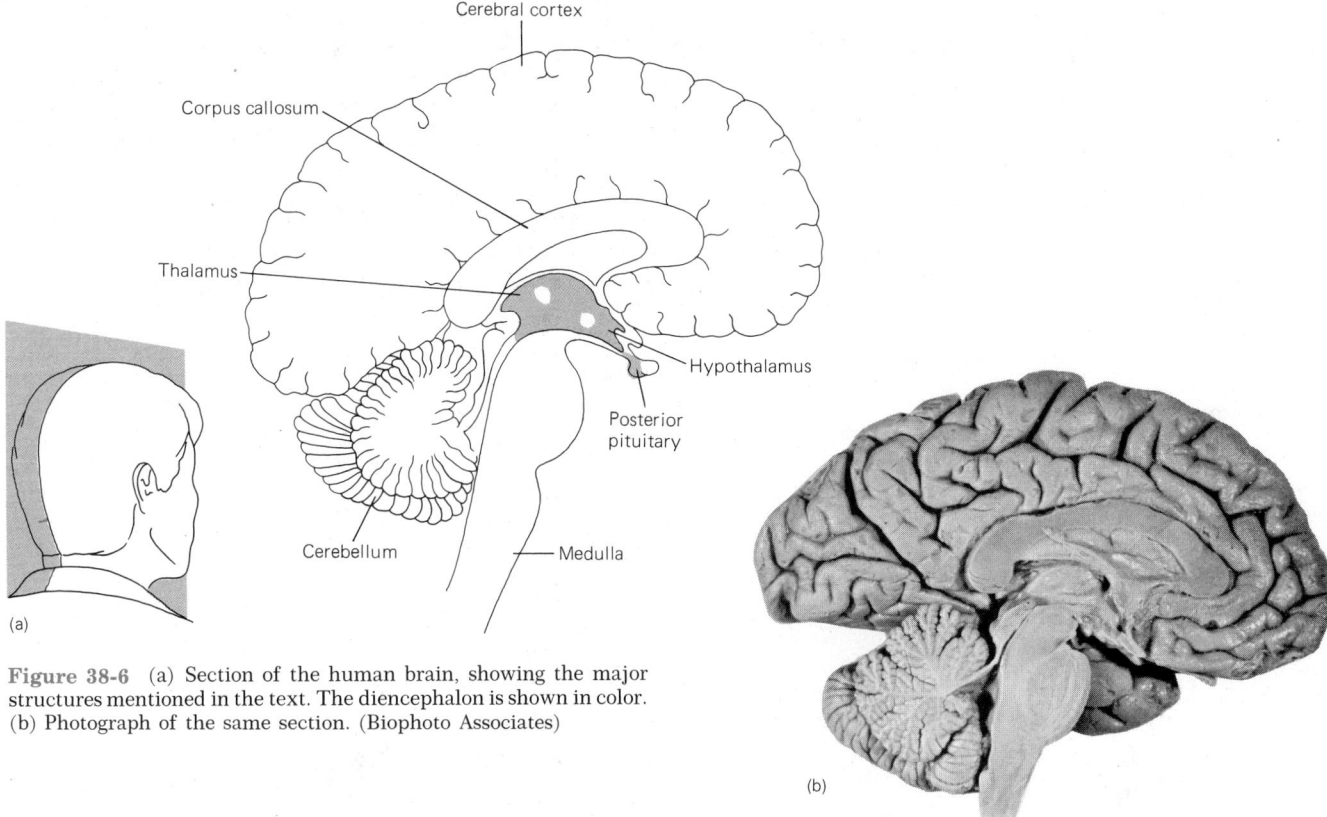

Cerebral cortex

Corpus callosum

Thalamus

Hypothalamus

Posterior pituitary

Cerebellum

Medulla

(a)

(b)

Figure 38-6 (a) Section of the human brain, showing the major structures mentioned in the text. The diencephalon is shown in color. (b) Photograph of the same section. (Biophoto Associates)

The functions of the different regions in the cerebral hemispheres have been investigated in various ways. By delivering electric shocks to different areas of the brains of laboratory animals or human patients, experimenters can map the activities of each area. Another method of investigation is to observe the effects of deliberate or accidental destruction of parts of the brain on its function. Such studies have shown that many activities are amazingly complex, involving cells in several areas of the brain. However, it has proved convenient to designate certain areas as being principally involved in certain functions. For example, there are primary sensory areas and primary motor areas of the cerebral cortex (Figure 38–7), which connect to one another by association pathways in the deeper layers of the cortex, and thus permit an appropriate response to a particular set of stimuli. Other areas of the cerebral cortex are involved in perception of visual or auditory stimuli and, in humans, in use of symbols and language.

In primitive mammals, each area of the surface of the cerebral cortex has specific sensory or motor functions. In more advanced mammals, especially pri-

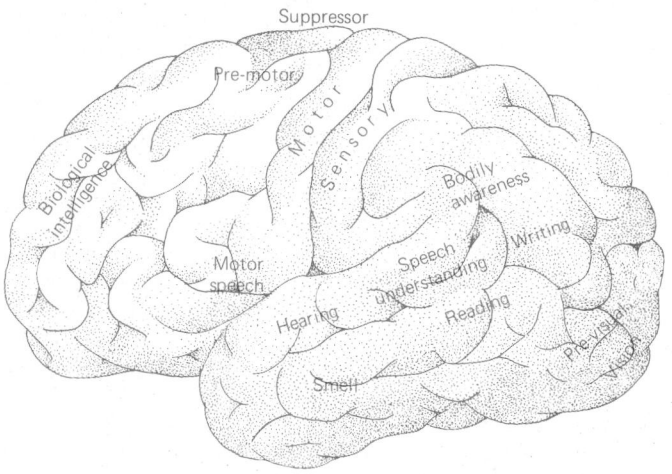

Suppressor

Pre-motor

Motor

Sensory

Biological intelligence

Bodily awareness

Motor speech

Speech understanding

Writing

Hearing

Reading

Pre-visual

Visual

Smell

Figure 38-7 Lateral view of the human cerebral cortex, showing functions assigned to several areas.

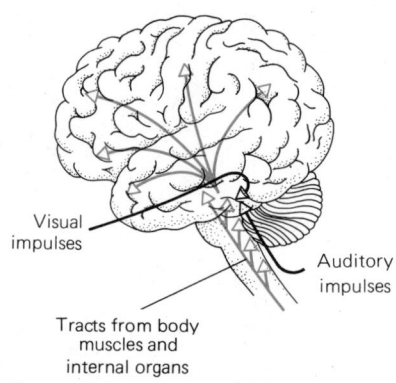

Visual
impulses

Auditory
impulses

Tracts from body
muscles and
internal organs

Figure 38-8 The pathways of sensory input via the reticular formation.

KEY o ON Retinal receptive field
(firing ganglion cell)

• OFF Retinal receptive field
(not firing ganglion cell)

This stimulus
impinging
on retina:

Causes these
cells to fire
maximally due
to arrangement
of input from
previous cells:

Retinal
ganglion
cells

Spot of light

Larger circles or
ovals of light with
various orientations

Lateral
geniculate
nucleus

Edges or corners of
various shapes

Simple
cortical
cells

Complex
cortical
cells

Movement or rotation of
various shapes and edges

Higher
centers

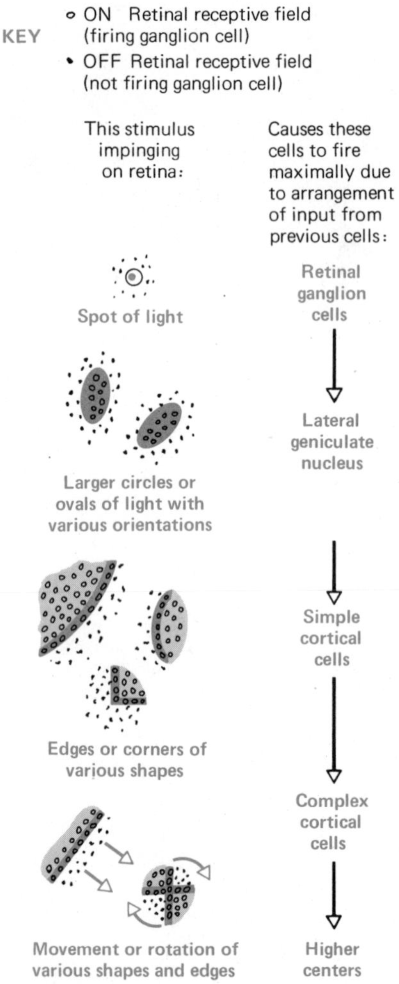

Figure 38-9 The receptive field of a retinal ganglion cell is arranged such that the ganglion cell responds maximally to a small spot of light. Retinal ganglion cells, in turn, are receptive fields for cells in the lateral geniculate nucleus, and so on. At each level, the receptive field is so organized that it permits the detection of more inclusive information.

mates, however, large areas of the cerebral hemispheres have no known specific function. Most researchers believe that these areas are important to the versatility of behavior so characteristic of higher mammals, and the capacity for abstract thinking. Personality may be influenced by these "unassigned" cells; it used to be a standard procedure to remove parts of the frontal lobes (frontal lobotomy) to alleviate psychopathologies or personality disorders. It is intriguing to speculate on the complexity of cellular organization that makes human beings think like human beings, that allows us to use language and to manipulate abstract concepts, but understanding of brain function at this level still lies far in the future.

Reticular Formation

One interesting subsystem of the brain is the **reticular formation** (Figure 38-8). This large, irregular collection of neurons and processes extends from the medulla through the base of the midbrain into the thalamus. Cells in the formation receive input from all types of sensory fibers entering the brain. Output from the formation goes back down the spinal cord, where it amplifies or reduces incoming activity. The reticular formation is the part of the nervous system most responsible for general levels of lethargy or liveliness. It is clearly advantageous for the whole nervous system to be more sensitive and react more readily when we cross a street or write a term paper than when we go to sleep, and the reticular formation provides this generalized control.

The **reticular activating system** extends from the reticular formation to the cerebral cortex. It acts as a filter that determines which sensory information reaches the level of consciousness. Cells in this system may induce arousal. You may have experienced a sudden waking from sleep, not knowing what woke you, but certain that something did. Whatever the stimulus was, it made its way into the reticular activating system, which flashed an "all points" arousal message throughout the brain.

Sensory Analysis

The processing of sensory information coming into the brain has been particularly well studied in the vertebrate visual system. This system serves to illustrate the complexity of information processing as sensory stimuli pass through higher and higher centers in the brain. It is sobering to realize that the visual system has received so much experimental attention because of its simplicity.

The **retina** is a thin layer of tissue in the eye; it contains many light-sensitive receptor cells. Groups of these receptor cells synapse with ganglion cells, also located in the retina. The receptor cells that synapse with a particular ganglion cell are called its **receptive field**, and these cells are arranged in two concentric circles. The receptors in the inner circle have excitatory* synapses with the ganglion cell, whereas the cells in the outer circle have inhibitory* synapses with the ganglion cell. Thus, when a spot of light falls only on the center circle, the excitatory input causes the ganglion cell to fire at a high frequency. As the spot grows bigger, more and more inhibitory input reaches the ganglion cell, and it fires less often. Each ganglion cell receives input from a large number of retinal cells, and each retinal cell sends information to a number of ganglion cells, making excitatory synapses with some ganglion cells and inhibitory synapses with others. Thus any light stimulus will cause a group of ganglion cells to respond, some weakly, some strongly.

The next higher center in the visual system is the **lateral geniculate nucleus** (Figure 38-9). A cell here has a receptive field consisting of a group of ganglion cells that receive input from a circle or oval of light shining on the retina. At the next level, **simple cortical cells** receive input from cells in the lateral geniculate nucleus such that edges or corners of light of certain size and orientation shining on the corresponding spot of the retina cause the greatest response in the simple cortical cells. Higher, more complex cortical cells receive input mainly from other cortical cells arranged in precise patterns. Thus, a group of "moving edge detector" cells,

---

* Excitatory Depolarizing the postsynaptic membrane and enhancing the chances of the postsynaptic cell's firing an action potential.
*Inhibitory Reducing the tendency of the postsynaptic cell to fire an action potential.

all lined up in a certain orientation in the cortex, will fire in response to an edge moving across the retina in a particular orientation. Movements of other kinds are the most effective possible stimuli for other complex cortical cells. There may be other, even higher-order cells in the visual cortex that respond to more abstract features of visual stimuli and allow recognition of various kinds of visual patterns. Thus each receptive field in the retina is **mapped** or **projected** to particular cells in the visual cortex via an orderly series of neurons connecting them. This specificity is necessary so that the brain can make a correct interpretation of the image falling on the retina, but just how the brain interprets the shape, motion, and color that we see is still a mystery.

Other sensory systems besides the visual system also relay information to the cerebral cortex (in mammals) for higher-order integration or decisions.

Motor Pathways

In the cerebral cortex, incoming sensory stimuli fire association neurons that connect with the motor pathways that will eventually cause appropriate action by effectors.

Motor commands from the cerebral cortex do not directly stimulate muscle contraction or gland secretion. As in sensory pathways, the command or message from the cortex may cross anywhere from three to thousands of synapses, allowing a great deal of filtering or modification of the message on its outward path. Starting from the motor cortex, axons may pass directly to spinal motor neurons, or cortical axons may first terminate in various ganglia, which, in turn, send messages to spinal motor neurons.

38-D The Spinal Cord

The spinal cord is a rope of nervous tissue that extends from the base of the hindbrain to the end of the vertebral column (Figure 38–10). Like the brain, it contains a central fluid-filled cavity. Again, the gray matter of the cord consists of cell bodies; the surrounding white matter is made up of bundles of myelinated axons carrying information to and from the brain and other parts of the spinal cord. The spinal cord is a relay system carrying information between the brain and the peripheral nervous system. It is also the seat of the many **spinal reflexes** that allow the body to make quick responses.

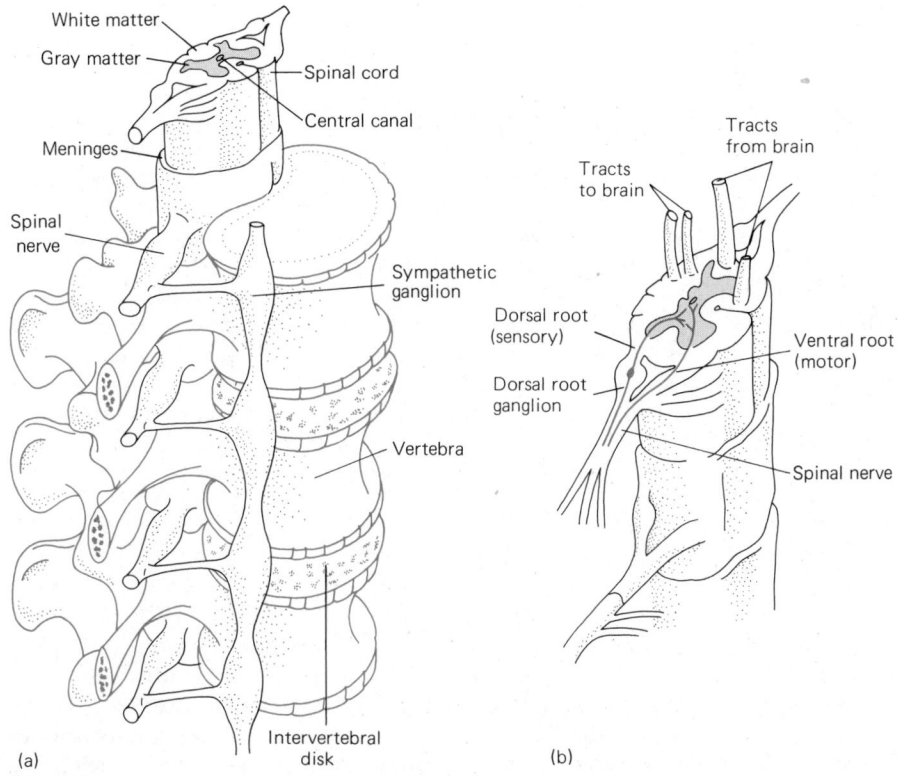

Figure 38-10 The human spinal cord. (a) With the vertebral column (colored). Paired spinal nerves protrude through the spaces between vertebrae. (b) Some nerve pathways of the spinal cord.

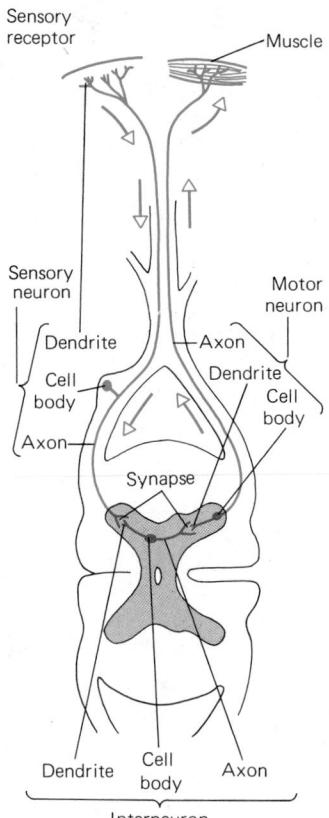

Figure 38-11 Simple reflex arc containing two synapses and three neurons: sensory neuron, interneuron, and motor neuron.

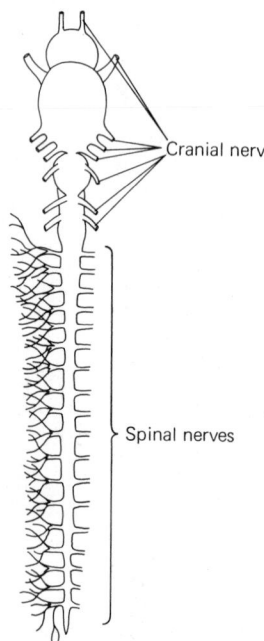

Figure 38-12 The nervous system of a fish. Paired cranial nerves arise from the brain, and spinal nerves from the spinal cord. Notice that the areas served by adjacent spinal nerves overlap.

Reflex Arcs

We have all learned to perform many complex activities; however, our responses to certain stimuli are simple and unvarying. For instance, it is hard to hold your lower leg still when the doctor hits you below the knee with a little rubber mallet. The knee jerk and the rapid withdrawal of a hand that touches a pin or a hot stove are controlled by **reflex arcs**, pathways of a few neurons each, under little control from higher centers. The simplest type of reflex arc has only two neurons, one sensory and one motor. More commonly, reflex arcs have one or more interneurons between the two (Figure 38-11). A reflex pathway saves time because it has a small number of synapses. The fewer synapses between receptor and effector, the sooner the effector responds to a stimulus.

The organs that regulate physiological homeostasis are controlled by more or less simple reflexes of the autonomic nervous system. Heart rate, breathing, dilation of the pupils, and digestion are all controlled in this way. More complicated activities, such as posture control, locomotion, sexual behavior, and defensive strategies, which are governed mainly by the somatic nervous system, also involve numerous reflex arcs.

The spinal cord plays an important role in the integration of reflex behavior. For instance, a "spinal" animal, one whose brain has been destroyed or removed, still shows reflexes. If a piece of acid-soaked paper is applied to the back of a spinal frog, one leg will come up and kick it away; the behavior is repeated no matter how many times the paper is placed on the skin. This response, involving many muscles working in a coordinated fashion, clearly demonstrates one of the chief characteristics of a reflex: fidelity of repetition. A frog with a brain might make the response two or three times, but eventually the higher centers would intervene and the frog would do something else—perhaps hop away.

The complexity of a spinal reflex varies with the number of neurons involved. The scratch reflex and the reflexes involved in walking use sensory information entering at many parts of the spinal cord. More complex reflexes may even involve brain structures such as the medulla and midbrain. Most reflexes involve motor signals to many muscles. For example, if you raise your arm to protect your face, muscles in your back must adjust their contraction or you would lose your balance.

Reflexes are advantageous because it saves both time and energy for an animal to perform such routine actions as walking and breathing without involving higher centers of the nervous system.

38-E   Plasticity in the Nervous System

The opposite of a reflex behavior pattern is a **plastic** behavior pattern, subject to modification by many neurons. For instance, the interaction of thousands of different synapses determines whether or not you immediately leap out of bed when the alarm clock rings, and for this reason the response may not follow hard upon the stimulus. **Plasticity** is the ability to learn, that is, to change the relationships between various parts of the neural circuitry in the central nervous system. If a frog's eyes are surgically rotated 180° in their sockets, the frog will perceive its prey to be in the direction exactly opposite from where the prey really is, and the frog will jump in the wrong direction, trying to catch its prey, until it starves to death. A mammal, which has more plastic behavior, can compensate for such a shift in its visual world in a relatively short time. You have made such a change in learning to comb your hair while looking in a mirror. The flexibility of the mammalian brain is thought to be due to the organization of the association neurons.

38-F   Cranial and Spinal Nerves

The peripheral nervous system of a vertebrate consists of paired nerves that branch from the central nervous system. In reptiles, birds, and mammals, 12 pairs of **cranial nerves** connect the brain with various structures, mostly in the head and neck; fish and amphibians have only the first 10 pairs (Figure 38-12). The thickest cranial nerves are the first, second, and eighth, which carry only sensory information coming to the brain from the major sense organs—the nose, the retinas

TABLE 38-2 **THE CRANIAL NERVES OF HUMANS**

NUMBER	NAME	ORIGIN OF SENSORY FIBERS	EFFECTOR INNERVATED BY MOTOR FIBERS
I	Olfactory	Olfactory mucosa of nose (smell)	None
II	Optic	Retina of eye (vision)	None
III	Oculomotor	Proprioceptors of eyeball muscles (muscle sense)	Muscles that move eyeball (with IV and VI); muscles that change shape of lens; muscles that constrict pupil
IV	Trochlear	Proprioceptors of eyeball muscles (muscle sense)	Other muscles that move eyeball
V	Trigeminal	Teeth and skin of face	Some muscles used in chewing
VI	Abducens	Proprioceptors of eyeball muscles (muscle sense)	Other muscles that move eyeball
VII	Facial	Taste buds of anterior part of tongue	Muscles of the face; submaxillary and sublingual salivary glands
VIII	Auditory	Cochlea (hearing) and semi-circular canals (senses of movement, balance, and rotation)	None
IX	Glossopharyngeal	Taste buds of posterior third of tongue, lining of pharynx	Parotid salivary gland; muscles of pharynx used in swallowing
X	Vagus	Nerve endings in many of the internal organs— lungs, stomach, aorta, larynx	Parasympathetic fibers to heart, stomach, small intestine, larynx, esophagus
XI	Spinal accessory	Muscles of shoulder (muscle sense)	Muscles of shoulder
XII	Hypoglossal	Muscles of tongue (muscle sense)	Muscles of tongue

of the eyes, and the cochlea (hearing) and vestibule (sense of equilibrium) of the ear. The other cranial nerves are **mixed nerves**, carrying both motor and sensory information. The longest cranial nerve is the tenth, the vagus, which innervates (supplies nervous connections to) many internal organs of the chest and upper abdomen, and is part of the autonomic system. Table 38-2 lists the 12 pairs of cranial nerves found in humans, with their functions.

In addition to the cranial nerves, human beings also have 32 pairs of **spinal nerves**, which branch out from the spinal cord between adjacent pairs of vertebrae. Each spinal nerve leaves the spinal cord in two parts: a **ventral root**, which contains the axons of motor neurons, and a **dorsal root**, which contains sensory neurons. Along each dorsal root lies the **dorsal root ganglion**, containing the cell bodies of the sensory neurons. The dorsal root ganglia are some of the few places where cell bodies are found outside the CNS. The axons of these sensory neurons continue into the spinal cord and synapse with other neurons (see Figures 38-10 and 38-11). Severing the dorsal root of a spinal nerve results in loss of sensation in the areas it serves. Cutting a ventral root paralyzes the corresponding muscles. This is not as serious as it sounds, because adjacent spinal nerves have considerable areas of overlap.

The dorsal and ventral roots join outside the cord and run a short way together as a spinal nerve. The nerve soon splits into three branches, each containing some sensory and some motor fibers. One branch serves the skin and muscles of the back, another serves the skin and muscles of the front of the body, and the third serves internal organs.

38-G The Autonomic Nervous System

The autonomic nervous system, found in all vertebrates, is a system of nerves that controls most of the body's homeostasis. The autonomic system regulates the heartbeat and controls contraction of the muscles in the walls of the blood vessels and the digestive, urinary, and reproductive tracts. Autonomic nerves also stimulate glands to secrete mucus, tears, and digestive enzymes.

There are two main parts of the autonomic nervous system: the sympathetic and the parasympathetic systems. Each has a different function. The **sympathetic system** dominates in time of stress: it initiates the **"fight or flight" reaction**— increases in blood pressure, heartbeat rate, breathing, and blood flow to the muscles and skin and decreases in the flow of blood to the digestive organs and kidneys. These changes ensure an adequate oxygen supply for muscular exertion, and increased blood flow to the skin allows the body to lose muscle-generated heat. In contrast, the **parasympathetic system** acts as a counterbalance by stimulating the opposite reactions in these organs. When the parasympathetic system is more active, digestion and elimination are promoted.

The ganglia of the sympathetic and parasympathetic systems are organized differently, in keeping with their different functions. The sympathetic ganglia lie just outside the spinal cord in the thoracic (chest) and upper lumbar (back) areas (Figure 38–13). They connect with each other as well as with the spinal cord, so that a stimulus to any part of the sympathetic system is quickly transmitted to all parts. Thus, in time of stress or danger, the entire body is quickly galvanized for action. By contrast, the ganglia of the parasympathetic system lie near the effector organs. Their contact with the central nervous system is by way of some of the cranial nerves and sacral (pelvic area) spinal nerves. The parasympathetic ganglia do not connect with each other, and thus each effector is controlled independently.

Another difference between the sympathetic and parasympathetic systems is in the chemical transmitters released at the synapse with the effector. Although acetylcholine (see Figure 37–15) is the transmitter at the ganglion in both sympathetic and parasympathetic systems, it is released at the synapse with the effector only in the parasympathetic system. In the sympathetic system noradrenalin is the transmitter at the effector.

Although the autonomic nervous system can carry out its function automatically, it is by no means completely independent of the animal's voluntary control. For example, it is possible to decide to stop breathing, at least for a short time. Recent studies have also shown that humans and animals can be trained to change their heart rates, blood pressures, and digestive reflexes voluntarily. Any voluntary control that endangers life, however, quickly disturbs homeostasis of the brain tissue, resulting in unconsciousness. When this happens, the autonomic system takes over again and restores normal functions.

38-H Memory

We still know very little about learning and memory, two of the more complex of the brain's functions. What is memory, the brain's information store? For many years scientists destroyed various parts of the brains of trained animals to find out where memory is located. They concluded that memory is nowhere and everywhere.

There are two kinds of memory: short-term and long-term. When we take lecture notes, we use short-term memory, remembering what the lecturer has said, with any luck, just long enough to write it down. If information is to be stored for any length of time, it must be transferred to the long-term memory where it may remain, much of it in subconscious form, for life. For many years, some researchers thought that long-term memory was laid down as RNA or protein molecules, because RNA and protein are synthesized when information is stored in the long-term memory, and because drugs that inhibit RNA synthesis destroy the memory. However, these experiments are inconclusive because RNA and protein synthesis are needed indirectly for many kinds of cell activity. You may have heard of experiments in which trained worms were fed to untrained worms that, supposedly, took over the memory of the worms they had eaten and so became instantly trained on

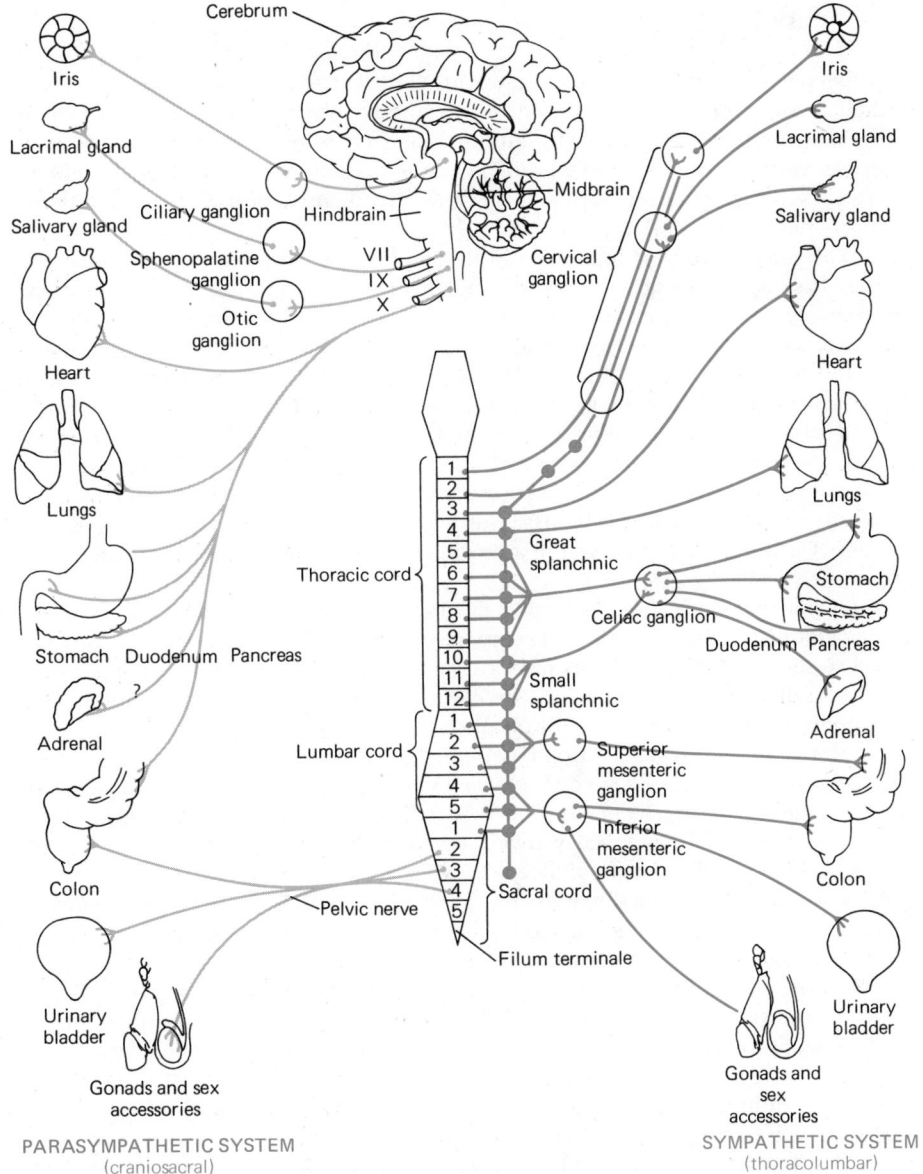

PARASYMPATHETIC SYSTEM
(craniosacral)

SYMPATHETIC SYSTEM
(thoracolumbar)

Figure 38-13 The structure of the autonomic nervous system, and the organs innervated by each of its branches. Each parasympathetic neuron shown ends at a ganglion in its target organ (heart, lungs, etc.). Here it synapses with postganglionic neurons, not shown in the diagram, which lie within the organ.

the strength of their meal. Most people now mistrust these feeding experiments since behavioral experimentation has grown more sophisticated. The net result is that we know almost as little about memory as we did a hundred years ago. The best guess is probably that memory is stored in the form of electrical impulses constantly whizzing around the myriad circuits in our brains.

New synapses form and old ones deteriorate in the vertebrate brain throughout life. This is why human thought and behavior are constantly adaptable, although young brains are more adaptable than old. The neurons in our brains die at the rate of about 10,000 a day; since no new neurons are formed after early childhood these can never be replaced. This does not produce total mental atrophy after the age of 30, however; if you live to be 100, less than 0.5% of your brain will die before you do—barring strokes, which can kill areas of the brain by cutting off the blood with its supply of needed food and oxygen.

Sleep is another mysterious nervous function. Most mammals and birds sleep, although many of them keep more reflexes active than humans do, and remain upright while they sleep. Reptiles, amphibians, and fishes also show periods when they are very unresponsive to stimuli. Sleep is undoubtedly related to the equally mysterious daily periods of activity known as circadian rhythms, which all animals display (see Section 41-E).

With all the progress in biology in the last 50 years, we still have no idea why we must sleep each night, nor why sleeping has such a profound effect on our temper, alertness, and emotional stability. To make things more complicated, the period of sleep required varies considerably from one individual to another. Guinea pigs and human beings with brain damage have lived for years without sleep, so sleep is obviously not necessary to survival. Since a sleeping animal is highly vulnerable to predators, sleep must have some powerful counteracting selective advantage.

At the moment, all we can do is describe the physiological changes that occur during sleep. Heart rate and blood pressure drop, breathing becomes more shallow, and body temperature drops slightly. However, the temperature of the big toes rises about 5°C; this change was used in a recent study as the first reliable indicator that a subject was asleep.

The most obvious changes during sleep are seen in the electrical activity of the brain (Figure 38–14). Electrical "brainwaves" can be detected by way of electrodes taped to the head. The patterns of brain activity recorded from a sleeping person show four different stages of sleep. The most striking of these is **REM** or **paradoxical sleep**, which occurs every 80 to 120 minutes.

REM is an acronym for "rapid eye movements." During REM sleep the brain activity resembles that of someone who is awake, although, in fact, the muscles are more relaxed and the sleeper harder to awaken than at any other stage. People awakened from REM sleep nearly always recall dreams, although dreaming can also occur during the other stages of sleep. REM sleep seems to be the sleep stage most crucial to our psychological well-being; people deprived of REM sleep become extremely tired, and they compensate for loss of REM sleep by increasing periods of REM sleep on subsequent nights.

Studies on the biochemistry of sleep have shown that the concentrations of neurotransmitters in the brain are different in waking and in different kinds of sleep. For instance, noradrenalin, dopamine, and acetylcholine are present in higher concentrations in the brain of someone who is awake, whereas the concentration of serotonin rises in the brain of a sleeping subject. Cause and effect are difficult to distinguish in this situation. For instance, it may be that an increase in the level of serotonin puts one to sleep. (There are reports that injection or ingestion of the amino acid tryptophan, a precursor of serotonin, induces sleep.) On the other hand, the higher levels of serotonin during sleep may be due to increased activity of neurons that produce serotonin.

It seems likely that sleep permits the restoration of biochemical functions depleted by the day's activity and also permits the processing and reorganization of information already present in the nervous system, but how and why these things happen will probably take a long time to discover.

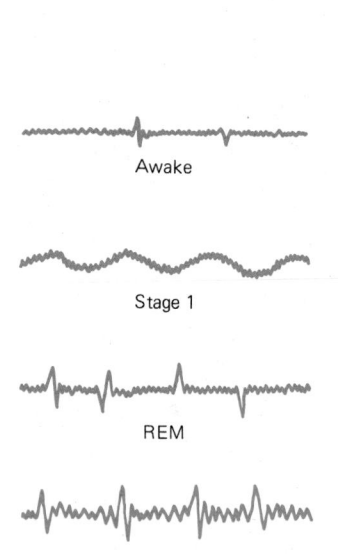

Figure 38-14 Parts of an electroencephalogram showing electrical activity of the brain of someone awake and in three different stages of sleep. Each recording covers 20 seconds.

Awake

Stage 1

REM

Stage 4

SUMMARY

During the evolution of the nervous system, progressive cephalization has resulted in the formation of a brain and major sense organs at the anterior end of the body. The nervous system of a vertebrate consists of the brain and spinal cord, which together compose its central nervous system, and the peripheral nervous system in the rest of the body.

The vertebrate brain has three major parts: the forebrain, midbrain, and hindbrain. During evolution, the vertebrate brain has increased in size and complexity. Some parts of the brain have retained their primitive functions, while others have taken on new functions as body structure and behavior have become more complex, and as intelligence has increased. The function of the brain is to "make decisions." Using information coded as patterns of action potentials coming from the external or internal environment via the sense organs, the brain produces a set of directions coded as another set of action potentials that cause the effector organs to respond. Information passes through various levels of organization in both motor and sensory areas as the brain determines appropriate responses to sensory input.

The spinal cord is primarily a relay station connecting the brain with the peripheral nervous system, although there are spinal reflexes in which motor and sensory components interact through the spinal cord without input from the brain.

The peripheral nervous system is divided into the somatic nervous system, which largely serves the muscles under conscious control, and the autonomic nervous system, which carries motor impulses to the muscles and glands of the internal organs, under little conscious control.

Memory and sleep are complex functions of the nervous system about which very little is known.

OBJECTIVES

From your study of this chapter, you should be able to:

1. Define the terms cephalization, ganglion, nucleus, brain, spinal cord, central nervous system, peripheral nervous system, autonomic nervous system, nerve, spinal nerve, cranial nerve.

2. Give the derivation (from forebrain, midbrain, or hindbrain) and known functions of the cerebral hemispheres, cerebellum, medulla, thalamus, hypothalamus, and tectum (colliculi in mammals).

3. Discuss the differences in relative size and function of the structures listed in Objective #2 between the brain of a fish and the brain of a human being.

4. Describe the reticular formation in vertebrates, and explain the relationship of the reticular activating system to arousal.

5. Briefly outline how visual information is processed and projected in the vertebrate brain.

6. Draw a labeled diagram of a simple reflex arc, and describe how it works.

7. Discuss how plasticity in a nervous system is related to learning and behavior.

8. Outline the major structural and functional differences between the parasympathetic and sympathetic nervous systems.

SELF-QUIZ

1. Regulatory control of deep body temperature, osmoregulation, thirst, and hunger occur in the:
   a. anterior colliculi
   b. hypothalamus
   c. thalamus
   d. cerebellum
   e. tectum

2. In looking at the evolution of the vertebrate brain from fish to humans, the greatest increases in size are seen in the:
   a. medulla and cerebellum
   b. cerebellum and tectum
   c. tectum and cerebral hemispheres
   d. cerebral hemispheres and cerebellum
   e. medulla and thalamus

3. The reticular formation:
   I. acts as a filter for sensory input
   II. regulates mechanisms of arousal
   III. affects spinal cord activity via a feedback system

   a. I only
   b. II only
   c. I and II
   d. I, II, and III

4. Plasticity in a nervous system is reflected in an animal's:
   a. degree of ability to learn
   b. degree of inability to learn
   c. constancy of response to repeated presentation of the same stimulus
   d. randomness of response to repeated presentation of the same stimulus
   e. repetition of the same behavior pattern in response to various different stimuli

5. Label the parts of the reflex arc shown in the figure below.

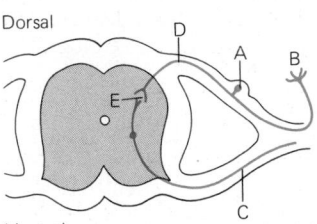

(Quiz continues on page 626.)

6. The parasympathetic nervous system:
   a. arises from the cranial nerves and the sacral spinal cord
   b. arises from thoracic and lumbar segments of the spinal cord
   c. has no synapses between the spinal cord neuron and effector
   d. has direct connections between adjacent ganglia without passing through the CNS
   e. is composed of sensory neurons running parallel to the motor neurons of the sympathetic system

7. The major transmitter between neuron and effector in the sympathetic system is _____, while in the parasympathetic system it is _____.

QUESTIONS FOR DISCUSSION

1. What possible disadvantages are there in the evolutionary trend toward cephalization of the nervous system? What advantages?

2. Viewed on an evolutionary time scale, the size of the cerebral cortex has virtually "exploded" to the relatively large size in modern humans. Can this expanded layer of cells account for the differences between apes and humans? Is intelligence a matter of number of synapses? (Scientists do not yet know the answer!)

3. What would happen to our basic behavioral responses if the ganglionic organization of the sympathetic and parasympathetic nervous systems were reversed?

4. During embryonic development the nervous system forms all the neurons an animal will ever have. After birth, neurons never divide to form new neurons. What is the adaptive advantage of this state of affairs? What are some drawbacks?

REFERENCES AND FURTHER READING

Some of the references listed for Chapter 37 will also be of use here.

Current Research on Sleep and Dreams. DHEW Publication No. (ADM) 75–244. Washington, DC: U.S. Government Printing Office, 1973.

DiCara, L. V. "Learning in the autonomic nervous system." *Scientific American,* Jaunary 1970. Experiments with rats and humans suggest that training can modify "involuntary" responses.

Eccles, J. C. *The Understanding of the Brain.* New York: McGraw-Hill, 1977. An eminent authority's short introduction to the activities of the human brain.

Guillery, R. W. "Visual pathways in albinos." *Scientific American,* May 1974. Explains how genetic defects in albino animals were used to help study visual pathways in the brain.

Hubel, D. H. "The visual cortex of the brain." *Scientific American,* November 1963. Studies on hierarchies of visual processing in cats.

Jouvet, M. "The states of sleep." *Scientific American,* February 1967. Experiments with cats to determine what structures and chemicals in the brain govern sleeping patterns; speculates on the function and evolution of sleep.

Konorski, J. *Integrative Activity of the Brain: An Interdisciplinary Approach.* Chicago: University of Chicago Press, 1967. A fascinating book dealing with brain function as evidenced by disorders in neurological patients.

Luria, A. R. "Functional organization of the brain." *Scientific American,* March 1970. Describes lines of evidence being used to try to determine what functions are performed in various areas of the cerebral cortex.

Nichols, J. G., and D. van Essen. "The nervous system of the leech." *Scientific American,* January 1974. Studies of a nervous system with comparatively few brain cells to determine how the patterns of cell organization dictate the animal's behavior.

Pribram, K. H. "The neurophysiology of remembering." *Scientific American,* January 1969. Presents evidence that a memory is stored in scattered areas of the brain rather than in one particular spot.

Routtenberg, A. "The reward system of the brain." *Scientific American,* November 1978. The role of the hypothalamus and drugs in learning.

Scientific American. "The Brain." September 1979 issue. The articles in this issue cover various aspects of brain structure and function.

Sperry, R. W. "The great cerebral commissure." *Scientific American,* January 1964. A fascinating account of how the function of the corpus callosum was demonstrated through experiments on animals; well-illustrated.

Stein, D. G., and J. J. Rosen, Eds. *Learning and Memory.* New York: Macmillan, 1974. A collection of scientific papers, with comments and questions by the editors to guide the reader's study of the subject.

CHAPTER 39

SENSE ORGANS

Sense organs permit an animal to detect objects and events in its own body and in the world around it. Information collected by sense organs is passed to the nervous system, which determines and initiates an appropriate response.

There is a popular myth that we have five senses. In fact, we have more than a dozen different types of sense organs, which monitor conditions both outside and inside our bodies. Internal sense organs detect and report changes in conditions such as body temperature, osmotic relationships, and pH. Without this information an animal could not maintain its physiological homeostasis. Sense organs that report sights, sounds, and chemicals in the outside world are used in feeding, in finding mates, in avoiding enemies, and in making other adaptive responses to the environment.

Our studies of other animals' sense organs are inevitably biased by our own senses, which give us a lopsided perception of the world. Humans are creatures of vision and, to a lesser extent, of hearing. It is hard for us to empathize with, say, an earthworm, for whom neither line nor color exists.

Many animals have sense organs far more sensitive than ours, and some respond to stimuli that our senses cannot detect at all. Bats flying about in total darkness can avoid obstacles, and homing pigeons can find their way over hundreds of miles of unfamiliar country with their eyes covered by opaque contact lenses. Until recently, people were baffled by these powers, which humans could not duplicate without sophisticated electronic equipment. Now we know that bats emit ultrasonic* squeaks and detect their faint echoes with such acute hearing that they can skirt telephone wires and catch flying moths on the darkest nights, and that pigeons can sense the earth's magnetic field. We cannot imagine what it would be like to have such senses. In terms of perception, we live in a different world from that of the worm burrowing in the flowerbed or the pigeon strutting on the sidewalk. Nevertheless, there are many similarities in the types of sense organs found among different animals, and in how these organs work.

In this chapter we shall consider how sense organs work and then survey the various types of sense organs found among animals.

39-A   Sense Organs and Their Functions

A **sense organ** is one or a collection of **receptor cells**, which may be sensory neurons or cells closely associated with sensory neurons (Figure 39–1). Receptor cells respond to stimuli by producing electrical activity in the nervous system. A **stimulus** is some form of energy; various receptors respond to energy in the form of pressure, light, electric current, chemical changes, osmotic potential, and heat. All of these stimuli will cause changes in the membrane of almost any cell. Even in unicellular organisms, however, there is specialization, with different areas of the cell peculiarly sensitive to specific forms of energy. In higher animals each receptor cell is specialized so that it reacts maximally to certain forms of energy.

An animal responds to a stimulus in a three-step process:

1. Receptor cells are changed by the energy of a stimulus in such a way that the stimulus energy is converted into electrical activity, which can travel in the nervous system.

2. The nervous system may alter the electrical potentials it receives from the sense organs. It then passes the resulting signals to the appropriate effectors.

3. **Effectors**, usually muscles and glands, contract or secrete chemicals, producing suitable responses to the information passed to the nervous system from the sense organs.

In this chapter, we shall consider only the processing of stimuli by receptors, and the entrance of the resulting information into the nervous system.

Stimuli

A sense organ reacts to a stimulus, which is some form of energy. Receptor cells may be classified by the form of stimulus energy to which they react: **mechanoreceptors** detect mechanical energy in the form of movement, pressure, or tension*; **photoreceptors** detect light energy; **thermoreceptors** detect heat; **chemoreceptors** detect chemicals; and **electroreceptors** detect electrical energy (Table 39–1).

A sense organ may filter incoming energy so that a modified form of the stimulus reaches the actual receptor cell. For instance, the human eye contains various **accessory structures** (see Figure 39–18). One of these, the **iris**, controls pupil size and thus changes the amount of light entering the eye. The **lens** adjusts the focus of the light, and it also filters out ultraviolet light so that these wavelengths do

---

* Ultrasonic At sound waves of a frequency higher than the human ear can detect.
* Tension Pull or stretch.

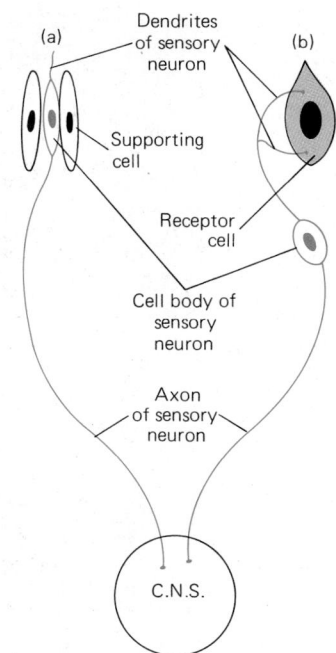

Figure 39-1 A sensory neuron conveys information from the sense organ to the central nervous system (CNS). Non-nervous cells are also important. (a) In an organ such as a vertebrate olfactory receptor (responsible for smell), the sensory dendrite is the actual receptor, and non-nervous cells serve merely to support it. (b) In sense organs such as vertebrate taste buds, a non-nervous receptor cell actually detects the stimulus, and its reaction depolarizes the membranes of dendrites of the sensory neuron ending on its surface.

TABLE 39-1 **CLASSIFICATION OF SOME RECEPTORS BY STIMULI**

GENERAL NAME	*EXAMPLES*	*EFFECTIVE STIMULUS*
Mechanoreceptors	Pacinian corpuscles (vertebrate connective tissue)	Deep touch and vibration
	Meissner's corpuscles	Touch on skin
	Proprioceptors	Position of parts of body
	Joint receptors	Angular movement in joint
	Golgi tendon organs	Stretch of a tendon
	Muscle spindles	Degree of muscle contraction
	Hair cells	
	Statocysts in invertebrates	Gravity
	Utriculus and sacculus in vertebrates	Gravity
	Cochlea	Airborne sound waves
	Semicircular canals	Acceleration
Photoreceptors	Ommatidia of arthropods	Wavelength of light
	Retina of vertebrates	Wavelength of light
Thermoreceptors	Pit organs of pit vipers, and labial organs of some boas; nerve endings in the tongue of mammals, skin of mammals and birds, and many invertebrates	Increasing and decreasing infrared radiation
	Krause's end bulbs	Cold on skin
Chemoreceptors	Taste buds and olfactory organs of vertebrates	Unknown features of the chemistry of specific molecules in air or water
	Chemoreceptors of invertebrates	
Electroreceptors	Organs in the skin of some elasmobranchs* and bony fish; ampullae of Lorenzini in elasmobranchs	Electric currents in surrounding water

* Elasmobranchs Sharks, skates, and rays.

Figure 39-2 Different eyes see different things. The human eye sees a marsh marigold with no markings (left). An insect sees big patches on the same flowers (right) because its eyes react to ultraviolet light energy, whereas ours do not. (Biophoto Associates)

not reach the receptor cells in the **retina**; animals such as honeybees, which do not have ultraviolet filters in their eyes, can perceive and react to ultraviolet light (Figure 39–2). It is not clear whether vertebrates other than humans can see ultraviolet light or not.

Transduction

The most important role of a receptor is to convert, or **transduce**, the energy of a stimulus into electrical energy—the only form of energy that can be transmitted by the nervous system. The stimulus may depolarize* the membrane of the receptor to produce a **generator potential**, an electrical potential whose magnitude is proportional to the intensity of the stimulus.

The first direct evidence for a generator potential was found by Bernard Katz in 1950. Katz studied the sensory neuron of a vertebrate **muscle spindle**, a receptor that detects the degree of stretch or contraction in a muscle. When recording electrodes were placed close to the dendrites of the sensory neuron near the muscle spindle, Katz recorded a generator potential. The size of the generator potential reflected the degree of stretch in the muscle (Figure 39–3). If the generator potential in the sensory neuron was large enough, it exceeded the threshold and induced the neuron to fire action potentials. The larger the generator potential, the greater the frequency of action potentials. Thus, the frequency of the action potentials fired by

---

*Depolarize Decrease the electrical potential difference across.

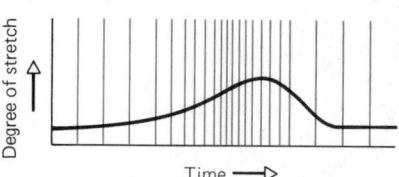

ANIMAL BIOLOGY

Figure 39-3 Recording from the sensory neuron of a muscle spindle, a stretch receptor in vertebrate muscle. The colored lines represent action potentials in the sensory neuron's axon. The black line shows the degree of stretch in the muscle. As the muscle is stretched, the frequency of sensory impulses increases and then falls again as the stretch subsides.

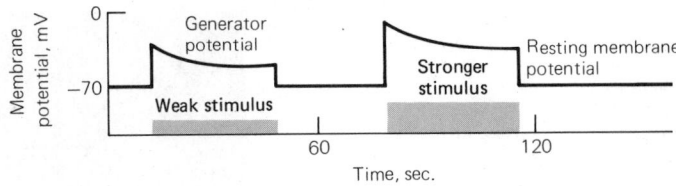

Figure 39-4 Generator potentials from a crayfish stretch receptor subjected to two stimuli (colored bars) of the same duration but different intensity. This is a slowly adapting receptor; the generator potential decays only very slowly as long as the stimulus remains.

the sensory neuron was proportional to the degree of stretch of the muscle (up to a point), and the sensory neuron conveyed precise information about the extent of muscle contraction to the central nervous system.

In some sense organs, the receptor consists merely of the dendrites of the sensory neuron. In others, the receptor is a separate, non-nervous cell close to the sensory neuron's dendrites. In either case, transduction must cause a generator potential that exceeds the sensory neuron's threshold and fires action potentials in its axon before information will be transmitted to the central nervous system.

Adaptation

If a constant stimulus is applied to a receptor for any length of time, the frequency of action potentials in the sensory neuron decreases as time passes. This diminishing response to a constant stimulus is called **adaptation**. Adaptation may result from the receptor's producing a smaller generator potential, or from the sensory neuron's becoming less responsive to stimulation with time, or both.

Both receptors and sensory neurons may adapt slowly or rapidly to a constant stimulus, and there are advantages to possessing cells with both kinds of responses. A rapidly adapting receptor allows an animal to ignore an unchanging stimulus. Touch receptors in the skin adapt rapidly to a constant stimulus and fire rapidly again only if the stimulus changes; for example, when we dress in the morning we notice the contact between our clothing and skin. This sensation rapidly disappears, and we are not distracted by the feel of our clothes all day. However, a fly crawling down the neck causes our touch receptors to respond immediately. Receptors that adapt rapidly detect stimuli that usually change rapidly. Visual and sound receptors adapt rapidly, and animals are consequently much more sensitive to changing sights and sounds than to constant ones.

In slowly adapting receptors, action potentials in the sensory neuron continue for as long as the stimulus lasts (Figure 39–4). Slowly adapting receptors provide information about the intensity of a stimulus for as long as the stimulus is present. Examples in humans include receptors for pain and cold.

Adaptation occurs not only in receptors and sensory neurons but also in the central nervous system. Indeed, adaptation at all levels in the nervous system is important in determining an animal's reaction to a particular stimulus.

39-B Mechanoreceptors

Mechanoreceptors respond to movement. Transduction in mechanoreceptors is poorly understood; a stimulus stretches the membrane of the receptor, and this somehow changes the permeability of the membrane and causes a generator potential. Some mechanoreceptors adapt rapidly to a constant stimulus, and so they are admirably suited to detect changes in pressure—that is, movement—in their surroundings. Examples are the receptors at the bases of hairs. Other mechanoreceptors adapt slowly. The pressure receptors that constantly monitor blood pressure in the large arteries adapt slowly, and their rate of firing reflects the actual blood pressure very accurately.

Vertebrates have many different types of mechanoreceptors. Human touch receptors are concentrated on the tongue, lips, face, and fingertips, reflecting the importance of the head and of the manipulative fingers in the higher primates. We will consider only a few of the more important types of mechanoreceptors here.

Figure 39-5 Hair receptors on the back of a caterpillar. (Biophoto Associates)

Proprioceptors

The central nervous system continually monitors the position and movements of parts of the body. The necessary information comes from **proprioceptors**, mechanoreceptors in the joints and muscles. Vertebrates have three main types of proprioceptors: **joint receptors**, which detect angular movement in the ligaments that hold the bones of a joint together; **Golgi tendon organs**, which determine stretch in the tendons that hold muscles to bones; and **muscle spindles**, which detect muscle movement.

MUSCLE SPINDLES

A muscle spindle consists of several modified **intrafusal muscle fibers** separated from the surrounding ordinary (extrafusal) muscle fibers by a capsule of connective tissue (Figure 39–6); a muscle's stretching also stretches the intrafusal muscle fibers embedded in it. The dendrites of two types of sensory neurons lie in contact with the intrafusal fibers. One type of sensory neuron first alerts the central nervous system to a change in the stretch or contraction of the muscle, and then adapts and sends information about the degree of stretch. Meanwhile, the second type of sensory neuron has been continuously signaling the degree of stretch.

The sensory neurons of a muscle spindle form part of a reflex arc that sends motor signals back to the same muscle and cause it to return to its previous state of contraction; in this way muscles responsible for control of posture, for example, can be regulated to keep the body steady.

The muscle spindle itself receives signals from motor neurons (gamma fibers) that cause parts of the spindle to contract. This changes the sensitivity of the spindle to the degree of muscle stretch, so that sometimes only large changes in stretch are detected, whereas at other times even small changes are signaled. The sensitivity of many sense organs is controlled by the central nervous system in this way.

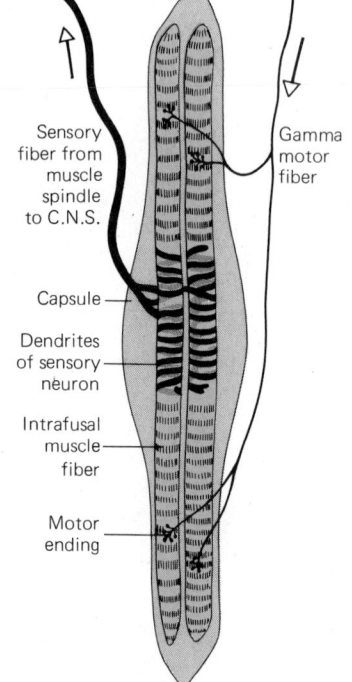

Sensory fiber from muscle spindle to C.N.S.

Gamma motor fiber

Capsule

Dendrites of sensory neuron

Intrafusal muscle fiber

Motor ending

Figure 39-6 A vertebrate muscle spindle. A muscle may contain hundreds of spindles. When the muscle stretches, it stretches the intrafusal fibers; this stimulates the sensory dendrites, and initiates impulses that travel to the central nervous system (CNS). The intrafusal muscle fibers contract when stimulated by the gamma fiber. This changes the spindle's sensitivity to stretching of the muscle.

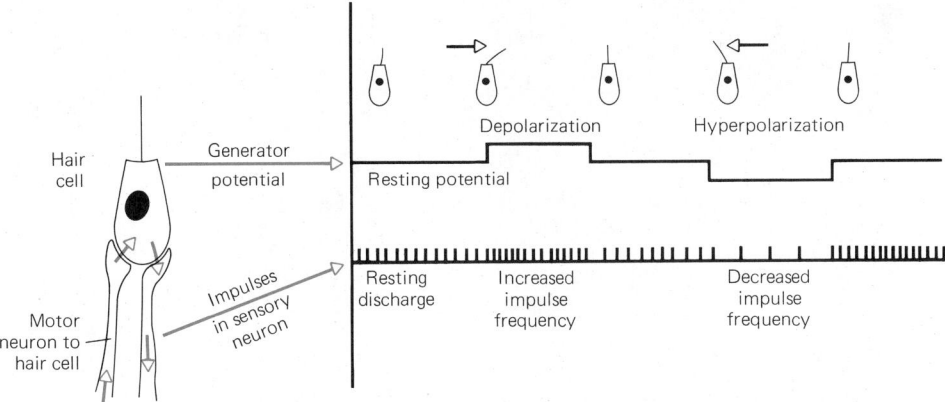

Figure 39-7 A vertebrate hair cell responds to deflection in either of two directions. When the cell is not stimulated, it produces a resting discharge of action potentials in the sensory neuron. The frequency of these action potentials can be altered by the effect of the motor neuron on the cell so as to make the cell more or less sensitive to stimulation. When the hair is bent in one direction (black arrow pointing right), the cell produces a depolarizing generator potential, which increases the discharge frequency in the sensory neuron. When the hair is bent in the opposite direction, the resulting hyperpolarization of the cell reduces the frequency of firing in the sensory neuron. The frequency of discharge in the sensory neuron informs the central neurons which way the hair is bent.

Hair Cells

In vertebrates, the mechanoreceptors that detect gravity or vibrations in air or water (sounds) are **hair cells** bearing fine projections like cilia, into which the dendrites of sensory neurons protrude (Figure 39–7). Typically, when a hair cell is bent in one direction, the sensory neuron fires more frequently; when it is bent in the other direction, the sensory neuron fires less frequently. Bending in either direction thus changes the pattern of action potentials sent to the central nervous system. We will consider a few examples of hair cells and the way they work.

STATOCYSTS

All gravity receptors have essentially the same structure. They consist of a cavity that is lined with hair cells, and that contains dense particles such as grains of calcium carbonate or sand (Figure 39–8). The particles are pulled downwards by gravity and stimulate the hair cells on which they lie. By detecting which hair cells are firing at any time, the animal knows where "down" is and so determines how its body is oriented.

Gravity receptors in invertebrates are usually called **statocysts**. A classic experiment on the function of statocysts was done by Kreidl in 1893. Kreidl used shrimp, whose statocysts are located in chambers at the bases of the antennae. The lining of the chambers is part of the exoskeleton, so that when a shrimp molts, it loses the lining of the statocyst as well as the sand grains inside the chamber. The shrimp must then pick up more grains of sand with its claws and place them in its statocyst chambers. Kreidl waited until his shrimp molted, and then placed iron filings in their tank instead of sand. The shrimp refilled their statocysts with these filings, and swam around quite normally. Kreidl then overcame the force of gravity by holding magnets above the shrimp. The filings were attracted upwards toward the magnets, and the shrimp, responding to the pressure of the filings on the hair cells, promptly began to swim upside down in response to their new perception of gravity.

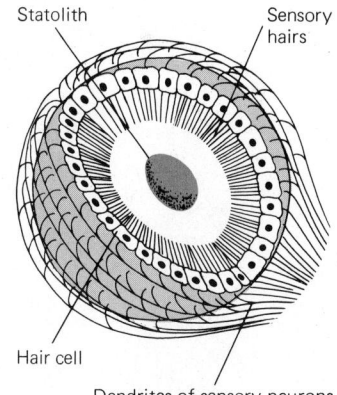

Figure 39-8 The statocyst of a crustacean. The statolith is a particle, and it moves around under the influence of gravity. Wherever it comes to rest, it stimulates the underlying hair cells of the statocyst lining.

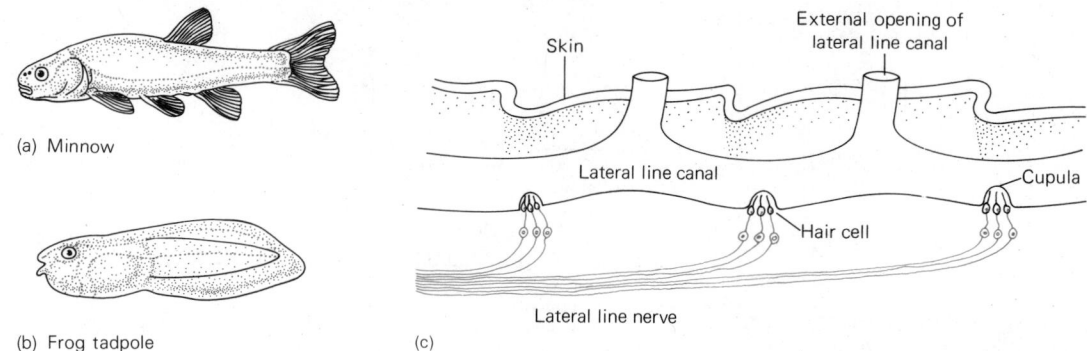

(a) Minnow

(b) Frog tadpole

(c)

Skin

External opening of
lateral line canal

Lateral line canal

Cupula

Hair cell

Lateral line nerve

Figure 39-9 The lateral line system. (a, b) Colored dots show the distribution of openings of the lateral line system. (c) Longitudinal section through the lateral line organ. The lateral line canal connects with the water outside the fish via openings in the skin. The actual pressure receptors are hair cells embedded in cupulae lining the canal. Dendrites from sensory neurons synapse with the hair cell bodies and detect their depolarization.

LATERAL LINE ORGANS

The **lateral line organs** of the members of the classes Agnatha, Chondrichthyes, and Osteichthyes and of larval amphibians are hair cell pressure receptors. The lateral line consists of a row of water-filled canals or tunnels extending along the animal's side and onto its head (Figure 39–9). The canals are lined with hair cells, which are stuck together by a gelatinous glue to form clumps called **cupulae** (singular: **cupula**) (Figure 39–9). When the water in a lateral line canal moves, it moves a cupula and distorts the hair cells, generating a signal in the associated sensory neurons. The lateral line organ is a remarkably sensitive system by which the animal can detect such things as the direction and force of water currents, the movements of other animals in the water, and pressure waves bouncing off stationary objects nearby.

LABYRINTH OF THE EAR

The anterior end of the lateral line organ evolved into the **labyrinth** found in the inner ears of all vertebrates (see Figure 39–12). Mechanoreceptors in different parts of the labyrinth detect the direction of gravity and the acceleration* of the head.

Two chambers, called the **sacculus** and **utriculus**, house a vertebrate's gravity detectors. Each chamber contains **otoliths**, crystals of calcium carbonate embedded in a gelatinous secretion atop a tuft of hair cells (Figure 39–10). The hair cell tufts in the two chambers lie in different planes, allowing the animal to tell the direction of gravity when the head is in any position. They also detect linear acceleration, changes in speed when the body is moving in a straight line.

*Acceleration   A change in the speed or direction of motion.

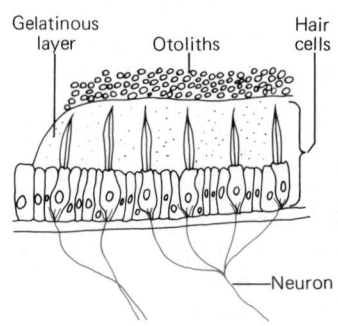

Gelatinous
layer

Otoliths

Hair
cells

Neuron

Figure 39-10 Sensory cells in the utriculus and sacculus have their tips embedded in a gelatinous material; particles of calcium carbonate (otoliths) lying on the gelatinous material respond to gravity, deflecting the hair cells as they do so.

Figure 39-11 (a) Semicircular canals of vertebrates are arranged in three planes at right angles to each other. Angular acceleration of the head is interpreted by combining the movements detected in each plane. (b) Each semicircular canal contains a cupula, located in an expanded chamber. Hair cells embedded in the gelatinous material of the cupula are bent by movement of the fluid. Bending the hair cells one way increases the rate of action potentials in the nerve fiber; bending the opposite way lessens the rate of action potentials.

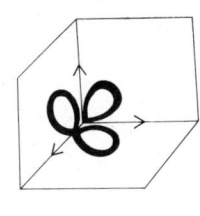

(a)

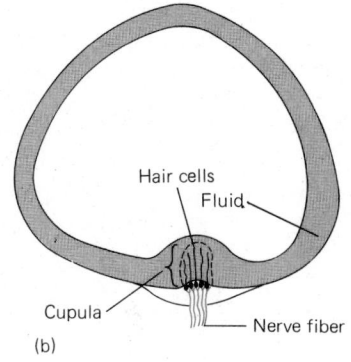

(b)

The three **semicircular canals** detect angular (rotational) acceleration; they are the sense organs of equilibrium. Since the canals lie in three different planes, they can detect acceleration of the head in any direction. The term "semicircular" is something of a misnomer; each canal is actually a hollow ring filled with fluid. At one point in the ring is a cupula, which operates like a swinging door across the canal. The tips of hair cell mechanoreceptors are embedded in the cupula of each semicircular canal.

The semicircular canals can detect changes only in the speed or direction of rotation of the head; they do not react if the head is moving at a constant speed or in a straight line. Travelling in a car at constant speed in one direction gives no sense of movement; only if the car turns do the semicircular canals relay a sensation of motion. This is due to the inertia of the fluid in the canals. When the head is stationary, or is moving at a constant speed or in a straight line, the fluid does not move with respect to the canals. When the head changes direction, however, the fluid tends to keep going in the same direction, and so it pushes the cupula into a new position (Figure 39–11).

HEARING

The **cochlea** of the inner ear contains another set of mechanoreceptor hair cells, which detect pressure waves, or sound. The cochlea lies in a fluid-filled cavity, and its hair cells respond to pressure waves in fluid that is inside the cochlea. In terrestrial vertebrates, accessory structures in the outer and middle ear transform the stimulus of sound waves in air to pressure waves in the cochlear fluid. In the human ear, for example, airborne vibrations strike the **tympanic membrane**, or eardrum. Vibration of the tympanic membrane moves three small bones that span the cavity of the middle ear. The third bone presses against the **oval window** of the cochlea. Vibration of the oval window in turn moves the fluid inside the cochlea (Figure 39–12).

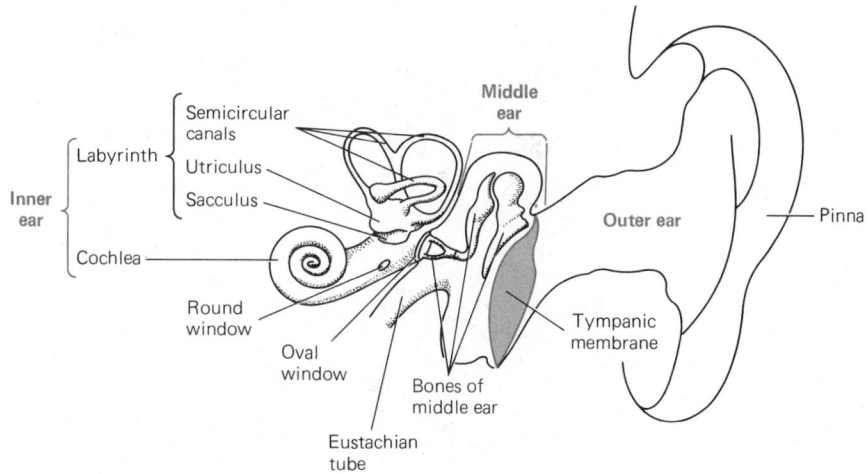

Figure 39-12 The human ear. Sound is transmitted from the outer ear, through the three bones in the air-filled middle ear, to the fluid of the cochlea in the inner ear. The inner ear also contains the labyrinth, where gravity and acceleration are detected.

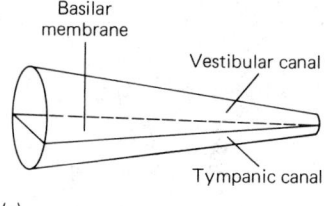

Basilar
membrane

Vestibular canal

Tympanic canal

(a)

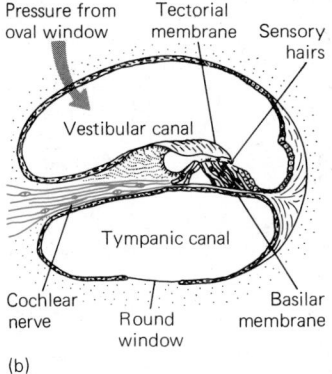

Pressure from   Tectorial
oval window   membrane   Sensory
                                hairs

Vestibular canal

Tympanic canal

Cochlear   Round   Basilar
nerve   window   membrane

(b)

Figure 39-13 (a) The cochlea unrolled to show its basic structure, a tapering tube divided into compartments by the basilar membrane. (b) Cross section through the cochlea in the inner ear of a mammal.

The entire cochlea is shaped like a snail shell; it is a coiled tube of increasing diameter. The **basilar membrane** stretches across the space inside the cochlea along its entire length (Figure 39–13), dividing the cochlea into the **tympanic** and **vestibular** canals. The **organ of Corti** consists of hair cells sandwiched between the basilar membrane and the **tectorial membrane**; these hair cells synapse with neurons of the auditory nerve.

Pressure on the oval window moves the fluid in the vestibular canal, causing it to distort the basilar membrane so that the hair cells of the organ of Corti move. The membranes of these cells depolarize, initiating generator potentials, which may become action potentials.

We can ordinarily distinguish three characteristics of a sound: pitch, volume, and tone quality. Pitch is a function of frequency; low frequency vibrations produce a sensation of low pitch, and high frequency vibrations produce a sensation of high pitch. At frequencies below 60 cycles per second, the entire basilar membrane moves and produces action potentials in the auditory nerve in synchrony with the rhythm of the sound. The rhythm of the spikes conveys information about the pitch of the sound. Frequencies above 60 cycles per second cause the basilar membrane to vibrate unequally along its length. Each frequency produces a maximum vibration at one point, where the width of the organ of Corti is tuned to the frequency. Thus the brain "knows" the pitch of the sound because it knows the location of the sensory neurons that are firing (Figure 39–14). The cells that vibrate most inhibit firing of neighboring cells, so that the brain receives a clear sensation of the true pitch.

The volume or loudness of a sound is a function of the amplitude of its vibrations (Figure 39–15). High-amplitude vibrations in air produce high-amplitude oscillations of the cochlear fluid and basilar membrane, producing more intense stimulation of the hair cells and a greater frequency of action potentials in the auditory nerve.

Figure 39-14 Each pitch of sound above 60 cps (cycles per second) causes a maximum vibration of the basilar membrane in a different part of the cochlea; high-pitched sounds are detected near the base of the cochlea, low-pitched sounds near the apex. As a result, different sensory neurons fire most rapidly in response to sounds of different frequencies, conveying information about the pitch of a sound to the brain (λ = wavelength = 1/frequency).

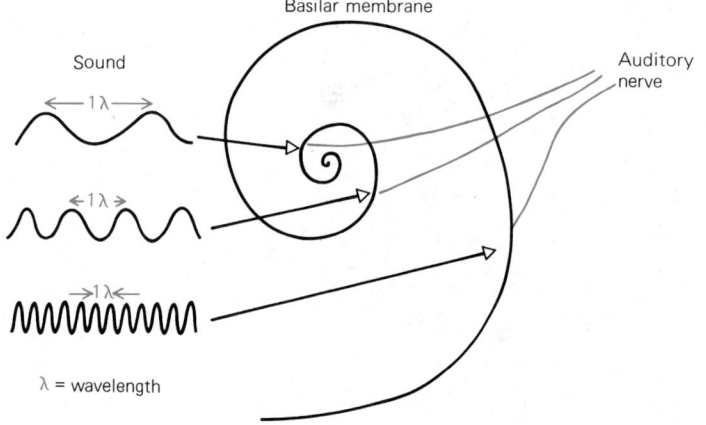

Basilar membrane

Sound

Auditory
nerve

λ = wavelength

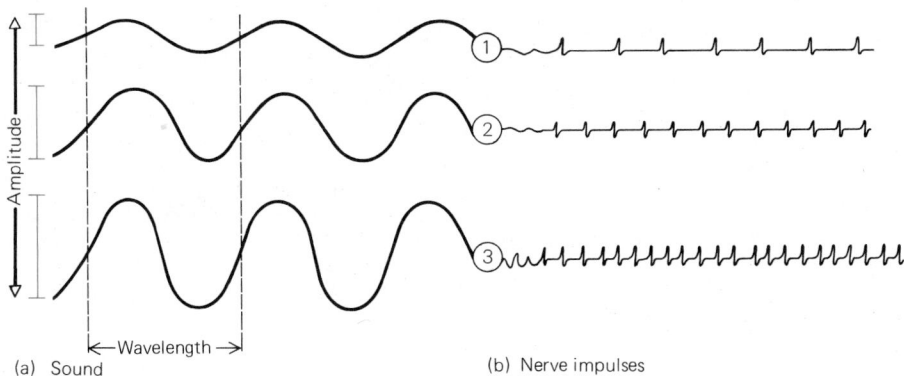

(a) Sound   (b) Nerve impulses

Figure 39-15 Sound of different amplitude (loudness) (a) translated into impulses in the auditory nerve (b). The brain interprets an increased rate of action potentials (1 to 3) as an increase in loudness. The louder the sound, the greater the amplitude of its vibrations; wavelength stays the same. The rate of action potentials in the auditory nerve fibers increases as the amplitude of the sound increases (up to a point, beyond which sounds seem to be of the same [deafening] magnitude).

The **tone quality** of a sound is a function of the frequency and amplitude of its harmonics. **Harmonics** are vibrations at integer multiples of the main pitch of the sound. If a violin, a piano, and a clarinet play a note at the same pitch and volume, each produces a characteristic tone quality because of the differences in the loudness of its various harmonics; each harmonic pattern stimulates a different combination of hair cells in various regions of the cochlea.

39-C Photoreceptors

Almost all animals have **photoreceptors** that transduce light energy using **pigments**, colored molecules that undergo chemical changes when they are struck by light rays. In platyhelminths and annelids, photoreceptors are merely **eyespots** that detect light as opposed to darkness (Figure 39-16). **Eyes** detect more detail; in an eye, light from the visual field projects an image onto light-sensitive receptor cells.

There are three basic types of eyes (Figure 39-17). The simplest image-forming eye, found in many invertebrates, is analogous to a pinhole camera. Relatively little light can enter the eye through the tiny aperture. The type of eye found in vertebrates and cephalopods (e.g., octopuses, squids) contains a lens, which can focus incoming light. The third type of eye is the multifaceted compound eye of some arthropods. Such an eye contains many closely packed individual units, the ommatidia, each receiving light from a narrow area of the visual field.

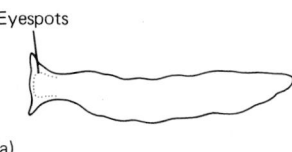

(a)

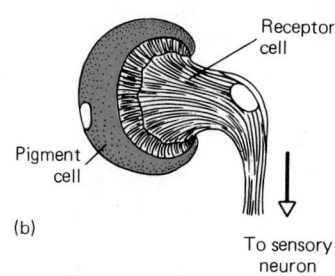

(b)

Figure 39-16 An eyespot. (a) Like many invertebrates, this planarian worm has many eyespots. (b) Section through an eyespot. The receptor cell (there may be more than one) is covered by one or several pigment cell(s) where light is absorbed, initiating changes that cause the cell membrane to depolarize.

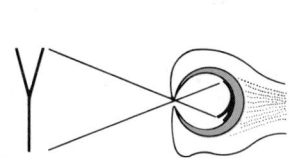

(a) Most molluscs

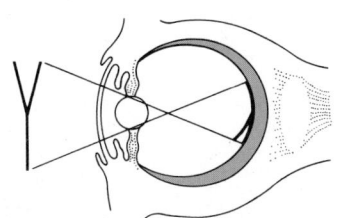

(b) Most cephalopods and vertebrates

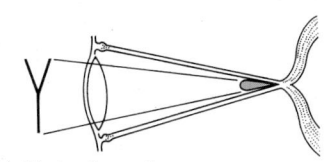

(c) Most arthropods

Figure 39-17 The three main plans of eyes. The position of the receptor cells is shown in color. The object in front of the eye produces an image, shown inside the eye. (a) A lensless "pinhole camera" type of eye found in some molluscs. (b) A vertebrate eye, with an adjustable lens that can change the focus. Some molluscs, such as some bivalves and cephalopods, have a very similar eye. (c) A crustacean eye, containing a lens but only one or two receptor cells. In insects, thousands of these individual units are organized into a single eye (see Figure 39-21).

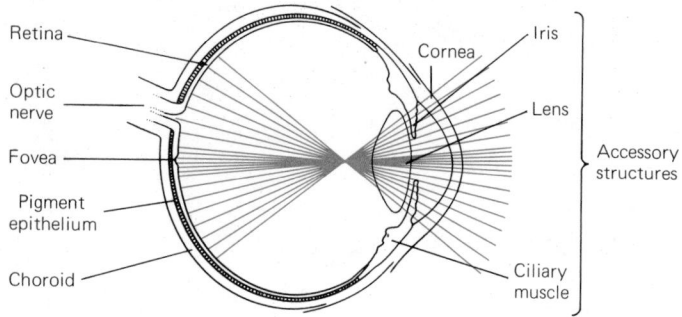

Figure 39-18 The human eye, sliced at its equator and viewed from above. The retina contains the receptor cells and their associated neurons, which send messages to the brain via the optic nerve. The choroid layer contains blood vessels, and the layer of dark pigment epithelium absorbs light which has passed through the retina.

In vertebrates, the curved cornea focuses light entering the eye (Figure 39–18); the focus is adjusted by the lens, which is somewhat elastic and changes shape when it is pulled by muscles around its edges. The iris is a diaphragm with a central opening, the pupil, whose size can be changed to regulate the amount of light passing through the lens. The **retina** is a delicate layer containing the photoreceptors and their associated neurons. Visual information passes to the brain by way of the optic nerve.

The actual photoreceptors of the retina are the rods and cones (Figure 39–19); they are not neurons but specialized receptor cells, apparently evolved from hair cells. Each receptor synapses with the dendrites of a sensory neuron. A remarkable feature of the vertebrate retina is that the sensory neurons lie in front of the rods and cones, and light must pass through a layer of nerve cells before it reaches the photoreceptors.

The images detected by way of the rods are coarse and poorly defined because the associated neurons may receive input from many rods. However, the rods and their nervous connections respond to light that is too dim to stimulate vision by way of the cones. Neurons carrying information from cones tend to keep the information from different cones separated, and this gives good resolution of detail in bright light; the cone system is also much better at interpreting color than is the rod system. Cones are especially densely packed in an area of the retina called the **fovea**; we tend to move our eyes so that the image of the object we want to see most clearly falls on the fovea. The fovea contains only cones, and most of the other cones of the retina are nearby. Rods are most concentrated in a ring about 20° from the fovea; their density is lower toward the edges of the fovea and toward the periphery of the retina.

Figure 39-19 Cross section of the retina of a vertebrate. The receptor cells are pigment-filled rods and cones, which synapse with bipolar neurons. These synapse with ganglion cells, neurons whose cell bodies lie in the retina and whose axons form the beginning of the optic nerve. Light must pass through several layers of cells before it reaches the receptors.

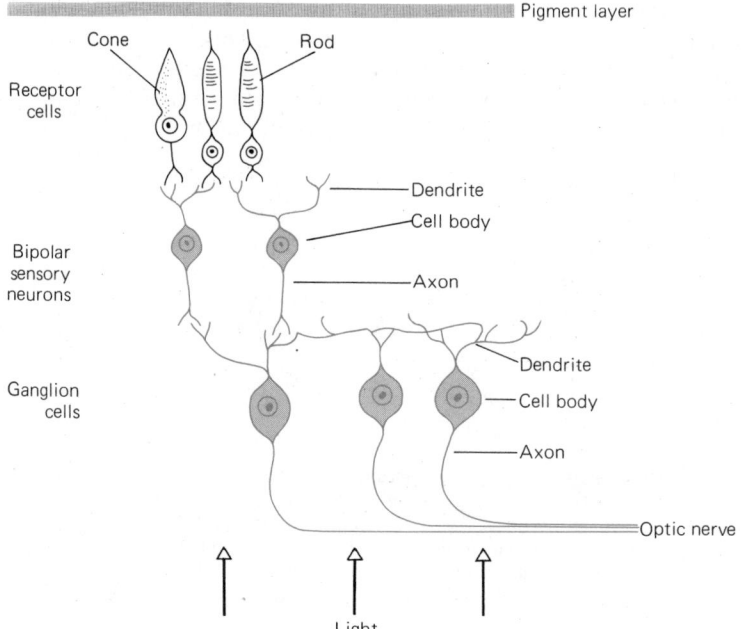

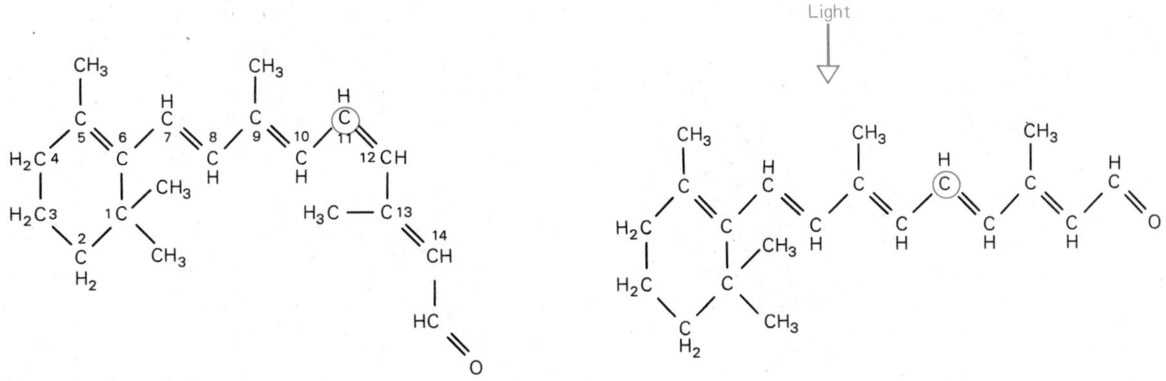

(a) 11-*cis* Retinaldehyde

(b) All-*trans* Retinaldehyde

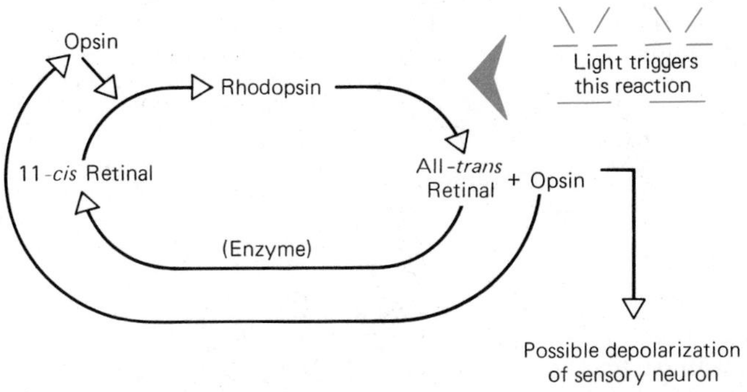

(c) The visual cycle

Figure 39-20 (a) and (b) The two isomers of retinaldehyde, a visual pigment found in many animals. Light causes the change from one form to another. The molecule changes shape by twisting around the 11-carbon atom. (c) Diagram of the visual cycle.

Electromagnetic wavelengths shorter than about 10 nm contain so much energy that they break bonds in molecules. Wavelengths longer than about 1000 nm contain so little energy that they cause little or no change when they hit a molecule, and so cannot be absorbed by pigments. Plants absorbing light for photosynthesis and animals using light for vision both have carotenoid pigments that absorb wavelengths of about 300 to 800 nm, between these two extremes. As far as we know, animals cannot synthesize carotenoids and must obtain their visual pigments directly or indirectly from plants. β-carotene (see Figure 8–7), a common plant carotenoid, can be broken down into two molecules of vitamin A, which in turn is used to synthesize retinaldehyde, a component of the visual pigment in vertebrate rod cells.

Retinaldehyde can exist as two isomers, the all-*trans* and 11-*cis* forms (Figure 39–20). 11-*cis* retinaldehyde combines with a protein, **opsin**, to form the visual pigment **rhodopsin**. When light strikes rhodopsin, it causes the 11-*cis* form to become all-*trans* retinaldehyde, and this change in shape causes the retinaldehyde molecule to break away from the opsin. This initiates a sequence of events that somehow leads to depolarization of the sensory neurons that synapse with the rod cells. The all-*trans* retinaldehyde is then converted back to the 11-*cis* isomer by an enzyme, the retinaldehyde recombines with opsin, and rhodopsin is re-formed, completing the sequence of reactions called the **visual cycle**.

All vertebrates seem to have at least a sprinkling of cone cells in their retinas, but many of them do not seem to respond to different colors. Although little is known about the chemistry of color detection, recent experiments have shown roughly what happens. An animal with color vision has two or more types of cone cells, each producing a different pigment; in primates, including humans, there are three types. Light of different wavelengths (colors) affects the different pigment molecules differently. By comparing the levels of excitation in cones containing different pigments, the visual system can obtain a relatively precise measure of the wavelength of incoming light. Various degrees of color blindness result if one or more of the cone pigments is missing.

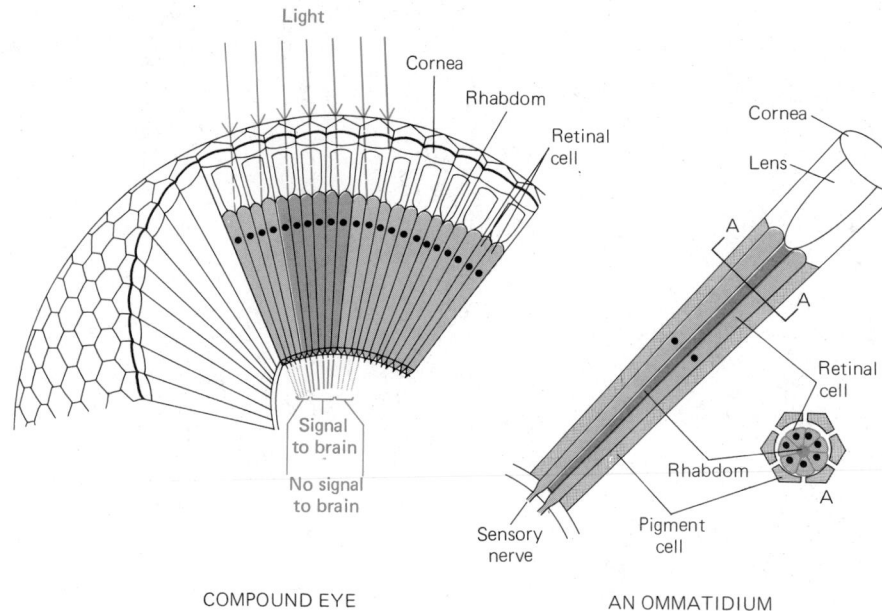

Light
Cornea
Rhabdom
Retinal cell
Signal to brain
No signal to brain
COMPOUND EYE

Cornea
Lens
A
A
Retinal cell
Rhabdom
A
Pigment cell
Sensory nerve
AN OMMATIDIUM

Figure 39-21 The compound eye of insects and some other arthropods is composed of many units, called ommatidia. Each ommatidium consists of a cornea, a lens, and a light-sensitive rhabdom surrounded by retinal cells, which transmit the sensory stimulus. Pigment cells around each ommatidium prevent light from passing into surrounding ommatidia. Only light rays that are parallel to the rhabdom will excite the retinal cells.

The compound eyes of many arthropods look very different from the vertebrate eye; compound eyes may be composed of as many as 20,000 **ommatidia** (Figures 39–21, 39–22). Each ommatidium has a lens-like structure that focuses incoming light rays onto a receptor called a rhabdom, made up of parts of the membranes of several cells. In the eyes of bees, flies, and many other diurnal insects, opaque walls around the receptor cut out light coming in from the side, and each receptor detects only a narrow beam of light. Such an eye appears to present a rather crude mosaic picture of the outside world. Its main advantage is that it permits an animal to detect rapid movement with a sensitivity about five times that of the human eye. In arthropods that are active in dim light, such as moths and lobsters, the pigment can move out of the way so that it does not block light from passing between ommatidia. The function of these compound eyes is not well understood.

Many invertebrates (and some vertebrates) can detect polarized light. This ability is used for determining compass direction and so for navigation by some animals: the light in the sky is polarized in different directions relative to the sun and to the observer.

Figure 39-22 A fly's big compound eyes allow it to view most of its surroundings at once and to detect slight movements as new patterns of ommatidia become excited. Numerous hair-like pressure receptors, which respond to air movement or touch, are also shown in this picture. (Biophoto Associates)

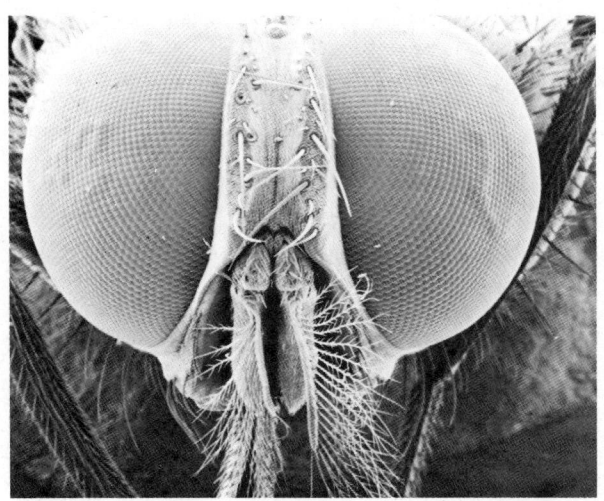

39-D Thermoreceptors

Heat, or infrared radiation, is another portion of the electromagnetic spectrum that most animals can detect, by using thermoreceptors. Since temperature influences the rate of all biochemical reactions, every living cell responds to temperature changes, and this makes it difficult to locate the specific receptors responsible for behavior patterns such as moving towards or away from heat. Temperature detectors may be very sensitive. Blood-sucking insects and ticks can distinguish temperature changes as small as 0.5°C in their search for warm-blooded hosts.

The most thoroughly studied thermoreceptors are the pits in the faces of crotalid snakes ("pit vipers") and the labial organs of some boas. These snakes use their heat-detecting organs to locate warm-blooded prey. A pit viper can detect the body heat generated by a small mammal at distances of up to about 1 metre. There are pits on both sides of the head, and when the snake moves its head until the temperatures detected on each side are the same, it is facing its prey directly.

Mammals have thermoreceptors consisting of free nerve endings. These are scattered over the surface of the body, particularly on the tongue. There are also thermoreceptors that detect internal body temperature in the hypothalamus of the brain. Information from internal and external thermoreceptors is integrated in the hypothalamus to produce appropriate behavior such as shivering or sweating.

39-E Chemoreceptors

Chemoreceptors detect various chemicals. Even primitive organisms such as bacteria and protists can detect chemicals in their surroundings; they avoid noxious chemicals in the water and swim toward substances such as food and oxygen.

In all invertebrate and most vertebrate chemoreceptors, the receptor is the sensory neuron itself. Chemoreceptors that are not neurons occur only in vertebrate taste organs (and possibly in some chemoreceptors in the circulatory system).

While we have relatively few external receptors for detecting chemicals in the world around us, our bodies are full of internal receptors for chemicals such as nutrients, oxygen, carbon dioxide, hormones, and neurotransmitters. Internal chemical reception is covered in other chapters; in this section we consider only receptors that detect chemicals in the external environment.

External chemoreceptors in humans are responsible for the senses of smell **(olfaction)** and taste **(gustation)**. The number of possible tastes is believed to be only four: sweet, sour, bitter, and salt. The "taste" of more complicated flavors is in fact a combination of olfactory and gustatory sensations. The important role of smell in the sensation of "taste" is evident during a bad cold; with a stopped-up nose, every dish tastes like cardboard.

The structure of the receptors and the organization of the nervous connections involved in olfaction and gustation are much simpler than those for vision or hearing. The vertebrate tongue is covered with numerous bumps, or **papillae**. In the grooves around the sides of a papilla lie the clusters of cells known as **taste buds**. A taste bud (Figure 39–23) consists of sensory hair cells, whose tips project to the edge of the epithelium, and supporting nonsensory cells. A single bipolar neuron sends branches of its dendrite to all of the hair cells in several taste buds.

Figure 39-23 Chemoreceptors involved in the sense of taste in humans. (a) The distribution of taste buds sensitive to sweet, bitter, sour, and salt on the tongue. (b) A taste bud in the tongue.

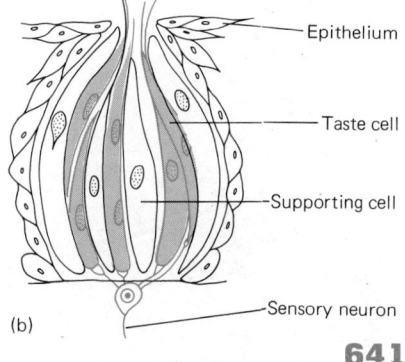

Epithelium

Taste cell

Supporting cell

(b) Sensory neuron

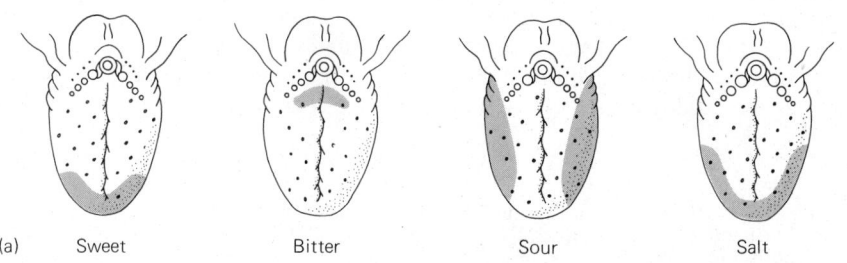

(a) Sweet Bitter Sour Salt

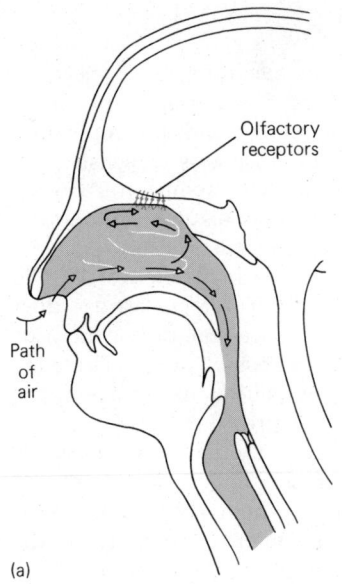

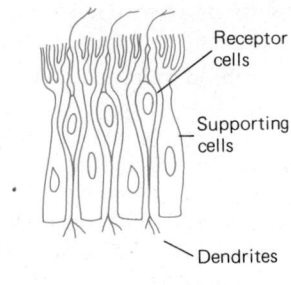

(b)

(a)

Figure 39-24 (a) Olfactory receptors are found in the top of the nasal cavity in a region of spongy tissue. (b) Chemoreceptors in the nose are free dendrites of neurons in the olfactory epithelium.

The olfactory endings of a vertebrate lie in the upper reaches of the nasal cavity and are quite densely packed; a dog has up to 40 million such endings per square centimetre. The dendrite of the olfactory sensory neuron extends beyond the epithelial cells in the nose, and its end is fringed with processes, where the initial chemoreception takes place (Figure 39–24). Since the olfactory epithelium of the nose is not on the main pathway to the lungs, air must travel a circuitous route to reach the olfactory endings. Sniffing aids the sense of smell by moving the air more rapidly into the upper recesses of the nose, where the olfactory endings lie. Here, contact between the olfactory chemoreceptor and the appropriate chemical creates a generator potential. Presumably the chemical combines with a specialized macromolecule on the surface of the neuron, and this combination makes the macromolecule change its shape and render the cell leaky to sodium.

Some snakes and lizards have specialized areas of chemoreception in the **organ of Jacobson** in the roof of the mouth. The animal's tongue flicks out, gathers chemicals from the environment, and transfers these chemicals to the organ (Figure 39–25).

Among invertebrates, insects have the best-studied chemical senses. Insect chemoreceptors may be located on the mouthparts, legs, and antennae (Figure 39–26). Chemoreceptors of some insects can distinguish between sugars and amino acids, between sodium and other ions, and between many very similar organic compounds.

39-F   Electroreceptors

In a number of bony and cartilaginous fishes, parts of the lateral line organ have become modified to detect electric currents in the surrounding water. Every living organism generates weak electrical fields, and the ability to detect these fields may permit a fish to capture prey or to avoid predators. Such an ability is especially valuable in turbulent, deep, or murky water, where the senses of vision and olfaction are of little use. Some fish generate electrical fields and then use their electroreceptors to detect how surrounding objects distort the field; this allows the fish to navigate in the muddy rivers where they live (see Essay, Chapter 40).

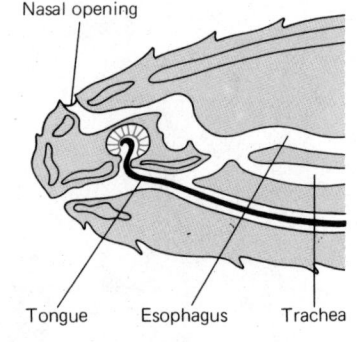

Figure 39-25 Chemoreception in the organ of Jacobson (colored) of a lizard. A lizard or snake shoots out its tongue to pick up chemicals in the environment. The tongue is then drawn back into the mouth and inserted into the organ of Jacobson, which is lined with chemoreceptor cells.

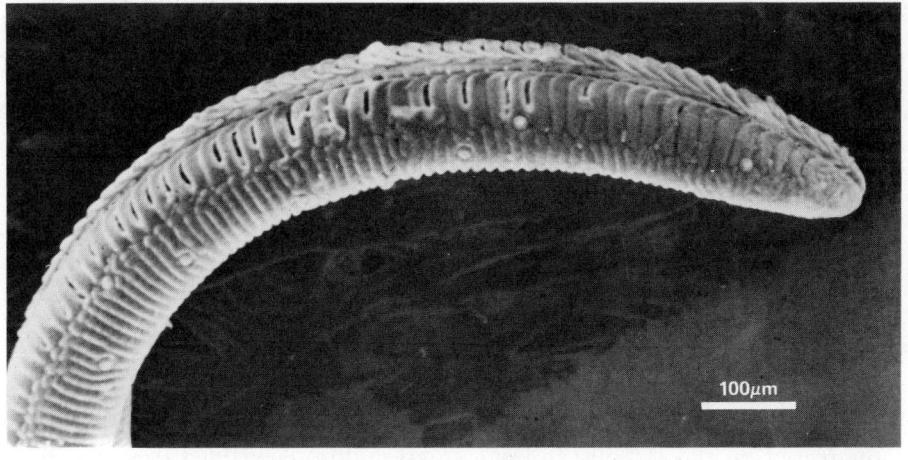

(a)

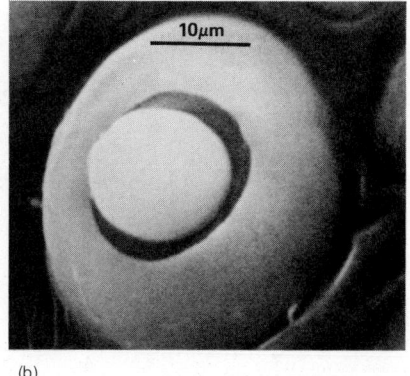

10μm

(b)

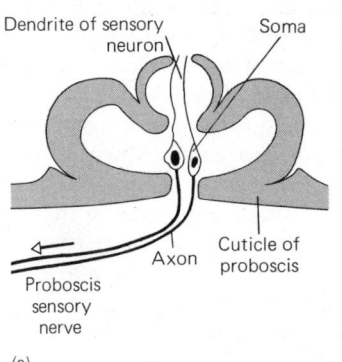

Dendrite of sensory neuron

Soma

Axon

Cuticle of proboscis

Proboscis sensory nerve

(c)

Figure 39-26 (a) Scanning electron micrograph of the tip of the proboscis of a black swallowtail butterfly. The button-like structures are chemoreceptors. (b) Closeup of one chemoreceptor. (c) A schematic cross section of the chemoreceptor. The hair-like dendrites of sensory neurons are the actual receptors.

SUMMARY

In order to produce adaptive responses, both behavioral and physiological, an animal must be able to detect changes inside its body and in the world around it. Sense organs are collections of cells specialized to react to particular forms of energy by producing electrical impulses in the nervous system. Sense organs may detect mechanical stimuli, light, heat, chemical changes, or electric current.

The actual receptors in sense organs may be the dendrites of sensory neurons or specialized non-nervous cells. Accessory structures may modify the external stimulus before it reaches the receptor. Adaptation at the level of the receptor, sensory neuron, or central nervous system changes the signal transmitted or received in the presence of a constant stimulus. The central nervous system usually has feedback control of the sensitivity of sense organs.

Mechanoreceptors detect mechanical distortions such as pressure, touch, sound, muscle stretch, movement of joints, and blood pressure. Hair cells are mechanoreceptors found in statocysts, lateral line organs, and the labyrinth and cochlea of the inner ear. Movement of the hair-like processes depolarizes or hyperpolarizes the hair cell membrane, and this potential change is transmitted to dendrites of a sensory neuron that synapses with the receptor.

Photoreceptors in eyespots, eyes, and ommatidia detect light by means of pigments whose structure is changed by electromagnetic radiation of appropriate wavelength. This photochemical reaction depolarizes the receptor cell membrane and causes a generator potential in an adjacent sensory neuron. Thermoreceptors are sensory neurons that respond directly to electromagnetic radiation in the infrared region.

Chemoreceptors are usually dendrites of sensory neurons, although the taste buds, and possibly some internal chemoreceptors, of vertebrates contain non-nervous receptor cells.

Electroreceptors are found in some bony and cartilaginous fishes; they detect electric currents generated by organisms.

The particular collection of sense organs in each animal species gives its members a unique perception of their bodies and of their environment.

Essay: Moths and Bats

Sense organs are vital in the perennial battle between predators and their prey. One of the most thoroughly studied systems of predator-prey sense organs is found in nocturnal bats and some of the moths they eat.

It is a warm spring night. As a bat awakes, the receptors in her empty stomach stimulate her to go in search of food. As she flies, she utters four or five short chirps per second, far too high-pitched for us to hear. The bat then listens for the echo of each chirp. Since the echoes are very faint, it seems strange that the bat is not deafened to them by the cry she made a split second earlier. But every time she makes a cry, an accessory structure moves in her middle ear, and the three bones that conduct sound from the tympanic membrane to the inner ear slide aside so that they no longer transmit sound to the cochlea. When the cry is finished, the three bones slide back into place, and the bat can hear the echo of her cry bouncing off nearby objects. Since she can tell both the time elapsed between uttering the cry and hearing its echo, and the direction of the echo, the bat can use the echoes of her cries as a sonar system to avoid obstacles and to detect other moving objects. Eventually the bat hears echoes which she identifies as coming from a fat noctuid* moth.

The moth is looking for a mate, using the sensitive chemoreceptors on his antennae to detect her scent. He also has very simple ears, with only two receptor cells apiece, but the cells are sensitive to the high-pitched sound of a bat. One receptor is sensitive to faint sounds; the other responds only to loud sounds. In effect, the first receptor detects distant sounds and the second detects nearby

---

* Noctuid Member of the Noctuidae, a family of night-flying moths with sound-detecting cells.

---

sounds. As the bat approaches, her cries stimulate the moth's "long-distance" auditory cells more and more strongly. If the bat approaches from the moth's left, his left-hand long-distance cell receives a stronger stimulus, and reacts more intensely. The moth's defensive reaction is to turn as he flies, until the stimulus intensity is the same in both ears. When this is the case he will be flying either directly away from or directly toward the bat. He can tell whether the bat is ahead of him or behind him because his ears are located at the rear of his thorax,* just in front of his abdomen, and the sound com-

---

* Thorax The part of an insect's body between the head and the abdomen.

---

ing from certain directions is deflected as the wings pass over the ears during flight. The information about differences in sound intensity, combined with information about wing position, give the moth an accurate indication of where the bat is. If the bat is still far away, the moth may escape by turning and flying away. However, when the bat is very close, the "loud" cells in the moth's ear begin to fire. This signals the moth's brain that immediate evasive action is necessary. He may execute a loop, or may close his wings in a crash dive. Sometimes these tactics succeed, sometimes not. The battle of the sense organs continues night after night, with survival as the stake, a strong selective pressure for sensitive and efficient sense organs.

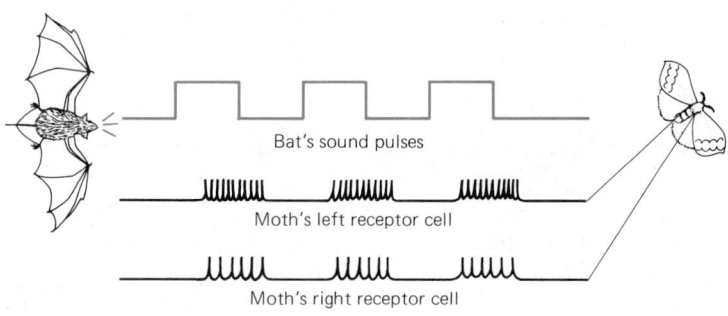

(a)

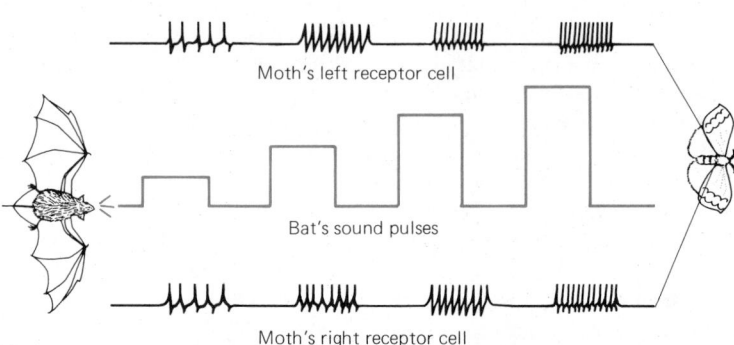

(b)

Figure 39-A Interactions between a bat and the moth it pursues. The nerve impulses recorded from the moth's right and left receptor cells are shown with the pulses of sound produced by the bat. (a) When a bat approaches from the left, the receptor on the left of the moth receives more intense stimulation and fires at a more rapid rate than the receptor on the right. (b) As the bat approaches the moth directly from the rear, the sounds emitted by the bat become louder (from the moth's point of view), and the moth's receptor cells fire at a greater rate.

OBJECTIVES

From your study of this chapter, you should be able to:

1. Write a short paragraph discussing the general importance of sense organs to an animal.

2. Name the five general types of receptor cells, and name some sense organs in which each is found.

3. Define transduction and generator potential, and explain the importance of each.

4. Define sensory adaptation; state the difference between rapidly adapting and slowly adapting receptors, and explain the relative advantages of each.

5. State what a proprioceptor is, and briefly describe the importance of the muscle spindle to the body.

6. Describe the roles of mechanoreceptors in the inner ear; explain how the ear recognizes differences in pitch, tone, and volume of sounds.

7. Sketch and label the eye of a vertebrate and that of an arthropod.

8. Name the two types of vertebrate photoreceptors, state their function, and describe the role of rhodopsin in vision.

SELF-QUIZ

Match the type of receptor on the right with the sense organs listed on the left:

___ 1. Nose
___ 2. Retina
___ 3. Pit organ
___ 4. Muscle spindle
___ 5. Semicircular canals

a. chemoreceptors
b. photoreceptors
c. electroreceptors
d. mechanoreceptors
e. thermoreceptors

6. The sensory neuron transmits patterns of action potentials that indicate (choose two):
   a. the type of stimulus received
   b. the intensity of stimulus received
   c. the duration of the stimulus
   d. the source of the stimulus
   e. the significance of the stimulus

7. A receptor that detects the position of parts of the body is called:
   a. a proprioceptor
   b. a hair cell
   c. a mechanoreceptor
   d. a muscle spindle
   e. a statocyst

8. Draw the basic structure of a vertebrate eye, and label the accessory structures that modify the stimulus, and the site of transmission of the stimulus to the central nervous system. Also name the two main types of receptors in the human eye and indicate their distribution.

9. Would a rapidly adapting or a slowly adapting receptor be more likely to respond to a stimulus in the manner shown in the figure below?

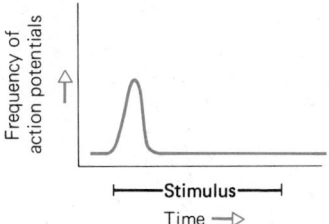

Choose one word from each pair in parentheses:

10. The visual cycle involving rhodopsin occurs in photoreceptors called (cones/rods). These receptors and their associated neurons are primarily used for vision in (bright/dim) light, and they are (good/poor) discriminators of color.

QUESTIONS FOR DISCUSSION

1. Name the two kinds of sensory information needed for the body to maintain equilibrium, and the anatomical structure(s) involved in detecting each type of information.

2. As tilt of the head increases, how do the nerve impulses convey the sensation of increased tilt to the brain?

3. Why do we say that sense organs detect changes in an animal's internal or external environment rather than just detecting the state of the environment?

4. In humans, the ability to taste phenylthiourea (PTU) is inherited as a dominant trait. Can the theory of taste presented in this chapter account for the genetic evidence?

5. Why do some people like certain foods that other people do not?

6. Why do people with vitamin A deficiencies have "night blindness?"

7. What visual problems would be experienced by people whose eyes lacked rods? Whose eyes lacked cones?

8. Why is it that you can sometimes see a dim star out of the corner of your eye, but the star seems to disappear if you look straight at it?

REFERENCES AND FURTHER READING

Bekesy, G. von. "The ear." *Scientific American*, August 1957. A description of the normal structure and functioning of the ear and how age, damage, and disease may affect hearing.

Gamow, R. I., and J. F. Harris. "Infrared receptors of snakes." *Scientific American*, May 1973.

Held, R. "Plasticity in sensory-motor systems." *Scientific American*, November 1965. Experiments showing that motor feedback plays an important role in how we perceive stimuli.

Hodgson, E. S. "Taste receptors." *Scientific American*, May 1961.

Horridge, G. A. "The compound eye of insects." *Scientific American*, July 1977. A description of the compound eye and of adaptations of the eyes of different insects to different ways of life.

Parker, D. E. "The vestibular apparatus." *Scientific American*, November 1980. Describes how the labyrinth interacts with other sense organs in detecting position and motion, and how experiments in space stations can extend our knowledge.

Roeder, K. D. "Moths and ultrasound." *Scientific American*, April 1965. Explains the anatomy of the moth's acoustic sensory cells and how these cells gather information to help a flying moth evade hungry bats.

Rushton, W. A. H. "Visual pigments and color blindness." *Scientific American*, March 1975. Describes how experiments with color-blind people have helped to elucidate both normal and abnormal color vision.

Wehner, R. "Polarized-light navigation by insects." *Scientific American*, July 1976.

CHAPTER 40

MUSCLES AND SKELETONS

An animal detects things with its sense organs, processes the information in its nervous system, and reacts with its muscles and glands. Muscles are part of the body's system of **effectors**: cells, tissues, or organs specialized to respond to stimuli detected by other cells. This definition excludes structures like nematocysts,* which both detect and react to stimuli, and organelles like cilia, which are merely parts of a cell, although both might be considered effectors under a less restrictive definition.

Muscles and glands are the most widespread effectors, but electric organs and light-emitting organs also respond to information gathered by sense organs:

RECEPTORS $\xrightarrow{\text{CONDUCTION}}$ CNS $\xrightarrow{\text{CONDUCTION}}$ EFFECTORS
e.g., eyes,   sensory   brain,   motor   muscles,
muscle   neurons   spinal   neurons   glands, etc.
spindles   cord

In this chapter we shall consider muscles and electric organs as effectors; endocrine glands are considered in Chapter 41.

Figure 40-1 Luigi Galvani, an eighteenth century scientist who discovered that electrical stimuli can cause muscles to contract. He invented a machine that produced electric sparks. One day the machine accidentally stimulated a freshly dissected frog's leg in Galvani's laboratory. (Smithsonian Institution)

---

*Nematocysts Stinging structures of cnidarians (jellyfish, sea anemones, etc.) (see Figure 27–7).

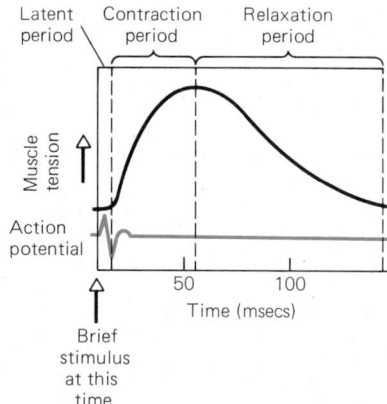

Latent period | Contraction period | Relaxation period

Muscle tension

Action potential

50   100

Time (msecs)

Brief stimulus at this time

Figure 40-2 Electrical activity (colored) and mechanical activity (black) recorded from a skeletal muscle fiber. When a skeletal muscle is electrically stimulated, it responds by firing an action potential (electrical activity) and by contracting with a twitch (mechanical activity).

40-A   Properties of Muscle Tissue

Muscle tissue has two distinguishing characteristics: contractility and electrical excitability (Figure 40-2). Excitability is also a property of neurons (see Chapter 37), and in both muscle and nervous tissue, it is due to energy stored as an asymmetrical distribution of ions across the plasma membrane, which results in an electrical potential difference across the membrane. A muscle membrane depolarizes in response to a chemical neurotransmitter* released at a **neuromuscular junction** (the synapse between a neuron and a muscle); in most types of muscle the membrane may also depolarize spontaneously, without stimulation by a neurotransmitter.

Although both nervous and muscular tissue transmit electrical impulses, most muscles also contract. There is no clear line between neurons and muscles on this basis, however, because some muscle cells in the vertebrate heart are specialized to conduct electrical impulses and cannot contract.

40-B   Types of Muscle

Vertebrates have three types of muscle: **cardiac muscle**, which makes up the heart; **smooth muscle**, which lines the walls of many of the internal organs; and **skeletal** or **striated muscle**, which is responsible for locomotion and change of position (Figure 40–3); striated muscle is normally the only type of muscle under the animal's voluntary control.

We shall consider all three types of vertebrate muscle, starting with smooth and cardiac muscle, which are least well understood, and working up to skeletal muscle, which has been studied most intensively. Then we shall see how skeletal muscle interacts with the skeleton to produce locomotion of the body or movement of its parts.

---

* Neurotransmitter Chemical that travels across the synaptic cleft from one cell, which has just fired an action potential, to the other cell, whose membrane potential is altered by the neurotransmitter.

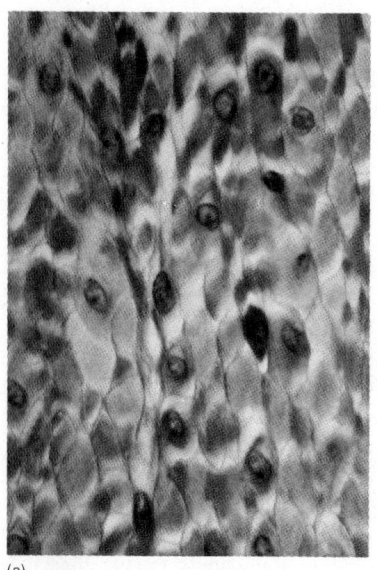

(a)

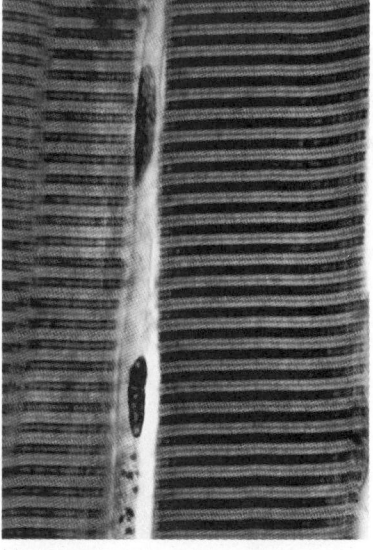

Nucleus

Intercalated disc

(b)   (c)

Figure 40-3 Types of muscle. (a) Smooth muscle. (b) Cardiac muscle. (c) Skeletal (striated) muscle. (a, c, Biophoto Associates; b, Carolina Biological Supply Company)

TABLE 40-1 DISTRIBUTION AND FUNCTIONS OF MUSCULAR TISSUES AND ORGANS IN THE HUMAN BODY

FUNCTION	STRUCTURES	MUSCLE TYPE
Circulation	Heart	Cardiac
	Arteries, veins	Smooth
Excretion	Walls of renal pelvis, ureter, bladder	Smooth
	Internal sphincter between bladder and urethra	
	External sphincter near exit from body	Striated
Digestion	Tongue	Striated
	Muscles of jaw, pharynx	
	External sphincter of anus	
	Lining of gut (esophagus, stomach, intestines)	Smooth
	Internal sphincter of anus	
Ventilation	Diaphragm	Striated
	Intercostal muscles	
Ejaculation	Walls of genital ducts	Smooth
	Skeletal muscles at base of penis	Striated
Parturition (childbirth)	Uterine wall	Smooth
	Cervix	
	Abdominal muscles	Striated
	Diaphragm	
Heat production	Skeletal muscles (exercise, shivering)	Striated
Maintenance of posture Change of position Locomotion	Skeletal muscles	Striated

Smooth Muscle

Smooth muscle is found in many internal organs, including the walls of the arteries, veins, digestive tract, bladder, and reproductive organs. The most common function of smooth muscle is to exert pressure on the contents of the space it surrounds. Smooth muscle constricts blood vessels, moves food along the gut, and expels urine from the bladder, semen from the seminal vesicle, or a baby from the uterus.

Smooth muscle is made up of sheets of muscle cells, each packed with contractile proteins. Neurons can often be seen between the muscle cells, but these end in fine branches rather than in the definite synapses visible between neurons and striated muscle (Figure 40-4).

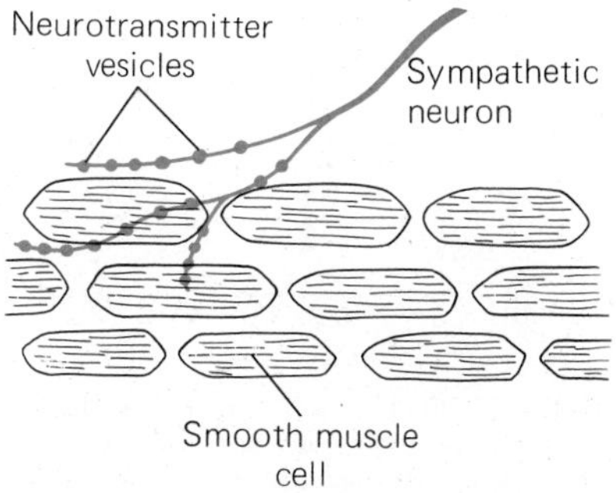

Some smooth muscles contract spontaneously, without nervous stimulation; others contract only in response to stimulation by nerves or hormones. Most smooth muscle is **innervated** (supplied with nerves) by the sympathetic* nervous system, in which noradrenalin is the neurotransmitter released at the effector. Single sympathetic neurons may activate many different cells in a smooth muscle because the neurons branch out in an indefinite pattern and release noradrenalin from vesicles scattered along the last few micrometres of the axon. From these vesicles the transmitter diffuses across to the muscle and stimulates the muscle cells. Because noradrenalin is almost the same molecule as the hormone adrenalin, it is not surprising that smooth muscle also reacts to adrenalin arriving from the adrenal glands by way of the bloodstream.

Smooth muscle produces a gradual contraction of variable force. A smooth muscle cell's electrical activity is not an all-or-none phenomenon, but a graded response, so that the extent of contraction depends on how much the cell is depolarized. Contraction of smooth muscle is often sustained for long periods of time without fatigue.

Figure 40-4 Diagram of a sheet of smooth muscle cells with the ending of a sympathetic neuron.

---

*Sympathetic nervous system Part of the autonomic nervous system, which governs internal bodily functions (Section 38-G).

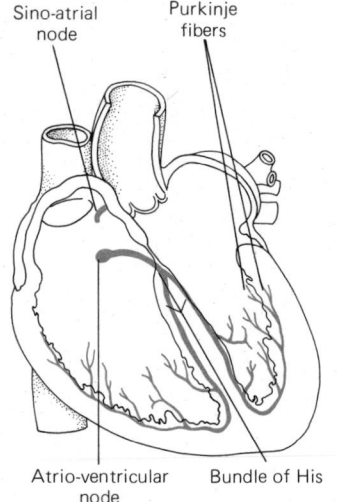

Sino-atrial node

Purkinje fibers

Atrio-ventricular node

Bundle of His

Figure 40-5 The electrical conduction system of the human heart.

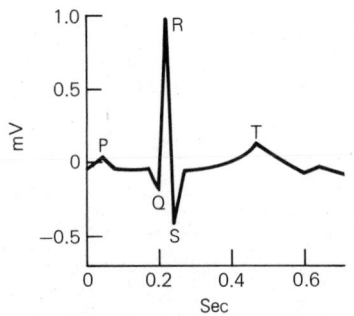

Figure 40-6 An electrocardiogram (EKG). The waves of electrical excitation that spread through the heart and cause it to beat are also conducted throughout the body in the body fluids. These currents can be monitored by electrodes attached to the skin. An EKG is used to determine whether the heart's electrical activity is normal. The P wave is produced by depolarization of the atria, Q, R, and S by depolarization of the ventricles. T is a result of ventricular repolarization.

Cardiac Muscle

Cardiac muscle occurs only in the vertebrate heart, which is made up largely of muscle cells arranged in long columns of fibers.

Nowhere is the similarity between nerve and muscle tissue seen more clearly than in cardiac muscle. An electrical impulse spreads all over the membrane of any muscle cell, but in the heart some cells (e.g., those of the **Purkinje fibers**) are so specialized to conduct impulses that they have lost their contractile proteins and behave much more like neurons than like muscle cells.

The noncontractile conducting cells of the heart are part of the mechanism that controls the heartbeat. Each beat is initiated by spontaneous electrical activity of the heart's pacemaker, the **sinoatrial node** in the wall of the right atrium (Figure 40-5). (How the pacemaker cells fire in their rhythmic and spontaneous pattern is a complete mystery.) From the sinoatrial node, the impulse must spread to all parts of the atrium, and then to the **atrioventricular node**, located near the atria on the partition between the two ventricles. From here the impulse spreads rapidly through the ventricular walls, triggering simultaneous contraction throughout the ventricles (Figure 40-6).

There are no nerves to carry the impulses from the pacemaker to the rest of the heart; the depolarization travels by way of the membranes of the heart muscle cells. The impulse travels from one muscle cell to another electrically, rather than by way of chemical transmitters such as those at most synapses between neurons.

Any cardiac muscle cell can pass its electrical activity (which is much like an action potential in an axon) to any adjacent cell, but some cells are specialized so that transmission between them is particularly fast. Such cells are joined together by modifications of their plasma membranes called **gap junctions**, which contain tiny pores through which ions, and therefore electric current, can pass from one cell to another. There is very little electrical resistance between cells linked by gap junctions, and the depolarization travels rapidly from one cell to the next. A column of cells connected by gap junctions conducts an impulse faster than a similar column without them.

Gap junctions occur between many cells in the heart. However, they are most common between cells in the Purkinje fibers, the cells that conduct impulses throughout the walls of the ventricles. Gap junctions are arranged in such a way that a single impulse leaving the pacemaker reaches different parts of the heart at precise times regardless of distance from the pacemaker. Thus, for instance, two parts of the ventricles that lie at different distances from the pacemaker both contract at the same time.

The innervation of the heart is complex and poorly understood. The main nerve to the heart is the tenth cranial nerve, the **vagus**. It contains branches of the sympathetic system, whose activity speeds up the heart, as well as branches of the parasympathetic system, whose impulses slow the heart rate. However, if all of the nerves to the heart from the central nervous system are removed, a vertebrate can survive in an apparently normal condition, and its heartbeat alters with the changing demands of exercise just as it does in an intact animal. This remarkable situation is possible because, unlike skeletal muscle, cardiac muscle contracts without nervous stimulation (even single cells contract spontaneously if they are isolated from the rest of the heart). The rate and force of heartbeats are governed partly by hormones that reach the heart through the bloodstream, and partly by reflexes in a system of nerves that lie completely within the heart and work even though they are not in contact with the rest of the nervous system.

Contraction in cardiac muscle is basically similar to that in other muscles, but there are some differences in detail. For instance, cardiac muscle must clearly be highly resistant to fatigue if it is to beat regularly throughout an animal's lifetime. Not surprisingly, cardiac muscle has abundant mitochondria and a relatively enormous blood supply, which guarantees adequate oxygen.

Skeletal Muscle

The muscles that attach to, and move, the skeleton of a vertebrate are called **skeletal** or **striated** (striped) or **voluntary** muscles. Most of the research on vertebrate muscle has been done on this readily available tissue.

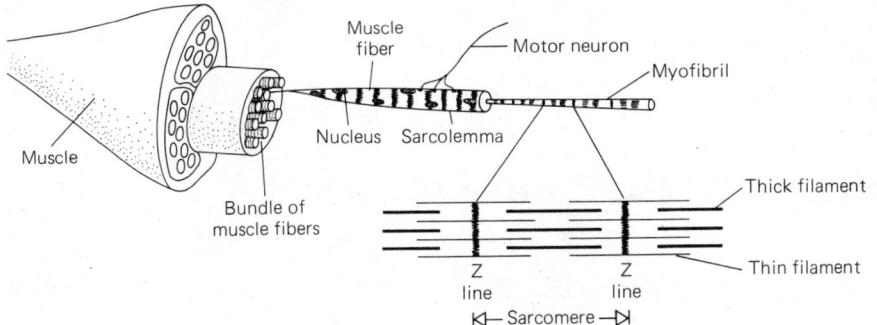

Figure 40-7 Anatomy of a vertebrate skeletal muscle. Sarcomeres are arranged end to end, single file, forming a myofibril. Bundles of myofibrils make up a syncytial muscle fiber, in which all the fibrils and nuclei are enclosed within a common sarcolemma. A muscle consists of dozens or hundreds of bundles of muscle fibers.

A skeletal muscle, such as the biceps in the upper arm, is made up of numerous **muscle fibers**, each of which runs the entire length of the muscle. Each skeletal muscle fiber originates from the fusion of several embryonic muscle cells called **myoblasts**; hence a fiber is a **syncytium**, a structure containing many nuclei that are not separated from one another by plasma membranes. The nuclei of the original cells are arranged along the outside of the fiber (see Figure 40–3c). Each muscle fiber is innervated by one (or more) motor neuron, whose axon may have terminal branches forming neuromuscular junctions with many fibers. The axon ends close to the muscle fiber's **sarcolemma**, a membrane analogous to the plasma membrane of a single cell, which encloses the whole muscle fiber. Several to over a hundred muscle fibers may be innervated by a single motor neuron.

Each striated muscle fiber consists of a bundle of **myofibrils**; each myofibril is made up of units called **sarcomeres** strung single file along the length of the fiber. A sarcomere is the part of the myofibril between two adjacent **Z lines**, protein-containing structures extending across the myofibril (Figure 40–7).

40-C Muscle Contraction

Sliding Filament Mechanism

In the 1950s A. F. Huxley and R. Niedegarde found that the Z lines of skeletal muscle fibers move closer together when a muscle contracts. H. E. Huxley and Jean Hanson suggested that this was because each myofibril is made up of filaments that slide along each other and mesh together more closely as the muscle contracts (Figure 40–8). This **sliding filament** theory of muscle contraction has been borne out by a large body of experimental work.

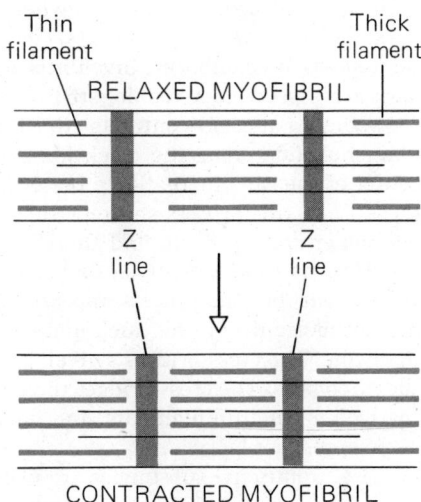

Figure 40-8 Sliding filament model of muscle contraction. The Z lines are attached to the thin filaments. The muscle contracts when the thick and thin filaments slide past each other, reducing the distance between adjacent Z lines. The filaments do not shorten: the myofibril contracts because all its sarcomeres contract.

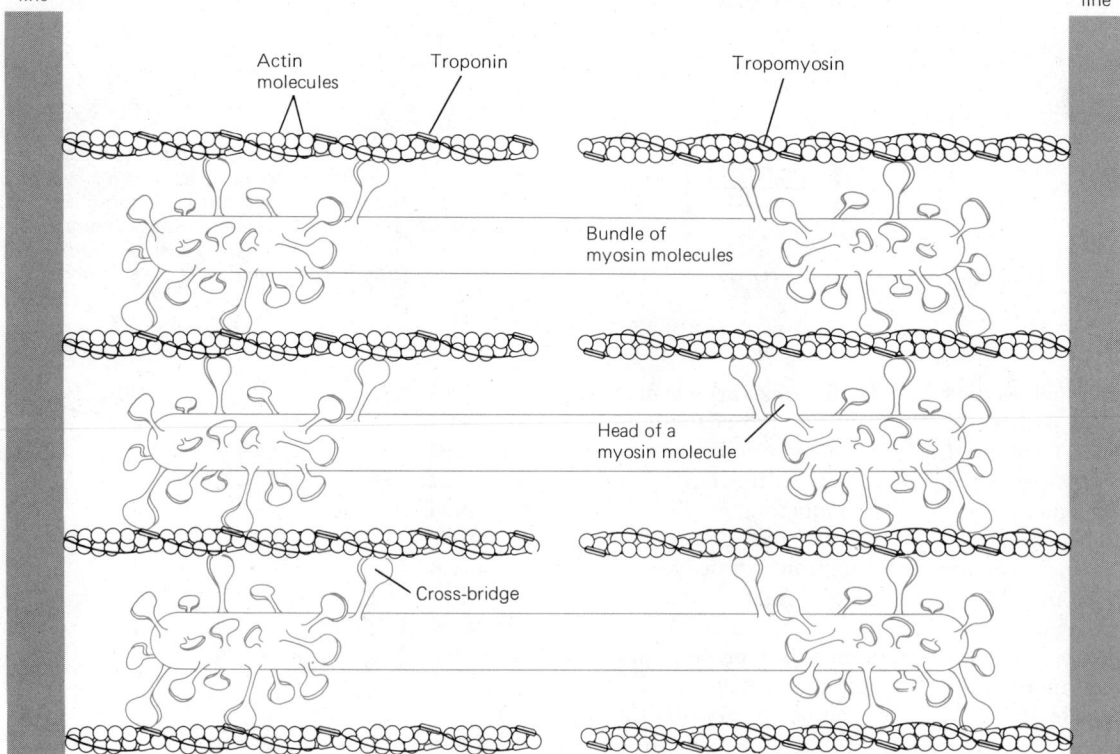

Figure 40-9 Molecular structure of thick and thin filaments. The filaments interact chemically when the heads of the myosin molecules attach to actin molecules, forming cross bridges between the filaments. The heads then swivel, causing the muscle to contract.

Muscle contraction involves two kinds of protein filaments. Attached to each side of a Z line are **thin filaments**, made of a twisted double strand of the protein **actin**, with smaller amounts of the proteins **troponin** and **tropomyosin** (Figure 40–9). The thin filaments extend less than halfway to the center of a sarcomere from the Z lines at either end. When the muscle is at rest, the thin filaments overlap somewhat with **thick filaments**, which lie in the middle of the sarcomere. The thick filaments consist of many molecules of the protein **myosin**, twisted into a rope-like structure with the "heads" of the molecules sticking out at the edges of the filament (Figure 40–9). The area where the thick and thin filaments overlap appears dark under the microscope. A skeletal muscle fiber appears striped because the sarcomeres of neighboring myofibrils are lined up side by side, so that their light and dark areas produce a visible, striped pattern extending across the fiber.

When a muscle contracts, the filaments stay the same length, but the thin filaments slide along the thick filaments, and their free ends move closer to the center of the sarcomere. Since the other ends of the thin filaments are attached to the Z lines, this process shortens the sarcomere. Shortening each sarcomere shortens the entire myofibril, and therefore the whole muscle.

How do the filaments slide past each other? Not all of the details have been worked out, but the process appears to work something like this: the heads of the myosin molecules in the thick filaments form cross-bridge attachments to the thin filaments. The cross-bridges swivel, pushing the thin filaments toward the center of the sarcomere. The cross-bridges then detach, swivel back to their original positions, reattach to the thin filaments, and push them still further toward the center of the sarcomere.

Two additional substances are needed for this sequence of events. One is ATP, which provides energy for the swiveling of the cross-bridges; the other is calcium ion (Ca²⁺). In the swiveling process, an ATP molecule binds to the head of each myosin molecule. The myosin-ATP complex is then ready to form a cross-bridge with actin molecules in the thin filaments. In a resting muscle, however, cross-bridges cannot form, apparently because the binding site on the actin molecules is blocked by tropomyosin molecules.

The role of calcium is to remove this barrier. When the muscle is stimulated to contract, calcium ions are released into the sarcomere from nearby storage areas and bind to troponin molecules. This causes the tropomyosin molecule attached to a calcium-troponin complex to move down into a groove in the thin filament, exposing the binding sites on the actin molecules. Now the ATP-myosin complex can bind to the actin. The ATP hydrolyzes, releasing energy that causes the cross-bridge to swivel and push the free end of the thin filament toward the center of the sarcomere. Then, a new ATP binds to the myosin head, which detaches, returns to its original position, reattaches to the thin filament, and swivels once more, driving the thin filament still closer to the center of the sarcomere. Since there are many cross-bridges and their movements are not synchronous, the thin filament cannot slip back while individual myosin heads are detached. During contraction, when abundant ATP and calcium are present, each cross-bridge breaks and reforms some 50 to 100 times a second, so that the thin filament swings smoothly past the thick filament, moving through a distance equivalent to the lengths of many individual cross-bridges.

One thing that puzzled researchers for a long time is that ATP is needed for a muscle to relax as well as for it to contract, whereas we might ordinarily expect one reaction to be active and its reverse passive. The phenomenon of **rigor mortis** makes it clear that this is not the case. After death, as ATP disappears from the body, the muscles lose their ability to contract or to relax. In fact, they become locked in whatever position they occupied when the ATP ran out. ATP is needed for muscle relaxation because the actin-myosin complex, formed when the muscle contracts, is chemically stable until a new ATP binds with the myosin molecule and breaks the cross-bridge.

Control of Contraction

A muscle contracts if sufficient ATP and Ca²⁺ are present within the sarcomere. Not surprisingly, research has shown that calcium is released into the sarcomere when a muscle is activated and removed when it relaxes.

A skeletal muscle is normally activated by the arrival of an action potential at the neuromuscular junction, the synapse between the motor neuron and the muscle fiber. The neurotransmitter acetylcholine released from the nerve ending depolarizes the sarcolemma. Depolarization spreads over the surface of the fiber and reaches the **T system**, a series of **transverse tubules**, indentations of the sarcolemma. The T system comes into intimate contact with the membranes of the endoplasmic reticulum, which in skeletal muscle is called the **sarcoplasmic reticulum** (Figure 40–10). The channels of the sarcoplasmic reticulum act as a reservoir where calcium ions are stored out of contact with the filaments. Depolarization of the

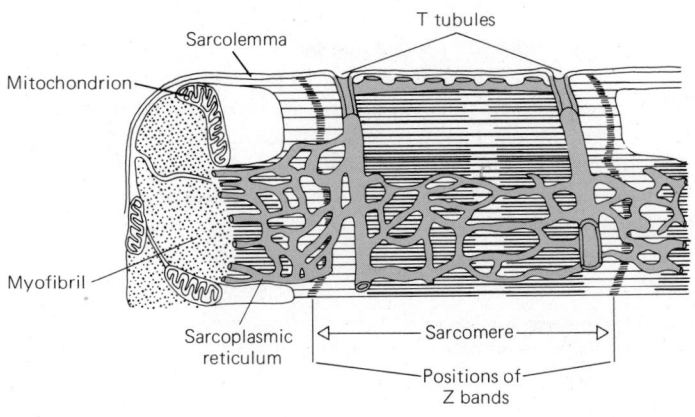

Sarcolemma
T tubules
Mitochondrion
Myofibril
Sarcoplasmic reticulum
Sarcomere
Positions of Z bands

Figure 40-10 T system and sarcoplasmic reticulum in part of a muscle fiber. The sarcoplasmic reticulum is a reservoir of calcium ions. When the sarcolemma is depolarized, the depolarization spreads down the membranes of the T tubules to the sarcoplasmic reticulum, which releases calcium ions into the myofibril.

sarcolemma and transverse tubules increases the permeability of the sarcoplasmic reticulum to calcium ions, allowing calcium to leak into the cytoplasm and initiate contraction. Calcium is continually pumped back into the channels of the sarcoplasmic reticulum by active transport, but as long as stimulation continues, calcium leaks out faster than it can be pumped back in, and the concentration of calcium around the troponin molecules remains high. As soon as the stimulation stops, the membrane of the sarcoplasmic reticulum is no longer permeable to calcium, and the calcium pump moves most of the Ca^{2+} ions back inside the reticulum. The tropomyosin then moves out and blocks the binding sites on the actin once more, and the fiber relaxes.

Graded Response of an Intact Muscle

Our everyday experience tells us that muscles do not always contract or relax completely. They contract to different degrees as the situation demands. Various mechanisms for producing this graded response are found in different animals.

In arthropods,* there are inhibitory* synapses onto some muscles as well as the more usual excitatory synapses (see Section 37-F). In this situation, the degree of contraction depends upon the balance between inhibitory and excitatory impulses acting on the muscle at any one time.

Most invertebrate skeletal muscles consist of fibers that can produce graded responses. A sequence of impulses in the motor nerve produces ever greater depolarization of the muscle membrane; this is reflected in an increase in contraction of the muscle fiber with time, and muscle tension builds up slowly. Muscle fibers of this type, known as "slow" fibers, are also found in vertebrates, where they produce graded contractions in a similar manner. They are common in fish, birds, and reptiles. Most mammalian skeletal muscle fibers, however, are of the "fast" or "twitch" variety. They cannot produce a graded contraction, but respond to stimulation in an all-or-nothing manner. The main advantage of this type of muscle is that it contracts faster than "slow" muscle, and so permits rapid movement. The flight muscles of insects contract even more rapidly than mammalian "fast" muscles.

The possession of "fast" fibers makes producing a graded muscle response more complicated. Each "fast" fiber is innervated by a branch of a motor neuron. The neuron and the fibers it innervates are known as a **motor unit**. Since each skeletal muscle is made up of hundreds of motor units, the degree of contraction of the whole muscle can be altered by controlling the number of motor units that are active at any one time. Sustained contractions are possible if the motor units fire in turn. For example, in the posture muscles, which work for long periods of sitting or standing, several different fibers are active at any one time. These then relax while others contract. The result is that the entire muscle remains partially contracted, but no one fiber need stay contracted until it exhausts its supply of ATP. The disadvantage of this system is that it requires many more motor neurons than does a "slow" muscle system, since many different motor units are necessary. This is one reason that mammals have so many more central and peripheral neurons than do lower vertebrates or invertebrates.

Tetanus and Fatigue

A single impulse from its motor neuron causes a vertebrate "fast" fiber to produce an all-or-none twitch, after which it relaxes. Depolarization of the muscle membrane is followed by a **refractory period**, during which the membrane cannot be electrically stimulated. All the electrical events in the muscle, however, take a short time compared to the mechanical sequence of contraction and relaxation (Figure 40–11). Normally, a muscle is stimulated, not by one impulse in the motor neuron, but by a sequence of impulses. Under such conditions, the muscle is electri-

* Arthropods   Crabs, barnacles, insects, and so on.
* Inhibitory   Making the postsynaptic membrane require more excitatory input before it will fire an action potential.

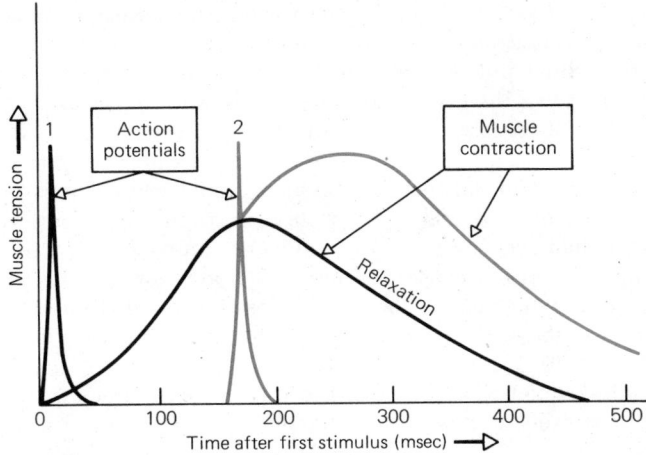

Figure 40-11 A muscle contracts more strongly in response to more than one action potential. A single action potential (1) causes the black contraction/relaxation curve. If a second action potential (2) arrives before the muscle has relaxed from the first one, the muscle contracts further (colored curve) before it relaxes.

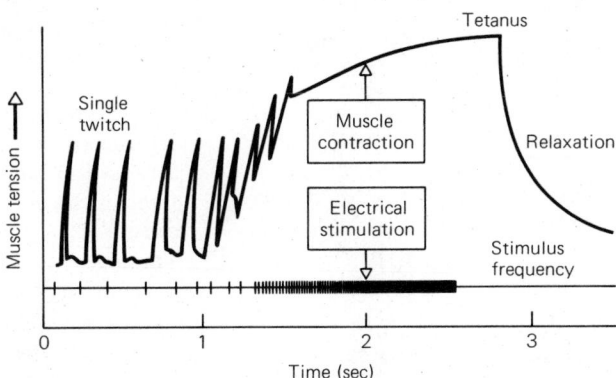

Figure 40-12 Tetanic contraction. The muscle's motor neuron is stimulated electrically. When the rate of stimulation is low, the muscle contracts in individual twitches. As the rate of stimulation increases, the muscle does not have time to relax fully between stimuli. Finally, a smooth, sustained, tetanic contraction occurs.

cally stimulated before it has relaxed from the previous impulse, and the individual twitches fuse into a smooth contraction called **tetanic contraction**, or **tetanus**. A tetanic contraction is smooth and steady and, furthermore, induces a fiber to produce more force than it does during a single twitch (Figures 40–11, 40–12); how it does this is not understood.

A skeletal muscle that is stimulated to contract repeatedly for a long period will enter a phase where it no longer relaxes between contractions. Eventually, it becomes unable to respond to further stimulations, and it gradually returns to its resting length. This phenomenon, known as **fatigue**, occurs only in striated muscle. Complete fatigue seldom occurs in an intact organism, but it is easily induced in a muscle that has been removed from the body.

40-D How Muscles and Skeletons Interact

Muscular contraction can move parts of the body only because muscles work against skeletons. A **skeleton** may be defined as anything on which a muscle exerts force. Some structures that act as skeletons according to this definition are not part of what is usually described as an animal's skeleton. For instance, the muscles that surround a blood vessel work against the walls of the vessel, and against the blood itself, when they contract and reduce the diameter of the vessel. When the muscles of the uterus contract during childbirth, they could do no useful work if there were no baby to push against. Muscles that cause movement inside the body shorten so little, and exert so little force when they contract, that only their attachment to nearby cells gives them enough leverage to operate. The muscles used in locomotion, on the other hand, may contract forcefully and dramatically, and they usually exert their force on what we generally think of as a skeleton: a hydrostatic skeleton, such as the fluid-filled cavities of a cnidarian or annelid; an exoskeleton, such as the cuticle of an insect or the shell of a snail; or an endoskeleton, such as the bones of a vertebrate.

Antagonistic Muscles

In systems that lack hard skeletons, including soft-bodied invertebrates and the internal organs of vertebrates, muscles are arranged in circular and longitudinal sheets. Contraction of longitudinal muscles, which run lengthwise along the body (or organ), makes it shorter and wider. Contraction of circular muscles, which run

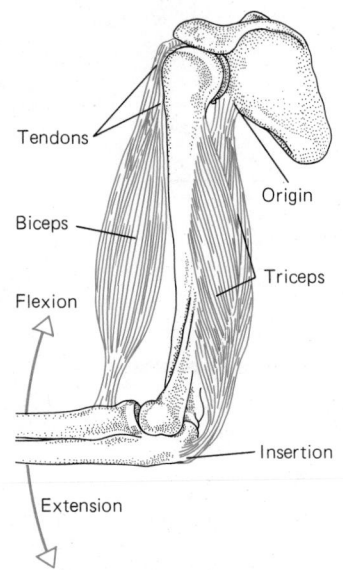

Figure 40-13 Skeletal muscles in the human upper arm. Each muscle is attached to the skeleton by tendons at its insertion (the end attached to the bone that moves) and origin (the end attached to the less mobile bone).

Labels on figure: Tendons, Biceps, Flexion, Origin, Triceps, Insertion, Extension

around the body (or organ), makes it longer and narrower. Such **antagonistic** action of two sets of muscles with opposite effects controls all movements. The antagonistic action of longitudinal and circular muscles against the contents of the alimentary canal moves food down the gut in most animals. Similar forces exerted against the fluid-filled coelom of an earthworm permit the worm to move (see Figure 28–6).

Where muscles are attached to hard skeletons, such as the exoskeletons of arthropods or molluscs or the endoskeletons of vertebrates, the muscles may be arranged in bundles rather than sheets. Antagonistic muscles run parallel across a joint in the skeleton. The muscle that causes the joint to stretch out is called an **extensor**, and its **antagonist**, which causes the joint to close up, is called a **flexor**. The biceps is the main flexor across the human elbow joint, and the triceps is its antagonistic extensor (Figure 40–13).

Exoskeletons and endoskeletons may differ in the way muscles attach to the skeleton. In vertebrates most skeletal muscles are attached at a single point; however, the greater relative surface area of an invertebrate exoskeleton allows the attachment of muscle fibers all along the skeleton of a particular limb. The great strength of an insect relative to its size results partly from the variety of different types of muscle attachment possible with an exoskeleton.

Reciprocal Inhibition

Locomotory muscles are so powerful that if a flexor and its antagonistic extensor were both to contract strongly at the same time, they could easily break a bone. **Reciprocal inhibition**, a process involving a reflex arc from each muscle to its antagonist, prevents this from happening. Muscle spindle receptors detect how far the muscle is contracted (see Section 39-B). The sensory neurons from the spindles in one muscle connect with interneurons* in the spinal cord. The interneurons in turn hyperpolarize* the motor neurons to the antagonistic muscle and so reduce their activity. Thus any stimulus that causes a muscle to contract also inhibits the contraction of its antagonist, and the joint moves smoothly.

This is the simplest reflex involved in locomotion. Much more complex reflexes come into play when we walk. For instance, reflexes from the opposite limb ensure that one leg moves after another. Such reflexes are even more complicated in animals that have many legs or that use their tails for balance.

40-E   The Vertebrate Skeleton

The skeleton of a vertebrate consists of two major portions. The **axial** skeleton runs along the axis of the body and includes the skull, other bones in the head, the backbone, ribs, and tail. The **appendicular** (limb) skeleton includes the bones in the limbs and in the girdles that attach the limbs to the backbone; the pectoral girdle includes the collarbones and shoulder blades, and the pelvic girdle includes the large, fused hip bones (ilium, ischium, and pubis) (Figure 40–14).

The bones and cartilages in our bodies permit the locomotory muscles to perform their allotted tasks, but they serve many other functions as well. First, they provide a framework that supports the body against the pull of gravity; land vertebrates must have skeletons more substantial than those of vertebrates that live supported by the buoyancy of water. Part of the skeleton also provides protection for delicate internal organs. The skull serves the dual functions of providing anchorage for the muscles of the face and jaw as well as protecting the brain and major sense organs. Many other parts of the skeleton also have both anchoring and protective functions; the backbone surrounds the spinal cord, and the pectoral girdle protects the heart.

---

* Interneurons   Neurons in the central nervous system that are neither sensory nor motor, but transmit information between other neurons.
* Hyperpolarize   Increase the membrane potential of.

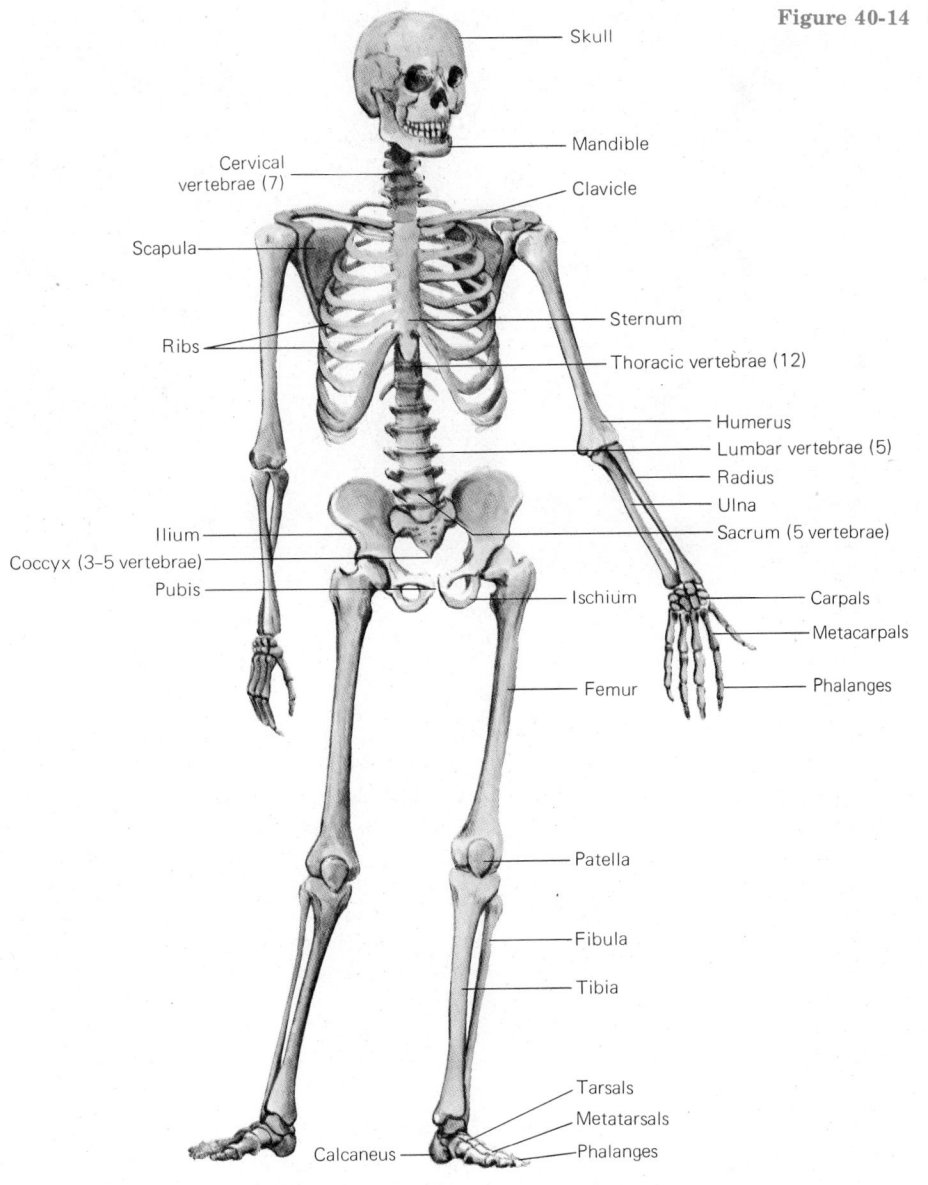

Figure 40-14 Some major bones of the human skeleton.

Skull

Mandible

Cervical
vertebrae (7)

Clavicle

Scapula

Sternum

Ribs

Thoracic vertebrae (12)

Humerus

Lumbar vertebrae (5)

Radius

Ulna

Ilium

Sacrum (5 vertebrae)

Coccyx (3–5 vertebrae)

Pubis

Ischium

Carpals

Metacarpals

Femur

Phalanges

Patella

Fibula

Tibia

Tarsals

Metatarsals

Calcaneus

Phalanges

Joints in the Vertebrate Skeleton

Joints between bones in the skeleton can be of many different types and degrees of rigidity. At one extreme are the **sutures**, wiggly lines in the skull where bones meet at interlocking projections that hold them very tightly together. Most bones, however, are joined to each other by **ligaments**. An extremely flexible joint in females is the one between the two bones in the front of the hip girdle, which are held loosely together by a ligament that stretches considerably and allows the bones to separate during childbirth. The joints of the limbs must permit smooth movement in various directions. The ends of the bones in these joints are covered by a smooth elastic sheet of connective tissue, and the whole joint is surrounded by a sac filled with **synovial fluid**, which lubricates the joint (Figure 40–15).

A skeletal muscle is attached to the skeleton either directly or by way of a **tendon**. In some cases—for instance, with the muscles that move our fingers—the tendons may be almost as long as the muscle.

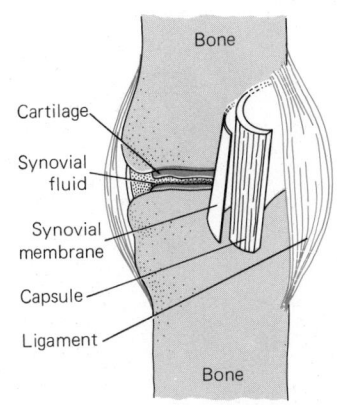

Bone

Cartilage

Synovial
fluid

Synovial
membrane

Capsule

Ligament

Bone

Figure 40-15 Structure of a joint. Cartilage cushions the ends of the bones. The space between the bones is filled with fluid, kept in place by a membrane. A tough, fibrous capsule surrounds the membrane. Ligaments bind the bones together and limit the movement between them.

**CHAPTER 40
MUSCLES AND
SKELETONS** 657

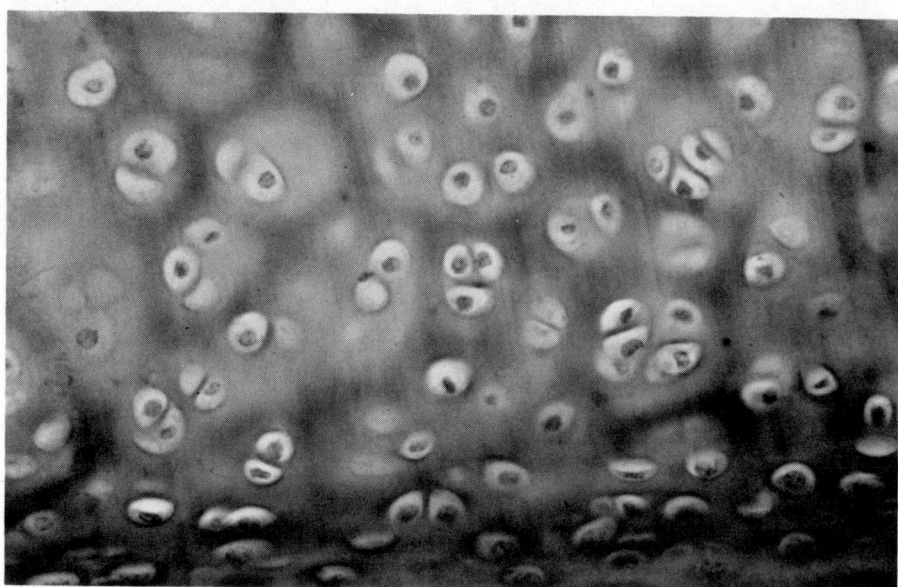

Figure 40-16 Cartilage. (Biophoto Associates)

40-F   Connective Tissue

Various types of **connective tissue**, composed mainly of intercellular* substances secreted by scattered cells, make up the bones, cartilages, tendons, and ligaments. In these structures, the intercellular substance contains fibers of a tough, elastic protein, **collagen** (which is also the main ingredient of gelatin). Bone is hard and brittle because it has mineral deposits in addition to these fibers. Cartilage has a great deal of a firm intercellular jelly surrounding the protein fibers and cells (Figure 40–16), while tendons and ligaments are mostly composed of fibers, with very few cells.

Cartilage

Unlike bone, cartilage receives no blood supply; its cells rely on diffusion of nutrients from capillaries in surrounding tissues. Cartilage makes up the entire skeleton in members of the class Chondrichthyes* and in all early vertebrate embryos. During embryonic development of vertebrates with bony skeletons, minerals are deposited in most of this embryonic cartilage, a blood supply and other typical features of bone appear, and the cartilage becomes bone. Cartilage persists in the adult only in areas where flexibility is necessary. These areas include:

1. The ends of the bones where they form synovial joints.
2. The discs between the vertebrae in the backbone.
3. The ends of the ribs where they join the breastbone.
4. The rings that thicken the walls of the trachea (windpipe) and keep it from collapsing.
5. The larynx, or "voice box" in the throat, at the anterior end of the trachea.
6. The external ear.
7. The eustachian tube, which connects the throat to the middle ear.
8. The tip of the nose.

---

* Intercellular Between cells.
* Chondrichthyes Sharks, skates, and rays.

Bone

Although the mineral matter of the dried human skeleton weighs only about 5 kg, the skeleton of a living human being is much heavier because bone is a living tissue that contains cells, blood vessels, nerves, fluid, and fat deposits.

Bone varies in structure depending upon its position and function in the body. At one extreme is **spongy bone**, composed of an irregular network of mineralized bars, and at the other is **compact bone**, composed of tubular units called Haversian systems (Figure 40–17).

The hard part of bone is made up of organic matter and inorganic salts. Most of the organic matter consists of collagen fibers that give bone most of its ability to withstand pull (tension). The inorganic salts are mainly a complex phosphate of calcium and smaller quantities of other ions. These give bone its considerable ability to withstand compression and side-slippage.

Distributed throughout a bone are the cells that lay down the collagen and mineral deposits. These are obviously necessary in a growing bone, but they also last into adult life and are responsible for repair and replacement of broken bone and for the formation of **calluses**, which develop at points of pressure on a bone.

NON-SKELETAL FUNCTIONS OF BONE

The spaces in bone are filled with blood vessels and **marrow**, a soft tissue with a number of different functions. Some marrow is primarily a fat depot; some produces red blood cells, platelets, white blood cells, and antibodies of the immunological system.

Bone in vertebrates is also involved in regulating the concentration of calcium ions in the blood. The calcium phosphate in bone is in equilibrium with that in the surrounding extracellular fluid. Thus, if the calcium level in the extracellular fluid rises, some of it is deposited in bone, and if the calcium concentration in the extracellular fluid falls, calcium from the bones dissolves into the fluid. The chemical equilibrium between solution and deposition determines the general level of calcium in the extracellular fluid and in the blood.

Fine tuning of the calcium level in the body fluids is under the control of hormones. **Parathyroid hormone**, secreted by the parathyroid glands, which lie behind the thyroid gland in the neck (see Figure 41–1), stimulates the cells in bone to dissolve additional calcium out of the mineral deposits and release it into the blood. A drop in the blood calcium level stimulates the parathyroid glands to produce this hormone; calcium is then released from the bone to make up the deficit.

If the blood calcium level rises, the hormone **calcitonin**, secreted by the thyroid gland, stimulates the bone cells to increase deposition of minerals, so that the excess calcium is removed from the blood.

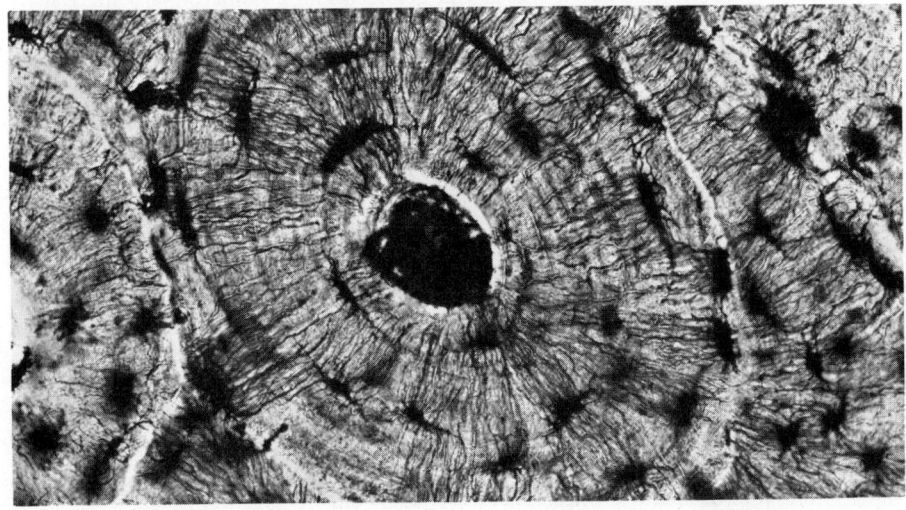

Figure 40-17 A cross section of compact bone. The dark oval in the middle of the picture is a Haversian canal, a channel for blood vessels and nerves. The many light-colored layers wrapped around the Haversian canal are mineral deposits, laid down by living cells that inhabit the small, dark, spider-shaped spaces. (Carolina Biological Supply Company)

Essay: Electric Fish

The usual function of a muscle is to move part of the body. In some vertebrates, however, skeletal muscles have been modified for another function: producing electrical discharges. Electric organs have evolved independently in several groups of fishes, including the Chondrichthyes (cartilaginous fishes) and several different families of the Osteichthyes (bony fishes).

In the electric ray *Torpedo* (class Chondrichthyes), the electric organ, which can produce a powerful electric shock, consists of a row of parallel muscle fibers (known as **electroplaques**) that have lost their contractile proteins and consist of little more than large neuromuscular junctions (Figure 40–A). Each electroplaque is innervated on the same side as all the others by a motor neuron. The electric organ works much like a battery with the electroplaques coupled in series. The motor neurons fire synchronously, and the depolarizations of the electroplaques sum to produce voltages up to 50 volts with several amperes of current. *Torpedo* uses its electrical discharges to stun its prey or to discourage invaders. Why it doesn't electrocute itself is still a mystery.

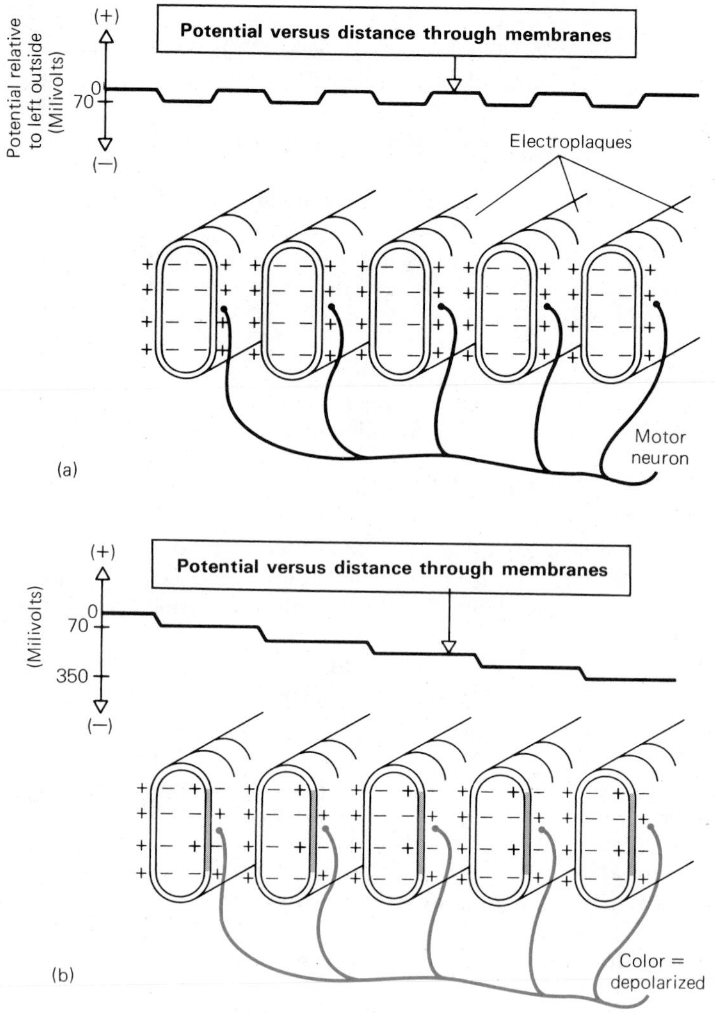

A number of different fish, both Chondrichthyes and Osteichthyes, can detect electric currents in the water around them. The electroreceptors of fish are not neurons, but cells that probably evolved from mechanoreceptors in the lateral line organ; the receptors are sunk in a pit or canal. The voltage of the stimulus determines the frequency of impulses in the sensory nerve.

Cartilaginous fish have receptors called **ampullae of Lorenzini** embedded in a gelatinous substance at the bottoms of pits in the lateral line system. The function of these organs has been debated for many years be-

Figure 40-A Some of the electroplaques of an electric ray. The graphs above the diagrams show potential versus distance from left to right through the membranes. (a) At rest, electrical potentials on the right and left hand membranes of an electroplaque cancel each other out, for a total of zero potential across all the membranes shown. (b) When the motor neurons fire, they depolarize the right hand side of each muscle cell (colored) to zero. Each of the left hand membranes has the normal resting potential of −70 millivolts across it; with the right hand membranes depolarized the left hand membranes sum (as in a battery) to create a total potential across all the membranes of −350 millivolts in the five electroplaques shown.

cause they are sensitive to small changes in pressure, osmotic potential, and temperature. Recent work shows that they are even more sensitive to weak electric fields, such as those generated by every living organism. Dogfish sharks can use this sensory ability to find prey, such as worms that they cannot see or smell, from a distance of many centimetres.

Members of some families of freshwater bony fish have electric

organs that produce weak electrical discharges. H. W. Lissman has established by behavioral experiments that these fish use their electric organs for orientation. The electrical discharges last only a few milliseconds and are repeated regularly several times a second. Each discharge creates an electric field around the animal. The fish detects the shape of the field through its electroreceptors (Figure 40–B). Any object in the

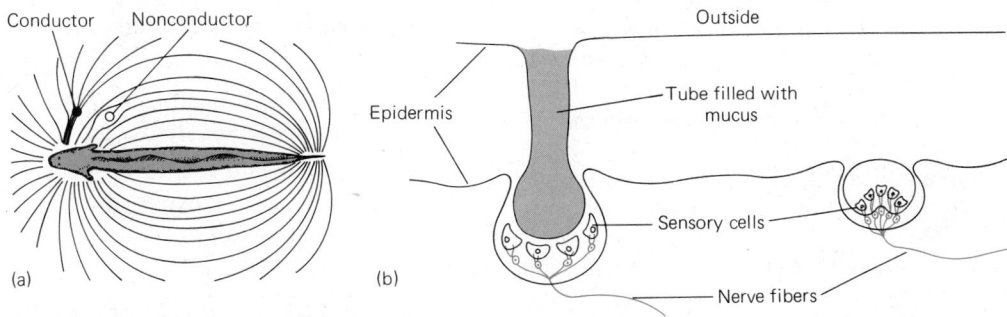

Figure 40-B (a) The electric fish *Gymnarchus* generates an electric field with its tail and detects the field with electroreceptors which are most concentrated in the head. The field is distorted by nearby objects. (b) The two types of electroreceptors in the skin of *Gymnarchus*.

electric field with an electrical conductivity different from that of the water around it distorts the field, and so is detected by the fish. The system is phenomenally sensitive. By rewarding a fish with food, Lissman showed that it could distinguish between glass and metal rods 2 mm in diameter and detect currents of about 3×10^{-15} amp.

In the turbulent muddy waters where they usually live, sight and smell are of little use to these weakly electric fish. Their electroreceptors, in conjunction with their electric organs, provide them with a navigation system whereby they can move around freely and find their food.

SUMMARY

Most of the movement in an animal's body is due to the action of its muscles. A muscle works by shortening so that it pulls against the skeleton or against adjacent tissues.

The contractile mechanism of muscle has been worked out in detail only for vertebrate skeletal muscle, although it appears to be similar in all muscle. In the presence of ATP and calcium ions, filaments of the protein myosin form cross-bridges to filaments of the protein actin; the cross-bridges then swivel and move the filaments past each other, shortening the distance between the Z lines of the sarcomere.

Contraction of a muscle is stimulated by depolarization of the muscle membrane, usually induced by a neurotransmitter released from the axon of a motor neuron or, in the case of smooth or cardiac muscle, by hormones in the body fluids. The arrangement of the neurons that innervate a muscle allows for graded muscle contractions.

The vertebrate skeleton supports the body, protects internal organs, and permits the locomotory muscles to move the animal. Pairs of antagonistic muscles move bones back and forth at joints. Reciprocal inhibition prevents simultaneous strong contraction by both members of a pair of antagonistic muscles.

The vertebrate skeleton is made up of relatively flexible cartilage and harder bone. Spaces in the bone are filled with marrow, which stores fat or forms cells, and with the cells that deposit or release calcium. Bones store or release calcium to maintain equilibrium with the body fluids. Hormones maintain a fine tuning of the blood/bone calcium balance.

From your study of this chapter, you should be able to:

1. Name the three types of muscles found in the vertebrate body, and give the location, innervation, and cellular organization of each type.

2. Describe how electrical impulses are conducted through the heart.

3. Draw a sarcomere from a skeletal myofibril and describe the interaction of its thick and thin filaments during muscular contraction.

4. List the four proteins involved in contraction of skeletal muscle, explain the role of each, and state how calcium and ATP interact with these proteins in the process of contraction.

5. Describe the sequence of events that takes place (on a chemical and cellular level) from the time a nerve impulse arrives at the neuromuscular junction in a skeletal muscle, through contraction of the muscle fiber, to subsequent relaxation of the fiber.

6. Explain why rigor mortis occurs in muscles after death.

7. Explain how the strength and duration of contraction of a skeletal muscle are controlled.

8. Define tetanus and fatigue in skeletal muscle.

9. List three functions of the skeleton as a whole and two non-skeletal functions of the bones.

10. Describe the structure of bone.

11. Describe the role of bone in maintaining the body's circulating calcium supply and the two ways in which bone releases calcium into the bloodstream.

12. Define cartilage, ligament, tendon, joint, flexor, extensor, antagonistic muscles, and reciprocal inhibition.

SELF-QUIZ

1. For each phrase below, give the type(s) of muscle that show the characteristic.
   ___ a. Syncytial
   ___ b. Innervated by autonomic nervous system
   ___ c. Can contract without nervous stimulation
   ___ d. Typically found in sheets rather than in bundles
   ___ e. Unicellular organization

   I. Cardiac
   II. Skeletal
   III. Smooth

2. Indicate which of the following items would be found at each of the places indicated on the diagram below:

   actin   tropomyosin
   calcium   troponin
   mitochondria   Z line
   myosin

   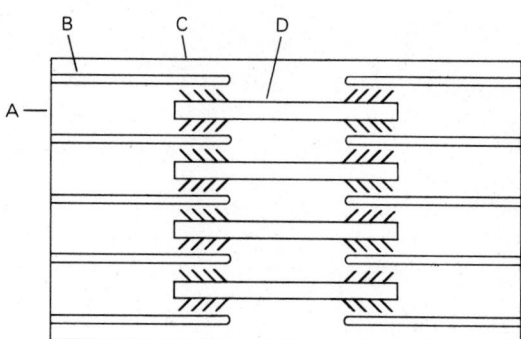

3. Below is a list of events that take place from the time an action potential arrives at a neuromuscular junction, through the contraction and subsequent relaxation of the muscle fiber. Arrange these events in correct chronological order.
   a. Myosin heads detach from the thin filaments.
   b. Ca^{2+} binds with troponin.
   c. Thick and thin filaments slide, lengthening the distance between Z lines.
   d. Hydrolysis of ATP causes the myosin heads to swivel.
   e. Thick and thin filaments slide, shortening the distances between Z lines.
   f. The calcium pump returns calcium ions into the sarcoplasmic reticulum.
   g. Tropomyosin molecules move out and cover the actin binding sites.
   h. Excitation of the T system causes the sarcoplasmic reticulum to release Ca^{2+}.
   i. Myosin-ATP complexes form cross-bridges to the thin filaments.
   j. Tropomyosin molecules move into the grooves of the thin filaments, exposing the binding sites of the actin.
   k. Electrical excitation spreads into the T system tubules.
   l. An action potential in a motor neuron arrives at the neuromuscular junction.
   m. Acetylcholine is released from the motor neuron.
   n. Electrical excitation begins to spread through the sarcolemma.
   o. Acetylcholine is received by the sarcolemma.

4. Rigor mortis is due to:
    a. leakage of calcium out of the sarcoplasmic reticulum after death
    b. lack of ATP
    c. death of the motor neuron
    d. loss of electrical potential difference across the sarcolemma
    e. cooling of the body after death

5. The extent of contraction of a mammalian skeletal muscle is controlled by:
    a. interaction of excitatory and inhibitory nervous input to individual fibers
    b. contraction of some sarcomeres in a muscle fiber but not others
    c. contraction of some entire muscle fibers while others remain relaxed

    d. reciprocal inhibition
    e. all of the above

6. Which of the following is *not* a role of bones?
    a. maintenance of simple chemical equilibrium between dissolved and deposited calcium without intervention of living cells
    b. dissolving of calcium deposits carried out by living bone cells in response to a hormone
    c. production of red blood cells
    d. production of white blood cells
    e. production of hormones to regulate calcium levels in body fluids
    f. storage of fat
    g. providing attachment sites for muscles
    h. protection of internal organs

QUESTIONS FOR DISCUSSION

1. If you are a Perry Mason fan, you know that it takes a variable length of time for rigor mortis to set in after death. What are some reasons for this variation?

2. Rigor mortis lasts for several hours and then disappears. Why does it eventually go away?

3. When you carry a full cup of hot coffee across a room, are your muscles exhibiting twitch or tetanic contraction?

4. During the time that a muscle is in tetanic contraction, what are the motor neurons that innervate the muscle doing?

5. Why doesn't a muscle relax between the arrival of nerve impulses when it is in tetanic contraction?

6. What is the advantage of having cartilaginous tissue in each area of the human body where it is characteristically found? (See Section 40-F.)

7. What are some advantages of having muscles attached to bones via tendons rather than directly?

8. Comment on the biological validity of the sayings "I can feel it in my bones" and "dry as a bone."

9. The disease rickets is characterized by bending of the bones due to lack of calcium deposits to keep them stiffened into the proper shape. Why is it advantageous for the bones to give up these calcium deposits even though doing so results in permanent deformation of the skeleton?

10. Would you expect levels of parathyroid hormone or of calcitonin to be elevated in a victim of rickets? In a pregnant woman? In her fetus?

11. Members of the class Chondrichthyes have skeletons composed entirely of cartilage. How might these vertebrates regulate the level of calcium in their body fluids?

REFERENCES AND FURTHER READING

Hildebrand, M. "How animals run." *Scientific American*, May 1960. Discusses adaptations of the skeleton and muscles that increase speed of running in mammals, notably horses and cats.

Huxley, H. E. "Mechanism of muscle contraction." *Scientific American*, December 1965. Describes experimental techniques used to isolate muscle proteins and study their structure. Many photographs of the isolated protein fibers.

Lissman, H. W. "Electric location by fishes." *Scientific American*, March 1963.

McLean, F. C. "Bone." *Scientific American*, February 1955. A brief and clear description of the structure, growth, and physiology of bone.

Murray, J. M., and A. Weber. "The cooperative action of muscle proteins." *Scientific American*, February 1974. Recounts the experimental unraveling of the roles of the four muscle proteins, ATP, and calcium in muscle contraction.

Smith, D. S. "The flight muscles of insects." *Scientific American*, June 1965. An interesting comparison of the structure and function of two types of insect flight muscle with each other and with the striated muscle tissue of vertebrates.

CHAPTER
41

ANIMAL HORMONES

Wherever there is division of labor, there must also be an exchange of information that will produce coordination. On a construction site, dozens of people may perform different jobs. They must know what their coworkers are doing and interact appropriately with one another to build the structure properly. In any multicellular organism, the jobs to be done are divided among many different types of cells. Like the different workers on a construction site, these cells must exchange information and work together if the organism is to survive.

The nervous system and the body's chemical messengers work together in coordinating the activities of all the different cells in the body. Originally, hormones were defined as chemical messengers (such as testosterone and insulin) produced by ductless endocrine glands (such as the testis and pancreas, Figure 41–1) that exert their effects some distance from where they are produced. Testosterone, for instance, is produced by the testes but, among other effects, causes an increase in the size of the muscles. We now know, however, that animals have dozens of chemical messengers in addition to those produced by endocrine glands. If we define a chemical messenger as a substance produced by one cell that affects another cell, the definition includes:

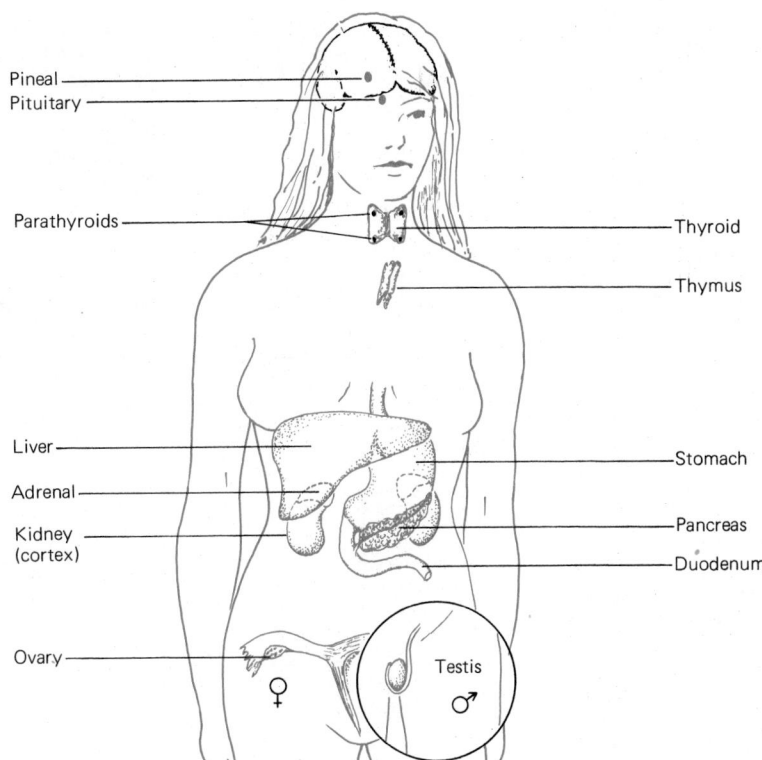

Figure 41-1 The locations of the major endocrine glands in humans.

1. Hormones produced by endocrine glands.

2. Hormones and transmitters produced by neurons. In Chapter 37 we encountered the neurotransmitters, such as acetylcholine and norepinephrine (noradrenalin), which carry nerve impulses across synapses. Some neurons, however, respond to nervous stimulation by releasing neurohormones (such as oxytocin, which induces labor during childbirth) from their axons into the bloodstream.

3. Other assorted internal chemical messengers and local hormones. This is a ragbag collection of chemical messengers not produced by endocrine glands or neurons. It includes histamine, which participates in inflammatory and immune reactions (Sections 33-D and 34-K); local hormones, such as those produced by gut-lining cells that regulate digestion (Figure 31–8); chalones, inhibitors of mitosis produced by dividing cells (Chapter 11); and prostaglandins, lipids with a variety of effects.

4. Pheromones. These are chemical signals produced by one animal that affect the behavior of another individual of the same species. The term was originally applied to the sex attractants of insects, but we now know of many other pheromones that affect animals' interactions.

Chemical messengers may act independently of, or in conjunction with, the nervous system to carry out two basic kinds of regulation. First, they may contribute to the body's physiological homeostasis; when certain conditions within the body are altered, hormones that reverse the changes are released. Second, hormones play a role in adaptive reactions to events outside the body. For instance, hormones ensure that an animal will be in reproductive condition when environmental conditions are most favorable for the survival of the young. It is in this type of control that the nervous and hormonal systems interact most obviously. An animal can detect environmental events only through its sense organs, and the nervous system must convey this information to the glands that produce adaptive hormonal changes. In the vertebrates, the hypothalamus of the brain is the part of the nervous system that communicates with the pituitary gland of the endocrine system, which in turn regulates most of the other endocrine glands.

In this chapter we shall consider the hormones produced by endocrine glands and by nervous tissue, as well as a few of the local hormones that we have not previously encountered. We shall see examples of how hormones contribute both to

TABLE 41-1 SOME VERTEBRATE PITUITARY HORMONES, THEIR SOURCES AND ACTIONS†

HORMONE	WHERE SECRETED	CHEMICAL NATURE	ACTIONS; STIMULATES:
Releasing factors	Hypothalamus	Polypeptides, etc.	Release of hormones from anterior pituitary
Oxytocin	Hypothalamus via posterior pituitary (=neurohypophysis)	Polypeptide	1. Uterine contractions in mammals 2. Milk production in mammals
Vasopressin (=ADH, antidiuretic hormone)	Hypothalamus via posterior pituitary	Polypeptide	1. Water resorption by nephron tubules 2. Increase in permeability of skin to water in amphibians
Adrenocorticotropic hormone (ACTH)	Anterior pituitary (=adenohypophysis)	Polypeptide	Secretion of corticosteroids by adrenal cortex
Thyrotropin (TSH, thyroid-stimulating hormone)	Anterior pituitary	Glycoprotein*	Secretion of hormones by thyroid
Follicle-stimulating hormone (FSH)	Anterior pituitary	Glycoprotein	Production of gametes in both sexes
Luteinizing hormone (LH)	Anterior pituitary	Glycoprotein	1. Secretion of sex hormones by gonads in both sexes 2. Ovulation in females
Prolactin (=LTH, luteotropic hormone)	Anterior pituitary	Protein	1. Mammary gland growth and lactation in mammals 2. Maintenance of corpus luteum* in mammals 3. Migration to water in amphibians 4. Reproductive functions in birds
Growth hormone	Anterior pituitary	Protein	1. Body growth in reptiles and mammals 2. Increased blood sugar in mammals

† Where the class of vertebrates in which a hormone acts is not specified, the action applies to members of all classes as far as we know.

homeostasis and to an animal's responses to environmental stimuli. Then we shall see some of the roles played by pheromones in influencing the behavior of different individuals of the same species.

41-A Hormones

What sort of evidence shows that a hormone exists and what it does? Most hormones have been discovered by observing animals that lack the corresponding endocrine gland. For instance, an amphibian tadpole without a thyroid gland never undergoes metamorphosis into a frog: it is reasonable to hypothesize that a hormone produced by the thyroid causes metamorphosis. This can be tested by transplanting a thyroid gland into a tadpole that lacks one. Since this operation is followed by metamorphosis, the theory is strengthened.

More convincing evidence for the hormone can then be produced by breaking down thyroid tissue into various chemical fractions and showing that some of these, but not others, produce metamorphosis after they are injected into a tadpole. Finally, it may be possible to isolate the pure hormone from the thyroid gland.

Such experiments are much harder when the endocrine gland involved, such as the vertebrate **pituitary gland**, produces many hormones (Table 41–1). In such situations, the most common approach is to take blood from an animal that is thought to contain the hormone and transfuse it into an animal that does not. For instance, the vertebrate pituitary was long suspected of producing **vasopressin**, a

---

*Glycoprotein Protein bonded to carbohydrate.
*Corpus luteum "Yellow body" that forms in an egg follicle of an ovary after it has ruptured and released a ripe egg.

TABLE 41-2 SOME VERTEBRATE HORMONES, THEIR SOURCES AND ACTIONS

HORMONE	WHERE PRODUCED	CHEMICAL NATURE	STIMULATES:
Thyroxin	Thyroid	Iodinated* amino acid derivative	1. Growth and metabolism 2. Metamorphosis in amphibians
Calcitonin	Thyroid	Polypeptide	Decrease in blood calcium by uptake of calcium into bones
Parathyroid hormone	Parathyroids	Polypeptide	Increase in blood calcium by release of calcium stored in bones
Insulin	Pancreas	Polypeptide	Decrease in blood sugar
Glucagon	Pancreas	Polypeptide	Increase in blood sugar
Gastrin	Stomach	Not known	Secretion of HCl by stomach
Adrenalin (epinephrine)	Adrenal medulla	Catecholamine (see Figure 41-3)	1. Dilation of blood vessels 2. Increase in blood pressure 3. Increase in blood sugar
Noradrenalin (norepinephrine)	Adrenal medulla	Catecholamine	1. Same as adrenalin 2. Also serves as a neurotransmitter
Corticosterone, cortisol, etc.	Adrenal cortex	Steroids (see Figure 41-4)	Metabolism of carbohydrate, protein, fat
Aldosterone	Adrenal cortex	Steroid	1. $Na^+ - K^+$ retention by kidney 2. Sex drive
Chorionic gonadotropin	Placenta	Glycoprotein	Maintenance of all body functions necessary for pregnancy in mammals
Progesterone	Corpus luteum of ovary	Steroid	1. Maintenance of uterine endometrium in mammals 2. Enlargement of breasts during pregnancy in mammals
Estrogens (estradiol, estrone, etc.)	Ovary	Steroids	Initiation and maintenance of sexual maturity in female mammals (affects uterus, vagina, mammary glands, skeleton, sexual behavior, metabolism, etc.)
Testosterone	Testis	Steroid	Initiation and maintenance of sexual maturity in male mammals; necessary for sperm production (affects voice, musculature, skeleton, metabolism, sexual behavior, etc.)
Melatonin	Pineal	Amine	Involved in light-regulated control of reproduction

water-conserving hormone, if the animal became dehydrated. To test this, blood from a dehydrated dog was transfused into a dog that had an adequate water supply. Urine production in the second animal decreased; it returned to normal only when the transfusion was stopped. This suggested that the blood of the first dog did in fact contain an antidiuretic* hormone. Showing that the hormone comes from the pituitary is more difficult. However, such an assumption would be strengthened if chemical analysis showed that the hormone is found only in the pituitary or in the bloodstream.

This method works well for a hormone with a reasonably long life and a strong influence on the body's activities. It is much more difficult—often impossible—to confirm the existence of a suspected hormone that seems to be destroyed very rapidly, or that is so diluted by transfusion that it has no demonstrable effect on the recipient animal.

Feedback Control of Secretion

Many of the hormones listed in Tables 41-1 and 41-2 control the composition of the body fluids or the rate of metabolism. For these homeostatic mechanisms to function properly, the hormones must be secreted into the bloodstream only when they are needed. Their secretion is under **negative feedback** control, whereby a process automatically limits itself.

---

*Iodinated Having atoms of iodine attached.
*Antidiuretic Opposing the tendency to excrete water.

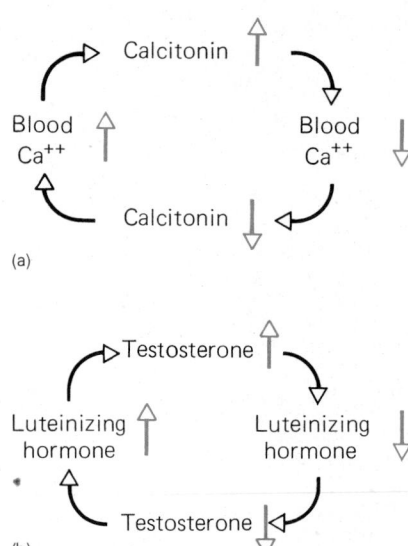

(a)

(b)

Figure 41-2 Two examples of feedback control: (a) between a hormone and calcium; (b) between two hormones. In both cases, each of the two substances serves to regulate the level of the other in the body fluids. Colored arrows indicate increases or decreases in levels of the substances in the body fluids.

A thermostat controls a familiar negative feedback system. The thermostat responds to a drop in temperature by turning the furnace on. When the thermostat detects that the temperature has risen to the set level, it turns the furnace off. As a biological example, a rise in the level of calcium in the blood causes secretion of the hormone **calcitonin** by the **thyroid gland**. Calcitonin causes cells in the bones to absorb calcium from the blood, decreasing the calcium level; this decrease, in turn, shuts off calcitonin secretion. This negative feedback loop is one of the control systems ensuring that the level of calcium in the blood remains constant.

A hormonal feedback loop may involve two or more hormones, instead of a hormone and some other substance (e.g., calcium). This is how hormones secreted by the pituitary gland control the release of other hormones. For instance, a male vertebrate must have the right levels of both **testosterone** and **luteinizing hormone (LH)** in the blood for the testes to produce sperm and function properly. LH stimulates testosterone secretion, but testosterone inhibits the secretion of LH. Thus, if the level of testosterone in the bloodstream rises, it inhibits the secretion of LH, and less testosterone is produced until the testosterone level falls low enough that the inhibition of LH is turned off. LH secretion then rises again, stimulating the secretion of more testosterone. Thus, a rise or fall in the level of either hormone is automatically corrected by way of the other (Figure 41–2).

Feedback control loops act at different speeds, depending on the number of hormones involved in the chain and the time that each takes to act. The system that controls the secretion of digestive enzymes in the stomach and intestine acts very rapidly in response to food in the stomach, whereas the feedback loop that controls fertility in women acts much more slowly; the menstrual cycle, one complete pass through a feedback control loop, takes about 28 days to complete (see Figure 36–5).

Chemistry of Hormones

Most hormones can be classified either as proteins (or amino acids, polypeptides, or glycoproteins) or as lipids of some sort.

The chemical nature of the hormone(s) secreted by a particular gland depends upon the embryonic and evolutionary origin of the gland. For instance, endocrine glands that develop from the embryonic endoderm* produce protein or polypeptide hormones. Among the endodermal glands are the thyroid, which evolved from part of the gut in early chordates,* the closely associated parathyroids, and glands in the stomach wall and in the pancreas. The other glands that secrete polypeptide hormones are derived embryonically from nervous tissue; these include the pituitary and hypothalamus in the vertebrate brain, and various invertebrate neurons that secrete hormones.

Related to the polypeptide hormones are the **catecholamines—adrenalin** and **noradrenalin**—also known as **epinephrine** and **norepinephrine**—from the **medulla** (internal cells) of the **adrenal glands**, at the anterior ends of the kidneys. Amines, as their name implies, are amino acid derivatives (Figure 41–3). One of the catecholamines, norepinephrine, is identical with the neurotransmitter produced by sympathetic neurons that synapse with effectors.

A number of hormones are lipids, synthesized from acetyl CoA* or cholesterol (Figure 41–4). Invertebrate hormones in this class include **juvenile hormone** of insects (Section 12-H). Vertebrate lipid hormones are the prostaglandins and steroid hormones. Steroid hormones include the **cortical steroids—cortisol** and its relatives—which are synthesized in the cortex (outer layers) of the adrenal glands, as well as the sex hormones. Steroid hormones are synthesized in endocrine glands derived from the embryonic mesoderm.

You should note that although we speak of "testosterone" or "estrogen" as if each were a specific molecule, there are actually dozens of slightly different versions of these and other hormones. Different estrogens occur in different species and also

---

*Endoderm   The innermost of the three germ layers in the early embryo of higher animals (see Section 36-G).

*Chordates   Animals with notochords and pharyngeal gill slits at some stage of development (see Chapter 29 for early chordates).

*Acetyl CoA   Acetyl coenzyme A, a versatile molecule that can donate its (2-carbon) acetyl group to synthesis of fatty acids or to the Krebs cycle, where it is broken down to yield energy (see Figure 7–7).

Figure 41-3 Some of the smaller proteinaceous hormones: (a) two posterior pituitary hormones which differ by only two
amino acids; (b) amines; and (c) thyroid
hormones, characterized by their content
of the iodine-containing amino acid thyro-
nine.

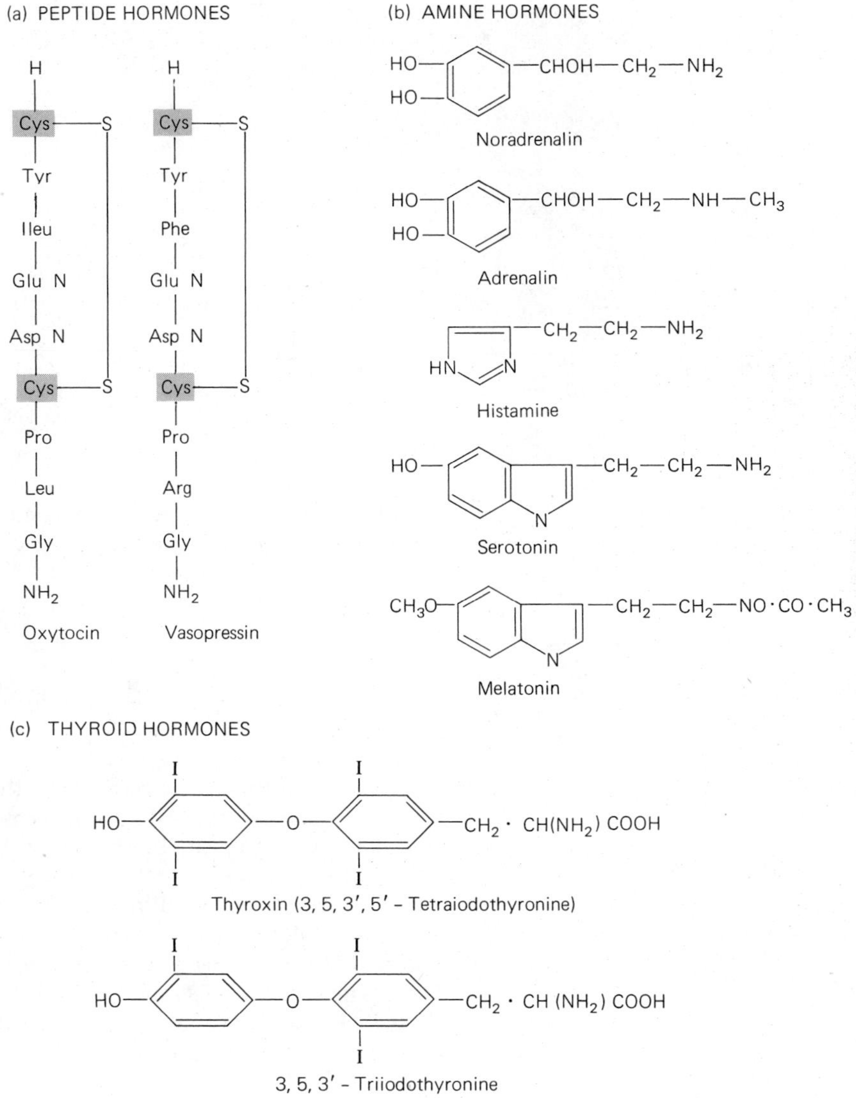

(a) PEPTIDE HORMONES

Oxytocin   Vasopressin

(b) AMINE HORMONES

Noradrenalin

Adrenalin

Histamine

Serotonin

Melatonin

(c) THYROID HORMONES

Thyroxin (3, 5, 3', 5' – Tetraiodothyronine)

3, 5, 3' – Triiodothyronine

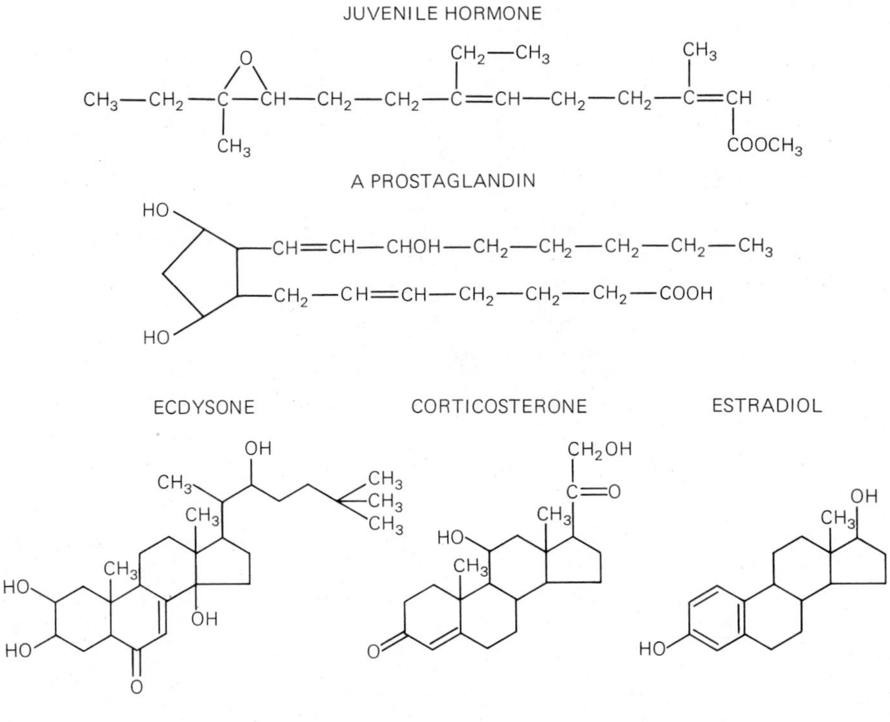

JUVENILE HORMONE

Figure 41-4 Some lipid hormones.

A PROSTAGLANDIN

ECDYSONE   CORTICOSTERONE   ESTRADIOL

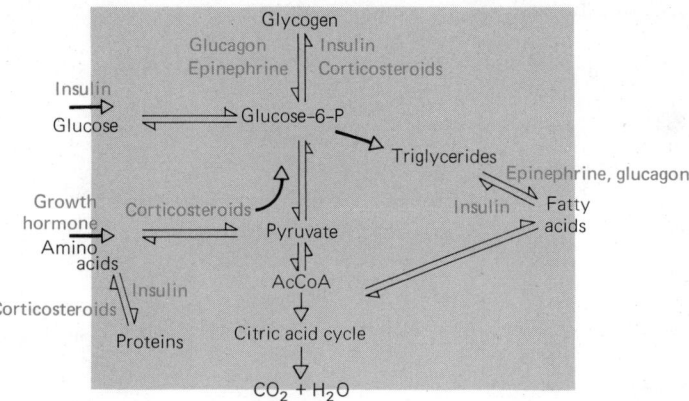

Figure 41-5 The effects of hormones on intermediary metabolism. The shaded box represents a cell. Hormones are shown in color next to arrows indicating the reaction they promote. For instance, insulin promotes passage of glucose into a cell, and epinephrine promotes the conversion of glycogen into glucose-6-phosphate. Hormones have different effects on different cells. The hormone effects shown here are typical of liver and muscle cells.

within individuals of one species. It now seems clear that each molecule has different effects and is a different hormone from its fellows, although we lump them all together and call them "testosterone" or "estrogen" because they are similar chemically.

Action of Hormones

Because they are released into the bloodstream or extracellular fluid, hormones reach most of the cells in the body. However, many of them are specific, influencing only their destined "target cells"; other cells do not react to them. For instance, although **insulin** travels throughout the body, only certain types of cells will respond to its presence by taking up glucose. Similarly, the reaction of the target tissue to a hormone differs at different times. Injection of LH will not make the ovaries of a sexually immature girl release an egg, since the ovarian cells cannot yet respond to the hormone as they will in later life. Such findings suggest that target cells carry receptor molecules that recognize and bind particular hormones, while other cells do not. In several cases, this model has been confirmed.

When a hormone is applied to a target cell, many changes may occur within the cell. The activity of various enzymes may increase or decrease, and permeability of the plasma membrane to specific substances may alter (Figure 41–5). Many hormones increase the activity of certain genes and thus alter the types of messenger RNA and proteins synthesized by the target cell (Figure 41–6). This was first shown

Figure 41-6 Model for the action of lipid and steroid hormones, which alter transcription. These hormones are transported to the nucleus, where they stimulate transcription of specific genes. In contrast, most protein hormones probably do not enter the cell but attach to a receptor at the plasma membrane and act via adenyl cyclase, an enzyme in the plasma membrane, which stimulates cyclic AMP production.

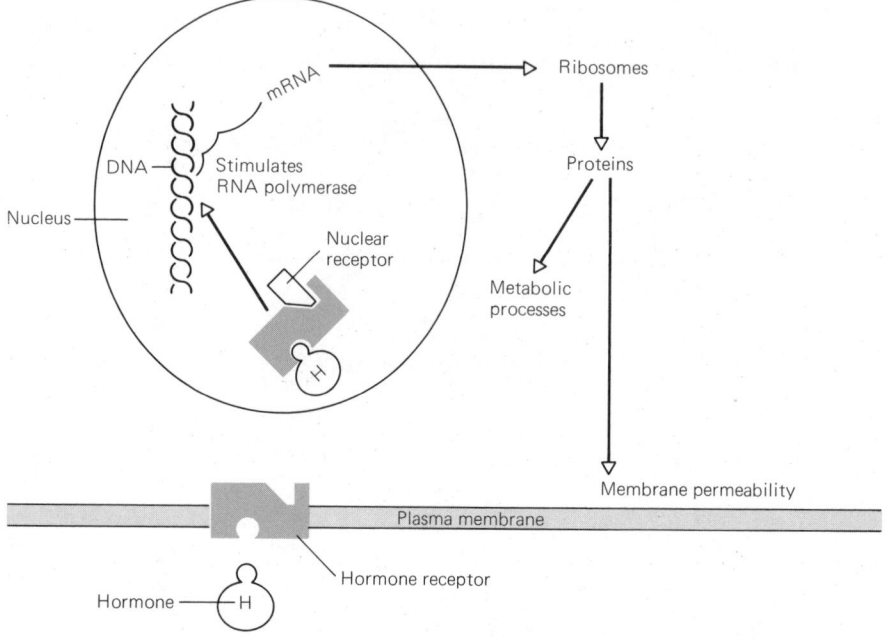

by P. Karlson and J. Edstrom in 1961 in studies on **ecdysone**, the hormone that controls molting in insects; it has since been shown that hormones affect gene activity in many vertebrate systems as well.

Plainly, if a hormone alters genetic activity and protein production, response to the hormone will be relatively slow. This is the case with the sex hormones, which may take days or weeks to exert their effects. Other hormones, such as insulin and epinephrine, can affect tissues within minutes and probably do not act by altering gene activity and protein synthesis.

In general, most lipid hormones seem to affect the genetic activity of their target cells, while proteinaceous hormones affect membrane permeability and enzyme action.

CYCLIC AMP

In the 1960s, Earl Sutherland showed that some hormones act on target tissues via an intermediary, identified as the nucleotide adenosine-3′,5′-monophosphate, commonly known as **cyclic AMP** (Figure 41-7). Sutherland showed that the hormone **glucagon** and the catecholamine hormones increase the level of an enzyme in liver cells if, and only if, the cells contain cyclic AMP. This has turned out to be a very important finding; most of the polypeptide and protein hormones are now known to act by way of cyclic AMP, and it is possible that some of the steroid hormones do too.

Sutherland proposed that a hormone binds to a specific protein receptor on the target cell surface, and that this stimulates activity of a membrane enzyme, adenyl cyclase, which catalyzes the production of cyclic AMP from ATP (Figure 41–8). The cyclic AMP, in turn, works as a "second messenger," activating enzymes, altering membrane permeability, and even initiating protein synthesis, depending on the type of cell. In other words, just as its precursor, ATP, acts as a universal energy-donating molecule in many reactions, so cyclic AMP acts as the common intermediary molecule that initiates all of the target cell's various responses to its specific hormone. The cyclic AMP molecule is admirably suited to this task because it is made from ATP, a molecule found in all living cells.

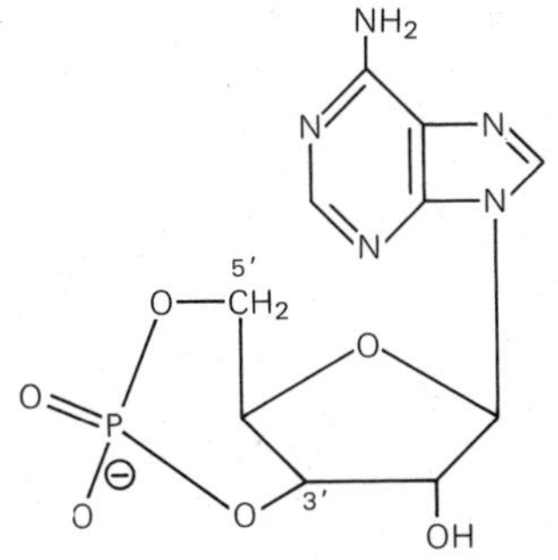

Figure 41-7 Adenosine-3′,5′-monophosphate, or cyclic AMP.

41-B Hormonal and Nervous Control

It takes longer for a hormone to act on its target cells than for a neuron to activate an effector. Not only may the hormone cause relatively slow reactions in the target tissue, but, since target tissues react only when the appropriate hormones are present at particular concentrations, it may also take time for a hormone to reach its target organ in sufficient quantity to have an effect.

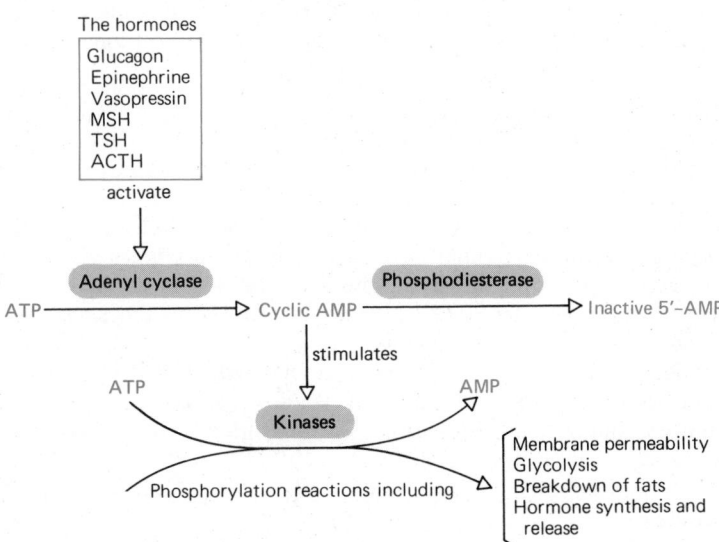

Figure 41-8 The role of cyclic AMP. Various hormones stimulate adenyl cyclase to produce cyclic AMP from ATP. Cyclic AMP stimulates the activity of kinase enzymes, which mediate the initial phosphorylation steps of various reactions. Enzymes are in colored ovals.

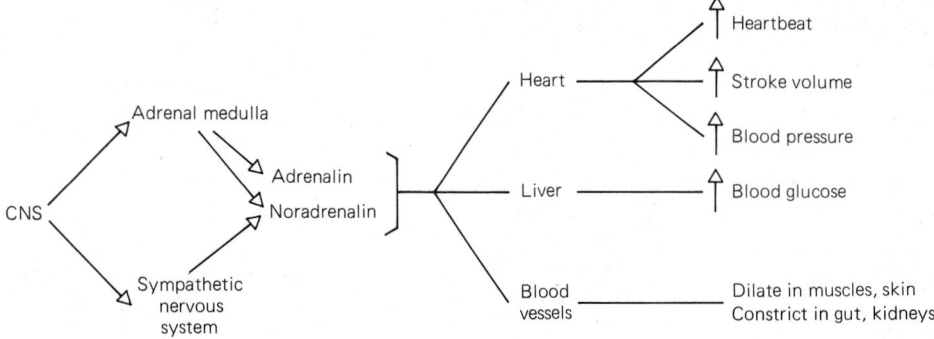

Figure 41-9 A few of the changes involved in the "fight or flight" reaction, mediated by both the nervous system and hormones.

By having nervous and endocrine control systems that react at various speeds, the body is equipped with mechanisms to suit a variety of occasions. The nervous system enables an animal to escape from an enemy or to withdraw from a painful stimulus in a fraction of a second. At the other extreme, slow-acting hormones can keep the body pregnant for many months. Hormones are more suitable than nerves for controlling long-term changes involving many different organs, whereas nerves are more suitable for rapid reactions involving relatively few organs. Often the two systems acting together can control a situation more efficiently than either of them acting alone, as in the example below.

Fight or Flight

A hot, red face, perspiring hands, and a rapidly beating heart commonly precede a stage appearance or an important exam. These symptoms are part of the "fight or flight" reaction, which prepares the body to meet stress or danger. We shall mention only a few aspects of this reaction here (Figure 41–9).

When a vertebrate senses danger or stress, the central nervous system (CNS) stimulates the adrenal medulla to release the hormones adrenalin and noradrenalin into the bloodstream. In addition, much of the sympathetic nervous system is activated, releasing more noradrenalin. Adrenalin and noradrenalin cause the heart to beat faster and to increase the volume of blood pumped per stroke; this raises the blood pressure and circulates the blood more rapidly. In addition, glucose is released into the bloodstream by the liver, raising the blood sugar level. Blood vessels in the muscles and skin dilate, increasing the blood supply to these organs; this prepares the muscles for action by supplying the muscles with extra oxygen and glucose. The increased blood supply to the skin permits the body to lose the heat generated by muscular activity. At the same time, constriction of the vessels supplying the kidneys and alimentary canal reduce their supply of blood.

Nervous control evokes these reactions very rapidly in time of danger; hormones provide a backup that can maintain the response for a long period.

The Hypothalamus–Pituitary Connection

Most of the endocrine glands in the vertebrate body are controlled, directly or indirectly, by the brain. The interface between the nervous and endocrine systems lies in the connections between the hypothalamus and the pituitary gland.

The **hypothalamus** is a small area of the forebrain lying under the cerebral cortex; it receives input from all parts of the brain (see Figure 38–6). Stimulation of various cells in the hypothalamus elicits sensations and behaviors such as sex drive, pleasure, rage, fear, satiation, hunger, and thirst.

The hypothalamus synthesizes some hormones which it does not release into the bloodstream. Rather, these hormones travel down secretory neurons* from the hypothalamus to the posterior lobe of the pituitary gland, where they are released

---

*Secretory neurons Nerve cells that secrete substances that travel in the body fluids and act as hormones or regulatory substances.

into the blood. In addition, the hypothalamus produces hormones called **releasing factors**, whose function is to control the release of other hormones from the anterior lobe of the pituitary gland (Figure 41–10). Secretory neurons in the hypothalamus discharge the releasing factors into a portal system,* which carries them to the anterior pituitary (Figure 41–11). Each releasing factor causes the anterior pituitary to release a specific hormone into the bloodstream.

The hypothalamus is the body's single most important control center. Messages from sensory neurons, and the chemistry of the surrounding cerebrospinal fluid,* provide the hypothalamus with a continuous flow of information about the state of the body. The hypothalamus reacts to these stimuli by producing activity in the autonomic nervous system,* by initiating behaviors such as feeding, and by controlling the pituitary's secretion of hormones.

The **posterior pituitary gland**, or **neurohypophysis**, releases polypeptide hormones manufactured in the hypothalamus. The best-known of these are **vasopressin**, a water-conserving hormone in terrestrial vertebrates, and **oxytocin**, which induces muscular contractions by the uterus and by the ducts of the mammary glands in mammals. The posterior pituitary releases many other hormones, but very little is known about them.

Many of the hormones secreted by the **anterior pituitary** (or **adenohypophysis**) induce other endocrine glands in various parts of the body to secrete their particular hormones.

All of the anterior pituitary hormones are proteinaceous. They include **lipotropin**, which mobilizes stored fats from fat cells in the adipose tissue, and **thyrotropin** and **corticotropin**, which cause hormone secretion by the thyroid and adrenal cortex, respectively. **Melanocyte-stimulating hormone (MSH)**

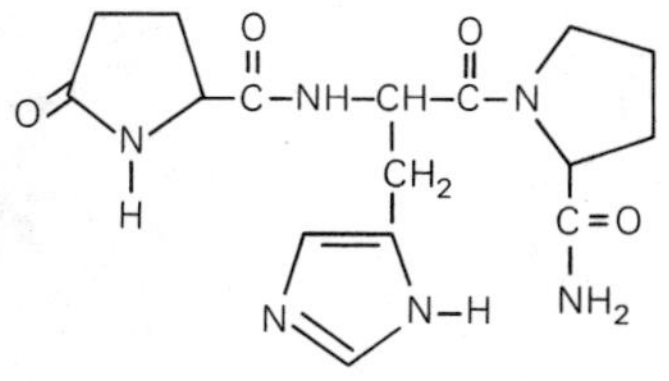

---

*Portal system   A series of blood vessels that carries blood between two sets of capillary beds without first returning to the heart.

*Cerebrospinal fluid   A clear fluid, derived from the blood, that bathes and cushions the brain and spinal cord.

*Autonomic nervous system   The part of the vertebrate nervous system over which the animal usually has no control; composed of the sympathetic and parasympathetic systems (see Section 38-G).

Figure 41-10   Thyrotropin-releasing hormone (TRH), one of the small regulatory peptide hormones that are produced in the hypothalamus and cause the release of hormones from the anterior pituitary. TRH causes the anterior pituitary to release thyroid-stimulating hormone (TSH, also called thyrotropin) and TSH, in turn, stimulates the thyroid gland.

Figure 41-11   The interrelations of hypothalamus and pituitary in the vertebrate brain.

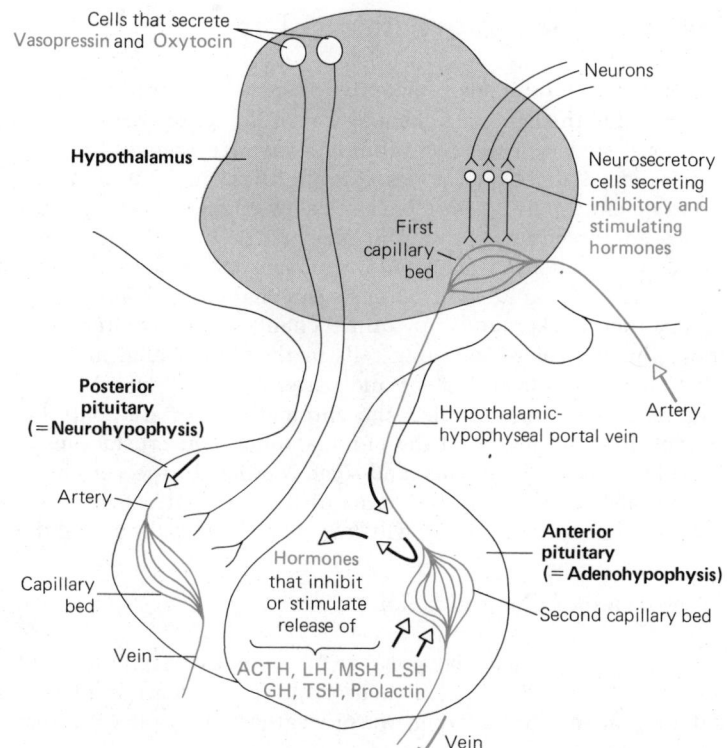

Cells that secrete Vasopressin and Oxytocin

Neurons

Hypothalamus

Neurosecretory cells secreting inhibitory and stimulating hormones

First capillary bed

Posterior pituitary (=Neurohypophysis)

Hypothalamic-hypophyseal portal vein

Artery

Artery

Anterior pituitary (=Adenohypophysis)

Capillary bed

Hormones that inhibit or stimulate release of

Second capillary bed

Vein

ACTH, LH, MSH, LSH GH, TSH, Prolactin

Vein

causes pigment dispersal in the color cells in the skins of vertebrates that can change their color, and affects sexual maturity and general health in birds and mammals. The anterior pituitary also produces follicle-stimulating hormone (FSH) and luteinizing hormone (LH), which are necessary for hormone production by the gonads in all vertebrates; **growth hormone**, which is essential for normal growth of young vertebrates; and **prolactin**, the hormone for milk production in mammals.

41-C Prostaglandins

Prostaglandins are a family of lipids discovered in human semen in the 1930s; we now know that they occur in most vertebrate tissues. Prostaglandins caused considerable excitement in endocrinological circles in the 1970s and are now beginning to find medical uses. Although we know a lot about their chemistry, however, we still know very little about the natural functions of prostaglandins in the body. The chemical structure of a prostaglandin is shown in Figure 41–4; the 5-carbon ring and two "tails" are found in all members of the group.

Different prostaglandins have a wide range of actions when they are applied to different tissues. Some stimulate smooth muscle to contract, and others cause it to relax; some cause constriction of capillaries, and others cause dilatation. Prostaglandins are probably involved in the inflammatory response to infection and are definitely involved in many different aspects of reproduction.

Some prostaglandins increase the cyclic AMP content of a number of tissues including lung, spleen, liver, and adipose tissue. It seems likely that the prostaglandins stimulate the activity of adenyl cyclase (Figure 41–8), but how they do this is not known. One view is that prostaglandins act as transducers between hormone receptors and adenyl cyclase, in which case they would be vital to the activity of many hormones reaching a cell via the blood, but this is far from proven. We know very little about the normal action or synthesis of these local hormones; they are nevertheless already being administered as contraceptives, muscle relaxants, and antiviral agents. Drugs that inhibit the synthesis of prostaglandins reduce some kinds of pain, but since we do not know the functions of most prostaglandins, the effect of inhibiting them is completely unknown and possibly unwise.

41-D Hormones and Animal Life

Animals exhibit many adaptive responses to information about the outside world sent to the nervous system by way of the sense organs. Although most of these reactions, such as eating or running away, are brought about by motor neurons, some are mediated by hormones. It is not surprising that most of the reactions to the environment involving hormones are slow changes such as occur when an animal comes into breeding condition in the spring.

Other hormonally mediated reactions, such as those involved in color change, are surprisingly rapid. Many animals can change color so that they are camouflaged against their backgrounds. An animal changes color by altering the size of its **chromatophores**, color-containing cells in the skin, which may be of various colors. This changes its overall color and pattern (Figure 41-12). In cephalopods, some bony fish, and some reptiles, the chromatophores are controlled by the nervous system. Hormones control the chromatophores in crustaceans, cartilaginous fish, some bony fish, amphibians, and some reptiles. In the vertebrates on this list, the pattern of light reaching the retina of the eye controls the release of melanocyte-stimulating hormone (MSH), which in turn changes the size of the chromatophores.

Environmental Control of Reproduction

There is great selective pressure for animals to reproduce under conditions that are likely to favor survival of their offspring. For most animals this means birth or hatching in the spring, when warm weather and a plentiful food supply offer the

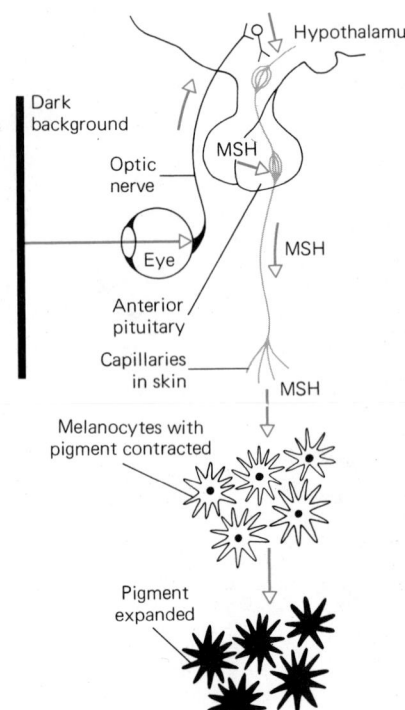

Figure 41-12 The mechanism of color change in a frog placed on a dark background. The tone of the background is perceived via the eyes, which initiate a chain of nervous and hormonal events. Finally, MSH from the anterior pituitary stimulates the pigment in melanocytes in the skin to spread out in the cell, darkening the skin. Colored arrows show the path of information flow.

best possible conditions. What triggers the various changes that bring the animal into breeding condition? Breeding often involves dramatic alterations in the anatomy, physiology, and behavior of an animal. The gonads of many animals shrivel after the annual breeding season, and must absorb food, grow, and produce gametes anew each year (Figure 41–13). Many animals must eat voraciously and produce large, yolky eggs, or they may have to accumulate fat reserves that will last them while they care for the young. Behavior usually changes in ways that improve the chances of finding and accepting a mate.

Experiments have shown that animals come into breeding condition in response to environmental conditions such as increasing daylength, temperature, rainfall, food, and the like.

Animals detect the external stimuli that induce breeding in various ways. Temperature and rainfall are detected by receptors in the skin; light may be detected by the eyes or by the pineal eye, the "third eye" that lies in the middle of the top of the head in many lower vertebrates. The appropriate stimulus causes hormone production by way of the hypothalamus in vertebrates. The hypothalamus does two things: first, it stimulates the pituitary to release hormones that cause the gonads to grow and produce sex hormones, which in turn cause growth of other reproductive organs; second, the sex hormones feed back to the hypothalamus, which then initiates reproductive behavior by sending out the appropriate signals in the nervous system.

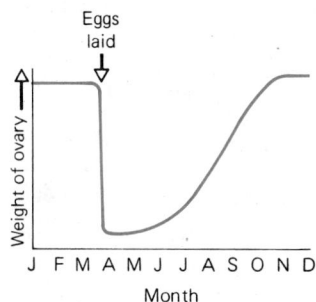

Figure 41-13 The weight of the ovaries of a toad (*Bufo bufo*) at different months of the year. Ovarian growth and egg laying are under hormonal control.

41-E   Biological Rhythms

Reproductive cycles are examples of rhythmic or cyclical events in an animal's life. A number of other cycles are evident in the physiology and behavior of eukaryotic organisms.

Circadian Rhythms

Circadian ("about a day") **rhythms** have a period of about 24 hours. They are very common in eukaryotes, and they affect many physiological processes. For instance, in most vertebrates the metabolic rate, body temperature, blood sugar level, composition of the urine, general level of nervous activity, and many other functions alter regularly in a 24-hour cycle (Figure 41–14).

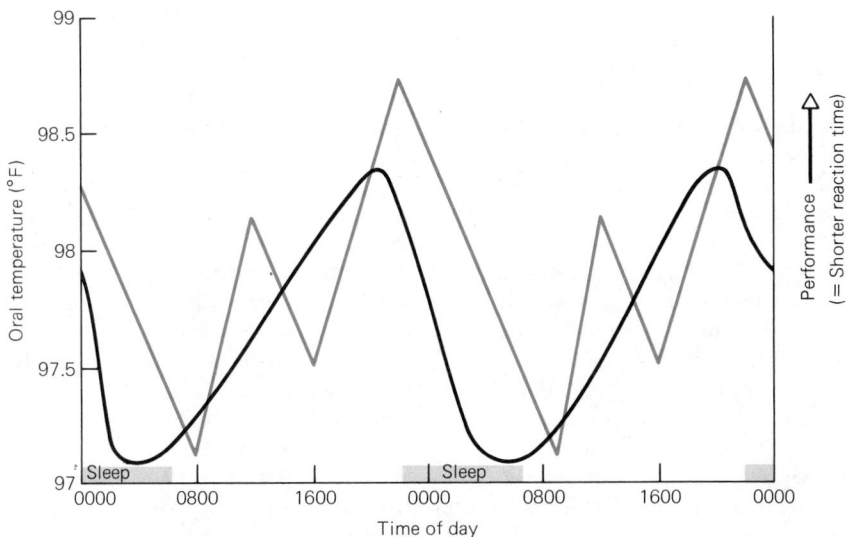

Figure 41-14 Circadian rhythms of oral temperature (black line) and reaction time (colored) in humans. (Reaction time is the time it takes to react to a stimulus when you are told to react as fast as possible, e.g., press a button when you see a red light.)

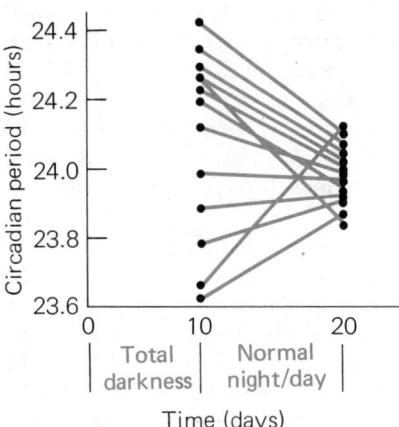

Figure 41-15 This experiment showed that the circadian period of a hamster is closer to 24 hours when the animal is exposed to normal light and dark cycles than when it is kept in total darkness. Each pair of dots joined by a line represents a single animal. The hamsters were kept in total darkness for 10 days, and then the length of the daily cycle of sleep and activity was measured. The hamsters were then returned to a normal day/night regime, and their cycles were measured again 10 days later. At this time the animals were closer to a 24-hour cycle than when they were in total darkness. (After Pittendrigh and Dunn, *Science* 186:548–50, November 1974. Copyright 1974 by the American Association for the Advancement of Science)

Because 24 hours is the period of the earth's rotation and of one light-dark cycle, it might seem obvious that daily cycles would be controlled by the onset of light or of dark. Is this the case, or is the rhythm **endogenous** ("built into" the animal)? To answer this question, investigators cut an animal off from environmental cues that might tell it the time of day, and see if the circadian rhythm persists. Under conditions of constant temperature, humidity, and light intensity, the animal still maintains circadian rhythms, but the cycles are no longer exactly 24 hours. The animal's endogenous rhythm must be related to an "internal clock."

In its normal environment, an animal has a cycle of exactly 24 hours. The cycle the animal maintains without environmental cues is always the same for any individual, although it may be between 22 and 26 hours long (Figure 41–15). This suggests that the cycle itself is endogenous, but is usually driven, or "entrained," by the 24-hour cycle of light and dark. This is further shown by the fact that it is possible to reset the circadian clock. If a vertebrate is kept in a controlled environment where the light is switched on only during what is really night, night becomes day and vice versa for the animal, and its daily rhythm soon becomes reset exactly 12 hours later (or earlier). Many zoos do this to **nocturnal** animals, which are usually active only at night, so that they become active during the day when visitors want to see them. These animals are exposed to bright light at night, and to a dim red light during the day. Their internal clocks are set so that the animals act as though it is night during what is really daytime.

Circadian rhythms may be remarkably persistent. For example, a reptile continues to show daily fluctuations in body temperature even when it has been kept at a constant temperature and light intensity for months. Another intriguing feature of circadian rhythms is their ability to adjust to changes in temperature. The rates of biochemical processes usually vary with temperature, and we assume that circadian rhythms are due to some biochemical process(es). Hence we would expect that the rhythm would be slower, and it would therefore take longer to complete each cycle, when the temperature is lowered. When the temperature of the environment is lowered, the animal's rhythm does slow down, but within a few days the rhythm adjusts to the new temperature and resumes its natural frequency. How this temperature compensation is made is an interesting puzzle.

Why did circadian rhythms evolve in the first place? Night and day always follow each other on a 24-hour cycle, so that there is a perfectly good environmental stimulus available to trigger daily cycles. Why should an organism also have an internal cycle of its own? The answer is probably that its internal rhythm permits an animal to anticipate regular daily events before environmental cues appear. This ability is valuable in many ways. It permits a bat to start hunting at the time of day when its insect prey will also be active, without wasting the energy it would take to check on the insect life of the area every few hours; it permits a reptile to start raising its body temperature before dawn so that it is already prepared to be active at daybreak.

Annual Rhythms

In addition to endogenous circadian rhythms, some animals have endogenous yearly cycles. Examples have been found among both vertebrates and invertebrates. Many mammals continue to hibernate at roughly the right time even if they are deprived of environmental cues that could tell them the time of year. Deer kept under constant conditions continue to shed their antlers at the same time every year. Several species of birds start migratory behavior at the same time of year even if they are deprived of environmental cues. In this last case, the adaptive advantage of the annual rhythm is probably that it permits a bird to return north for the breeding season even though it receives few cues of seasonal change in the relatively constant tropical environment of its wintering area.

Biological Clocks

The existence of circadian and annual rhythms begs the question "How does it work?" Organisms must plainly have internal "clocks" of some sort that keep track of time independently of environmental stimuli. Such a clock is very valuable. It provides an animal with a time sense that allows it to tell the season of the year by detecting the length of time between sunrise and sunset. It also permits animals to navigate using the sun as a compass. (The clock is necessary because the sun is at different points on the compass at different times of the day.)

Biological clocks have been studied in some detail in the reproductive cycle of female white rats. The cycle starts with the release of LH from the pituitary, an event that must be due to the rat's biological clock since it does not depend on environmental factors. The clock is set so that LH release can occur only between certain hours of the day. In addition, LH release can occur only on days when the level of estrogen in the blood is sufficiently high. The interaction between the animal's clock and the estrogen level in the blood controls the reproductive cycle and ensures that females come into breeding condition every four or five days, depending on the strain of rat (in the five-day strain the level of estrogen in the blood builds up more slowly).

In this particular case, the clock seems to be physically located in neurons in the hypothalamus. If part of the hypothalamus is destroyed, a female rat comes into permanent breeding condition, although the pituitary acts normally and the gonads produce their hormones as usual.

It seems likely that animals' biological clocks and pacemakers may turn out to be the result of oscillations (perhaps of membrane properties) in the electrical activity of specialized neurons, but we are still far from understanding how they work.

Although the nature of the endogenous clock is not understood, in at least one case we know how the clock is reset by the normal 24-hour cycle of light and dark. In chickens, the pineal gland releases the hormone melatonin (Table 41–2) in a daily rhythm. The activity of the enzyme acetyltransferase, needed to synthesize melatonin, is reduced in the presence of light, and so the amount of melatonin in the blood falls at daybreak every day. It is not yet certain how the chicken pineal gland detects light, but in many mammals the message that there is light reaches the pineal gland via the eyes. Light does not cause the rhythm of melatonin release: if the pineal gland is removed, or if the chicken is kept in darkness, acetyltransferase activity continues to rise and fall. The normal daily cycles of light and dark do, however, reset the clock. Interactions between the level of melatonin and the chicken's other internal rhythms might then influence reproductive activities. The final outcome is that chickens lay eggs more frequently when the days are long than when days are short.

HONEYBEE

$CH_3 \cdot CO(CH_2)_5 CH = CH \cdot COOH$

LEAF ROLLER MOTH

$CH_3 \cdot CH_2 CH = CH(CH_2)_{10} - O - COCH_3$

DOUGLAS-FIR TUSSOCK MOTH

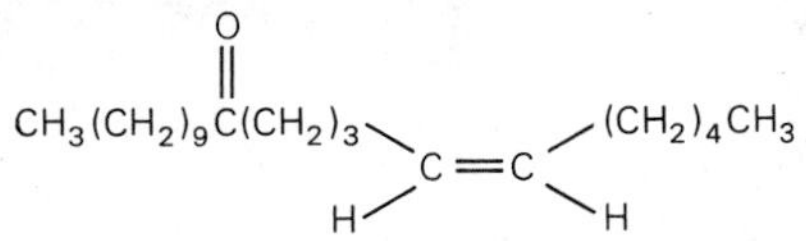

$$CH_3(CH_2)_9 \overset{\overset{\displaystyle O}{\|}}{C}(CH_2)_3 \underset{H}{\overset{}{C}} = \underset{H}{\overset{}{C}} (CH_2)_4 CH_3$$

Figure 41-16 Three insect sex attractant pheromones. Sex attractants can be used to trap insects for use in scientific research. Researchers are searching for ways to use these chemicals for the control of insect pests like the leaf roller and tussock moth.

41-F Pheromones

A hormone is a chemical that carries information within the body of an animal; a pheromone is a chemical that travels outside the body, carrying information to other members of the same species. The first pheromones described were sex attractants from insects (Figure 41–16). In many species of moths, beetles, cockroaches, and flies, the female releases a chemical that attracts the male. He finds his mate by flying or crawling up the odor gradient toward her.

Many vertebrates, and particularly mammals, use pheromones in urine or feces, or from special scent glands, to mark trails and territories. When a dog urinates on a fire hydrant, he is depositing a pheromone that tells other dogs that the hydrant is part of his territory. Pheromones also accelerate reproductive maturity in a number of species and permit members of one sex to distinguish which members of the opposite sex are in breeding condition.

The pheromones used to mark territories or attract a mate produce immediate effects on the nervous system, physiology, and behavior of the animal that receives the pheromone. There are other pheromones that act more slowly and have longer-lasting effects. For instance, if many female mice are caged together, the estrus (sexual receptivity) periods of all of them are interrupted by periods in which the females are infertile because the normal estrous cycle does not occur. Similarly, the odor of a strange male will terminate the pregnancy of a newly fertilized female mouse. In this case, the pheromone responsible comes from the male's urine. It is received via the female's olfactory organs and triggers nervous activity in the hypothalamus. The hypothalamus causes the release of pituitary hormones that reduce steroid output by the ovaries, and this reduction means that the uterus does not receive the hormones needed to permit implantation and pregnancy. The newly fertilized eggs cannot implant, and the pregnancy aborts.

Pheromones may also synchronize reproduction within a group. In desert locusts, any sexually mature male produces a pheromone that speeds up sexual development in immature members of both sexes. This system ensures that most of the locusts in the area reproduce at the same time: so many young locusts hatch at once that predators can kill only a small fraction of them, and many more survive than would be the case if they hatched over a longer period, giving predators time to eat one brood before turning their attention to the next.

The best evidence for human reproductive pheromones comes from anecdotal evidence that when numbers of women live together, in dormitories or similar situations, their menstrual cycles eventually synchronize. Other than this, there is no convincing evidence for human pheromones, despite the large number of reports on the subject that appear periodically.

In sexual systems involving pheromones, a single chemical determines whether or not an animal will reproduce. For this reason, a slight change in the pheromone might have profound evolutionary effects. For example, there are two pheromone "races" of the European corn borer moth. Males in Italy, the Netherlands, and the eastern United States respond to a pheromone mixture containing 96% *trans*-11-tetradecenyl acetate and 4% of the isomer* *cis*-11-tetradecenyl acetate. Elsewhere in Europe and North America, males are attracted to the same compounds but in the

*Isomers Molecules containing the same atoms but in different arrangements.

reverse proportions. Both forms of the insect occur together in Pennsylvania, where they do not interbreed; they can therefore be regarded as separate species. There is a good chance that these species arose in the same area in Europe, before both were introduced accidentally into North America. Speciation would have required only two independent mutations, one altering a *trans* receptor protein to a *cis* receptor in the male, and the second causing the female to produce the *cis* instead of the *trans* isomer of the pheromone.

In the insect societies of bees, ants, and termites, pheromones organize not just the reproduction but also the behavior and social structure of a colony, as we shall discuss in more detail in Chapter 42.

SUMMARY

The nervous and hormonal systems carry messages that travel between an animal's cells and coordinate their activities. Hormones may act on a wide range of cells over a long time; neurons affect specific cells briefly.

Animals have three types of chemical messengers: proteinaceous and lipid hormones, produced by endocrine glands or secretory neurons, and carried all over the body by the blood; local hormones such as histamine, chalones, and prostaglandins, which usually act near the cells that produce them; and pheromones, which carry information between different individuals of the same species.

Each hormone has only limited effects because receptors for a hormone are restricted to specific target cells. Many hormones activate production of cyclic AMP, which acts as a relay messenger inside the cell, triggering the cell's preset response to the hormone.

Hormone secretion is usually controlled by negative feedback in response to some disturbance in the body or to the level of another hormone in the blood. Hormones and the nervous system often interact with one another to control both long- and short-term aspects of an animal's response to a stimulus. The interaction between the two systems occurs mainly in the hypothalamus. The hypothalamus receives nervous signals from the sense organs via the rest of the brain, and also detects changes in the blood chemistry. It initiates appropriate responses by way of the nervous system and by way of the pituitary gland, which releases hormones responsible for the maintenance and activity of many of the body's other endocrine glands.

Hormones are involved both in homeostatic mechanisms within the body and in many of an animal's responses to its environment. Reproductive cycles and color changes may be mediated by hormones.

Animals have endogenous biological clocks that tell them the time of day; the clocks may be driven to some extent by environmental stimuli. Although it is clear that an animal's clock interacts with its nervous and endocrine mechanisms, how the clock works is still not understood.

OBJECTIVES

From your study of this chapter, you should be able to:

1. State what a hormone is, and explain how hormones may be identified and their action determined.

2. Describe the process of feedback control of hormone production, and diagram it using a specific example.

3. Describe what is known about how hormones affect their target organs.

4. Explain the differences in the control exerted by the nervous system and by the endocrine system.

5. Describe the role of cyclic AMP in a cell's response to a hormone.

6. Describe the "fight or flight" response, listing six physiological changes that occur, and explaining the role of

the nervous and endocrine systems in the response; discuss the roles played by adrenalin and noradrenalin in the response.

7. Describe the relationship between the hypothalamus and both the anterior and the posterior pituitary gland.

8. Explain the role of the pituitary gland in the body's endocrine system.

9. Give two examples each of hormonal responses that maintain homeostasis and of hormonal responses to conditions in the animal's environment.

10. Define circadian rhythm, give an example of a biological clock, and tell what is known about their properties.

11. Define pheromone, and explain the biological roles of pheromones.

679

SELF-QUIZ

1. Which of the following techniques would be *least* likely to help elucidate the function of a hormone produced by an endocrine gland?
   a. removal of the gland and subsequent analysis of what functions are lost
   b. transplantation of the gland into an animal that lacked the gland
   c. transfusions of blood from an animal lacking the gland into an animal that has the gland and observation of its effects
   d. observing effects of gland extract on various tissues grown in culture
   e. observing the condition of animals with a tumor that causes the gland to be overactive

2. All of the following commonly serve as signals stimulating hormone secretion *except:*
   a. conditions outside the body
   b. rising levels of another hormone
   c. rising levels of the hormone in question
   d. falling levels of the hormone in question
   e. falling levels of another hormone

3. Hormones are known to cause all the following changes in target cells *except:*
   a. changes in genetic makeup
   b. changes in permeability
   c. changes in metabolic rate
   d. increase in cyclic AMP concentration
   e. synthesis of different messenger RNA

4. An advantage to having the endocrine system as well as the nervous system involved in the "fight or flight" response is:
   a. the endocrine system responds faster
   b. the endocrine response usually lasts longer
   c. the endocrine system is tuned more precisely to the degree of need
   d. the endocrine system affects only the target organs whose response is needed to meet the emergency
   e. response by the endocrine system frees the nervous system to think of a way out of the situation instead of simply maintaining the body in an alert state

5. Information from internal or external receptors may initiate a response from the endocrine system by passing through the part of the brain known as the _____. This area connects with the anterior pituitary via _____ and with the posterior pituitary via _____. The pituitary stimulates various glands to secrete hormones by _____.

QUESTIONS FOR DISCUSSION

1. Using the secretion of thyroxin as an example, diagram the feedback control system for hormone production.

2. Thyroxin levels are generally at about 100 units. A patient has only 80 units in the bloodstream. Normal therapy for this situation is to inject the extra 20 units. Why might this therapy *not* be effective?

3. It's spring! Time to go on a diet so you won't bulge too much at the beach this summer. If you go on a low sugar diet, your body will metabolize proteins and polysaccharides to sugar. Explain how the endocrine glands and hormones accomplish this, including what will happen to the glands and hormones when sugar levels return to normal.

4. Draw a diagram to show the glands and hormones that control growth. If the growth hormone has no known feedback mechanism, what sort of control mechanism might determine when growth stops?

5. What does the biological clock have to do with the price of eggs?

REFERENCES AND FURTHER READING

Binkley, S. "A timekeeping enzyme in the pineal gland." *Scientific American,* April 1979. Includes a discussion of what we know and don't know about endogenous clocks in vertebrates.

Guillemin, R., and R. Burgus. "The hormones of the hypothalamus." *Scientific American,* November 1972. Recounts how the hypothalamus–pituitary links were discovered and the means of isolating and characterizing hypothalamic hormones.

Le Baron, R. *Hormones: A Delicate Balance.* New York: Pegasus, 1972. A delightfully written layman's account of (mainly human) hormones and their functions.

McEwen, B. S. "Interaction between hormones and nerve tissue." *Scientific American,* July 1976. Experiments on nervous–endocrine interactions and on neurosecretion and its control.

O'Malley, B. W., and W. T. Schrader. "The receptors of steroid hormones." *Scientific American,* February 1976. Discusses the molecular effects of steroid hormones on target cells.

Pastan, I. "Cyclic AMP." *Scientific American,* August 1972. Describes the roles cyclic AMP plays in regulating metabolism in organisms as different as bacteria and vertebrates.

Pike, J. E. "Prostaglandins." *Scientific American,* November 1971. Outlines the discovery of these elusive local hormones, their fascinating range of biological effects, and the outlook for using these molecules for medical purposes.

Saunders, D. S. "The biological clock of insects." *Scientific American,* February 1976. Presents hypotheses as to how internal rhythms operate, and outlines work to distinguish between them.

Turner, C. D., and J. T. Bagnara. *General Endocrinology,* 6th ed. Philadelphia: W. B. Saunders, 1976.

CHAPTER 42

BEHAVIOR

We are prone to conclude that a dog is "ashamed" when it puts its tail between its legs and sneaks into a corner after a spanking, and "happy" when it wags its tail. This type of thinking is called **anthropomorphism**, ascribing human emotions to animals. At the opposite extreme, we may say that a bird sings from instinct, because it is incapable of behaving intelligently. The view of human behavior as intelligent and that of other animals as instinctive is reflected in the tendency to ascribe actions of which we are ashamed to "animal instincts." Neither of these approaches to animal behavior gives much useful insight into why animals behave as they do. Recent research has attempted to study animal behavior with as little bias as possible from our human prejudices.

Because much behavior is so complicated, many workers have looked for simple behavior pattens that they hope will reveal the essential features of more complex activities. We are now finding out how certain individual neurons function in locomotion and escape reactions of invertebrates such as cockroaches, crayfish, the sea slug *Aplysia*, and leeches, but the picture is still far from complete.

Figure 42-1   Why do animals behave as they do? Is this porpoise leaping for joy, to escape predators, or what? (U.S. Navy)

In many ways, natural selection acts more directly on behavior than on anything else: dozens of different behavior patterns may distinguish the individual that reproduces from the one that does not, and all of an animal's anatomical and physiological adaptations are useless if the animal does not feed itself, escape from predators, and find a mate.

An animal's genes determine the range of characteristics it can develop, but just which traits develop depends upon interactions between the genes and the environment. In this chapter we shall consider how behaviorists may try to disentangle the genetic and environmental influences on an animal's behavior, and the kinds of selective pressures that have produced the varied behavioral repertoires of different animals.

42-A   Proximate and Ultimate Causes

A frog is sitting in the grass when a fly buzzes past. Zip! the frog's tongue flicks out and pulls the fly into the frog's mouth. How and why does the frog do this? The question "how" can be answered by a description of the frog's sensory, nervous, and muscular systems. The stimulus of a moving fly in the visual field sends impulses along sensory neurons to the central nervous system, which in turn activates and directs the tongue muscles used to catch the fly.

The question of "why" the frog catches the fly is different because it can be answered on two levels. The immediate, or **proximate**, reason is that the behavior pattern results from a nervous reflex. Seeing a fly activates a reflex system that results in the frog's striking at the fly. However, there is also an evolutionary or **ultimate** answer to the question "why?" That particular behavior pattern exists because it has been selected for during the course of evolution.

Three main ultimate causes, or selective pressures, have brought about the behavior patterns we see today:

Figure 42-2   A black-headed gull removing an egg shell.

1. Ultimately, an animal's behavior patterns will be selected for as they contribute to its reproductive success. It is occasionally possible to see how (why) a behavior pattern has evolved by showing that deviation from this behavior is disadvantageous. For instance, many birds remove the empty egg shell from the nest after the young has hatched. Niko Tinbergen showed that adult black-headed gulls that did not remove the shells lost more chicks to predators: the white inside of the egg shell allowed certain predators to discover the otherwise camouflaged nest and chicks (Figure 42-2).

2. On a more short-term basis, behavior patterns must allow an animal to solve immediate problems. Hungry animals must feed, and hunted animals must escape predators if they are to survive and reproduce.

3. Sights, sounds, and other stimuli in its environment continually bombard any animal. Behavioral adaptations permit an animal to detect stimuli that are important to survival or reproduction, and then to carry out behavior patterns appropriate to those stimuli. Mechanisms for discriminating between stimuli and for ensuring that an animal completes a behavior pattern are crucial parts of any animal's behavioral makeup.

Let us begin our attempt to disentangle proximate and ultimate causes of behavior by considering some direct experimental demonstrations that genes and the environment do indeed affect an animal's behavior.

42-B Genes and Environment

Most genes influence behavior because they control an animal's anatomy and physiology and therefore determine how it can and does respond to its environment. In laboratory populations of the fruit fly *Drosophila*, it is possible to select a mutant that differs from the other flies by only one gene and to see if that gene affects its behavior. Seymour Benzer has found single genes that cause a fly to do such things as court members of the same sex, copulate for much longer than normal, or follow a 19-hour rhythm of activity instead of the usual 24-hour cycle. These behaviors result from mutations that cause some abnormality in the sense organs or in the nervous or muscular systems.

Environment affects behavior in two main ways. In the short run, animals perform many behavior patterns only when they are induced to do so by environmental stimuli; in the long run, the environment influences gene expression in the development of many behavior patterns. Even reflex behaviors do not develop normally in restricted environments. A. H. Riesen reared newborn chimpanzees in the dark until they were 40 months old; when they were brought into the light, they showed no eye-blink reflex. The reflex finally appeared after the chimpanzees had been in the light for about five days. The presence of light is necessary for the development of this reflex.

One reason physical characteristics are easier to study than behavior is that most of the interactions between genes and environment that determine physiology and anatomy are complete by the time the animal is sexually mature. In animals with complicated nervous systems and musculature, genes and environment may interact and alter an animal's behavior throughout its life. We shall consider the development of behavior in immature animals in more detail in Section 42-F.

Figure 42-3 A lacewing takes off when prodded by the stick at the bottom of the picture. Many behavior patterns are immediate responses to environmental stimuli. (Biophoto Associates)

The idea that a behavior pattern is either instinctive or learned (or, more often, a combination of the two) is pervasive among behaviorists and in everyday life. **Instinctive** or **innate** behavior is genetically programmed in the "wiring" of the nervous system and is difficult or impossible to alter. **Learned** behavior is acquired or eliminated as a result of experience.

In the 1940s, disagreement over whether most of the behavior of birds and mammals is instinctive or learned (the so-called "nature-nurture" controversy) resulted in a deep rift between two schools of behaviorists. One school consists largely of American experimental psychologists. The tools of their trade are Skinner boxes* and mazes; they are almost exclusively interested in learning, particularly in such questions as how much animals can learn and how reward and punishment affect learning. This school has demonstrated that the behavior of many animals is highly susceptible to modification or learning. The best-known member of this school is the American B. F. Skinner, after whom Skinner boxes are named.

The findings of the experimental psychologists counterbalanced the influence of ethology, originally a European school of thought. **Ethology** is the study of animal behavior in nature; it emphasizes the evolution and adaptive value of behavior patterns. The founding fathers of ethology are Konrad Lorenz and Niko Tinbergen; in 1973 Lorenz, Tinbergen, and Karl von Frisch became the first field behaviorists to win the Nobel Prize.

The idea that much of animal behavior is instinctive came from ethologists' observations that members of the same species tend to show identical behavior patterns in the wild. "Instinct" is a difficult concept, however, because it is hard to define except by negatives: instinctive behavior develops without the animal's having to learn it. Such negative definitions are notoriously difficult to use. Furthermore, the only possible experiment to determine whether or not a behavior pattern is instinctive is to deprive the developing animal of as many environmental stimuli as possible and see if the behavior pattern still appears. And even if the pattern does appear under such circumstances, it may still not be instinctive; the experimenter might merely have failed to remove the stimuli that permit the animal to learn the behavior.

Adaptive Value of Learned and Innate Behaviors

One way to make sense of the diversity of behavior is to consider the selective advantages of the two extremes: innate and learned behavior.

Innate behavior is produced perfectly the first time it is performed. This is valuable if the cost of an initial mistake is high. For instance, kittiwakes are sea birds that nest on narrow ledges; the chicks stand still from the moment they hatch, whereas related herring gull chicks move around. This innate behavior (or non-behavior) pattern of kittiwakes is clearly adaptive because a false step means death to a kittiwake chick. There is no room for learning.

There is also selection for a behavior to become innate if the stimulus is always the same and if the same response is always appropriate. If the song of the male is always the same, the female who reacts without having to learn the song saves time and energy. There is no penalty for reacting automatically, and the reward may be finding a mate rapidly.

Energy saving is an important advantage of innate behavior. Learning may involve considerable expenditure of time and energy, and there is clearly a selective advantage to the animal that does not waste energy learning responses that are sure to be required frequently and without variation.

With all of these advantages of innate behavior, then, why are so many behavior patterns learned? In particular, why are vital behavior patterns, such as recognizing a mate or learning to fly, so often partly learned? Learning may be important because it gives the animal the flexibility to adapt to a changing environment by acquiring new behavior patterns as they become appropriate, or by responding in

---

*Skinner boxes Boxes used in experiments on an animal's ability to discriminate between stimuli and to learn. Typically, the animal is placed in the box, presses knobs in response to stimuli, and is rewarded with food.

new ways to old stimuli. For animals such as vertebrates, which have relatively long lifespans and so experience changing environmental conditions, this flexibility often means the difference between life and death.

Learning is also necessary whenever a stimulus differs for individual members of a species. For instance, every mobile animal with a home base must learn to find that home. No amount of genetic programming will permit a crab to find its own burrow among all the holes on a sandy beach. Differences among local conditions also probably select for the existence of a learned local dialect component in many bird songs. Local populations of animals also evolve genetic adaptations that make them better suited to survive in their own habitat than in any other; it is advantageous for their genes to include the ability of the animals to learn that they are members of a particular population and to choose mates within it, since interbreeding with other populations is likely to produce offspring that are genetically adapted to neither locality. In addition, many social animals live in environments where the relationships between individuals change constantly, and such relationships must almost always be learned.

In general it seems that learning plays a large part in the development of behavior patterns evoked by stimuli that are local or changeable. Innate behavior is appropriate where the stimulus is always the same, where speed of reaction is important, and where the cost of an initial mistake is high.

A species' way of life also determines whether its members evolve learned or innate behavior patterns. Consider a solitary wasp, which hatches alone, develops as a larva and matures with practically no interaction with other members of her species. The behavior by which she finds a male, mates, builds a nest, and lays her eggs must be largely innate in order for her to perform each action perfectly the first, and perhaps the only, time in her life. On the other hand, a social animal such as a cat can learn much of its behavior from observing other members of its group. It would, however, be an enormous oversimplification to say that the behavior of an insect is innate, and the behavior of a mammal is learned. In any group of animals above the annelids, both types of behavior are vitally important and frequently impossible to disentangle. Even the solitary wasp learns to search for food, to find her way back to her nest, and many other behavior patterns during her short life (Figure 42–4). Similarly, mammals have many innate behavior patterns.

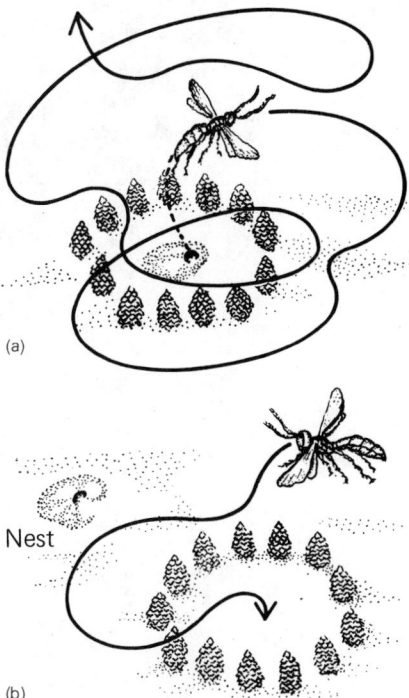

Figure 42-4 Learning by an invertebrate. A digger wasp finds her nest by visual landmarks. (a) The wasp makes an orientation flight over a nest entrance that the investigator has surrounded with pine cones. (b) She returns to the center of the ring of cones, which the investigator has moved in her absence.

42-D The Neurophysiological Basis of Behavior

At a physiological level, a behavior pattern is the action of an animal's effectors* in response to a stimulus activating its receptors.* Between receptor and effector lies the nervous system, which determines what information travels from one to the other. In many ways, the nervous system is still the "black box" of behavior. The stimulus that goes in and the behavior that comes out can often be defined, but precisely what goes on inside the nervous system is, in most cases, a mystery. However, the characteristics of a behavior pattern must reflect the organization of the nerve cells that control it.

Vertebrate Reflexes and Behavior

A reflex is the simplest model of the neural basis of behavior. The few neurons of a reflex pathway process a stimulus in such a way that a predictable behavior pattern always results.

Reflexes and more complex behavior patterns share a number of properties that result from the way neurons operate. Both show **latency**, a time delay between stimulus and response. Latency is due to the time necessary for transduction in the sense organ, conduction through the nervous system, and excitation of the effector.

A common property of behaviors is that the stronger the stimulus, the shorter the latency (Figure 42–5). This is true of the flexion reflex by which a dog withdraws its leg in response to a painful stimulus, and of more complex behaviors. Chaffinches (birds), presented with objects that elicit their alarm call, respond most quickly to the stimulus that is known on other grounds to be the strongest.

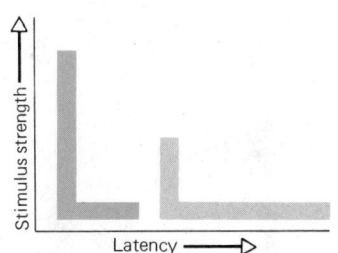

Figure 42-5 Relationship between stimulus strength and latency in the flexion reflex of a dog. Latency is the time that elapses between stimulus and response. If the dog's foot is pricked with a pin, the stronger the prick the faster the dog withdraws its leg.

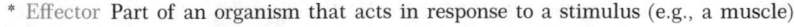

* Effector Part of an organism that acts in response to a stimulus (e.g., a muscle).
* Receptor The part of an organism that detects a stimulus.

The central nervous system may add together stimuli arriving from different sense organs or at different times, a phenomenon known as **summation**. In the early 1900s, the great physiologist Sir Charles Sherrington gave a clear example of this in the scratch reflex elicited by tickling a dog's back. Sherrington showed that two stimuli, each of which was too weak to produce a scratching response by itself, could add to one another and elicit scratching if they were applied simultaneously several centimetres apart. Similarly, it has been shown that male rats respond sexually to a combination of visual, tactile, and olfactory stimuli from a receptive female. Young males will not respond unless at least two of these stimuli are present. These properties of behavior patterns clearly arise from the properties of the neurons involved.

Fixed Action Patterns

A striking feature of the behavior of any animal is its repertoire of **fixed action patterns** or **rhythmic behaviors**—acts, involving the use of many muscles in a precisely timed sequence, that are always performed in an essentially identical pattern. Reflexes are the simplest fixed action patterns, but more complicated activities such as locomotion, sound production, breathing, and feeding also fall into this category. Analysis of rhythmic behaviors has concentrated mostly on invertebrates, which use many fewer neurons than vertebrates do to achieve similar behavior patterns. In some invertebrate systems, every neuron involved in a particular behavior pattern has been identified, a level of analysis that is still unthinkable in any vertebrate.

Studies of invertebrate neurophysiology suggest that fixed action patterns differ from other behaviors in two ways: they are controlled by very few neurons in the central nervous system, and they can occur without feedback from the sense organs, although such feedback is available and is often used.

By contrast, consider what happens when you pick up a glass of water. This behavior, which is not a fixed action pattern, is a complicated series of interactions between the sense organs and muscles in the arm. The sense organs signal how much the water is slopping about and how far and how fast the glass is rising. These sensory messages reach thousands of neurons in the central nervous system, and hundreds of motor neurons respond, controlling the muscles in the arm so that the glass rises steadily and the water does not spill. Hundreds of central neurons and continual feedback from the sense organs are necessary for this behavior pattern.

With a fixed action pattern, this is not necessarily the case. For example, the fixed action patterns by which a crayfish flexes its abdomen when it swims are induced by the activity of single, identifiable neurons in the central nervous system. At least some fixed action patterns are "programmed" into the nervous system. They are always produced in identical fashion because only one or a few control cells trigger all of the motor neurons for the entire behavior pattern.

Fixed action patterns stop and start under the influence of stimuli just as other behaviors do, but unless it is stimulated to stop, a fixed action pattern runs its course automatically even if some parts of the behavior are inappropriate at the time. A dog hiding a bone under the living room rug moves as though covering it up with earth. Similarly, most dogs turn around several times before they lie down to sleep, although there is no grass to be trampled down.

Some fixed action patterns are innate; others are learned. The cockroach's escape reaction is innate, but a rat's pressing a lever for food is a fixed action pattern that has been learned. Each rat presses the lever with a characteristic gesture. One uses a fist, another one finger, and each uses its own gesture time after time. Similarly, how you hold your pen, walk, play the piano, or ride a bicycle is a learned fixed action pattern that is very conservative and unique to you.

The selective advantage of fixed action patterns programmed into the nervous system is probably that they reduce the number of neurons used in a relatively complex task that must be performed perfectly and often. Some stereotyped behaviors, such as locomotory and escape movements, must be performed perfectly to work at all. Others save energy; the muscular movements of writing or feeding need not be worked out with sensory feedback every time they are performed. All animals seem to be able to "write" programs for fixed action patterns into the nervous system during embryonic development and, in many cases, in later life.

(a)

(b)

Figure 42-6 (a) Walking is a fixed action pattern. These emperor penguins need little feedback from their sense organs to stroll across the ice. (b) Landing is *not* a fixed action pattern. This blue tit needs all the information it gets from its superb binocular vision to stop flying and balance on this branch. (a, U.S. Navy; b, Biophoto Associates)

Sign Stimuli

What sorts of stimuli trigger behavior patterns? In the 1940s Tinbergen was studying male three-spined sticklebacks, which sported the red belly characteristic of these fish in breeding condition. Every time a red truck drove past a nearby window, all of the fish made frantic attempts to swim through the glass of their tanks toward the truck, as if they would attack it. A male stickleback in reproductive condition will also attack other breeding males.

What stimulus provoked this attack behavior? Tinbergen presented various models to the sticklebacks to find out. When he showed wooden models of sticklebacks to males in reproductive condition, they attacked crude models with an eye and a red belly in preference to lifelike models without the red belly (Figure 42–7). The red belly of the male stickleback in reproductive condition is thus the sign stimulus that triggers the fixed action pattern of attack by another breeding male. A **sign stimulus** (also known as a **releaser**) is that portion of the total stimulus which releases a particular behavior pattern.

A sign stimulus does not invariably provoke its particular fixed action pattern; the animal's physiological condition and nervous system filtering may interfere. Thus, only when they are in breeding condition, with high levels of the hormone testosterone, do male sticklebacks attack other fish with red bellies.

A stimulus may be filtered out in the nervous system so that it never produces a response. A stimulus hierarchy in the nervous system ensures that an animal reacts to stimuli in a particular order. For instance, the sign stimulus of food will normally stimulate feeding behavior in a hungry mouse. If the mouse sees a cat about to pounce at the same time that it sees the food, however, the food stimulus will become ineffective, or be filtered out, while the mouse avoids the cat. The hierarchy may change with the physiological state of the animal. In general, stimuli that elicit escape behavior take precedence, but this can alter; a starving animal may ignore stimuli that would normally send it scuttling for safety.

 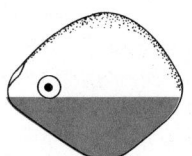 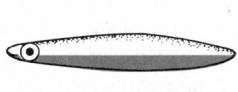

Figure 42-7 The sign stimulus for attack by a male stickleback. Niko Tinbergen found that male sticklebacks in breeding condition do not attack the life-like model stickleback lacking a red belly, but they will attack either of the crude models with a red undersurface. The eye is also necessary. The presence of an eye is often necessary for an animal to identify an object as another animal.

(a) (b) (c)

Figure 42-8 Models used in testing for releasers. Young herring gull chicks peck at the parent's bill; this induces the parent to feed the chick. (a) A life-like (though flat) model of the parent's head releases fewer pecks by a newborn chick than (b) a model in which the bill is longer and thinner than normal; this model is less effective than (c), a model that is long and thin and emphasizes the contrast between bill color and bill-patch.

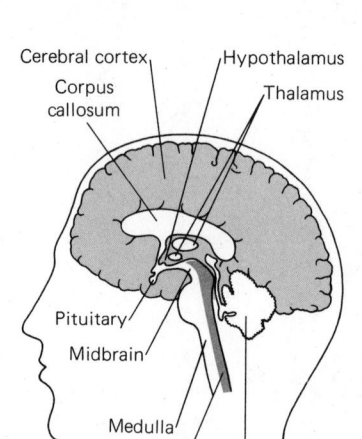

Figure 42-9 Position of important structures in the human brain.

An interesting extrapolation of the theory of how sign stimuli work is that it is possible to produce a model called a **supernormal stimulus**, which provokes a behavior pattern more effectively than does the normal stimulus. For instance, herring gull chicks peck at the stimulus provided by a red spot on the parent's bill. (This induces the parent to feed the chick with regurgitated fish.) When models of the stimulus were tested, it was found that a bar with big red and white stripes provoked more pecks from young chicks than did a realistic model of the bill (Figure 42-8); the bar was a supernormal stimulus for the fixed action pattern.

Drive and Motivation

A particular stimulus may evoke a different response in the same animal at different times. For example, an animal that sees food will eat it if it is hungry but may ignore the food if it has just eaten. Something inside the animal, which we may call **motivation** or **drive**, is different at these two times. Since different behaviors are appropriate at different times even when the stimulus is the same, variations in motivation help to ensure that an animal's behavior changes to fulfill its short-term needs.

Research on mammals has cast an interesting light on the interaction between stimuli and motivation. Stimuli reach the reticular formation of the brain, where they may produce a state of arousal (Figure 42-9). This changes the animal's responsiveness to subsequent stimuli.

In the vertebrate brain, the hypothalamus is the area most highly implicated in the control of motivation. Attack, escape, and sexual behavior can be evoked by electrical stimulation of certain parts of the hypothalamus. The hypothalamus seldom acts alone to determine motivation; its activities can be modified by input from higher and lower centers in the brain and by hormone levels.

42-E   Learning

Learning produces adaptive changes in an individual's behavior as a result of its experiences. It occurs in so many different ways that we have to classify them somehow, although there is no evidence that the classification used here bears any relationship to the largely unknown physiological basis of learning.

Habituation is the loss of old responses. Animals may learn not to respond to stimuli that are repeated frequently and are unimportant to them; young animals often show alarm behavior at a variety of stimuli, most of which they rapidly learn to ignore. Habituation is advantageous in increasing the animal's reaction to new stimuli, which stand out against the background of stimuli to which the animal has become habituated.

Conditioned reflexes are behavior patterns evoked by a previously neutral stimulus which an animal has learned to associate with the stimulus that normally elicits the reflex. The Russian physiologist Ivan Pavlov showed that there is a reflex which causes hungry dogs to secrete saliva when they see food. Pavlov rang a bell when he showed food to the dogs, and after several trials the dogs would salivate when the bell was rung even though he stopped showing them food. The dogs had learned to respond to the new stimulus, the bell, to which they had not previously responded. Pavlov called this the **conditioned stimulus**. A conditioned response to a negative stimulus, or punishment, can be formed in the same way.

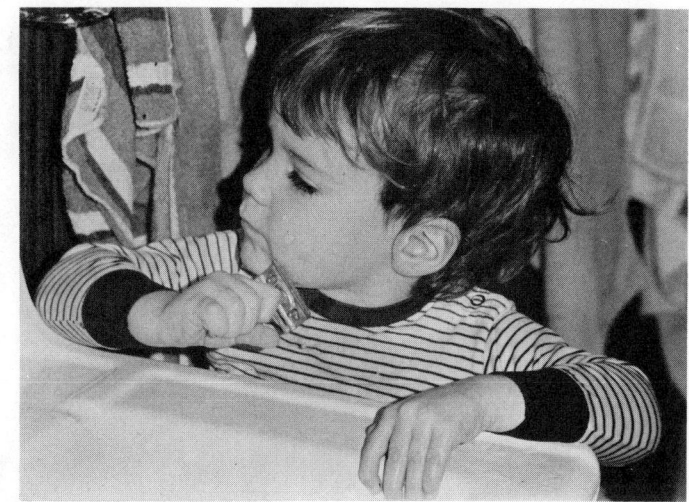

Figure 42-10 Animals learn many things by trial and error. Social animals, in particular, also learn things by imitation.

Trial and error learning is what its name implies. An animal's spontaneous movements may by chance produce a reward, and the animal learns by trial and error to repeat that behavior pattern. The reward may often be the "pleasure" of performing an action more accurately than before. Trial and error is probably the most appropriate category for the learning of new motor skills. Young mammals and birds perfect their prey-catching movements, and humans learn to play the piano, by a trial and error form of practice.

All of these types of learning are varieties of **associative learning**, first described by Pavlov. **Reinforcement** (reward or punishment) is a central feature of associative learning, and associative learning usually occurs most readily where stimulus, response, and reinforcement occur fairly close together. Modern experiments have shown, however, that dogs and other animals can learn to avoid a particular type of food if the food makes them sick many hours after they have eaten it. Another characteristic of associative learning is its improvement with repetition. Associative learning permits an animal to generalize from the learned stimulus to other stimuli, or to discriminate the stimulus to as fine a level as the sense organs will permit, if such discrimination is rewarded. If rewards for a conditioned response cease, the response becomes reduced and eventually disappears.

Latent learning occurs without any obvious reward or punishment; it is learning that produces no obvious behavior at the time it occurs. This often happens during exploratory behavior. A recently fed animal may give no sign that it has noticed a new food source until it later returns to feed there.

Insight learning is a form of reasoning that draws on the results of past experiences to arrive at the solution of a novel problem. The classic example of insight in animals came from the work of Wolfgang Köhler on chimpanzees. Presented with a bunch of bananas too high to reach, they would pile up boxes to make a stand from which they could reach the bananas (Figure 42–11). Reasoning of this sort has been shown in many mammals and in some birds, although it is often difficult to distinguish from other forms of learning.

42-F Development of Behavior

An animal's behavior, like its anatomy and physiology, forms during its development, through the interaction between its genetic makeup and its environment.

There are often critical periods when a particular environmental influence must be present if a particular behavior pattern is to appear. Torsten Wiesel and David Hubel sewed one eye of a kitten shut just after birth. When they opened it some months later, the kitten was found to be permanently blind in that eye because the neurons in the visual cortex of the brain had developed abnormally. If the eye of a normal adult cat is sewn shut for a year, its vision is unaffected. There is thus clearly a critical period during its life when a kitten's eye must be exposed to light if its visual nervous pathways are to develop normally.

Figure 42-11 Insight learning. In Köhler's famous experiment, a chimpanzee was left in a room with a number of boxes and a bunch of bananas hanging from the ceiling. After a period (perhaps of thought) the chimpanzee piled the boxes on top of one another, climbed up, and reached the bananas.

Figure 42-12 A jackdaw. (Biophoto Associates, N.H.P.A.)

Another example of a critical period has been found in studies of **imprinting** in young animals. Goslings (young geese) and ducklings learn to follow their parents, and to respond to their parents' signals, during a critical period after they hatch. Konrad Lorenz found that young birds would follow him as if he were their mother if they saw him rather than their mother during the critical period.

Many animals learn what their future mates will look like by a similar process of sexual imprinting during a critical period. (Lorenz had a tame jackdaw (Figure 42–12) that unfortunately became sexually imprinted on him before he understood how the process worked. It caused Lorenz great inconvenience by stuffing regurgitated worms into his ear during its "courtship feeding.")

Imprinting is sometimes involved in the development of social behavior in mammals. The Harlows' primate research group has shown that rhesus monkeys must be exposed to the social life of members of their own species for normal sexual and maternal behavior to develop. The same is true of human beings. However, with both rhesus monkeys and humans, abnormal behavior, which may result from

Figure 42-13 Sexual partners are often determined by imprinting. The large bird is a male Lady Amhurst pheasant that was reared by a bantam hen. He is inappropriately displaying his sexual allure to an unappreciative bantam. (Biophoto Associates, N.H.P.A.)

learning inappropriate social behavior as a juvenile, can be reversed under the influence of associative learning at a later age.

The interactions between genetics and environment involved in development of mature behavior patterns have been studied intensively in the case of bird song. Mark Konishi and Fernando Nottebohm deafened male white-crowned sparrows (Figure 42–14) at various stages of their development by removing the cochlea of the inner ear. If this is done to a very young bird, it will never sing a song at all, producing instead only a series of disconnected notes. In order to produce even the inherited song of its species, the bird has to be able to hear itself. (Human beings who are deaf from birth also have great difficulty learning to speak.)

Peter Marler and M. Tamura showed that male white-crowned sparrows reared in isolation sing only the inherited song of their species, whereas in the wild, white-crowned sparrows learn the distinctive dialect of their own local population by listening to adult birds singing. To complicate matters, the birds do not learn the dialect when they sing it, but many months earlier. A bird must be exposed to the dialect during a critical period when it is about 3 months old if it is to produce the dialect when it first begins to sing, at the age of one year. And even if it has heard the dialect during the critical period, it will never sing this dialect correctly if it is deafened before it has also sung the dialect. Once a bird has sung the full dialect song, however, deafening has no effect on its further performance (Figure 42–15). This example shows that many factors may be involved in the development of a normal adult behavior pattern.

One further conclusion from the work on bird song is that an animal inherits a tendency to learn some behavior patterns but not others. White-crowned sparrows learn the dialects of their own species, but exposing them to the songs of other (even closely related) species during the critical period does not make them learn the songs of these other species. Birds treated in this way end up with a song sounding like that of the completely isolated male.

Animals are genetically predisposed to learn the behavior patterns of their own species. Tinbergen showed that herring gulls learn to recognize their own chicks but not their own eggs; presumably chicks stray from the nest and have to be retrieved, whereas eggs do not. Guillemots, a related species that build no nests but lay their eggs on open cliffs, do recognize their own eggs and retrieve them; eggs not in nests may roll around, and it is clearly adaptive to learn which ones to retrieve, ensuring that parents are perpetuating their own genes.

The environmental stimulus necessary to the normal development of a behavior pattern may be very precise, or it may be rather general. Rats or mice that have been picked up and returned to the nest once or twice a week in their youth mature more quickly, in behavioral terms, than do those never handled. This is a rather unspecific stimulus to the maturation of behavior.

Figure 42-14 A white-crowned sparrow. (Cornell Laboratory of Ornithology)

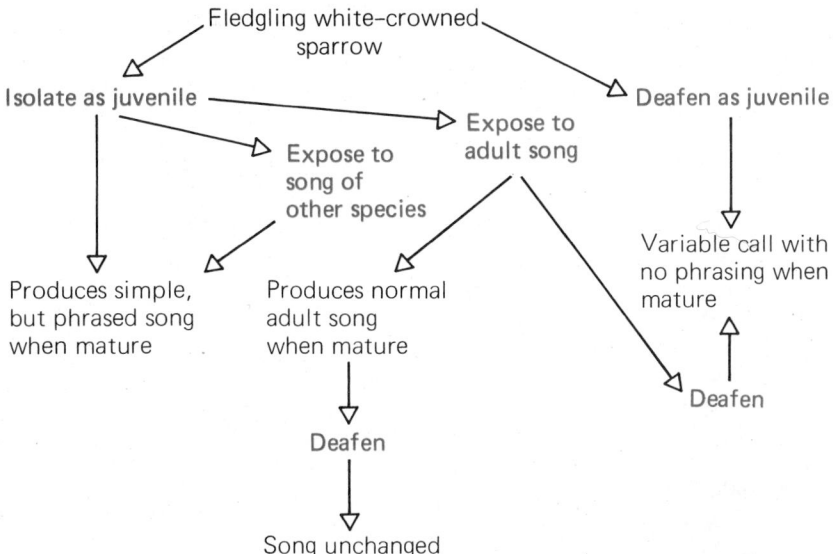

Figure 42-15 Summary of findings of Marler, Tamura, and Konishi on song development in white-crowned sparrows.

Figure 42-16 Territorial behavior. Two families of Brent geese threaten each other across the boundary between their territories. (Biophoto Associates, N.H.P.A.)

42-G   Territorial Behavior

Now that we have seen something of the evolutionary origins and physical basis of behavior, we go on to consider particular behavior patterns that illustrate these ideas.

Many animals defend **territories**, areas where they raise their young or where they have a monopoly on the food resources. A territory holder attacks and drives away other members of the same species. Aggression in territorial behavior must be very precisely controlled. It is to an animal's advantage to defend the territory with a minimum of attack behavior, since every attack carrries the risk that the attacker will be injured or spotted by predators. Animals have evolved several features that minimize damage during territorial encounters. For instance, fighting is infrequent because there are "rules" about who wins encounters between two individuals.

Consider a male thrush defending a territory before the female arrives in the spring. The male is most aggressive near the center of his territory. As he moves toward the boundary, his attacks on a trespassing neighbor become less violent, until he reaches a point at which he is as likely to escape as to attack when he sees another male thrush. This point marks the boundary of his territory. When two neighbors meet at the boundary of their territories, they both act as if they have conflicting escape and attack motivations; these tendencies are manifested as conflict behaviors.

Conflict behavior usually contains elements of the two conflicting tendencies (in this case movements toward escape and toward attack), as well as containing movements apparently unrelated to the issue at hand, such as preening* or pecking at the ground (Figure 42–17). In many species, patterns of conflict behavior appear

Figure 42-17 Conflict behavior of the type known as displacement behavior. A gull involved in a territorial clash violently pulls up clumps of grass. The bird acts as if it is caught in a conflict between tendencies to attack and to flee. Instead of doing either, it engages in apparently irrelevant displacement activity—pulling up grass. A more placid form of grass-pulling is part of its nest-building behavior.

---

*Preening   Maintaining the feathers by oiling and smoothing them with the bill.

Figure 42-18 An emperor goose at his nest threatens an intruder. (Biophoto Associates, N.H.P.A.)

to have evolved into ritualized **threat** displays that are directed toward intruders. The physiological stress induced by opposing behavioral tendencies is often evident in threat behavior. For instance, erect hair or feathers, a result of activity of the sympathetic nervous system, is a common manifestation of threat behavior in mammals and birds. Threat is obviously more advantageous than actual fighting in that it does not injure the animal. In the case of a threat display between two animals (which is effectively a ritualized fight), an experienced observer can predict which animal will win by deciding which animal incorporates more attack movements in its display. The loser will eventually move away from the winner.

42-H   Conflict and Courtship

Most animals, even those that live in social groups, maintain a minimum **individual distance** from one another. For example, swallows sitting on a telephone wire are always a certain minimum distance apart. The invasion of individual distance is a threat, and the invading animal is usually attacked. The conflicting tendencies to attack and to permit another animal to come close enough to mate are often evident in **courtship behavior**, the behavior patterns that precede mating in most animals. The courtship displays of many species seem to have evolved from such conflict behavior.

In a well-studied example of courtship behavior, the male black-headed gull attracts a female to his territory. She alights near him, and both gulls adopt a series of postures that resemble, but are slightly different from, the characteristic threat display of the species. If neither bird attacks the other, both display appeasement gestures, which imply that the hostility between them has lessened. Eventually the female flies off, but she may return many times, and each bird will display fewer threatening and more appeasement gestures with each visit. Eventually the greeting ceremony ceases entirely, and the male feeds the female. After this, copulation can

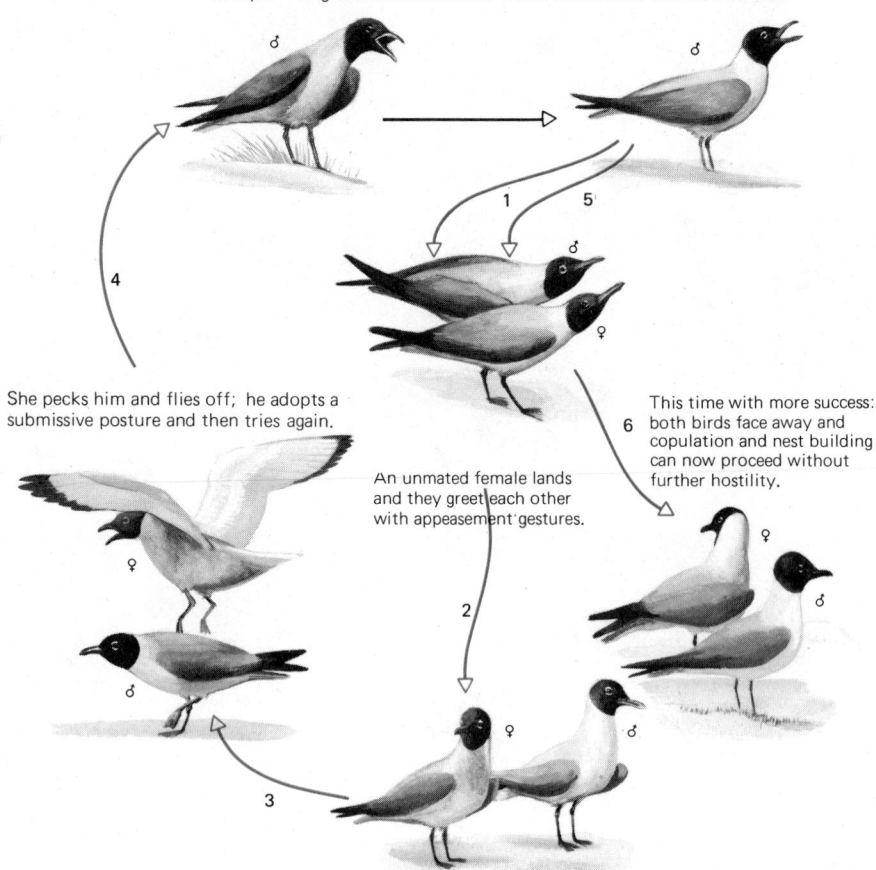

Oblique — long call intimidates other males and attracts unmated females.

She pecks him and flies off; he adopts a submissive posture and then tries again.

An unmated female lands and they greet each other with appeasement gestures.

This time with more success: both birds face away and copulation and nest building can now proceed without further hostility.

He turns his head away with bill down in another appeasement gesture. But she does not reciprocate; her head still points forward and her wings are raised in an aggressive posture.

Figure 42-19 A courtship sequence in the black-headed gull. Follow the arrows around from the top.

occur and a permanent pair bond forms (Figure 42–19).

In their initial encounters, the birds are displaying behavior that reveals three conflicting tendencies—to attack, to flee, and to stay together. The behavior patterns that result from the conflict have evolved into an elaborate courtship ritual.

42-I Migration and Homing

Many animals have remarkable navigational ability, sometimes traveling over hundreds of miles of land and sea. Migrating animals can do many things that we cannot do ourselves. A Manx shearwater, which had never been more than 10 miles from home, was removed from her nest on an island off the coast of Wales, flown to Boston, and released (Figure 42–20). She was back on her nest before the letter announcing her release reached the observers in Wales. To perform an equivalent feat, such as sailing from Boston to Wales, a human being would have to spend hours learning to use compass and sextant* to cross the ocean, and would still need a map to find the nest on the other side. Birds, monarch butterflies, fish, and salamanders all perform equivalent journeys without mechanical aids and with little or no learning.

Many animals, like horses, dogs, cats, and humans, orient themselves by landmarks that they learn and recognize visually. Many animals can also find their way around by using their chemoreceptors.* Dogs can follow long and complicated scent

---

* Sextant A navigational instrument used to measure the altitude of the sun or stars above the horizon, a measurement from which one's position on the earth's surface can be calculated if one also knows the time.

* Chemoreceptors Sense cells that respond to chemical stimuli.

Figure 42-20 Manx shearwaters. (Biophoto Associates, N.H.P.A.)

trails in unfamiliar territory; moths find mates and ants find their nests by following odor gradients. A dramatic case is that of the salmon, which hatches in a freshwater stream and matures hundreds of miles away in the ocean. Seven years later, when the time comes to spawn, each salmon finds its way back to the very stream in which it hatched. If its olfactory* organs are plugged or destroyed, a salmon can no longer distinguish between chemicals in experimental situations, nor can it find its spawning stream. It is unclear whether the remembered odor of its birthplace guides the salmon from the time it first enters a river or merely directs it finally to the right stream.

Many animals can move in a specific compass direction. Von Frisch showed that bees can tell direction from the sun, and Gustav Kramer showed that birds can orient themselves in a specific compass direction in the same way. The sun appears to move from east to west during the day; thus to maintain a constant compass direction by reference to the sun, a bird must also know the time of day. Animals have internal clocks that control their circadian rhythms (see Section 41-E); by exposing birds to artificial daylight that began and ended 6 hours (one-fourth of a day) later than natural daylight, it proved possible to "clock-shift," or reset, the internal clocks of migrating birds. The birds then interpreted the sun's position incorrectly and oriented themselves in a compass direction that was 90° (one-fourth of a circle) clockwise away from the correct direction for migration (Figure 42–21).

*Olfactory Pertaining to the sense of smell.

(a) LIGHT AND DARK DURING CLOCK-SHIFTS

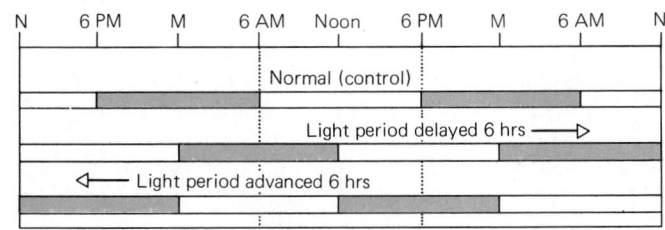

(b) PATHS OF ANIMALS RELEASED AFTER EXPOSURE
TO ARTIFICIAL NIGHT AND DAY

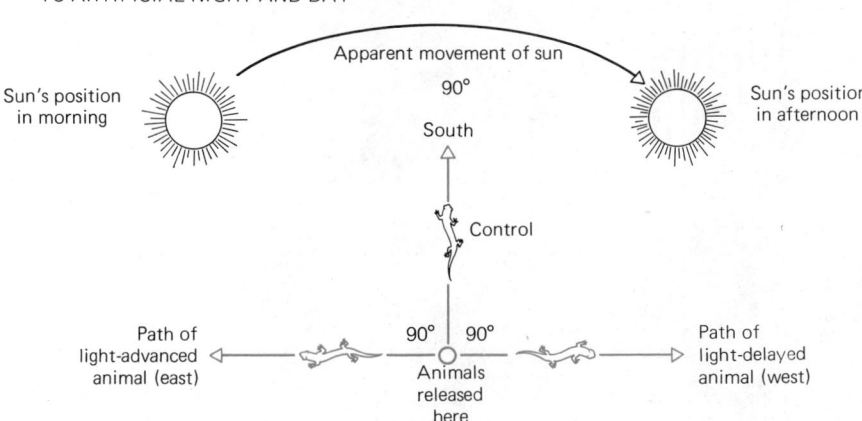

Figure 42-21 Effect of clock-shifting on compass orientation in migrating salamanders (in the Northern Hemisphere). The animals were captured as they migrated south and exposed to artificial light and dark (a) which reflected normal night and day or a "day" in which "daylight" was advanced or delayed by six hours. (b) When they are released, the control animal sets off south as expected; the experimental animals go east or west, which is 90° away from south on the compass. This is evidence that an internal clock (which can be reset by artificial light and dark) and the sun are involved in orientation. The light-delayed animal's internal clock tells it that the time is 8 a.m. when the animal is released at 2 p.m. To travel south, the animal keeps the sun on its left and therefore actually sets off to the west, since the animal is mistaken (clock-shifted) as to the time of day. (Kraig Adler, *Photochem. Photobiol.*, vol. 23, p. 288, 1976 and Pergamon Press)

Figure 42-22 A flock of several different species of terns migrating up the coast of South Africa. (Biophoto Associates, N.H.P.A.)

These experiments confirmed that by combining information from the direction of the sun and their internal clocks, birds can point themselves in a particular compass direction. It was later shown that nocturnal migrants can also use star patterns to find a compass direction.

The ability to fly on a constant compass setting using the sun or stars, however, does not explain the ability of the Manx shearwater to cross the Atlantic or of a homing pigeon to return to its loft after it has been released in unknown country. In order to get from A to B using a compass you have to know whether A is north, south, east, or west of B and how far away it is. This is called a **map sense** because it means that you must know the relative positions of A and B on a hypothetical map. Many animals obviously have a map sense, but we have not the slightest idea what it is.

When pigeons are prevented from using the sun compass (they are fitted with opaque contact lenses and clock-shifted!), they can find their way home by using magnetic cues; presumably they can detect the earth's magnetic field. Navigation by magnetic cues is not affected by clock-shifting, but it can be upset by attaching small magnets to the pigeons. Honeybees can also orient themselves with respect to the earth's magnetic field. Many questions remain unanswered: how does an animal detect a magnetic field? The earth's magnetic field can obviously give a crude compass direction, as it does in a mechanical magnetic compass. Can it also supply an

Figure 42-23 Pigeons are famed for their homing ability. One hospital in England saves time and money by using them to fly blood samples 20 kilometres to the laboratory. Here its proud owner displays the bird that won the 1979 France to England race. (Biophoto Associates, N.H.P.A.)

animal with a map sense? We do not know. If not, an animal's map sense—its ability to find out where on earth it is—remains as mysterious as ever.

Studies of how pigeons navigate are greatly complicated by the redundancy of their navigational system; pigeons have an innate ability to find their way home using any one of several cues. Learning also plays a part, and experienced birds reach home faster than naïve ones. Inexperienced birds use a simple hierarchy of navigational cues. If the sun is shining, they use a sun compass, so that if they have been clock-shifted, they fly off in the wrong direction. Experienced birds are less likely to be fooled by a clock shift. They appear to cross-check the information from their magnetic and sun compasses, which will give conflicting information if the birds have been clock-shifted. In this situation an experienced bird goes to sleep in the nearest tree until the effect of the clock-shift wears off; it then flies straight home. It is clear that even extremely complex innate behavior patterns can be modified by experience, even in a bird as notoriously hard to educate as a pigeon.

Although much has been learned about animal navigation and orientation in the last two decades, we are still far from understanding this remarkable collection of behavior patterns.

42-J Social Behavior (Sociobiology)

We define ''social'' as involving cooperation between members of the same species. Some animals have very little contact with members of their own kind, but in many species, some degree of interaction with others is apparent. This may be of limited duration, as it is in flocks of chickadees and red-winged blackbirds that forage together in the winter, or it may last throughout the animal's life and be vital to its survival.

It is often difficult to tell whether particular groups of animals are truly social. For instance, gray herons spend much of their lives alone or with their mates (Figure 42–24). Occasionally, however, these birds spend days, and even weeks, together feeding. Is this feeding group social? Do the birds communicate with each other or help each other with the fishing? Do they take turns standing guard and warning each other against predators? It turns out that the herons are all in the same place only because each of them has found food there. They do not interact any more than human diners in a restaurant do. Thus it is not safe to assume that a group of animals constitutes a society. A group of animals is truly social only when there is considerable interaction among the individuals. Examples of extremely social animals are honeybees, humans, and wolves, which form cooperative, long-lived societies upon which the individual's very life depends.

Figure 42-24 A gray heron at its nest. (Biophoto Associates)

Communication

All animals that live in societies communicate with other individuals; those animals with more complex societies tend to have more elaborate methods of communicating. Every communication involves action by a communicator and reception by another individual. Human beings, for instance, use sound and hearing (when we speak, clap, or laugh), and visual stimuli and vision (advertising posters, dressing up, shaking a fist) among our means of communication. Birds, like humans, have highly developed vision. It is not surprising that they communicate largely by movement and color (also by sound and hearing). Because our sense of smell is poor, we pay little attention to the chemical communication so common in other animals. Many mammals, such as dogs, mark their territories, determine another animal's mood, find their mates and food, and, for all we know, communicate in many other ways, by scent. (We do not even have a common word, equivalent to ''blind'' or ''deaf,'' for lacking the sense of smell.)

A pheromone (see also Section 41-F) is a chemical whose function is communication among members of the same species. A better definition might be that it is a chemical that affects the behavior of another member of the species, because we can never discover the ''intent'' of an animal of another species and have to judge a pheromone's function by noting whether the chemical influences behavior. We shall consider the importance of pheromones in honeybee societies later in this section.

(a)

(b)

(c)

Figure 42-25 The life of a honeybee. (a) A worker stands on her head and fans pheromone from her scent gland through the hive. (b) Workers tending larvae (white objects). (c) A swarm. (Biophoto Associates, N.H.P.A.)

Communication by sound can be highly elaborate even if it does not involve language, which is the most important aspect of our own sound communication. Crickets, frogs, and mosquitoes, among others, produce sounds in the mating season that have two effects—they tell a listening individual whether or not the sound-maker is a member of the same or a different species, and they permit the sexes to find each other.

Alarm calls, pheromones, speech, and courtship displays have presumably evolved primarily for their value in communicating with other individuals. Other signals which animals produce, such as the electric currents emitted by electric fish and the sonar signals of cetaceans, are also used for communication, although they probably originally evolved under other pressures—for navigation equipment in the case of sonar and of electric organs that permit their owners to detect objects in the water around them.

Honeybee Societies

Many insects are more or less social, but honeybee (and ant) societies are the most elaborate and widely studied. The unit of social organization is a family of related individuals. A society of honeybees (*Apis mellifera*) typically consists of a reproductive female (the queen) and her daughters (and sometimes her sons).

A honeybee hive may contain 80,000 individuals, each with its own job. Labor is divided so that the queen lays eggs, drones (males) produce sperm needed to fertilize the eggs, and workers (sterile females) carry out different tasks in sequence during their lives: tending larvae, cleaning the hive, and foraging for food. The tasks a bee performs, the number of queens produced, and the founding of a new hive are all organized by pheromones. It is not at all obvious what selective pressure could lead to the evolution of sterile workers who devote energy to raising another bee's (the queen's) offspring. The genetics of this peculiar situation are explored in Section 18-F.

The queen mates once, stores the sperm, and uses it to fertilize the thousands of eggs she lays during her life of seven years or more. Her eggs hatch into larvae, cared for by the workers. The diet fed to a larva determines whether it develops into a queen or a worker. A new worker usually first serves as a nurse, preparing cells for eggs and feeding the larvae after they hatch. After about two weeks the worker becomes a house-bee, cleaning, secreting wax for the honeycomb, and guarding the hive. After this she forages outside the hive for the remaining five or six weeks of her life.

As we would expect from such a complex society, honeybees communicate extensively. Karl von Frisch, who shared a Nobel Prize with Tinbergen and Lorenz, found that foraging bees returning from a successful trip "dance" on the honeycomb, recruiting other bees to harvest a new good food source, and telling them where to find it (Figure 42-25). Pheromones permit bees to identify their own hive and serve as alarm signals. A pheromone produced by the queen prevents the workers from producing any more queens and ensures that all the female larvae are fed so that they develop as workers. This continues until the hive is overcrowded, when the queen stops producing that particular pheromone, and the workers start to raise new queens. The queen also lays unfertilized eggs, which develop into drones, one of whom may furnish sperm for the queen of a new hive.

Eventually the old queen may leave, for only one queen can survive in a hive. As she leaves, the queen secretes a swarming pheromone, which attracts many of the workers and keeps them with her. The swarm lands somewhere and may remain several days while scouts search for a new site. The scouts return to "dance" a description of the location of a possible new nest. The intensity of her dance conveys the scout's impression of the merits of the site. Other workers go to inspect the sites, and finally a consensus emerges when all the scouts are dancing for one site. The swarm then flies to the new site and settles in. This method of making a decision impresses us by its resemblance to the way we like to think humans act.

Vertebrate Societies

Like insect societies, most vertebrate societies consist of genetically related individuals. Unlike insect societies, however, all members of the vertebrate society are fertile, and competition to reproduce is the main principle determining the social system. A typical vertebrate society consists of a leader (usually a male), his mate(s), and their descendants.

Most vertebrates learn a much greater proportion of their behavioral repertoires than do insects; how does this difference affect the social behavior of the two groups? Member of vertebrate societies can identify each member of the group individually, whereas insects probably cannot. There is usually a dominance hierarchy,

Figure 42-26 All the sea lions in this photograph belong to the harem of one bull. (Biophoto Associates, N.H.P.A.)

Figure 42-27 Nearly all primates, like these long-tailed macaques, live in social groups with dominance hierarchies. Grooming behavior, possible only between social animals, removes parasites and contributes to the health of all members of the group. (Biophoto Associates, N.H.P.A.)

or "pecking order," which ensures that dominant individuals have first choice of desirable but limited commodities such as food, shelter, or mates. Individuals may fall in rank as the result of age or disability; in one baboon troop, the top male changed five times in two years. An individual's role in the society is largely determined by its position in the hierarchy, and so individual ability has more effect on an individual's role in a vertebrate than in an insect society. Position in the hierarchy is not determined solely by an individual's size or fighting ability, however. In many species (probably most primates), having a mother of high social status gives one an initial boost up the social ladder.

Threat displays (Section 42-G) and related behavior patterns are important in maintaining dominance hierarchies. A dominant individual displaces a subordinate by threat behavior; a subordinate responds with **appeasement** gestures, which inhibit other animals from attacking.

The evolutionary advantage of a social hierarchy is probably that it reduces the deleterious effects (such as injury from fights) of the inevitable competition between related individuals living in the same area. The society ensures that at times when resources (such as food) are in short supply, some individuals will get all the food they need to survive instead of the whole group becoming half-starved and likely to die, as happens with honeybees. When members of a group are related, such apparent altruism* will be selected for because individuals carry many of the same genes, and an individual that starves while a relative lives to reproduce is actually contributing to the survival of many of his or her own genes in future generations.

Animal societies, like anything else biological, have survived because they enhance reproductive success. A wolf or an elephant seal, for instance, can raise more offspring as a member of a society than as an individual. Cooperation in hunting, defense, tending the young, and so forth, may all be activities that contribute to reproductive success of the society's members.

---

* Altruism Behavior that contributes to the welfare of another animal at the expense of the actor (see Section 18-F).

SUMMARY   The genes that an animal inherits determine the range of behavior patterns it can develop. In addition, most behavior patterns, innate or learned, will not develop normally if an animal is not exposed to the appropriate environmental conditions.

The proximate reason that an animal behaves in a particular way is that it has been exposed to environmental stimuli that induce the behavior pattern while it was in the appropriate physiological state. Ultimately, behavior patterns that must be produced perfectly at the first exposure to the stimulus are usually innate. Learning requires time and energy and is reserved for behavior that must be flexible in meeting local or changing conditions. Many behavior patterns, both innate and learned, become programmed into the nervous system as fixed action patterns that may be triggered by sign stimuli and controlled by a small number of neurons with minimal sensory feedback.

Animals are always exposed to a variety of stimuli, which may or may not evoke a response. Action or inaction is determined by factors such as the animal's physiological state and its conscious or unconscious hierarchical ranking of stimuli. Conflict behavior, frequently visible in courtship and territorial displays, is one possible outcome of mutually exclusive behavioral tendencies.

Most animals seldom or never cooperate with other members of their own species, but true societies have evolved in some species of insects and of vertebrates. A society usually consists of genetically related individuals. Communication between individuals is most highly developed in social animals. Vertebrate societies are characterized by hierarchies that determine an individual's access to limited resources. The society provides an individual with protection, and its members cooperate in various aspects of their lives.

OBJECTIVES

From your study of this chapter, you should be able to:

1. Explain the theoretical difference between innate and learned behavior, and give two examples of each.
2. Explain what is meant by stereotyped, or fixed action pattern, behavior, and give two examples. Describe the neurophysiological characteristics of fixed action patterns.
3. List the selective advantages of innate and learned behavior and of fixed action patterns.
4. Describe what is meant by the filtering and summation of stimuli.
5. Give examples of motivation or drive, sign stimuli, and supernormal stimuli.
6. Explain and give examples of how the reticular formation and hypothalamus of the brain influence motivation for behavior patterns.

7. Describe the characteristics of territorial behavior, and give an example of such behavior.
8. Describe conflict behavior, and explain why it is thought to have played a role in the evolution of threat displays and courtship behavior.
9. Distinguish between habituation, conditioning, trial and error learning, insight learning, and imprinting.
10. Summarize what is known about migration and homing in animals.
11. Compare and contrast the societies of honeybees and vertebrates.
12. Describe the functions of threat and appeasement behavior in the maintenance of a dominance hierarchy.

SELF-QUIZ

From the list of types of behavior patterns below, choose the one exemplified by each of the following situations.

___ 1. A male cardinal attacks any other male cardinal that tries to come into your back yard.

___ 2. A puppy rolls on its back when a strange adult dog growls at it.

___ 3. A cat meeting a strange (and not overly large or fierce) dog arches its back, fluffs up its fur, and hisses.

___ 4. A student in a typing class makes fewer errors on the tenth homework assignment than on the first.

___ 5. When someone jumps out at you from behind a door, it takes a few milliseconds before you emit a piercing scream.

___ 6. Your signature looks the same every time you write it.

___ 7. Newly hatched ducklings follow a windup toy as if it were their mother.

___ 8. A male stickleback attacks a cardboard model of a fish with an eye and a red belly more intensely than it attacks a model with a red belly but no eye.

___ 9. A skilled musician can play a tune after hearing someone hum a few bars even if he has never heard the tune before.

a. appeasement
b. dominance
c. imprinting
d. insight
e. conditioned reflex
f. fixed action pattern
g. territoriality
h. threat
i. latent learning
j. trial and error
k. summation
l. latency

(Quiz continues on page 702.)

10. Courtship behavior is said to show conflict because:
    a. the two mates fight a lot
    b. the mates cannot immediately agree on a nest site
    c. the mates are sexually attracted to each other but do not normally permit another animal to get as close as copulation demands
    d. the mates must choose each other from a large number of members of the opposite sex
    e. the mates are in competition for food in a territory of limited size

11. The part of the nervous system that is strongly implicated in the control of drives for such behavior patterns as feeding, drinking, and sexual behavior is the:
    a. pituitary
    b. sympathetic system
    c. parasympathetic system
    d. reticular activating system
    e. hypothalamus

12. Which of the following is *not* true of fixed action patterns?
    a. They may be triggered by sign stimuli.
    b. They are initiated by one or a few neurons.
    c. They cannot be extinguished.
    d. They exhibit latency.
    e. They can be learned or innate.

QUESTIONS FOR DISCUSSION

1. How can one distinguish learned and innate behavior patterns? Are deprivation studies an adequate test?

2. Despite everything you have read in this chapter, you probably still think that most of an insect's behavior is innate, and that most of a human's or chimpanzee's is learned. Can you justify this position?

3. What neurophysiological mechanisms might produce filtering of stimuli so that many stimuli are ignored but the key stimulus "comes through" and evokes a particular behavior pattern?

4. What are some of the possible selective advantages to defending a territory, an activity that consumes a lot of time and energy and increases the risk of injury?

5. Why does a behavior pattern have a shorter latency with increasing intensity of stimulus?

6. William Dilger studied the nest-building behavior of parakeets of the genus *Agapornis*. He crossed members of a species that carries nest-building material in its beak with members of a species that carries its nesting material tucked under its tail feathers, and observed the behavior of the hybrid offspring. These offspring showed hybrid behavior and usually dropped the material whether they carried it in their beaks or their feathers. One particular bird tried to build a nest 48 times and failed. On the 49th try it was successful. What does this tell you about whether nest-building behavior in this genus is innate or learned?

7. What experimental techniques would you use to determine which part(s) of the nervous system governed a certain type of behavior, such as threat, feeding, or sexual behavior?

REFERENCES AND FURTHER READING

Adler, K. "Extraocular photoreception in amphibians." *Photochemistry and Photobiology* 23:275, 1976. Includes amphibian compass orientation.

Alcock, J. *Animal Behavior.* Stamford, CT: Sinauer Associates, 1975. Excellent behavior text.

Dilger, W. C. "The behavior of lovebirds." *Scientific American*, December 1962. The genetic basis of behavior in lovebirds.

Hasler, A. D., and J. A. Larsen. "The homing salmon." *Scientific American*, August 1955.

Keeton, W. T., and A. Gobert. "Orientation by untrained pigeons requires the sun." *Proceedings of the National Academy of Science, U.S.A.*, 65:853–856, 1970.

Keeton, W. T. "Magnets interfere with pigeon homing." *Proceedings of the National Academy of Science, U.S.A.*, 68:102–106, 1971.

Lorenz, K. Z. "The evolution of behavior." *Scientific American*, December 1958. An ethologist's view of behavior.

Lorenz, K. Z. *King Solomon's Ring.* London: Methuen, 1942. Delightfully written autobiographical account of life with animals.

Tinbergen, N. "The evolution of behavior in gulls." *Scientific American*, December 1960.

Tinbergen, N. "The curious behavior of the stickleback." *Scientific American*, December 1952.

von Frisch, K. *The Dancing Bees.* Translated by D. Ilse. New York: Harcourt Brace, 1953. Experiments that showed how bees learn and communicate.

PART SIX

PLANT BIOLOGY

CHAPTER 43

STRUCTURE AND GROWTH OF VASCULAR PLANTS

Most vascular plants are nutritionally self-sufficient, or **autotrophic**; they use sunlight and carbon dioxide from the atmosphere, and water and minerals from the soil, to make all the organic molecules they need. The structure of plants is adapted to their autotrophic way of life. A typical vascular plant has a **root system** that absorbs water and minerals from the soil and a **shoot system** made up of one or more stems with leaves. Stems hold the leaves up where they can intercept sunlight, and the vast surface area of the leaves allows them to gather a great deal of light energy, which they store in food molecules.

Plants have **indeterminate growth**: in a favorable environment, a plant can increase in size and add new parts throughout its life. Most of the plant body consists of mature, differentiated cells, with specialized functions. Under normal circumstances, these cells will never divide again. A plant's ability to grow indefinitely is due to the presence of **meristems**, tissues whose cells retain the capacity to divide and produce new cells. Some of these cells differentiate and become new parts of the plant; others remain meristematic.

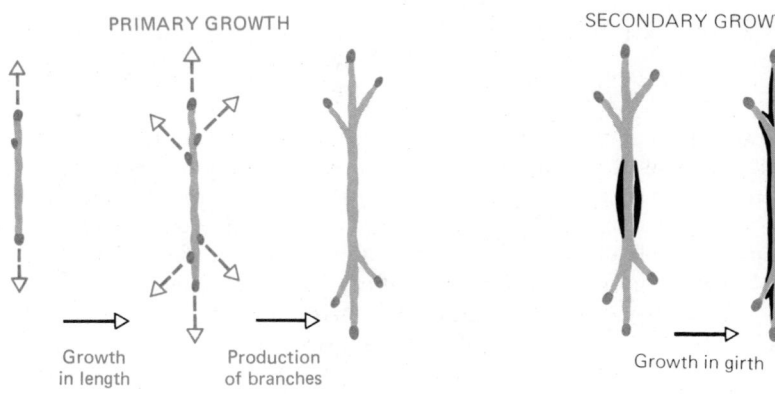

PRIMARY GROWTH   SECONDARY GROWTH

Growth in length   Production of branches   Growth in girth

Figure 43-1 Comparison of primary and secondary growth. Primary growth is mainly growth in length and production of new branches; secondary growth produces growth in girth.

The growth of plants can be divided into two aspects (Figure 43–1). **Primary growth** is principally growth in length of the shoots and roots, and production of new root and shoot branches, or **laterals**. **Secondary growth** accounts for almost all the growth in girth, or thickness, of the stems and roots produced by primary growth. The increased thickness contributed by secondary growth strengthens the plant and provides the support necessary for new primary growth, both in height and in spread. Although all vascular plants exhibit primary growth, only some are capable of secondary growth.

The structure of a plant depends on which parts grow, how long and how thick they become, where laterals arise, and how much the laterals grow with respect to the other laterals and the main shoot or root. Ultimately, the size and shape of a plant reflect the activities of its meristems, which lay down the cells for each part.

In this chapter we shall follow the growth of a plant seedling and see how its primary structure is produced by the activities of primary meristems. Then we shall go on to see how secondary tissues are added to the primary plant body through the action of secondary meristems.

43-A The Bean Seed

The "skin" of a bean seed is the tough protective **seed coat**, wrapped tightly around the embryo (Figure 43–2). Within this coat is the embryonic bean plant. The two "halves" of the bean, the **cotyledons**, are the embryo's "seed leaves"; the presence of two cotyledons shows that the bean is a member of the **dicotyledons (dicots)**, the group of flowering plants whose embryos have two seed leaves. The cotyledons of a bean are extremely large because they have absorbed the food supply that will nourish the growing bean seedling. However, the seeds of many plant species do not store their food within the cotyledons.

If we separate the cotyledons with care, we can see that they are part of the embryo, attached to its tiny main axis. The part of the axis below the cotyledons' place of attachment is called the **hypocotyl**. Below the hypocotyl is the **radicle**, which will become the seedling's first root. Above the cotyledons is the **plumule**, including the future shoot tip and the two **first foliage leaves**.

The bean embryo has two meristems, one at the tip of the plumule, between the first foliage leaves, and one at the tip of the radicle. Since each of these meristems is located at a tip, or apex, of the plant, they are called **apical meristems**. Cell division in these meristems, followed by growth and differentiation of some of the cells formed, results in growth of the root and shoot systems of the plant.

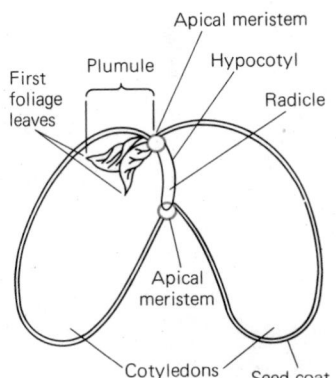

Figure 43-2 A bean seed split open to reveal the embryo. Food is stored in the bean embryo's seed leaves, the cotyledons. The apical meristem of the shoot is hidden between the first foliage leaves.

CHAPTER 43 STRUCTURE AND GROWTH OF VASCULAR PLANTS   **705**

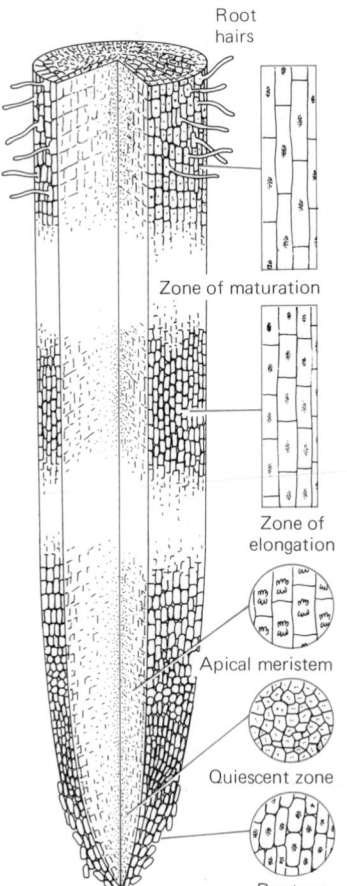

Root
hairs

Zone of maturation

Zone of
elongation

Apical meristem

Quiescent zone

Root cap

Figure 43-3   A growing root tip. The colored circles and boxes show what the cells in each area look like through the microscope.

Primary Growth of Roots: Growth in Length

The first part of the bean embryo to begin growth is the radicle. In the root apical meristem, at the tip of the radicle, there is a central **quiescent zone**, where little or no cell division occurs (Figure 43–3). Around this region, cells divide and produce more cells. The cells produced toward the tip of the root become part of the **root cap**, a thimble of cells that protects the interior cells from damage.

The cells produced on the other side of the apical meristem, toward the main part of the plant, become arranged in definite strands. Eventually these cells stop dividing and begin to elongate. Finally, they stop growing and mature as specialized cells performing particular functions. These mature cells form the root's **primary tissues**, that is, the tissues laid down by the activity of the apical meristem. Meanwhile, the apical meristem has continued to divide and produce new cells, which follow the same sequence of division, growth, and development. So at any one time, cells in different parts of the root tip are of different ages and are thus going through different stages of development; the farther back from the root tip, the older the cells.

The actual growth of the root tip through the soil is due to the elongation of the cells in the **zone of elongation** (Figure 43–3); thus the root tip is literally pushed through the soil. As the root passes between soil particles, the outermost cells of the root cap may be rasped off by the abrasive edges of the particles. These broken cells leave a slimy coating on the soil particles that allows the rest of the root to slide along more easily. Cells lost from the root cap are replaced by the continuous division of cells in the apical meristem.

In the **zone of maturation**, the cells have attained their full size and are becoming differentiated. Here the outermost cells become **epidermal** cells, which fit together tightly and form a protective layer, one cell thick, over the outside of the root. Some of the epidermal cells become **root hairs** by forming extensions that grow out among the soil particles. Root hairs anchor the plant in the soil and increase the surface area for absorption of water and minerals.

It is extremely important that root hairs form in the zone of maturation rather than in the zone of elongation. If root hairs were formed where cells were still elongating and pushing through the soil, the delicate extensions growing out sideways between soil particles would be pulled off.

Primary Structure of Roots

A cross section of a mature primary root, through a region above the zone of maturation, shows the tissues that eventually differentiate from the cells laid down by the apical meristem (Figure 43–4). The outermost layer is the **epidermis**, which includes the root hair cells. The **cortex**, which is usually several cells thick, lies just inside the epidermis. The cells in the cortex are of a type called **parenchyma**; they are rather rounded in cross section, with relatively thin cell walls and with many spaces between cells. The cells may contain **amyloplasts**, organelles that store starch. Parenchyma cells like those in the root cortex usually make up the bulk of the primary tissue of a plant.

The single layer of cells at the inner edge of the cortex is called the **endodermis** ("inner skin"). The endodermis plays a role in controlling the movement of substances between the root cortex and the interior of the root (see Section 45-C). Just within the endodermis is another single layer of cells, the **pericycle**. Cells in the pericycle do not differentiate, but remain meristematic, capable of division as needed.

Within the pericycle are the **vascular tissues**, the tissues that conduct substances for long distances through the plant. There are two types of vascular tissue: **xylem**, which conducts water and minerals taken in through the roots up to the stems and leaves, and **phloem**, which conducts food from the shoot system to the living root cells. Conducting cells of the phloem are among the first to become differentiated; they provide food for the growth and differentiation of cells near the root tip. Conducting cells in the xylem and phloem are extremely long and narrow, like sections of a pipe. The details of their structure and function will be presented in the next chapter; for the present, note that the vascular tissue usually occupies

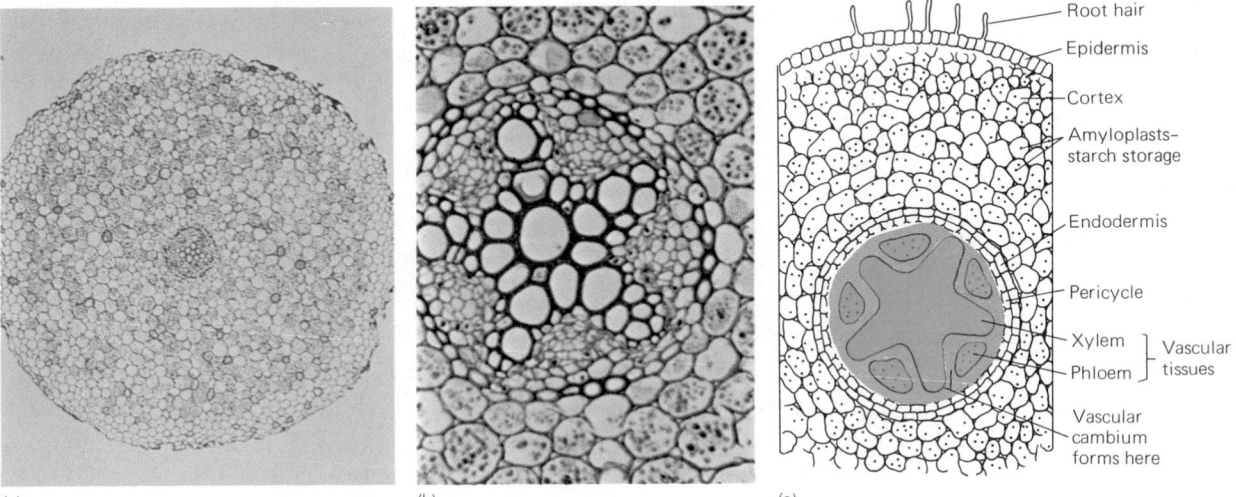

(a)　　　　　　　　　　(b)　　　　　　　　　　(c)

Figure 43-4 The tissues of a mature primary root. (a) Cross section of a buttercup (*Ranunculus*) root. (b) Closeup of the center of the root in (a). (c) Drawing to show the various tissues in the primary root (the cortex has been drawn smaller in proportion to the vascular tissues in the center). (Carolina Biological Supply Company)

the core of the root. In cross sections of dicot roots, there is commonly a central "star" of xylem, with arms extending outward; the phloem forms in pockets between these arms of xylem.

Primary Growth of Roots: Production of Laterals

The first root of a bean seedling grows quickly and soon begins to produce laterals. Cells in the pericycle, which remained meristematic when their neighbors differentiated, begin to divide and form a new apical meristem. As the innermost cells begin to elongate and mature, they push the new meristem out through the endodermis, cortex, and epidermis, and on into the soil (Figure 43–5). Once a new branch root has formed, its structure and growth are just like those of the first primary root discussed above. The new primary roots, in turn, may give rise to more branch roots.

There are two basic types of root systems (Figure 43–6). A **taproot system** has one main root, the **taproot**, which is by far the longest and thickest root in the entire plant. The taproot does have side branches, but these are distinctly shorter and thinner than the taproot. The taproot of an old grape vine may reach 15 metres down and obtain water far beneath the soil surface.

Fibrous root systems are highly branched, with extensive growth throughout a large volume of soil. They consist of several large roots of about equal size, instead of just one.

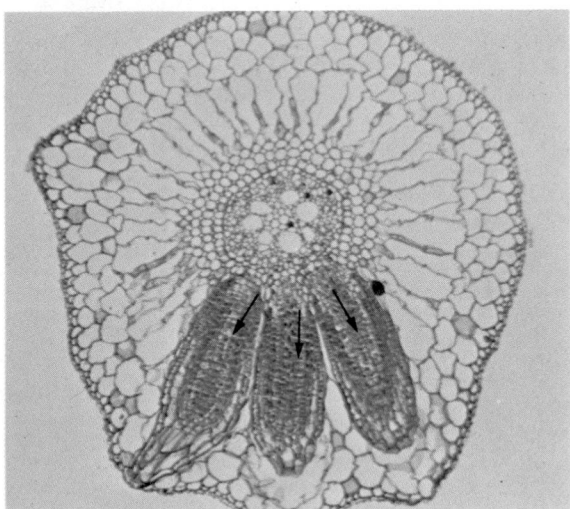

Figure 43-5 Origin of lateral roots in a water hyacinth. Three branch roots are growing out from the center of the root (arrows). (Carolina Biological Supply Company)

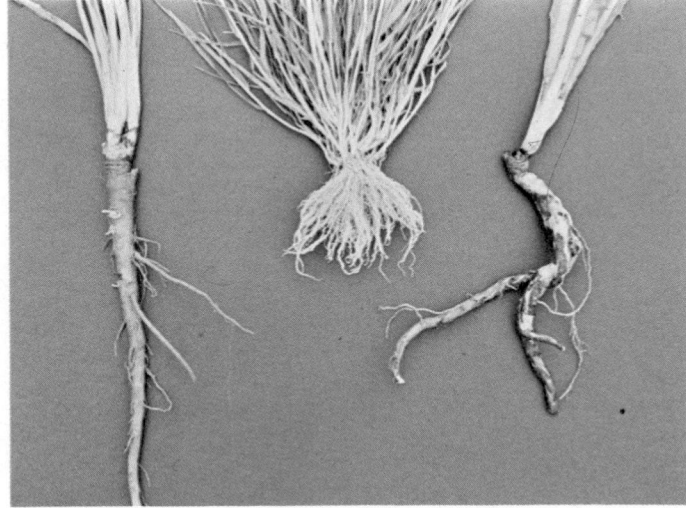

Figure 43-6 Two types of root systems. Left, a taproot system; center, a fibrous root system. What do you suppose happened to the plant on the right? (Biophoto Associates)

Not all new roots arise as lateral outgrowths of previously existing roots. **Adventitious roots** are formed by meristems growing from other parts of the plant. For example, a mature corn plant has many roots that arose adventitiously from the base of the stem. Many climbing plants form tiny roots on the undersides of their stems as they climb up brick walls or the stems of larger plants. Some plants, such as African violets, can also form roots from the undersides of leaves that are placed on moist soil. Cuttings of many plants form adventitious roots if the stems are placed in water.

Functions of Roots

The preceding discussion of roots points out their four main functions:

1. Roots anchor the plant in the soil. Taproots are especially effective anchors, since they are long, and so thick and tough that it is difficult to break them.

2. Roots absorb water and minerals from the soil. Fibrous root systems often expose more surface area to the soil than do taproot systems.

3. Roots transport water and minerals up to the shoot system. The central core of xylem is responsible for this function. The endodermal layer to some extent controls the passage of substances into or out of this central core of vascular tissue.

4. Roots store food. The root cortex is the primary food storage area in some plants. In many other kinds of plants, the root stores only a small supply of food for itself.

43-C   Stems

Primary Growth of Stems: Growth in Length

Once the growing bean embryo has established roots in the soil, the shoot begins to grow. The apical meristem of the shoot lies between the first foliage leaves, at the tip of the stem axis. This apical meristem is similar to that of the root tip, but it has no quiescent zone nor any structure similar to the root cap.

During the initial stage of shoot growth, the hypocotyl forms a loop that pushes up through the soil and then straightens, pulling the cotyledons free of the soil (Figure 43–7). Meanwhile, the apical meristem remains hidden and protected be-

Figure 43-7 Stages in the growth of a bean seed into a young plant.

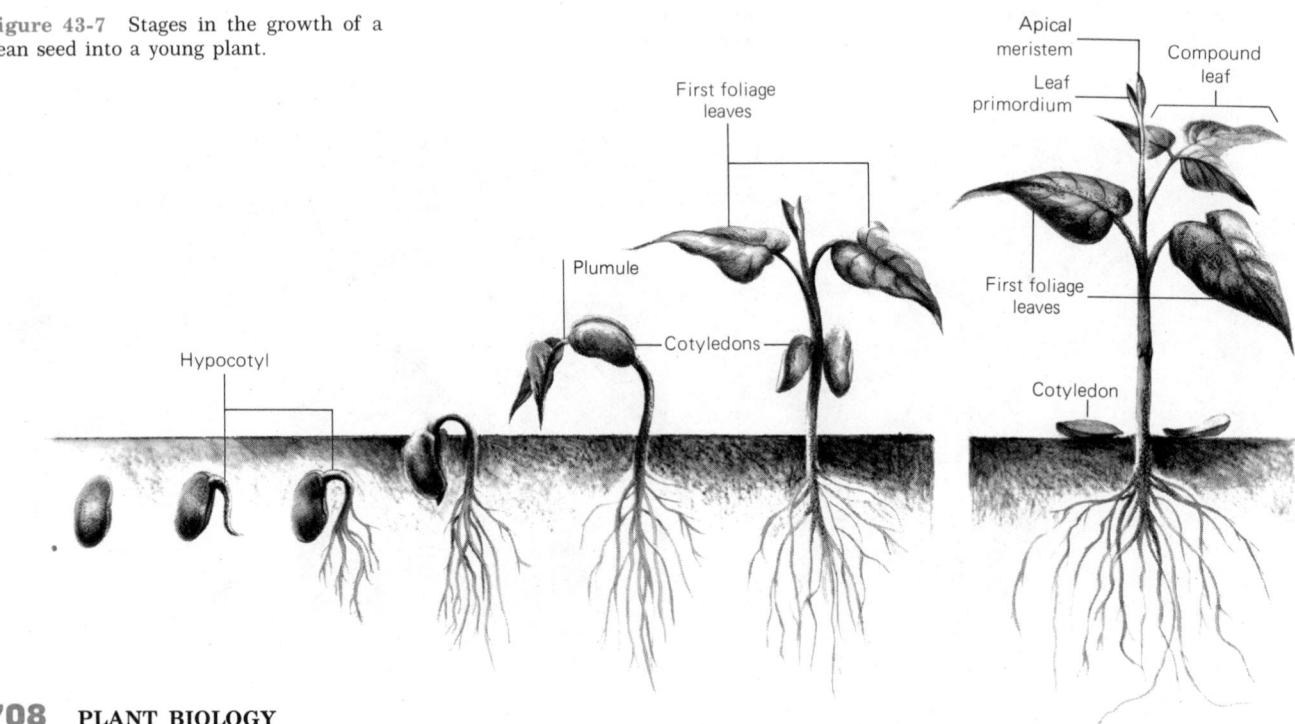

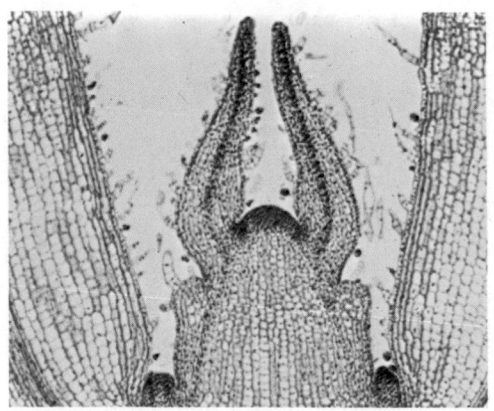
(a)

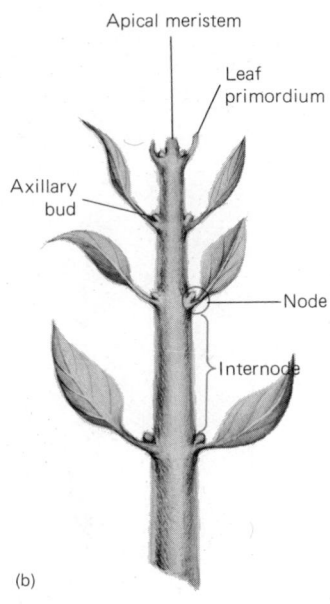

Apical meristem

Leaf primordium

Axillary bud

Node

Internode

(b)

(c)

Figure 43-8 Growth of a shoot tip. (a) Longitudinal section of a *Coleus* shoot tip, showing the apical meristem flanked by new leaf primordia. Note the axillary buds in the axils of the next older pair of leaves, which are still differentiating. (b) Drawing to assist in identifying the structures shown in part (a). (c) Axillary buds developing into new shoots on a privet twig; note that one bud is still dormant in the third node from the top. (a, Carolina Biological Supply Company; c, Biophoto Associates)

tween the cotyledons. Once the cotyledons are above the ground, they spread apart, revealing the plumule. The first foliage leaves, which were fully formed in the bean embryo, have already begun to expand. When they are exposed to sunlight they become green and expand more rapidly. Until these leaves can support the plant by their photosynthesis, the bean seedling continues to use the food stored in the cotyledons. The cotyledons themselves are leaves, and when they are exposed to light they too become green and carry out some photosynthesis. Eventually, the cotyledons shrivel away as their food supply is used up, and they fall off.

The apical meristem consists of a little mound of rapidly dividing cells. As in the root tip, the cells just below the shoot apical meristem elongate and then mature into differentiated cells of the primary plant stem. Growth in both length and diameter by cells in the shoot may last for longer than it does in the root, so that growth in the shoot can be detected quite a distance behind the growing tip of the apical meristem.

Primary Growth of Stems: Production of Laterals

As the apical meristem of the shoot grows, it lays down not only the cells for the primary tissues of the stem, but also **leaf primordia**, which develop into leaves (Figure 43-8). The parts of stems where leaves are attached are called **nodes**; **internodes** are the stretches of stem between two adjacent nodes. In the **axil** of each leaf, that is, in the angle formed between the base of the leaf and the stem, a small patch of cells remains meristematic; these patches produce **axillary** (axilla = armpit) or **lateral buds**. The meristems of these axillary buds remain dormant for a time.

A new shoot branch arises when an axillary bud becomes active and produces more cells, some of which enlarge and mature to form the primary tissues of a branch (Figure 43-8c). These branches produce leaves with axillary buds, just as the main stem does.

Some of the shoot system's axillary buds become apical meristems of new branches and some remain dormant. Picture what the plant would look like with all the axillary buds growing into new branches; the plant would soon become impossibly complex, with many more branches than it could support, and with a great deal of crowding among the leaves for a place in the sun. Normally, most of the axillary buds of a plant are repressed by hormones from the apical meristems. This phenomenon, called **apical dominance**, will be discussed in Chapter 47.

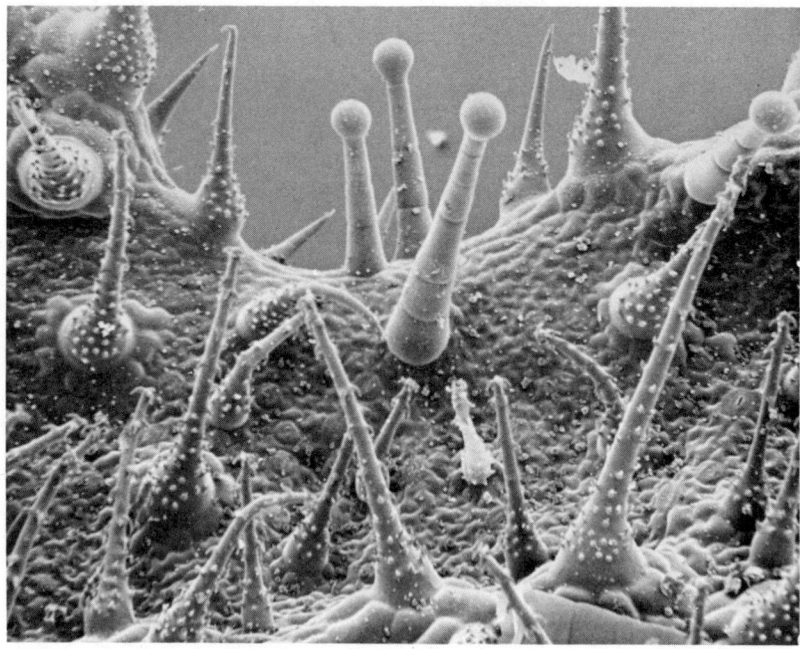

Figure 43-9 The epidermis of a South American nettle leaf has two kinds of protective hairs: One kind is pointed, with hooked ends, and the other kind secretes droplets of irritating fluid. (Biophoto Associates)

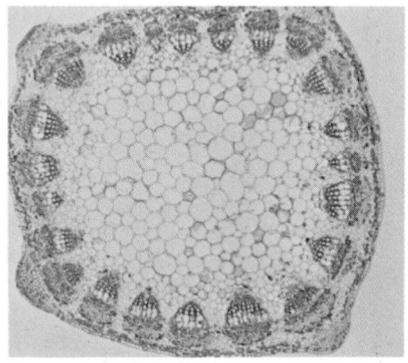

(a)

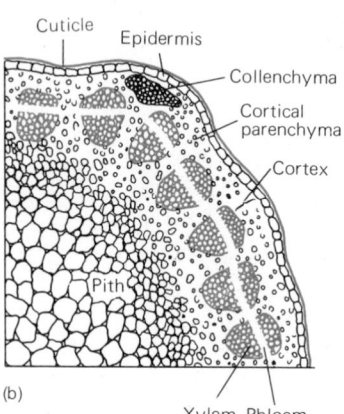

(b)

Figure 43-10 Tissues of a mature primary stem of a dicot. (a) Cross section of a mature primary stem of alfalfa (*Medicago*). (b) Drawing pointing out the various tissues in (a). (Carolina Biological Supply Company)

Structure of the Mature Primary Stem

Cells laid down by the apical meristem of the shoot divide, enlarge, and differentiate to become the primary tissues of the stem. The epidermis of the stem is usually just one cell thick. The epidermal cells must fit tightly together to prevent both loss of moisture and damage by invading fungi or insects. A further adaptation that reduces moisture loss is the **cuticle**, a layer of waxy substances secreted by the epidermal cells. This forms a waterproof covering over the surface of the stem. The cuticle of the stems is continuous with a layer of cuticle that covers the leaves.

The epidermal layer may have other protective functions. Epidermal hairs of tomato plants, for instance, secrete a juice that irritates any animals brushing against them. Many plants have nonsecretory epidermal hairs. The hook-shaped hairs of beans entangle small insects, causing them to stop feeding while they struggle to free themselves (Figure 43-9).

The next layer inside the epidermis is the cortex (Figure 43-10). The outer layer of the cortex may contain cells called **collenchyma**. These cells are shaped like columns, with thickened ridges of strengthening material forming edges that run from the top of the cell to the bottom. Bundles of collenchyma may form discernible ridges on the surface of a plant; the ribs of celery are composed mainly of collenchyma. Most of the cortex is made up of parenchyma; again, as in the root, these cells have a rounded cross section, with thin walls and many intercellular spaces. Some cells in the cortex may contain chloroplasts and carry out photosynthesis; in some cases cells of the cortex have amyloplasts. The **pith** in the center of the stem is also composed of parenchyma cells. In this respect, stems differ from roots, in which the center is usually occupied by the cylinder of vascular tissue. As the stem continues to grow, the cells in the outer parts may enlarge and expand so much that the pith is pulled apart, and the stem becomes hollow.

The primary vascular tissue in the stem occurs as a ring of discrete vascular bundles rather than as the central core found in roots. Generally the xylem is located in the part of the bundle toward the inside of the stem, and the phloem lies toward the epidermis. Each vascular bundle in the stem is partly surrounded by thick-walled **fiber** cells, which provide extra support. The walls of the conducting cells in the xylem may also be especially thick. The extra strengthening material in the vascular bundles, along with the collenchyma, helps to support the stem in the air, which gives much less support than the soil that surrounds the roots.

Stems usually lack the endodermis and pericycle found in roots.

Blade

Midrib

Petiole

Figure 43-11 Basic structure of a leaf. (William Camp)

Functions of Stems

The foregoing description shows that stems perform four main functions:

1. Stems support structures of the shoot system. The collenchyma and vascular tissue of stems form a tough framework that is often very difficult to break. Added to this is the turgor* of the parenchyma cells in the pith and cortex, which keeps the stem firm. With this tough, resilient structure the stem can hold the leaves up to the sunlight they need for photosynthesis.

2. Stems transport substances between the roots and leaves. The vascular tissue of stems is continuous with that of the roots and leaves. The xylem transports water and minerals taken in by the roots up to the leaves and to the living cells of the stem, and the phloem carries food from the leaves or from storage areas to the living and growing parts of the plant, some of which are not capable of photosynthesis.

3. Stems produce food. Some stems are green and carry out photosynthesis. In most plants this supplements the photosynthesis carried out by the leaves, but in plants such as cactuses, the stems are the main photosynthetic organs.

4. Stems store substances. Some stems contain a large number of amyloplasts. The familiar potato is an enormous underground shoot with a large stem specialized for storage. Other stems may store lesser amounts of food. Some stems, such as those of some cactuses, may store large quantities of water.

43-D Leaves

A leaf begins as a leaf primordium laid down by the activity of an apical meristem. When mature, the leaf usually has three main parts: the **petiole**, or stalk; the **midrib**, the main vein of vascular tissue; and the photosynthetic **blade**, which is usually broad and flattened (Figure 43–11).

The broad, flat shape of leaves gives a maximum amount of surface area per unit of volume. Surface area is important both for gathering the energy of sunlight and for exchanging gases with the atmosphere. The comparatively small volume makes the leaves very lightweight; heavy leaves would require more supporting tissue in the stems and roots.

---

*Turgor   Internal pressure that results from being filled with fluid.

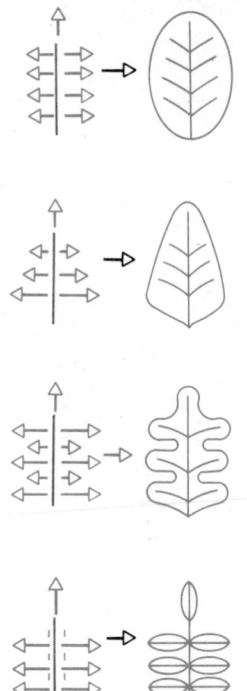

Figure 43-12 Leaf shape is determined by rate of growth and division of cells along different parts of the original leaf axis.

Leaf shape varies a great deal among plants, and is often used as a convenient way to identify plants. What makes leaf shape different in different species of plants? The growth pattern of cells in a developing leaf is genetically determined, although in some species the environment influences which of several possible patterns a leaf follows. A leaf begins as a slender rod of meristematic cells (Figure 43–12). If the cells near the edges of the developing leaf keep dividing in a uniform manner, the leaf grows in a smooth, rounded shape. On the other hand, if cells in some areas stop dividing while cells in other areas do not, an irregular shape results, with lobes or teeth alternating with indentations. In extreme cases, the cells in some areas near the midrib do not divide at all, and the mature leaf is composed of several discrete parts, or **leaflets**, connected only by the midrib, and not by any areas of leaf blade. Such leaves are called **compound leaves**, and to the uninitiated each leaflet appears to be a separate leaf.

How can **simple leaves**, that is, leaves with undivided blades, be distinguished from the leaflets of compound leaves? The vital clue is the bud in the axil. A leaf has an axillary bud at its base; a leaflet does not, but there should be an axillary bud where the structure to which the leaflet is attached joins the stem (Figure 43–13).

Our bean plant has both simple and compound leaves. The first foliage leaves are simple, somewhat heart-shaped leaves, each with an axillary bud in its axil. The leaves formed by the plant subsequently, however, are compound leaves, with three leaflets each (see Figure 43–7).

The bean plant also shows two different types of leaf arrangement. The two first foliage leaves are **opposite**; that is, there are two leaves at the node, at a 180° angle to each other. The compound leaves formed later are **alternate**, with only one leaf per node, and with the leaves at successive nodes growing out from the stem at different angles. A third type of leaf arrangement, not seen in bean plants, is the **whorled** arrangement; here several leaves all arise from the same node, at well-spaced angles around the circumference of the stem. The arrangement of leaves in opposite, alternate, or whorled patterns is often used as an identifying characteristic of plants (Figure 43–14).

Structure of Leaves

As in the primary root and stem, the leaf's outer surface, both top and bottom, is covered by the epidermis; the epidermal cells of leaves, like those of stems, secrete a protective waterproof cuticle. Although the cuticle retards the loss of water to the atmosphere, it also impedes the passage of gases from the atmosphere into the plant. Since the leaves need carbon dioxide from the air for photosynthesis, the secretion of a continuous layer of cuticle would solve one problem but create another.

In fact, however, the cuticle is not continuous; it is interrupted at intervals by **stomata** (= mouths; singular: **stoma**). Each stoma is a pore between a pair of **guard cells** (Figure 43–15). The size of the pore can be controlled by the guard

Figure 43-13 Simple leaves versus compound leaves. Axillary buds occur where the leaf joins the stem. Thus, (a) is a plant with simple leaves (American beech), whereas (b) has compound leaves, with the leaf blade forming several separate leaflets (shagbark hickory).

(a) Simple leaves

(b) Compound leaves

Figure 43-14 Arrangements of leaves. (a) Opposite; (b) alternate; (c) whorled.

cells; when the stoma is open, more carbon dioxide can enter the plant from the atmosphere, and also more water vapor can escape into the atmosphere; closing the stoma reduces the exchange of these gases between the interior of the leaf and the atmosphere. Most dicot leaves have stomata only in the lower epidermis.

Interestingly, a study of maple leaves from different sections of the city of Montreal revealed that the more polluted the air, the fewer stomata were present in a given amount of leaf surface area. Leaves from trees in an unpolluted suburban area had 10 times as many stomata per unit area as leaves from a very polluted industrial section.

Stomata are not limited to leaves; they are found in the epidermal layers of stems, flower parts, and so forth. All parts of a plant need oxygen for respiration, and some also need carbon dioxide for photosynthesis.

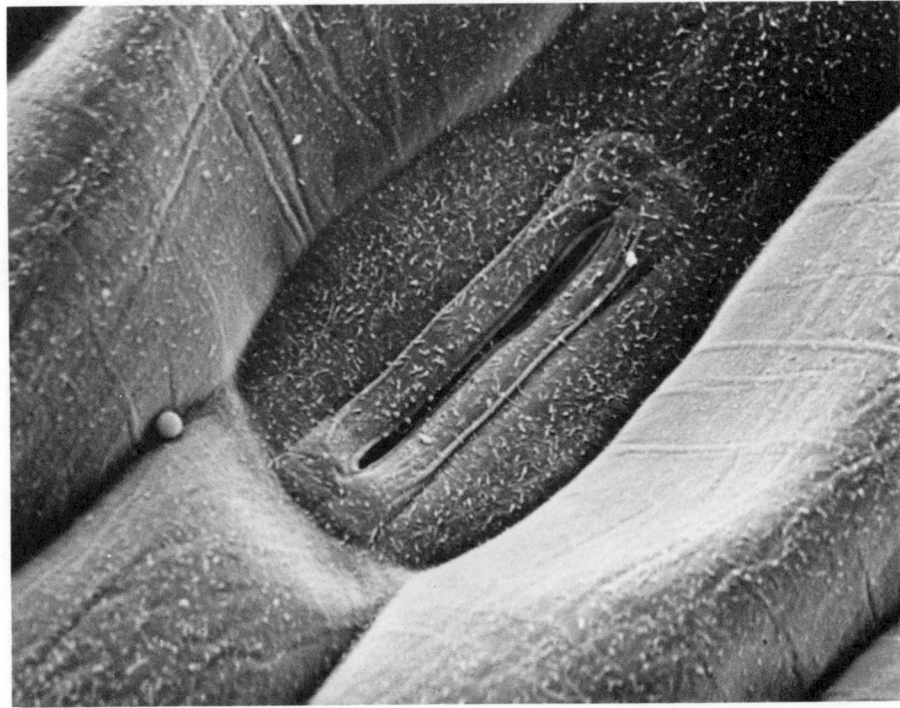

Figure 43-15 A slitlike stoma between two guard cells. The white flecks are part of the waxy, waterproof cuticle. (Biophoto Associates)

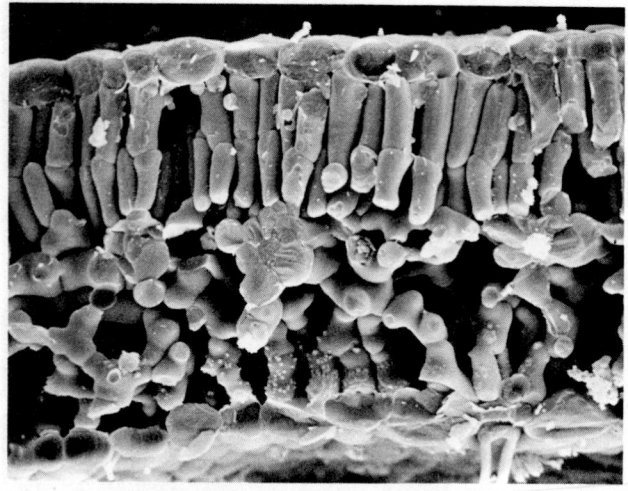

(a)

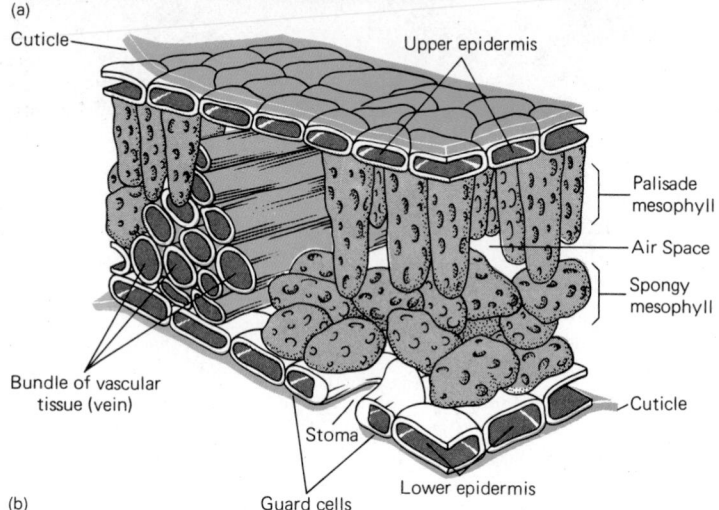

Cuticle

Upper epidermis

Palisade
mesophyll

Air Space

Spongy
mesophyll

Bundle of vascular
tissue (vein)

Cuticle

(b)

Stoma

Guard cells

Lower epidermis

Figure 43-16 Cross section of a leaf. (a) Scanning electron micrograph; (b) drawing to assist in interpreting part (a). (a, Biophoto Associates)

Between the epidermal layers of the leaf are two layers of **mesophyll**, the photosynthetic tissues that make up the bulk of the living matter in the leaf (Figure 43–16). The cells of the mesophyll layers are modified forms of parenchyma, with many chloroplasts. Just beneath the upper epidermis is the **palisade mesophyll**. Its cells are long and thin, and stand on end, perpendicular to the upper epidermis. There may be one or several layers of palisade cells, depending on the species. Below the palisade mesophyll is the **spongy mesophyll**. Its name comes from the presence of numerous air spaces between the cells. These air spaces communicate with the atmosphere via the stomata, and thus provide for circulation of gases to the cells of the mesophyll.

The leaf also has vascular tissue, running through the mesophyll. The vascular tissue brings water to the leaves from the roots, and carries away the products of photosynthesis. It also provides a framework that stretches the delicate photosynthetic tissues out where they can intercept the rays of the sun and the gases in the atmosphere (see Figure 43–11). The strands of vascular tissue are commonly called the **veins** of the leaf. Various species of plants differ greatly in the pattern of veins, or **venation**, in their leaves.

43-E Secondary Growth

Secondary Growth of Stems

Secondary growth is growth in diameter due to addition of new cells. It is most familiar in woody, perennial (living for many years) dicots and gymnosperms. It is also found to some extent in annual (living just one season) dicots, such as alfalfa and sunflowers, which produce additional thickening and support as they increase in height.

Figure 43-17 Secondary growth of a woody stem. (a) Sequence of events in the formation of secondary tissues: secondary (2°) xylem, secondary phloem, and cork. (b) Cross section of a basswood (*Tilia*) stem at the end of three years of secondary growth. (Carolina Biological Supply Company)

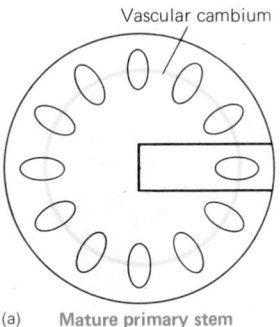

(a) Mature primary stem

Cork
Cortex
Phloem
Vascular cambium
Pith
Primary xylem
Secondary xylem

(b)

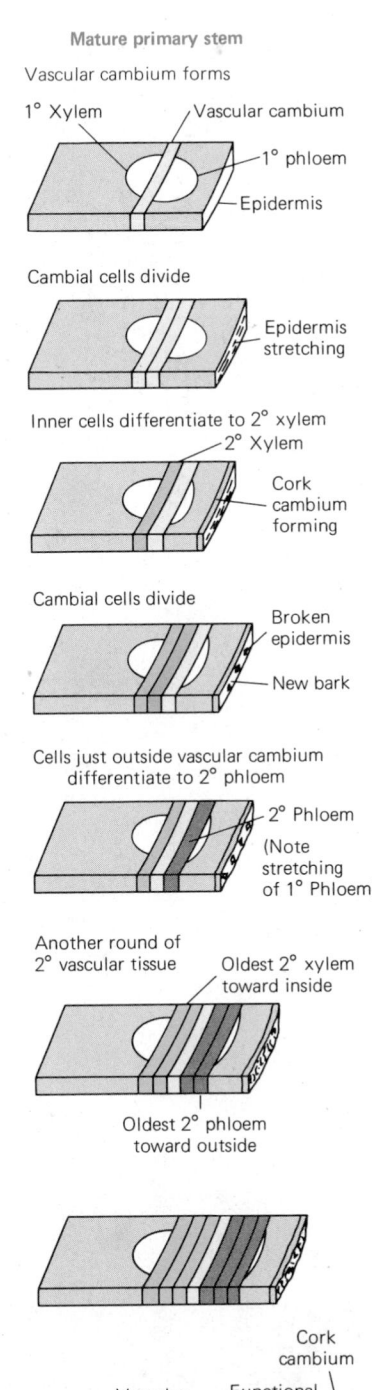

Mature primary stem

Vascular cambium forms
1° Xylem
Vascular cambium
1° phloem
Epidermis

Cambial cells divide
Epidermis stretching

Inner cells differentiate to 2° xylem
2° Xylem
Cork cambium forming

Cambial cells divide
Broken epidermis
New bark

Cells just outside vascular cambium differentiate to 2° phloem
2° Phloem
(Note stretching of 1° Phloem)

Another round of 2° vascular tissue
Oldest 2° xylem toward inside
Oldest 2° phloem toward outside

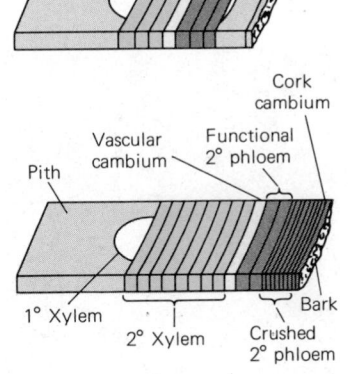

Cork cambium
Vascular cambium
Functional 2° phloem
Pith
1° Xylem
2° Xylem
Crushed 2° phloem
Bark

Stem with much secondary growth

Secondary growth begins with the formation of a **vascular cambium** among the primary tissues. The vascular cambium is known as a **lateral meristem**; it divides to produce cells that mature into **secondary vascular tissues**. In the stem, one or more rows of cells that separate the primary xylem from the primary phloem in the vascular bundles become meristematic and form part of the vascular cambium. Additional cells between the vascular bundles also become meristematic, and thus the vascular cambium can be seen in a cross section of the stem as a continuous ring of tissue, with the xylem and pith on the inside and the phloem, cortex, and epidermis on the outside (see Figure 43–17a).

Once a continuous ring of vascular cambium has formed, it begins to divide, producing new cells. The cells produced inside the ring of vascular cambium differentiate into **secondary xylem** tissue; most of these cells have very thick cell walls. Secondary xylem tissue is also called **wood**. As the vascular cambium produces new wood on the inside of the ring, the plant rapidly grows in diameter. The vascular cambium divides and adds more cells to the ring, and thus it continues to surround the most recently formed ring of secondary xylem completely. Meanwhile, the phloem tissue that was outside the vascular cambium will have been stretched and crushed by the expansion of the inner part of the plant. The cells produced just outside the vascular cambium differentiate as **secondary phloem** tissue, which can carry on food transport in place of the primary phloem that may be destroyed. As more secondary xylem forms, the first secondary phloem is destroyed in its turn, and more secondary phloem is added just outside of the vascular cambium (that is, immediately to the inside of the previously formed secondary phloem) (Figure 43–17).

As the secondary xylem and phloem are added by the vascular cambium, increasing the diameter of the stem, the cortex and epidermis, which originally lay outside the vascular tissues, are also stretched and destroyed. This would leave the internal tissues of the stem unprotected, were it not for the formation of a new protective outer layer of secondary tissue. When the vascular cambium is formed, cells in the cortex form another lateral meristem, the **cork cambium**. It divides and produces new cells to the outside; these cells become impregnated with a water-

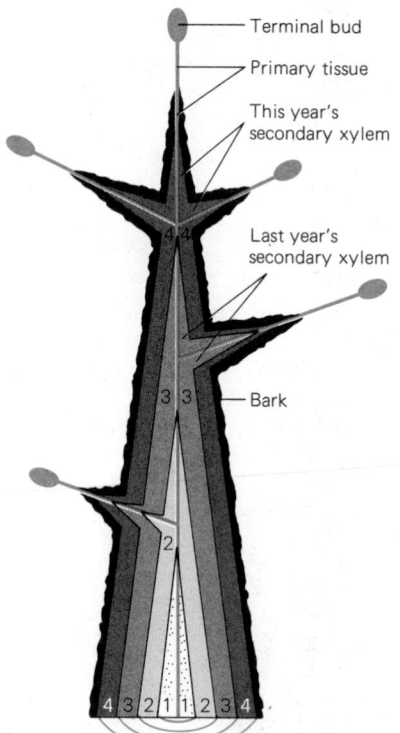

Terminal bud
Primary tissue
This year's secondary xylem
Last year's secondary xylem
Bark

Figure 43-18 Longitudinal section through a (fat) four-year-old tree, showing the relationship between primary and secondary tissues. 1 = first year's secondary growth, 2 = second year's, and so on.

proof waxy material and then die, forming a protective layer of **cork**, or **outer bark**, on the outside of the tree. The bark often bears distinctive markings called **lenticels** (Figure 43–19); these are areas where the cork cambium produces loosely arranged cells that permit exchange of gases between the atmosphere and the living tissue of the plant.

Thus, in the cross section of a stem that has well-developed secondary growth, we would find these tissues, starting from the center of the stem and working outward: pith, primary xylem, secondary xylem, vascular cambium, secondary phloem, the crushed remains of primary phloem and cortex (these would be present but

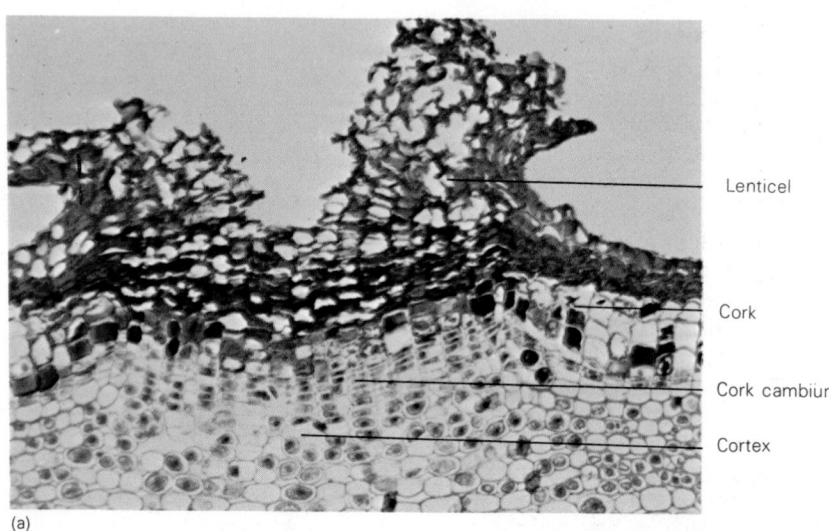

Lenticel

Cork

Cork cambium

Cortex

(a)

(b)

Figure 43-19 (a) A lenticel in the bark of an elder (*Sambucus*) twig. (b) The thick bark of a giant sequoia can withstand insects and fire, allowing the tree to survive for thousands of years. (a, Carolina Biological Supply Company)

TABLE 43-1 SUMMARY OF PRIMARY AND SECONDARY GROWTH OF A WOODY DICOT STEM

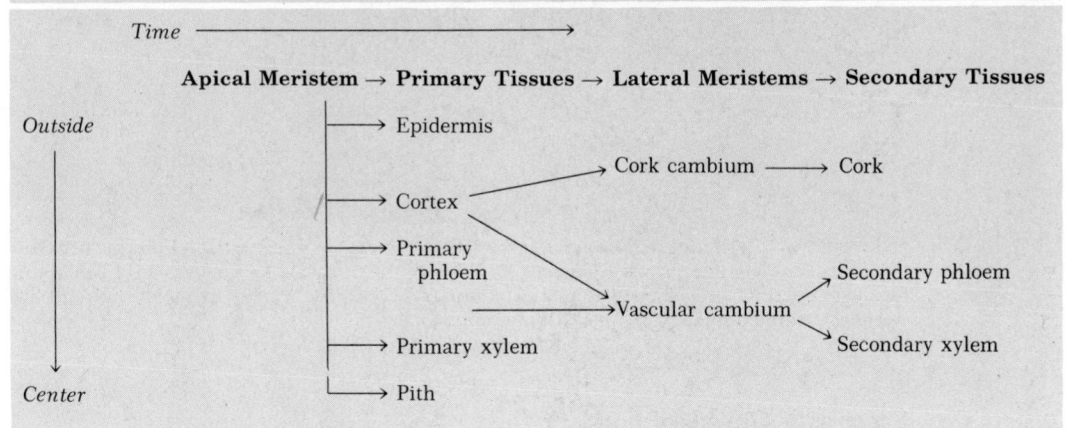

Time ⟶

Apical Meristem → Primary Tissues → Lateral Meristems → Secondary Tissues

Outside

⟶ Epidermis

⟶ Cork cambium ⟶ Cork

⟶ Cortex

⟶ Primary phloem

⟶ Vascular cambium ⟶ Secondary phloem

⟶ Primary xylem

Center

⟶ Pith

Secondary xylem

perhaps not identifiable), cork cambium, and cork. The primary layer of epidermis has by this time ruptured and sloughed away. When the bark is peeled off a tree, it breaks at the layer of delicate undifferentiated cells in the region of the vascular cambium. The trunk of a tree is almost entirely secondary xylem, with a very slender column of pith and primary xylem in the center; the remainder of the above-listed tissues, that is, those outside the vascular cambium, are all in the part of the trunk commonly called the bark.

Structure of Twigs

The finer branches of a woody plant are the **twigs** that bear the leaves, and the flowers and fruits in season. In woody plants that grow in areas where there is a season of dormancy, the apical and axillary meristems pass this season as **terminal buds** and **lateral buds**, respectively. Usually the buds are surrounded by a cluster of **bud scales**—tough, modified leaves impregnated with a waterproof substance (Figure 43-20). The bud scales protect the delicate tissues within.

The bud contains a very short stem bearing rudimentary leaves, and possibly flowers, formed during the previous year. When the bud breaks dormancy, the bud scales are shed, and the leaves and flowers within enlarge tremendously; at the same time the cells in the internodes of the new length of stem elongate, spreading the leaves in the sunlight.

The stem soon completes its primary growth; it then forms its vascular and cork cambia, which in turn produce secondary xylem and phloem, and a layer of bark. Meanwhile, the apical meristem at the end of the twig may have been producing new stem growth and new leaves in addition to those originally contained within the bud. All of the leaves have axillary buds in their axils; some of these axillary buds may grow out and form lateral twigs, but most remain dormant. At the end of the growing season, the buds become enclosed in their protective armor of bud scales. Near the base of each leaf, now grown old, cells divide and enzymes digest some of the cell wall material, forming a zone of weakness. Eventually the leaf falls off, leaving a **leaf scar**. A layer of corky material seals the leaf scar, protecting the twig from loss of water and from attack by insects or pathogens.

The extent of growth during the previous season can be determined by measuring the distance from the base of the terminal bud back to the nearest circle of **bud scale scars**, the marks left by the bud scales of the preceding winter. Such rings of bud scale scars indicate the positions of the terminal bud during its dormant periods.

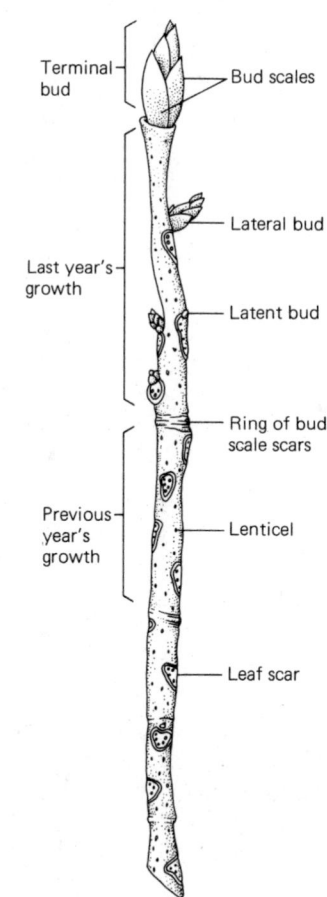

Terminal bud

Bud scales

Lateral bud

Last year's growth

Latent bud

Ring of bud scale scars

Previous year's growth

Lenticel

Leaf scar

Figure 43-20 External features of a hickory (*Carya*) twig.

**CHAPTER 43
STRUCTURE AND
GROWTH OF
VASCULAR PLANTS 717**

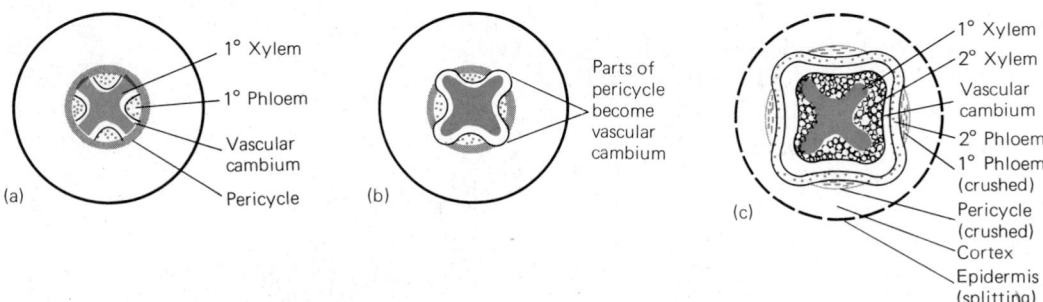

Figure 43-21 Beginnings of secondary growth in a root. (a) Vascular cambium forms between the primary xylem and primary phloem. (b) Parts of the pericycle become vascular cambium, forming a continuous layer. (c) Vascular cambium lays down secondary xylem unevenly, and the vascular cambium quickly assumes a more circular outline.

Secondary Growth of Roots

Secondary growth in roots is similar to that in stems. Vascular cambium forms as a layer of tissue between the primary xylem and primary phloem. However, due to the arrangement of primary vascular tissue in roots, these layers of vascular cambium are roughly U-shaped in cross section (Figure 43-21a). The vascular cambium is continued between these arcs by cells of the pericycle (Figure 43-21b). Thus the vascular cambium is again continuous, but in a complicated shape rather than in the roughly circular ring seen in cross sections of stems. Quite soon, however, the vascular cambium assumes a more circular form, because secondary xylem is added more rapidly to the areas between the arms of primary xylem (Figure 43-21c).

The roots of a tree taper off rather rapidly in diameter just a short distance from the base of the trunk; they may then run for some distance with little change in diameter. Most trees that have taproots as seedlings or saplings develop a more branching root system later on. Typically, there are many large roots spreading in all directions not far below the surface of the soil. The longest roots might be up to twice as long as the longest stem, but most of the major roots are about the same length as the shoot system.

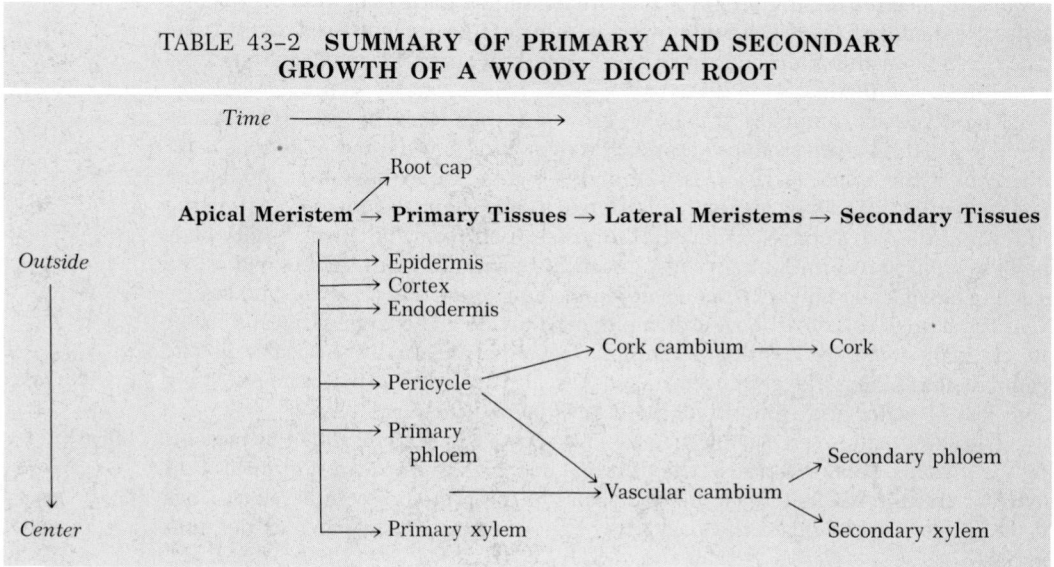

TABLE 43-2 **SUMMARY OF PRIMARY AND SECONDARY GROWTH OF A WOODY DICOT ROOT**

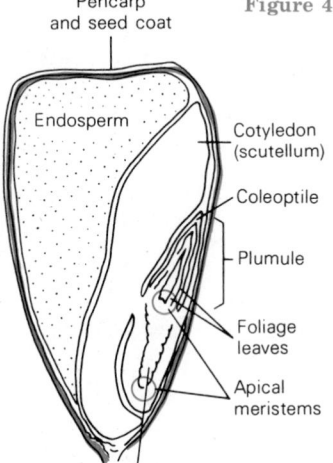

Figure 43-22   Structure of a corn kernel.

Pericarp
and seed coat

Endosperm

Cotyledon
(scutellum)

Coleoptile

Plumule

Foliage
leaves

Apical
meristems

Hypocotyl

Figure 43-23   Growth of a corn seedling. The coleoptile grows upward, while the roots grow down. Note the root hairs on the root at the right. (Biophoto Associates)

43-F   Monocotyledons

The monocotyledons are separated from the dicotyledons on the basis of the number of cotyledons (seed leaves) possessed by the embryo; in monocotyledons, there is only one cotyledon (Figure 43–22). There are other differences between the two groups of plants in the structure of their embryos.

The corn kernel is convenient for studying the structure of a monocot seed. A kernel of corn is actually a one-seeded fruit. The **pericarp**, a part of the fruit, is closely attached to the seed coat, so that the two peel off together as the "skin" of a kernel of corn. In the corn seed, the **endosperm**, a nutritive tissue, is separate from the embryo, whereas in beans the food-rich endosperm has been absorbed into the cotyledons. (The distribution of endosperm is not consistently different between monocots and dicots, however.) The corn embryo digests the food stored in the endosperm and uses it for growth until it becomes established as an independent, photosynthetic plant. The plumule of the corn embryo consists mainly of developing leaves, wrapped above the apical meristem. A tough sheath, the **coleoptile**, protects the leaves and the apical meristem from injury as the seedling pushes up through the soil. Once the coleoptile has penetrated into the light above the ground, the leaves expand greatly, rupture the tip of the coleoptile, and grow on out.

Monocots also differ from dicots in the structure of the primary plant body and the arrangement of its tissues. In cross section, monocot roots may show many more "arms" in the star-shaped xylem (corn has 20 to 40), and there may be an area of pith in the center of the root. In monocot stems, there are numerous small scattered vascular bundles (Figure 43–24), rather than the single ring of bundles found in dicots. Because of this arrangement of bundles, there is no clear demarcation between "pith" and "cortex" regions in monocot stems. In the leaves of monocots, the

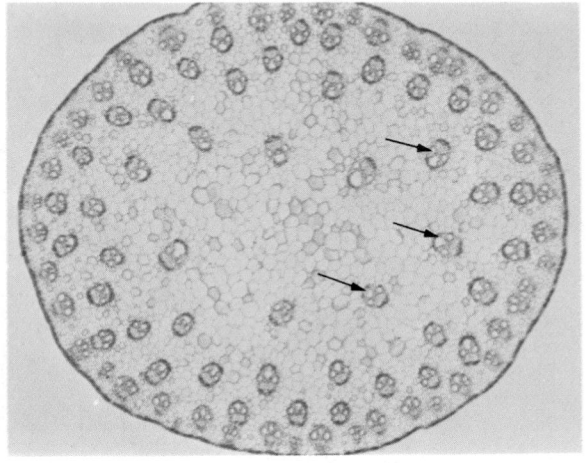

Figure 43-24   Cross section of a mature primary stem of a corn plant (*Zea mays*). Note that the vascular bundles are scattered throughout the stem (arrows), rather than being arranged in a single ring as in dicots (compare with Fig. 43-10). (Carolina Biological Supply Company)

TABLE 43–3 COMPARISON OF MONOCOTYLEDONS AND DICOTYLEDONS

CHARACTERISTIC	MONOCOTS	DICOTS
Seed leaf (cotyledon)	One	Two
Flower parts	In 3's or multiples of 3, or irregular	4, 5, their multiples, or irregular
Leaf venation	Parallel	Netted or fan-like
Vascular tissue	Bundles scattered through stem; usually no cambium	Single ring of bundles or continuous ring; cambium in woody forms
Form	Mostly herbs; few trees, e.g., palms	Herbs, shrubs, and trees
Examples	Lilies, grasses, grains, orchids, irises, onions, palms, crocuses, daffodils, etc.	Oaks, maples, legumes, roses, mints, squashes, daisies, walnuts cactuses, violets, buttercups, poppies, etc.

principal veins of vascular tissue run parallel to each other, rather than in the fan-like or net-like pattern common in dicots. Monocot leaves often stand more or less upright, and stomata are plentiful in both epidermal layers of the leaf.

Another difference between monocotyledons and dicotyledons is their mode of growth. Besides the apical meristems, monocots have **intercalary meristems**, which produce new cells at the bases of the leaves. As a result, a leaf can grow after its top has been damaged, and repair the loss to some extent. A familiar example is the growth of grass in a lawn after each mowing. This mode of growth keeps the developing cells that need the most nutrients close to the roots that supply some of

(a)

(b)

Figure 43-25 Monocots and dicots. Monocot features include (a) leaves with parallel venation and (b) flower parts in multiples of three. (c) The netlike leaf venation and flower parts in fives mark the primrose as a dicot. (a, William Camp)

(c)

these needs. These young, succulent cells have not yet toughened to maturity; their position protects them to some extent from being eaten by hungry grazing animals, scorched by the sun, or desiccated by wind.

Monocots are mostly herbaceous (soft-bodied, non-woody) annuals, or perennials with leaves that die back to the ground after each growing season. Monocots lack secondary tissue and lateral meristems; the few examples of monocots that grow in the form of trees, such as palms and Joshua trees, achieve height and longevity by various means unlike the type of secondary growth described previously.

SUMMARY

A vascular plant grows in one place, making its own food and competing with other plants for the basic resources of sunlight, water, and minerals. The body of a vascular plant is made up of roots, stems, and leaves, all containing vascular tissue, which conducts water, minerals, and food rapidly from one part of the body to another. This body plan is adapted to obtaining water and sunlight efficiently. The plant's indeterminate growth pattern enables it to produce new organs throughout its life; if it is successful in exploiting its environment, it can continue to grow and obtain more and more of the resources it needs.

A seed consists of a protective covering, food supply, and plant embryo with a rudimentary root, stem, and leaves.

Early in life, all cells of the plant can divide. Later most cells mature, specialize, and lose their capacity to divide. Cells that retain the ability to divide are found in meristems at specific locations in the plant body.

Primary growth is principally growth in length and production of new root and shoot branches. The tips of stems and roots grow longer through the activity of apical meristems. Stem branches arise from axillary buds, located at leaf nodes. An axillary meristem thus becomes the apical meristem of a new branch. Lateral roots arise from the pericycle tissue inside the mature primary root. This produces a new apical meristem of a branch root.

Secondary growth is growth in girth by production of secondary tissues. Secondary xylem adds strength and increases the capacity for water conduction to the leaves; secondary protective tissue (cork) replaces the epidermis that has been destroyed by expansion of the tissues interior to it.

Monocotyledons and dicotyledons differ in the number of cotyledons found in the embryo (one versus two, respectively), in the arrangement of vascular tissue, especially in the stems and leaves, and in the arrangement of leaf stomata. In addition, very few monocots have secondary growth, whereas secondary growth is found even in some annual dicots.

OBJECTIVES

From your study of this chapter, you should be able to:

1. Explain how the basic structure (roots, stems, and leaves) and growth pattern of vascular plants are adapted to their functions.

2. List four functions of the root system, and compare the advantages and disadvantages of fibrous roots versus taproots in accomplishing these functions.

3. List the functions of the stem and leaves of a vascular plant.

4. List or point out the parts of a bean seed and of a kernel of corn; state the function or future fate of each part.

5. State what is meant by primary and secondary growth in vascular plants, and describe the sequence of events in primary growth of a root or stem and in secondary growth of a stem.

6. Define the term meristem, and state the role of each of the following in the growth of a plant: apical meristem, zone of elongation, zone of maturation, root cap, terminal or apical bud, axillary bud, intercalary meristem, pericycle, vascular cambium, cork cambium.

7. Sketch a cross section of a leaf and of a mature primary root and stem, and label the following structures if present: epidermis, root hairs, cuticle, stomata, guard cells, endodermis, pericycle, vascular tissue, pith, cortex, collenchyma, palisade and spongy mesophyll, air spaces; give the function of each part.

8. Distinguish between simple and compound leaves, and between alternate, opposite, and whorled arrangements of leaves; define node and internode.

9. Briefly contrast the way branches are formed in roots versus stems.

10. Explain how to tell the age of a twig, and identify the external features of a twig, including: bud scales, lateral or axillary bud, terminal bud, leaf scars, bud scale scars, lenticels.

11. List or recognize differences in structure and growth between monocotyledons and dicotyledons.

SELF-QUIZ

1. A fibrous root system is apt to perform which function better than a taproot system?
   - a. absorption
   - c. food storage
   - b. anchorage
   - d. transport

2. One difference between a bean seed and a kernel of corn is that:
   - a. a bean seed has a seed coat but a kernel of corn does not
   - b. only the bean has two cotyledons; the corn has one
   - c. the bean contains stored food; the corn does not
   - d. the bean embryo has leaves, while the corn embryo lacks leaves
   - e. the bean embryo has two apical meristems; the corn embryo has only one

3. Primary growth of a tree:
   - a. occurs through the activities of apical meristems
   - b. occurs through the activity of a vascular cambium
   - c. occurs through the activity of the root cap
   - d. occurs only in the first year of the tree's life
   - e. occurs in stems, but generally does not occur in roots

4. Arrange the following events during the growth of a shoot in proper order:
   - a. cell division
   - b. cell maturation
   - c. cell elongation

5. Lateral roots arise from:
   - a. root hairs
   - b. pericycle
   - d. endodermis
   - c. cork cambium
   - e. axillary buds

6. Secondary xylem and phloem are laid down by:
   - a. apical meristems
   - b. axillary meristems
   - d. cork cambium
   - c. vascular cambium
   - e. intercalary meristems

7. Which of the following would *not* secrete a cuticle?
   - a. leaf epidermis
   - b. stem epidermis
   - c. root epidermis

8. One year's growth in length of a young woody shoot is the distance between successive:
   - a. rings of bud scale scars
   - b. leaf scars
   - d. branches
   - c. axillary buds
   - e. any of the above

9. For each characteristic listed below, tell whether it is characteristic of monocotyledons, dicotyledons, or both:
   - a. stomata on both leaf surfaces
   - b. roots with root caps
   - c. parallel venation in leaves
   - d. many vascular bundles scattered throughout stem cross section
   - e. vascular cambium

10. Horace and Hermione visited a forest on their honeymoon. Horace selected a tree that was 10 m tall and 30 cm in diameter. He carved their initials into its bark 1.5 m above ground level. On their tenth anniversary, Horace and Hermione return to the forest; the tree is now 12 m tall and 33 cm in diameter. Their initials are now:
    - a. 1.5 m above ground level
    - b. 2 m above ground level
    - c. 3.5 m above ground level

QUESTIONS FOR DISCUSSION

1. What is the advantage of the root hairs' developing their extensions in the zone of maturation of the root, rather than in the apical meristem or the zone of elongation?

2. What is the advantage of the arrangement of xylem in a star-shaped pattern in the root, rather than its being internal to the phloem as it is in the shoot?

3. Since the cotyledons of a bean seedling become photosynthetic, one might expect them to be valuable organs of the plant even after their stored food is used up. Why might it be adaptively advantageous to the plant to shed them soon after it becomes well established?

4. Leaf structure varies from one species to another, and it is often closely correlated with the habitat of the plant. In what type of habitat would you expect to find each of the following modifications of leaf structure?
   - a. more than one cell layer in the epidermis
   - b. extra thick layers of cuticle
   - c. little or no cuticle
   - d. little or no cuticle; little or no xylem; no stomata
   - e. large air spaces in the mesophyll, and stomata in the upper epidermis instead of in the lower epidermis

REFERENCES AND FURTHER READING

Cronshaw, J. "Support and protection in plants." *Topics in the Study of Life*. New York: Harper and Row, 1971. A good summary of basic plant structure.

Elias, T. S., and H. S. Irwin. "Urban trees." *Scientific American*, November 1976.

Esau, K. *Plant Anatomy*, 2d ed. New York: John Wiley and Sons, 1965. A reliable reference on plant structure, illustrated with numerous drawings and photomicrographs.

Pillemer, E., and W. Tingey. "Hooked trichomes: a physical plant barrier to a major agricultural pest." *Science* 193:482, 1976.

Weier, T. E., C. R. Stocking, and M. G. Barbour. *Botany: An Introduction to Plant Biology*, 5th ed. New York: John Wiley and Sons, 1974. An especially well-illustrated botany text.

Wilson, B. F. *The Growing Tree*. Amherst: University of Massachusetts Press, 1970. Focuses on structure and growth of trees.

Wilson, C., W. Loomis, and T. Steeves. *Botany*, 5th ed. New York: Holt, Rinehart and Winston, 1971. A very clear introduction to this and many other aspects of plants.

CHAPTER

44

TRANSPORT IN VASCULAR PLANTS

The root endings of an oak tree may be hundreds of feet from its leaves. The roots and leaves are connected by the vascular tissues, which transport water and minerals from the roots to the leaves, and conduct food from the leaves to the roots, as well as to growing buds, flowers, and fruits. Vascular tissue—the wood in the roots and stems, and the veins in leaves—also provides strength and support.

Having an efficient transport system permits the tree to have parts specialized for different functions, such as water-gathering, energy-gathering, and reproduction. Furthermore, the support provided by vascular tissue permits the tree to exploit a much larger volume of soil and air, and so increases its ability to compete with other plants.

How do materials move hundreds of feet to the various parts of a tall tree? To answer this question, we must study the structure of the vascular tissues, **xylem** and **phloem**, and the activities of nearby cells.

723

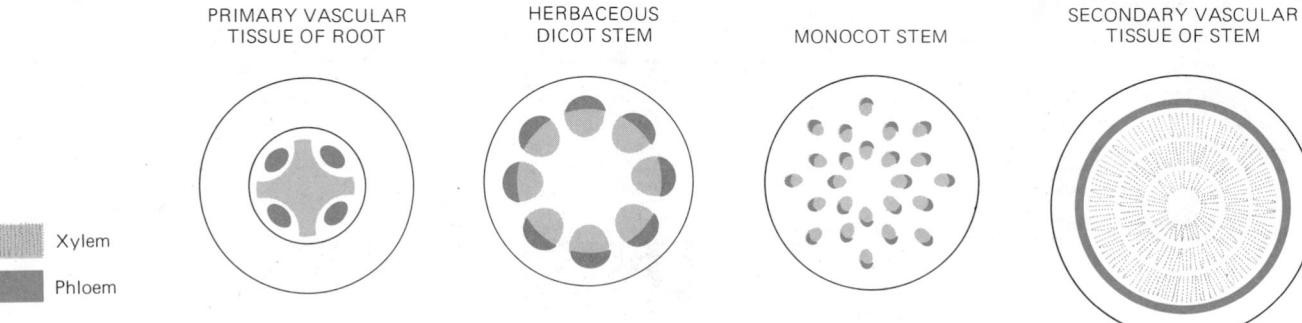

PRIMARY VASCULAR
TISSUE OF ROOT

HERBACEOUS
DICOT STEM

MONOCOT STEM

SECONDARY VASCULAR
TISSUE OF STEM

Xylem

Phloem

Figure 44-1 Arrangement of the vascular tissues, xylem and phloem, in various plant parts. Primary vascular tissue is located in the center of the root, and in discrete bundles in stems. The bundles are arranged in a single ring in dicots, and are scattered throughout the cross section of the stem in monocots. Secondary vascular tissues form in concentric rings in woody stems.

44-A Functions of Xylem and Phloem

The arrangement of vascular tissue differs in different plants and in different parts of the same plant (Figure 44–1). In both roots and stems of most vascular plants, the xylem is located toward the interior of the plant, and the phloem toward the exterior. What is the evidence that xylem and phloem are transporting tissues, and how do we know what is transported by each?

Young children enjoy coloring white flowers by placing their cut stems in a solution of food coloring, and waiting for the dye to rise into the flower. When the stem of the flower is held up to the light, strands of color can be seen inside. Under the microscope, a section of the stem shows that the dye is inside large cells in the xylem tissue; the rest of the cells in the stem retain their original color. This is evidence that xylem transports water upward in a plant.

In 1679, Marcello Malpighi, an Italian scientist, performed a girdling experiment to determine the functions of xylem and phloem. **Girdling** consists of removing the bark in a complete ring around the trunk of a tree. This removes the phloem, which is attached to the inner bark, but leaves the secondary xylem, or **wood**, intact. After this treatment, Malpighi found that a swelling appeared in the bark just above the stripped area; the fluid that exuded from this swelling was sweet (we now know that it contained sucrose). The leaves appeared to be unaffected for days or months; eventually, however, the leaves wilted and then died, and the entire tree was soon dead (Figure 44–2).

From these observations, Malpighi concluded that phloem transports food, such as the sugar in the liquid exuded from the bark, to the roots. Without this supply of

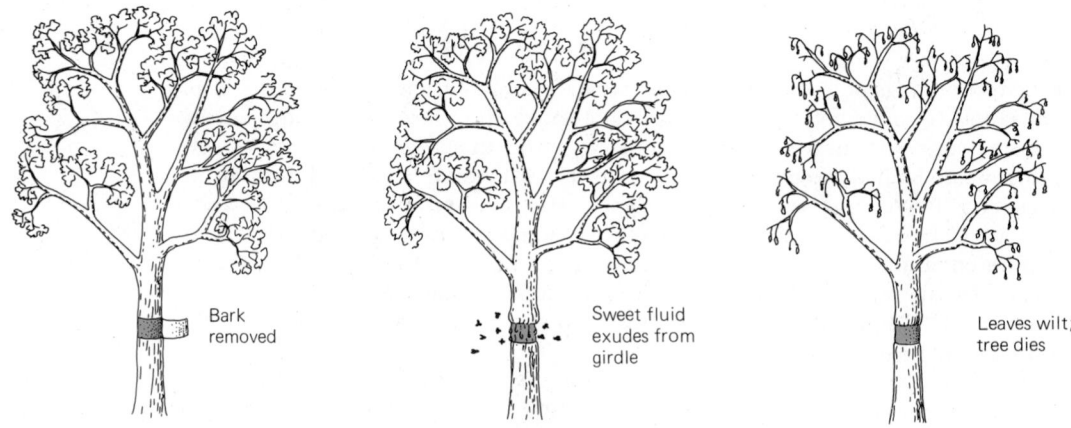

Bark
removed

Sweet fluid
exudes from
girdle

Leaves wilt;
tree dies

724 **PLANT BIOLOGY**   Figure 44-2 A girdling experiment performed by Malpighi.

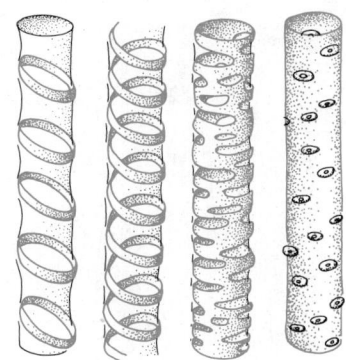

Figure 44-3 Secondary thickenings (color) in xylem cells vary greatly in extent.

food, the roots died after they had used the food reserves stored below the girdle. Since the leaves remained healthy for a time, Malpighi concluded that xylem transports water to the leaves; without a continuous supply of water, the leaves would have wilted and died in just a few hours.

So it appeared that the vascular tissues divide the task of transport, with the xylem moving water from the roots to the shoot system, and phloem moving food from leaves to roots. Other girdling experiments have since shown that phloem also conducts food to growing buds, flowers, and fruits.

44-B   Structure of Xylem

Xylem is a complex tissue; that is, it contains many different types of cells. Only some of these conduct the mixture of water and solutes called **sap**. Both conducting and nonconducting cells may contribute strength to the plant body. The structure of xylem shows many adaptations that enhance its dual functions of transport and support.

Right after they are formed by cell division, most plant cells lay down **primary cell walls**, composed of cellulose, pectin, and other polysaccharides. When they have expanded to their mature size, the cells that will become xylem conducting cells lay down **secondary thickenings**, which strengthen the cell walls. These secondary thickenings vary from disconnected rings to extensive **secondary cell walls**, which cover the surface of the cell almost completely (Figure 44–3). The secondary thickenings are composed of cellulose and **lignin**, a tough, complex organic compound that makes wood woody; in general, the more secondary thickening in the walls of its conducting cells, the more woody the plant.

A very important feature of the conducting cells of xylem is that they die once their cell walls are complete. The cell contents disintegrate, leaving a strong, hollow cylinder filled with sap. Since the conducting cells are stacked on top of one another, sap can travel in a more or less straight line up the plant.

The conducting cells of primitive vascular plants are **tracheids**, long, extremely thin cells with slanting end walls. The wood of a gymnosperm such as a pine is made up almost entirely of tracheids (Figures 44–4, 44–5b). The heavy secondary cell walls slow the passage of sap from one tracheid to the next. Areas called **pits** in the walls of tracheids permit more rapid passage of sap from cell to cell. The pit is not a hole in the wall; rather, it is an area where there is little or no secondary cell wall material between neighboring cells. Pits occur both in the side walls and in the end walls of tracheids; this permits water to move **laterally**, or sideways, as well as upwards in the plant, and permits adjustment of the water supply to different sides of the tree.

In addition to tracheids, the secondary xylem, or wood, of a pine also contains living parenchyma cells. Some of these parenchyma cells form **resin canals** (see Figure 44–7b). **Resin** is a sticky, pungent fluid secreted by cells lining the canals; it inhibits the growth of certain pathogenic (disease-causing) organisms and so helps protect an injured tree from disease. Resin may be collected and distilled to make turpentine. It is also the raw material for pitch and tars, used to protect wood from rotting.

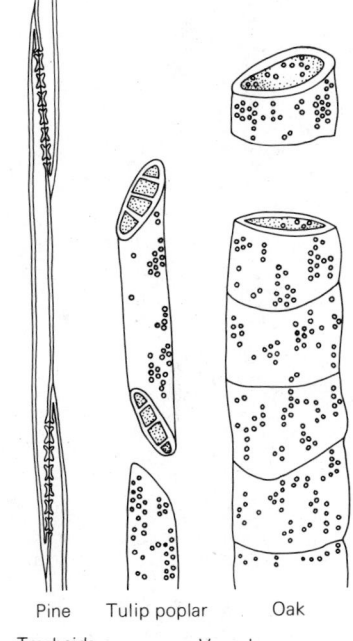

Pine Tulip poplar Oak
Tracheids   Vessels

Figure 44-4 Conducting cells found in xylem. Tracheids occur in angiosperms as well as in lower vascular plants. Vessels are characteristic of angiosperms. Vessel members are thought to have evolved from tracheids; they are generally shorter and wider than tracheids, with end walls perforated or absent.

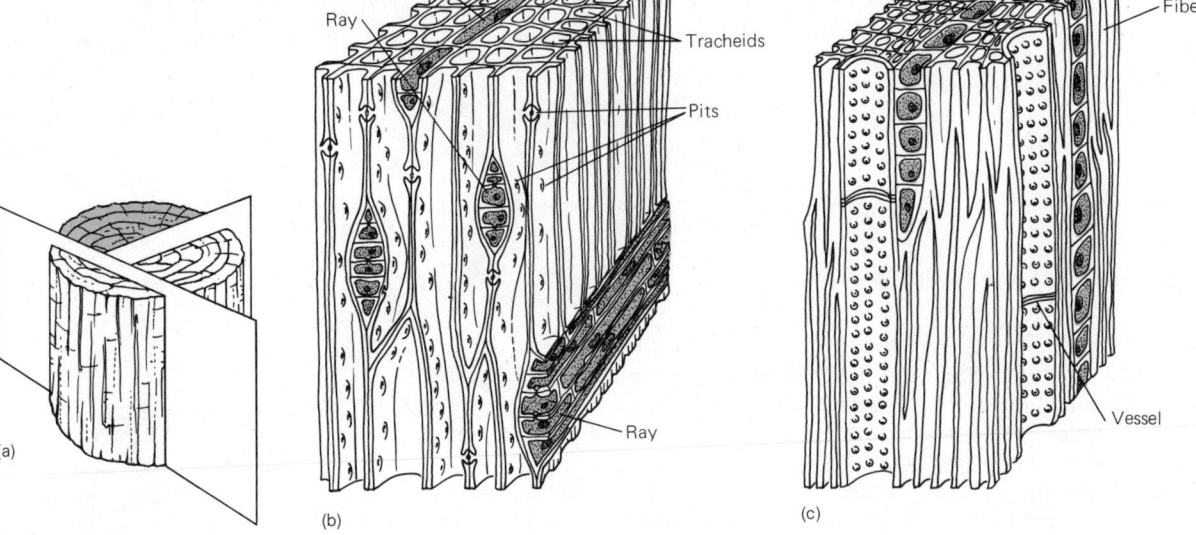

(a)

(b)

(c)

Figure 44-5 Structure of xylem. (a) Shows the planes in which logs are cut to produce the sections shown in (b) and (c). The colored section of the log in (a) is enlarged in parts (b) and (c). (b) The secondary xylem (wood) of pine. (c) The secondary xylem of an angiosperm. Note that ray cells in both kinds of wood are elongated for lateral conduction.

Pine wood also contains other living parenchyma cells arranged in radial formation, or **rays**, running out from the center of the tree. The living ray cells conduct laterally, moving materials out to the living tissues in the vascular cambium, phloem, and cork cambium. They may also serve as depots for food storage.

A major evolutionary advance of angiosperms was further specialization of cells in the xylem. Some conducting cells have literally made an evolutionary breakthrough; their last act in life is to digest parts of their end walls to form real holes, not just thin pit areas (see Figure 44–4). This allows sap to flow directly from one cell to the next. In lower angiosperms, the end walls are perforated; in more advanced forms, they are entirely absent, and the cells are like sections of pipe.

Along with the loss of material from the end walls of conducting cells came a progressive shortening and widening of the cells, increasing the diameter of the pipelines from the roots to the leaves. This greater diameter reduces the frictional drag on sap passing through xylem and permits the sap to travel faster. The hollow xylem tubes of angiosperms (and of a few gymnosperms) are called **vessels**; the individual cells in the vessels are called **vessel members** or **vessel elements**.

Some of the wood of an angiosperm such as an oak is made up of vessels (Figure 44–5), but most of it is composed of other types of cells. Among these are **fibers**— long, thin cells with thick cell walls. It is not hard to imagine that these cells evolved from tracheids by an exaggeration of the secondary cell wall thickenings, whereas the vessel members evolved in another direction and lost parts of their cell walls

Figure 44-6 The wood of a hazel (*Corylus*) tree as it appears under the scanning electron microscope. (Biophoto Associates)

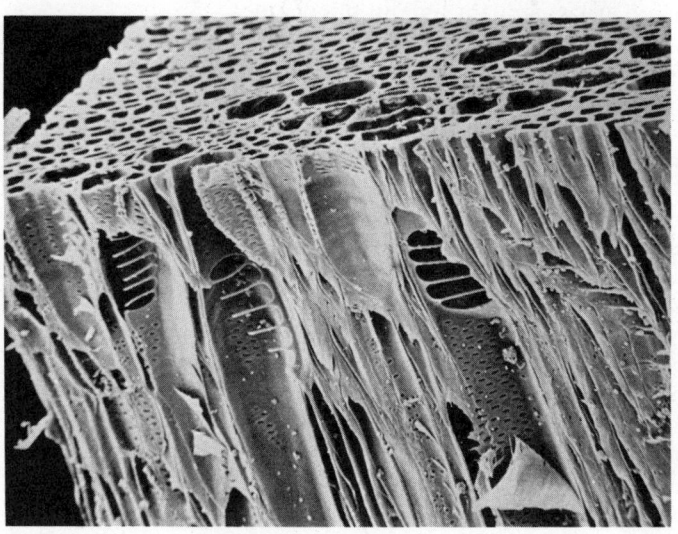

TABLE 44-1

SUMMARY OF CELL TYPES IN XYLEM TISSUE

CELL TYPE	DESCRIPTION	FUNCTION
Tracheids	Long, thin, thick-walled, dead; thin areas (pits) in walls let contents into adjoining cells.	Transport of sap
Vessel members	Shorter and wider than tracheids, thick walls with holes, which let contents into next cell; dead	Transport of sap
Fiber and sclereid cells	Fibers long, thin, thick-walled, often dead; sclereids variable shape, thick-walled, dead	Strengthening
Parenchyma	Relatively thin-walled, relatively unspecialized, living cells	Lateral conduction; starch storage

completely. Like vessel elements, some fibers die, leaving strong walls, which help support the tree. Angiosperm xylem may also contain irregularly shaped **sclereids**, or stone cells. Sclereids are similar to fibers in that they secrete a thick secondary cell wall and then die. Like gymnosperms, angiosperms have xylem rays made up of living parenchyma cells, and many also have tracheids as well as vessel members.

Changes in the Xylem of Woody Plants

The xylem of a woody plant changes throughout its life. As we saw in Chapter 43, the vascular cambium adds a new ring of xylem outside the existing xylem each year. These **annual rings** are an eye-catching feature of the wood of some trees (Figure 44-7a). In certain kinds of trees, the inner part of an annual ring, the **spring wood**, is light in color. In the spring there is usually a great deal of moisture available to the tree; under such good growing conditions, the tree produces a relatively high concentration of the growth hormone auxin, and the new cells in the xylem grow rapidly and reach a large size before they die. The summer is usually drier; less auxin is produced, and the tracheids and vessel members grow more slowly, and so are smaller when they die. **Summer wood** appears darker than spring wood because cell walls occupy a greater proportion of its area. Annual rings can be used to determine the age of a tree or branch; there is usually one annual ring for each year of age.

Figure 44-7 Annual rings. (a) Cross section of a pine tree trunk. Note the annual rings of alternating light and dark wood. (b) Under the light microscope, a cross section of pine wood shows the differences in tracheid cell diameter between spring and summer wood. The cells at the bottom of the picture are the oldest. (a, William Camp; b, Biophoto Associates)

(a)

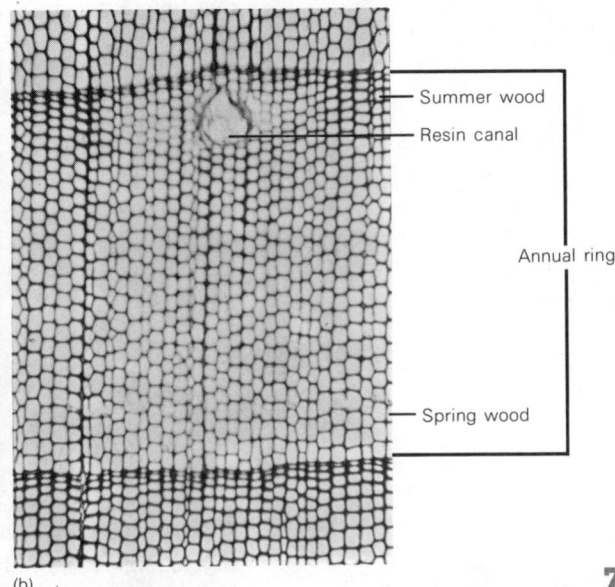

(b)

Summer wood

Resin canal

Annual ring

Spring wood

Figure 44-8 The cliff palace at Mesa Verde, Colorado, built between 1200 and 1300 A.D. The pattern of annual rings in the building timbers suggests that a long drought forced the inhabitants to abandon their intriguing cliff towns. (National Park Service)

Annual rings are sometimes used to date archaeological finds, or to determine the climate at some time in history. A wide band of growth in an annual ring reflects a good growing season, while a narrow band is formed in a poor year. By comparing the pattern of tree rings in building timbers with the rings in large, old trees in the same areas, archaeologists have been able to determine when the structures were built (Figure 44–8).

Another change in xylem with age is the formation of heartwood. The annual rings of the latest several years' xylem are called **sapwood** because they actually conduct the sap in trees. Interior to the sapwood is the **heartwood**, older xylem that no longer conducts sap. The heartwood is often plugged by deposits of waste material. Sometimes it rots away, leaving a hollow tree that is still alive because its sapwood and phloem are still functioning.

44-C Transport in the Xylem

Because there is no living matter in the conducting cells of xylem, it looks as though xylem conduction must depend on physical phenomena rather than on the activities of living cells. The tubular shape of tracheids and vessels, and the stacking of the conducting cells in columns, is obviously of advantage in the transport of sap. Of equal importance, however, is the physical nature of water, the main constituent of the sap that travels through the vessels.

Properties of Water

Let us briefly review some of the properties of water that will be important in discussing how it moves through the xylem of a tree:

1. **Diffusion and osmosis**. Water molecules tend to move down a water potential gradient. **Water potential** is a measure of the energy of water, and it depends mainly on two factors: the osmotic potential, due to the presence of solutes, and the hydrostatic or turgor pressure of water in the cells pushing outward against the cell walls (see Section 5-F):

water potential = osmotic potential + turgor pressure

The osmotic potential is zero for pure water, and it is negative for water containing solutes; thus the negative osmotic potential of a cell's cytoplasm opposes the positive value of the cell's turgor pressure. When the magnitudes of the osmotic potential and turgor pressure are equal, the two cancel out, and the water potential is zero; there is no net movement of water into or out of the cell. As we shall see, however, the osmotic potential, and hence the water potential, can be lowered (made more negative) either by addition of solutes or by removal of some of the solvent, water. Water tends to move into areas with lower water potential.

2. **Evaporation**. Some of the molecules of water escape from the liquid and enter the gaseous state (water vapor).

3. **Cohesion**. Water molecules stick together with considerable strength because of the numerous, weak hydrogen bonds between them.

4. **Adhesion**. Water molecules adhere to polar substances, such as the components of plant cell walls.

5. **Capillarity**. The cohesive and adhesive nature of water causes it to move into small spaces lined with polar molecules, even against the pull of gravity. Thus water flows up the sides of tracheids or vessels, and water in the center of the tube is pulled up because it sticks to the water that is attracted to the sides of the cells. The smaller the bore (internal diameter) of the tube, the higher water will rise in it (Figure 44–9). Calculations indicate that water moving by capillarity in the most slender tracheids can reach a maximum height of about five feet, not high enough to account for the rise of sap in many plants.

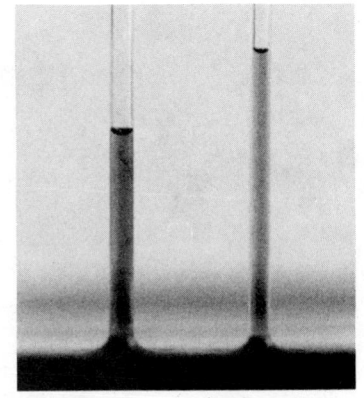

Figure 44-9 When the end of a glass capillary tube is touched to the surface of water, water automatically rises into the tube. The smaller the bore of the tube, the farther the water will rise.

Root Pressure

If sap cannot climb on its own, it must either be pushed by some force from the bottom of the plant, or be pulled from the top of the plant. The push from below is called root pressure. Root pressure can be demonstrated in some plants by cutting the plant off near the ground and sealing the stump into a glass tube; sap rises into the tube (Figure 44–10). The **root pressure**, or the force with which sap is pushed up into the stem of the plant, can be calculated from the final height of the fluid in the tube.

Root pressure depends on active transport in living root cells, which lowers the water potential. The root cells accumulate ions from the soil solution by active transport and move them toward the center of the root. This accumulation of ions lowers the water potential of the sap in the xylem, and water follows, building up hydrostatic pressure in the xylem vessels. The endodermal layer surrounding the vascular tissues of the root forms a barrier that prevents the sap from flowing out of the root back to the soil; thus there is nowhere for the sap to go but up.

Because roots use a great deal of energy for active transport, they must have an adequate oxygen supply. When the soil contains ample water and oxygen, root pressure may be high. If, at the same time, the atmosphere is saturated with water vapor, water does not vaporize readily from the leaves, and sap forced up from the

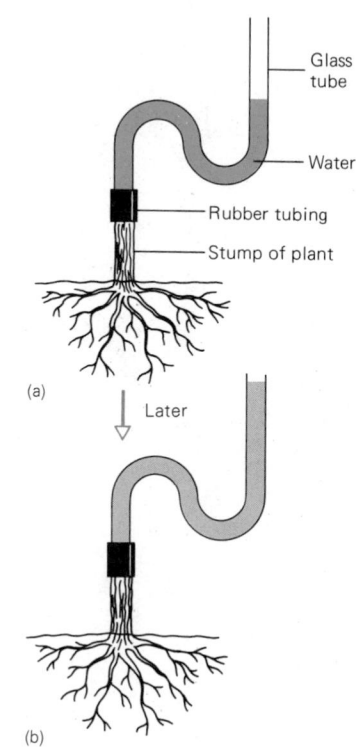

Figure 44-10 Demonstration of root pressure.

Figure 44-11 Guttation. (Biophoto Associates)

roots may be exuded from the tips of the leaves. This **guttation** occurs only in rather short plants, where the tips of the leaves are comparatively close to the root endings, the source of the root pressure (Figure 44–11).

However, the greatest root pressures measured are sufficient to push sap only a few feet. Thus, root pressure alone can account for xylem conduction only in plants that are at most a foot or two tall. Root pressure has not been found at all in gymnosperms, which include the tallest trees known.

Transpiration Pull

This evidence forces us to look at a third mechanism for xylem transport, the pull from above, now known as **transpiration pull**. In 1727 Stephen Hales, an English clergyman, demonstrated that **transpiration**, the evaporation of water from leaves, can pull sap up through the xylem of a plant. Hales cut sets of similar leafy branches, removed different amounts of leaves from some of them, and left the rest intact. He then set each branch in a container with a measured amount of water and observed that the amount of water removed from the container was roughly proportional to the area of leaf surface on the branch (Figure 44–12). Hales decided that it was some activity of the leaves, rather than merely the activities of the xylem "capillaries" discovered by Malpighi, that caused sap to rise in the stem of a plant.

In another experiment, Hales found that this movement depended on the leaves' being dry and exposed to the air. A branch with its leaves immersed in water could not "perspire" (or transpire, as we would say today) and thus water did not move through the branch at an appreciable rate; when the leaves were allowed to dry and stand in the air, however, they "perspired" a great deal and pulled water up into the tube (Figure 44–13).

Start

Finish

Figure 44-12 Stephen Hales determined that water movement depends on the leaves by measuring the rates of water absorption by branches with differing amounts of leaf surface.

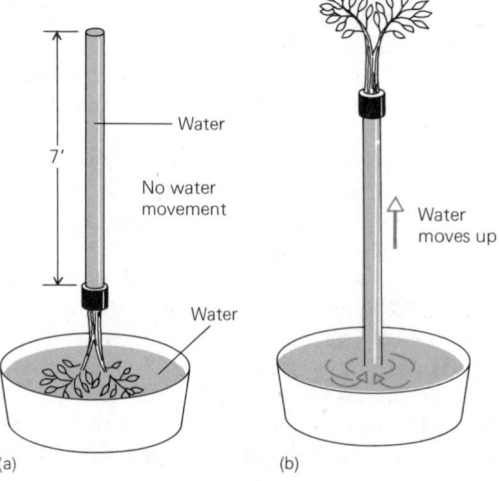

Figure 44-13 In this experiment, Hales showed that water moves through a plant only when its leaves are dry. A glass tube seven feet long was attached to a branch and filled with water. (a) No water moved through the branch when the leaves were immersed in water, even though the water in the tube was being pulled down by the force of gravity. (b) However, when the leaves were dry the branch would pull water up the tube against the pull of gravity.

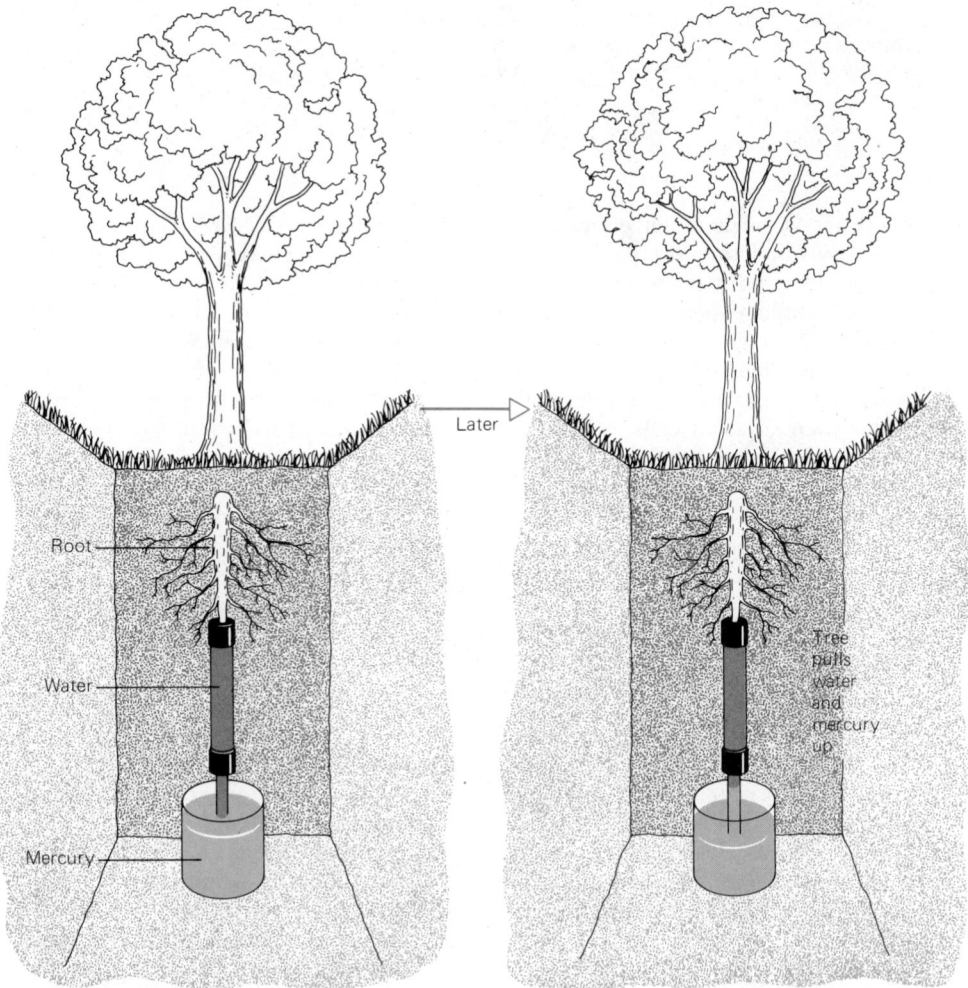

Later

Root

Water

Mercury

Tree pulls water and mercury up

Figure 44-14 Hales measured the pull exerted by a transpiring tree by attaching a tube filled with water to the cut root of a pear tree. He then placed a vessel of mercury in contact with the water and measured how far the mercury was pulled up into the tube as the tree withdrew water.

In yet another experiment Hales dug a deep hole next to a pear tree, exposing a root. He attached a large tube of water to the root, and below this tube he fixed a smaller tube filled with mercury. The lower end of this tube was immersed in a container of mercury (Figure 44–14). As the root drew water from the large tube, the water adhered to the mercury and pulled it up. By measuring the rise of the mercury, Hales could calculate the pull exerted on the water by the tree. He observed that the water and mercury were pulled up further on bright, sunny dry days than on cloudy or damp days, and that the level of mercury fell at night. These were exactly the results to expect if the evaporation of water from the leaves were indeed the force that pulled water up through the tree.

Although Hales clearly demonstrated transpiration pull, the role of water cohesion was pointed out by Henry Dixon and John Joly, who did similar experiments in the 1890s and published convincing arguments for the theory.

ANATOMICAL BASIS OF TRANSPIRATION PULL Leaves inevitably lose water, in the gaseous form of water vapor, because they must exchange gases with the atmosphere. Evaporation increases with the amount of surface area exposed to the air, and in a "typical" plant the leaves account for most of the exposed area. The leaf's waterproof cuticle helps to reduce evaporation, but since it is relatively impermeable to gases, it also prevents the leaf from obtaining carbon dioxide for photo-

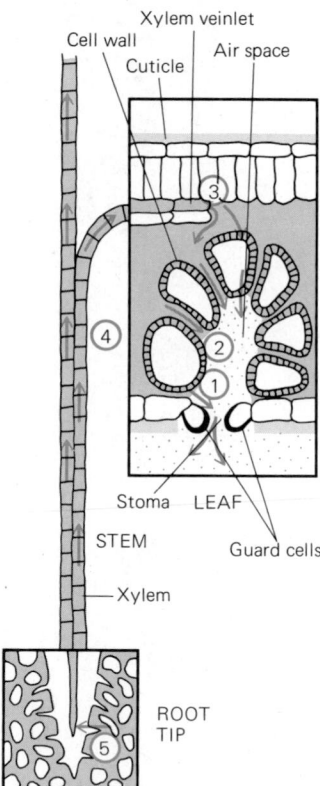

Cell wall
Xylem veinlet
Cuticle
Air space

Stoma   LEAF
STEM
Guard cells

Xylem

ROOT
TIP

Figure 44-15 Movement of the "transpiration stream" through a plant. (After Weier, Stocking, and Barbour, *Botany* 4th ed., p. 229. New York: John Wiley & Sons)

synthesis through much of its surface. **Stomata** are pores that allow leaves to reduce water loss and still exchange carbon dioxide and oxygen (and inevitably, water vapor) with the atmosphere.

Each stoma opens into an air space surrounded by a loose collection of parenchyma cells (Figure 44–15). Water evaporates into the air space from the spaces in the parenchyma cell walls and diffuses out through the stoma. The evaporation of water lowers the water potential in the cell wall spaces; the lost water is replaced by water from adjacent cells, cell walls, and intercellular spaces. This water in turn pulls the water behind it, and so forth, until the replacement water comes from the end of a xylem veinlet. This is replaced by water that it pulls up behind it, and thus the entire water column in the thin xylem pipeline, all the way down to the ends of the xylem in the roots, creeps up a little. Eventually the pull reaches root cells which take in soil water to replace that lost through the leaves.

Transpiration seems to be a necessary evil, the price plants pay for having leaves, which are very efficient structures for obtaining sunlight and carbon dioxide. The cuticle, guard cells, and internal air spaces seem to have evolved as a means of limiting water loss through the leaves.

COHESIVE STRENGTH OF WATER Water follows a pressure gradient from the soil, in which water may be at a low positive pressure, to the plant, in which pressure rapidly decreases and (under most circumstances) becomes negative. Under such conditions, water in the xylem is under **tension**, that is, a pull or negative force. Tensions in large trees may "stretch" the water in the xylem so much that it sucks the walls of the xylem in, and the tree's circumference shrinks measurably.

The cohesive strength that keeps water molecules together must be able to withstand tensions increasing by one bar† (1000 kg/cm-sec^2) for every 10.2 metres in height. In addition, the walls of the xylem exert a frictional resistance that at least doubles the pull required to raise the water column. On days when transpiration is especially rapid, or when little water is available from the soil, the tension on the water in the xylem column increases even more.

Because water in the xylem conducting cells is under tension, the actual cohesion of water as it is pulled up is difficult to measure. Cutting or puncturing the xylem to take internal measurements lets air rush in as the water column snaps apart, just as the two ends of a stretched piece of elastic snap when it is cut. When air enters the xylem column, it breaks the cohesion of the water, preventing further movement of sap by transpiration pull through the affected tracheid or vessel.

RATE OF TRANSPIRATION Hales estimated that, weight for weight, the daily water intake of a plant is 17 times that of a human being. This huge intake is necessary because the plant transpires so much water to the atmosphere. On a warm, dry day Hales measured the transpiration of a one-metre-tall sunflower plant as 0.9 litre in 12 hours. On dry nights, this rate decreased to less than 0.1 litre in 12 hours, and on nights with dew, the plant gained water by absorbing the dew through its leaves.

The rate of transpiration for any particular plant depends on many factors. Structural features of the plant, such as the density and distribution of the stomata, the thickness of the cuticle, and the geometry of the leaves with respect to each other and to the ground, all influence the rate of transpiration.

---

† 1 bar is about 1 atmosphere = 14.8 lb/in^2. The seeming disparity is due to the fact that the pound is a force, whereas the kilogram is a mass, which must be multiplied by the acceleration of gravity to obtain a measure of force. (Force = Mass $\times$ Acceleration)

Environmental factors that affect transpiration include the availability of water in the soil, the intensity of sunlight, and the relative humidity, temperature, and wind speed of the air. The availability of soil water determines how much water the plant can "afford" to lose by transpiration. Sunlight stimulates photosynthesis and causes the plant to open its stomata and take up carbon dioxide more rapidly; thus the plant loses water more rapidly in bright sunlight. Low humidity increases transpiration because the air can pick up a great deal more water vapor before it becomes saturated. Furthermore, the higher the air temperature, the more water vapor air can gain before it reaches the saturation point. If the air is still, the water vapor transpired by a leaf remains near the leaf, forming a protective layer of saturated air that retards further evaporation from the leaf. However, if there is a breeze this layer of still, saturated air blows away, and the leaf loses more water.

Plants can regulate their transpiration according to environmental conditions. This regulation is performed largely by the guard cells, which open and close the stomata.

OPERATION OF GUARD CELLS A stoma is surrounded by two guard cells (Figure 44–16). The ability of a pair of guard cells to control the size of the opening of the stoma depends on their peculiar structure. The cell walls are thicker next to the stomatal opening, and thinner on the ends and sides away from the stoma.

Basically, guard cells open the stoma by accumulating solutes, lowering their osmotic potential, and hence their water potential, so that water enters and the guard cells become turgid. Because of their unique structure, the guard cells do not swell evenly all around. The thin parts of the guard cell walls, at the ends and the outsides, stretch easily, while the thick part next to the opening stretches very little. As a result, the guard cells assume a curved shape. As the ends of the guard cells swell and push against each other, the thick inner walls are pulled apart, opening the pore.

To visualize this situation better, consider the guard cells as elongated balloons. Some long balloons do not blow up to be uniformly long, but instead bend because of a difference in the thickness in the rubber. The thicker parts of the balloon do not expand as much as the thinner parts, and as a result, the shape of the balloon turns out bent.

How do the guard cells lower their osmotic potential? At one time, the fact that guard cells are the only epidermal cells with chloroplasts was regarded as evidence that the guard cells increased their concentration of solutes by manufacturing sugar. More recently, however, a better correlation has been found between the opening of the guard cells and their accumulation of potassium ions. But if the guard cells operate by gaining salts instead of sugar, why do they have chloroplasts? It seems that the crucial role of guard cell chloroplasts may be the immediate production of ATP (see Section 8-C) rather than the ultimate production of sugar; ATP could supply the energy needed for active transport of potassium ions into the cell.

STOMA CLOSED

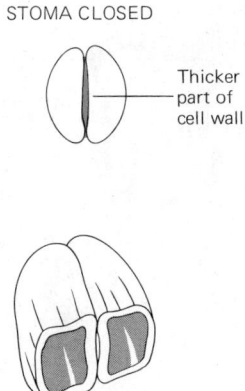

Thicker part of cell wall

STOMA OPEN

Thicker part of cell wall

Figure 44-16 Opening and closing of the stoma depends on turgor pressure in the guard cells. The guard cells have thicker walls on the side facing the stoma; thus when the cells swell, the inner part of the cell wall can expand less than the end and outer portions, and the stoma opens.

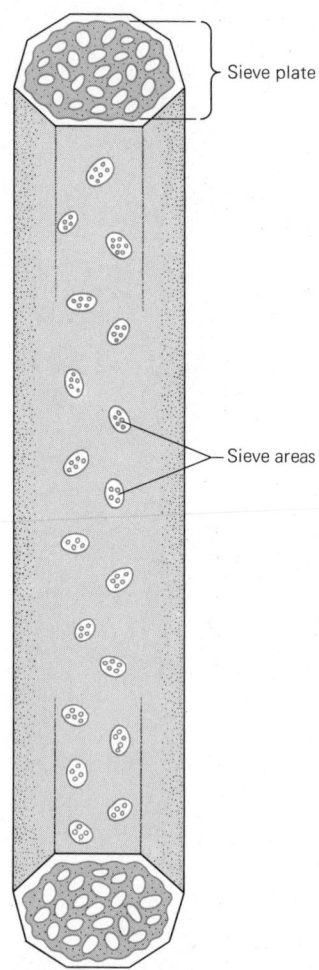

Figure 44-17 An angiosperm sieve tube member. The top and bottom walls are slanting.

The stomata open when the concentration of carbon dioxide in the guard cells is low. Under normal circumstances, this happens in the daytime, when the plant's photosynthesis is using up carbon dioxide. However, the stomata will also open in the dark if the carbon dioxide content of the surrounding air is artificially lowered. It is not yet known how the low carbon dioxide level triggers the active transport of potassium into the guard cells.

Water loss by transpiration sometimes exceeds the ability of the xylem to deliver water. When the water potential in the leaves becomes too low, the hormone abscisic acid is quickly synthesized. This hormone causes the guard cells to lose their turgor, and the stomata close until the delivery of water through the xylem catches up with the needs of the leaves.

44-D Phloem Structure

In addition to xylem, vascular plants have another transport tissue, the phloem. Phloem is responsible for the conduction of various substances, especially organic compounds, throughout the plant from sites of production to sites of use or storage. In most plants, the phloem lies near the xylem.

Like xylem, phloem contains many types of cells. In the more primitive vascular plants, including gymnosperms, the conducting cells of the phloem are the **sieve cells**. Angiosperms have evolved more specialized phloem conducting cells, called **sieve tube members**, stacked one on top of the other in long columns called **sieve tubes**. The sieve cells of lower vascular plants are longer and narrower than the sieve tube members of angiosperms. Both types of cells, however, have special **sieve areas**, consisting of numerous pores in their cell walls where substances can be exchanged with neighboring cells. Sieve areas in the end walls of sieve tube members have evolved as **sieve plates** with rather large pores (Figure 44–17).

The conducting cells of phloem, unlike those of xylem, contain living cytoplasm. Before a cell becomes able to conduct, it loses its nucleus and most of its other organelles. In addition, there is a great enlargement of its **plasmodesmata**, strands of cytoplasm that pass through the sieve areas and sieve plates and connect neighboring cells. Strands of a material called **P-protein** pass through the cytoplasm both within the cell and in the plasmodesmata between cells.

After the conducting cells have lost most of their own metabolic machinery, they are believed to be maintained by the metabolism of adjacent cells that remain intact. In primitive vascular plants, neighboring parenchyma cells perform this function. In angiosperms, unequal division of a single mother cell produces a large cell that becomes a sieve tube member and a smaller cell that becomes the **companion cell** for that sieve tube member (Figure 44–18). The companion cell is believed to contribute to the metabolism of its sieve tube member.

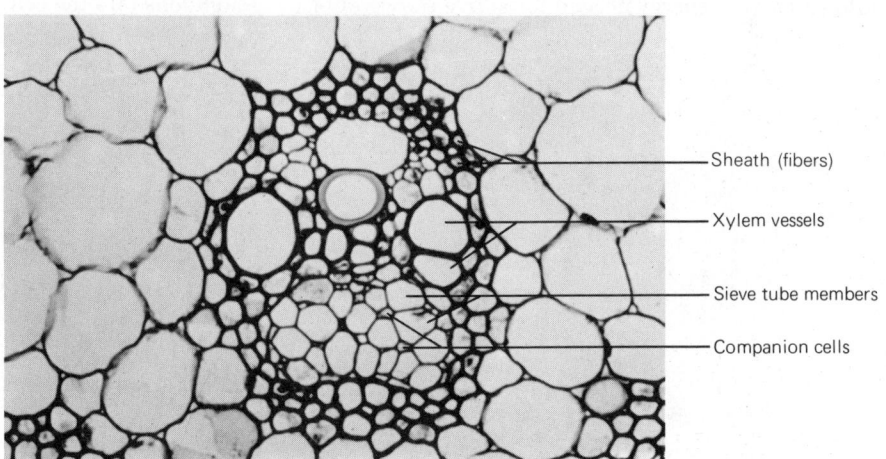

Figure 44-18 Cross section through a vascular bundle of a corn stem. (Carolina Biological Supply Company)

The phloem of woody plants contains extensions of the xylem rays (see Section 44-B). As in xylem, living parenchyma cells in the rays conduct materials laterally and store food.

Phloem also contains nonliving **phloem fibers**, similar to the fibers found in xylem. These fibers strengthen the phloem and are the only cells in the phloem that have secondary walls.

44-E Transport in Phloem

Plant physiologists are still debating how transport occurs in the phloem. The problem is difficult to investigate because phloem is extremely delicate; even slight disturbances to living phloem cause the sieve plates to become plugged and stop transport altogether.

Since phloem is too sensitive to be studied directly, investigators have devised a number of indirect methods of observation. Data obtained by several methods can then be organized into a theory that accounts for all the observations.

One much-used method is tracing the movement of radioactive materials through a plant. A substance containing a radioactive isotope is applied to the roots or leaves; radioactive carbon may even be applied in gaseous form (CO_2) to be taken in through the stomata and incorporated into sugar, which then travels through the phloem. The movement of a substance in a large tree may be measured by following the radioactivity with a Geiger counter. A more commonly used method, which works well on small plants, is autoradiography. After the plant has had time to take up and transport the radioactive substance, the plant is pressed flat and placed on a sheet of photographic film. Particles emitted by the radioactive substance expose (blacken) the film. The plant thus "takes its own radioactivity portrait" (see Figure 44-21). Autoradiography allows experimenters to find out where the plant sends a particular substance, by what route (individual sieve tubes can often be distinguished in autoradiographs), and how fast.

Another technique for the study of phloem transport is to use aphids, tiny insects that feed on plants (Figure 44-19). An aphid's mouthparts form a long tube. The aphid can insert its mouthparts into a plant in such a way that the end of the

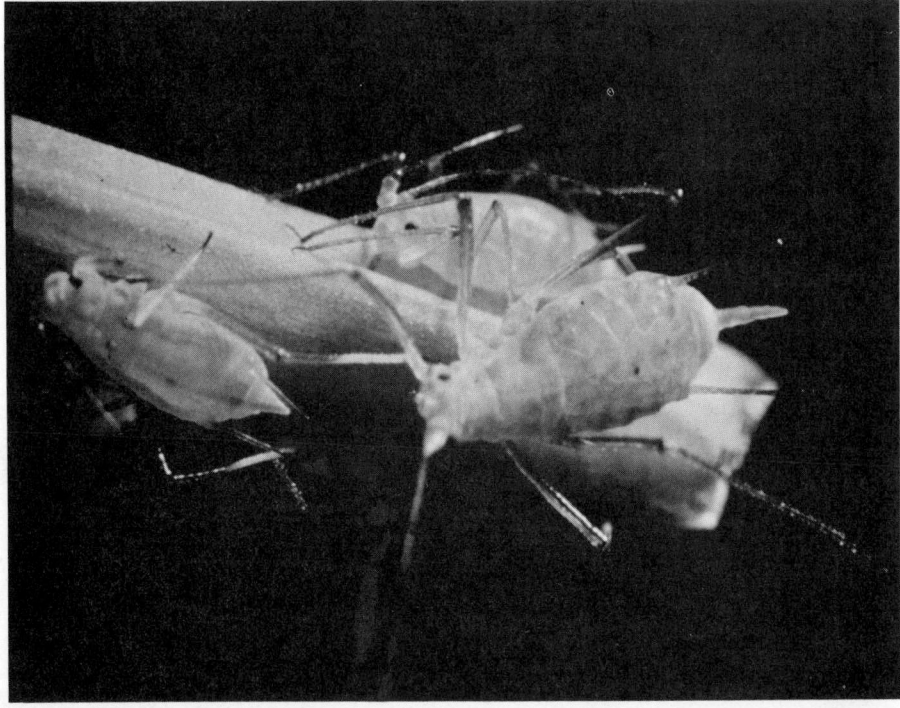

Figure 44-19 Aphids, insects that suck the fluid from conducting cells in the phloem, are useful tools in studying how the phloem works. (Gill Renard)

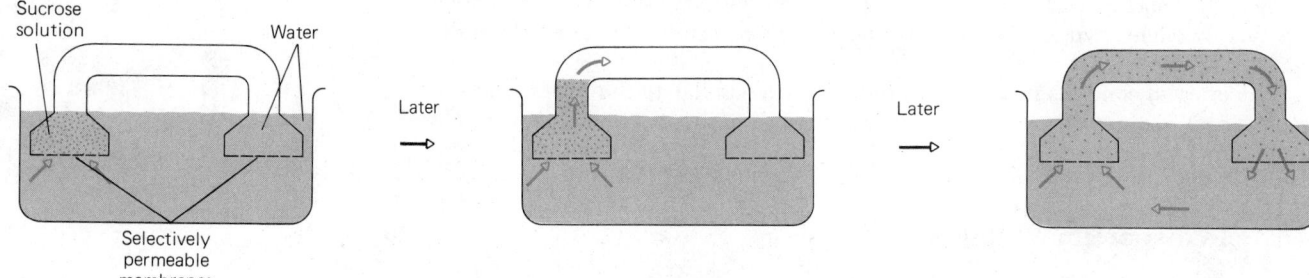

Sucrose solution

Water

Later →

Later →

Selectively permeable membranes

Figure 44-20 Munch's model of mass flow in the phloem. The selectively permeable membrane is permeable to water but not to sucrose. Colored dots represent sucrose; colored arrows represent movement of water or solution.

tube taps a single sieve element. The contents of the sieve element are under considerable pressure, and so the fluid in the phloem oozes into the tube and on through the aphid's gut. It flows with such force that feeding aphids often have a drop of fluid, called "honeydew," on their posterior ends. Since passage through the aphid changes the composition of honeydew, researchers seldom work with intact aphids. After the aphid has started to feed, it is anesthetized with carbon dioxide and severed from its mouthparts. The fluid that oozes out of the phloem through the mouthparts can then be collected and analyzed. By using several aphids on different parts of the plant, an investigator can introduce substances at one point and study their speed and direction of travel.

Other methods of studying phloem include tracing dyes that move through the phloem system, using chemicals that inhibit the activities of the phloem, applying heat to small areas, and studying the microscopic structure of the tissue.

Any theory of phloem transport must provide a reasonable explanation for these observations:

1. The contents of conducting cells in the phloem are under pressure, and a huge volume of fluid passes through any particular cell in a very short time.

2. In intact plants, substances may travel at speeds ranging from 30 to 200 cm per hour. Higher rates have been measured in badly damaged phloem. The extremely high rates in disturbed phloem reflect the release of the tremendous pressure normally exerted by the contents of the phloem.

3. The direction of flow in a particular sieve element may reverse from time to time.

4. Neighboring sieve elements may conduct in opposite directions at the same time.

5. The small size of the pores in the sieve plates must hinder the flow of materials from cell to cell.

6. The conducting cells contain living cytoplasm, and if the cells are killed, transport in that part of the plant stops; however, inhibiting phloem metabolism with poisons or low temperature does not prevent transport.

Over the course of the years, a number of theories of phloem transport have been advanced. These include activated diffusion and cytoplasmic streaming, which are both felt to be too slow to account for the observed rates of transport. A third theory proposes a form of osmosis driven by an electrical membrane potential. Still another theory is that substances spread by a rapid surface-interface flow at the boundary between two different media within the cells. Recently, there has been a spate of new theories, based on evidence that P-proteins have some properties like the contractile protein actin, which is involved in cytoplasmic streaming, amoeboid movement, and muscle contraction (Section 40–C).

By far the most widely accepted theory was advanced by Ernst Munch in 1926; it is variously called the **mass flow**, **solution flow**, or **pressure flow** theory. Munch proposed a physical model that behaved very much like the phloem system in a living plant (Figure 44-20). Two chambers were connected by a length of tube, and each chamber had a membrane permeable to water but not to sucrose. One chamber was filled with plain water, the other with a sucrose solution. When the two chambers were immersed in plain water, water immediately began to flow into the chamber with the sucrose solution, which had a lower water potential. The pressure in this chamber increased as water entered, forcing some of the solution to rise into the connecting arm and flow across to the other chamber, taking some of the sucrose along. As sucrose solution arrived at the second chamber, it forced some of the water there to flow out through the membrane into the water bath. The flow continued until the sucrose was equally concentrated in the two chambers; at that point, the two chambers had equal water potentials, and the flow ceased.

If sucrose could be continuously replenished in the first chamber and continuously removed from the second chamber as it arrived, the water potential gradient would be maintained, and the flow would continue indefinitely. Such conditions are indeed found in the phloem system of a plant. The source of sucrose could be the leaves or other photosynthetic parts of the plant or, alternatively, storage areas where starch is broken down to sucrose. The **sucrose sink**, where sucrose is consumed as it arrives, could be any cell that needs energy. Storage tissues are also sucrose sinks, where sugar molecules are converted into starch granules, which have less osmotic activity.

In Munch's model, the water bath represents the xylem, the source of the water that enters the phloem solution by osmosis. Thus, water arriving at the end of a xylem vessel in a leaf might flow immediately into a neighboring phloem conducting cell. Here the water would increase the pressure, and this would force the solution through the phloem until it arrived at a tissue that removed the dissolved nutrients. The water would then be forced out of the phloem by the pressure of the more concentrated fluid moving behind it, and drawn once more into the xylem.

How does sugar enter the phloem in the first place? It is now believed that sucrose moves into phloem conducting cells by active transport, which accumulates the sugar in fairly high concentrations (10 to 25%). There are no further membranes to cross; strands of cytoplasm connect the cells to each other through the sieve plates. Once arrived near a hungry cell, the sucrose would be removed from the phloem and stored or metabolized.

This theory agrees well with the observations on the speed and pressure at which substances move through phloem. It can also account for observed changes in direction of flow. Flow proceeds from an area of high pressure to one of lower pressure. The pressure is lowest in tissues that withdraw nutrients most actively. At a particular time, the roots may be creating the lowest pressure in their end of the phloem system. But suppose that a bud or a fruit then starts to grow more quickly, and begins to withdraw nutrients faster, making the pressure in its end of the phloem system even lower than that in the root. This situation would reverse the flow, and more sugar would be carried to the tissue where the demand for nutrients had been increased. Demands from two areas of the plant could account for the simultaneous flow of solution in different directions in different parts of the phloem system.

44-F   Distribution of Substances

The two transport tissues, xylem and phloem, between them carry the various substances that must be moved from one part of the plant to another. Water and minerals taken up by the roots are transported by xylem. The movement of minerals in the xylem is not always correlated with the movement of water. Both water and minerals may be removed from the xylem by living cells in any part of the plant.

Normally, the vast bulk of water moved to the leaves of a plant is lost to the atmosphere. However, some water becomes part of the cytoplasm; some serves as a raw material for photosynthesis, and some moves into the phloem, but this water eventually returns to the xylem.

(a)   (b)

Figure 44-21 Autoradiographs of two bean plants, showing movement of radioactive phosphorus (^{32}P) from older parts of the plant to younger parts. The roots of the living plants were placed in a solution containing radioactive phosphorus. After plant (a) had had time to take up and transport the phosphorus, it was autoradiographed. The darkest areas represent the parts of the plant with the most radioactive substance; note that the radioactive material was most concentrated in the roots and youngest leaves. Plant (b) was removed from the solution containing ^{32}P at the same time as plant (a), but was allowed to continue growing for a time in nonradioactive solution before it was autoradiographed. Note that the radioactive phosphorus has been moved to the younger leaves, which had not yet formed at the time that the plant was in the radioactive solution. (Susann and Orlin Biddulph)

Some minerals also eventually make their way into the phloem. As leaves grow old, their minerals may be transported back into the plant for use in younger leaves or for storage over the winter. Phosphorus, potassium, and nitrogen especially are reclaimed; iron may also be found in the phloem as it is recycled under certain circumstances. Calcium is a mineral that does not move once the xylem has delivered it to the leaves. Minerals that move out of the leaves through the phloem may eventually make their way back into the xylem, and sucrose moved to storage areas by the phloem may appear in dilute solutions in the xylem when the sap "rises" in the plant the following spring. Thus, it appears that plants may have a slight and very slow "circulation" in their vascular systems, not just one-way transport.

The main substance transported by the phloem is sucrose; it may be present in concentrations of up to 25%. Other sugars and small polysaccharides are transported in lower concentrations. When these sugars arrive at the roots, some of them are used by the root cells in their respiration, or are stored as food reserves; other sugars are combined with ammonium (NH_4^+ or nitrate (NO_3^-), taken in through the roots, to form amino acids. These amino acids then enter the xylem or the phloem for transport to parts of the plant where they can be used. Phloem also carries nucleotides, hormones, and various other organic compounds.

A particular organ of the plant may need some of the substances transported in the phloem but not others. How is the movement of substances in the phloem determined? As we have seen, sucrose is the main solute in the phloem, and movement of the phloem solution is thought to follow a gradient based on solute concentration. Therefore it is not surprising that the phloem solution always moves down a sucrose gradient, regardless of the concentration gradients of the other solutes present. For example, if a solution of phosphorus is applied to the roots of a plant, and a weaker solution of radioactive phosphorus is applied to the leaves, radioactive phosphorus can be traced from the leaves to the roots with the flow of sucrose, even though phosphorus is already more concentrated in the roots than in the leaves. Likewise, if a substance is applied to the leaves of a plant when little sucrose is being transported out of the leaves, it will move out of the leaves more slowly than if it were applied to the leaves during a time of active sucrose export. Thus, only when the leaf is exporting sugar will substances move quickly out through the phloem to other parts of the plant.

Practical Applications

While the problem of transport in the vascular tissues of plants is an intriguing intellectual puzzle, it is of more than academic interest. Knowing how plants handle various minerals and organic molecules can be helpful in the management of plants and plant pests.

For example, leaves can absorb some substances placed on their surfaces in solution, and move them into the phloem, which transports the substance throughout the plant. This is the basis of **foliar feeding**, or fertilizing with a nutrient solution sprayed onto the leaves. Other substances must be worked into the soil because only the roots will absorb them.

Systemic pesticides, watered into the soil or sprayed onto the leaves, are absorbed and distributed throughout the plant. Systemic pesticides combat internal pests, such as leaf miners, insect larvae that live entirely inside leaves and never come into contact with sprays applied externally. These pesticides are useful to protect ornamental plants but cannot be used on food crops because they are toxic to humans as well as to insects.

Topical pesticides are not absorbed into the plant, but remain on the leaf surfaces and protect the plant by killing leaf-chewing insects or invading fungi. Because these pesticides stay put, they do not contaminate fruits produced after the spraying nor roots used as food.

44-G   Root Grafts and Transport Between Trees

Roots, like stems, grow in diameter by adding secondary vascular tissue. If two roots grow near each other, their secondary tissues may touch and eventually merge to form a **root graft**. The two roots can exchange materials through the graft. If the roots are from two different trees (of the same species), each tree can then affect the health and well-being of the other.

The effects of root grafting were observed during the droughts of the 1950s and 1960s. In plantations of red pine that had been started in the 1930s, clumps of trees in poor, shallow soil were dying. This was not surprising, because there was not enough water in the soil to support them. However, all of the trees in a clump were dying at once, rather than the weaker or more spindly trees dying first, followed by the larger, healthier trees. In these clumps, unseen root grafts formed an underground network linking the vascular systems of the trees to each other, and the available water supply was shared among the trees. This was not a conscious act of altruism, of course; the movement of sap in the xylem, as usual, was merely following water potential gradients, and moved throughout the clump of trees. When the entire soil volume had been depleted of its water, all of the trees died.

Similarly, when these plantations were due for thinning, the foresters girdled the trees that they wished to remove. This was supposed to kill the unwanted trees. However, thanks to the root grafts, the roots of the girdled trees were not killed by being disconnected from the phloem of their own shoot systems. Food produced by their neighbors was transported through the phloem to the root grafts, where some of the food followed the pressure gradient and ended up in the girdled trees' roots.

Girdling will kill trees of many species even when root grafts are present, because fungal diseases and woodboring insects invade through the wounded tissue of the girdle. However, in red pines and other trees that produce protective resins, the exposed wood is protected from these pests and the wound soon heals over. The tree continues to produce new wood above the girdle, and continues to grow new branches and needles. At last, after a decade or more, the tree becomes too top-heavy, and it will break at its weak point, the girdle, during a high wind or after a heavy snowfall.

SUMMARY   The conducting cells of xylem and phloem show structural adaptations for their rapid transport of substances. Tracheids, found in the xylem of both gymnosperms and angiosperms, have pits in their walls, while vessel cells, found only in angiosperms, may have their end walls perforated or absent. Conducting cells in the phloem have sieve areas in their cell walls; the sieve tube members of angiosperms have well-developed sieve plates in their end walls.

In xylem, a series of dead, tubular cells conducts water and dissolved minerals upward from the roots to the leaves, flowers, and fruits. Pressure in the roots pushes sap up a short distance in the stems of some plants. In most cases, however, much more upward movement results from the transpiration of water vapor from the leaves, which in turn creates a pull from the top of the water column; water follows a water potential gradient from the soil, through the roots and stem, and out through the leaves. The cohesion of water and its adhesion to the walls of the xylem conducting cells is crucial to the transport of water to the top of tall trees. Plants control transpiration to some extent by opening and closing the stomata.

Phloem transport is still poorly understood. The conducting cells of phloem form continuous tubes filled with living cytoplasm. The mass flow theory of phloem transport suggests that a high concentration of sugar in the phloem cells of leaves creates a low osmotic potential; water moves into the phloem from the xylem. This uptake of water creates hydrostatic pressure, which pushes the phloem contents along from one cell to the next. Phloem distributes sugars, amino acids, hormones, and some minerals to the roots, fruits, and growing buds.

An understanding of transport mechanisms in plants, as well as a knowledge of the pathways followed by various substances within the plant body, is important in the planning of fertilization, pest control, and hormone treatment programs in modern agriculture.

OBJECTIVES

From your study of this chapter, you should be able to:
1. Name the two conducting tissues in plants, give their functions, and list substances commonly transported by each.
2. Describe the girdling experiments, and explain what they show about the functions of conducting tissues in trees.
3. Sketch or describe the adaptations of tracheids, vessel elements, sieve cells, and sieve tube elements that suit these cells for their roles in transport.
4. Outline the path by which water moves from the soil through a plant to the atmosphere, referring, as appropriate, to the molecular mechanisms and plant structures that make this movement possible.
5. Explain how (1) the root pressure mechanism and (2) the transpiration pull mechanism of xylem transport are believed to work; describe experiments that support each of these theories, and discuss the relative importance of each mechanism in the total conduction that takes place

in the xylem of plants; predict how changes in the environment would affect each process.
6. Explain how the structure of the leaf represents a compromise between the plant's need for water and its need to obtain sunlight and carbon dioxide.
7. Sketch two guard cells and the stoma between them and describe in a few sentences the mechanism by which guard cells open the stomatal pore; state the advantage to the plant of being able to open and close its stomata.
8. Sketch or identify the following features of a woody stem: annual ring, spring wood, summer wood, heartwood, sapwood, secondary xylem, bark, rays, and the approximate position in which you would find vascular cambium and secondary phloem. State the functions of each of these features.
9. Explain how transport in phloem may occur by mass flow in the living plant, and predict how changing relevant conditions will modify the transport occurring in the phloem.

SELF-QUIZ

1. The main solute transported by phloem is:
   a. glucose
   b. pyruvate
   c. sucrose
   d. starch
   e. amino acids

2. The girdling experiments performed by Malpighi supported the theory that:
   a. water moves in a tree by the root pressure mechanism
   b. water moves in a tree by a transpiration-cohesion mechanism
   c. xylem is primarily responsible for conducting water from the roots to the leaves
   d. phloem is primarily responsible for conducting water from the roots to the leaves

3. List two functions of xylem tissue, and describe at least one adaptation of cells found in the xylem that contributes to the performance of each function.

4. The movement of water up through a tree trunk depends on:
   a. the high boiling point of water
   b. exclusion of air molecules from the sap solution
   c. the vapor pressure of water
   d. attraction between water molecules
   e. low osmotic potential in the sap

5. Would the rate of transpiration increase, decrease, or remain the same under the following conditions?
   _____ a. high humidity
   _____ b. increased turgor pressure in the guard cells
   _____ c. increased light
   _____ d. increased wind

6. Would root pressure increase, decrease, or remain the same under the following conditions?
   _____ a. high humidity
   _____ b. watering dried-out soil
   _____ c. darkness

7. Xylem that is not conducting water is called:
   a. heartwood
   b. sapwood
   c. spring wood
   d. summer wood
   e. rays

8. Which of the following is *not* necessary to the operation of the mass flow theory as it is presently understood?
   a. ATP
   b. root pressure
   c. intact membranes in conducting cells
   d. difference in osmotic potential in different parts of the plant
   e. constant production or release of sugar molecules

QUESTIONS FOR DISCUSSION

1. In 1936, Bruno Huber performed an experiment in which he inserted thin heating wires into the xylem of a tree. He placed a thermocouple (a sensitive heat-detecting device) farther up the stem and timed how long it took before the heated sap passed the thermocouple. Huber found that the sap moved slowly at night. In the morning, the sap movement speeded up first in the twigs; later, the sap began to rise more quickly in the trunk further down the tree. Do these results support the root pressure or the transpiration pull theory of xylem transport? Justify your answer.

2. Refer to the photographs below. These three cuttings were taken from the same plant at the same time. Stem 1 was placed in a container of water immediately. Stem 2 was left lying on the table for half an hour, and then a two-inch length was cut off the bottom end. It was then placed in a container of water. Cutting 3 was also left lying on the table for half an hour, after which time it was put into a container of water without further treatment. The photographs were taken $2\frac{1}{2}$ hours after the cuttings were taken from the plant. What differences can you see among the three stems? What accounts for the results?

3. What is the best time of day to cut flowers? Justify your answer.

4. We have seen in this chapter that cohesion of water keeps a column of water traveling up through the xylem of a tall tree; how did the water reach the top in the first place?

5. Why should the ground around evergreen plants be watered thoroughly before the ground freezes for the winter?

6. Contrast the conditions affecting transpiration that are experienced by a houseplant in the winter and in the summer.

7. Account for the fact that the stomata are generally closed at night and open during the daytime (except when the plant is losing too much water via transpiration).

8. Why does a bean leaf have more stomata in the lower epidermis than in the upper epidermis? Why are the stomata of a corn leaf located in both epidermal layers?

9. What does wilting indicate about the movement of water in a plant? What does it indicate about the water content of the soil?

10. What is the relative importance of the activities of roots and leaves in the transport of water in a vascular plant? Would a plant with healthy roots but no leaves be able to carry on transport in the xylem? Would a plant with healthy leaves be able to carry out transport in the xylem if the roots were killed?

11. The virus disease known as "beet yellow" is transmitted from plant to plant by aphids. Why does the disease spread through the plant rapidly and kill it quickly?

12. Large brown algae known as kelps have a system of "sieve filaments" whose cells look remarkably like the sieve cells of vascular plants; these sieve filaments transport carbohydrates from the blades (photosynthetic parts) to the stipe (stalk) and holdfast of the plant. Why do these kelps have phloem-like tissue but no xylem-like tissue?

13. Predict what happens when foresters try to thin a red pine planting in which there are extensive root grafts by injecting poisons into the unwanted trees.

REFERENCES AND FURTHER READING

Biddulph, S., and O. Biddulph. "The circulatory system of plants." *Scientific American*, February 1959. An interesting account of experiments showing how various substances move in plants.

Cohen, I. B. "Stephen Hales." *Scientific American*, May 1976. A biographical sketch of the life and experimental work of an energetic pioneer in plant physiology.

Evert, R. F. "Transport of food substances in plants." *Topics in the Study of Life*. New York: Harper and Row, 1971. Describes studies of anatomy, development, and translocation directed to understanding how substances are transported in the phloem.

Fritts, H. C. "Tree rings and climate." *Scientific American*, May 1972. How tree ring patterns are used to analyze the climate of bygone times.

Jensen, W. A., and F. B. Salisbury. *Botany: An Ecological Approach*. Belmont, CA: Wadsworth, 1972. Especially good presentation of methods for measuring water cohesion and modern experiments on transport.

Ray. P. *The Living Plant*, 2d ed. New York: Holt, Rinehart and Winston, 1972. Plant physiology from a modern, quantitative viewpoint.

Stone, E. L., J. E. Stone, and R. C. McKittrick. "Root grafting in pine trees." *New York's Food and Life Sciences Quarterly* 6(2):19, 1973.

Zimmermann, M. H. "How sap moves in trees." *Scientific American*, March 1963. Outlines various methods used to try to elucidate how transport occurs in xylem and phloem.

CHAPTER 45

SOIL, ROOTS, AND PLANT NUTRITION

Most terrestrial plants are autotrophic; that is, they make their own food using inorganic substances obtained from the environment. All living organisms, including plants, need the same major nutrients. Land plants obtain carbon, in the form of carbon dioxide from the air, via their stomata (see Section 43-D). The wide array of mineral nutrients and the huge quantities of water needed to keep a plant healthy, however, are absorbed by its roots.

In this chapter we shall see how roots take in water and minerals and how mineral nutrients are used by a plant. We shall also study special nutritional adaptations that allow some plants to live in poor soils or with no soil at all.

45-A Nutritional Requirements of Plants

Plants need nitrogen, phosphorus, potassium, calcium, magnesium, sulfur, and iron in order to grow well. Farmers have long fertilized their soil by adding these elements in either inorganic or organic form. All except iron are needed in relatively

Medium lacking nitrogen: poor growth

(a)

Medium lacking iron: extreme chlorosis, poor growth

(b)

Medium lacking micronutrients: short internodes at tops

(c)

Nutritionally complete medium

(d)

Figure 45-1 Sunflower seedlings grown for four weeks in water with various nutrient contents. (Carolyn Eberhard)

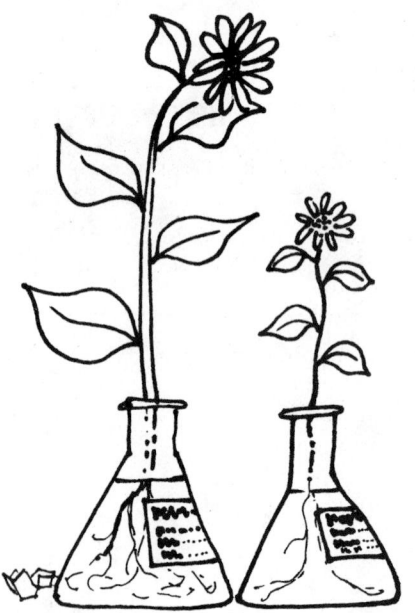

"I took the precaution of packing some sandwiches before we came." (Rosemary Smith)

large quantities (Table 45–1), and so are referred to as **macronutrients** (macro = large). Nitrogen, phosphorus, and potassium are required in the greatest quantities; commercial fertilizers are rated by the percentage of each of these elements they contain, e.g., a "5-10-5" fertilizer contains 5% nitrogen, 10% phosphorus, and 5% potassium, by weight.

Besides macronutrients, all plants need a variety of **micronutrients**, minerals that are used by plants in small amounts. For instance, whereas nitrogen is applied to the soil at the rate of several hundred pounds per acre per year, the treatment for molybdenum-deficient soils in Australia is 2 ounces of MoO_3 per acre, applied once every 10 years.

Table 45–1 indicates the roles of various mineral nutrients in plants. Since many minerals are vital components of the most important biological molecules, an inadequate supply of any one of several minerals may result in rather general deficiency symptoms, such as **chlorosis** (paleness) and poor growth. On the other hand, the deficiency symptoms of some nutrients may be quite specific. Zinc, for example, is believed to be necessary for the production of the plant hormones called **auxins**, which cause elongation or enlargement of cells during growth. If a plant lacks zinc, its cells do not grow to full size, the leaves are small, the cells of the internodes* do not elongate, and the plant grows in the form of a rosette.* (Many species of plants normally grow as rosettes even when well-nourished.)

Some plants have special adaptations that require additional nutrients. For example, legumes (pea family) need cobalt, a component of the vitamin B_{12} used by their symbiotic root nodule bacteria in the process of nitrogen fixation.* Similarly, sphenopsids* and some grasses require silicon, which they incorporate into an abrasive material that both strengthens their stems and wears down the teeth of grazing animals.

---

* Internodes Portions of a stem between sites of leaf attachment.
* Rosette A flat-growing plant with leaves arranged all around the very short stem, e.g., dandelion, saxifrages.
* Nitrogen fixation Conversion of gaseous nitrogen (N_2) to ammonia (NH_3).
* Sphenopsids A group of lower vascular plants (horsetails, see Section 26-E).

TABLE 45-1 **THE ROLES OF MINERAL NUTRIENTS IN PLANTS**

NUTRIENT	ABSORBED FORM	LB/ACRE REMOVED BY WHEAT CROP IN ONE SEASON†	DEFICIENCY SYMPTOMS	ROLE IN PLANT
Macronutrients:				
Nitrogen	NO_3^-, NH_4^+	76	Mild: older leaves yellow Severe: stunting	Component of proteins, nucleic acids, chlorophyll, some hormones, secondary chemicals (see Section 51–C)
Potassium	K^+	42	Various, general	Not well understood; probably necessary for cell division and protein synthesis in meristems,* production of chlorophyll, synthesis and transport of carbohydrate, reduction of nitrates to ammonium
Phosphorus	$H_2PO_4^-$	14	Various, general; poor roots, slow maturation	Component of nucleic acids, ATP, phospholipids, activated sugars
Sulfur	$SO_4^=$	11	Chlorosis, poor roots	Component of some amino acids, coenzyme A*
Calcium	Ca^{++}	12	Meristem death; abnormal cell division; breakdown of membrane structure	Component of middle lamella* of cell walls; ties up waste products, especially oxalic acid, as insoluble salts; involved in transport of sugars and amino acids and in membrane permeability
Magnesium	Mg^{++}	8	Chlorosis	Constituent of chlorophyll; important cofactor of enzymes, e.g., in glycolysis
Micronutrients:				
Iron	Fe^{+++}	0.7	Chlorosis, appearing first in youngest leaves	Needed for chlorophyll production; part of cytochromes and ferredoxins (electron transport) and of some enzymes
Boron	$HB_4O_7^-$	0.3	Darker color, abnormal growth, malformations, flowering inhibited, meristem death	Mostly unknown; pollen germination; regulation of carbohydrate metabolism
Zinc	Zn^{++}	0.2	Small leaves, short internodes	Synthesis of tryptophan (= precursor of auxins); component of some enzymes
Manganese	Mn^{++}	0.5	Mottled chlorosis	Activates citric acid cycle enzymes; involved in evolution of O_2 during photosynthesis
Chlorine	Cl^-		Small leaves, slow growth	Evolution of oxygen during photosynthesis; other roles not understood
Molybdenum	$HMoO_4^-$		Same as nitrogen deficiency	Part of enzymes for nitrate reduction and nitrogen fixation; possibly other functions
Copper	Cu^{++}	0.03	Lowered protein synthesis	Component of some enzymes; component of cytochromes

† From Weier, Stocking, and Barbour, *Botany*, 4th ed., 1970, John Wiley & Sons.

Nutrient requirements are usually determined by growing plants in a solution of water containing all the known nutrients. If plants grow poorly when provided with all of the known required nutrients, then it is assumed that yet another nutrient must be supplied for normal growth.

In practice, there are several difficulties in determining the nutritional requirements of plants. For instance, the proportions of the various nutrients to each other are important. In addition, the presence of some nonessential elements may reduce the toxicity of other elements; if silicon is provided, a plant can tolerate higher

---

* Meristems Regions of dividing cells in growing areas of the plant.
* Coenzyme A An important transfer molecule in respiration and in fat metabolism (see Figure 7–12).
* Middle lamella The shared partition between the cell walls of adjacent plant cells.

levels of magnesium without developing symptoms of magnesium poisoning, but silicon itself is not a necessary nutrient for most plants. Plants can also sometimes use one nutrient in place of another that is in short supply; the ions Ca^{++}, Mg^{++}, and Mn^{++} can substitute for one another in some reactions.

Some nutrients are needed in such small amounts that they are almost always present in sufficient quantities as impurities in the chemicals used to make up nutrient test solutions. These have been added to the list of required nutrients by using triply-purified salts and triple-distilled water to identify elements needed in only the most minute quantities.

Still another problem is the mobility of elements in plants. Nutritional studies generally use seedlings started under good nutritional conditions. These seedlings may have stored nutrients that permit them to grow in a normal-appearing fashion even after they have been moved to a solution that lacks certain nutrients. Many nutrients are quite mobile in plants, and can be moved from an older to a younger part of the plant.

Nutrient deficiencies of plants may be due to low concentrations of nutrients in the soil, but deficiencies may also occur in soils where the nutrient is present but unavailable to plants. Another cause of nutrient deficiency is genetic defects that result in an inability to use the nutrients that are available. One strain of soybeans is unable to convert Fe^{+++} to Fe^{++}, a step necessary in the transport of iron from the roots to the shoot system. The roots of these plants can obtain iron from the soil, but the shoot systems show iron deficiency symptoms. Likewise, some tomato plants are unable to transport boron from their roots to their shoot systems. Their shoot systems show the boron deficiency disease known as "brittle stem." Other plants have lost the ability to reduce* nitrate (NO_3^-) or sulfate (SO_4^{-2}), reactions that are necessary before the plant can form amino acids.

45-B Soil

The minerals used by plants come ultimately from the rock particles that make up much of the soil. Soils can be classified by the average size of their rock particles. The finest particles give rise to **clay**, larger particles to **silt**, and still larger particles to **sand**.

The size of soil particles influences the soil's capacity to hold **soil water**. Some of the water that falls on a soil will drain away under the pull of gravity, carrying dissolved nutrients along. The water that stays in the soil is held in three main ways. First, some of it forms a film around the soil particles. Second, wedges of water are held between the soil particles by capillary attraction. These films and wedges together make up the **capillary water** of the soil. Third, water may be held by **imbibition**, that is, by the formation of **colloidal systems** in which water becomes tightly bound up in the structure of particles of clay or organic matter (Figure 45–2). When the soil loses water, capillary water wedges are most easily removed, followed by the films around the soil particles, and finally by the imbibed water, which is tightly bound.

---

* Reduce Bond hydrogen atoms to.

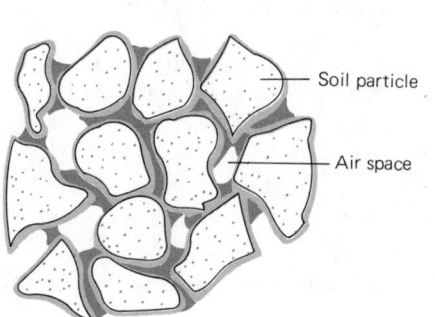

▭ Water film around soil particles
▮ Water wedges in smallest spaces between soil particles
▱ Imbibed water, tightly bound in soil particles

— Soil particle

— Air space

Figure 45-2 Soil water is held in three ways: as capillary films around individual soil particles, as capillary wedges in small spaces between soil particles, and as imbibed water incorporated into the colloidal structure of particles of clay and organic matter.

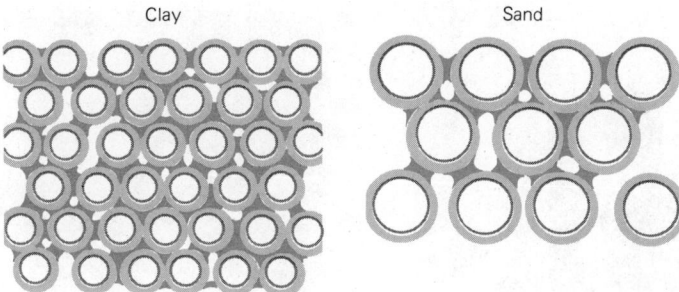

Clay   Sand

Figure 45-3 Comparison between water held by clay soils (small particles) and water held by sandy soils (larger particles).

Clay soils can hold more water than sandy soils, on all three counts (Figure 45–3). They have finer particles with more surface area to hold surface films; the spaces between particles are smaller, and thus have higher capillary attraction and tend to fill up more completely (see Section 44-C); and clay particles are colloidal and so can imbibe water. Sandy soils are often very well drained, whereas clay soils tend to become waterlogged. Yet clay holds its capillary water so strongly that plants growing in clay often cannot obtain enough water. Thus the best soil for most plants is one that contains a mixture of particle sizes.

The most fertile soils also contain a great deal of organic matter: dead plants or plant parts, dead animals, animal excrement, and dead microorganisms. Such organic matter holds soil water by imbibition. It swells with water after a rain and then gradually gives up some of this water as the soil dries out. This alternate swelling and shrinking helps to keep the soil loose, allowing roots to grow through it easily. Organic matter on the top of the soil insulates the soil from the heat of the sun and reduces evaporation from the soil. More importantly, the organic matter in soil represents a reservoir of nutrients that are released slowly by decomposition.

Living organisms are another important constituent of soil. Soil residents include microorganisms such as bacteria, fungi, algae, and protozoans, and larger organisms, including worms and insects, in addition to the roots of plants. Many soil organisms break down organic matter into its inorganic constituents, slowly releasing minerals that plants can absorb. Soil organisms also affect conditions such as pH and the oxidation state of minerals. Some bacteria can fix nitrogen, converting gaseous nitrogen (N_2) dissolved in the soil water to ammonium (NH_4^+). Nitrifying bacteria can use fixed nitrogen to make nitrites (NO_2^-), or nitrates (NO_3^-)(see Section 22-F). The burrowing of larger organisms mixes and breaks up the soil, allowing air to penetrate.

Oxygen is a very important component of soil. If the soil spaces are filled with water and contain no air, aerobic soil organisms, including plant roots, will die of asphyxiation. In waterlogged soil, only anaerobic organisms can make a good living. In general, anaerobic organisms cause changes that are unfavorable to plant growth. This makes it doubly important for plants to live in well-aerated soil.

The Soil Solution

Like all other living cells, root cells can absorb substances only when they are dissolved. The soil water acts as a solvent for various molecules and ions.

Some of the minerals in the soil water become tightly bound to soil particles. Clay particles, for instance, have negatively charged surfaces and so bind positively charged ions. Only the small percentage of positively charged mineral ions that remains dissolved in the soil solution is available to the roots of plants.

Acid soils tend to be deficient in nutrients because their hydrogen ions displace other positively charged ions from the surfaces of the soil particles. The displaced mineral ions are then readily washed, or **leached**, out of the soil by percolation and runoff. Some necessary plant nutrients, such as iron, are generally more available in acidic conditions and less available under neutral or basic conditions. Calcium and nitrogen, on the other hand, are more available in neutral soil. Most plants do best at a soil pH of 6.0 to 7.5, although there are plants adapted to growing at higher or lower pH ranges.

Figure 45-A Soil erosion: Water washes 8 tons of soil per acre per year off this farm in Wisconsin. Soil particles, pesticides, and fertilizers washed into waterways cause water pollution. (U.S.D.A. Soil Conservation Service)

Figure 45-B The better practice: Terraced fields on this hillside in Nepal prevent soil erosion. Where land has been passed down from parent to child for generations, the soil is better preserved and often more fertile than in the great agricultural areas of North America. (U.S. Agency for International Development)

Essay:
Soil Erosion

Soil erosion is both an agricultural and an ecological problem; it occurs when wind or water remove soil from one location, which is often agricultural land, to another location, usually a river or non-agricultural land. In the United States, an average of 9 to 12 tons of soil are lost from each acre of agricultural land every year by erosion (Figure 45–A).

In many areas, the soil is now only a few inches deep. The Great Plains, which once had more than six feet of topsoil, have lost about two feet of soil through erosion in a hundred years. Many areas in the eastern United States and in California have lost essentially all their soil through erosion and have become uneconomical to farm. Because soil is produced very slowly, by the weathering of underlying rock and by the decomposition of organic matter, soil replacement is slow.

Soil erosion is largely a problem of newly developed farmland. It is very rare in Europe and Asia, where the same land has often been farmed for 2000 years or more. In such a situation, farmers must prevent erosion and put back into the soil what they take out of it, or their children may starve (Figure 45–B). In newly agricultural countries, on the other hand, people have traditionally moved on when they have exhausted the soil in one area, and the incentives to preserve the soil have been few.

Soil erosion contributes to problems other than the loss of agricultural land. Soil washing into streams and rivers carries fertilizer and pesticides with it and contributes largely to water pollution. The soil particles may also block out much of the light needed by photosynthetic organisms in streams and lakes.

Soil erosion can be prevented by simple measures. First, land should not be left unplanted for long, since plant roots hold soil in place. Terracing and contour plowing prevent soil from washing down slopes, and planting windbreaks checks erosion by high winds (Figure 45–C). Also, since organic debris can bind a great deal of water, soil with a high content of organic matter is less likely to be washed away; fertilizing land with manure instead of chemical fertilizers is a considerable deterrent to erosion, although this practice is almost obsolete in Western countries.

The soil erosion problem is largely economic: an individual farmer can profit more by planting crops instead of rows of windbreak trees, by cultivating land that is flooded periodically, or even by leaving land without plant cover for a period. Today we face the difficult task of designing economic incentives that will encourage farmers to conserve soil. The alternative may well be a drastic drop in our ability to grow food for our still-increasing population. It has been theorized that loss of topsoil from agricultural land brought about the fall of the magnificent civilization of the Mayas, a thousand years ago; much of that land has still not redeveloped soil of the original quality. Many agricultural areas today seem to be facing the same fate.

(a) (b)

Figure 45-C Wind erosion and part of the cure. (a) When the wind stopped, this road in Idaho was covered with soil, which was knee-deep in some places. (b) Windbreaks of willow prevent soil erosion on this farm in Michigan. (U.S.D.A. Soil Conservation Service)

Soil Treatments

Various treatments improve the condition of soil for growing plants. One of the oldest agricultural practices is **tilling** the soil, that is, turning it over before planting seeds, and turning over the soil around the plants as they grow. Tilling has two main purposes. First, it mixes up the nutrients and loosens soil particles, making it easier for the roots to penetrate the soil. Second, it gives the crop plants, which are left untouched, a competitive advantage over the weeds, which are deliberately disturbed to damage their root systems. Another effect of tilling is the introduction of oxygen into the soil. This stimulates growth because roots need oxygen to take up minerals from the soil solution. The additional oxygen also permits aerobic microorganisms to break down organic material in the soil more rapidly. The latter effect is not necessarily beneficial to the plants; if the nutrient supply is already adequate, accelerating the release of minerals into the soil solution may promote leaching, as the plants cannot take up the released nutrients fast enough.

Fertilization is another common agricultural practice. Organic fertilizers, such as manure, have the advantage of improving the texture of the soil as well as providing a supply of nutrients that are released slowly by decomposition. Their disadvantage is that they are heavy and expensive to spread, and that their nutrient content varies. For this reason chemical fertilizers that provide precise quantities of nutrients are more often used in modern agriculture. The main disadvantage of chemical fertilizers is that they are easily abused. Improperly applied, they waste money, pollute water supplies, and damage plants. If the soil solution becomes too concentrated, plants can no longer take up water from the soil by osmosis, and instead may actually lose water. Such "fertilizer burn" may dehydrate and kill the plant.

Of the mineral elements needed by plants, nitrogen is most often deficient. Plants need nitrogen in large quantities because it is a major constituent of proteins and nucleic acids. Most of the nitrogen in the world exists in the form of nitrogen gas in the air, which plants cannot use. Plants can absorb nitrogen only in the form of nitrate (NO_3^-) or ammonium (NH_4^+). Nitrogen may be applied to the soil as a salt of either of these, as ammonia (NH_3) gas, or in a form such as urea ($H_2N-CO-NH_2$), which is broken down by microorganisms to release ammonia. Ammonia reacts with soil water to produce ammonium:

$$H_2O + NH_3 \rightleftharpoons NH_4^+ + OH^-$$

water + ammonia $\rightleftharpoons$ ammonium + hydroxide

A problem with any form of nitrogen is that it is also available to microorganisms, which may convert the nitrogen into forms that plants cannot use. Nitrogen fixation by other soil organisms usually does not completely offset this effect. Another problem is that nitrate, a negatively charged ion, leaches out of the soil easily because it is not attracted to soil particles as positively charged minerals are.

It would seem preferable to apply positively charged ammonium to the soil, but this too has its drawbacks. Ammonium can be taken up readily by plants and combined with carbohydrates to form amino acids, whereas nitrates must first be converted to ammonium. Ammonium may therefore give plants too much nitrogen too fast and allow them to outgrow their carbohydrate supply if photosynthesis cannot keep up with protein synthesis. Since nitrogen is readily lost from the soil in all of the ways described above, nitrogen fertilizers must usually be applied several times during the growing season to provide a steady supply of nitrogen.

Soil is often treated to adjust its pH to a better level for the intended crop. Lime is applied to acid soils to raise the pH, whereas sulfur or ammonium sulfate is applied to alkaline soils. These treatments have other effects besides changing the pH; for example, lime provides calcium, and ammonium sulfate contributes nitrogen. Changing the pH may also have an effect on the solubility of various minerals in the soil, and hence on their availability to plants. For example, if the pH of an acid soil is raised too much, it may cause iron to precipitate out of the soil solution. One way to overcome such an effect is to apply **chelated iron**—iron bound to organic molecules so that it cannot be precipitated—along with lime. Other minerals whose solubility is decreased in the presence of lime are copper, manganese, and zinc. However, lime increases the solubility of some minerals, notably phosphorus and molybdenum. Thus it is important to determine the pH of the soil carefully and then plan exactly how much of a chemical to apply to correct the situation.

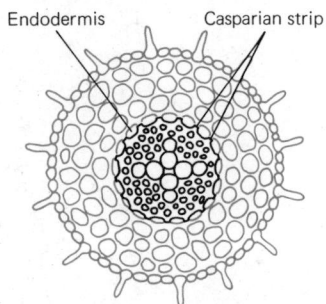

Endodermis   Casparian strip

(a) **ABSORPTIVE SURFACE AREA**

Available for uptake of water and
minerals across plasma membranes

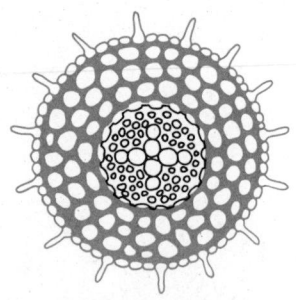

(b) **OUTER PART OF APOPLAST**

Volume of root occupied
by soil solution

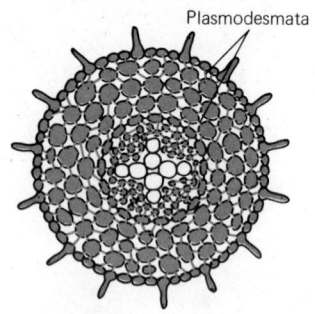

Plasmodesmata

(c) **SYMPLAST**

Living cytoplasm of cells is
connected throughout the root

Figure 45-4 Uptake of substances from soil solution by roots. (a) Substances may be taken up through the plasma membranes of cells in the epidermis, the cortex, and outward-facing parts of the endodermis (color). (b) The part of the apoplast outside the endodermis (color) consists of pores in the cell walls of the epidermis, cortex, and outer face of the endodermis, as well as spaces between these cells. All of these spaces are freely penetrated by the soil solution. (c) The symplast consists of all the living matter in the root (color); the cytoplasm of the living cells is interconnected by plasmodesmata. The membranes of these living cells are selectively permeable to various minerals. Once taken up by a cell, substances can move throughout the symplast from one cell to another, and eventually can be passed to the vascular tissues, which transport them to other parts of the plant.

45-C Absorption by the Roots

A continuous "transpiration* stream" moves from the roots of a plant, up through the xylem, and out through the stomatal pores in the leaves. Water lost through the leaves is continuously replaced as the soil solution is taken up through the surfaces of root cells. The outer surface area of a root is increased by root hairs, epidermal cells with extensions that grow out between soil particles. But the epidermal layer is only part of the root surface exposed to the soil solution. The soil solution moves freely into the **cell wall spaces** between the cellulose fibers of the cell walls and on into the spaces between the cells of the root cortex. Thus a plant can take up water and minerals through the entire surface area of the plasma membranes of the epidermis and cortex (Figure 45-4a).

The system of cell wall spaces and intercellular spaces in the root forms a functional unit, consisting of spaces filled with nonliving material, called the **apoplast** (apo = apart from; plast = living material). The apoplast in the epidermis and cortex is virtually continuous with the external soil solution (Figure 45-4b).

The apoplast is divided into outer and inner areas, separated by the endodermal layer. The cells of the endodermis have a rather peculiar structure. Each cell is shaped like a brick, and the cells form a layer that is like a cylindrical brick wall extending up and down the root, separating the epidermis and cortex from the interior of the root. A **Casparian strip**, composed mainly of an impermeable waxy material, **suberin**, runs right around each cell in the endodermis (Figure 45-5). This material impregnates the cell wall and extends beyond the outside of the cell wall until it meets the Casparian strips of the adjacent endodermal cells. Since the Casparian strips are impermeable to water and dissolved solutes, no substances can pass through the pores of the cell walls from the outside of the endodermis to the inside, or vice versa.

All substances that travel through the endodermis, therefore, must pass through the cytoplasm of the endodermal cells themselves. Thus the endodermis lives up to the translation of its name, "inner skin," by serving as a selective barrier. Substances in the cortex must pass through the living cells of the endodermis before they can enter the vascular tissue and be transported to the rest of the plant. The importance of this barrier can be dramatized by allowing a root to become plasmolyzed.* In such a case, rather than shrinking uniformly all around, like most plant cells, the endodermal cells remain in contact with the Casparian strips and shrink away only from their inner and outer cell walls.

The apoplast occupies only about 10% of the total root volume, but it greatly increases the area of contact between the soil solution and the living root cells. These living cells form a second functional entity, the **symplast** (sym = together with; plast = living matter), a continuous system of cytoplasm that extends throughout the root. Although each cell is encased in its own cell wall, the cells are connected by **plasmodesmata**, strands of cytoplasm that extend between neighboring cells (Figure 45-4c). Plasmodesmata are estimated to take up about 1% of the total surface area of the root cells.

---

* Transpiration Loss of water by evaporation through pores (stomata) in the shoot system of a plant.

* Plasmolyzed Shrunken due to loss of water under conditions where water is scarce or concentration of external solutes is high.

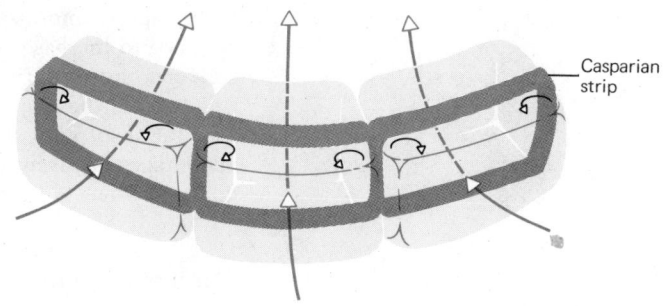

Figure 45-5 Structure of the endodermis. The cells are arranged like bricks fitting tightly against neighboring endodermal cells. The Casparian strip is a layer of watertight "mortar" impregnating the cell walls in a continuous strip, preventing passage of the soil solution from the exterior to the interior of the root by way of spaces in the cell walls; any substance that reaches the interior of the root has passed through the plasma membrane and living cytoplasm of (at least) the endodermal cells.

Casparian strip

In 1938, Alden Crafts and Theodore Broyer proposed the symplast theory of mineral uptake in the roots of plants. According to this theory, all the cells in the epidermis and cortex, as well as the external surfaces of the endodermal cells, absorb minerals from the soil solution in the apoplast. Minerals are then transported through the symplast, which continues right through the endodermal layer, and into the living cells around the conducting elements of the xylem. Thus minerals move toward the interior of the root through both the apoplast and the symplast, although eventually they must pass through the symplast if they are to reach the xylem and, from it, the rest of the plant. The apoplast provides for rapid movement of materials, whereas the living symplast, with its large surface area of selectively permeable membranes, exercises some degree of control over the kinds and amounts of substances taken in.

Some mineral ions, such as calcium and magnesium, are believed to enter root cells by diffusion across the plasma membranes. Most minerals, however, are taken in by active transport; ions absorbed in this way include potassium, nitrate, phosphate, and sulfate. Such active transport is halted in the absence of oxygen or in the presence of substances known to block membrane carrier molecules.

Once minerals have passed through the living endodermis, they must leave the symplast, either by diffusion or by active transport, and move through the apoplast of the root interior until they reach the dead conducting elements of the xylem. From here they are transported upward to the rest of the plant.

The movement of minerals into the root cells decreases the cells' osmotic potential, and hence their water potential (see Section 44-C). The water potential may be further lowered if the shoot system is exerting transpiration pull. Water moves into the root by osmosis, following the water potential gradient.

45-D   Nutritional Adaptations

Some plants have adaptations that allow them to live in places where the soil is unsuitable for the growth of most plants, or where there is little or no soil. We shall examine a few of these adaptations briefly.

Mycorrhizae

An adaptation that enables some plants to live in poor soil is the formation of **mycorrhizae**, symbiotic associations between the roots of higher plants and some soil fungi. The fungus grows in or around the root and penetrates between, and sometimes into, the cells of the root cortex. This seems to be a nutritional adaptation; related species of plants that grow on soils with better nutrient content do not form mycorrhizae. Trees are the best studied forms of higher plants that form mycorrhizae. Trees by definition live in habitats where there is enough rainfall to support large plants, and this means that there is also enough rainfall to leach nutrients out of the soil.

The classic study of mycorrhizae was done by A. B. Hatch. In comparisons of pine seedlings that had formed mycorrhizae with those that had not, Hatch found that the seedlings with mycorrhizae absorbed nitrogen, phosphorus, and potassium more rapidly and also grew much more rapidly. Some mycorrhizal fungi are also believed to increase the solubility of soil nutrients by exuding substances into the soil, or to increase the rate of conversion of nutrients into forms that the plant can utilize. It is these special activities of the fungi, rather than the increased surface area for absorption added by their filaments, that are believed to be the important advantages of mycorrhizae to the plant. In many mycorrhizae studied, the fungus does not seem to be nourished to any extent by the plant.

Carnivorous Plants

Acid bogs are habitats where nutritional conditions are poor for most plants. Acidity retards the growth of bacteria that could release nutrients from organic matter, and it also makes the nutrients that are released very soluble, so that they are easily leached away. Thus many plants that grow in such an area cannot obtain an adequate supply of nutrients through their roots. Some of these plants have "turned the tables" on some members of the animal kingdom by becoming carnivorous. The bodies of animals contain relatively high concentrations of protein, a good source of nitrogen, as well as minerals that plants need. A carnivorous plant, like most other plants, obtains most of its carbohydrates by photosynthesis, and indeed carnivorous plants experience little competition for sunlight because few non-carnivorous plants can survive in these habitats.

There are three basic types of carnivorous plants: active traps, such as the Venus's flytrap; pitfall traps, such as the many varieties of pitcher plants; and flypaper traps, such as the sundews. All three have certain features in common. The trap mechanism is formed from modified leaves, supplied with nectar glands that exude substances attractive to insects; in most other plants, nectar glands are confined to the flowers. The leaves of carnivorous plants also have glands that produce digestive enzymes. A third feature common to carnivorous plants is the modification of leaf hairs in such a way that they aid in capturing prey (Figures 45–6, 45–7, and 45–8).

Once the insect has been trapped and digested, the leaves absorb nitrogen and minerals from its body. The plant absorbs most of its minerals across the surface of the leaves. The roots are small and serve mainly to anchor the plant and to absorb water.

Besides structural adaptations, flytrap and sundew plants also show growth adaptations; rapid changes in hormonal levels when an insect is caught make the cells enlarge in a manner that allows a more secure grip on the prey. After the insect has been digested, these plants "reset" their traps by another round of rapid differential growth. Because only a limited amount of growth is possible, these types of plants can capture only a few insects with each leaf before its cells have grown too much to be of further use. As some leaves become too large, they are shed and new traps are grown. Pitcher plants, however, can reuse the same pitchers over and over, and can even trap many insects at once in each leaf.

Some Other Nutritional Adaptations

Some plants inhabit soils with high concentrations of heavy metals, such as the slag heaps of mines. In Wales, a species of the bent grass *Agrostis* grows well on soil containing tailings from copper mines. This species does not exclude the potentially poisonous copper; when grown in solutions containing copper, it takes up as much of the metal as do plants of related species that are killed by these levels of copper. Copper-tolerant plants are immune only to copper, not to all heavy metals. This strongly suggests that the ability to tolerate the heavy metal is genetic, and involves the synthesis of a chelating agent that inactivates copper within the plant. Such tolerance is doubly beneficial to the plant. Not only does it permit the plant to thrive in soils where other species would perish, so that it encounters little competition from other plants, but also it makes the plant toxic to herbivorous animals, for whom the levels of copper are just as toxic as they are for most plants. Prospectors make use of plants' abilities to tolerate high concentrations of heavy metals; certain types of vegetation indicate likely places for mining particular metals.

(a)

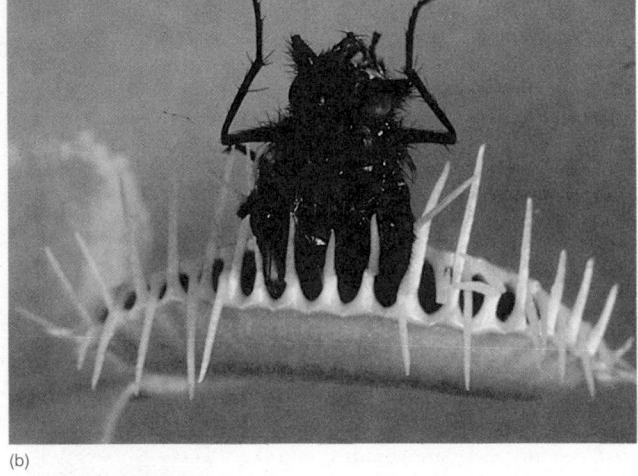

(b)

Figure 45-6 Venus's flytrap, an endangered carnivorous plant native only to the Carolinas. (a) An entire plant; the red lining of the bilobed leaves attracts insects. Sensitive hairs on the inner surface of the leaves respond to an insect's jostling by starting a wave of depolarization similar to a nerve impulse. This causes rapid changes in the water content of certain leaf cells; the leaf folds up quickly and imprisons the prey. Hairs along the edges of the leaf form a cage around the prey. (b) Part of a fly protruding from a trap. Glands on the trap's inner surface secrete enzymes that digest away the insect's soft parts. The rest of the fly blows away when the trap reopens, ready for another victim. (a, Carolina Biological Supply Company; b, Biophoto Associates)

Figure 45-7 Pitcher plants, pitfall types of carnivorous plants. (a) The leaves of pitcher plants are tubular; a hood over the opening prevents entry of rain. Nectar-secreting glands on the lip of the pitcher attract insects. Just beyond these glands is a slick area where unwary insects slip and plunge to a pool of digestive juices below. Downward-pointing hairs inside the pitcher cause the prey to skid into the pitcher and prevent them from crawling back out. (b) A leaf cut in half to show its prey. (a, Biophoto Associates; b, Carolina Biological Supply Company)

(a)

(b)

(a)

Figure 45-8 Sundew, a flypaper-type carnivorous plant. (a) Each leaf is covered with hairs that secrete glistening, sticky droplets attractive to insects. (b) Once an insect becomes entangled, nearby hairs grow towards it and hold it more firmly. Again, digestive enzymes are secreted and the digestion products absorbed. (Carolina Biological Supply Company)

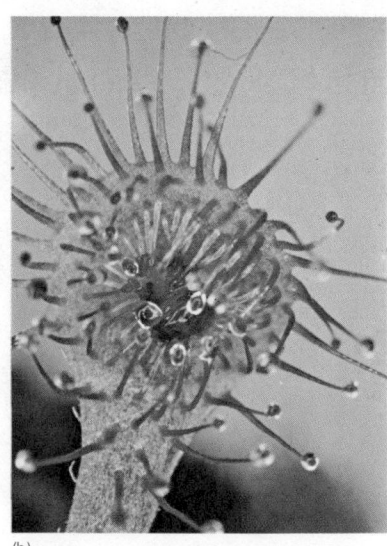

(b)

753

Figure 45-9 An epiphyte growing on the trunk of a tree. The bases of the leaves form pockets that catch water and dust from the air; these are absorbed into the plant. (Biophoto Associates)

Another group of plants with special nutritional adaptations are **epiphytes**, plants that grow on other plants. Some epiphytes catch water and minerals from the air in pockets at the bases of leaves (Figure 45–9). Others grow on trees whose rough bark catches dust and organic debris and forms a shallow layer of soil.

An interesting nutritional adaptation is seen in mistletoes, which can carry on photosynthesis to meet their needs for organic molecules, but which are parasitic on their tree hosts for water and minerals. The roots of mistletoes invade the vascular tissue of the host tree and extract sap from it (Figure 45–10). Mistletoe is so well adapted to this way of life that if the host tree is girdled,* the leaves of the tree die for lack of nutrients and water before the mistletoe does!

45-E Food Storage

Nutrition involves not only obtaining and using nutrients, but also storing them for use in future growth and reproduction. Most plants do not store much food in the leaves, where it is made, but move it to other parts of the plant.

The food storage organs of a plant include roots, such as those of radishes, carrots, and rutabagas; underground stems, such as those of potatoes; and some-times even underground leaves, such as those of onions and other bulbs. One adap-

---

*Girdled Stripped of bark in a complete ring around the trunk or a branch (see Section 44-A).

Figure 45-10 Mistletoes on host trees. (a) The mistletoe sends root-like haustoria (a term borrowed from the parasitic fungi) into the tissues of the host, where they absorb water and minerals from the host plant. The leathery leaves reduce transpiration, and the mistletoe grows slowly; these adaptations reduce the stress it exerts on the host, which must live if the mistletoe itself is to survive. (b) Dwarf mistletoe (yellow) on a pine branch. (a, Carolina Biological Supply Company)

(a) (b)

tive advantage of such underground storage may be that it is difficult for many herbivores to locate and obtain the food when it is underground. In addition, underground storage organs are less vulnerable to freezing or desiccation.

During reproduction, plants move their food reserves from storage areas to the developing fruits and seeds. The food in a seed nourishes the young embryo as it germinates, before it becomes nutritionally self-sufficient. Storage of food in fruits is an adaptation to dispersal of the seed by animals in some cases. The animal is attracted by the nutritious fruit, and eats it. The seed is usually adapted to passing through the animal's digestive tract unharmed.

Plants store inorganic nutrients as well as organic foods. Each autumn, before it drops its leaves, a tree digests many proteins and pigments, and moves nutrients such as nitrogen, phosphorus, and magnesium back into the trunk before the leaves are shed. Some elements, such as calcium, stay in the leaves and must be recycled by the action of soil microorganisms.

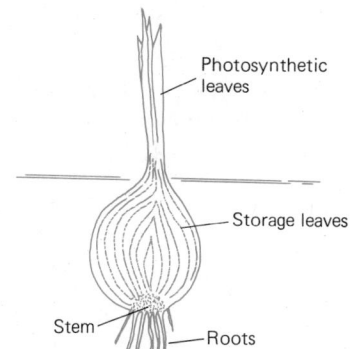

Figure 45-11 Food storage in an onion.

SUMMARY

The minerals required by plants are divided into two groups, macronutrients and micronutrients, depending on the quantities needed. Some nutrients are needed only by plants with special adaptations. The nutritional requirements of plants are difficult to assess because different concentrations or proportions of nutrients, and the ability of plants to use one nutrient in place of another, may change the outcome of experiments.

Nutrient deficiencies of plants may result from lack of nutrients in the soil, a soil pH which makes nutrients unavailable to plants, unfavorable proportions of one nutrient to another, or genetic deficiencies that make the plant unable to handle nutrients properly.

Plants take up most of the water and minerals they need from the soil. The nature of the soil is determined mostly by the type of rock from which it is derived; rainfall, organic matter, soil organisms, and oxygen are also important in determining the quality of the soil. All of these components of soil interact and affect the availability of water and minerals in the soil solution. Agricultural practices such as tilling, fertilizing, or liming can improve the soil to meet the needs of crop plants.

The soil solution moves freely into the outer region of the apoplast, between the cells of the root cortex. The endodermis forms a barrier between the soil solution and the rest of the plant. The symplast of the root absorbs minerals and moves them through the endodermis to the xylem, which transports them to the rest of the plant. Water enters the symplast by osmosis, following the minerals accumulated mostly by active transport.

Plants growing in nutrient-poor conditions may show a variety of adaptations that enable them to acquire a better supply of nutrients. Many trees form mycorrhizal associations with soil fungi, enabling them to take advantage of the fungus's superior ability to absorb nutrients and convert them into usable form; carnivorous plants are adapted to capturing the nutrients in the bodies of insects; epiphytes capture rain or dust from the air with their leaves; mistletoes tap the vascular systems of their host plants for water and nutrients.

Plants can store both organic and inorganic nutrients for future use in growth or reproduction.

OBJECTIVES

From your study of this chapter, you should be able to:

1. List or recognize the macronutrient and micronutrient elements needed by plants.

2. List and explain several reasons why a plant might show symptoms of nutrient deficiency.

3. List the main components of soil, and explain their role in plant nutrition.

4. Discuss several ways in which minerals become available to plants and ways in which they become unavailable.

5. Define the terms apoplast, symplast, epidermis, root hair, cortex, endodermis, and Casparian strip, and explain the role of each in the uptake of water and minerals by the plant.

6. Define the terms mycorrhiza and epiphyte; explain the nutritional adaptations found in each.

7. Explain how carnivorous plants are adapted to capturing animals and why these adaptations are advantageous to such plants.

SELF-QUIZ

1. Match the following components of the soil with their role in plant nutrition:

___ Living organism

___ Organic matter

___ Oxygen

___ Rock particles

___ Water

  a. ultimate source of most of the minerals available through the soil

  b. used in breakdown of organic molecules to release energy and minerals

  c. dissolves minerals and carries them into roots

  d. provides food for fungi and bacteria in soil

  e. releases minerals bound in organic molecules

2. Which of the following would *not* make minerals more available to plants?
   a. increasing the rainfall in a wet, forested area
   b. raising the pH of a very acid soil
   c. tilling a packed-down or waterlogged soil
   d. introducing fungi that can form mycorrhizae into a soil that lacks them

3. The intercellular spaces and cell wall spaces of the root are collectively known as the (a)_____. Its function is (b)_____. The soil solution is prevented from penetrating the entire plant by the presence of the (c)_____ in the (d)_____ layer of root cells. Substances are absorbed by the living cells of the __(e)__ and _____ layers of the root, which together are part of the (f)_____.

4. From the list below, pick the three nutrients needed by plants from the soil in highest amounts.

boron	magnesium	phosphorus
calcium	manganese	potassium
copper	molybdenum	sulfur
iron	nitrogen	zinc

5. List at least three reasons why the following statement is untrue: plants will be well-nourished as long as the mineral nutrients they need are present in the soil in the necessary quantities.

6. Which of the following adaptations would *not* be seen in carnivorous plants?
   a. extensive root system
   b. nectar glands
   c. glands that secrete digestive enzymes
   d. trapping hairs
   e. specialized leaf shape

QUESTIONS FOR DISCUSSION

1. Determining what nutrients plants need would be much easier if a healthy plant were analyzed for its chemical makeup. However, this gives a less accurate picture of plant requirements than the more laborious method described in Section 45-A. Why is chemical analysis of the plant not a good indication of the plant's nutritional needs?

2. Why is vegetation often sparse in soils with large numbers of pebbles and boulders?

3. What are the advantages of organic rather than inorganic fertilizers? What fallacies are involved in statements that organic fertilizers are more nutritious for plants than are inorganic fertilizers?

4. Higher plants may supply organic foods to fungi with which they form mycorrhizae. Why might it be advantageous to the plant to release organic compounds that stimulate the growth of non-symbiotic microorganisms in the soil?

5. Boron-deficient soils are improved by the application of sodium tetraborate at a recommended rate of 20 to 50 pounds per acre. A study of mineral uptake by a wheat crop in one growing season showed that the plants took up only 0.3 pounds of boron per acre. If these values are typical, what happens to the rest of the boron that is applied to the soil?

6. Examine Table 45–1 and see if you can guess why potassium and chlorine are highly mobile in the plant, while calcium, iron, manganese, boron, zinc, and molybdenum are mostly immobile.

7. From looking at Table 45–1, can you tell why the symptoms of molybdenum deficiency are like those of nitrogen deficiency?

8. How do carnivorous plants obtain minerals other than nitrogen? Do you think they can obtain them from the swampy soil where they live? How would you go about showing what is the main source of these other elements for carnivorous plants?

REFERENCES AND FURTHER READING

Bowling, D. J. F. *Uptake of Ions by Plant Roots*. London: Chapman and Hall, 1976. A clearly written review of experimental evidence that sheds light on how roots obtain nutrients; extensive bibliography.

Epstein, E. *Mineral Nutrition of Plants: Principles and Perspectives*. New York: John Wiley and Sons, 1972. Clearly written overview of how plants obtain and use minerals, including genetic and ecological aspects. Cites studies, with bibliography.

Epstein, E. "Roots." *Scientific American*, May 1973. Compares the form of root systems in different types of plants as well as how roots absorb minerals from the soil.

Heslop-Harrison, Y. "Carnivorous plants." *Scientific American*, February 1978. Briefly describes the various types of carnivorous plants known and presents recent experimental and photographic evidence on how prey is captured and digested.

Lloyd, F. E. *The Carnivorous Plants*. Waltham, MA: Chronica Botanica, 1942. The classic work on carnivorous plants.

Stefferud, A., Ed. *Soil: The 1957 Yearbook of Agriculture*. U.S. Department of Agriculture. Washington, DC: U.S. Government Printing Office, 1957. A compilation of interesting and useful information about soil characteristics and their effects on plant growth.

Weier, T., C. R. Stocking, and M. G. Barbour. *Botany: An Introduction To Plant Biology*, 4th ed. New York: John Wiley and Sons, 1970. A well-illustrated botany text.

CHAPTER 46

REPRODUCTION IN FLOWERING PLANTS

In the last few chapters we have studied the organization and function of plants and seen how plants grow and acquire more resources from the environment. Ultimately, natural selection ensures that the plant's resources are channeled effectively into reproduction.

There are two kinds of reproduction in flowering plants. **Vegetative reproduction** is an extension of the kinds of growth we have already seen in Chapter 43. It gives rise to new individuals that are genetically identical to the parent, and thus perpetuates gene combinations that are well adapted to the local environment. Individuals of the favorable genetic combination quickly spread throughout the immediate area where the parent plant is growing.

Sexual reproduction is more complex. It involves the production and growth of a group of structures making up the flower, the production and fertilization of gametes, and development of the embryo, seed, and fruit. The chief advantage of sexual reproduction is that it forms new genetic combinations in each generation. In addition, it produces seeds, which can disperse over a wide area, and which are protected against adverse environmental conditions that might kill the parent plant.

In Chapters 25 and 26, we studied the basic plant life history, in which the diploid (2N) sporophyte generation alternates with the haploid (N) gametophyte generation:

sporophyte (2N)	$\xrightarrow{\text{meiosis}}$	spore (N)	$\longrightarrow$	gametophyte (N)	$\longrightarrow$	gamete (N)	$\xrightarrow{\text{fertilization}}$	zygote (2N)	$\longrightarrow$	sporophyte (2N)
flowering plant		megaspore microspore		embryo sac pollen tube		egg nucleus sperm nuclei				embryo

(You may wish to refer back to this diagram as you read.)

In our brief study of the angiosperm (flowering plant) life history in Chapter 26, we saw that the sporophyte generation is dominant and the gametophyte is very much reduced: the sporophyte is the familiar plant of garden, field, or forest, and the gametophytes are only minute parts of the flower. The male gametophyte consists of the pollen grain and the tube that grows from it, and the female gametophyte is hidden within the female flower parts.

In this chapter we shall look first at sexual reproduction—the events from flower formation to establishment of the embryo of the next sporophyte generation—and then at vegetative reproduction. We shall see that the study of both kinds of reproduction is important in our attempts to improve the strains of plants that we grow for our own use.

46-A Flowers

We have seen in Chapter 43 that the shoot system of a plant grows as its apical meristems produce new cells. Some of these cells enlarge and differentiate into new tissues, whereas others remain meristematic and continue to divide. When it is time to flower, however, the cells in the meristem differentiate and become parts of the flower. Although the meristem is lost in the process, meristematic characteristics soon arise again in the zygote(s) of the next generation formed within the flower. Furthermore, the meristem that produces a flower may lie near another meristem

(a)

(b)

Figure 46-1 Flower structure. (a) *Trillium,* with its flower parts arranged in threes, is a monocot. Three green sepals and three showy red petals surround the six stamens, which end in v-shaped anthers. In the center of the flower, the three arms of the stigma and six angles of the ovary indicate that the compound pistil forms from more than one carpel. (b) Marsh marigolds are dicots (note the five petals). A wreath of stamens surrounds a central group of simple pistils, each made up of a single carpel.

that remains meristematic and grows after the flower has withered and set seed; in this case, we must look carefully to see that the new meristem is not in fact a direct continuation of the old.

Meristems differentiate and become flowers in response to hormonal changes. In Chapter 47 we shall see that the length of the light period or of the dark period a plant receives each day may trigger a change in hormone balance and initiate flowering; these responses may be modified by temperature, moisture, or other factors. In other plants, flowering occurs when the plant reaches a certain stage of maturity; for example, tomatoes begin flowering when the plant has produced a certain number of leaves. Desert plants may respond to a heavy rain by flowering.

The new hormone balance in a meristem makes it develop into a different type of shoot. A flower is really a stem with highly modified leaves and very short internodes. In a typical flower there are four types of modified leaves; like vegetative leaves, they mature and differentiate in order, from the base of the shoot to the tip. The outermost, basal leaves are the **sepals**, which are often green; they develop first and protect the other parts of the flower maturing inside the bud. The next parts inside the sepals are the **petals**, which are often large and showy, with brightly colored patterns that attract animal pollinators. Inside the petals lie the **stamens**, each consisting of a stalk, or **filament**, bearing an **anther**, a chamber where pollen grains develop. In the center of the flower are one or more **carpels**, modified leaves that contain ovules (Figures 46–1 and 46–2). A flower may have one carpel, or several separate carpels, or several carpels fused to one another forming a single structure. The term **pistil** refers either to a single independent carpel (**simple pistil**), or to the structure formed by the fusion of several carpels (**compound pistil**) (Figures 46–1 and 46–2). Each carpel has three parts: the **stigma** receives pollen and often secretes a sticky substance that allows the pollen to stick to it; a **style**, or stalk, connects the stigma to the third part, the **ovary**, which encloses one or more **ovules** (Figures 46–1 and 46–2c).

Many variations on these typical flower parts are found among the quarter-million species of flowering plants. Lilies, for example, have petals and sepals that are indistinguishable from one another except by position; the six ''petals'' of a lily are really three petals and three sepals (Figure 46–2a). In plants that are pollinated by wind, the sepals and petals are often reduced or absent; this allows greater exposure of the anthers as they give up their pollen, and of the stigmas as they receive it. Several plants produce separate male and female flowers (e.g., corn and members of the squash family, including cucumbers and pumpkins), while others have separate male and female (or staminate and ovulate) plants; examples include spinach, willows, some hollies, and hemps. There are also many plants in which

(a)

(b)

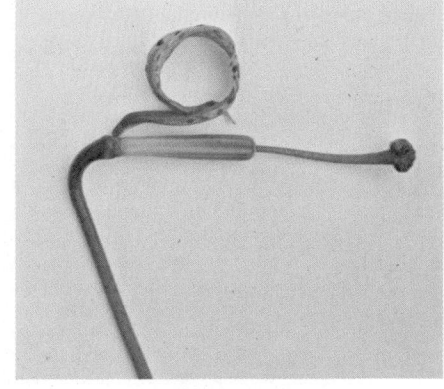
(c)

Figure 46-2 Structure of a lily flower. (a) The petals and sepals are similar in appearance; six stamens surround the pistil. (b) Some sepals and petals removed to show how the stamens attach to the flower stalk. The stamens are composed of white filaments bearing dark anthers, which produce pollen. (c) The pistil is composed of the brown stigma, red style, and green ovary. The three-lobed structure of the stigma and ovary indicates that the pistil is compound, formed of three fused carpels. (William Camp)

(a)

(b)

structures near the flowers act as parts of the "flower." The "petals" of dogwood and poinsettias, for example, are really modified leaves that surround clusters of small, inconspicuous flowers (Figure 46-3).

46-B Pollen

In plants meiosis gives rise to **spores**, rather than to gametes, which are the products of meiosis in animals. The haploid spores grow into haploid **gametophytes**, which in turn produce the gametes that take part in fertilization (see Section 25-G). In flowering plants there are spores of two sizes, microspores and megaspores, which give rise to male and female gametophytes, respectively. Pollen is the (immature) male gametophyte.

Pollen is formed in chambers within the anthers (Figure 46-4). Each **microspore mother cell** in these chambers divides by meiosis to form four haploid cells, which are called **microspores** (Figure 46-5).

Each microspore develops into a male gametophyte. In this process, the haploid microspore nucleus divides once by mitosis, producing two identical haploid nuclei. Before it can complete its development, the male gametophyte must be deposited on the stigma of a flower. It makes this journey in the form of a **pollen grain**, an immature male gametophyte enclosed in a protective wall.

Just as leaf and flower structures vary from one kind of plant to another, so too do the shape and pattern of the pollen grain wall. In fact, experts can easily place a particular pollen grain into the proper genus by its distinctive cell wall pattern (Figure 46-6). The wall is resistant to strong acids and bases and to intense heat; indeed, pollen grains may persist for millions of years and become incorporated into rock formations or peat deposits. Since the sculpturing of the walls of pollen grains is so characteristic of each plant genus, palynologists—people who study pollen— can trace the history of the vegetation of an area by examining the pollen in layers of rock or peat deposits.

Once the pollen grains have become encased in their walls, the anther splits open and frees the pollen.

Figure 46-4 Cross section of a flower bud. Pollen (stained red) is produced in chambers in the anthers; the central ovary contains two ovules. (Biophoto Associates)

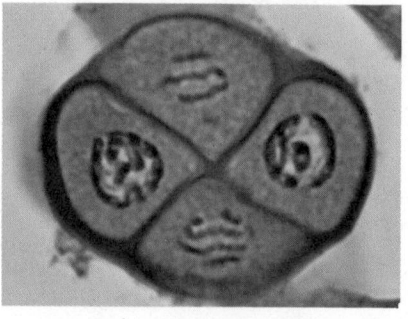

Figure 46-5 A tetrad of microspores formed by meiosis of a microspore mother cell in the anther of a lily. Each microspore will develop into a pollen grain. (Carolina Biological Supply Company)

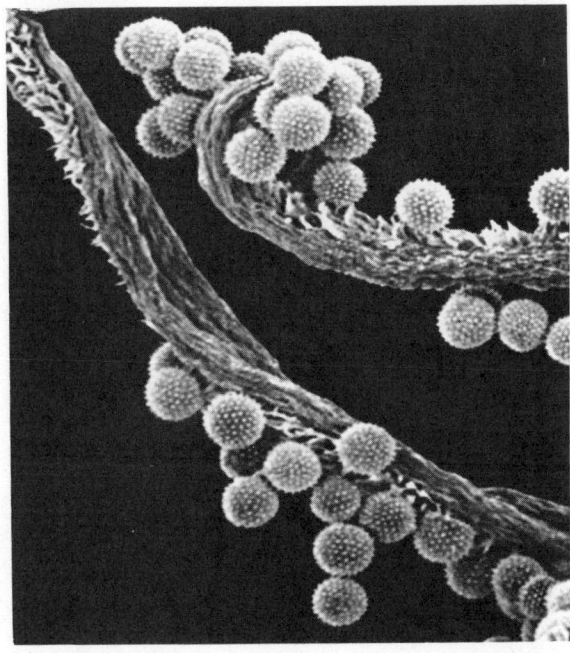
(a)

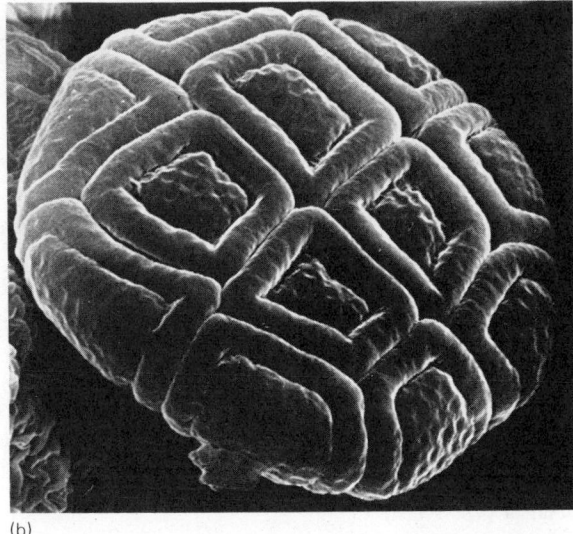
(b)

Figure 46-6 Scanning electron micrographs of pollen grains. (a) Pollen of hollyhock on strands of tissue from the anther. (b) A pollen grain of mimosa. (Biophoto Associates)

Pollination

Pollination is the transfer of pollen to the stigma. Pollen may simply fall from the anther onto the stigma of the same flower, resulting in **self-pollination**. Some flowers, such as peas and their relatives (see Figure 13–1), are so constructed that their stamens and pistils are completely enclosed within the petals, ensuring a high percentage of self-pollination.

Although many types of flowers do self-pollinate, it is often genetically desirable for plants to be **cross-pollinated**, that is, to receive pollen from another individual of the same species. Many plants have adaptations that ensure cross-pollination. For example, a flower's carpels may mature only after its anthers have shed their pollen. The existence of separate male and female plants or flowers, mentioned above, is probably due to selective pressure for cross-pollination.

Pollen, of course, cannot move on its own power; plants rely on wind or animals as agents of pollination. From a plant's point of view, pollination by animals may have various advantages over pollination by wind. First, wind pollination is wasteful because much of the pollen never reaches another flower; the plant may be able

Figure 46-7 Wind-pollinated plants must release a great deal of pollen. (Biophoto Associates)

Figure 46-A The red markings on this poppy flower are "nectar guides," which lead pollinators to the center of the flower. Note the compound pistil. (William Camp)

Figure 46-B A wild bee liberally dusted with pollen. (Biophoto Associates, N.H.P.A.)

Essay: Coevolution of Flowers and Their Pollinators

Much of the evolutionary success of flowering plants has been attributed to the fact that they evolved after terrestrial animal life had become well established. Flowering plants and animals have exerted strong selective pressures on one another, and each has shaped the evolution of the other in many ways. Pollination systems offer many fascinating examples of such **coevolution**.

The most important pollinating animals are the bees. A flower enjoys several advantages in being pollinated by bees. Bees are widely distributed and numerous. Bees also work very hard at visiting flowers because many bees depend entirely on the food they obtain from flowers, both to nourish themselves and to feed the larvae. The behavior of bees also makes them highly desirable pollinators; bees quickly learn to tell the different types of flowers apart, and they are faithful to one particular kind of flower for long periods. Bees are also available throughout the growing season, and they can remain active even at very low environmental temperatures, which immobilize most other insects.

Various species of the order Lepidoptera (butterflies and moths) are important flower pollinators in all parts of the world. Since these insects rely on nectar only as a supplementary food for their short-lived adult stage, however, they are not as effective pollinators as bees. Most moths are nocturnal, and the flowers that depend on them for pollination tend to be of pale colors visible in dim light. Some flowers, such as *Nicotiana* (a member of the tobacco family), produce scent only at night, when the moths that pollinate them are active. Flowers pollinated by butterflies, on the other hand, are more likely to have bright colors that stand out by day.

Although many butterflies and moths feed on nectar from more than one species of flower, they concentrate on one flower at a time. Thus, a hawk moth feeds only on, say, toadflax for as many as five days, and then switches to feeding on nothing but bedstraw. This faithfulness is plainly advantageous to both flowers and insects. The flower benefits because the insect is likely to convey pollen to another flower of the same species. The pollinator benefits by "keying in" on certain cues provided by the flower; the insect

Figure 46-C A butterfly sucking nectar through its tubular proboscis. (Biophoto Associates)

to save energy on pollen production if it is pollinated by an animal. Second, wind pollination is very inefficient for a plant that does not live in dense populations; if the nearest neighbor of the same species is far away, there is a good chance that no pollen will reach its stigmas. By contrast, an animal that visits only one kind of plant carries pollen directly from one individual to another of the same species. Many flowers have evolved structures such that only one species of animal can pollinate them, and these flowers enjoy highly specific transfer of pollen from one individual to another.

Animal pollinators are attracted, first, by some type of reward—usually a sweet nectar—and second, by an advertisement, such as the odor, shape, or color of a flower—preferably all three—that catch the animal's attention. The reward is so located that the animal cannot reach it without at the same time acquiring a load of pollen. All of this has a cost: the animal-pollinated flower must invest energy in making its nectar and its large, showy petals, even though it need not make the prodigious quantities of pollen required for successful wind pollination.

Animals that serve as pollinators include insects—bees, butterflies, moths, wasps, flies, and beetles—and vertebrates such as birds, bats, and even a South African mouse!

can then find more flowers of that species efficiently and ignore the cues from competing "restaurants" (just as some people key in on the "Golden Arches"!).

Many species of birds feed on nectar and supplement their diet with insects. However, birds often pierce the sides of tubular flowers and so obtain the nectar without picking up pollen, a situation that is disadvantageous to the flower. This may be one of the selective pressures that led to the evolution of flowers shaped in such a way that birds can reach the nectar more conveniently from a position where they also brush against the pollen.

The 300 species of humming-birds are the largest group of bird pollinators. They nearly always feed while in flight, hovering in front of a flower and using their long bills and tubelike tongues to suck up the nectar deep within the flower. Flowers pollinated by hummingbirds usually have long stigmas that pick up pollen from the bird's head. In tropical areas, particularly, the length of the bird's bill and the depth of the flower trumpet dictate considerable specificity, making any one species of hummingbird able to feed only on certain species of flowers.

Flowers have evolved several refinements that ensure quality pollination. Some of these adaptations ensure attractiveness to the preferred pollinators, the bees. Nectar with a high concentration of sugar often also has a high concentration of amino acids and some lipids, and this nutritious nectar tends also to contain alkaloids (see Figure 51–5), substances that are highly distasteful to adult butterflies, but that bees will tolerate.

Plants need to ensure not only that their pollinators are faithful to their species, but also that the pollinators visit the flowers frequently. Flowers produce limited quantities of nectar, making it necessary for the pollinator to visit many flowers, and to revisit each flower again and again. Another adaptation that ensures frequent visits is the distinctiveness of a flower; if the flower looks and smells different from other flowers that bloom at the same time in the same area, the pollinator can find the flower easily and discover more flowers of the same species quickly, so that it does not waste time and energy hunting around.

There is also strong selection for different plant species to bloom at different times, so that each species receives the attentions of pollinators in its turn, rather than having all flowers competing for attention during a brief period. Such staggered blooming also provides pollinators with a steady food supply throughout the growing season.

Figure 46-D Part of this orchid flower resembles the rear end of a female bee. Deluded male bees mistakenly copulate with the flower and pick up pollen, which they carry to other orchids of the same species. This adaptation ensures that the flowers are visited frequently and faithfully. (Biophoto Associates, N.H.P.A.)

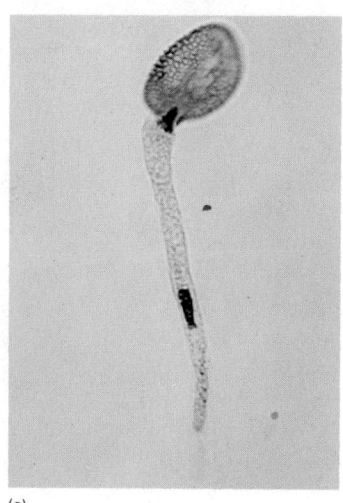

(a)

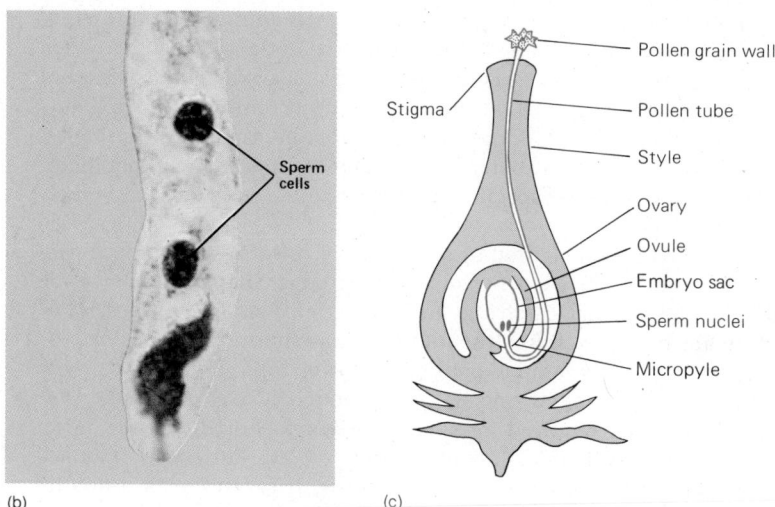

(b)   (c)

Figure 46-8 Events from pollination to fertilization. (a) A lily pollen grain germinated to form a pollen tube. (b) Two sperm nuclei at the tip of the pollen tube, ready to be released for fertilization. (c) A pollen grain normally grows from the stigma to the embryo sac, where it releases two sperm nuclei. (a, b, Carolina Biological Supply Company)

Pollen Maturation

A pollen grain is an immature male gametophyte; it completes development after it has landed on the stigma. The protective coat of the pollen grain ruptures and produces a **pollen tube**, which digests its way through the surface of the stigma (Figure 46–8a). Pollen can be grown in the laboratory in a solution containing sugar, and the sweet, sticky tip of the stigma probably provides it with nourishment in nature.

The wall of the pollen grain contains glycoproteins that must be compatible with proteins in the stigma if the pollen is to grow. Pollen will usually not germinate on the stigma of flowers of a different species. Many kinds of plants have a system of compatibility genes such that pollen containing the same genes as those of the stigma is prevented from completing its development. This mechanism assures that the flower does not fertilize itself and maintains genetic diversity in the population.

If the pollen and stigma are compatible, the pollen tube grows along a chemical gradient (the exact chemical seems to depend on the kind of flower), down the style toward the ovule(s) in the ovary at the base of the pistil. Many pollen grains may land and produce pollen tubes in the same style.

When the pollen tube nears or reaches the ovule, one of its nuclei divides by mitosis to form two **sperm nuclei** (Figure 46–8b) (the other nucleus from the original pollen grain disintegrates during the growth of the pollen tube). The pollen tube grows into the ovule through a tiny pore, the **micropyle**, and releases its two sperm nuclei (Figure 46–8c). The micropyle then closes, preventing the entry of any more pollen tubes. If there are more ovules, other pollen tubes enter through their micropyles. The traffic problem that must exist in the style of a cantaloupe flower is fearful to contemplate!

46-C Preparation of the Ovule

Before the pollen tube arrives at the micropyle, a series of changes inside the ovule has produced an **embryo sac**, a mature female gametophyte. Here again the process begins with the formation of spores by meiosis. In the ovule, a **megaspore mother cell** undergoes meiosis to produce four **megaspores**. In most species of flowering plants, as in female animals, three of the four cells formed during meiosis disintegrate. Only the fourth megaspore, the one nearest the micropyle, survives.

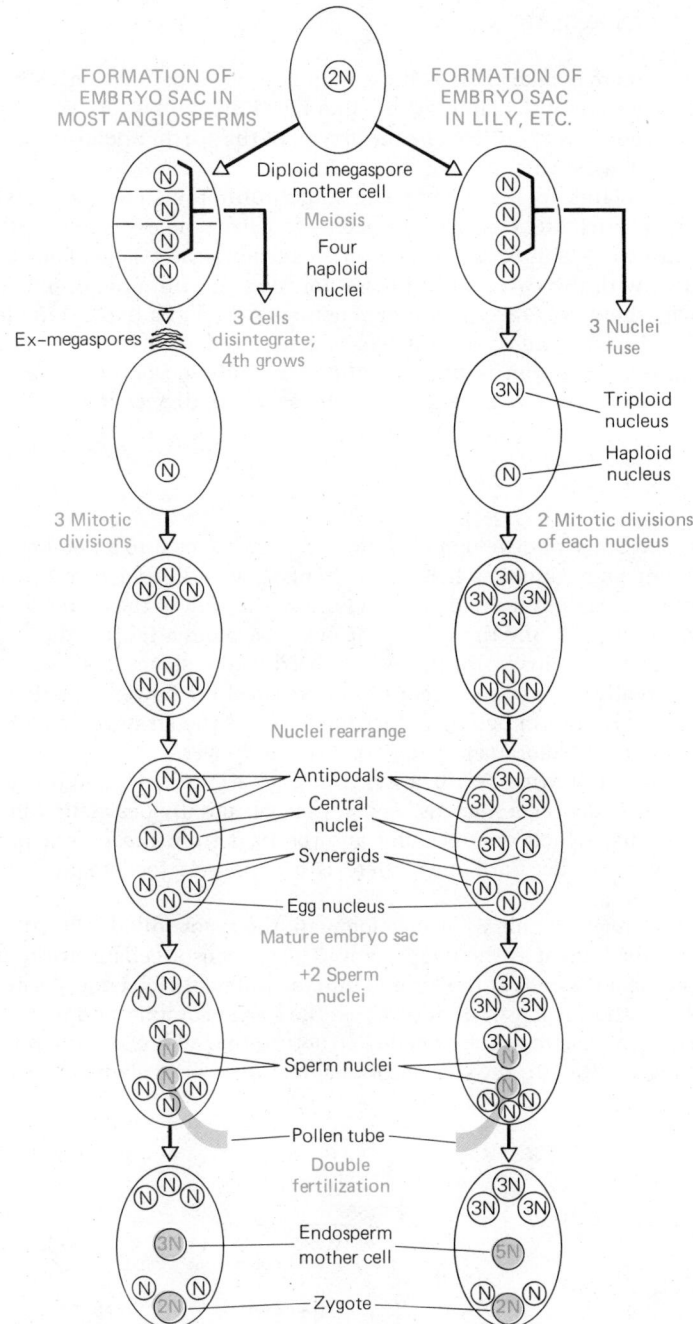

(2N)

Diploid megaspore mother cell

Meiosis

Four haploid nuclei

Ex-megaspores

3 Cells disintegrate; 4th grows

3 Nuclei fuse

(3N) Triploid nucleus

(N) Haploid nucleus

3 Mitotic divisions

2 Mitotic divisions of each nucleus

Nuclei rearrange

Antipodals

Central nuclei

Synergids

Egg nucleus

Mature embryo sac

+2 Sperm nuclei

Sperm nuclei

Pollen tube

Double fertilization

Endosperm mother cell

Zygote

Figure 46-9 Development of the female gametophyte from the megaspore mother cell. Most angiosperms show the pattern depicted on the left; the more complicated lily pattern, shown on the right, is often studied in the laboratory because of its convenience and the large size of the structures involved.

This megaspore enlarges greatly by absorbing nutrients, including the remains of its sister megaspores, and its haploid nucleus goes through three mitotic divisions, producing eight identical haploid nuclei. The enlarged, elongated eight-nucleated structure so formed is called the embryo sac. The eight nuclei arrange themselves with three at the end of the embryo sac near the micropyle, three at the opposite end, and two in the middle (Figure 46–9). The female gametophyte is now ready to be fertilized.

46-D   Fertilization

Fertilization may take place as little as an hour after pollination, as in barley, or as much as several months later, as in witch hazel, which flowers and is pollinated in the late fall but is not fertilized by the arrival of the sperm nuclei at the micropyle until the following spring.

Flowering plants are unique in possessing **double fertilization**, in which both sperm nuclei participate. One sperm nucleus fertilizes the egg nucleus, the central one of the three nuclei next to the micropyle, forming the zygote. The other sperm nucleus fuses with the two central nuclei, forming an **endosperm** nucleus that is triploid (3N, where N is the amount of genetic material in a haploid nucleus of the species). (In lilies the endosperm nucleus is pentaploid, or 5N.) The adaptive value of this second fertilization is unclear, although the endosperm tissue that arises from this nucleus has a very important role, as we shall see shortly.

46-E   Development of the Seed and Fruit

In the next stage of development, the zygote develops into an embryonic plant, and the parent plant supplies it with nutrients that will help it to become established as an independent individual. In addition, the wall of the ovule develops into a protective **seed coat**, and the wall of the ovary becomes a **fruit**. Other parts of the plant near the flower may also develop into fruit-like structures; for example, a strawberry is really an enlarged **receptacle**, the tip of the flower stalk that holds the carpels; its ''seeds'' are in botanical fact the fruits of the strawberry plant, each of them having arisen from a separate carpel of the flower.

Immediately following fertilization, the zygote enters a period of dormancy. Meanwhile, the endosperm nucleus becomes active, and divides many times to form endosperm tissue, which enlarges and absorbs food from the parent plant. Thus when the zygote breaks dormancy, there is a supply of food ready for it in the endosperm tissue.

In the first stage of embryonic development, the zygote divides by mitosis, forming a line of cells known as the **suspensor**. The suspensor cells near the micropyle elongate and push the cells at the far end into the nutrient-rich endosperm. Soon the cell at the tip of the suspensor divides across the suspensor axis, and two subsequent cell divisions give rise to an eight-celled structure that will eventually become the new plant (Figure 46–10); the suspensor disintegrates as the plant embryo develops.

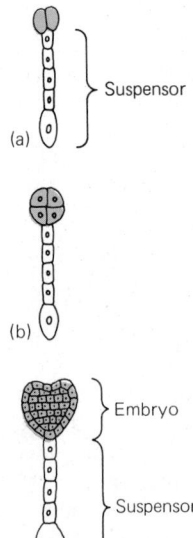

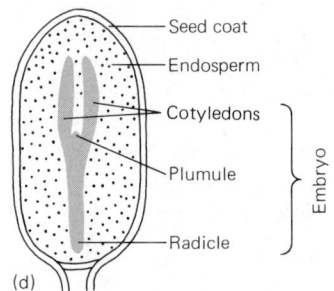

Figure 46-10 Early development of the embryo. (a) A series of cell divisions gives rise to a string of cells (suspensor); the end shown at the top here is pushed into the area of the nutritive endosperm. The end cell divides to form an embryo (b, c). (d) At the completion of embryonic development, the embryo has a recognizable axis, with apical meristems at each end, and its cotyledons have developed. The seed coat has developed from the wall of the ovule, and there is a food supply for the new plant (in this case, the endosperm). (See Figures 43-2 and 43-22 for different seed structures.)

The mature embryo has a **radicle**, which will develop into the primary root of the new plant, a **plumule**, which will develop into the shoot system, and one or two **cotyledons**, depending on whether it is a monocotyledon or a dicotyledon (see Table 43–3).

During the growth of the embryo, the endosperm continues to absorb food from the parent plant, and it may persist as a reserve food supply for the embryo or may be completely absorbed into the embryo as the seed matures. The wall of the ovule, which is part of the parent plant, becomes larger as the embryo grows, and usually hardens to form the protective seed coat as the seed matures.

Outside the seed coat, the wall of the ovary also enlarges and absorbs more nutrients to form a fruit. Fruit growth is initiated by the pollen tube's release of tiny amounts of the hormones auxin and gibberellin (to be discussed in Chapter 47). Soon the developing seed begins to produce its own hormones, which continue to stimulate growth of the fruit. The presence of these growth substances, or the activity they stimulate, enhances the transport of cytokinins, hormones that stimulate cell division, from the parent plant to the growing fruit. Under the influence of these three growth-promoting hormones, the fruit grows by both cell division and cell enlargement.

Most types of plants do not set seed and develop fruits unless their flowers have been pollinated and fertilized. In some species, however, spraying the flowers with the proper concentration of auxin or auxin plus gibberellin induces the production of seedless fruits. Some fruits, such as cultivated strains of bananas and pineapples, develop naturally without fertilization and are therefore seedless. Humans have increased the populations of these economically desirable plants by vegetative propagation, and the plants have thus become successful through artificial selection.

Not all fruits are fleshy and delicious, however. Many are rather minimal protective layers; when the seeds are ripe, the fruit ruptures along its lines of weakness and releases the seeds. Pods of peas, beans, and milkweeds are examples of such fruits.

Figure 46-11 Bananas are hexaploid, and their chromosomes cannot undergo the delicate process of meiosis properly. The nutritious fruits develop without fertilization, and the plants readily propagate vegetatively. (William Camp)

(a)

(b)

(c)

Figure 46-12 Fruit dispersal. (a) The familiar winged fruits of dandelions disperse by blowing in the wind. (b) The unfortunately equally familiar burdock bur hooks onto clothing or fur and hitches a ride. (c) Raspberry seeds are dispersed in bird and animal droppings; the fruit changes color as it ripens, attracting animals who eat the fruit and eliminate the seed undamaged. (a, b, Biophoto Associates)

46-F   Dispersal of Seeds and Fruit

Once seeds mature, they are ready for dispersal. Small, lightweight milkweed seeds and dandelion fruits have parachute-like tufts of fiber that enable them to float on the breezes for dispersal through the air. Larger seeds have a distinct competitive advantage because they contain more food to supply the embryo's needs until it becomes established. However, wind dispersal cannot move a large seed far from the parent plant. An animal, on the other hand, may carry even a large seed a

Figure 46-13 Defense against seed predators: a large, nutritious seed (right) may be covered with a prickly husk (left). (Biophoto Associates)

considerable distance. The importance of animal dispersal is shown by the large energy investments of many plants in adaptations that promote dispersal by animals. Some fruits or seeds have hook-like extensions that attach to the feathers, fur, or clothing of passing animals or people, which give the seed a free ride to a new home.

The most common adaptation, however, is protection of seeds in indigestible seed coats; the seeds are surrounded with a tasty, nutritious fruit that an animal will eat. The seeds then pass unharmed through the animal's gut and are deposited with a small pile of organic fertilizer. In fact, the seeds of some plants will not grow unless their seed coats have been eroded to some extent by passage through an animal's digestive system.

Fruits are usually protected from being eaten before the seeds are ripe for dispersal; an unripe fruit is often distasteful and may even contain toxic chemicals. When the fruit is ripe, its chemical composition changes so that it tastes good. The fruit may also change color, a visual signal that it is now ready to be eaten (Figure 46–12c).

Seeds and Seed Predators

Because seeds contain the food supply for an embryonic plant, they also make good food for animals. There is therefore strong selection for adaptations that protect the seeds from predation. The types of adaptations that are effective depend on the main types of predators eating the seeds. An undiscriminating predator gobbles up every seed it finds until it is full. Plants with this type of predator usually produce many small seeds, providing a better chance for some of the seeds to survive.

A discriminating predator maximizes its food intake for the energy it expends; it may attack plants that have the most seeds in a fruit, or the seeds that are largest or easiest to chew. Plants attacked by such predators usually enclose their seeds in a hard covering, such as a nutshell, that discourages the predator; producing smaller seeds works only if the seeds become smaller than those of another species to which the seed predator might switch.

Another adaptation is **seed masting**, the simultaneous release of seeds by all the trees of the same species in an area, at intervals of two years or longer; this makes seeds available to predators for a minimum time period. Beech trees and some oaks show this characteristic, but the most impressive examples are bamboos. Part of a bamboo stand in India was collected and sent to botanical gardens in the United States and Britain at the beginning of the nineteenth century. The plants grew vegetatively for 130 years afterwards, and then the bamboo stands on all three continents produced seeds in the same year! The advantage of seed masting to the plant is that most of the time there are no seeds to support the growth of large populations of seed-eating animals. When seeds are finally shed, there are so many of them that a small population of seed predators cannot eat them all, and some seeds escape to produce the next generation of plants.

46-G Germination

Many seeds enter a period of dormancy or metabolic inactivity after they have formed. As a seed enters dormancy, it steadily dries out, until its water content may be less than 5% of its total weight. In many kinds of seeds, dryness seems to be the main factor assuring the seed's **viability**, or ability to break dormancy and grow into a new plant after an extended period of time. Many commercial seed suppliers now dry their seeds thoroughly and wrap them in moisture-proof foil packets.

The viability of seeds varies from species to species, and from individual to individual within a species. The record for viability is held by certain legume seeds stored as study specimens in the National Museum in Paris, France. In the 1930s some of these seeds were tested and found to have retained their viability for over a century! In contrast, sugar maple seeds live less than a week. These seeds do not have a dormant period, and their viability is better if they are not allowed to dry out very much.

In order to **germinate**, or begin growing into a new plant, seeds must be supplied with water. However, the seed coat is often so thick or impermeable that its contents cannot absorb water until the seed has had some special treatment; the seed coat may need to be partially digested by animals or decomposer organisms, or abraded by contact with soil particles, before water can enter. Most seeds also require oxygen, and many require particular temperature and light conditions before they will germinate. Furthermore, germination requirements may vary not only from species to species, but also from individual to individual. Thus a plant does not have all of its seeds "in one basket"; germination may be spread over months or years, and at least some seeds are likely to find conditions that favor their survival.

Germination of seeds seems to be associated with an increase in the growth-stimulating gibberellin hormones. As a seed breaks dormancy, its gibberellin content may increase. Application of gibberellins at appropriate concentrations can often overcome the special light or temperature requirements of a seed. Thus, lettuce seeds that are normally induced to germinate by exposure to light can be germinated by applying gibberellins instead. Likewise, many seeds, such as some types of rye and wheat, require a period of cold (**vernalization**) in order to germinate. Again, gibberellins can substitute for the cold treatment. It appears that these environmental stimuli somehow change the physiology of the seed and result in production or activation of gibberellin. In some species, other plant hormones are needed for germination.

Germination has been best studied in barley. The barley embryo secretes gibberellin, which induces the **aleurone**, a tissue layer between the seed coat and the endosperm, to secrete a variety of enzymes. These enzymes break down starch and other stored food in the endosperm layer, making it available for absorption by the developing embryo. As the embryo uses up its food, the dry weight of the seed decreases. When the embryo becomes well-enough established to make its own food through photosynthesis, the dry weight of the seedling begins to increase again. The seedling then grows as described in Chapter 43.

46-H Breeding Programs

Probably since the beginnings of agriculture, humans have been breeding plants selectively in order to improve the characteristics that make them useful to us. An obvious starting point in such a program is choosing plants that look good in the field and collecting seeds from them to plant for the next generation. For thousands of years, this was the only available method of improving plants, but it nevertheless produced strains of cultivated plants that were strikingly different, especially in their high food yield, from their wild ancestors. More control of the process can be gained by carefully cross-pollinating plants selected for various desirable characteristics, in an attempt to obtain offspring that combine all of the desired characteristics in the same plant. Our modern knowledge of genetics now guides the selection of parental strains for such breeding programs.

The genetics of some crop plants are fairly well understood. For example, corn, one of the most important crops in the United States, has been intensively studied; many of its genes have been analyzed and mapped. The popular strains of hybrid sweet corn are produced by carefully planned crosses between a number of strains, each bred for particular traits. It may take several generations of crosses to produce the seed that is sold for growing hybrid sweet corn, and these crosses must be made anew each year in a continuous breeding program, since the hybrids do not breed true. Thus seed for hybrid corn must be purchased from breeders each year; planting seeds saved from hybrid ears results in a motley assortment of characteristics rather than the desired genetic uniformity.

Figure 46-14 The hybrid tomato plant on the left was bred to bear more and larger fruits than the standard tomato plant on the right. (W. Atlee Burpee Co.)

Corn is a fairly easy crop to use in genetics programs. Male and female flowers are borne in separate, very large, clusters; thus corn plants are easy to manipulate in carrying out a desired pollination. In addition, corn has only 10 pairs of chromosomes, with a relatively limited number of possible combinations. Corn also has the advantage of being an annual plant, so that the success of any particular cross can be judged within a year. In addition, since corn is so widely grown, there is a large number of plants from which desirable new characteristics can be chosen.

Other plants pose difficulties. For example, a breeding program to produce new and better strains of apples must wait 4 to 10 years until planted seeds grow into mature trees and bear fruit. Furthermore, only about 1% of such trees bear fruit that is even equal to varieties of apples already available; most carry new genetic

Figure 46-15 "Sugar bush"—a type of watermelon that grows on space-saving bushy vines ("normal" watermelon vines may sprawl over several metres) and also has desirable flavor, color, and disease resistance—took almost 20 years to develop through selective breeding. (W. Atlee Burpee Company)

combinations that are inferior to old varieties. Any new tree that looks promising must be screened for another 10 years before it is ready for marketing . . . or the woodpile.

Even in annual plants, it may take many years to develop new varieties if the genetics of the species has not been worked out well (Figure 46–15).

46-I Vegetative Reproduction

The new genetic combinations of sexually produced offspring are often less suited to the environment than is the genetic makeup of the parent plant. Furthermore, even seeds with favorable genetic combinations have high mortality from predation and from dispersal to unfavorable habitats. Thus it is advantageous for plants that are well adapted to their environments to spread by vegetative growth and cover the surrounding area with more individuals of the same genetic makeup. This forms a **clone**, a population of individuals with identical genotypes.

Vegetatively produced offspring remain attached to the parent plant for a time. Consequently, they are larger than sexually produced offspring, and each one repre-

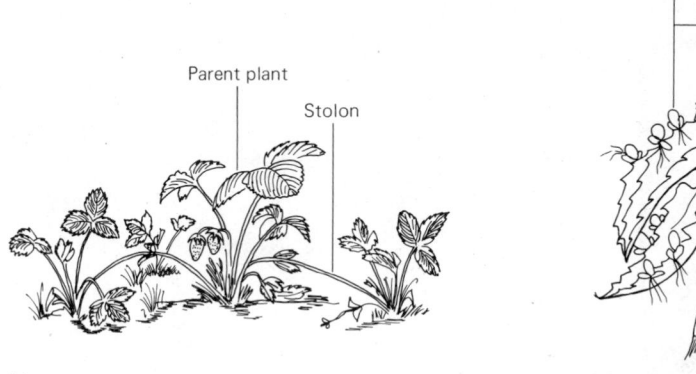

Plantlets

Parent plant

Stolon

(a)   (b)

Figure 46-16 A few modes of vegetative reproduction. (a) Stolons of strawberry plants begin as slender stems that arch away from the parent plant and form roots and leaves of a new plant where they touch the ground. (b) The leaves of the Kalanchoe bear tiny plantlets, complete with roots, which eventually fall off and grow on their own.

sents a larger energy investment than a sexually produced seed. However, the vegetatively produced offspring is almost always an evolutionary success, whereas the tiny seeds are risky investments. Thus vegetative reproduction is often more advantageous to a plant than sexual reproduction is.

Vegetative propagation may be by means of horizontal stems or roots that send up new shoot crowns, with attached roots, at intervals; by **stolons**, which grow up and out from the plant and eventually touch ground again and produce the roots and shoot of a new individual, as in strawberries (Figure 46–16); or by **layering**, in which a woody branch may produce new roots, and hence a potential new individual, at a point where it touches moist soil, as in blackberries and raspberries.

Dandelions, hawkweeds, and many grasses reproduce by **agamospermy**. In this process, seeds develop from ovules in the flower, and so reproduction appears to be a normal sexual procedure. Closer study, however, shows that meiosis and fertilization do not take place. In some agamospermous plants pollination may occur, but it is only a stimulus to the ovule to develop into a seed, and the pollen contributes no genetic material to the offspring. In other species, agamospermy is an emergency method of reproduction when pollination does not occur. The ubiquitous and hardy dandelions and hawkweeds are obviously well fitted to meet the various conditions found in their environments, and asexual reproduction has replaced sexual reproduction in these plants.

Vegetative propagation is often desirable from the human as well as from the plant point of view, and we use it to reproduce many economically important plants such as strawberries and potatoes. Many plants reproduce vegetative parts underground, and can be helped along in their spread. The bulbs of daffodils, tulips, and onions, for instance, divide into two or more pieces at intervals; the pieces can be dug up, separated, and replanted, to increase the spread of the plant.

Potatoes are an important crop produced by vegetative means. A potato is an underground stem, or **tuber**, and by digging up potatoes, cutting them into pieces, each with an "eye" (bud) and replanting them, a large number of offspring can be obtained from each potato plant. Commercially, potato plants are grown from "seed potatoes," that is, tubers of good quality, rather than from "true seed" produced by potato flowers; potato plants grown from true seed are almost inevitably sorry affairs. However, breeders do grow plants from true seed in an attempt to produce new strains with characteristics that presently available potatoes lack, such as resistance to certain diseases and pests and an ability to form tubers when grown in tropical climates.

Many plants do not normally reproduce vegetatively, but will do so if properly manipulated. Home gardeners often root cuttings to produce new plants. *Coleus* is

Figure 46-17 A successful graft in an apple tree. Scion (above) and stock (below) formed a bulging callus as they grew together and sealed themselves into a single unit several years before this picture was taken. Eventually the callus will disappear as new tissue grows smoothly around it.

perhaps the most popular plant propagated by rooting cuttings. Geraniums, ivy, and other house plants can also be grown in this way. African violets and other plants with fleshy leaves can often be grown from a single leaf placed on moist soil until it roots. Some plants can be rooted more successfully with the use of solutions of plant hormones, available from most plant nurseries and florists.

Another means of propagating plants vegetatively is by **grafting**. In this process, a **scion**, a twig or bud of a desirable plant, is attached to a **stock**, the root system or stem of another plant, which has had a twig or bud of corresponding size removed from the graft area. The scion and stock are wrapped or sealed together, with their vascular cambia* as close to each other as possible. The cut area of each plant produces undifferentiated callus cells, which soon produce new cambial strands that form a vascular connection between the stock and the scion. Each part of the graft retains its genetic identity, but the two do interact physiologically, by the exchange of nutrients and hormones.

To be successful, a graft must be between plants of the same or closely related species. The most common use of grafts is in the production of fruit trees, grape vines, or rose bushes in such a way that the desirable fruits or flowers of the scion are coupled with a sturdy, pest-resistant rootstock of another strain of the same species. Thus, for example, all the Red Delicious, Golden Delicious, and McIntosh apple trees now in existence are derived by grafting, starting from single trees that originated as fencerow "volunteers" that happened to have desirable fruits. Dwarf fruit trees are produced by grafting scions onto rootstocks of related species; thus dwarf pears are grown by grafting onto quince roots. It is now possible to purchase small, manageable apple and pear trees with five or six grafted branches, each bearing fruits of a different variety.

---

*Vascular cambium Tissue that produces secondary xylem and phloem, transport tissues that must form a physiological link between the two members if the graft is to be successful.

A more recent application of grafting is the development of trees for planting along city streets. One variety of honey locust tree has a rootstock that is remarkably tolerant of salt runoff from sidewalks and nitrogenous wastes from promenading dogs. The scion comes from another variety of honey locust that produces a straight, aesthetically pleasing trunk and bears very few seed pods; seed pods are undesirable to humans because of the mess they make on sidewalks.

An exciting recent development in vegetative propagation is the culturing of meristematic cells on growth medium under laboratory conditions to produce adult plants. The technique has the advantage that it can produce more replicas of a desirable plant quickly, since only a small lump of meristematic cells is needed to start each new plant, rather than the large, leafy cutting that was formerly required. The method also reduces the amount of space needed for maintaining starter plants. Cells kept in a flask or two in the laboratory can substitute for acres of plants formerly kept as sources for cuttings (Figure 46–18). This method was developed using chrysanthemums, but it is applicable to a number of other plants as well.

Thus, vegetative propagation of plants, either by allowing them to spread by their natural method, or by rooting, dividing, grafting, or meristem propagation, allows us to make many exact replicas of plants that have desirable genetic combinations, combinations that originally arose as a result of selective breeding or even as a result of unsupervised sexual reproduction.

Figure 46-18 Plantlets grown in tissue culture in the laboratory from meristem cells of a chrysanthemum. (Robert W. Langhans)

Plants flower in response to specific cues that differ greatly among the various species, and each of the quarter-million species of angiosperms has its own distinctive flower structure. In all this diversity, however, we can find a basic unity in the structure and function of flowers.

A flower is an abbreviated shoot, in which all of the cells differentiate into parts of the flower stalk or its modified leaves, the flower parts. Certain cells divide by meiosis, and the resulting haploid cells develop into the haploid male gametophytes, or pollen grains, and female gametophytes, or embryo sacs. Double fertilization forms a zygote and an endosperm mother cell. The endosperm mother cell divides and develops into the endosperm, which absorbs food from the parent plant, and the zygote soon develops into the embryo of a new plant. The parent plant, besides contributing food to the new embryo, also protects it and its food supply in a seed coat derived from the wall of the ovule, which in turn is surrounded by the fruit, derived from the wall of the ovary, another parental structure.

Animals may aid in the reproduction of plants by distributing pollen among flowers, or by carrying seeds to new locations. In many plants one or both of these functions are carried out by the wind.

Many seeds enter a period of dormancy following their release from the parent plant. Eventually the seed germinates in response to environmental cues and establishes itself as a new plant.

Sexual reproduction results in individuals with new combinations of genetic characters. In many cases, these combinations are less desirable than those of the parents, from the point of view either of humans or of the plants. Many plants have some means of asexual reproduction, which perpetuates a particularly favorable combination of genes unchanged, in addition to, or instead of, sexual reproduction. Humans propagate many plants vegetatively by artificial means such as rooting and grafting.

OBJECTIVES

From your study of this chapter, you should be able to:

1. Define the following terms, and use them properly: sepal, petal, stamen, anther, carpel, pistil, style, stigma, ovary, ovule, pollen, pollen tube, embryo sac, endosperm, seed coat, cotyledon, fruit, pollination, fertilization, germination.

2. Name the parts of a flower, and state the role of each in reproduction of the plant.

3. Explain what is meant by the terms megaspore, microspore, and male and female gametophytes, and tell where each is found in flowering plants.

4. Explain how pollination and fertilization take place in flowering plants.

5. Name the three main parts of a seed, and explain how each develops.

6. List four factors that may be required for germination of a seed.

7. Explain the advantage of asexual, or vegetative, reproduction to a plant, and list some ways in which plants may propagate asexually; explain why humans use vegetative propagation of plants, and name some ways in which plants can be propagated by human manipulation.

SELF-QUIZ

1. Pollen is produced in the:
   a. carpel
   b. anther
   c. pistil
   d. stigma
   e. pollen tube

2. A female gametophyte would be found in the:
   a. ovule
   b. stigma
   c. endosperm
   d. fruit
   e. seed

3. A flower part whose primary role is protection is the:
   a. stamen
   b. ovary
   c. sepal
   d. seed coat
   e. embryo sac

4. True or False. The terms pollination and fertilization can be used interchangeably.

5. The seed coat develops from the:
   a. sepals
   b. ovary wall
   c. ovule wall
   d. endosperm
   e. fruit

6. Which of the following is *not* required for the germination of seeds?
   a. certain temperature conditions
   b. oxygen
   c. water
   d. light
   e. none of the above

7. Grafting is used to propagate plants because:
   a. it is faster than growing seeds
   b. it maintains a desired set of genetic characteristics
   c. it combines the genetic characteristics of two desirable strains of plants
   d. healthy plants will graft by themselves, so that they reproduce profusely
   e. a plant can produce many more scions than seeds

QUESTIONS FOR DISCUSSION

1. Why do banana plants put so much energy into producing fruits that contain no seeds?

2. Plants given large amounts of fertilizer, especially fertilizer containing much nitrogen, often flower poorly or not at all, and do not accumulate food reserves; instead they engage in vigorous vegetative growth. Is there an adaptive advantage to this?

3. Some plants, such as dandelions and hawkweeds, have lost the ability to reproduce sexually but still produce flowers and set seed by development of the ovule without meiosis or fertilization. What is the advantage to this system over a more orthodox means of vegetative reproduction?

4. What are some advantages to the plant in having its flower parts differentiate in sequence rather than all at once?

REFERENCES AND FURTHER READING

Echlin, P. "Pollen." *Scientific American*, April 1968. Many interesting facts and illustrations, plus a discussion of how pollen develops.

Elias, T. S., and H. S. Irwin. "Urban trees." *Scientific American*, November 1976.

Heinrich, B. "The energetics of the bumblebee." *Scientific American*, April 1973. A detailed study of the relationships between bumblebees and the flowers they pollinate, viewed in terms of the influence of energy expenditure on evolution of adaptations.

Koller, D. "Germination." *Scientific American*, April 1959. Describes germination requirements of various species, showing how widely these requirements may vary.

Langhans, R. W., E. D. Earle, and S. R. Bush. "Chrysanthemum micropropagation." *New York's Food and Life Sciences Quarterly* 7(2):3–7. Describes procedures for meristem propagation of chrysanthemum plants.

Proctor, M., and P. Yeo. *The Pollination of Flowers*. Glasgow: William Collins Sons, 1973.

Wilson, C. L., W. Loomis, and T. Steeves. *Botany*, 5th ed. New York: Holt, Rinehart and Winston, 1971. Contains well-illustrated discussions of sexual and vegetative modes of reproduction.

CHAPTER 47

PLANT HORMONES

In Chapter 43 we saw that a plant grows as meristematic cells divide and produce cells that grow, differentiate, and mature. But how is this growth controlled? Why does one cell differentiate to become an epidermal cell while another becomes part of a xylem vessel? Why does the root grow down into the soil and the shoot grow up into the air? What determines whether a lateral bud forms a new branch this year, or next year, or never? What makes the vascular cambium produce enough new secondary xylem to support all the branches and leaves? Why do all the cherry trees in an orchard bloom at the same time in the early spring, and why do their fruits ripen over a short season in early summer? Why do all the thistles in a field bloom at once in the late summer? And why do coffee trees bear flowers, unripe fruits, and ripe fruits simultaneously during every season of the year?

Every aspect of the production, differentiation, growth, and maturation of a plant or of its parts is regulated by chemicals, the plant hormones. In this chapter we shall introduce the plant hormones that are known and give some examples of their functions and their usefulness to humans.

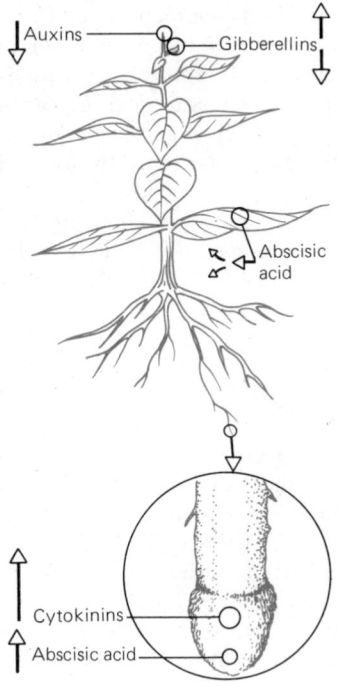

Figure 47-1 Sites of hormone production in a plant. Auxins are produced in growing apical meristems of the shoot, and small amounts (not indicated) are also made in the apical meristems of the roots; gibberellins are produced in young leaves; cytokinins are produced in the apical meristems of the roots; abscisic acid is produced in older leaves and in the root caps; ethylene (not shown) is produced in many areas when the concentration of auxin exceeds a certain threshold. Arrows indicate direction in which each hormone is transported.

47-A Plant Hormones

Hormones are chemical messengers produced in one part of an organism and transported to other parts, where they exert an effect even though they are present in very small concentrations.

Whereas animals often have definite endocrine organs whose sole function is the production of hormones, the tissues or organs that produce plant hormones are usually also responsible for some other function (Figure 47–1). In addition, a particular plant hormone may be produced by different tissues at different stages in the plant's life. In animals, a hormone often acts quite specifically on particular target tissues, and will not affect other tissues of the body; however, plant hormones seem to affect virtually any tissue in the plant. (In both plants and animals, however, different tissues may respond differently to a particular hormone, depending on the tissues' stage of development and on the concentration of the hormone.) Plant hormones do not maintain homeostasis, as do many animal hormones; instead they usually affect differentiation and growth by causing permanent, irreversible changes in the activities of cells. These changes include the division, elongation, and death of cells. Although no purely homeostatic hormones have been discovered in plants, this does not necessarily mean that none exists.

How do plant hormones influence differentiation and growth? Applying a hormone to part of a plant is often followed by changes in synthesis of RNA or proteins, or changes in the activity of certain enzymes. However, it is not yet clear whether the hormones themselves bring about these changes, or whether the hormones act indirectly, by causing changes that in turn lead to the observed activities.

A cell's differentiation is determined by the environment inside the cell, which in turn is affected by the activities of adjacent cells. The cell's environment also includes the rest of the plant. Growth of new parts that change the size and shape of a plant is regulated by hormonal messages passing between different parts of the plant. Thus the root system remains in physiological balance with the shoot system, and the position of a branch or leaf or root determines where other branches or leaves or roots are added to the plant body.

A plant must also respond to stimuli in its physical environment. Growth responses to environmental gradients are known as **tropisms**. The response of a plant to gravity, for example, is called **geotropism**. In most plants, the roots are **positively geotropic**, tending to grow downward, while the shoot is **negatively geotropic**, growing away from the center of the earth, or up. Such responses are under the control of hormones. Hormones also influence the plant's response to seasonal changes.

The plant hormones identified and studied to date can be divided into five groups: auxins, gibberellins, and cytokinins are generally considered to be stimulatory, inducing cell division or cell growth; abscisic acid and ethylene usually inhibit these processes and instead produce dormancy or aging.

47-B   Auxins

The first plant hormones to be discovered were the **auxins**. They were detected because of their role in **phototropism**, the growth of a plant in response to light. If a plant is placed so that light falls on it from one direction, the stem bends in the zone of elongation and the shoot grows toward the light. When the shoot has grown so that it receives light uniformly on all sides, it resumes its growth straight upward.

Oat seedlings are favorite subjects for the study of phototropism. A sheath, the **coleoptile**, covers the first leaves of the oat seedling. As long as the coleoptile is in darkness, it grows upward, but if it is exposed to light from one side, it shows positive phototropism and bends toward the light. Light stimulates growth of the leaves within the coleoptile; the expanding leaves quickly rupture the coleoptile and grow out into the light where the phototropic response of the coleoptile has positioned them.

Charles Darwin studied this phototropic response by covering different parts of oat coleoptiles and subjecting them to light from one direction. Although the growth response takes place in the region of elongation behind the coleoptile tip, Darwin found that covering this area had no effect on the phototropic response. However, when he covered or removed the coleoptile tip, the phototropic response did not occur. He concluded that phototropic bending in the zone of elongation was controlled by the tip (Figure 47-2).

How did the tip of the coleoptile exert this control? At the time, animal physiologists knew that animal responses to the environment were controlled by two means: nerve impulses and hormone production. Could either of these be working in plants? Although plants were known to have no nervous tissue, they might have

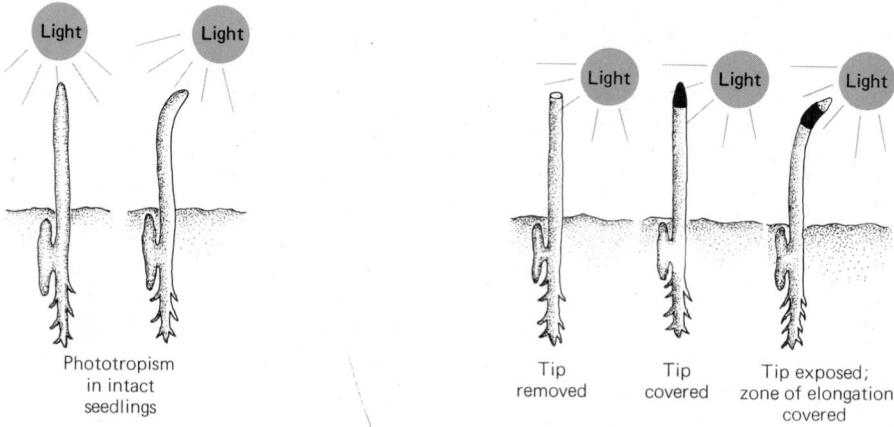

Figure 47-2 Phototropism in the coleoptile of an oat seedling. (Left) The coleoptile grows towards light. (Right) Covering or removing the coleoptile tip destroys the phototropic response, whereas covering the zone of elongation, where bending actually occurs, has no effect.

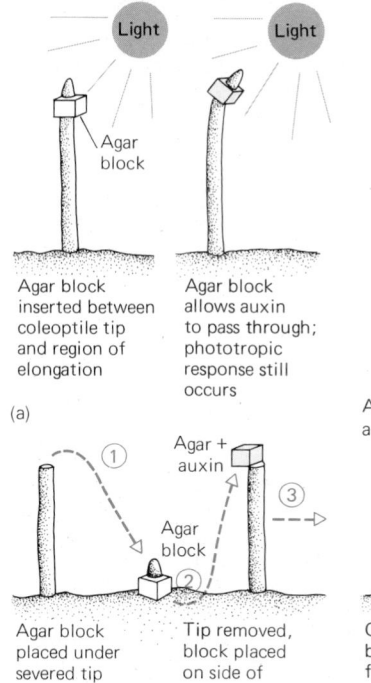

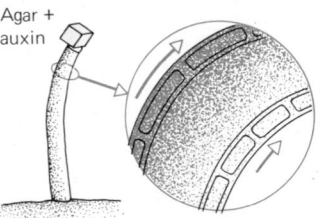

Figure 47-3 Chemical control of phototropism. (a) An agar block can transmit the agent responsible for phototropism from the coleoptile tip to the zone of elongation. (b) This agent can be collected from the tip and later transferred to a second plant, showing that electric current is not the active agent. Bending is due to greater elongation of cells on the side away from the light, that is, on the side with a higher concentration of auxin.

Light

Light

Agar block

Agar block inserted between coleoptile tip and region of elongation

Agar block allows auxin to pass through; phototropic response still occurs

(a)

Agar + auxin

Agar block

Agar block placed under severed tip

Tip removed, block placed on side of second coleoptile

(b)

Agar + auxin

Coleoptile bends away from side of auxin application

Greater concentration of auxin on left stimulates more elongation; coleoptile bends toward right

some sort of electrochemical impulse that could pass through the plant, and chemical transport was certainly possible.

In 1926, Frits Went, then a graduate student, devised an experiment to distinguish between the two hypotheses. It had already been found that inserting a block of agar* between the severed tip of a coleoptile and the zone of elongation did not interfere with the phototropic response. Presumably, the impulse or hormone could pass through the agar. Went had the idea of making the transfer a two-step process, from tip to block and then from block to coleoptile (Figure 47–3). To do this, he first allowed a severed coleoptile tip to sit on a block of agar for a while; when he placed this block on one side of a freshly decapitated coleoptile, the coleoptile grew and bent away from the side on which the block was placed. Since an electrical impulse would not be expected to remain in the agar block until it was placed on the coleoptile, Went had provided evidence for a chemical messenger. Furthermore, if he placed two coleoptile tips instead of just one on an agar block, the block produced twice as much bending when it was later placed to one side of a headless coleoptile. The chemical responsible for the phototropic response was named auxin.

Bending occurs only when different concentrations of auxin are applied to different areas of the coleoptile sheath. If the concentration of auxin is uniform across the stump, the entire coleoptile grows straight up. Up to a certain point, the greater the concentration of auxin applied, the greater is the elongation of the coleoptile. These results suggest that the phototropic response in the intact seedling is due to a higher concentration of auxin on the dark side of the coleoptile than on the side toward the light; this would be expected to produce the greater elongation that was in fact observed on the dark side. Thus auxins must either be destroyed on the lighted side of the coleoptile, or transported to the shady side in response to the stimulus of light falling on the lighted side. By applying radioactively labeled auxin to plants, investigators have established that auxin is transported from the lighted to the shaded side, where it stimulates increased cell growth—hence the increased bending seen in the phototropic response.

---

* Agar   A gelatin-like substance extracted from certain algae.

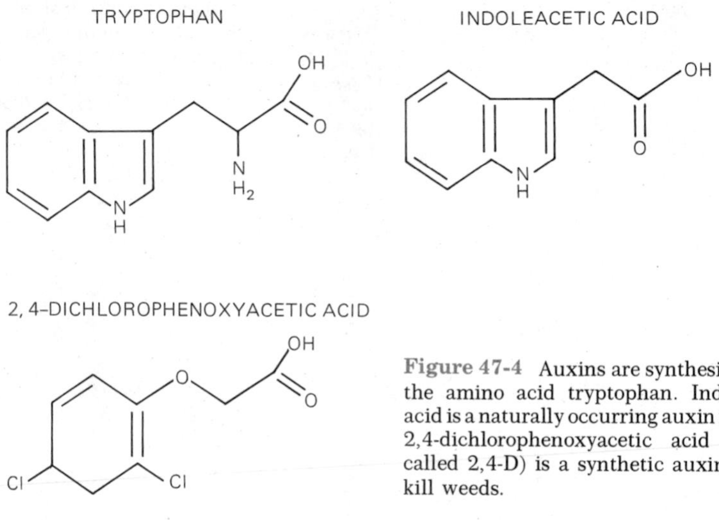

TRYPTOPHAN

INDOLEACETIC ACID

2, 4-DICHLOROPHENOXYACETIC ACID

Figure 47-4 Auxins are synthesized from the amino acid tryptophan. Indoleacetic acid is a naturally occurring auxin in plants. 2,4-dichlorophenoxyacetic acid (usually called 2,4-D) is a synthetic auxin used to kill weeds.

Since these early experiments, the auxin involved in the phototropic response has been isolated and identified as **indoleacetic acid** (Figure 47–4). It is derived from the amino acid tryptophan, and is produced mainly in the tips of coleoptiles and in growing apical meristems of the shoot system; a very small amount is produced in root tips. Elongating cells lose the ability to produce auxin.

Auxins from the shoot are transported primarily toward the roots. Auxins are moved slowly from one cell to another rather than being transported quickly through the vascular tissues.

Auxins have been found to affect many plant activities besides phototropism. Auxins are responsible for **apical dominance**, the inhibition of the growth of lateral buds exerted by the apical meristem. High concentrations of auxins promote production of ethylene, another plant hormone. Auxins induce cell division in the cambium, formation of adventitious and lateral roots, and development of female flower parts and fruits.

Synthetic auxins, such as 2,4-D, are used as weed killers in lawns. Different plants are sensitive to these chemicals in different ranges of concentrations. Dicots are much more sensitive to low concentrations of auxins than are monocots; thus, since most "weeds" in lawns are dicots, and grasses are monocots, application of certain dosages of these synthetic auxins can cause the dicots literally to grow themselves to death. They grow abnormally, becoming grossly deformed before they finally die. Monocots are not visibly affected. To be most effective, weed killers must be applied when plants are most sensitive to them, during active growth.

During the Vietnam conflict, airplanes sprayed "agent orange," a herbicide containing synthetic auxins, on forests and crops. This had a drastic effect on the ecology of the countryside, and a contaminant in the spray was found to cause malformations of developing fetuses. After this contaminant was detected in drinking water and in fish, one of the few sources of protein in the Vietnamese diet, the spraying was abandoned.

Auxins are also used to counteract the effects of hormones that promote the dropping of fruit from trees. Spraying auxins on pear and apple trees is a standard practice to make the trees retain and ripen more of their fruits.

Biochemical studies have produced at least a partial picture of how auxins stimulate growth. Auxins cause the plasma membrane to transport hydrogen ions into the cell wall. This makes the fluid in the cell wall spaces more acidic, which in turn increases the activity of enzymes that loosen the crosslinks between cellulose fibers in the cell wall. As the cell takes up water from the dilute sap of the plant, it expands and pushes the cell wall outward. This expansion starts in the first half hour after auxin is applied. After this period, the auxin also continues to promote growth by speeding the synthesis of proteins needed for growth.

One response to the environment that is mediated by auxins is the division of cells in the vascular cambium in response to mechanical disturbance of a tree. Trees grown in greenhouses or held up by guy wires do not toss in the wind; they are rather spindly compared to trees grown outdoors with no support. However, if trees growing in greenhouses are periodically shaken, their trunks grow thicker than those of control trees growing under the same conditions and not shaken. Increased production of xylem, which strengthens and thickens the tree trunk, is stimulated by auxins. It is theorized that the living cells in the rays of the tree trunk may be able to detect mechanical stimuli and somehow cause more auxin to be distributed to the areas of stress. The auxin would then induce cell division in the cambium, which would increase the amount of xylem present in the area, thus thickening and strengthening that part of the tree.

Apical Dominance

Many types of plants exhibit apical dominance, in which the growing apical meristem represses the growth of lateral buds. If the apical meristem is cut off and a solution of auxin is placed on the cut stump, growth of the lateral buds continues to be inhibited. If the apical meristem is cut off and the stump is treated with plain solvent, however, axillary buds soon grow and form new lateral branches (Figure 47–5).

If auxins inhibit the growth of meristems, how can the apical meristem continue to grow when it is in fact the source of auxin? A clue to the answer can be found by performing a variation on the apical dominance experiment just described. If auxin is not placed on the cut stump of the apical meristem quite soon after the meristem has been removed, the lateral buds break dormancy and begin to grow as the apical meristems of new lateral branches. Once they have started this activity, they cannot be rendered dormant again by application of auxin. Thus the effect of the hormone must depend on the physiological state of the tissue that receives it. If the tissue is active, auxin has no effect; if it is dormant, continued exposure to auxin keeps it dormant.

It is also suspected that another plant hormone, cytokinin, may have some effect in the removal of apical dominance. Virtually all of the cytokinin produced in the roots of a plant is transported to the growing apical meristem(s). However, if this cytokinin "sink" is removed, the amount of cytokinin going to the lateral buds

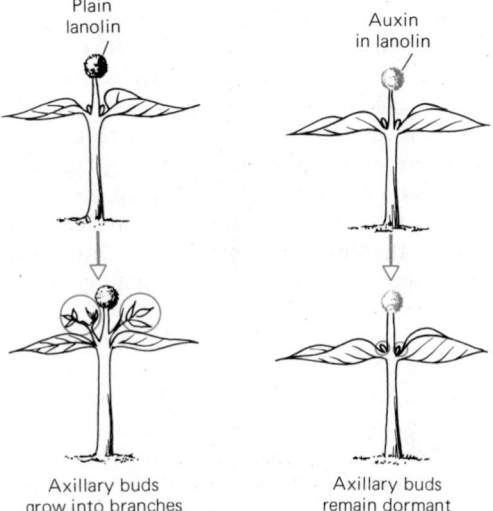

Plain
lanolin

Auxin
in lanolin

Axillary buds
grow into branches

Axillary buds
remain dormant

Figure 47-5 Demonstration that apical dominance is mediated by auxin. (Left) Removal of the apical bud permits growth of axillary buds, but if the apical meristem is replaced with an auxin solution (right), the axillary buds are still inhibited from growing. (Why is plain lanolin applied to the plant on the left?)

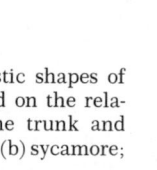

Figure 47-6 The characteristic shapes of trees of various species depend on the relative amounts of growth of the trunk and branches. (a) shagbark hickory; (b) sycamore; (c) American elm.

increases. Thus it may be the higher concentration of cytokinin, perhaps in combination with the lower concentration of auxin, that releases the lateral buds from inhibition.

Along similar lines, it is known that food and water are transported preferentially into areas of the plant where auxin concentration is high. Thus it is possible that the lateral buds fail to grow due to starvation rather than repression; removal of the auxin-producing apical meristem would reduce concentrations of auxin at the tip of the plant and permit more nourishment to be transported to the lateral buds, which could then begin to grow and produce their own auxin, causing them, in turn, to receive more nutrients from the transport system.

It is difficult to design experiments to show whether the inhibition of lateral buds comes from too much auxin, from too little cytokinin, from the wrong combination of auxin and cytokinin, from starvation, or from some combination of these.

By looking at the plants around us, we can easily see that the degree of apical dominance varies a great deal in the plant world (Figure 47–6). Some of this variation is due to genetic differences, which give rise to differences in physiology from one species or variety of plants to another. Environment, too, can affect apical dominance; a beech tree in a forest will be tall and slender, with short, thin branches, whereas another beech tree, growing in the open, will be wide and spreading, and a third growing at the edge of the wood will have large branches on the open side, and small, thin branches on the shady side. These variations in shape suggest that light may somehow counteract apical dominance; this allows the tree to grow new branches into any openings in the forest canopy.

47-C Gibberellins

Another important group of plant hormones is the **gibberellins**. The first one known, discovered by Kurosawa in Japan, was actually a product of a fungus, *Gibberella fujikoroi*, which causes "foolish seedling" disease of rice. Rice seedlings

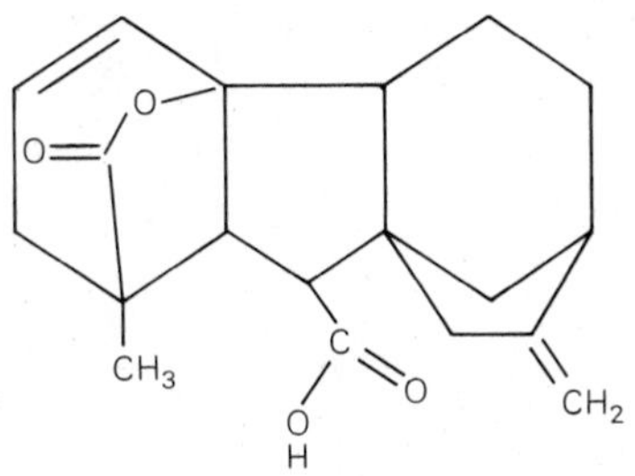

Figure 47-7 Structure of gibberellic acid, a gibberellin.

with the disease grow abnormally fast, but are spindly and unhealthy and seldom yield fruit. The substance responsible for the rapid growth was isolated from cultures of the fungus and named **gibberellic acid** (Figure 47–7). (Commercial production of gibberellins is still carried out by raising the fungus in large vats.) The search for similar substances in plants has revealed that many chemically related compounds, collectively known as gibberellins, are produced not only by fungi but also by vascular plants and by brown and green algae.

One of the important roles of gibberellins in higher plants is to stimulate elongation in stem cells. Dwarf or miniature varieties of plants are often genetic mutants that do not produce gibberellins; they can be stimulated to grow to the same size as normal varieties by application of gibberellins at appropriate concentrations. Application of gibberellins to normal plants, on the other hand, has little effect (Figure 47–8).

Other effects of gibberellins include stimulation of leaf growth, especially in monocots, inhibition of root formation, stimulation of cell division at the stem apex, and stimulation of the development of male flower parts. Gibberellins also seem to stimulate the production of auxins, as well as making plants more responsive to auxin treatment. Gibberellins are also involved in the responses of plants to their environment. Some plants that require cold in order to flower or germinate will respond after gibberellin treatment even if they have not experienced cold. However, flowering occurs in a different sequence depending on whether cold or gibberellin was the initial stimulus. Cold-treated plants form flower buds before the flower stalk elongates, whereas gibberellin-treated plants form elongated stems, which then form flower buds.

Gibberellins are produced in young leaves, mainly in the chloroplasts. From here they are distributed throughout the plant, and they affect virtually every part.

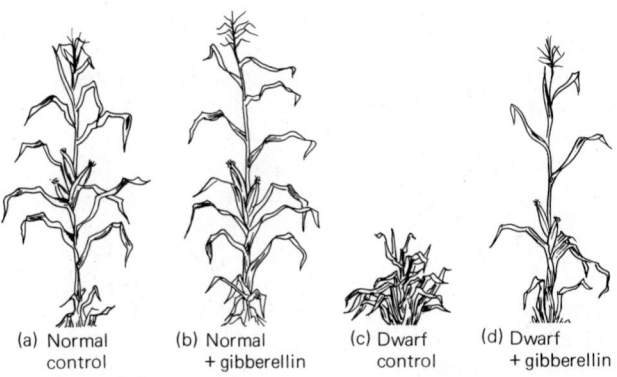

(a) Normal control

(b) Normal + gibberellin

(c) Dwarf control

(d) Dwarf + gibberellin

Figure 47-8 Effect of gibberellic acid on normal and dwarf strains of corn plants.

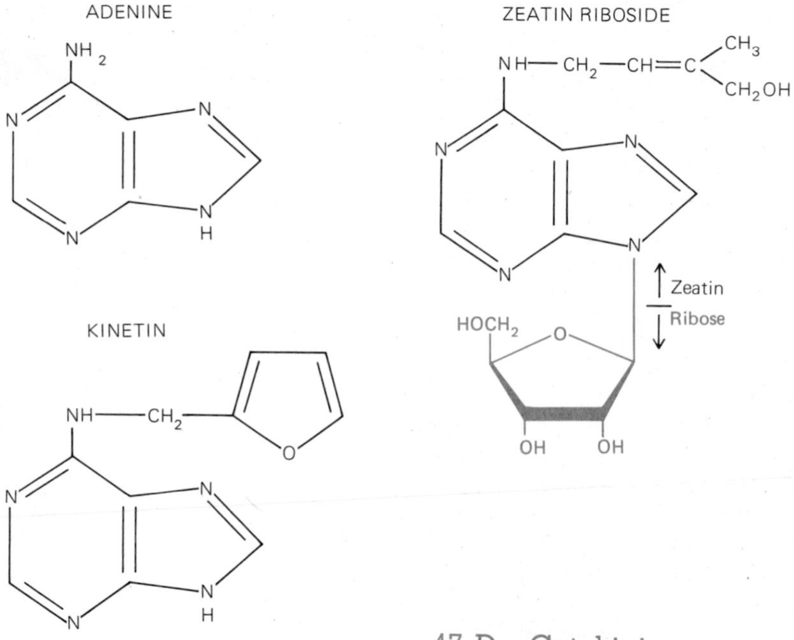

ADENINE

KINETIN

ZEATIN RIBOSIDE

Figure 47-9 Cytokinins are variations on the structure of adenine. Kinetin is a synthetic cytokinin widely used in hormone experiments. Two naturally occurring cytokinins are zeatin (black portion of zeatin riboside), first identified in young corn seeds, and zeatin riboside, isolated from coconut milk after over a quarter-century of painstaking work.

47-D Cytokinins

The first known cytokinin was a component of coconut milk. This compound proved difficult to isolate, but whole coconut milk became a standard additive to cultures of plant tissue in the laboratory because it contained a substance that made plant cells divide when they grew in tissue culture.

Cytokinins were eventually isolated from coconut milk as well as from immature organs of corn and from various other sources. Cytokinins are basically variations on the nitrogenous base adenine (Figure 47–9). They are found as part of certain transfer RNAs, molecules that transport amino acids to ribosomes during protein synthesis; their presence seems to be necessary for the binding of these transfer RNAs to ribosomes. However, it seems clear that this does not account for all the known effects of cytokinins.

Cytokinins are produced in the roots and transported through the xylem to the leaves and the shoot meristems. Apple trees have especially high concentrations of cytokinins in the xylem sap during the spring. The effects of cytokinins on intact plants are consistent with their observed stimulation of cell division in laboratory cultures: cytokinins are involved with the growth of the plant and with the production of fruits and seeds. They also seem to promote transport of food to tissues that contain them, but this may be either a cause or an effect of growth promotion.

Application of cytokinins to lateral buds releases them from apical inhibition, and since the cytokinins come from the roots it has been proposed that the lower buds on a shoot may break dormancy owing to the accumulation of more cytokinins and the reduction in the amount of auxins reaching them from the apical meristem. Cytokinins also prevent the onset of dormancy and slow the process of aging, or **senescence**, in cut leaves or fruits. Indeed, cytokinins are often sprayed on holly that is cut early for the holiday season.

Cytokinins and other hormones may interact in various ways. For example, the relative concentrations of cytokinins and auxins applied to cells in tissue culture determine how the cells differentiate. Without cytokinins, cells grow in size but do not divide; a certain level of cytokinins is necessary for cell division to occur. With high concentrations of both cytokinins and auxins, the tissue continues to grow as a **callus**, or lump of undifferentiated tissue. If the concentration of auxin is higher compared to the cytokinin concentration, the cultured tissue forms roots; if the concentration of cytokinin is higher compared to the auxin concentration, the tissue forms shoots with leaves. Presumably, similar interactions between these hormones in the intact plant govern the differential switching on and off of gene activity that results in the differentiation of the various tissues and organs.

47-E Abscisic Acid

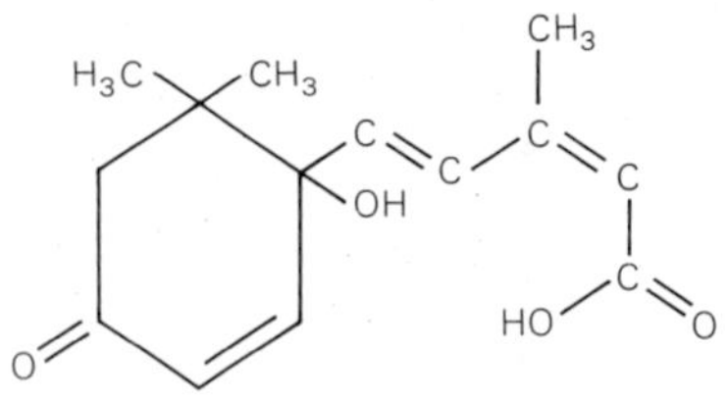

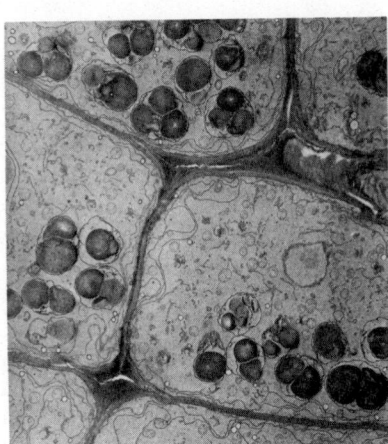

Auxins, gibberellins, and cytokinins have many growth-stimulating roles. In contrast, abscisic acid is chiefly a growth inhibitor.

Abscisic acid was discovered almost simultaneously by Frederick Addicott, working with cotton bolls,* and by P. F. Wareing, working with sycamore leaves. In both cases, the substance isolated was found to accumulate in tissue as it aged and to counteract the effects of auxin when it was applied to auxin-treated coleoptiles. This inhibitory hormone was named **abscisic acid (ABA)** (Figure 47–10).

Figure 47-10 Molecular structure of abscisic acid.

Abscisic acid is produced in mature leaves and transported to the rest of the plant. When its concentration is high compared to the levels of gibberellins and cytokinins, at the end of the growing season, ABA induces the apical meristem to stop dividing, and the newest leaf primordia form into protective bud scales around the tip instead of becoming normal photosynthetic leaves. Application of abscisic acid to germinating seeds may inhibit them from further growth and make them dormant once again. ABA maintains the dormant condition of twigs and seeds until it is leached away by water or overcome by production or application of a stimulatory hormone, usually gibberellin.

An important role of abscisic acid during the growing season is the closing of stomata, by causing potassium to leave the guard cells, during periods of water shortage (Section 44-C).

There is growing evidence that abscisic acid also mediates the geotropic response in roots. This role was formerly attributed to auxins, which are still thought to play a minor role in root geotropism (auxins do in fact mediate the negative geotropism—upward growth—of shoots).

Roots are positively geotropic, growing toward the center of the earth. Abscisic acid is synthesized in cells in the core of the root cap and transported toward the shoot system. If the root is growing straight down, abscisic acid is distributed evenly to all sides of the root, and the root continues to grow straight down. However, if the root is horizontal, the hormone is distributed so that its concentration is higher on the lower side of the root. Since abscisic acid inhibits growth, the cells on the upper side of the root grow more than the cells on the lower side, and the root bends in the region of elongation until it is growing straight down.

A root's response to gravity depends on starch-storing plastids, the **amyloplasts**, of root cap cells. Because of their high density, these amyloplasts tend to accumulate at the bottoms of the cells (Figure 47–11); they are called **statoliths** by analogy to the particles that respond to the pull of gravity in the sense organs of many animals (see Figure 39–8). It is not clear how the statoliths' response to gravity results in a differential distribution of abscisic acid in the root.

---

* Cotton bolls Pods containing seeds and cotton fibers, the commercially valuable parts of the cotton plant.

Figure 47-11 Electron micrograph of cells in the root cap of the primary root of a corn seedling. The dark-stained amyloplasts have fallen towards the bottoms of the cells. (Barrie Juniper)

47-F Ethylene

The final known plant hormone is **ethylene.** Unlike the plant hormones discussed above, ethylene is a gas at normal temperatures and pressures. Thus, whereas other hormones travel inside the plant to their target tissues, ethylene travels not only within the plant but also through the air, from its site of production to its target tissue, which may be part of the same plant or of another plant. For example, ethylene emitted by apples inhibits the sprouting of potatoes stored in the same bin.

The existence of a gas that influences plant physiology has been known for a long time. Nineteenth-century growers of greenhouse plants knew that some component of the gas used in gas lighting was responsible for the premature withering of blossoms, and it was a common practice of mango and pineapple growers to light fires near the trees because something in the smoke would synchronize the flowering and ripening of the crop. Once the component responsible for these effects had been identified as ethylene (Figure 47–12), it was possible to study the effects of ethylene by applying this gas to plants or parts of plants in airtight enclosures and observing its effect on growth.

The production and effect of ethylene seem to be most closely tied to auxin. After auxin concentration has exceeded a certain level, it stimulates production and release of ethylene; ethylene in turn counteracts the effect of auxins. Thus production of ethylene by lateral buds as they receive auxins seems to play some part in the inhibition of their growth in apical dominance. Likewise, in the roots of some plants, very low auxin concentrations are sufficient to promote ethylene production, which in turn slows root growth, whereas stem tissue does not produce ethylene until the auxin concentration reaches much higher levels, and so stems are stimulated at auxin concentrations that would inhibit roots.

Other interesting effects of ethylene are its roles in the senescence of plants and in the ripening of fruits. The development of a fruit is stimulated by auxins, gibberellins, and cytokinins. The level of auxins builds up and then drops; this is followed by production of ethylene. Ethylene stimulates the activity of some enzymes, including enzymes that convert starch and acids of the unripe fruit to sugars, and pectinase, an enzyme that softens the fruit by breaking down pectins in the cell walls.

The release of ethylene has a positive feedback effect: the more ethylene that is produced, the more the fruit is stimulated to produce additional ethylene. Hence the entire fruit ripens at once, and if there are many fruits in an area, the first to begin ripening stimulates ripening of its neighbors. By the same token, overripening is also contagious, and it is quite true that "one bad apple spoils all the good ones." Apples are now stored in refrigerated, airtight rooms in an atmosphere enriched with carbon dioxide, which somehow inhibits the action of ethylene. Ethylene finds uses, however, in the production of tropical fruits for far-away markets. In this case, rather than letting the fruit ripen on the trees and risk loss by overripening in transit, growers pick fruits such as bananas, pineapples, and citrus fruits when they are still green and ripen them by applying ethylene after they reach their destination.

Senescence

Flowering, setting of seeds and fruit, and germination of seeds are events in a plant's life history that occur predictably in response to environmental or genetic factors. **Senescence**, the preparation of all or part of the plant for death, is also an integral part of its life history. A wheat plant turns yellow, dries up, and dies after it has set seed; a plum tree drops its fruits during a short period in early summer and loses all its leaves during the fall.

Senescence is not degeneration of a plant due to starvation or cold; it is under hormonal control, probably by ethylene. In annual plants, the hormonal changes that initiate senescence of the whole plant often seem to be triggered by the setting of seed.

In perennial plants, various parts such as leaves, fruits, and withered flower

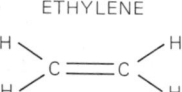

ETHYLENE

Figure 47-12 Molecular structure of ethylene, a gas that acts as a plant hormone.

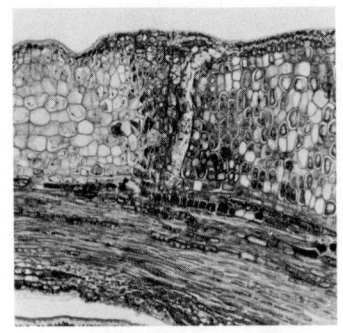

Figure 47-13 An abscission layer has formed in a line running down from the top center of this picture. (Carolina Biological Supply Company)

parts senesce and drop off each year. This process is known as **abscission**. Although its name suggests a link with abscisic acid, abscission seems to be controlled primarily by ethylene. Auxins may also speed abscission if applied after the process is under way, although they are inhibitory if applied before abscission begins.

The abscission of leaves has been studied most extensively. Leaves may senesce and drop as a regular part of the life of a plant; in *Coleus*, for instance, the oldest leaves are shed as new ones appear at the top of the plant. Abscission may also follow injury to a leaf, and wholesale abscission is a response of many plants to the onset of winter.

Before abscission, several changes take place. Nutrients may be withdrawn from the leaf; enzymes that degrade the leaf tissue become active, and a series of cell divisions may take place in the **abscission zone**, where the leaf joins the stem. In this region enzymes break down the cell walls, creating a zone of weakness (Figure 47–13). Finally, the weight of the leaf, perhaps assisted by the wind, causes this zone to break apart, and the leaf drifts away. The leaf scar is quickly healed by the deposition of a corky, waterproof seal that prevents the loss of water or entry of pathogens (disease-causing organisms).

47-G Photoperiodic Control of Flowering

To many people, the parade of flowers blooming one after the other symbolizes the passing of the seasons. What would summer be without the fragrance of roses, or autumn without the rich colors of goldenrod and purple asters?

We now know that the predictable flowering of many plants is due, at least in part, to the changing length of the days. As the year passes, the days become longer in the spring and shorter again in the fall. Many plants exhibit **photoperiodism—** the response to changes in daylength by making appropriate physiological changes, such as flowering. Although this might seem obvious, the influence of daylength on the physiology of plants was discovered a surprisingly short time ago.

The first study that showed daylength to be an important cue in the flowering of plants was done in 1920 by W. W. Garner and H. A. Allard. They were investigating two phenomena. One was the behavior of a mutant variety of tobacco known as "Maryland mammoth." These plants reached heights over 3 metres in the field, but did not flower. However, if cuttings were rooted and grown during the winter in greenhouses, the plants would flower even when they were much smaller. The second peculiarity was the flowering of "Biloxi" soybeans. Biloxi soybeans planted at different times before June 1 all flowered and produced beans in early September, despite their different ages.

In both cases, it seemed that some cue from the environment must determine when the plant flowered. Garner and Allard experimented with the effects of different temperatures and light intensities on growth and flowering, but found they had no effect. They were reluctantly forced to consider another hypothesis, that plants could respond to varying daylengths, which seemed silly because it meant that plants would have to be able to tell time. Garner and Allard artificially changed the daylength by lighting the greenhouse at night to give plants longer days, or by

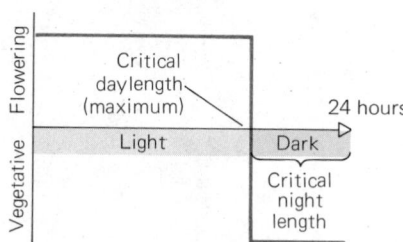

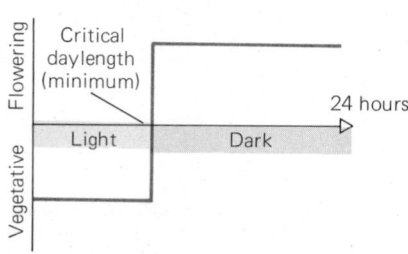

Figure 47-14 Comparison of requirements for flowering in short-day and long-day plants. Short-day plants will flower when periods of light (daylengths) shorter than the critical daylength are administered during 24-hour days; long-day plants flower only when the daylength is longer than the critical daylength. The value of the critical daylength varies according to species, and the terms short-day and long-day do not imply that the dark period must be longer or shorter than the period of light.

placing the plants in dark cupboards to shorten the days. They found that the plants would flower only if they were exposed to light for less than a certain amount of time, the **critical photoperiod**, each day. Because a maximum photoperiod could not be exceeded if these plants were to flower, they came to be called **short-day plants**. Later experiments showed that some types of plants required a certain minimum length of photoperiod before they would flower; these came to be known as **long-day plants**. Examples are spinach, radish, and barley. Note that the definition of long-day and short-day plants does not imply that all long-day plants require periods of daylight that are longer than those required by any short-day plant; the definitions simply refer to whether there is a maximum or minimum daylength in the range of proper photoperiods (Figure 47–14). There are also many **day-neutral** plants, such as tomatoes and cucumbers, which begin to flower at a certain stage of growth and continue to flower and bear fruit until they freeze to death.

In 1938, K. C. Hamner and J. Bonner found that the cocklebur, a short-day plant, would flower whenever it received more than nine hours of darkness, regardless of the length of the light period. Furthermore, exposing short-day plants to a brief flash of light during the dark period suppressed flowering, even though the critical photoperiod had not been exceeded. Thus, it is not the length of the day, but rather the length of the night, or dark period, that governs flowering. Short-day plants should really be called "long-night" plants. However, by the time this was discovered, the term "short-day" was firmly established. In nature, of course, long nights automatically go with short days; if there is light for eight hours, there will necessarily be darkness for the other 16.

In many plants, factors other than daylength influence the flowering response; for example, many plants require a particular light/dark regime but produce more flowers, or flower for a longer time, when they are also given the proper temperature, or day/night temperature changes, or moisture, or a sequence of short days after long days or vice versa. Usually the light and dark periods must also be correct for several days running; seldom does a plant respond to a single proper light/dark sequence.

The florist industry has benefited from studies of the light/dark regimes needed to induce flowering in various species of plants. We can now obtain chrysanthemum plants in full bloom for Mother's Day and present carnations to our Valentines.

Phytochrome

The ability of plants to respond to varying photoperiods indicates that they must be able to measure time—but how? Because the response of any organism to light is mediated by pigment molecules, which change when they absorb light, it is reasonable to expect that pigments play a part in plants' photoperiodic responses.

In order to find out what pigment activates the photoperiodic responses of plants, light of different wavelengths was shined on different plants, to discover

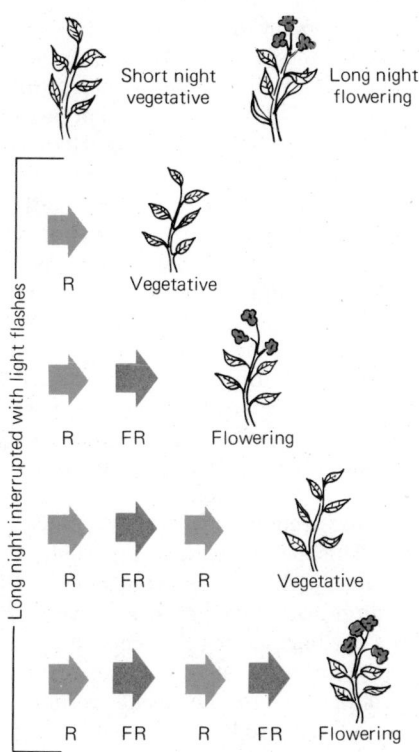

Figure 47-15 Response of a short-day plant when its critical dark period is interrupted with flashes of light. A flash of red light (R) will suppress the flowering response normally initiated by a regime of dark periods of critical length; following a flash of red light with a flash of far-red light (FR) reverses the effects of the red flash, and the plant enters the flowering phase. When a train of flashes is given, the plant always responds to the last flash received.

which wavelengths promoted the change. Pigments that absorb the effective wavelengths could then be isolated from the plant.

When different wavelengths of light were used to interrupt the dark period, it was found that red light of about 660 nm was most effective in inhibiting the flowering of short-day plants. There was also a rather curious effect when far-red light (730 nm) was used for the flash; this wavelength had no effect, and the plants acted as though no light were given in the dark period. Furthermore, when a flash of far-red light was given immediately after a flash of red light, the plants again acted as though they had received no light during the dark period! If successive flashes of red and far-red light were given to the plants during their dark period, the plants always responded to the last wavelength received; when red was the last wavelength given, the plants did not flower, but when far-red was given last, the plants flowered, regardless of how many doses of red light they had received (Figure 47–15). On the other hand, long-day plants held on short-day photoperiods could be made to flower if they were given a flash of red light during the dark period. This could be reversed by far-red light, and re-reversed by red light as described above.

To explain this, researchers hypothesized that the pigment involved, which they named **phytochrome**, has two different forms. When it is illuminated with red light, phytochrome switches to a form called P_{fr}, or phytochrome that can absorb far-red light. If it is then illuminated with far-red light, P_{fr} switches back to its original form, $\mathbf{P_r}$, or phytochrome that can absorb red light (Figure 47–16).

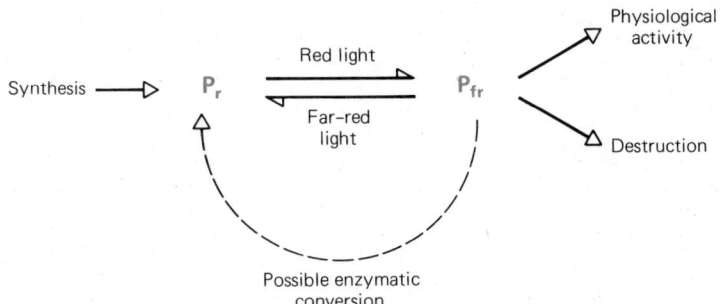

Figure 47-16 The interconversion of phytochrome between the active (P_{fr}) and inactive (P_r) forms.

Phytochrome is present in very small quantities, but researchers finally succeeded in isolating it. The molecule consists of a nonpigmented protein and a much smaller light-absorbing pigment unit. It now appears that phytochrome is synthesized in the P_r form, which can be converted to P_{fr}, the physiologically active form, by exposure to light (red light is most effective).

During the daytime, plants receive both red and far-red light from the sun, and phytochrome molecules cycle rapidly back and forth between the two forms. While in the P_{fr} form, phytochrome is slowly destroyed. When the sun goes down, about half of the plant's phytochrome is left in the P_{fr} form, which is slowly degraded during the night. In short-day plants, there appears to be a need for relatively high levels of P_{fr} during the early part of the night, because exposing a plant to far-red light (converting its phytochrome to P_r) at the beginning of the dark period inhibits flowering. As the night progresses, P_{fr} is destroyed, and the plant seems to need to remain in the dark for a certain period after P_{fr} has been reduced to a low level in order to initiate flowering; giving red light late in the dark period (converting P_r to P_{fr}) also inhibits flowering. In summary, flowering in short-day plants seems to rely on the proper sequence of two processes, the first occurring when P_{fr} levels are high and the second later, when P_{fr} levels are low.

This flower-initiating sequence probably involves the interaction of phytochrome with the "internal clock." Plants, like other eukaryotic organisms, have **endogenous circadian rhythms** (circa = around; dies = a day), which change in a regular pattern that repeats roughly every 24 hours when the plant is in constant light or constant darkness. In nature, the rhythm repeats almost exactly every 24 hours because the clock is reset daily by external cues such as sunrise, sunset, or both. The periodicity of these changes is practically independent of temperature, which normally has a profound effect on the rates of biochemical reactions. We do not yet know exactly what the elusive clock is.

Phytochrome's role seems to be to reset the clock. Phytochrome is associated with the plasma membrane and with some membranes inside the cell. When a plant is exposed to alternating flashes of red and far-red light, the electrical potential across the plasma membranes changes; this suggests that the change from P_r to P_{fr} somehow alters membrane permeability to electrically charged particles. All of this could reset the clock by initiating some phase of chemical activity; the response to photoperiod would then depend on the interaction between the internal rhythms, which seem to have phases favored by light and phases favored by darkness, and the state of phytochrome. This outline is very tentative, and there are still many unanswered questions.

The outcome of all this seems to be changes in the hormonal state of the plant. It can be shown that flowering is induced by hormones produced in one part of a plant and transported to other parts; in some kinds of plants, if only a single branch, or even a single leaf, of a plant is exposed to the appropriate light regime, the plant responds by flowering even though the part that flowers never received the proper light treatment (Figure 47–17). Similarly, grafting* a plant that received the proper

---

*Grafting   Cutting two plants and placing them so that their vascular tissues are in contact. Transport occurs between the two, and the wounded area heals over, making the two into a single physiological unit (Section 46-I).

Figure 47-17 Flowering of short-day plant exposed to long-day treatment. A single leaf of the plant is darkened to give it the short-day treatment normally required for flowering; hormones produced in this leaf are transported to the rest of the plant and induce flowering.

treatment to another plant causes the second plant to flower even though it has not received the proper light treatment. Some, but not all, long-day plants flower without the long-day treatment if appropriate concentrations of gibberellin are applied. The question of what internal chemical changes initiate flowering is still open, and many researchers feel that the answer in most cases may be a yet-undiscovered flowering hormone, which has already been named "florigen"!

SUMMARY

Five groups of plant hormones are known. Auxins, gibberellins, and cytokinins are generally growth-promoting; abscisic acid is often growth-inhibiting, and ethylene promotes ripening of fruits and senescence of leaves. Probably no single effect can be attributed to any one of these hormones alone; rather, the interactions of these hormones govern a plant's growth, so that different parts, such as the leaves and roots, remain in anatomical and physiological balance with each other. Hormones are also involved in a plant's response to its environment. They enable the plant to respond appropriately to the direction of light, gravity, or prevailing winds, and to changes in daylength that signal changes in the seasons.

How a plant responds to a particular hormone depends on the tissue that receives the hormone, the concentration of the hormone, the presence and concentration of other hormones, the age and physiological state of the tissue, and environmental factors such as temperature, light, or light/dark cycles. Since plant hormones can change a cell's synthesis of RNA or proteins, a plant's response to a particular hormone is greatly influenced by its genetic makeup, and different species, varieties, or even individual plants will respond to the same treatment in different ways. This has given rise to a large body of conflicting experimental information that defies neat generalizations. However, we are learning that these five groups of plant hormones play a rich repertoire of roles in the diverse genetic and environmental settings of plants.

OBJECTIVES

From your study of this chapter, you should be able to:

1. Name or recognize the five plant hormone groups (auxins, gibberellins, cytokinins, abscisic acid, and ethylene); list parts of the plant where each is synthesized; and discuss the general biochemical roles of plant hormones.

2. Explain the role of auxins in phototropism and apical dominance, and list or recognize at least two other roles of auxins.

3. List or recognize two main effects each for gibberellins, cytokinins, abscisic acid, and ethylene.

4. List five or more factors that may influence how a plant responds to application of a particular hormone.

5. Explain what is meant by the terms short-day, long-day, and day-neutral plants; give the evidence that photoperiodic flowering responses are controlled by changes in phytochrome.

SELF-QUIZ

1. In phototropism, auxins:
   a. promote growth of cells
   b. stimulate differential growth of cells in different sides of the plant
   c. inhibit growth of cells
   d. inhibit cell division
   e. absorb stimuli and signal the direction of light or gravity to the plant

Match the hormones listed below to their effects.

___ 2. Promotes ripening of fruits

___ 3. Initiates cell division in tissue culture

___ 4. High concentrations stimulate ethylene production

___ 5. Substitutes for long days in the flowering of some long-day plants

___ 6. Responsible for geotropic response in roots

___ 7. Counteracts the effects of auxin

___ 8. Promotes onset of dormancy

a. abscisic acid
b. auxin
c. cytokinin
d. ethylene
e. gibberellin

9. In a short-day plant growing in a home garden, phytochrome is normally switched from one form to the other by:
   a. red and far-red light
   b. sunlight
   c. activation by gibberellin or abscisic acid
   d. different electrical potentials in the plant's plasma membranes
   e. measuring the length of the dark period between light periods

10. A long-day plant is one that:
   a. requires more than 12 hours of light in order to flower
   b. increases in height when it flowers
   c. needs a certain minimum length of photoperiod in order to flower
   d. is not affected by temperature in its flowering response
   e. will not flower if its dark period is interrupted by a flash of light

QUESTIONS FOR DISCUSSION

1. Some trees, such as the weeping willow, may grow roots sideways a dozen metres or more in response to abundant water, such as water in a sewer line or septic system. This response is called hydrotropism. Propose a mechanism by which this tropism might work.

2. The fungus that causes the "foolish seedling" disease in rice appears not to need the gibberellin it produces for its own growth. Why might it be selectively advantageous for the fungus to be secreting this substance?

3. Why are apples, oranges, and grapefruit sold in plastic bags with holes in them rather than in unperforated bags? Why does it often turn out that produce packaged in market trays with clear wrap is too soft on the underside, which you could not see when you picked it out in the store?

4. The leaves of a forest canopy absorb more of the red light than of the far-red as sunlight passes through them. Many plants of the forest floor produce seeds that germinate only in the early spring, before the canopy leafs out. Propose an explanation for the timing of this germination, and explain its adaptive advantage.

5. In some species of trees, individuals growing near streetlights become dormant later in the fall than do other individuals. How could you account for this?

6. Exposure to a flash of light during the dark period can change the plant's subsequent flowering response (or lack of it); however, interrupting the light period with an interval of dark has no effect. How can you explain this?

REFERENCES AND FURTHER READING

Bidwell, R. G. S. *Plant Physiology*, 2d ed. New York: Macmillan Publishing Co., Inc., 1979. Especially interesting for its thoughtful discussions.

Galston, A. W., P. J. Davies, and R. L. Satter. *The Life of the Green Plant*, 3d ed. Englewood Cliffs, NJ: Prentice-Hall, 1980. A clear, well-illustrated plant physiology text.

Juniper, B. E. "Geotropism." *Annual Review of Plant Physiology* 27:385–406, 1976. A review of the available evidence on geotropism, especially in roots.

Overbeek, J. van. "The control of plant growth." *Scientific American*, July 1968. A brief history of the discovery of the plant hormones (except ethylene).

Wareing, P. F., and I. D. J. Phillips. *Control of Growth and Differentiation in Plants*, 2d ed. Elmsford, NY: Pergamon Press, 1978.

CHAPTER 48

THE BIOSPHERE

In the next four chapters we shall study ecology, the branch of biology that examines the interactions of organisms with their environments. Ecologists study the patterns of distribution and abundance of organisms in nature, how these patterns are maintained in the short run, and how they change during the course of evolution. The science of ecology has gradually emerged from the descriptive field of natural history since the middle of the nineteenth century. The word "ecology" was coined by the German biologist Ernst Haeckel in 1869. It is based on the Greek word *oikos*, meaning "house" or, more loosely, "habitat." The study of ecology has gained impetus recently with the realization that human activities have a profound effect upon the living world and that this in turn affects us by altering our own environment.

Life on earth requires, among other things, water, a source of energy (usually light from the sun), and various nutrients, including CO_2 and minerals that have been released from rocks by weathering. Suitable combinations of these ingredients are found only in a narrow layer covering the surface of the earth. The **biosphere**, as this layer is called, extends about 8 kilometres up into the atmosphere, where insects and the spores of microorganisms and plants may be found, and as much as 8 kilometres down into the depths of the sea. Living organisms are not distributed uniformly through the biosphere; few organisms live in the polar regions, while many live in tropical rain forests.

In this chapter, we consider the question, "Why are organisms where they are?" This question arose out of two general patterns observed by Western naturalists as they explored the world and catalogued its life. First, each newly discovered area contained previously unknown species of organisms, and the list of known species has grown steadily. Second, in spite of the ever-increasing numbers of known species, there are only a few basic types of **communities**, that is, groups of organisms living in the same place. Walking through a tropical forest in South America, we would find tall trees with large leaves and fruits, festooned with immense climbing vines, and we would see colorful butterflies and birds flitting through the gloomy shade. A tropical forest in Africa would look very much the same, but the particular species of trees and vines, and of butterflies and birds, would be different. Other types of communities—desert, shrubland, grassland, or tundra—look much the same wherever they occur. Why are such similar communities of organisms found in different parts of the world?

48-A Climate and Vegetation

If we look at a map of the world showing the kinds of communities in different places (see Figure 48–5), we find that communities of the same type have similar climates. Climate is the main factor determining the type of plants that can grow in the area; soil type also has an influence. The kinds of plants present, in turn, determine the appearance of the community.

Climate depends basically on the sun. Near the equator, the sun's rays strike almost vertically, and this gives tropical plants much more of the sun's energy than is enjoyed by plants outside the tropics, which receive oblique rays (Figure 48–1). Because of the tilt of the earth on its axis, in nontropical areas the seasons vary at different times of the year, while in the tropics there is little seasonal difference in daylength and temperature.

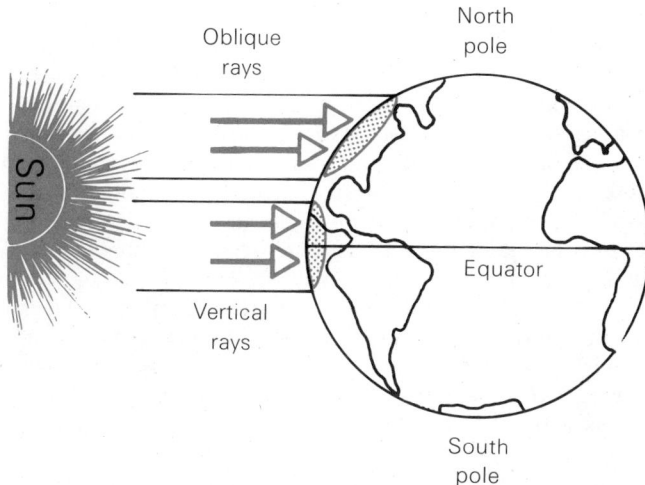

Figure 48-1 A beam of solar radiation striking the earth's surface at high latitudes is spread over a wider area, and is therefore less intense at any one point, than a similar beam striking near the equator.

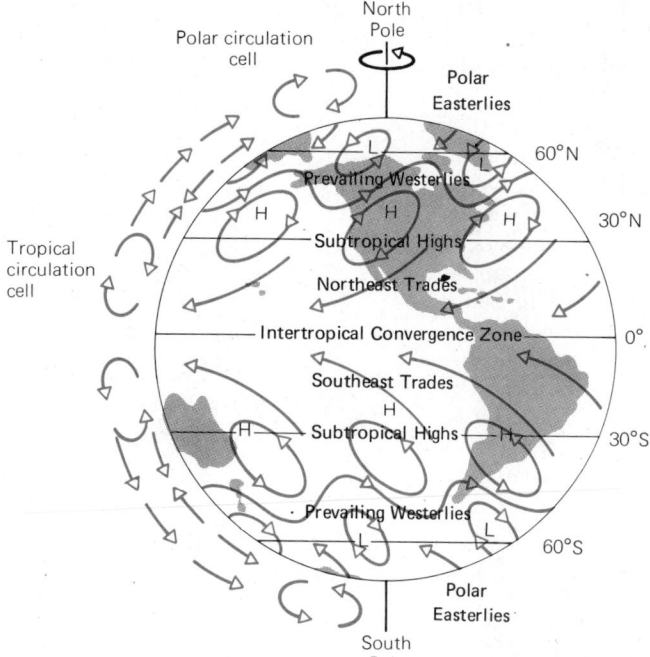

North
Pole

Polar circulation
cell

Polar
Easterlies

60°N

Prevailing Westerlies

30°N

Subtropical Highs

Tropical
circulation
cell

Northeast Trades

Intertropical Convergence Zone

0°

Southeast Trades

Subtropical Highs

30°S

Prevailing Westerlies

60°S

Polar
Easterlies

South
Pole

Figure 48-2 General circulation of air within the lower atmosphere. Heated air rises at the equator and flows poleward at high altitude, causing low pressure and frequent calms in equatorial regions (the "doldrums"). Some of the cooled poleward-moving air falls at about latitudes 30°N and 30°S, heating up as it contracts and forming regions of high pressure. One of these "subtropical highs," the Azores High, is usually situated over the Atlantic off the coast of northwest Africa; another is generally present over the eastern Pacific, off the coast of California. Some of the descending air returns towards the equator as the northeast or southeast trades, completing the tropical circulation cells. A second pair of circulation cells (the polar cells) occurs at high latitudes, giving rise to the "polar easterlies." Circulation in mid-latitudes is more complex, and the prevailing wind direction is westerly. At the surface, poleward movement of tropical air and movement of polar air towards the tropics occurs around cyclones (low pressure areas, often associated with storms) and anticyclones (spinning masses of high pressure air, usually broken off from the polar air mass). At high altitudes the winds are generally westerly, reaching 320 kilometres per hour (200 mph) or more in the undulating polar jet stream (at 9000 to 12,000 m), which influences the location and movement of the polar front beneath it.

The sun is responsible not only for the amount of light available for photosynthesis, but also for the general temperature. Tropical climates, receiving near-vertical sunlight throughout the year, have fairly steady, high temperatures. In other areas, the temperature varies roughly with the amount and intensity of sunlight at different seasons.

The other important component of climate is moisture, and this depends on sunlight and temperature. Warm air holds more moisture than cool air, and as air cools some of its moisture may condense as rain, snow, or dew. Air heated at the equator rises, expanding and cooling as it mounts higher into the atmosphere, and releasing some of its moisture as it does so; the result is the steamy rains of tropical jungles. The air moves on, at high altitudes, both north and south from the equator, and eventually sinks to earth again, becoming warmer and soaking up more moisture as it does so. The descent of this dry air creates the world's great deserts (Figure 48–2). Still further north and south, in the temperate latitudes that include most of the United States and Europe, swirling winds pull masses of air, sometimes from warm tropical areas, sometimes from frigid polar regions, giving us varied weather patterns and keeping us faithful to the evening news to see what the weather map has in store for the morrow (maybe).

In 1889 C. Hart Merriam, a young naturalist surveying the biology of part of Arizona, noticed that San Francisco Mountain showed a variation in vegetation from base to peak similar to that seen when traveling further and further north (or south) from the equator (Figure 48–3). Since temperature varies with altitude as well as with latitude, Merriam concluded that the type of vegetation in an area is determined by its temperature.

We now know that this conclusion was oversimplified, for moisture plays a role just as important as temperature. Heavy rainfall is needed to support the growth of large trees, while progressively lighter rainfall supports communities dominated by small trees, shrubs, grasses, and finally scattered cactuses or other desert plants; in extreme cases, lack of rainfall results in total lack of plants.

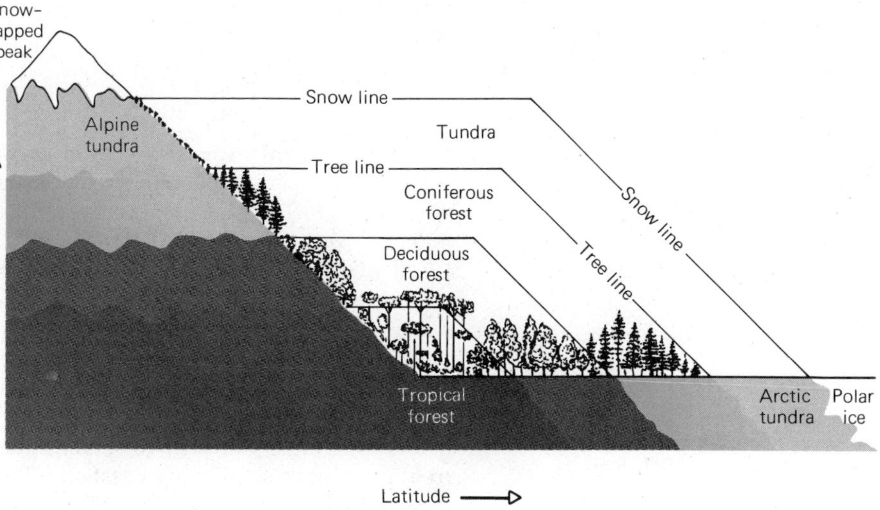

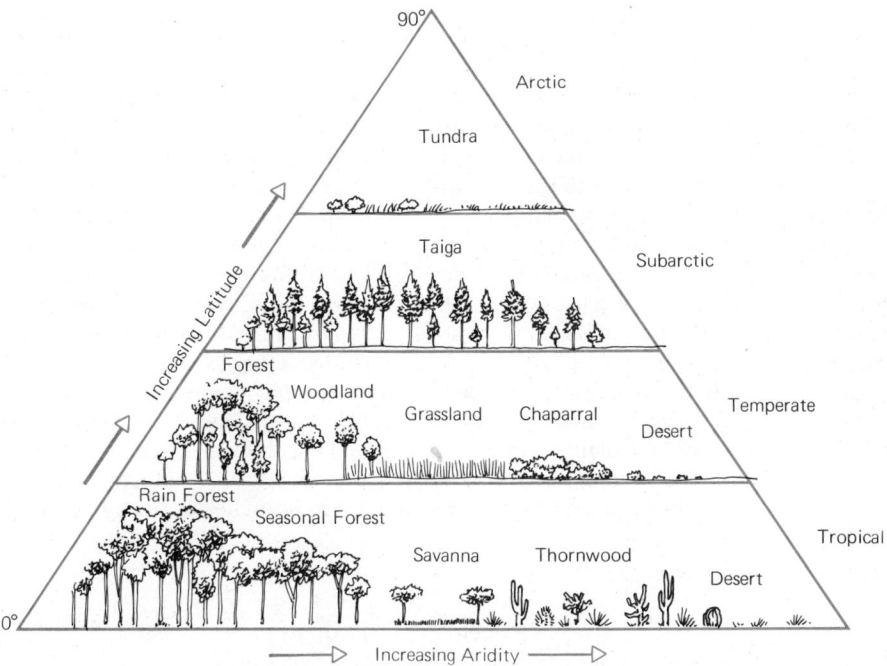

Figure 48-3 Belts of vegetation at successively higher altitudes often correspond to those at successively higher latitudes because similar temperatures favor similar types of plants. Vegetation type is also influenced by moisture. The example shown here is of a transect (cross section) through communities with abundant precipitation. The horizontal axis is greatly compressed relative to the vertical axis.

Biomes

Biomes are the earth's major kinds of terrestrial communities; each is recognizable by the characteristic structure of its dominant vegetation. Ecologists differ as to the number of biomes that should be recognized. One scheme may distinguish between two or more biomes that another scheme lumps together as a single biome. This confusion reflects the fact that biomes seldom have sharp boundaries. Instead they gradually merge into one another, forming gradients, or **ecoclines**, of changing community type in relation to gradients of changing climate and soil type.

Here we lump the world's terrestrial biota into 12 major biomes (see Figure 48–5). Starting at the equator, we shall discuss them in order of the increasing latitudes (i.e., decreasing temperatures) at which they occur at low altitudes. Within the tropical and temperate categories, the biomes can be arranged along ecoclines of increasing aridity (Figure 48–4). The arctic and subarctic biomes are more uniform, reflecting the cold, dry winter winds and the short growing season, which limit plant growth in these communities.

Figure 48-4 Simplified scheme of the major terrestrial biomes, arranged along ecoclines of increasing aridity at different latitudes, illustrating the predominant influence of moisture and temperature on the structure of plant communities.

Tropical Rain Forest

Tropical rain forest occurs in areas where high, fairly constant rainfall and temperatures permit plants to grow throughout the year. In such areas, a month with less than 10 cm of rain is considered dry, and annual precipitation may exceed 400 cm.

The soils in most rain forests are highly leached (depleted of minerals because the rain water dissolves the minerals and carries them away). These soils are often waterlogged, and the surface layer of organic litter is very thin. Temperature and moisture are ideal for decomposer organisms, so that organic matter falling to the forest floor is quickly decomposed. The minerals released are rapidly taken up again by the plants, and so almost the entire nutrient pool of the forest is locked within the bodies of living organisms. To obtain minerals from the shallow, soggy soil surface, trees must have shallow root systems. The tall, shallow-rooted trees are often buttressed for more support. Mycorrhizal associations* are common; they permit trees to obtain some of their nutrients in organic form directly from the litter, avoiding the nutrient loss that occurs when organic matter is decomposed to a form that can leach out of the soil. So efficient are these forests at retaining nutrients that water running off the forest floor into a stream is frequently as free of inorganic solutes as distilled water.

If a tropical rain forest is cut down and burned, the heavy rain causes intensive leaching of the suddenly mineralized* nutrients. Except in some young, nutrient-rich volcanic soils (e.g., in Indonesia and Central America), the cleared land rapidly loses its fertility, and sustained agriculture is impossible.

Natural tropical rain forest is the richest of all biomes, in that it has the greatest diversity of species for a given area. The number of tree species, for example, may exceed 100 per hectare.* The dominant plants are tall trees with slender trunks that branch only near the top, covering the forest with a dense **canopy** of leathery evergreen leaves that shed water rapidly. The general canopy height is usually between 35 and 50 m, but it may be punctuated at intervals by "giant" trees, reaching up to 60 m. The dense canopy permits as little as 0.1% of the incident sunlight to reach the forest floor. Since few plants grow in this permanent twilight, the lower levels of a tropical rain forest are fairly open and easy to walk through. Thick "jungle" occurs only in open areas, along river banks or in clearings created by fallen trees, where sunlight can reach the ground (Figure 48–6a). There is no definite flowering season: some trees are in flower and some in fruit at most times of the year.

The tall trees provide surfaces where many other plants grow. **Lianas** are large woody climbers, using trees to support their rapidly growing, flexible shoots. Once established in the canopy they can be long-lived, for if the supporting tree dies, lianas may remain fastened to the crowns of other trees, with their stems hanging down like ropes. **Epiphytes**, plants growing with their roots anchored to the surfaces of other plants, include a great variety of orchids, bromeliads, cactuses, and ferns. These plants have a variety of adaptations for storing water, permitting them to survive occasional dry periods. Bromeliads, for example, have water-absorbing scales that take up water collected in funnels formed by the leaf bases. Ferns, which cannot tolerate dehydration, produce their own soil by acting as traps for dust and organic litter. The organic matter trapped by epiphytes may amount to several tons per hectare; in addition to conserving moisture for the roots of epiphytes, it provides a habitat for many other plants and for some animals. Epiphytes disperse by wind-borne spores (ferns), by dust-like seeds (orchids), or by berries (cactuses, bromeliads) that are eaten by birds and later deposited on tree branches in the birds' droppings.

---

* Mycorrhizal associations Associations between plant roots and soil fungi, in which the plant receives nutrients absorbed from the soil by the fungi.
* Mineralized Released from organic compounds as soluble inorganic ions.
* Hectare 10,000 square metres, about 2.5 acres.

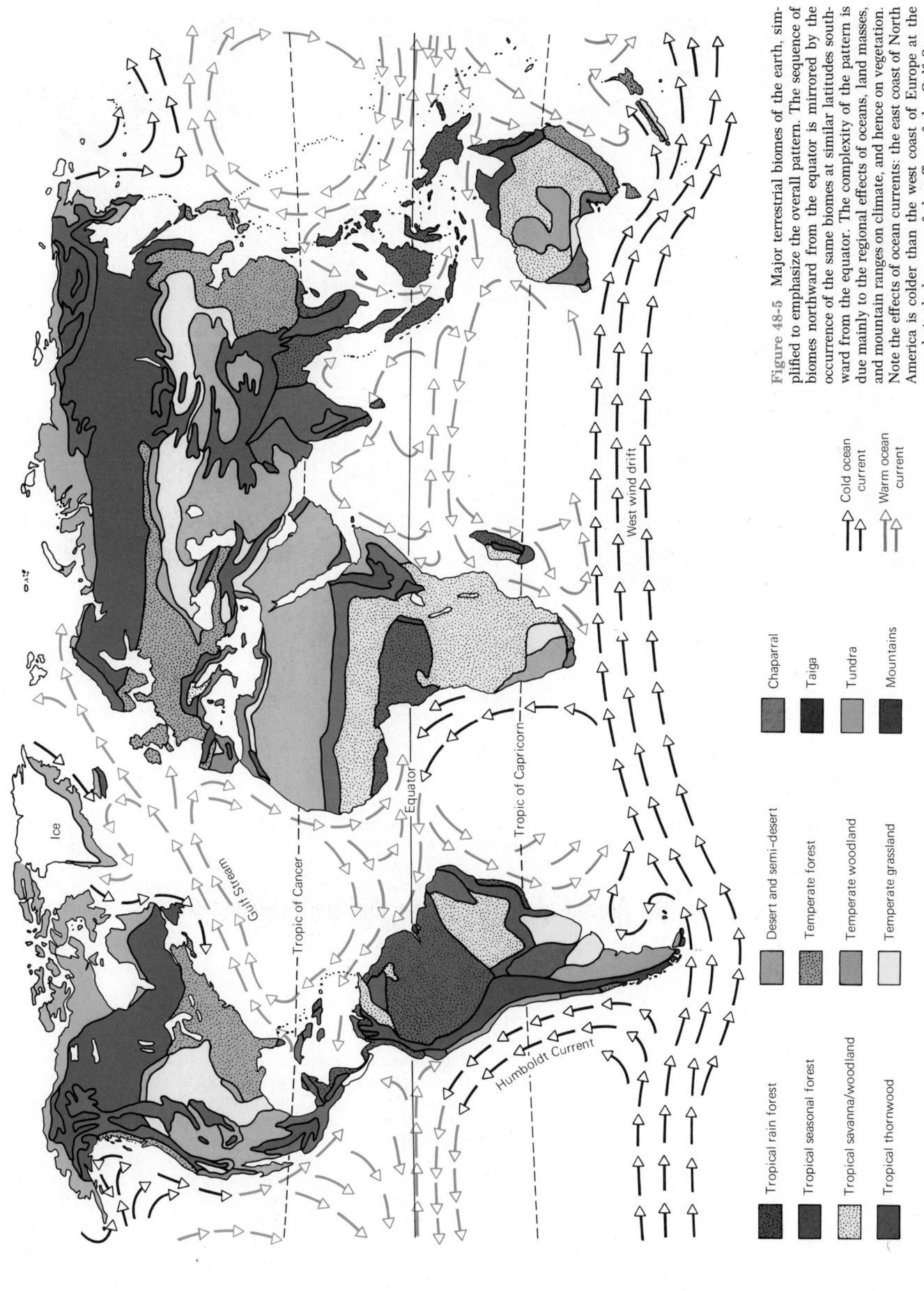

Figure 48-5 Major terrestrial biomes of the earth, simplified to emphasize the overall pattern. The sequence of biomes northward from the equator is mirrored by the occurrence of the same biomes at similar latitudes southward from the equator. The complexity of the pattern is due mainly to the regional effects of oceans, land masses, and mountain ranges on climate, and hence on vegetation. Note the effects of ocean currents: the east coast of North America is colder than the west coast of Europe at the same latitude because of the effect of the Gulf Stream.

Chaparral

Taiga

Tundra

Mountains

Cold ocean current

Warm ocean current

West wind drift

Ice

Gulf Stream

Tropic of Cancer

Equator

Tropic of Capricorn

Humboldt Current

Desert and semi-desert

Temperate forest

Temperate woodland

Temperate grassland

Tropical rain forest

Tropical seasonal forest

Tropical savanna/woodland

Tropical thornwood

(a)

(b)

(c)

(d)

Figure 48-6 Tropical rain forest. (a) Tree ferns, tangled vines, and tall, skinny trees (top left) characterize the perennial wet twilight of this patch of rain forest where some of the larger trees have been cut down, allowing understory plants to grow. (b) A rain forest feels uncannily like home because so many of the plants are familiar to us as house plants. (c) Thousands of species of epiphytes, like this orchid, live high in the canopy. (d) Most of the animals, like this tree frog from Central America, live high up in the treetop canopy. (Biophoto Associates, N.H.P.A.)

The animal life of rain forests is also exceedingly rich. Birds, butterflies, beetles, and frogs exhibit an almost bewildering diversity of striking color patterns (Figure 48–6). Since most of the plant food is high up in the canopy, most of the animals, including the mammals and reptiles, are arboreal.

Sadly, most of us will never see the incredible beauty of a tropical rain forest. Developing nations are clearing these forests at great speed, using the wood mainly as fuel for cooking. At the present rate, nearly all of the rain forests will be gone by the year 2000.

Tropical Seasonal Forest

Further away from the equator, climates tend to have distinguishable seasons, with precipitation becoming concentrated during part of the year and an increasingly pronounced dry season. Tropical rain forest grades more or less sharply into **tropical seasonal forest**. Canopy heights are lower, and the proportion of deciduous* trees increases as precipitation decreases and the length of the dry season increases. At one extreme, where there is only a short dry season, tropical seasonal forest is largely evergreen and is quite similar to rain forest. At the other extreme, with a pronounced and lengthy dry season, all of the trees may be deciduous and

* Deciduous   Losing the leaves during one season of the year; not evergreen.

Figure 48-7 Mist swirls over a tropical seasonal forest in Brunei. (Biophoto Associates)

lose their leaves during the dry season. Tropical seasonal forests include the monsoon forests of India and Southeast Asia; they also occur in South and Central America, the West Indies and northern Australia (Figure 48–7).

Tropical Savanna

Tropical savanna extends over large areas, often in the interiors of continents, where rainfall is insufficient to support forests, or (it is thought in some cases) where the development of forests is prevented by recurrent fires. Typical savanna consists of grassland dotted with scattered small trees or shrubs, such as acacias (Figure 48–8). Some savannas are entirely grassland, while others, sometimes referred to as tropical broadleaf woodlands, contain many trees.

(a)

(b)

Figure 48-8 Savanna. (a) Zebras in Tsavo National Park, Kenya. (b) In the dry season, thousands of square kilometres of savanna are parched and brown. (a, Biophoto Associates; b, Paul Feeny)

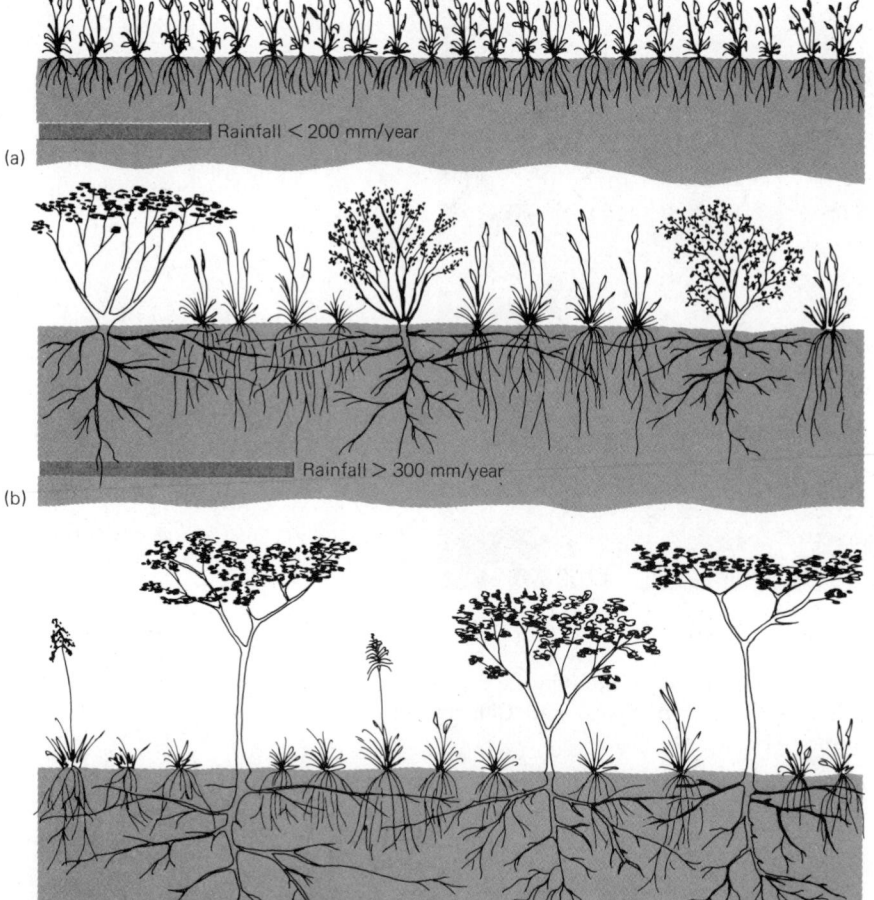

(a)

Rainfall < 200 mm/year

(b)

Rainfall > 300 mm/year

(c)

Rainfall > 400 mm/year

Figure 48-9 Relative proportions of trees and grasses in a savanna are influenced by rainfall (colored bars). In (a), where annual rainfall is 200 mm or less, the roots of the grasses use up all of the water available during the rainy season. If annual rainfall reaches 300 mm or more (b), not all of the water is used up by grasses and some is available to permit scattered shrubs or trees to survive the dry season. At annual rainfalls above 400 mm, trees come to predominate, forming a tree savanna or (at higher rainfall levels) tropical woodland. (After H. Walter, *Vegetation of the Earth*, New York: Springer-Verlag, 1973)

The proportion of trees in a savanna reflects competition between trees and grasses for water. Where rainfall is light, the roots of grasses are able to absorb all of it during the wet season; woody plants cannot survive because no water is available during the dry season. As rainfall increases, the grasses are unable to absorb it all, leaving water available for scattered trees. Where rainfall is sufficient to support a woodland, the canopy shade inhibits the development of grasses and the competitive relationship is reversed (Figure 48–9).

Savannas are most extensive in Africa, where they support a rich fauna of grazing mammals, such as zebras, wildebeest, and gazelles. The spectacular migrations of some of these species are related to shifting patterns of local rainfall that permit the growth of the young, nutritious foliage of grasses.

Tropical Thornwood

Tropical thornwood occurs in many regions too dry to support forest, but with at least a short rainy season each year. It often merges into savanna or seasonal forest on the one hand, and on the other, as the climate becomes increasingly dry, it merges into semidesert scrub, with smaller shrubs spaced further and further apart, or into desert. Spiny species of *Acacia* and other trees of the pea family often dominate thornwoods of the Americas and Africa. Many of the plants in a tropical thornwood lose their small leaves during the long dry season, and their growth and reproduction take place entirely during the wet season.

Although many tropical areas support lush natural vegetation, they are unsuited to the agriculture that produces most human food. As one ecologist pointed out recently, the main reason the world's developing countries are poverty-stricken is that they have inherited poor real estate (tropical forest and savanna); soil and climate alike are unsuited to the intensive agriculture needed to feed populations that have roughly quadrupled in the twentieth century.

48-C Desert

Deserts generally occur in regions having a rainfall of less than about 20 cm per year. Typical hot deserts are found at latitudes of about 20 to 30° north and south, where dry air from the equator falls from the upper atmosphere, warming as it is compressed at lower altitudes. Because it contains little water vapor, the atmosphere over a desert is a poor insulator, and although days can be very hot, nights are often cold because the ground radiates heat rapidly.

The Sahara desert, stretching across Africa from the Atlantic coast to the Red Sea, is the largest hot desert in the world. Hot deserts occur also in the southwestern United States and northwestern Mexico, the west coast of South America from Chile to Peru, southwest Africa, and central Australia. Desert areas with less than 2 cm of rain per year support little life of any kind, and the terrain is dominated by rocks and sand. Less extreme areas, including parts of the Sahara, have highly specialized plants, many of them annuals that grow, bloom, and set seed in the few days when water is available. Most desert perennials, such as the American cactuses, are **succulents**, plants that store water in their tissues, or small woody shrubs that shed their leaves during the dry season (Figure 48–10). Desert animals have adaptations that restrict the loss of water through their skin and lungs and in their urine and feces. Many are nocturnal, avoiding the desiccating heat of the day by burrowing into the cooler soil.

(a)

(b)

(c)

Figure 48-10 Desert. (a) Where there is more than 2 cm of rain a year, as in this area of Nevada, specialized desert plants can grow. (b) A strawberry hedgehog cactus in flower in Arizona. (c) A desert-colored horned viper from the Namib Desert. (a, William Camp; b, c, Biophoto Associates, N.H.P.A.)

(a)

(b)

Figure 48-11 Temperate deciduous forest (a) in
spring, (b) in autumn. Animals of temperate forests
include (c) a barred owl, and (d) a wood mouse.
(a, Paul Feeny; d, Biophoto Associates)

(c)

(d)

48-D Temperate Biomes

Temperate Forest

North and south of the tropics and their adjacent deserts lie the world's temperate regions, so-called because their climate is typically moderate in terms of temperature. The **temperate forest** biome occurs in regions of abundant rainfall. The
composition of temperate forests, the proportion of deciduous versus evergreen
trees, and the spacing and height of the trees, depend largely on the seasonal
distribution of precipitation, the severity of the winters, the nature of the soil, and
the frequency of fires. Three major categories of temperate forest can be distinguished: deciduous forest, evergreen forest, and rain forest.

Temperate deciduous forests occur in moderately humid inland climates
where precipitation occurs throughout the year, but where winters are cold, restrict-
ing plant growth to the warm summers. Most of the trees lose their leaves in the fall,
and thus they lose little water by transpiration* in the winter when their roots could
not replace it from the frozen soil. Broad-leaved deciduous trees, such as beeches,
oaks, hickories, and maples, dominate this kind of forest; there is also a well-
developed understory of shrubs and herbaceous plants on the forest floor (Figure
48–11). The soil is rich in minerals and organic matter.

*Transpiration Evaporation of water through pores in leaves.

Figure 48-12 Mist is an important source of moisture for these coast redwoods in the temperate rain forest on the California coast. (Paul Feeny)

Mammals typical of North American deciduous forests include white-tailed deer, chipmunks, squirrels, opossums, raccoons, and foxes. Wolves, black bears, bobcats, and mountain lions roamed widely until they were largely eliminated by human activities. As winter draws near, up to three-quarters of the birds may migrate south, insects enter diapause,* and many of the mammals hibernate. In the spring, herbaceous plants such as skunk cabbage, violets, Solomon's seal, and *Trillium* produce their leaves and flower before the tree canopy has leafed out and cut off most of the light from the forest floor.

Temperate evergreen forests occur over wide areas where conditions favor needle-leaved conifers or broad-leaved evergreens over deciduous trees. These conditions include poor soils and a high frequency of droughts and forest fires. Various species of pine can grow in poor soils and have specialized adaptations for surviving fire. In some pines, for example, the cones open and their seeds germinate only when exposed to temperatures of several hundred degrees. The seeds thus germinate in areas that have just been burned. In the United States, temperate evergreen forests include impressive stands of ponderosa and other pines in the west, as well as the extensive pine forests of the southern states. These are now prime areas for commercial timber operations. Elsewhere in the world, temperate evergreen forests occur in eastern Asia, in southern Chile, in New Zealand, and in Australia, where forests are dominated by various species of *Eucalyptus*.

Temperate rain forests occur in cool maritime (near the sea) climates with abundant winter rainfall and summer cloudiness or fog. They include the forests of giant trees along the Pacific coast of North America, stretching from the mixed coniferous forest of the Olympic Peninsula of Washington to the coastal redwood forests of Oregon and northern California (Figure 48–12). The heights of these

---

*Diapause Period when an animal suspends activity, accompanied by a drop in metabolic rate.

forests may reach 60 to 90 m, with some trees over 100 m, the tallest in the world, rivaled only by *Eucalyptus regnans* in the temperate rain forests of eastern Australia. Although there is little rainfall in California in summer, the foliage of redwoods can absorb water from the frequent fogs. Other temperate rainforest types include the coastal Sitka spruce forests extending north to Alaska, forests of Southern Hemisphere conifers in New Zealand and Chile, and various forests at higher elevations on tropical mountains.

Temperate Woodland

The **temperate woodland** biome occurs in climates too dry to support forests, yet with sufficient moisture to support more than grassland. It includes a range of communities, from small trees with an almost full canopy cover, to open woodland with scattered trees. The dominant tree species may be conifers, or evergreen angiosperms with tough, thick leaves, or deciduous trees. Pygmy conifer woodlands of piñon pine and juniper cover extensive areas of the American west, between the grassland and semidesert biomes at lower elevations and the pine forests at higher elevations. Oak woodlands are common in central California, and evergreen oak and oak-pine woodlands are extensive in the southwestern states and in Mexico. Temperate woodlands occur in Asia, around the Mediterranean, and in temperate regions of the Southern Hemisphere.

Temperate Shrubland

The **temperate shrubland** biome is best represented by the **chaparral** communities that occur in all five regions of the world with a Mediterranean climate: coastal California and Chile, the Mediterranean region, southern Australia, and the southern tip of Africa (Figure 48–13). These areas have moderately dry, maritime climates with little or no summer rain. The shrubs are mainly angiosperms with leathery leaves, ranging in height from 1 to about 5 m. They are often distinctly aromatic, with volatile and inflammable terpenes* in their leaves. Fires are frequent and pose a constant threat to residents of Santa Barbara and other cities located within this biome. After fires, the dominant shrubs regrow from surviving tissues near the ground.

---

*Terpenes   Organic compounds, including the essential oils (see, for example, Figure 51–5).

Figure 48-13   Chapparal in California. (Paul Feeny)

(a)

(b)

Figure 48-14 Temperate grassland. (a) Prairie in South Dakota. (b) Buffalo, one of the many species of ungulates native to grasslands. (a, Paul Feeny; b, William Camp)

Temperate Grassland

Temperate grassland, known variously as prairie (North America), steppe (Asia), pampas (South America), or veldt (South Africa), covers extensive areas in the interiors of continents where there is not enough moisture to support forest or woodland (Figure 48–14). Scattered shrubs may occur, often in depressions or watercourses where more water is available. Overgrazing of grasslands in the southwestern United States has permitted invasion by shrubs, such as mesquite. When the grasses are grazed, water losses due to transpiration cease, leaving enough water in the soil to support woody plants.

Although grassland vegetation forms only a single layer, many plant species may be present. The persistence of many prairie wildflowers is dependent on periodic fires. There are fewer species of birds in temperate grasslands than in deciduous forests, probably because of the low structural diversity of the vegetation. Mammals of North American prairies include small burrowing species, such as prairie dogs and ground squirrels (rodents), and large grazing herbivores such as bison and pronghorn antelope. Because of the rich deep soil that underlies many temperate grasslands, these regions, including the midwestern United States and the Ukraine in Russia, have become prime areas for sustained and highly productive agriculture. The original prairie of North America is now represented only by a few tiny, scattered, unfarmed remnants.

Temperate Desert

Temperate desert or semidesert occurs in regions too dry to support grassland, often in the rain shadows of mountain ranges (Figure 48–15). Cool semidesert

Figure 48-15 Temperate desert in the mountains of northern Iran. Summers are hot as in hot deserts, but winters are cold. (Paul Feeny)

occupies much of the Great Basin east of the Cascade and northern Sierra Nevada mountain ranges in the western United States. In these regions, large areas are dominated by sagebrush (*Artemisia* sp.), interspersed with perennial grasses. Typical animals include jack rabbits, sage grouse, and various pocket mice and kangaroo rats. Cool temperate semideserts also occur in central Asia, South America (Patagonia and the Andes) and Australia. True deserts, too dry for almost any kind of life, are scarce in temperate regions, although such areas do occur in central Asia, in the Andes and other arid mountains, and locally within other areas of semidesert.

48-E Taiga

The term "taiga" derives from a Russian word meaning "primeval forest." The **taiga** biome is dominated by subarctic needle-leaved forest, consisting mainly of gymnosperms—spruces, pines, and firs—that can survive extreme cold in winter. As in temperate evergreen forests and woodlands, trees in the taiga tend to be further apart than the trees of a deciduous forest, and light penetrating to the forest floor is used by an extensive ground cover of shrubs, mostly of the Ericaceae (the blueberry and heather family). The taiga, or **boreal forest**, as it is sometimes called, stretches in almost unending monotony in a giant circle through Canada and Siberia (Figure 48–16). This monotony, due to the low diversity of tree species, is occasionally interrupted by extensive areas of bog or "muskeg."

Much of the precipitation in the taiga falls as snow, and in the winter, many of the resident animals grow white fur or plumage that blends with the white background. Animals characteristic of the North American taiga include moose, wolverines, wolves, lynx, spruce grouse, gray jays, crossbills, and (in summer) many species of warblers (birds).

(a)

(b)

(c)

Figure 48-16 (a) Taiga covers these mountains in northern Idaho. Animals of the taiga include (b) the gray jay, and (c) the snowshoe hare. (b, William Camp)

48-F Tundra

The **tundra**, a treeless biome, occurs far north in the arctic regions, where winters are too cold and dry to permit the growth of trees (Figure 48–17). In many areas the deeper layers of soil remain frozen as **permafrost** throughout the year, and only the surface layer of soil thaws during the summer. The number of different species of organisms is low. Another important characteristic of the tundra is that decomposition occurs very slowly because the ground is so cold for most of the year. Because of the low rate of decomposition, the shallow soil, and the slow growth rate of the plants, tundra takes a long time to recover when it is destroyed. Thus tundra is peculiarly vulnerable to destruction by humans. This is why conservationists are so concerned about the effects of running oil pipelines through the tundra.

Tundra vegetation is dominated by sedges, grasses, mosses, lichens, and dwarf woody shrubs. Bogs are common because the permafrost retards drainage. The largest animals of the tundra are caribou in North America and reindeer in Europe and Asia. Lemmings, ptarmigans, snowy owls, arctic foxes, and wolverines are also typical. Hordes of mosquitoes, deerflies, and blackflies breed in the wet spots during the brief arctic summer. These insects contribute to the food available for a variety of migratory birds, including various plovers and sandpipers, snow buntings, longspurs, and horned larks, which nest in the tundra.

Neither taiga nor tundra occurs at sea level in the Southern Hemisphere because the continents do not extend far enough south. Antarctica harbors only a very scanty flora* and fauna* around its edges.

---

* Flora   Plant life (often includes photosynthetic protists such as diatoms).
* Fauna   Animal life (including any animal-like protists present).

(b)

(c)

(a)

Figure 48-17 Tundra and its inhabitants. (a) The treeless landscape; (b) a reindeer; (c) a dwarf willow, one of the small shrubs that can survive in the thin soil. (Biophoto Associates)

(a)

(c)

(b)

(d)

Figure 48-18 Alpine tundra and grassland in the North American Rocky Mountains. (a) Tundra above the tree line at about 3,500 m. (b) An alpine meadow. (c) Mountain avens. (d) Elk. (a, Paul Feeny; c, d, Biophoto Associates, N.H.P.A.)

A variety of **alpine grasslands**, **alpine shrublands**, and **alpine semideserts** are found on mountains, between the timberline and higher regions where nothing can live (Figure 48–18). This resembles arctic tundra in many ways, although nights are cool throughout the year in alpine areas, whereas they are warm during the brief summer in arctic regions. In northern temperate mountains, alpine meadows cover extensive areas above the timberline. These meadows are dominated by sedges and grasses, interspersed with dwarf willows, heaths, and other shrubs. Alpine cushion plants increase at higher and drier sites. Many plants cultivated in rock gardens, such as gentians, saxifrages, and edelweiss, are alpine species. Alpine meadows in North America are inhabited by mountain sheep, mountain goats, grizzly bears, marmots, and pikas. Many of the larger animals migrate to lower elevations during the winter, and all organisms, like those of the tundra, are adapted to take advantage of the short growing season.

Alpine grasslands also occur above the timberline on tropical mountains, often grading into alpine shrublands and sometimes extending almost to the snow line. The *paramo* of the South American Andes is partly alpine grassland, though other *paramo* communities have a distinctive flora of rosette shrubs and tussock grasses. Communities of similar structure, but widely different in evolutionary origin, occur in the alpine zones of African mountains. African alpine communities also contain heaths, and these shrubs dominate the alpine shrublands of the Himalayas. In New Zealand, tussock grasslands extend from the alpine zone down to more temperate elevations; such grasslands also occur on sub-Antarctic islands.

Strictly speaking, the term biome is used only to refer to communities on land. However, there is a great array of different aquatic communities, both marine and freshwater, which, like biomes, exhibit similarities wherever in the world they occur. We shall consider temperate freshwater lake communities in Chapter 49. Here, we restrict ourselves to a brief consideration of life in the sea.

Like the land, the sea can be divided, rather artificially, into various zones, based on the prevailing physical conditions and the different types of organisms that these conditions support. As on land, temperature and the intensity of sunlight determine what grows where; lack of water is not a problem in the sea, but there are areas where a shortage of minerals dissolved in the water limits ocean life. We shall consider several different types of marine community—remembering, as on land, that our division of organisms into communities is artificial and that in real life these communities may merge into each other.

Littoral Zone

Along the seacoasts, many kinds of plants and animals thrive in the **littoral zone**, the area between high and low water marks, where they are submerged for part of the day. The organisms in the littoral zone must be adapted to withstand desiccation when the tide is out, whether by having a waterproof covering, by hiding or burrowing in moist places, or by retiring into a tube or shell. There are three main types of littoral zone: rocky, sandy, and muddy shores, which support very different communities (Figure 48–19).

Mudflats occur in places such as estuaries and bays, where the water moves slowly enough to deposit sediment. Algae cover the mud and make up the food of numerous burrowing molluscs, worms, and some crustaceans. When the tide is low, most of the animals avoid desiccation and escape from predaceous shorebirds by burrowing in the mud.

Sandy beaches are less stable than mudflats, for sand shifts constantly and dries out more rapidly than mud when the tide is out. Microscopic algae coat the sand grains and provide the food for some burrowing worms and specialized crustaceans, but most of the tiny protists and crustaceans that live between the sand grains eat marine plankton stranded when the tide goes out.

Neither muddy nor sandy shores provide much foothold for sessile animals or anchored seaweed. These forms are largely restricted to rocky shores, which support

Figure 48-19   (a) Rocky and sandy shore on the coast of Maine. (b) Sea urchins and algae in a rocky littoral zone. (a, Paul Feeny)

(a)

(b)

a large variety of plant and animal life, each specialized to its own level with respect to the low water mark. The rocks provide anchorage for large and small algae; the algae as well as the rocks provide anchorage for numerous animals, both sessile and motile. Since the water crashes onto the rocks, everything must be firmly anchored. Large algae have tough holdfasts; sessile animals such as barnacles secrete a layer of cement that holds them to the rocks; cnidarians, echinoderms, and molluscs hold on by suction. Motile animals such as crustaceans anchor themselves firmly to rocks or seaweeds by their legs, or hide in crevices.

Vertebrates are few in the littoral zone, although a number of birds come in at low tide to scavenge or to prey on burrowing invertebrates and scuttling crustaceans.

The **sublittoral zone** occupies the **continental shelves**—the edges of continents—extending from low tide mark to a depth of about 200 metres. In the sublittoral zone, temperature fluctuates less, and wave movement is less violent, than in the littoral zone; in addition, mineral nutrients are readily available, washed from the land by rivers. Sublittoral areas are among the most densely populated on earth.

Coral Reefs

Coral reefs are found only in the tropics. Most of a reef lies in the sublittoral zone, although its top may be exposed at low tide and therefore in the littoral zone. A reef may form as an atoll,* as a fringe around an island, or as an offshore barrier reef. Coral reefs are restricted to warm oceans where the water temperature seldom falls below 21°C. Corals are cnidarians that live in association with photosynthetic dinoflagellates.* The reef itself is made up of calcareous material, secreted by the coral animals, and by red and green algae. Since photosynthetic organisms are so important to their formation, coral reefs are found only in clear, shallow (less than 100 m deep) water, where there is enough light for photosynthesis. The reef acts physically like a rocky shore in providing anchorage for algae and sessile animals. Since most of the reef is usually under water, the water movement is usually less than on a rocky shore, and a great variety of fish and swimming invertebrates can find shelter within the crevices of the reef (Figure 48–20).

*Atoll   A doughnut-shaped island with a seawater lagoon in the center. An atoll forms around an island that later sinks or erodes away.
*Dinoflagellates   One-celled organisms with two flagella each (Section 23-D).

Figure 48-20 A coral reef. (a) Snappers patrolling their territory on a Caribbean reef. (b) Hydroids (cnidarians) on a reef. (Steven Webster)

(a)

(b)

(a)

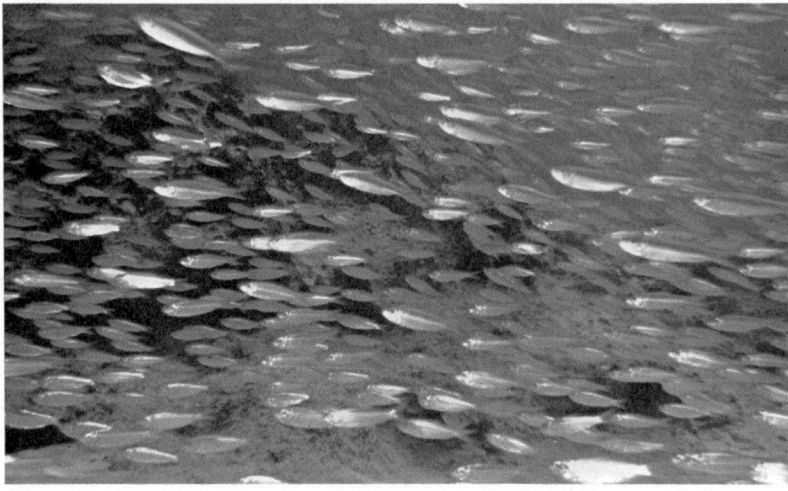
(b)

Figure 48-21 Animals of the nekton. (a) A baby green turtle struggles towards the sea soon after it has hatched; these turtles live for decades as members of the nekton, but they lay their eggs ashore. Human beings have brought them to the brink of extinction by stealing the eggs. (b) A school of fish. (a, Biophoto Associates, N.H.P.A.; b, Steven Webster)

The Open Ocean

The open ocean can be divided into the top 100 metres or so, where photosynthesis can occur, and the ocean depths. In the surface waters live the **plankton**, drifting protists, plants, and animals not powerful enough to swim against the ocean currents. They can, however, control their vertical position in the water, thereby moving from one current to another, for currents flow in different directions at different depths in many parts of the ocean. Fish and similar large animals in the ocean make up the **nekton**, creatures that can swim in any direction. Nektonic organisms feed mainly on plankton, and so their presence in any area depends largely on whether the water contains enough mineral nutrients to support a large population of phytoplankton (photosynthetic plankton). Because mineral nutrients are scarce in many areas of the open ocean, the world's major fisheries lie on continental shelves, which receive minerals washed down rivers, and in parts of the open ocean where currents carry minerals up from the bottom.

Seventy-five percent of the oceans' water lies more than 1000 metres deep. For many years, people assumed that there was little life in the depths of the ocean because it was too dark for photosynthesis to occur. Improved diving techniques that permit sampling at depths of more than 6000 metres have, however, revealed fascinating communities on the ocean floor, and many dives have turned up organisms hitherto unknown to science.

Bacteria are ubiquitous on land but few and far between in shallow sea water; an accidental experiment revealed that they are more numerous in deeper water. In 1968 a submersible* sank in 1500 metres of water off Nantucket. The crew escaped but left their sandwiches behind in plastic food containers. Several weeks later, the sandwiches were rescued—with a population of bacteria, the only organisms that were small enough to get into the containers. It is now clear that the ocean floor, at all depths, supports a vast population of decomposer bacteria that live on dead organisms or feces falling from the surface layers above them or on dead members of the deep-sea community. Larger organisms also feed on falling carcasses, or on the decomposers, and filter-feeders strain food out of the water.

---

*Submersible   A submarine specially built for deep-sea work.

We all know that coastal California is predominantly covered, not by chaparral, but by farms, roads, and buildings, and we all know why: human civilization has disturbed the natural communities, clearing the vegetation to make room for human affairs and their adjunct parking lots. So, when we say that climate determines the type of community in an area, we mean the community that would exist if the area were left alone long enough, rather than what may actually exist there. The community that forms if the land is left undisturbed, and that perpetuates itself as long as no disturbances occur, is called the **climax community**. In reality the climax community may vary somewhat within the same biome, usually in relation to such factors as soil type, soil drainage, and the frequency of fires.

When a climax community is disturbed, either by human activities or by natural means such as floods or fires, it begins a slow process of returning to its original state by a process known as succession. **Succession** is a progressive series of changes that ultimately produces a climax community.

Primary Succession

Primary succession occurs when the terrain is initially lifeless or almost so—when a new island rises out of the sea, or when a glacier retreats or a mountainside caves in, leaving a pile of rocks. Primary succession is usually very slow because it starts without any soil. Consider an area of barren rock created by a landslide. Water seeping into cracks in the rock may freeze, expanding and breaking the rock into chunks. The surface of the rock is also weathered by the beating of wind and rain. Carbonic acid, formed when atmospheric CO_2 dissolves in water, helps to dissolve minerals in the rock fragments, providing nutrients. Lichens adapted to exposed conditions may spread over the rock surface; they produce organic acids, which further dissolve the rock. Dead lichens also contribute organic matter to the forming soil, and mosses may gain a hold in even a thin layer of lichen remains and rock dust. As the mosses break up the rock more and add their own dead bodies to the pile, the seeds of small rooted plants can germinate and grow. The process continues along similar lines, with progressively more rock broken up into soil and progressively larger plants moving in, until the climax community becomes

Figure 48-22 Primary succession: mosses colonize rocks at the edge of a stream. (Biophoto Associates)

Figure 48-23　Secondary succession in an old field in New York. (Paul Feeny)

established; it may take thousands of years for the soil and the climax vegetation to develop fully.

Primary succession also occurs as a lake or pond fills up with silt and fallen leaves and the shoreline creeps toward the center of the lake. Gradually the lake turns into a marsh and then into dry land, eventually colonized by plants of climax species from surrounding communities.

Secondary Succession

Secondary succession is the series of community changes that takes place in disturbed areas that have not been totally stripped of their soil and vegetation. Although it may take a hundred years or more for the climax stage to return during secondary succession, the process is nevertheless much faster than primary succession because soil already exists.

A familiar example of secondary succession in the New England area is "old field succession," by which abandoned farms return to the climax deciduous forest (Figure 48–23). When a farmer stops cultivating the land, grasses and weeds quickly move in and clothe the earth with a carpet of green: wild carrot, black mustard, and dandelions. The "pioneers" of newly available habitats, these plants grow rapidly and produce seeds adapted to dispersal by wind or animals over a relatively wide area. Soon taller plants, such as goldenrod and perennial grasses, move in. Because these newcomers shade the ground and their long root systems monopolize the soil water, it is difficult for seedlings of the pioneer species to grow. But even as these tall weeds choke out the sun-loving pioneer species, they are in turn shaded and deprived of water by the seedlings of pioneer trees, such as pin cherries, dogwoods, sumac, and aspens, which take longer (2 to 5 years) to become established but command the lion's share of the resources once they reach a respectable size. Succession is still not complete, for the pioneer trees are not members of the species that make up the mature climax forest; after 5 to 30 years, slower-growing oak, maple, beech, and hickory trees will eventually move in and take over, shading out the saplings of the pioneer tree species. After perhaps a century or two, the land is covered with a mature climax forest.

Secondary succession can thus be considered as the gradual return of a climax community to equilibrium following disturbance. If the disturbance has resulted in

(a)

(b)

Figure 48-24 Fugitive species. (a) Fireweed in a
forest clearing. (b) A tomato hornworm caterpillar.
(a, Biophoto Associates; b. Paul Feeny)

major damage to the soil, as by erosion or leaching of nutrients, the successional
end-point may be an impoverished climax. In the eastern United States, so much
topsoil has been lost because of poor farming practices that it may take thousands of
years for nutrient levels in the present second-growth forests to build up again by
slow release from the parent rock.

In any tract of land, we can always find at least small patches that are undergoing succession following disturbance—a spot where a large tree has fallen, leaving a
light gap where pioneer weeds can move in and begin a miniature old field succession; a small pond filled in with dead leaves and gradually developing the species
composition of the surrounding forest community; a slope where a landslide has
occurred; or a burned forest. The existence of various patches undergoing succession ensures that there is a steady supply of **fugitive** plants, the fast-growing, here-
today-and-gone-tomorrow weeds (Figure 48–24). These species have seeds that can
spread over appreciable distances, carried by wind or by animals. In addition, the
seeds of many of these fugitive plants are adapted to live for long periods in a
dormant state, germinating when a disturbance provides the proper conditions,
such as increased light.

Animals, as well as plants, may be fugitive species (Figure 48–24). Insects that
specialize in eating a particular plant species may travel far and use their keen
senses to smell out new patches of their food plant some distance away. Some of our
agricultural pest problems stem from the fact that most crop plants originated as
fugitive species, depending on their sparse distribution and their nomadic habits
(never in the same place for many seasons in a row) to protect them from their
insect predators. By planting fields exclusively to one crop year after year, farmers
create a paradise for such fugitive animals as cabbage worms and cucumber beetles,
which no longer have to spend energy to find food and have nothing to do but eat
and multiply.

Repeated disturbance will prevent an area from returning to the climax state.
Suburban homeowners spend considerable time, energy, and money creating a continuous series of disturbances that maintains a lawn of a few species of short
grasses, constantly interrupting the old field succession of tall weeds, shrubs, and
light-loving tree seedlings that will take over if vigilance is relaxed. Many areas in
the Midwest were prevented from reverting to climax forest by native Americans
who deliberately set fires after they found that herds of buffalo, which live on the
prairie, could be hunted more efficiently than the solitary white-tailed deer of the
young forests.

Succession occurs because of progressive changes that make the environment
less favorable for the species that are present and more favorable for colonization by
others. Some of the changes are purely physical, like the silting in of a lake or the

weathering of rock, but many are caused by the organisms themselves. As succession proceeds, the supply of available nutrients in the soil declines as minerals become increasingly locked up in living organisms. The community's production of new organic matter (through photosynthesis), and the total weight of all the organisms in the community, both increase during succession, leveling off as the climax is approached.

Fire-Maintained Communities

Fires, set by lightning, storms, or human activities, occasionally sweep through large areas of taiga and northern temperate forest, burning tree crowns and destroying entire communities of animals and plants. Burned areas undergo secondary succession. In the spruce-fir forests of the Rocky Mountains, for example, burned areas are rapidly colonized by wind-borne seeds of fireweed (see Figure 48–24a), which grows and clothes the slopes with its purple flowers in summer, until it is displaced by spruces and other trees (Figure 48–25). Fires in northern or mountain regions do not often occur in the same place and thus do not prevent communities from reaching the normal climax pattern characteristic of their biome.

(a)

(b)

Figure 48-25 Periodic fires determine the vegetation in many types of places. (a) A grassland fire. (b) Fast-growing conifers are replacing those destroyed by a forest fire several years earlier. (a, Biophoto Associates; b, William Schlesinger)

In many other communities, fire is sufficiently frequent to determine the nature of the climax vegetation. Such communities include chaparral, temperate grassland, and many southern and western pine forests in the United States. Seedlings and saplings of deciduous trees are especially susceptible to fire, whereas many pines are relatively resistant and so are favored in regions where fires occur often. Many pines are adapted to survive, and even to exploit, fires. The seedlings of longleaf pine in the southern United States, for example, remain as grassy tufts for the first six years or so of their lives. Meanwhile, they accumulate large food reserves in their roots, with the result that during about their seventh year they can grow extremely rapidly. Fires in the southern forests where longleaf pine grows are light ground fires, very different from the intense crown fires of northern forests, and longleaf pine saplings can survive ground fires both when they are in their young "grass" stage and when they are more than about 6 metres tall. Their peculiar growth habit permits them to go through the middle stage, when they are most vulnerable to fire, in a very short time. Grasses also readily regenerate after fires that would kill trees; thus recurrent fires may prevent grassland or savanna from turning into woodland.

If fires are prevented in a fire-adapted pine forest, deciduous trees may become established. In addition, dead wood and litter build up on the ground, and so when a fire eventually does occur, it is more severe than usual, destroying not only any deciduous colonizers, but also the pines and other species. Odd though it may seem at first, frequent burning is essential for the preservation of many natural communities.

48-I Why Are Organisms Where They Are?

In our study of succession, we have seen that species may be widespread if they have efficient dispersal mechanisms, such as seeds that float on the wind or are carried by animals. However, there is a limit to how far an organism can travel over territory that is unsuitable for its survival. If we return to the question that opened our discussion, "Why are organisms where they are?" we find that at least two factors determine the answer: climate, and the organisms' ability to disperse to areas where the climate is appropriate.

When we look again at our African and South American rain forests and repeat the question, "Why are there different species in these two areas?" our answer is

(a)

(b)

Figure 48-26 Convergent evolution. (a) Cactuses in Arizona; (b) euphorbs in the Canary Islands. (Biophoto Associates)

that, although each area has a climate suitable for growth of the plants and animals of the other area, they are separated by long stretches of ocean, an impassable barrier to dispersal. There are exceptions: for example, small animals called rotifers form cysts that can be blown almost anywhere in the world, but the animals can live only in very restricted types of environments; hence the rotifer species in a marble cemetery urn in Pennsylvania may be the same as that in a marble urn in a South African cemetery, but different from that in the granite urn on the next grave!

Because most species cannot travel between continents, there are different species in the tropical rain forests of the different continents. The similarities of form and color of species in each place result from **convergent evolution**—the evolution of similar adaptations suited to similar environments. For instance, the advantage of being able to conserve water in desert habitats has led to the evolution of plants with thick, water-storing stems, as well as spiny leaves that deter animals from using the stems as the source of their own water, in deserts all over the world (Figure 48–26).

Two main patterns can be seen in a worldwide survey of the distribution of organisms:

1. Different areas of the world are inhabited by different species of plants and animals.

2. Terrestrial communities in different parts of the world can be divided into a fairly small number of categories, or biomes, on the basis of vegetational structure. These biomes are worldwide and are not restricted to single continents.

While the actual species found in an area depend on the area's evolutionary history, the biome depends mainly upon rainfall, temperature, and soil type. Similar changes in biomes occur with increasing altitude and increasing latitude.

The richest biome is tropical rain forest, where high temperatures and rainfall permit plants to grow throughout the year. Most of the plant and animal life is found in the canopy among the broad evergreen leaves of trees. Decomposition is rapid and the soil is poor; at any one time most of the nutrients are locked in the bodies of living organisms.

In temperate deciduous forest the soil is much richer in nutrients because the trees lose their leaves in the fall, creating a litter layer that decomposes only slowly. Deciduous forest is an important biome of North America and Eurasia in areas with warm, moist summers and cold winters.

Where the soil is poor or fire is frequent, temperate evergreen forest replaces temperate deciduous forest. Further north, both are replaced by taiga, a biome dominated by gymnosperms adapted to growing in sparse soil and to resisting extreme cold and water loss in winter. The taiga stretches around the world in the subarctic region.

North of the taiga and above the timberline lie the tundra and alpine grasslands, dominated by cold-resistant woody shrubs or by sedges and grasses, depending on the soil type and the amount of moisture in the soil.

Grasslands receive more rainfall than deserts and less than deciduous forests. Grasslands occur in the drier interiors of continents in the Americas, Asia, and Australia. Shrubs and trees may be scattered among the tall grasses. Grasslands are replaced by semidesert scrub or by small woody shrubs in areas where there is too little water for grasses to grow.

Deserts have hot days, cold nights, and very little rainfall. Their plant life is mainly annuals with very short growing seasons and succulent perennials adapted to the low rainfall.

The distribution of marine organisms is determined by water temperature and by the availability of light and minerals. The littoral and sublittoral zones are well supplied with both light and minerals and support dense communities of life. Coral reefs are specialized sublittoral communities found only in tropical waters. In the open ocean, the availability of light for photosynthesis restricts plankton to the upper layers of the water, but scarcity of nutrients in these layers may limit the numbers of organisms. Larger nektonic organisms are found primarily where their planktonic food is abundant. Dead organisms from the surface layers of the ocean supply food for a community of bacteria and other organisms that live on the deep-sea floor.

Although the climate of an area determines the composition of its climax community, patches of the area are always in various stages of ecological succession as a result of disturbances of the climax community. Organisms adapted to living in the unstable communities of early successional stages have effective dispersal mechanisms and perpetuate themselves by continuously colonizing new habitats as they arise in the surrounding climax community.

The limits of their dispersal ability prevent most organisms from colonizing all possible habitats, and so in different parts of the world we find similar communities inhabited by similar species, which have arisen by the convergent evolution of similar adaptations to similar environments.

OBJECTIVES

From your study of this chapter, you should be able to:

1. Describe the major patterns of air movement in the lower atmosphere, and explain how they affect precipitation at different latitudes.

2. List the three main factors that determine the distribution of biomes.

3. Describe why gradients of vegetation are similar with increasing altitude and increasing latitude.

4. State the conditions under which you would expect to find each of the following biomes, and list the type(s) of plants characteristic of each: tropical rain forest, tropical seasonal forest, tropical savanna, tropical thornwood, desert, temperate forest, temperate woodland, temperate shrubland, temperate grassland, taiga, tundra.

5. List three different types of littoral habitat and the types of organisms found in each.

6. Describe the conditions under which a coral reef may be formed.

7. List the main factors that determine the distribution of life in the oceans; describe the ocean surface and ocean bottom communities, and tell how organisms in these communities acquire food.

8. Define the following: community, climax community, fugitive species.

9. Distinguish between primary and secondary succession, outline an example of each, and explain why succession occurs; comment on the extent to which succession leads to a single type of climax community in a given biome.

10. List three examples of major community types that are maintained by fire. Account for the survival of plants in communities where fires occur frequently.

11. Explain why widely separated areas with similar climate usually contain similar species.

12. Define the term "convergent evolution," and give examples.

SELF-QUIZ

1. Rising air tends to become (warmer/cooler) and to (gain/lose) moisture.

2. One reason that the effect of increasing altitude on vegetation is similar to the effect of increasing latitude is that:
   a. temperature decreases with both latitude and altitude
   b. mountains decrease the angle of sunlight
   c. there are always clouds over mountaintops
   d. it is hard for plants to disperse up the sides of mountains
   e. both are higher up in the biosphere

3. The distribution of biomes is determined mainly by:
   a. soil, water, and temperature
   b. water, succession, and altitude
   c. wind, altitude or latitude, and water
   d. fires, convergent evolution, and moisture
   e. sunlight, water, and the height of the tallest plants

4. Which of the following has a vegetation structure with only one level?
   a. tropical rain forest
   b. taiga
   c. grassland
   d. shrubland
   e. desert

5. Which of the following communities would have trees?
   a. taiga
   b. littoral zone
   c. shrubland
   d. tundra
   e. plankton

6. A biome with high temperature, high rainfall, and poor soil is:
   a. shrubland
   b. coral reef
   c. semidesert scrub
   d. tropical rain forest
   e. temperate evergreen forest

7. A community with no living green plants is:
   a. a rocky shore
   b. the plankton
   c. a mud flat
   d. the deep ocean floor
   e. a coral reef

8. Colonization of an abandoned stone quarry would be an example of (primary/secondary) succession.

9. A pond in a deciduous forest becomes filled in with rock particles and dead leaves, creating soil. List, in order, the types of vegetation that would be seen as this area undergoes ecological succession, and name the climax community that would eventually result.

10. A(n) (early successional/climax) community would have a high proportion of fugitive species.

11. The American prairies and the Asian steppes do not have the same species of grasses because _____. However, both are inhabited primarily by grasses because _____.

QUESTIONS FOR DISCUSSION

1. What biome do you live in?

2. The 30° N latitude line runs through southern Louisiana and northern Florida as well as through desert country in Mexico and Texas. Why is the area in Louisiana and Florida not desert like the area in Mexico and Texas?

3. Why is it proving difficult to carry out large-scale "agribusiness" farming in vast tracts of land cleared of their tropical rainforest vegetation?

4. Why does secondary succession slow down as it proceeds?

5. How can frequent fires increase the species diversity of a region?

6. Why does a light gap contain some species of animals and plants that differ from those in surrounding climax forest?

REFERENCES AND FURTHER READING

International Wildlife Series: This Fragile Earth. Part I: "Doomed jungles?" by Peter Gwynne. July–August 1976. Part II: "The island dilemma," by Mariana Gosnell. September–October 1976. Part III: "Mountains besieged," by Edward R. Ricciuti. November–December 1976. Part IV: "Shifting sands," by Frederic Golden. January–February 1977. Part V: "Margin of Life," by Robert Allen. March–April 1977. Part VI: "The living sea," by Arthur Fisher. May–June 1977.

Jannasch, H. W., and C. O. Wirsen. "Microbial life in the deep sea." *Scientific American,* June 1977.

Richards, P. *The Life of the Jungle.* New York: McGraw-Hill, 1970. Short and beautifully illustrated introduction to the tropical rain forest, by a leading authority.

Strahler, A. N. *The Earth Sciences,* 2d ed. New York: Harper and Row, 1971. Excellent introduction to the physical processes that shape the earth's climate and geography.

Walter, H. *Vegetation of the Earth in Relation to Climate and the Eco-Physiological Conditions.* Translated from 2d German edition by Joy Wieser. London: The English Universities Press Ltd.; New York, Heidelberg, Berlin: Springer-Verlag, 1973. Despite its formidable title, this book is short and readable; probably the best account of world vegetation zones and the conditions that determine what biome occurs where.

Whittaker, R. H. *Communities and Ecosystems,* 2d ed. New York: Macmillan, 1975. A moderately advanced textbook on communities by a leader in the field, with a good section on biomes.

CHAPTER 49

ECOSYSTEMS AND COMMUNITIES

In Chapter 48 we saw that a tropical rain forest or a prairie may cover thousands of square kilometres. For convenience's sake, ecologists usually study smaller units— for example, a hillside, a lake, or a field. The value of considering such units was recognized in 1887 by Stephen Forbes, biologist for the Illinois Natural History Survey,when he wrote:

> A lake . . . forms a little world within itself—a microcosm within which all the elemental forces are at work and the play of life goes on in full, but on so small a scale as to bring it easily within the mental grasp . . . If one wishes to become acquainted with the black bass, for example, he will learn but little if he limits himself to that species. He must evidently study also the species upon which it depends for its existence, and the various conditions upon which these depend.

Nowadays, we would call Forbes's lake—or any other manageably small unit, with more or less distinct boundaries—an **ecosystem**. Forbes's writing points out the other characteristics of an ecosystem: it consists of the **community** of all the differ-

Figure 49-1 A lake forms a partly isolated ecosystem enclosed by its banks.

ent organisms living in the area, along with their physical environment. These all interact and change one another, so that the study of an ecosystem is indeed a complex undertaking. For convenience, we usually regard an ecosystem as an isolated unit, but in fact, things invariably move from one ecosystem to another, as when soil and leaves wash from a forest into a lake, or birds migrate between their summer and winter homes.

Not all ecosystems are natural; a space station, an aquarium, and a pot of houseplants are artificial ecosystems. A farm is often considered as an ecosystem because we must recognize the interactions between crop plants, fertilizers, pesticides, soil, climate, and the natural flora and fauna in order to manage the farm effectively.

An ecologist views interactions between organisms and their environment on two different time scales. The events observed and measured in ecological studies show what is happening here and now, in **ecological time**: rainfall, plant growth, animals eating plants, and so forth. Ecologists seek to understand the short-term consequences of such events to ecosystems and to their resident organisms. But these events also represent evolution in action, and so they can be viewed on the scale of **evolutionary time**. Every environmental event affecting organisms is potentially a selective force that may shape the course of their evolution. Each time an owl catches a mouse, it not only reduces the number of mice but also selects for those mice that are better at avoiding capture by owls.

In this chapter we shall first consider the flow of energy and cycling of nutrients in ecosystems and then examine some characteristics of the communities of organisms that inhabit various ecosystems. In Chapter 50 we discuss populations, also in ecological time. We shall reexamine ecology from an evolutionary point of view in Chapter 51.

49-A The Basic Components of Ecosystems

The sun is the ultimate source of energy for almost all ecosystems. During photosynthesis, green plants "trap" solar energy and use it to convert CO_2 and water into carbohydrates (see Chapter 8). Plants also incorporate other inorganic nutrients, such as nitrate, phosphate, and ions of various metals, into proteins, nucleic acids, pigments, and so forth. Because they are **autotrophs**, able to make their own organic food molecules from inorganic substances, green plants are called the **producers** of all the food in the ecosystem.

Figure 49-2 In some ecosystems, such as this heather moor in Scotland, remains of plants accumulate faster than they are decomposed. The resulting peat can be cut in blocks, stacked to dry as in this picture, and burned as fuel. (Paul Feeny)

Plants inevitably die. Their remains do not usually accumulate in the ecosystem (coal, oil, and peat deposits are an exception [Figure 49–2]) but instead are broken down by **decomposers**, organisms that acquire their food molecules from nonliving organic material (Figure 49–3). In the process of extracting energy and nutrients from this nonliving material, decomposers release some of the nutrients back into the ecosystem, where they are again available to producers. Nutrients are thus cycled through the ecosystem and may be used again and again in the same small area.

Figure 49-3 The main producers on land are (a) grasses and shrubs, (b) trees such as this coconut palm. Decomposers include (c) mushrooms, the reproductive structures of decomposer fungi, and (d) many millipedes, like the one shown here, which feed on decaying plant material. (a,b,c, Biophoto Associates; d, Paul Feeny)

(a)

(b)

(c)

(d)

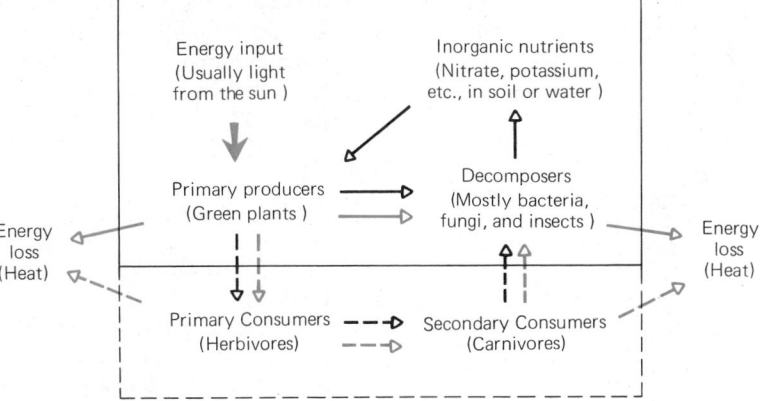

Figure 49-4 General pattern of energy flow (colored arrows) and nutrient cycling (black arrows) in an ecosystem. The solid line encloses the minimal components required for a self-sustaining ecosystem. In addition to these, most ecosystems contain primary, secondary, and even higher levels of consumers. Energy is continuously lost from the ecosystem in the form of heat produced by metabolism.

Energy, by contrast, is not cycled but is continuously lost from an ecosystem. Most organisms would soon die if the sun's energy were cut off for any length of time. (An interesting exception may be organisms in some unusual ecosystems in deep ocean trenches where the entire source of energy is geothermally heated water and H$_2$S seeping from cracks in the sea floor. In these ecosystems, certain bacteria can derive energy from the H$_2$S by **chemosynthesis**, providing the food for a variety of clams, sea anemones, crabs, and other species.)

The basic requirements for a self-sustaining ecosystem are **abiotic**, or nonliving, nutrients, together with producers, decomposers, and a source of energy (Figure 49–4). In most natural ecosystems, however, some of the green plant material is eaten by plant-eating animals, collectively known as **herbivores** or **primary consumers**. In turn, either these may die and pass directly to the decomposers, or a proportion of them may be eaten by **carnivores**, or **secondary consumers** (Figure 49–5). There may be tertiary or even quaternary consumers, carnivores that

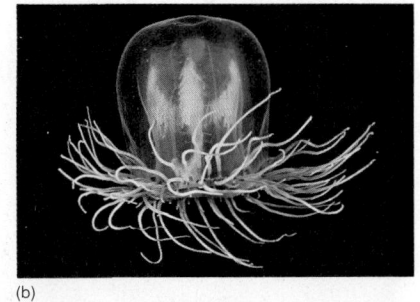

Figure 49-5 Consumers are the herbivores and carnivores of an ecosystem. (a) A Masai giraffe eating a prickly acacia. (b) A carnivorous scyphozoan jellyfish. (c) An eagle feeding her nestling. (a,c, Biophoto Associates, N.H.P.A.; b, Langdon Quetin)

feed on the secondary and tertiary consumers, respectively. All consumers, as well as most decomposers, are **heterotrophs**, feeding on organic material produced by other organisms. Rotting logs, piles of dung, and other ecosystems that contain only heterotrophs can obviously not sustain themselves indefinitely (see the Essay on page 845).

49-B   Food Webs

A **food chain** is a series of organisms, each of which provides the food supply for the next. Examples are the passage of energy from leaves to caterpillars to chickadees to hawks, or from dead organic matter to bacteria to fly maggots to parasitic wasps. Food chains are rarely isolated sequences; they are usually interconnected with one another, and the overall pattern is called a **food web** (Figure 49–6). Food webs may be very complex.

We are often warned that human interference, particularly with food webs we do not understand, can have unforeseen results. Ecologist Lamont Cole investigated one such situation in the 1950s. The World Health Organization tried to eliminate malaria from Borneo by spraying with the insecticide DDT. The spray did indeed kill the mosquitoes that carry malaria, but there was a snag. The spray turned out to weaken the cockroaches, which are bigger than mosquitoes and more resistant to DDT. Geckoes (insect-eating lizards) that ate the cockroaches suffered nerve damage from the DDT; their reflexes became slower, and many more of them than usual

Figure 49-6   A food web in a salt marsh. (Modified from *Ecology and Field Biology*, 2nd Edition, Robert Leo Smith. Copyright © 1974 Robert Leo Smith. Reprinted by permission of Harper & Row, Publishers, Inc.).

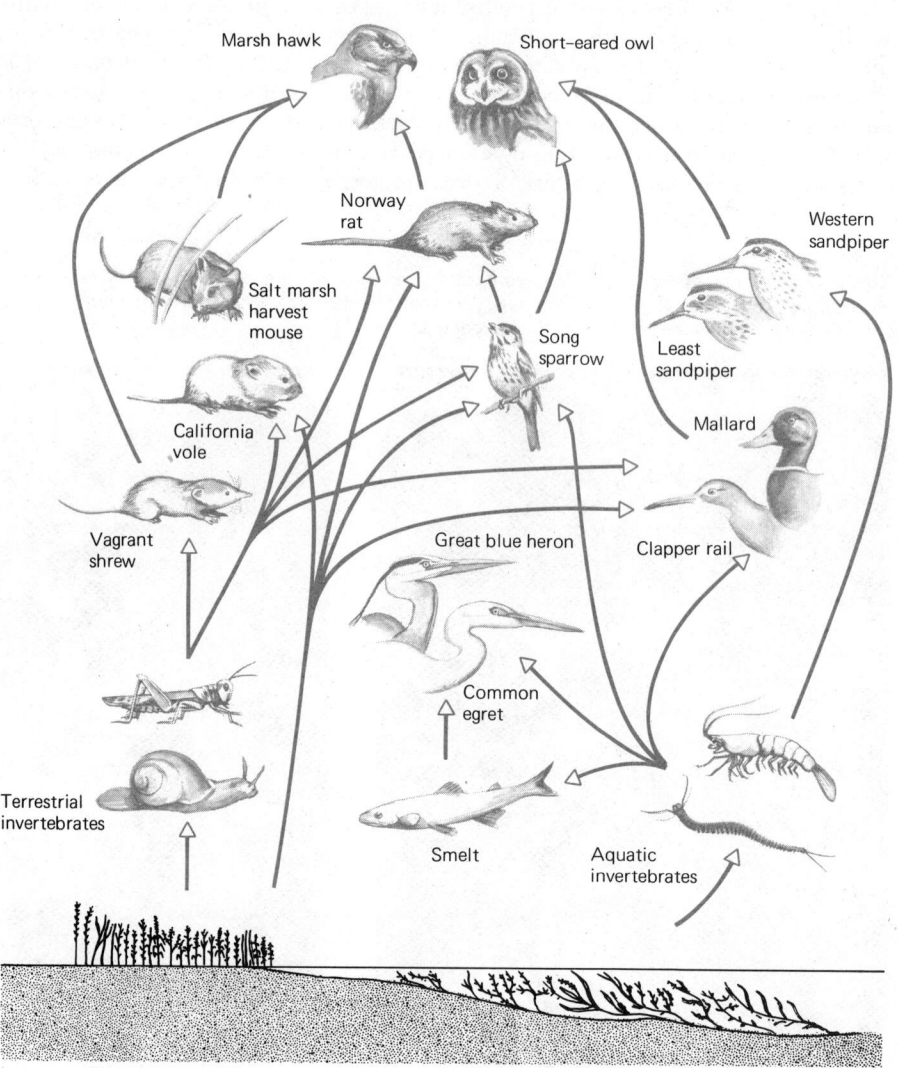

Marsh hawk

Short-eared owl

Norway rat

Western sandpiper

Salt marsh harvest mouse

Song sparrow

Least sandpiper

California vole

Mallard

Vagrant shrew

Great blue heron

Clapper rail

Common egret

Terrestrial invertebrates

Smelt

Aquatic invertebrates

Terrestrial plants

Aquatic plants

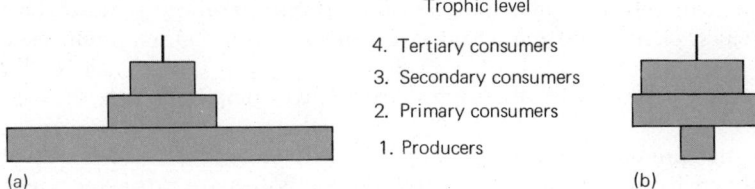

Trophic level

4. Tertiary consumers
3. Secondary consumers
2. Primary consumers
1. Producers

(a)   (b)

Figure 49-7 Pyramids of numbers showing how many individuals exist at each trophic level in (a) grassy fields in the northern United States and (b) a temperate deciduous forest in England. The "pyramid" for the forest is partly inverted because each tree supports a great many herbivorous insects. The third trophic level in the forest contains almost as many individuals as the second trophic level because it includes insect parasites. A single herbivorous insect may serve as food for several parasites.

were caught and eaten by cats. Because most of their gecko predators were now gone, caterpillars eating the local thatched roofs multiplied unchecked, and the roofs started to collapse. In addition, the cats were soon dying of DDT poisoning; rats moved in from the forest, and with them came rat fleas carrying the bacteria that cause the plague. Most people would rather have malaria than plague, and so the WHO stopped spraying DDT and, in an attempt to remedy the damage already done, parachuted large numbers of cats into the jungle—an expensive lesson in the importance of understanding a food web before you start pulling out its strands!

The **trophic level** to which an organism belongs is an indication of how far it is removed from plants in the food chain (Figure 49–7). Green plants constitute the first (producer) trophic level; the second trophic level contains the plant-eating animals (primary consumers), and higher trophic levels are made up of carnivores (secondary consumers, and so forth). Trophic levels are functional categories, useful in discussing energy flow and nutrient cycling in ecosystems. However, an organism cannot always be assigned to any one trophic level. Thus some plants, such as sundews, are carnivores as well as autotrophs. A tadpole eats diatoms or other vegetable material, while the adult frog is carnivorous. Many mammals, such as foxes, wolves, and humans, are **omnivores**, organisms that belong to several trophic levels because they eat not only other animals but also plant food.

Because energy is lost at each step, there are seldom more than five trophic levels in an ecosystem. A **pyramid of numbers** can be used to depict the progressively smaller numbers of organisms at successive trophic levels (Figure 49–7). Sometimes these pyramids are inverted, for example, when a single tree is attacked by many individual insects (Figure 49–7b). The trophic levels of an ecosystem may also be represented as a **pyramid of biomass**, showing the total dry weight of living material present at each level (Figure 49–8). These can also sometimes be inverted when the energy turnover (see Section 49-E) in a lower trophic level is unusually high (Figure 49–8b).

49-C   Primary Productivity

Primary productivity is the rate at which energy is stored as organic matter by photosynthesis. It is usually expressed in terms of energy stored (e.g., $Kcal/m^2/year$) or of biomass gained per unit area per unit time. Not all of the organic matter fixed during primary productivity accumulates as plant biomass; about half of it is metabolized in the plant's own respiration. **Gross primary productivity** is the total rate of photosynthesis, including the organic matter that is

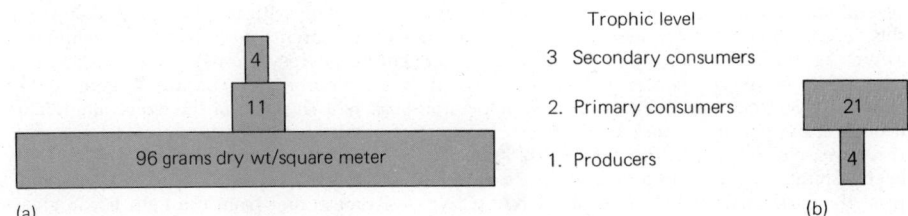

Trophic level

3 Secondary consumers
2. Primary consumers
1. Producers

(a)   (b)

Figure 49-8 Pyramids of biomass at different trophic levels for (a) a lake in Wisconsin and (b) the English Channel. The turnover of phytoplankton was so fast during the English Channel study that the standing crop of phytoplankton present at any one time was actually less than that of the zooplankton feeding on it.

829

almost immediately used up in plant respiration. **Net primary productivity** is the rate at which plants store energy or organic matter not used up in their respiration.

Net Productivity = Gross Productivity − Respiration Rate

The net productivity over time appears as plant growth and is available for consumption by heterotrophs. Note that productivities are rates; the **yield** of an agricultural crop, by contrast, is a measurable weight or volume of harvestable material. It represents a certain fraction of the crop's net productivity multiplied by the length of the growing season and the area of crop. Nor should productivity be confused with **standing crop**, which is the amount of biomass present at any one time. In some situations productivity is actually higher at lower standing crop. For instance, a field or lawn that is mowed regularly or grazed by cattle (i.e., low standing crop of grass) can have a higher net productivity than it would have if just left to grow (higher standing crop).

How is productivity measured? The most obvious way of measuring net productivity—determining the yield of biomass per area over a certain time period—is often difficult in natural communities. One problem with such a measurement is the estimation of how much of the yield is being consumed by heterotrophs during the period of measurement. Less direct methods are usually used. The overall equation for photosynthesis and respiration is:

$$6CO_2 + 6H_2O + Energy \underset{\text{Respiration}}{\overset{\text{Photosynthesis}}{\rightleftharpoons}} C_6H_{12}O_6 + 6O_2$$

Net productivity is directly proportional to the rate of production of oxygen and to the rate of depletion of CO_2. Measurement of either permits calculation of net productivity. If we can, in addition, measure respiration in the absence of photosynthesis, by shutting off the light source, then gross productivity can also be estimated,

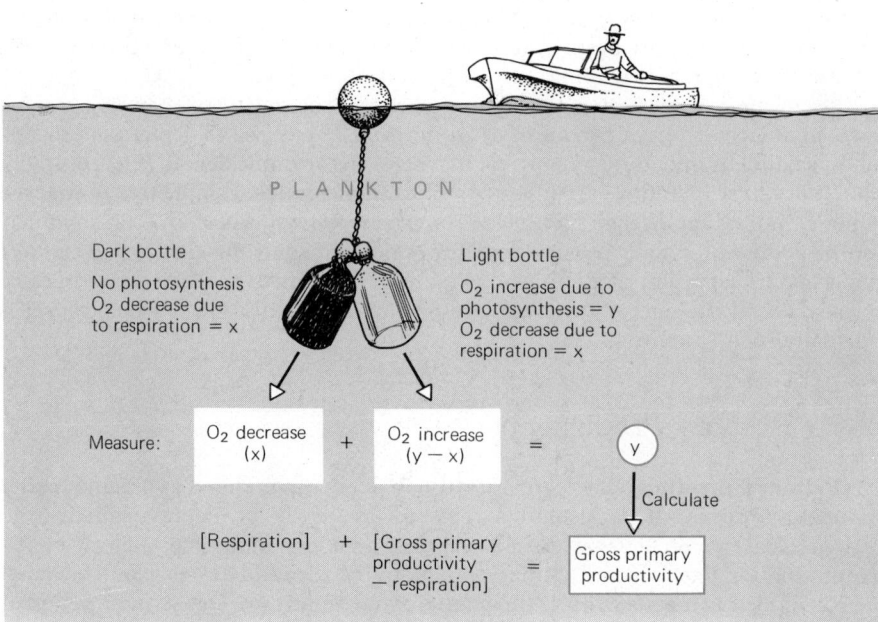

Figure 49-9 Measuring productivity. "Light-and-dark-bottle" method for measuring gross primary productivity in aquatic ecosystems. A pair of bottles is suspended at the desired depth for several hours. The bottles have been filled with water (complete with its plankton) taken from this depth. The bottles are identical except that one is transparent (permitting photosynthesis as well as respiration) and the other is opaque (preventing photosynthesis). Later, back in the laboratory, the oxygen concentration in each bottle at the end of the experiment is measured chemically and compared with that in water samples taken at the start of the experiment. The decline of oxygen in the dark bottle gives a measure of respiration by both plants and animals; the increase of oxygen in the light bottle indicates the net gain of photosynthetic productivity over respiration; respiration is assumed to be the same in both bottles. Adding "Respiration" from the dark bottle to "Gross primary productivity—Respiration" from the light bottle gives a figure for the oxygen produced by photosynthesis, which can be used to calculate gross primary productivity in terms of fixed carbon or energy.

TABLE 49-1 **NET PRIMARY PRODUCTIVITY OF MAJOR ECOSYSTEMS**[1]

ECOSYSTEM TYPE	NET PRIMARY PRODUCTIVITY ($g/m^2/year$)	
	Range	Mean
Tropical rain forest	1000–3500	2200
Temperate evergreen forest	600–2500	1300
Temperate deciduous forest	600–2500	1200
Boreal forest (taiga)	400–2000	800
Savanna	200–2000	900
Temperate grassland	200–1500	600
Tundra and alpine communities	10– 400	140
Desert and semidesert scrub	10– 250	90
Extreme desert	0– 10	3
Cultivated land	100–3500	650
Lake and stream	100–1500	250
Open ocean	2– 400	125
Upwelling zones	400–1000	500
Continental shelf	200– 600	360
Algal beds and coral reefs	500–4000	2500
Estuaries*	200–3500	1500

[1]From R. H. Whittaker, *Communities and Ecosystems*, 2d ed., Macmillan, 1975.

since gross productivity = net productivity + respiration rate (Figure 49-9).† Measuring the productivity of terrestrial plants presents considerable problems, but it can be done by enclosing plants or parts of plants in large transparent tents, and comparing changing CO_2 levels of the air inside these tents by day and by night.

Estimates of the net primary productivity of regional ecosystem types (biomes) are shown in Table 49–1. These values represent, on average, about 50% of the corresponding gross primary productivities. As might be expected, productivity generally increases from arctic regions towards the tropics, reflecting the increasing temperature and annual solar radiation. Where water is scarce, as in savannas and deserts, or where nutrients are lacking, productivity is low no matter how much sunlight reaches the area. In arid climates, net primary productivity increases almost linearly with increasing precipitation (Figure 49–10). Intensive agriculture, using special crop varieties, irrigation, and fertilizers, and planting two or more crops on the same land per year, can achieve net productivities as high as those of any naturally occurring terrestrial vegetation. Because crop plants are maintained in peak condition, with no senescent* plant material, respiration losses may be as low as 20%, or even less, of gross productivity.

Marine productivity suffers from a constant drain of nutrients as plant and animal remains sink below the sunlit layer of the sea where photosynthesis can occur. Thus productivity in the open oceans is generally low except in regions of upwelling, where nutrients are brought up to the surface layers by ocean currents, and in the relatively shallow waters of the continental shelf.

---

† This discussion of productivity disregards the effects of photorespiration (Section 8-G), which must be included in a more rigorous experiment.

* Estuaries Shallow areas where rivers meet the sea.

* Senescent Undergoing changes in preparation for death.

Figure 49-10 Net primary productivity increases with mean annual precipitation. As would be expected, the relationship is most striking in ecosystems that are most arid, where productivity is most likely to be limited by the availability of water. (After H. Lieth, *Human Ecology* 1, 303, 1973)

"That's *not* what I meant by measuring the net productivity of the shore." (Rosemary Smith)

Coral reefs, in contrast to the oceans surrounding them, are among the most productive ecosystems. They are very efficient at extracting nutrients from the water and at recycling these nutrients.

The annual net primary productivity of the whole biosphere is approximately 170 billion tons (dry weight) of organic matter. Of this total, 115 billion tons are produced on land and 55 billion tons in the oceans (despite the fact that the oceans occupy about 70% of the earth's surface). Humans harvest about 1.2 billion tons per year as plant food.

These are enormous quantities of material, but for several reasons, they actually represent the use of only a tiny fraction of incoming solar radiation. First, much of the solar energy reaching the earth is reflected at once back into space (causing the earth to "shine" like other planets). Most of the shortwave ultraviolet radiation is absorbed by the ozone layer of the atmosphere. Other wavelengths are also absorbed or scattered to some extent, so that even at noon on a clear summer day only about 67% of the incoming light reaches the earth's surface. The proportion is usually much less, due to scattering or absorption by dust, water vapor, clouds, and so forth. Second, of the radiation reaching the ground on a clear day, only 45% is in the visible region of the spectrum used in photosynthesis. Third, photosynthesis uses primarily the red and blue wavelengths; much of the green light is not absorbed by plants. Under some conditions, a barley plant may convert as much as 14% of the visible light energy it receives to net biomass. However, the average gross and net primary productivities over the globe represent use of only 0.6% and 0.3%, respectively, of the radiant energy reaching the earth's surface in the visible part of the spectrum.

49-D   Secondary Productivity

Secondary productivity is defined as the rate of formation of new organic matter, as growth and reproduction, by heterotrophs. Of the net primary production available in temperate forests, herbivores (e.g., caterpillars, aphids, deer) eat only about 1 to 3%; in some other communities, such as old fields, 10% or more of the vegetation may be eaten. Not all of this becomes secondary productivity, however. The gut of a caterpillar feeding on herbaceous plants absorbs only about half of the leaf material eaten; the rest is egested as feces. Furthermore, of the food absorbed, about two-thirds is used in respiration, so that only about 15% of ingested food appears as secondary productivity (Figure 49–11). Grasshoppers in a Tennessee field were found to convert only about 4% of the food they ate into secondary productivity (Figure 49–12). An average figure is about 10% for animals; this would be somewhat lower for herbivores and higher for carnivores, because a given weight of meat contains more of the nutrients an animal needs than the same weight of plant food.

Even if the organisms at each trophic level were able to find, capture, and eat all of the net productivity from the previous trophic level, the tertiary consumers would receive only about $1/10 \times 1/10 \times 1/10 = 1/1000$th of the energy present in the original producers in their food web. It is clear from this that in times of food scarcity, eating meat is a luxury for omnivores, including human beings. By adopting a vegetarian diet, it is possible to skip the energy losses at one trophic level so that more people could be supported by a given area of land.

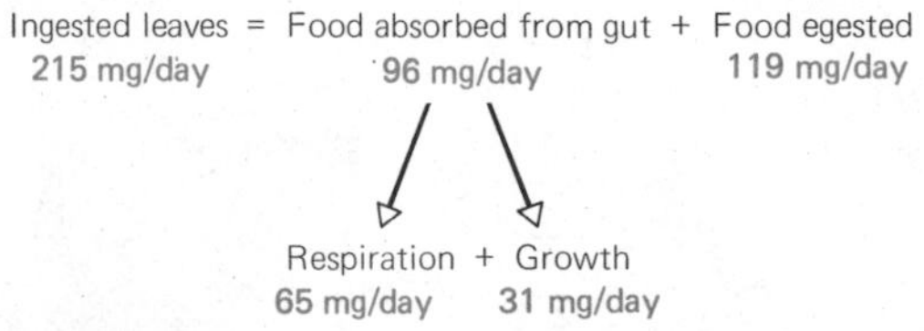

Ingested leaves = Food absorbed from gut + Food egested
215 mg/day   96 mg/day   119 mg/day

Respiration + Growth
65 mg/day   31 mg/day

Absorption efficiency = $\frac{96}{215} \times 100$ = **44.7%**

Growth efficiency = $\frac{31}{215} \times 100$ = **14.4%**

Figure 49-11 Food budget of black swallowtail butterfly caterpillars feeding on leaves of wild carrot plants (in dry weight). (Courtesy of J. Mark Scriber and Paul Feeny)

1 m² of field for 1 yr.

Figure 49-12 Productivities at producer, herbivore, and carnivore trophic levels measured in a Tennessee field. Figures are for a square metre of field in one year. Considerably less than 10% of the food (net productivity) available at one trophic level is converted into net productivity at the next highest level. The figures are 2.2% for plants to herbivores (colored) and 6.7% for herbivores to carnivores. Thus more than 90% of potential food is lost at each transfer from one trophic level to the next. (From Van Hook, Robert I., Jr., "Grassland Energy and Nutrient Dynamics" (*Ecological Monographs* 41:1–26). Copyright 1971 by The Ecological Society of America).

d. Net carnivore productivity: 0.4 g (0.15% of a; 6.7% of c)

c. Net herbivore productivity: 6 g (2.2% of a; 20.7% of b)

b. Every year herbivores eat: 29 g (10.7% of a)

a. Net primary productivity (plants): 270 g (dry weight)

49-E Pyramids of Energy

The flow of energy through an ecosystem can be represented in the form of a **pyramid of energy**, which shows the total amount of incoming energy for successive trophic levels (Figure 49–13). Unlike pyramids of numbers or biomass, however, pyramids of energy must always be "right side up," because some energy is always lost as heat in going from one trophic level to the next.

In a mature ecosystem, one that is not increasing in biomass, solar energy input is in approximate balance with total community respiration. The rate of energy loss is the same as the rate of energy storage, though lost energy is in a more degraded and therefore less useful form (infrared heat rather than visible light). This is more or less true for the biosphere as a whole, since it is not heating or cooling appreciably and the first law of thermodynamics must be obeyed (see Section 6-A).

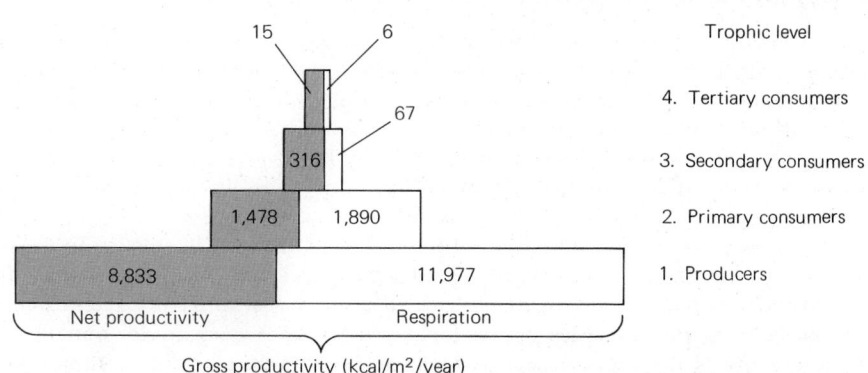

Trophic level

15   6

67

4. Tertiary consumers

316

3. Secondary consumers

1,478   1,890

2. Primary consumers

8,833   11,977

1. Producers

Net productivity   Respiration

Gross productivity (kcal/m²/year)

Figure 49-13 Pyramid of energy flow for Silver Springs, a river ecosystem in Florida, showing total energy input at each trophic level and its division into net productivity and respiration. (From Odum, H. T., "Trophic Structure and Productivity of Silver Springs" (*Ecological Monographs* 27: 55–112). Copyright 1957 by The Ecological Society of America)

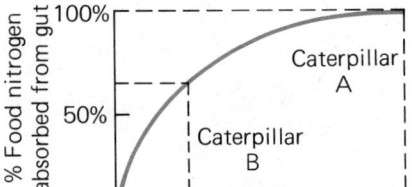

Figure 49-14 Hypothetical illustration of how feeding efficiency may be sacrificed to increase the rate of acquisition of a limiting nutrient. Not all plant nitrogen is easily digestible. Caterpillar "A" retains food in its gut until all the nitrogen is assimilated (100% efficiency). Caterpillar "B" assimilates only the easily-digested nitrogen and excretes the remaining food. Though less efficient, "B" can digest four gut-loads of food in the time it takes "A" to digest one. Caterpillar "B" ends up assimilating nitrogen (and therefore growing) at a faster rate.

There are several reasons why ecosystems seldom have more than four or five trophic levels. First, not all of the food available at one trophic level is actually eaten by animals in the next level. This results partly from the difficulty of finding food, and from the inedibility of some of this food (e.g., bones, tree trunks, toxic plants). In addition, the abundance of many organisms is limited, by predators, disease, weather, and so forth, to levels below those that could be supported by their food supply.

Second, not all of the food eaten is actually useful. For instance, the growth of herbivores is often limited by the availability of nutrients such as nitrogen in their food. They may excrete and "waste" large amounts of energy-rich plant material in order to eat enough food to extract the amount of nitrogen they need, and they do not do even this with maximum efficiency. Since nitrogen is present in food in forms of varying digestibility, it pays a caterpillar, for example, to excrete one meal after assimilating the easily digestible nitrogen from it, and then to eat another meal, rather than taking the time to digest the first meal fully (Figure 49–14). Rates and efficiencies cannot be maximized simultaneously. The compromise in most animals, as in machines like cars, tends to be toward maximizing rate (or power output) rather than efficiency.

A third and important source of energy loss is metabolism. Large quantities of energy must be spent in the maintenance and repair of body tissues, in the performance of normal functions such as feeding, excretion, and circulation, and in behavior. The respiration of glucose, the major body fuel, is only about 50% efficient, and more energy is lost as the ATP formed during respiration is used to drive metabolic reactions.

Because of all of these losses of energy from one trophic level to the next, there is not enough energy left to support higher trophic levels. A wolf may have to travel 30 kilometres a day to find enough food to eat, and a tiger requires a home range of up to 250 square kilometres. An animal that fed on wolves or tigers would have to be extremely large, and it would have to cover a wide hunting area to try to find enough of its widely scattered prey. It is not energetically feasible to try to harvest the small amount of food energy available in the highest trophic level.

49-F   Cycling of Mineral Nutrients

Although the productivity of an ecosystem may be limited by the supply of sunlight or availability of water, in many cases it is limited instead by the availability of inorganic nutrients.

Living organisms require six elements in relatively large quantities: carbon, hydrogen, oxygen, nitrogen, phosphorus, and sulfur. Elements needed in smaller amounts include sodium, potassium, manganese, calcium, iron, magnesium, chlorine, iodine, cobalt, and boron. These elements are present in rocks, usually as salts. They are released, by erosion and weathering, into soil, rivers, lakes, and the oceans. Some elements, such as nitrogen and oxygen, are also present in the atmosphere. The movements of nutrient elements through the biosphere, or through any particular ecosystem, by the physical processes of erosion, deposition, evaporation, and precipitation, and by the biological processes of uptake and release, are called **biogeochemical cycles**. They are called cycles because nutrient elements, unlike solar energy, may be used over and over again by living systems. On the average, every breath you inhale contains several million atoms once inhaled by Plato—or by any other person in history who lived for 65 years.

Nutrients are sometimes recycled rapidly through living systems, as in grasslands, where the above-ground vegetation dies back each year and its nutrients are made available again the following season. In other cases, nutrients spend many years apart from the activities of the biological world. For example, remains of marine organisms may sink to the bottom and be incorporated into sedimentary rocks that are uplifted and exposed to erosion and nutrient release only after millions of years. Every nutrient element has a somewhat different fate, depending on its physical and chemical properties and on its role in living organisms; we shall illustrate the concept of nutrient cycling with only a few simplified examples.

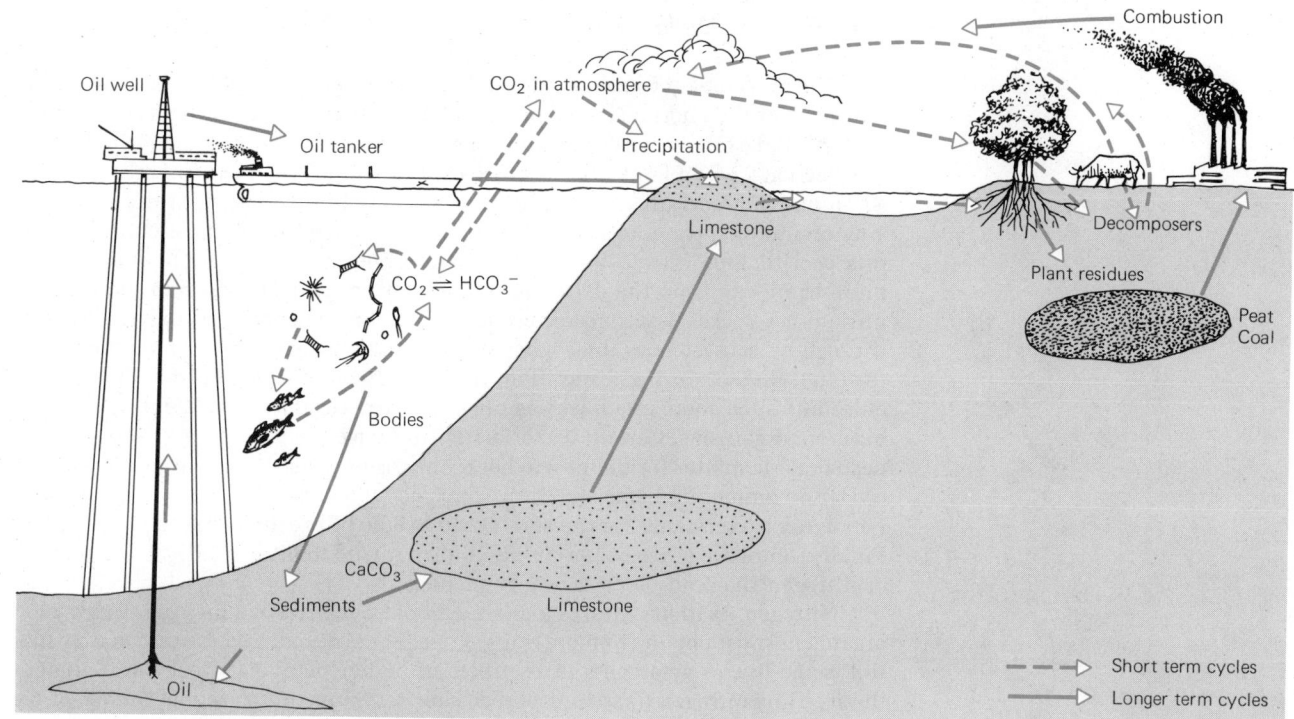

Figure 49-15 Simplified carbon cycle. Carbon passes through the processes indicated by dashed arrows more rapidly than through those indicated by solid arrows.

The Carbon Cycle

In terrestrial ecosystems there is a short-term cycling of carbon between living organisms and the atmosphere, as a result of photosynthesis and respiration (Figure 49–15). Most of the carbon fixed as organic matter is rapidly broken down and released back into the atmosphere as CO_2. In aquatic ecosystems, the corresponding exchange is between living organisms and CO_2 or bicarbonate (HCO_3^-) dissolved in the surrounding water. In some ecosystems, however, carbon may accumulate either as undecomposed organic matter, as in the peaty layers of bogs and moorlands, or as insoluble carbonates, which accumulate in bottom sediments in aquatic systems. Such carbon has entered a cycle of longer duration, and may take a long time to be released. Deposits of coal, oil, and natural gas represent part of the net production of a bygone era. These fossil fuels are organic compounds that were buried before they could be decomposed, and were subsequently transformed by time and geological processes. As they are burned, their carbon is released back into the atmosphere as CO_2. Sedimentary carbon (for example, in the insoluble calcium carbonate, $CaCO_3$, of various seashells) eventually turns into sedimentary rocks such as limestone and dolomite. After millions of years, these rocks may be lifted above sea level and exposed to erosion, releasing their carbon as carbonate (CO_3^{2-}) and bicarbonate into streams and rivers: "hard" water has usually flowed through limestone at some point, picking up carbonate, which may accumulate as "fur" in kettles when the water is boiled.

The carbon balance of the biosphere as a whole is moderated by exchange of CO_2 between the atmosphere and the oceans. The CO_2 content of the atmosphere is about 0.03% but is steadily increasing (from 312 to 330 parts per million in the last 20 years), owing both to the burning of fossil fuels and to the cutting down and clearing of forests (which is proceeding at a rate of 0.5 to 1% of the remaining forest area per year in the tropics). The rate of CO_2 increase in the atmosphere, however, is less than half of what would be expected from these human activities. The remainder is almost certainly being absorbed by the oceans, which thus act as a global "buffer" for CO_2, moderating its concentration in the atmosphere and accounting for the basic stability of the global carbon budget.

The Nitrogen Cycle

The movement of nitrogen through an ecosystem differs from that of carbon in that most of the enormous pool of molecular nitrogen gas is available only to those bacteria and cyanobacteria that can fix gaseous nitrogen by reducing it to ammonia (see Section 22-F). Plants must acquire this vital nutrient either directly from nitrogen-fixing microorganisms or indirectly from the limited pool of nitrogen found as ammonium or nitrate ions in soil water, rivers, lakes and oceans. Other inputs to this pool include nitrate formed from atmospheric nitrogen by lightning, ammonium or nitrate from the decay of organic matter, and nitrate from the erosion of nitrate-rich rocks. A second difference in the movements of nitrogen and carbon through an ecosystem is that, whereas most organisms release carbon dioxide into the environment during respiration, the conversion of nitrogen from organic molecules into such inorganic forms as nitrates involves steps carried out by a series of different organisms. Certain bacteria and fungi use the proteins and amino acids in dead organic matter for their own dietary nitrogen, releasing the excess as ammonia (NH_3) or ammonium ions (NH_4^+), some of which is again taken up by plants. Nitrifying bacteria, also in the soil, oxidize some of the ammonia to nitrite (NO_2^-). Finally, there are other bacteria that oxidize nitrite to nitrate (NO_3^-), another form of nitrogen that can be absorbed by plants.

Nitrogen fixation requires a great deal of energy, which nitrogen-fixing microorganisms must obtain from organic matter. Some decompose organic matter in the soil, some live as symbionts in the roots of higher plants, and some are photosynthetic. Thus nitrogen fixation proceeds slowly relative to the rate of uptake of fixed nitrogen by plants. Consequently, despite the fact that nitrogen makes up 78% of the atmosphere, nitrogen is often the limiting nutrient in plant growth. Nitrogen is thus one of the elements most commonly applied in crop fertilizers.

Although many ecosystems, such as tropical rain forests, are remarkably efficient at retaining their nitrogen, all terrestrial ecosystems continually lose some nitrogen as nitrate and organic matter dissolved in ground water and streams. Nitrate leached out of the soils of the Andes mountains, and deposited where the streams carrying it dried up in deserts, formed the immense deposits of nitrate rock in Chile. Nitrogen is returned to the atmospheric "reservoir" by denitrifying bacteria living in the anaerobic mud of fertile lakes, and in bogs, in estuaries, and on parts of the sea floor. The input of usable nitrogen into the biosphere is also being increased by the industrial manufacture of ammonia-based fertilizers from atmospheric nitrogen, using catalysts and large amounts of fossil fuel energy (Figure 49–16).

The Phosphorus Cycle

Phosphorus is a major constituent of ATP, DNA, RNA, and biological membranes, and many animals need large quantities of this element for shells, bones, and teeth. Since phosphorus almost never occurs as a gas, its cycle, unlike the **atmospheric cycles** of nitrogen or carbon, is a **sedimentary cycle** (Figure 49–17). When rocks are eroded by weather, minute amounts of phosphorus dissolve, usually as phosphate (PO_4^{3-}), and so become available to plants. Unlike nitrogen, which must be released from organic compounds before plants can take it up again, much of the phosphorus excreted by animals is in the form of phosphate, which is immediately reusable by plants. Thus the cycling of phosphorus in terrestrial ecosystems is usually very efficient, although small amounts are continually lost downstream and to the oceans. Since some phosphate salts are not very soluble in water, the oceans continually deposit phosphate minerals in their sediments. The principal reason for the infertility of the open oceans is their low phosphorus content.

At the level of the biosphere, the phosphorus cycle is, in the short run, a one-way flow—from rocks to land ecosystems to the ocean and finally to ocean sediments. Apart from trivial amounts deposited on land in the droppings of oceanic birds, the only natural way for phosphorus to return to land is via immensely slow geological processes in which sea floor sediments may again become terrestrial rocks. However, terrestrial ecosystems are able to retain much of their phosphorus

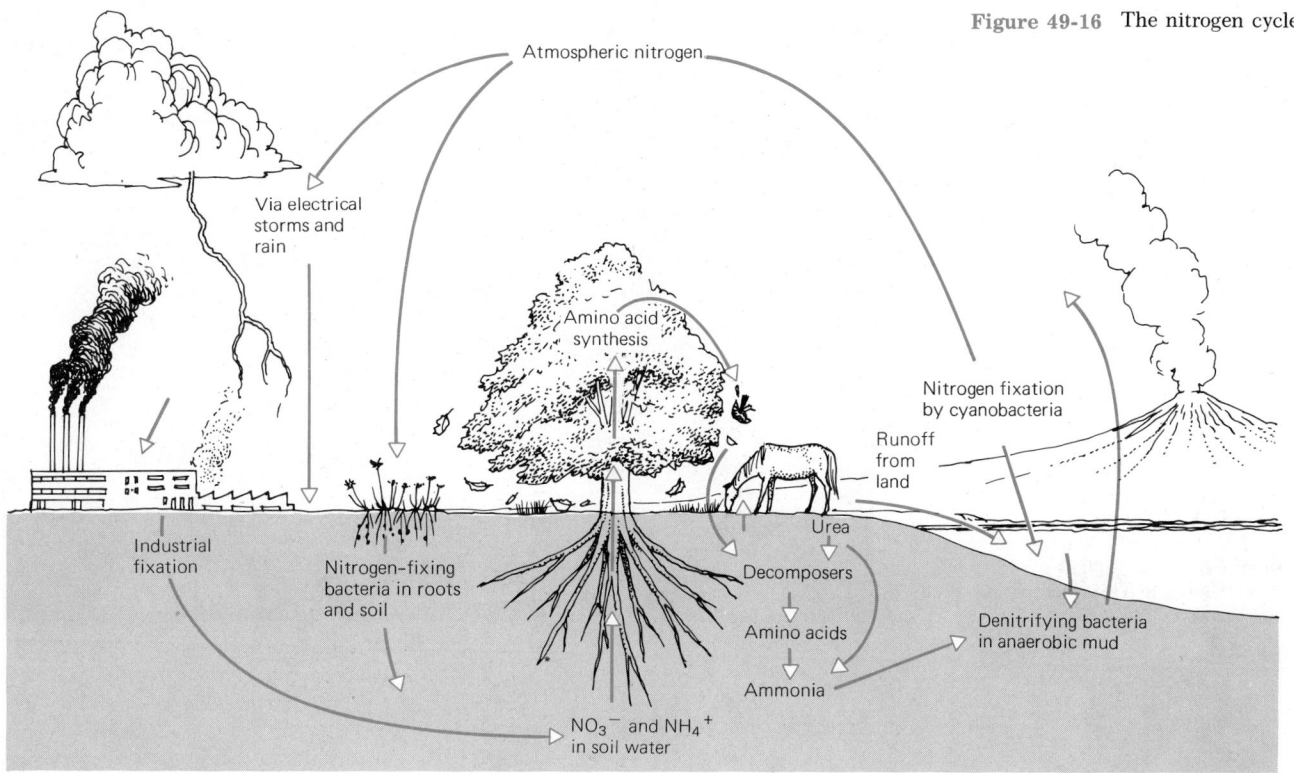

Figure 49-16 The nitrogen cycle.

for short-term cycling, since both organic humus and inorganic soil particles absorb phosphate ions, providing a local reservoir of this element in the soil and helping to promote a steady supply of it into and out of the living organisms of the ecosystem. Soil erosion robs the ecosystem of its phosphorus, and it may take thousands of years to recoup this loss through the weathering of rocks. Phosphate is much in demand as a fertilizer, but the richest and most easily mined deposits of phosphate, such as those near Tampa, Florida, are being depleted rapidly.

Figure 49-17 The phosphorus cycle.

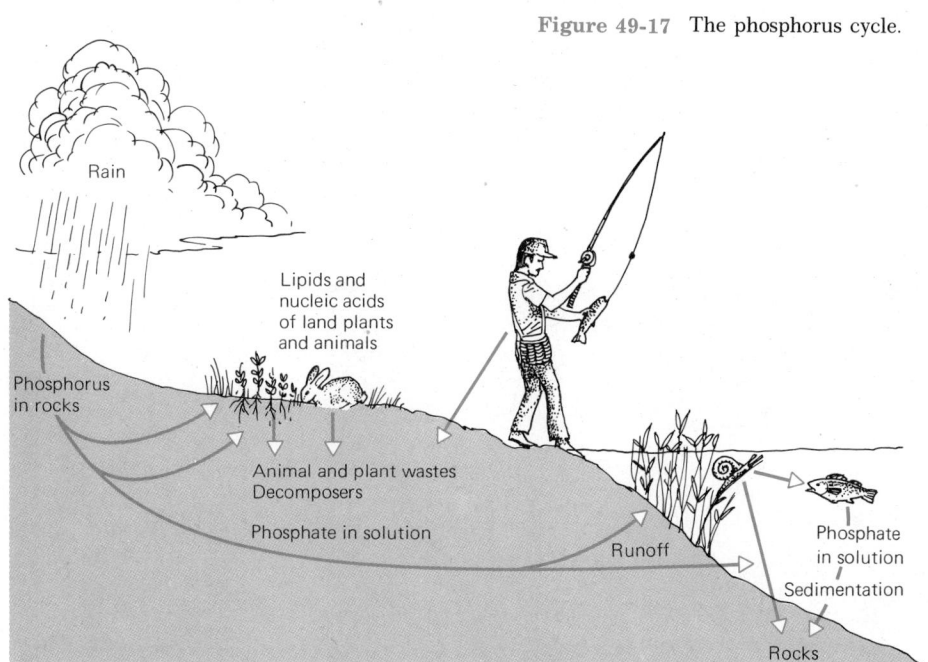

Figure 49-18 "V-notch" weir at the lower end of a deforested watershed at the Hubbard Brook Experimental Forest. The rate of water run-off from the watershed is measured by the height of water in the "V" (very low when this picture was taken). (Peter Marks)

An Experimental Ecosystem

Nutrient cycles in natural ecosystems, and the effect of human activities on such cycles, are now being studied experimentally on a relatively large scale in several parts of the world. One such ecosystem is the Hubbard Brook Experimental Forest in the White Mountains of central New Hampshire. This large watershed, in a region of temperate deciduous forest, has been studied continuously since 1963 by Herbert Bormann, Gene Likens, and their colleagues. The forest consists of a group of valleys, each with its own tributary creek running down the middle.

The first project in the Hubbard Brook ecosystem was to measure the inputs and outputs of water and nutrients from undisturbed forest. To do this, concrete "V-notch" weirs (Figure 49–18) were built across the creeks at the bottoms of six valleys. Since the weirs were anchored in impermeable bedrock, all of the water leaving each watershed (apart from that lost by evaporation) had to flow over a weir, where its flow could be measured and its nutrient content analyzed. Precipitation gauges, placed throughout the watersheds, were used to measure the input of rain and snow and their dissolved nutrients. Data from these undisturbed watersheds revealed that the forest is extremely efficient at retaining nutrients. Inputs of nutrients in precipitation approximately balanced the nutrient outputs in stream flow, and both were small relative to the total amounts of nutrients present.

Figure 49-19 An aerial view of part of the Hubbard Brook Experimental Forest in winter, showing the deforested watershed (unbroken white snow) and two adjacent watersheds. On the left of the picture is another watershed in which trees had been cut down in horizontal strips. (Robert Pierce, U.S. Forest Service)

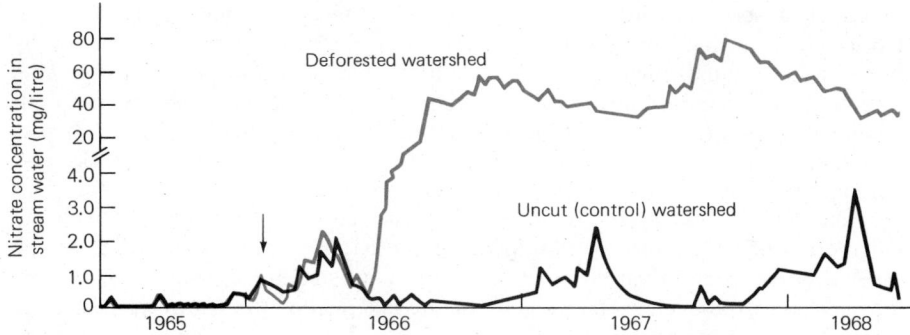

Figure 49-20 Accelerated loss of nitrate in stream water in one of the watersheds at the Hubbard Brook Experimental Forest after it was deforested. The date of deforestation is shown by the arrow. Note the change in scale on the y axis, needed to keep the lines for control and experimental watersheds on the same graph. (From Likens et al., "Effects of Forest Cutting and Herbicide Treatment on Nutrient Budgets in the Hubbard Brook Watershed-Ecosystem" (*Ecological Monographs* 40:23–47). Copyright 1970 by The Ecological Society of America.

The investigators next examined what happens to an ecosystem's nutrients when a forest is cut down. In the winter of 1965–1966 they cut down all of the trees and shrubs in one of the six sub-watersheds, leaving them where they fell and preventing regrowth by spraying with herbicides (Figure 49–19). Dramatic effects became obvious almost immediately. In comparison with undisturbed (control) watersheds nearby, water runoff through the weir in the deforested watershed increased by 40%; this water would ordinarily have left the ecosystem by transpiration (evaporation from plants via pores in their leaves). More important, the total rate of loss of inorganic nutrients dissolved in the stream water increased six- to eightfold. Loss of nitrate was particularly striking (Figure 49–20). Other experiments at Hubbard Brook have revealed that nutrient losses are reduced if the forest is cut in horizontal strips, leaving strips of standing trees, rather than being clear cut.

Measurements at Hubbard Brook and elsewhere have also revealed that rain in the eastern United States has become increasingly acidic, with a pH averaging 4.0 to 4.2, and sometimes reaching a pH of 3 or even lower. The acidity of rain increases from western states towards the east because prevailing winds are westerly, and because sulfur-containing industrial pollutants, which react with water vapor and form acids, accumulate in the atmosphere as an air mass travels across the country. Acid rain dissolves paint and stone from buildings, causing great economic damage, and it also affects natural ecosystems. A 1979 survey in the Adirondack Mountains of upstate New York showed that the pH of hundreds of ponds and lakes was less than 5, and that 264 of these had completely lost their fish populations. Water from melting snow is also very acidic, and it may kill the year's crop of fish eggs and hatchlings if it lowers the pH of the lake water during the breeding season.

49-G Lake Ecosystems

As an example of an ecosystem, we shall examine a lake in a little more detail, considering the physical characteristics of the lake and the community of organisms within it.

All ecosystems have structure, not only in horizontal and vertical dimensions but also in time, reflecting changes at different hours or seasons. Because of their sharp boundaries with surrounding land and air, lakes provide clear examples of ecosystem structure.

As light passes down through the waters of a lake, it is gradually absorbed and scattered. Much of it is used in photosynthesis by phytoplankton near the surface. Here the color of the light becomes green, partly because the chlorophyll of the algae selectively absorbs blue and red wavelengths. Further down there occurs a certain depth (which depends on the clarity of the water) at which oxygen production by photosynthesis equals the respiratory consumption of oxygen by all organisms. This is known as the **compensation depth** (Figure 49–21). The region above this depth is known as the **euphotic zone**, in which more organic matter is formed than is consumed. In the region below the compensation depth, decomposition is the predominant process.

Zonation of a lake according to light intensity has a profound influence on the distribution of organisms. Shallow parts of the euphotic zone, referred to as the **littoral zone**, contain various rooted aquatic plants such as water lilies, rushes, and sedges. In the open surface waters of the lake, called the **limnetic zone**, the dominant organisms are the floating algae, herbivorous rotifers and small arthropods, and carnivorous animals.

Beneath the limnetic zone, the chief input of energy is dead organic matter that falls from above. Various kinds of decomposers, fish, and invertebrates consume this dead matter or one another. On the lake bottom, the decomposers use up dissolved oxygen as they convert dead organic matter back into inorganic nutrients. In lakes with highly productive surface waters, so much dead matter falls to the bottom ooze, or **benthic zone**, that little dissolved oxygen is left in the water near the bottom. Under these conditions the dominant decomposers are anaerobic bacteria. In better-oxygenated waters, the benthic region supports a variety of flatworms, protists, clams, crustaceans, and insect larvae. Most of these species can tolerate severe oxygen deficiency.

Aquatic life is richest in the littoral zone around the lake edges, where sunlight penetrates to the bottom and supports rooted aquatic plants. The shallow waters are sometimes covered with duckweeds, and a variety of animals, such as insect larvae, rotifers, crustaceans, hydras, snails, and flatworms, wander among these plants. Closer to the shore are plants such as pond lilies, bullrushes, and cattails. Dragonflies and damselflies lay their eggs on the stems of these plants just below the water line, and the surrounding waters teem at certain seasons with tadpoles, fish fry, leeches, and insects such as diving beetles and water boatmen. Here pickerel, sunfish, and other fish find plenty of food and shelter among the plants.

Temperature has a profound influence on the seasonal activities in a lake. Water has the peculiar property of being most dense at 4°C. As a result, water at 4°C sinks beneath water that is either warmer or colder. In winter, therefore, water

Figure 49-21 The structure of a lake ecosystem.

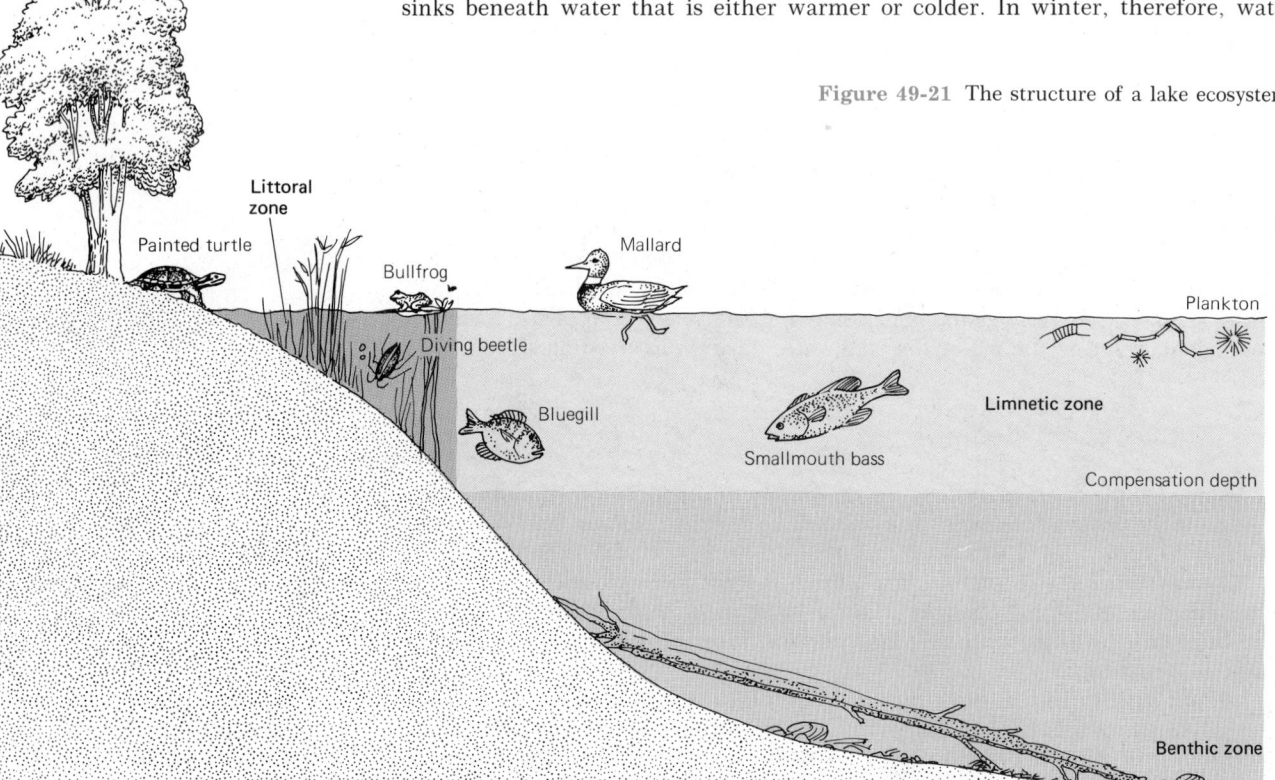

cooled below 4°C rises above water at 4°C, and at 0°C the surface water freezes. Under the ice the water remains at 0 to 4°C, with perhaps a slight rise at the bottom due to heat coming from the underlying bedrock; plants and animals survive in the water under the ice. In spring, the ice melts and the sun warms the surface waters, which sink as they approach 4°C, forcing colder water below to rise to the surface. This **spring overturn** is important because it brings nutrients from the bottom to the surface and carries oxygen down to the deeper waters.

You may have been swimming in the summer in a lake where the surface layer is lovely and warm, but if you let your toes sink they encounter much colder water. This is because the sun-warmed surface waters stay on top and do not mix with the denser, colder water underneath. As a result, the lake becomes **thermally stratified**, with an upper layer (**epilimnion**) of warm water, which may reach 25°C at the surface, and a deeper layer (**hypolimnion**), which never warms much above 4°C. Between the two there is a narrow region of very rapid temperature drop, the **thermocline**. During the summer, the depth of the epilimnion increases to as much as 20 m. In the fall the thermocline begins to rise again as the air and surface water temperatures fall, eventually leading to the **fall overturn**. Once again nutrients are recharged at the surface and oxygen is mixed into the deep waters. Trout fishermen know that spring and fall are the times when trout can be found in cold, rising surface waters. Trout need oxygen-rich water, and since warm water holds less oxygen than cold water, trout spend the summer in the deeper, colder waters.

Lakes in temperate regions go through two overturns each year, as described above. In the Arctic and in parts of the tropics, the water reaches 4°C only once a year, in midsummer or midwinter, and there is thus only one overturn.

We can divide lakes into categories based on their production of organic matter. **Eutrophic** ("good food") lakes are relatively shallow and rich in organic matter and nutrients; oxygen depletion occurs in the hypolimnion during the summer, owing to the heavy accumulation of dead organic matter and the rapid rate of oxygen use by aerobic decomposers. These lakes have a rich production of organic matter. In contrast, **oligotrophic** ("few food") lakes are usually characterized by

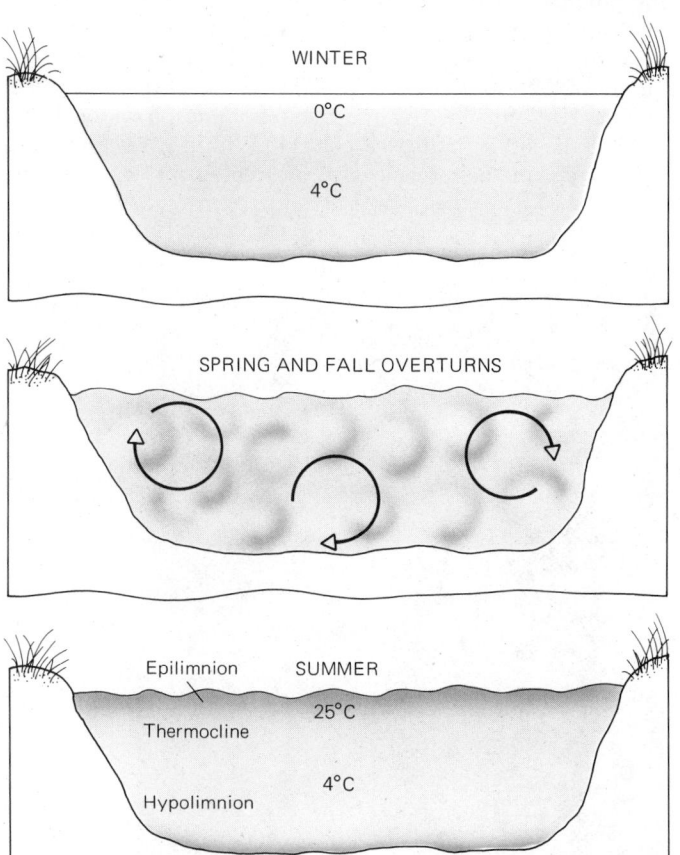

Figure 49-22 Seasonal stratification of temperature in a temperate lake.

WINTER

0°C

4°C

SPRING AND FALL OVERTURNS

Epilimnion   SUMMER

25°C

Thermocline

4°C

Hypolimnion

greater depth, steeper sides, and a poorly developed littoral zone. These lakes are poor in nutrients such as phosphorus, calcium, and nitrogen; hence there are few organisms, and therefore very little organic material, in such lakes. Their water is usually very clear, with the deep waters always containing oxygen. In the normal course of events, a lake ages as it is steadily filled in with sediment and organic matter. It usually becomes more eutrophic as it ages, and will eventually turn into a bog or marsh and finally into dry land. For a deep oligotrophic lake, such an aging process may take millions of years.

Pollution of Lakes

Broadly defined, **pollution** is an undesirable change in the physical, chemical, or biological characteristics of an ecosystem. Ecological research, along with research in the health-related sciences, has revealed that many human activities once thought to be innocuous actually create pollution. Human activities that affect lakes, such as the dumping of wastes or the building of power plants, are prime examples. Pollution cannot be avoided entirely, but it can be minimized. This requires an understanding of its effects on ecosystems, both natural and artificial.

Some forms of pollution in lakes are fairly obvious. Chlorinated hydrocarbons, such as DDT and the polychlorinated biphenyls (PCBs), or toxic heavy metals such as mercury, are a clear danger to natural ecosystems, just as they are to humans. Many of these substances are concentrated in living tissues because organisms have no means to excrete them. They accumulate and are passed on, at successively higher concentrations, to predators higher in the food chain. Their effects are therefore particularly damaging to carnivores such as hawks and fish-eating sea birds; for example, DDT is known to cause female birds to lay eggs with thinner shells, resulting in breakage before the egg hatches, as well as various deformities of the developing young.

Less likely forms of pollution, at first sight, are nutrient enrichment and thermal pollution. Phosphorus, for instance, is one of the elements most likely to limit the productivity of an ecosystem. Nowhere has this been demonstrated more dramatically than in lakes. David Schindler divided a small oligotrophic lake in Manitoba into two halves by suspending a plastic sheet across a narrow neck in the middle of the lake and anchoring the bottom of the sheet securely to the lake's rocky bottom. He then added a large amount of phosphate fertilizer to one half of the lake, leaving the other half as a control. Within a few weeks the fertilized half had become opaque as a result of a massive bloom* of plankton (Figure 49–23).

*Bloom Rapid increase in population.

Figure 49-23 The effect of phosphorus on the productivity of a lake. This lake in Manitoba was divided in two by plastic sheeting across the narrow neck in the middle of the photograph. Phosphorus was added to the half of the lake in the upper part of the photograph. Several weeks later, the phosphorus-fertilized half of the lake was opaque as a result of massive plankton bloom; the lower part of the lake remained clear and oligotrophic. (David Schindler)

When phosphorus is added in a single dose, as in this experiment, the effects are not long-lasting because phosphorus is continually removed from the system by deposition in bottom sediments or by stream outflow. On the other hand, if phosphorus is added continuously, as it often is in the form of sewage, phosphate-containing detergents, or fertilizer runoff from agricultural land, the productivity of a lake will remain high and the process of eutrophication will be hastened, changing the character of the lake, often irrevocably. Such changes have fundamentally altered Lake Erie, which is relatively shallow. Even deep oligotrophic lakes, such as Lake Tahoe on the California-Nevada border, have become noticeably more eutrophic in the last 20 years as a result of pollution. Rich productivity in the surface waters of such lakes provides more debris in the deeper regions, and the aerobic decomposers often cannot obtain enough oxygen to decompose all of this debris. Nutrient enrichment therefore leads to oxygen starvation and anaerobic conditions at depth.

Oligotrophic lakes with clear water are much more appealing and useful than eutrophic lakes covered with algae, clogged with weeds, and stinking from the by-products of anaerobic bacteria. For this reason, there has been strong pressure to ban the use of detergents containing phosphorus and to speed up the installation of tertiary sewage treatment plants, which remove organic and inorganic molecules from the water. In some cases, these measures have been able to bring lakes back towards a more oligotrophic condition.

Thermal pollution results from the addition of excessive amounts of heat to a lake. The production of organic matter in the limnetic zone of a lake depends, among other things, on the temperature of the water in this zone during the summer, and on the duration of the warm period. If extra heat is put into the lake—as for example in the effluent from a power station cooling plant—organic production is increased because the water becomes warmer and because the period of thermal stratification is extended. Greater production of organic matter means more work for the aerobic decomposers in the hypolimnion. A lengthening of the time between the spring and fall overturns may even cause the hypolimnion to use up all of its oxygen before the fall overturn replenishes the supply, leading to the same kind of anaerobic conditions that result from nutrient enrichment.

49-H   Community Structure in Space and Time

The open waters of lakes and oceans contain communities with a loosely defined structure. Their vertical distribution depends on the penetration of light and heat to different depths, and their horizontal distribution is influenced by winds and currents. In terrestrial communities, by contrast, structure is profoundly influenced by the organisms themselves, and especially by plants. The rooted immobility of plants produces a relatively permanent horizontal structure and the height of trees adds a vertical dimension to the community. Unlike many planktonic communities, terrestrial communities modify their physical environments—the distribution of light and moisture, the formation of soil and the cycling of nutrients, and even the velocity of the wind.

Vertical structure is most obvious in forest communities. The canopy trees in temperate forests intercept 50% or more of the sunlight reaching the community, and also absorb some wavelengths more than others as light passes through the leaves. Below the canopy there are usually many smaller trees, including immature canopy trees and mature trees of other species. Less than 10% of the initial sunlight may reach the next level down, the shrubs. Beneath the shrubs there is usually a layer of low-growing, soft-stemmed herbs that receives only 1 to 5% of the original sunlight striking the forest. Mosses and creeping herbs may provide yet another layer of vegetation close to the ground. Vertical structure continues down into the soil. The roots of different plants extend to different depths, and the soil itself is made up of various layers, or **horizons**, from the leaf litter on down. Here we have gradients not of light, but of plant nutrients and water. The vertical structure of a forest is obvious to us; to a grasshopper, the vertical structure of a meadow might be equally striking.

Animals, as well as plants, live a more-or-less vertically stratified existence within communities. For example, caterpillars are generally found on leaves above the ground, ground beetles in the litter at the soil surface, and earthworms at various depths below ground level. Some warblers (small birds) hunt for insects primarily in the canopy layer of the forest, others spend appreciable time in the intermediate layers, and yet others, such as the ovenbird, usually hunt amidst the leaf litter on the ground.

The two-dimensional horizontal structure of communities is characterized by patchiness. Patchiness may result from the irregular distribution of light gaps or other disturbed areas in various stages of ecological succession.* Patchiness may also result from the uneven distribution of light, moisture, and nutrients within a community and from minor differences in topography that affect these characteristics. Rattlesnakes, which raise their body temperatures by sunbathing, frequent sunny, south-facing slopes (in the Northern Hemisphere). Droppings from birds or mammals may release nutrients that favor the establishment of one plant species over another. Patches are usually randomly distributed.

A rare example of almost regular distribution of patches is found where the dispersion of birds and mammals is influenced by territorial behavior. Outside the breeding season, however, the distribution of even these animal species is influenced by the patchy occurrence of seeds, fruits, or other food.

Community structure changes more or less rhythmically over short time periods in response to the earth's relationship to the sun, which causes daily and seasonal cycles, and to the moon, which influences tidal cycles. As an example of a diurnal rhythm, different predators hunt the flying insects in a forest canopy at different times of day: flycatchers and swallows by day, nighthawks at dusk, and bats at night. Wildflowers on the forest floor in California are visited by pollinating insects primarily during the few minutes each day that sunflecks pass over them; the sunflecks make the flowers more conspicuous and raise the local temperature, making it advantageous for the ectothermic* insects to follow the sunflecks (Figure 49–24). During the summer months in much of North America, the community structure steadily changes as one species of herb after another sends up flowers and then sets and drops it seeds.

Seasonal effects are especially marked in communities exposed to long, severe winters or dry seasons. Most birds leave northern communities in winter, migrating south to join other communities and returning the following spring. Many mammals hibernate, and insects enter diapause, a resting stage in which the metabolic rate is very low. Annual plants spend the winter as seeds, while deciduous shrubs and trees shed their leaves. Decomposer activities slow down or cease. The activity of the community is restricted to evergreen plants, which can continue photosynthesis, and foraging by a few hardy endotherms such as chickadees, woodpeckers, mice, and deer.

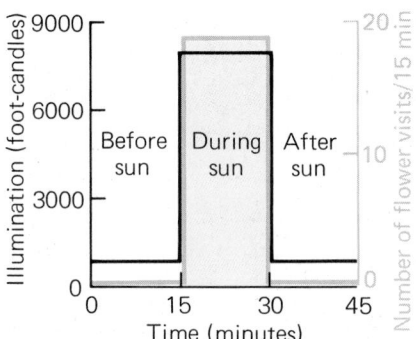

Figure 49-24 Visits by pollinating insects to violets on the floor of a redwood forest are restricted to about 15 minutes each day when a sunfleck passes over the plant. (After A. J. Beattie, *Madroño* 21:120, 1971)

49-I Species Diversity

Communities differ in their **species diversity**, the number of species they contain. The simplest measure of such diversity is the number of species in a given area. Using this measure, a two-species community with 98 individuals of one species and two individuals of the other would have the same diversity as a two-species community with 50 individuals of each species, and yet the characters of the two communities would be very different.

For this reason, ecologists often use more complex measures of diversity that take account of the relative abundances of the species as well as the actual number of different species. Such an approach usually reveals that there are a few **dominant** species, which are especially abundant or obvious in the community, and a host of species that are rare or inconspicuous. (Some rarer species may nonetheless play significant roles in the community; a small number of bees may have a profound influence on plant pollination, or a beetle that specializes in eating one kind of plant may keep it from crowding other species of plants out of the community.)

We might expect that one kind of decomposer would become so efficient at digesting and absorbing dead matter that it would drive competing species of decomposers to extinction, or that a species of "super-predator" would gradually monopo-

---

* Succession Series of changes that eventually produce a climax community (see Section 48-H).
* Ectothermic Regulating body temperatures by behavioral means.

Essay: An Ecosystem at the Drop of a Pat

An ecosystem need not be as large as a lake or a field—it can be as small as a cow pat. Two Finnish scientists, H. Koskela and I. Hanski, studied the structure and succession of the beetle community in cow dung in fields and in pine and spruce forests throughout southern Finland. The beetle community they studied consisted of 62,500 individuals belonging to 179 species. Within the community two basic feeding types were distinguished: the coprophages, or dung-eaters, and the carnivores, or predators.

According to Koskela and Hanski, the succession of a dung pat community progresses in three stages. The first stage, characterized by rapid changes in species composition, is dominated by the copro-phages, most of them species with highly efficient mechanisms for dispersal and thus with exceptional ability for finding freshly deposited "habitats." Colonization takes place so rapidly that species diversity peaks on the second day after deposition. The second stage, the stable phase, lasts longer than the first stage and is characterized by the arrival and increase of carnivorous species. The peak for carnivores occurs between the third and eighth days, shortly after their prey (the coprophages) reach their peak density. The last and longest stage, the final phase, is not characterized so much by species replacement as it is by gradual disintegration and disappearance of the pat due to the actions of weather and other abiotic factors.

Succession in a cow pat differs in some very fundamental ways from, for example, secondary succession from old field to forest. It is, for one thing, a heterotrophic succession; that is, the energy sources are greatest as the habitat opens for colonization and decrease continually as succession proceeds. In contrast to forest successions, both total biomass and species diversity are highest in the early stages. Another contrasting feature is that in a cowpat succession, predaceous species (carnivores) outnumber prey species (coprophages), a situation that appears to run counter to all the predictions of trophic balance in community ecology. There are several possible explanations for the strange pattern—that carnivores may have alternate food sources outside the cowpat community, that number of individuals rather than number of species is the critical parameter in maintaining trophic balance, or that carnivorous species are numerous but are restricted in distribution by habitat preferences (say, for dung in pine forests rather than fields). To distinguish the hypotheses, however, requires further study in what is obviously a fertile field for research.

(Contributed by May Berenbaum)

lize all available prey. Competition between species for limited resources sometimes does lead to the extinction of one of them, but obviously this is not always the case. What maintains species diversity? What prevents one or more species in a trophic level from eliminating the others through competition? One answer is that when species are in potential competition for a food supply, they may subdivide this resource in some way. Different insectivorous warblers, for instance, forage at different heights in a forest. Each warbler species is a **specialist** at feeding on insects at a particular range of heights above the ground; flycatchers, nighthawks, and bats feed on the same general food resource (flying insects), but at different times of the day or night. An important reason for species diversity within a particular habitat, therefore, is specialization in the use of limited resources.

A second contribution to the overall species diversity of a region is the variety of habitats it contains, each characterized by its own diversity. In a climax community,* species diversity may vary with gradients of elevation, soil type, and so forth. Another cause of diversity is the creation of different habitats within a region by periodic disturbance. Light gaps in a forest may be inhabited by different species of birds and insects from those in the nearby climax forest. In forests the climax plants are usually efficient competitors, and if it were not for frequent disturbance they would be able to monopolize the entire area and reduce diversity significantly. The great diversity of Australian tropical forest trees results largely from subtle habitat differences due to disturbances. Some of these occurred a very long time ago, but their flora is still distinct from surrounding vegetation.

Predation is also important in maintaining species diversity. Robert Paine studied this phenomenon in an intertidal community on the rocky coast of Washington. The community contains mussels, barnacles, and other species that feed by filtering water when the tide is in, and other species such as limpets that move around on the

---

*Climax community   The community that eventually comes to occupy an area if it is left undisturbed.

rocks as they graze attached algae. All are fed on by the top carnivore of the food web, a sea star called *Pisaster*. When the sea stars were removed from the rocks in an 8 × 2 m area, Paine found that *Balanus* barnacles settled successfully over most of the area and took up about 80% of the available space within three months. Later, however, the barnacles were gradually crowded out by two species of mussels that occupy the rock space more efficiently. After a year or so, the experimental area was dominated by the mussel *Mytilus californianus*, and species diversity of the community had dropped from 15 species to 8. Thus, in the natural community, the sea star predator maintains diversity by preventing some of its prey species from excluding the others by competition for space. Grazing by generalized herbivores can sometimes have similar effects. Several long-term experiments, in which voles,[*] rabbits, or larger mammals have been excluded from areas of grassland by fencing, revealed increased dominance by certain grasses or herbs and a reduction in overall species diversity with time.

Specialized herbivorous insects can contribute to community diversity. In tropical forests, almost all of the seeds and seedlings of some leguminous tree species die if they remain in the vicinity of their parent tree, because they are attacked by insects. Only if seeds are dispersed some distance from their parent tree do they have a reasonable chance of escaping the insects near the parent tree. This sort of predation probably contributes to the enormous diversity of tree species in tropical forests. Species that become common are most easily discovered by enemies, which quickly drive their food plants back to rarity.

Species Turnover in Communities

A 1968 survey of the number of species of birds breeding on each of the nine Channel Islands off the coast of southern California showed that the total number of species on each island had changed little since a similar survey 50 years earlier; however, the species composition had changed markedly. On San Nicolas Island, for example, there were 11 species of birds in 1917 and the same number in 1968, but only five of these species were the same. Six species had become extinct on this island, but six other species had colonized it.

The number of species on an island results from a balance between colonization and extinction. Species that colonize islands tend to have good means of dispersal, either by flying under their own power, by hitching a ride on a drifting log or on the

---

[*] Voles   Small rodents that look like short-tailed mice.

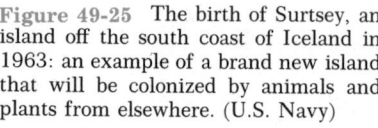
Figure 49-25 The birth of Surtsey, an island off the south coast of Iceland in 1963: an example of a brand new island that will be colonized by animals and plants from elsewhere. (U.S. Navy)

feathers of a flying bird, or by floating as seeds or spores through the air. Good colonizers also tend to be species that do not require highly specialized conditions to live and reproduce—a plant that self-pollinates, or that reproduces vegetatively, is more likely to establish a new population than is a plant that requires the presence of other individuals and of particular insect pollinators for cross-pollination. Colonizers also tend to have high reproductive rates, allowing them to establish a thriving population quickly. Islands are usually colonized by a disproportionate number of fugitive species (see Section 48-H), which tend to possess these characteristics. As the island fills up, organisms continue to arrive, but the rate of successful colonization declines because the established species are already exploiting the resources that newcomers need.

Island species are especially susceptible to extinction. This is partly because the population of each species is relatively small, and random events may kill all the individuals. Furthermore, island communities tend to have low species diversity, and often contain few predators; this can increase the likelihood of one species' crowding out some of its neighbors, or at least reducing their populations to such low numbers that they become extinct through random processes.

The immigration and extinction curves for any island should cross each other, and the point of intersection gives the island's "equilibrium" number of species (Figure 49–26). According to this model it is only the *number* of species that will remain the same. The *identity* of the species present may change, as some become extinct and others become established.

The number of species on an island at equilibrium depends on the island's distance from the source of species and on the island's size and diversity of potential habitats. Fewer species can reach an island that is far from the mainland, so it will end up with fewer species than an island closer in. Larger islands have more species at equilibrium, partly because they intercept more dispersing organisms, but probably mostly because they permit species to maintain larger population sizes, which are less susceptible to random extinctions.

Daniel Simberloff and Edward Wilson tested these predictions experimentally. First, they counted the number of species, mainly insects, on six small mangrove islands in Florida. Next they completely exterminated all of the arthropods on these islands by enclosing the islands in enormous plastic tents and fumigating them (Figure 49–27). They then removed the tents and sampled the fauna regularly. Within six months, the number of species on each island had returned to approximately the number present before fumigation, and remained at about the same level

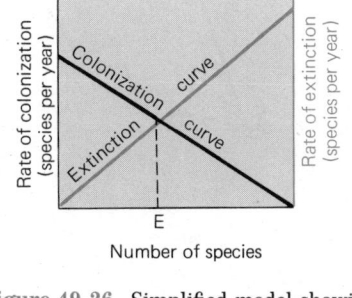

Figure 49-26 Simplified model showing how colonization by new species and extinction of existing species interact to determine the number of species on an island. As the number of species present increases, the rate of colonization by new species is expected to decrease and the rate of extinction of species already present is likely to increase. The point where the curves cross corresponds to an equilibrium number of species (E), at which extinctions balance colonizations.

Figure 49-27 Scaffolding completely encloses a small mangrove island in the Florida Keys, prior to enclosure with plastic sheeting. After elimination of the island's insect fauna by fumigation, tent and scaffolding were removed and the process of recolonization of the island monitored periodically. (Daniel Simberloff)

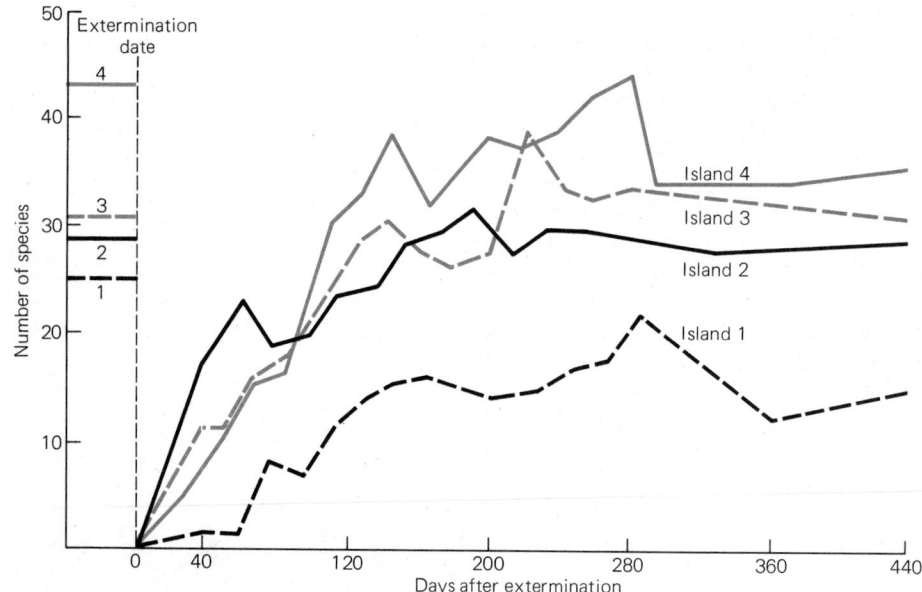

Figure 49-28 Recolonization of four small mangrove islands following extermination of their insect fauna. The numbers of species on each island before fumigation are indicated on the left. (From Simberloff and Wilson, "Experimental Zoogeography of Islands" (Ecology 51:934–37). Copyright 1970 by the Ecological Society of America)

for as long as observations were continued (Figure 49–28). As predicted, however, the species compositions were not identical to those before the experiment, and species turnover continued on each island, with extinction approximately balanced by immigration.

Species turnover is not restricted to island communities. Turnover is slow in most communities on continents because population sizes are usually large and local extinction is easily offset by immigration from nearby areas. Small, isolated communities such as mountain tops, bogs, and lakes, however, may have appreciable species turnover rates. An understanding of island communities promises to be a valuable tool in the design of parks and nature reserves, artificial "islands" where decisions must be made as to what area of habitat should be preserved and whether more species will be saved if the park contains one large or several smaller pieces of the habitat.

SUMMARY

An ecosystem consists of a community of organisms that are dependent on one another in various ways, plus their physical environment. Two of the most important factors determining the biomass and makeup of an ecosystem are its productivity and nutrient cycles. The productivity of an ecosystem is determined by the availability of light, water, and minerals for photosynthesis. Energy passes from one trophic level to another, with approximately 90% being lost at each step. In consequence, the biomass that an ecosystem can support at each successive trophic level declines rapidly, and energy flow is, for all intents and purposes, one-way.

Nutrients cycle through an ecosystem. They are taken in by organisms as inorganic substances and may remain as minerals or be incorporated into organic molecules. Nutrients may pass along the food chain for a time, but eventually they are once again released into the environment as inorganic substances. An ecosystem may be very efficient at conserving and recycling its nutrients. The availability of nutrients often limits the productivity of an ecosystem.

Lakes are convenient ecosystems to study because their boundaries are clearly defined. The distribution of organisms in a lake is determined by the depth of the water and by the distribution of light and nutrients. The seasonal mixing of different layers of water is vital to the lake's nutrient cycling. Nutrient pollution may promote the growth of some organisms at the expense of others, leading to an imbalance that may destroy the character of the ecosystem. Thermal pollution may increase the metabolic rates of some organisms while making oxygen less available for respiration. It may also change the pattern of mixing in the lake so that oxygen and nutrients are not replenished on schedule.

Each community has a structure, determined by the distribution of organisms and by the height and shape of the plants. Community structure may change in response to seasonal, diurnal, and tidal cycles. Species diversity is maintained by the ability of potentially competing species to subdivide limited resources (such as food) within habitats, and to specialize in using different habitats. Species diversity may also be maintained by predation, which can prevent one prey species from eliminating others by competition.

The study of islands shows that there is species turnover in communities, with new species moving in and resident species becoming locally extinct. The balance between immigration and extinction establishes an equilibrium number of species, which varies from one community to another depending on the community's size, diversity of habitats, and distance from the source of colonists.

OBJECTIVES

From your study of this chapter, you should be able to:

1. Define the following words and use them in context: ecosystem, community, producer, (primary and secondary) consumer, herbivore, carnivore, decomposer, autotroph, heterotroph, food chain, food web, trophic level, biomass, productivity, eutrophication, oligotrophic.

2. Define gross and net primary productivity, and discuss factors that affect them.

3. Outline energy flow through an ecosystem and give rough estimates of the energy loss between adjacent trophic levels.

4. Describe a simplified nutrient cycle for carbon, nitrogen, and a non-atmospheric element such as phosphorus, and point out the important differences between them.

5. Describe the structure of a lake ecosystem, and explain how thermal and nutrient pollution hasten eutrophication.

6. Describe what is meant by community structure, and give some examples of changes in community structure in space and time.

7. Explain how the species diversity of a community is affected by interspecific competition, predation, and the number and types of habitats it contains.

8. Define species turnover, and explain how species diversity may be affected by a community's area (size) and degree of isolation from similar communities.

SELF-QUIZ

1. Using the items listed below, diagram a food web; indicate the trophic level of each organism.

deer	herbivorous insect
soil microbes	spider
shrub	sparrow
wolf	hawk

2. The annual primary production of any ecosystem is greater than the annual increase in biomass of the herbivores in that ecosystem because:
   a. plants are more efficient in converting energy input to mass than are animals
   b. energy is lost during each energy transformation
   c. there are always more plants than there are plant eaters
   d. woody plants live much longer than most herbivores

3. Of the total amount of energy that passes from one trophic level to another in a food chain, about 10% is:
   a. transpired
   b. "burned" in respiration
   c. stored as body tissue
   d. reradiated in the form of heat
   e. passed out in the feces

4. Nutrient cycles may involve:
   a. movement of the nutrient from the organism to the atmosphere
   b. movement of nutrients into the soil
   c. limitations on the number of organisms in the ecosystem due to shortage of some nutrients
   d. loss of the nutrient from the ecosystem due to abiotic forces
   e. all of the above

5. Wolves and lions may be said to occupy the same trophic level because:
   a. they both eat primary consumers
   b. they both utilize their food with about 10% efficiency
   c. they both live on land
   d. they are both large mammals
   e. they both have a wide range of dietary items

6. Eutrophication
   a. is a non-reversible process
   b. is often caused by excessive phosphorus input
   c. is caused by inhibition of algal blooms
   d. decreases the productivity of a lake
   e. need never happen to a lake if human beings take proper precautions to avoid detergent, sewage, and thermal pollution in the watershed.

7. State whether the species diversity of a community would be likely to increase or decrease under each of the following conditions:
   ___ a. increased frequency of disturbances
   ___ b. extermination of a predator that preys on many other species
   ___ c. partitioning of a shared resource among competing species

8. Species turnover is likely to be especially fast on islands that are:
   a. far from another land mass
   b. small
   c. occupied by climax communities
   d. extinct volcanoes

849

QUESTIONS FOR DISCUSSION

1. Is more energy lost from an ecosystem when an herbivore eats a plant or when a carnivore eats an animal? Why?

2. Is there one or more than one food web in any ecosystem?

3. How is the flow of energy in an ecosystem linked to the flow of nutrients? How do energy and nutrient flow differ?

4. The productivity of an ecosystem increases during the course of ecological succession. Why is this so?

5. Robert MacArthur found that the number of bird species in forests is not correlated with plant species diversity but is correlated with the amount of layering of foliage at different heights. Can you account for these findings?

6. How can frequent fires increase the species diversity of a region?

7. Which are likely to be the better colonizers of islands, dandelions or oak trees? Why?

8. Suppose you are a member of a congressional committee developing legislation that would set aside 100,000 acres of land within a particular region for a biological preserve or national park. The committee is faced with the question of whether to recommend government purchase of one large area or several smaller areas. What would your advice be if the prime objective was to preserve (1) an endangered large mammal population (such as the Texas red wolf), (2) as many species as possible, (3) as many local habitats as possible, and (4) the best possible compromise of these?

REFERENCES AND FURTHER READING

Bell, R. H. V. "A grazing ecosystem of the Serengeti." *Scientific American*, July 1971. An excellent study on partitioning of food by herbivores, showing how different specializations are coupled to the animals' physiology and behavior.

Bormann, F. H., and G. E. Likens. "Nutrient cycles of an ecosystem." *Scientific American*, October 1970. A short account of the Hubbard Brook watershed study of nutrient cycles.

Corliss, J. B., and R. D. Ballard. "Oases of life in the cold abyss." *National Geographic*, October 1977. Description of the chemosynthetic ecosystem discovered in the ocean depths.

Forbes, S. A. "The lake as a microcosm." *Bulletin of the Illinois Natural Historical Survey* 15:537, 1887. Forbes's original description of a lake as an ecosystem.

Gosz, J. R., *et al.* "The flow of energy in a forest ecosystem." *Scientific American*, March 1978.

Janick, J., C. H. Noller, and C. L. Rhykerd. "The cycles of plant and animal nutrition." *Scientific American*, September 1976.

Koskela, H., and I. Hanski. "Structure and succession in a beetle community inhabiting cow dung." *Ann. Zool. Fennici* 14:204, 1977. The reference for Dr. Berenbaum's essay at the end of this chapter.

Likens, G. E., F. H. Bormann, R. S. Pierce, J. S. Eaton, and N. M. Johnson. *Biogeochemistry of a Forested Ecosystem.* New York: Springer-Verlag, 1977. The detailed story of the Hubbard Brook studies.

Likens, G. E., *et al.* "Acid rain." *Scientific American*, October 1979.

Odum, E. P. *Fundamentals of Ecology*, 3d ed. Philadelphia: W. B. Saunders, 1971.

Pough, F. H., and R. E. Wilson. "Acid precipitation and reproductive success of *Ambystoma* salamanders." *Water, Air, and Soil Pollution* 7:307, 1977. A description of one deleterious effect of acid rain.

Schindler, D. W. "Eutrophication and recovery in experimental lakes: implications for lake management." *Science* 184:897, 1974.

Scientific American: The Biosphere, September 1970 issue. A collection of articles relating to energy and nutrient flows, including in order of appearance:

Hutchinson, G. E. "The biosphere."
Oort, A. H. "The energy cycle of the earth."
Woodwell, G. "The energy cycle of the biosphere."
Penman, H. L. "The water cycle."
Cloud, P., and A. Gibor. "The oxygen cycle."
Bolin, B. "The carbon cycle."
Delwiche, C. C. "The nitrogen cycle."
Deevy, E. S., Jr. "Mineral cycles."
Brown, L. A. "Human food production as a process in the biosphere."
Singer, S. F. "Human energy production as a process in the biosphere."
Brown, H. "Human materials production as a process in the biosphere."

Simberloff, D. S., and E. O. Wilson. "Experimental zoogeography of islands—a two year record of colonization." *Ecology* 51:934, 1970. The story of experiments in which mangrove islands were fumigated to kill their animal populations and their recolonization followed.

Whittaker, R. H. *Communities and Ecosystems*, 2d ed. New York: Macmillan, 1975.

CHAPTER

50

POPULATIONS

In the two previous chapters we have considered the importance of energy flow, nutrient cycling, and climate to the survival and distribution of organisms. In this chapter, by contrast, we examine populations, the basic biological units of ecosystems, and discuss the interactions that affect their size and distribution in ecological time.

A **population** is all of the members of a species that occupy a particular area at the same time. Examples are the bluegill population of a pond, the deer population of Vermont, and the human population of the United States. As a group, a population has a number of characteristics that are not typical of its component individuals. For instance, each population has a characteristic niche (Section 50-A), size, density, dispersion, and age structure.

The numerical size of a population at any one time depends upon the balance between birth and immigration,* which add individuals to the population, and death and emigration,* which remove them:

---

* Immigration Taking up (more or less permanent) residence in a new area.
* Emigration Leaving an area of residence for some other place.

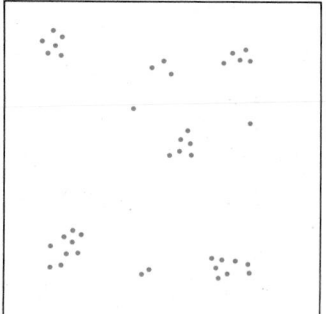

(a)

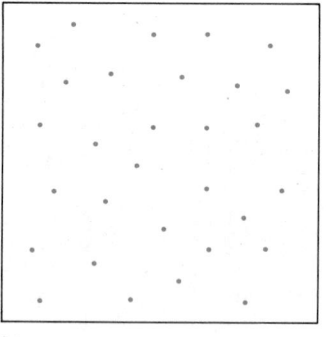

(b)

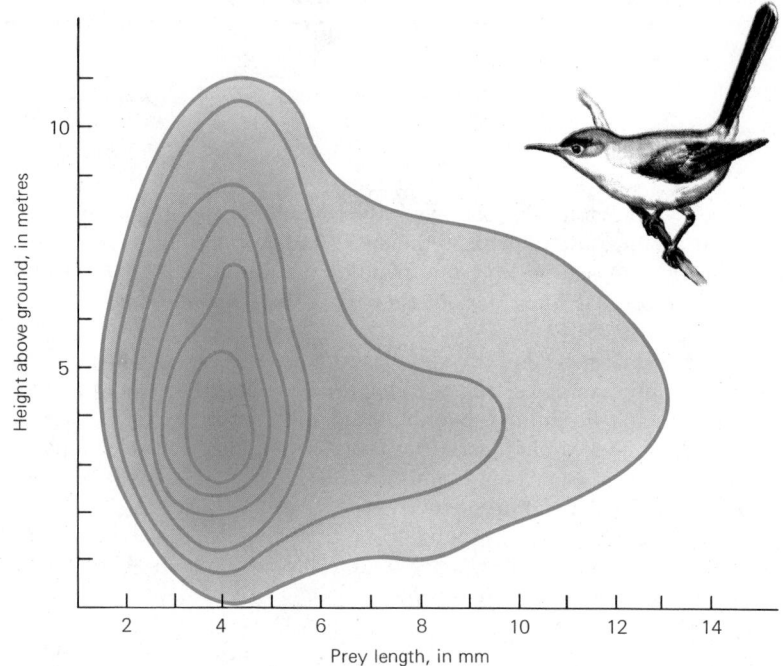

(c)

Change in population size = (Births + Immigration) − (Deaths + Emigration)

If the term (Birth + Immigration) exceeds the term (Death + Emigration), the population will grow; if the terms are equal, the population will remain the same size; and if the second term exceeds the first, the population will decrease.

Population **density** is the number of individuals of a population per unit area or volume—the number of oak trees per hectare,* or of *Daphnia** per cubic metre of water, for example. **Dispersion** is the pattern of population distribution; individuals may be distributed randomly or in clumps (Figure 50–1). A third possibility, regular distribution, is much less common in nature. The **age structure** of a population is the proportion of individuals of each age, and as we saw in Chapter 15, populations also have various genetic characteristics.

50-A   Habitat and Niche

The **habitat** of an organism or population is the place where it lives. Every habitat is characterized by particular ranges of temperature, humidity, soil or vegetation structure, food type, competitors, predators, and other factors that make up the organism's environment. The **niche** of a population or species is its functional role in an ecosystem. Using a human analogy, the niche is the species' profession or way of life, whereas the habitat is where this way of life is carried on—its address. The responses of a population to various characteristics of its habitat are part of its way of life and, therefore, of its niche.

To illustrate niche, let us consider a study of the blue-gray gnatcatcher, a small insectivorous bird, in oak woodlands in California. Two "dimensions" of the gnatcatcher's niche were measured: the lengths of insects caught by gnatcatchers and the heights at which the birds foraged in the trees. When plotted together as axes on a graph, they reveal the bird's foraging activities in two dimensions (Figure 50–2). A third axis could be added to the graph, representing perhaps the distribution of foraging activity over 24 hours, or its variation with temperature. This would give us three dimensions. Any organism's niche has many more than three components.

---

* Hectare   10,000 square metres (about 2.5 acres).
* *Daphnia*   A small, freshwater arthropod.

Figure 50-1 Dispersion. The individuals in most populations are (a) spaced randomly or (b) clumped. (c) In some populations, such as populations of some territorial birds during the breeding season, individuals are more evenly spaced.

Figure 50-2 Two niche dimensions of the blue-gray gnatcatcher in oak woodlands in California. The contour lines, moving in towards the middle, indicate increasing frequencies of capture of prey (mostly insects) in the trees. Prey are captured most frequently between 3 and 5 m above the ground, and prey about 4 mm long are captured more often than smaller or larger prey. (After Whittaker, Levin and Root, *American Naturalist,* 107. Copyright 1973, University of Chicago Press)

Consider the niche of a cottontail rabbit, a common inhabitant of rural and suburban areas. The rabbit is a herbivore; to live, it must eat plants, and it eats some plants more than others. It tramples and digs burrows in the soil, which alters the soil; it produces feces, little ecosystems in themselves; it drinks water and eats salt and other minerals, removing these from wherever they are found and depositing them as urine or dung; it attracts flies that feed on its feces; it may be eaten by foxes or hawks, and so forth. Every interaction of the rabbit with its environment is part of its niche and determines where it can live and what other organisms can coexist with it.

50-B Population Growth

The size of a population of mice in a field or of violets in a woodlot seems, at first sight, to vary little from year to year. Is this really the case? Surely these organisms produce so many offspring that their populations could increase greatly from one year to the next. What limits the size of natural populations?

To answer these questions, it is convenient first to find out how rapidly populations could increase if nothing stopped their growth.

The Russian ecologist G. F. Gause studied the growth of populations of the protist *Paramecium caudatum*. Every few hours a well-nourished paramecium divides to form two new individuals. Gause set up tubes containing ample bacteria for food and introduced one paramecium into each. He then followed the growth of the paramecium populations. If nothing checked its growth, a population showed **exponential growth**; that is, as time went on, the number of individuals added in each time period kept increasing. When the population size is graphed on a linear axis, this exponential growth plots as a curve that grows steeper and steeper; when the population size is plotted on a logarithmic axis, the exponential growth plots as a straight line (Figure 50–3). The equation describing this growth is

$$\frac{dN}{dt} = r_m N$$

where N is the number of individuals in the population, $\frac{dN}{dt}$ is the rate of change in numbers with time, and r_m is the **innate capacity for increase** or **biotic potential** of the population. That is, r_m is the maximum rate of population growth, per individual, that is achieved when food and space are superabundant and there is no interference from members of other species.

Biotic potential is as much an adaptation of an organism as is size or shape. It is much more difficult to measure, though, because optimum conditions for growth almost never occur except under artificial laboratory conditions. What we see in nature is the outcome of interaction between a population's biotic potential and various environmental circumstances that restrict its actual growth rate. This **actual rate of population increase** at any time is given the symbol r. In most natural populations, r (the difference between the birth rate and death rate per individual per unit of time) varies continually in response to interactions between a population and its environment.

Exponential growth occurs in nature when a population has a superabundant supply of resources. History records many cases of exponential growth by species imported, intentionally or accidentally, into areas where resources were available and natural enemies or competitors were lacking. For example, dandelions, starlings, and house sparrows introduced into the United States underwent dramatic population explosions. Similar population explosions occur when bacteria invade the intestinal tract of a newborn animal or when decomposers invade a freshly dead animal or plant. Exponential growth does not necessarily mean that the population is growing at its biotic potential. The human population started growing exponentially in the mid-eighteenth century, although women were not bearing infants as frequently as is biologically possible. In addition, many people who could have reproduced did not. Most human populations increase at a rate of 3% or less per year ($r = 0.03$ per year).

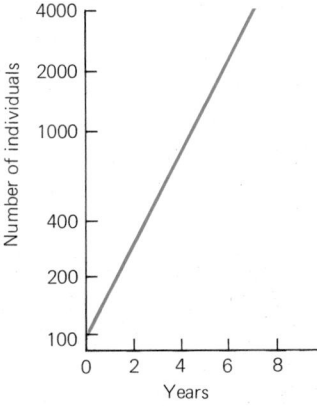

(a)

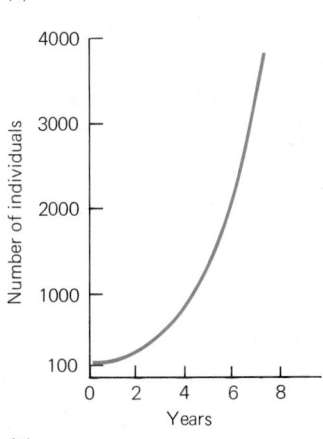

(b)

Figure 50-3 Exponential (= geometric) growth of a hypothetical population expressed on (a) a logarithmic scale, (b) a linear scale. The equation describing these graphs is

$$\frac{dN}{dt} = r_m N$$

where $\frac{dN}{dt}$ = Increase of population size per unit time

r_m = Innate capacity for increase

N = Population size

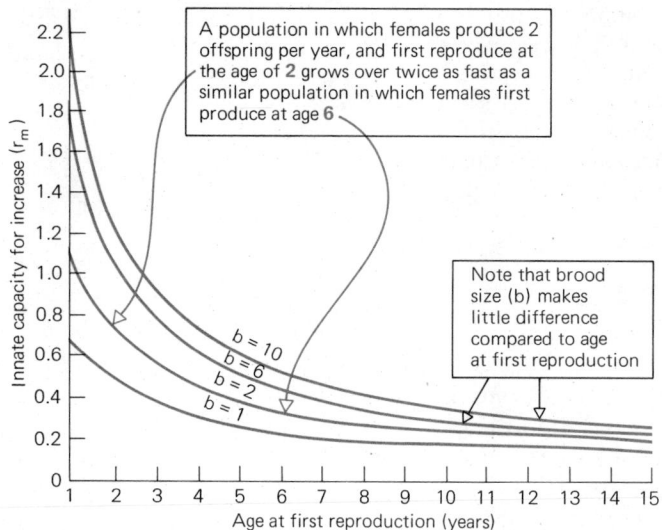

A population in which females produce 2 offspring per year, and first reproduce at the age of 2 grows over twice as fast as a similar population in which females first produce at age 6

Note that brood size (b) makes little difference compared to age at first reproduction

Figure 50-4 The effect of females' age at first reproduction on the innate capacity for increase (biotic potential) of a population. This is a population of long-lived organisms in which young are produced in one brood or litter once a year. b = the number of young produced per year. (After Cole, 1954)

Factors Affecting Biotic Potential

The biotic potential differs from one species to another; bacterial populations can grow faster than populations of oak trees. An individual's rate of reproduction can be increased in any or all of three ways: (a) by producing a large number of offspring each time it reproduces, (b) by having a long reproductive life, or (c) by reproducing as early in life as possible. Of these three factors, the last is by far the most important. A bacterium neither lives for a long time nor produces many offspring each time it reproduces. Its reproductive potential is higher than that of a dog because most bacteria can reproduce within an hour after being formed by cell division, whereas a dog is not able to reproduce until it is about 6 months old. The shorter the generation time of a species, the higher its reproductive potential (Figure 50–4). Lamont Cole, who drew attention to the significance of age of first reproduction, showed that a woman who bears one child annually for 3 years, starting at age 13, contributes as much to the growth rate of a human population as a woman who bears five children but starts at age 30.

In the case of organisms with equal generation times, the number of offspring produced determines which has the higher potential for population growth. Thus the population of a plant that produces 100 seeds a year can potentially grow faster than the population of a plant that produces only 10 seeds a year. The longer the pre-reproductive period, however, the less the effect of the number of progeny on biotic potential. Cole calculated that the biotic potential of a human population in which women lived forever, producing a child every year starting at age 20, would be only twice that of a population in which women bore only five children altogether, one each year from the age of 20!

Reproductive Strategies and Survivorship

The number of offspring produced and the age of first reproduction are part of the evolved **reproductive strategy** of a species. The term strategy in this context does not, of course, imply conscious planning by the organism; rather, natural selection has favored organisms with different reproductive performances under different circumstances. There are two extremes in reproductive strategies. At one extreme, parents may produce many small individuals, and provide each with a very small amount of food and parental care. At the other extreme, an organism may produce very few offspring but invest a great deal of energy in each one before it becomes independent of the parents. As we might expect, the rate of survival among the young tends to be higher in the second case. Where many small offspring are produced and receive no care, there is extremely high mortality among the young. Because they have produced so many offspring, however, the parents have a good

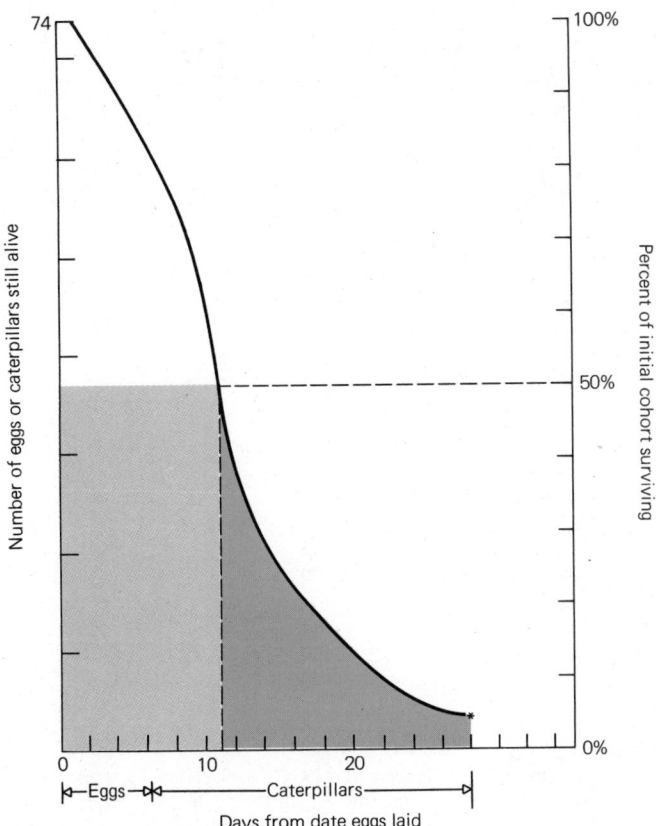

Figure 50-5 Survivorship curve for eggs and caterpillars of the black swallowtail butterfly on wild carrot plants in a New York hayfield. The initial cohort of 74 eggs was found by following female butterflies and placing flagged stakes nears the plants on which they laid eggs. These were then visited daily to record their fate before and after they hatched into caterpillars. Data are unavailable beyond May 28 because the large caterpillars wander from the plants before pupating. 50 percent of the insects had died by Day 11; that is, an egg laid on Day 1 had only a 50 percent chance of surviving to Day 11. Most of the deaths are due to predation by spiders and insects. (Data courtesy of Paul Feeny)

chance that some of their offspring will survive to become parents of the next generation.

By following the fate of young individuals throughout their lives, we can construct **survivorship curves** that describe mortality in relation to age (Figure 50–5). To construct a survivorship curve, we start with a **cohort** of many individuals newly added to the population, and follow them to determine the age of death of each member of the cohort. The study is completed when the last individual dies. We then plot number of survivors against ages. Survivorship curves are usually adjusted to portray what would have happened if the initial cohort had been some convenient round number like 100 or 1000. By labeling the vertical axis from 0 to 100, the survivorship curve can be used to estimate the probability (in %) that an individual of starting age will reach any particular age, x (Figure 50–5).

In a Type I survivorship curve, most individuals survive for a long time, and die as a result of the diseases of old age (Figure 50–6). This type of survivorship curve is usually associated with a reproductive strategy in which the parents devote considerable energy and care to their offspring. A "perfect" Type I curve never occurs because there is always some early mortality. Most human populations in developed nations approach a Type I survivorship curve after the first year of life, with its high infant mortality from genetic or developmental defects or birth accidents. A person who survives the first year of life is likely to live for another 60-plus years (Figure 50–7).

In a Type III survivorship curve, most individuals meet their demise at an early age, as eggs or larvae. This type of survivorship is associated with a "lay them and leave them" reproductive strategy, where large numbers of offspring are produced. It is characteristic of many species of invertebrates, bony fish, plants, and fungi.

The Type II curve falls between Types I and III. There is again an initial period of high mortality due to defective genes or accidents during development, birth, or hatching. However, once past this critical period, an individual is just as likely to die at any subsequent age; the chances of dying or being killed are equal throughout life. This type of curve is typical of several birds and of human beings exposed to poor nutrition and hygiene. In this case, reproductive strategy also lies somewhere

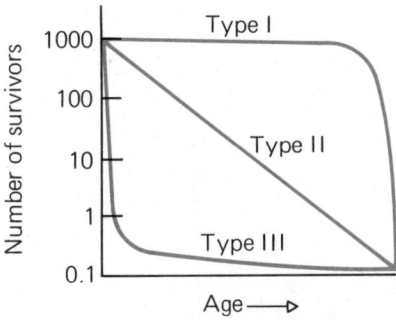

Figure 50-6 Main types of hypothetical survivorship curves (note logarithmic scale on vertical axis). Survivorship curves for real populations may fall between these curves. Type I and Type II curves are rarely, if ever, exactly as shown, because mortality among very young individuals is nearly always higher than that of older individuals.

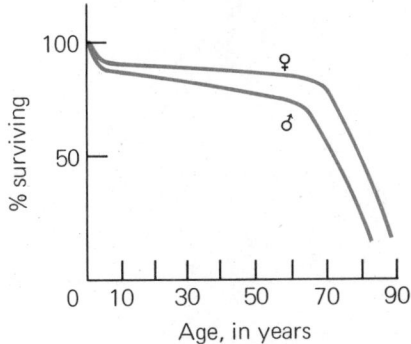

Figure 50-7 Survivorship curves for Americans based on the 1970 census data. Human survivorship approximates a Type I curve.

TABLE 50-1 LIFE TABLE FOR THE TOTAL POPULATION OF THE UNITED STATES, 1975. Note the high infant mortality. A child aged 1 year has a higher life expectancy than a 6-month-old baby.

AGE INTERVAL	OF 100,000 BORN ALIVE		AVERAGE REMAINING LIFETIME
Period of life between two exact ages stated in years	Number living at beginning of age interval	Number dying during age interval	Average number of years of life remaining at beginning of age interval
0–1	100,000	1,606	72.5
1–5	98,394	275	72.7
5–10	98,119	175	68.9
10–15	97,944	176	64.0
15–20	97,768	498	59.1
20–25	97,270	671	54.4
25–30	96,599	658	49.8
30–35	95,941	724	45.1
35–40	95,217	991	40.4
40–45	94,226	1,522	35.8
45–50	92,704	2,348	31.4
50–55	90,356	3,488	27.1
55–60	86,868	5,077	23.1
60–65	81,791	7,198	19.4
65–70	74,593	9,064	16.0
70–75	65,529	12,113	12.8
75–80	53,416	14,408	10.2
80–85	39,008	14,465	8.0
85 and over	24,543	24,543	6.2

Data courtesy of the U.S. Department of Health, Education and Welfare, National Center for Health Statistics.

between the other two. Parents usually produce more offspring than members of Type I populations, a situation selected for by the higher mortality rate of the offspring.

Survivorship within a population can also be represented in the form of a **life table**, a summary of the mortality operating on cohorts of each age group (Table 50–1). Life tables for human populations are used by life insurance companies to predict how much longer people of a given age are likely to live; this determines the price of insurance for people of various ages. Life tables for populations of animals and plants are useful aids for summarizing and analyzing the effects of different sources of mortality acting on populations (see Section 50-F).

50-C Population Forecasts

It is often useful to be able to predict how the size of a population will alter in the future. To do this, we need information about survivorship, age structure (Figure 50–8), and reproductive rates.

A **stationary** population remains the same size over time. In a stationary population, the number of individuals added by birth and immigration must be equal to the number removed by death and emigration in the same period. However, such a population is not necessarily stable; a **stable** population is one with an unchanging age structure. A stable population may be steadily increasing, decreasing, or stationary.

Whether a population is stationary, increasing, or decreasing can be found from life tables for female survivorship that also include the female fertility in each age

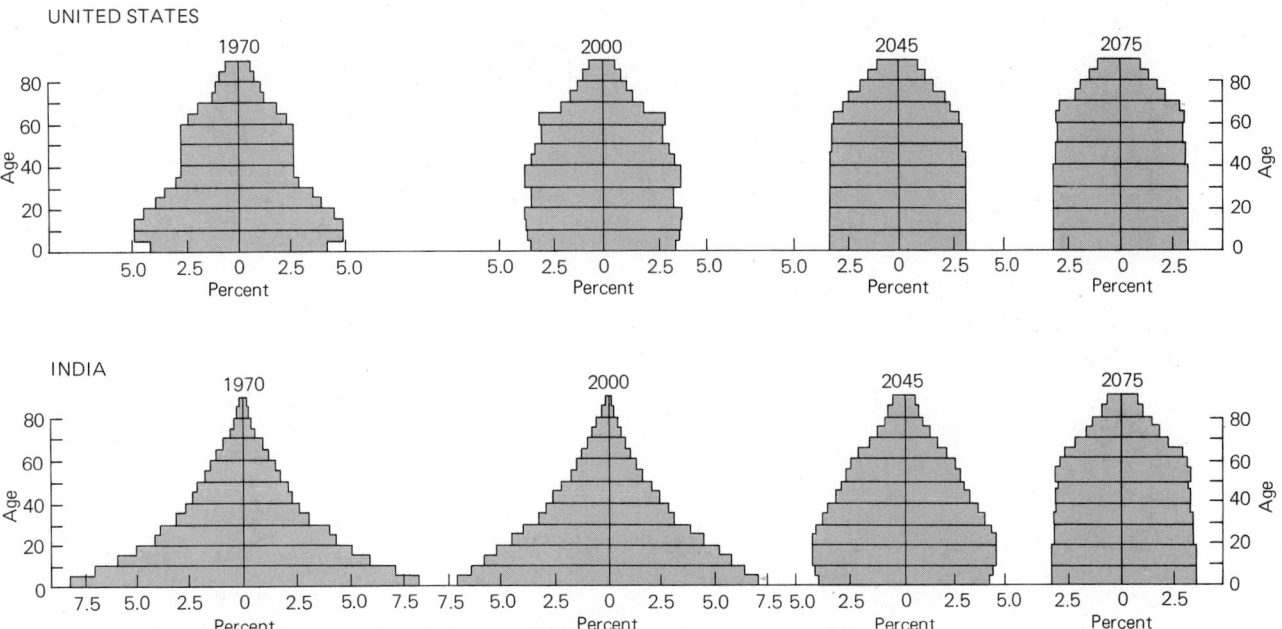

Figure 50-8 Age structures of populations. The percentage of the population in each age group is shown. These are projections of the changing age structures in India and the United States as population growth slows down. The age structure for India in 1970 is characteristic of a very rapidly growing population. The age structure projected for 2075 in both countries is characteristic of a stationary human population. (Population Reference Bureau, Inc.)

category. By multiplying the number of females remaining alive at each age by the average number of female offspring produced by a female at that age, we find the expected total number of female offspring produced by all females from the original cohort to reach that age. We can then add these totals for all age groups to find the total number of female offspring produced (Table 50–2). If this number exceeds the number of females in the original cohort, the population is increasing; if it is lower, the population is declining.

TABLE 50-2 **SURVIVORSHIP AND FERTILITY TABLES FOR WOMEN IN THE UNITED STATES, 1967**

AGE PERIOD (x)	AVERAGE NUMBER OF FEMALES (FROM INITIAL COHORT OF 100) SURVIVING TO MIDDLE OF AGE PERIOD x (l_x)	AVERAGE NUMBER OF FEMALE OFFSPRING PER FEMALE DURING AGE PERIOD x (m_x)	TOTAL NUMBER OF FEMALE OFFSPRING PRODUCED DURING AGE PERIOD ($l_x \times m_x$)
0-9	97.75	0	0
10-14	97.52	0.0062	0.60
15-19	97.30	0.1656	16.11
20-24	96.98	0.3544	34.36
25-29	96.61	0.3278	31.66
30-34	96.13	0.1934	18.59
35-39	95.41	0.1039	9.91
40-44	94.34	0.0259	2.44
45-49	92.75	0.0017	0.16
50-over	—	0	0
		Number of daughters born to 100 women	113.82

Data courtesy of U.S. Department of Health, Education and Welfare, National Center for Health Statistics.

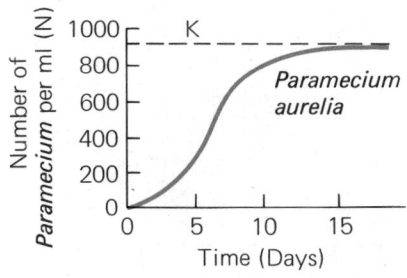

Figure 50-9 Growth of a *Paramecium aurelia* population in a culture medium to which Gause added a constant supply of bacteria as food each day. The population grew exponentially at first, but levelled out as numbers approached the carrying capacity, K. These data are described rather well by the logistic equation for population growth:

$$\frac{dN}{dt} = r_m N \left(\frac{K - N}{K}\right)$$

where N = population numbers (or density)

$\frac{dN}{dt}$ = change of population numbers or density per unit time (i.e., population growth rate)

r_m = innate capacity for increase

K = carrying capacity

The term $\frac{(K - N)}{K}$ indicates how much of the resources are still available to the population. When N is much less than K, the term $\frac{(K - N)}{K}$ approximates 1 and the equation becomes $\frac{dN}{dt} = r_m N$ (the equation describing exponential growth). As N approaches K, the term $\frac{(K - N)}{K}$ approaches zero, and $\frac{dN}{dt}$ (i.e., growth rate) also becomes zero.

50-D Carrying Capacity

No population can grow exponentially for long. Gause found that his *Paramecium* populations eventually stopped growing at a population ceiling (Figure 50–9). **Carrying capacity (K)** is the number of individuals of a particular species that a particular environment can support indefinitely. When a population reaches this size it may stabilize, with fluctuations above and below the carrying capacity. The curve described as a population reaches its carrying capacity, following exponential growth, is called a **logistic curve** (Figure 50–9). Carrying capacity is determined by a multitude of factors, including predation, competition, and climatic conditions. All of the factors that limit a population are collectively known as the **environmental resistance** to population growth. Since the factors that create environmental resistance are many and varied, it is clear that the carrying capacity of any area for a population may vary considerably over time.

50-E Regulation of Population Size

The number of individuals in a natural population varies with time, in some cases dramatically. If the size of a population declines too drastically, the population may become extinct. Local populations of many species do become extinct quite frequently, but may later be re-established by immigration from other populations. Small populations on islands or in restricted patches of suitable habitat are espe-

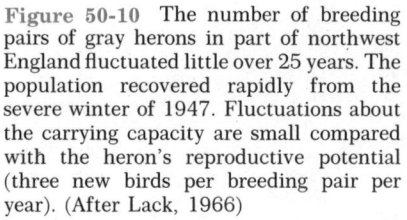

Figure 50-10 The number of breeding pairs of gray herons in part of northwest England fluctuated little over 25 years. The population recovered rapidly from the severe winter of 1947. Fluctuations about the carrying capacity are small compared with the heron's reproductive potential (three new birds per breeding pair per year). (After Lack, 1966)

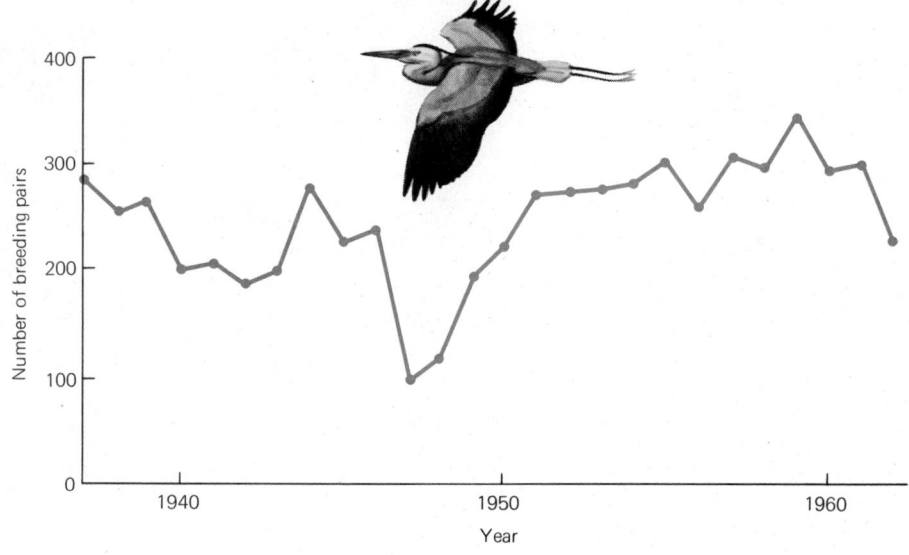

cially prone to extinction, as are populations exposed to occasional extremes of weather. For example, butterfly populations in some of the high mountain valleys of Colorado are occasionally exterminated by freak midsummer snowstorms.

In spite of fluctuations, however, an outstanding feature of most large populations is that their average sizes change relatively little over the years, and certainly considerably less than one might expect from their biotic potentials (Figure 50–10). This suggests that population sizes are regulated in such a way that small populations grow fast, larger populations grow more slowly, and still larger populations decline.

What kinds of mechanisms could bring about such ecological homeostasis? To bring about such regulation, at least some of the mortality factors affecting the population would have to be **density-dependent**, killing a larger proportion of the individuals in the population as population density increases (Figure 50–11). Several kinds of mortality can act in a density-dependent manner. The two most important are competition for resources and the action of predators and parasites. A density-dependent decrease in the number of offspring produced will also contribute to population regulation.

If some factors that limit population growth are density-dependent, are there any sources of mortality that are **density-independent**, killing a proportion of the population that is not correlated with its density? Bad weather is often cited as an example. A hurricane, a severe winter, or a drought may kill all of the individuals in a population, no matter what the population density. The action of bad weather is not always totally independent of population density, however, since some individuals may be able to shelter from it: if the number of shelters is limited, all of the members of a small population can find shelter, whereas only a fraction of a larger population will be protected.

It would be convenient if we could pinpoint one or two factors and say that they determine the size of a particular population. However, the sizes of natural populations are often affected by many different factors whose interactions can be complex. We shall later cite some examples where a particular cause of death has actually been shown to limit the size of a population.

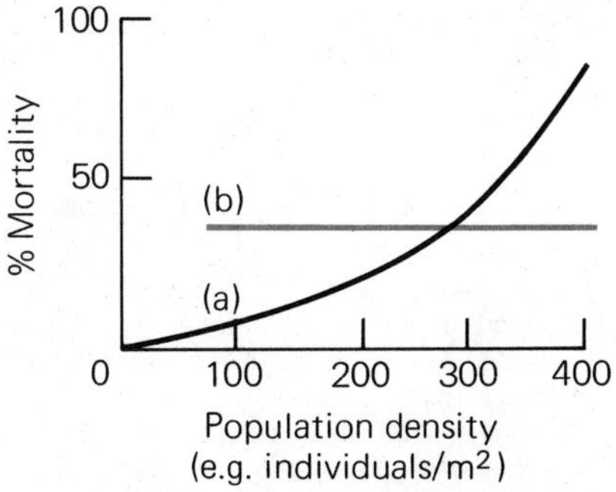

Dispersal and Habitat Patches

All environments are made up of **patches**, local areas of habitat, but patches are perceived differently by different species. An area of forest may represent one large patch to chickadees or deer but dozens of different patches (individual trees) to insects or fungi.

The size, distribution, and longevity of habitat patches and the sizes of the local populations they contain have important effects on population abundance and distribution. If patches are relatively permanent and the populations within them are large (for example the chickadees of a large forest), then the population may survive indefinitely within the patch. When patches are ephemeral* or their populations small, however, local populations within the patches frequently become extinct. Most of the herbaceous* plant species in a particular forest clearing will inevitably become extinct when the clearing grows back to forest. These short-lived patches are colonized by populations that grow rapidly, perhaps exponentially, and then "crash" to local extinction. The dispersal (emigration and immigration) of members of the species to new patches, however, may be such that the size of the population over a large area containing many patches is fairly constant over time.

Intraspecific Competition

Competition is a density-dependent factor that helps to regulate the size of most populations. **Competition** occurs when two or more organisms attempt to exploit a limited resource, such as food or space. Competition between individuals of the same species (**intraspecific competition**) is very common in nature. It is true that in some bird species the males and females have different beak lengths, enabling

Figure 50-11 Hypothetical curves describing the action of (a) density-dependent mortality: an increasing proportion of the population is killed by the mortality factor as population density increases; (b) density-independent mortality: the same proportion is killed, regardless of population density (though of course higher *numbers* of individuals are killed at higher densities).

---

* Ephemeral Lasting only a short time.
* Herbaceous With non-woody stems.

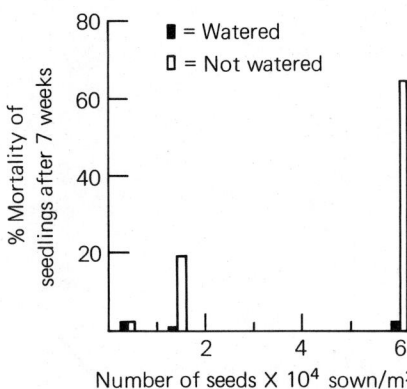

Figure 50-12 The effect of seed density on survival of white clover seedlings. Unshaded bars represent the mortality of seedlings that were not watered after the 18th day. (After J. L. Harper, *Society for Experimental Biology Symposium* 15:1–39, 1961, Cambridge University Press)

them to feed on different insect prey, and in many fish and butterfly species, the young feed on food different from that of adults. Generally, however, members of the same species need the same resources and are thus bound to compete for them, except when colonizing a new patch, or when predation, weather, or some other factor keeps the population small in comparison to the available resources. Gause's *Paramecium* population in the laboratory was clearly limited by the density-dependent action of intraspecific competition as the population leveled off. How does competition manifest itself in nature? When does intraspecific competition limit the size of natural populations? Let us consider some examples.

In one experiment, seeds of white clover, *Trifolium repens*, were planted at three different densities. Half of the plants at each density were watered throughout the experiment, but the other half were watered only for the first 18 days. After seven weeks, the densities of the surviving seedlings were measured. Among the seedlings that were watered regularly, mortality was low regardless of density. Among the seedlings deprived of water, however, the proportion of seedlings killed was three times greater in the high density than in the intermediate density plots, dramatic evidence of density-dependent mortality (Figure 50–12).

In the kind of competition exemplified by this experiment, each individual uses the resources without regard to other individuals competing with it. Such **scramble competition** is what might happen among humans competing for a limited supply of movie tickets if we did not adhere to the convention of forming a line. In nature, scramble competition is a common form of competition among organisms incapable of elaborate behavioral conventions, and among organisms whose resources are patchy or ephemeral. Scramble competition occurs among the maggots of blowflies that infest the carcasses of dead mammals. The first female flies that find a freshly dead carcass lay their eggs, and most of their larvae will have enough food to reach pupation. As the population builds up, however, there comes a time when the food runs out and most of the remaining maggots die.

Instead of competing directly for a limiting resource, as in the scramble type of competition, individuals of many vertebrate and some invertebrate species may compete indirectly for social dominance or for a territory. In **contest competition**, as this is called, the successful individuals are assured of an adequate supply of the limiting resource, whereas the losers may end up with nothing. Possession of a territory may be of no value in itself; however, it guarantees the territory-holder an adequate supply of some ultimate limiting resource such as food, space, or a mate.

Competition for territories that guarantee their owners access to shelter and food is common in higher vertebrates. Paul Errington studied muskrats in Iowa marshes for about 25 years. During this time, a given 270 acres of favorable marsh always contained about 400 muskrats. This remarkable stability was shown to result from the density-dependent action of competition. Males compete for territories that guarantee a suitable area, free from interference, where they can find food, mate, and raise their young. Animals in their territories are comparatively safe from predation because they know their territories well, and the territories have ample cover for escape. Unsuccessful males are forced into unfavorable marginal areas of marsh where they, with their mates and progeny, suffer enormous mortality due to overcrowding, inadequate food, predation, and interference from other muskrats. Density-independent factors are not effective in regulating muskrat populations: after a major catastrophe due to bad weather the populations are reduced considerably, but the high reproductive potential soon brings them back up again to the carrying capacity. Errington concluded that muskrat populations are limited by intraspecific competition.

David Lack, a noted English ornithologist, was of the opinion that most bird populations are limited by intraspecific competition for food. This seems to be true also for populations of many mammals and other vertebrates. Among invertebrates that exploit patchy and ephemeral resources, the significance of intraspecific competition is less clear. Although it undoubtedly sets an upper limit to population growth within any particular patch, such as a carcass or a stand of food-plants, overall population levels for a wide area may be limited by the rate at which new patches are formed and discovered. Intraspecific competition for light, nutrients, or water is important in many plant populations.

Interspecific Competition

Interspecific competition, competition between individuals of two (or more) different species for the same resource, may occur when the ecological niches of the species overlap. Mathematical models developed from the logistic equation during the 1920s predicted that if individuals of either or both of the two species compete intensely, then one species will always eliminate the other. Only if the two species compete very little does the model predict continued coexistence and sharing of the resource. In the early 1930s Gause set out to test these predictions by raising two species of *Paramecium* in the same culture. He chose two species that have similar food requirements and hence are likely to be strong competitors. In accordance with the predictions, Gause found that one species always eliminated the other. Which species was the victor depended upon the particular conditions in the culture (Figure 50–13).

Impressed with these and other similar results, Gause formulated what is now generally known as the **competitive exclusion principle**, which states that "complete competitors cannot coexist" or, put another way, "no two species can occupy the same niche." This was once known as "Gause's Principle" but the idea was familiar to several people before his time.

It often looks as though several species can coexist while competing strongly for the same resources. Many different species of trees live under similar conditions in the same forest. Lake Baikal in Siberia contains 239 species of gammarid* crustaceans. Several species of warblers feed on the same kinds of insects in the same areas of forest. Whenever such cases have been examined in detail, however, it has always been found that the niches are not identical; the species invariably partition the resources in some way. Thus Robert MacArthur found that wood warblers in northeastern forests forage for insects in different parts of the trees, reducing their competition (Figure 50–14). David Lack found that two very similar sea birds inhabiting the coast of Britain partition the fishing such that one of the species, the shag, specializes on bottom-dwelling fish and the other, the cormorant, takes only fish that swim above the bottom. Apparent exceptions to the competitive exclusion principle thus turn out not to be exceptions after all.

How common is the competitive elimination of one species by another in nature? In undisturbed communities it is hard to say. It is difficult or impossible to tell what happened in the past because any "losing" species are no longer around. Furthermore, the elimination of one species by another may be happening, but so slowly that we do not notice it. Extinction due to competitive exclusion can sometimes be demonstrated dramatically, however, when species are introduced into new areas. The large-scale extinctions of marsupial* mammals in South America

---

*Gammarid   Belonging to a family of arthropods whose members swim on their sides and feed on dead organic matter.
*Marsupial   Mammal whose young are born quite early in development and complete their development attached to a nipple in the mother's marsupium, or pouch.

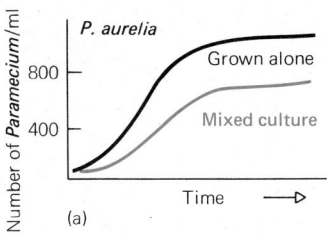

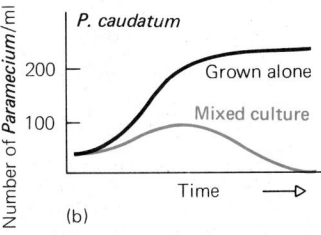

Figure 50-13 Gause's experiment with two species of *Paramecium*. Both species grew well by themselves in culture tubes with daily changes of water and inputs of bacterial food. When placed together under these conditions, *P. aurelia* (a) survived, while *P. caudatum* (b), the larger and slower-growing of the two species, always declined to extinction. If the water was not changed, waste could build up, and the competitive outcome was invariably reversed.

Figure 50-14 Several species of warbler in the genus *Dendroica* hunt for insects in coniferous trees in the same New England forests. Each usually forages in a different part of the trees (shaded), thus reducing competition for food.

BLACKBURNIAN WARBLER   BAY-BREASTED WARBLER   MYRTLE WARBLER   **861**

are believed to have resulted in large part from competitive exclusion by placentals*
from North America when the Central American land bridge formed between the
two continents about 3 million years ago. A more recent case is that of a unique
species of giant tortoise that lived on Abingdon Island in the Galapagos Archipelago,
off the northwest coast of South America. In 1962 an expedition was sent to this
island to check on the status of this historic species, which had been known to
Darwin. A search of the island showed that the species was extinct, although the
expedition found the remains of tortoises that could not have been dead for more
than a year or two. The reason for the extinction was clear. The tortoises' food
plants had been completely consumed by goats, introduced onto the island by a party
of fishermen in 1957.

Interspecific competition can also restrict the abundance and distribution of
natural populations in ecological time. An example is competition between two
species of barnacles, *Balanus* and *Chthamalus*, on the rocky coast of western Scotland. *Chthamalus* occupies the upper part of the intertidal zone and *Balanus* a
lower zone, with little overlap between them. The planktonic larvae of both species
settle on rocks in both zones. However, *Balanus* cannot survive in the upper zone
because it is less tolerant than *Chthamalus* to prolonged exposure at low tide. By
removing *Balanus* larvae as they settled on experimental pieces of slate in the lower
zone, Joseph Connell showed that *Chthamalus* survives in the lower zone when
Balanus is absent. *Balanus* grows faster than *Chthamalus*, so that when both
species compete for space on the rocks, *Balanus* grows over *Chthamalus* and crowds
it out. Thus the lower limit of the zone occupied by *Chthamalus* is determined by
competitive exclusion by *Balanus* (Figure 50–15). The **realized niche** actually
occupied by *Chthamalus* is a more restricted set of environmental conditions than
the **fundamental niche** that *Chthamalus* could occupy if *Balanus* were absent.
Competition that restricts the niche, and hence the population size, is probably very
common in nature. Although each species is uniquely adapted to exploit a certain
range of conditions, there is usually a broader range of conditions from which it can
be excluded by competition.

Predation

Predation, like competition, is a density-dependent factor that can regulate population size. The usual impression of a predator is of something like a lion or a
wolf—a free-living carnivore that kills and devours individuals of various prey
species for food. A parasite, by contrast, is commonly visualized as something like a
fungus or a flea—an organism that lives in or on the body of a larger organism, its
host, from which it derives food. There are so many exceptions to this neat classification, including parasites that kill their hosts and large herbivores that do not kill
the plants they graze, that it is often convenient to consider **predation** as any case
where individuals of one species exploit those of a **prey** species for food.

---

* Placentals Mammals that carry the young in the mother's uterus, where they receive food and
oxygen via the placenta, until a fairly advanced stage of development.

Figure 50-15 On the rocky coast of
Scotland the vertical distribution of
Chthamalus barnacles is limited by competitive exclusion. *Chthamalus* larvae settling from the plankton are crowded out by
growing *Balanus* barnacles. The upper
limit of *Balanus* is restricted by its limited
tolerance of exposure at low tide.

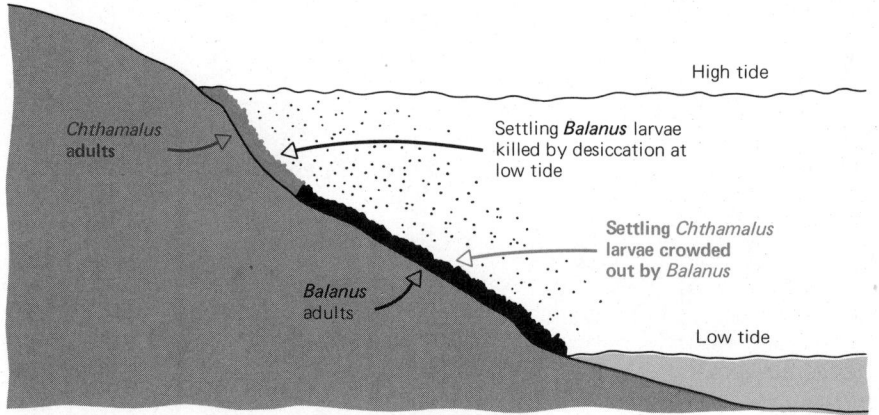

High tide

Chthamalus
adults

Settling *Balanus* larvae
killed by desiccation at
low tide

Settling *Chthamalus*
larvae crowded
out by *Balanus*

Balanus
adults

Low tide

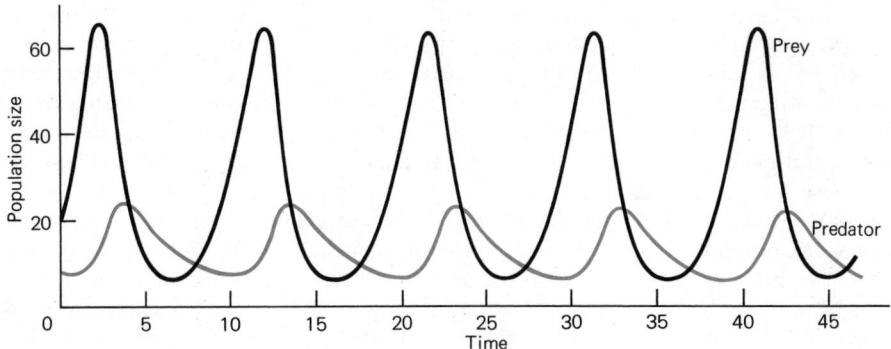

Figure 50-16 Oscillating population sizes of predators and prey as predicted by mathematical models for simple two-species systems.

Predation is a major source of mortality to most natural populations, but can it limit or regulate the size of prey populations? This question has been debated for a long time and the answer seems to be "sometimes." Early mathematical models, derived from the logistic equation, predicted that a simple system containing one predator species and one prey species should oscillate, with cycles in the abundance of predator and prey. As prey density increased, the model predicted that predator numbers would follow. The higher numbers of predators would then take an increasing proportion of the prey, causing the prey population to decline. Finally the predator population would also decline, whereupon the cycle would start again (Figure 50–16).

The tireless Gause set out to test these predictions by introducing two protistan species into tubes containing a culture medium. He found that the population of the prey, a species of *Paramecium*, grew rapidly at first but then declined to extinction as numbers of the predator, *Didinium*, increased; deprived of food, the *Didinium* population then also declined to extinction. Gause was able to prevent extinction and to produce oscillations in the predator and prey populations only by adding new individuals of each species to the culture periodically. More recently, Carl Huffaker in the United States tested the same predictions with two species of mites,* one of which preys on the other. Individuals of both species were placed on a set of trays containing a regular array of oranges, the food of the prey species. In early tests, Huffaker, like Gause before him, found that the predator always eliminated the prey. Later, however, he laid down lines of Vaseline between some of the oranges to provide barriers across which the mites could not move freely. Under these conditions, which simulate a patchy rather than a homogeneous environment, the predator and prey populations did indeed oscillate for several months, more or less as the mathematical model predicted (Figure 50–17). The heterogeneous environment provided refuges for the prey mites so that while some groups were being discovered and devoured by the predator mites, other groups of prey, free from predators for a time, could thrive. With some modifications, therefore, the mathematical predictions can be fulfilled in laboratory populations. But what happens in the real world?

From California comes an impressive demonstration of the power of a small beetle to control the numbers of its prey, in this case an attractive yellow-flowered

---

*Mites Small arthropods in the subphylum (Chelicerata) that includes spiders. Mites have eight legs, and the body is not divided into two parts as it is in spiders.

Figure 50-17 Stable oscillations of predator and prey mite species in a laboratory experiment simulating a heterogeneous (patchy) environment. The prey mite, *Eotetranychus*, fed on oranges, spaced in an array on a large tray. The predator mite, *Typhlodromus*, fed on *Eotetranychus*. (After C. B . Huffaker, *Hilgardia* 27:343–383, 1958. Division of Agricultural Sciences, University of California)

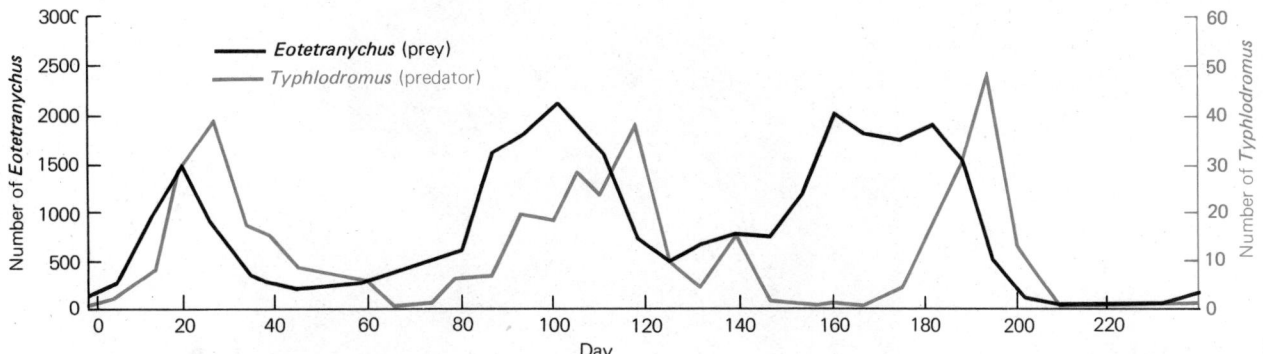

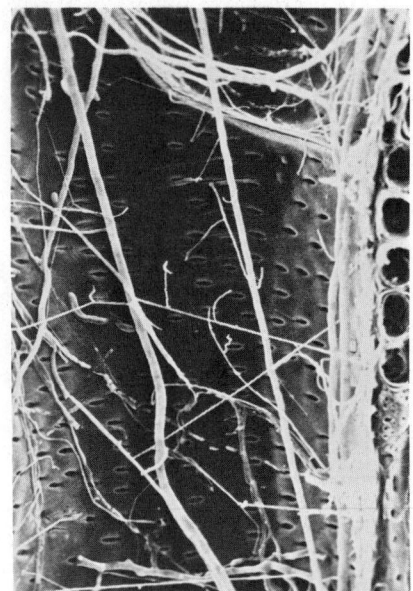

Figure 50-18 Dutch elm disease. The branching fibers are hyphae of a pathogenic fungus growing inside a xylem vessel of an elm. (Biophoto Associates)

perennial* herb called perforated St. John's wort or Klamath weed (*Hypericum perforatum*). This plant is a native of Europe and Asia and was first reported in the United States near Lancaster, Pennsylvania, in 1793. By 1900, it had appeared along the Klamath River in northern California. Not only is this plant toxic to cattle, but also it is an aggressive competitor on grazed range land, eliminating desirable grasses and herbs. By the late 1940s it occupied over 2 million acres in California, Oregon, and nearby states, covering 80% of the ground in many areas and making it useless for ranching. In Europe, St. John's wort is attacked by several host-specific insects, including two species of leaf beetles of the genus *Chrysolina*. These species had shown promise as biological control agents in Australia, where the introduction of St. John's wort had also created a problem. Supplies of beetles were obtained from Australia and were let loose in California in 1945 and 1946. They flourished; by 1959 Klamath weed had been reduced to less than 1% of its former abundance, chiefly due to the enormous damage inflicted by the beetle larvae. Both the beetles and the plant persist in the United States, but at low overall densities. The plant survives best in shady areas, where the beetles are less effective. Without knowing the history of this predator–prey interaction it would be hard to realize how effective these small and now rather rare beetles are at population control. This is why examples of biological control are so dramatic as evidence of the importance of predation.

Pathogens provide further examples of the ability of specialized parasites with a high biotic potential to bring down the numbers of their hosts. A familiar example is the rapid decline of the American elm tree in much of eastern and central North America because of Dutch elm disease. The damage is done by a fungus that is carried from one tree to another by beetles. A virus is responsible for the chestnut blight that has all but eliminated another tree, the American chestnut, from the temperate deciduous forests of North America. There is no question that specialized predators and parasites are capable of limiting population numbers, and there is little doubt that this kind of control is widespread in nature. In many cases the prey or host survives at low densities, rather like the prey mites in Huffaker's orange experiment, showing that the control exerted by the enemies is density-dependent.

The role of generalized predators, those that feed on many prey species, in limiting population size is less clear. Some examples of prey limitation by generalized predators are known. The numbers of mussels in the rocky intertidal zone of Washington state, for instance, are kept down by a predatory sea star. The survival rate of young sockeye salmon in a lake in British Columbia increased threefold after predatory squawfish and trout had been removed by gill netting. In many field studies, however, the role of generalized predators has been shown to be subsidiary to other limiting factors, such as food supply.

An example of such a study comes from Isle Royale, a forested island, 45 miles long and up to 9 miles wide, in Lake Superior. The island has populations of about 600 moose and of about 20 timber wolves, which feed on the moose and on many types of smaller prey. It was once thought that the size of the moose population was controlled by the wolves. However, more recent studies have strongly suggested that the number of moose on Isle Royale is limited by their supply of the micronutrient

---

* Perennial Living for many years.

Figure 50-19 A female moose.

sodium. The wolves kill mainly ailing and feeble moose, the young and the old, whereas healthy adult moose can outrun wolves or stand their ground and fight; a single blow from the leg of a healthy adult moose can be fatal to a wolf. The predators seem to be taking mainly moose that would have died anyway.

After mountain lions (cougars) and wolves were eliminated from much of their former range in the United States, there was a noticeable increase in the populations of deer. This led many to believe that the deer populations had been limited by the large carnivores. Such a conclusion is now much less secure, however, since the rise in deer populations is also correlated with other human activities that have resulted in an increase in secondary growth vegetation,* a food favored by deer. The deer populations of old may have been limited by food rather than by predators.

It is not surprising that clear-cut cases of population limitation by generalized predators are hard to come by. Such species are able to exploit any prey species that becomes abundant in their habitat for some reason. By concentrating on such a prey species for a while they may be able to slow and perhaps stop a population outbreak. This is more likely when the prey, such as the mussels attached to a rock surface, have no means of escape. Many generalized predators are unlikely to be able to prevent an outbreak, however, because their biotic potential is low relative to that of their prey.

50-F   Winter Moth Caterpillars on English Oak Trees: A Case History

One of the best-understood examples of population regulation in nature is that of the winter moth, a small brown moth whose green "inchworm" caterpillars feed on oak leaves in the spring. This was studied for 20 years by George Varley and George Gradwell in Wytham Wood, a biological reserve in central England.

Every spring, the fresh young leaves of oak trees in England are attacked by the caterpillars of many different species of moths. Varley and Gradwell found that the abundances of the different species of caterpillars changed from year to year, sometimes dramatically (Figure 50–20); these fluctuations tended to be synchronous for most species. However, in spite of the large fluctuations, the abundance of each species stayed within a certain range over the course of several years. The lower the abundance of any particular species in one year, the more likely it was to increase in

---

* Secondary growth vegetation   Plants that grow as cleared and abandoned land returns to its original climax state (deer browse on twigs and buds of shrubs and small trees).

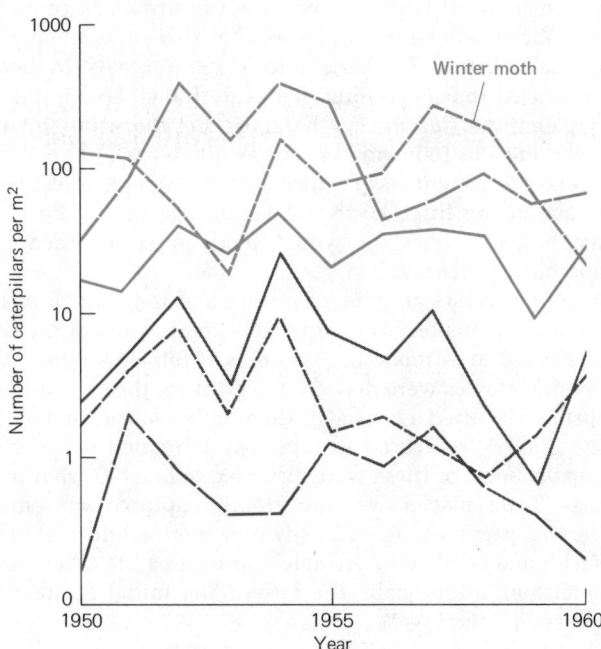

Figure 50-20 Stability of population densities of caterpillar species on English oak trees in the spring. In spite of annual fluctuations, the density of each species remains within a certain range, suggesting density-dependent regulation. (After G.C. Varley, *Proceedings of the 18th Annual Session of the Ceylon Association for the Advancement of Science,* 142–156, 1962)

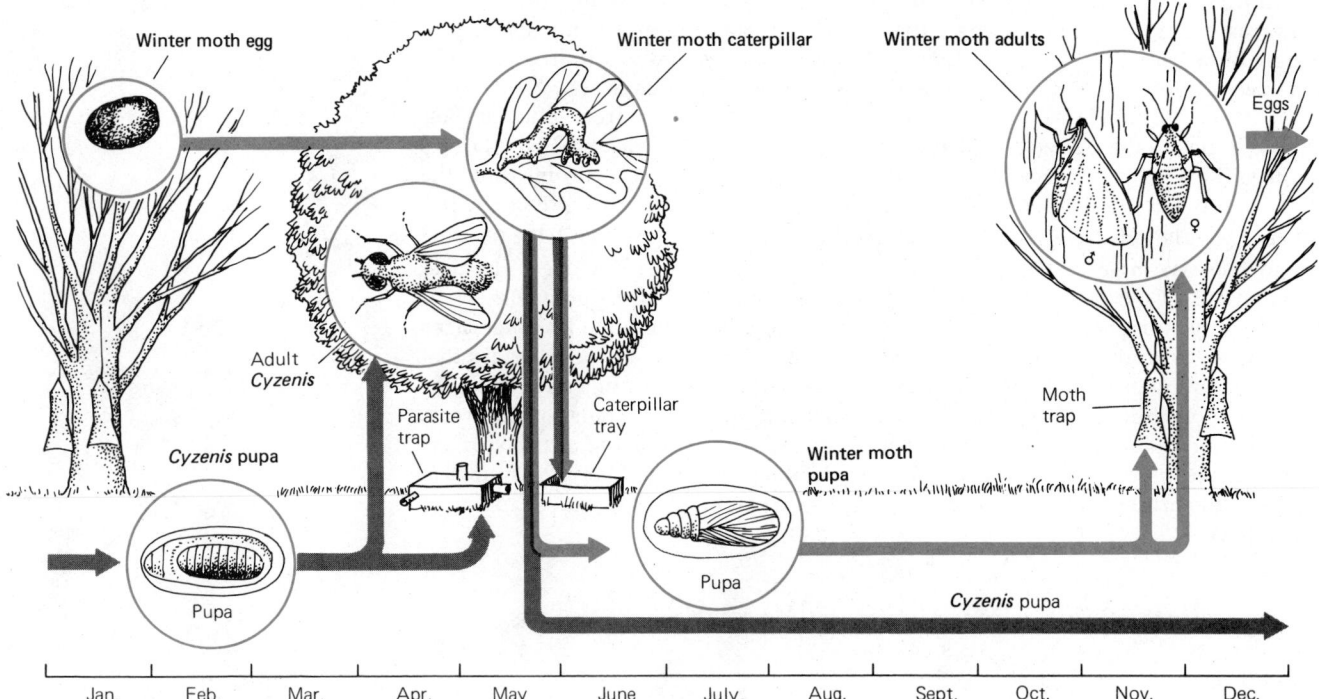

Winter moth egg

Winter moth caterpillar

Winter moth adults

Eggs

Adult *Cyzenis*

Parasite trap

Caterpillar tray

Moth trap

Cyzenis pupa

Winter moth pupa

Pupa

Pupa

Winter moth pupa

Cyzenis pupa

| Jan. | Feb. | Mar. | Apr. | May | June | July. | Aug. | Sept. | Oct. | Nov. | Dec. |

Figure 50-21 The life history of the winter moth (colored arrows) and *Cyzenis*, one of its parasites. The diagram also shows stages in the life history at which the animals are trapped so that they can be counted. (After Varley, Gradwell, and Hassell, 1973)

numbers the following year. This suggested some kind of density-dependent regulation of population size.

Varley and Gradwell picked the winter moth, the most abundant species, to study in detail. Females of this species are wingless; they emerge in November and December from pupae in the soil litter and climb the trunk of the nearest tree, often the oak they fed on as a caterpillar. After mating on the way up (the males are winged and can fly), the females lay about 150 eggs each on twigs in the tree canopy. The following spring, the eggs hatch at about the same time that the oaks' leaf buds begin to open. The tiny caterpillars crawl into the opening buds and feed on the developing leaves. After about three weeks, when the caterpillars are full grown, they spin down to the ground on silk threads, crawl into the soil and remain there, as pupae, until late fall when the life history begins again (Figure 50-21).

Varley and Gradwell found that two parasitic insects kill a number of winter moths each year. The first, called *Cyzenis*, is a fly. In the spring, each female *Cyzenis* lays an egg in every winter moth caterpillar she can find. When the winter moth has pupated, the *Cyzenis* egg hatches, eats the winter moth pupa and itself pupates, emerging the following spring as another *Cyzenis* fly (Figure 50-21). The other parasite is a small wasp called *Cratichneumon*, whose females seek out winter moth pupae in the litter layer and lay an egg in any they discover. The egg, in turn, hatches, consumes the winter moth pupa and then pupates, later to emerge as another *Cratichneumon*.

The investigators made a census of the populations of the winter moth and of its major parasites every year, at several points in the life history. Tree trunk traps were used to estimate the numbers of female moths climbing the trunks each fall; several females were dissected to estimate the average egg load. Leaf samples in the spring permitted a census of the number of caterpillars feeding, and metal trays on the ground were used to intercept a fraction of the fully fed larvae dropping to pupate. Some of these were dissected to find out what proportion contained *Cyzenis* eggs. Traps placed over the ground captured any emerging *Cratichneumon* and *Cyzenis* parasites, as well as winter moth adults (Figure 50-21). From these data, Varley and Gradwell were able to prepare a life table for the winter moth each year, describing numerically the fate of the initial number of eggs as the season progressed (Table 50-3).

LIFE HISTORY STAGE	NUMBER STARTING STAGE (per m^2)	MORTALITY FACTOR	NUMBER KILLED (per m^2)	NUMBER SURVIVING (per m^2)	k VALUE
Adult (females) Climbing trees in 1955	**4.39**†				
Egg Number of females × 150	658	Winter disappearance	561.6	**96.4**	$k_1 = \log(658) - \log(96.4) = 0.84$
Caterpillar Successfully hatched	**96.4**	*Cyzenis* parasite	**6.2**	90.2	$k_2 = \log(96.4) - \log(90.2) = 0.03$
	90.2	Other parasites	**2.6**	87.6	$k_3 = \log(90.2) - \log(87.6) = 0.01$
	87.6	Protozoan disease	**4.6**	83.0	$k_4 = \log(87.6) - \log(83.0) = 0.02$
Pupa	83.0	Predators (shrews, etc.)	**54.6**	28.4	$k_5 = \log(83.0) - \log(28.4) = 0.47$
	28.4	*Cratichneumon* parasite	**13.4**	15.0	$k_6 = \log(28.4) - \log(15.0) = 0.27$
Adult (total)	15.0				
Adult (females) Climbing trees in 1956	**7.5**				

TABLE 50–3 LIFE TABLE FOR THE WINTER MOTH FOR 1955–1956

† Figures in boldface were measured by census; the other figures in this table are derived from them.
Modified from Varley, Gradwell, and Hassell, 1973.

The life table data were used to determine the factors responsible, first, for the major year-to-year fluctuations in the population of winter moths, and, second, for the apparent control of numbers over several years. Varley and Gradwell used a method known as **key factor analysis**. For each source of mortality, they assigned a "k value," a measure of the "killing power" of that mortality (Table 50–3). It is calculated as the logarithm of the number of winter moths present before the particular mortality, less the logarithm of the number of survivors.

$$k = \log \text{ (Initial Number)} - \log \text{ (Number Surviving)}$$

A great convenience of expressing numbers as logarithms is that successive mortalities are then additive rather than multiplicative, so that total mortality $K = k_1 + k_2 + k_3 +$ etc. (just as $\log 20 = \log 4 + \log 5$, while $20 = 4 \times 5$). When annual k values for each source of mortality from one year to the next were plotted, one of them, k_1, was found to be strongly correlated with winter moth abundance (Figure 50–22), while none of the others showed any correlation. Thus k_1 is the "key factor" reflecting population change from one year to the next. It is due to the "winter disappearance" of eggs that fail to show up as caterpillars feeding on the leaves. Varley and Gradwell found that the death of these individuals is due to the weather. Each year only a fraction of winter moth eggs hatches during exactly the right two- or three-day period when the oak buds are opening. Caterpillars that hatch out too soon cannot penetrate the closed buds; caterpillars that hatch too late are unable to start feeding on the rapidly toughening leaves. Instead of starving where they are, caterpillars that get the timing wrong disperse on silk threads, carried by the wind. They have a slight chance of landing on an oak tree with buds just opening. The time at which oak buds open is determined by temperature and other weather changes in the early spring. Only occasionally do winter moths nearly all hatch at just the "right" time.

Winter disappearance accounts for about 90% of winter moth mortality, on the

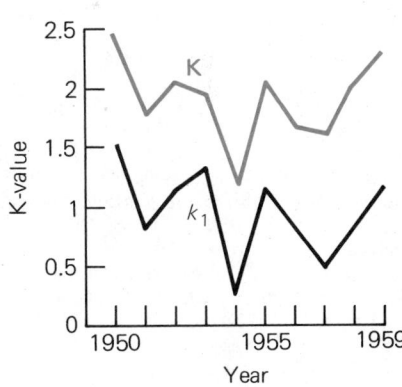

Figure 50-22 Synchronous fluctuations of K (total mortality) and k_1 (winter disappearance) of winter moths. (After Varley, Gradwell, and Hassell, 1973)

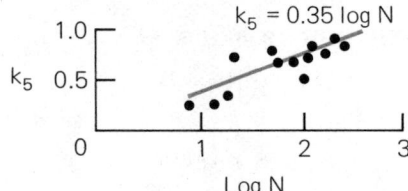

Figure 50-23 Density-dependent action of predation by shrews and beetles on winter moth pupae. Each dot represents data from one year. N = number of winter moth pupae per m². (After Varley, Gradwell, and Hassell, 1973)

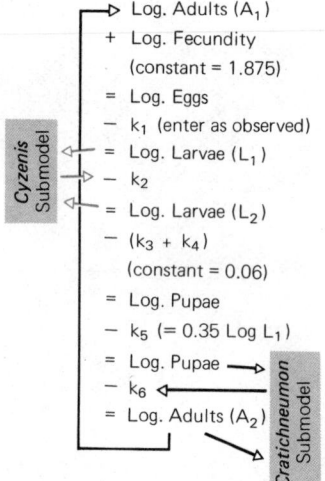

Figure 50-24 Simplified population model for winter moth. (After Varley, Gradwell, and Hassell, 1973)

Figure 50-25 Annual densities of winter moth and its *Cyzenis* and *Cratichneumon* parasites from 1950 to 1968. Observed values are shown in black; values calculated from the model described in the text are shown in color. (After Varley, Gradwell and Hassell, 1973)

average. It explains two of the early findings, namely the large fluctuations from one year to another and, since the same weather problem affects most other species in the same way, the synchrony in these fluctuations among species. It does not explain the apparent homeostasis* of population regulation, however, which requires some form of density-dependent mortality.

To determine which factors represented density-dependent mortality, Varley and Gradwell plotted each k value over the years against the abundance or density of the winter moth population on which it acted. If a source of mortality is density-dependent, its k value should be directly proportional to the logarithm of the size of the population on which it acts (see Figure 50–11). Such a relationship was found for the mortality for winter moth pupae in the soil as a result of predation by shrews and beetles (Figure 50–23). Parasitism by *Cyzenis* (k_2) and by *Cratichneumon* (k_6) was also density-dependent, although in these cases the parasitism was proportional to winter moth density the *previous* year (which determines the number of parasites in the following year's population). Mortality due to protozoan disease (k_4) and to the action of certain other parasites (k_3) and of bird predators was found to be trivial and not density-dependent. Winter disappearance, as expected, was also found to be density-independent. Thus the only mortality capable of regulating the local winter moth population is that due to *Cyzenis*, *Cratichneumon*, and the predators of pupae in the soil.

The final stage in the study was the construction of a mathematical model, based on the life tables and key factor analysis, to try to simulate the population dynamics of the winter moth and its parasites. If all the important factors were taken into account, and measurements and analyses were suitably accurate, it should have been possible to construct a mathematical model that could not only describe the population dynamics found, but also predict future changes in population density.

The model developed by Varley, Gradwell, and Michael Hassell is shown in Figure 50–24. Starting with the number of adults at the beginning of a generation, the model calculates the effects of each source of mortality, acting in sequence, ending up with the predicted number of adults available to start the next generation. Sub-models, not shown here, calculate the predicted densities of the two parasites, *Cyzenis* and *Cratichneumon*, available for the next generation. The only thing that the model cannot predict is the weather; this source of mortality, as k_1, must be entered into the model at each generation. The success of the model is clear from Figure 50–25. Entering into the model only the starting densities of winter moths, *Cyzenis* and *Cratichneumon* in 1950, with the k_1 values for each year, the model (in the form of a computer program) generated annual values of the densities of each species over 18 years that were remarkably similar to the densities actually found.

This case history illustrates a number of important features of population stud-

---

* Homeostasis Condition of "staying the same"; that is, fluctuating only within narrow limits.

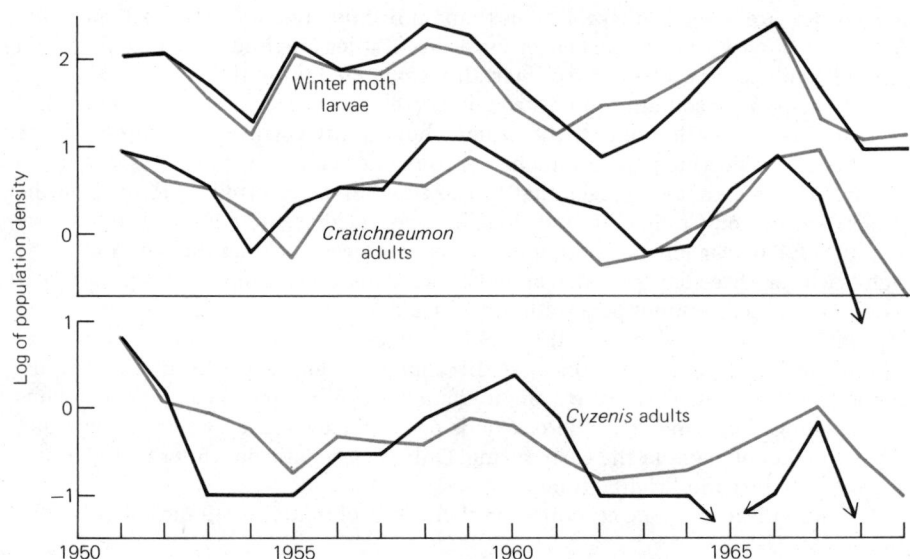

ies. It shows how the density-dependent action of predators and parasites can regulate a population over a number of years, even though the mortality they cause is considerably less than that due to the unpredictable density-independent effects of weather. The study also demonstrates the usefulness of making and analyzing life tables that can later be applied to the preparation of models. The success of a model gives greater confidence that all of the important factors have been taken into account. Moreover, models can be used to predict future population changes, perhaps of major economic interest. Finally, this study emphasizes the critical importance of understanding the natural history of all the species involved, something that usually requires long-term studies. Varley and Gradwell's study was initially frustrating because their early models failed to work. Only after eight years did the critical importance of *Cratichneumon* parasitism of pupae in the soil become apparent.

50-G Human Population

Human population growth is ultimately responsible for most of our other ecological problems, such as our steadily diminishing energy supply and the pollution of our air and water. The population explosion began when early tribes started to obtain food from agriculture instead of from hunting and gathering, a changeover that had dramatic effects on human history.

For most of our evolutionary history, human beings were **hunter-gatherers**, killing wild animals and collecting plants for food. It is estimated that as recently as 2000 years ago, farmers had occupied only about half the land suitable for agriculture. By the time of Columbus, about 500 years ago, hunters occupied much less than half of the earth's land area; they were located mainly in parts of the Americas, Australia, and Africa. Today, very few hunting populations remain.

Agriculture is the process of breeding and caring for animals and plants that are used for food and clothing. Its adoption by human societies can be traced back at least 10,000 years. It has long been thought that agriculture originated east of the Mediterranean and spread from this area to the rest of the Old World, and that it had a separate origin in America. Fossils of domesticated dogs dating from 11,000 years ago have been found in Iraq, and cultivated plants date back to at least 9000 years ago in the same area. Now that archeological excavations have become more widespread, however, it is clear that agriculture originated, probably independently, in many different places at about the same time (Figure 50–26).

Figure 50-26 Areas where plants and animals were probably first domesticated. Some species were probably domesticated independently in two or more areas and these are shown in both places.

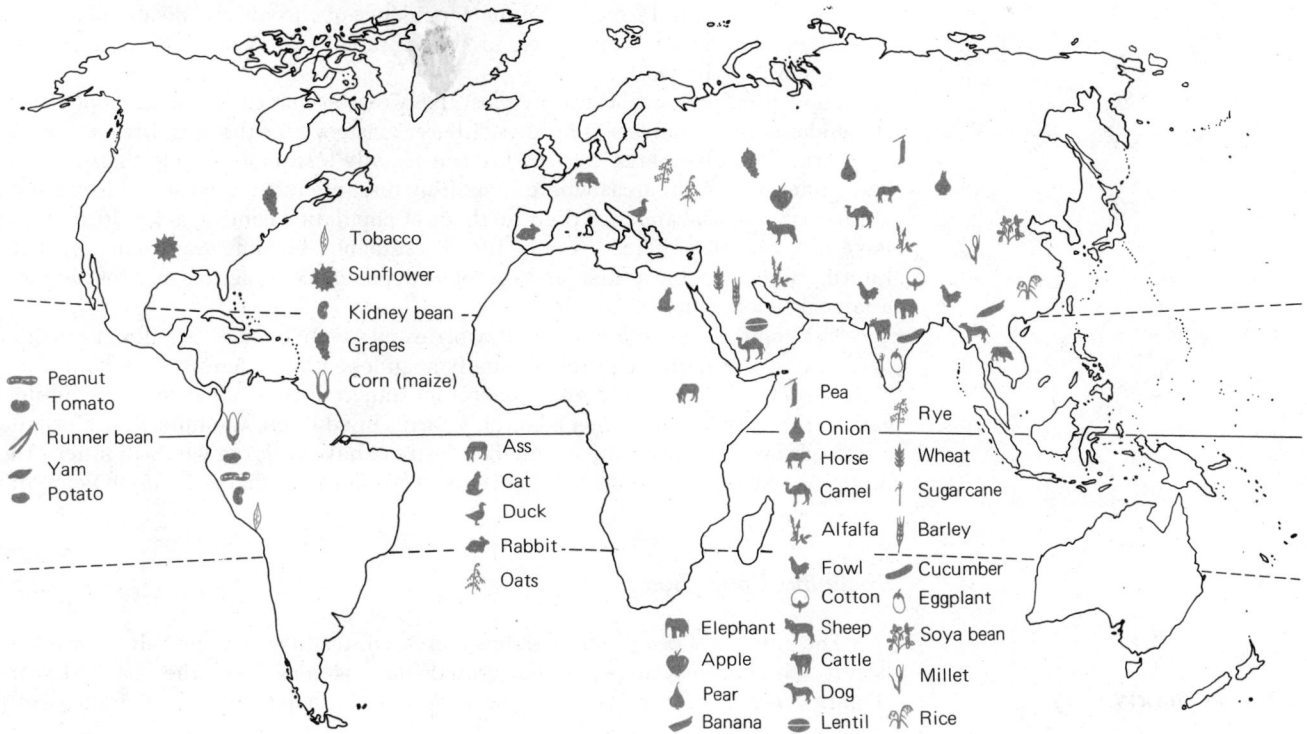

Tobacco
Sunflower
Kidney bean
Grapes
Corn (maize)

Peanut
Tomato
Runner bean
Yam
Potato

Ass
Cat
Duck
Rabbit
Oats

Elephant
Apple
Pear
Banana

Pea
Onion
Horse
Camel
Alfalfa
Fowl
Cotton
Sheep
Cattle
Dog
Lentil

Rye
Wheat
Sugarcane
Barley
Cucumber
Eggplant
Soya bean
Millet
Rice

Figure 50-27 A Kalahari bushman beside a partly constructed hut in the family camp. These people are one of the last surviving groups of hunter-gatherers. Even today, they survive without agriculture. (Biophoto Associates, N.H.P.A.)

The changeover from hunting and gathering to agriculture has had such a dramatic impact on human societies that it is often known as the **agricultural revolution**. The advantages of agriculture are not immediately obvious. For instance, hunter-gatherers do not face the constant battle with pests, droughts, and famines that beset all agricultural communities. Studies in southern Africa during a drought showed that farmers starved while the population of hunter-gatherer Bushmen in the Kalahari desert remained stable in size and the people well-fed. This is probably because most hunter-gatherer populations stay well below the carrying capacities of their territories. This is not because of excessive mortality; the people make conscious efforts to keep their population size down by such practices as abstention from sexual intercourse, abortion, infanticide, late marriage, and late weaning. Furthermore, these hunter-gatherer populations have a more balanced diet than that of most farmers, and their incidence of chronic and disabling diseases is no higher; their life expectancies are comparable with those of agricultural peoples in most of the world.

Population pressure seems the most likely reason that early human populations surrendered their wandering hunter-gatherer existence for the hard life on a primitive farm. In a given area, agriculture can usually feed more people than hunting and gathering can. In areas where something disrupted the social structure in such a way that people abandoned their methods of population control, agriculture might have become the preferred way of life. Presumably, farming was combined with hunting and gathering at first, and complete dependence on agriculture for food was a later development.

The population explosion, which worries us so much today, is a direct result of agriculture. Population control is usually abandoned by agricultural communities. This is partly because children, who are not important food collectors in a hunter-gatherer society, are useful as labor on a farm. In addition, the inheritance of land and goods becomes more important. The desire to have children who will inherit the property and care for their aged parents is a recurrent theme in mythology and literature.

Declining Death Rates

The number of people on earth has increased steadily since agriculture was first begun, but the greatest population growth has taken place in the last 200 years (Figure 50–28). This is primarily because the death rate has been dramatically

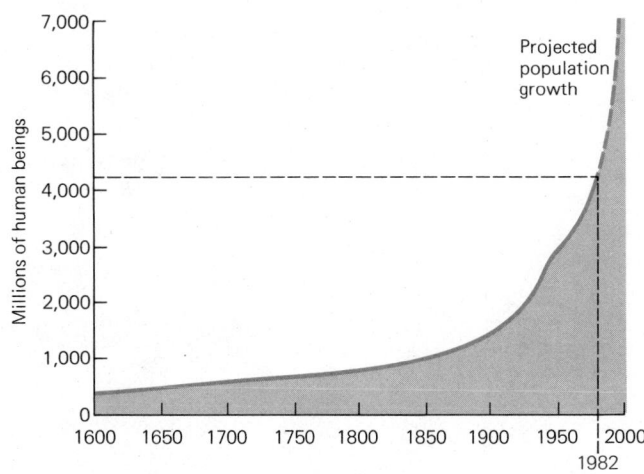

Figure 50-28 Growth of the world human population in the past 400 years. The dotted line shows the approximate world population in 1982.

reduced in most countries. The discovery of antibiotics in the twentieth century has made a slight contribution to the decline of the death rate, but much more important were the earlier and less spectacular improvements in nutrition and hygiene. Studies on a Navajo reservation in Arizona showed that the presence of a modern hospital within easy reach saved only one life in 10 years, whereas a district nurse who instructed parents on infant care reduced the infant mortality in the first year of life to 6 deaths instead of the 15 expected on the basis of the infant mortality rate before the arrival of the nurse. In developing countries, most deaths are due to respiratory and digestive tract infections in infants, and such deaths are easily reduced without expensive medical care. The United States has a high infant mortality rate. This is because the quality of health education and medical care for the poor is much worse than that received by the poor in most other developed countries. The infant mortality among poor white and nonwhite Americans is nearly 30 per 1000 live births—about twice as high as for all whites.

Life expectancy is the average number of years that a group of newborn babies is expected to survive. Not surprisingly, life expectancy is higher in wealthier countries whose people receive better health education and better medical care. Life expectancy in Switzerland, most of Europe, and North America, for example, is now more than 70 years; the average for Latin America and for East Asia (which includes China and Japan) is 63 years, and for most of the other Asian countries and nearly all of Africa, less than 45 years.

The Demographic Transition

Historically, it has been found that a decline in a country's birth rate eventually follows a reduction in its death rate. This decline in the birth rate is called the **demographic transition**. The single factor most clearly correlated with a lowered birth rate is an increase in the level of education and employment of women. The spread of knowledge that lowers the death rate is usually part of a general program to improve the educational level. Educated women find that they need not bear a large number of children to ensure that a few survive, and they also learn contraceptive techniques. In addition, women find that they can contribute to the family's increasing prosperity by holding a job and by spending less time and energy on raising children. This is usually attractive to women, even in countries where religious doctrine and tradition dictate large families.

The Population Explosion

The demographic transition has always taken from one to three generations to spread through the population in any country. During its progress, death rates are low, but birth rates remain high, and so the population grows enormously. Currently the population of the world is growing at the rate of about 60 million people a

TABLE 50-4 **PER CAPITA GROSS NATIONAL PRODUCT IN 1978**

TABLE 50-4 **PER CAPITA GROSS NATIONAL PRODUCT IN 1978**
(The gross national product is divided by the number of people in the country.)

COUNTRY	PER CAPITA GNP (IN U.S. $)	DOUBLING TIME† (IN YEARS)	COMMENTS
Kuwait	11,510	18	Oil! Immigrants contribute largely to population growth
Switzerland	8,050	173	
U.S.A.	7,060	116	
Japan	4,460	63	
Libya	5,080	18	Highest per capita GNP in Africa
U.S.S.R.	2,620	77	
Argentina	1,590	53	
Albania	600	28	Europe's highest birth rate and lowest per capita GNP
China	350	41	
Bolivia	320	27	Lowest per capita GNP and one of the highest birth rates in Latin America
Pakistan	140	24	One of the highest per capita GNPs in Southeast Asia
Nepal	110	30	One of the lowest per capita GNPs and highest birth rates in Southeast Asia
Yemen	210	24	Lowest per capita GNP in the Middle East
Burundi	100	33	One of the world's highest birth rates

† Doubling time in this table is the time it will take the country to double its population if rate of population increase remains the same as it was in 1978.
Figures from *World Population Growth and Responses*, Population Reference Bureau, 1978.

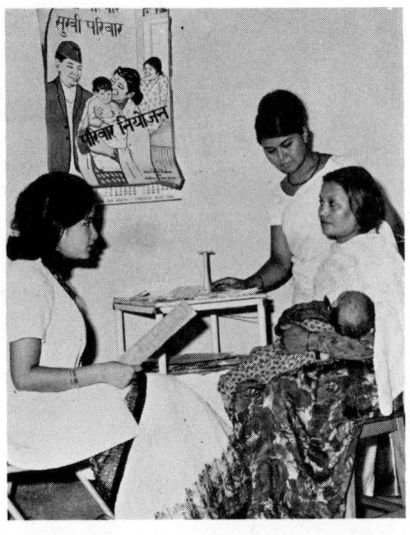

Figure 50-29 The demographic transition in progress. A doctor talks to a patient at a family planning center in Nepal; the wall poster emphasizes the pleasures of wanted instead of unwanted infants. In a "saturation" campaign to do something about its disastrous population problem, Nepal makes use of clinics, door-to-door visits, and a scheme whereby every village storekeeper will eventually sell condoms and contraceptive pills and keep 40% of the proceeds as an incentive to sales. (U.S. Agency for International Development)

year; we are gaining the equivalent of the population of New York City every six months. The world population has grown from about 500 million in 1650 to nearly 5 billion in 1981, and it is expected to approach 8 billion by the year 2000. The **doubling time**, the time it takes for the population to double in size, has decreased rapidly in recent years. It took more than a hundred years for the world population to double from 1 to 2 billion; it doubled from 2 to 4 billion in less than 70 years. Table 50-4 shows that the doubling time of a population varies greatly from one country to another, and that population growth is fastest in those countries that are poorest and can least afford it.

Recently the world population has begun to grow more slowly. Between 1965 and 1974, world birth rates dropped appreciably and, for the first time in more than 50 years, declined faster than death rates. Nevertheless, in 1975 the world population increased by 1.8%, and in Africa the birth rate was still increasing. Even if birth rates are reduced until they equal death rates by the year 2000, the world's population will grow past 8 billion people. This is because of the age structure of the population: more than half the people in the world are under 25, and even if they have only two children per couple, the number of people on earth will increase vastly.

Starvation

Since food production has not kept up with the population increase, the population explosion has resulted in widespread starvation. About 20,000 people die of starvation every day, and at least 10 million children in the world are so **malnourished** (poorly fed) that their lives are in danger.

Starvation means death from lack of food, but most people who are inadequately fed do not actually die because they take in too few calories to sustain life, but because their malnourished bodies have little resistance to diseases that would not be fatal to the properly fed. Most malnourished people get about as many calories as they need, but their food is deficient in essential amino acids and vitamins. Amino acid (protein) deficiencies are particularly damaging. Protein deficiency dis-

Figure 50-30 Agriculture in California's Central Valley. (U.S.D.A.)

eases such as kwashiorkor also lead to mental retardation, particularly when they occur in young children.

Malnourishment is, therefore, the more accurate term, but this word loses much of its force when we are told that half the teenagers in the United States are malnourished (which merely means that they consume more calories and fewer vitamins than would be ideal, not that they are likely to die from a common cold or an intestinal parasite).

Human starvation is not a necessary condition at the moment. The world's farmers produce enough food calories, proteins, and vitamins to keep more than 5 billion people in good health. The trouble is that food, and the income to buy food, are unevenly distributed between the rich and the poor. The problem is, therefore, more economic and political than biological. We can learn something of the solutions that will or will not work, however, by examining the biology of the situation.

The Efficiency of Agriculture

Food production depends ultimately on photosynthesis. However, crops seldom suffer from lack of sunlight; lack of moisture, nutrients (especially nitrogen and phosphorus), and temperature nearly always limits the growth of crops to a fraction of their photosynthetic potential. In nearly all parts of the world, the most productive agriculture occurs on land that is artificially irrigated (Figure 50–30). Very few areas have both good soil and enough rain, since adequate rainfall quickly leaches* nutrients out of the soil.

Plants can use nitrogen only in certain forms, mainly nitrates and ammonium ions. In nature, these forms are produced by soil bacteria as they decompose plant and animal matter or as they "fix" gaseous nitrogen from the atmosphere. Before about 1940, farmers increased the nitrogen supply to their crops by spreading manure and by planting legumes,* whose roots contain nitrogen-fixing bacteria. Recently, advanced agriculture has switched almost completely to nitrogenous fertilizers made by combining natural gas with atmospheric nitrogen. This method can provide much more nitrogen per acre than traditional practices. In Indonesia, the increase in rice yields from 1.8 to 2.7 metric tons per hectare was attributed mainly to the nitrogen supplied in inorganic fertilizers.

The world's farms can be loosely divided into subsistence farms, which provide the family's food and sometimes a small crop that can be sold for cash, and "modern" farms, where crops are raised to be sold. Modern farms are much less self-sufficient, since they need seed, fertilizer, equipment, credit, and marketing facilities from the rest of society. Modern farms are much more productive in terms of output per person per year, but they produce less food per unit area, partly because the plants must be far enough apart to permit machinery to move between the rows.

---

* Leaches   Dissolves minerals and washes them out of the topsoil.
* Legumes   The plant family to which peas belong; includes clover, alfalfa, and so on.

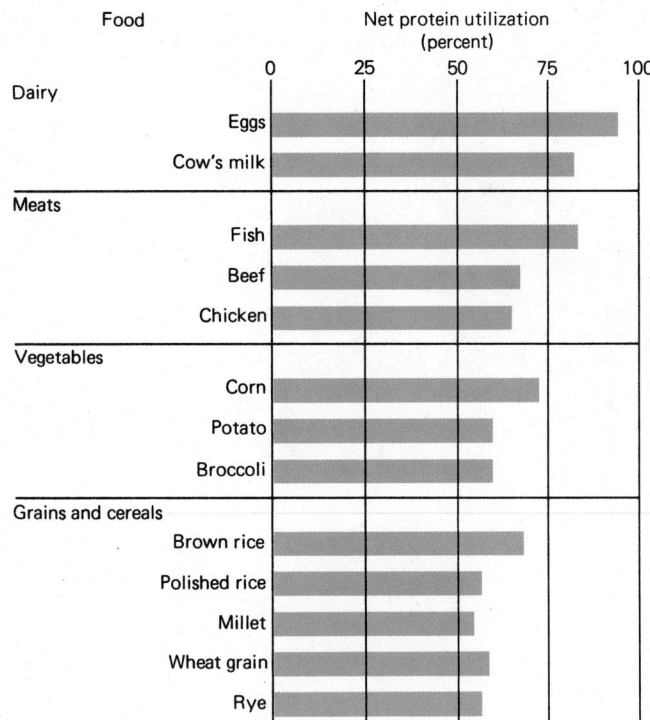

Food

Net protein utilization
(percent)

Dairy
Eggs
Cow's milk

Meats
Fish
Beef
Chicken

Vegetables
Corn
Potato
Broccoli

Grains and cereals
Brown rice
Polished rice
Millet
Wheat grain
Rye

Figure 50-31 Protein "quality" of various food sources. Net protein utilization is a measure of how much of the protein in the food is used by the body. The higher the percent utilization, the more closely the amino acid composition of the food approximates the amino acid requirement of the human body. Foods deficient in one or more essential amino acids have low utilization values.

A subsistence farmer, or a vegetable gardener, can produce much more per year out of the same area by planting one crop between the rows of another, and by planting more than one crop per year. Thus there is nothing inherently wrong with labor-intensive small farms. In many parts of the world, land has been subdivided by inheritance to the point where each family has a small holding of as little as one acre. Merging these into large modern farms would throw even more of the rural poor out of work. It would be more efficient to supply programs that increase the productivity of subsistence farms until they produce as much food as they possibly can.

People feed themselves most efficiently when they eat plants or plant parts. From 30 to 40% of a plant's net primary productivity* can usually be harvested, and 70 to 80% of such a harvest can be digested by human beings. The protein content of plants varies from about 6 to 20%. Since a plant must expend more energy on synthesizing protein than on synthesizing carbohydrates, and since this protein synthesis also reduces the amount of energy that a plant stores, it is more efficient to grow and eat plants with a lower protein content, as long as enough of the amino acids essential to human health are present. Cereals produce around 10% protein, which is nutritionally adequate for an adult human (Figure 50-31). This is why world agriculture is now so dependent on cereals (Figure 50-32).

Eating animals is much less efficient than eating plants. The animals must eat plants first, and animals convert only 10 to 15% of their food calories into calories that are available to their predators (see Sections 49-D, E). In no case is the efficiency of conversion higher than the 25% attained in milk and egg production. As societies become wealthier, their consumption of animal products increases. This makes it even more difficult for food production to surpass population growth as the number (if not the proportion) of people with a reasonable income increases. On the other hand, since some grazing land is unsuitable for growing plant crops, growing meat in these areas is the most effective way to use this land to increase the human food supply.

In both traditional and more developed societies, the energy used in processing, distributing, and cooking food is greater than the energy used to produce the food in the first place. For instance, about twice as much energy goes into cooking a kilogram of rice in rural India as was invested in producing it. In the United States, it

---

* Net primary productivity The amount of energy or matter stored by plants during photosynthesis minus that used in respiration.

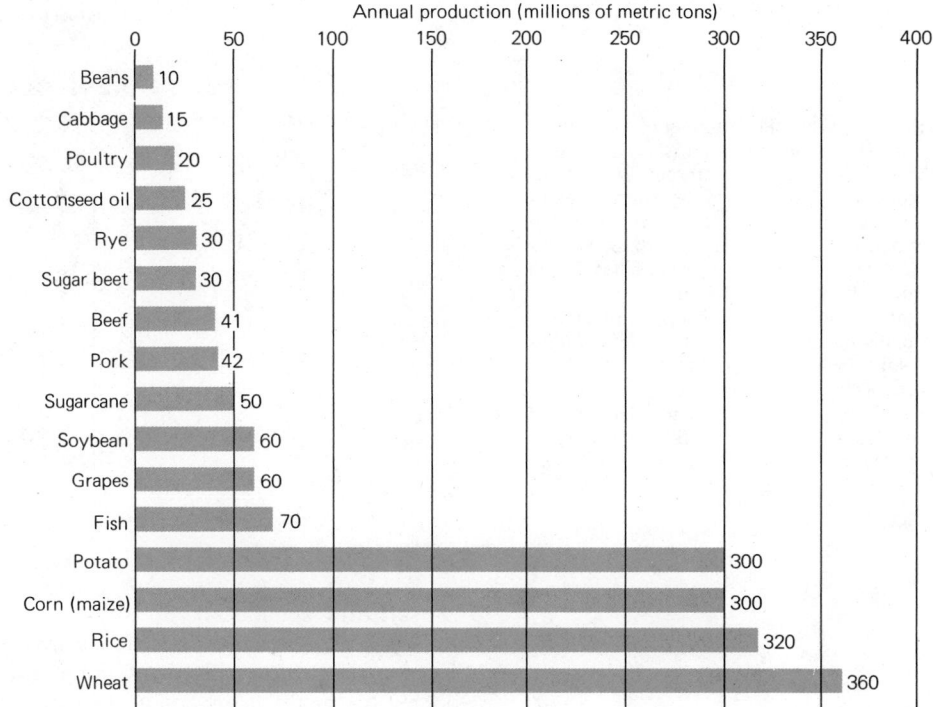

Figure 50-32 Annual world production of major food sources. Note the overwhelming importance of the major cereals.

Annual production (millions of metric tons)

Beans	10
Cabbage	15
Poultry	20
Cottonseed oil	25
Rye	30
Sugar beet	30
Beef	41
Pork	42
Sugarcane	50
Soybean	60
Grapes	60
Fish	70
Potato	300
Corn (maize)	300
Rice	320
Wheat	360

has been estimated that each calorie on our dinner tables has cost 9 calories to put there. Half a calorie represents investment on the farm; the rest represents the cost of processing, packaging, distribution, and cooking. (Packaging was the single largest contributor to the increase in retail food prices in the United States between 1967 and 1977.)

Thus it seems that the most effective way to solve the world's food problems is for hungry nations to increase their production of crops with the lowest possible adequate protein content, and to encourage more efficient agriculture among small farmers, who make up about half the population and occupy more than half of the land in many countries. National leaders are beginning to realize the political dangers of a large, unemployed, starving population, who flock to crowded urban slums. Food production began to increase dramatically in countries such as India, Mexico, and China in the 1950s, although most of Africa has not experienced any agricultural gains.

The Green Revolution

The production of wheat in Mexico increased more than eightfold from 1950 to 1970; twice as much land was brought into production, while the yield per acre quadrupled. In the same period, India doubled its production of grain. These spectacular increases, hailed as the "green revolution," resulted from intensive efforts to breed new strains of plants, to educate farmers in new growing techniques, and to change the economic structure of agriculture in developing nations.

By the late 1960s, food production in Mexico began to level off. It has now become clear that this is because the agricultural revolution reached the larger farms, but had no effect upon the less prosperous farmers, who work 80% of Mexico's agricultural land but produce only 55% of its crops. Small farmers did not plant the new strains of "miracle wheat" because the education, transportation, irrigation, money, and good soil needed for growing improved strains of crops are all more readily available to large farmers. Large farmers can take more risks; they can borrow money to tide them over a crop failure, whereas the subsistence farmer cannot borrow to buy the fertilizer even for one year's crop, much less survive a crop failure. As a result, the green revolution has often worsened the plight of small farmers, whose produce is more expensive than crops produced by modern practices.

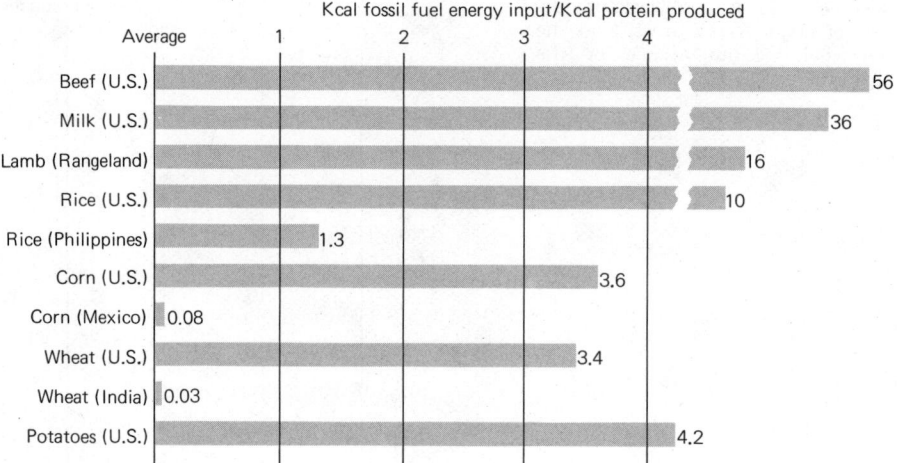

Figure 50-33 The number of kilocalories of fossil fuel energy used to produce each kilocalorie of edible protein for various crops in various parts of the world. The difference in energy cost reflects the degree to which agriculture is mechanized. Rice is cultivated largely by manual labor in the Philippines, but by machine in the United States. The enormous energy cost of producing meat and milk in the United States results from the fact that animal feed is produced by machinery and an animal converts only 10 to 15% of the plant food it consumes into meat or milk. (Data courtesy of David Pimentel)

The Fossil Fuel Shortage

Fossil fuel, such as oil and coal, has formed from fossil plants that lived millions of years ago. Most countries today are completely dependent upon fossil fuels, not just for the niceties of civilization but also for feeding their populations (Figure 50–33). At the moment the end of the supply of fossil fuels would mean the end of civilization as we know it. The trouble is that the earth contains finite amounts of fossil fuel. Whether we exhaust the supply tomorrow or in 300 years' time, the disappearance of these fuels will drastically reduce the earth's carrying capacity for humans unless we succeed in developing some new source of energy in the meanwhile. Extracting fossil fuel from the earth also becomes more expensive every year as the superficial supplies of oil, coal, and gas that can be extracted cheaply are used up.

Nuclear fission was once hailed as the energy supply of the future, but it is now clear that even if nuclear power plants can be made acceptably safe, the supply of uranium fuel (another non-renewable resource) is much more limited than was once thought.

The sun and hydrogen bombs produce heat and light by the process of nuclear fusion. Harnessing of human-controlled nuclear fusion for useful energy on earth is not technologically possible at the moment, and best estimates are that we shall not have usable energy from nuclear fusion power plants for at least another century.

Attention has recently turned to producing energy from sources that are essentially renewable—wind and sunlight. Both have been used in a minor way throughout human history. Many people doubt, however, that they can be developed to supply enough energy in time to make much difference as the fossil fuel supply dwindles.

The energy situation is such that unless people reduce their consumption of fossil fuels and develop new energy sources with unprecedented speed, energy shortages alone will cause widespread reductions in the standard of living in industrialized countries as the price of fuel rises, and widespread starvation in developing countries, since the production of food depends upon fossil fuel for fertilizers and transportation.

Pollution

All organisms expel their waste products into the environment around them. These waste products include carbon dioxide, dead leaves, feces, and inedible parts of plants and animals. Although waste products make the environment less favorable for the organism that produced them, in a balanced ecosystem, one organism's wastes are another's food and drink, and so waste does not accumulate. **Pollution** results when human waste products are not destroyed as fast as they are produced and so accumulate, making the environment less hospitable to human life, and often to other organisms as well.

Figure 50-34 Pollution. A factory burning fossil fuel without using adequate filters such as scrubbers. The smoke contains particulate matter and sulfur, making the air dangerous to breathe and causing acid rain to fall as much as thousands of miles downwind of the factory. (Peter Brussard)

The modern pollution crisis dates from the industrial revolution with its use of fossil fuels for energy. Polluted air is air that damages the health of people who breathe it. Until about 1950, most air pollution was caused by burning coal. London "pea-souper" fogs, happily extinct since the passage of the Clean Air Act in 1952, were a mixture of fog with the sulfur and particulate matter produced by burning coal; they caused thousands of deaths. Modern "smogs" are caused largely by the combustion products of gasoline from cars. In cities such as Los Angeles and Tokyo, smog frequently renders the air unfit for human consumption. It has been calculated that a nonsmoker who lives in New York City suffers lung damage from air pollution that is equivalent to that suffered by someone who smokes a pack of cigarettes a day but breathes the less polluted air of rural New York State. Since it is impractical to filter air as people breathe it, air pollution must be controlled at its source.

Since about 1950, pollution from solid waste has become a much more acute problem than it ever was before. Part of this is due to the population explosion, which has resulted in a greater total amount of solid wastes. Our higher standard of living in a "throw-away" society has also increased the amount of waste produced per person. The use of fossil fuel, this time as the raw material for the manufacture of many artificial polymers* collectively known as "plastics," has also contributed to solid waste pollution.

Modern artificial polymers are synthesized from long chains of carbon atoms, present as organic polymers in fossil fuel. Polyester, polyethylene, vinyls, rayons, and acetates are just some of the artificial polymers used in clothing, containers, skis, bottles, artificial arteries, and thousands of other items. Plastics have replaced animal and vegetable products in many areas of our lives. Substances like paper, cotton, cardboard, and leather are produced from plants and animals and so are biodegradable; that is, they can be broken down into harmless minerals, carbon dioxide, and water by decomposer organisms. Most artificial polymers cannot be broken down by any microorganisms, so that they are apparently with us forever.

Since artificial plastics and polymers are based on carbon chains, however, there is plainly hope that microorganisms that can break them down will one day evolve, and experiments are being conducted to encourage this.

Like other forms of pollution, water pollution results from overpopulation and intensive use of technology. Until recently, human populations were sufficiently dispersed that people could dump wastes into streams and rivers; the wastes would be diluted and decomposed into harmless components before the water was used by the next person downstream. Nowadays, human populations are too dense for natural processes to take care of the wastes, and sewage treatment facilities must be operated to make water usable once again. Even so, the quality of water for drinking, washing, recreation, and manufacturing has deteriorated in many areas in recent years.

---

* Polymers Large molecules made up of subunits that are smaller molecules similar or identical to one another.

Essay: Private Interest and Public Welfare

The main difficulty in our attempts to solve ecological problems such as overpopulation and pollution is the conflict between the short-term welfare of individuals and the long-term welfare of society. Garrett Hardin has called this phenomenon "the tragedy of the commons." He illustrates it with the case of commons in medieval Europe. A common was grazing land that belonged to a whole village; any member of the community could graze cows and sheep there. It was in the interest of each individual to put as many animals on the common as possible, to take advantage of the free animal feed. However, if too many animals grazed on the common, they eventually destroyed the grass; then everyone suffered because no one could raise cattle on it. For this reason, common land was eventually replaced by individually owned, enclosed fields. In this situation, the owner is careful not to put too many cows on one patch of grass, because overgrazing this year means that fewer cows can be supported next year.

In the same way, if we, as individuals or as corporations, knew we should pay directly for the overpopulation and the pollution each of us causes, we would each have fewer children and contribute less to pollution. An example comes from laws in the United States, which made employers liable for on-the-job injuries. When a firm knows that injuries will cost it money, it takes steps to ensure greater safety on the job so that fewer injuries occur. In the case of ecological problems, however, it is not usually possible to assign responsibility directly to the people who cause the problem. Future generations will pay most of the price for the fact that we have too many children now, and our individual contributions to water pollution cannot be easily distinguished.

If I decide not to use my car and not to use phosphate detergents, I can make little difference to the overall levels of air and water pollution. Furthermore, I shall be at a disadvantage compared with my neighbors because I cannot take a job that would require me to use my car, and I probably would not get the job anyway because of my "tattletale gray" clothes! Thus, although incentives for individual action are effective ways of getting most things done, environmental problems can be solved only by effective action by the government, an agency that, at least in theory, can look beyond the immediate interests of individuals and plan for the long-term welfare of society.

A government can act in three main ways. First, since educated people can understand "the tragedy of the commons" and do not really want to make the earth uninhabitable for future generations, programs to educate the public about the problems are vital. Second, probably the most effective way for a government to reduce pollution, for example, is to apply incentives (usually in the form of tax reductions) to those who pollute less and to levy a "pollution tax" against individual and corporate polluters. Third, when all else fails, government must resort to regulation, banning peculiarly dangerous practices and products altogether. Regulation becomes necessary in two main cases: first, when a practice is so damaging that no financial value can be placed upon it (it is not enough to permit a firm to market thalidomide or exterminate the blue whale and tax them for it afterwards), or when it is impossible to assign responsibility with any degree of justice (how should the price for sulfur discharged into the air by a coal-fired power station be distributed between the utility that owns the station and the consumers who use its electricity?).

Thus the solution to our ecological problems lies in modifications of individual and collective behavior sponsored by a government that can enforce compliance of everyone concerned. In a democratic society, this can only be done with the support and consent of a well-informed public that understands the biology of organisms and the intricate web that connects us to all living things.

SUMMARY

Populations of various species of organisms are the basic biological units of ecosystems. Given ideal environmental conditions, the number of individuals in a population increases geometrically at its innate capacity for natural increase (r_m). This biotic potential is determined mainly by the age of the (female) parent at first reproduction, but is also influenced by the number of individuals produced at each reproductive event and by the parent's reproductive lifespan. A population seldom, if ever, reproduces at its r_m; environmental resistance slows population increase to r, the actual rate of increase (or decrease), even when the population is growing exponentially. Factors that limit population growth may be density-dependent, for example, predation or competition for resources, or density-independent, for example, bad weather. When growth of a population ceases, under the influence of one or more of these factors, the size of the population may fluctuate at approximately the carrying capacity of the environment for that species.

Survivorship curves for members of a population reflect the population's reproductive strategy. At the two extremes, the members of a population may produce many small offspring and give them no parental assistance in subsequent survival, or may produce a few large offspring that are nourished and trained by the parents. Survivorship curves are the basis for constructing life tables, useful in predicting future changes of population size.

Competition may play an important part in regulating the size of a population. In scramble competition, all members of a species have equal opportunity to share a resource; in times of abundance, the population may increase dramatically, but all may suffer when there is not enough of the resource to go around. In contest competition, the strongest or most aggressive members of a species acquire enough of the resource they need, leaving the weaker members to survive if they can on what is left. If two species compete strongly for the same resource, the competitive exclusion principle states that one species will become extinct in the face of the superior competitive ability of the other. However, in both intraspecific and interspecific competition, specialization may partition a resource and allow it to be shared.

Predation is another factor that may cause high mortality in a population. Many specialized predators and parasites are known to keep their prey species at low density. Generalized predators tend to prey on the most abundant prey available until its level is considerably reduced (by predation or by other factors). Generalized predators have been shown to limit the population sizes of prey species in a few cases.

It is clear that many factors influence the size of a population, and that the main factor actually controlling population size may vary from population to population and from time to time for the same population.

The invention of agriculture permitted people to live in a population more dense than that possible with a hunter-gatherer economy. The number of people in the world has grown even more rapidly in the last few hundred years as our understanding of nutrition and hygiene has increased. During the industrial revolution, human beings began to switch from wood to fossil fuels as the main source of energy, and this change has accelerated the population explosion. Today the population of many countries is growing faster than is their food production. Food production can be increased dramatically by the techniques of the "green revolution." Among the disadvantages of switching from traditional to modern agricultural techniques are that modern techniques widen the economic gap between small and large farmers, increase a nation's dependence on dwindling supplies of fossil fuel, and involve the use of pesticides and herbicides that may become dangerous pollutants.

The widespread use of fossil fuel and the population explosion are the main factors responsible for the pollution of air and water that threatens human health, especially in industrialized countries.

OBJECTIVES

From your study of this chapter, you should be able to:

1. Define population, habitat, niche, exponential growth, innate capacity for increase (biotic potential, r_m), actual rate of increase (r), logistic growth, carrying capacity (K), competition, and predation.

2. Draw and interpret any or all of the three main kinds of survivorship curves; relate the survivorship curve for a species to its reproductive strategy.

3. Explain what a life table is and why it is useful.

4. Distinguish between a stable population and a stationary population.

5. Distinguish between density-dependent and density-independent mortality factors; give an example of each.

6. Explain how a species that occupies short-lived patches of habitat could have a relatively constant overall population size over a wide area.

7. Describe the competitive exclusion principle, and discuss its validity.

8. Describe an example of successful biological control of a pest species.

9. Contrast the effects of specialized predation and generalized predation on population sizes of prey species, citing evidence to support your argument.

10. Write a paragraph summarizing your understanding of the factors that limit the sizes of populations and your assessment of their relative importance.

11. Explain why the agricultural revolution is believed to have been responsible for a dramatic increase in the human population.

12. Explain what is meant by the demographic transition, and give some reasons why it occurs; explain why the human population is still growing even though its birth rate is declining.

13. Contrast the efficiencies of subsistence and modern farms, and of vegetarianism and carnivory.

14. Describe the *biological* basis of the green revolution, and explain its impact on human society, economics, and ecology. Explain why the green revolution has fallen short of original expectations.

15. Explain how agriculture, use of fossil fuels, and the manufacture and use of artificial polymers contribute to pollution.

SELF-QUIZ

1. Draw a curve showing the long-term growth of a population of bacteria on a nutrient medium in a Petri dish.

2. A population can grow exponentially:
   a. when food is the only limiting resource
   b. when first invading a suitable and previously unoccupied habitat
   c. only if there is no predation
   d. only in the laboratory

3. Which of the following does *not* directly affect biotic potential?
   a. a female's age of first reproduction
   b. carrying capacity of the environment
   c. length of time a female is fertile
   d. average number of offspring per brood or litter

4. The leveling off of an S-shaped curve describing population growth is caused by:
   a. the carrying capacity
   b. competitive exclusion
   c. environmental resistance
   d. a change in the biotic potential

5. If a population exceeds the carrying capacity of the environment:
   a. it will evolve adaptations to avoid a population crash
   b. its numbers will probably decrease rapidly
   c. its food supply will increase in the next generation
   d. the mean number of young per individual will increase

6. Which of the following would be *least likely* to act as a density-dependent limiting factor in a population of mice?
   a. parasitism
   b. buildup of waste products
   c. predation
   d. unfavorable climate

7. A female elephant bears one offspring every two years. Which type of survivorship curve would you expect elephant populations to show: I, II, or III?

8. Studies on the competitive exclusion principle suggest that:

   a. two species may share the same resource only if they are both strong competitors for it
   b. two species that appear to be sharing the same resource are probably specializing so that each uses only a particular part of the resource
   c. if two species are sharing the same resource, neither is capable of exploiting the part of the resource used by the other
   d. no resource can be used indefinitely by more than one species

9. For each statement listed below, tell whether it is more likely to be true of a situation involving a specialized predator, a generalized predator, both, or neither.
   a. Predation keeps populations of both prey and predator at very low overall levels.
   b. The biomass of the predator population is smaller than that of the prey population.
   c. Predation does not have much influence on the size of the prey population(s).
   d. Predation accounts for a significant fraction of mortality in the prey species.

10. The agricultural revolution is believed to have been responsible for a dramatic increase in the human population. Which of the following was *not* a factor in this increase?
    a. Food became more concentrated and thus easier to obtain.
    b. Many methods of birth control were abandoned.
    c. Improved medical knowledge increased life expectancy.
    d. Larger amounts of food could be obtained by fewer people.
    e. People could accumulate possessions, and wanted more children to pass them to.

11. The demographic transition in a country is correlated with:
    a. women's liberation
    b. education for women
    c. the green revolution
    d. the industrial revolution
    e. improved hygiene and nutrition
    f. improved treatment for illnesses

QUESTIONS FOR DISCUSSION

1. Paul Ehrlich has said, "It is quite possible that the penalty for frantic attempts to feed burgeoning populations in the next decade may be a lowering of the carrying capacity of the entire planet." What does this mean? Do you think he is right? Why?

2. Why has the human population grown so rapidly in the last few centuries? What are some of the lines of evidence that indicate that the human population has already exceeded its carrying capacity?

3. How does a project such as filling in a marsh for a housing development or building a four-lane highway affect the populations of organisms in an area?

4. Give some examples of populations that have grown exponentially in nature. What generalizations can be made about the conditions necessary for exponential growth?

5. Why does rate of population growth slow as the carrying capacity is approached? Does the current rate of increase in human population tell us anything about the value of the planet's carrying capacity for humans? Can the earth be considered as an ephemeral "patch"?

6. The 1970s saw a decline in the birth rate in the United States. Why is the population of the United States still growing?

7. Consider two women born in the same year, each of whom will give birth to twin girls as her only children. However, one woman (A) will have her twins at age 18, the other (B) at age 36. Each daughter will have twin daughters at the same age her mother gave birth and so on. All mothers will die at age 72.
   a. How many descendants does A have when she dies?
   b. How many descendants does B have when she dies?
   c. Construct a graph to show the growth of populations A and B.
   d. How do the rates of increase compare in the two populations? (Find a *numerical* answer if you can!)
   e. What are some factors that contribute to an increase in the age of the first reproduction in humans?

8. People worry about overpopulation of a number of species. What are some of these species? Why should we worry about their overpopulation?

9. People also worry about underpopulation of some species. What are some of these species? Should we worry about possible extinction of some species?

10. Historians often view time as a pendulum that swings back and forth, passing through the same phases repeatedly. What recent trends of modern society are reminiscent of the hunter-gatherer way of life?

11. It has been argued that advances in human civilization can be traced to exploitation of new sources of energy. Major steps forward (or backward if you are a pessimist) included the addition of meat to the herbivorous diet of our hominid ancestors, the taming of fire, and the use of fossil fuels. Other advances along the way have been domestication of beasts of burden and harnessing of wind and water power. Explain how each of these has affected the ecology of human beings.

REFERENCES AND FURTHER READING

Cole, L. C. "The population consequences of life history phenomena." *Quarterly Review of Biology* 29:103, 1954. The paper in which Cole showed how much more a female contributes to the growth of a population if she breeds as young as possible.

Colinvaux, P. A. *Introduction to Ecology*, 2d ed. New York: John Wiley & Sons, 1978. A very readable general ecology text, including extensive discussion of population regulation.

Hazen, W. E., Ed. *Readings in Population and Community Ecology*, 2d ed. Philadelphia: W. B. Saunders, 1970. Several classic papers are reprinted here, including studies by Huffaker (mites), MacArthur (warblers), and Connell (barnacles) discussed in this chapter.

Holloway, J. K. "Weed control by insect." *Scientific American*, July 1957. The fascinating story of the biological control of Klamath weed.

Krebs, C. J. *Ecology: The Experimental Analysis of Distribution and Abundance*. New York: Harper and Row, 1972. A textbook of general ecology with a particularly clear and comprehensive treatment of population ecology.

Lack, D. *Population Studies of Birds*. New York: Clarendon Press, 1966. A very readable book about population regulation, full of interesting tidbits on the natural history of birds by a famous ornithologist.

Langer, W. "Checks on population growth 1750–1850." *Scientific American*, February 1972. An interesting account of how human populations in Europe were held in check before the advent of modern contraception.

Root, R. B. "The niche exploitation pattern of the blue-gray gnatcatcher." *Ecological Monographs* 37:317, 1967.

U.S. Department of Health, Education and Welfare, *Vital Statistics of the United States, 1975*. A series of pamphlets containing vast amounts of statistical information about the human population of the United States.

Varley, G. C., G. R. Gradwell, and M. P. Hassell. *Insect Population Ecology*. Oxford: Blackwell Scientific Publications, 1973. A readable introduction to insect population dynamics, including the winter moth studies.

Wilson, E. O., and W. H. Bossert. *A Primer of Population Biology*. Stamford, CT: Sinauer Associates, 1971. A short, fairly mathematical book on population biology.

CHAPTER
51

EVOLUTIONARY ECOLOGY

In the last two chapters, we have discussed factors—such as energy flow, nutrient cycles, competition for resources, and predation—that determine the structures of populations and communities from day to day, in ecological time. In the long run, in evolutionary time, these factors have acted as selective pressures and have determined how the members of any population or community have evolved.

Every organism has adaptations resulting from four main categories of selective pressure:

1. The physical environment
2. The need for nourishment
3. Predation
4. The need to reproduce

Figure 51-1 Queen Anne's lace, a plant defended against herbivores by the toxic chemicals it contains.

We have studied many adaptations to the physical environment in our survey of the physiology of various organisms, and we have seen many reproductive adaptations in Chapters 18, 36, and 46. In this chapter, we shall concentrate on adaptations in the second and third categories—adaptations that permit organisms to obtain food more efficiently and to avoid predation.

An organism can acquire and use only a limited amount of energy in a lifetime. The more energy it expends on one adaptation, the less is available for other adaptations or for reproduction. The balance of selective forces dictates that energy will be spent on a particular adaptation only insofar as that adaptation increases reproductive success, on the average, over the average success rate from spending the same energy and nutrients in some other way.

When we examine particular adaptations of organisms, we can often deduce the selective pressures that have been at work. For example, when we find a plant expending large amounts of energy producing chemicals that are toxic to herbivores, we may reasonably guess that herbivores have been important agents of selection in the evolution of that plant species. We shall examine this and some other rather dramatic examples of adaptations in this chapter.

51-A   How to Avoid Being Eaten

Most organisms have adaptations that reduce their risk of being eaten by other organisms. These adaptations are of two general kinds, defense and escape.

Thorns and spines are plant defenses against vertebrate predators. Thorns on stems are not much defense against small insects, as any rose gardener knows, but many plants have spiny or hairy leaves that discourage these insects. For instance, varieties of potatoes and beans with hairy leaves are much more resistant to leaf hoppers, a major insect pest, than are varieties with smooth leaves (Figure 51-2). A coating of silica, the raw material of sand and glass, makes an effective defense for grasses and horsetails*; leaf toughness or stringiness and cells containing sharp, needle-shaped crystals known as raphides may also induce herbivores to look elsewhere for a meal.

Physical defenses can sometimes be **induced**, that is, produced or elaborated

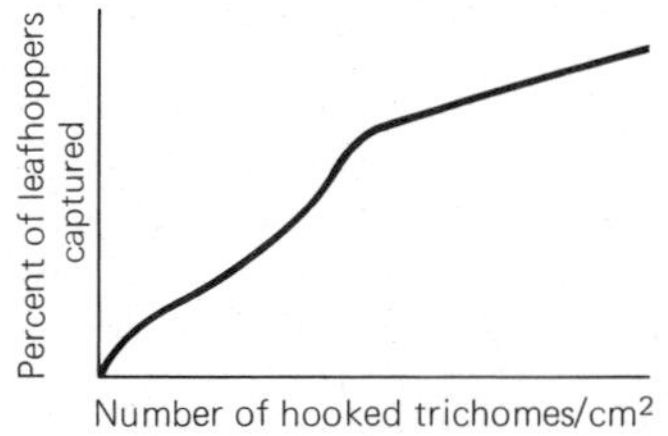

---

* Horsetails   Members of a group of primitive vascular plants (Section 26-E).

Figure 51-2 The percentage of 20 leaf-hopper nymphs captured by hooked trichomes on leaves of various bean varieties after 24 hours increased with the density of trichomes. (Data courtesy of E. A. Pillemer and W. M. Tingey)

Percent of leafhoppers captured (vertical axis)

Number of hooked trichomes/cm² (horizontal axis)

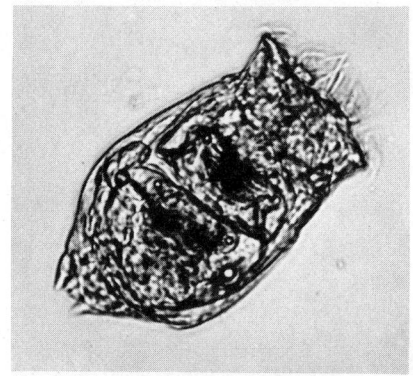

(a)

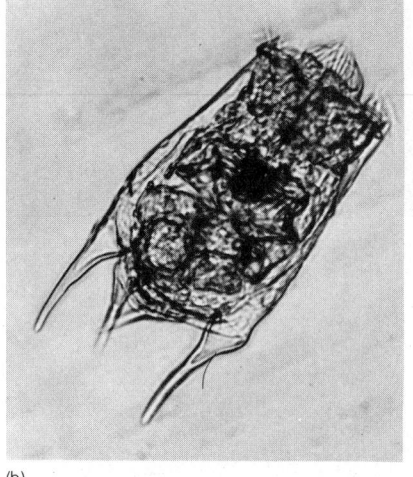

(b)

Figure 51-3 Eggs of planktonic rotifer *Brachionus calyciflorus* develop into individuals with a pair of long spines when the predatory rotifer *Asplanchna* is present. (a) *Brachionus* with no long spines (b) *Brachionus* with long *Asplanchna*-induced spines. (John J. Gilbert)

only when they are needed. For instance, the small planktonic rotifer* *Brachionus* has short spines that protect it against some predatory species of zooplankton* but not against *Asplanchna*, another rotifer. However, *Brachionus* can detect a chemical that *Asplanchna* releases into the water, and this stimulus causes the next generation of *Brachionus* to develop with longer spines, making the animal too big to be eaten by *Asplanchna* (Figure 51–3). Such induction has the advantage that energy is spent on defense only when it is actually needed.

51-B   Nutrient Exclusion

Plants in general must be very apparent to herbivores. Why, then, do they dominate the landscape even in environments with large populations of herbivores? The general explanation seems to be that, for many different reasons, all that is green is not necessarily edible. For example, a plant may be a less than ideal food for a herbivore due to **nutrient exclusion**: many plants exclude from their tissues nutrients that they themselves do not need (or need only in small quantities), but that are required in larger quantities by animals. It is probably no coincidence that plants contain much lower quantities of nitrogen, sodium, and iron than are required by herbivores; this forces herbivores to consume large quantities of food in order to extract sufficient nutrients, or to find these nutrients in some other way. Many plants are also deficient in amino acids, such as methionine, that are essential for animals. The practice of breeding corn for higher lysine content, which makes it more valuable for human food, may backfire if it also makes corn a much more suitable diet for insect pests.

51-C   Chemical Defenses

Plants

Everyone has heard of such poisonous plants as poison ivy and deadly nightshade. The poisons in these plants represent only a fraction of an enormous array of natural products known as **secondary compounds**, whose distribution in nature is sporadic, as opposed to the **primary compounds** (proteins, carbohydrates, lipids, nucleic acids) that are found in every living organism. As recently as the 1950s most biologists thought that secondary compounds were waste products of metabolism. In 1959, however, Gottfried Fraenkel argued convincingly that at least some secondary compounds of plants represent defenses against herbivores. Although this

---

*Rotifer   Small freshwater animal with a mouth surrounded by cilia (Section 27-H).
*Zooplankton   Animals floating in the surface layers of water.

Figure 51-4 This scanning electron micrograph shows the hooked spines that make goosegrass difficult for a small herbivore to walk over, let alone eat. (Biophoto Associates)

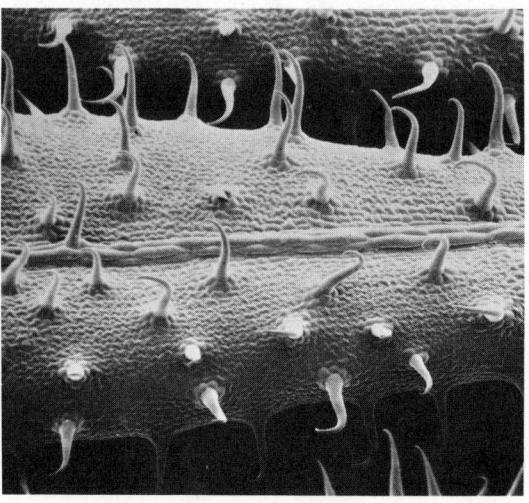

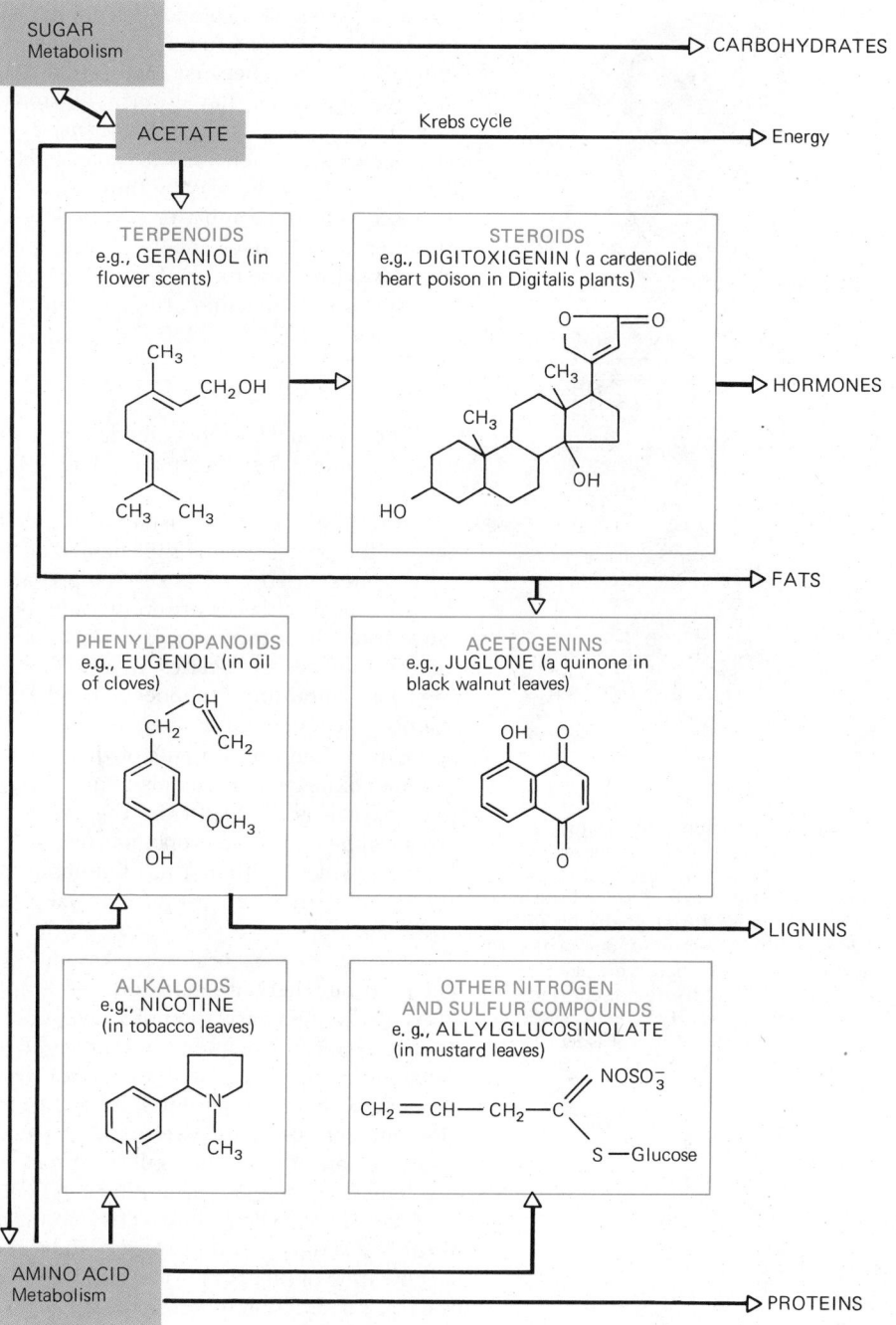

Figure 51-5 The major groups of secondary compounds (in colored outlines) and their relationships to primary metabolism. (Redrawn from Whittaker and Feeny, *Science* 171:757, 1971)

has been shown to be true beyond doubt in only a few cases, it seems likely that all land plants contain chemical defenses of one sort or another. Not all of these defensive compounds are toxic to humans, but this does not mean that they are harmless to other animals as well. It seems to be a world in which "one man's meat is another man's poison."

Studies of plant metabolism have revealed that secondary compounds are produced in side branches or extensions of the main metabolic pathways that produce primary compounds, such as proteins and fats. Acquiring the capacity to produce a defensive chemical probably involved the evolution of only a few extra enzymes that diverted some intermediate molecule from a main metabolic pathway and altered it slightly. Several thousand secondary compounds are known, all of them members of only six major groups of chemicals, derived in various ways from acetate or from common amino acids (Figure 51–5).

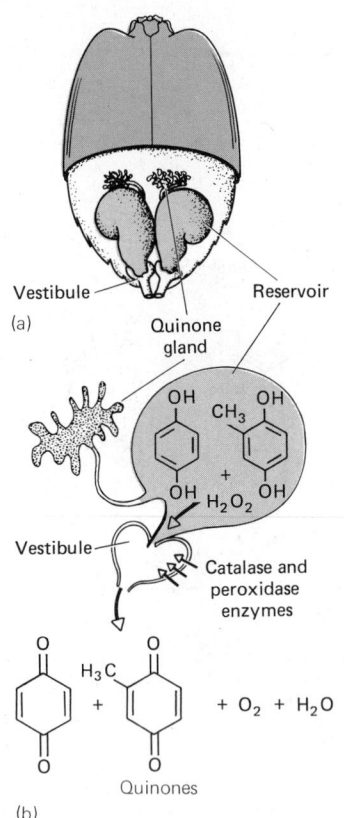

Vestibule

(a)

Quinone
gland

Reservoir

Vestibule

Catalase and
peroxidase
enzymes

Quinones

(b)

Figure 51-6 Defense apparatus of the bombardier beetle (*Brachinus*). (a) Dorsal view of a beetle with part of its back removed to reveal a pair of structures joining at a rotatable turret at the tip of the abdomen. (b) Diagram of one structure showing its gland, reservoir containing aqueous solution of hydroquinone, methylhydroquinone, and H_2O_2, and vestibule (reaction chamber). (After Eisner, 1970)

Many secondary compounds not only deter or poison herbivores but also inhibit the establishment of fungi, bacteria, and viruses, thus serving several defensive functions at once. Because many secondary compounds are also toxic to the plants that produce them, they are usually stored in special glands or ducts, or are complexed with sugars in the form of inactive **glycosides**; these can be hydrolyzed by special enzymes when a tissue is damaged, releasing the active defensive compound. Plants, like animals, may be induced to produce chemical defenses when they are damaged; when an apple or potato is peeled, for example, tannins are formed by oxidation of precursors in the tissues and turn the damaged area brown. Tannins complex with proteins, such as the enzymes liberated by the hyphae of fungi, and thus serve as a first line of defense until a protective skin can reform.

Animals

Many animals, such as monarch caterpillars and skunks, contain toxic chemicals that defend them against predation. But whereas plants with chemical defenses can tolerate some damage to their tissues before their toxin has its effect on a herbivore, animals must usually defend themselves without being seriously damaged in the process. Thus the chemical defenses of animals are often in the form of sprays, bites, or stings that reach a predator before the predator can inflict damage.

Chemical defenses are most common and diverse among the arthropods, and range from the stings of scorpions, bees, and wasps to the defensive secretions of many millipedes and beetles. Bombardier beetles spray predators with a hot (100°C) secretion containing quinones.* The beetles have a special gland made up of two chambers. One contains phenols, precursors of the quinones, as well as hydrogen peroxide. When the contents of the chambers are mixed, an enzyme in the second chamber oxidizes the phenols to quinones in an exothermic* reaction. At the same time, oxygen is released from the hydrogen peroxide, and the pressure of this gas provides the explosive propellant for the spray (Figure 51-6). Most insect defenses are less elaborate than this, but many ants, wasps, and predatory bugs* have venoms that they can use to paralyze their prey or to defend themselves against attackers.

Among the vertebrates, snakes and skunks are well known for their chemical defenses, but there are many other examples. **Tetrodotoxin** is a nerve poison that prevents the depolarization of nerve and muscle membranes. It was first found in Japanese puffer fish, and investigators have since found it in the Asian goby fish, an American newt, and some Costa Rican frogs. Many frogs and toads secrete toxins in the mucus that covers their skin. Some of these toxins are incredibly poisonous. A few micrograms of **batrachotoxin**, from the frog used by Colombian Indians to make poison arrows, will kill a human being; tetrodotoxin is equally poisonous.

The energy "cost" of producing chemical defenses is suggested by two sorts of evidence. First, when the selective pressure of herbivores on a chemically defended plant is relaxed—in other words, when defensive chemicals become unnecessary—a population of plants may lose its ability to manufacture the chemicals. Members of the plant population that lose the genes for producing the chemicals then have more energy for reproduction than the plants that retain the genes have, and the gene pool changes as a higher proportion of each generation is plants without the genes that produce the defensive chemicals. Second, in plant species that encounter fluctuating selection pressures, polymorphism* may exist, with some members of the population able to produce defensive chemicals and others lacking this ability. For example, the leaves of some clover plants produce compounds that release cyanide when the leaves are damaged; other clover plants in the same population may lack this defense. When many herbivores are present, the defended clover is at a selec-

---

*Quinones   Compounds derived from quinone, a 6-carbon unsaturated ring compound with two keto groups.
*Exothermic   Heat-releasing.
*Bugs   Members of the insect order Hemiptera; e.g., water striders, backswimmers.
*Polymorphism   Simultaneous presence in a population of two genetically different forms of a trait at frequencies higher than could be maintained by recurrent mutation (see Section 15-F).

Figure 51-7 Chemically defended animals. (a) A puffer fish. (b) A tropical frog. (a, Biophoto Associates; b, Christine Boake)

(a)   (b)

tive advantage, but when the clover is relatively free from predation, those plants that are not putting their energy into manufacturing defenses can out-reproduce the defended plants.

Animals sometimes obtain their chemical defenses at second hand. Nudibranchs, which feed on cnidarians, manage to move the complete and undischarged nematocysts* of their prey into their own skins, where the nematocysts are redeployed against potential predators of the nudibranchs (Figure 51–8). Many insects have adaptations that permit them not only to tolerate toxins in their food plants, but also to reuse these toxins for their own defense. For example, monarch butterflies are toxic because they contain cardiac glycosides* from the milkweed plants they ate as larvae (Figure 51–9). Cardiac glycosides can also be synthesized by animals, such as toads, for defensive purposes—an example of the convergent evolution of chemical defense.

---

*Nematocysts   Stinging structures (see Figure 27–7).
*Cardiac glycosides   Sugar-linked steroids that cause the heart to beat faster, sometimes leading to death of the animal.

Figure 51-8 Nudibranchs (sea slugs), such as these *Cyprioma* grazing on a coral, preserve the stinging cells from their hydroid prey and use them for their own defense. The purple dye seen here is secreted by many molluscs and was used to dye the robes of Roman emperors. (Steven Webster)

Figure 51-9 A monarch caterpillar feeding on milkweed. (William Camp)

Figure 51-10 An aposematically colored nudibranch. Orange and yellow, black and white, arranged in stripes, are common aposematic patterns because they are highly visible. (Biophoto Associates, N.H.P.A.)

51-D Aposematic Coloration

Many animals with effective defenses, such as wasps, skunks, and poisonous fish and frogs, are strikingly colored (Figure 51–10). This **aposematic coloration** warns potential predators to stay away and not attack them.

Aposematic coloration protects best against predators that can learn, especially higher vertebrates. Jane and Lincoln Brower showed that a toad stung once by a bumblebee will not touch anything looking remotely like a bumblebee for a long time afterwards. They found that the same is true of the Florida scrub jay, an insectivorous bird. Dogs will seldom tangle with a skunk more than once.

The only defense of some aposematic animals is that they are distasteful to most animals. Ladybird beetles and cinnabar caterpillars are examples. Here the learning ability of the predator plays a vital role in the success of the aposematic coloration, since some individuals will be killed by inexperienced predators. If the predator did not learn, after one or two trials, to leave cinnabar caterpillars alone, many caterpillars would be killed even if they tasted so nasty that the predator spat them out.

Aposematic coloration is not absolute. By altering their behavior, many animals can be conspicuous at some times and hidden at others. A moth that is invisible against a similarly colored background, for instance, may flash a highly visible eye spot when disturbed (see Figure 51–18). Some animals use conspicuous colors and behavior in functions other than protection, such as in courtship or in warning other members of the species against danger.

Aposematic Sound

Bats produce ultrasonic squeaks as they hunt moths at night. Dorothy Dunning and Kenneth Roeder found, to their surprise, that some moths respond to bat sounds by flexing their legs, so that they produce ultrasonic clicks. This seemed peculiar since the bats can undoubtedly hear the clicks and so locate the moths. Dunning found that bats refuse to eat moths that click, whether they are presented alive or dead. Presumably the moths are protected by a distasteful chemical and have evolved the ability to identify themselves, allowing bats to learn to avoid them. This is an example of **aposematic sound**—the acoustic equivalent of aposematic coloration.

(a)

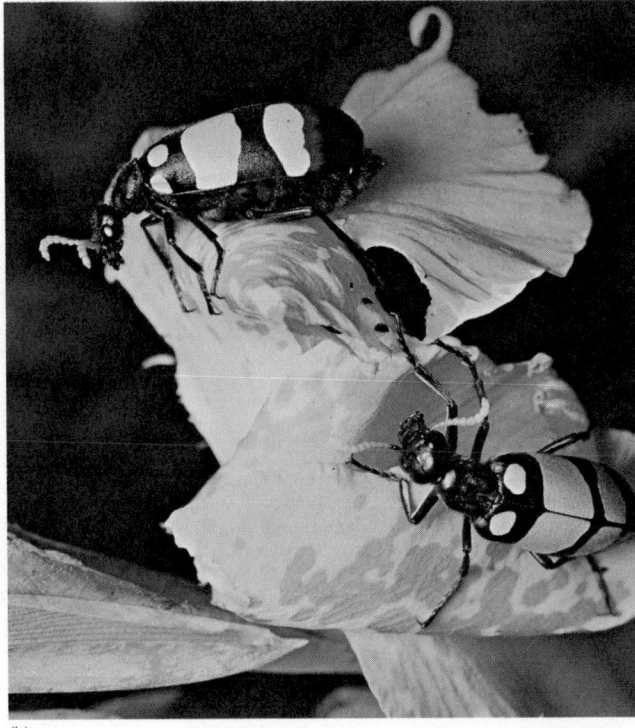

(b)

Figure 51-11 Mullerian mimicry. Many different noxious insects, such as the hornet (a) and blister beetles (b), bear similar color schemes. (Biophoto Associates, N.H.P.A.)

51-E Mimicry

Mullerian Mimicry

During the course of evolution, members of several different well-protected species have come to resemble one another, a phenomenon known as **Mullerian mimicry** (Figure 51–11). For instance, a number of different species of hymenopterans (the bees and wasps) have black and yellow stripes. The advantage of Mullerian mimicry is that it reduces the numbers of each species that are killed while a predator learns to avoid a given aposematic coloration. Suppose, for example, that a bird will avoid any insect with yellow and black stripes after it has been stung by two such insects. It is more advantageous for each of the striped prey species if the bird, in learning this lesson, kills only one member of one species and one of another. If the bird had to learn to avoid any insect with yellow stripes or green dots or red circles, because all species of stinging insects looked different, many more insects would be killed during the predator's learning process, and aposematic coloration would be much less valuable.

It may seem logically obvious that this is the selective pressure behind the evolution of Mullerian mimicry, but it is usually extremely difficult to prove; maybe so many different Hymenoptera have yellow stripes, not because there is an advantage in resembling one another, but because it is biochemically easier for them to produce yellow stripes than, say, green dots. In later sections we shall consider some of the factors that can help us decide whether or not a resemblance is true mimicry.

Batesian Mimicry

If a bird may learn to avoid eating all insects with black and yellow stripes, then there might be enormous advantage to a tasty, unprotected species if it too had black and yellow stripes. **Batesian mimicry** is the resemblance of a **mimic**, an unprotected species, to a **model**, an aposematically colored, protected species. There is a family of flies whose members mimic bees and wasps. You yourself have probably

(a)

(b)

Figure 51-12 Batesian mimicry. (a) The monarch butterfly (larger) is protected from birds by the cardiac glycosides it contains. The viceroy (smaller) is edible but birds avoid it because they mistake it for the monarch which it resembles. (b) This fly is a very close mimic of a bee. (a, Biophoto Associates, N.H.P.A.)

reacted to Batesian mimicry by moving carefully away from such a harmless fly wearing black and yellow stripes (Figure 51–12).

Mimicry is often very precise, down to details of anatomy and behavior. Most flies are actually quite different from bees and wasps. Flies have only one pair of wings, whereas other winged insects have two pairs. Most flies also lack the distinct "waist" of the body between the thorax and the abdomen that is found in bees and wasps, but flies that mimic hymenopterans usually look as though they do have this constriction because the body either has an actual constriction or is colored to look as if it does. A mimic fly's wings may also be so large that they look like two pairs (Figure 51–12). These flies are often found hovering over flowers, with rapidly beating wings, just where one would expect to find bees.

EXPERIMENTAL EVIDENCE FOR BATESIAN MIMICRY

The Browers showed that Batesian mimicry does indeed afford its owners protection from insectivorous birds and toads. Many of their experiments were done with the large orange and black monarch butterfly, *Danaus plexippus*, which is mimicked by an unrelated butterfly, the viceroy, *Limenitis archippus*. The viceroy is a little smaller, but otherwise very hard to distinguish from the monarch. The Browers found that a Florida scrub jay, raised in captivity, which had never seen either butterfly before, would eat the viceroy quite cheerfully. After one or two mouthfuls of a monarch, however, the bird would not even peck at a monarch, and after it had learned to avoid the monarch, it would not eat any more viceroys either. The Browers later discovered that monarchs contain cardiac glycosides, which are toxic to vertebrates (Section 51-C).

No protection in nature is ever absolute. Since there is always an evolutionary arms race, as it were, between predator and prey, any animal will always have predators that can learn how to avoid its sting or that have evolved the biochemical pathways to cope with its toxic chemicals. Although most birds avoid wasps and bees, shrikes and bee-eaters have behavior patterns that permit them to eat Hymenoptera without being stung. Although most herbivores will not eat milkweed, monarch caterpillars do, and they are not harmed by the cardiac glycosides. Evolving the ability to eat a defended species is very advantageous because it opens up a food source for which there may be little competition.

BATESIAN MIMICRY IN NATURE

How can we decide if two similar-looking animals are examples of Batesian mimicry, and which species is the model and which the mimic? One (tedious) way would be to perform a series of experiments, but other clues may often be used. First, a model usually looks and behaves like closely related species, whereas a mimic may be enormously different from its relatives. Thus if an insect looks and acts like a wasp, but on close examination turns out to be a fly, it is a fair guess that it is a mimic.

(a)

(b)

Figure 51-13 Two examples of camouflage. (a) This katydid's resemblance to a leaf and twigs protects it from predators. (b) The pink bulge on this coral is a nudibranch that eats the coral and is almost invisible to potential predators. (Biophoto Associates, N.H.P.A.)

Atypical behavior is another good clue to mimicry. There are black-and-yellow-striped flies that, if you touch them, will rapidly curve their abdomens around and stab your hand. In fact they have no sting, so the gesture is meaningless, but the fly looks so much like a wasp about to sting that you are likely to shake it off your hand without harming it. Most moths fly by night and rest by day, but moths that mimic butterflies have reversed this behavior pattern and fly by day with the butterflies they mimic. Similarly, the juvenile form of a South African lizard that mimics a noxious beetle moves in a way more characteristic of beetles than of lizards.

Lastly, we must consider a rather controversial point about the theory of mimicry. We might expect always to find fewer mimics than models in any particular area because the mimic is parasitic on the model's reputation; that is, the existence of the mimic detracts from the protection given the model by its aposematic coloration. If many of the prey attacked by a predator are palatable Batesian mimics, the predator is likely to attack more members of the model species before it finally learns to avoid prey with that color pattern. Thus, as we might predict, models are usually more common than mimics in any one area. Sometimes, however, this is not the case. Jane Brower showed that the viceroy butterfly gained some protection from its resemblance to the monarch even when she offered more mimics than models (60% mimics to 40% models) to Florida scrub jays. It is clear, however, that the more common the mimics, the less valuable the aposematic coloration to the model.

51-F Escaping Predation

So far we have considered various ways in which organisms can defend themselves against predation. Escape is the other major way of reducing losses to predation; escape may include hiding, fleeing, or finding safety in numbers. Each has its drawbacks: in hiding, an individual loses time that could have been spent feeding; an animal that flees must expend energy in developing long legs or wings and the muscles to work them.

Camouflage

Many organisms that hide lurk under rocks or leaves or in burrows, but others have coloration that hides them even in plain sight. **Camouflage** is a means of disguising things so that they are difficult to perceive (Figure 51–13). The camouflage of many animals (and plants) protects them from discovery by their predators or prey. For example, it is difficult to see a green frog in the grass on the bank of a pond or a trout against the gravel of a streambed. Experiments have shown that camouflage is in fact of selective advantage to its owner. We have studied one such case, the evolution of melanism in moths (see Section 16-C); moths the same color as their background are less likely to be eaten by birds than moths that contrast with their background.

(a)　　　　　　　　(b)　　　　　　　　(c)

Figure 51-14 Silhouette, eye, and bulk are the main features that permit us to recognize an animal.

Since vision is the dominant sense in humans, visual camouflage is very obvious to us and so has been the most studied. Other ways of hiding are advantageous in interactions with animals to whom other senses are more important. Some animals can camouflage their smell or the sounds they make. We shall concentrate here on protection against visually oriented animals.

We recognize an animal by three main visible features: its silhouette, its eye, and its bulk, or appearance of being rounded. The silhouette in Figure 51–14 might be almost anything. Adding an eye and ears makes the silhouette recognizable as an animal. The importance of bulk is clear from the fact that we would not confuse even a perfectly colored cardboard cutout of a cow with a real cow because it would not have any of the shadows that the bulk of a real cow would make.

An animal's camouflage must disguise its silhouette, its bulk, and its eyes. Bulk is nearly always disguised by **countershading** (Figure 51–15). If an object is the same color all over, its underside appears darker when light falls on it from above and makes it appear rounded. The vast majority of animals have light-colored bellies and dark backs; light falling from above makes them look uniformly colored and therefore flat. A camouflaged gun or plane is also painted with a pattern that countershades it. Interestingly, animals that habitually live upside down, such as the three-toed sloth, the backswimmers (aquatic bugs), and some caterpillars, have inverted countershading, with light backs and dark bellies.

Camouflage often involves coloration that disrupts the silhouette (Figure 51–16). In such **disruptive coloration**, some parts of the body appear the same color and intensity as the normal background, while others contrast with it strongly. Under these conditions, some parts of the object stand out whereas others seem to disappear. The result is an incoherent pattern of splotches rather than a recognizable animal.

Camouflaging the eye is important for two reasons. First, where there is an eye, there is an animal. Second, an eye is always near the brain, one of an animal's most

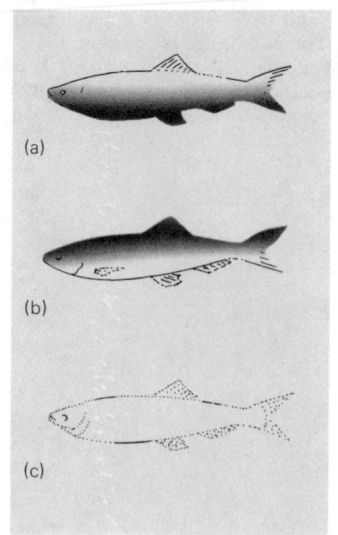

Figure 51-15 Countershading. (a) Appearance of a uniformly-colored object lit from above. (b) Appearance of a countershaded object lit uniformly from all sides. (c) Appearance of a countershaded object lit from above.

Figure 51-16 Disruptive coloration. When it rests against a black background, the black border of this butterfly becomes invisible, leaving only a serrated green area which does not remind a predator of a butterfly. (Biophoto Associates)

(a)

(b)

vital organs. The eyes may be disguised in various ways (Figure 51–17). In addition, many animals have false eyes on some other part of the body. Eyespots on the wings of moths (Figure 51–18) are quite common. Their function is not completely clear. They certainly distract attention from the moth's head, and it may be that a bird seeing these spots is fooled into thinking this is not a moth but a much larger animal, such as might possess two large eyes so far apart.

Camouflage is utterly useless unless appropriate behavior patterns also evolve. For instance, a "leaf" wandering up and down a twig is apt to be noticed by a predator. Most butterflies are not camouflaged; they fly by day and are visible whatever their coloration. Camouflage is much more common in moths, most of which are nocturnal; during the day they rest motionless and camouflaged on some appropriate surface.

FUNCTIONS OF CAMOUFLAGE

What is the advantage of visual camouflage? First, **cryptic coloration**—coloration that hides an animal against its background—might have been selected for in animals with numerous predators that hunt by sight. If this is true, we would expect that animals with few predators would be less likely to be cryptically colored. This is

Figure 51-18 When this *Polyphemus* moth is disturbed, it moves its upper wings sideways, revealing remarkably realistic eyespots on its lower wings. (Biophoto Associates, N.H.P.A.)

in fact the case. Large birds such as swans and sea gulls, which have few predators, are conspicuous, but their small, vulnerable young are cryptically colored (Figure 51–19). The highly vulnerable eider duck, incubating her eggs on an open nest, is also cryptically colored.

Plants, too, can be camouflaged. Mistletoes in Australian forests, for example, mimic the leaf shapes of their plant hosts in a most remarkable fashion. This probably makes the mistletoes less apparent to the butterflies and perhaps to some mammals that feed on them.

Escape by Numbers

Another adaptation that allows organisms to escape predation is the presence of so many potential food items that the predators cannot possibly eat them all. An organism may produce so many offspring that some are almost bound to survive. The vast numbers of offspring produced by planktonic crustaceans, for example, compensate for losses caused both by unwanted dispersal and by predation by fish and plankton-feeding whales. The nymphs of the periodic cicada of the southern United States take 17 years to grow, living in the soil and feeding mainly on fluids from the roots of trees. At the end of the 17 years, all of the nymphs in the population emerge at the same time, climb up on vegetation, telephone poles, and other structures, shed their exoskeletons, and emerge as winged adults. Then they mate, the females lay their eggs, and the next generation develops over another 17 years. These periodic aggregations are so dense that the ground may be covered inches deep with dead cicadas and the discarded exoskeletons of the nymphs. The adaptive value of this life history seems to be escape from predation. During the outbreaks,

(a)

(b)

(c)

Figure 51-19 Cryptic coloration is common in animals that remain in sight of potential predators for any length of time. (a) This moth spends the day resting on a tree trunk. (b) A turnstone sitting on her nest. (c) A gull chick in the nest. (b, Biophoto Associates, N.H.P.A.; c, William Camp)

predators such as blue jays and rodents stuff themselves, but cannot make a significant impact on the cicadas. In the intervening years there is no food supply available to permit predators to build up to population levels that could inflict significant losses. Moreover, the length of the cycle may make it particularly difficult for parasites or predators to evolve synchronous life cycles, which might make it possible for the predators to emerge every 17 years also.

A third form of escape is unpredictability. Fugitive species*—for example, early successional plants and insects (see Section 48-H)—may escape their predators in space or in time, at least for long enough to produce offspring that can disperse to other patches of habitat. Populations of the black swallowtail butterfly in Costa Rica rely on a year-round supply of ephemeral patches of open habitat in which their food plant, *Spananthe,* survives. The first brood of butterfly larvae to appear in a newly formed patch does very well, but successive broods in the same patch have progressively higher mortality as predators and parasites discover the patch and accumulate.

51-G Adaptations for Obtaining Food

There are three main ways of feeding: staying in a good spot and letting the food come, moving around and hunting for food, or parasitizing another organism's ability to obtain food. All three are common.

Roughly half of all animals are hunters of one sort or another, ranging from mountain lions and giraffes that roam over many square kilometres to nematodes that hunt bacteria in a few cubic centimetres of soil. Hunting can be divided into two different types, depending on whether organisms actively search out and pursue their prey, or instead sit in a likely spot and wait for prey to come close. Active searchers include the majority of vertebrates and many invertebrates.

Active hunters often have behavioral adaptations that permit them to maximize the amount of food eaten for the amount of energy spent in searching for food. It may be energetically advantageous to pass up the opportunity to catch a small or less nutritious prey if there is a good chance of finding a larger or more nutritious prey during the time that would be spent chasing and consuming the small prey. Chickadees abandon searching for insects in a particular spot if their feeding rate drops below a certain rate of success that they remember from the immediate past. They may be able to consume more food if they fly to another site where the food density is likely to be higher. The size of the territory defended by a male hummingbird is inversely proportional to the density of nectar-producing flowers within it. Other

---

* Fugitive species Species that occupy an area for only a short time.

Figure 51-21 Aggressive mimicry. A female *Photuris* firefly eating a *Photinus* male, which she has attracted by mimicking the pattern of light flashes produced by female *Photinus*. (James E. Lloyd)

bird species, feeding on flowers that are widely dispersed, show a different behavior, called **traplining**. They follow a particular route over and over, visiting the same flowers along the way and drinking the nectar the flowers have produced since the last visit.

Animals that hunt actively can increase their feeding success by forming **search images**. A search image predisposes a predator to discovering a particular type of prey, with the result that such prey is captured more often than would be predicted from its abundance in the habitat. Because a search image gives a predator fewer decisions to make, its overall success rate is higher (it need not examine and identify every object to see if it is good to eat, but merely scan to find items of a certain type). We ourselves form search images; during a quest for a particular item in a supermarket, it is easy to overlook another item that may also be on the shopping list.

Camouflage and mimicry are common among "sit and wait" predators, affording advantages in both offense against prey and defense against predators (Figure 51–20). Some praying mantids, for example, look remarkably like twigs, leaves, or even flowers; unsuspecting insects come close enough for the mantid to strike and capture them, while predators often overlook the mantids. Certain spiders emit a chemical compound that mimics the sex attractant pheromone* produced by females of a moth species; this attracts male moths of that species, which the spider pounces on and eats. Likewise, the light flashes of fireflies are part of their courtship behavior; males are attracted to a female by her pattern of flashes. But females of the genus *Photuris* have turned this fact to predatory advantage. They produce flashes that mimic those produced by females of another genus, *Photinus*. Thus, *Photinus* males are attracted to the *Photuris* female, who promptly eats them (Figure 51–21). When she is in a more romantic mood, the *Photuris* female flashes in the courtship pattern that attracts *Photuris* males.

51-H   Competition

Competition for food itself acts as a selective force, favoring those individuals that can exploit food resources for which there is little competition. Intraspecific* competition for food may result in a sexual dimorphism* in which the two sexes eat different kinds of food. Female and male black-capped chickadees hunt for insects in different parts of trees, the males on the trunk and inner branches, the females on

---

*Pheromone   Chemical released by one member of a species which influences the behavior of another member of the species.
*Intraspecific   Within a species.
*Sexual dimorphism   Dimorphism = two forms; in sexual dimorphism, the two sexes look (or in this case, act) different.

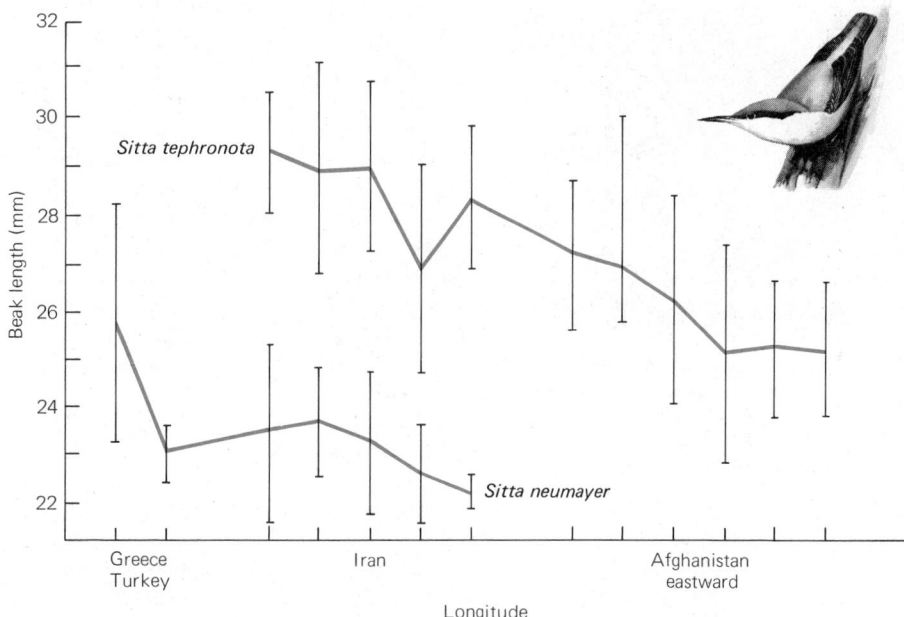

Figure 51-22 Character displacement in two species of rock nuthatch (*Sitta*) that occupy similar habitat (open rocky areas). Where the ranges of the two species overlap (in Iran), beak length of *S. neumayer* is smaller and that of *S. tephronota* larger than in areas where they occur alone. Vertical bars represent the range of beak lengths of each site. (After C. Vaurie, *Proceedings of the Tenth International Ornithology Congress,* 163–166, 1950)

the outer branches and twigs. In other species, individuals of different ages may feed on different foods; the use of different foods by larvae and adults of many insect species is a major advantage of their kind of life history.

Competition between members of different species leads, in ecological time, either to the competitive exclusion of one species by another, or to the subdivision of the resource (see Section 50-E). In evolutionary terms, such competition exerts selection for individuals that can avoid the shared resource. The divergence of two competing species in a region of overlap (**character displacement**) is illustrated by two species of nuthatches whose ranges overlap in northern Iran. Elsewhere in their ranges, the beaks of the nuthatches are similar in size, indicating that they feed on food of similar size. In Iran, however, the beak size of one species is reduced in size, while that of the other species is larger than it is elsewhere, showing that the two species have diverged in their food habits, reducing competition, in the area where they exist together (Figure 51–22).

The migrations of many bird and mammal species can be viewed as a consequence of competition, both interspecific and intraspecific. Formerly tropical species, which can move north to exploit the abundant summer food supply, avoid the competition for food in tropical habitats during their breeding period. Such migrations presumably evolved gradually, with the earliest individuals moving only a relatively short distance north.

Organisms also have a variety of adaptations that enable them to deter competitors of the same or other species. Many higher plants release chemicals that prevent the growth or germination of other plants. Such effects have been studied in shrubs of the chaparral in California. These plants produce a number of toxic substances that are washed from their leaves into the soil by rain, inhibiting the germination of seeds in the soil. This protects the shrubs from competition for space and light. Similarly, it is notoriously difficult to grow anything under a walnut tree. The leaves of walnut trees produce a substance called juglone, which washes onto the ground and there inhibits the growth of other plants.

51-I Coevolution Between Predators and Prey

Coevolution is the evolutionary change that occurs in two or more different species as a result of their action as selective pressures on one another. The cases of defense and competition that we have considered so far in this chapter are examples of coevolution. For instance, a herbivore selects for improved defenses in any plant it eats; this in turn may select for the ability of the herbivore to tolerate the improved plant defense. Parasites evolve adaptations that make them better at finding

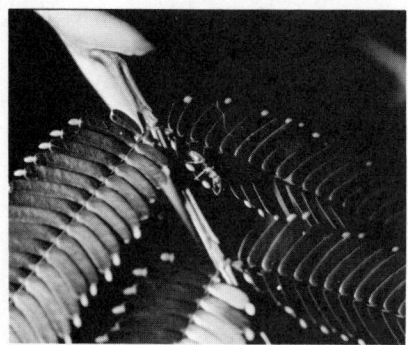

Figure 51-23 An Acacia ant (*Pseudomyrmex nigrocincta*) feeding on the yellowish-brown "Beltian bodies" on an *Acacia cornigera* plant. A nectar gland is also visible on the stem below the ant. (Paul Feeny)

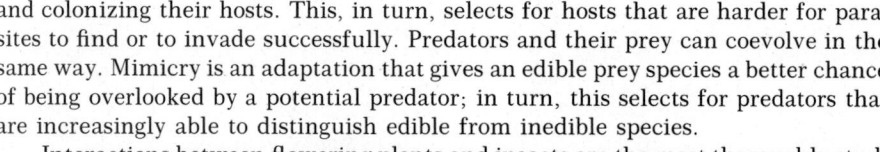

and colonizing their hosts. This, in turn, selects for hosts that are harder for parasites to find or to invade successfully. Predators and their prey can coevolve in the same way. Mimicry is an adaptation that gives an edible prey species a better chance of being overlooked by a potential predator; in turn, this selects for predators that are increasingly able to distinguish edible from inedible species.

Interactions between flowering plants and insects are the most thoroughly studied examples of coevolution between plants and animals. Insects are the most important animal pollinators of flowers, and many insects are specialized so that they feed on the leaves, seeds, or other parts of only one or a few species of plants. Also, insects' behavior patterns are relatively simple compared with those of vertebrates. Thus insects and flowering plants exhibit relatively simple, easily studied patterns of coevolution. Here we examine two examples of coevolution between plants and insects that have been investigated experimentally.

Mutualism with Ants

Acacias are leguminous* trees found in tropical areas throughout the world. Many acacias have spines that protect them against herbivorous mammals, and some Central American acacias are protected by a mutually beneficial relationship, or mutualism, with the ants that live on them. Why ants? Ants are very abundant in tropical areas; thousands of species are known, and many more have not yet been described. Many ants have potent stings, and they are social insects: an ant can recruit other members of its colony for cooperation, whereas a non-social insect usually acts alone.

Ant acacias have several structures that benefit the ants; these include swollen, hollow thorns in which the ants live, nectar glands on their leaves, and Beltian bodies (Figure 51–23). **Beltian bodies** are swollen, nutrient-rich leaf tips, which the ants cut off and feed to their larvae. The ants also collect nectar secreted by glands on the stems of the acacias.

It is advantageous for the acacia tree to have tenant ant colonies because the ants reduce the damage that herbivores do to the tree. Daniel Janzen found that when he removed the ants from an ant acacia, the tree usually died after a few months. Fungus grew on it, and its shoot tips were eaten so that it grew more slowly and became overgrown with vines. When ants are present, they react to anything that touches the tree. They remove dust, fungal spores, pollen grains, and spider webbing. They destroy the seedlings of other plants that sprout under their tree and sting other insects or mammals that try to eat the tree (Figure 51–24).

A few species of insects have evolved defenses that permit them to survive on an acacia tree that is guarded by ants. Some of these insect species seem totally immune to ant stings and ignore the ants; others can pick up the ants and throw them off the tree; still others have hard cuticles that an ant's sting cannot penetrate.

The coevolution of ant acacias and their insect populations probably went something like this: ants invaded an acacia and fed on the leaf parts and nectar of the tree. The ants also removed other insects, allowing the acacia to grow faster, and making it able to shade out other plants and to produce more offspring. Acacias that were more attractive to ants reproduced more rapidly. The availability of food and shelter, in turn, exerted selection on the ants to protect the tree with increasing efficiency. Of the insects that fed on the tree before the ants arrived, most species were expelled, but a few evolved defenses against ant attacks.

Acacia species that are not defended by ants have cyanides and other protective compounds in their leaves. These seem to have been lost in the ant acacias, presumably because they are no longer needed in the presence of ants.

Cabbages and Butterflies

When you cook cabbage or broccoli or mustard greens or almost any other member of the **crucifer** family, it gives off the characteristic odor of mustard oils—the group of secondary compounds characteristic of this family of plants. In the

(a)

(b)

Figure 51-24 (a) The ground around this young ant-defended acacia in Costa Rica is kept bare by the ants, thus protecting the tree from competing plants and from fires in the dry season. (b) This army-worm caterpillar, placed on an ant acacia, was stung to death in minutes. Also seen here are the hollow swollen thorns in which acacia ants make their nests. (Paul Feeny)

*Leguminous Belonging to the family of plants that includes peas, clover, and so on.

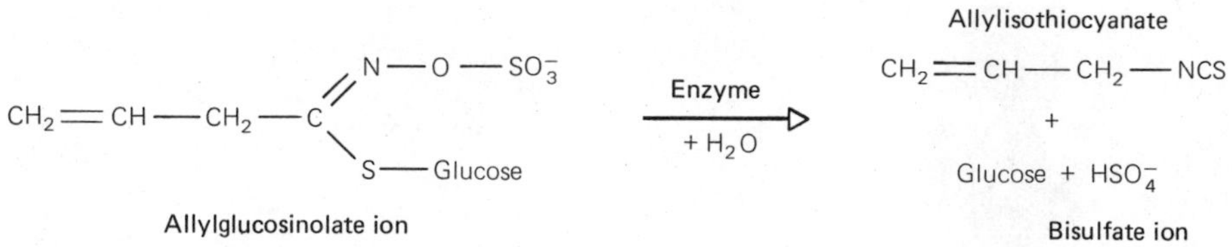

Figure 51-25 When tissues of crucifer plants are damaged, glucosinolates come into contact with enzymes that hydrolyze them to volatile and toxic mustard oils. Allylisothiocyanate, formed from allylglucosinolate, is responsible for the odor of cooked cabbage and the pungent taste of horseradish.

intact plant, the mustard oils are usually bonded to sugar molecules to form glucosinolates. The sugar molecules make the mustard oils much less toxic, allowing them to be stored in the plant without damaging the tissues they are defending. When a cell is damaged—as it is when an insect bites into it—an enzyme cleaves off the sugar molecule, which is analogous to pulling the pin on a hand grenade, and the mustard oil is released (Figure 51-25).

How toxic are mustard oils? They are plainly not very poisonous to humans or to the cabbage white butterfly caterpillars that sometimes wipe out entire plantings in the home garden. The main difficulty in answering this question is that insects usually do not eat anything except their normal food plants. You cannot take a caterpillar from an oak tree and plunk it on a cabbage leaf to see if it will be poisoned, because the caterpillar will not eat the cabbage. In an experiment to get around this problem, black swallowtail butterfly larvae, which normally feed on plants of the carrot family, were raised on some rather special carrot leaves. These leaves were cultured in solutions containing various concentrations of allylglucosinolate (Figure 51-25). The larvae were therefore feeding on their usual carrot diet, plus secondary compounds from a family of plants that the larvae do not normally eat.

As the concentration of allylglucosinolate in the carrot leaves increased, the larvae ate more slowly, but lost weight much more rapidly than would be expected from their feeding rate. At glucosinolate concentrations that occur naturally in crucifer plants, the larvae lost large quantities of fluid in their feces and soon died. Clearly, then, glucosinolates can be an effective defense against insects that do not normally attack the plants that contain them. These compounds are also toxic to a variety of fungi and bacteria, and thus seem to defend crucifer plants in several ways.

51-J Counteradaptations to Secondary Chemicals

Since every plant is in fact eaten by some herbivore, it is clear that herbivores have evolved counteradaptations that permit them to eat some species of plants that are toxic to most other animals. For instance, cabbage butterfly caterpillars can grow normally on crucifer leaves in which glucosinolate levels have been increased, by culturing, to more than 10 times the concentrations found in nature. Much of the detoxification of poisonous compounds is thought to be done by the battery of special enzymes in the guts of insects.

Some of the counteradaptations that insects have evolved against the chemical defenses of plants are remarkably sophisticated. For instance, the seeds of some tropical legumes contain canavanine, an amino acid that is an analog of arginine (Figure 51-26). In most animals, canavanine can replace arginine during protein synthesis. This produces proteins that do not function properly, and so canavanine is extremely toxic. The larvae of some Costa Rican beetles, however, eat legume seeds that contain canavanine. The arginyl-tRNA synthetase enzyme of these larvae is more specific than it is in other animals and will join only arginine, and not canavanine, to the tRNA that carries arginine to the ribosomes for protein synthesis. As a result, the beetle does not synthesize useless proteins that contain canavanine but instead uses the canavanine as a source of dietary nitrogen.

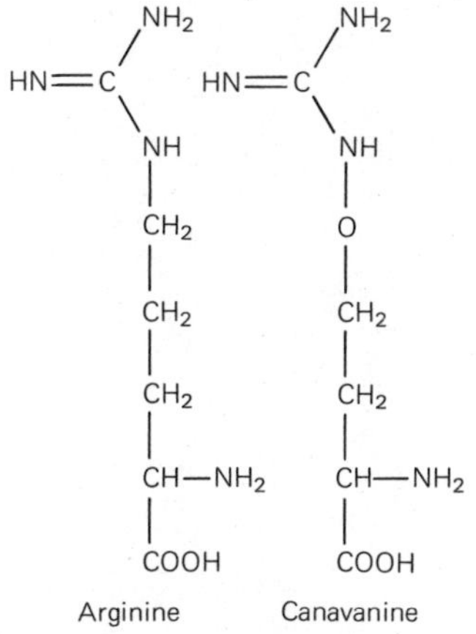

Figure 51-26 The rare amino acid canavanine is similar to arginine, a common amino acid in the proteins of all organisms.

Figure 51-27 Trapping cabbage flea beetles with mustard oil traps. The glass vial on the left contains a 1 percent aqueous solution of allylisothiocyanate. Many beetles have been attracted by this odor and remain stuck to the glue lining the inside of the carton. The vial on the right is a control, containing water only, and has caught almost no beetles. (Paul Feeny)

51-K Secondary Compounds That Attract Herbivores

Many insects use secondary compounds given off by their food plants as cues that help them to find the plants. For example, some flea beetles can pick out their cruciferous food plants from a field containing many other crops. In order to show that these beetles are indeed finding their food by scent, the smell must be separated from the plant. When this is done, traps containing solutions of pure mustard oils and a sticky glue to catch the beetles will attract just as many flea beetles as will a real plot of crucifers, whereas few beetles are caught in traps containing the solvent and glue alone.

Another insect that moves towards the scent of mustard oil in behavioral experiments is the wasp *Diaretiella rapae*. This insect is not a herbivore, but a parasite of an aphid that feeds on crucifers. Moving towards mustard oils permits it to find its host. Experiments have shown that the aphid is almost free from parasitism by *D. rapae* when it is on a non-crucifer, such as a beet plant. When it is on a crucifer it is heavily parasitized. Aphids feeding on beets smeared with mustard oil are more heavily parasitized than aphids feeding on ordinary beets.

These experiments are merely a few examples of work that has shown that a variety of animals living on or near crucifers are organized by the secondary compounds, the mustard oils, that protect crucifers from being eaten by many other herbivores. The same compound in a plant can plainly act as both an attractant and a repellent. Thus mustard oils are feeding stimulants to flea beetles, and are oviposition (egg-laying) stimulants to the cabbage white butterfly, who lays her eggs on crucifers. Both are attracted by the mustard oils that crucifers produce. On the other hand, the mustard oils repel those herbivores that cannot eat crucifers efficiently.

51-L Interference with Hormone Control

Some plants produce secondary compounds with structures similar to insect hormones. An overdose of the insect molting hormone **ecdysone** or similar compounds, produced by the common polypody fern and by some conifers and flowering plants, speeds up some phases of insect development so much that other phases, which are not much influenced by the hormone, cannot keep up. The result is a misshapen individual that soon dies.

The **juvenile hormone** of insects normally prevents metamorphosis before the insect is mature. If the gland that produces this hormone is removed from an insect, the insect pupates before it has grown to its full size. On the other hand, if juvenile hormone is added to a nymph* or a pupa, the adult never forms. The presence of

---

*Nymph Immature stage of some insects in which there is no pupa; a nymph is a small, sexually immature, and wingless version of the adult.

substances similar to juvenile hormone in plants was discovered in an interesting way. A Czechoslovakian scientist came to Harvard as a visiting researcher, bringing with him some insects that he had been raising for several generations in his laboratory in Europe. When he attempted to raise them at Harvard, the insects grew into bigger nymphs than usual, but never became adults. An extensive investigation revealed that the paper towels placed in the bottom of the insect cultures were the culprits. All American newspapers and journals had the same effect on the insects. The paper manufacturers traced the paper back to balsam fir trees, which, when tested, also prevented maturation. The active substance was extracted from a few hundred paper towels, and tested on a variety of insects. The substance had a juvenile hormone-like effect only on members of the family Pyrrhocoridae. Apparently, other groups of insects have juvenile hormones with somewhat different chemical structures.

The family Pyrrhocoridae includes several destructive pests; we can surmise that somewhere along the evolutionary line they may have exerted selective pressure against the balsam fir and some other conifers, such as the eastern hemlock, Pacific yew, and larch, which also contain this compound.

Several plants are now known to produce secondary compounds that act as insecticides because they interfere with the hormones that control insect growth and metamorphosis. One such compound, found in the bedding plant *Ageratum*, was named **precocene** because it causes premature metamorphosis in larval milkweed bugs. Female milkweed bugs that metamorphose prematurely are all sterile. Precocene also prevents the development of the ovaries in some adult insects; it is insecticidal because it suppresses the secretion of juvenile hormone.

SUMMARY

Every organism has adaptations that permit it to survive in its physical environment, to obtain nourishment, to avoid predation, and to reproduce. Selective forces often exert their effects in conflicting directions, and the adaptations seen in organisms represent compromises of the amount of time, nutrients, and energy that can be allotted to counteracting various pressures. In general, more energy will be channeled into adaptations that meet stronger selective pressures.

On the other hand, some adaptations may meet more than one selective pressure; for example, territoriality may permit an animal to construct or defend a suitable nest or burrow where it is protected from the elements, to reserve a food supply free from competition from other members of its species, to become familiar with good spots to hide or to stand and fight off predators, and to raise a family in comparative peace.

Plants defend themselves from being consumed by herbivores in a number of different ways: by physical defenses such as spines, tough cuticles, or silica; by nutrient exclusion; or by toxic chemicals. Plants expend considerable energy in producing these defenses.

Some adaptations preclude others: a cryptically colored animal cannot become an active daytime predator; an animal cannot become a Batesian mimic unless it can exploit a food source in the vicinity of its model; and a species that obtains food by lurking and pouncing cannot defend a large territory against others of its species without becoming obvious to its prey.

Some animals avoid being eaten by defending themselves, either with weapons that allow them to fight or with chemicals that make them unpleasant or poisonous. Other animals escape from predators; means of escape include hiding, camouflage, fleeing, unpredictability, or congregating in such large numbers that no predator could possibly consume all of the individuals present.

Coevolution between different species is universal. The best-studied examples come from the evolutionary interactions between flowering plants and the insects that feed on them and pollinate them. For example, acacia trees provide food and shelter for ants, which protect the tree from herbivores and from competing plants.

OBJECTIVES

From your study of this chapter, you should be able to:

1. Explain the advantages and disadvantages to a plant of producing secondary compounds.

2. Describe experiments to show that aposematic coloration and Mullerian and Batesian mimicry confer protection on their owners.

3. Describe the main differences between Mullerian and Batesian mimicry.

4. Explain why a Batesian mimic is considered "parasitic" on its model's reputation.

5. List the four main principles of camouflage, and give an example of each.

6. Describe the importance of appropriate behavior to the adaptive advantage of protective coloration.

7. Describe how you would determine whether the pattern of coloration of an animal means that the animal is cryptic, aposematic, a Batesian model, a Batesian mimic, or a member of a Mullerian mimicry complex.

8. Tell what is meant by the following terms, and explain the importance of each to an animal's ability to find food: traplining, search image, camouflage, mimicry.

9. Tell how you would design and carry out an experiment to show that a particular physical or chemical characteristic of a plant is an effective defense against herbivores; describe such an experiment that has already been done.

10. List the benefits of mutualism to acacias and to the ants that live on them.

11. Tell how you would design and carry out an experiment to demonstrate that a particular secondary compound may function as a feeding attractant, oviposition stimulus, or repellent for herbivores; describe such an experiment that has already been done.

SELF-QUIZ

1. The quills of a porcupine are an example of:
   a. avoiding predation by defense
   b. avoiding predation by escape
   c. a chemical deterrent to predation
   d. Batesian mimicry
   e. crypsis

2. True or False? A Batesian mimic is considered parasitic on its model's reputation because the existence of a mimic is detrimental to members of the model species.

3. Which of the following is true of a Batesian mimicry complex but not of a Mullerian mimicry complex?
   a. The species in the complex are found in the same area at the same time.
   b. Resemblance between species in the complex is as detailed as possible.
   c. Protection of the species involved depends on the predator's ability to learn.
   d. The species are aposematically colored.
   e. Appropriate behavior patterns enhance the effects of similar coloration.

4. Pretend you are a birdwatcher seeking the shy and elusive Connecticut warbler, a small woodland bird. You wish to be as inconspicuous as possible. Which of the following will probably *not* help you to achieve your aim?
   a. wearing clothing of a mottled green and brown pattern
   b. painting a strip of brown mascara across your face from ear to ear, taking care not to get it in your lovely brown eyes
   c. wearing a hat with leaves or twigs attached to it
   d. wearing a dark shirt or jacket and light-colored slacks
   e. sitting or standing perfectly still

5. A black bug with bright red-orange triangles on its wings lives on plants that have gray-green leaves and dull pink flowers. From this information, you conclude that the bug's coloration is an example of:
   a. crypsis
   b. aposematic coloration
   c. disruptive coloration
   d. countershading

6. Insect species A and B look very much alike. Uninitiated toads quickly learn to leave insects of species A alone. The same toads are then offered species B. The toads don't even try to catch it. Species B is:
   a. a Batesian model
   b. a Batesian mimic
   c. a Mullerian mimic
   d. not enough information given to answer the question

7. Design experiments that would help you find the correct answer to Question 6 above.

8. For each adaptation below, tell whether it would be more advantageous to an animal that was an active hunter (H) or a sit-and-wait predator (S).
   a. camouflage
   b. mimicry
   c. traplining
   d. search image

9. Which of the following is a disadvantage to a plant in producing secondary compounds?
   a. Herbivores that can detoxify them use them to "home in" on the plant for feeding or egg laying.
   b. The chemical damages the plant.
   c. The chemical will repel pollinators.
   d. The chemical may prevent the growth of other plants nearby.

10. Many conifers combat insect herbivores by producing juvenile hormone. This is an effective defense because:
    a. juvenile hormone is toxic to insects

b. juvenile hormone prevents larvae from maturing into adults and producing another generation of herbivorous larvae

c. juvenile hormone keeps the insects from digesting their food efficiently

d. insects are repelled by the odor of juvenile hormone, thinking that there are already too many young insects present

11. An advantage of being a herbivore that feeds only on one or a few related plant species is:
   a. only a few detoxifying enzymes are needed
   b. it is easy to obtain a well-rounded diet
   c. it is easy to find enough food
   d. there will be no other herbivores competing for the food source

QUESTIONS FOR DISCUSSION

1. Explain the advantage to an animal of having its defensive chemicals in the form of a spray, sting, bite, or noxious body coating rather than as an integral part of its body, as seen in chemical defenses of plants.

2. Why is Batesian mimicry so rare? What would happen if every non-defended species mimicked a defended one?

3. Are flowering plants the only land plants that have evolved defenses against animals?

4. Richard Southwood has found a correlation between the abundance of members of various plant groups over geological history and the number of specialist herbivores associated with each group. Can you explain this?

5. Vincent Dethier has stated that a host plant probably exerts a greater selective pressure on a herbivorous insect specialized to feed on it than the insect exerts on the plant. What are the selective pressures that are likely to be acting (1) on the insect, (2) on the plant? Do you agree with Dethier?

6. *Dentaria* is a crucifer that lives in shady woods instead of in the sunny open fields where most crucifers are found. Few herbivores are observed feeding on *Dentaria* in its woody habitat. What would you expect to happen if *Dentaria* were planted in a field? Why?

7. Why might some insects not be very susceptible to man-made insecticides?

8. Why do insect populations that have become resistant to insecticides lose their resistance if insecticide spraying is discontinued for a long time?

REFERENCES AND FURTHER READING

Barlow, A. B., and D. Wiens. "Host-parasite resemblance in Australian mistletoes: the case for cryptic mimicry." *Evolution* 31:69, 1977.

Brower, L. P. "Ecological chemistry." *Scientific American,* February 1969. The story of insects that are unpalatable because they eat toxic plants and of a blue jay's response to eating the insects.

Ehrlich, P., and P. H. Raven. "Butterflies and plants." *Scientific American,* June 1967. Cites numerous examples of coevolution between butterflies and various species of plants.

Eisner, T. "Chemical defense against predation in arthropods." In: *Chemical Ecology,* E. Sondheimer and J. B. Simeone, Eds. New York: Academic Press, 1970. Excellent illustrated account, including the bombardier beetle story.

Huey, R. B., and E. R. Pianka. "Natural selection for juvenile lizards mimicking noxious beetles." *Science* 195:201, 1977.

Janzen, D. H. "Coevolution of mutualism between ants and acacias in Central America." *Evolution* 20:249, 1966.

Lloyd, J. E. "Aggressive mimicry in *Photuris* fireflies: signal repertoires by femmes fatales." *Science* 187:452, 1975.

Pillemer, E. A., and W. M. Tingey. "Hooked trichomes: a physical plant barrier to a major agricultural pest." *Science* 193:482, 1976.

Rehr, S. S., P. P. Feeny, and D. H. Janzen. "Chemical defence in Central American non-ant-acacias." *Journal of Animal Ecology* 42:405, 1973.

Rosenthal, G. A., *et al.* "Degradation and detoxification of canavanine by a specialized seed predator." *Science* 196:658, 1977. The canavanine story.

Whittaker, R. H., and P. P. Feeny. "Allelochemics: chemical interactions between species." *Science* 171:757, 1971.

ANSWERS TO SELF-QUIZZES

Chapter 2
SOME BASIC CHEMISTRY

1. basic; decreased
2. a.
3. c.
4. covalent
5. covalent
6. a.
7. a.

Chapter 3
BIOLOGICAL CHEMISTRY

1. c.
2. H_3C-
3. d.
4. Check your answer using Figure 3-7(b).
5. c.
6. b.
7. d.
8. a.

Chapter 4
STRUCTURE AND FUNCTION OF CELLS

1. n.
2. e.
3. b.
4. d.
5. h.
6. i.
7. m.
8. o. (m.)
9. l.
10. j.
11. all
12. animals, some plants, some prokaryotes
13. plants, prokaryotes
14. animals, plants (prokaryote DNA is not organized into chromosomes)
15. animals, plants
16. e.
17. b.

Chapter 5
HOW THINGS ENTER AND LEAVE CELLS

1. lower, more
2. water enters; level of solution in tube rises; meanwhile glucose leaves; water follows; at equilibrium, the solutions in the tube and beaker will be equally concentrated and will stand at the same level.
3. d.
4. hypotonic, a.
5. a.
6. larger

Chapter 6
ENERGY AND LIVING CELLS

1. photosynthesis
2. reduced
3. substrate-level
4. c.
5. e.

Chapter 7
FOOD AS FUEL

1. c.
2. c.
3. a. CO_2, ATP, NADH + H^+, $FADH_2$, oxaloacetic acid, CoA
   b. ATP, NAD^+, CO_2, CH_3CH_2OH
   c. NAD^+, FAD, H_2O
4. a.
5. true
6. true

Chapter 8
PHOTOSYNTHESIS

1. c.   8. b.
2. a.   9. e.
3. c.   10. c., f.
4. d.   11. a.
5. e.   12. d.
6. a., f.   13. c.
7. f.   14. a.

Chapter 9
DNA

1. a.
2.

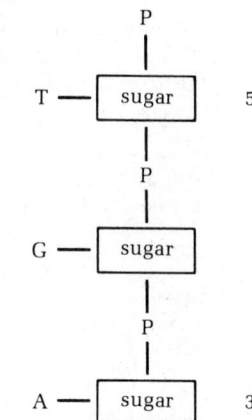

3. d.
4. c.
5. d.
6. **Similarities:**
   1. In both, the DNA forms a double helix, with A = T and C = G.
   2. Both have semiconservative replication.
   Differences:
   1. Eukaryote DNA is complexed with protein in chromosomes; prokaryote DNA is not complexed with protein.

904

2. Prokaryote DNA is circular; eukaryote DNA is linear.

7. c.

Chapter 10
PROTEIN SYNTHESIS

1. a. U-C-C-G-G-A-C-G-A-A-U
   b. A-C-C-G-U-C-G-A-U-G
   c. A-A-A-U-G-C-G-U-G-G
2. a. Met—His—Arg—Arg—Pro—Ile—Val
   b. Met—Phe—Leu—Lys—Gly—Arg (always begin with the sequence AUG, which occurs in the second, third, and fourth nucleotides shown)
3. mRNA: A-U-G-U-U-C-A-U-G-A-A-C-A-A-A-G-A-A
   amino acid sequence: Met—Phe—Met—Asn—Lys—Glu
4. a. point mutation
   b. the second amino acid would be Leu instead of Phe, and the fourth amino acid would be Lys instead of Asn
   c. frame-shift mutation
   d. the second amino acid would be Leu instead of Phe, and the chain would terminate after this amino acid, because the next codon is now a *Stop* codon.
5. e.
6. a. DNA double-stranded; RNA single-stranded
   b. DNA nucleotides contain deoxyribose; RNA nucleotides contain ribose
   c. DNA contains thymine nucleotides; RNA contains uracil nucleotides
7. a.

Chapter 11
CELL REPRODUCTION

1. b.   5. b.
2. 3; 6   6. d.
3. b.   7. b.
4. c.

Chapter 12
CELL DEVELOPMENT AND AGING

1. d.
2. a.
3. a.
4. b.
5. a.

Chapter 13
MENDELIAN GENETICS

1. a. $\frac{3}{4}$ tall: $\frac{1}{4}$ short
   b. all tall
   c. $\frac{1}{2}$ tall: $\frac{1}{2}$ short
2. $\frac{1}{2}$ axial: $\frac{1}{2}$ terminal
3. a. dumpy recessive to normal, which is dominant
   b. both heterozygous for dumpy wings
4. 40
5. a. 250
   b. 125
6. a. Sniffles: homozygous dominant (colored)
      Whiskers: heterozygous for albino
      Esmeralda: homozygous recessive (albino)
   b. $\frac{3}{4}$ colored: $\frac{1}{4}$ albino
   c. $\frac{1}{2}$ colored: $\frac{1}{2}$ albino
7. Mate his dog to bitches known to carry the trait. If any pups show it, the dog is heterozygous for the allele. If none of a large number of pups shows it, the dog is probably homozygous normal.
8.

a.

	DH	Dh	dH	dh
Dh	DDHh	DDhh	DdHh	Ddhh
dh	DdHh	Ddhh	ddHh	ddhh

b.

	DH	Dh
Dh	DDHh	DDhh
dh	DdHh	Ddhh

c.

	DH	Dh	dH	dh
dh	DdHh	Ddhh	ddHh	ddhh

   d. $\frac{1}{2}$
9. a. stamens: straight dominant to incurved; petals (red vs. streaky): can't tell from information given
   b. stamens: both parents heterozygous; petals: one heterozygous, one homozygous recessive, but no indication which is which from information given.
   c. red × red and streaky × streaky: if red is dominant, red × red will produce some streaky progeny, and vice versa

10. a. $\frac{3}{16}$
    b. $\frac{3}{16}$
    c. $\frac{9}{16}$
11. a. 480
    b. 160
    c. 40
12. a. $\frac{1}{8}$
    b. $\frac{1}{8}$
    c. $\frac{3}{8}$
13. let T^A = crosswise stripes
    T^L = lengthwise stripes
    1 crosswise: 2 plaid: 1 lengthwise

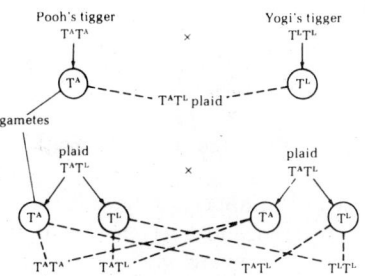

14. a. genotype: SsRR′ phenotype: straight roan
    b. straight red ($\frac{1}{8}$)
       straight roan ($\frac{1}{4}$)
       straight white ($\frac{1}{8}$)
       curly red ($\frac{1}{8}$)
       curly roan ($\frac{1}{4}$)
       curly white ($\frac{1}{8}$)
15. a. clover patch: $\frac{1}{2}$ roan, $\frac{1}{2}$ white
       alfalfa field: $\frac{1}{2}$ red, $\frac{1}{2}$ roan
       cornfield: $\frac{1}{4}$ red, $\frac{1}{2}$ roan, $\frac{1}{4}$ white
    b. it doesn't matter; $\frac{1}{2}$ the calves will be roan in any case
16. a. $\frac{1}{4}$
    b. $\frac{1}{2}$
    c. $\frac{1}{4}$
    d. $\frac{3}{16}$
    e. $\frac{1}{8}$
17. a. $\frac{1}{2}$
    b. $\frac{1}{4}$
    c. $\frac{1}{4}$
    d. 1
18. a. $\frac{1}{4}$
    b. $\frac{1}{2}$
    c. $\frac{1}{16}$
    d. $\frac{1}{4}$
19. $\frac{1}{2}$
20. If the dog is heterozygous, the

probability that one pup will not show retinal atrophy is $\frac{3}{4}$ and the probability that 9 pups will not show the trait is $(\frac{3}{4})^9 = .075$. Subtracting .075 from 1 gives .925, or 92.5% certainty that the dog is homozygous dominant.

21. a. $\dfrac{\overline{MV}}{mv}$

   b. 3%

22. a. $\dfrac{\overline{SM}}{sm}$

   b. $\underbrace{\underbrace{s \quad v \quad m}_{3} \text{ or } \underbrace{s \quad m}_{7} \quad \underbrace{v}_{3}}_{7}$

   c. SsVv × ssvv

   d. SsVvMm × ssvvmm (with any linkage in the heterozygote)

23. The loci appear to be unlinked (different chromosomes), but in fact they are just over 50 map units apart on the same chromosome.

24. a. First of all, her own chance of having the trait is $\frac{2}{3}$. Possible genotypes from her heterozygous parents are $1DD : 2Dd : 1dd$. Since she is obviously not dd, there are three choices left, two of which are carriers. Her chances of having an afflicted child are thus: 1/10,000 of marrying someone with the allele × $\frac{2}{3}$ that she has it × $\frac{1}{4}$ that her child will be dd = 1/60,000.

   b. $\frac{1}{24}$; $\frac{2}{3}$ chance she has it × $\frac{1}{2}$ chance grandparent passed it to spouse's parent × $\frac{1}{2}$ chance spouse's parent passed it on to spouse × $\frac{1}{4}$ chance of homozygous recessive child if they are both heterozygous.

25. a. .21; .42 that chromosome with eye and muscle traits will be in gamete × .5 that Kryptonite sensitivity allele will also be present.

   b. .04; .08 chance of crossing over to get X-ray and normal muscles on same chromosome × $\frac{1}{2}$ to get chromosome with Kryptonite sensitivity in same gamete.

26. $\frac{9}{16}$ purple: $\frac{7}{16}$ white

Chapter 14
INHERITANCE PATTERNS AND GENE EXPRESSION

1. a. $\frac{2}{3}$
   b. sell cows bearing these calves and try a new bull

2. a. $\frac{1}{4}$
   b. $\frac{1}{2}$ normal: $\frac{1}{2}$ brachydactylic

3. a. yellow mice are heterozygous
   b. "yellow" allele is lethal in the homozygous condition, early in embryonic development (2 : 1 ratio)
   c. homozygous yellow dies as an early embryo and is resorbed by the uterus
   d. carry out yellow × yellow matings and examine contents of females' uteri early in pregnancy to detect defective embryos

4. a. heterozygous for an allele that is lethal in the homozygous condition
   b. they die before hatching

5. a. $I^A i$
   b. $I^B i$
   c. $I^A i$ or $I^A I^B$
   d. $I^B i$
   e. $I^A I^A$ or $I^A i$ or $I^A I^B$
   f. $I^A i$
   g. $I^A i$ or $I^B i$ or ii
   h. $I^A i$
   i. $I^B i$

6. Yes, John Smith is really Tom Jones! A baby with blood type M cannot have a parent with blood type N. Also, a parent with blood type AB (Ms. Jones) cannot have a child with blood type O.

7. a. $\frac{1}{2}$ normal: $\frac{1}{4}$ chinchilla: $\frac{1}{4}$ Himalayan
   b. $\frac{1}{2}$ chinchilla: $\frac{1}{4}$ Himalayan: $\frac{1}{4}$ albino
   c. $\frac{3}{4}$ chinchilla: $\frac{1}{4}$ albino

8.

$\frac{1}{2}$ barred males: $\frac{1}{2}$ unbarred females (remember male birds are ZZ and females ZW [see Figure 14-8]).

9. $\frac{1}{2}$ the sons hemophiliacs: $\frac{1}{2}$ the sons normal.
$\frac{1}{2}$ the daughters heterozygous carriers: $\frac{1}{2}$ the daughters homozygous normal.

10. when the mother is either a hemophiliac or a carrier

11. a. Red-green color blindness in humans is carried on the X chromosome. Since males have only one X chromosome, any male who receives the allele for color blindness will be colorblind. A color-blind female will have two X chromosomes bearing the allele. The frequency of this allele among females is thus $(\frac{1}{12})^2 = \frac{1}{144}$
   b. mother homozygous recessive color-blind; father normal allele + Y.
   c. $\frac{1}{2}$
   d. $\frac{1}{2}$

12. 2 female offspring: 1 male.

13. No; since the baldness trait is carried on the autosomes, a man could inherit the trait from either parent who carried it, and from any of his four grandparents who passed it on to the appropriate parent.

Chapter 15
POPULATION GENETICS

1. e.
2. d.
3. a. stabilizing
   b. The normal allele would gradually increase, and the sickle allele would decrease, but not to zero; the sickle allele would remain at low levels in the heterozygous condition.
   c. $30\frac{1}{4}$% homozygous normal $49\frac{1}{2}$% heterozygous $20\frac{1}{4}$% homozygous sickle (Allele frequencies are $0.10 + \frac{1}{2}(0.90) = 0.55$ for the normal allele and $\frac{1}{2}(0.90) = 0.45$ for the sickle allele; in the next generation, the frequency of homozygous normal is $(0.55)^2 = 0.3025$; of heterozygous is $2(0.55)(0.45) =$

0.4950; and of homozygous
sickle is $(0.45)^2 = 0.2025$)
4. a.
5. c.
6. d.
7. true
8. a.
9. a. increase
   b. increase
   c. increase
   d. increase
10. c.

Chapter 16
NATURAL SELECTION AND EVOLUTION

1. c.   3. e.
2. d.   4. b.

Chapter 17
SPECIATION

Identification of dichotomous key specimens (Figure 17-1):
   I. Chilopoda
  II. Arachnida
 III. Crustacea
  IV. Insecta
   V. Crustacea
  VI. Diplopoda
1. a. morphological
   b. biological
   c. biological
   d. morphological
   e. morphological
   f. biological
2. true
3. g. prezygotic
4. a. prezygotic
5. e. postzygotic
6. b. prezygotic

Chapter 18
EVOLUTION AND REPRODUCTION

1. b.
2. favorable; unfavorable
3. b.
4. c.
5. d.
6. a.
7. c.

Chapter 19
ORIGIN OF LIFE

1. b.
2. 1. organic monomers
   2. proteinoids
   3. fermentation
   4. water-splitting photosynthesis
   5. aerobic respiration
   6. intracellular organelles
3. e.
4. 1. Addition of O_2 to the atmosphere made it oxidizing rather than mildly reducing.
   2. An ozone layer formed.
   3. As a result of increased food production, the earth could support a greater number of organisms.
5. d.

Chapter 20
CLASSIFICATION OF ORGANISMS

1. d.
2. d.
3. b.
4. b.

Chapter 21
VIRUSES

1. d.
2. c.
3. a. lytic (true of lysogenic when it has become lytic)
   b. lytic
   c. lysogenic
   d. lysogenic
   e. both
4. true

Chapter 22
MONERA

1. d.   5. true
2. a.   6. true
3. a.   7. false
4. c.

Chapter 23
PROTISTA AND THE ORIGIN OF MULTICELLULARITY

1. eukaryotic, unicellular
2. a. iii.
   b. i.
   c. ii.
3. nutrients; light and warmth
4. e.

Chapter 24
FUNGI

1. c.   3. b.
2. a.   4. e

5. b.   8. a.
6. a.   9. a.
7. c.

Chapter 25
LOWER PLANTS

1. Phaeophyta
2. Chlorophyta
3. Phaeophyta
4. Rhodophyta
5. d.
6. c.
7. a.
8. a.
9. a.
10. b., d.
11. b., d., e., c., a.

Chapter 26
HIGHER PLANTS

1. c.
2. d.
3. d.
4. d.
5. c.

Chapter 27
THE LOWER INVERTEBRATES

1. Cnidaria   5. c., d., h.
2. Porifera   6. a., c., d., h.
3. bilateral   7. b., e.
4. a., d., h.   8. b., f., g.

Chapter 28
SOME HIGHER INVERTEBRATES

1. d.   9. b.
2. a.   10. b.
3. h., j.   11. g.
4. c.   12. a.
5. b., g.   13. i.
6. f.   14. d.
7. a.   15. e.
8. a.

Chapter 29
THE ORIGIN OF VERTEBRATES

1. d.
2. a., b., d.
3. e.
4. d.
5. b.

Chapter 30
VERTEBRATE ANATOMY AND EVOLUTION

1. a.
2. feathers
3. a.
4. Osteichthyes
5. 1. support
   2. dehydration of body and of eggs laid outside the body
   3. extraction of O_2 from (dry) air rather than water
6. Reptilia
7. Reptilia
8. true

Chapter 31
ANIMAL NUTRITION AND DIGESTION

1. f.
2. a.
3. a., b., c.
4. a., c.
5. d.
6. b.
7. f.
8. a.
9. c.
10. e.
11. b.
12. c.
13. a.

Chapter 32
GAS EXCHANGE IN ANIMALS

1. right
2. a.
3. a.
4. b.
5. a.
6. c.
7. a.
8. a.
9. a. right
   b. left
   c. left
   d. left
10. a.

Chapter 33
ANIMAL TRANSPORT SYSTEMS

1.

cnidarian	earthworm	insect	fish	mammal
x	x	x	x	x
	x		x	x
				x
	x	x	x	x

2. b.
3. c., d.
4. d.
5. b.
6. cell in intestine → extracellular fluid → lymphatics → thoracic duct → vena cava → right atrium → right ventricle → pulmonary artery → lung capillaries → pulmonary vein → left atrium → left ventricle → aorta → artery → arteriole → capillary → extracellular fluid → cell in fatty tissue
7. c.
8. a.
9. a. ectotherms
   b. both
   c. endotherms

Chapter 34
DEFENSES AGAINST DISEASE

1. e.
2. a.
3. f.
4. d.
5. d.
6. e.
7. b.
8. c.
9. c.

Chapter 35
ANIMAL EXCRETION

1. a.
2. b. (it is a mammal)
3. a. low tide
   b. active transport of salts into its body from its environment or from its urine (or both)
   c. ammonia
4. a.
5. a.
6. b.
7. d.
8. Loop of Henle: e.
   Renal artery: a.
   Proximal convoluted tubule: b.
   Glomerulus: d.
   Distal convoluted tubule: c.
9. b.
10. decrease/c.
11. increase/increase

Chapter 36
SEXUAL REPRODUCTION AND EMBRYONIC DEVELOPMENT

1. j.
2. i.
3. c., e.
4. b.
5. a.
6. f.
7. h.
8. f. → j. → g. → i. → a. → h. → b.
9. a.
10. f.
11. d.
12. b.

13. a.
14. c.
15. a.
16. C
17. C
18. N
19. O
20. C
21. G, N, O
22. G
23. a.
24. e.

Chapter 37
NEURONS

1. d.
2. a. action potential
   b. action potential
   c. action potential
   d. local response
   e. local response
3. a.
4. a. A neurotransmitter substance carries information between two neurons in that it is released as a result of the arrival of an action potential at the presynaptic membrane of one neuron, and its arrival at the postsynaptic membrane stimulates electrical activity in the postsynaptic cell.
   b. Vesicles store neurotransmitter molecules in the presynaptic terminal and release them when an action potential arrives at the terminal.
   c. On combining with neurotransmitter molecules, receptor molecules change the permeability of the postsynaptic membrane to ions, resulting in an EPSP or IPSP.
   d. Enzymes that destroy neurotransmitter molecules in effect "turn off" the signal brought across the synapse by the neurotransmitter.
5. d.
6. d.

Chapter 38
THE VERTEBRATE NERVOUS SYSTEM

1. b.
2. d.
3. d.
4. a.
5. A. soma of sensory neuron

B. dendrite of sensory neuron
C. axon of motor neuron
D. axon of sensory neuron
E. synapse between sensory and motor neuron
6. a.
7. noradrenalin
acetylcholine

Chapter 39
SENSE ORGANS

1. a.
2. b.
3. e.
4. d.
5. d.
6. b., c.
7. a.
8. check your drawing against Figure 39-18; accessory structures: iris, lens; transmission of stimulus to brain is via the optic nerve; receptor types: rods (throughout retina except fovea) and cones (fovea and adjacent parts of retina).
9. rapidly adapting
10. rods; dim; poor

Chapter 40
MUSCLES AND SKELETONS

1. a. II
   b. I, III
   c. I, III
   d. III
   e. I, III
2. A. Z line
   B. actin, tropomyosin, troponin
   C. mitochondria, calcium
   D. myosin
3. lmonkhbjidefgac
4. b.
5. c.
6. e.

Chapter 41
ANIMAL HORMONES

1. c.
2. c.
3. a.
4. b.
5. hypothalamus; the bloodstream; secretory neurons; releasing its hormones, which travel in the blood to other, target, glands.

Chapter 42
BEHAVIOR

1. g.   7. c.
2. a.   8. k.
3. h.   9. d.
4. j.   10. c.
5. l.   11. e.
6. f.   12. c.

Chapter 43
STRUCTURE AND GROWTH OF VASCULAR PLANTS

1. a.
2. b.
3. a.
4. a., c., b.
5. b.
6. c.
7. c.
8. a.
9. a. monocotyledons
   b. both
   c. monocotyledons
   d. monocotyledons
   e. dicotyledons
10. a.

Chapter 44
TRANSPORT IN VASCULAR PLANTS

1. c.
2. c.
3. function: transport of sap
   adaptation: tubular, dead and hollow, pits or open ends

   function: support of plant body
   adaptation: secondary wall thickenings
4. d.
5. a. decrease
   b. increase
   c. increase
   d. increase
6. a. remain the same
   b. increase
   c. remain the same
7. a.
8. b.

Chapter 45
SOIL, ROOTS, AND PLANT NUTRITION

1. living organism: e.
   organic matter: d.
   oxygen: b.
   rock particles: a.
   water: c.

2. a.
3. (a) apoplast
   (b) increases surface area for absorption of minerals and water from soil solution
   (c) Casparian strip
   (d) endodermal
   (e) epidermal, cortical, outer surfaces of endodermal
   (f) symplast
4. nitrogen, phosphorus, potassium
5. a. proportions of minerals to each other may not be correct
   b. minerals may be bound to soil particles and unavailable to plants ·
   c. genetic defects of plants may prevent absorption or utilization of nutrients
6. a.

Chapter 46
REPRODUCTION IN FLOWERING PLANTS

1. b.
2. a.
3. c.
4. false
5. c.
6. e. (but not all seeds require all factors listed)
7. b.

Chapter 47
PLANT HORMONES

1. b.   6. a.
2. d.   7. d.
3. c.   8. a.
4. b.   9. b.
5. e.   10. c.

Chapter 48
THE BIOSPHERE

1. cooler, lose   5. a.
2. a.   6. d.
3. a.   7. d.
4. c.   8. primary
9. (1) short fugitive plants
   (2) taller, perennial plants
   (3) shrubs and pioneer trees
   (4) climax trees
10. early successional
11. they are separated by great distances; they have similar climates, which exert similar selective pressures (in this case, nothing much taller than

grasses can grow with so little rainfall).

Chapter 49
ECOSYSTEMS AND COMMUNITIES

1.

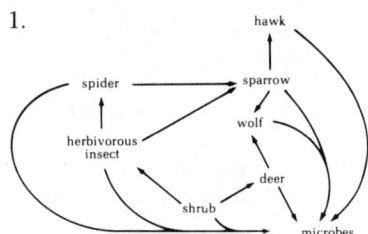

decomposers: soil microbes
producer: shrub
primary consumers: deer, herbivorous insect
secondary consumers: wolf, spider, (sparrow)
tertiary consumers: hawk, (sparrow)
(sparrow feeds at several possible trophic levels)

2. b.
3. c.
4. e.
5. a.
6. b.
7. a. increase
   b. decrease
   c. increase
8. b.

Chapter 50
POPULATIONS

1.

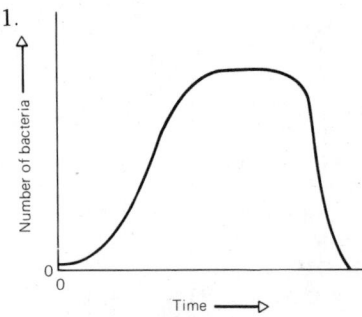

The population grows exponentially, levels off, and then declines rapidly to extinction in this "closed" environment. Mark yourself $\frac{1}{2}$ off if you forgot to label the axes.

2. b.
3. b.
4. c.
5. b.
6. d.
7. I
8. b.
9. a. specialized
   b. both, in most cases
   c. likely to be true of generalized predator; also occurs in specialized predation
   d. specialized
10. c.
11. b.

Chapter 51
EVOLUTIONARY ECOLOGY

1. a.
2. true
3. b.
4. d.
5. b.
6. d.
7. i. Offer an insect of completely different appearance (C) to toad to see if it is hungry. If it eats C, try Experiment #ii. If not, repeat offering B and C later. If it then refuses B but eats C, go on to Experiment ii. Experiment ii: Offer another uninitiated toad insects of species B, followed by members of species A. If it eats B at first but then starts to refuse both A and B, B is probably a Batesian mimic. (Be sure to test for hunger by offering C when A or B is refused.) If it starts to refuse B before it is offered any A's, A and B are probably Mullerian mimics.
8. a. S
   b. H
   c. H
   d. H
9. a.
10. b.
11. a.

INDEX/GLOSSARY†

†**Boldface** page numbers indicate pages on which the index item is defined; *italicized* page numbers indicate pages containing illustrations. Glossary terms, printed in boldface, are defined here as well as in the text. Some biological terms defined here are used throughout the text and may not have specific page references.

Agnatha, 446–447, 450
Agricultural revolution, **870**
Agriculture, **869**–870, 873–875
    productivity, 831
Agrostis, 752
Air bladders Air-filled sacs providing
    buoyancy
    in algae, *378,* **379**
    in fish (swimbladders), **510**
Air, breathing, 502
Air movements and climate, *798*
Air pollution, 877
Air sacs (birds), *463,* **510,** 532
Air spaces in leaves, *117*
Alanine, 37
Albinism, 208–209
Albugo, 359
Alcohol, 591
    and brain, 607
    in diet, 480
    enzyme induction, 170
Alcohol group, *28*
Alcoholic fermentation, 97–98
Aldehyde group, *28*
Aldosterone, **568,** 569, 667
Aleurone, **770**
Alfalfa, *710*
Algae Photosynthetic organisms with a
    one-celled or simple multicellular
    body plan, 376–386, *813*
    Chlorophyta, 376, 377, *381–384*
    classification, 376
    economic uses, *380–381*
    life histories, 382–384
    Phaeophyta, 376, 377, *378–381*
    Rhodophyta, 376, *377–378,* 380–381
    unicellular (Protista), 347–350
Alginate, **381**
Alkaline Having a (basic) pH of more
    than 7, 25
Alkaloid A member of a class of
    compounds which are usually basic;
    most contain one or more nitrogen
    atoms in a cyclic structure, and
    many are toxic, 763, 885
Allantois A sac that grows out of the
    embryonic gut in higher vertebrates;
    stores the embryo's nitrogenous
    waste until hatching in reptiles and
    birds; forms the main embryonic
    respiratory surface and part of the
    placenta in mammals, **457,** 590
Allard, H. A., 789
Alleles Portions of DNA that carry the
    information for contrasting forms of
    the same genetic trait (e.g., blue
    eyes and brown eyes), **187**
    allele frequency vs. genotype
        frequency, *226*
    multiple, 210–211
    neutral, 237–238
Allergen, **549**
Allergy, 548–549
Alligators, 460
Allopatric Living in different areas, **258**
Allopatric speciation, 262–264
Allylglucosinolate, *899*
Alpine communities, *812,* 831
Alternate leaf arrangement, **712,** *713*
Alternation of generations 383–384, 390
Altitude and vegetation, *798–799*
Altitude sickness, 509

Altruistic behavior, **283**–287
Alveoli, *502,* **503,** *520*
Amanita, 371
Ambient temperature Temperature of
    the surroundings
Ambystoma, *175,* 456
Ameba, *see* Amoeba
Ametabola, 435
Amino acid Small organic molecule
    containing both a carboxyl and an
    amino group bonded to the same
    carbon atom; amino acids are the
    monomers from which polypeptides
    are made, *28,* 33–35
    abbreviations, 140
    analyzing sequence in proteins, 148–149
    essential, **478**
    formation under prebiotic conditions,
        294, 295
    invariant, **150**
    m. w., 42
    production in plants, 115
    synthesis in plants, 738
Amino acid analyzer, 148
Amino acid sequencer, 148
Aminoacyl attachment site, ***144***
Amino group, *28*
Aminopeptidase, *485*
Ammonia (NH$_3$)
    excretion of, *556, 557, 558, 559*
Ammonites, fossil, *2*
Ammonium (NH$_4^+$), 749
Amniocentesis, **217**
Amnion A fluid-filled sac that surrounds
    the embryo in reptiles, birds, and
    mammals, **457,** 590
Amniotic egg, **457,** 559
Amoeba, 345, 347, 350
    feeding in, *70*
Amoeboid movement Motion
    accomplished by the flowing of
    cytoplasm in one direction, bulging
    the plasma membrane out so that
    the rest of the cytoplasm follows
    along, *346,* 587
Amphetamine, 606, 607
Amphibia Class of vertebrates containing
    frogs, toads, salamanders, newts,
    and their relatives, *450,* 455–456
    brain, *615*
    circulation, *518–519*
    color change, *674*
    excretion, 557, 558, 559
    gas exchange, *503,* 519
    metabolic rate, 515
    metamorphosis, 57, 177–178
    regeneration of limbs, *175–176*
Amphineura, *425, 426*
Amphioxus, 446
Amphiploidy, **264**
Ampullae of Lorenzini, **452,** 629, **660**
Amylase An enzyme that breaks glucose
    monomers from starch molecules,
    485, 486
Amyloplasts, **55, 706,** *707,* 711, **787**
Anabaena, 332
Anacondas, 460
Anaerobic 1. Without oxygen; 2. Not
    requiring molecular oxygen for
    extraction of energy from food
    (respiration)

Anaphase, *158, 161*
Anaphylactic shock, **548**
Anemia A deficiency of hemoglobin or of
    red blood cells, resulting in a
    deficiency of oxygen supply to the
    tissues, 479, **525**
Anesthetics, 498
Angiosperms Flowering plants, 391, 400–
    403
    reproduction, 757–777
    wood, *725, 726*
Angiotensin, angiotensinogen, **568,** 569
Angstrom unit (Å) 10^{-10} m
Animal cells, *50*
    water relations, *72*
Animalia (kingdom), 311, 313, *315,* 316, 406
Animals
    chemical defenses, 886–887
    evolution, *407–408*
    marine, 408
Anion A negatively charged ion
Anisogamy, **278**
Annelida, *421–424*
    circulatory system, *516*
    metabolic rate, 515
    nervous system, *613*
Annual rhythms, 677
Annual rings, **727,** *728*
Anoplura, 435
Antagonistic muscles, 655–**656**
Antelope, 266
Anterior At or toward the front end
    of an animal
Anterior pituitary, 673
Anther, *758,* **759**
Antheridium, **360,** *362*
Anthocyanins Pigments of red, blue, or
    purple hue, commonly found in
    vacuoles of plant cells, 56
Anthozoa, 411, *413*
Anthropoids Monkeys, apes and humans,
    which make up the suborder
    Anthropoidea, **471,** *472*
Anthropomorphism Attribution of human
    characteristics to other animals, **681**
Antibiotics Substances that destroy
    microorganisms, 339
    and evolution, 249
    and fungi, 366
    and protein synthesis, 147
    reduction in human deaths, 339, 871
    resistance in bacteria, 339
    source, 335
Antibody Protein produced by the body
    that reacts with an antigen (foreign
    protein or other foreign substance),
    and thus defends the body, **527,**
    542–544, 546–548
    genetics, 547–548
Anticodon, ***144,*** 145, 146
Antidiuretic hormone, **568,** 666
    See also Vasopressin
Antifreeze, 24
Antigen A substance that binds to an
    antibody, **540,** 542, *543*
    ABO, 210, **527**
    MHC, 210, **542**
Antiparallel (Of nucleic acids)
    Arrangement of two strands side by
    side, but with the 3′ end of one next
    to the 5′ end of the other and vice
    versa, 127

Ants and acacias, 287, 898
Anura An order of amphibians, containing the frogs and toads, 455–456
Aorta The large blood vessel that carries blood from the heart toward the rest of the body, *520, 521,* 522
Apes, 471
uric acid, 558
Aphids, 735–736, 742
parasitism of, 900
Aphrodite, 422
Apical dominance, **709, 782,** 783–784
Apical meristem Area of dividing cells at the root and stem tips of a plant, **705,** *706,* **709,** *719*
Apoda, 455
Apoplast, **750,** 751
Aposematic coloration Conspicuous coloration of an animal that is dangerous or distasteful to eat, *888*
Aposematic sound, 888
Appeasement behavior, **700**
Appendicular skeleton, **656,** *657*
Appendix, 488
Appetites for specific nutrients, 493
Apple maggot, 265
Apples, 788
Apple trees, 771, 774, 786
Aquatic Of water (fresh or salt)
Aqueous Containing water
Aqueous solution Solution in which water is the solvent
Arachnida, 430, *431–432*
Arber, Werner, 131
Arboreal Living in trees
Arcella, 351
Archenteron, **586**
Arcyria, 372
Archaeolemur, 473
Archaeopteryx, 461
Archenteron, **586**
Architeuthis, 429
Arenicola, 423
Arginine, 899
Aristotle, 243, 291
Armyworm caterpillar, 898
Artemisia, 810
Arteriole, **519,** *521*
Artery Blood vessel that carries blood away from the heart, **517,** *519,* 522
hardening, *480*
renal, *564*
Arthritis, rheumatoid, 548
Arthropods Members of the phylum of animals with jointed appendages and stiff external skeletons, e.g., crabs, lobsters, barnacles, insects, *430–437*
Arachnida, 430, *431–432*
Chilopoda, *430*
Crustacea, 430, *432–433*
Diplopoda, *430*
Insecta, 430, *433–437*
key to classes, *257*
nervous system, *613*
Artificial insemination, **579**
Artificial pollination, 186
Artificial selection, 244, *245,* 265
Artiodactyls, 465, *470*
Asbestos, 551
Ascaris, 183, 417

Ascidian A marine invertebrate chordate (sea squirt), *445*
Ascomycetes, 359, *362*
Ascophyllum, 379
Ascorbic acid, 478, 479, 494, 496
Ascus, *362*
Asexual reproduction versus sexual reproduction, 273–275
Aspergillus, 362, 370
Aspirin, 367
Asplanchna, 884
Association (nervous system), **616,** 617, 618
Associative learning, 688–**689**
Asteroidea, 440, *441*
Asymmetron, 446
Atherosclerosis, *480*
Athiorhodaceae, 302–303
Athlete's foot, 364
Atmosphere of early earth, 293, 294
Atmospheric nutrient cycles, **836**
Atmospheric pressure The pressure exerted by the weight of the overlying air; since there is a thinner layer of air over a mountain than over a low-lying area, the atmospheric pressure is lower at high altitudes
Atoll, 814
Atom, *17–18*
Atomic mass, **18**
Atomic number, 18
ATP, *see* **Adenosine triphosphate**
ATPase An enzyme that synthesizes or hydrolyzes ATP, *96,* 104, *105, 113*
Atrioventricular node, **650**
Atrium (pl., **atria**) A chamber; e.g., a chamber that receives blood as it enters the heart, *518, 520*
amphioxus (gill slits), **446**
Atropine, 606
Auditory nerve, 621
Aulacodiscus, 348
Aurelia, 515
Australopithecus, 472, *473*
Autoimmunity, **548**
Autonomic nervous system The part of the vertebrate nervous system over which the animal usually has no control; composed of the sympathetic and parasympathetic systems, **614,** 622, 623
Autopolyploid, **264**
Autoradiography Technique whereby the position of a radioactive element introduced into the system is determined from the fact that radiation emitted by radioactive substances exposes (blackens) photographic film that is placed next to them, 735, *738*
Autosomes, **212**
Autotroph Organism that can make its own food molecules from inorganic constituents, **302**
bacteria, 335–336
origin of, 302–303
Auxins Plant hormones that stimulate cell enlargement, 727, **780–784,** 785, 786, 787
and ethylene, 788
in fruit development, 767

Avery, O.T., 124
Aves The class of vertebrates that contains the birds, **450,** *461–464*
See also Birds
Avian Pertaining to birds
Axial skeleton, **656,** *657*
Axil, **709,** *712*
Axillary bud, **709,** *712*
Axolotl, 178
Axon Extension of a neuron that carries nerve impulses to the next cell(s) in line, 595, **596,** 597, 602–603, *604*
Azotobacter, 337

β-carotene, *108–109,* 639
β chain of hemoglobin, evolution, 151
β-globin, *143*
Babies, *see* Infants
Baboons, 282, 472
Bacillus (pl., **bacilli**) Rod-shaped bacterium, *334*
Background radiation, 164
Bacteria
cell structure, *62, 63*
chemical composition, 42
chemiosmosis, 83–85
chemosynthetic, 336, 337
classification, 313, 333–335
control of protein synthesis, 168, *169,* 170
defense against phages, 131
denitrifying, 836
disease caused by, 330, 335, 338–339
DNA structure, 130
drug resistance, *249,* 339
flora of animals, 337–338
and food, 339–340
fossils, 319
genetic recombination (conjugation, transduction, transformation), 331
Gram staining, 334
habitats, range, 333
heterotrophic, 337–338
marine, 815
nitrifying, 337, 836
nitrogen-fixing, 336–337, 836
photosynthetic, 335–336
purple non-sulfur 302–303
reproduction, 331–332
spore formation, 170
structure, *330–331*
sulfur, 303, 336
toxins, **338,** 339, 340
Bacterial restriction enzymes, *131–133*
Bacterial transformation, **123–124**
Bacteriophage, *see* Phage
Bacteroides, 337
Balanus, 846, 862
Baldness, 215
Baleen, 468
Balsam fir, chemistry, 901
Bamboo, 769
Banana, 767, 788
Bar Measure of pressure; 1 bar = 1000 kg/cm-sec^2
Barbary rock apes, *283*
Barbiturates, 607
Bark, **716**
Barley, 766, 790, 832
germination, 770
Barnacles, 433, 846, 862

Brood size, *854*
  selection for, 247
Brower, Jane and Lincoln, 888, 890, 891
Brown algae, *see* Phaeophyta
Broyer, Theodore, 751
Brucellosis, 339
Bryophyta Embryophytes lacking
  vascular tissue: mosses, liverworts,
  and hornworts, *391–393, 401*
Budding Production of a smaller
  individual attached to its parent; the
  new individual may later become
  independent, or may remain
  attached, forming a colony, **273**
Budget,
  energy, 13
  food, *832*
Bud scale, **717**, *787*
Bud scale scar, **717**
Buffalo, *809*
Buffering of blood, 508
Buffon, Georges-Louis Le Clerc de, 243
Bugs Members of the insect order
  Hemiptera, 435
Bulbs, 773
Bullough, W. S., 157
Bundle of His, *650*
Bundle-sheath cells, **117**
Burdock, 768
Burkitt's lymphoma, 326, 550
Burnet, Sir Frank MacFarlane, 549
Burns, Barry, 498
Bursa of Fabricius, **541**
Bushmen, 870
Buttercup root cross-section, *707*
Butterflies, *4*, 435
  courtship, *261*
  disruptive coloration, *892*
  genetics, *231, 234, 236*
  molting, *434*
  pollination by, *762*
  speciation, 266–267
    See also Armyworm, Cabbage
    white, Fritillary, Monarch, *Papilio,*
    Queen, Swallowtail, Viceroy
Butyric acid, *31*
B vitamins, 88, 479

C_3 photosynthesis, **114**, *115*
C_4 photosynthesis, *117–118*
Cabbage flea beetle, 900
Cabbage white butterfly, 899
Cabbage, white rust of, 359
Cachalot, toothed, 429
Cactus (family Cactaceae), *118, 401*, 711,
  805, 820
Caecilians, 455, 456
Caecum (pl., caeca), **483**, *488*
Caffeine, 591
Caimans, 460
Caisson disease, 533
Calcaneus, *657*
Calcareous Composed of or containing
  calcium carbonate ($CaCO_3$)
Calciferol, 479
Calcitonin, **659**, 667, **668**
Calcium
  active transport of, 85, 96
  in cell division, 157–158
  control of level in blood, 659, 668
  in muscle contraction, 653, 654

and plant nutrition, 745, 746, 751
  role in animals and dietary sources, 481
Calico cats, 173
Calimyra fig, *260*
Callus, **659**
Caloric value (of foods), 477
Calorie 1 Calorie = 1000 calories = 1
  kilocalorie or kcal. A calorie is the
  amount of heat needed to raise one
  gram of water from 14.5°C to
  15.5°C. The energy values of foods
  are expressed in Calories
Calvin cycle, **114**, *115*
Calvin, Melvin, 114
CAM (Crassulacean acid metabolism), 118
Cambium Meristematic tissue that
  produces new cells which increase
  the diameter of a woody stem or
  root, **715**, *716*, 718
Cambrian Period, 317, *450*
Camptosaurus, 457
Canavanine, *899*
Cancer, 542, 549–551
  cells, *587*
  and estrogen, 578
  radiation treatment, 164
  skin, 164
  viruses, 325, 326, 550
Canines The pointed teeth next to the
  incisors in mammals, **465**, *488, 489*
Canopy, **800**, *843*
Capillaries The smallest blood vessels in
  a circulatory system; exchange of
  substances between the blood and
  the extracellular fluid takes place
  across the thin walls of the
  capillaries, **517**, *519, 522, 555*
  transport of substances across walls of,
  70
Capillarity Movement of fluid into a
  narrow tube or space based on
  attraction of the walls of the space
  for the molecules of the fluid, **23**
  of water, **729**
Capillary water, **746**, *747*
Capsule (prokaryotic cell), **330**
Carbohydrates A class of compounds
  whose members have the general
  formula $(CH_2O)_n$ and contain at least
  one double-bonded oxygen, **29–31**,
  42
  conversion to fat, 98
  energy yield, 100
  storage, 100
Carbon, *18*, 27–28
  carbon-14, *18*, 114
  vs. silicon, 15, *66*
Carbon cycle, *835*
Carbon dioxide, 21, *28*
  in atmosphere and oceans, 835
  excretion, 555
  production in fermentation, 97
  in respiration, *93, 94*
  transport in blood, 507
  and photorespiration, 116
  effect on photosynthesis, 116
  use in photosynthesis, *114*, 115
Carbon fixation, **110**, *114–115*, 117–118
Carbonic anhydrase, **507**
Carboniferous period (Mississippian +
  Pennsylvanian), 317, *454*

Carboxylase An enzyme that carboxylates
  (adds a carboxyl group to) a substrate
  molecule
Carboxyl group A functional group
  (COOH) consisting of a carbon atom
  double-bonded to an oxygen atom
  and single-bonded to another oxygen
  atom, which in turn is bonded to a
  hydrogen atom, *28*
Carboxypeptidase, *485*
Carcinogens Cancer-causing agents, **550**,
  551
Carcinus, 570
Cardiac glycosides Sugar-containing
  chemicals that cause the heart to
  beat faster, sometimes leading to
  death of the animal, 887
Cardiac muscle The type of muscle that
  makes up the vertebrate heart, **648**,
  650
Cardiovascular disease, **480**
Carnivora, 465, *468*
Carnivores Animals that eat other
  animals, **476**, *488, 489*
Carnivorous plants, 752, *753*
Carotenoids Accessory photosynthetic
  pigments that usually appear yellow,
  orange, or brown, **108–109**, 332,
  349, 350
  in algae, 377, 378, 381
  in land plants, 388
Carotid artery, **520**
Carpals The bones of the wrist, *657*
Carpel A female flower part, **403**, *758*,
  759
Carrageenan, **381**
Carrier, of genetic trait, **208**
Carrier molecules (plasma membrane),
  68, 69
Carrots, *167*
Carrying capacity (K) Number of
  individuals of a species that the
  environment can support
  indefinitely, *858*
Cartilage 1. Tissue composed of scattered
  cells surrounded by tough, flexible
  intercellular protein fibers, *658*; 2.
  Skeletal element composed of
  cartilage, 658
Carya (hickory), *712, 717*
Casparian strip, **750**, *751*
Catalyst, **36**
Catecholamines A group of chemicals
  containing an amine group, **668**
  See Also **Adrenalin,** Noradrenalin,
  Dopamine
Caterpillars, selective pressures on, 248–
  249
Cation A positively charged ion
Cats
  color, *173*, 214–215
  development of visual pathways, 689
  genetics, 214–215
  skeleton, *464*
Cattle
  antibiotics in feed, 249
  disease, 350, 351
  genetics, 202, 206, 218
  selective breeding, 244, *245*
Caudal Toward the tail
cc Cubic centimetre; equal to one
  millilitre

consisting of DNA and proteins and carrying genetic information, 52, *53*, *130–131*, **155**
abnormalities, in newborns, 217
bands, *171–172*
condensation, 158
homologous, 155, **156**, ***192–193***, *194, 195, 196, 197*
human, *155*, 156
lampbrush, ***172***
mapping, ***198***
number in some organisms, 156
polytene (giant), *171–172*
puffs, *171–172*
replication, **155**
structure, *130–131*
X and Y, 156
Chrysanthemums, 775
Chrysolina, 864
Chrysophyta Diatoms and their relatives, *347, 348, 350, 376, 391*
relation to brown algae, 378
Chthamalus, 862
Chylomicrons, **487**
Chyme, **485**
Chymotrypsin, 149, *485*
Cichlid fish, 267
Ciliates, *347, 352–353*
Ciliophora, *347, 352–353*
Cilium (pl., cilia), **58–59**
locomotion using, 346
Cinnabar caterpillars, 888
Circadian rhythms Cycles of about 24 hours in the physiology and behavior of eukaryotic organisms, **675–677**, 792
and navigation, 695
Circular muscle, 656
Circulatory systems, 514, 516–524
closed, ***516***
comparative vertebrate, 517–519
connection to other fluid compartments, 555
double, *518*
earthworm, *516*
fish, *518*
insect, *517*
mammalian, **517**
mollusc, 536
open, **517**
single, *518*
Citric acid, *94*
Citric acid cycle, *see* Tricarboxylic acid cycle
Citrus fruits, 788
Citrus scale insect, 249
Cladocerans, 430
Clam, *427–428*, *501*
Class, in taxonomy, **310**, 311
Classification of organisms, 308–316
algae, 376–377
bacteria, 333–335
binomial nomenclature, 309–310
five-kingdom system, 313–316
methods, reasons, problems, 311–312
Protista, 345–346
taxonomic system, 308–309, 310–311
Claviceps, 371
Clavicle, 657
Clay soil, **746**, *747*
Cleavage A series of cell divisions of the zygote, ending with formation of the blastula, *583–585*, 590

Cleidoic egg, ***457***, 559
Clever, U., 171
Cliff dwelling, *728*
Climate and vegetation, *797–799*, *801*
Climax community, **816**
Cline A series of interconnected populations extending from one area to another, with gradual, continuous changes from one population to the next, ***227–228***, *258*
Clitoris, *573*, *574*, *576*
Cloaca The vestibule, found in most vertebrates, into which urine, feces, and sperm or eggs are discharged before they leave the body, **558**
Clock-shifting, *695*
Clone A population of cells or individuals descended from one original cell or individual by asexual propagation, and hence genetically identical, **274, 772**
carrot, *167*
frog, *168*
mouse, *168*
human, **181**
Closed circulatory system, **516**
Clostridium, 324, 337, 339, 340
Clotting of blood, **525**
Clover, 860, 886
Club fungi (Basidiomycetes), *362–364*
Club mosses, *393, 394, 396, 401*
Cnidaria Jellyfish, *Hydra*, corals, sea anemones, etc.; simple animals with only two well-developed layers of cells, only one opening into the gastrovascular cavity, tentacles, and stinging nematocysts, 409, *410–413*, *814*
excretion, 562
feeding and digestion, *482–483*
metabolic rate, 515
nerve network, 612, *613*
transport, *515*
CNS, *see* **Central nervous system**
CoA, *see* Coenzyme A
Coacervate droplets, **296, 297,** *298*
Coadapted gene complexes, 231–233
Cobalamin, 479
Cobalt, 481, 744
Cobra venom, 36
Cocaine, 606
Coccidiosis, 351
Coccus (pl., cocci) Sphere-shaped bacterium, ***334***
Coccyx, *657*
Cochlea, **635**–637
Cocklebur, photoperiodism, 790
Cockroach
circulation, *517*
escape reaction, 602, 686
metabolic rate, 515
Coconut, 160, 826
milk, 786
Codein, 607
Codons, table of, ***140***
initiation, 141, 145
stop, 141, *146*
Coelenterates Old name of a phylum whose members are now placed in the phyla Cnidaria *(see)* and Ctenophora (a small group not covered in this book)

Coelom A body cavity lined with mesodermal tissue, in which the internal organs are suspended, **420**, *421*
development, 589
Coenocyte Structure in fungi or plants in which many nuclei occupy a mass of cytoplasm, without being separated by plasma membranes, **160, 357**
Coenzyme Organic molecule that must be present for an enzyme to act, **38**, 88
Coenzyme A, 88, 93, *94*
Coevolution Evolution of two or more species whose members exert selective pressures on one another, **762, 897–898**
and hormone analogs, 900–901
and mustard oils, 898–899, 900
and pollination, 762
and secondary compounds, **884–886**, 898–901
Cofactor Ion or molecule that must be associated with an enzyme for the enzyme's proper functioning, **38**
Cohesion The sticking together of molecules of the same substance, **23**
of water, **729**, 732
Cohort, 855
Colchicine, **159**
Cold sores, 325, 326
Cold treatment, 770, 785
Cole, Lamont, 828, 854
Coleoptera, 435, *436*
Coleoptile The sheath covering the embryonic shoot in a monocot seed, ***719***, **780–781**
Coleus, 773, 789
shoot tip longitudinal cross-section, *709*
Collagen A structural protein that forms fibers; common intercellular component of connective tissue, **658**
Collecting duct, 564, **565,** *566,* 567, 568
Collembola, 435
Collenchyma, **710,** 711
Colliculi (brain), 615, **616**
Colobus monkeys, 488
Colon (large intestine), *484*, **485,** 487
Colonization, 846–848
Colony A group of more-or-less independent individuals attached to one another, 353
Coloration, protective, 891–894
Color blindness, 639
red-green, 214
Color change, 609, *674*
bony fish, 454
cephalopods, 428
Color vision, 639
Column chromatography, **148**
Commensalism Close association between members of different species in which one member benefits and the effect on the other is neutral or unknown, **368**
Communication
between cells, *60*
between social animals, 697–700
Community All the organisms living in a particular habitat, *797*, 824, *843–848*

climax, **816**
in dung, 845
fire-maintained, 819–820
structure, 843–844
Companion cell, *734*
Compensation depth, *840*
Competition, 845, 859–862
Competitive exclusion Name given to the idea that two species with the same niche cannot exist together in the same time and place, *862–863*
Complement reaction, **545**
Compositae A family of flowering plants including daisies, asters, zinnias, and other plants with flower heads made up of many small flowers
Compound, **20**
Compound eye, *637*, *640*
Compound leaf, *708*, **712**
Compound microscope, *46*
Concentration The proportion of one substance found in the total of a mixture of several substances; may be given in terms of proportion of molecules, of weight, and so on. Concentration is symbolized: [sugar] = concentration of sugar
Concentration gradient The change in concentration of a substance over a distance, *67*
Condensation reaction Chemical reaction in which two molecules become covalently bonded by removing an –H from one and an –OH from the other, with the removed atoms forming a molecule of water, *30*, *31*
Condensed heterochromatin, **172**
Condenser lens, *46*
Conditioned reflexes, **688**
Condom, 579
Cone, pine, *397*
Cones in eye, *638*, *639*
Conflict behavior, **692–693**
Conidiospores, *362*
Conifers Cone-bearing gymnosperms (including junipers and yews, whose reproductive structures do not resemble cones, as well as the pines, spruces, etc.), *391*, *397*, **399**, *400*
evolution of, 269
Conjugation, bacterial, **331**
Connective tissue Tissue composed of scattered cells and much intercellular material secreted by the cells, *61*, **658**–659
Connell, Joseph, 862
Conservative Changing very little during the course of evolution, **377**
Consumer Organism that eats another organism, **827**
Contest competition, **860**
Continental drift, **268**–269
Continental shelf communities, *814*, 831
Contraception, **578**
Contractile vacuoles, **346**, *349*, *353*, **562**
Control, in an experiment, **6**
Conus arteriosus, *518*
Convergent evolution Evolution of similar features by unrelated organisms, **351**, *379*, **821**
Copepods, 432–433

Copper
hydroxide and fungi, 367
and plant nutrition, 745, 749
role in animals and dietary sources, 481
tolerance in plants, 752
Coprolite, *316*
Copulation, **575–576**
Coral reefs, *814*, 831, 832
algae, *378*, *381*
Corals, *413*, 814
Cork, *716*
Cork cambium, **715**, *716*
Cormorants, 861
Corn (maize)
breeding, 770–771
chromosome number, 156
dwarf, *785*
genetics, *211*
heterozygote advantage, *232*
and human nutrition, 884
hybrid, *232*, 770–771
kernel, *719*
leaf cross-section, *117*
nutritional value, 478
source of cytokinins, 786
stem cross-section, *719*
Corn borers, 678
Cornea, *638*
Coronary artery, **520**
Corpus callosum, **616**, *617*
Corpus luteum "Yellow body" that forms in an egg follicle of an ovary after the follicle has ruptured and released an egg, *576*, **577**, 667
Corpus striatum, *615*, **616**
Correns, Carl, 185
Cortex Layer lying just inside the outermost boundary (epidermal or epithelial) layer of a stem, root, kidney, brain, etc.
root, **706**, *707*
stem, *710*, 716
Corti, organ of, **636**
Cortical steroids, **668**, *670*
Corticosterone, *667*, *669*
Cortisol, *667*, **668**
Cortisone, *33*
Corylus, *726*
Corynebacterium diphtheriae, *324*
Cotton, 118
Cotyledon A seed leaf of a plant embryo, **705**, *708*, *709*, *719*
Countercurrent exchange
in gills of fish, 500–*501*
heat, *532*, 533
rete mirabile, 510
Countercurrent multiplier, kidney, **567**
Countershading, *892*
Courtship behavior, **693–694**
Covalent bond, *19*–**20**
Cow, *see* Cattle
Cowper's glands, **575**
Crabgrass, *117*
Crabs, *431*, 433
Crafts, Alden, 751
Cramps, muscle, 556
Cranial nerves, **620**–621
Crassulaceae, 118
Crassulacean acid metabolism, 118
Cratichneumon, *866*, 868
Cravings (food), 493
Crayfish, 686

Crepidula, 213
Cretaceous period, 317, 458, 464, 471
Crick, Francis, 126, 127, 139
Crinoidea, 440, *441*
Crocodilia Order of reptiles including crocodiles, alligators, etc., 458, *460–461*
Crop, *483*, *490*
Cross-fertilization Fertilization of one plant by sperm nuclei from another plant; not to be confused with cross-pollination, deposition of pollen on stigmas of flowers on another plant
Crossing over (genetic), **162**, *196–197*
Crossover value, **197**
Cross-pollination The transfer of pollen from male flower parts of one plant to female flower parts of another, **186**, 761
Crotalids, 641
Crucifers, **898–899**, 900, 903
Crustacea, 430, 431, *432–433*
color change, 674
Cryptic Camouflaged; inconspicuous against the background
Cryptic coloration, **893**, *894*
C-terminal The end of a peptide chain with a free (not part of a peptide bond) carboxyl group, *34*
Cucumber, 790
Cud, 489
Cupula (pl., cupulae), **634**, *635*
Curare, 606
Cutaneous veins, **519**
Cuticle Layer of waxy waterproof substance secreted on the outer surface of an organism, *389*, **710**, 712, *713*, *714*
arthropod, **431**
Cuttlefish, 428
Cyanide, 56, 101, 886
Cyanobacteria, *106*, *307*, 329, 332–333, 336
relationship to other organisms, *376*, 378, *391*
Cyanogen, 295
Cyanophyta Former name for cyanobacteria *(see)*
Cycads Palmlike gymnosperms, *391*, **398**
Cycles, nutrient, 834–837
Cyclic AMP, **157**–158
and cell division, **157**–158
and hormones, **671**
and slime molds, 372
Cyclic electron flow in photosynthetic membranes, *113*
Cyclic GMP, 157–158
Cycloheximide, 147, 366
Cyclostomes, *447*
Cynognathus, *268*
Cyprioma, *887*
Cysteine, *33*, *34*, 35
Cystic fibrosis, 208
Cystitis, 338
Cyst A dormant organism within a resistant covering; stage in which some organisms pass through adverse conditions, *335*, 346
Cytochrome An electron carrier molecule consisting of a protein and a porphyrin ring (see Figure 32-15) and containing a metal ion, **95**

cytochrome *c* evolution, 150–*151*, 237
cytochrome *P450*, 498
in photosynthesis, 112
Cytochrome chain, *see* Electron transport chain
Cytokinesis Division of a eukaryotic cell in two, following nuclear division, *158*, **159**
Cytokinins Plant hormones that stimulate cell division, *786*
and apical dominance, 783
in fruit development, 767
Cytoplasm, effects on nucleus, 176
Cytoplasmic streaming Flow of cytoplasm within a cell or between adjacent cells, 58
Cytosine, *39*, *127*
Cytoskeleton, **52**, 58
Cyzenis, *866*, *868*

2,4-D, 782
Dalmatian dogs, 558
Dandelion, *275*, *768*, *773*, 853
Danaus, *890*
See also Monarch butterfly
Dark reaction, see footnote, page 110
Dart, Raymond, 472
Darwin, Charles, 242, 243, 244, *252–253*
and phototropism, 780
Darwin, Emma, 253
Darwin's finches, *244*
Dausset, Jean, 542
Day-neutral plants, **790**
D&C, 580, *see* Dilation and curettage
DDT, 828, 842
Deamination Removal of an amino ($-NH_3$) group, *556*, **557**
Death rates, human, 871–872
Decapods, *433*
Decarboxylation Removal of a carboxyl ($-COOH$) group from a molecule
Deciduous Of plants that lose their leaves during one season of the year; not evergreen
Decomposer, 337, **826**
fungi, 364–365
Deer, *281*, 465, *475*
Deer populations, 865
Defecation, **485**
Defenses
of animals, 884, 886, 887, 888–895
of fungi, *371*
of plants, 883–886, 897–899, 900–901
Deforestation, *839*
Degeneracy of genetic code, **141**
Dehydration synthesis reaction, 31
See also **Condensation reaction**
Deletion mutation, 147
Demerol, 607
Demographic transition, **871**
Denaturation of proteins, 39
Dendrites Extensions of a neuron that receive impulses from stimuli or from other neurons, 595, **596**
Dendrobates, 456
Denitrifying bacteria, 836
Density-dependent mortality factors, **859**, *868*
Density-independent mortality factors, 859
Dentaria, 903
Denticle, *451*

Deoxyribonuclease, 485
Deoxyribonucleic acid, **39**, 42, 122–135
base content, 125
base ratios in, 125, 126
content of cells, 124–125
as genetic material, 123–125
historical, 123–127
m. w., 42
origin as genetic material, 300–301
and radiation, 164
recombinant, 131, *132*, 133
repetitious, 131, 144
replication, 127, *128*, *129*, 130
structure, *126–127*
Deoxyribose, structure, *40*, **41**, *137*
Depolarize Decrease the electrical potential difference across, *598*, *599*
Dermaptera, 465
Desert, **805**, 809–810, 831
Desiccation Drying out
Desmid, *381*
Desmosome, *50*, **59**
Detergents, 32
Dethier, Vincent, 903
Detritus Bits of dead organic matter
Deuteromycetes, 359, 364
Deuterostomes Animals in which the mouth forms some distance from the embryonic blastopore (as opposed to protostomes, in which the blastopore forms the mouth). Include echinoderms and chordates, **420**
Development
animal, **572**, 577, 582–591
insect embryo, *160*
plant, 706–709, 766
Devonian Period, 317, *450*, 454
De Vries, Hugo, 185
Dexedrine, 607
Dexter cattle, *206*
Diabetes insipidus, 568
Diabetes mellitus, 568
Diapause A resting condition in which the metabolic rate is very low, 807
Diaphragm (birth control), 579
Diaphragm (breathing), *502*, **503**
Diaretiella, 900
Diastolic pressure, **521**
Diatoms, 314, 347, *348*, 350
Dichlorophenoxyacetic acid (2,4-D), 782
Dichotomous Forking into two
key, 257
Dicot, *see* Dicotyledon
Dicotyledon Member of the group of flowering plants whose embryos have two cotyledons, **705**, 720, 782
Dictyosome, **57**
Dictyostelium, 372
Didinium, 352, 863
Diencephalon, 615, **616**, *617*
Diet,
and cardiovascular disease, 480
fad, 478
and population problems, 874–875
Differentiation, cellular, **166**, 168–179, **583**
in prokaryotes, 168–170
Diffraction of light and resolving power, 47
Diffusion, **67**

facilitated, **68**–69
Digestion, **476**, 482–490
in birds, *490*
extracellular, **482**
in herbivores, 488–489
in humans, 484–487
intracellular, 70, **482**
in invertebrates, 482–483
in mammalian carnivores, 488, 489
regulation of feeding, 492–493
ruminant, 488–489
Digestive enzymes, 485–486
Digestive systems, 482–485, 488–490
human, 484
Digger wasp, 685
Digitalis, 885
Dihybrid cross, **190**–191
Dihydroxyacetone, 29
Dihydroxyacetone phosphate, 91
Dikaryon Fungus or part of fungus in which each cell contains two haploid nuclei, **357**
Dilation and curettage, **580**
Dilger, William, 702
Dimethyl ether, *21*
Dimorphism, sexual, 235, *280*
Dinitrophenols, 85
Dinoflagellates One-celled organisms with two flagella each, 347, *348*, 349
Dinosaurs, 269, *457*–458, 460
Dipeptide, **34**
Diphosphoglycerate, 91
Diphtheria, 324, 338
Diploid Containing twice the number of chromosomes found in a gamete; having paired homologous chromosomes, **156**
Diplopoda, *430*
Diptera, 435
Directional selection, **230**
Disaccharide A molecule formed by the condensation of two simple sugars (monosaccharides), *30*, **31**
Disease
bacterial, 330, 335, 338, 339
defenses of human body, 338
fungal, *365–367*
insect vectors, 436
protozoan, 350, 351
viroids, 326
virus, 325
Dispersal, 846–847
of seeds and fruit, 768–769
Dispersion, **852**
Disruptive coloration, **892**
Disruptive selection, **230**
Dissociation, 24–25
Distal In a position or direction away from the point of an appendage's attachment to the body, *see* **Proximal**
Distal convoluted tubule, *564*, **565**
Distance, individual, **693**
Disulfide bond, in protein structure, *34*, **35**
Diurnal Active during the daytime
Diurnal rhythms in communities, 844
Diving beetles, 505
Diving mammals, *533*
Division A taxon of plants or fungi equivalent to a phylum in the animal kingdom, 358

Dixon, Henry, 731
Dizygotic twins Twins arising from two different zygotes (fertilized eggs); nonidentical twins
DNA, *see* Deoxyribonucleic acid
DNA ligase, **130**
DNA polymerase I, III, **130**
  m.w., 42
Dobzhansky, Theodosius, 257
Dog
  behavior, 686, 688, 694
  genetics, 201
  mongrel, 232
  sense of smell, 642
  species, 256
Dogfish, embryo, *8*
Dolphin, *468*
Domestication of other species, *869*
Dominance (of a gene), *187*, 205
  evolution of, 232
  incomplete, *192*
Dominance hierarchy, 699–700
Dominant allele Allele expressed in the heterozygote, *see* Dominance of a gene
Dominant generation Larger, more conspicuous stage in the life history of a species with alternation of generations, **393**
Dominant species, 844
Dopamine, *605*
Dormin, *see* Abscisic acid
Dorsal Toward the back, or uppermost surface of an animal
Double bond, *19*
Double circulation, **518**
Double fertilization, 765, **766**
Double helix, 126–*127*
Doubling time, 872
Dover, White Cliffs of, 351
Down's syndrome, 217
Drive, **688**
Drosophila
  adaptive radiation in Hawaii, 267
  artificial selection, 265
  behavior (genetics), 683
  chromosome number, 156
  courtship, 271, *280*
  genes, 131, *196*, *198*, 201, 202
  genetics, 214, 232
  mate discrimination, 288
  polytene chromosomes, 171–172
  sex determination, 212–213
  sex linkage, 214
Drug resistance in bacteria, *249*
Drugs
  and cell reproduction, 158
  effect on neurons, 606, 607
  excretion, 556
Duck, *8*
Duck-billed platypus, 466
Dugesia, 414
Dung communities, 845
Dunning, Dorothy, 888
Duodenum, *484*, **485**
Dutch elm disease, 367, *864*
Dwarf plants, 785
Dysentery, 338, 339

Eagle, 827
Ear, 634–637

Earth, age of, 3
Earthworm, *279*, *421*, *422*, *423*–*424*
  circulatory system, *516*
  excretion, *559*, *563*
  metabolic rate, 515
  nervous system, *613*
Ecdysone, 178, **671**
  analogs made by plants, **900**
  chemistry, *669*
  and chromosome puffing, *172*, 183
ECF, *see* Extracellular fluid
Echidna, 466
Echinodermata, 439–*443*
Echinoidea, 440, *442*–443
Echolocation, 644
Ecocline, **799**
Ecological succession, **816**–820
Ecological time, **825**
Ecology Study of the relationships of organisms with other organisms and with their physical environment, 796–797
  and photosynthesis, 117–*118*
Ecosystem All of the organisms present in a particular area, together with their physical environment, 824–843
  components, 825–828
Ectoderm Outermost of the three germ layers of the embryonic gastrula, giving rise to skin, nervous system, and associated structures, **420**, *421*, **586**, *587*, 589
Ectoparasites, **416**, 424
Ectothermy Regulation of body temperature by behavioral means, **529**, 530–*531*
Edema Swelling due to retention of excessive amounts of fluid in parts of the body
Edentata, 465
Edstrom, J., 671
Effectors Muscles and glands which effect an animal's response to a stimulus, **595**, **628**, **647**
Efferent arteriole, **564**
Efficiency vs. rate, 834
Egestion, **483**, **556**
Egg, 278, **572**, *581*, 582
  cleidoic (reptiles and birds), 457, 559
  cytoplasm, effects on nucleus, 176
  fish, amphibian, 453
  follicle, 574, 576, 581
  formation of, 581–582
  insect, development, *160*
  sperm complementarity, 572
Ehrlich, Paul, 880
Ejaculation, **576**
EKG (Electrocardiogram), 650
Elasmobranchs Cartilaginous fish with jaws; the sharks, skates, and rays. Another name for Chondrichthyes, *451*–452
  See also **Chondrichthyes**
Elder (lenticel), *716*
Electrical potential (neuron membrane), **597**
Electrical synapse, 605
Electric current, 77
Electric eel, 454
Electric fish, 660–661
Electric organs (fish), 452, 454, *660*
Electrocardiogram, 650

Electrochemical (potential) gradient, 82
Electroencephalogram, *624*
Electromagnetic spectrum, *164*
Electron, **17**
Electron charge clouds, **18**
  See also Bonds between atoms
Electronegativity, **19**
Electron microscope, 46, 47, **48**
Electron orbital, **17**–*18*
Electron shell, **17**–*18*
Electron transport chain (system), **82**–83
  in cellular respiration, 95–96
  in photosynthesis, 104, 110, 111, *112*, *113*
Electrophoretic separation A technique that separates substances, using the fact that they move at different rates (depending on their size and electrical charge) when subjected to an electric current, 234
Electrophorus, 454
Electroplaques, **660**
Electroreceptors, 452, 454, **628**, 629, 642, 660–661
Elements, chemical, 16, 17
  found in animals, 17
  trace, 17
Elephants, *469*
Elk, *812*
Elm, 367, *784*, *864*
Embolism, 536
Embryo
  animal, *589*, *590*
  bean, *705*
  chick, *584*, *588*, *589*
  corn, *719*
  dogfish (shark), *8*
  fish, amphibian, *453*
  insect, *160*
  plant, 766–767
Embryonic development
  animal, **572**, *577*, 582–591
  cleavage, *583*–585
  flowering plants, 766–767
  gastrulation, 585–587
  nervous system, **614**
  neurulation, 588–589
  organogenesis, 589–591
Embryonic induction, **587**
Embryophytes, 390–391
Embryo sac, **764**–765
Emergent properties, **297**
Emigration Leaving an area of residence for some other place
Emu, 268
Endemic (adj. often used as a noun) Peculiar to a particular population or locality, where it originated
Endergonic reaction Reaction in which the energy of the products is greater than the energy of the reactants, **79**
Endocrine gland A gland whose hormone secretion enters the body fluids directly rather than being transported to its site of action through a duct, *664*, *665*
Endocytosis Engulfing of a particle by a cell, **70**
Endoderm The innermost of the three germ layers in the early embryo of higher animals, **420**, *421*, **586**, 589, 668
Endodermis, **706**, *707*, *750*, *751*

Endogenous rhythms, **676**
Endometrium, **577**, *590*
Endoparasite, **416**
Endopeptidase, **485**
Endoplasmic reticulum, *50, 56*, **57**
Endorphins, **607**
Endoskeleton Internal skeleton
Endosperm Nutritive tissue in a seed, 160, **403**, *719*, *765*, **766**, 767
Endospore formation (bacteria), **170**
Endostyle, **446**
Endothecia, 359
Endothermy Maintenance of a particular body temperature by physiological regulation of heat gain and loss, **462**, 465, **529**, 531–532
Endotoxins, **338**
Energy
  in chemical reactions, 79–81
  flow vs. flow of matter, *78, 827, 833*
  free, **76**
  kinetic, **77**
  potential, **77**
  transfer between trophic levels, 834
  transformation, **77**
    See also **Free Energy**
Energy of activation Energy input required before a chemical reaction can proceed, **37**
Energy budget, 12–13
Energy intermediates Molecules (or gradients) that store energy temporarily and donate it to energy–requiring reactions, 80–81
  electrochemical gradients as, 85
Energy levels of electrons, *18*
Energy pyramids, 883
Engelmann, T. W., 107
Enterogastrone, **486**
Enterokinase, **486**
Entomology The study of insects, **437**
Entomophthora, 359
Entosphenus, **447**
Entropy, **76**
Environment An organism's physical and biological surroundings (often includes conditions within the organism's own body)
  sea, 408
Environmental resistance, **858**
Enzyme activity, 36–39
  effects of pH, 38–39
  effects of temperature, *39*
  and hormones, 670–671
Enzyme induction, 168–*169*, 170–171
Enzyme-substrate complex, 37–38
Eocene epoch, 317
Eotetranychus, *863*
Ephedra, *399*
Ephemeral Lasting only a short time
Ephemeroptera, 435
Ephestia, 233
Epicotyl, *see* Plumule
Epidermis Layer of cells covering the outside of the body
  animal, 538
  embryonic development, **586**
  plant, **62**, *706*, *707*, *710*, *712*, *714*
Epidermophyton, 359
Epididymis, **574**, *575*
Epiglottis, **485**
Epilimnion, *841*

Epinephrine (= **Adrenalin,** *see*)
Epiphytes Plants that grow on the surface of larger plants, which are not harmed by having the smaller plant growing there, **754**, **800**, *802*
Episome, **339**
Epithelial tissue, *61*
Epstein-Barr virus, 326, 550
Equilibrium Balance
Equilibrium point of a reaction, **22, 36,** 79
Equisetum, 393–394, 396
Erection of penis, 575
Ergot, *371*
Ericaceae Plant family including azaleas, rhododendrons, heathers, blueberries, cranberries, etc., 810
Errington, Paul, 860
Erythroblastosis fetalis, **527**
Erythrocytes, **524**–525, 526, 527; *see* **Red blood cells**
Erythromycin, 335
Erythronium, 310
Erythropoietin, **526**
Escape by numbers, 894–895
Escape reaction (cockroach), 602, 686
Escherichia coli, 337
  active transport, 69
  electron transport, 101
  in intestinal flora, 337–338
  movement of flagellum, 85
  phages of, *322, 323*
  protein synthesis control, 168–*169*
  and recombinant DNA, *132*–133
  recombination in, 331
  size, 47
Esophagus, *484*, **485**
Essential amino acids, **478**
Ester group, *28*
Estivation (also **Aestivation**) Period of dormancy during the summer, 456, **530**
Estrogen, 171, *576*, **577**, *578*, *667*, *669*
  and cancer, 551
Estrous cycles (rodents), 677, 678
Estuaries, 831
Ethanol (ethyl alcohol), *21*, 97
Ethology, **684**
Ethylene, 782, **788**–789
Eubacteria, *334*
Eucalyptus, 400
Euchromatin, *172*–173
Eugenol, 885
Euglena, 313, 347, 349
Euglenophyta, 347, *349*, *376*, *391*
Eukaryotic Having a nuclear membrane surrounding the genetic material, and with other membrane-bound organelles in the cytoplasm
Eukaryotic cells, 49–60
  compared with prokaryotic cells, 63
  origin, 344, 345
Eumycophyta, 359–364
Euphorbiaceae (euphorbs), 118, *820*
Euphotic zone, *840*
Euphydryas, 236
Eustachian tube, 635
Eutrophic Of a body of fresh water rich in nutrients and hence in living organisms
Eutrophication Process in which debris accumulates in a lake, making it

richer in nutrients and hence in organisms, until eventually it fills in and becomes dry land
Eutrophic lake, 841, 843
Evaporative cooling Reduction of temperature by escape of the fastest-moving (that is, warmest) water molecules as water vapor, **23**, **532**
Evolution 1. Process by which organisms come to differ from generation to generation; 2. Change in the gene pool of a population from generation to generation, 10–11, **222**, *226*–231, 232, **241**–242, 244–249
  and adaptations, 12
  causes of, 226–231
  convergent, *see* **Convergent evolution**
  of cytochrome *c*, 150–*151*
  of hemoglobin, 151
  history of theory, 242–244, 252–253
  human, *471*–473
  protein, 150–151
  rate of, 237
  and reproduction, 272–290
Evolutionary success Degree to which an individual contributes genes to the gene pool of future generations, **11**, 247
Evolutionary time, **825**
Evolutionary tree
  of animals, *407*
  of plants, *391*
  of proteins, *151*
Excitatory Tending to cause depolarization of the postsynaptic membrane and hence enhancing the chances of the postsynaptic cell's firing an action potential, **598**, *599*, **604**
Excretion, 554–571
  and egg, 559
  energy expenditure, 557, 558, 562
  excretory organs, 562
    of invertebrates, *562*–563
    of vertebrates, *563*–568
  nitrogenous wastes, 556–559
  and osmoregulation, 560–562
  substances excreted, 555–556
Exercise, effects on circulation, 523–524
  lactic acid fermentation during, 98
Exergonic reactions Reactions in which the energy of the products is less than the energy of the reactants, **79**
Exocytosis Fusing of a vesicle membrane with the plasma membrane, followed by opening of the vesicle to the outside of the cell, thus releasing the vesicle contents to the outside of the cell, **70**
Exon, *143*
Exopeptidase, **485**
Exoskeleton External skeleton, outside the rest of the body, 431
  of cell, 431
Exothermic Heat-releasing
Exotoxins, **338**
Experiment, **5**
Experimental psychology, 684
Exponential population growth, *853*
Extensor, *656*
Extinction of populations, 847
Extracellular Outside a cell

Franklin, Rosalind, 126
Free energy Energy that is available or usable, **76**
Frequency-dependent selection, **236**
Fritillary population explosion, 236
Frogs, *455–456*, 802
color change, *674*
embryonic development, *585*, *586*
fertilization of egg, *583*
mating times, *259*
metamorphosis, 177–178
nuclear transplantation, *168*
poisonous, *886*, *887*
reproductive isolation, *259*
tadpole, *11*
Frond A leaf, usually highly divided (usually applied to ferns or palms), **394**, *395*
Fructose, *29*, 30
Fructose bisphosphate, *90*
Fructose-6-phosphate, 90
Fruit Structure formed from the ovary of a flower around one or more seeds, **403**, 766–767
dispersal, *768–769*
drop, 782
ripening, 788
Fruit fly, *see* Drosophila
Fruiting bodies Rather large, prominent reproductive structures of some fungi and myxobacteria, *335*, **358**, *363*, *364*, 370, *372*
FSH, *see* Follicle stimulating hormone
Fucoxanthin, *109*, 350, 378
Fucus, *378*, *379*, *382*
Fugitive species Species that occupy an area for only a short time, *818*
Functional groups, chemical, *28*, **29**
Fungi, 356–374
Ascomycetes (sac fungi), 359, *362*
Basidiomycetes (club fungi), 359, *362–364*
classification, 358–364
as decomposers, 364–365
Deuteromycetes (imperfect fungi), 359, 364
diseases, 359, *365–367*
as food, 369–*370*
kingdom, 313, *314*, 316, 356, 358
life and times, 357–358
Myxomycophyta (slime molds), 359, 371–372
Oomycetes (water molds), 359, *360*
poisonous, 371
symbiosis, 368
Zygomycetes, 359, *361*
See also Lichen, Mushroom, **Mycorrhiza,** names of individual fungi
Fungicide, **365**, 366, 367
Fusarium, 359

γ-aminobutyric acid, *605*, 607
γ rays, 164
G_1 and G_2 periods, *156*, 157, 158
Gaffkya, 334
Galapagos Islands, *243*, 244
Galapagos marine iguana, *458*
Galapagos tortoises, 862
Gall bladder, *484*, **491**
Gall midges, 183

Gallstones, 491
Galvani, Luigi, *647*
Gametangium, *361*
Gametes Eggs and sperm genetics, *188*, *191*
Gametic wastage, 260–261
Gametophyte Haploid plant that produces haploid gametes by mitosis, **383**, **390**, *395*, 396
of flowering plants, 758, *760*, 764–765
See also Algae, Bryophyta, Fern allies, Ferns, Gymnosperms
Gamma fibers, *632*
Gammaridae A family of crustacea (arthropods) whose members swim on their sides and feed on dead organic matter, 267, 861
Ganglion (pl., **ganglia**) A group of neuron cell bodies, 613
Gap junction A junction between two cells across which the electrical resistance is low, indicating that charged ions can travel across the junction from one cell to the other, **60**, 605, 650
Gap periods, *156*, 157, 158
Garner, W. W., 789
Garstang, Walter, 445
Gas exchange, 497–505
air vs. water, 499
regulation, 508–509
respiratory pigments, **505–507**
Gastrin, *486*, 667
Gastrocoel The hollow in the embryonic gastrula that becomes the lumen of the gut, *586*
Gastrointestinal tract, *484*
Gastropoda, *425*, *426–427*
Gastrovascular cavity A cavity in the body that serves for both digestion and distribution of food, *411*, *482*, *515*
Gastrula, *586*
Gastrulation, *585–587*
Gause, G. F., 853, 858, 861, 863
Gavial, 460
Gecko, *458*, *459*, 828
Geese, *692*, *693*
Gefter, Malcom, 130
Gemmae cups, *392*
Gene, **137**, 204
effect of hormones on, *670–671*
inhibitory, **168**, *169*
modifier, **232**
neutral, 237–238
operator, **168**, *169*
overlap, 141
regulatory, **168**, *169*
splicing, 131–133, *143*
structural, **168**, *169*
Gene expression, *137*, **166**, 168–174, 177–179, 204–220
Gene fixation, **229**
Gene flow Transfer of genes between one more-or-less isolated population and another, **227**–228
Gene pool All of the genes present in a population of organisms, **222**, 241–242
Genera Plural of genus
General Sherman tree, 397
Generation time Time elapsed from production (birth) of a new

individual to production of its first offspring, *854*
Generator potential, **598**, **630**–*631*
Gene regulation, 168, *169*, *170*
Genetic code, 138–141
origin of, 300–301
Genetic drift, **228**, 234
Genetic engineering, **133**
Genetic load, **233**
Genetic recombination, **162**, **274**
Genetics,
and behavior, 683
blending theory, 185, 244
and meiosis, 192–*194*
Mendelian, 184–203
and probability theory, 198–200
viruses, 141
See also Gene expression, Population genetics
Genetic variation, 234–237
Genital Of the reproductive system
Genome The total of an individual's genetic material
Genotype The particular genes present in an individual, some of which may not be expressed in the phenotype (*see* **Phenotype**); usually refers to only one or a few gene pairs, **187**, *188*
Genotype frequency vs. allele frequency, *226*
Gentian violet, 367
Genus, **309**, 310, 311
Geologic timetable, 317
Geotropism, **780**, 787
Geraniol, 885
German measles, 591
Germ cells Cells that give rise to reproductive cells (eggs, sperm, spores), 156
Germination, 769–**770**
Germ layers The three layers of cells in the embryonic gastrula, **586**, 587
See also **Ectoderm, Endoderm,** and **Mesoderm**
Gestation Period during which young are carried within the mother's body before birth, **578**, 591
Geyser, *12*
Giant fibers (nerve), 602
Giant sequoia, 716
Giant squid, 429
Giant tortoise, 862
Gibberella, 784, 794
Gibberellins Plant hormones that stimulate cell enlargement, 784–785, 787, 794
in fruit development, 767
in germination, 770
Gibbons, 471
Gilbert, Walter, 131
Gills, **500**–502
aquatic insects, 505
heat loss, 531
Gill slits, pharyngeal, **444**, *446*, *452*
Gin, 401
Ginkgo, *391*, 398, *399*
Giraffe, *827*
Girdling Stripping off the bark in a complete ring around the trunk or branch of a woody plant, **724–725**, 739

Gizzard Part of the stomach or gut modified as a heavy-walled grinding chamber, **462, 483, 490**

Glaciations, Pleistocene, *263*

Glans (of the penis), 575

Glial cells, **596**–597

Gloecapsa, 332

Glomerulus, **564**, 565

Glossopteris, 268

Glucagon, 667, *670,* **671**

Glucose A hexose (6-carbon) monosaccharide sugar, *29, 30*
  energy yield from, 96–97
  facilitated diffusion of, 68–69
  m. w., 42

Glucosinolates, *899*

Glutamate (glutamic acid), *605*

Glutamine, *33*

Glyceraldehyde, *29*

Glyceraldehyde phosphate, *see* Phosphoglyceraldehyde

Glycerol, *32*

Glycine, *33*

Glycogen A storage polysaccharide made up of glucose monomers and commonly found in animals, *30,* **31**
  in metabolism, *99*

Glycolipid, **31**

Glycolysis, 89, *90–92,* 97, 98, 99

Glycoprotein Protein bonded to carbohydrate, **36,** *51, 52*

Glycosidases, **485**

Glycosides, 886

Gnatcatcher, *852*

Goby fish, *886*

Golden algae, 347, 350

Goldfish, temperature acclimation, 529

Golgi complex Series of membrane-enclosed sacs that store secretory products of a cell, **57**

Golgi tendon organs, **632**

Gonads Organs that produce gametes, **414**

Gondwanaland, 268–269

Gonorrhea, 249, 339, 593

Gonyaulax, 347, 349

"**Good**" **species** Group of interbreeding organisms reproductively isolated from other similar groups, 265, 266

Goosegrass, *49, 884*

Gorilla, 473

Gout, 215, 558

Gradwell, George, 865, 866

Grafting 1. Attaching a cut stem or bud of one plant to the rooted stem of another, with the transport tissues of the two in contact. Transport occurs between the two, and the wounded area heals over, making the two into a single physiological unit, **774;** 2. A similar operation in which the tissue of one animal is incorporated into the body of another individual

Graft rejection, 541

Gram, Hans Christian, 334

Gram molecular weight, **21**
  See also **Mole**

Gram-negative bacteria, **334,** 335, 338

Gram-positive bacteria, **334**

Gram stain, 334

Grass, *402, 720,* 773

Grasshopper, *432*
  food utilization, 832

nervous system, 613

Grassland, *809,* 812, 831

Gravity detection
  in animals, *633, 634*
  in plants, 787

Gray heron, *697, 858*

Gray jay, *810*

Gray matter Nervous tissue made up of unmyelinated neuron cell bodies and processes, **616,** *619*

Green algae (Chlorophyta) Group of plants containing both unicellular forms and multicellular forms with simple body plans, 376, 377, *381–384*

Green plants Photosynthetic organisms, including some members of the kingdoms Monera and Protista as well as most members of the kingdom Plantae

Green revolution, 875

Green sulfur bacteria, 303

Grizzly bear, 309

Gross National Products, per capita, 872

Gross primary productivity Total amount of organic matter produced by photosynthesis in a given area during a given period of time, **829**–831

Ground pine, 393, 394, 396, 401

Growth curves (population), *853, 858*

Growth hormone, 666, *670,* **674**

GTP *see* Guanosine triphosphate

Guanine, 39, *127*

Guanosine triphosphate (GTP), 41, 80

Guard cells Two cells surrounding a stoma, or pore, in the epidermis of a plant, **389, 712,** *713, 714*
  structure and operation, 733

Guillemots, 691

Guinea worm, 417

Gulls, *682, 684, 688,* 691, 692, *693–694, 894*

Gullet, *349, 352, 353*

Gurdon, John, 176

Gurtner, Gail, 498

Gustation Tasting, *641*

Guttation, **730**

Gymnarchus, 661

Gymnosperm Non-flowering plant that produces seeds, e.g., pines, redwoods, cycads, *Ginkgo,* 391, *397–400,* 401

Habitat The physical area where an organism lives, *852*

Habituation, **688**

Haeckel, Ernst, 796

Hagfish, 447

Hair cells, 629, *633–636,* 638

Hairs
  caterpillar, *538*
  plant, 710, 884

Haldane, J. B. S., 294, 296

Hales, Stephen 730, 731, 732

Hamilton, William, 284

Hamner, K. C., 790

Hamsters, 229, 676

Hanski, I., 845

Hanson, Jean, 651

Haploid Containing the number of (unpaired) chromosomes found in a gamete; equal to half the number of

chromosomes found in a body cell of most higher plants and animals, *156*

Haploid organisms, 161

Hardin, Garrett, 878

Hardy, Alister, 349

Hardy-Weinberg Law, 223–226, 228

Hares, *467, 810*

Harlow, 690

Harmonics, **637**

Harris, Henry, 142

Harris tweed, 369

Harvestmen, 430

Harvey, Moses, 429

Hatch, A. B. (mycorrhizae), 751

Hatch, M. D. (C4 photosynthesis), 117

Hatch-Slack pathway, 117–118

Haustorium (pl., **Haustoria**) An extension of a fungus into a living plant cell, **357**

Haversian systems, *659*

Hawkweeds, 773

Hazel wood, 726

Hb *see* **Hemoglobin**

HCG, *see* Human chorionic gonadotropin

HDL, *see* High-density lipoprotein

Head-foot, *425*

Hearing, 635–637
  in moths, *644*

Heart, *520*
  beating, *521, 650*
  earthworm, 516
  insect, 517
  sea urchin, 536

Heart attack, 480

Heartwood, **728**

Heather moor, *826*

Hectare 10,000 square meters, or about 2.5 acres

Heliozoan, *351*

Helium, *17,* 18

Helix, double, 126–127

Heme Iron-containing group in hemoglobin, myoglobin, and cytochromes, *505*

Hemimetabola, 435

Hemiptera, 435

Hemocoel Body cavity containing a fluid that acts as the transport medium in many invertebrates, **431,** 517

Hemocyanins, 506

Hemoglobin Respiratory pigment that carries oxygen in the blood of vertebrates and various invertebrates, **505**–507, 508
  $\alpha, \beta, \delta, \epsilon$ chains, 151
  in amphibian metamorphosis, 177
  evolution of, 151
  fetal, 151
  m. w., 42
  sickle cell, *207–208*

Hemolymph Fluid that fills the body cavity and acts as blood in animals with open circulatory systems, **431, 517**

Hemolysis The bursting of blood cells

Hemophilia, 214, 219

Hemopoeitic tissue, **540**

Hemorrhage Bleeding, e.g., from a ruptured blood vessel

Hemorrhoids, **520**

Hemotoxylin stain, 49

Hens-and-chicks, 118
Hepatic portal system and vein, **491, 526**
Hepatitis, 325
Herbaceous With non-woody stems
Herbivore An animal that eats plants or parts of plants, **476,** *488–489,* **827**
Hermaphroditic Containing both male and female organs, **279, 414**
Hernia A split in a sheet of muscle such as that of the abdominal wall or diaphragm, 574
Heroin, 607
Herons, *697,* 895
  population, 858
Herpes viruses, 325–326, 550
Hershey, Alfred, 124
Heterochromatin, **172**–*173*
Heterocysts, *332,* **333,** 336
Heterospory Production of spores of two distinct sizes, which give rise to gametophytes of the two sexes: microspores to male gametophytes, megaspores to female, **398**
Heterotroph Organism dependent on other organisms for its organic (food) molecules
Heterozygote Individual with two different alleles of a gene, one at each of the loci for that gene; adj., **heterozygous, 187**
Heterozygote advantage (superiority), **232**–233
Hexosaminidase, 208
Hexose, **29**
  formation in photosynthesis, *115*
Hg Chemical symbol for mercury: air pressure is commonly measured as the height to which it can push a column of mercury. Air at sea level has a pressure of 760 mm Hg = 1 atmosphere
Hibernation, 456, 491, 530
  and mitochondria, 85
Hickory, *784*
  leaf, *712*
  twig, *717*
High altitude, human adaptations to, 509
High-density lipoprotein, 480
Himalayan rabbit, *216*
Hindbrain, **614**–616
Hippocampus, **616**
Hirudinea (leeches), 422, 424
Hirudo, 424
Histamine, **525,** 549, 665, *669*
Histones Basic proteins, attached to nearly all eukaryotic chromosomes, which may play a role in the regulation of transcription, *130*–131, 172
Hodgkin, A. L., 597
Hodgkin's disease, 553
Holdfast, **378**
Holly, 786
Holmes, Oliver Wendell, 338
Holometabola, 435
Holothuroidea (sea cucumbers), 440, *443*
Homeostasis The maintenance of conditions inside the body within the narrow limits required for life, **66, 554**–555
Homeothermy Maintenance of a constant, high body temperature

and metabolic rate, **529**–530, 531–532
Homing behavior, 694–697
Homing pigeons, *696*–697
Hominids Humans and their direct ancestors: members of the family hominidae, with large brains, small teeth, and bipedal locomotion, 311, **472**
Hominoids Humans and apes, large tailless primates
Homo erectus, 473
Homologous 1. (Of chromosomes) Of the same origin and containing the same kinds of genetic information, 162, **192**–*193, 194, 195, 196, 197;* 2. (Of structures, etc.) Originating from the same structure in ancestral forms (e.g., a bird's wing and a seal's front flipper are homologous structures)
Homoptera, 435
Homo sapiens, 311, *473*
Homozygotes Individuals having alleles for the same form of the gene at both loci for that gene; adj., **homozygous, 187**
Homozygous recessive Having two recessive alleles, one at each locus for the gene, **187**
Honeybee, 286–287, *698*–699
Honeycreeper, *229*
Hoofed mammals, 465, 469–470
Hooke, Robert, 45
Hooker, Joseph Dalton, 244
Hormones Substances that are produced in one part of the body and specifically influence certain activities of cells in another part of the body, **779**–780
  animal, 664–680
  chemistry, 668–670
  and color change, *674*
  and cyclic AMP, 671
  detection of, 666–667
  and digestion, *486*
  feedback control of, 667–668
  and the hypothalamus, 672–673
  local, 665
  mode of action, 670–671
  and nervous system, 671–674
  insect, *see* Ecdysone, Juvenile hormone
  and menstrual cycle, 576–577
  plant, 778–793
    See also Abscisic acid, **Auxin, Cytokinin,** Ethylene, **Gibberellins**
  reproductive, *576*–578
  roles, 664–666, 674–677
  vertebrate, 666, 667
Hornet, *889*
Horses, *477*
  evolution of, 267
Horsetails, 391, *393–394,* 396
Hot springs organisms, *12*
House sparrows, 853
Hubbard Brook experimental forest, *838–839*
Hubel, David, 689
Huber, Bruno, 741
Huffaker, Carl, 863
Human beings, taxonomic classification, 311

Human body, substances exchanged with environment, 555
Human chorionic gonadotropin, **577**
Human evolution, *471–473*
Human genetic traits, 202
Human life history, *382*
Human mating systems, 282, 288
Human population, 869–878
  death rates, 871–872
  demographic transition, **871**
  explosion, 871–872
  and food, 874–876
  life expectancy, **871**
  pollution, **876**–877, 878
  starvation, 872–873
Humerus, *657*
Hummingbirds, **530,** 895–896
  as pollinators, 763
Humoral immune response, **541,** 542–544
Hunter-gatherers, **869**–870
Huntington's chorea, 240
Hutton, James, 243
Huxley, A. F., 597, 651
Huxley, H. E., 651
Hyaluronic acid, 587
Hybrid Offspring of a mating between genetically different individuals selection against, 266–267
Hybrid breakdown, inviability, sterility, **260**
Hybrid corn, 770–771
Hybrid embryos (frogs), 585
Hybrid vigor, **232**–233
Hybridization of ribosomes, **340**
Hydra, 273, 412
  nerve net, 612, *613*
  transport in, *515*
Hydration of sodium ions, *24*
Hydraulic pressure Force exerted by a fluid
Hydrochloric acid in stomach, 486
Hydrogen atom, *17*
  molecule, *19*
  size, *47*
Hydrogen bond Weak bond between two molecules or two parts of the same molecule due to the attraction of a hydrogen with a partial positive charge to an oxygen or nitrogen with a partial negative charge, **20**
  between DNA base pairs, *127*
  in protein structure, 35
  in water, 22
Hydrogen chloride, *20*
Hydrogen cyanide, 296, 299
Hydrogen gas, *19*
Hydrogen ion reservoir
  in photosynthesis, 104, *105,* 111–*113*
  in respiration, 95–96
Hydrolase, **485**
Hydrolysis The breaking apart of a macromolecule by addition of the components of a water molecule into each of the covalent bonds linking the monomers, **30,** 31
Hydrophilic Able to dissolve in water, 31
Hydrophobic Unable to dissolve in water; nonpolar, 31
Hydrophobic interaction, 34
Hydrostatic pressure Pressure exerted by confined fluid, *71*

Hydrostatic skeleton, *423*
Hydrotropism, 794
Hydroxyproline, 494
Hydroxytryptamine, *605*
Hydrozoa, *411–412*
Hygiene, 338, 339, 871
Hymen, **573**
Hymenoptera The order of insects that includes bees, wasps, ants, etc., 287, 435
   sex determination in, 213
   social, *286–287*, 698–699
Hypericum, 864
Hyperpolarize Increase the membrane potential of, **598**, *599*
Hypersensitivity reaction, **548**
Hypertension, **480**
Hypertonic (Used of a solution) Having a higher osmotic pressure (and therefore a lower osmotic potential) than a solution to which it is being compared, **72**
Hypha One of the threadlike structures that makes up the body of a fungus, *357*
Hypocotyl, **705**, *708*
Hypolimnion, *841*
Hypothalamus A part of the brain, responsible for monitoring internal conditions in the body and initiating behaviors that tend to maintain physiological homeostasis, 615, **616**, *617*, 641
   and biological clocks, 677, 678
   and hormones, 666, 668, **672–673**
   and motivation, 688
Hypothermia, **533**
Hypothesis, *5*
   alternate, 128
Hypotonic (Used of a solution) Having a lower osmotic pressure (and therefore a higher osmotic potential) than a solution to which it is being compared, **72**
Hypoxia, **509**

Ice, *22*, *23–24*
Ice fish, 512
Iguana, marine, 458
Ileum, *484*
Ilium, *657*
Imbibition, **746**
Immigration Movement of new individuals into an area
Immune response, 537, 541
   cell-mediated, **541–542**
   humoral, **541**, 542–544
   primary and secondary, *544*
Immune system, 540–549
   and allergies, 548
   anatomy, 540–541
   complement reaction, 545
   failure of, 548
   and vaccination, 546
Immunity to disease, 544–546
Immunoglobulins **542–544**, 546–548
   E, 549
Immunological memory, 544–545, 546
Immunological tolerance, 549
Impala, *282*
Implantation, **577**, 590

Imprinting, *690*–691
Inborn errors of metabolism, 208–209
Incest taboos, 233
Incisors Chisel-shaped cutting teeth found in the center of the lower and upper jaws in mammals, **465**, **484**, *488*
Incomplete dominance (genetics), *192*
Independent assortment, law of, **190**–*191*, *194*
Indeterminate growth, **704**
Index of relatedness, **284**–287
India, population, *857*
Individual distance, **693**
Indoleacetic acid, **782**
Induction 1. The initiation of protein synthesis by the presence of an inducer substance, 168, *169*, *170–171*; 2. Conversion of one type of tissue into another as a result of the tissue's contact with a particular stimulus, 183, **587**, *884*
Industrial melanism, 244–246
Infants
   lungs of premature, 503
   Rh complications, 527
Infectious drug resistance transfer, **339**
Inflammation, **525**, 674
Information coding in neurons, 608
Information flow, origin of (genetic), 300–301
Ingest To take into the body through the mouth, 313
Inguinal canal, 574, 575
Inhibition of a neuron, **598**, *599*
Inhibitory gene, **168**, *169*
Inhibitory postsynaptic potential, **604**
Inhibitory synapse, 604–605
Innate behavior, **684–685**
Innate capacity for increase (r_m), **853**
Innervate Supply nervous connections to
Insecticides, 607
Insectivora, 465, 467
Insectivores Animals that eat insects
Insects, 430, 433–437
   anatomy, 434
   chemoreception, 642, *643*
   circulatory system, *517*
   control of, 436
   embryo, *160*
   excretion, 558, *559*, 563
   eyes, *640*
   in fossil record, 317
   gas exchange, *504–505*, *509*
   life history, *292*, 435
   major orders of, 435
   metabolic rate, 515
   metamorphosis, *178–179*
   molting, 671
   nervous system, *613*
   pheromones, 678
   societies, *286–287*, 698–699
Insertion mutation, 147
Insight learning, **689**
Instinct vs. learning, **684–685**
Insulin A small protein hormone, one of whose functions is to regulate cellular uptake of glucose from the blood, *34*, 69, 667, *670*
   effect on liver, 491
   m. w., 42
   recombinant DNA and, 133

Integration (nervous system), **613**
Intercalary meristem, **720**
Intercalated disks (heart), *648*
Interferon, **539**
   recombinant DNA, 133
Intermediary metabolism Total of all of a cell's enzyme-mediated reactions, *99*
Interneurons Neurons in the central nervous system that are neither sensory nor motor, but transmit information between other neurons, 595
Internodes Portions of a stem between sites of leaf attachment, **709**
Interphase, *156*, *158*
Intertidal Between tidemarks; covered by water at high tide and exposed to the air at low tide
Intervening sequence of DNA, **142**, *143*
Intestine, *484*, **485**, 486, 487
Intracellular Inside a cell
Intrafusal muscle fibers, **632**
Intraspecific Within one species
Intrauterine device, 579
Introgression, **265**
Intromittent organ Male body part used to introduce sperm into body of female (penis, claspers, etc.)
Intron, **142**, *143*
Invariant amino acids, **150**
Iodine, 481
Invertebrate An animal that lacks a backbone (e.g., earthworm, snail)
In vitro Latin = "in glass," (e.g., test tubes), as opposed to "*in vivo*," in the intact living organism
Ion, *19*
   dissociation, *23*, *24*
Ionic bond, **19**, *23*
Ionizing radiation, **164**
Iris (of eye), **628**, *638*
Irish moss, 381
Irish potato famine, 366
Iron, in animals and dietary sources, 481
Iron deficiency in plants, *744*, 745
Iron-sulfur protein, 95
Ischium, *657*
Island communities, 846–848
Isle Royale, 864
Isoelectric point The pH at which the charged groups on a molecule sum to electrical neutrality; charged groups combine with or dissociate from H^+ or OH^- groups in the solution depending on the pH, 234
Isogamy, *276*, **277**
Isolating mechanisms (reproduction), **259–262**
Isomers Molecules containing identical numbers and types of atoms, but with these atoms arranged differently
Isoptera, 287, 435
Isotonic (Used of a solution) Of the same osmotic pressure and osmotic potential as a solution to which it is being compared, **72**
Isotopes Different forms of atoms of an element, differing in the number of neutrons in the atomic nucleus. Heavy isotopes are those with more neutrons in the nucleus; a

origin of, 291–306

Life cycles, *see* Life histories

Life expectancy, human, **871**

Life functions, 354

Life histories of plants, *382–384*

Life tables, **856,** 867

Ligament, *657*

Light, *106*
  effect on organisms, 639
  and photorespiration, 116
  and photosynthesis, *107–109,* 110, 111,
    112, 113, 116 (rate)

Light microscope, *46–48*

Light reaction Outmoded term; see
  footnote page 110

Light trap, **109**

Lignin, **55, 725**

Likens, Gene, 838

Lilac, diseased, *366*

Lily, *720, 759*
  embryo sac development, *765, 766*
  pollen, *760, 764*

Limb girdles, *455, 457, 462, 464*

Limbic structures, **616**

Limenitis, 890, *see* Viceroy butterfly

Limestone, 835

Limeys, 494

Limnetic zone, *840*

Linkage, genetic, *193, 194,* **195–198, 232**
  sex, 213–215

Linnaeus, Carolus, 256, 308, *309,* 310,
  311, *313,* 429

Lions, 468, 477, *803*

Lipase, **485**

Lipids A large class of organic molecules
  including fats, waxes, oils, steroids,
  carotenes, *31–33*
  hormones, 668–671
  sizes, 42

Lipopolysaccharide Molecule composed
  of lipid and sugar polymers

Lipoprotein, **31,** 480

Lipotropin, **673**

Lissman, H. W., 660

Lister, Joseph, 338

Littoral zone, *813,* **840**

Liver, *484,* 491, 526
  glucose uptake, 68
  regeneration, 175

Liverworts, 391–392

Lizards, *458, 459, 531, 642*

Llama, 509

Loading pressure (= loading tension),
  506

Lobsters, *433*

Local response, **598–600**

Locus (pl., **loci**) The position on a
  chromosome that is occupied by an
  allele for a particular gene, e.g., the
  hemoglobin beta chain locus may be
  occupied by the allele for normal or
  sickle hemoglobin, *193*
  linked loci, *193, 194,* **195,** *196, 197*
  multiple loci, **211**

Locusts, 278, 436, 678

Locust trees, 775

Logistic curve, *858*

Long-day plants, *790–793*

Longitudinal muscles, 656

Longleaf pine, 820

Loop of Henle, *564,* **565,** *566–567*

Lorenz, Konrad, 684, 690

Lorenzini, ampullae of, **452,** 629, **660**

Loris, *472*

Low-density lipoprotein, 480

LSD, 371

LTH, *see* Prolactin

Lugworm, 422, 423

Lumen Space in the center of a tube

Lungfish, 268, 559

Lungs, *502–504*

Luteinizing hormone, *576,* **577,** *578,* 666,
  668, 674

Luteotropic hormone, *see* Prolactin

Lycoperdon, 359

Lycopsida Primitive vascular plants: club
  mosses and ground pines, 391, *393,*
  394, 396, 401

Lyell, Charles, *243,* 244, 252

Lymphatic system, 528, **541,** 555

Lymph nodes, 540, *541*

Lymphocytes, 540, **541,** 542, 543, 544,
  545, 548

Lymphoid tissue, **540**

Lyon, Mary, 173

Lyonization, *173–174*

Lysenko, T. D., 250

Lysergic acid diethylamide, 371, 607

Lysine, 884

Lysis Bursting of a cell, **72**

Lysogenic cycle (virus), **327**

Lysosome Membrane-bound organelle
  filled with hydrolytic enzymes, *50,*
  57, *70, 543*

Lysozyme, **538**

Lystrosaurus, 268

Lytic phages, **323**

Macaque, *700*

MacArthur, Robert, 850, 861

MacLeod, C. M., 124

Macromolecule, **27**

Macronucleus, *352, 353*

Macronutrients Nutrients needed in
  relatively large amounts, **477–478**
  of plants, **744,** 745

Macrophage, 540, **543,** *545*

Macroscopic Visible to the unaided eye

Magnesium
  in animals, 481
  in chlorophyll, *108*
  in plant nutrition, 745, 746, 751

Magnetic sense, 696

Maize, *see* Corn

Major histocompatibility complex, **542**

Malaria, 208
  control, 828

Male, **277**
  evolutionary role, 279–280

Male pattern baldness, *215*

Malnourishment, **872–873**

Malpighi, Marcello, 724–725, 730

Malpighian tubules, *434,* **563**

Malthus, Thomas Robert, 244

Maltose, *30*

Mammalia, 311

Mammals Warm-blooded animals, with
  fur or hair, whose young are
  nourished by milk from the
  mammary glands of the female
  parent, e.g., humans, rabbits, *450,*
  464–473
  brain, *615*

circulation, 519–524
  efficiency of food conversion, 490
  egg-laying, 466
  evolution, 269, 464–466, 471–473
  excretion, 557, 558, 559, 561–562
  in fossil record, 317
  marsupial, *460*
  metabolic rate, **499,** 515
  orders of, 465
  osmoregulation, *561–562*
  placental, *467–470*
  sodium conservation by, 569

Mandible, *657*

Manganese, 113, 481
  in plant nutrition, 745, 746, 749

Mango, 788

Mangrove islands, *847–848*

Manning, Aubrey, 265

Mantle Sheet of tissue surmounting the
  visceral mass of a mollusc and
  secreting the shell, if present, **425,**
  426, 427, 428

Manure, 749
  and mushrooms, 374

Manx shearwater, 694, *695*

Map distance, (genetics), **198**

Maple
  distribution of stomata, 713
  polymorphism, 236
  seeds, 770

Mapping of chromosomes, **198**
  of receptive fields, 618–619

Map sense, **696**

Margulis, Lynn, 340

Marine Living in the sea

Marine bacteria, 815

Marine communities, *813–815,* 831

Marine iguana, 458

Mark-release-recapture, **246**

Marler, Peter, 691

Marsh fritillary, 236

Marsh marigold, *630, 758*

Marsupials Mammals whose young are
  born quite early in development and
  complete their development
  attached to a nipple in the mother's
  marsupium, or pouch, 269, 465,
  466, 480
  extinctions in South America, 862

Mass flow (phloem), **736–737**

Mass vs. weight, 21

Mast cell, 540, **545,** 549

Masting (seed) **769**

Mate selection, 280–281

Mating systems, 281–283

Mating types The equivalent of sexes in
  fungi and some bacteria, **358**

Matter and energy flow, 78

Matthaei, Heinrich, 140

Maturation, 592

Mayas, 748

Mayr, Ernst, 257

McCarty, M., 124

Mechanoreceptors, **628,** 629, 631–637

Median The value that falls between the
  lowest and the highest 50% of
  individual measurements

Medicago stem cross-section, *710, see*
  Alfalfa

Medulla of brain, **614,** *615,* 617

Medusa, *411, 412*

Megaspores Large spores, produced by

meiosis in some plants, and giving rise to female gametophytes, *397, 400, 403, 758,* **764,** *765*

Meiosis A series of nuclear divisions that reduces the amount of genetic material in each resulting daughter nucleus to half that in the original nucleus, **154,** 155, 160–*161,* 274
and genetic patterns, 192–194, 194–198

Meissner's corpuscles, 629

Melanin Black pigment that gives dark color to organisms, 164
genetics, 211, 215, 216

Melanism Dark coloration, **232**
industrial, 244–246

Melanocyte stimulating hormone, 671, **673,** 674

Melatonin, 667, *669,* 677

Membrane, *see* **Plasma membrane**

Membrane potential, 597–*598,* 599

Memory, 622–623

Memory, immunological, 544–545, 546

Menadione, 479

Mendel, Gregor, 185–186, 187, 188, 189, 190

Mendelian genetics, 184–191, 200

Meninges, **614,** *619*

Meningitis, **614**

Menstrual cycle, **576**–577, 678

Mental retardation, from phenylketonuria, 209

Meristems Regions of dividing cells in growing areas of the plant
apical, *706, 709*
intercalary, **720**
lateral, **715**
propagation using, *775*

Merriam, C. Hart, 798

Mesa Verde, 728

Mescaline, 607

Meselson, Matthew, 128

Mesentery A sheet of mesodermal tissue that suspends the internal organs in the coelom

Mesoderm The middle one of the three embryonic germ layers of the gastrula; gives rise to most of the muscles, heart, kidneys, gonads, etc., *420, 421,* **586,** *589, 668*

Mesoglea The "middle glue" layer, containing few, scattered cells, between the outer and inner layers of a cnidarian, **410,** *421*

Mesophyll, *117, 714*

Mesosaurus, 268

Mesosome, **63, 330**

Mesozoic era, 269, 317

Messenger RNA The molecule that carries genetic information from DNA to the ribosome, where the information is used to determine the order in which amino acids are joined to form a polypeptide, **138,** 140, 142, *143,* 144, 145, 146
in egg cell, 585, 586

Metabolic heat Heat released by chemical reactions in the body, **529**

Metabolic pathways of cells, *99*

Metabolic rate The rate of the total of an organism's biochemical reactions; usually measured as the rate of oxygen consumption by respiration,

since respiration produces the energy needed for the other biochemical process, *499–500,* 515

Metabolism, origin of, 297–299

Metacarpals, *657*

Metameric segmentation, *421*

Metamorphic rocks, **316**

Metamorphosis The radical change in shape, physiology, and behavior that occurs when a larva becomes a very different-looking adult, 177–179
amphibian, 57, 177–178, 666
ascidian, 445
insect, 178–179

Metaphase, *158, 161*

Metaphase plate, **158**

Metastasis, **550**

Metatarsals, *657*

Methadone, 607

Methane, 21

Methanol, 28

Methionine, 145

Methyl group, 28

Mexico, agriculture, 875

MHC antigens, 210, **542**

Mice, *467, 806*
behavior maturation, 691
genetics, 201, *207,* 218, 547
nuclear transplantation, 168

Microcephaly, 209

Microfilaments, **58,** 70

Microfossils, 319

Microglia, 597

Micrometre (μm) 10^{-3}mm = 10^{-6}m

Micron, *see* **Micrometre**

Micronucleus, **352,** *353*

Micronutrients Nutrients needed in relatively small amounts, **477,** 478–479, 481–482
of plants, **744,** 745

Microorganisms Unicellular or simple many-celled organisms (e.g., bacteria, fungi)

Micropyle, **764**

Microscopy, *46–49*

Microspheres, **297,** 298, *300*

Microspores Small spores, produced by meiosis in some plants, and giving rise to male gametophytes, *397, 400,* 403, 758, **760**

Microsporum, 359

Microtubules, **58,** *59, 158, 159*

Microvilli, *487*

Midbrain, *615,* 616

Middle lamella The shared partition between the cell walls of adjacent plant cells, **55,** 159

Midge, 171

Midrib, **711,** 712

Migration, 694–697
and genetics, 235

Mildew, *358, 359, 366*

Milkweed, 887

Millardet, Pierre, 367

Miller, Stanley, 294, 296

Millet, *117*

Millipedes, *430, 826*

Mimicry, *889*–891, *896*

Mineralization Release of soluble inorganic ions from organic compounds, 747

Mineral nutrients, **478,** 481–482

needed by plants, 743–746

Miocene epoch, 317, 472

Miscarriage, **577**

Missing link, 472

Mississippian epoch, 317, *450*

Mistletoes, *754, 894*

Mitchell, Peter, 82

Mites Small arthropods in the subphylum (Chelicerata) that includes spiders. Mites have eight legs, and the body is not divided into two parts as it is in spiders, 431, 432
predation experiments, 863

Mitochondrion (pl., mitochondria), 54–55, 2–93
ATP synthesis, *83–85, 95–96*
genetic code in, 141
origin of, 340
and photorespiration, 116
size of, 47

Mitosis Series of events that results in the division of one cell nucleus into two nuclei identical to the original one, **154,** 155, *158, 159*

Mitotic spindle, **158**

Mixing in lakes, 840–*841*

MN blood groups, human, 219

Modification enzymes, 131

Modifier genes, **232**

Molars (teeth), **465,** *484, 488,* 489

Molar solution, **21**

Molds, slime, 371–372

Molds, water, 359, *360*

Mole Gram molecular weight; the number of grams equal to the molecular weight of a substance (e.g., a mole of water = 18g). Equal numbers of moles contain equal numbers of molecules regardless of the molecular weight of a substance, and so molar measurements are used when comparing the ratios of numbers of molecules of various substances present, **21**

Molecular biology Study of the molecular basis of inheritance, **123**

Molecular formula, **21**

Molecular weight, **21**

Molecule, 20–**21**

Molluscs Soft-bodied animals with a muscular head-foot and a mantle, which usually secretes a shell, e.g., snails, clams, squids, *425–429*
circulation, 536
metabolic rate, 515

Molting Shedding of skin, exoskeleton, or feathers, *431, 434,* 671

Molybdenum, 481, 744, 745, 749

Monarch butterfly, *886, 887, 890*

Monera Kingdom containing prokaryotic organisms, 313, *314,* 316, 329
fossils, 319

"Mongolism," 217

Mongrels, 232

Monkeys, *471, 472, 473,* 700
development of behavior, 690

Monoclonal antibodies, **546**

Monocotyledon Member of the group of flowering plants whose embryos have only one cotyledon, 719–721, 782

Monoculture Agricultural planting of

only one crop in an extensive tract of land, **366**
Monod, Jacques, 168
Monoecious (of plants) Having both male and female organs in the same individual, **279**
Monogamy The mating of one male with one female, either for life or for the duration of one breeding season, **281**, 282
human, 288
Monoglyceride Molecule composed of a glycerol molecule covalently bonded to one fatty acid molecule
Monohybrid cross, *187, 188,* **189**
Monomers Small molecules that may become joined together to form large (macro) molecules; e.g., amino acids are the monomers that make up polypeptides, **27**
Mononucleosis, 325, 553
Monosaccharides Simple sugars, with formulae given by $(CH_2O)_n$, e.g., glucose, ribose, **29**
Monotremes Egg-laying mammals, 465, 466
Monozygotic twins Twins arising from the same fertilized egg; identical (genetically) twins, 122
Mons Veneris, *573, 574*
Moor, *826*
Moose, *864*
Morel, *370*
Morph Form, variety, **235–236**
Morphine, 607
Morphology Structure, anatomy
Mortality factors, **859**, 867–868
Mortality rate, human, 871
Mosquito larva, *505*
Mosses, club, *393, 394, 396,* 401
Mosses, true, *391–392,* 401, *816*
cells, *13*
Moss, Spanish, *401*
Moth, *893, 894*
genetics, 232, 233, 234
hearing, *644*
pollination by, *762*
Motile Able to move from place to place
Motivation, **688**
Motor neuron (= motoneuron), **595**, 654
Motor pathways in nervous system, 619
Motor unit, **654**
Mouse, *see* Mice
Mountain avens, *812*
MSH, *see* Melanocyte stimulating hormone
Mucous glands, **486**
Mudflats, 813
Mules, 257
Mullerian mimicry, *889*
Multicellular Composed of more than one cell
Multicellularity, origin of, 353–354
Multiple alleles, **210**
Multiple loci, **211**
Multiple sclerosis, 553
Munch, Ernst, 736
Muscarine, 606
Muscle fiber, *651*
Muscles
antagonistic, 655–656
contraction of, *651–655*

distribution in human body, 649
fermentation in, *98*
glucose uptake by, 68–69
interaction with skeleton, 655–656
regeneration of, *176*
reciprocal inhibition, **656**
Muscle spindle, **630, 632,** 656
Muscle tissue, **61,** *648–651*
Mushrooms, *9, 363, 370, 371, 826*
Muskrats, 860
Mustard oils, 898–899, 900
Mutagenic agents (mutagens) Agents (e.g., chemicals, certain kinds of radiation) that cause permanent change in the genetic material, 147, **205,** 550
Mutation Inheritable change in the genetic material (DNA)
and evolution, **227**
frameshift, point, **147**
and gene expression, 204, **205,** 207
and radiation, 164
rate, 205
reasons for recessiveness, 205
Mutualism Close association between members of different species that benefits both, **368,** 898
Mycelium The body of a fungus, **357**
Mycobacterium, 335
Mycology Study of fungi
Mycorrhiza Association between a fungus and the roots of a higher plant; the fungus takes up mineral nutrients from the soil and passes them to the plant but any benefit to the fungus is still unclear, **368,** 751–752, 800
Myelin sheath Layers of fatty, insulating wrapping around the axons of some neurons, 595, **597,** 602–603, *604*
Myoblasts, *176,* **651**
Myofibrils, *651*
Myoglobin Oxygen-storing molecule in muscles, 151, **533**
Myosin, **652**–653
Myotomes Segmentally arranged blocks of muscle, *444*
Mytilus, 846
Myxamoeba, **372**
Myxobacteria, 335
Myxomycetes, 359, *372*
Myxomycophyta, 359, 371–372

NAD, NADH *see* Nicotinamide adenine dinucleotide
NADH dehydrogenase, *see* Nicotinamide adenine dinucleotide dehydrogenase
NADP, NADPH *see* Nicotinamide adenine dinucleotide phosphate
Nagana, 350
Naiads Immature aquatic stages of some insects, 435
Nanometre (nm) 10^{-9}m
Naris (pl., **nares**) Nostrils, **561**
Nasal gland (birds), **561**
Nathans, Daniel, 131
Natural selection Differential reproduction of genotypes, **229**–231, **242,** 244–249
for brood size, 247
frequency dependent, 236

for industrial melanism, 244–246
introduction to, 10–13
selective neutrality, 237
types of, *230–231*
Nature-nurture controversy, 684
Nauplius larva, 430
Nautilus, 425, 428
Navigation, 695–697, 698
electric, 660–661
Nectar guides, *762*
Necturus, 178
Negative feedback Mechanism whereby the change detected in some condition stimulates compensating physiological activity that brings the condition back to within its normal range, **667**–668
Negative pressure breathing, **503**–504
Neisseria, 339
Nekton, **815**
Nematocysts Stinging structures characteristic of cnidarians, *411*
Nematode Roundworm, 409, *417*
Neomycin, 335
Neoplasm, *see* Tumor
Neoteny The evolutionary trend in which metamorphosis occurs progressively later in the life history relative to sexual maturity until metamorphosis is eliminated and the previously larval form is now the adult; also applied to the retention of only some features of the ancestral larva in the adult form of its descendants, 178
Nephric capsule, **564**
Nephridium, *421,* **422,** 425, **563**
Nephron, **563**–567
Nereocystis, 379
Nerve gases, 607
Nerve impulse, **601**–603
Nerves, **614**
cranial, **620**–621
effect on regeneration, 176
spinal, *619,* 620, **621**
Nervous system, 612–626
autonomic, 622, *623*
embryonic development of, *614*
See also Neurulation
evolution of, 612, 613, 615
and hormones, 671–674
information coding in, 607–*608*
invertebrate (evolution), *613*
learning (plasticity), 620, 622–623
organization (vertebrate), 614
peripheral, 620–621
sensory analysis, *618–619*
sleep, 624
somatic, **614**
Nervous tissue, **61**
Nest-building behavior, 702
Net primary productivity, **830**
Neural tube, **588,** 589
Neurectoderm, **586,** *588, 589*
Neuritis, 479
Neuroglia (glial cells), 596–597
Neurohormones, 609, 672–673
Neurohypophysis, **673**
Neuromuscular junction, **603,** *604,* 653
Neurons Nerve cells, **594**–611
action potential, 600–603
classification, 613
and drugs, 606, 607

inhibition of, **598**, *599*, 604–605
local response in, 598–600
number of connections, 596
numbers of, 612
refractory periods, *602*
resting potential, 597–598
speed of impulse propagation, 602–603
structure, *595–596*
summation in, **600**, *601*
synapses between, *603–605*
transmitters, *605–607*
Neuroptera, 435
Neurosecretion, 609, 672–673
Neurospora, 197
Neurotransmitter Chemical that travels across the synaptic cleft from a neuron that has just fired an action potential to another cell, which tends to become excited to or inhibited from responding as a result of the arrival of the neurotransmitter, **603**, *604*, *605*, 607
Neurotrophic effect, *176*
Neurula, *588*
Neurulation, **588–589**, *590*
Neutral genes (alleles), 237–238
Neutron, **17**, 164
Neutrophil, *540*
Newborn, immunity of, 546
Newt Aquatic amphibian with four legs and a tail in the adult stage, 455
Nexus junctions, **60**
Niacin, 88, 479
Niche The way of life of a species; includes the habitat, food, nest sites, and so on that it needs in order to survive, *852–853*, 862
Nicolson, Garth, 51
Nicotiana, 762
Nicotiana tabacum chloroplast, *105*
Nicotinamide adenine dinucleotide, 88, 89
  in fermentation, 97, 98
  in glycolysis, *91*
  in respiration, *93*, *94*, 95, *96*
Nicotinamide adenine dinucleotide dehydrogenase, 208
Nicotinamide adenine dinucleotide phosphate, 88
  in photosynthesis, 110, 111, *112*, *114*, *115*
Nicotine, 606, 885
Niedegarde, R., 651
Night blindness, 479
Nirenberg, Marshall, 140
Nitrate NO_3^-, 749
Nitrifying bacteria, **337**, 836
Nitrobacter, 337
Nitrogen 15 and DNA replication, 126–127
Nitrogenase, **336**
Nitrogen cycle, 836, *837*
Nitrogen deficiency in plants, *744*, 745, 749, 751
Nitrogen fixation Conversion of gaseous nitrogen (N_2) to ammonia (NH_3), **333**, 336–337, 836
Nitrogen mustards, 158
Nitrogenous bases, 39, 125, 138
Nitrogenous wastes, 556–560
  and habitat, 558–560
  production of, 556–558
Nitrosomonas, 337

Noble rot, 370
Noctiluca, 347, 349
Noctuids, 644
Nocturnal Active at night
Node, leaf, **709**
Node of Ranvier, 595, **597**, *603*
Nomenclature, binomial, 309–310, *311*
Noncyclic electron flow, 111–113
Nondisjunction Failure of homologous chromosomes or of sister chromatids to separate; if this occurs during meiosis, one of the resulting nuclei will have an extra copy of the chromosome, while another will have one chromosome too few, **217**
Nonpolar Electrically symmetrical
Nonpolar bonds, **20**
Noradrenalin, *605*, 606, 607, 667, **668**, *669*, 672
  effect on smooth muscle, 649
Norepinephrine (= Noradrenalin, *see*)
Nori, *380*
Nose, *642*
Notochord Elastic rod dorsal to the gut in all chordate embryos; in most adult chordates (i.e., vertebrates), the notochord is replaced by vertebrae, which form around it, **443**, *444*, *446*, *449*
  embryonic development, *586*, 589
Nottebohm, Fernando, 691
N-terminal The end of a peptide chain with a free (not part of a peptide bond) amino group, *34*
Nuclear area, *63*
Nuclear envelope, *see* Nuclear membrane
Nuclear membrane, **53**
  in mitosis, 158, 159
Nuclear pores, *53*
Nuclear power, 876
  sources, 164
Nuclear transplantation, **168**, 176
Nucleic acids Class of macromolecules, made up of nucleotide monomers, that contains the genetic information of organisms; DNA and RNA, 39–41, 42
  and radiation, 164
Nucleocytoplasmic interactions, 176
Nucleolus, *52*, **53**, 158, 159
Nucleoside, structure, *40*, **41**
Nucleosome, **130**
Nucleotide Monomer unit that makes up nucleic acids; consists of a nitrogenous base, a pentose sugar, and a phosphate group, *39–41*, 42
Nucleus 1. The more-or-less central part of an atom, consisting of proton(s) and (except in most hydrogen atoms) neutrons, **17**; 2. The membrane-bound area of a eukaryotic cell that contains the cell's chromosomes, *53*; 3. A cluster of neuron cell bodies in the brain, **614**
Nudibranch, 425, *427*, 887, *888*, 891
Nuthatch, 897
Nutrient cycles, 827, 834–839
Nutrient deficiency diseases, 478, *479*
Nutrient deficiency in plants, *744–745*
Nutrient exclusion, **884**
Nutrition

animal, 477–482, *492–493*
and human population, 871, 872–873, 874–875
plant, 743–746
Nymph Immature stage of some insects that have no pupal stage; a nymph is a small, sexually immature, and wingless version of the adult, 435
Nystatin, 335, 366

Oak-feeding insects, *248*
Oat seedlings, 780–781
Obesity, 85, 478
Objective lens, *46*
Ocean communities, 815
  productivity, 831
Octaploid plants, 165
Octopus, 425, 428
Ocular lens, *46*
Odonata, 435
Odonthalia, 377
Oedogonium, 277, 279
Oil, use of fossils in finding, 351
Oils, *32*
Okazaki fragments, **130**
Old field succession, *817*
Oleic acid, *31*
Olfaction, 629, **641**, *642*
Olfactory Pertaining to the sense of smell
Oligocene epoch, 269, 317
Oligochaeta, 422, *423–424*
Oligodendroglia, **597**
Oligotrophic Of a body of fresh water that contains few nutrients and few organisms; lakes, 841, 843
Omasum, **489**
Ommatidium, **640**
Omnivore Animal that eats both plants and animals, **476**, 829
Oncogenic virus Virus that causes cancer, **326**
Onion, 755
Oocyte, **581**, 590
Oogamy Form of sexual reproduction in which the egg is large and nonmotile and the sperm is small and motile, **277**
Oogenesis, **581**–582, 590
Oogonium (pl., oogonia), **360**, 581
Oomycetes, 359, *360*
Oparin, Alexander, 294, 296, 297, 298
Open circulatory system, **517**
Operator region of gene, **168**, *169*
Operculum 1. Common covering of all the gills on each side in bony fish, *451*, **453**; 2. Structure used to seal the shell shut in some snails
Operon, **168**, *169*
Ophiuroidea, 440, *442*
Opiates, 607
Opisthorchis, 415
Opium, 607
Opossum, *466*
Opposable thumb, 472
Opposite leaf arrangement, **712**, *713*
Opsin, **639**
Optic lobe (brain), 616
Optic nerve, 587, 621, *638*
Optic vesicle, 587
Orangutans, 471
Orchids, *763*, 802

to and from the muscles, sense organs, internal organs, and so on, **614**, 620–622

Perissodactyls, 465, *470*

Peristalsis, **484**

Peritoneum, **563**

Periwinkles, 425

Permafrost, **811**

Permeability of plasma membrane, 68, 73

Permian period, 268, 317

Pernicious anemia, 479

Peroxisomes, **116**

Pesticides
  application of, 739
  resistance to, 436

Pests, insect, 436

Petals, *403*, 758, **759**

Petiole, **711**

Peziza, 359

P$_{fr}$, 791

PGA (Phosphoglycerate), 91, *114*, 115

PGAL (Phosphoglyceraldehyde), 91, *114*, 115

pH Measure of how acidic or basic a solution is, on a scale of 0 to 14 (0 = very acidic, 14 = very basic, 7 = neutral); pH = −log [H$^+$], **25**
  of body fluids, 25
  and chloroplasts, 113
  effect on enzyme activity, *38–39*
  of stomach, 25

Phaeophyta, 376, 377, *378–381*, *391*

Phages, **124**, *321*, **322**, *323–324*
  evidence for DNA as genetic material, *124*
  genetics of, 141
  messenger RNA, 143
  ϕX174, 141

Phagocyte, 70, **538**, *543*

Phagocytosis, **70**, *543*

Phalanges, 657

Phallus, *314*

Pharynx Part of gut just behind mouth in many animals, **483**, **484**

Pheasant, *690*

Phenols, 886

Phenotype The sum total of expression of an organism's genes, influenced by the presence of other genes and by the environment. Not all the genes in the genotype are expressed in the phenotype; presence of genes that are not expressed cannot be detected without special biochemical techniques or knowledge of pedigree, *187*, *188*

Phenylalanine, *33*, 209

Phenylketonuria (PKU), 208–209

Phenylpropanoids, 885

Phenylpyruvic acid, *209*

Phenylthiourea, ability to taste, 645

Pheromone Chemical released by one member of a species which influences the behavior of another member of the species, 678, 697–698

Philodendron, 118

Phloem Tissue in plants that conducts food from sites of synthesis or storage to sites where food is used or stored, **723**, 734–737, 738
  in cross-sections of plants, **706**, *707*, *710*, *715*

Phosphate bonds, high energy, **80**

Phosphate group, *28*

Phosphate pollution (detergent), *842–843*

Phosphatidic acid, *32*

Phosphoenolpyruvate, *92*, 117

Phosphoglyceraldehyde, *91*, *114*, 115

Phosphoglycerate, *91*, *114*, 115

Phospholipids, **32**

Phosphorus
  role in animals and dietary source, 481
  as plant nutrient, 745, 751
  transport in plants, *738*

Phosphorus cycle, *836–837*

Phosphorylation Addition of a phosphate group

Photinus, 896

Photochemical reaction, **109**, 110, 111, 112

Photoperiodism in plants, **789–793**

Photoreceptors, **628**, 629, **637–640**

Photorespiration, 116–117

Photosynthesis, 103–120
  bacterial, 335–336
  C$_3$, *114–115*
  C$_4$, 117–118
  carbon fixation, *114–115*
  chemiosmosis, *113*
  crassulacean acid metabolism, 118
  cyclic electron flow, 113
  dark reaction, see footnote page 110
  ecological aspects, 117–118
  electron flow, *111–113*
  light absorption, 111
  light reaction, see footnote page 110
  light utilization, 106–109
  origin of, 302–303
  overview, 109–111
  oxygen production, *112*
  rate, factors controlling, 115–116
  and respiration, 78
  sulfur bacteria, 120

Photosynthetic phosphorylation, **111**, *112–113*

Photosynthetic pigments, *106–109*
  in algae, 377
  in cyanobacteria, 332
  in embryophytes, 390
  in protists, 347

Photosynthetic reaction center, **109**

Photosystems I and II, **109**, *112*, *113*

Phototropism, **780**

Photuris, 896

Phycobilins Pigments found in Cyanobacteria and in Rhodophyta; major ones are phycocyanin and phycoerythrin, *108–109*, 332, 378

Phycocyanin, *109*, 332, 378
  See also **Phycobilins**

Phycoerythrin, *108–109*, 332, 378
  See also **Phycobilins**

Phycology Study of algae

Phyletic, *see* **Phylogeny**

Phyllobates, 456

Phylogeny Line of evolutionary descent (adj., **phylogenetic** or **phyletic**), **312–313**

Phylum, in taxonomic hierarchy, **310**, 311

Physarum, 372

Physiology The processes by which an organism carries out its various biological functions; how an organism works

Phytochrome, *791*

Phytophthora, 359, *365*, 366

Phytoplankton Plants floating in the upper layers of a body of water, **333**, 348

P$_i$ Abbreviation for an inorganic phosphate group

Pickles, 339

Pieris, 310

Pigeon
  homing, *696–697*
  skeleton, *462*

Pigment Molecule that differentially absorbs various wavelengths of visible light and so appears colored
  photosynthetic, *106–109*
  respiratory, **505–507**
  visual, **637**, *639*, 640

"Pill, the," *578*

Pilobolus, 359

Pilus, **331**

Pine
  cones, 397
  and fire, 820
  life history, *400*
  and mycorrhizal fungi, 751–752
  pollen, 390
  wood, 725, 726, 727

Pineal eye, 675

Pineal gland, *665*, 667, 677

Pineapple, 767, 788

Pine cone, 397

Pinnipedia, 465, *469*

Pinocytosis, **70**

Pinus longaeva (bristlecone), 397

Pipa, 456

Pisaster, 846

Pistil, 758, **759**, 762

Pisum sativum, *185–186*, *187*, *189*, *190–191*

Pit (in xylem), **725**, *726*

Pitch
  from trees, 725
  of a sound, *636*

Pitcher plants, 752, *753*

Pith, **710**, *715*

Pit organs (heat detection), 641

Pituitary gland, 615, *617*, *665*, **666**, 668, *673*

Pituitary hormones, *666*, 673–674, 678

Pit vipers, 460, 641

PKU, *see* Phenylketonuria

Placenta Organ in mammals in which blood capillaries from mother and fetus lie close together and through which substances pass between the two bloodstreams, **467**, **577**, **590**, 591, 667

Placentals Mammals that carry the young in the mother's uterus, where they receive food and oxygen via the placenta, until a fairly advanced stage of development, *467–470*
  evolution, 269

Plague, the, 338

Planarians
  excretion and osmoregulation, 562
  eyespots, **637**
  metabolic rate, 515
  nervous system, *613*
  regeneration, *175*
  transport, *516*

Plankton Organisms floating in the top layer of water, **815**

Plant cells, *50, 55–56*
  water relations, *73*
Plant hormones, *778–793*
Plant kingdom, *313, 315*
  evolution and classification, *376–377, 387–391*
Plants
  adaptations to land, *387–390*
  breeding, *770–772*
  defenses against herbivores, *883–886, 897–899, 900–901*
  evolutionary tree, *391*
  food storage, *754–755*
  as human food, *380–381, 401*
  life histories, *382–384, 392, 396, 400, 403*
  nutrition, *743–746*
  reproduction in flowering, *403, 757–777*
  selective pressures on, *375–376*
  structure and growth, *704–721*
  support, *73*
  See also Vegetation and **Biome**
Plasma, blood, **524**
Plasmalemma, *see* **Plasma membrane**
Plasma membrane The membrane bounding the outside of a cell's cytoplasm, *51–52*
  effect of hormones on, *670–671*
  exchange of particles, *69–70*
  permeability, *68*
  transport across, *68–69*
Plasmid, *132*
Plasmodesma (pl., **plasmodesmata**) Strand of cytoplasm passing from one plant cell to its neighbor, **60, 750**
Plasmodium, 208
Plasmodium of slime molds, *372*
Plasmolysis Shrinkage of a cell due to loss of water under conditions where water is scarce or concentration of external solutes is high, **73**
Plasmopara, 367
Plasticity Ability to change
  evolutionary, **377**
  in learning, **620**
Plastics, 877
Plastids, *55, 56*
Platelets, **524, 525**
Plate tectonics, **268**
Plato, 241, 310
Platyhelminthes Flatworms, e.g., planarians, tapeworms, flukes, 409, *414–416,* 417
  digestion, *483*
  excretion, *562*
  nervous system, *613*
Platypus, 466
Pleistocene epoch, 317, 472, 473
  glaciations, *263–264*
Pliocene epoch; 317
Plumaria, 377
Plumule, **705,** *708, 719*
Pogo, 76
Poikilothermy Condition of having a body temperature which is usually close to that of the environment, **529**
  and metabolic rate, *500*
Poinsettia, *760*
Point mutations, **147**
Poison arrow, 456, 886

Poisonous fish, 454
Poisonous frogs, 456
Polar Electrically asymmetrical
Polar body, **581**
Polar bonds, **20**
Polarized light, 640
Polio, 553
Polio virus, 325
Pollen, 121, *390,* **398,** *760–764*
Pollen cone, 397
Pollination Deposition of pollen on or near the female parts of a gymnosperm or angiosperm, *402–403,* **761–763**
Pollution, **876–877,** 878
  of lakes, *842–843*
Polyandry Mating system in which a female mates with more than one male, **282**
Polyanthus, *264*
Polychaeta, *422–423*
  regeneration, *176*
Polygamy Mating system in which an individual may have more than one mate, **281,** *282*
Polygenic characters, **211,** 237
Polygyny Mating system in which one male mates with more than one female, **282**
Polymer Large molecule made up of subunits that are smaller molecules similar or identical to one another, **27**
  artificial, 877
Polymorphism Simultaneous presence in a population of two genetically different forms of a trait at frequencies higher than could be maintained by recurrent mutation, **234–236**
Polyoma virus, 326
Polyp, *411, 412*
Polypeptide A polymer composed of many amino acid monomers, **34**
Polyphemus, 893
Polyploidy Possession of more than 2N sets of chromosomes, **264**–265
Polypodium, 395
Polypody fern, 900
Polyporus, 359
Polyribosomes, **146.** *147*
Polysaccharide A macromolecule made up of many subunits which are simple sugars, *30,* **31**
Polysomes, **146,** *147*
Polytene chromosomes, *171–172*
Pongo, 465
Poppies, 531, 762
Population All members of a species living in a particular area and making up one breeding group, **851–881**
  age structure, **852**
  carrying capacity, 858
  characteristics of, **852**
  forecasts, *856–858*
  growth, *853, 858*
  human, *869–877*
  models, *863, 868*
  regulation, *858–865*
  stable, **856**
  stationary, **856**

  winter moth, *865–869*
Population explosion, *871–872*
Population genetics, 222–240
  coadapted gene complexes, *231–233*
  founder effect, **229**
  gene fixation, **229**
  gene flow, **227**–228
  genetic drift, **228**–229
  genetic load, **233**
  Hardy-Weinberg law, **223**–226
  heterozygote advantage, **232**–233
  kin selection, **284**–285, **286**–287
  natural selection, **229**–231
  neutral alleles, **237**
  polymorphism, **235**–236
  unit of selection, 237
  variation in nature, *233–237*
Porifera, *409–410*
Porphyra, 380
Porpoise, 682
Portal system A series of blood vessels that carry blood between two sets of capillary beds without first returning to the heart
  hepatic portal system, **525**
  hypothalamus-pituitary, *673*
Porter, Keith, 57
Portuguese man-of-war, 411
Positive pressure breathing, **503,** 504
Posterior At or toward the rear end of an animal
Posterior pituitary, *673*
Postsynaptic Receiving stimuli from a nerve cell across a small gap, or synapse, *603,* 604
Postsynaptic potential, 598
Postzygotic isolating mechanisms, **260**
Potassium
  active transport of, 69
  in animal nutrition
  in plant nutrition, 745, 751
Potassium permanganate, 367
Potatoes, 711, 773, 788
Potatoes, late blight of, *365,* 366
Potato famine, Irish, 366
Potential energy, **77**
Powdery mildew, *358, 366*
P-protein, **734,** 736
P_r 791
Prairie, *809*
Praying mantid, 896
Prebiotic Before life arose
Precambrian Before the Cambrian period, which began about 600 million years ago, 317
  fossils, 319
Precocene, **901**
Predation, *862*
Predators Animals that capture other animals for food
Pregnancy, *577–578, 590–591*
  risks, 578, 579
  smoking and anesthesia, 498
Prehensile Grasping
Premature birth, 591
Pressure flow (phloem), **736**–737
Presynaptic Pertaining to part of the neuron that sends a signal across a synapse; the receiving cell is postsynaptic
Presynaptic membrane, *603, 604*
Prezygotic isolating mechanisms, *259–261*

Primary cell wall, **55, 725**

Primary consumer, *827*

Primary growth Growth in length and production of new stem and root branches in plants, **705**, *706*, *707*, *708–709*, *717*, *718*

Primary immune response, *544*

Primary productivity The amount of food manufactured from inorganic substances by photosynthesis and chemosynthetic organisms, **829–832**

Primary structure (protein), **34**

Primary succession, *816–817*

Primary tissues Plant tissues that differentiate from cells laid down by the apical meristem, **706**, *707*, *710*, *719*

Primates The order of mammals that contains monkeys, apes, humans, and so on, 311, 465, 467
evolution, *471–473*
societies, 282, 283, 700

Primitive Showing features believed to have arisen early in evolution

Primrose, 720

Primula, 264

Probability
and Mendel, 185
of origin of life, 293
rules of, 198–200

Proboscidea, 465, *469*

Proboscis, butterfly, *643*

Producers Photosynthetic and chemosynthetic organisms, **825**, *826*

Productivity, **829–833**

Products of a chemical reaction, **21**

Progeny Offspring

Progesterone, 33, 171, 576, **577**, *578*, 667

Proglottids, *416*

Projection of sensory information, 618–**619**

Projector lens, 46, **48**

Prokaryotes Organisms composed of one or more cells, which lack a nuclear membrane separating the DNA from the cytoplasm and without other membrane-bound organelles
differentiation, 168, *169*, 170
evolutionary success, 332
fossils, 319

Prokaryotic cells, 62–**63**, 330–331
DNA structure, *130*
vs. eukaryotic, 63
reproduction of, 331–332

Prolactin, 666, **674**

Proline, *34*

Promoter area of genetic material, **142**

Prophase, *158*, *161*

Proprioceptors, 629, **632**

Prosimians, **471**, *472*

Prostaglandins, 665, *669*, 674

Prostate gland, **575**

Prosthetic group Chemical group that is covalently bound to a protein and necessary in order for the protein to function properly, **38**

Proteases, **485**

Protective coloration, *889–894*

Protein A functional unit made up of one or more polypeptides, 33, *34*–**39**
amino acid sequence determination, 148–149

conversion to fat, 98

energy yield from, 100

evolution of, *150–151*

functions, 36

repressor, **168**

respiration of, 98

sizes, 42

structure, *34–35*

Protein deficiency disease, 478

Proteinoids, **296**, 297

Protein quality of foods, 874

Protein synthesis, *145*, *146*, *147*
and canavanine, 899
effects of antibiotics on, 147
effects of hormones on, 670–671
effect of mutations on, 147
origin of, 300–301
regulation of, 168–171
transcription, **138**, *142–144*
translation, **138**, *145–146*

Prothrombin, 479, **525**

Protista Kingdom containing one-celled eukaryotic organisms, 313, *314*, 315, 344–347, *348–353*

Proton, *17*

Protonema, *392*

Protostomes An evolutionary line of animals in which the embryonic blastopore becomes the mouth. Includes Platyhelminthes, Annelida, Arthropoda, and Mollusca, **420**

Protozoa, **344**, *345–347*, *350–353*

Proximal In a position or direction near the point at which an appendage attaches to the body, *see* **Distal**

Proximal convoluted tubule, *564*, **565**

Pseudocoelom, **417**

Pseudogenes, **144**

Pseudomyrmex, 898

Pseudoplasmodium, **372**

Pseudopodia, *346*

Psilocybin, 607

Psychology, experimental, 684

Pterodactyls, 221

Pteropsida, 394, 396

Pubis, 657

Puccinia, 359

Puddling, *493*

Puffballs, 359

Puffer fish, 454, 886, *887*

Puffins, *463*

Puffs, chromosome, *171*, *172*

Pulmonary artery/vein, **519**, **520**

Punnett, Reginald Crundall, 189

Punnett Square, **189**, *191*

Pupa Stage between larva and adult in insects with complete metamorphosis (holometabolous insects), 435

Pupation 1. Time of entry into the pupal stage; 2. State of being in the pupal stage

Purine Member of a group of nitrogenous bases chemically related to purine (e.g., adenine and guanine, found in DNA and RNA), 39

Purkinje fibers, **650**

Puromycin, 147, 366

Purple non-sulfur bacteria, 302–303

Purple sulfur bacteria, 303

Pus, **525**, 545

Pyramid
of biomass, *829*

of energy, *833*

of numbers, *829*

Pyridoxine, 479

Pyrimidine Member of a group of nitrogenous bases chemically related to pyrimidine, *39*

Pyrrhocoridae, 901

Pyrrophyta, 347, *348*, 349, *376*, *391*

Pyruvate (pyruvic acid), *92*, *93*, *97*, *98*, *99*

Pythons, 457, 460

QB phage, genetics of, 141

Quadrupeds Animals that walk on four legs, **457**, *464*

Quaternary period, 317

Quaternary structure of proteins, **35**

Queen Anne's lace, *883*

Queen bee, 287, *698–699*

Queen butterfly, *261*

Quiescent zone, **706**

Quinones Compounds derived from quinone, a 6-carbon unsaturated ring compound with 2 keto groups, *886*

r The actual rate of increase of a population, **853**

Rabbits
genetics, *216*, 219
sodium conservation, 569

Raccoon, *468*

Radial cleavage, *584*

Radial symmetry, **413**

Radiation and cell division, *164*

Radicle, **705**, *706*

Radioactive substances
in human health, 164
in phloem transport, 735, *738*

Radiolarians, *351*

Radish, 790

Radius, *657*

Radula, **426**

Rainfall, 798, *799*, 804

Rain forest, 800, *802*, *807–808*, 831

Ranunculus root cross-section, *707*

Ranvier, node of, *595*, **597**, *603*

Raphides, *883*

Rapid eye movements (REM), 624

Raspberry, 768, *773*

Rats
birth control, 578
mating behavior, 686
maturation of behavior, 691
osmoregulation, 561
reproductive cycle, 677

Rattlesnakes, 844

Ray (fish), 451–452

Ray (plant), **726**

RBC, *see* **Red blood cells**

Reactants, **21**

Reactions, chemical, **21**–22
coupled, 80
redox, **80**

Recent epoch, 317

Receptacle, **766**

Receptive fields (visual), **618**

Receptors Sense cells or organs that detect stimuli, changes in the animal's external or internal environment, **628–631**

Salmonella, 340
Salt A substance that yields neither hydrogen nor hydroxyl ions when it dissociates in water, **25**
   appetite for, 493
   conservation of, 569
   regulation and excretion of, 556, 560–562
Saltatory conduction, *603*
Saltbush, 12
Salt deficiency, 493
Salt gland (birds), **561**
Sambucus, 716
Sampling error, 228
Sand, **746,** *747*
Sandy beaches, *813*
San Francisco Mountain, 798
Sanger, Frederick, 131, 141, 148, 149
Sap Mixture of water, minerals, etc., conducted in xylem tissue of plants, **725**
Saprobe Organism using nonliving organic matter for food, **335,** 337
Saprolegnia, 359, *360*
Sapwood, **728**
Sarcodina, 347, 350–*351*
Sarcolemma, **651,** 653
Sarcomeres, **651**
Sarcoplasmic reticulum, *653*–654
Saturated fats, **31,** 32
Sauerkraut, 339
Saurischia, 460, *see* Dinosaurs
Sauternes, 370
Savanna, *803*–*804,* 831
Scale
   shark, *451*
   reptile, *459*
Scale insects, 249
Scanning electron microscope, 47, **48**
Scapula, *657*
Scarlet fever, 324, 338
Scavengers Animals that eat dead organisms or organic matter
Schindler, David, *842*
Schistosoma, 415
Schizophyta, *see* Bacteria
Schleiden, Matthias, 45
Schwann cell, **597**
Schwann, Theodor, 45
Science, 2
   experiments, **5**–6
   limitations of, 7
   and religion, 7
   research, 6–7
   and society, 4, 6–8
Scientific method, 5–6
Scion, **774**
Sclereids, **727**
Scolex, *416*
Scorpions, 431, *432*
Scouring rushes, *see* Sphenopsida
Scramble competition, **860**
Scrotum, **574,** *575*
Scurvy, 479, 494
Scutellum, *719*
Scyphozoa, 411, *412,* 827
Sea, as cradle of life, 554
Sea anemones, *10, 315, 413*
Sea cucumber, *443*
Sea lilies, *441*
Sea lions, 699
Sea mouse, *422*

Search image Image used by an animal seeking something (usually food) with a particular visual appearance; said to exist when the animal ignores different-looking items that in fact are equally valuable as food and that are eaten by the animal at other times, 236, **896**
Sea slug, *see* Nudibranch
Sea squirts, *445*
Sea stars, *441,* 846
Sea urchins, *442,* 536, *813*
Seawater, salt composition, 597
Seaweed Multicellular algae in marine habitats; some members of the Rhodophyta, Phaeophyta, and Chlorophyta, 376–386
   economic uses, *380*–*381*
Sebaceous gland, *538, 539*
Secondary cell wall, **725**
Secondary compounds Organic compounds that occur in some organisms but not others; i.e., chemicals other than proteins, carbohydrates, lipids, nucleic acids, 884–886, 898–901
Secondary consumers, **827**
Secondary electrons, **48**
Secondary growth Growth in girth in plants, **705,** 714–718
Secondary immune response, **544**
Secondary productivity, **832**
Secondary sexual characteristics, **573**
Secondary structure of proteins, **34**
Secondary succession, 817–818
Secondary tissue (in plants), *715*–718
Secondary xylem, **715**
Second filial generation, *187,* **188**
Secretin, *486*
Secretion (urine formation), **564,** 565
Secretory neurons, *672*–673
Sedimentary nutrient cycle, **836**
Sedimentary rocks, **316,** 835
Sedimentation, 316–318
Seed Dispersal unit of gymnosperms and angiosperms, consisting of a seed coat, embryonic plant, and food supply, **397**
   bean, *705*
   corn, *719*
   development, *766*
   germination, 769–**770**
   pine, *400*
   viability, **769**–770
Seed coat Outer covering of a seed, developed from the outer layers of the ovule, **705,** *719,* **766,** 769
Seed dispersal, 768–769
Seed masting, **769**
Seed predation, 769
Segmented worms, *see* Annelida
Segregation, law of, *188, 191, 193, 194*
Selection, 229–231, **242**
   frequency-dependent, **236**
   level of action, 237
Selective breeding, 244, *245*
   of plants, 770–772
Selective permeability Property of allowing some substances to pass through more easily than others, **68**
Selenium, 481
Selfishness and altruism, 286

Self-pollination The transfer of pollen from male flower parts to female flower parts on the same plant, *186,* **761**
SEM (scanning electron microscope), 47, 48
Semen, **575**
Semicircular canals, **635**
Seminal vesicles, **575**
Seminiferous tubules, **574,** *580*–581
Semmelweiss, 338
Senescence in plants, **786,** *788*–789
Sense organs, 627–646
   adaptation, *631*
   role, 628
   transduction, 630–631
   types, 628–630
Sensory Having to do with detection of changes in the external or internal environment of an organism
Sensory coding, 630–631
   auditory, *636*–637
   visual, *618*
Sensory neuron, **594,** 628, *629*
Sepal, *403,* **758,** *759*
Sepia, 428, 429
Septum (pl., **septa**) Partition
   in annelids, *421*
   in fungi, 357
Sequoia, 397
Sequoiadendron, 397
Serotonin, **525,** 624, *669*
Serpulid, *423*
Serum, **524**
Serum agglutinins, 210
Sessile "Sitting"; not moving from place to place
Seta (pl., **setae**) Bristles, *422*
Seventeen-year locust, *see* Periodic cicada
Sevin, 607
Sewage treatment, 333, 843
Sex change, in *Crepidula,* 213
Sex chromosomes, 212–215
Sex determination, 212–213
Sex differentiation of embryo, 212
Sex-influenced genes, **215**
Sex linkage, **213**–215
Sexual dimorphism Dimorphism = "two forms;" in sexual dimorphism, the two sexes look or act different, *280*–281
Sexual intercourse, **575**–576
Sexual reproduction, 572–578, 580–582
   vs. asexual, 273–275
   in flowering plants, **757**–772
   gametogenesis, 580–582
   human, anatomy and physiology, 573–578
   origin of, *276*–278
Sexual selection, **231**
Shade leaves, 118
Shade plants, *118*
Shag, 861
Sharks, 451–452
   embryo, 8
   excretion and osmoregulation, *561*
Sheep, *10, 470*
   food choice, 492
Sherrington, Sir Charles, 686
Shigella, 338, 339
Shivering, 532

protective covering when conditions are unfavorable for growth
algal, **383**
bacterial, 170, **332**
fungal, **357**
Sporophyte Diploid plant that produces haploid spores following meiosis, **383, 390**
moss, *391*
Sporozoa, 347, 351
Spring wood, **727**
Squamata Order of reptiles including lizards and snakes, 458–460
Squid, 425, *428*
axon, blood, salts, 597
circulatory system, 536
giant, 429
locomotion, *502*
nitrogenous wastes, 559
Stabilizing selection, **230**
Stable population, **856**
Stahl, Franklin, 128
Staining of specimens, 48–49
Stamen, *402, 403, 758,* **759**
Standing crop The biomass of organisms actually present at any one time, 830
St. Anthony's fire, 371
Staphylococcus, 337, 339
Starch A polysaccharide made up of glucose monomers, commonly found in plants, 30, **31**
size, 42
Starfish, *see* Asteroidea
Starling, 853
brood size, 247
Starvation, *491,* **872–873**
Stationary population, **856**
Statocyst, **633**
Statolith, *633,* **787**
Stearic acid, 32
Stem cell, *540*
Stems, 708–711, *719*
Steppe, 809
Sterilization 1. Any more-or-less permanent change that prevents an animal from reproducing sexually, 579; 2. Destruction of living organisms by applying heat, ultraviolet radiation, poisonous chemicals, etc., 164
Sternum Breastbone, *462,* **463,** 657
Steroids, **33,** 668–671, 885
and protein synthesis, *170, 171*
Steward, Frederick C., 167
Sticklebacks, *687*
Stigma Tip of female flower part, usually sticky, allowing pollen to adhere to it easily, *402, 403, 758,* **759**
Stimulus (pl., **stimuli**) A form of energy (chemical, electrical, thermal, light, mechanical, etc.) in the external or internal environment of an organism, to which the organism may respond, **628**
intensity coding, 630–631
Stinkhorn, *314,* 359
Stipe The "stalk" of a multicellular alga or of a mushroom, *378,* **379**
St. John's wort, 864
St. Kilda wren, *262*
Stolons, *773*
Stoma (pl., **stomata**) Pores in the outer

layer of a plant that permit gases to be exchanged between the plant and the air, *117,* **389,** *712, 713, 714*
regulation of opening, 733–734
role in sap movement, *732, 733*
Stomach, *484,* **485,** *486,* 667
of sheep, *489*
Storage of food in body, 100
Storage proteins, 36
Strawberry, *766, 773*
Streptococcus, 324, 330, 339
Streptomyces, 335
Streptomycin, 147, 158, 335
Striated muscle, *648,* 650–655
Strobilus (pl., **strobili**) Cluster of leaves bearing sporangia, *393, 394*
Stroma, **105,** 114
Stromatolites, 302, 319
Structural formula, **21**
Structural gene A section of chromosome that carries information determining the sequence of amino acids in a polypeptide or protein, **168,** *169*
Structural proteins, **36**
Style, *403,* **759**
Suberin, **750**
Substrate 1. Reactant in an enzyme-mediated chemical reaction; 2. Solid underlying surface, e.g., a rock in the ocean bed
Substrate-level phosphorylation, *82*
in glycolysis, *91, 97*
in respiration, *96, 97*
in tricarboxylic acid cycle, *94, 95, 97, 98*
Succession Process in which the environment becomes less favorable for existing inhabitants and more favorable for members of other species, which eventually come to dominate the community, and which may later be displaced by still other species, 816–820
in dung, 845
Succulents Plants that store water in fleshy stems or leaves, *805*
Sucrose A disaccharide consisting of a glucose and a fructose residue; table sugar, *30;* m.w., 42
Sugar, 29–31
formation in photosynthesis, *115*
Sugarcane, *117*
Sugar maple, *236,* 770
Sugar phosphate backbone (DNA), *126*
Sulfanilamide, 366
Sulfide bacteria, 336
Sulfonamides, 366
Sulfur, *481, 745, 751*
Sulfur bacteria, 120, 336
Summation
in behavior, **686**
of local potentials, **600,** *601*
Summer wood, **727**
Sun, 797–798
radiation from, *106*
Sunbathing, *531*
Sundew, *752, 753*
Sunflecks, 844
Sunflower, *744*
transpiration, *732*
Sun leaves, *118*
Sunlight, fate of, 832

Sun plants, 118
Supernatant ("Floating above") The liquid above solid matter in the bottom of a tube, beaker, etc., *54*
Supernormal stimulus, **688**
Suppressor T lymphocyte, 544
Surface area to volume ratio, *353*
Surface tension of water, **23**
Surfactant, **503**
Surtsey, *846*
Survivorship curves, **855**
Suspensor, **766**
Sutherland, Earl, 671
Suture (joint), **657**
Swallowtail butterfly, 492, 895, 899
caterpillar survivorship, 855
genetics, *231*
Swan, *8, 284*
Sweat glands, 532, 538
Sweating, 532
Sweet peas, 203
Swimbladder, *451, 453, 454,* **510**
Swivelase, 130
Sycamore tree, 784
Symbiont Organism that lives in close association with a member of another species (usually refers to a microorganism living in relationship with a larger host organism), **350**
in digestive tract, 488, 489
Symbiosis Intimate association between members of different species, **368**
Symmetry, pentaradial, *440, 441, 442*
Symmetry, radial vs. bilateral, *413*
Sympathetic nervous system That portion of the nervous system which prepares the body to meet stressful or dangerous situations, **622,** *623,* 672
Sympatric Living in the same area, **258**
Sympatric speciation, **264–266**
Symplast, **750,** 751
Synapse, **595,** *603–605,* 610
electrical, 605
inhibitory, 604–605
reciprocal, 610
Synapsis The lining up of sets of sister chromatids in the early stages of meiosis, **162**
Synaptic cleft, knobs, vesicle, *603, 604*
Syncytium A "cell" in an animal that contains more than one nucleus within one plasma membrane, **160,** *651*
Synovial fluid, **657**
Synthesis period, 156, 157
Syphilis, 335
Systemic pesticides, **739**
Systolic pressure, **521**

Tachyglossus, 466
Tactile Pertaining to the sense of touch
Tadpole, *11, 168*
Taenia, 416
Taiga Northern coniferous forest, **810,** 831
Tamura, M., 691
Tannins, 248
Tapeworms, *414, 416, 417*
Taproot system, **707,** 708
Target cells, 670

Tars, 725
Tarsals, *657*
Tarsier, *472*
Taste, sense of, 641
Taste bud, *629*, **641**
"Tasters," 645
Tatum, Edward, 136, 331
Taxon (pl., **taxa**) One of the hierarchical categories into which organisms are classified, e.g., species, order, class, **310**
Taxonomy Study of the classification and identification of living organisms, 310–313
Taylor, J. H., 130
Tay-Sachs disease, 208
TCA cycle Former abbreviation for tricarboxylic acid cycle; should no longer be used
T-cells, *see* T lymphocytes
Tectorial membrane, *636*
Tectum, *615*, **616**
Teeth
  carnivore, *488*, 489
  herbivore, *488*
  human, *484*
  sea urchin, *442*
  shark, *451*
Telencephalon, *615*, **616**
Teleosts, *451*, **453**
Television, 164
Telophase, *158*, *161*
Temperate biomes, *806–810*
  productivity, 831
Temperate desert, *809–810*
Temperate forest, **806**–808
Temperate grassland, *809*
Temperate phages, **324**
Temperate shrubland, *808*
Temperate woodland, *808*
Temperature
  and enzyme activity, *39*
  and lakes, 840–841, 843
  and metabolic rate, *500*
  and photorespiration, 116
  and rate of photosynthesis, *116*
Temperature regulation, 510, 528–532, 533
Template A pattern or mold
Temporal Of time
Temporal summation, **600**, *601*
Tendon, *656*, **657**
Terminal bud, **717**
Termites, 287, 350, 435
Terns, *696*
Terpenes, 808
Terpenoids, *885*
Terracing, *747*
Terrestrial Of land
Territorial behavior, 280, *692*
Territoriality, 860
Tertiary period, 317
Tertiary structure of protein, **34**, 35
Test Protective covering, 445
Test cross, **189**, *190*
Testis, **574**, *575*, *580–581*, *665*, 667
  development of, 590
Testosterone, *33*, **576**, 667, **668**
  and behavior, 687
Tetanic contraction, **655**
Tetanus (disease), 338
Tetracycline, 147, 158, 335, 366

Tetrad Foursome consisting of two sets of two (=4 altogether) linked sister chromatids lined up at synapsis, *161*, **162**, *193*
Tetraploid, **156**
Tetrapods "Four-footed" vertebrates; amphibians, reptiles, birds and mammals, **455–473**
Tetrapyrrolic ring, *108*
Tetrodotoxin, **454**, **886**
Tetrose, *29*
T even phages, *322*, *323*
Thalamus, *615*, **616**, *617*
Thalassemia, 208
Thalidomide, 591
Theory, scientific, **6**
Thermal conductivity of water, **23**
Thermal pollution Discharge of excessive amounts of waste heat into the environment, 843
Thermal stratification, *841*
Thermochemical reaction, **109**, 116
Thermocline, *841*
Thermodynamics, **76**, **78**
Thermoreceptors, **628**, 629, 641
Thermoregulation, 510, **528**–*532*, 533
Thiamine, 479
Thiorhodaceae, 303
Thistle, *402*
Thoracic duct, **528**, *541*
Thorax Part of the body between the head and the abdomen
Thornwood, 804–805
Threat behavior, **693**, 700
3′ end, 5′ end The 3′ end of a nucleotide strand is the one at which the 3′ carbon of the sugar of the last nucleotide is not bound to the phosphate of another nucleotide. At the 5′ end, the phosphate bound to the 5′ carbon of the sugar is not bound to another nucleotide, *126*
Thrombin, **525**
Thrombocyte, **524**, 525
Thromboplastin, **525**
Thrombus, **536**
Thylakoids, **104**–*105*
Thymine, *39*, *127*, *138*
Thymus gland, 541, *665*
Thyroid gland, *665*, 667, **668**
  in amphibian metamorphosis, 666
Thyroid stimulating hormone, *see* Thyrotropin
Thyrotropin, 666, 671, **673**
Thyroxin, *178*, 667, *669*, 680
Thysanura, 435
Tibia, *657*
Ticks, 431, 432
Tiger, 834
Tight junction, 50, 60
Tilia stem cross section, *715*
Tilling, *749*
Time and origin of life, 293
Tinbergen, Niko, 682, 684, 687, 691
Tissues Groups of cells that perform a particular task in an organism; e.g., blood, cartilage, xylem, *61*, 62
  See also names of tissue types
T_L, *506*
T lymphocyte, *540*, **541**–542, 544, 548
Toads, 455
  defenses, 887

reproductive isolation, 259
Tobacco, 789
  genetics, 201
  mosaic virus, *325*
Tocopherol, 479
Tomato hornworm, *818*
Tomato plants, 118, 710, *771*, 790
  genetics, 201
Tone quality of a sound, **637**
Tongue, *641*
Tonoplast Membrane enclosing the central vacuole of a plant cell, 50, **56**, 73
Tools, 472, 473
Tonsils, 540
Topical pesticides, **739**
Torpedo, 660
Torpor, **529**–*530*
Tortoise, 862
Toxins Poisons
  bacterial, **338**, 339, 340
  See also names of specific toxins
Toxoplasma, 351
Trace minerals, **481**, 482
Trachea 1. Vertebrates: windpipe; tube that conducts air from the pharynx to the lungs, **502**; 2. Insects: tubes that conduct air from the outside throughout the interior of the body, *434*, *504–505*, 509, **517**
Tracheids, **725**, *726*, *727*
Tracheophyta, *393–403*
Tragedy of the commons, 878
Transcribe Synthesize RNA using a DNA template
Transcription, **138**, *142–144*
  control in eukaryotes, *170–174*
  control in prokaryotes, *168–170*
  in embryonic development, 176
Transduction 1. Conversion of the energy of a stimulus into electrical energy which can be transmitted by the nervous system, **630–631**; 2. Transfer to genetic material from one bacterium to another by a phage, **324**, 331
Transfer RNA (tRNA) RNA molecules that transport amino acids to ribosomes during protein synthesis, **138**, *144*, 146
Transformation, bacterial, **123–124**, 331
Transformation by tumor viruses, **326**
Transfusion, 527
Translation The assembling of amino acids in a sequence specified by a molecule of mRNA, **138**, *145–146*
Translocation, genetic, **217**
Translocation (in protein synthesis), *146*
Transmission electron microscope, *46*, 47, 48
Transpiration Loss of water by evaporation through pores (stomata) in the shoot system of a plant, **730**–734, 741
Transplants, nuclear, **168**
Transplants, organ, 542
Transport, active, *see* **Active transport**
Transport in animals, 514–523
Transport in plants
  of minerals, 737–738
  by phloem, 735–737
  by xylem, 728–734

seeds or for flowering of some types of plants, **770**
Verril, A. E., 429
Vertebrae, *449*, *619*
Vertebrates Animals with backbones, e.g., fish, humans, 449–473
　characteristics, 446
　comparison of brains, *615*
　evolution, 449, *450*
　origin of, 445, 446–447
　societies, *699-700*
　　See also Agnatha, **Amphibia, Aves, Chondrichthyes, Mammals, Osteichthyes, Reptiles**
Vesicle Membrane-enclosed sac, holding secretory products, enzymes, etc., in a cell, **57**
Vessel elements, vessel members, vessels (in xylem), *725*, **726**, *727*
Vestigial Much reduced in size and nonfunctional
Viability of seeds, **769**–770
Vibrio, 338
Viceroy butterfly, *890*
Victoria, Queen, 214
Vietnam, defoliation, 782
Villi, *487*
Vinblastine, 159
Vincristine, 159
Vinegar, 339
Viper, *805*
Virchow, Rudolf, 45
Viroids, **326**
Virulent Extremely damaging
Virus Particle composed of nucleic acid and protein, and exhibiting some properties similar to those of living organisms, 321–328
　and cancer, 326, 550
　disease, 325–326
　DNA as genetic material, *124*
　Epstein-Barr, 326, 550
　genetics, 141
　herpes, 325–326, 550
　lysogenic, *324*
　lytic, *323*
　messenger RNA, 143
　φx174, 141
　QB, 141
　RNA *324–325*
　structure, *322–323*
Vision, 638–640
　color, 639
　information processing, 618
　in primates, 471
Visual cycle, **639**
Visual pigment, **637**, *639*, 640
Vitamin A, 639
Vitamin C, 496
Vitamins, **478**–479, 489
　See also B vitamins, names of specific vitamins (p. 479)
Viviparity Condition of giving birth to young rather than laying eggs, **452**
Vole Small rodent resembling a short-tailed mouse, *828*, 846
Volume, of a sound, 636, 637

Voluntary muscle, *648*, **650**–*655*
von Frisch, Karl, 695, 699
von Linne, Carl, *see* Linnaeus, Carolus
Vulva, **573**, 576

Waddington, Charles H., 251
Wallace, Alfred Russel, 242, 243, 244, 252, 253
Wall pressure, **72**
Walnut tree, 897
Walruses, *469*
Warblers Group of small insectivorous birds
　adaptive radiation, speciation, 263–*264*
　competition and niche, *861*
Wareing, P. F., 787
Warning coloration, *888*
Wasp, solitary, 685
Water, *22–24*
　cohesion of, 732
　properties, 729
　as respiratory medium, 500
Water hyacinth root, *707*
Water lily, *315*, *401*
Watermelon, 772
Water molds, 359, *360*
Water pollution, 333, *842–843*, 877
Water potential A measure of the energy of water, determined by the opposing forces of osmotic potential and turgor pressure, **729**
Water-soluble vitamins, **478**
Water vascular system, **440**
Watson, J. D., 126, 127
Wavelength Light and sound may be considered as traveling in wavy lines. The distance between adjacent peaks of the line is the wavelength, symbolized by: λ (lambda)
Waxes, **32**
WBC, *see* White blood cell
W chromosome, *212*
Weaning, 286
Weather, 797–798
Weaverbird, 280
Weed killer, 782
Wegener, Alfred, 268
Weight, ideal body, 492
Weight vs. mass, 21
Went, Frits, 781
Wessells, Norman, 172
Whales, 7, 468
Wheat, 770
Wheat production, 875
Wheel animals, *see Rotifera*
White blood cells Cells in the blood that are involved in defending the body against foreign organisms and substances, *173*, *208*, **524**, *525*, *540*
White-crowned sparrow, *691*
White matter Nervous tissue consisting of myelinated axons, **616**, *619*
Whittaker, Robert, 313, 314
Whooping cough, 338
Whorled leaf arrangement, **712**, *713*
Wiesel, Torsten, 689

Wigglesworth, Vincent, 178
Wild type Showing the normal phenotype of wild members of the species for the trait in question, **196**
Wilkins, Maurice, 126, 127
Willow, 794, *811*
Wilson, Edward, 847
Wilson, H. V., 409
Windbreaks, 747
Wine, 97, 99, 370
Winter moth, 865–869
　natural selection, 248–249
Witch hazel, 766
Wolf, 834
Wolffia, 400
Wolves on Isle Royale, 864
Wood Secondary xylem, **715**, **724**–728
Wood alcohol, 28
Woodland, temperate, **808**
Woodpecker, *464*
World map, *801*
Wren, *262*
Wright, Sewall, 228

X chromosome, 156, *173–174*, *212–215*
　inactivation, *173–174*, 215
X-ray diffraction of DNA crystals, 126
X rays, 164, 551
Xylem Tissue that conducts sap from the roots to the leaves of plants, **723**–734
　in cross sections of plants, **706**, *707*, *710*, *715*
　mechanisms of transport in, 728–734

Y chromosome, *155*, 156, *212–213*
　human, role, 213
Yeast, 97–98, 99, 362
Yellow sulfur butterflies, 266–267
Yersinia, 338
Yield, **830**
Yogurt, 339
Yohimbine, 607
Yolk, 573, 584
Yolk sac, *8*, **457**

Z chromosome, *212*
Zea stem cross section, *719*
　See also Corn
Zeatin (riboside), *786*
Zebra, *470*, *803*
Zinc, 481
　as plant nutrient, 744, 745, 749
Z lines, **651**, *652*
Zone of elongation, **706**
Zone of maturation, **706**
Zooflagellate, 347, *350*
Zoomastigina, 347, *350*
Zooplankton Animals floating in the surface layers of a body of water, **408**
Zoospores, *276*, *360*, **383**, *384*
Zygomycetes, 359, *361*
Zygospore, *361*
Zygote A fertilized egg, *155*, *156*, *572*, **582**
　importance in life history, 384

GEOLOGIC TIME SCALE

Years Ago (millions)	Era	Period*	Epoch*	Climate and Physical Events
	CENOZOIC	Quaternary	Recent	4 ice ages; rise of Sierra Nevada
			Pleistocene	
2.5		Tertiary	Pliocene	Rise of Panama; cool; extinction of many species
7			Miocene	Rocky Mountains rise further
26			Oligocene	Rise of Alps and Himalayas
38			Eocene	Mild to tropical weather
53			Paleocene	Most continental seas disappear
65	MESOZOIC	Cretaceous		Rise of Rockies reduces rainfall to their east
135		Jurassic		Much of Europe covered by sea. Breakup of Pangaea
195		Triassic		Large areas arid and mountainous. Appalachians rising
225	PALEOZOIC	Permian		Appalachians rising; glaciation in Southern Hemisphere
280		Carboniferous	Pennsylvanian	Land low, covered by shallow seas and swamps; subtropical climate
			Mississippian	
345		Devonian		U.S. largely low and sea-covered; Europe mountainous
395		Silurian		Continents flat; mountains beginning to rise in Europe
430		Ordovician		Mild climate; shallow seas cover continents, which are flat
500		Cambrian		
600	PRECAMBRIAN	Precambrian		Planet cooling, shallow seas, many mountains rise

(The Earth is about 4½ billion years old)

*Remember the periods and epochs (beginning with the earliest) with a mnemonic such as: Penguins Can Only Swim Deeply, Carried Past The Jewel-Crusted Pyramids Ensconced On Mildly Pettish Palfreys.